89
87
85
83
48
46
44
42
ONT.
ONTARIO
LAKE SUPERIOR
ISLE ROYALE
KEWEENAW
HOUGHTON
ONTONAGON
BARAGA
MARQUETTE
GOGEBIC
IRON
ALGER
LUCE
CHIPPEWA
SCHOOLCRAFT
MACKINAC
DICKINSON
DELTA
MENOMINEE
MACKINAC IS.
BOIS BLANC IS.
DRUMMOND IS.
BEAVER IS.
FOX IS.
MANITOU IS.
EMMET
CHEBOYGAN
CHARLEVOIX
PRESQUE ISLE
LAKE HURON
LEELANAU
ANTRIM
OTSEGO
MONT-MORENCY
ALPENA
BENZIE
GRAND TRAVERSE
KALKASKA
CRAWFORD
OSCODA
ALCONA
WISCONSIN
LAKE MICHIGAN
MANISTEE
WEXFORD
MISSAUKEE
ROS-COMMON
OGEMAW
IOSCO
ARENAC
CHARITY IS.
MASON
LAKE
OSCEOLA
CLARE
GLADWIN
HURON
OCEANA
NEWAYGO
MECOSTA
ISABELLA
MIDLAND
BAY
TUSCOLA
SANILAC
MUSKEGON
MONTCALM
GRATIOT
SAGINAW
KENT
OTTAWA
IONIA
CLINTON
SHIA-WASSEE
GENESEE
LAPEER
ST. CLAIR
ALLEGAN
BARRY
EATON
INGHAM
LIVINGSTON
OAKLAND
MACOMB
VAN BUREN
KALAMAZOO
CALHOUN
JACKSON
WASHTENAW
WAYNE
ONT.
BERRIEN
CASS
ST. JOSEPH
BRANCH
HILLS-DALE
LENAWEE
MONROE
LAKE ERIE
INDIANA
OHIO
MICHIGAN
0 10 50
Scale of Miles
UMMZ-1957-WLB

MICHIGAN FLORA

Frontispiece

Rubus parviflorus (Thimbleberry)
First discovered by Thomas Nuttall at Mackinac Island in 1810, and disjunct between the northern Great Lakes region and western North America (see map 462).

(Photo by E. G. Voss in Emmet County, June 28, 1984.)

MICHIGAN FLORA

A guide to the identification and occurrence of the native and naturalized seed-plants of the state.

Part II

DICOTS (Saururaceae–Cornaceae)

Edward G. Voss

Cranbrook Institute of Science
Bulletin 59 *and*
University of Michigan Herbarium
1985

Library of Congress Catalog Card No. 85-72083

ISBN 87737-037-0

BULLETIN 59
Second printing 1998, with
corrigenda and addenda on pp. 725–727

Designed by William A. Bostick

Composition by Edwards Brothers, Inc.

Color printing by Edwards Brothers, Inc.

Printed and bound by Edwards Brothers, Inc.

Edited by Christine E. Bartz

To the memory of my parents, who for so long
shared and encouraged my love of nature:
KATHERINE G. VOSS (1904–1973)
DAVID O. VOSS (1898–1978)

Preface

A FLORA is a census, a basic resource like soil surveys, topographic maps, and vegetation (or land resource) maps, essential to the wise stewardship and management of our natural heritage.

The first part of this Flora, published in 1972, dealt with 723 species of gymnosperms (Gymnospermae) and monocots (Liliopsida) then known to grow (or to have grown) in Michigan outside of cultivation. The present part covers the first half or more of the dicots: nearly 1000 species in over 80 families, including most of our trees; willows, dogwoods, and many other shrubs; such large groups as the pink, buttercup, mustard, rose, legume, and carrot families. Altogether, this part presents the "apetalous" and "polypetalous" families of dicots: those with flowers lacking petals or with the petals separate from each other. The third and final part will cover the "sympetalous" families: those with united petals, i. e., families from the heaths to the composites in the traditional Englerian sequence. The division is thus the same as in the three volumes of Gleason's *New Britton and Brown Illustrated Flora*.

A key to the dicot families would be premature at this time, as it is not yet certain what species in the various families of Part III will have to be covered, and the breadth of variation in each family will govern the construction of a workable artificial key. Such a key will be included in Part III.

Much introductory material on history, occurrence of plants in Michigan, and use of keys appeared in Part I and mostly is not repeated now. As before, the keys are strictly artificial, designed solely to facilitate identification and not to indicate natural relationships. Any flora has its own special emphases, which are stressed to the neglect of other aspects. Here, the stress is on identification and on known distribution (past and present) and habitat of plants in Michigan. It will thus inevitably indicate what is *not* known about plants in the state: what species do not grow here, where those which do occur have not yet been collected, and which are most unsatisfactorily distinguished from each other. This is a *local* flora, to provide local information (which indeed may also be useful over a much wider area); it is not a comprehensive treatise to present general information on the kinds of plants that occur in Michigan. A flora is essentially an inventory—and this one aims to be complete for the state—but there is no need for every flora to contain repetitive descriptions, statements of total range, and additional material which is in fact often copied from yet other (and sometimes unreliable) sources.

Parts I and II will not be revised before completion of Part III. In the meantime, persons collecting additional state or county records are urged to supply specimens to the University of Michigan Herbarium or at least to report what they deposit elsewhere. All herbaria, especially those outside the state, cannot be examined exhaustively again to locate records relevant to a revision (which may take the form of a one-volume compact edition for more convenient field use).

Acknowledgments

The indebtedness of an author of a work such as this is immense: to collectors of the specimens upon which it is based; to authors of other floras, monographs, and notes; to specialists who have annotated herbarium specimens over the years; to colleagues who have made helpful suggestions, whether brief or extensive, and who have facilitated examination of collections.

Preparation of a flora requires familiarity not only with the species included but also with the geography of the area and the collectors who have made specimens. Many place names—of which lakes are the most obvious—have been duplicated in the state and too many collectors have provided less than adequate data with their specimens. Information accumulated on the lives and travels of 19th century botanists has helped a great deal with interpreting their collections. A narrative account has presented much of the story (Voss 1978) and greatly expands on the brief historical survey of plant collecting given on pp. 10–15 of Part I. Collectors' field notes, annotated manuals, and correspondence have all helped to provide data when labels were incomplete or unclear.

Since the brief history in Part I, recent collections of major size or importance in Michigan include those of Robert H. Read in Alger County (Pictured Rocks National Lakeshore); Russell Garlitz in Alpena County and others nearby; Heidi Appel and Judy Gendlin in Antrim County; F. Glenn Goff especially in Bay and Ottawa counties; William R. Overlease in Benzie County; Timothy S. Mustard in Mason and Oceana counties (Manistee National Forest); Brian T. Hazlett in Leelanau County (Fox Islands and portions of the Sleeping Bear Dunes National Lakeshore including Manitou Islands) and Mason County (Nordhouse Dunes); N. William Easterly in Monroe County; Don Henson in Schoolcraft and other Upper Peninsula counties; John Merkle (trees) and later John V. Freudenstein, both especially in the east-central Lower Peninsula; and several general collectors about the state: Roy E. Gereau, William T. Gillis, Richard K. Rabeler, A. A. Reznicek, Paul W. Thompson, and James R. Wells—among others. Some of these collections were made with support from the Hanes Fund and/or in conjunction with environmental consulting or evaluation projects, or explicitly to improve the maps for this Flora. The herbaria at Andrews University and the Morton Arboretum are rich in recent Berrien County collections; those at Alma College and Central Michigan University, in collections from the central Lower Peninsula. Collectors at Northern Michigan University have concentrated on Marquette County.

The growth of herbaria in Michigan since this project began in 1956 has been enormous. Several institutional herbaria now rich in records had few if any specimens 30 years ago. The herbaria of Michigan State University and the Cranbrook Institute of Science have expanded twice into new quarters and that of the University of Michigan once during this period. The total number of specimens is now quite overwhelming. Unfortunately time has not made it possible to annotate every Michigan specimen in some families in some herbaria, although I hope that most serious errors in identification have been corrected and that no county records have been overlooked (unless, of course, specimens were unfiled, misfiled, or away on loan when herbaria were examined). At times, it seemed as if herbarium material were accumulating faster than I could check it, so great has been the surge of collecting in recent years. But essentially all Michigan specimens in the following herbaria have been checked (for families in this volume) and complete data permanently filed for many if not all of them. These are all in Michigan unless otherwise indicated, and the symbols in parentheses are the standard ones as listed in *Index Herbariorum* (Holmgren et al. 1981).

Albion College, Albion (ALBC)
Alma College, Alma (ALM)
Andrews University, Berrien Springs (AUB)
Central Michigan University, Mount Pleasant (CMC)
Cornell University, Ithaca, New York (CU & BH)
Cranbrook Institute of Science, Bloomfield Hills (BLH)
Ford Forestry Center, Michigan Technological University, L'Anse (MCTF)
Harvard University, Cambridge, Massachusetts (GH & A)
Isle Royale National Park, Houghton (IRP)
Michigan State University, East Lansing (MSC)
Missouri Botanical Garden, St. Louis, Missouri (MO)
Morton Arboretum, Lisle, Illinois (MOR)
Northern Michigan University, Marquette (NM)
University of Michigan, Ann Arbor (MICH)
University of Michigan Biological Station, Pellston (UMBS)
University of Notre Dame, Notre Dame, Indiana (ND & NDG)
Wayne State University, Detroit (WUD)
Western Michigan University, Kalamazoo (WMU)

The above institutions were visited for examination of specimens and, often, selection of specimens for loan, except for Alma College (specimens kindly supplied on loan by Ronald O. Kapp), Cornell University (Michigan specimens, omitting *Rubus* subg. *Rubus*, generously selected and loaned by Peter A. Hyypio), and Isle Royale National Park (specimens added or needing rechecking since initial examination of both monocots and dicots in 1958 supplied by a succession of Park naturalists from Robert M. Linn to Robert A. Janke and Bruce E. Weber). Dicots from several herbaria (e. g., MGR, SEN) listed in Part I were examined when that part was prepared and no further checking has been done. I regret that time simply did not permit checking several additional herbaria both in and out of the state.

Selected specimens, ranging from a few to many, have been received on loan from the following herbaria, usually in response to specific requests:

Academy of Natural Sciences, Philadelphia, Pennsylvania (PH)
Field Museum, Chicago, Illinois (F)
Glen Oaks Community College, Centreville, Michigan (GOCC—not in Index Herb.)
Iowa State University, Ames, Iowa (ISC)
New York Botanical Garden, Bronx, New York (NY)
Ohio State University, Columbus, Ohio (OS)
University of Wisconsin, Madison, Wisconsin (WIS)

I am deeply grateful to the curators of the above herbaria together with their supporting staff for the high priority they have given my visits and loans and their unfailing cooperation in making available the specimens which are the foundation of this Flora. A special word of gratitude goes to those who had no official duties at several institutional herbaria but whose searching and filing and assorted volunteer efforts greatly increased the efficiency and productivity of my use of collections: Ruth MacFarlane (MCTF), Richard K. Rabeler (MSC), Anne M. Richards (NM), Kerry S. Walter (NY), and Anna L. Weitzman (GH/A).

The collections of the Grand Rapids Public Museum, formerly on long-term loan to Aquinas College (AQC), are now on indefinite deposit at the University of Michigan. The private herbaria of Leroy H. Harvey and F. J. Hermann have been presented to the University of Michigan. Private herbaria from which at least selected specimens have been made available for Part II include those of Kenneth W. George, W. R. Overlease, Anne M. Richards, and E. G. Voss. A few specimens from herbaria additional to those thus far cited have been seen and recorded when students or colleagues have had them on loan for other purposes.

This part of the Flora includes many groups of exceptional taxonomic difficulty. More than ever, I have had to draw on the wisdom of others, while trying to avoid the universal problem that specialists in one group often wish those who treat *other* groups to do more "lumping." I have no taxonomic specialty myself and have tried to take as equal an approach as possible in defining taxa in all groups—impossible as such a task may be when the processes of evolution have not created equal taxa. Consequently, it has not been possible to follow all advice—sometimes contradictory—received from colleagues and correspondents, but I am nevertheless extremely grateful for the generous help of all those who looked at portions of my draft manuscripts and offered few to numerous comments. These comments saved me from many errors (if not all!) of fact or judgment and improved the final results. If I have failed to utilize all information supplied in print, in correspondence, or in personal discussion with my fellow botanists, it is not necessarily because of any doubt concerning the significance of their inquiries—but because other characters, or alternative taxonomic positions, or deferral of some judgments have seemed to provide more useful results for identification purposes. Usually reference is made to alternative positions or to relevant literature which will offer more details.

Those whose criticisms were especially extensive and helpful are acknowledged again at the appropriate places in the text. Chief among these are Harvey E. Ballard, Jr., who has conquered *Viola*; and Warren H. Wagner, Jr., on whom I have relied for key characters and identifications in *Quercus*. Burton V. Barnes and W. H. Wagner, co-authors of the revised *Michigan Trees,* have been helpful with other woody plants as well. A. A. Reznicek, my colleague in the University of Michigan Herbarium, has suffered much from my countless interruptions to seek his advice, a problem he could have avoided only by being less well versed on the local flora and relevant literature. Other authorities on certain families and genera who have examined portions of the manuscript, in early or late stages and in detail or more superficially, include the following: Susan G. Aiken, George W. Argus, Ernest O. Beal, David E. Boufford, Lincoln Constance, Arthur Cronquist, Thomas O. Duncan, Shirley T. Graham, Peter C. Hoch, Emil P. Kruschke, Donald H. Les, Ernest J. Palmer, James B. Phipps, Richard K. Rabeler, Peter H. Raven, Reed C. Rollins, C. Marvin Rogers, Jonathan D. Sauer, Warren L. Wagner, and Dennis W. Woodland.

Numerous individuals have called attention to interesting collecting sites, unusual species, or useful information (including data on obscure localities given on labels). My thanks go to all: students and colleagues, citizens of Michigan and visitors, botanists both amateur and professional. Many students and other part-time assistants over the years have given essential aid in the assembling, recording, and organizing of records. In the terminal phases of Part II, when herbarium records, maps, illustrations, and text were being pulled together, the intelligent, efficient, and loyal assistance of John V. Freudenstein and Kevin E. Bradtke deserves a special word of appreciation.

Preparation of this Flora as an identification manual began in 1956 under a five-year grant from the Faculty Research Fund of the Horace H. Rackham School of Graduate Studies of the University of Michigan. It has been carried on as a part of the research program of the University Herbarium, with increasing support for travel, hourly assistance, and other expenses of producing the manuscript from the Clarence R. and Florence N. Hanes Fund established by bequest of Mrs. Hanes in 1966. All those who use the Flora will share my gratitude to the trustees of the Hanes Fund (and to Mr. and Mrs. Hanes) for their encouragement and aid. They have made possible a more thoroughly documented and illustrated work than would otherwise have been achieved, with the publication subsidized to bring it within the reach of more users.

The continued support of the University of Michigan Herbarium, under directors Rogers McVaugh and Robert L. Shaffer during preparation of Part II, has enabled me to make the concentrated effort which this project requires.

ILLUSTRATIONS

The number of dicot families and genera, compared to Part I of the Flora, has made it impossible to continue the aim of illustrating all of them—especially since quite a number are represented in our flora only by an introduced species, perhaps collected as a waif long ago. Nevertheless, a diverse representation of species, genera, and families is included, within the constraints of space and availability of suitable drawings and photographs. Some species rarely illustrated in floras for eastern North America are depicted, along with many familiar and attractive wildflowers. But well known cultivated species which occasionally or rarely spread from gardens and plantings are often not illustrated.

The color illustrations are all made from 35 mm transparencies of plants in Michigan. The source (county or island) of each is indicated, along with the last name of the photographer. I am indebted to the several skilled photographers among my friends who responded to my invitation to submit slides to supplement or supplant my own, and from whose offerings selections were made—often not easily!: Frederick W. Case, Jr. (Saginaw, Michigan), Susan R. Crispin (Lansing, Michigan), John V. Freudenstein (Saginaw, Michigan), T. L. Mellichamp (Charlotte, North Carolina), Jon D. Monroe (Ithaca, N.Y.), Michael R. Penskar (Chelsea, Michigan), James R. Wells (Bloomfield Hills, Michigan), and Gary R. Williams (Glen Ellyn, Illinois).

The line drawings were mostly not made from Michigan plants. They have been borrowed from previously published sources, which it is a pleasure to acknowledge as follows. In most cases, original drawings were made available, improving the quality of reproduction.

Figure 1, the map of Michigan counties, is used through the courtesy of the University of Michigan Museum of Zoology.

Figures 3, 4, 5, 6, 7, 8, 12, 13, 14, 17, 18, 20, 21, 22, 23, 24, 25, 26, 27, 28, 29, 114, 115, 116, 171, 173, 175, 176, 177, 179, 181, 186, 189,

192, 193, 195, 221, 222, 224, 248, 267, 270, 273, 274, 275, 279, 280, 283, 284, 285, 305, 306, 349, and 351, by Elizabeth Dalvé with the assistance of Elizabeth King, are reprinted from *The Woody Plants of Ohio: Trees, Shrubs, and Woody Climbers, Native, Naturalized, and Escaped* by E. Lucy Braun, copyright © 1961 by the Ohio State University Press, by permission of the Ohio State University Press and the Ohio Academy of Science, which kindly lent the original drawings for reproduction.

Figures 9 and 10, by Thomas Cobbe, are reprinted from Billington's *Shrubs of Michigan* (Bull. 20, revised 1949); Figures 42, 66, 98, and 164, by Ruth Powell Brede, are reprinted from Smith's *Michigan Wildflowers* (Bull. 42, revised 1966); both published by Cranbrook Institute of Science.

Figures 11, 15, 30, 32, 113, 180, 208, 276, and 277, by Sarah Phelps and Janice Glimn Lacy, are from Otis' *Michigan Trees,* revised by Barnes and Wagner, published by the University of Michigan Press (1981), and are used by permission of the Press.

Figure 16, by Carol Ann Kanter and Hazel M. Hartman, is from Viereck and Little's *Alaska Trees and Shrubs* (Handb. 410, 1972); Figures 54, 61, 68, 69, 87, 93, 135, 266, 293, 321, and 347, by Regina O. Hughes, are from Reed's *Selected Weeds of the United States* (Handb. 366, 1970); both published by the U. S. Department of Agriculture.

Figures 33, 56, 112, 178, 210, 220, 249, and 315 are from vols. 1 and 2 of *Iconographia Cormophytorum Sinicorum,* published in 1972 in Beijing.

Figures 31 and 182, by Ronald A. With, are from *Shrubs of Ontario,* by James H. Soper and Margaret L. Heimburger, published by the Royal Ontario Museum (1982) and are used by permission of the authors and publisher.

Figures 37, 57, 58, 59, 70, 72, 74, 75, 78, 81, 83, 84, 85, 104, 105, 119, 123, 134, 137, 138, 140, 143, 151, 154, 161, 165, 202, 207, 209, 211, 212, 213, 218, 235, 241, 252, 255, 257, 291, and 335, by F. Schuyler Mathews, are from Beal's *Michigan Weeds* (Bull. 267, 1911); Figure 203 is from Darlington, Bessey, and Megee's *Some Important Michigan Weeds* (Spec. Bull. 304, 1940); both published by the Michigan Agricultural Experiment Station.

Figures 39, 41, 50, 53, 91, 103, 144, 264, 307, and 311 are from Mason's *A Flora of the Marshes of California* (1957) and are used by permission of the publisher, the University of California Press.

Figures 76, 107, 111, 118, 126, 152, 157, 242, 265, 286, 290, and 295, by Isaac Sprague, are from *The Genera of the Plants of the United States,* by Asa Gray (1848–1849) and are used with permission from the director of the Harvard University Herbaria.

Figures 62, 92, 117, 122, 125, 254, 256, 260, 328, 334, 343, 345, 348, and 350 were prepared for the generic flora of the southeastern United States under the direction of Carroll E. Wood, who generously made them available; fig. 62 is used with permission of the Journal of the Arnold Arboretum,

© 1981 President and Fellows of Harvard College; most of the others also appeared earlier in the same journal.

Figures 102, 106, 120, 121, and 225 are from *Wildflowers of the Great Lakes Region* by Roberta L. Simonds and Henrietta H. Tweedie (Chicago Review Press, 1983), and are used with the gracious permission of the authors.

Figure 141 was drawn for me by Edward M. Barrows and previously appeared in *The Michigan Botanist* 6: 48 (1967).

Figures 231 and 232 are from *Rhodora* 38: 188 (1936); Figures 258 and 259 are from *Rhodora* 43: plates 657C and 660B (1941).

Figures 288 and 289 are from the *Transactions* of the Wisconsin Academy of Sciences, Arts and Letters 58: 314 (1970) and are used with permission from the University of Wisconsin Herbarium.

Figure 314 is from the *Annals* of the Missouri Botanical Garden 69: 853 (1983) and is used with permission of David Boufford and the Garden.

All of the remaining line illustrations are reproduced by permission of the New York Botanical Garden from H. A. Gleason's *New Britton and Brown Illustrated Flora of the Northeastern United States and Adjacent Canada* (Vol. 2, 1952).

E. G. V.

Ann Arbor
February 22, 1985

University of Michigan Herbarium

Contents

Introductory Section

Introduction *Page* 3
The Basis of this Flora *Page* 3
Is It an Established or Recognizable Species? *Page* 7
Taxonomy and Nomenclature *Page* 8
References *Page* 10

Taxonomic Section

Using this Flora: Keys, Style, Abbreviations *Page* 25
  Keys *Page* 25
  Scope of Part II *Page* 26
  Style *Page* 26
  Illustrations *Page* 27
  Abbreviations and Symbols *Page* 27
Magnoliopsida (Dicots) *Page* 29
Glossary *Page* 683
Index *Page* 693
Corrigenda *Page* 725
Addenda *Page* 727

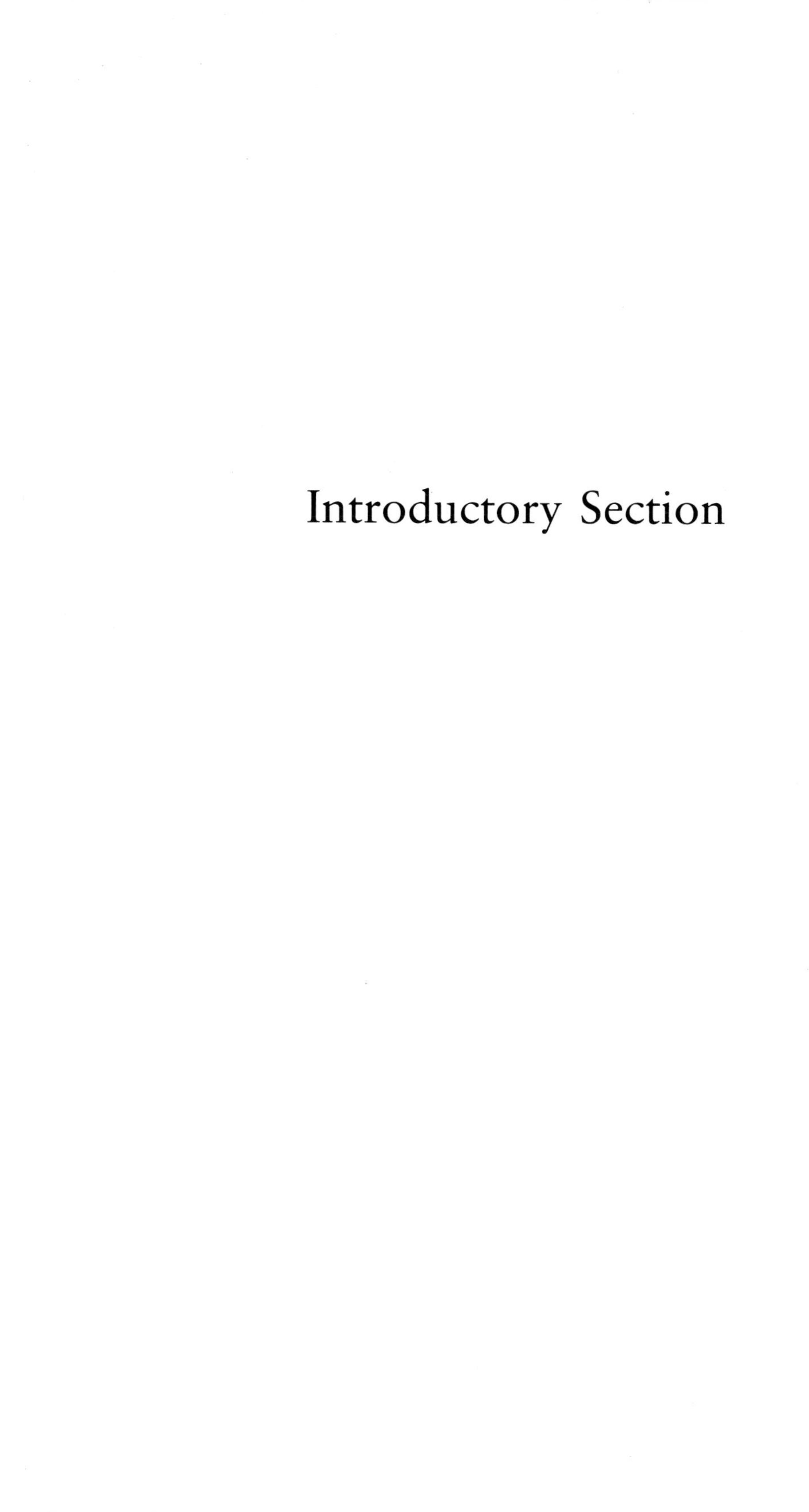

Introductory Section

Introduction

Interest in the plant life of Michigan appears strong. Photographers, berry-pickers, hikers, naturalists of diverse sorts, hunters and fishermen, "wild foods" devotees, and citizens who simply love beauty, all appreciate our natural heritage. Concern for natural areas and protection of rare species is widespread—though far from universal. The Michigan Natural Features Inventory, initiated by The Nature Conservancy and the Department of Natural Resources, gathers data on the occurrence of rare communities and organisms, and monitors their welfare. Under Michigan's Endangered Species Act of 1974, as amended, some 200 vascular plant species are officially listed as threatened or endangered in the state.

Sorting and classifying plants and other creatures is an activity as old as civilization. Primitive classifications were based on the life-and-death qualities of edibility and other uses. Plant classification is now more scientific, and identification can be a leisure-time activity. Most people enjoy knowing some of the plants and animals around them, just as knowing other neighbors makes one feel more at home. For many, a lifelong hobby may develop from outdoor pursuits. The purpose of the *Michigan Flora* is to help interested persons to expand their knowledge of the plants with which we share this part of the planet: how to identify them, where they grow in terms of both geography and habitat, and what still needs to be found out about them. While this kind of study brings pleasure to many "amateur" botanists (who often know more than the "professionals"), it is essential to those who must prepare assessments of environmental impact from developments, determine whether threatened or endangered species grow in a project area, identify a berry consumed by a child, decide whether to make jelly from a wild fruit, look up information on a plant producing an allergic reaction, or label a prize photograph.

THE BASIS OF THIS FLORA

It is tragic when any flora, whether for a local region or a larger area, fails to take into account the labors of previous workers, and I have tried to distill the literature and the collections (specimens and their labels) representing 175 years of botanical work in the state. Published literature is often of great help in understanding the status of plants with less than complete specimen labels, but identifications are often erroneous. Reports are frequently not documented by specimens at all; and, conversely, many species have been collected in the state but never attributed to it in publications.

The *Michigan Flora*, therefore, is based primarily on Michigan specimens—not on literature, and not on specimens from elsewhere. One often reads that a flora has been "compiled." Insofar as that word is applicable to the present work, it generally means "compiled from original data," although of course existing keys and monographs have often been enormously helpful in suggesting what diagnostic characters to look for on our specimens. For species rare in Michigan, both native and introduced, measurements have sometimes had to be based on material from elsewhere (preferably adjacent areas), and occasionally an extreme measurement is indicated in square brackets to note that a much larger or smaller figure is given in some literature. The range of variation in a species as it occurs *here* has been deemed most important.

Likewise, habitat information is based on what collectors have said on labels of Michigan specimens (adjusted with some common sense), on my own field experience throughout the state, and on statements in original local literature. The intent, as discussed in Part I, is to give an *impression* as to representative kinds of places where each species grows, not to catalog all possible sites and not to conform to any rigid predetermined classification of habitats—to which a collector writing a label doubtless did not conform in the first place. For many weeds and introduced species, as well as interesting native ones, a few words are given about their history in the state, going back, in some instances, to the first Geological Survey of Michigan. The First Survey, under the direction of Douglass Houghton, was responsible from 1837 to 1840 for botanical and zoological investigations as well as geological, and a list of the plant collections of 1838 was published the next year by the legislature—the first significant list of plants in the state (see McVaugh 1970).

The maps, as explained at length in Part I, are based *only* on specimens personally examined by me, not on reports in literature, correspondence, manuscripts, or conversation. There is no fully satisfactory (not to mention diplomatic) way to distinguish the reliable from the unreliable in such reports. Even when there can be no doubt that a report is based on an accurate identification of an existing specimen (and in conformity with the taxonomic dispositions herein adopted), there may still be doubt as to interpretation of the label. Many specimens cited in literature, or merely mapped, turn out to have been erroneously attributed to a county, or even state. Several Michigan communities (Lake, Mason, Oscoda, St. Joseph, Schoolcraft) are not located in the county of the same name; many township, lake, and river names occur repeatedly about the state; the Detroit Zoo is not in the same county as Detroit, nor is Isle Royale Mine on Isle Royale. Easy sources of geographic error abound!

The distinction should be made clear between a "report" and a "record" as these terms are used here. A report is a statement in a publication, or letter, or even oral communication; it is not necessarily supported by a specimen (or at least not by a correctly identified one or by one examined in the preparation of this Flora). A record is based on an actual specimen, checked and noted for this Flora—in short, a *verified* report.

In addition to the 83 counties in Michigan, seven islands or island groups in the Great Lakes are mapped separately as they are distinct enough geographically and/or phytogeographically from the mainland counties to which they are politically attached. These are shown on the map in Fig. 1, and are listed below:

Charity Islands (Arenac County)
Beaver Islands (Charlevoix County) [The entire group.]
Drummond Island (Chippewa County) [Including a few small adjacent islands, but not islands of the St. Mary's River.]
Isle Royale (Keweenaw County) [The entire archipelago comprising the National Park.]
Fox Islands (Leelanau County)
Manitou Islands (Leelanau County)
Mackinac, Round, and Bois Blanc Islands (Mackinac County) [But all other islands in the Straits of Mackinac have been included with the mainland.]

A dot in a county (or island) means that one or more specimens have been examined from that mapping unit, with complete label data recorded and filed by species and county in the University of Michigan Herbarium, where they can be examined by anyone interested in the basis for any dot: How many collections are there? When were they made? From what precise localities or habitats? By whom? Where can the specimens be found and re-examined? What specialists, if any, have checked them? All specimens for which such complete data were recorded have been rubber-stamped "Noted, 19—, Michigan Flora Project." Not even a rough count has been kept, but the total number examined (with most of them fully recorded) for Part II must approach 100,000.

Absence of a dot from a county (or island) means only that no specimen has been seen in the herbaria examined (as listed in the Acknowledgments). It does not necessarily mean that the species does not grow there. (I may even have seen it there myself, but for some reason made no specimen.) The farther away the nearest county with a record is, the more likely it is that the absence of a dot means the absence of the species. If the specialized

1. Map of Michigan, showing counties and major islands in the Great Lakes.

habitat required by some species (e. g., limestone outcrop, shoreline sand dune) does not occur in a county, the absence of the species is also to be expected. The commonest and most familiar plants, including many weeds, trees, and escapes from cultivation, which everyone assumes are well known, are often the ones with least complete maps, while the rare and interesting species are often much more thoroughly mapped.

Some localities, especially lakes, straddle a county line or refer to more than one county (e. g., "Keweenaw Peninsula") and if a collector failed to indicate the county on a label there may be no way to assign it. Records from such localities have not been mapped unless there is no other record from either of two counties; then the dot has been placed on the county line.

IS IT AN ESTABLISHED OR RECOGNIZABLE SPECIES?

Two problems relating to whether or not to include a species in this Flora are especially acute in the dicots compared to the monocots. First is the status of escapes from cultivation. Collectors all too often fail to record on their labels whether garden species were considered by them to be established outside of cultivation. There is nothing wrong with making herbarium specimens of cultivated plants, for future reference, but they should be clearly labeled as such. Likewise, specimens collected from sites where they were not planted (spreading by seeds or even suckers), as along roadsides, should also be clearly labeled as to apparent status. It is not the intention of this Flora to include records of species growing where they were planted, no matter how long they have survived. Collectors are not even consistent in using a word like "established," which could mean merely "still alive" rather than connoting "reproducing on its own" as botanists usually use that word. When a label says "garden," did the collector mean that the specimen was a weed in the garden or that it was planted there?

I have tried to admit (and hence include in keys and maps) only those species and county records which appear reasonably to represent plants growing spontaneously (not planted). Our common weeds were all once new to the state, represented by only a few individuals. Who is to say which casual waifs will never again be found in the state and which will become well established—or how long one should wait before reaching such a conclusion? Records of waifs and escapes are included, regardless of date. (Some native plants may also never again be found in the state, and indeed some are already assumed to be extinct here.) Attention is called to some dubious records. A few trees, shrubs, and ornamentals which may persist or spread slightly from cultivation are sometimes only mentioned in the text but are omitted from the keys (especially if specimens were discovered too late to remodel the keys satisfactorily).

The other special problem involves groups of particular taxonomic difficulty because they seem to be in a state of active evolution—or at least to be reproducing asexually (as apomicts) and/or hybridizing extensively. Included in this category are such notorious problem genera as *Amelanchier, Crataegus, Rubus,* and *Viola*. The practical solution is to recognize some "complexes" which include several named "species" and then to move on without delaying to solve the problems of evolution. These complexes are nothing more than groups of similar, obviously related plants for which the taxonomic relationships are imperfectly understood—or which fit imperfectly into the mold of orthodox floristic treatment. Species are not all uniform in their distinctness, and in these problem genera the matter of definition is nearly insoluble—especially in a local flora. For such groups I have elected to offer ad hoc treatments that will enable placement of most specimens and to suggest sources for further information if one cares to define taxa more narrowly.

TAXONOMY AND NOMENCLATURE

My taxonomic approach is frankly "conservative" or "traditional" if there is any need to label it. Anyone whose tastes prefer genera defined more narrowly (in, say, *Polygonum, Arenaria,* or *Euphorbia*) or broadly (as for *Pyrus*) may do so without affecting in any way the information presented on identifying characters, distribution, or habitat. Likewise, assigning all of our plants to a particular variety or subspecies will not affect the information provided. Similarly, if one prefers to "lump" species here treated as distinct, information can be combined. If, however, one desires to recognize more infraspecific taxa, or to "split" species, the maps may prove disappointing although the text may give some idea of general range for possible segregates. I consider too many named infraspecific taxa to be of little or no taxonomic significance. Such taxa are mentioned, as in Part I, only if they seem to be particularly distinctive (including those often recognized as separate species) or if they are accepted in current manuals *and* the type locality is in Michigan. Numerous names for which the type came from Michigan are not mentioned if ignored by recent manuals and applicable to no entity that appears significant. Only in *Rubus* and *Crataegus* are *all* names typified by Michigan collections accounted for.

Hybrids were neither keyed nor mapped in Part I and essentially the same practice is continued here except for a very few which are more abundant than their parents (*Rhus* ×*pulvinata*) or are often treated as species (*Drosera* ×*anglica, Spiraea* ×*vanhouttei, Viola* ×*primulifolia*). Of course, many species are of hybrid origin with increase of chromosomes, and these as usual are treated as good species.

The sequence of families, as in Part I, follows the traditional Englerian system, as exemplified by *Gray's Manual* (Fernald 1950). I follow this sequence not because it is now considered natural, but because (like the traditional tribes of the Gramineae) it is based on easy-to-remember "key" characters and makes possible ready comparisons among the manuals useful in this region which use the same system. Within each family and genus the sequence of genera and species, respectively, is the same as the sequence in which they are numbered in the keys. This system facilitates making further comparisons beyond those stated in the keys.

There is much room for differences of opinion in taxonomy, and I have indicated alternative dispositions for some of our taxa, but it is not possible to call attention to all opinions on our plants. When differing treatments exist in reliable manuals and monographs, I have pragmatically based the disposition of our specimens on the one which seems to deal most satisfactorily with the material at hand, preferring whenever possible to retain familiar concepts and names in the interest of communication.

While there are no rules for defining a species or circumscribing a genus, there are rules governing the application of scientific names, once the taxonomist has decided what needs a name. These are laid out in the *International Code of Botanical Nomenclature* (Voss, Greuter, et al. 1983). Sometimes names necessarily change because of a change in classification. If one decides, for example, that certain St. John's-worts differ at generic level from the rest, they must be called *Triadenum* rather than be retained in *Hypericum*. If one decides that our common species in this group is different from the longer-styled acute-sepaled plants of the Coastal Plain, it is to be called neither *Hypericum virginicum* nor *Triadenum virginicum*, but *Hypericum fraseri* or *Triadenum fraseri* (or *T. virginicum* ssp. *fraseri*). Any of these names *can* be correct for our plants, given a certain opinion on their classification. If one decides that our goldthread is the same species as the one in Asia, it must be called by the older name for the latter, *Coptis trifolia*, and not the later *C. groenlandica*. If the differences are thought to warrant, it may be called *C. trifolia* ssp. *groenlandica* or *C. trifolia* var. *groenlandica*. All these different taxonomic opinions are inevitably reflected in different names. On the other hand, names are sometimes discovered to be contrary to the Code, perhaps because they were not published in the proper form (invalid) or because they violate one of the other rules of nomenclature, such as not being the oldest applicable name. More rarely, a change in the Code itself requires a change of name. Sometimes names of families and genera have been conserved by an International Botanical Congress, i. e., preserved in the interests of stability and uniformity, to avoid problems such as lack of priority, variant spellings, or diverse applications.

If a name (or the author cited for it) differs in this Flora from that in other sources, even though there is no change in classification, it can usually be assumed that investigation has shown some error to be corrected. Widespread synonyms are cited herein (whether taxonomic or nomenclatural in nature), especially ones employed in Fernald (1950) or Gleason (1952). A brief indication is often given of the reason for using a relatively unfamiliar name, and a few nomenclatural problems are discussed at greater length elsewhere (Voss 1985). Further elucidation of the principles of nomenclature is given briefly in Part I (pp. 26–29) and more fully in an excellent recent book (Jeffrey 1977).

Common names are governed by no code, and as for the monocots are given when they seem truly to be in common use and not manufactured merely to add to the nomenclatural burden. Common names which are consistent with taxonomic correctness are written as separate words (red clover) and those which are not are hyphenated (bush-clover; mountain-ash). When a common name applies to all of our species in a genus, it is given after the generic name except when there is only one species in the genus. Common names applied to a single species in our flora are given after the scientific name of that species.

REFERENCES

For general identification of vascular plants in the northeastern United States and adjacent Canada, the standard manuals are still Fernald's 8th edition of *Gray's Manual* (1950), Gleason's three-volume *New Britton and Brown Illustrated Flora* (1952), and Cronquist's unillustrated condensation of the latter as Gleason and Cronquist's *Manual of Vascular Plants* (1963, currently under revision). These should be helpful for family keys, further descriptive information or illustrations, indication of total range, and alternative keys when the present work proves unsatisfactory. Smith's well illustrated *Michigan Wildflowers* (1966) includes many of the more conspicuous species but is not a complete flora. Braun's *Woody Plants of Ohio* (1961) is beautifully illustrated and very helpful especially in the southern part of Michigan. Otis' classic *Michigan Trees* has been completely rewritten by Barnes and Wagner (1981) and is excellent. Fassett's *Spring Flora of Wisconsin* is now in a well deserved fourth edition (1976) and works very well for plants blooming before mid-June. Swink and Wilhelm (1979) provide original keys and other information for plants of the Chicago area, including southwesternmost Michigan (Berrien County). A fine new flora for Manitoulin Island, Ontario—only a short distance east of the eastern end of Michigan's Upper Peninsula—includes keys for some critical groups and much additional information on identification and habitats (Morton & Venn 1984).

Several parts have appeared in *Contributions to a Flora of New York State*, edited by Mitchell. A number of volumes are now published in Mohlenbrock's *Illustrated Flora of Illinois*. These are all helpful in our region, as far as they go. The various papers in a series comprising the generic flora of the southeastern United States, under the general direction of Wood, are well illustrated, full of general information and bibliographic references, with keys to genera; an index to the first 100 parts has been published (Wood 1983).

Since the discussion in Part I on the occurrence and distribution of plants in Michigan, several publications have dealt with groups of particular phytogeographic, ecological, or other special interest in our area. These include threatened and endangered species (Beaman 1977; Beaman et al. 1985),* arctic-alpine species (Given & Soper 1981), western disjunct species (Marquis & Voss 1981), and halophytes (Catling & McKay 1981). Attention should also be called to a detailed directory of over 150 nature preserves, totaling over 250,000 acres, in Michigan (Crispin 1980).

During the preparation of this volume, several works have appeared in print which have been more or less heavily drawn upon for information on circumpolar and introduced species. These include the English translation of the *Flora of the U.S.S.R.* (Komarov et al.) and Czerepanov's supplement updating it (1973); the complete *Flora Europaea* (Tutin et al. 1964–1980); *Hortus Third* (Bailey Hortorium 1976); and Part II of Hultén's *Circumpolar Plants* (1971), which offers a worldwide perspective on many of our species. The most readily consultable checklist for North America thus far is the substantial volume by Kartesz and Kartesz (1980) although it has no bibliographic data; a more recent, bulky, and cryptic list—but more up to date in some respects—was published as the "National List of Scientific Plant Names" by the U.S. Department of Agriculture in 1982. Little's revised *Checklist of United States Trees* (1979) gives full synonymy and bibliographic information for names of trees.

All of the works mentioned above have been constantly consulted, and much helpful information has been derived from them. They are cited in full below, but not each time they are mentioned on subsequent pages. Besides all these general works and series which the serious reader may want to consult, there is a vast literature dealing with particular families, genera, and species. Some of the most useful of these are listed under "References" throughout this Flora. Included are works helpful in identification of species in our area and understanding taxonomy and variation, as well as some reports of special interest on aspects of natural history and distribution of certain plants, particularly when they represent observations made in or adjacent to the Great Lakes region. Some references less locally oriented but with excellent bibliographies are also among the selected references cited.

*Lists of taxa officially accepted as threatened or endangered in Michigan, revised periodically as required by the state's Endangered Species Act, are available from the Department of Natural Resources in Lansing.

Obviously, many additional titles could be listed, but those included will aim the reader toward further information.

Also included in the citations below are a few references of special importance, repeated from those given in Part I, such as the exemplary floras for Indiana (Deam 1940) and Missouri (Steyermark 1963), sources on Michigan botanical history, and some of the early lists for the state or parts of it (Cole 1901; Beal 1905). These references are repeated since they are frequently referred to in the text. Cronquist (1981) presents a comprehensive technical survey of families on a worldwide basis and their organization into higher categories according to modern ideas on plant systematics, with extensive documentation.

Bailey Hortorium. 1976. Hortus Third A Concise Dictionary of Plants Cultivated in the United States and Canada. Macmillan, New York. xiv + 1290 pp.

Barnes, Burton V., & Warren H. Wagner, Jr. 1981. Michigan Trees. Univ. Mich. Press, Ann Arbor. 383 pp.

Beal, W. J. 1905 ["1904"]. Michigan Flora. Rep. Mich. Acad. 5: 1-147.

Beaman, John H. 1977. Commentary on Endangered and Threatened Plants in Michigan. Mich. Bot. 16: 110–122.

Beaman, J. H., et al. 1985. Endangered and Threatened Vascular Plants in Michigan. II. Third Biennial Review Proposed List. Mich. Bot. 24: 99–116.

Braun, E. Lucy. 1961. The Woody Plants of Ohio Trees, Shrubs, and Woody Climbers Native Naturalized, and Escaped. Ohio State Univ. Press, Columbus. 362 pp.

Catling, P. M., & S. M. McKay. 1981. A Review of the Occurrence of Halophytes in the Eastern Great Lakes Region. Mich. Bot. 20: 167–179.

Cole, Emma J. 1901. Grand Rapids Flora. A. Van Dort, Grand Rapids. 170 pp.

Crispin, Susan R. 1980. Nature Preserves in Michigan, 1920–1979. Mich. Bot. 19: 99–242.

Cronquist, Arthur. 1981. An Integrated System of Classification of Flowering Plants. Columbia Univ. Press, New York. 1262 pp.

Deam, Charles C. 1940. Flora of Indiana. Dep. Conservation, Indianapolis. 1236 pp.

Deam, Charles C. 1953. Trees of Indiana. 3rd ed. Dep. Conservation, Indianapolis. 330 pp.

Fassett, Norman C. 1976. Spring Flora of Wisconsin. 4th ed., rev. by Olive S. Thomson. Univ. Wisconsin Press, Madison. 413 pp.

Fernald, Merritt Lyndon. 1950. Gray's Manual of Botany. Ed. 8. Am. Book Co., New York. lxiv + 1632 pp. [Some corrections in later printings.]

Given, David R., & James H. Soper. 1981. The Arctic-Alpine Element of the Vascular Flora at Lake Superior. Natl. Mus. Canada Publ. Bot. 10. 70 pp.

Gleason, Henry A. 1952. The New Britton and Brown Illustrated Flora of the Northeastern United States and Adjacent Canada. N. Y. Bot. Gard., New York. 3 vol. [Minor corrections in later printings.]

Gleason, Henry A., & Arthur Cronquist. 1963. Manual of Vascular Plants of Northeastern United States and Adjacent Canada. Van Nostrand, Princeton. li + 810 pp. [Some corrections in later printings.]

Hanes, Clarence R., & Florence N. Hanes. 1947. Flora of Kalamazoo County, Michigan. Vascular Plants. [Authors], Schoolcraft, Mich. 295 pp.

Holmgren, Patricia K., Wil Keuken, & Eileen K. Schofield. 1981. Index Herbariorum Part I The Herbaria of the World. 7th ed. Regnum Vegetabile 106. 452 pp.

Hultén, Eric. 1971. The Circumpolar Plants. II Dicotyledons. Sv. Vet-akad. Handl. IV. 13(1). 463 pp.

Jeffrey, Charles. 1977. Biological Nomenclature. 2nd ed. Edward Arnold, London. 72 pp.

Kartesz, John T., & Rosemarie Kartesz. 1980. A Synonymized Checklist of the Vascular Flora of the United States, Canada, and Greenland. Univ. North Carolina Press, Chapel Hill. xlviii + 498 pp.

Komarov, V. L., ed., et al. 1968– . Flora of the U.S.S.R. Israel Program for Scientific Translations, Jerusalem. Vol. 1–21 + 24 [of 30]. [Supplemented by S. K. Czerepanov. 1973. Additamenta et Corrigenda ad "Floram URSS" (tomi I-XXX). Nauka, Leningrad. 667 pp.]

Little, Elbert L., Jr. 1979. Checklist of United States Trees (Native and Naturalized). U.S. Dep. Agr. Agr. Handb. 541. 375 pp.

Marquis, Robert J., & Edward G. Voss. 1981. Distributions of Some Western North American Plants Disjunct in the Great Lakes Region. Mich. Bot. 20: 53–82.

McVaugh, Rogers, Stanley A. Cain, & Dale J. Hagenah. 1953. Farwelliana: An Account of the Life and Botanical Work of Oliver Atkins Farwell, 1867–1944. Cranbrook Inst. Sci. Bull. 34. 101 pp.

McVaugh, Rogers. 1970. Botanical Results of the Michigan Geological Survey under the Direction of Douglass Houghton, 1837–1840. Mich. Bot. 9: 213–243.

Mitchell, Richard S., ed. [1978]– . Contributions to a Flora of New York State. N.Y. State Mus. Bull. 431, 435, 442, 446, 451 [to date]

Mohlenbrock, Robert H. 1967– . The Illustrated Flora of Illinois. Southern Illinois Univ. Press, Carbondale.

Morton, J. K., & Joan M. Venn. 1984. The Flora of Manitoulin Island. 2nd ed. Univ. Waterloo Biol. Ser. 28. 106 pp.

Schwarten, Lazella, & Harold William Rickett. 1958. Abbreviations of Titles of Serials Cited by Botanists. Bull. Torrey Bot. Club 85: 277–300.

Smith, Helen V. 1966. Michigan Wildflowers. Cranbrook Inst. Sci. Bull. 42, revised. 468 pp.

Steyermark, Julian A. [1963]. Flora of Missouri. Iowa State Univ. Press, Ames. lxxxiii + 1725 pp.

Tutin, T. G., V. H. Heywood, N. A. Burges, D. H. Valentine, S. M. Walters, & D. A. Webb, eds. 1964–1980. Flora Europaea. Cambridge Univ. Press, Cambridge. 5 vol.

U. S. Department of Agriculture. 1982. National List of Scientific Plant Names. Soil Conserv. Serv. SCS-TP-159. 2 vol.

Voss, Edward G. 1972. Michigan Flora. Part I Gymnosperms and Monocots. Cranbrook Inst. Sci. Bull. 55 & Univ. Mich Herb. 488 pp.

Voss, Edward G. 1978. Botanical Beachcombers and Explorers: Pioneers of the 19th Century in the Upper Great Lakes. Contr. Univ. Mich. Herb. 13. 100 pp.

Voss, E. G., W. Greuter, et al. 1983. International Code of Botanical Nomenclature Adopted by the Thirteenth International Botanical Congress, Sydney, August 1981. Regnum Vegetabile 111. 472 pp.

Voss, Edward G. 1985. Nomenclatural Notes on Some Michigan Dicots. Mich. Bot. 24: 117–124.

Wood, Carroll E., Jr. 1983. Indexes to Papers 1 to 100 Published as Parts of the Generic Flora of the Southeastern United States. Jour. Arnold Arb. 64: 547–563.

Taxonomic Section

A. **Salix pyrifolia**
Monroe Luce Co.

E. **Chenopodium capitatum**
Monroe Emmet Co.

PLATE 1

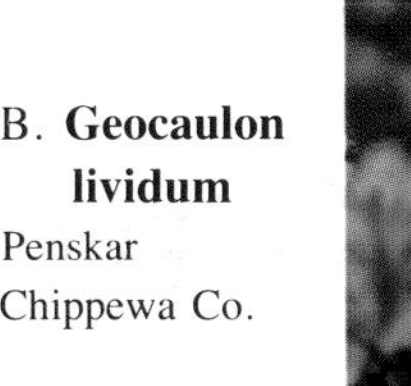

B. **Geocaulon lividum**
Penskar
Chippewa Co.

D. **Asarum canadense**
Wells Wayne Co.

C. **Arceuthobium pusillum**
Voss Emmet Co.

Sagina nodosa
s Isle Royale

F. **Claytonia virginica**
Wells Oakland Co.

E. **Aquilegia canadensis**
Wells Emmet Co.

C. **Nymphaea odorata**
Williams Schoolcraft Co.

B. **Nelumbo lutea**
Crispin Monroe Co.

D. **Clematis occidentalis**
Case Keweenaw Co.

PLATE 2

F. **Anemone multifida**
Voss Emmet Co.

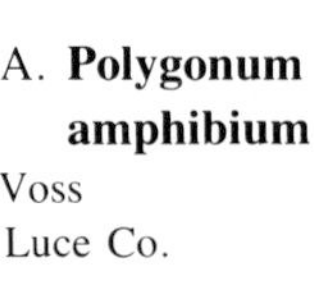

A. **Polygonum amphibium**
Voss
Luce Co.

D. **Jeffersonia diphylla**
Wells Oakland Co.

A. **Ranunculus hispidus**
Wells Oakland Co.

PLATE 3

B. **Caltha palustris**
Wells Oakland Co.

F. **Sarracenia purpurea**
Case Alger Co.

E. **Corydalis aurea**
Voss Emmet Co.

C. **Caulophyllum thalictroides**
Voss Arenac Co.

D. **Mitella diphylla**
Wells Leelanau Co.

F. **Saxifraga virginiensis**
Voss Isle Royale

PLATE 4

A. **Sanguinaria canadensis**
Mellichamp Washtenaw Co.

E. **Mitella nuda**
Case Emmet Co.

B. **Drosera linearis**
Case Chippewa Co.

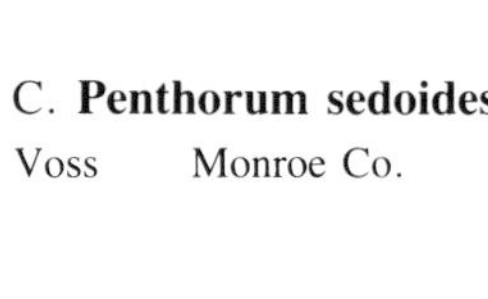

C. **Penthorum sedoides**
Voss Monroe Co.

PLATE 5

E. **Dalibarda repens**
Freudenstein Crawford Co.

C. **Rosa acicularis**
Voss Emmet Co.

D. **Sorbus decora**
Voss Luce Co.

B. **Rubus acaulis**
Voss Schoolcraft Co.

A. **Ribes triste**
Voss Emmet Co.

G. **Euphorbia polygonifolia**
Voss Emmet Co.

F. **Geranium bicknellii**
Voss Emmet Co.

B. **Nemopanthus mucronatus**
Voss Emmet Co.

A. **Empetrum nigrum**
Voss Isle Royale

PLATE 6

F. **Hypericum kalmianum**
Voss Emmet Co.

C. **Impatiens capensis**
Wells Oakland Co.

E. **Triadenum virginicum**
Voss Berrien Co.

D. **Hibiscus moscheutos**
Voss Monroe Co.

C. **Opuntia fragilis**
Wells   Marquette Co.

E (above) and F (below)
Oplopanax horridus
Voss   Isle Royale

A. **Hudsonia tomentosa**
Voss   Chippewa Co.

B. **Viola pedata**
Williams   Berrien Co.

PLATE 7

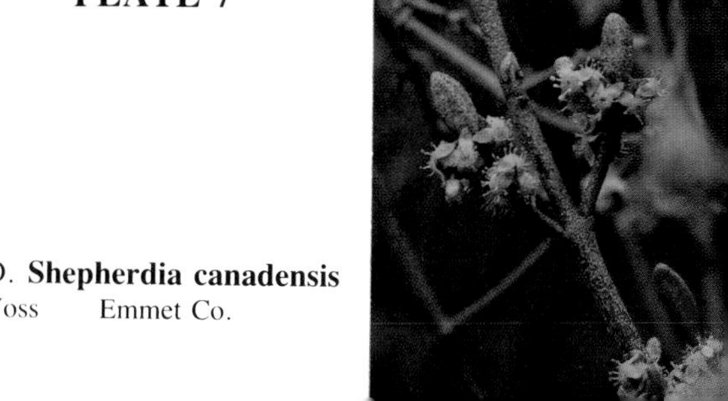

D. **Shepherdia canadensis**
Voss   Emmet Co.

A. **Epilobium hirsutum**
Voss
Cheboygan Co.

E (above) and F (below)
Cornus canadensis
Mellichamp Emmet Co.

PLATE 8

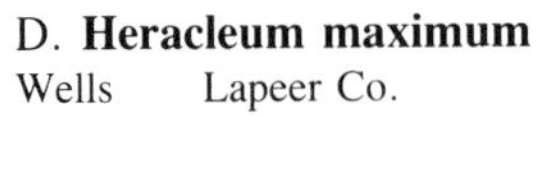

D. **Heracleum maximum**
Wells Lapeer Co.

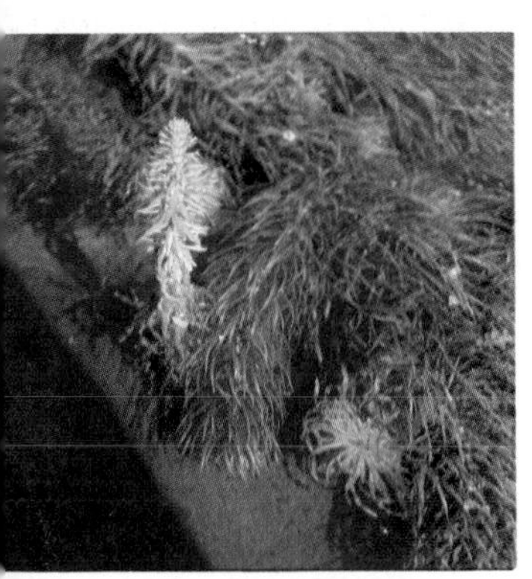

B. **Hippuris vulgaris**
Voss Emmet Co.

C. **Angelica atropurpurea**
Voss Emmet Co.

USING THIS FLORA: KEYS, STYLE, ABBREVIATIONS

"The naming of cats is a difficult matter," wrote T. S. Eliot. Plants are often as difficult. But the naming of plants becomes easier with experience, common sense—and a practical guide. It should not be necessary to repeat here the full section on identification and keying presented in Part I (pp. 30–35). Stressing a few points may nevertheless be helpful.

Keys

The keys are always strictly dichotomous: two choices ("leads") in contrasting form. The two contrasting leads bearing the same number are a "couplet." Always read *both* leads *completely* before deciding which applies best to an unknown plant. If that lead does not give you a name, drop to the next couplet below it, and continue until a name is reached. The characters considered best for identification or easiest to determine are generally stated first. In lieu of descriptions in the text (which could be very repetitive), descriptive material is presented in the keys insofar as it offers good contrasts. This style makes for longer keys but more reliable ones than those that depend on a single character (which can often be too easily misinterpreted, or absent, or ambiguous—and then there is no other to fall back upon). The user must often make a judgment about which lead applies best, and if that one fails to produce a satisfactory answer, one must go back to the "fork in the road" and try the alternative lead.

Some variable or difficult taxa may be included more than once in a key, or the key may direct one to go forward or backward to another couplet; it is thus often possible to use easier characters than if the taxon had to be keyed down at a single point. Once a name has been reached in the keys, do not assume that either you or the key is perfect. Check further information or figures in the text, or consult descriptions or illustrations in another work. Look at distribution maps and statements about habitat to see if they make sense for your specimen. If possible, compare specimens with authoritatively identified ones in a herbarium.

When identifying unknown plants, it will help to use a good lens (magnifying about 12×) and good light. A dissecting microscope with strong illumination is best, and frees the hands for manipulation. But a hand lens, bright sunlight, sharp fingernail and pencil, can accomplish wonders in the field. A scale graduated in millimeters (or preferably half millimeters) is essential. It never pays to guess at such points as measurements and ratios.

Remember that the best characters for field and herbarium identification may not be the most important features from the standpoint of technical plant classification or evolution. The curious botanist using a key, however, does not always have available the chromosome number, underground parts, both flowers and fruit simultaneously, chemical data, large population samples, related species for relative comparison, and other such very helpful information. Furthermore, the function of a local flora is not to accommodate all the variation of the taxa it includes insofar as this variation occurs in other parts of their ranges. It is precisely the restriction to local taxa and their local range of variation that makes a local flora useful in the area it is intended to serve. If it proves useful in other areas, so much the better, but that is secondary.

Scope of Part II

A question inherent in any partial flora, such as this one, is whether the plant one wishes to identify is in the book at all. In Part I of this Flora (pp. 44–47) is a discussion of the major groups of seed-plants, including several practical hints for distinguishing monocots from dicots. Any seed-plant not belonging in Part I is a dicot, of which Part II covers (with very few exceptions) the families in which the flowers lack petals or in which the petals are separate from each other. (The Pyrolaceae, for example, with separate petals, are obviously so closely related to the Ericaceae that they are to be included with the latter in Part III.)

The beginner with an unknown dicot to identify needs first to determine the family to which it belongs, using the family key in any convenient manual or text, if the family of an unknown plant (as may often be the case for the smaller families in this volume) is not apparent. The present volume is really a supplement, with local information, to the general manuals that deal with a broader area.

Style

For each family in this Flora, there is a key to genera; for each genus, a key to species. (When there is only one genus in a family in our area, or only one species in a genus, a key is of course not necessary.) Each species is listed under the same number as it bears in the key. Following information on its habitat in Michigan and any remarks on distribution, there may be comments on variation, additional (often more subjective) identifying characters, alternative classifications or names, and other matters of interest.

No special typography distinguishes the names of species not considered indigenous in the local flora. However, all species are considered native unless there is a statement to the contrary. Records of special interest (the rarest species, unusual forms, doubtful records, etc.) are often documented in the customary style, with collector's name and collection number (if any) in italics; the year of collection, if cited, following "in" and not italicized; the herbaria where specimens have been seen indicated by the standard symbols as given in the Acknowledgments (pp. xi–xii). Square brackets in giving locality data indicate information (usually the county) which was not included in the source (label or literature), but which has been added.

In describing plant parts in keys and descriptions, the standard units of the metric system are used. In describing locations (distances), habitats (depths), and trees of large size, where the English system is so widely used, it has generally been retained in the interest of clarity and familiarity.

Measurements of plant parts are based on dry specimens—the only way to gain a broad range from throughout the state. Fresh material may run a trifle larger, especially for delicate structures such as anthers or petals. The commonest range of conditions is stated, with unusual extremes at one or both ends added in parentheses. No particular statistical precision is to be assumed from these. An occasional extreme is given in square brackets, to indicate a published measurement considerably different from those obtained locally. "Ca." (about) is sometimes used when the measurements have been few or difficult to determine. The symbol ±, meaning "more or less," is frequently used as a space-saving way of saying "rather," somewhat," or "to a greater or lesser extent."

Illustrations

The scale of enlargement or reduction of a whole plant or inflorescence (or part of one) is given first in the legends, followed by indication of any details and their scale. Separate figure numbers are assigned to details only when they are on a different page from the main drawing or if they have been taken from a different source. The scales are only approximate, and usually are those stated (if any) in the sources from which the drawings were taken. On each page of figures, they are numbered in the same sequence as the species appear in the text. Occasionally, space has required that a species appear "out of order" on a different page, however, and it is then numbered in the sequence for that page.

When critical points mentioned in the keys, especially qualitative comparisons, are likely to be significantly clarified by an illustration, figure numbers have often been cited; but most figures are not cited in the keys. All figures and color illustrations are cited after the names of the species in the text.

Abbreviations and Symbols

Abbreviations for titles of periodicals basically follow Schwarten and Rickett (1958). Herbarium abbreviations are listed on pp. xi–xii. Generic names are abbreviated to the first letter when it is clear from the context what name is intended.

ca.	about (Latin: circa)
cf.	compare (Latin: confer)
cm	centimeter (see scale below, the numbered metric units)
dm	decimeter (= 10 centimeters; see scale below)
e. g.	for example (Latin: exempli gratia)
f.	form (forma)
i. e.	that is (Latin: id est)
m	meter (= 10 decimeters, 100 centimeters, 1000 millimeters)
mm	millimeter (= 0.1 centimeter; see scale below)
sens. lat.	in a broad sense (Latin: sensu lato)
sens. str.	in a narrow sense (Latin: sensu stricto)
sp.	species (singular)
spp.	species (plural)
ssp.	subspecies
TL	type locality; the locality from which a type specimen came (this is indicated by the abbreviation or by fuller discussion for all names mentioned when the type locality is in Michigan)
Tp.	Township
var.	variety (varietas)
±	more or less
×	in figure legends, scale of enlargement or reduction; otherwise the sign of a hybrid: before a specific epithet, × indicates that the binomial applies to a hybrid; between two binomials, it indicates a hybrid (sometimes putative) of that parentage

Magnoliopsida (Dicots)

SAURURACEAE — Lizard's-tail Family

A single genus in our region, represented by a single species, which occurs in eastern North America. A second species grows in eastern Asia.

REFERENCE

Wood, Carroll E., Jr. 1971. The Saururaceae in the Southeastern United States. Jour. Arnold Arb. 52: 479–485.

1. **Saururus**

1. **S. cernuus** L. Fig. 2 — Lizard's-tail

Map 1. Swamps (usually deciduous but sometimes cedar), floodplains, shallow water and mudflats at the borders of streams and ponds.

The base of the petiole surrounds the stem, leaving a scar which encircles the stem. The fruit consists of 3–4 (5) very rough nutlet-like 1-seeded carpels (± fleshy when fresh). The axis of the inflorescence, the short pedicels, the spatulate bracts (adnate to the pedicel or apparently at its summit), and often the bases of the main veins of the leaves are all ± loosely pubescent with many-jointed hairs.

SALICACEAE — Willow Family

KEY TO THE GENERA

1. Leaves with lanceolate to linear blades (1.5) 2 or more times as long as wide, the lowermost several pairs of lateral veins shorter than those toward middle of blade, scarcely prominent; buds each covered with a single scale; bracts ("scales") in ament entire; flowers with glands at base but no disc; stigmas 2, unlobed .... 1. **Salix**
1. Leaves with ovate to deltoid blades mostly less than twice as long as wide, the lowermost (or next to lowermost) pair of lateral veins equaling or exceeding all others in length and prominence (and with several smaller branches toward margin of blade); buds each covered with several overlapping scales; bracts ("scales") in ament coarsely toothed to lacerate or fringed; flowers with an oblique or symmetrical cuplike disc at base; stigmas consisting of 4 or more lobes .... 2. **Populus**

1. **Salix** Willow

The flower of a willow consists of a single pistil or (1) 2–7 (occasionally more) stamens apparently in the axil of a small scale or bract. These flowers are grouped in "catkins" or *aments*. Since the flowers are small and lack a perianth, the species are more difficult to describe than in many genera, with the added complications that the sexes are on separate plants and that in many species the flowers appear before the leaves. In addition, hybridization is not uncommon, so that the willows have a reputation for being unusually difficult to identify.

The general key below includes all species, and is based on characteristics of the flowers, fruit, and leaves. It includes fairly full descriptions of the species insofar as they differ from each other, and is intended to work with either staminate or pistillate plants, as long as leaves are present. In addition, supplementary keys are provided for staminate and pistillate plants of those species which may be found (at least sometimes) with the flowers or fruit well developed before the leaves appear. These keys will be helpful in many cases, but additional leaf collections from the same individual plant should be made for certainty. One should not expect to be able to identify incomplete or immature specimens of *Salix* any more readily than such specimens in other genera. A key to all the species of, say, *Panicum* or *Carex* would be unthinkable if based solely on leaves or solely on flowers. In a sense, *Salix* might be considered an easier genus than many, since the majority of specimens (except for vigorous new sprouts, which should not be collected for specimens) *can* usually be identified on the basis of flowers or leaves alone. But the best practice is to mark a plant and collect material from it at different seasons.

Identification of willows requires patience and good magnification of about 10× with adequate illumination. In several species, the presence or absence of copper-colored hairs on young foliage is a very helpful characteristic; do not assume that these are absent until a careful search has been made. (In some species, such as *S. fragilis* and *S. amygdaloides*, the presence of such copper hairs is not significant.) Similarly, examine the petioles of a number of leaves before concluding that glands are absent. Look at several leaves to gain an impression of shape and the nature of the margin. In short, do not base decisions on a single examination of a single part. In a few species, the sequence of flowering in the ament is distinctive. This is a characteristic best seen in partly developed staminate aments; fully mature or exceedingly young aments will not demonstrate the sequence. The filaments of older stamens may shrivel and shorten, making them falsely appear younger, so great caution must be exercised in using this character. In *S. discolor, S. humilis, S. myricoides,* and *S. cordata* the first well developed stamens with ripe anthers appear at the apex of the ament, those at the base developing subsequently.

The keys presented here attempt to use several characters whenever possible at each contrast, to give a better picture of the plants involved, and they treat several ambiguous or variable species at more than one place. The resulting keys are somewhat complex in places, with more cross-references than would look tidy, but they may lead more often to the correct identification than very brief keys. These were first drafted in 1958, based on examination and measurement of approximately 2000 specimens of *Salix* from Michigan, almost all of them annotated (many shortly before his death) by C. R. Ball, long the outstanding authority on willows. Dr. Ball's annotations, correspondence, suggestions, and published accounts were of very considerable help. However, the selection of "key characters" and design of the keys are original. After the text had been in manuscript for many years, some modifications were made based on the recent investigations by Robert D. Dorn and George W. Argus. Suggestions and identifications by by Dr. Argus are gratefully acknowledged. Specimens annotated by Ball, or by me prior to 1983, may bear names not now adopted (*S. glaucophylloides, S. interior, S. rigida, S. subsericea*).

The largest known trees of any species, wild or planted, in Michigan are specimens of *S. alba* and *S. nigra* each about 9 feet in diameter and the "national champion" of its kind.

REFERENCES

Argus, George W. 1965. Preliminary Reports on the Flora of Wisconsin. No. 51. Salicaceae. The Genus Salix—the Willows. Trans. Wisconsin Acad. 53: 217–272. [Includes very helpful illustrations.]

Argus, George W. 1980. The Typification and Identity of Salix eriocephala Michx. (Salicaceae). Brittonia 32: 170–177.

Dorn, Robert D. 1975. A Systematic Study of Salix section Cordatae in North America. Canad. Jour. Bot. 53: 1491–1522.

Dorn, Robert D. 1976. A Synopsis of American Salix. Canad. Jour. Bot. 54: 2769–2789.

Soper, James H., & Margaret L. Heimburger. 1982. Shrubs of Ontario. Royal Ontario Museum, Toronto. 495 pp. [*Salix*, pp. 14–85.]

Wagner, Warren H., Jr. 1968. Teratological Stamens and Carpels of a Willow from Northern Michigan. Mich. Bot. 7: 113–120.

GENERAL KEY TO THE SPECIES

1. Leaves green beneath (similar to upper surface, or ± yellow-green, glabrous or pubescent, but neither whitened nor glaucous nor with the surface hidden by dense white pubescence)

 2. Leaves ± entire, revolute, the upper surface with impressed (sunken) veins; capsules (and often other parts) white-tomentose; stamens 2, the filaments glabrous . ... 9. **S. candida**

 2. Leaves serrate to remotely denticulate (very young ones sometimes nearly entire), not revolute (though margins sometimes thickened), the upper surface smooth or with the veins slightly raised; capsules and young growth glabrous to silky, but not tomentose; stamens various

3. Leaf blades remotely denticulate (the greatest distance between teeth on mature leaves mostly 3–6 mm), mostly linear-oblong or narrowly lanceolate; stipules minute or none; stamens 2, the filaments densely pilose on lower half; capsules pubescent or glabrous . . . . 1. **S. exigua**

3. Leaf blades closely serrate (the teeth on mature leaves less–usually much less–than 3 mm apart), narrowly lanceolate to ovate or elliptic (except in *S. nigra*, with 4–7 stamens); stipules, at least on shoots, conspicuous in some species; stamens more than 2 or the filaments nearly or quite glabrous; capsules glabrous

4. Blades linear-lanceolate, acute to somewhat rounded at base, at maturity mostly 4–10 times as long as wide; stamens 4–7; petioles usually not glandular at the summit; capsules 3–4 (4.5) mm long, with styles less than 0.25 mm long (or rarely slightly longer) . . . . 2. **S. nigra**

4. Blades lanceolate or broader, rounded to subcordate at base, at maturity 1.5–6 times as long as wide; stamens 2 or if more the petioles glandular at the summit; capsules various

5. Petioles with 2 or more glands at or near junction with the blade; stamens (2) 3 or more (usually 5); year-old branchlets glabrous and usually shiny . . . . [go to couplet 22]

5. Petioles without glands (though margin of blade may be glandular-toothed nearly to its base); stamens 2; year-old branchlets often pubescent or at least puberulent above the nodes

6. Unfolding young leaves usually very reddish, moderately to very sparsely pubescent; older leaves usually somewhat whitened beneath, glabrous or nearly so, the marginal teeth tending to be crenulate and with glands absent or very small and ± sunken; stipes ("pedicels") 1–2.5 mm long, usually about equaling or exceeding the scales; styles ca. 0.5 mm long or shorter; young aments flowering from base to apex; year-old branchlets densely to sparsely (and locally) puberulent, rarely somewhat villous . . . . 3. **S. eriocephala**

6. Unfolding leaves, even if (as occasionally) reddish, mostly covered with dense long silky hairs; older leaves densely hairy to glabrate, definitely green beneath (at least under the hairs), the margins usually sharply (sometimes doubly) serrate, the teeth tipped with prominent enlarged glands; stipes under 1 mm long, shorter than the scales; styles usually 0.6–1.5 mm long; young staminate aments flowering from apex to base; branchlets ± villous-pubescent . . . . 4. **S. cordata**

1. Leaves whitened beneath, usually ± strongly glaucous and/or the lower surface completely hidden by dense white pubescence

7. Margins of foliage leaves (not always peduncular bracts) entire or sometimes (especially on sprouts) rather coarsely and irregularly or obscurely crenulate-serrate on apical half, often ± revolute; stamens 2 (or 1 in *S. purpurea*); capsules pubescent (except in *S. pedicellaris* & *S. myricoides*) [Note: All species included under this lead, except *S. pedicellaris*, are also in the supplementary keys to flowering specimens; those keys may be helpful to consult for material with aments but with leaves not yet fully mature.]

8. Leaves completely glabrous (or hairy when first emerging from bud, the hairs rapidly deciduous)

9. Aments on long or short branchlets ("peduncles") bearing green bracts or leaves; capsules glabrous, on stipes 1.2–4 mm long; filaments glabrous; anthers yellow; axillary buds at most 5 mm long (*S. pedicellaris*) or 7 mm (*S. myricoides*); leaves strictly entire or obscurely toothed

10. Aments up to 3 cm long, terminating leafy branchlets ca. 1–3 cm long; capsules strongly red-tinged (rarely bright green); style very short, not over 0.25

mm long; anthers ca. 0.5 (0.4–0.6) mm long; leaves strictly entire, slightly revolute, often ± rounded at apex; branchlets glabrous; stipules absent .... 5. **S. pedicellaris**

10. Aments at maturity 3–9 cm long (or the staminate, and rarely pistillate, as short as 2 cm), on short green-bracted branchlets; capsules yellow-brown to greenish; styles usually 0.6–1.5 mm long; anthers 0.6–1 mm long; leaves ± acute at apex, with at least a few obscure teeth; branchlets, especially of current year, usually ± densely pubescent; stipules often present .. 22. **S. myricoides**

9. Aments essentially sessile, at most with tiny green or brown bracts at base; capsules pubescent, the stipes various; filaments often pilose at base; anthers yellow to red or purplish; axillary buds up to 13 mm long on sprouts (though often under 5 mm, especially in *S. discolor* & *S. planifolia*); leaves ± obscurely toothed on about the apical third or half

11. Capsules less than 3 mm long, sessile; styles essentially none; filaments (and sometimes anthers) united into a single stamen pilose at the base; anthers less than 0.5 mm long, dark brown or purplish; stipules none; leaves often tending to be subopposite (especially on older twigs), thickish, purplish, ± oblanceolate or narrowly oblong; shrub escaped from cultivation......... 6. **S. purpurea**

11. Capsules (3.5) 5–10 mm long, sessile or on stipes up to 4.5 mm long; styles usually 0.5–1 mm or (in *S. planifolia*) longer; filaments distinct, glabrous or pilose at the base; anthers ca. (0.5) 0.6–1.2 mm long, yellow or sometimes flushed with reddish; stipules often present; leaves alternate, thin, the blades green (above), obovate (-oblanceolate) to elliptic; shrubs native .... [go to couplet 18]

8. Leaves pubescent on one or both surfaces, at least until well opened

12. Branchlets glabrous, usually ± strongly and extensively glaucous; leaf blades rather densely covered with lustrous hairs beneath when mature (glabrous or nearly so above), with narrowly but strongly revolute margins; stipules none . ... 7. **S. pellita**

12. Branchlets pubescent or glabrous, not (or occasionally slightly) glaucous; leaves and stipules various

13. Pubescence of leaves consisting of straightish, silky, and appressed hairs

14. Leaf margins ± revolute; capsules sessile or nearly so; styles 0.5–1.5 mm long; plant escaped from cultivation ....8. **S. viminalis**

14. Leaf margins at most thickened, not revolute; capsules distinctly stipitate (stipes 0.5–3.5 mm long); styles less than 0.5 mm long; plants widespread native shrubs .... [go to couplet 33]

13. Pubescence of leaves consisting of ± curled, tomentose, or woolly hairs (at most silky on youngest leaves just emerging from bud)

15. Under surface of leaf blade actually green, hidden by dense white- or gray-tomentose pubescence (rarely glaucous with sparser white tomentum); blades linear to lanceolate or narrowly oblong, (4) 5–12 (17) times as long as wide; ovary and capsule ± densely white-tomentose, on stipe up to 1 mm long; young branchlets usually with flocculent tomentum..............9. **S. candida**

15. Under surface of blades ± whitened or glaucous, seldom completely hidden by the pubescence; blades generally broader, usually less than 5 times as long as wide (rarely up to 6 times); ovary and capsule silky-pubescent, on stipe 1–6 mm long; young branchlets glabrous or puberulent to villous (not flocculent-tomentose)

16. Aments on short leafy branchlets; pubescence (even on young reddish leaves) whitish, with no admixture of red or coppery hairs; leaves usually retaining at least sparse pubescence throughout above when fully developed, rugose, with veins and veinlets impressed above and prominent beneath; scales of

aments rather narrow, whitish to pale brown, of uniform color or slightly darker at base or pink-tipped . . . . 10. **S. bebbiana**

16. Aments on short branchlets or sessile, but only rarely leafy-bracted (hybrids?); pubescence of unfolding leaves (especially above) in small or large part of copper-colored hairs; leaves becoming glabrous above when fully developed except often for whitish or coppery hairs on midrib, flat or (*S. humilis*) slightly rugose (a few coppery hairs occasionally present at maturity); scales of aments typically rather broad, ± obovate, very dark brown to black except for pale base

17. Year-old branchlets glabrous

18. Capsules ca. (3.5) 4–6 mm long, sessile or on stipes up to 1 mm; styles 0.7–1.5 mm long; leaf blades glossy (rarely dull) dark green above with ± parallel lateral veins; plant of Isle Royale . . . . 11. **S. planifolia**

18. Capsules (5) 6–10 mm long, on stipes (1) 1.5–4.5 mm; styles usually 0.5–1 mm long; leaf blades not glossy, with more irregular lateral veins; plant common throughout the state . . . . 12. **S. discolor**

17. Year-old branchlets pubescent or puberulent at least in small patches above the nodes [Note: Sterile specimens not safely distinguished; see text. If old aments are present, the following may help.]

19. Pistillate aments 0.5–2.5 (rarely to 4 or even 5.5) cm long; anthers red-purple to brown, 0.4–0.6 mm long (rarely larger); leaves often slightly rugose (and somewhat revolute-margined), the veinlets impressed above . . . . 13. **S. humilis**

19. Pistillate aments 3–12 (14) cm long; anthers yellow or flushed with reddish, (0.6) 0.7–0.9 (1.2) mm long; leaves flat or the main veins slightly raised above

20. Wood smooth beneath the bark; plant a common native shrub . . . . 12. **S. discolor**

20. Wood (at least of 2–3-year-old twigs) with distinct long ridges beneath the bark; plant a rare escape from cultivation . . . . 14. **S. cinerea**

7. Margins of leaves (at least mature foliage leaves) ± finely and distinctly serrate (the teeth rather distant in *S. exigua* & *S. fragilis*) almost or quite to the base of the blade; stamens 2 to several; capsule pubescent or glabrous

21. Petiole with prominent (sometimes stalked) glands or projections above at or near junction with the blade

22. Leaves lanceolate to linear-lanceolate, attenuate to apex; capsules 3–6 mm long; stamens 2–3

23. Blades of leaves usually ± silky, at least beneath, at maturity, with 6–10 (13) teeth per cm on the margin; capsules sessile or nearly so, 3–3.5 (4.5) mm long . . . . 15. **S. alba**

23. Blades of leaves glabrous at maturity (silky when very young), with 4–6 teeth per cm on the margin; capsules on short stipes (ca. 0.7 mm long), 3.5–6 mm long . . . . 16. **S. fragilis**

22. Leaves broadly lanceolate to ovate-elliptic with acute or acuminate apex; capsules 4–10 (12) mm long; stamens usually 5

24. Blades of mature leaves ca. 1.5–3.5 times as long as wide, green or slightly whitened beneath, acute to short-acuminate; petioles and young foliage glabrous; stipules usually present (falling early) on young sprouts; leaves or leafy bracts on the short flowering branchlets usually entire or nearly so; scales, especially in staminate aments, sparsely pubescent to glabrous except at base; capsules smoothish and 4–6 mm long, dehiscing after July 1 (or sometimes as early as May in southern Michigan); plant an introduced shrub or small tree rarely escaped . . . . 17. **S. pentandra**

24. Blades of mature leaves 2–6 times as long as wide, green to strongly glaucous beneath; leafy bracts of flowering branchlets finely toothed, like the foliage leaves; plants common native shrubs, differing from *S. pentandra* as described below (e. g., if leaves whitened beneath, then stipules always lacking, capsules larger, and scales pubescent to their tips; if green beneath, then tending to be long-acuminate on sprouts, often pubescent on petioles when young, the capsules ripening early)

25. Leaves somewhat whitened to strongly glaucous beneath, acute or short-acuminate, estipulate; petioles and young leaves glabrous; scales of both staminate and pistillate aments usually ± pubescent to their tips; mature capsules (6) 7.5–10 (12) mm long, ± granular-roughened or wrinkled throughout, dehiscing after July 1 (occasionally June 15–July 1 in southern Michigan or in dry or early seasons) .... 18. **S. serissima**

25. Leaves dark to pale green but not whitened beneath, often stipulate on sprouts and long-acuminate at maturity; petioles and young foliage often sparsely to ± heavily pubescent with copper-colored (sometimes whitish) hairs; scales usually with a prominent glabrate area on apical third or half; mature capsules 4–6.5 (7.5) mm long, smooth (except usually in wrinkled beak), dehiscing before June 15 (or as late as July 1 in northern Michigan or late seasons) . ... 19. **S. lucida**

21. Petioles without glands (or these occasionally obscure in *S. amygdaloides*)

26. Scales of aments yellowish, pilose at base and margins, glabrate on back, deciduous before ripening of capsules; flowers tending to be spaced in whorls, in rather slender, lax aments; capsules glabrous; stamens 2–7; plants native or introduced trees (when full-grown), with lanceolate, usually attenuate leaves, these glabrous to somewhat silky at maturity

27. Leaves with vein islets coarse or obscure; stamens 2 (3); capsules 3–6 mm long, sessile or on very short stipes .... [go to couplet 23]

27. Leaves with a very fine network of veins, forming tiny islets (ca. 4 per mm) clearly visible below; stamens 4–7; capsules 4.5–6 mm long, on stipes (0.8) 1–2.5 mm long ....20. **S. amygdaloides**

26. Scales (except in the otherwise distinctive *S. exigua* & *S. myricoides*) brown to black (pale at base), frequently pilose throughout, persistent; flowers generally crowded into thickish aments; capsules glabrous or pubescent; stamens 2; plants native shrubs, with leaves various

28. Surface of leaves actually green beneath, covered with dense whitish pubescence

29. Margins of leaves closely serrate, the teeth tipped with enlarged glands; capsules glabrous; filaments glabrous or nearly so; young aments appearing with or slightly before the leaves; scales dark-tipped, persistent .... 4. **S. cordata**

29. Margins of leaves rather remotely denticulate, the teeth often 3 mm or more apart, without enlarged glands; capsules often thinly silky; filaments ± densely pilose on basal half; aments developing after the leaves; scales yellowish, deciduous .... 1. **S. exigua**

28. Surface of leaves whitened or glaucous beneath, glabrous or retaining pubescence. [Note: If leaves are rather coarsely crenate-toothed and the specimen does not work well here, try couplet 6. Furthermore, all five species below are included in the supplementary keys to flowering material (although only the last two normally have aments before the leaves); those keys may be helpful to consult for material with aments.]

30. Leaf blades ± rounded (occasionally acute) to cordate at base, often over 1.5 cm wide at maturity, glabrous or retaining a little pubescence along the midrib; capsules glabrous; stipules often present and conspicuous, especially

on sprouts, ovate to reniform; young unfolding leaves at tips of branchlets usually reddish

31. Foliage with a definite "balsamic" (almost cosmetic or spicy) odor, which is usually persistent long after drying; mature leaves completely glabrous; year-old branchlets red and shiny, smooth and glabrous (the current year's branchlets glabrous to somewhat puberulent when very young); mature capsules on stipes 2–4 mm long, subtended by rather uniformly light brown scales .... 21. **S. pyrifolia**

31. Foliage without balsamic odor; mature leaves often ± pubescent on petioles and midribs; year-old branchlets usually puberulent above the nodes if not more densely pubescent (current year's branchlets frequently densely short-villous), not shiny and seldom red; mature capsules on stipes 1–3 (3.5) mm long, subtended by dark brown to black scales pale at the base

32. Young unfolding leaf blades at tips of branchlets often reddish, glabrous or usually with some copper-colored hairs mixed with whitish; mature leaves thick, usually very glaucous beneath; styles ca. 0.6–1.5 mm long; young staminate aments flowering from apex to base .... 22. **S. myricoides**

32. Young unfolding leaves ± strongly red, but only with whitish hairs (neither glabrous nor with coppery hairs); mature leaves thinner in texture, ± whitened to almost green beneath, not strongly glaucous; styles ca. 0.5 mm long; young staminate aments flowering from base to apex .... 3. **S. eriocephala**

30. Leaf blades acute to (rarely) rounded at base, nearly always less than 1.5 (rarely to 2) cm wide at maturity, glabrous to silky; capsules at least thinly pubescent; stipules rarely present (at most tiny and narrow); young unfolding leaves not really red, though often with coppery hairs and/or blackening in drying

33. Mature leaves glabrous or nearly so above (except for puberulent midrib), the underside ± silky, usually rather densely so, with the main lateral veins prominently rib-like; young leaves densely silky, without any coppery hairs; year-old branchlets puberulent (or sometimes glabrate), those of the current year usually densely so; capsules (2) 3.5–6.5 mm long, rather plump (ovoid) and rounded at tip .... 23. **S. sericea**

33. Mature leaves ± pubescent above or essentially glabrous on both sides, the lateral veins not prominent below; young leaves nearly always with few to many coppery hairs mixed with whitish ones (or practically glabrous); year-old branchlets glabrous (sometimes slightly glaucous) to occasionally slightly puberulent; current year's branchlets puberulent to glabrate; capsules 4.5–8.5 mm long, ± lanceolate .... 24. **S. petiolaris**

SUPPLEMENTARY KEYS TO FLOWERING MATERIAL WITH AMENTS WELL DEVELOPED BEFORE THE LEAVES

Do not try to use these keys on specimens either with very immature aments or with foliage leaves developed, as the species may not be included or the measurements may differ from those given here. (Developing leafy bracts may, however, occur on the flowering branchlets or "peduncles" in several species keyed below.)

These keys are only guides to typical material; extremes of variation may cause difficulty. It is always best to collect leaves from the same plant later in the season. Be sure to measure the most mature examples of any part described, and several of them. Stipes, in particular, may be shorter than the measurements given if the flowers are not fully developed. Measurements of styles are of the undivided portion. Measurements are based primarily on dry specimens, and anthers, in particular, may be slightly larger when fresh. Although some unfortunate exceptions occur, these keys are offered for whatever help they may provide.

Pistillate Specimens

1. Ovaries completely glabrous
  2. Year-old branchlets (but not necessarily new branchlets) completely glabrous, reddish, shining; bracts on flowering branchlets with conspicuous glandular margins; plants with a balsamic odor persisting in dry specimens; styles scarcely 0.5 mm long . . . . 21. **S. pyrifolia**
  2. Year-old branchlets ± densely puberulent or pubescent, at least in a small region above each node, scarcely ever both reddish and shining throughout (if smooth, the bracts not conspicuously glandular); flowers sweet-smelling, perhaps, but without persistent balsamic odor; styles ca. 0.5–1.5 mm long
    3. Longest styles ca. 0.5 mm long; stipes 1–1.5 (rarely 2.5) mm long, at most equaling to slightly exceeding the body (not hairs) of the subtending scales . . . . 3. **S. eriocephala**
    3. Longest styles (0.5) 0.7–1.5 mm long; stipes various
      4. Stipes very short, inconspicuous, scarcely 0.5 mm long (rarely to 1 mm), shorter than subtending scales . . . . 4. **S. cordata**
      4. Stipes 1–3 mm long, about equaling or exceeding the scales . . . . 22. **S. myricoides**
1. Ovaries pubescent
  5. Ovaries sessile or nearly so, even the most mature stipes not exceeding 1 mm, shorter than subtending scales (except often in *S. candida*)
    6. Styles obsolete or very short (to 0.4 mm)
      7. Ovaries up to 3 mm long; leaves (leaf scars) tending to be subopposite (especially on older twigs); growth of current year not at all begun or essentially glabrous . . . . 6. **S. purpurea**
      7. Ovaries up to 5.5 mm long; leaves always alternate; youngest branchlets and growth of current year (including bracts on flowering branchlets) just beginning at flowering time, finely silky-pubescent . . . . 23 **S. sericea** (but cf. also couplet 14)
    6. Styles distinct, 0.5–1.5 (2) mm long
      8. Pubescence of ovaries (and, usually, bud scales and young branchlets) definitely tomentose or woolly . . . . 9. **S. candida**
      8. Pubescence silky or velvety, of ± straight hairs
        9. Branchlets puberulent; plant escaped from cultivation . . . . 8. **S. viminalis**
        9. Branchlets glabrous; plant native, in northern Michigan
          10. Scales black; branchlets not glaucous; aments appearing strictly before the leaves . . . . 11. **S. planifolia**
          10. Scales brown; branchlets often very glaucous; aments usually appearing as leaves begin to expand . . . . 7. **S. pellita**
  5. Ovaries on stipes 1–5 mm long (except in very immature aments), in some plants exceeding the subtending scales

11. Styles 0.5–1 mm long
12. Wood smooth beneath the bark; plant a common native shrub...... 12. **S. discolor**
12. Wood (at least of 2–3-year-old twigs) with distinct long ridges beneath the bark; plant a rare escape from cultivation........................ 14. **S. cinerea**
11. Styles obsolete to 0.5 mm long
13. Scales subtending stipes straw-colored to pale brown, ± elongate, uniformly colored or pinkish at apex or slightly darker at base; stigmas often completely sessile; leafy bracts developing on flowering branchlets; aments 1–6 cm long; year-old branchlets usually pubescent (at least finely puberulent), sometimes glabrous .... 10. **S. bebbiana**
13. Scales pale to dark brown or black, usually rather broadly elliptic or obovate and paler at base; stigmas usually on short styles; bracts, aments, and branchets various
14. Leafy green bracts usually developing on flowering branchlets; aments 1–3 (3.5) cm long [Note: Mature leaves generally necessary for accurate identification.]
15. Year-old branchlets nearly always glabrous and shining, frequently a little glaucous; ovaries usually 2.5–7 mm long, lanceolate (the sides straight to concave, ± tapered to apex) .... 24. **S. petiolaris**
15. Year-old branchlets generally finely puberulent, at least in patches (sometimes completely glabrous); ovaries tending to be shorter (often under 3 mm), blunt, and ovoid (convex-sided) .... 23. **S. sericea** (and some pubescent specimens of *S. petiolaris*)
14. Leafy bracts not developing on flowering branchlets (at least not until foliage leaves are also present; rarely [hybrids?] earlier); aments 0.5–12 (14) cm long
16. Year-old branchlets glabrous .... 12. **S. discolor**
16. Year-old branchlets pubescent, at least in patches above the nodes
17. Aments 3–12 (14) cm long; mature stigmas often elongate (up to 5–8 times as long as thick and 1 mm long).......................... 12. **S. discolor**
17. Aments 0.5–2.5 (rarely to 4) cm long; stigmas shorter and relatively stouter .... 13. **S. humilis**

Staminate Specimens

1. Filaments (and sometimes anthers) united, forming a single stamen pilose at the base; leaves (leaf scars) tending to be subopposite (especially on older twigs).... .... 6. **S. purpurea**
1. Filaments 2, distinct (or rarely fused at base only), pilose or glabrous; leaves alternate
2. Flowering branchlets developing at most small brown bracts (occasionally green in *S. discolor*)
3. Anthers 0.6–1.2 mm long, averaging 0.7–0.8 mm or more; year-old branchlets often (not always) completely glabrous
4. Wood smooth beneath the bark; plant a common native shrub ...... 12. **S. discolor**
4. Wood (at least of 2–3-year-old twigs) with distinct long ridges beneath the bark; plant a rare escape from cultivation .... 14. **S. cinerea**
3. Anthers 0.3–0.7 mm long, averaging at most 0.6 mm (usually ca. 0.4–0.6 mm); year-old branchlets densely pubescent throughout to puberulent above the nodes only (or occasionally in yellow-anthered plants completely glabrous)
5. Anthers predominantly yellow (rarely yellow-brown) .... [go to couplet 8]
5. Anthers very dark, red-violet (or pale brown in age, but scales often strongly tinged with red-violet, placing such plants here)

6. Pubescence of branchlets largely or entirely flocculent-tomentose, white; pubescence of scales often ± wavy or curled .... 9. **S. candida**
6. Pubescence gray or sordid, villous to puberulent; hairs of scales ± straight.. .... 13. **S. humilis**

2. Flowering branchlets developing small to conspicuous leafy green (sometimes densely silky) bracts
   7. Anthers dark red-violet, 0.3–0.5 mm long; pubescence of branchlets, bracts, etc., mostly or entirely flocculent-tomentose, white .... 9. **S. candida**
   7. Anthers predominantly yellow (to yellow-brown), 0.4–1.1 mm long; pubescence of branchlets not flocculent-tomentose—usually rather gray or sordid, villous to puberulent, or none
      8. Anthers 0.6–1.1 mm long, averaging ca. 0.7 mm or more
         9. Mature scales whitish to pale brown, of uniform color or slightly darker at base or pink-tipped; year-old branchlets pubescent to glabrate .... 10. **S. bebbiana**
         9. Mature scales brown to black, usually paler at base (occasionally ± uniform medium brown or reddish); year-old branchlets pubescent or glabrous
            10. Year-old branchlets completely glabrous................... [go to couplet 16]
            10. Year-old branchlets densely pubescent to finely puberulent, at least in patches above the nodes
               11. Bracts of flowering branchlets densely silky-pubescent beneath, the margins with prominent enlarged ± spherical glands (sometimes hidden in the hairs); aments flowering from apex to base .... 4. **S. cordata**
               11. Bracts thinly silky or glabrous except at base, the margins with glands absent or scarcely enlarged and appearing merely as tips of low teeth; aments flowering variously
                  12. Filaments pilose toward base; bracts rather small and inconspicuous.... .... 24. **S. petiolaris**
                  12. Filaments glabrous; bracts small or, often, well developed and conspicuous
                     13. Scales with rather wavy or curled hairs; bracts of flowering branchlets not glandular-margined; aments flowering from base to apex......... .... 3. **S. eriocephala**
                     13. Scales with ± straight hairs; bracts often with small glandular-tipped teeth; aments flowering from apex to base .... 22. **S. myricoides**
      8. Anthers 0.4–0.7 mm long, averaging at most 0.5–0.6 mm
         14. Year-old branchlets densely pubescent throughout to puberulent at least in small patches above the nodes
            15. Bracts densely silky-pubescent beneath, the margins with prominent enlarged ± spherical glands (sometimes hidden by the hairs); aments flowering from apex to base .... 4. **S. cordata**
            15. Bracts thinly silky or glabrous except at base, the margins without glands; aments flowering from base to apex or from middle to both ends .... 23. **S. sericea***
         14. Year-old branchlets completely glabrous
            16. Branchlets rather strongly glaucous .... 7. **S. pellita**
            16. Branchlets not glaucous (or slightly so in *S. petiolaris*)
               17. Branchlets reddish and shiny; margins of bracts prominently glandular-toothed; aments flowering from base to apex; plants with persistent balsamic odor when dry .... 21. **S. pyrifolia**

*Some puberulent-branched specimens of *S. petiolaris* will run here, as will small-anthered specimens of *S. eriocephala* and also *S. viminalis*, a rare escape from cultivation. Pistillate material and/or leaves are necessary to distinguish these safely.

17. Branchlets greenish, yellowish, or dark, but seldom red and shiny (cf. *S. planifolia*); margins of bracts (if any) not glandular, or if obscurely so, the aments flowering from apex to base; flowers sweet-smelling at most, but plant without persistent balsamic odor

18. Anthers 0.6–1 mm long; aments flowering from apex to base . . . . 22. **S. myricoides**

18. Anthers 0.4–0.7 mm long; aments flowering from base to apex or from middle toward both ends

19. Scales black; aments sessile, without bracts on flowering branchlets, appearing strictly before the leaves; plant of Isle Royale . . . . . 11. **S. planifolia**

19. Scales brown; aments usually on bracted flowering branchlets and appearing as leaves begin to expand; plant widespread . . . 24. **S. petiolaris** (and some specimens of *S. sericea*, which usually has puberulent branchlets)

1. **S. exigua** Nutt. Fig. 3 Sandbar Willow

Map 2. Shores, dunes, stream margins, ditches, only rarely on dry ground; forms large dense thickets from extensive root systems.

Long known as *S. interior* Rowlee, which is now considered to be the same species as *S. exigua* of western North America. This is one of our most distinctive willows in its remotely toothed leaves. The leaves are ± densely silky and also tend to be broader in a form which usually occurs only on dunes and other sandy sites, f. *wheeleri* (Rowlee) E. G. Voss (TL: Black Lake, Cheboygan Co.)—named for Michigan's accomplished field botanist, Charles F. Wheeler (1842–1910).

2. **S. nigra** Marsh. Fig. 4 Black Willow

Map 3. A shrub to large tree, along rivers, streams, and lakes, and in other moist places.

The very fine reticulation of veinlets visible on the undersides of the leaves is similar to that in *S. amygdaloides*, which differs in having the leaves glaucous beneath and normally somewhat broader as well as estipulate. The hybrid of the two species, *S. ×glatfelteri* Schneider, is known from Michigan and should be expected where the two are found together. Dry specimens of *S. nigra* could be confused superficially with *S. petiolaris* if the

1. Saururus cernuus

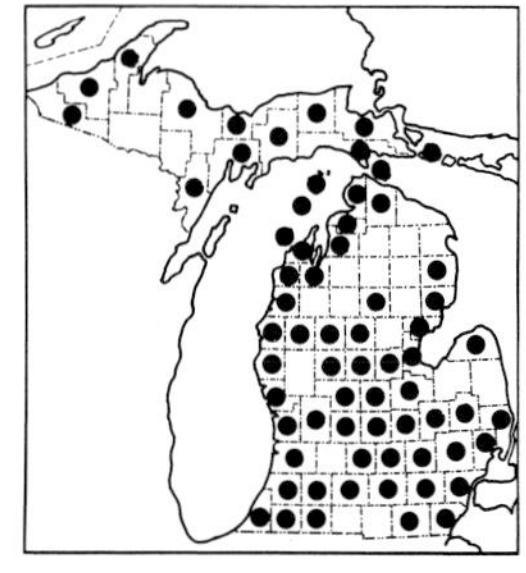

2. Salix exigua

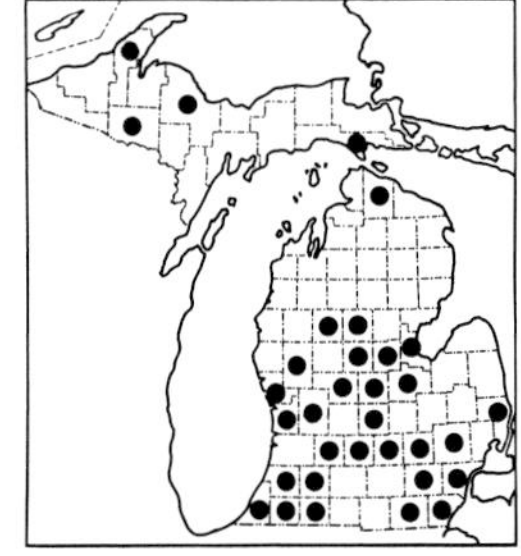

3. Salix nigra

strongly glaucous lower surface of the leaves in the latter were thought to have been lost in drying, but *S. petiolaris* lacks stipules, while young shoots of *S. nigra* are generally stipulate. See also comments under the next species.

3. **S. eriocephala** Michaux

Map 4. Swamps, shores, dunes, ditches and swales, stream banks, and other wet, sometimes calcareous sites; rarely upland.

This willow, once widely known as *S. cordata* Muhl. (whence the ambiguous common name "heart-leaved willow"), has more recently been well known as *S. rigida* Muhl., not now considered a distinct species (see Argus 1980). Narrow-leaved extremes, with the blades ± acute at both ends, superficially resemble *S. petiolaris*, but may usually be distinguished by the large, conspicuous stipules, especially on sprouts, and absence of any copper-colored hairs in the pubescence of young leaves (which are generally red in this species but not in *S. petiolaris*). Young pistillate specimens in which the leaves are only partly grown—particularly if they lack the usual red color of unfolding leaves—are sometimes confused with *S. nigra*, although the latter generally has more distinct small vein islets in the leaves. *S. nigra* is usually also distinctive in its very short or obsolete styles, capsules less dense in ament and ± whorled, and more persistent dark scales in the ament. (Staminate plants of these two species are readily distinguished by stamen number: 2 in *S. eriocephala* and 4–7 in *S. nigra*.)

4. **S. cordata** Michaux Fig. 5 Sand-dune or Furry Willow

Map 5. Almost completely restricted, in Michigan, to sandy shores and dunes along the Great Lakes; also on sandy shores of Douglas Lake, Cheboygan County. Overall range of the species is from Newfoundland to Hudson Bay, south to Maine and the Great Lakes.

Most Michigan plants have been segregated under the name *S. syrticola* Fern. [or were called in the past *S. adenophylla* Hooker], but the differences do not seem sufficiently consistent to warrant recognition of more than the

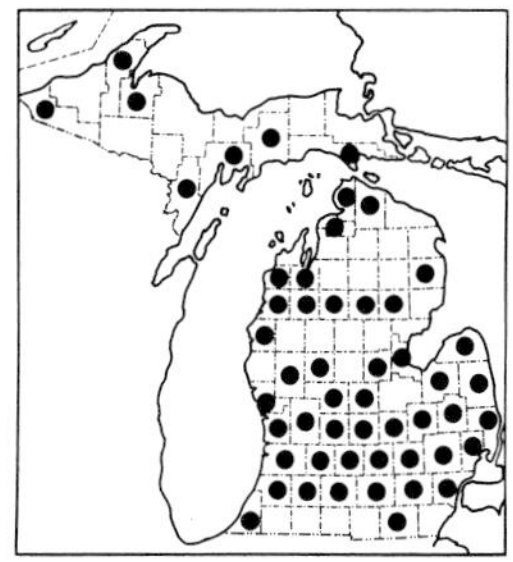
4. Salix eriocephala

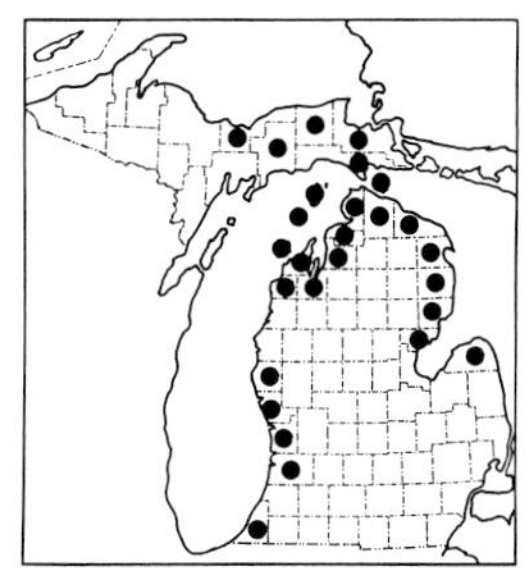
5. Salix cordata

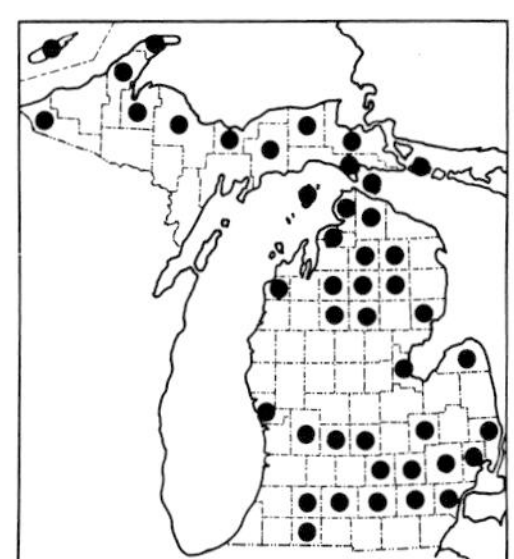
6. Salix pedicellaris

2. *Saururus cernuus* $\times ^1/_2$
3. *Salix exigua* $\times ^1/_2$
4. *S. nigra* $\times ^1/_2$
5. *S. cordata* $\times ^1/_2$
6. *S. pedicellaris* $\times ^1/_2$
7. *S. lucida* $\times ^1/_2$

single species. Normally rather densely silky, even on mature leaves, although occasional specimens have the foliage almost completely glabrous [var. *abrasa* Fern.].

5. **S. pedicellaris** Pursh   Fig. 6   Bog Willow

Map 6. Bogs, fens, and similar peatlands, as well as wet interdunal hollows.

A species very distinct in its complete lack of pubescence, strictly entire leaves, long-peduncled aments, and bog habitat. The leaves are strongly glaucous beneath when fresh, although the waxy layer may be lost in drying.

6. **S. purpurea** L.   Basket or Purple-osier Willow

Map 7. A European willow originally grown in this country for basket-making, locally well established as an escape from cultivation, especially on lake shores, riversides, and wet banks.

The leaves tend to be opposite or nearly so, except on vigorous new shoots. Herbarium specimens are sometimes confused with the preceding species; in addition to the habitat distinction and position of the leaves, *S. purpurea* has a distinctive purplish color to the foliage, the leaf margins are obscurely toothed toward the apex, and the axillary buds are (3) 5–13 mm long on sprouts.

7. **S. pellita** Schneider   Satiny Willow

Map 8. River banks, sandy shores, and (as at Isle Royale) hollows in rock outcrops. Known in Michigan only from a very few locations in the Upper Peninsula and apparently near Alpena.

The leaves have a distinctive lustrous and velvety appearance beneath, the hairs ranging from rather straight and silky to somewhat tomentose. Staminate material of this species has only very rarely been collected, and no pistillate specimens have been made in Michigan. A form with the leaves glabrescent beneath, and often with twigs scarcely if at all glaucous [f. *psila* Schneider] is frequent along the north shore of Lake Superior in Ontario and has been collected on the upper beach of Lake Superior in Luce County.

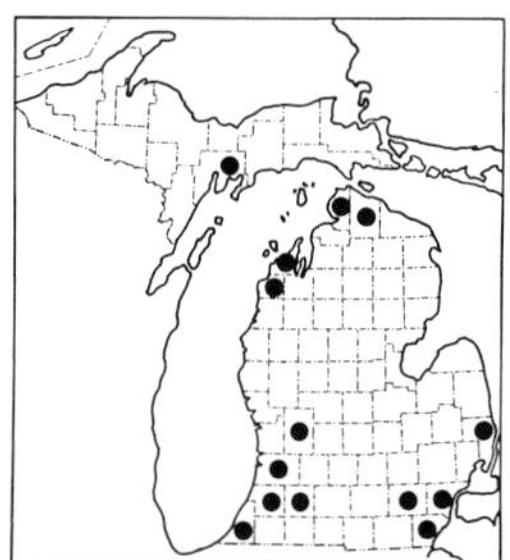
7. Salix purpurea

8. Salix pellita

9. Salix viminalis

8. **S. viminalis** L. Basket Willow; Common or Silky Osier

Map 9. Another species originally introduced from Europe for basket-making. Known in Michigan only as a presumed escape from cultivation at East Lansing (*Beal* in 1899, MSC).

Specimens of *S. pellita* f. *psila* might key here.

9. **S. candida** Willd. Sage or Hoary Willow

Map 10. Bogs, fens, beach meadows, tamarack swamps, stream borders, usually in open ± calcareous sites.

One of the more easily recognized species, distinctive in its narrow leaves and usually abundant white-tomentose pubescence. It hybridizes with many other species, the influence of this parent generally recognizable by presence of the characteristic tomentum. Plants with the leaves glabrous or nearly so (sometimes glaucous rather than green on the under surface) are known as f. *denudata* (Andersson) Rouleau, and occur sparingly throughout the state. F. C. Gates collected specimens illustrating both this form and the typical form from the same plant in Cheboygan County, the branches coming from portions of the shrub in flowing and quiet water, respectively. This observation suggests that the differences may in part, at least, result from ecological factors. Hybrids with *S. petiolaris* have been called *S.* ×*clarkei*, with Flint, Michigan, suggested as the "type locality," but that name has not been validly published.

10. **S. bebbiana** Sarg. Beaked or Bebb's Willow

Map 11. A common, often tall shrub (to 10 meters), especially in wet places: swamps, thickets, stream banks, shores, borders of wetlands, ditches.

This is not only one of our most common species but also one of our most variable ones, and evidently hybridizes with several others. The narrow pale scales of the aments and long stipes (up to 6 mm at maturity) are helpful characters in identification, in conjunction with the complete absence of any copper-colored hairs (though some may be a little brownish). The filaments are often pilose at the base, as in a number of other species. Aments are

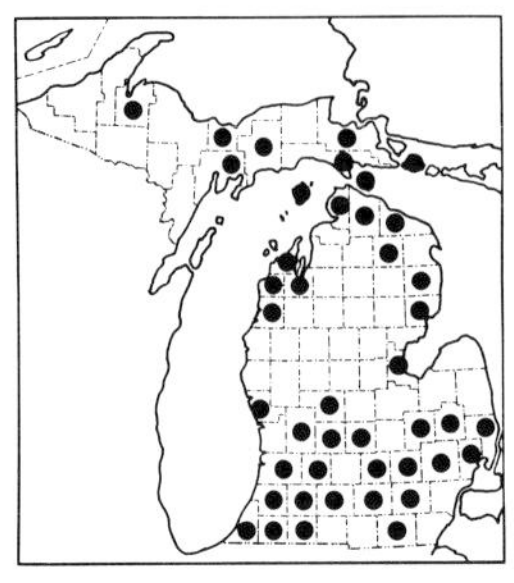

10. Salix candida

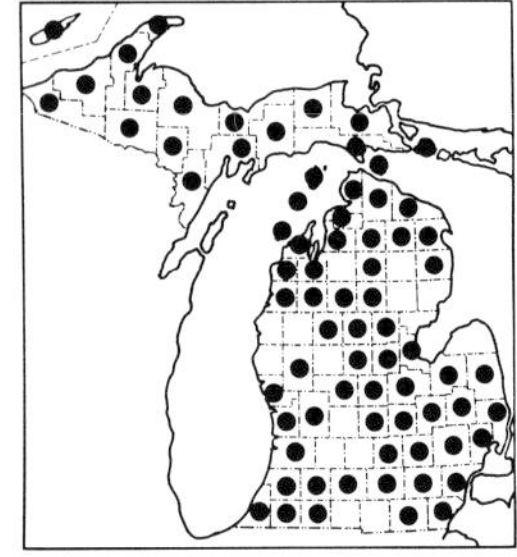

11. Salix bebbiana

12. Salix planifolia

occasionally found with both staminate and pistillate flowers, as well as many intermediate ones (Wagner 1968).

11. **S. planifolia** Pursh

Map 12. A boreal species ranging south in our area barely to Lake Superior, where it is frequent along the Canadian shore and apparently very local on rocky shores at Isle Royale.

This variable species is often confused with others, especially *S. discolor*, with which it probably hybridizes. In its most distinctive form with highly glossy leaves and red-purple twigs, it is quite readily recognized.

12. **S. discolor** Muhl. Pussy Willow

Map 13. Bogs, swamps, shores, stream banks, ditches, wet thickets and fields, rarely upland in sandy or rocky ground.

Another of our commonest but most variable species, often hybridizing. Plants with mature branchlets and foliage glabrous or nearly so [var. *discolor*] are fairly readily recognized. Pubescent plants [var. *latifolia* Andersson] are almost equally abundant throughout the state; they may sometimes closely resemble *S. humilis* (and may in part even be a result of hybridization with that species). Fortunately, the mature aments in var. *latifolia* tend to be longer than in typical var. *discolor*, in contrast to the short aments of *S. humilis*. Vegetative material is more of a problem to identify, the differences in leaf shape and texture being rather subtle. The leaves of *S. discolor* have more of a tendency to be elliptic; those of *S. humilis* run more consistently to an oblanceolate or narrowly obovate shape and also tend to be more rugose and pubescent beneath. The normally dry upland habitat of *S. humilis* in this region is a helpful contrast with the moist habitat most typical of *S. discolor*. A variant differing in its large and broadly obovate to elliptical leaves has been called *S. discolor* var. *overi* Ball and is known from Isle Royale. As in some other species, plants rarely occur with both staminate and pistillate flowers. See also comments under the next two species.

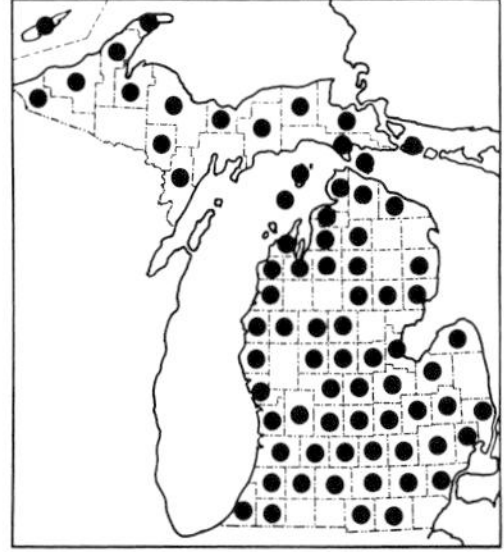
13. Salix discolor

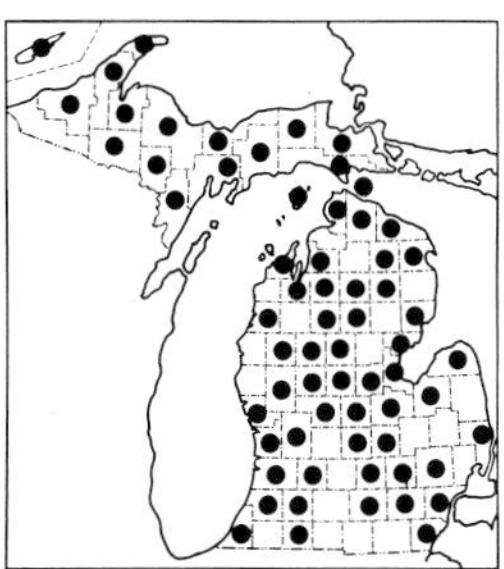
14. Salix humilis

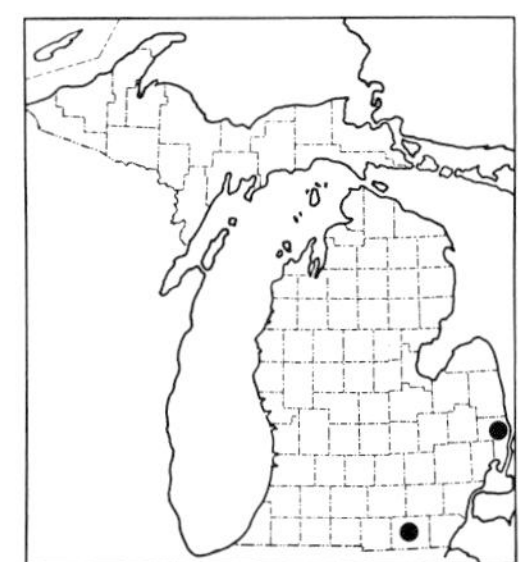
15. Salix cinerea

13. **S. humilis** Marsh. Upland or Prairie Willow

Map 14. Usually in dry sandy or rocky ground, often with jack pine and oak, but sometimes on moist sites.

In southernmost Michigan, plants smaller in every way (especially in the small, almost rotund aments) and also estipulate may be var. *microphylla* (Andersson) Fern., recognized by some authors as a distinct species, *S. tristis* Aiton [or *S. occidentalis* Walter; see Brittonia 36: 328–329. 1984]. Plants with very broadly obovate, densely pubescent leaves have been called var. *keweenawensis* Farw. (TL: [Clifton], Keweenaw Co.); such plants are found primarily in the Upper Peninsula but are hardly worth naming. Also of dubious merit is a narrow-leaved and glabrate extreme, var. *rigidiuscula* (Andersson) Robinson & Fern. It is occasionally collected, and is rarely completely glabrous or with slight pubescence only on the base of the midrib and petiole; in the key, such plants will probably run to *S. discolor*, from which they differ in their narrow shape and, usually, dense pubescence of at least the youngest branchlets. See also comments under *S. discolor*.

14. **S. cinerea** L. Gray Willow

Map 15. A Eurasian species, rarely found as an escape from cultivation along streams and river banks. C. K. Dodge collected it on the banks of the Black River at Port Huron in 1905, noting that it had persisted for more than 10 years.

This is very similar to the preceding two species, but the wood (at least of 2–3-year-old twigs) is prominently ridged. (Soak the bark as necessary to peel it carefully from the wood.)

Another cultivated Eurasian species, *S. caprea* L. (goat or florists' willow), has been collected only once in the state, from "among rocks" in Keweenaw County (*Farwell* in 1888, MICH), a somewhat dubious record. This is the "pussy willow" of the florists, and with smooth wood will run to *S. discolor* in the keys. Young staminate material is readily distinguished, however, by the very large, thick aments which flower from the base to the apex. The leaves are very broad (sometimes almost orbicular) and densely gray-tomentose beneath.

15. **S. alba** L. White Willow

Map 16. A tree introduced from Europe and locally naturalized, especially on shores and banks.

Plants with conspicuous yellow first-year twigs have been called var. *vitellina* (L.) Stokes, the golden willow, or var. *tristis* (Ser.) Gaudin (with "weeping" habit). This hybridizes with the next species, producing plants [*S.* ×*rubens* Schrank] intermediate in character or showing some of each species (e. g., the serrations of one with the pubescence or capsules of the other). Collections apparently intermediate or, at least, not assignable to either *S. alba* or *S. fragilis* have been seen from several counties in addition to those from which one or both species are mapped: Benzie, Eaton, Genesee, Gladwin, Ingham, and Tuscola.

16. **S. fragilis** L. Crack or Brittle Willow

Map 17. Another Old World native, like the preceding established in damp ground as on shores and banks.

The leaves or bracts on the short flowering branchlets of this species and the preceding are usually ± entire; and glands are generally conspicuous at the summit of the petiole on mature leaves, but usually not on younger ones. The branchlets of *S. fragilis* are very brittle (whence the name) at the base, breaking off in windstorms. Twigs broken in this way may root at the edges of rivers and streams.

Salix babylonica L., weeping willow, has been reported from the state, but is not hardy quite this far north. We have at most very few documented records of any weeping willow (i. e., one with long pendulous branches) escaping from cultivation in Michigan—though they may persist long after cultivation. However, most if not all "weeping willows" in this region are actually more hardy hybrids of *S. babylonica* with *S. fragilis*, if not cultivars of *S. alba* var. *tristis*. True *S. babylonica* has capsules shorter than either *S. alba* or *S. fragilis*, and the leaves are finely toothed. But the taxonomy of the cultivated "weeping willows" is confusing at best. (For further discussion, see Steyermark's *Flora of Missouri*, pp. 496–497. 1963.)

17. **S. pentandra** L. Bay-leaved or Laurel Willow

Map 18. A European species, occasionally escaped from cultivation and apparently established in fields and on dunes, streamsides, borders of woods, and other places. Trees 30–40 feet tall, much larger than usually found, are well established in Gogebic County along the Middle Branch of the Ontonagon River northeast of Watersmeet (*Brodowicz 261*, MICH).

Sometimes difficult to distinguish from the next two species (see full descriptions in the key), but with leaves typically paler beneath than in *S.*

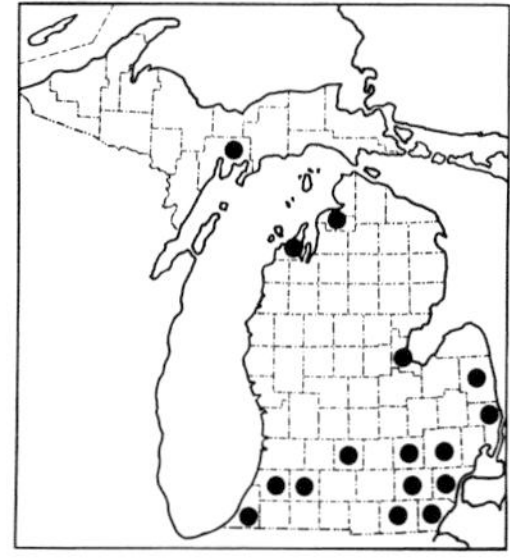

16. Salix alba

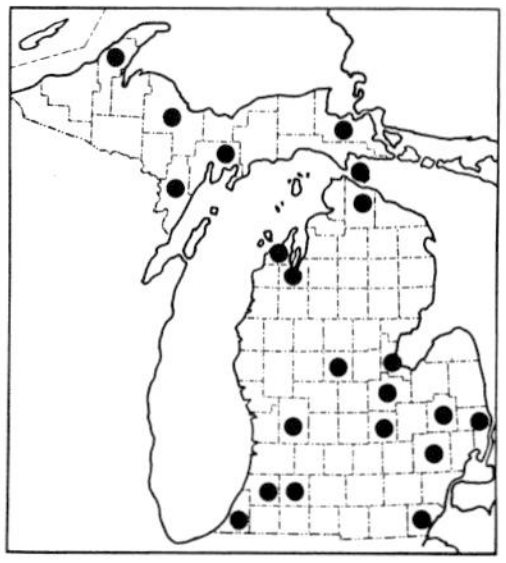

17. Salix fragilis

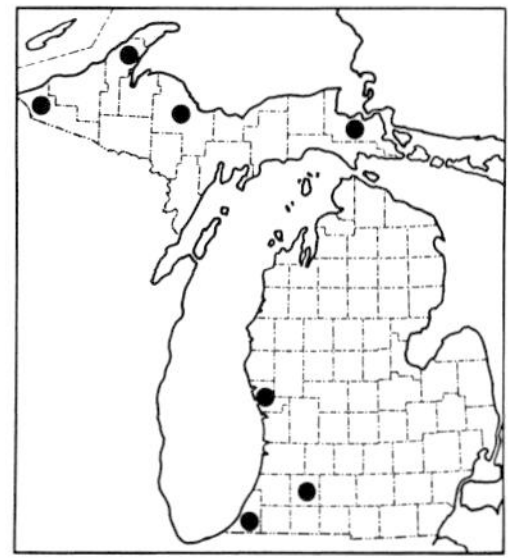

18. Salix pentandra

lucida and not long-acuminate, but not so glaucous as in *S. serissima*. The measurements in the key for mature capsules are based on cultivated and escaped Michigan plants, and are in accord with manuals; some European specimens have capsules as long as 8 mm and granular-roughened as in *S. serissima*.

18. **S. serissima** (Bailey) Fern. Autumn Willow

Map 19. Bogs (or fens), cedar and tamarack swamps, shores and streamsides, usually in ± calcareous places.

Frequently distinguished in manuals from the next species by shorter aments. Although this is a tendency, it is not sufficiently consistent to use as a key character. Mature pistillate aments of Michigan specimens of *S. serissima* range 2–5 cm in length; those of *S. lucida*, 1.5–5 cm. Staminate aments are ca. 2.5–3.5 cm in *S. serissima* and 1.5–6 cm in *S. lucida*.

19. **S. lucida** Muhl. Fig. 7 Shining Willow

Map 20. Shores and low dunes, swales, ditches, and wetlands generally.

The young branchlets and undersides of mature leaves are pubescent in var. *intonsa* Fern. Plants with leaves narrower than usual have been described as var. *angustifolia* Andersson. See also comments under the preceding two species.

20. **S. amygdaloides** Andersson Peach-leaved Willow

Map 21. Shores, stream banks, floodplains, and borders of marshes.

Specimens collected many years ago near Grand Rapids have both staminate and pistillate flowers in the same aments. See comments under *S. nigra*.

21. **S. pyrifolia** Andersson Plate 1-A Balsam Willow

Map 22. Bogs and conifer swamps, rather local and more often seen along roadsides and other clearings; boreal forest and rock outcrops (Isle Royale and northward).

The rather spicy balsamic fragrance of this plant, lasting as long as 100

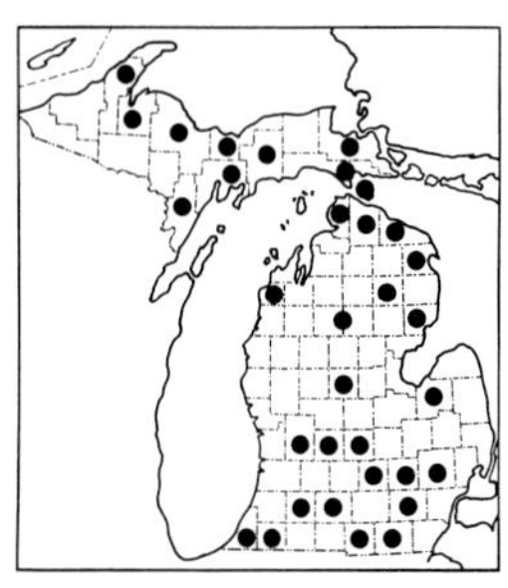

19. Salix serissima

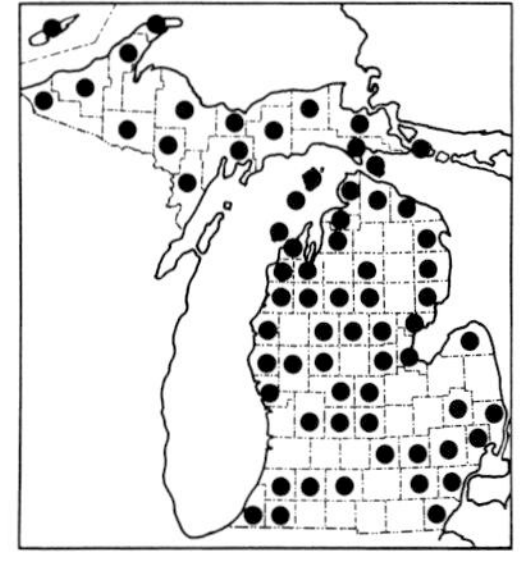

20. Salix lucida

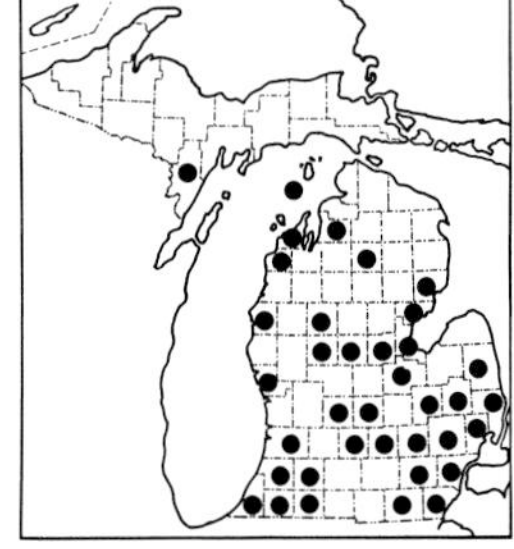

21. Salix amygdaloides

years in herbarium specimens, is notable. I can usually detect it from some little distance away—but not in fresh material.

22. **S. myricoides** Muhl. Blueleaf Willow

Map 23. Sandy shores, dunes, and damp interdunal hollows along the Great Lakes; rarely inland on ± calcareous shores and banks.

Dorn and Argus have both concluded that this older name applies to what has for many years been well known as *S. glaucophylloides* Fern. Plants with ± densely white-villous branchlets and the leaves retaining some pubescence represent var. *albovestita* (Ball) Dorn; this is the usual form in the state, strictly glabrous plants being relatively rare. Ball referred plants with small thick leaves (in part, at least, probably an ecological response) to *S. glaucophylloides* var. *brevifolia* (Bebb) Ball. Vegetative specimens with mature leaves are often difficult to distinguish from *S. eriocephala*, especially if one is not familiar enough with both species to evaluate the comparatively thick leaf texture and more heavily glaucous surface of *S. myricoides*. This species apparently hybridizes on the dunes with *S. cordata*.

23. **S. sericea** Marsh. Silky Willow

Map 24. Borders of lakes and streams, ditches, and in other low places.

Many published reports for this species, especially from the northern part of the state, are based on specimens of *S. petiolaris* with leaves pubescent above.

24. **S. petiolaris** J. E. Smith Slender or Meadow Willow

Map 25. Marshy and swampy ground, shores, ditches, and bogs, usually in ± open places.

The shoots, even when young, lack stipules (cf. *S. nigra*). See also comments under *S. eriocephala*. Plants with leaves permanently ± pubescent, at least above, have been called *S. subsericea* (Andersson) Schneider. Such plants are often considered to be hybrids between *S. sericea* and *S. petiolaris*; the fruit, especially, may be intermediate. However, this was maintained by Ball as a good species in view of the pubescence on the upper

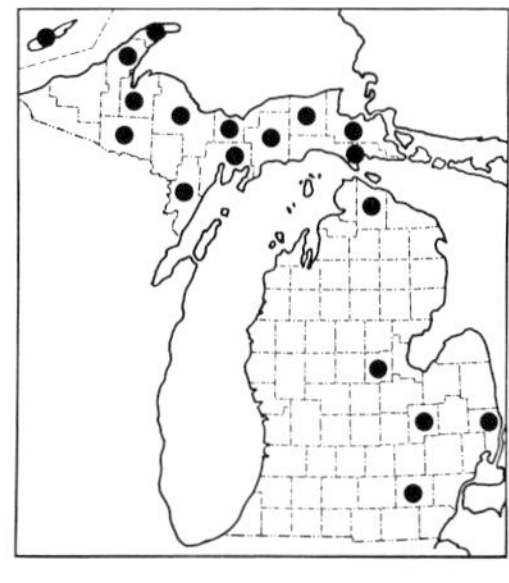

22. Salix pyrifolia

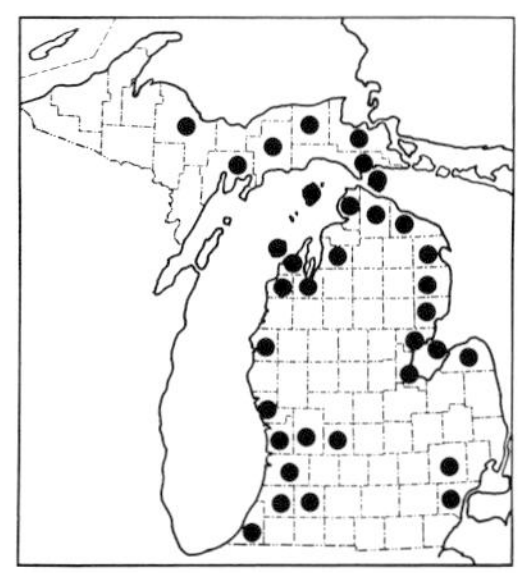

23. Salix myricoides

24. Salix sericea

surfaces of the leaves (as in neither supposed parent) and the failure of the leaves to blacken in drying (as they do regularly in *S. sericea* and often in typical *S. petiolaris*). Furthermore, we do not find *S. sericea* north of Saginaw Bay in the state, while "*S. subsericea*" is rather common throughout. The similarity in leaf shape suggests affinity with *S. petiolaris* as does the ± continuous range of variation in pubescence, although the capsules have a tendency to be shorter and plumper. As in Wisconsin (Argus 1965), Michigan plants called *S. subsericea* seem to belong with *S. petiolaris*.

2. **Populus** Poplar

There is a strong tendency in most species to form underground suckers, so that a grove of aspens or poplars may represent a single clone. Furthermore, leaf shape is quite variable and further complicated by hybridization in at least some species. So while normal typical material is distinctive and readily identifiable, specimens of suckers or sprouts, hybrids, or other atypical states may cause confusion. Barnes (1961) and Spies and Barnes (1982) discuss hybrids involving the two aspens and *P. alba* in Michigan.

The key is not specifically designed for use with specimens bearing aments before the appearance of any leaves, but sufficient data on flowers, buds, and habit are given in the key or text that incorrect identification of flowering material may usually be avoided.

REFERENCES

Barnes, Burton V. 1961. Hybrid Aspens in the Lower Peninsula of Michigan. Rhodora 63: 311–324.

Barnes, Burton V., & Kurt S. Pregitzer. 1985. Occurrence of Hybrids Between Bigtooth and Trembling Aspen in Michigan. Canad. Jour. Bot. 63: [in press].

Erlanson, Eileen Whitehead, & Frederick J. Hermann. 1928. The Morphology and Cytology of Perfect Flowers in Populus tremuloides Michx. Pap. Mich. Acad. 8: 97–110.

Spies, T. A., & B. V. Barnes. 1982. Natural Hybridization Between Populus alba L. and the Native Aspens in Southeastern Michigan. Canad. Jour. For. Res. 12: 653–660.

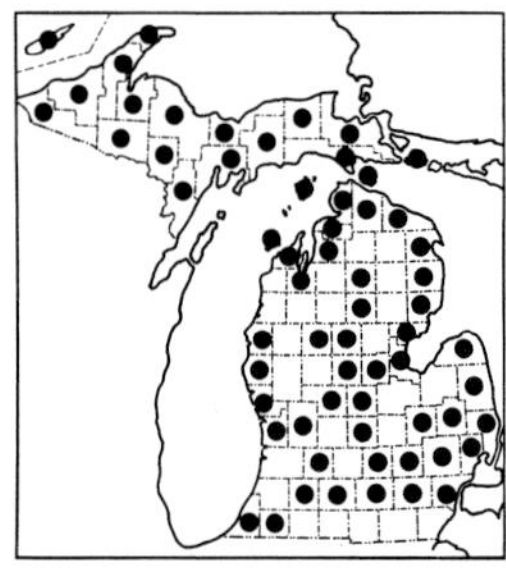

25. Salix petiolaris

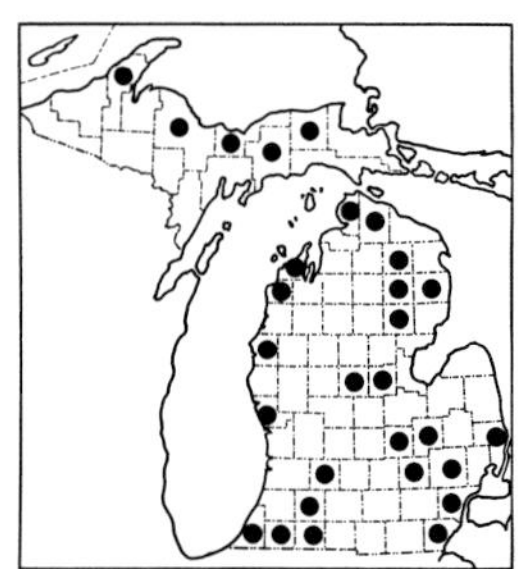

26. Populus alba

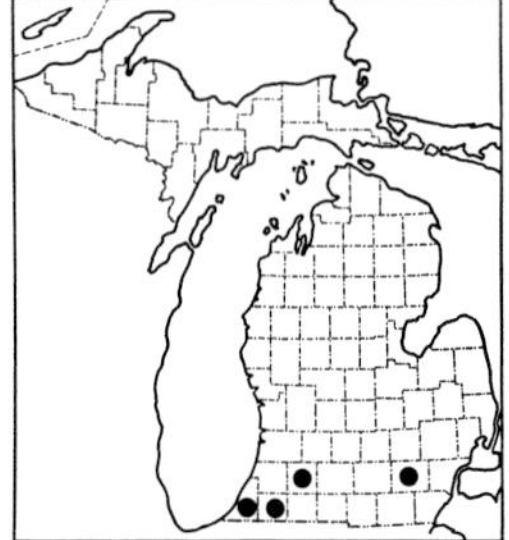

27. Populus heterophylla

KEY TO THE SPECIES

1. Mature leaf blades beneath and petioles densely felted with white pubescence; leaves of long shoots ± sinuate with 5 or fewer irregular rounded lobes on a side; trees spreading from cultivation. . . . . 1. **P. alba**
1. Mature leaves glabrous or nearly so (white-felted when young in nos. 2 & 7); leaves usually with more than 5 teeth on a side; trees native or spreading from cultivation
  2. Petioles terete (often somewhat grooved above)
    3. Outline of leaf blade usually slightly convex toward blunt or rounded tip; buds usually ± pubescent, not (or only slightly) glutinous; expanding leaves (and often edges of mature midribs beneath) densely cottony-tomentose; flowers on pedicels ca. 2–8 mm long . . . . 2. **P. heterophylla**
    3. Outline of leaf blade usually slightly concave toward very acute or short-acuminate tip; buds very strongly glutinous, varnished with a fragrant gum; expanding leaves glabrous or at most minutely puberulent; flowers sessile or nearly so in the aments (pedicels less than 2 (2.5) mm long) . . . . 3. **P. balsamifera**
  2. Petioles strongly compressed laterally, especially near blade
    4. Leaf blades strongly deltoid or diamond-shaped, the margins with a firm colorless border thicker than adjacent major veinlets, the incurved teeth with callous tip; stamens more than 15; scales in ament glabrous, fringed with many thread-like segments
      5. Glands prominent at junction of blade and petiole; larger blades deltoid, mostly (6) 7–12 (20) cm broad, the margin (especially when young) usually ± densely ciliolate with minute hairs (at least on the teeth) . . . . 4. **P. deltoides**
      5. Glands absent at junction of blade and petiole; larger blades rhombic (± diamond-shaped), usually less than 8 cm broad (but broader than long), the margin always glabrous . . . . 5. **P. nigra**
    4. Leaf blades orbicular to reniform or obscurely deltoid, the margins often with no colorless thin border or with one no thicker than adjacent main veinlets, the teeth with or without a glandular tip; stamens 12 or fewer; scales in ament pilose, fringed with only ca. 3–7 (10) linear or narrowly lanceolate segments
      6. Leaf margins closely crenulate-serrate with 15–35 or more (–70) teeth on a side; buds and young growth glabrous or nearly so; scales fringed with mostly 3–5 narrow segments . . . . 6. **P. tremuloides**
      6. Leaf margins closely undulate-dentate with fewer than 10 (12) large teeth on a side (except on large-leaved sprouts); buds and new growth (leaves, branchlets) ± pilose or tomentose with white to gray pubescence; scales fringed with mostly 5–7 segments . . . . 7. **P. grandidentata**

1. **P. alba** L. — White or Silver Poplar

Map 26. A native of Europe, widely cultivated and spreading by suckers, sometimes forming small thickets along roadsides, at old homesites, in fields, and even into woods, especially in sandy soil.

Staminate plants are apparently not in cultivated clones. It is possible that some of the material included here is *P. canescens* (Aiton) J. E. Smith, variously treated as a similar species, a variety of *P. alba*, or a hybrid clone of *P. alba* × *P. tremula*, the European aspen. Hybrids of *P. alba* × *P. tremuloides* have been reported from Gladwin, Livingston, Midland, and Tuscola counties, and of *P. alba* × *P. grandidentata* from Jackson, Livingston, Saginaw, and Washtenaw counties.

2. **P. heterophylla** L. Swamp Cottonwood

Map 27. Swamps, wet hollows, and shores—very rare in Michigan.

The flowers are nearly distinctive in the genus for their long pedicels, although those of *P. deltoides* may also elongate.

3. **P. balsamifera** L. Balsam Poplar; Hackmatack

Map 28. Sandy shores, dunes, and interdunal hollows; swamp forests and cedar bogs; often spreading vigorously in damp hollows, swamp borders, and cutover forests, forming large thickets.

The leaves of this species are rather shiny, especially when young, and rather strongly whitened or yellowish beneath. The margins of the blades, petioles, and bases of main veins are sometimes minutely puberulent, but the foliage is otherwise glabrous. Buds and twigs are reddish brown, in contrast with *P. deltoides*, in which they are yellowish brown. The base of the blade varies from rounded or merely truncate [var. *balsamifera*] to subcordate [var. *subcordata* Hylander]; the latter form seems especially characteristic of sand dunes, but a few specimens referred here may represent the pistillate clone widely cultivated as "Balm-of-Gilead" [sometimes called *P.* ×*jackii* Sarg., later named *P.* ×*gileadensis* Rouleau, if thought to be a hybrid of *P. balsamifera* with *P. deltoides*].

4. **P. deltoides** Marsh. Cottonwood

Map 29. Borders of rich woods, floodplains and other swamp forests, river banks, sand dunes; spreading to shores, fields, and other open usually damp places; sometimes planted.

5. **P. nigra** L. Lombardy Poplar

Map 30. The cultivar 'Italica' of the European black poplar is widely planted, a familiar tree of distinctive columnar shape resulting from the strongly ascending habit of the branches. Only staminate plants seem to be grown here, and they have been variously interpreted as a hybrid or freakish clone. This poplar spreads to some extent by root suckers, and also from rooting

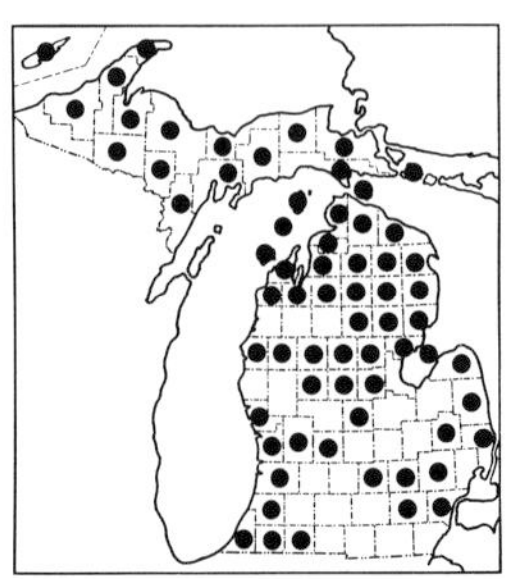

28. Populus balsamifera

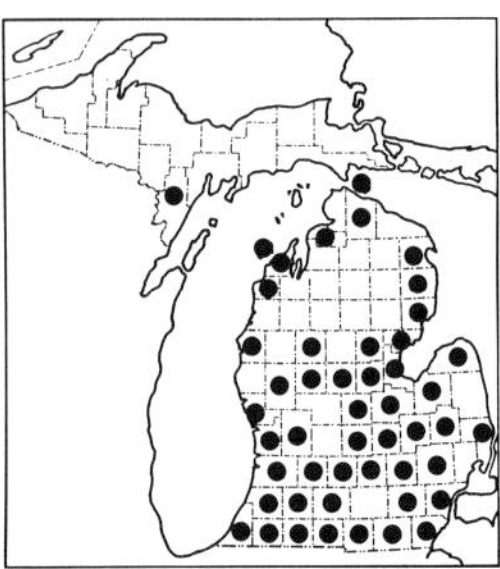

29. Populus deltoides

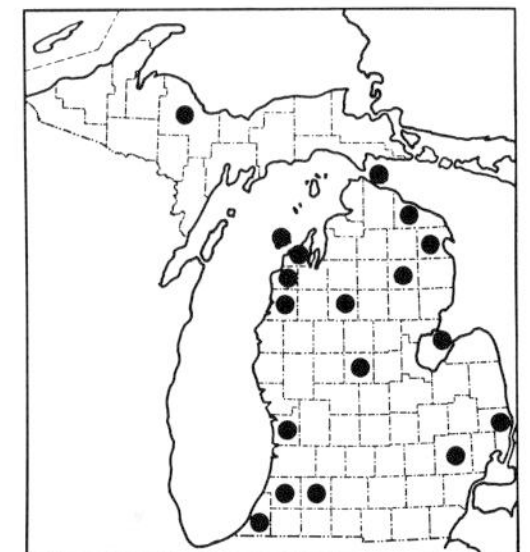

30. Populus nigra

of portions of the brittle branches. Locally established along roadsides, on shores and dunes, and in other disturbed sites.

A hybrid with *P. deltoides* [*P.* ×*canadensis* Moench] is often planted, as Carolina poplar; it has a few minute cilia on the leaf teeth, and may be expected to escape.

6. **P. tremuloides** Michaux Fig. 8 Quaking Aspen

Map 31. Nearly ubiquitous, like the next species, but tolerates and thrives in colder and wetter situations; one of the few deciduous trees of the boreal forest to the north of Michigan. Typical of swamps, with both conifers and hardwoods, but often on drier, even sandy sites; characteristically in clearings, after fire or logging.

This species hybridizes with the next, and hybrid clones are frequent [*P.* ×*smithii* Boivin] (see Barnes 1961; Barnes & Pregitzer 1985). First generation hybrids are rather easily recognized, with leaves intermediate between those of the distinctive parent species. Backcrosses (introgressants) are less readily identified. Hybrids also occur between our native aspens and *P. alba*. Perfect flowers, in contrast to the usual dioecious condition in this family, discovered in the Nichols Arboretum at Ann Arbor, have been described by Erlanson and Hermann (1928). Sometimes the bark of aspen is so light in color that the trees are mistaken for paper birch, especially by the casual tourist. See also remarks under the next species.

7. **P. grandidentata** Michaux Largetooth or Bigtooth Aspen

Map 32. A nearly ubiquitous tree of sandy soils, especially in dry cutover and burned areas, clearings, and woodlands, often with oaks and pines; persisting in mature deciduous forests. The largest tree of this species in the state is 3.5 feet in diameter and the largest of *P. tremuloides* (a "national champion") is also over 3 feet.

Easily distinguished from the preceding species by the coarsely toothed leaves, densely white-silky when first expanding, and pubescent bud scales. Extremely large leaves on young sprouts and suckers are quite different.

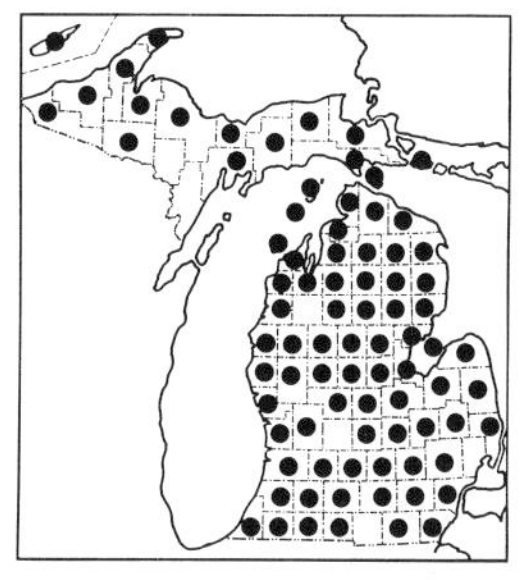
31. Populus tremuloides

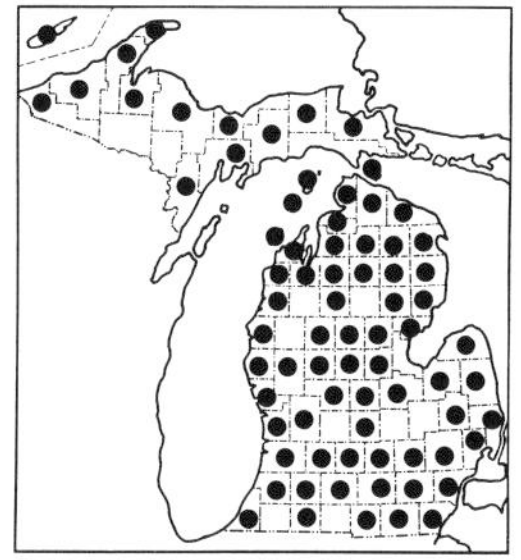
32. Populus grandidentata

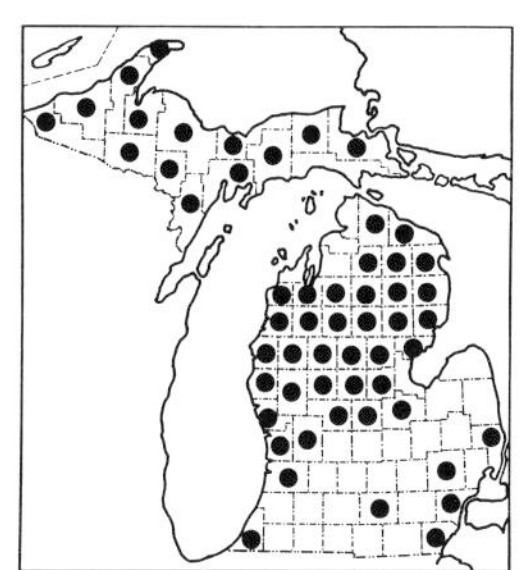
33. Comptonia peregrina

The aspens or "popples" are short-lived trees, rarely attaining the size of "champions." They are intolerant of shade but in their shade more desirable and lasting species, such as pine and fir, may get a sheltered start. The vast areas of aspen in Michigan have resulted from the abundant reproduction, chiefly sprouting from root suckers, which follows fire or cutting. Without frequent forest fires or cutting to maintain it, aspen will die of old age and be largely replaced by other trees. The wood is soft and light, but is valued commercially for pulp, boxes, cabin logs, and other uses. Aspens are important to wildlife, too, the young thickets providing cover and the foliage browse; the bark is a principal food of beaver.

MYRICACEAE — Bayberry Family

Our two species in this family are shrubs with a spicy-aromatic fragrance when bruised or crushed. The leaves and branchlets are ± pubescent in *Comptonia* and often so in *Myrica*. They also bear resinous orange dots, especially conspicuous in *Myrica*, in which the fruit and center of the staminate aments are rather heavily dotted. Myricaceae are among the few plants, apart from the Leguminosae, which bear nitrogen-fixing nodules on their roots. Both of our species may serve as alternate hosts for the sweetfern rust fungus [*Cronartium comptoniae* Arthur], which is a cause of stem cankers in jack pine and red pine, especially the former.

REFERENCES

Anderson, Gerald W. 1963. Sweetfern Rust on Hard Pines. U.S. Dep. Agr. Forest Pest Leafl. 79. 7 pp.

Elias, Thomas S. 1971. The Genera of Myricaceae in the Southeastern United States. Jour. Arnold Arb. 52: 305–318.

KEY TO THE GENERA

1. Leaves pinnatifid their entire length, stipulate (the stipules eventually deciduous); staminate aments cylindrical, usually becoming somewhat lax and recurved; pistillate aments globose, bur-like; nutlets smooth, shiny, ellipsoid to barrel-shaped, ca. 3.5–6 mm long, much surpassed by the long-acuminate bracts; shrub of dry ground . . . . 1. **Comptonia**
1. Leaves unlobed, entire except for a few teeth toward apex, estipulate; staminate and pistillate aments both short-cylindrical, rather cone-like, stiffly ascending; nutlets resin-dotted, flattened-ovoid, ca. 2.5–3 mm long (not including the pair of much swollen, adnate, acuminate bractlets), distinctly exceeding the broadly obtuse bracts; shrub of wet ground . . . . 2. **Myrica**

1. Comptonia

A single species, sometimes included in the next genus, in which case its name is *Myrica peregrina* (L.) Kuntze, not *M. aspleniifolia* L., which is often seen.

1. **C. peregrina** (L.) Coulter Fig. 9 Sweetfern
Map 33. Sandy plains and hills, usually in open woodland of oak, aspen, and/or jack pine; especially conspicuous and abundant from Clare northward on cutover pinelands.

2. **Myrica** Bayberry; Wax-myrtle
A single species in Michigan, sufficiently different from the bayberries that it is sometimes segregated as *Gale palustris* (Lam.) Chev.

1. **M. gale** L. Fig. 10 Sweet Gale
Map 34. Wet, especially calcareous places: bogs or fens, lake margins, beachpools and interdunal swales, streamsides, rock crevices along Lake Superior.

JUGLANDACEAE Walnut Family

REFERENCE

Elias, Thomas S. 1972. The Genera of Juglandaceae in the Southeastern United States. Jour. Arnold Arb. 53: 26–51.

KEY TO THE GENERA

1. Leaflets (9) 11–19, the terminal one usually about equal to those of lowest pair, or smaller; pith of year-old twigs becoming chambered (separating into thin plates and cavities, fig. 12); staminate aments sessile or nearly so .... 1. **Juglans**
1. Leaflets 5–9, the terminal one distinctly larger than those of lowest pair; pith of year-old twigs continuous; staminate aments peduncled in groups of 3 .... 2. **Carya**

1. Juglans

Both of our species are widely planted along roads and in farmyards, and it is sometimes not clear from herbarium labels whether certain specimens were collected from native trees—whether in a forest or allowed to persist in otherwise cleared sites. Records from the northern part of the state, particularly, have not been mapped unless they definitely represent trees believed not to have been planted.

The English walnut, *J. regia* L., is sometimes planted in Michigan. Its leaflets are nearly entire and glabrous. In our native species, the leaflets are toothed and ± pubescent, at least beneath (the rachis even more densely glandular-pubescent).

KEY TO THE SPECIES

1. Pith dark chocolate brown; fruit (also the nut within) ovoid-oblong ...... 1. **J. cinerea**
1. Pith pale (tan to cream); fruit globose .... 2. **J. nigra**

1. **J. cinerea** L. Figs. 12, 13 Butternut

Map 35. Stream banks and swamp forests, as well as upland beech-maple, oak-hickory, and mixed hardwood stands. The Alpena and Benzie county records are of individuals seeding in from planted trees.

There are some pubescence differences which will help to distinguish this species from the black walnut, even when the distinctive dark pith is not exposed. In *J. cinerea*, there is often a pad of dense small hairs extending transversely along the upper margin of the old leaf scars; in *J. nigra*, this pad is absent, although the circular area of bud pubescence is confusing, and some specimens are ambiguous. The underside of the leaflets in *J. cinerea* is ± densely covered with mostly stellate hairs, while in *J. nigra* the pubescence is sparser and mostly of simple hairs. The pubescence of *J. cinerea*, including that on the fruit, is more clammy than that of *J. nigra*. Persons familiar with the species in the field will also know differences in fragrance of the foliage and fruit and in stain from the husks—all difficult to express in words.

2. **J. nigra** L. Fig. 14 Black Walnut

Map 36. Like the preceding, grows in both lowland and upland woods. The Benzie County record is likewise of individuals seeding in from planted trees.

Unfortunately now rather scarce, this is a valuable tree for fine lumber as well as nuts.

2. Carya Hickory

Traditionally a difficult genus, the problems in part resulting from inadequate herbarium material; specimens of hickory should include not only foliage but also fully mature fruit (usually from late fall) and winter buds

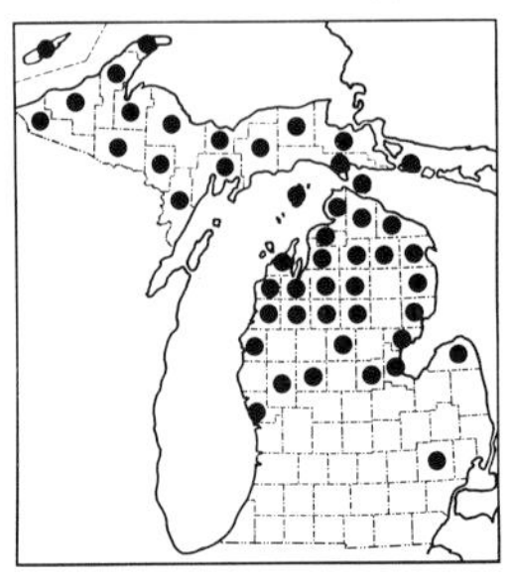
34. Myrica gale

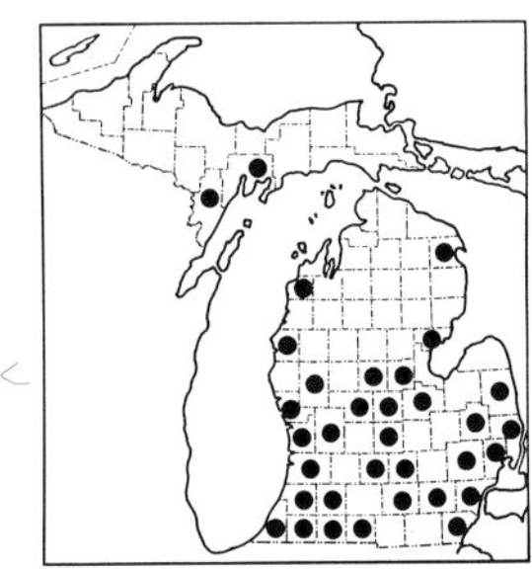
35. Juglans cinerea

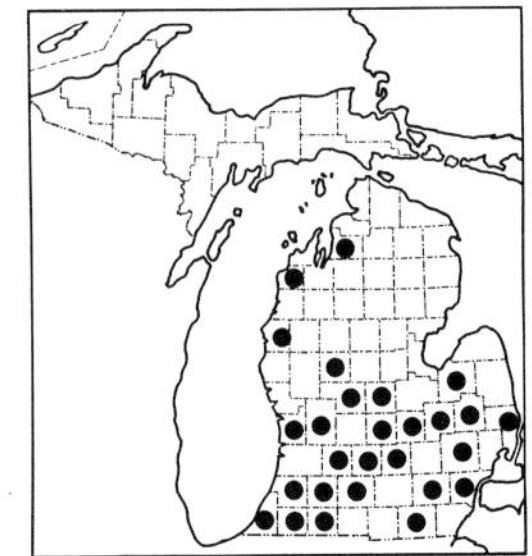
36. Juglans nigra

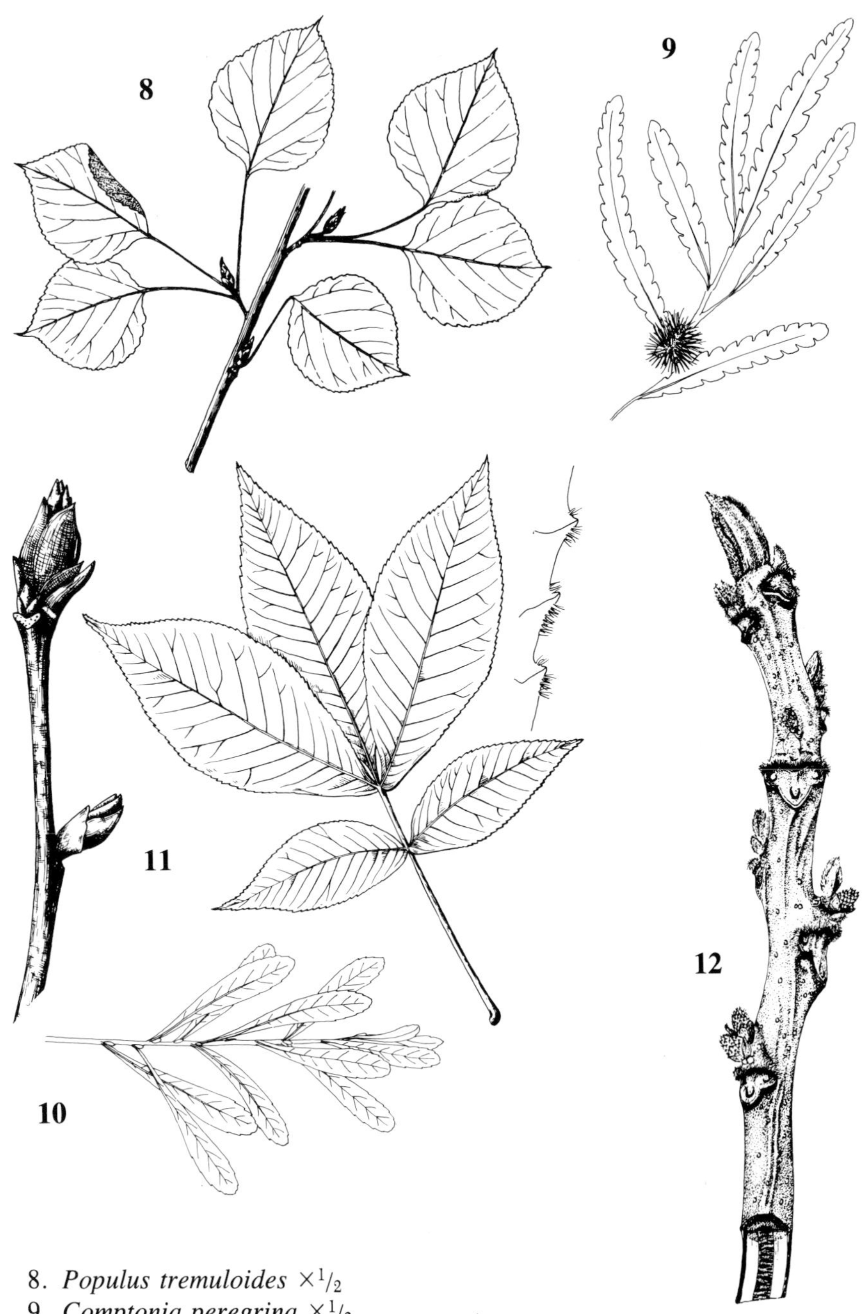

8. *Populus tremuloides* $\times^1/_2$
9. *Comptonia peregrina* $\times^1/_2$
10. *Myrica gale* $\times^1/_2$
11. *Carya ovata,* leaf $\times^1/_3$ (plus detail of margin); winter twig $\times 1$
12. *Juglans cinerea,* winter twig $\times 1$

as well as notes on bark. More data are desirable (see Deam's *Trees of Indiana*, 3rd ed., for good advice on collection and preparation of adequate specimens of *Carya* for serious study). In part, problems in hickories may result from hybridization.

Measurements of husk thickness in the key are based on the middle of the husk, not the thicker base (although there is sufficient variation that it usually makes little difference where one measures).

Hickory wood is well known for its combination of strength, toughness, and flexibility, although it is not durable in contact with the soil or when exposed to warping influences. It is widely used for tool handles, ladders, sporting goods, and fuel (especially in curing meats, to which the smoke gives a distinctive and desirable flavor). Hickories are not widespread enough in Michigan to constitute a major timber resource. The nuts, particularly of shagbark, are gathered for their sweet kernels. Pecan [*C. illinoinensis* (Wang.) K. Koch] is not native as far north as Michigan but is sometimes planted in the southern part of the state. Our native species are also planted northward in farmyards and along roadsides; such plants have been excluded from the maps.

Most of the Michigan specimens examined for this Flora have benefited from the annotations of Wayne E. Manning.

REFERENCES

Manning, Wayne E. 1950. A Key to the Hickories North of Virginia with Notes on the Two Pignuts, Carya glabra and C. ovalis. Rhodora 52: 188–199.

Manning, Wayne E. 1973. The Northern Limit of the Distribution of the Mockernut Hickory. Mich. Bot. 12: 203–209.

KEY TO THE SPECIES

1. Bud scales valvate, yellow (from dense granules), the 2–3 pairs leaving well separated glabrous scars; fruit small (1.8–2.7 cm long, including beak), the husks with prominent ridges or low wings at sutures above, not splitting all the way to the base, thin (ca. 1–1.5 mm thick); leaflets 7–9, never only 5, their margins glabrous or nearly so . . . . 1. **C. cordiformis**

1. Bud scales imbricate, at most with scattered yellow granules, more than 6, the scars crowded, usually ciliate when fresh; fruit various but scarcely if at all ridged at sutures and if as small as in the above, then the husk usually thick and/or splitting all the way to the base (only in some forms of *C. glabra*, with 5–7 leaflets, the fruit both small and with husks splitting only halfway to the base); leaflets 5–9

   2. Leaflets consistently 5, ciliate when young, retaining a dense tuft of small hairs on one or both sides of each tooth just below its apex (these tending to wear off on some leaflets by fall); fruit 2.1–4.1 (4.5) cm long, the husks 3–10 (12) mm thick, splitting all the way to the base at maturity; bark "shaggy" . . . . 2. **C. ovata**

   2. Leaflets 5–9, if 5 not ciliate, and never with dense tufts on teeth; fruit and bark various

3. Leaflets 5 (–7), glabrous or glabrate beneath or rarely pubescent especially on veins; larger winter buds (terminal or pseudoterminal) 5–10 mm long; rachis of leaves often glabrous; fruit 1.6–3.2 (3.5) cm long, the husks 0.6–4 mm thick . . . . 3. **C. glabra**

3. Leaflets 7 (–9) (sometimes a few small leaves with only 5), ± pubescent beneath; larger winter buds (6) 11–20 mm long; rachis of leaves glabrate to pubescent; fruit (3.5) 4–6 cm long, the husks 6–11 mm thick . . . . 4. **C. laciniosa**

1. **C. cordiformis** (Wang.) K. Koch Bitternut Hickory

Map 37. In beech-maple and mixed hardwood stands. The "national champion" bitternut is a tree 134 feet tall and nearly 4 feet in diameter in Cass County. Reported (without specimens) north to Alpena County.

This species is sometimes called "bitter pecan," the latter word alluding to the close relationship to the true pecan, and the former to the very bitter kernel of the nut. The bark is tight and close with interlacing ridges, not at all as in the "shagbark" hickories.

2. **C. ovata** (Miller) K. Koch Fig. 11 Shellbark or Shagbark Hickory

Map 38. Usually in dry upland woods, with oaks, but sometimes in lowland woods. The Benzie County record is of plants seeding in from planted trees; the Round Island tree presumably grew from a nut washed in from a cultivated tree.

Very variable, particularly in size of fruit and in amount of pubescence on leaves and young twigs. Several varieties have been named. The characteristic subapical tufts of hairs on the teeth of the leaflets are said to be absolutely distinctive in this species, so that if search on an old worn leaf in the fall reveals one or two teeth bearing a tuft, identification is positive. Very rarely there may be 7 leaflets rather than the usual 5.

The large loose elongate plates of bark characteristic of this species (and *C. laciniosa*) are best developed on older trees. Young trees may have smooth tight bark (as on one 1–2 dm in diameter, bearing large fruit, discovered by E. E. Sherff in Barry Co.). The husks may be slightly ridged along the sutures, but not so prominently as in bitternut.

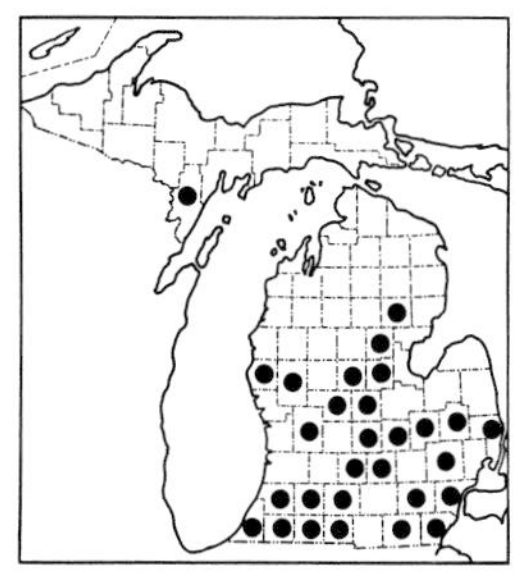
37. Carya cordiformis

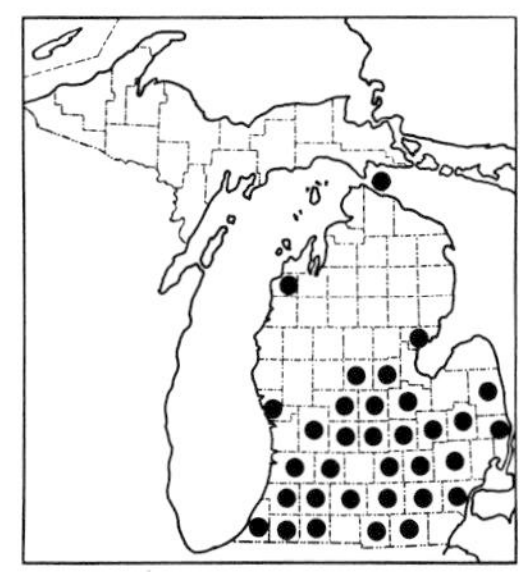
38. Carya ovata

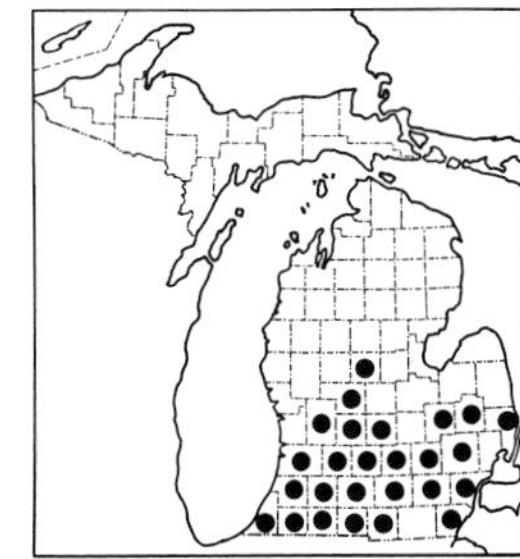
39. Carya glabra

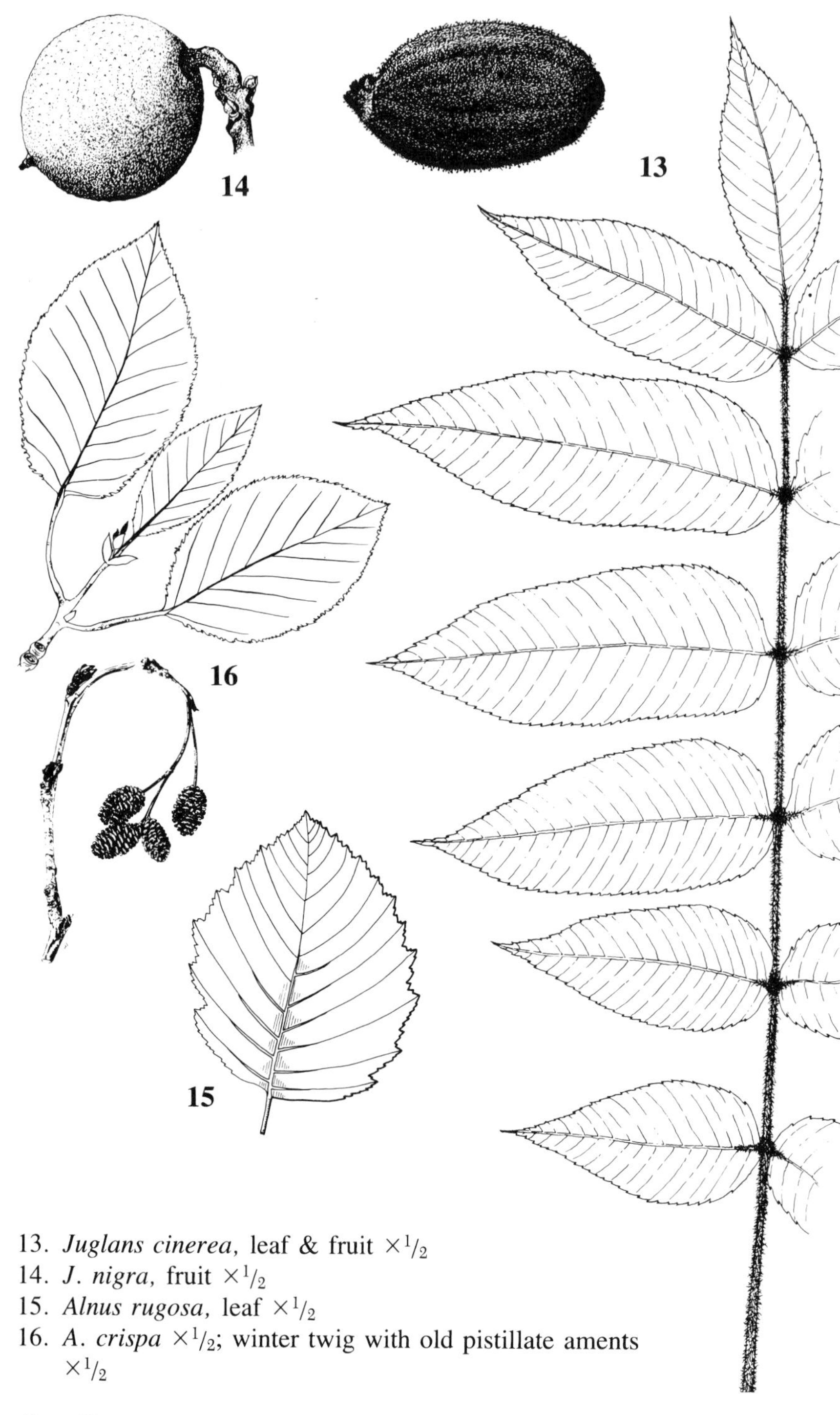

13. *Juglans cinerea,* leaf & fruit $\times ^1/_2$
14. *J. nigra,* fruit $\times ^1/_2$
15. *Alnus rugosa,* leaf $\times ^1/_2$
16. *A. crispa* $\times ^1/_2$; winter twig with old pistillate aments $\times ^1/_2$

3. **C. glabra** (Miller) Sweet — Pignut Hickory

Map 39. Usually in upland, often sandy, woods, associated with oaks.

The pignuts are here treated in a broad sense, in accord with the published (and private) opinions of several authorities. Others, however, maintain as distinct the red hickory or sweet pignut, *C. ovalis* (Wang.) Sarg. The distinction can be made with certainty only in the late fall, with fully mature fruit present. The husks split all the way to the base in *C. ovalis* and only halfway in *C. glabra*. Unfortunately, even this character, like others, is not fully reliable, for intermediates occur. *C. ovalis*, when older, characteristically has "shaggy" or scaly bark, while in *C. glabra* the bark remains tight and not scaly. There seems to be little correlation among the other characteristics used by various authors: presence of granular yellow dots on buds, shape of fruit and nut and angles on the latter, number of leaflets (said to be more often 7 in *C. ovalis*), distribution of pubescence and glandular dots, etc. The distribution of the two "species" is essentially the same. Available specimens indicate that the *ovalis* type is the usual one in Michigan and that true *glabra* is relatively rare, but many specimens cannot be positively referred to either, in the absence of mature or unambiguous fruit.

4. **C. laciniosa** (Michaux f.) G. Don — Kingnut or Shellbark Hickory

Map 40. Riverbanks and rich floodplain woods.

The stout branchlets and twigs of this species contrast (as do the very large buds and fruit) with the more slender (or smaller) ones of other species in the genus.

Another large-fruited, large-budded hickory, *C. tomentosa* (Poiret) Nutt., mockernut hickory, has been reported from southern Michigan. All efforts to document such reports with reliable specimens have failed (see Manning 1973). The mockernut has a more densely tomentose rachis in the leaves, darker brown twigs, and less compressed nuts than *C. laciniosa*. The latter also has shaggy bark on old trees, while in *C. tomentosa* the bark remains tight, ridged and furrowed.

BETULACEAE — Birch Family

Sometimes divided into two (or even three) families, but most modern opinion supports a single family of two (or three) tribes. The Betuleae include *Betula* and *Alnus*, the most primitive genera in the family; and Coryleae, the remaining genera (unless the Carpineae, including *Carpinus* and *Ostrya*, are further segregated). The name Betulaceae is conserved against Corylaceae if a single family is recognized.

These are all wind-pollinated, monoecious trees and shrubs, flowering before the leaves are mature. The staminate flowers are in ± elongate cylindrical aments in all genera. While the leaves are distinctive, they are hard to describe objectively, and the key must rely often on reproductive parts.

KEY TO THE GENERA

1. Pistillate flowers in dense elongate aments or cone-like inflorescences, with subtending deciduous or persistent woody bracts or scales less than 1 cm long; nut usually winged; scales of staminate ament somewhat peltate in appearance, usually ± obtuse or several-lobed (from adnate bractlets), with additional perianth parts and/or bractlets evident; anthers mostly glabrous, the separate locules 12 or 24 (the bifid stamens 6 or 12) per scale (in *B. pumila* often as few as 3 bifid stamens)
   2. Pistillate aments clustered (2 or more loosely racemose), old ones remaining on the plant all year, the scales at maturity persistent, woody, ± at right angles to rachis, glabrous, strongly wedge-shaped and truncate or obscurely 5-lobed (more definitely with 5 ± overlapping lobes when young) at apex; nut wingless (widespread species) or winged; staminate aments basically with 4 stamens per flower and 3 flowers per scale, thus with 12 stamens per scale, these slightly bifid, resulting in 24 barely separated anther locules . . . . 1. **Alnus**
   2. Pistillate aments solitary, the scales deciduous (or readily dislodged) at maturity, firm or only slightly woody, usually ± strongly ascending, ciliate or hairy (to glabrate in a bog shrub), strongly 3-lobed; nut with membranous wings; staminate aments basically with 2 stamens per flower and 3 flowers per scale, thus with 6 stamens per scale, these slightly to deeply bifid, resulting in 12 separate locules (in *B. pumila* sometimes only 3 or 4 stamens, hence 6 or 8 locules per scale) . . . . 2. **Betula**
1. Pistillate flowers in a short head or loose raceme-like ament, with subtending bracts becoming leafy, persistent, at least 1.5 cm long; nut wingless; scales of staminate ament acute to abruptly short-pointed, with or without partly adnate lobe-like bractlets but with no additional perianth; anthers with a few hairs at the tip, the separate locules either 8 or 16 or more, on 4 or 8 or more bifid filaments per scale
   3. Leaves with about 8 veins or fewer on each side of midrib, the blades up to twice as long as broad (but mostly ca. 1.5 times), with undulate or shallowly lobed (as well as toothed) margins; scales of staminate aments each with 4 bifid stamens (= 8 separate anther locules) and 2 small partly adnate bractlets; pistillate flowers few, in short (not over 6 mm) ovoid buds, concealed (with villous bractlets) by the bud scales except for protruding red stigmas (2 per flower); flowers appearing well before the leaves; fruit a large (ca. 1 cm) acorn-like nut hidden by involucral bracts . . . . 3. **Corylus**
   3. Leaves with about 10 or more veins on each side of midrib, the blades mostly about twice as long as broad or slightly longer, with margins only toothed (not lobed or undulate); scales of staminate aments each with 8 or more bifid stamens (= 16 or more locules) and no bractlets; pistillate flowers in a loose elongate terminal ament or drooping spike; flowers opening at about the same time as the leaf buds or later; fruit a much smaller nut, either enclosed in a sac-like bract or in pairs subtended by a leafy bract
      4. Plant a tall or medium shrub (usually with several crooked stems), the bark smooth, light gray, ± fluted or "muscular" in appearance; staminate aments solitary, covered by bud scales in winter, elongating in spring to 1.2–2 (2.7) cm long, the scales acute or broadly and obscurely acuminate but without any prolonged tip; nut with several strong ribs and a several-toothed crown at the apex, in pairs subtended by a few-toothed or -lobed bract which enlarges at maturity to 1.8–3 (4) cm long; leaves without glands on petioles, the blades with lateral veins only very rarely forked . . . . 4. **Carpinus**
      4. Plant a small tree with dark scaly bark and single erect trunk; staminate aments mostly in clusters of 2–3, naked in winter, elongating to 1–5 cm at full flowering in spring, the scales acuminate to a short prolonged tip; nut smooth (at most

obscurely ribbed toward end), without toothed crown (entire at apex), surrounded by a tubular (fused) bract which enlarges at maturity into an inflated ellipsoid sac (like a *Carex* perigynium) 1.5–2.5 cm long; leaves usually with sparse to dense stalked glands on petioles (these sometimes also on young branchlets) and lateral veins often branched (especially beyond the middle) . . . . 5. **Ostrya**

1. **Alnus** Alder

These familiar shrubs often form dense thickets, especially northward. Nodules on the roots harbor nitrogen-fixing microorganisms, a feature of relatively few plants other than the Leguminosae (cf. Myricaceae and *Ceanothus*).

Since it is more evident when the staminate aments (formed the previous season) are mature (shedding pollen) and they shrivel and fall afterwards, their season is stressed in the key, although the woody, persistent, cone-like pistillate aments have essentially the same time of floral maturity (in *A. crispa*, however, forming early in the spring).

Alnus serrulata (Aiton) Willd., smooth alder, has been collected in Indiana near the Michigan border and ranges southward. It might be expected in southwestern Michigan. It differs from *A. rugosa* in its more elongate, elliptical or slightly obovate, usually rather acute leaves with regular finely serrate margins (in this aspect, as also in slightly glutinous surface, resembling the northern *A. crispa*). In Ohio it occurs only in the eastern and southern portions of the state, and it extends across southern Indiana and Illinois, with isolated occurrences (now extirpated?) in northwestern Indiana.

REFERENCE

Furlow, John J. 1979. The Systematics of the American Species of Alnus (Betulaceae). Rhodora 81: 1–121; 151–248.

KEY TO THE SPECIES

1. Leaf blades all very broadly rounded to truncate or notched at apex, obovate to suborbicular in outline; plant an erect tree, rarely escaped from cultivation . . . . 1. **A. glutinosa**
1. Leaf blades all or mostly acute (barely obtuse) to short-acuminate at apex, roughly ovate to elliptical in outline; plant a tall native shrub, usually many-stemmed
   2. Staminate aments expanding to maturity in early spring, before the leaves begin to open; filaments almost fully adnate to the perianth, the anthers appearing sessile; leaves not glutinous, both finely serrate and more coarsely dentate or obscurely lobed, the appearance being one of teeth quite varying in size and regularity (fig. 15); nut not winged at maturity (but so thin when immature as sometimes to be misleading) . . . . 2. **A. rugosa**
   2. Staminate aments expanding at the same time as the leaves; filaments free, ca. 0.5–1 mm long; leaves glutinous (with shiny sticky dots and veinlets) especially beneath and when young, finely and regularly serrate but not basically dentate or lobed (fig. 16); nut with a broad membranous wing on each side . . . . 3. **A. crispa**

1. **A. glutinosa** (L.) Gaertner — Black Alder

Map 41. A Eurasian species, recorded as escaped along banks of rivers and lakes in the southeastern Lower Peninsula, and perhaps to be found elsewhere. Planted as far north as Baraga County.

Resembles *A. rugosa* in early flowering and nuts wingless or nearly so, but the young parts are very glutinous; the leaves are quite distinctive.

2. **A. rugosa** (Duroi) Sprengel Fig. 15 — Speckled Alder

Map 42. A common tall shrub in older zones of bogs, along lakes and streams, in extensive mucky swamps, and in all sorts of wetlands. This dominant of our "tag alder" swamps is taken for granted in Michigan, but it ranges south only into the northernmost parts of Illinois and Indiana and the northern half of Ohio.

This alder has often been included with the similar European *A. incana* (L.) Moench, of which it is perhaps better treated as ssp. *rugosa* (Duroi) Clausen. Plants with the mature leaves ± glaucous or whitened beneath occur throughout the state and have been called var. *americana* (Regel) Fern. [*A. incana* var. *glauca* (Michaux f.) Loudon]. Plants with the leaves green beneath also occur throughout, although dry specimens are not always clear. Both varieties have both glabrous and pubescent forms.

The pistillate aments are often attacked by a fungus (*Taphrina*) which causes hypertrophy; the resulting elongate proliferations from the ament are quite unlike the normal scales and are easily recognized as an abnormality.

3. **A. crispa** (Aiton) Pursh Fig. 16 — Green or Mountain Alder

Map 43. Forming thickets in mixed woods and clearings, on river banks, and along Lake Superior on upper beaches, sand bluffs, and rock outcrops. Often more abundant than the preceding species in the Lake Superior area, but very local elsewhere.

Sometimes included in the Eurasian *A. viridis* (Chaix) DC. as ssp. *crispa* (Aiton) Turrill. Plants with the leaves beneath, young branchlets, and peduncles ± pubescent have been called var. *mollis* (Fern.) Fern.

40. Carya laciniosa

41. Alnus glutinosa

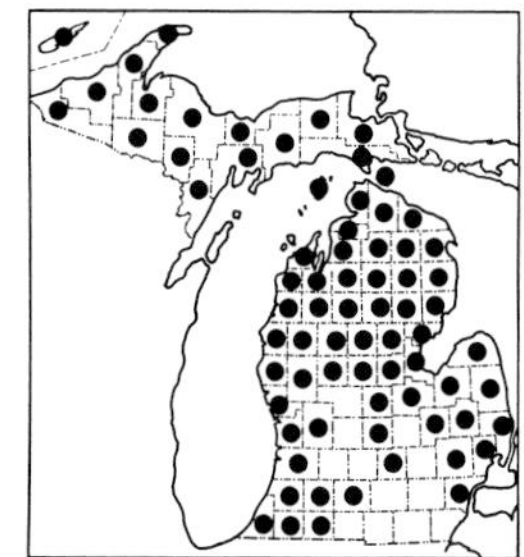

42. Alnus rugosa

2. **Betula** Birch

Familiar trees and shrubs of northern latitudes. The typically white and yellowish exfoliating (peeling) bark of white birch and yellow birch, respectively, makes many individuals readily recognizable in the field. However, the taxa are extremely variable, many individuals (especially when in the form of herbarium specimens) being less easy to recognize. Hybridization is rather frequent and is presumably a source of variability in addition to the natural lack of uniformity in the species.

REFERENCES

Barnes, Burton V., & Bruce P. Dancik. 1985. Characteristics and Origin of a New Birch Species, Betula murrayana, from Southeastern Michigan. Canad. Jour. Bot. 63: 223–226.

Brittain, W. H., & W. F. Grant. 1967. Observations on Canadian Birch (Betula) Collections at the Morgan Arboretum. V. B. papyrifera and B. cordifolia from Eastern Canada. Canad. Field-Nat. 81: 251–262.

Dancik, Bruce P. 1969. Dark-barked Birches of Southern Michigan. Mich. Bot. 8: 38–41.

Dancik, Bruce P., & Burton V. Barnes. 1972. Natural Variation and Hybridization of Yellow Birch and Bog Birch in Southeastern Michigan. Silvae Genet. 21: 1–9.

Dancik, Bruce P., & Burton V. Barnes. 1975. Leaf Variability in Yellow Birch (Betula alleghaniensis) in Relation to Environment. Canad. Jour. For. Res. 5: 149–159.

Grant, W. F., & B. K. Thompson. 1975. Observations on Canadian Birches, Betula cordifolia, B. neoalaskana, B. populifolia, B. papyrifera, and B. ×caerulea. Canad. Jour. Bot. 53: 1478–1490.

KEY TO THE SPECIES

1. Leaves with 8–12 pairs of distinct lateral veins; mature pistillate aments 12–21 (24) mm thick, sessile or nearly so (peduncles at most ca. 5 mm); bark of twigs with flavor of wintergreen; mature plant a tree with yellowish peeling or dark bark . . . . 1. **B. alleghaniensis**

1. Leaves with (2) 3–8 [9] pairs of distinct lateral veins; mature pistillate aments (5) 6–10 (12) mm thick, distinctly peduncled (the peduncles sometimes as short as 2 mm in bog shrub); bark of twigs without wintergreen flavor; mature plant usually a small to large many-stemmed shrub or a tree with white peeling bark

  2. Leaf blades 1–3.5 cm long (or up to 5 cm on sprouts), mostly ± obovate and rounded at the apex; wings of fruit narrower than the body; plant a bushy shrub with dark bark . . . . 2. **B. pumila**

  2. Leaf blades mostly 5–8 (10) cm long, ± ovate to deltoid, sometimes cordate, acute at apex; wings of fruit broader than the body; plant a tree with white bark (or often dark, especially when juvenile)

    3. Leaf blades completely glabrous on both sides, even when young, ± deltoid and long-acuminate; pistillate scales with middle lobe reduced, scarcely prolonged . . . . 3. **B. pendula**

    3. Leaf blades with at least some hairs in axils of lower lateral veins beneath, ± ovate, broadly cuneate to cordate at base, acute to short-acuminate; pistillate scales with middle lobe reduced to well developed and prolonged . . . . 4. **B. papyrifera**

1. **B. alleghaniensis** Britton Yellow Birch

Map 44. Grows in a diversity of habitats, from lowland hardwoods, especially in the southern part of the state, to hemlock-white pine-northern hardwoods forest of the north, becoming especially common in the western Upper Peninsula. The largest trees in the state are about 5 feet in diameter.

Reports of native *Betula lenta* L. (sweet, black, or cherry birch) and *B. nigra* L. (river birch) from Michigan should apparently all be referred to yellow birch. As is true of the other birches, the pistillate scales and bark of *B. alleghaniensis* are variable. Very old trees, as in virgin stands in the Upper Peninsula, may have dark bark so furrowed or checkered as not to resemble birch bark at all. Trees in swamps of the southern Lower Peninsula may have dark brown or blackish bark which peels very little if at all; such trees hardly resemble the widespread typical yellow birch with light yellow-brown, peeling bark, and have sometimes been erroneously referred to *B. lenta* (which differs in having smaller, nearly or quite glabrous and less densely ciliate scales in the pistillate aments). *B. lenta* and *B. alleghaniensis* both have a characteristic odor and flavor of wintergreen in the crushed bark of young twigs. (The wintergreen flavor is retained in dry specimens and can be detected by chewing on a bit of bark from specimens at least 100 years after they were pressed.) *B. nigra* lacks the wintergreen flavor and has leaves pale beneath with margins both toothed and undulate or shallowly lobed; the fruit ripens in late spring or early summer, with wings narrower than the body; the bark is pinkish, curly and tattered-looking, and the number of leaf veins is less than in *B. alleghaniensis*. *B. nigra* grows in wet places and is known in northwestern Indiana; it should be sought in southern Michigan, where it is sometimes planted but seems never to spread.

Yellow birch was long known as *B. lutea* Michaux, an illegitimate name. Hybrids with *B. papyrifera* may be less rare than usually supposed. Hybrids with *B. pumila* are described under that species below. The leaves on short lateral shoots are usually in pairs, whereas in *B. papyrifera* they are usually in 3's. The wood of yellow birch is one of the most valuable timbers in

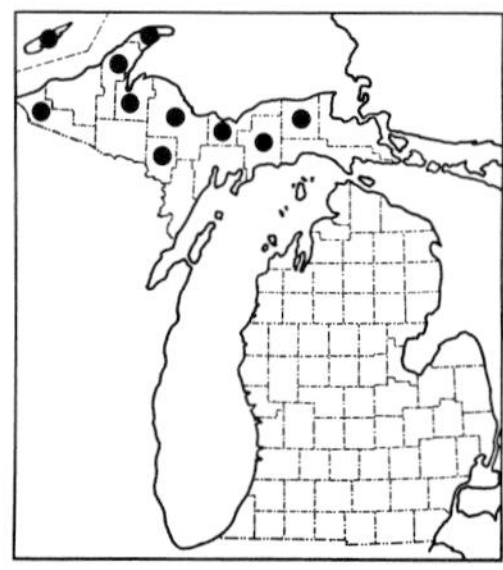
43. Alnus crispa

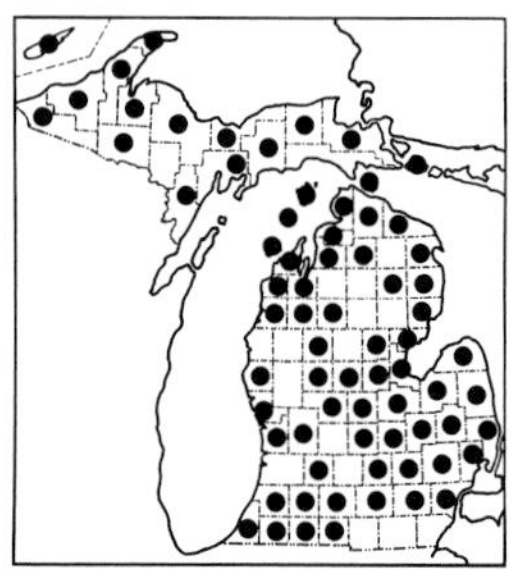
44. Betula alleghaniensis

45. Betula pumila

Michigan. It is strong and close-grained, polishes well, and is of special importance in furniture and veneers. Solid logs suitable for cutting veneer command a premium price.

2. **B. pumila** L. Fig. 17 Bog or Dwarf Birch

Map 45. Bogs and fens, conifer swamps, shrubby peatlands, stream borders; generally a calciphile.

Bog birch is a bushy shrub up to about 10 feet tall, but usually no more than half that large, the twigs and young leaves extremely variable: glabrous or pubescent (puberulent and/or with longer hairs), with warty glands absent, sparse, or prominent. The blades are often quite glaucous beneath and are rather thick in texture. Plants with neither glands nor pubescence have been called var. *glabra* Regel; those without glands but usually with conspicuous pubescence, var. *pumila* (which is the commonest variant in the southern Lower Peninsula); and those with both glands and some pubescence, var. *glandulifera* Regel (common in northern Michigan). Reports of the arctic-alpine diploid (2n = 28) *B. glandulosa* Michaux from this region should all presumably be referred to *B. pumila* var. *glandulifera*, but the classification of the dwarf birches is not yet completely satisfactory. The group is here considered a polymorphic complex; two or all three of the varieties may grow together, and all the characteristics vary considerably. The stamens in this species are often only 3 or 4 per scale.

Hybrids, beautifully intermediate, with the native tree species may be expected where the parents grow in proximity. Those with *B. papyrifera* seem more frequent, and have been called *B.* ×*sandbergii* Britton. They have leaves which are mostly acute, rather than obtuse, at the apex, with 5–6 (7) pairs of prominent lateral veins, in contrast with the (2) 3–5 pairs in *B. pumila*. Mature staminate aments may be as long as 5–6 cm, while in *B. pumila* they are only (0.8) 1–2 cm and in *B. papyrifera* usually 6–11 cm. The wings of the fruit in our hybrid material are a little narrower than the body, but resemble those of *B. papyrifera* much more than of *B. pumila*. Plants tend to be intermediate in size between the tree and shrub species and to have many stems with dark bark, scarcely if at all peeling. The hybrid produces pollen and seed copiously and thrives where well established. There is some question as to whether backcrossing and introgression are occurring or whether plants which resemble *B. papyrifera* more closely are merely F_1 progeny, the resemblance due to the presence of more chromosomes from the polyploid tree parent than from the tetraploid *B. pumila* (2n = 56). When the parents are as variable as they are, the hybrids must also be quite variable. *B.* ×*sandbergii* has been collected throughout the state, including a number of sites in the eastern Upper Peninsula and northernmost Lower Peninsula.

Hybrids with *B. alleghaniensis* have been named *B.* ×*purpusii* Schneider (TL: Clarks Lake [Jackson Co.]), and as in *B.* ×*sandbergii* are intermediate between the parents in stature and in shape and texture of leaves, which have (6) 7–9 (10) pairs of prominent lateral veins and usually an acute apex. The mature pistillate aments are much larger than those of *B. pumila*, being 10–12 or even as much as 18 mm thick. The wings on the fruit are much narrower than the body, as in both parents (but not in *B. papyrifera* or *B.* ×*sandbergii*). The bark of fresh twigs has the wintergreen flavor of yellow birch, but this may not persist on dry specimens. *B.* ×*purpusii* has been collected across the southern Lower Peninsula north to Clinton County. An octoploid derivative from this hybrid has been named *B. murrayana* Barnes & Dancik (TL: Third Sister Lake, Washtenaw Co.).

3. **B. pendula** Roth European White Birch

Map 46. Locally established, presumably as an escape from cultivation, in fields, swamps, bogs, and elsewhere. "Cut-leaved" forms are sometimes cultivated, but the collections referred here have normal leaves, with blades distinctly more acuminate than in *B. papyrifera* though not quite as caudate-acuminate as in the gray birch, *B. populifolia* Marsh. The latter is a close-barked (not exfoliating) small tree or tall shrub, chiefly of the northeastern states and adjacent Canada, which bears similar pistillate catkins of horizontally divergent scales with the lateral lobes much larger than the reduced terminal one. However, the pistillate scales are finely but densely pubescent all across both surfaces in *B. populifolia*, as well as being somewhat smaller (3–4 mm broad) than those of *B. pendula* (4–6 mm broad), which are usually glabrous or glabrate at least inside (adaxial surface). Specimens from Bear Lake bog, Ingham County, have scales as in *B. populifolia* but leaves less acuminate than in that species.

4. **B. papyrifera** Marsh. Fig. 18 Paper, White, or Canoe Birch

Map 47. Widespread in a diversity of habitats; especially characteristic after fire or other disturbance, when seedlings are often abundant. Frequently on dunes and associated with aspen on upland sites, but also in swampy places. Usually a relatively short-lived successional tree, growing in attractive clumps from stump sprouts that follow fire. But old trees may persist in the forest, as large as 2–4 feet in diameter, with rough bark.

This is a familiar and handsome tree, though seldom a large one, long famed as the source of birch bark used by Indians to construct canoes and many other objects. It is closely related to the white birches of Europe [*B. alba* L. or *B. pubescens* Ehrh. and *B. pendula*], forming part of a circumboreal complex. Often treated in the past as a variety of *B. alba*. White birch is extremely variable in leaf glandularity and shape and in size, shape, and pubescence of the pistillate scales—and even variable in bark, this being rich reddish brown on young trunks and even on some older ones being dark or pinkish rather than the usual chalky white.

Plants variously referred to *B. cordifolia* Regel or *B. papyrifera* var. *cor-*

difolia (Regel) Fern. have been distinguished chiefly by having cordate leaf blades; supposed differences in pistillate scales seem hardly consistent or significant against the enormous variability in the species. The work of Brittain and Grant in Canada indicates that *cordifolia* is a northern diploid element (2n = 28 or rarely 56) in contrast to typical *B. papyrifera* (2n = 70 or 84 or rarely 56). Cordate leaves are characteristic of the diploid but not restricted to it. It may be that some trees from northern Michigan are the diploid, however it should be treated taxonomically, but it seems best at present to interpret *B. papyrifera* in the widest sense to include all of our native white birches (which may, on investigation, reveal all four chromosome numbers).

Hybrids with *B. pumila* are rather frequent and are discussed under that species above. Hybrids with *B. alleghaniensis* also occur.

3. **Corylus** Hazel

The European filberts belong to this genus, and the fruits of our species provide edible nuts.

KEY TO THE SPECIES

1. Plants with scattered dark stiff gland-tipped bristles as well as very fine ± crinkly whitish hairs on petioles and young branchlets (and often upper leaf surfaces and involucral bracts); staminate aments on short woody peduncles or branchlets; involucre a pair of separate, broadly fan-shaped, sharply lacerate-toothed bractlets, scarcely twice as long as mature nut, which is often partly visible .... 1. **C. americana**
1. Plants without dark or gland-tipped bristles (although some whitish hairs, especially copious on involucre, are longer, stiffer, and straighter than in the preceding species); staminate aments sessile or nearly so; involucre a tubular prolongation of united bractlets, toothed at apex, ca. 2–5 times as long as the completely concealed nut .... 2. **C. cornuta**

1. **C. americana** Walter Hazelnut

Map 48. Dry woodlands, river banks, woods and thickets.

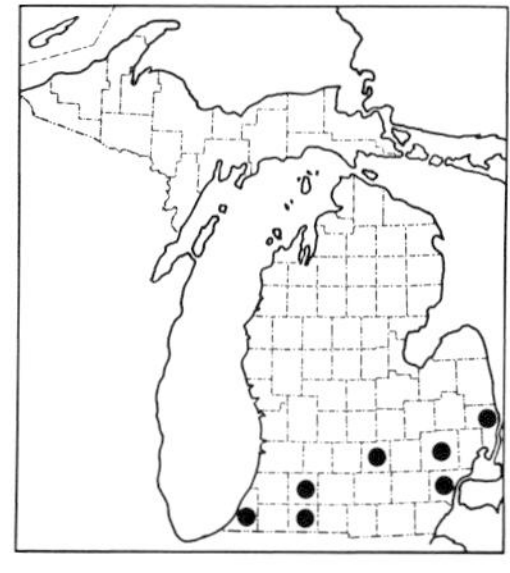
46. Betula pendula

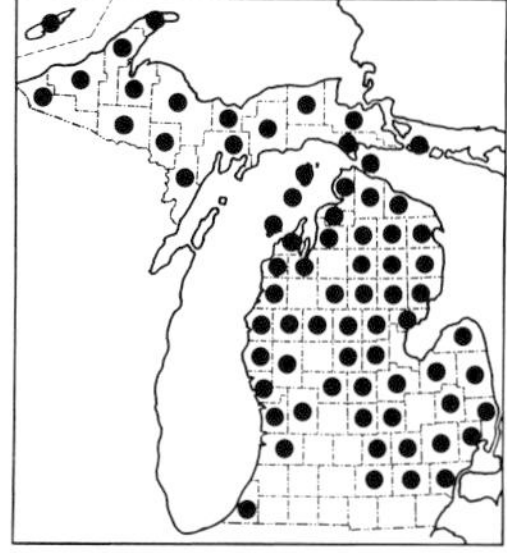
47. Betula papyrifera

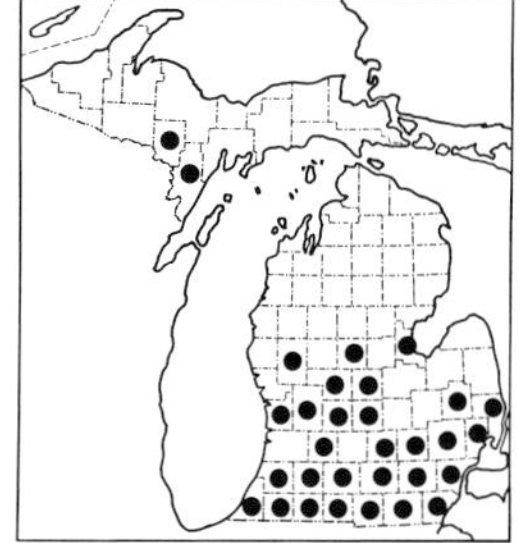
48. Corylus americana

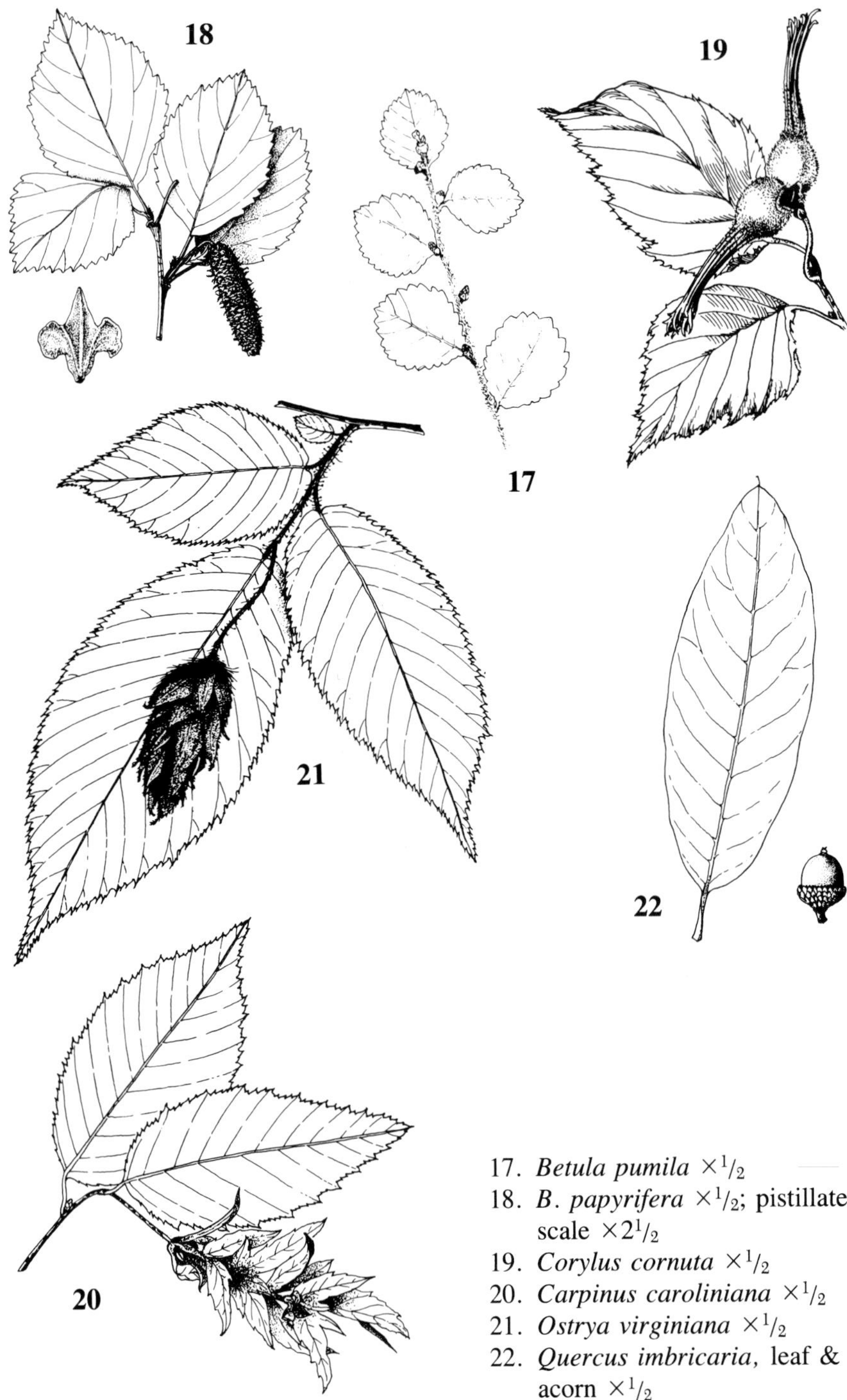

17. *Betula pumila* $\times^1/_2$
18. *B. papyrifera* $\times^1/_2$; pistillate scale $\times 2^1/_2$
19. *Corylus cornuta* $\times^1/_2$
20. *Carpinus caroliniana* $\times^1/_2$
21. *Ostrya virginiana* $\times^1/_2$
22. *Quercus imbricaria*, leaf & acorn $\times^1/_2$

2. **C. cornuta** Marsh. Fig. 19 Beaked Hazelnut

Map 49. A common understory shrub especially of borders and clearings throughout the northern hardwoods forest; also in dune thickets and (particularly in the southern part of its range) on river banks. Several collections (all before 1900) from Ingham County are presumably from plantings although only one label explicitly states "Cult. plants from Hersey [Osceola Co.]."

4. **Carpinus**

1. **C. caroliniana** Walter Fig. 20 Hornbeam; Blue-beech

Map 50. Swamp forests and stream banks; also upland woods, including aspen, oak-hickory, and rich beech-maple stands.

This is sometimes called ironwood, a name better reserved for *Ostrya*. The straight unbranched lateral veins of the leaf may suggest those of beech, and the leaves are also plicate in the bud; however, the margins are doubly serrate, while in beech the only teeth are at the ends of the veins. The nutlets are in pairs, subtended by a prominent green bract, much enlarged from the bractlet of the flower; the true subtending bract is deciduous. The smooth muscular-looking bark is an ideal field character.

5. **Ostrya**

1. **O. virginiana** (Miller) K. Koch Fig. 21 Ironwood; Hop-hornbeam

Map 51. A characteristic small tree of deciduous forests (oak, hemlock-hardwoods, and especially beech-maple). The largest tree in the state is 3 feet in diameter, but most are much more slender.

Sometimes (especially formerly) called leverwood, from the hardness and strength of the wood. The town of Ironwood in Gogebic County was named for a Mr. Wood whose nickname was "Iron"; it is thus only indirectly named for the tree.

The foliage resembles that of yellow birch (*Betula alleghaniensis*), but

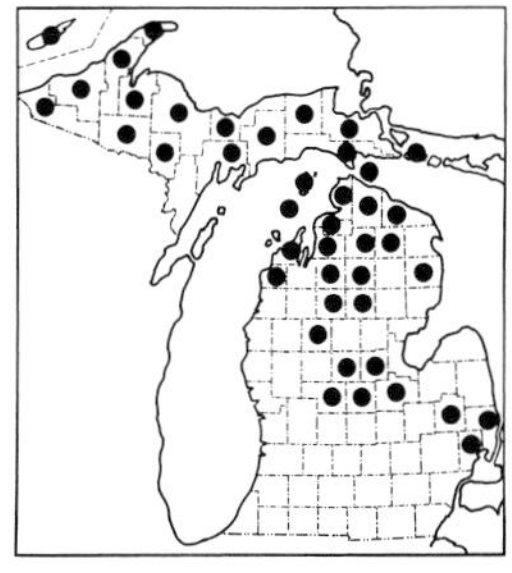

49. Corylus cornuta

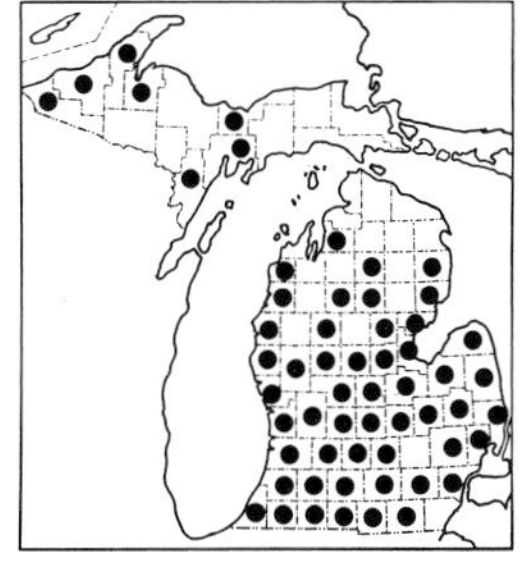

50. Carpinus caroliniana

51. Ostrya virginiana

as in *Carpinus* the leaves are always single, not in pairs, while in *B. alleghaniensis* the leaves on very short lateral shoots (not new branchlets) are generally paired. The stigmas tend to be more delicate and elongate and less brightly colored than those of *Carpinus* or *Corylus*. The nutlets are basically in pairs, but each is surrounded by a sac (modified bractlet) and the single true subtending bract is deciduous.

FAGACEAE — Beech Family

REFERENCE

Elias, Thomas S. 1971. The Genera of Fagaceae in the Southeastern United States. Jour. Arnold Arb. 52: 159–195.

KEY TO THE GENERA

1. Leaves entire to deeply lobed, or (3 species) if coarsely serrate, then finely stellate-pubescent over the lower surface; nut an "acorn," terete, seated (solitary) in a shallow to deep scaly cup which remains intact . . . . 1. **Quercus**
1. Leaves sharply and regularly serrate, glabrous to silky-pubescent (not stellate) along the veins beneath; nut slightly to strongly flattened on at least one side, enclosed (usually 2–3 together) in a prickly involucre (husk or bur) which splits into (2–) 4 valves
 2. Teeth of leaves at most 1.5 (2) mm long (measured along upper margin); flowers appearing in spring, at same time as the leaves, the staminate ones in a peduncled, pendent, globose ament; fruiting involucre ca. 2 cm long or less, bearing weak unbranched spines and normally containing 2 sharply 3-angled nuts; winter buds over 1 cm long; bark smooth and gray . . . . 2. **Fagus**
 2. Teeth of leaves mostly 2–5 mm long, acuminate and often somewhat incurved; flowers in midsummer, after leaves are fully developed, the staminate ones in elongate, stiff, essentially sessile aments; fruiting involucre over 2 cm long, densely covered with stiff and sharp branched spines and normally containing 2–3 somewhat flattened but not sharply angled nuts; winter buds less than 1 cm long; bark ridged and furrowed, brownish . . . . 3. **Castanea**

1. Quercus — Oak

Everyone knows an oak at sight, at least if it is bearing acorns, a distinctive type of fruit borne by no other genus in this region. However, distinguishing the species of oak is often difficult in the field and next to impossible in the herbarium. Collections for the latter should (but too rarely do) include typical mature foliage, winter buds, and fully mature acorns, as well as notes on bark and stature of the plant. (Only one of our species is a shrub, yet collectors almost always fail to record this essential key character on their labels.)

Oak leaves are extremely variable (in size, shape, and pubescence) from one part of the same tree to another, and from one tree to another of the same species. Leaves growing in the shade, as deep in the canopy or on lower branches, are often less deeply lobed than those receiving full sun, as on upper branches. Leaves on vigorous sprouts, as in the Salicaceae (and other groups), may appear quite unrecognizable. Add the fact that oaks frequently hybridize with other species of the same subgenus, producing first generation (F_1) progeny with ± intermediate characters (and apparently with some resultant introgression of characters following backcrossing), and one sees why the genus has a well deserved reputation for difficulty. Immature (e. g., flowering) specimens are usually impossible to name—at least without full information on the plant from which the juvenile material came. Young or abortive acorns should be avoided for identification.

Our oaks fall naturally into two subgenera, the white oaks (*Quercus* subgenus *Quercus*—long called subgenus *Lepidobalanus*) and the black or red oaks (*Quercus* subgenus *Erythrobalanus*). The latter group is readily recognized by the prominent bristle-tips of the lobes of the leaves (or, in the entire-leaved shingle oak, only at the tip of the leaf). Additional characters are included in the key and are discussed in other works. The serious student of oaks will want to become familiar with the plants in the field, including bark and general habit, and to study references such as those cited below. The very full descriptions and discussions, based on long field experience, by such authors as Deam and Palmer, together with specialized reports on variation and hybridization in our area, will add much to the brief account that can be given here.

Reports of the blackjack oak, *Q. marilandica* Muenchh., from Michigan are unsubstantiated. They appear to be based on aberrant leaves of other species (e. g., immature and other forms of *Q. velutina*) with a broad obovate shape.

Acorns, especially of the white oaks, are an important food for many mammals and birds and are sometimes used by humans. The higher tannin content of acorns of the red and black oaks makes leaching of them even more necessary to remove the bitterness. Oak wood is used for many purposes, including flooring, furniture, paneling, and other ornamental uses as well as in situations where resistance to decay is important, as in posts and pilings. It is also one of the best firewoods. "National champions" of five oak species and hybrids are recorded from Michigan, all of them 6 feet or more in diameter.

I have had relatively little field experience with this genus, and have relied heavily on the notes and observations of W. H. Wagner, who has generously helped me deal with the oaks—including annotation of many sterile or otherwise difficult herbarium specimens which his experience has enabled naming but which are insufficient to run with assurance through any key or to recognize as hybrids. Hybrids are not described here; they are to be expected near the parent trees (or where parent trees once grew) and are more easily recognized in the field than from inadequate herbarium material which is too easily misidentified as one of the parents.

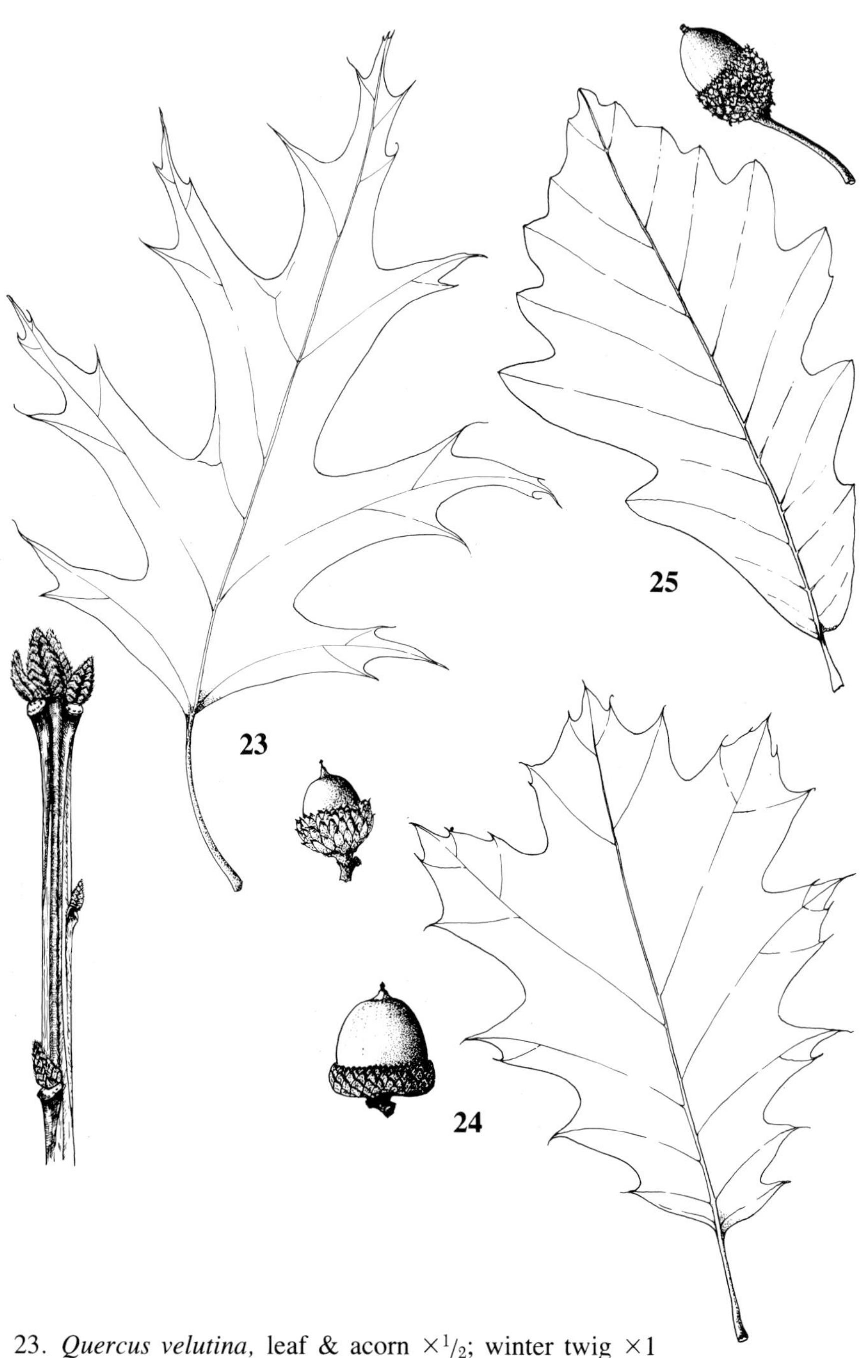

23. *Quercus velutina*, leaf & acorn $\times^1/_2$; winter twig $\times 1$
24. *Q. rubra*, leaf & acorn $\times^1/_2$
25. *Q. bicolor*, leaf & acorn $\times^1/_2$

REFERENCES

Deam, Charles C. 1953. Trees of Indiana. 3rd ed. Department of Conservation, Indianapolis. [*Quercus*, pp. 118–167.]

Hardin, James W. 1975. Hybridization and Introgression in Quercus alba. Jour. Arnold Arb. 56: 336–363.

Overlease, William R. 1975a. Populations Studies of Red Oak (Quercus rubra L.) and Northern Red Oak (Quercus rubra var. borealis (Michx. f.) Farw.). Proc. Pennsylvania Acad. 49: 138–140.

Overlease, William Roy. 1975b. A study of Variation in Black Oak (Quercus velutina Lam.) Populations from Unglaciated Southern Indiana to the Range Limits in Northern Michigan. Proc. Pennsylvania Acad. 49: 141–144.

Overlease, William Roy. 1977. A Study of the Relationship Between Scarlet Oak (Quercus coccinea Muenchh.) and Hill Oak (Quercus ellipsoidalis E. J. Hill) in Michigan and Nearby States. Proc. Pennsylvania Acad. 51: 47–50.

Palmer, Ernest J. 1948. Hybrid Oaks of North America. Jour. Arnold Arb. 29: 1–48.

Wagner, W. H., Jr., & D. J. Schoen. 1976. Shingle Oak (Quercus imbricaria) and Its Hybrids in Michigan. Mich. Bot. 15: 141–155.

KEY TO THE SPECIES

1. Leaf blades entire, unlobed, elliptic to oblong-lanceolate or slightly obovate . . . . 1. **Q. imbricaria**

1. Leaf blades toothed or lobed

2. Lobes and teeth of leaf prolonged into a distinct bristle; acorns ripening their second year (hence located on year-old twigs, below the leaves), densely pubescent on inside of the shell; bark dark and tight, often shiny, becoming furrowed (but not scaly) with age

3. Leaves with sinuses cut half or less the distance to the midrib (the longest lobes [measured along upper side] thus at most about equaling the uncut part of the blade)

4. Terminal winter buds dull whitish-hairy, angled, usually at least 6 mm long by September, longer at maturity; acorn about half covered by the cup, which has the upper scales ± loose and spreading; leaves glossy above, the petiole and/or lower portion of midrib above usually retaining some pubescence well into the summer; range in Lower Peninsula, especially southward . . . . 2. **Q. velutina**

4. Terminal winter buds shining reddish, glabrous or nearly so, terete, usually less than 6 mm long; acorn about a third or less covered by the cup, which has the upper scales appressed; leaves with blades dull above, the petiole and midrib above soon glabrous; range throughout the state . . . . 3. **Q. rubra**

3. Leaves with sinuses cut about two-thirds or more of the distance to the midrib (the longest lobes thus at least twice as long as the uncut middle part of the blade)

5. Cups of acorns usually gray-pubescent throughout, with uppermost scales loose and spreading, forming a fringe around the acorn; terminal winter buds angled, hairy, usually 6 mm long by September, longer when mature . . . . 2. **Q. velutina**

5. Cups of acorns glabrous or pubescent but with uppermost scales ± appressed, not forming a fringe; terminal winter buds terete or nearly so, glabrous at least on lower half, smaller

6. Fruit small: acorns not over 12 mm long, cup very shallow and saucer-like, not over 12 mm across; plant of low habitats (but often planted on drier sites) . . . . 4. **Q. palustris**

6. Fruit larger: nut longer and/or cup broader (and deeper); plants of upland (even dry and sandy) habitats . . . . 5. **Q. coccinea**

2. Lobes or teeth of leaf rounded or blunt, not bristle-tipped; acorns ripening their first year (hence located among the leaves), glabrous on inside of the shell; bark usually gray and flaky

7. Leaf blades completely glabrous and usually ± glaucous beneath at maturity . . . . 6. **Q. alba**

7. Leaf blades thinly to densely stellate-pubescent over some or all of their surface beneath at maturity

8. Leaves with the sinuses much deeper on lower half of blade (sometimes nearly to midrib); terminal buds pubescent; branchlets densely pubescent when very young, soon glabrate; older branchlets frequently with corky wings; acorn half or more covered by the deep and strongly fringed cup; range throughout the state except northwestern Upper Peninsula . . . . 7. **Q. macrocarpa**

8. Leaves more uniformly lobed or toothed (*Q. bicolor* often with a few larger or more irregular lobes near middle of blade); buds glabrous or nearly so; branchlets glabrous or nearly so, without corky wings; acorn about a third or less covered by the cup, which lacks fringing awns on upper scales; range limited to southern half of Lower Peninsula

9. All lateral veins of some leaves not ending in a tooth or lobe (some of them fading out or ending at a sinus); leaves often ± irregularly lobed (more shallowly than in *Q. macrocarpa*) and not bilaterally symmetrical; acorns (usually paired) on peduncles ca. (1) 3–7 cm long (peduncles longer than petioles) . . . . 8. **Q. bicolor**

9. All lateral veins (except sometimes the lowermost) of all leaves ending in a large tooth; leaves unlobed and essentially bilaterally symmetrical; acorns usually sessile or nearly so (peduncles if any shorter than petioles)

10. Plant a low colonial shrub of mostly open areas, ca. 1–2 m tall; leaf blades with (4) 6–8 (9) teeth per side . . . . 9. **Q. prinoides**

10. Plant an erect tree of forests; leaf blades with (7) 9–13 (15) teeth per side

11. Teeth of leaf acute, usually with distinct (even slightly prolonged) callus tip; acorn less than 2 cm long . . . . 10. **Q. muehlenbergii**

11. Teeth of leaf rounded, with at most an obscure callus tip; acorn at least 2 cm long . . . . 11. **Q. prinus**

1. **Q. imbricaria** Michaux Fig. 22 Shingle Oak

Map 52. Swamp forests and upland woods, but like many upland tree species in southern Michigan persisting largely along roadsides and edges of cleared fields.

The name is derived from use of the wood by early settlers to make shingles. The majority of shingle oaks retain their dried and brown leaves in the winter, a tendency much less commonly observed in other oaks except the scarlet oak. This is a very distinctive species in our area (there are other entire-leaved species elsewhere), the bristle at the tip of the leaf revealing its affinity with the black oak subgenus. Leaves with small lobes or bristles on the margin represent hybrids, for this oak hybridizes with several other species, including at least three in Michigan: *Q. rubra* [producing *Q.* ×*runcinata* (A. DC.) Engelm.], collected several times in Washtenaw County; *Q. velutina* [producing *Q.* ×*leana* Nutt.], also collected several times in

Washtenaw County; and *Q. coccinea,* collected in Jackson and Washtenaw counties. (See Wagner & Schoen 1976.)

2. **Q. velutina** Lam. Fig. 23 Black Oak

Map 53. Usually in dry sandy woods with other oaks and hickory or (in cutover areas) red maple and largetooth aspen.

Yellow or orange inner bark is a good character for this species, but is rarely noted on herbarium specimens. The scaly fringed aspect of the cup around the acorn and the large, hairy, angled winter buds are also diagnostic. The glossy leaves, often but not always lobed more than halfway to the midrib, will also help to distinguish black oak from red oak.

Black oak is at least as variable as the other species of its group. It has probably received genes from, and contributed them to, various populations through hybridization and introgression during postglacial time, as oaks were becoming established on the glacial sand plains of Michigan (as well as elsewhere) and some characters contributed features favorable for survival in certain environments. Many years ago, the late W. H. Camp, who had been studying oaks for some time, suggested to me that there was not much "pure *velutina*" in our area and that the *velutina—rubra—coccinea* complex was even more tangled to the south in Ohio. Overlease (1975b) concluded that at the northern limit of its range (in the northern Lower Peninsula of Michigan), black oak has smaller, narrower acorns, less covered by the cup (which is less evidently fringed), smaller winter buds, and lighter yellow inner bark than more southern plants. Many of the specimens from the northern Lower Peninsula which have been called *Q. velutina* seem better placed as hybrids if not *Q. coccinea* or *Q. rubra*. Hybrids with those two species and with *Q. imbricaria* are noted under the other parent.

3. **Q. rubra** L. Fig. 24 Red Oak

Map 54. Thrives in rich mesic woods, but also grows on sandy plains with jack pine and on rock outcrops in the Lake Superior region, rather scrubby in the latter situations but can be a magnificent tree on better soils.

52. Quercus imbricaria

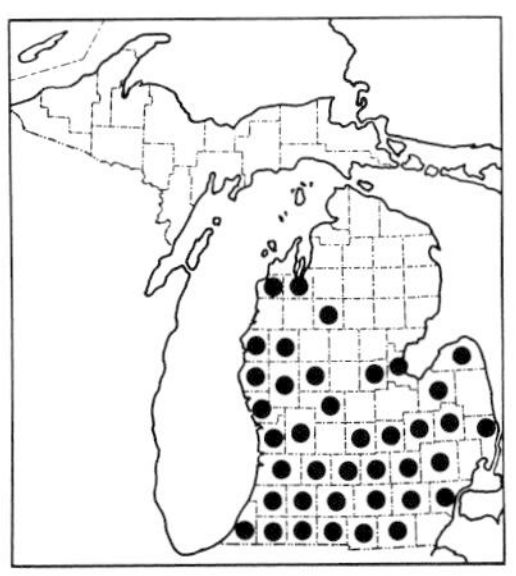
53. Quercus velutina

54. Quercus rubra

Ordinarily, red oak can be fairly easily recognized by the shallowly lobed, dull leaves, the lobes broadest at the base and ± tapering (not expanded) distally, the sinuses therefore narrowest at the base and widest at their open end (not narrowed again). The mature leaves are glabrous except usually for small tufts of pubescence persisting in the axils of the main veins beneath (as in the preceding species)—many descriptions of the leaves as completely glabrous notwithstanding. Most of our plants apparently have broad (to 2.5 cm), very flat, saucer-like cups beneath the large acorns (fig. 24) and have been referred to var. *rubra*. Plants with narrower acorns, the cup deeper and often with a suggestion of fringe around the acorn, are var. *ambigua* (A. Gray) Fern.—often given the later name of var. *borealis* (Michaux f.) Farw. But the acorn and cup differences may be the result of introgression from *Q. velutina* (see Overlease 1975a), with which this species hybridizes, producing *Q.* ×*hawkinsiae* Sudw., collected from Berrien County to Oscoda County. *Q. rubra* also hybridizes with *Q. imbricaria* (see above) and undoubtedly with *Q. coccinea*.

A few trees in Michigan with large, shallow-cupped acorns resembling those of *Q. rubra* (cups at least 2 cm broad) but the leaves glossier and more deeply lobed, suggesting very large leaves of *Q. coccinea*, have sometimes been identified as the more southern *Q. shumardii* Buckley. Material from Kalamazoo County was thus referred by E. J. Palmer. However, such trees seem more likely to be a form of *Q. rubra* or a hybrid involving that species. Their upland habitat is unlikely for a bottomland species like *Q. shumardii*, and other characters of leaves and fruit are inconsistent.

4. **Q. palustris** Muenchh. Fig. 26 Pin Oak

Map 55. Characteristic of low (periodically flooded) ground, including forests and edges of wet prairies; often planted and easily grown on upland sites.

According to Palmer (1942), besides the very small and distinctive acorns, the small, acute, nearly glabrous winter buds will serve to distinguish this species from *Q. coccinea*, although the buds are not very different from

55. Quercus palustris

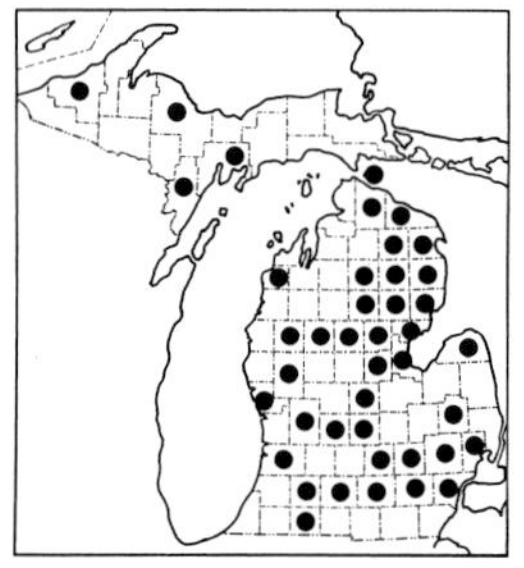

56. Quercus coccinea

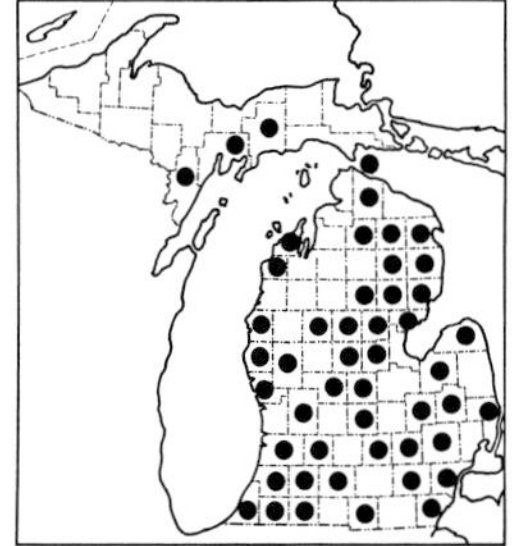

57. Quercus alba

those of other species (including *Q. ellipsoidalis,* here treated as part of the *Q. coccinea* complex). The leaves of *Q. palustris* are said to be usually recognizable by a combination of cuneate base of the blade (decurrent on the petiole) and conspicuous tufts of pubescence remaining until late in the season in the axils of the main veins and often along the midrib beneath. Our specimens, however, often have quite truncate leaf bases, but they do have prominent axillary tufts of pubescence (if not more along the midrib).

The habit of pin oak is so distinctive that it can be recognized in winter from a long distance. The trunk is very thick relative to the branches; the lowest branches are persistent and sweep downward.

5. **Q. coccinea** Muenchh. Fig. 27 Scarlet Oak

Map 56. Usually associated with other upland oaks and jack pine, on dry sandy soils.

Included in this complex are the plants often separated as *Q. ellipsoidalis* E. J. Hill, Hill's or northern pin oak, on the basis of various characters, including more pointed but less pubescent buds (the upper bud scales rather densely ciliate in *Q. coccinea*), narrower and more elongate acorns (longer than wide), more glossy and less pubescent cups (covering less of the acorn than in *Q. coccinea,* in which it may be half covered). There have long been differences of opinion as to the nature of the populations to which these names were originally applied, the circumscription of the populations to which they should now be applied, the significance of the characters, and their relation to possible hybridization. Those who studied the oaks extensively in field and herbarium over a period of years (e. g., C. C. Deam, E. J. Palmer, W. H. Camp) did not reach uniform conclusions. The characters used to separate the two supposed species are not well correlated, nor are there very clear discontinuities, and particular specimens have been assigned by different authorities to different species.

The best solution—short of a very long and detailed genetic analysis of the oaks—seems to be to treat these here as a single variable complex, doubtless of quite mixed nature. Some of the variability surely results from "contamination" by genes from *Q. velutina* and *Q. rubra*—and perhaps even other species over thousands of years. Overlease (1977) studied populations of *Q. coccinea* and *Q. ellipsoidalis* from southern Indiana to central Wisconsin and northern Michigan, concluding that Hill's oak was a northern small-fruited expression of scarlet oak, with the acorn less covered by the cup. While the extremes of these oaks are well marked, the broad zone of intermediacy in northern Indiana and Illinois and in southern Wisconsin and Michigan suggested to him no more significant differences than known in the similar north-south clines involving black oak and red oak. Small-fruited trees with yellow inner bark, sometimes identified as expressions of *Q. ellipsoidalis,* were felt by Overlease to belong more appropriately to the *Q. velutina* continuum.

Mature leaves are usually glabrous, except sometimes for tufts of pubescence in the axils of the main veins beneath. But a few specimens have leaves that have retained some pubescence on the petiole and midrib above, suggesting introgression from *Q. velutina*. More clearcut hybrids have angled buds (smaller than in *Q. velutina*) and/or a suggestion of fringe in the acorn cups. Some plants (or some branches, at least, collected for specimens) have rather shallowly lobed leaves resembling small ones of *Q. rubra,* but they are usually shiny above; when acorns are present, their small size, up to half covered by the cup, and the often densely ciliate apical portion of the small (frequently ± angled) winter buds, will further help to distinguish such specimens from *Q. rubra*. Only very rarely are acorn cups of *Q. coccinea* as broad as 2 cm.

The *Q. coccinea* complex constitutes the group often referred to as "scrub oaks," usually associated with very sterile, well drained, sandy plains and glacial deposits. Scarlet oaks tend to hold their rigid leaves much longer into winter than their near relatives.

Hybrids with *Q. imbricaria* are noted under that species. Hybrids with *Q. velutina* may be called *Q. ×palaeolithicola* Trel., a name narrowly applied to *Q. ellipsoidalis* × *Q. velutina* but available for the inclusive *Q. coccinea* × *Q. velutina,* for which no other binomial has been published; these hybrids are rather frequent throughout the overlapping ranges of the species in Michigan and even a bit beyond *Q. velutina,* to southern Cheboygan County.

6. **Q. alba** L. White Oak

Map 57. Oak-hickory, beech-maple, and mixed hardwood stands; in the northern Lower Peninsula, often with jack pine and other oaks on sandy plains.

A familiar tree to many people, the white oak is easily recognized by the rounded lobes and definite sinuses of the leaves, which are pale and completely glabrous beneath. (The dense pubescence on both surfaces of the young leaves is readily deciduous; rarely a few hairs may persist along the midrib beneath, but these are easily dislodged.) This is an important timber tree farther to the south and east, but it is not of so much commercial importance in Michigan.

Hybrids with several other species are known in the state: most often with *Q. macrocarpa* [producing *Q. ×bebbiana* Schneider], collected in Arenac, Cheboygan, Clare, Genesee, Kent, Lenawee, Montcalm, Oakland, Oscoda, St. Joseph, and Washtenaw counties; but also with *Q. bicolor* [producing *Q. ×jackiana* Schneider], mapped by Hardin from Huron and Shiawassee counties; with *Q. muehlenbergii* [for which no binomial is available, *Q. ×deamii* having been found to be of a different parentage], collected once in Huron County; and with *Q. prinoides* [producing *Q. ×faxonii* Trel.], mapped by Hardin from St. Clair County (see Hardin 1975). All are doubtless of wider occurrence than existing collections document. Reports of *Q. stellata* Wang. and *Q. lyrata* Walter from Michigan are apparently based on plants of *Q. alba* with unusually deeply lobed leaves some of which are ± minutely puberulent beneath, suggesting a little introgression from *Q. macrocarpa* (although there is little evidence of the latter in acorns or overall outline).

7. **Q. macrocarpa** Michaux Bur Oak

Map 58. Ranges from rich bottomlands and (especially northward) river banks or lake shores to dry sandy ± open upland woods or depressions in such woods. This was the characteristic scattered tree of savanas called "oak openings" in the early days of settlement in southern Michigan. Many fine large specimens of bur oak continue to grow as shade trees where they have escaped the ax, saw, and bulldozer, even in cities. Some of those in and near Ann Arbor are over 200 years old.

An occasional specimen of *Q. bicolor* may resemble this species if the leaves are larger and more deeply lobed than usual, but the glabrous buds, long peduncles, and more whitened undersides of the leaves in *Q. bicolor* should distinguish it. The large "mossy cup" acorns of *Q. macrocarpa* are sessile or on peduncles at most about 1 cm long, but even in the absence of this distinctive fruit the species can usually be easily recognized by the large leaves, much more deeply lobed on the basal half than on the apical half.

Hybrids with *Q. bicolor* [*Q.* ×*schuettei* Trel., later named *Q.* ×*hillii* Trel.] appear to be more frequent in Michigan than those with *Q. muehlenbergii* [*Q.* ×*deamii* Trel.]. *Q.* ×*schuettei* has been collected in Jackson, Kent, Lapeer, Lenawee, Menominee, Missaukee, Washtenaw, and perhaps Newaygo counties; *Q.* ×*deamii,* in Berrien, Genesee, Washtenaw, and perhaps Clinton counties. An apparent hybrid with *Q. prinoides* has been found in Livingston County.

8. **Q. bicolor** Willd. Fig. 25 Swamp White Oak

Map 59. Floodplains, swamp forest, and other poorly drained sites.

The undersides of the leaves in this species are usually more densely stellate-tomentose than in the next three, feeling rather velvety to the touch. The very whitish aspect of this surface on normal leaves is the basis for the epithet *bicolor*. However, shaded leaves or those on juvenile plants may be green and sparsely pubescent beneath. Especially on open-grown trees, the lower branches tend to hang downward and to persist when dead.

Hybridizes with *Q. alba* and *Q. macrocarpa,* as noted above. An apparent hybrid with *Q. prinoides* has been found in Livingston County.

9. **Q. prinoides** Willd. Dwarf Chestnut or Dwarf Chinquapin Oak

Map 60. Dry sandy hills and prairie-like sites.

Sometimes difficult to distinguish from the next species, which is often considered a variety of it. Collectors who neglect to record the habit of the plant on their label make determination of herbarium specimens difficult. The number of teeth is often said to overlap excessively with *Q. muehlenbergii,* but our material seems fairly well separated on this basis. In the field, young plants of *Q. muehlenbergii* can be easily distinguished by the absence of the woody rhizomes which characterize the colonial *Q. prinoides* and by the usual absence of fruit, the true shrubby species fruiting, of course, when quite short.

Hybridizes with *Q. alba* and doubtless with *Q. muehlenbergii.*

10. **Q. muehlenbergii** Engelm. Chinquapin or Yellow Chestnut Oak

Map 61. Beech-maple woods, stream banks, rich forests on stabilized dunes, often on ± calcareous sites.

Often included with the preceding species as var. *acuminata* (Michaux) Gl. The two probably hybridize, and *Q. muehlenbergii* also hybridizes with *Q. alba* and *Q. macrocarpa.*

The leaves are quite variable in shape, from narrowly elliptic-lanceolate to ± oblong to rather broadly obovate; the latter wide-leaved extreme is recognized sometimes as f. *alexanderi* (Britton) Trel. (TL: Birmingham [Oakland Co.]).

11. **Q. prinus** L. Rock Chestnut Oak

Map 62. Distributed in dry to mesic woods from the Appalachian region through eastern Ohio to southern Ontario and a single known colony in Michigan, in the Waterloo Recreation Area.

Often called *Q. montana* Willd. In *Q. prinus* the bark is dark (as in the red oaks), deeply furrowed on old trees, while in *Q. muehlenbergii* the bark is gray and flaky as in most white oaks.

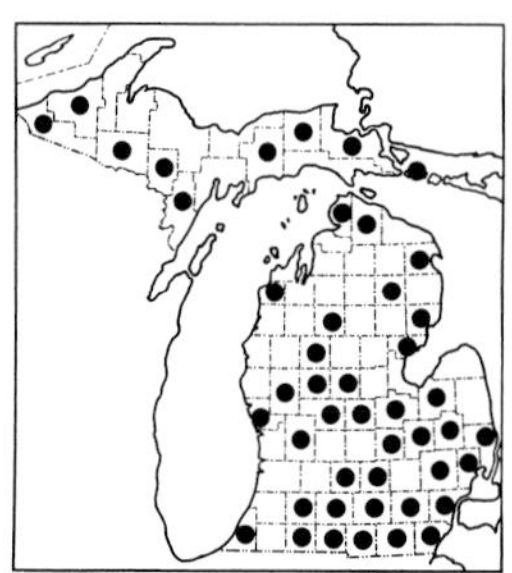
58. Quercus macrocarpa

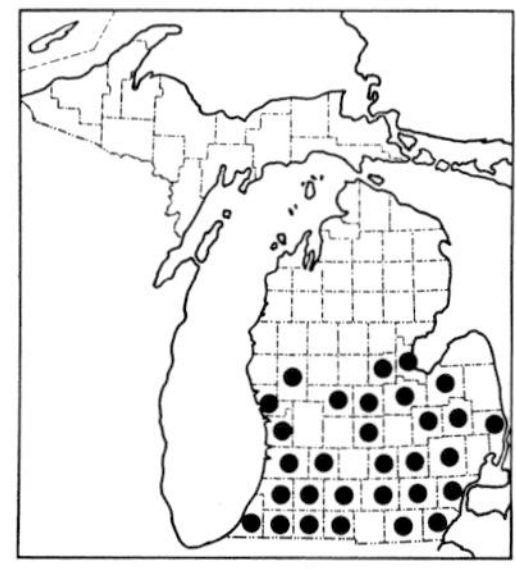
59. Quercus bicolor

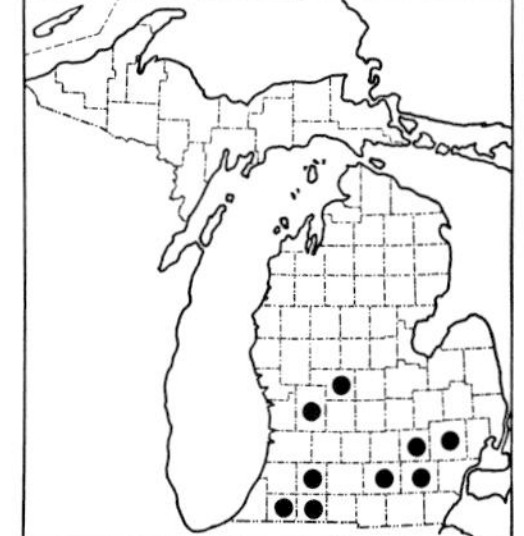
60. Quercus prinoides

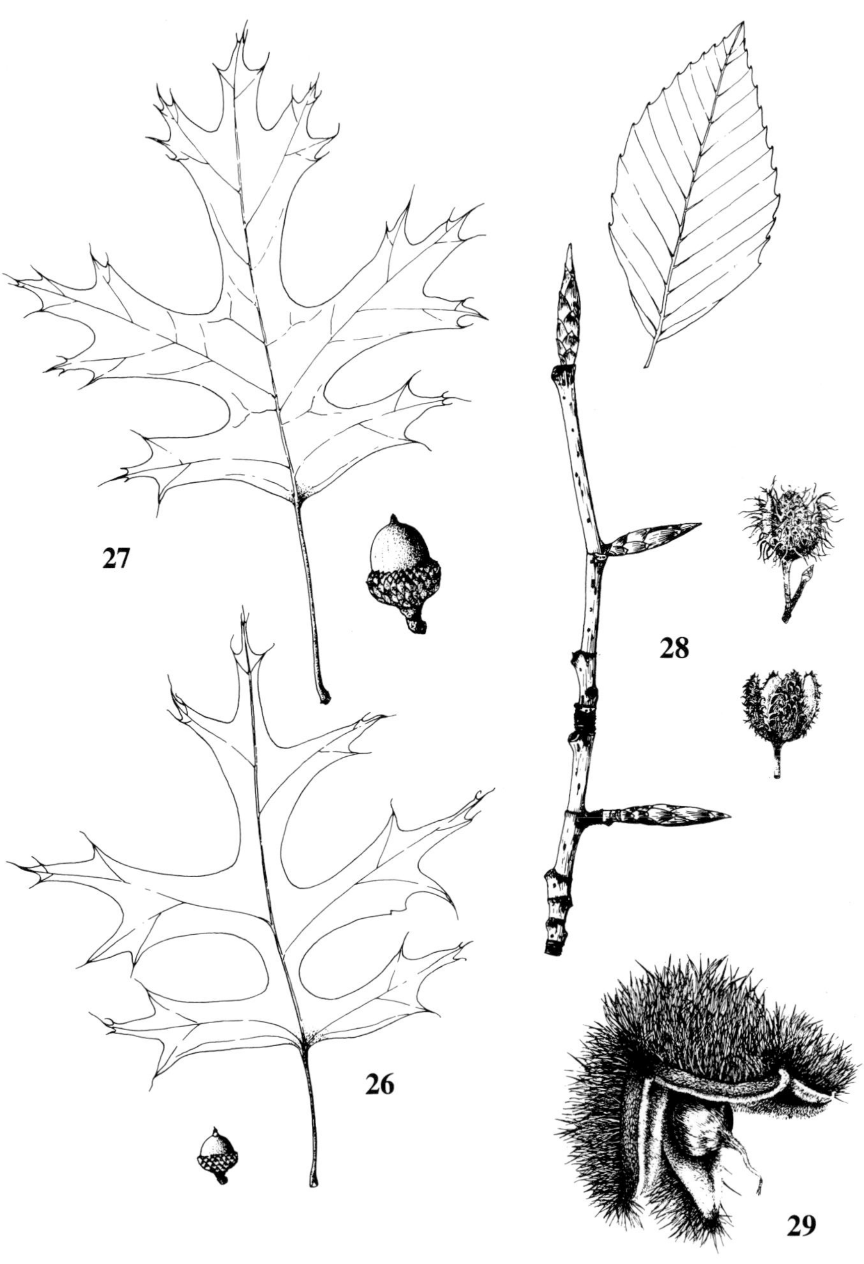

26. *Quercus palustris,* leaf & acorn $\times^1/_2$
27. *Q. coccinea,* leaf & acorn $\times^1/_2$
28. *Fagus grandifolia,* leaf & involucres (husks) $\times^1/_2$; winter twig $\times 1$
29. *Castanea dentata,* fruit & involucre $\times^1/_2$

2. Fagus

1. **Fagus grandifolia** Ehrh. Fig. 28 Beech

Map 63. Typical of beech-maple forests and hemlock-white pine-northern hardwoods. Thrives especially on islands and along the Lake Michigan shore where favored by moist winds.

We have a single species of beech, although studies of its variation by the late W. H. Camp indicated distinct races, of which the "red beech" and the northern "gray beech" meet and intermingle in Michigan, with some representation also of the southern "white beech" (see Deam, *Trees of Indiana,* 3rd ed., pp. 112–115. 1953). Michigan is at the western edge of the range of beech. It is absent from the westernmost Upper Peninsula, and extends southward in Wisconsin only near Lake Michigan.

The smooth tight steel-gray bark of this species, together with the distinctive sharp-pointed slender elongate winter buds (ca. 15–25 mm long when fully grown) make this a well known and easily recognized tree even in winter. Beech nuts, of which there are good crops only every few years, are an important food for wildlife as well as for the patient human. One quart of beech fruit, including burs (husks) will contain approximately one cup of nuts, which in turn will yield about one-half cup of shelled nutmeats. Our native American beech is said to be difficult to transplant; the European *Fagus sylvatica* L. is more often recommended as an equally handsome shade tree for planting.

The leaves of beech are especially attractive as they unfold, accordion-like, from the prominent buds in the spring; they usually retain some silky pubescence on the petiole and veins beneath.

3. Castanea

REFERENCES

Brewer, Lawrence G. 1982. The Present Status and Future Prospect for the American Chestnut in Michigan. Mich. Bot. 21: 117–128.

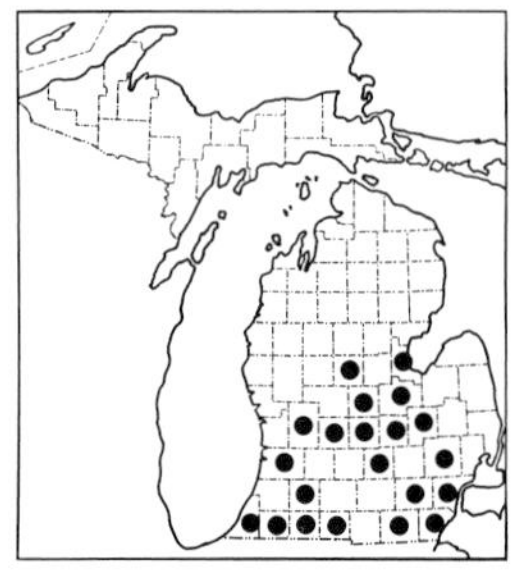

61. Quercus muehlenbergii

62. Quercus prinus

63. Fagus grandifolia

Thompson, Paul W. 1969. A Unique American Chestnut Grove. Mich. Academ. 1(3–4): 175–178.

1. **C. dentata** (Marsh.) Borkh. Fig. 29 Chestnut

Map 64. This formerly important tree has been nearly exterminated throughout its natural range by the chestnut blight, a parasitic fungus introduced from abroad about the turn of the century and first noted in Michigan about 1930. Stump sprouts, which do not grow to maturity, are known at some places, and a rare healthy tree of considerable size may survive in woods, as in Oakland County. Chestnut was originally found in Michigan only in forests (especially with oaks) in the southeasternmost Lower Peninsula, north to St. Clair County. Shrubby plants well established in a woods in Van Buren County are undoubtedly progeny from a tree once planted in the vicinity. Farther north, impressive stands of such reproducing trees are in Benzie, Leelanau, and Missaukee counties. Where planted along roadsides and in orchards beyond its original native range, some chestnuts have escaped the blight and have grown to a large size. Dozens of such trees are found in Michigan, some healthy and some only recently becoming afflicted, as far north as Emmet County and Beaver Island.

The leaves resemble large, elongate ones of beech in having straight lateral veins each terminating in a single tooth, but the teeth are longer and the veins more often glabrous or very sparsely hairy beneath.

ULMACEAE Elm Family

REFERENCE

Elias, Thomas S. 1970. The Genera of Ulmaceae in the Southeastern United States. Jour. Arnold Arb. 51: 18–40.

KEY TO THE GENERA

1. Leaves with prominent straight ± parallel lateral veins running into the principal teeth; flowers perfect, the perianth shallowly lobed (less than halfway to base); ovary flattened and winged; fruit clustered, a samara, winged on all sides, ripe in spring . . . . 1. **Ulmus**
1. Leaves with lateral veins curved and ascending, weaker and the branches anastomosing before the margin; flowers usually unisexual, the perianth very deeply lobed (nearly or quite to the base); ovary not flattened; fruit solitary, a drupe, ripe in the fall . . . . 2. **Celtis**

1. Ulmus Elm

Some Eurasian elms are planted as shade and ornamental trees. One of the most easily recognized is *U. glabra* Hudson, the Scotch or wych elm, with the longer side of the very scabrous and asymmetrical leaf blade so prominently lobed as to form an auricle which ± overlaps and hides the very short petiole; the samara is large and completely glabrous.

KEY TO THE SPECIES

1. Leaf blades less than 5 (7) cm long, the base only slightly if at all asymmetrical, the margin often simply serrate; samaras entirely glabrous; lateral buds not over 3 mm long; plant a shrub or small tree, locally established as an escape from cultivation . . . . 1. **U. pumila**
1. Leaf blades mostly longer than 7 cm, the base strongly asymmetrical and the margins doubly serrate; samaras pubescent on margins or sides (or both); lateral buds 4–10 mm or longer; plant a native tree
   2. Samaras eciliate, but pubescent on the sides over the nut (wings glabrous); tips of buds, stipules, and perianth covered with dense red-brown hairs; leaves very harshly scabrous above; petioles densely pubescent; flowers nearly sessile . . . . 2. **U. rubra**
   2. Samaras densely ciliate, glabrous on the sides or pubescent on both nut and wing; buds, stipules, and perianth glabrous to somewhat pubescent but with at most some red-brown hairs on bud-scales; leaves smooth to scabrous above; petioles glabrous to pubescent; flowers drooping on elongate pedicels (or racemose)
      3. Bark of older twigs not winged; samaras glabrous on the sides, ca. 9–14 mm long; mature leaves smooth to scabrous above . . . . 3. **U. americana**
      3. Bark of older twigs coarsely winged with corky ridges; samaras pubescent over entire surface, ca. 15–22 mm long; mature leaves very smooth and glabrous above (± pilose when young) . . . . 4. **U. thomasii**

1. **U. pumila** L. — Siberian Elm

Map 65. Native of Asia but widely planted and spreading from cultivation into vacant (and not so vacant) lots, sidewalk crevices, waste ground, shores, and woods. Evidently becoming quite aggressive in recent years.

This is a small, often bushy, rapidly growing tree. Sometimes called "Chinese Elm," but that name is best applied to *U. parvifolia* Jacq., which flowers in late summer or fall and has leaves with the apex (and teeth) less sharply acute than *U. pumila*.

2. **U. rubra** Muhl. — Slippery or Red Elm

Map 66. Floodplains, stream banks, and rich hardwoods. Seldom a large tree, but reported by Hanes as approaching 100 feet in height and 3 feet in diameter in Kalamazoo County.

For many years known as *U. fulva* Michaux. The inner bark is quite mucilaginous, especially when chewed—whence the common name. The samaras are often larger (11–18 mm long) than in *U. americana,* besides being eciliate and scarcely if at all notched at the apex (unlike the deeply notched ones of the next species). The rusty or red-brown pubescence of the buds and roughness on the undersides of the leaves are the most helpful characters for separating sterile specimens from scabrous-leaved ones of American elm.

3. **U. americana** L. Fig. 30 American or White Elm

Map 67. Characteristic of swamp forests, such as river floodplains, wet bogs, even cedar swamps, often with silver maple, and also in rich upland hardwoods.

Occasional plants, especially on young sprouts, produce leaves as harshly scabrous above as *U. rubra,* feeling like coarse sandpaper, but the leaves are more likely to be smooth beneath. Both this species and the preceding have a familiar and characteristic "vase-shape" whereas in the next the trunk generally extends straight into the crown before forking. Always susceptible to insect damage and disease (such as the viral elm phloem necrosis), our stately elms have suffered especially from the Dutch elm disease, first noted in Michigan in 1950 near Detroit and subsequently fatal to most trees in the southern part of the state; the disease is caused by an ascomycetous fungus, *Ceratocystis ulmi,* spores of which are carried by beetles and enter healthy tissue via wounds caused by feeding of the beetles on young shoots.

The outer bark of this species consists of alternating pale and dark brown layers, while in *U. rubra* the outer bark is solid brown.

4. **U. thomasii** Sarg. Rock or Cork Elm

Map 68. Mixed hardwood forests and low rich woods along rivers.

Sterile specimens often cannot be distinguished from smooth-leaved plants of the preceding species unless the characteristic corky-ridged bark is present. Furthermore, sprout leaves of rock elm can be scabrous above. The inflorescence has a ± elongate axis—almost racemose—whereas the flowers in our other species are more densely clustered.

2. Celtis Hackberry

REFERENCE

Wagner, W. H., Jr. 1974. Dwarf Hackberry (Ulmaceae: Celtis tenuifolia) in the Great Lakes Region. Mich. Bot. 13: 73–99.

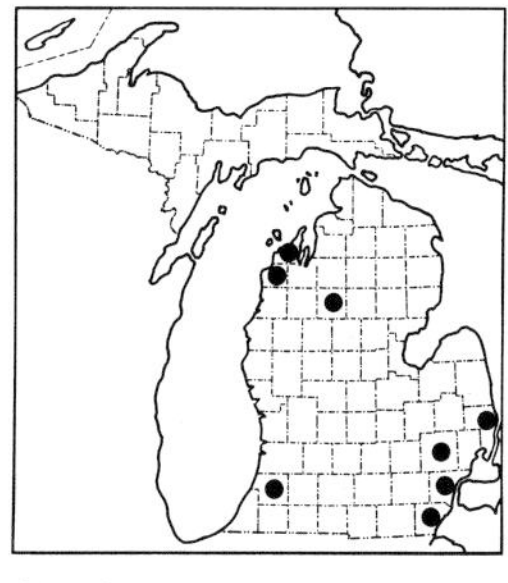
64. Castanea dentata

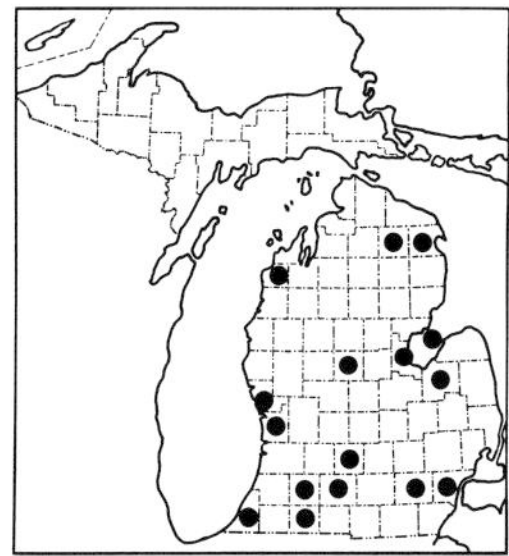
65. Ulmus pumila

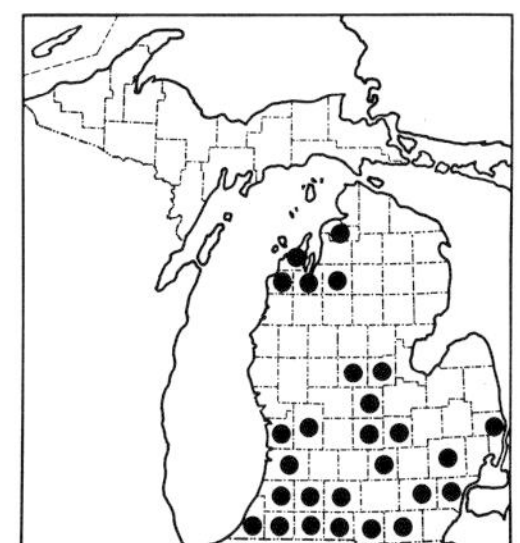
66. Ulmus rubra

KEY TO THE SPECIES

1. Fruiting pedicels ca. 10–20 mm long, at least 1.5 times as long as the subtending petiole; ripe drupes dark purple, coarsely and strongly wrinkled when dry; leaves mostly with acuminate apex and margins toothed nearly to the base . . 1. **C. occidentalis**
1. Fruiting pedicels ca. 3–6 mm long, about equaling the petioles or shorter; ripe drupes orange-red, remaining plump and smooth when dry; leaves mostly broadly acute to short-acuminate, the margins often entire or nearly so on basal third . . . 2. **C. tenuifolia**

1. **C. occidentalis** L. Hackberry

Map 69. River banks, stream valleys and ravines, rich moist woods, and rarely drier sites. Usually a medium-sized tree, especially at the northern edge of its range, which includes Michigan.

The leaves are quite variable in size and scabrousness. Hackberry often produces "witch's brooms" when disease causes proliferation of branch tips. Differences between this species and the next, including leaf shape and venation, are well shown by Wagner (1974).

2. **C. tenuifolia** Nutt. Fig. 31 Dwarf Hackberry

Map 70. Borders of woods, fields and fencerows, and open dryish sandy woods.

A shrub or small irregular tree to ca. 20 feet tall, with leaves tending to be smaller, more ovate, and more sparsely veined than in the preceding species.

MORACEAE Mulberry Family

A family chiefly of tropical and subtropical regions, including such edible fruits as figs, jack-fruit, and breadfruit, as well as other useful plants. Most species (including ours) are woody, with milky juice (latex), and produce a multiple fruit, i. e., a fruit derived from an entire inflorescence rather than a single flower, often including a ± fleshy common receptacle or axis of the inflorescence.

67. Ulmus americana

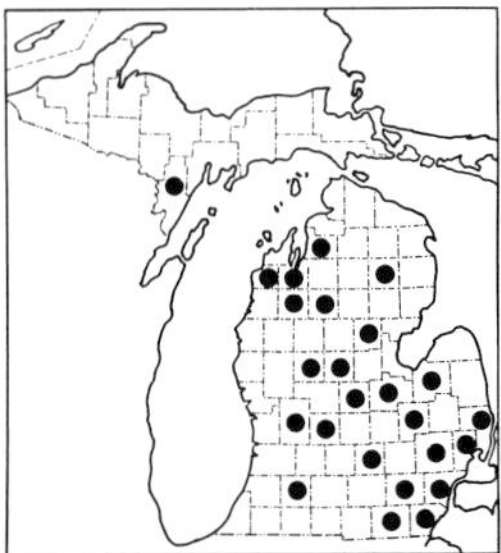
68. Ulmus thomasii

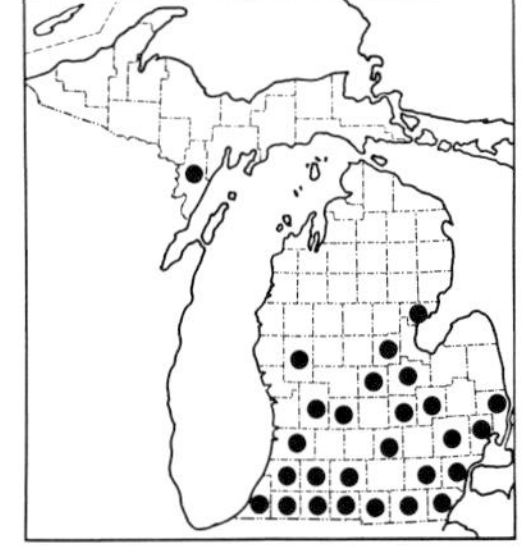
69. Celtis occidentalis

The common fig, *Ficus carica* L., has 3–5-lobed leaves; the flowers and nutlets are borne on the inner surface of the urn-like receptacle. An occasional waif may be found this far north (noted in Macomb and Washtenaw counties), but the stems die back to the ground each year; fruit, however, may be formed on first-year sprouts.

KEY TO THE GENERA

1. Leaves entire, unlobed, definitely pinnately veined; stems with axillary spines; fruit globose, green (to yellow when fully ripe), orange-like in external appearance but scarcely fleshy, nearly or fully 1 dm in diameter .... 1. **Maclura**
1. Leaves toothed, often lobed, ± palmately veined (a pair of strong lateral veins arising, with the midrib, from summit of the petiole); stems without spines; fruit short-cylindric, white to pink or purple, blackberry-like and juicy, ca. 3 cm long or smaller .... 2. **Morus**

1. Maclura

1. **M. pomifera** (Raf.) Schneider — Osage-orange

Map 71. Native from Arkansas to Oklahoma and Texas, but widely planted northward for hedges or living fences and rarely spreading in our region by root suckers or seeds. It is not always clear from herbarium specimens whether collections are from such barely established colonies or from planted trees, but there are definite reports of escape at least in Berrien, Kalamazoo, Kent, Oakland, and Washtenaw counties.

The wood is unusually hard and durable. The fruit ("hedge-apple") is quite inedible (except to some animals).

2. Morus — Mulberry

The fruit of both our species is relished by birds, which are responsible for the wide distribution and frequently almost weedy occurrence of the mulberries.

KEY TO THE SPECIES

1. Leaves glabrous (at most papillate-roughened) beneath except along the main veins, if lobed the lateral lobes obtuse or nearly so (fig. 32); fruit white to dark purple . ... 1. **M. alba**
1. Leaves pubescent beneath between the main veins, if lobed the lobes acute to acuminate; fruit dark purple .... 2. **M. rubra**

1. **M. alba** L. Fig. 32 — Russian or White Mulberry

Map 72. A native of China, where grown since antiquity, well known in the Orient as the source of food for silkworm caterpillars. Widely planted and escaped from cultivation in the eastern United States. Vacant lots, hedgerows, railroads, fields, woods, and streamsides.

2. **M. rubra** L. Red Mulberry

Map 73. Floodplains and riverbottom woods.

The leaves are sometimes rather scabrous above; when unlobed they are suggestive of *Tilia* in shape and size.

CANNABACEAE Hemp Family

The family name has been conserved as spelled here, but there are various spellings preferred by some botanists, including Cannabinaceae, Cannabidaceae, and Cannabiaceae. One solution would be to follow those authors who include this family in the Moraceae, from which it differs (at least in this region) in herbaceous habit, absence of latex, and opposite leaves. The leaves of all our species are very rough above, with minute resinous dots visible with a good lens, especially beneath.

REFERENCE

Miller, Norton G. 1970. The Genera of the Cannabaceae in the Southeastern United States. Jour. Arnold Arb. 51: 185–203.

KEY TO THE GENERA

1. Leaves palmately compound with lanceolate to oblanceolate leaflets, opposite on lower portion of the erect stem and alternate above; stems and petioles finely antrorse-hispid; achenes subtended by narrow bracts, not concealed . . . . 1. **Cannabis**
1. Leaves at most deeply lobed, all opposite on a climbing stem (vine); stems and petioles retrorse-scabrous; achenes concealed by broad, strongly overlapping bracts . . . . 2. **Humulus**

1. Cannabis

REFERENCE

Small, Ernest, & Arthur Cronquist. 1976. A Practical and Natural Taxonomy for Cannabis. Taxon 25: 405–435.

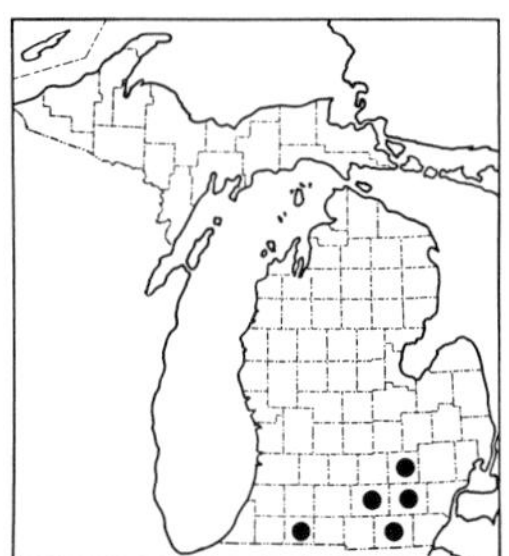

70. Celtis tenuifolia

71. Maclura pomifera

72. Morus alba

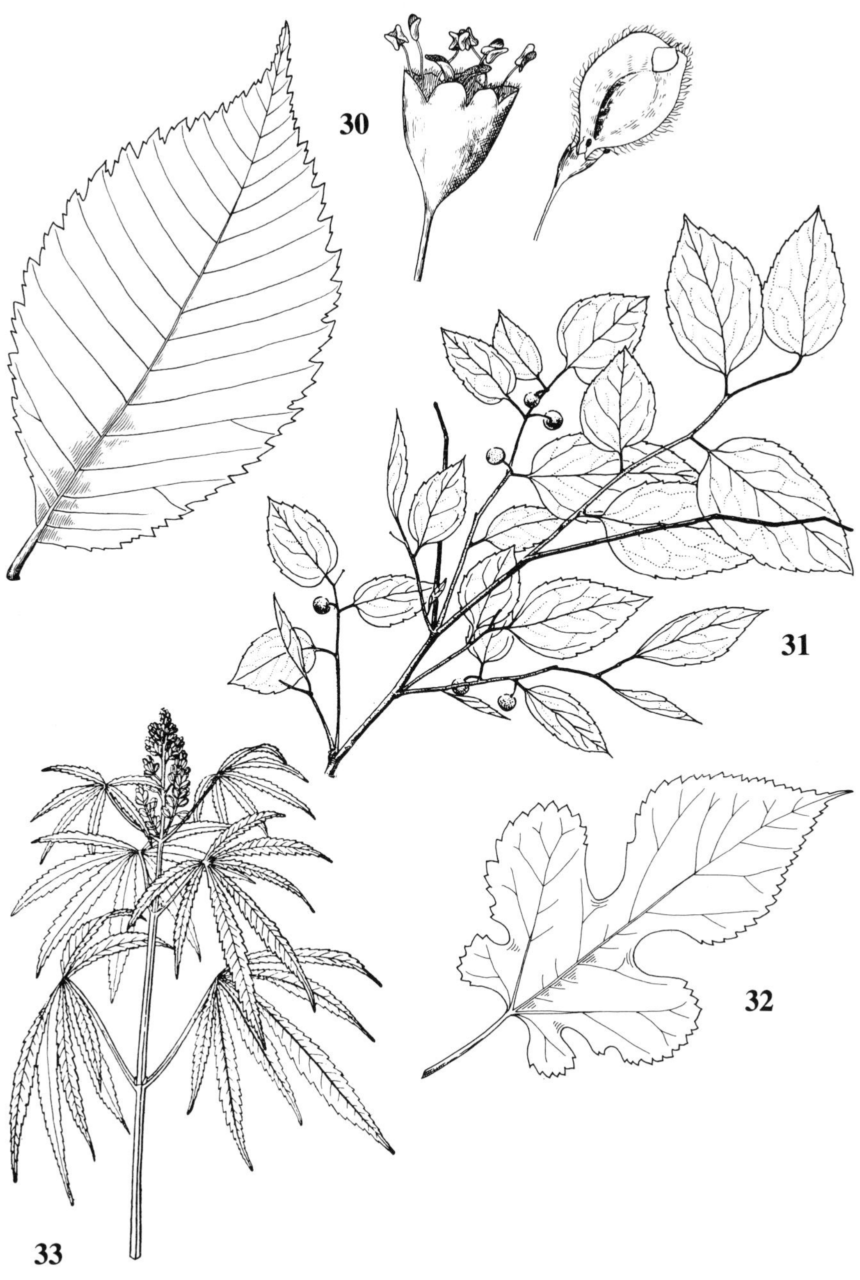

30. *Ulmus americana,* leaf $\times{}^1/_2$; flower $\times 6$; fruit $\times 2$
31. *Celtis tenuifolia* $\times{}^1/_2$
32. *Morus alba,* lobed leaf $\times{}^3/_4$
33. *Cannabis sativa* $\times{}^1/_2$

1. **C. sativa** L. Fig. 33 Hemp; Marijuana

Map 74. Surreptitiously cultivated, but spontaneous in many situations: roadsides, farmyards, vacant lots, old fields, stream banks; thrives on muck soils. Long established as a weed in waste places (collected in Michigan as early as 1837 by the First Survey). A native of Asia, this complex species has been cultivated since antiquity for the fiber, derived from the stem much as flax and used in ropes, nets, sailcloth, etc. Cultivation as a fiber plant and for the oil in the seeds has been encouraged as recently as World War II, and most wild plants represent escapes.

The species is also known as a drug plant, long used in Asia (source of bhang, hashish) and more recently gaining considerable notoriety elsewhere as a non-addictive stimulant. The resin of the plant, exuded especially in the pistillate inflorescences, is the source of the active principle. However, the strains cultivated for fiber are not the same as those particularly cherished for drug use. Growing of *Cannabis* is under strict federal control but like any weed it is difficult to eradicate by legislation.

2. Humulus Hops

REFERENCE

Small, Ernest. 1979. A Numerical and Nomenclatural Analysis of Morpho-geographic Taxa of Humulus. Syst. Bot. 3: 37–76.

KEY TO THE SPECIES

1. Main veins of leaf beneath with stiff, spinulose, simple hairs (pustulate-based); petiole about equaling or exceeding the leaf blade; larger blades usually 5–7-lobed; pistillate bracts narrowly acuminate, long-hispid, lacking resinous dots; anthers without resinous dots . . . . 1. **H. japonicus**
1. Main veins of leaf beneath glabrous to softly pubescent and/or with a few short stubby 2-hooked hairs; petiole shorter than the leaf blade; larger blades 3- (very rarely 5-) lobed; pistillate bracts (especially younger ones) broadly acute to obtuse or rounded, glabrous or with tiny hairs, usually with at least a few resinous dots basally; anthers usually with resinous dots . . . . 2. **H. lupulus**

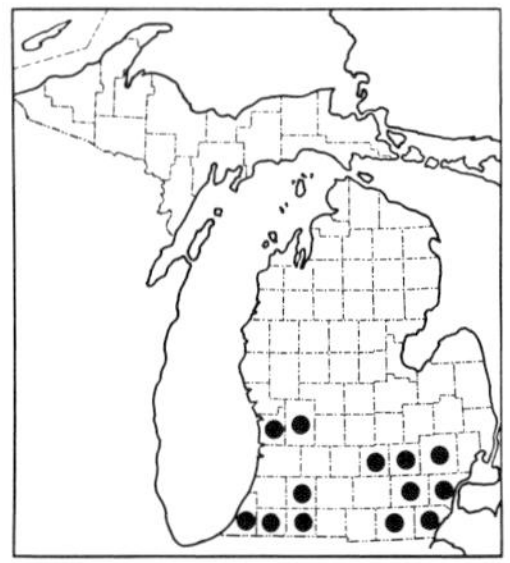

73. Morus rubra

74. Cannabis sativa

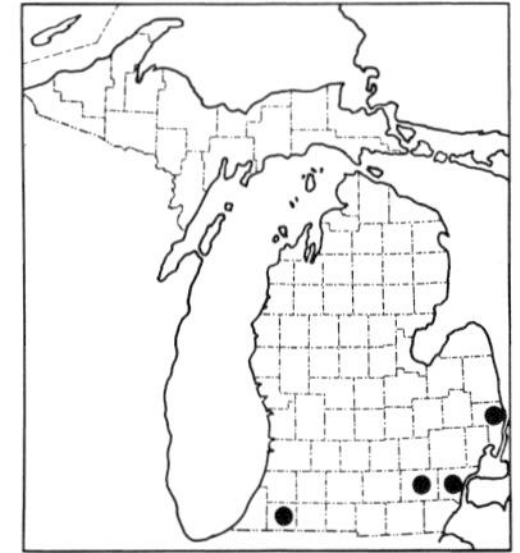

75. Humulus japonicus

1. **H. japonicus** Sieb. & Zucc. Japanese Hops

Map 75. A native of eastern Asia, sometimes cultivated as an ornamental annual vine. Collected in Michigan at several places, mostly half a century or more ago, as locally established in waste ground.

2. **H. lupulus** L. Common Hops; Hop

Map 76. Unlike the preceding, this species is a perennial vine, and consequently it persists and spreads long after cultivation at old homesites and similar places where it was once grown as an ornamental. Some of our records from floodplains, thickets, and lowlands doubtless represent native occurrences, but the original native range of hops in Michigan is obscure. The species was not reported (or collected) by the First Survey (1837–1841) and the first report seems to be from surveyor William A. Burt, who found the species in 1846 in T42N, R25W—in what now is southern Marquette County.

Most of our records probably represent plants persistent or spread from cultivation: the Eurasian var. *lupulus* as defined by Small, with the main leaf veins beneath glabrous or sparsely pubescent. However, there is a gradation to our native varieties, which are not themselves clearcut: var. *pubescens* E. Small has the midrib beneath densely pubescent and has hairs between the veins, such plants occurring in the southernmost Lower Peninsula; var. *lupuloides* E. Small is less pubescent, but not as glabrate as var. *lupulus*. Most plants of these varieties have on the midrib beneath a few of the broad-based, very short, stiff, 2-hooked barbs ("climbing hairs" of Small) which are especially characteristic of the petioles. American plants have sometimes been called *H. americanus* Nutt. However they are classified, they are not easily distinguished from Eurasian ones in our region, in part apparently the result of introgression.

URTICACEAE Nettle Family

The stinging hairs which bring some members of this family forcibly to our attention each consist of a long slender cell containing the irritating sub-

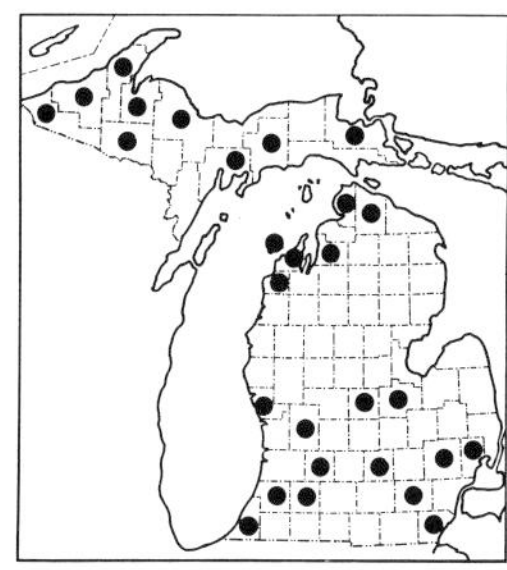
76. Humulus lupulus

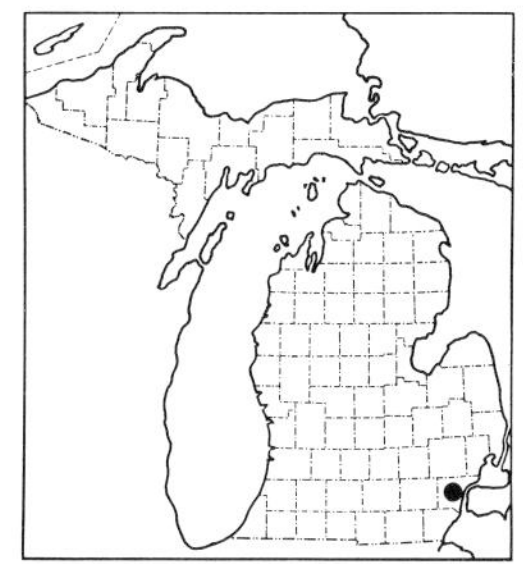
77. Parietaria diffusa

78. Parietaria pensylvanica

stance in a sac-like base. The bulbous tip of the hair is broken off upon contact, leaving a sharp point through which the irritant is injected when slight pressure on the hair compresses the sac at the base.

REFERENCES

Bassett, I. J., C. W. Crompton, & D. W. Woodland. 1974. The Family Urticaceae in Canada. Canad. Jour. Bot. 52: 503–516.

Miller, Norton G. 1971. The Genera of the Urticaceae in the Southeastern United States. Jour. Arnold Arb. 52: 40–68.

KEY TO THE GENERA

1. Leaves alternate
  2. Plant without stinging hairs (but not necessarily glabrous); leaves entire; tepals of pistillate flowers equal, fused at the base; style obsolete, remaining ± central and terminal as the fruit matures, the stigma a deciduous tuft of hairs; achene very shiny and glossy, terete . . . . 1. **Parietaria**
  2. Plant with stinging hairs; leaves prominently toothed; tepals of pistillate flowers separate, becoming unequal in fruit (2 larger, 2 smaller); style and elongate persistent stigma linear-subulate, becoming lateral as the fruit matures; achene ± dull, usually speckled until fully ripe, strongly compressed . . . . 2. **Laportea**
1. Leaves opposite
  3. Plants usually 1 m or more tall, with stinging hairs; tepals of pistillate flowers 4, separate, becoming unequal (2 larger, 2 smaller), ciliate to densely hispid with straight to curved bristles; stigma a deciduous tuft of hairs . . . . 3. **Urtica**
  3. Plants less than 1 m tall, without stinging hairs (pubescence, if any, minute and of hooked or ± appressed hairs); tepals of pistillate flowers 3, glabrous, and separate or completely united and bearing some hooked bristles; stigma linear
    4. Flowers in dense (± interrupted) elongate unbranched cylindrical spike-like inflorescences; achene (somewhat fleshy) completely enclosed by the perigynium-like perianth, which bears mixed straight and hooked bristles . . . . 4. **Boehmeria**
    4. Flowers in loose spreading branched clusters; achene evident, not hidden by the separate glabrous tepals . . . . .5. **Pilea**

1. Parietaria — Pellitory

KEY TO THE SPECIES

1. Leaves densely soft-pilose, the blades less than twice as long as broad; achenes black; bracts about equaling or shorter than the flowers; anthers ca. 1 mm long . . . . 1. **P. diffusa**
1. Leaves glabrous to only sparsely and minutely pilose, the blades mostly at least (2.5) 4–6 times as long as broad; achenes pale whitish to brown; bracts much exceeding the flowers; anthers less than 0.5 mm long . . . . 2. **P. pensylvanica**

1. **P. diffusa** Mert. & Koch

Map 77. A native of Europe, collected once at Detroit (*Farwell 8574* in 1929, MICH, BLH).

A much-branched, spreading or decumbent perennial.

2. **P. pensylvanica** Willd.

Map 78. Moist to dry woods, weedy sites, and gravelly shores.

A ± erect, weak annual, little if at all branched.

2. Laportea

1. **L. canadensis** (L.) Wedd. — Wood Nettle

Map 79. Floodplains, stream banks, springy banks and ravines, low spots in hardwood forests, sometimes in wet cedar swamps.

The stinging hairs are sometimes scarcely noticeable, particularly if plants are somewhat wilted. Ordinarily, however, they are entirely too potent—and with a delayed action effect which may call them to one's attention too late for him to escape comfortably from the midst of a colony.

3. Urtica — Nettle

REFERENCES

Bassett, I. J., C. W. Crompton, & D. W. Woodland. 1977. The Biology of Canadian Weeds. 21. Urtica dioica L. Canad. Jour. Pl. Sci. 57: 491–498.

Hermann, F. J. 1946. The Perennial Species of Urtica in the United States East of the Rocky Mountains. Am. Midl. Nat. 35: 773–778.

Woodland, Dennis W. 1982. Biosystematics of the Perennial North American Taxa of Urtica, II. Taxonomy. Syst. Bot. 7: 282–290.

KEY TO THE SPECIES

1. Inflorescences shorter than the petioles subtending them; leaves deeply serrate (teeth often twice as long as broad), with blades less than 4 cm long, ± obtuse; stipules less than 3 mm long; plant annual, a rare waif . . . . 1. **U. urens**
1. Inflorescences exceeding the petioles; leaves with shorter teeth and usually larger blades, narrowly acute; stipules longer; plant perennial, widespread . . . . 2. **U. dioica**

1. **U. urens** L.

Map 80. A European species, collected in waste ground at Ironwood in 1919 (*Bessey 3099,* MSC).

2. **U. dioica** L. — Stinging Nettle

Map 81. Swamps and (less often) marshes, damp disturbed woods and thickets, ditches and weedy areas, lake shores and river banks.

Typical European plants are said to be usually if not always dioecious and to have broader ovate and definitely cordate leaf blades which bear stinging hairs on both surfaces. Such plants are represented in Michigan only by a Houghton County collection described by Farwell as *U. gracilis* var. *latifolia* (TL: Lake Linden; *Farwell 8513* in 1929, BLH, MICH). Most if not all of our plants are monoecious (though this is not always obvious), have narrower, smoother leaves, and are the native var. *procera* (Willd.) Wedd. [or ssp. *gracilis* (Aiton) Selander, including *U. gracilis* Aiton and *U. viridis* Rydb.]. This nettle is extremely variable and the systematic and geographic significance of the variation has not been agreed upon by botanists, although most have concurred with Hermann's conclusion that American plants cannot be separated at the species level from European ones.

4. Boehmeria

B. nivea (L.) Gaudich. is the important fiber plant, ramie, a native of China but widely cultivated in warmer climates.

1. **B. cylindrica** (L.) Sw. Fig. 34 — False Nettle

Map 82. Floodplains and swamps (both deciduous and coniferous), less often in marshes or in upland deciduous woods.

There is considerable variability in amount of pubescence on the stems (sometimes glabrate, often with hooked and straight hairs) and in roughness of leaves.

5. Pilea — Clearweed; Richweed

Smooth watery-appearing annuals with translucent stems, often forming dense colonies where abundant seeds have been deposited in a previous season. A quite different-appearing species, with leaves entire and less than 1.5 cm long, *P. microphylla* (L.) Liebm., was collected as "apparently spontaneous" on the Michigan State University campus in 1958 (*H. W. Joyner*, MSC). This is a native of tropical America, known as a greenhouse weed but surely not hardy in our climate.

KEY TO THE SPECIES

1. Mature achenes 0.7–1 (1.1) mm wide, pale yellowish or greenish to light brown, often marked with darker ± elongate but scarcely raised speckles .... 1. **P. pumila**
1. Mature achenes mostly 1.1–1.4 mm wide, deep purple- or olive-black with narrow pale margins, warty on the sides with low irregular raised bumps........2. **P. fontana**

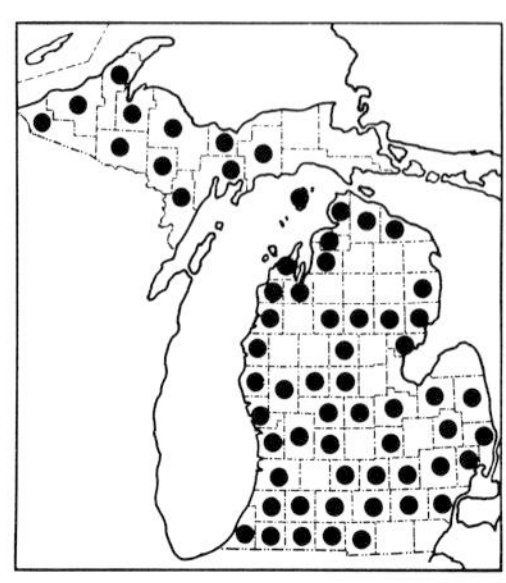
79. Laportea canadensis

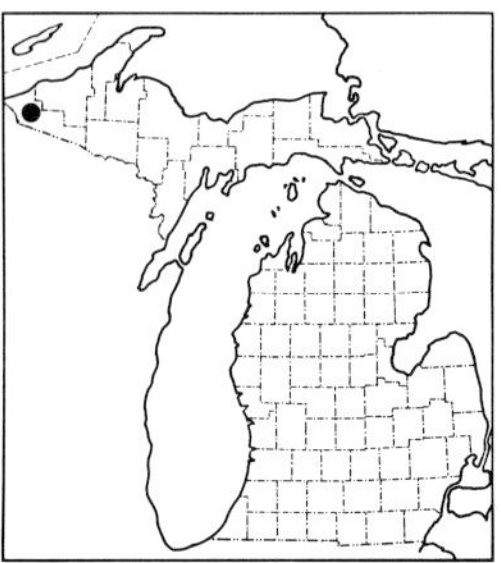
80. Urtica urens

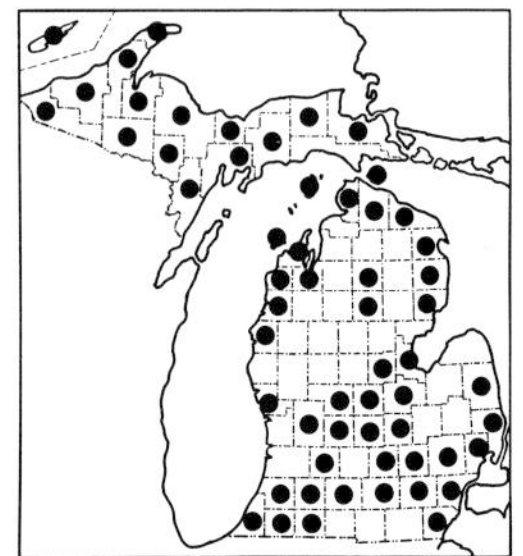
81. Urtica dioica

1. **P. pumila** (L.) A. Gray

Map 83. Low rich woods and floodplains; stream banks (and seasonally dry beds).

2. **P. fontana** (Lunell) Rydberg

Map 84. River banks, lake margins, swamps, marshes, usually in quite wet places.

Only the fruit can be relied upon to distinguish this species from the preceding, although there is a reputed (but not very clear) tendency for the petioles to be shorter in this species. Some authors unite this with the preceding as a variety or even form of *P. pumila*.

SANTALACEAE — Sandalwood Family

Largely a tropical family, often of woody plants, including the sandalwoods; represented in our area by two inconspicuous herbaceous species with extensive somewhat woody rhizomes. Both are hemiparasitic, i. e., in addition to bearing green leaves they are apparently always attached (by means of modified roots, or *haustoria*) to some other plant. Piehl (1965) reports that in the Great Lakes region alone over 200 species (herbaceous and woody, vascular and non-vascular) have been determined to serve as hosts of *Comandra umbellata*. This is the largest and most diverse natural host range known for a root-parasitic flowering plant. Little harm to the host plant seems to result from the parasitism.

Both species also serve as alternate hosts for the canker-producing comandra blister rust fungus (*Cronartium comandrae*), which in this region infects jack pine, although not as a serious enemy. The life cycle is similar to that of the better known white pine blister rust. The leaves of many plants, especially of *Comandra umbellata*, appear almost golden in late summer because of the rust infection.

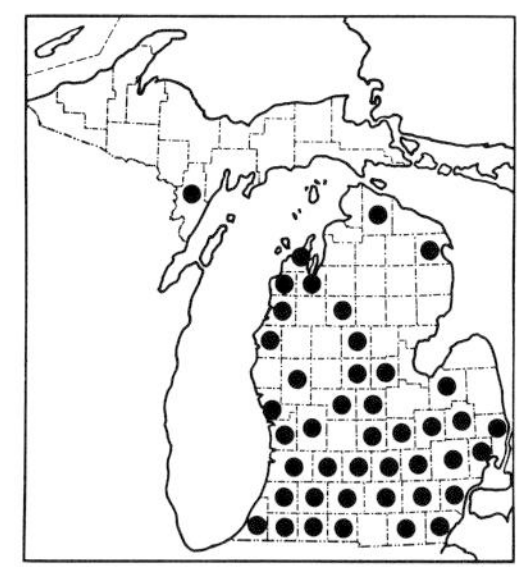

82. Boehmeria cylindrica

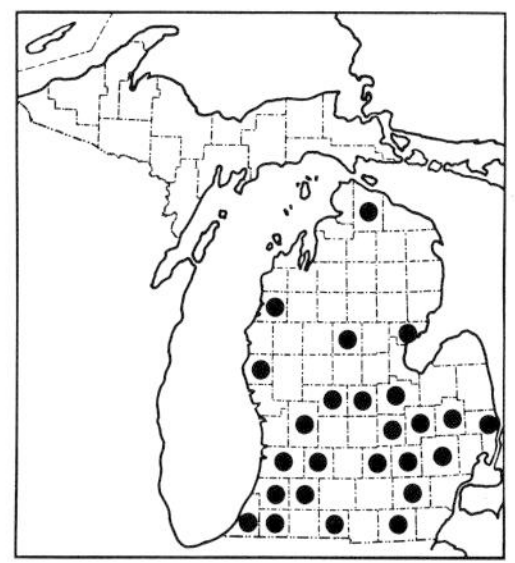

83. Pilea pumila

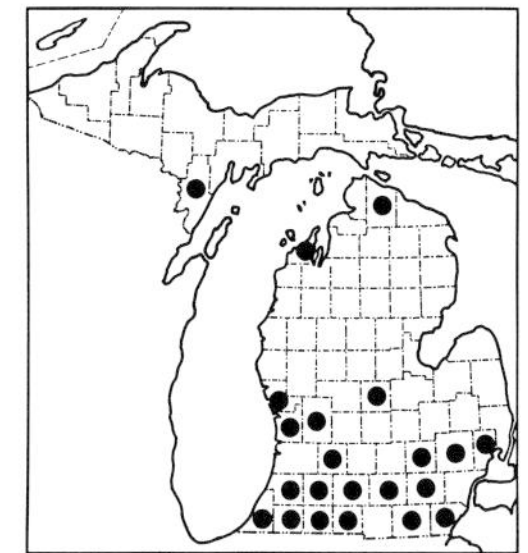

84. Pilea fontana

REFERENCES

Mielke, J. L., R. G. Krebill, & H. R. Powers, Jr. 1968. Comandra Blister Rust of Hard Pines. U.S. Dep. Agr. Forest Pest Leafl. 62, rev. 8 pp.

Piehl, Martin A. 1965. The Natural History and Taxonomy of Comandra (Santalaceae). Mem. Torrey Bot. Club 22(1): 1–97. [*Geocaulon* also discussed, on p. 41.]

KEY TO THE GENERA

1. Flowers greenish brown or purplish, in 2–3-flowered cymules on peduncles in the axils of middle or upper leaves; fruit a juicy orange to red drupe ca. 7–12 mm in diameter; cymules usually with only 1 flower perfect, the style less than 0.5 mm long; sepals deltoid, less than 1.5 times as long as broad .... 1. **Geocaulon**

1. Flowers white (rarely pinkish), bright green basally, all or mostly in a terminal inflorescence of few to many cymules; fruit a dry or slightly fleshy green or yellowish drupe less than 6 mm in diameter; cymules with all flowers perfect, the filiform style ca. 1.5–2.5 mm long; sepals usually becoming ca. twice as long as broad, or longer .... 2. **Comandra**

1. Geocaulon

1. **G. lividum** (Richardson) Fern. Plate 1-B

Map 85. Parasitic on a diversity of gymnosperms and angiosperms, usually on sandy (or rock) ridges or dunes at the borders of conifer thickets and woods, mostly not far from the shores of the Great Lakes; rarely in bogs or fens with *Chamaedaphne* and *Thuja*.

Sometimes included in the next genus, as *Comandra livida* Richardson. Vegetatively resembles some shade forms of *C. umbellata* rather closely and may be growing with it. Often overlooked when the inconspicuous axillary flowers are in bloom, but quite striking when bearing the ripe bright orange-red fruit, which is said to be edible (though I do not find it tasty).

2. Comandra

1. **C. umbellata** (L.) Nutt. Bastard-toadflax; Star-toadflax

Map 86. Sandy, gravelly, and rocky sites, often calcareous; shores, open woods (oak, aspen, jack pine) and clearings, dunes and ridges, sometimes in damper ground or at the borders of lakes and marshes.

In such characters as leaf size, depth of rhizome, shape of inflorescence, and sepal length, this is an extremely variable species, especially in our area, where some or all plants have sometimes been referred to a segregate, *C. richardsiana* Fern., which Piehl has been unable to justify as a distinct entity at any rank. All of our plants are treated as the typical variety or subspecies of *C. umbellata*; others occur from the Great Plains westward.

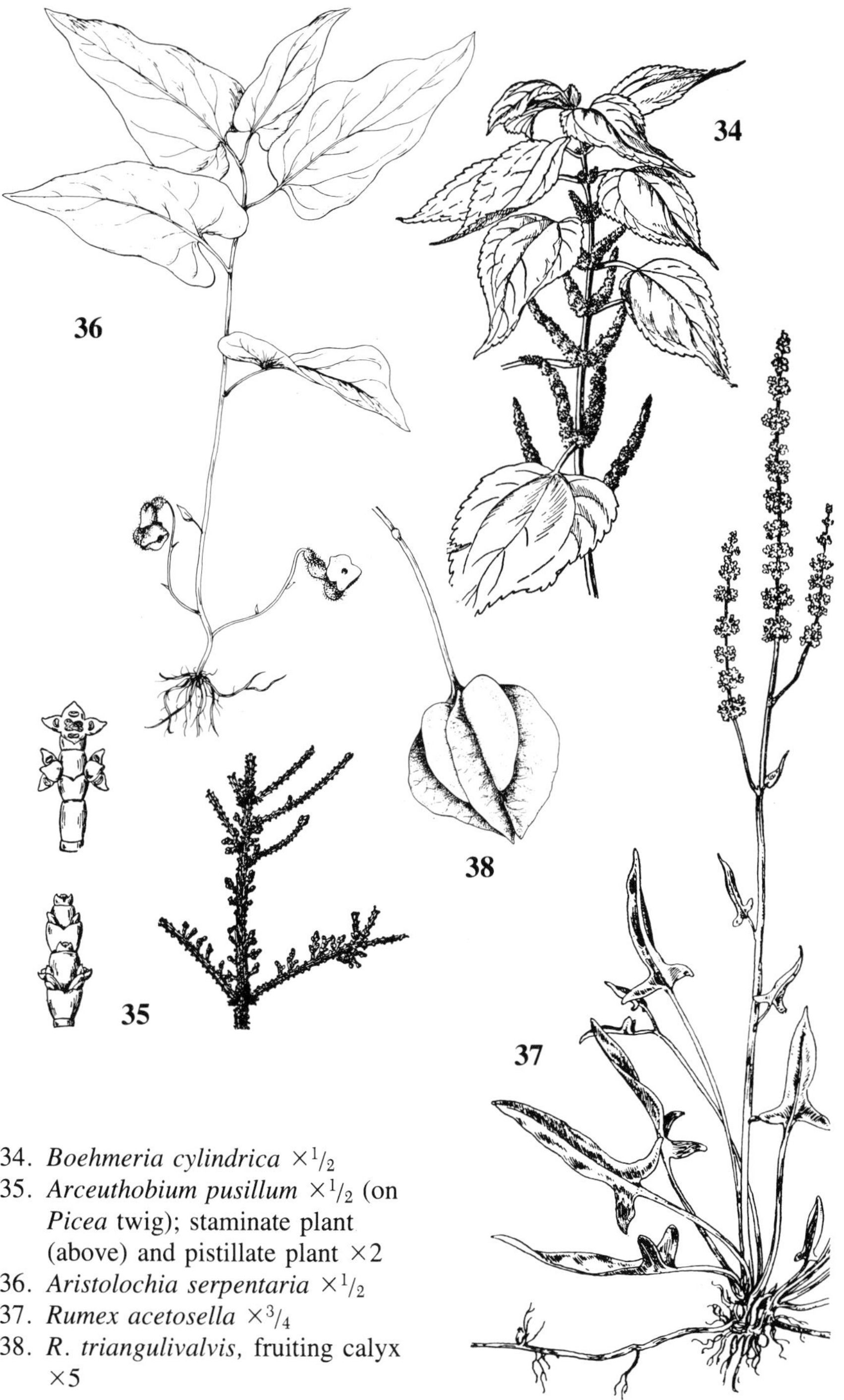

34. *Boehmeria cylindrica* ×1/2
35. *Arceuthobium pusillum* ×1/2 (on *Picea* twig); staminate plant (above) and pistillate plant ×2
36. *Aristolochia serpentaria* ×1/2
37. *Rumex acetosella* ×3/4
38. *R. triangulivalvis,* fruiting calyx ×5

The common mistletoe used in this country for decoration at the Christmas season is a member of this family, *Phoradendron serotinum* (Raf.) Johnston [long known incorrectly as *P. flavescens* (Pursh) Nutt.], a native of the southern states. The true European mistletoe is *Viscum album* L. Our non-green, essentially leafless dwarf mistletoe looks quite different from these familiar species. Until fairly recently, all these genera were usually included in the Loranthaceae, the "showy mistletoes," now restricted to plants with larger flowers as well as embryological and other differences.

REFERENCE

Kuijt, Job. 1982. The Viscaceae in the Southeastern United States. Jour. Arnold Arb. 63: 401–410.

1. Arceuthobium

In western North America, from southern Alaska to Mexico, there are several larger and more conspicuous species of this genus than ours; only a species of the Himalayas is smaller. These are important parasites on gymnosperms and are of serious concern to foresters.

REFERENCES

Hawksworth, Frank G., & Delbert Wiens. 1972. Biology and Classification of Dwarf Mistletoes (Arceuthobium). U.S. Dep. Agr. Agr. Handb. 40. 234 pp.

Ostry, Michael E., & Thomas H. Nicholls. 1979. Eastern Dwarf Mistletoe on Black Spruce. U.S. Dep. Agr. Forest Insect & Disease Leafl. 158. 7 pp.

Ostry, Michael E., Thomas H. Nicholls, & D. W. French. 1983. Animal Vectors of Eastern Dwarf Mistletoe of Black Spruce. U.S. Dep. Agr. For. Serv. Res. Pap. NC-232. 16 pp.

Voss, Edward G. 1967. Michigan Mistletoe. Mich. Audubon Newsl. 14(6): 5.

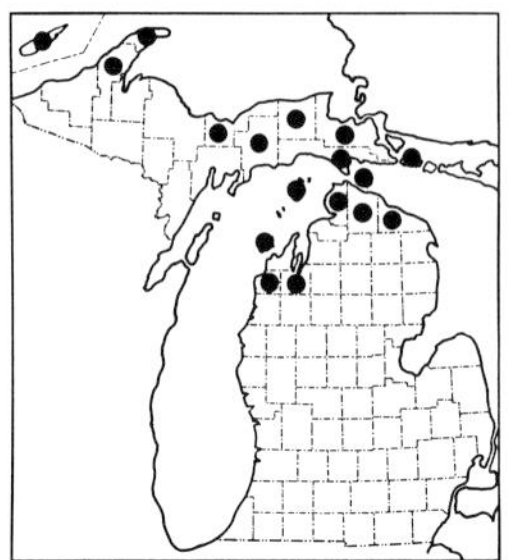

85. Geocaulon lividum

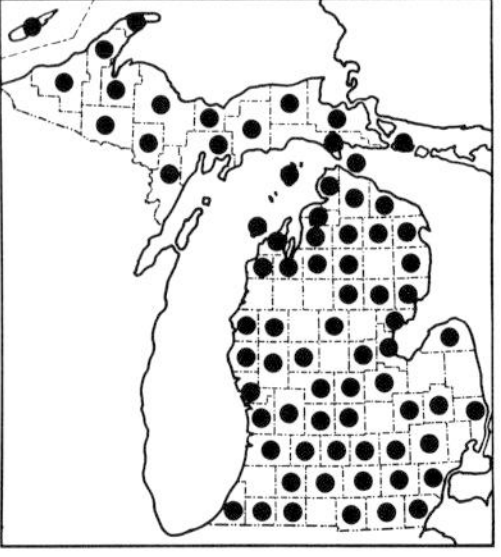

86. Comandra umbellata

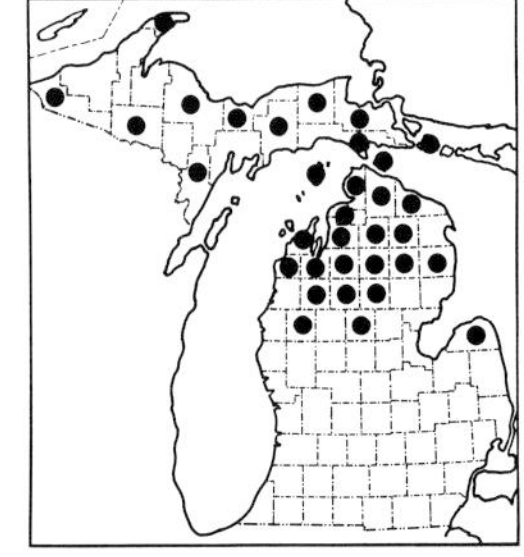

87. Arceuthobium pusillum

1. **A. pusillum** C. H. Peck Plate 1-C; fig. 35 Dwarf Mistletoe

Map 87. A tiny parasite, the shoots seldom over 1 cm high, on the branches of spruce (reported rarely on tamarack and pines elsewhere). The commonest host is black spruce (*Picea mariana*) and hence it is usually found in bogs, where it is rather frequent in Michigan, often locally abundant. In the vicinity of the Straits of Mackinac, however, as also in Door County, Wisconsin, it also parasitizes white spruce (*P. glauca*) in thickets and at borders of woods, especially on dunes, near the Great Lakes shore (Drummond Island, Bois Blanc Island, Beaver Island, and Wilderness State Park in Emmet County). Thus far, the southernmost stands of spruce in Michigan seem to be free of mistletoe, which is only in the northern part of the state, whence it ranges northward to Hudson Bay (but not as far north as spruce).

Our dwarf mistletoe is one of the first plants to bloom in April or May, often at about the same time as red maple is in flower. The sexes are separate, and pollination is primarily by insects. The minute perianth of the pistillate flower is 2-toothed; the staminate flower is only slightly more noticeable, with 3 (–4) small lobes, each bearing a yellow anther. The 1-seeded berry, ca. 3 mm long, ripens in the fall and the seed is expelled by rapid contraction of the fruit, traveling perhaps as much as 6 to 12 meters or more. Wind and even birds or squirrels, aided by a viscous coating on the seed, may play minor roles in dispersal. Germination of the seed and penetration of the host by the rapidly growing fungus-like absorptive tissues of the plant ordinarily occur the following spring. Plants do not flower until the fourth season and consequently fertile material will not be found on younger branches of the host.

The most conspicuous effect on the host is to produce the deformity known as a "witch's broom." The dwarf mistletoes are the only flowering plants which induce this unusual growth on the part of other plants, but since many parasitic fungi cause witch's brooms, the presence of such an aberration of growth is no guarantee that mistletoe is present. Anyone looking for the parasite, however, should watch for these deformities of spruce as a handy indicator. In old, well established infections, some of the host trees may

88. Aristolochia macrophylla

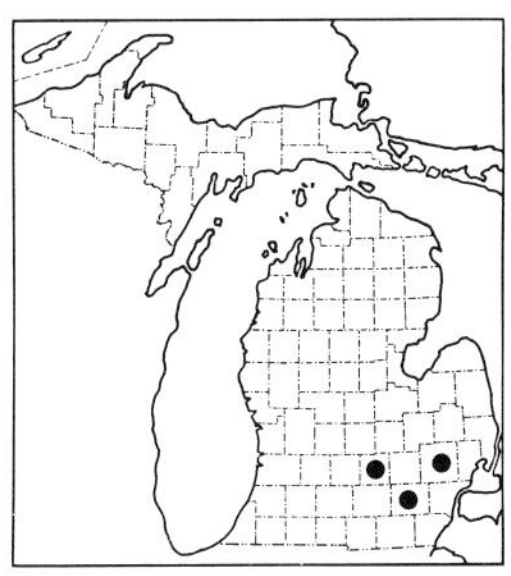

89. Aristolochia clematitis

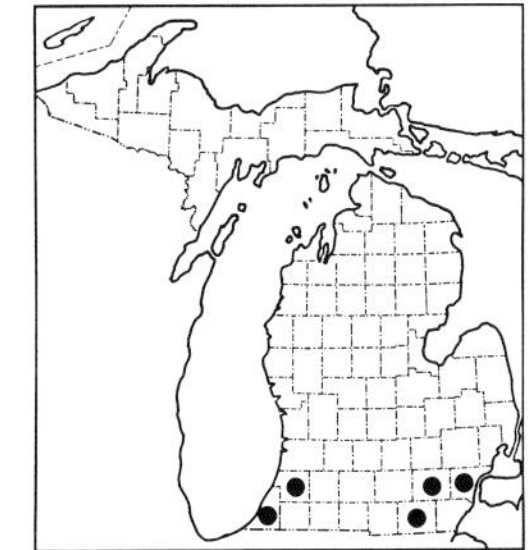

90. Aristolochia serpentaria

already be dead, their skeletons starkly revealing the witch's brooms. Fire was formerly a natural means of control.

It is surprising that a plant so widespread in northern bogs and producing such striking deformities should not have been reported from Michigan in any literature before 1904. Either its inconspicuous habit successfully concealed it from early botanists or else it has been spreading in recent times. That the latter may at least in part be the case is suggested by the fact that despite the considerable attention botanists have paid the Keweenaw Peninsula, dwarf mistletoe was not collected in the "Copper Country" until 1965.

ARISTOLOCHIACEAE — Birthwort Family

KEY TO THE GENERA

1. Stem erect (herbaceous or a woody vine), with odor none or not ginger-like; leaves alternate; perianth bilaterally symmetrical, with elongate and inflated calyx tube; stamens 6, the anthers sessile and adnate to the stigma .... 1. **Aristolochia**
1. Stem prostrate, with strong odor of ginger when freshly bruised; leaves in pairs; perianth regular, the tube very short, neither curved nor inflated but cup-shaped; stamens 12, with short filaments inserted on the ovary .... 2. **Asarum**

1. **Aristolochia** — Birthwort

REFERENCE

Pfeifer, Howard W. 1966. Revision of the North and Central American Hexandrous Species of Aristolochia (Aristolochiaceae). Ann. Missouri Bot. Gard. 53: 115–196.

KEY TO THE SPECIES

1. Plant a high-twining woody vine with flowers ca. 3 cm or more long; leaves usually more than 8 cm broad .... 1. **A. macrophylla**
1. Plant herbaceous, not twining, with flowers less than 3 cm long; leaves (except the largest) less than 8 cm broad
   2. Flowers axillary; perianth ± straight, yellow; principal leaf blades about as broad as long or broader; plant escaped from cultivation, in ± disturbed places .... 2. **A. clematitis**
   2. Flowers on peduncles arising at the base of the stem; perianth S-shaped, purplish; principal leaf blades distinctly longer than broad; plant native, in woods .... 3. **A. serpentaria**

1. **A. macrophylla** Lam. — Dutchman's Pipe; Pipe Vine

Map 88. Said to have escaped from cultivation near Ypsilanti (*Farwell 1095* in 1891, BLH; see Rep. Mich. Acad. 6: 206. 1905). Native in woods well to the south of Michigan.

Sometimes called *A. durior* Hill, a name of uncertain application.

2. **A. clematitis** L. Birthwort

Map 89. A native of Europe and Asia Minor, once cultivated as a medicinal plant, only very locally established on river banks and roadsides.

3. **A. serpentaria** L. Fig. 36 Virginia-snakeroot

Map 90. Barely ranges as far north as Michigan, where it is local in oak-hickory or less often mesic woods.

2. Asarum

1. **A. canadense** L. Plate 1-D Wild-ginger

Map 91. Rich, moist deciduous woods, especially on banks, much less often in cedar swamps; very local or absent in the eastern Upper Peninsula and northern Lower Peninsula, elsewhere common.

A familiar wildflower, blooming in the spring with flowers hiding at ground level, but the broad reniform leaves persisting through the summer. The pronounced ginger-like aroma of the plant, particularly the bruised prostrate stems or shallow rhizomes, readily reminds one of the true ginger, a tropical monocot in the Zingiberaceae.

Extremely variable in the length, shape, and reflexing of the sepals, but the varieties or species which have been proposed on these differences freely intergrade and seem unworthy of formal recognition. Flowers in which the calyx lobes have no tips or tips at most 3 mm long have been referred to var. *reflexum* (Bickn.) Robinson and apparently range north to Grand Traverse County. Flowers with prolonged acuminate to caudate tips—sometimes as long as 2 cm—represent typical var. *canadense* [including var. *acuminatum* Ashe] and are found throughout the state, with the longest and narrowest tips found on western Upper Peninsula plants.

Aberrant plants with 4 sepals have been found in Washtenaw County on several occasions and doubtless occur elsewhere. This tetramerous form was even named as a new species, *A. ypsilantense* Walpole (TL: Lefurge Woods [Washtenaw Co.]).

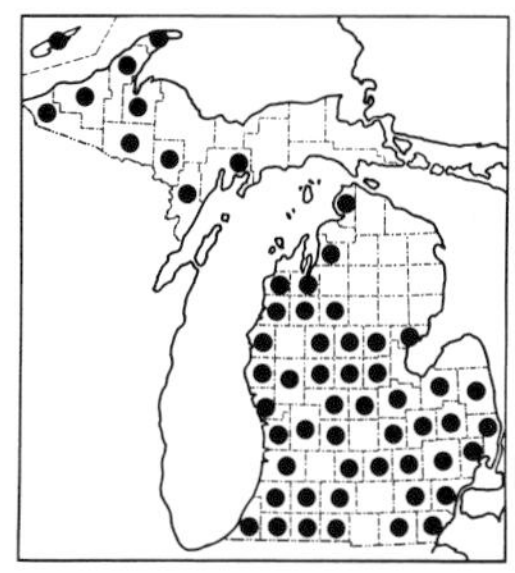

91. Asarum canadense

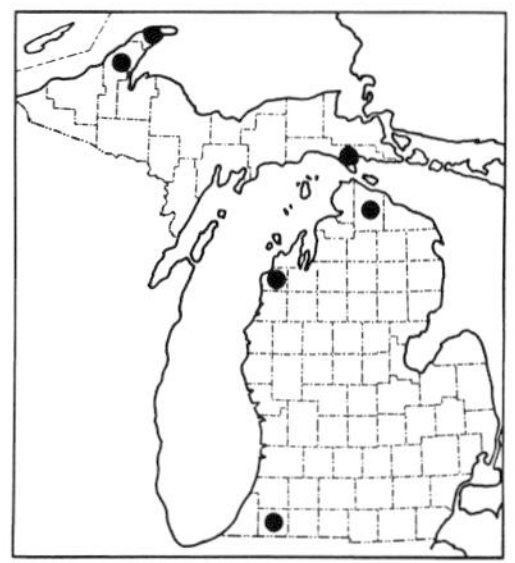

92. Rheum rhaponticum

93. Rumex acetosella

POLYGONACEAE — Smartweed Family

All members of this family in our area are easily recognized by the presence of a stipular sheath, or *ocrea,* which surrounds the stem above the attachment of each leaf. The similar reduced structure in the inflorescence is called an *ocreola*. The leaves are alternate and the nodes usually enlarged.

REFERENCES

Graham, Shirley A., & C. E. Wood, Jr. 1965. The Genera of Polygonaceae in the Southeastern United States. Jour. Arnold Arb. 46: 91–121.

Mitchell, Richard S., & J. Kenneth Dean. [1978]. Polygonaceae (Buckwheat Family) of New York State. New York St. Mus. Bull. 431. 81 pp.

KEY TO THE GENERA

1\. Achenes winged on the angles, becoming much longer than the perianth (ca. 3–5 times as long); stamens 9; stigmas conspicuous: knobby, subsessile and crowded at the summit of the ovary; leaves palmately veined; plant a stout perennial cultivated for its edible petioles and rarely escaped . . . . 1. **Rheum**

1\. Achenes not winged (though the perianth may be), no more than twice as long as the perianth (or longer in the annual *Fagopyrum*); stamens 8 or fewer; stigmas various (feathery, minute, and/or on distinct styles); leaves and habit various

2\. Tepals 6, greenish or reddish, scarcely petaloid, the 3 inner (but not the outer) ones enlarging in fruit and concealing the achene; stigmas a feathery tuft; plants in some species dioecious or polygamous and hence some flowers entirely staminate . . . . 2. **Rumex**

2\. Tepals 4–5, white to red and ± petaloid at least along the margins, uniform in size or the outer ones larger; stigmas usually not feathery and plants mostly with perfect flowers

3\. Pedicels with a swollen joint near the middle (but not far above the sheathing ocreolae), solitary in each ocreola, the inflorescence thus composed of slender racemes, appearing jointed because of the overlapping ocreolae; leaves not over 1 (1.1) mm wide; plant a delicate-looking annual . . . . 3. **Polygonella**

3\. Pedicels usually jointed near the summit (if at all), often crowded, the inflorescence various; leaves at least (1.5) 2 mm wide; plants usually stouter, annual or perennial

4\. Outer tepals winged or keeled in fruit, or plant somewhat twining or vine-like, or both; leaves ovate-cordate to broadly sagittate . . . . . . 5. **Polygonum** (couplet 11)

4\. Outer tepals not winged or keeled; plant not twining; leaves various

5\. Leaf blades about as broad as long, ± triangular-cordate; ripe achene 3-sided and much exceeding the perianth; styles 3, small but slender, several times as long as thick; plant a smooth branching annual with the uppermost inflorescences crowded in a ± corymbiform panicle . . . . 4. **Fagopyrum**

5\. Leaf blades usually much longer than wide, rarely if ever cordate (if triangular, the stem prickly); achene 2- or 3-sided, included in the perianth or slightly exserted; styles 1–3, in most species scarcely if at all longer than thick (elongate in *P. viviparum* & *P. virginianum*); plants of various habit, with flowers in axillary clusters, spikes or dense spike-like inflorescences, or small heads . . . . 5. **Polygonum**

1. **Rheum**

1. **R. rhaponticum** L. Rhubarb; Pie Plant

Map 92. An Old World species, widely cultivated for its tasty petioles, flavored principally with malic acid. However, the leaf blades, in contrast to the stalks, are poisonous and may even be fatal if eaten. Plants persist long after cultivation and may spread a little to roadsides and old fields. The species can hardly be said to be established, but doubtless occurs outside of present cultivation more widely than the map suggests.

Hortus Third adopts the name *R. rhabarbarum* L. for the garden rhubarb.

2. **Rumex** Dock

Half- or full-grown fruit is desirable, sometimes essential, for certain identification of species.

Reports of *R. conglomeratus* Murray from the state are based on immature material of *R. obtusifolius*; specimens of *R. crispus* and *R. altissimus* have also been misidentified as *R. conglomeratus*.

REFERENCES

Löve, Áskell, & Doris Löve. 1957. Rumex thyrsiflorus New to North America. Rhodora 59: 1–5.

Rechinger, K. H., Jr. 1937. The North American Species of Rumex. Field Mus. Nat. Hist. Bot. Ser. 17: 1–151.

Sarkar, Nina Marie. 1958. Cytotaxonomic Studies on Rumex Section Axillares. Canad. Jour. Bot. 36: 947–996.

KEY TO THE SPECIES

1. Leaf blades mostly hastate or sagittate with acute basal lobes, ± pleasantly acid to the taste; plants dioecious
  2. Outer tepals not at all reflexed; inner tepals of pistillate flowers closely covering the achene, or smaller, but no larger . . . . 1. **R. acetosella**
  2. Outer tepals strongly reflexed; inner tepals of pistillate flowers much expanded in fruit, at least twice as broad as the achene
    3. Inflorescence with branches at most branched once again and ± loosely flowered; tepals green to pinkish . . . . 2. **R. acetosa**
    3. Inflorescence repeatedly branched and densely flowered; tepals bright pink . . . . 3. **R. thyrsiflorus**
1. Leaf blades tapered, rounded, or subcordate at the base, the basal lobes, if any, not acute and the taste not sour; plants monoecious or with perfect flowers
  4. Margins of inner tepals with elongate teeth or spines at maturity (teeth much longer than broad)
    5. Swellings ("grains") well developed on midribs of all 3 inner tepals; plant an annual, of damp shores and marshy ground, with leaf blades tapered at the base . . . . 4. **R. maritimus**
    5. Swellings well developed on midrib of only 1 inner tepal; plant a stout perennial of waste ground generally, with subcordate or cordate leaf blades. . 5. **R. obtusifolius**

4. Margins of inner tepals entire to shallowly crenulate or toothed
6. Well developed swellings ("grains") present on at most 1 of the 3 inner tepals at maturity
7. Inner tepals with no grains well developed, distinctly cordate and reniform . . . . 6. **R. longifolius**
7. Inner tepals with a bulbous grain on 1 midrib (sometimes an obscure grain on others), scarcely cordate (except in the rare *R. patientia*) and usually as long as broad or longer
8. Leaves entire, flat, those of the main stem with short axillary shoots (branches or tufts of leaves); swollen joint ca. 1 mm or less from base of all fruiting pedicels . . . . 7. **R. altissimus**
8. Leaves crenulate-toothed, ± crinkly ("crisped") on the margins, without axillary shoots; swollen joint more than 1 mm from base of at least some fruiting pedicels
9. Inner tepals 3.5–5 mm long at maturity; petioles with a narrow groove on each side of midvein above . . . . 12. **R. crispus**
9. Inner tepals 6.5–8 mm long; petioles ± flat above . . . . 8. **R. patientia**
6. Well developed grains present on all 3 inner tepals at maturity (fig. 38)
10. Leaves entire, flat, those of the main stem with short axillary shoots (branches or tufts of leaves); swollen joints on fruiting pedicels 1 mm or less (usually 0.5 mm) from the base
11. Pedicels equaling or shorter than the perianth, ± curved, the inflorescence rather dense . . . . 9. **R. triangulivalvis**
11. Pedicels at least 2.5–3 times as long as the perianth, sharply deflexed and straight most of their length, the inflorescence in distinct whorls . . . . 10. **R. verticillatus**
10. Leaves crenulate-toothed and usually ± crinkly ("crisped") on the margin, without axillary shoots; swollen joints on at least some fruiting pedicels more than 1 mm from the base or not evident at all
12. Pedicels with joint not swollen, nearly or quite invisible; fruit ripe in September (rarely as early as late August in Upper Peninsula); base of swollen grain distinctly above base of tepal midrib . . . . 11. **R. orbiculatus**
12. Pedicels with a distinctly swollen joint on lower third (fig. 41); fruit ripe in June and July (rarely as late as August as far north as Isle Royale); base of swollen grain even with base of tepal midrib or bulging below it . . 12. **R. crispus**

1. **R. acetosella** L. Fig. 37 Sheep or Red Sorrel

Map 93. A pesky Eurasian weed, spreading by long horizontal roots (often mistaken for rhizomes) and forming large colonies. Almost ubiquitous, especially in dry sandy open places (disturbed ground, waste places, roadsides, old fields, stabilized dunes); invading woods and better soils.

The tepals range from greenish yellow to red and both extremes of color may be found on plants of both sexes. The sour leaves are a pleasant nibble and may be cooked as a vegetable.

2. **R. acetosa** L. Garden or Green Sorrel

Map 94. A fibrous-rooted native of Eurasia, sometimes cultivated for greens and rarely established. Collected ("rare") from waste ground at Bay City in 1897 (*Bradford,* MSC).

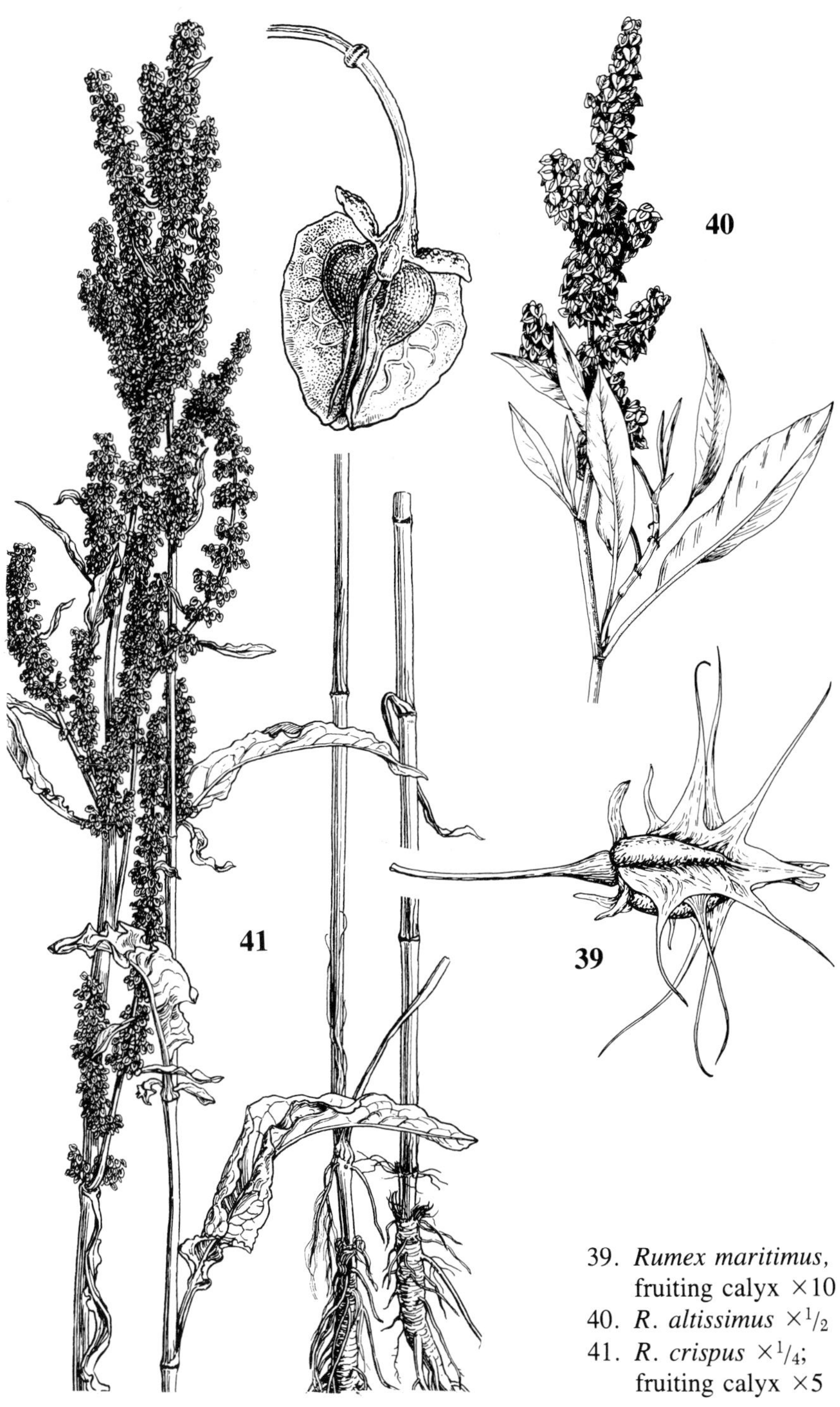

39. *Rumex maritimus*, fruiting calyx ×10
40. *R. altissimus* ×$^1/_2$
41. *R. crispus* ×$^1/_4$; fruiting calyx ×5

3. **R. thyrsiflorus** Fingerh.

Map 95. Another Eurasian introduction, similar to the preceding but more densely flowered and deeply tap-rooted. Collected in waste ground near the docks in St. Ignace in 1948 (*McVaugh 9297,* MICH, UMBS, MO, BLH), and by now apparently spread along the nearby expressway (*Voss 15642* in 1983, MICH).

4. **R. maritimus** L. Fig. 39

Map 96. Marshy ground, dikes, shores, and ditches.

Plants of our region have been separated from those of Eurasia and the Atlantic and Pacific coasts of North America as var. *fueginus* (Phil.) Dusén (see Mitchell in Brittonia 30: 293–296. 1978).

5. **R. obtusifolius** L. Bitter Dock

Map 97. A native of Europe, found in disturbed, usually moist ground on floodplains, along logging roads and borders of woods, in clearings and fields, around buildings. Known in Michigan since the First Survey (1838).

Material too young for the inner tepals to have enlarged and developed teeth may generally be placed by the fact that this is our only species of *Rumex* in which the leaf blades, at least the lower ones, are broadly and definitely subcordate to cordate and the inflorescence is dense. Hybrids with *R. crispus* are known (see below). The young leaves of all our docks can be eaten as greens, but this one becomes bitter early in the season.

6. **R. longifolius** DC.

Map 98. A native of northern Europe and Asia, collected several times (1901–1960) in moist ground at Isle Royale; other collections are apparently from waste places.

Long known in manuals as *R. domesticus* Hartman, a later name.

7. **R. altissimus** Wood Fig. 40

Map 99. River banks and also roadsides, not common.

Sometimes there is a second poorly developed (or rarely fully developed)

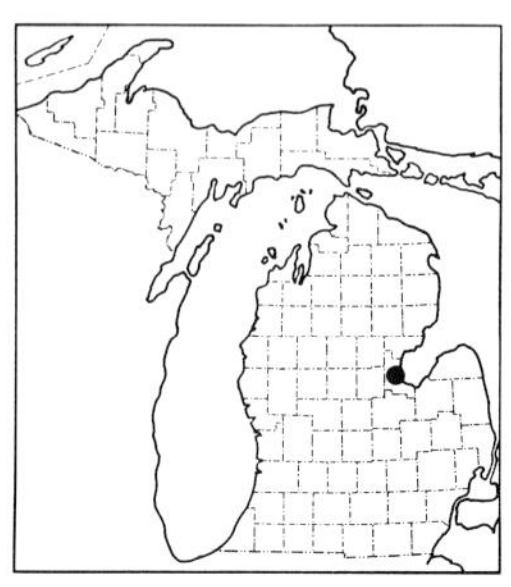

94. Rumex acetosa

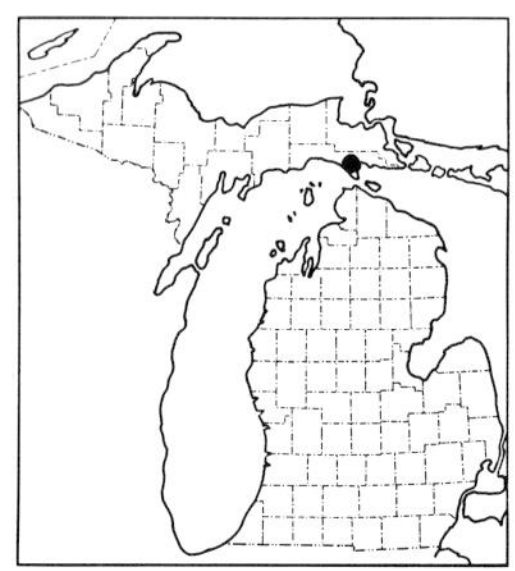

95. Rumex thyrsiflorus

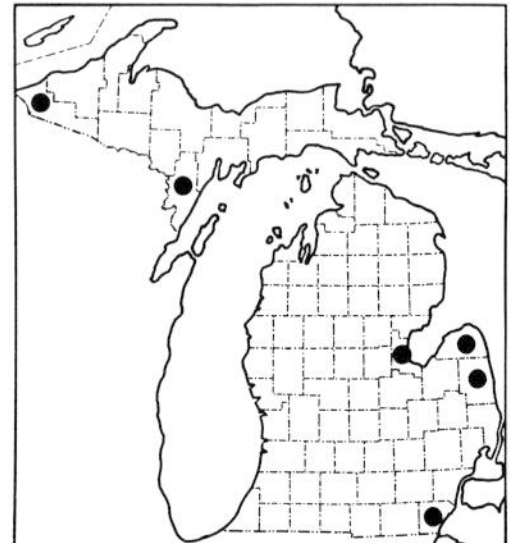

96. Rumex maritimus

grain on one of the inner tepals. One collection, from Cheboygan County (*Gates 10533*, F, MO), even has grains on all 3 tepals but has been determined by Rechinger as *R. altissimus*; it has large tepals and ovate-lanceolate leaves. Such specimens and those too immature to show clearly that only one of the inner tepals is forming a well developed grain can usually be distinguished from *R. triangulivalvis,* which also has flat entire leaves with axillary shoots, by the more ovate-lanceolate leaf blades (ca. 4–5 times as long as wide, or broader). The mature fruiting perianth is ca. 4–5 (6) mm long; in *R. triangulivalvis,* it is ca. 3–3.5 (4) mm long.

8. **R. patientia** L. Patience

Map 100. Native of Eurasia, collected by Wheeler along a railroad near Portland (in 1890, MSC; in 1892, GH).

The mature inner tepals are more deeply cordate than those of *R. crispus,* in this respect resembling those of *R. longifolius.*

9. **R. triangulivalvis** (Danser) Rech. f. Fig. 38

Map 101. Sandy or marshy shores, ditches, clearings, and waste ground.

The leaf blades are narrowly lanceolate, the best developed ones ca. (5) 6–10 times as long as broad (cf. *R. altissimus*). Rechinger restricted *R. mexicanus* Meissner to a larger-fruited species of Mexico and the Southwest,

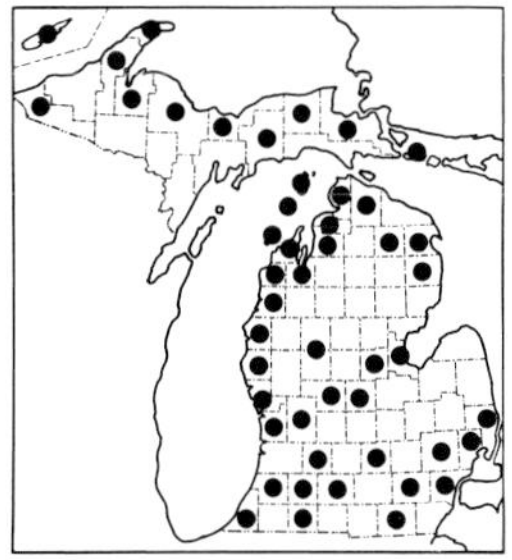

97. Rumex obtusifolius

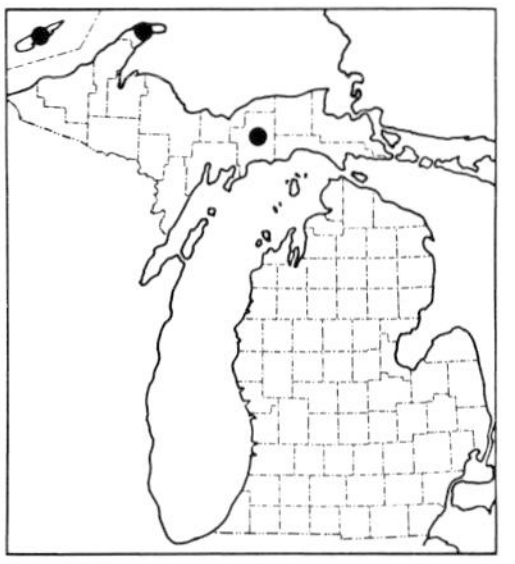

98. Rumex longifolius

99. Rumex altissimus

100. Rumex patientia

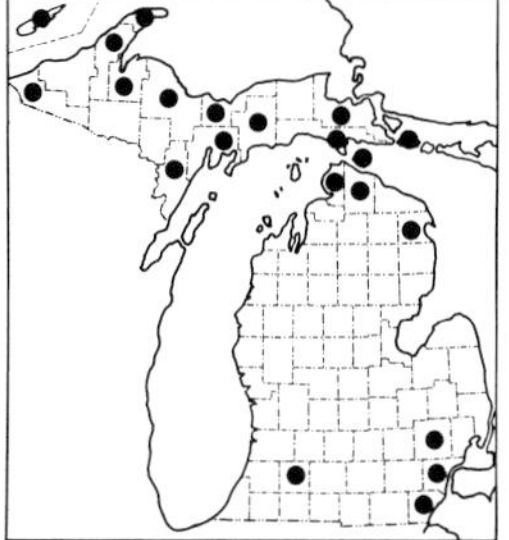

101. Rumex triangulivalvis

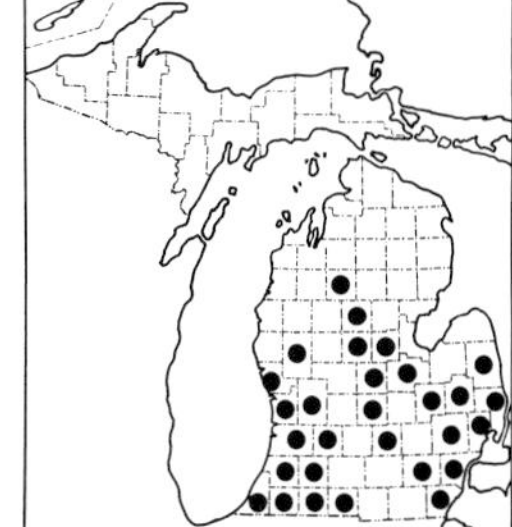

102. Rumex verticillatus

which Sarkar has shown to be tetraploid; these authors have been followed by Mitchell in referring all material from our region to the morphologically very similar *R. triangulivalvis,* though it has often been called *R. mexicanus.*

10. **R. verticillatus** L. Water Dock

Map 102. Swamp forests, river and lake margins, marshes, mudflats, often in shallow water.

Rather easily identified even when young by the very long pedicels in distinct whorls and the flat leaves.

11. **R. orbiculatus** A. Gray Great Water Dock

Map 103. Damp to very wet ground or shallow water of peatlands, river margins, marshes, ponds, swales, and ditches.

One of our most distinctive species in its very late fruiting, often extremely large stature in wet places, raised position of the grain on the tepals, and absence of a swelling at the joint in the pedicel. Even young flowering plants of *R. crispus* and related species have the joint evident. There is also a subtle difference in leaf venation, the main lateral veins in *R. orbiculatus* continuing prominent farther from the midrib than in *R. crispus* (and others), in which they soon become weak, branched, and anastomosed. The basal leaves may be over a meter long.

Long known under the name *R. britannica* L.

12. **R. crispus** L. Fig. 41 Curly or Sour Dock

Map 104. A European native, well known as a common and troublesome tap-rooted weed across North America. Collected by Dennis Cooley as early as 1837 in Michigan, and now found throughout in both moist and dry ground, thriving in disturbed places, along roadsides and also streamsides, in swampy clearings and woods, on shores and in fields.

Many of our plants have a well developed grain on only 1 of the 3 inner tepals [f. *unicallosus* Peterm.]. The inner tepals vary from entire to shallowly toothed, both types of margin often being found on the same plant—sometimes even on the same fruit.

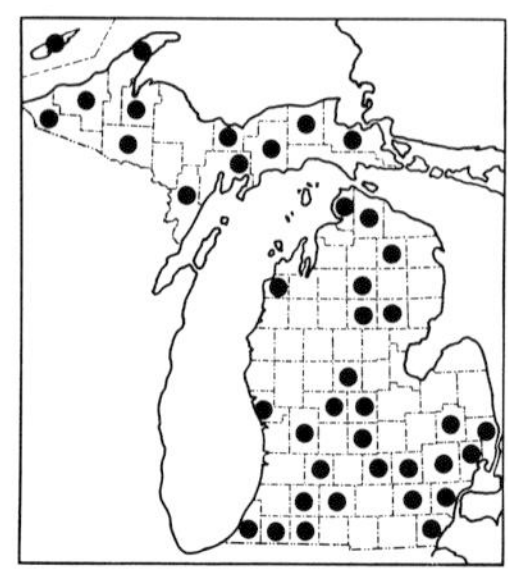

103. Rumex orbiculatus

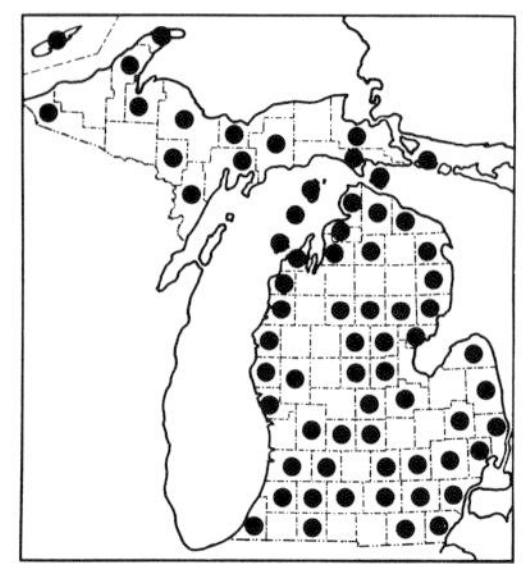

104. Rumex crispus

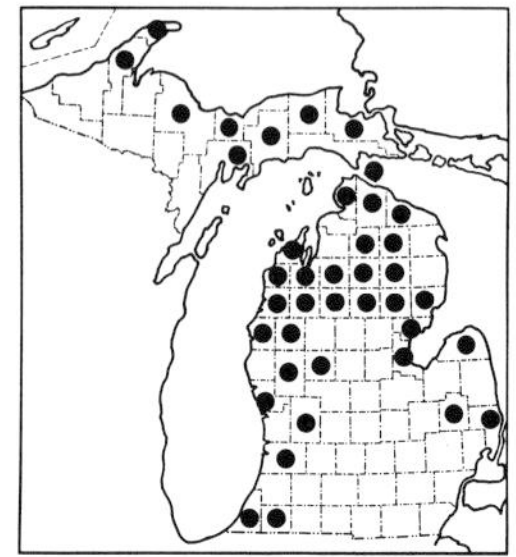

105. Polygonella articulata

Hybrids with *R. obtusifolius* [*R. ×pratensis* Mert. & Koch] occur occasionally. The inner tepals have sharp teeth, but shorter than in *R. obtusifolius*, and the plants are sterile, thus differing from *R. stenophyllus* Ledeb., a fertile species with similarly toothed tepals (see Löve & Bernard in Rhodora 60: 54–57. 1958). Apparent hybrids have been seen from Berrien, Houghton, Ingham, and Macomb counties and North Manitou Island (Leelanau Co.), and were reported from Chippewa County by Rechinger (1937, p. 149).

3. Polygonella

1. **P. articulata** (L.) Meissner Fig. 42 Jointweed

Map 105. Sandy banks, plains, and dunes, often in somewhat disturbed places. Common on the jack pine plains and also from Whitefish Point to the Huron Mountains along Lake Superior, where plants are relatively large and showy when blooming late in the summer.

The perianth varies from white to red. Until the plants bloom (chiefly August–September), the peculiar segmented appearance of the immature inflorescences, with closely overlapping ocreolae but no buds yet evident, often arouses curiosity in those not familiar with the species.

4. Fagopyrum

F. tataricum (L.) Gaertner, an Asian species introduced in Europe and North America, was collected at Corunna, Shiawassee County, but may well have been a cultivated plant there; the collection is undated but was made ca. 1880–1890. In this species the perianth is less than 2 mm long and the achenes are rough or tuberculate on the sides; in our common buckwheat, the perianth is ca. 2–4 mm long and the achenes are smooth.

1. **F. esculentum** Moench Fig. 43 Buckwheat

Map 106. A common cultivated plant of Asian origin, frequently escaped to waste ground, such as roadsides, fields, and railroads, but hardly well established.

The achenes are the source of buckwheat flour, and buckwheat honey is made from the flowers—but not by man. Sometimes known as *F. sagittatum* Gilib., a name not validly published.

5. Polygonum Smartweed; Knotweed

The genus is here treated in a broad sense, as by Graham and Wood and by Mitchell. Opinions of authorities differ widely as to which segregates, if any, should be recognized as distinct genera, for these often rely upon only a single character for diagnosis, and lines between the segregates are not as clearcut on the basis of assemblages of characters as they superficially appear. The smartweeds (species 16–25) are often separated as *Persicaria,* with the knotweeds (species 3–8) remaining in *Polygonum,* but this distinction is not maintained in *Flora Europaea* (see Webb & Chater in Feddes Repert. 68: 187–188. 1963). Species 10–11 are sometimes separated as *Reynoutria,* with distinct, feathery styles, and species 12–14 as *Fallopia* (which antedates the more often used *Bilderdykia*), with nearly sessile capitate stigmas (but cf. *P. cilinode*!); Shinners (Sida 3: 117–118. 1967) united all these under *Reynoutria* (the later name *Tiniaria* has been used but lacks priority). The alternative names, if these and some other segregates are recognized, are indicated for the species here treated, as they may be useful in consulting other recent (or future) works.

REFERENCES

Fassett, Norman C. 1949. The Variations of Polygonum punctatum. Brittonia 6: 369–393.

Hedberg, Olov. 1946. Pollen Morphology in the Genus Polygonum L. s. lat. and Its Taxonomical Significance. Svensk Bot. Tidskr. 40: 371–404.

Löve, Áskell, & Doris Löve. 1956. Chromosomes and Taxonomy of Eastern North American Polygonum. Canad. Jour. Bot. 34: 501–521.

Mitchell, Richard S. 1968. Variation in the Polygonum amphibium Complex and Its Taxonomic Significance. Univ. Calif. Publ. Bot. 45. 54 pp.

Mitchell, Richard S. 1971. A Guide to Aquatic Smartweeds (Polygonum) of the United States. Virginia Polyt. Inst. Water Resources Center Bull. 41. 52 pp.

Styles, B. T. 1962. The Taxonomy of Polygonum aviculare and Its Allies in Britain. Watsonia 5: 177–214.

KEY TO THE SPECIES

1. Stem and petioles with retrorse prickles; leaves hastate or sagittate (with acute basal lobes)
   2. Leaves hastate, the basal lobes divergent; achenes 2-sided . . . . 1. **P. arifolium**
   2. Leaves sagittate, the basal lobes with tips parallel; achenes 3-sided . . 2. **P. sagittatum**
1. Stem and petioles without prickles; leaves various
   3. Flowers 1–4 at a node, sessile or pediceled in the axils of foliage leaves or bracts; leaf blades jointed at the base, less than 2 (2.4) cm broad; summit of ocrea silvery white, becoming lacerate-shredded; plants annual
      4. Flowers in the axils of bracts which (except for the lowermost) are less than twice as long, the inflorescence thus appearing to be a remotely flowered slender spike; plants stiffly erect with leaves mostly linear (rarely all nearly elliptical) and very sharply acute
         5. Flowers and fruit erect; leaves longitudinally folded or grooved (W-shaped in section) with ciliolate margins; fruiting perianth ca. 3–3.5 mm long . . . . 3. **P. tenue**
         5. Flowers and fruit becoming strongly reflexed; leaves flat or with margins revolute but glabrous; fruiting perianth ca. 3.5–4.5 mm long . . . . 4. **P. douglasii**
      4. Flowers in the axils of foliage leaves which mostly are at least twice as long; plants erect to ascending or prostrate, with leaves acute to obtuse or blunt
         6. Mature achene with the puncticulate background not disrupting the very shiny surface aspect; plant erect with ascending branches and narrowly elliptic acute leaves . . . . 5. **P. ramosissimum**

6. Mature achene dull or sparkling with puncticulate surface breaking up the smooth shiny aspect—rarely glossy; plant erect to spreading or low and prostrate with leaves various (narrowly elliptic and acute almost only when plant is prostrate)
7. Outer tepals scarcely if at all keeled (rarely slightly keeled), not exceeding the inner ones; plants with at least the lower branches spreading, the whole plant often prostrate and mat-forming; leaves (1.5) 2–8 (15) mm broad, often narrowly elliptic, acute .... 6. **P. aviculare**
7. Outer tepals becoming prominently keeled (hence boat-shaped) and mostly exceeding the inner ones, often forming a definite beak-like projection beyond the achene (fig. 45); plants ± erect; leaves (at least the widest) (6) 8–18 (24) mm broad, elliptic to obovate, usually obtuse to rounded at apex
8. Tepals fused less than half their length (not as far as widest portion of fruit), their margins yellowish (or a few rosy) .... 7. **P. erectum**
8. Tepals fused at least half their length (slightly beyond widest portion of fruit), their margins whitish .... 8. **P. achoreum**
3. Flowers (or bulblets in *P. viviparum*) numerous in peduncled terminal or axillary spikes, racemes, or panicles, often densely crowded; leaves not jointed at base of blade, in some species over 2.5 cm broad; summit of ocrea tinged with brown, shattering at maturity but not shredding; plants annual or perennial
9. Inflorescence a narrow, elongate, remotely flowered spike-like raceme (flowers solitary at nodes), the main (terminal) one 15–36 (47) cm long; styles 2, soon elongating and becoming indurated, persistent, hooked at the tip; leaf blades (at least the larger ones) (4) 4.3–9 (10) cm broad, tapered at the base .... 9. **P. virginianum**
9. Inflorescences shorter, often densely flowered (flowers clustered at nodes); styles 1 or 3, obsolete or very short (except in the boreal *P. viviparum*), deciduous, not hooked; leaves narrower or cordate to rounded at base
10. Outer tepals keeled to narrowly or broadly winged, at least toward apex at maturity; plant becoming a stoloniferous twining vine or the larger leaves at least 6 cm broad
11. Plants stout (main stem ca. 8–25 mm thick), erect, with leaves at least 6 cm broad and blades truncate to shallowly cordate at base
12. Leaf blades cordate or subcordate, usually (at least the larger ones) over 15 cm long, with acute to short-acuminate apex .... 10. **P. sachalinense**
12. Leaf blades truncate at the base (often with a small lobe or angled shoulder at each outer angle plus a slight expansion of the blade at summit of the petiole, resulting in a shallowly 3-lobed aspect), less than 15 cm long, abruptly short-acuminate .... 11. **P. cuspidatum**
11. Plants with slender stems (up to 3 mm), becoming twining at least in part (young shoots ± erect), with leaves less than 6 (very rarely as much as 9) cm broad and blades deeply cordate with acute to rounded lobes
13. Nodes mostly with a ring of reflexed bristles; achene smooth and glossy; styles distinctly separate, divergent; outer tepals ± obscurely keeled .... 12. **P. cilinode**
13. Nodes glabrous; achene granular or outer tepals definitely keeled or winged; styles obsolete, the stigmas nearly sessile
14. Achene with sides granular or cellular in appearance; outer tepals ± papillose or roughened, narrowly keeled (wing less than 0.5 (0.6) mm broad); fruiting perianth less than 5 (6) mm long .... 13. **P. convolvulus**
14. Achene smooth and glossy; outer tepals smooth, ± conspicuously winged (wing over 0.5, usually 1, mm broad); fruiting perianth (6) 7–10 (11) mm long (including ± prolonged winged summit of pedicel) ...... 14. **P. scandens**

10. Outer tepals not keeled or winged; plant not at all vine-like; leaves less than 6 cm broad (except in *P. orientale*)

15. Inflorescence solitary, terminal, usually at least in part converted to sterile bulblets; cauline leaves no more than 3; basal leaves present, arising (with the stem) from a short thickened corm-like rhizome; style longer than the ovary . . . . 15. **P. viviparum**

15. Inflorescences solitary (*P. amphibium*) or several, without bulblets; cauline leaves numerous; basal leaves absent and stem not from a corm-like base; style shorter than ovary

16. Inflorescences 1 (−2), terminal, thick and densely flowered; leaves in some forms floating and in some forms subcordate (and in some, neither). . . . . 16. **P. amphibium**

16. Inflorescences few to many, axillary as well as terminal (if some plants with only 1–2, these very slender—less than 5 mm thick—and elongate); leaves neither floating (may be submerged) nor subcordate

17. Ocreae without bristles (very rarely with a few minute cilia)

18. Peduncle with few to dense distinctly stalked glands; inflorescences at least 9 mm thick, erect; outer tepals with veins becoming obscure toward margin; achenes 2.3–3.3 (3.7) mm broad . . . . 17. **P. pensylvanicum**

18. Peduncle with glands sessile or (usually) none; inflorescences rarely as thick as 9 mm, usually nodding; outer tepals with veins becoming strong and forked with recurved anchor-like branches toward margin (fig. 50); achenes (1.3) 1.5–2.2 mm broad . . . . 18. **P. lapathifolium**

17. Ocreae with at least a fringe of bristles on the margin (often with additional, usually appressed, hairs or bristles)

19. Peduncles and ocreae with spreading hairs

20. Hairs, at least on peduncles, gland-tipped; achenes ca. 1.5–1.7 mm broad; principal leaves lanceolate, up to 2.5 cm broad; inflorescences rather loosely flowered, less than 1 cm thick . . . . 19. **P. careyi**

20. Hairs not gland-tipped; achenes ca. 3–3.5 mm broad; principal leaves ovate, at least (2.5) 5 cm broad; inflorescences densely flowered, usually at least 1 cm thick . . . . 20. **P. orientale**

19. Peduncles and ocreae with hairs appressed or none

21. Tepals covered with glistening yellowish glandular dots (fig. 49) (especially noticeable on dry specimens); plant sharply peppery to taste (reaction may be briefly delayed)

22. Achene smooth and glossy; ocreae not swollen; tepals white-tipped or entirely green . . . . 21. **P. punctatum**

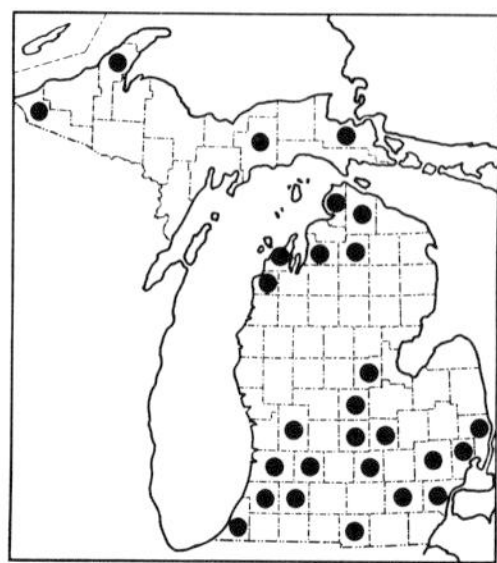

106. Fagopyrum esculentum

107. Polygonum arifolium

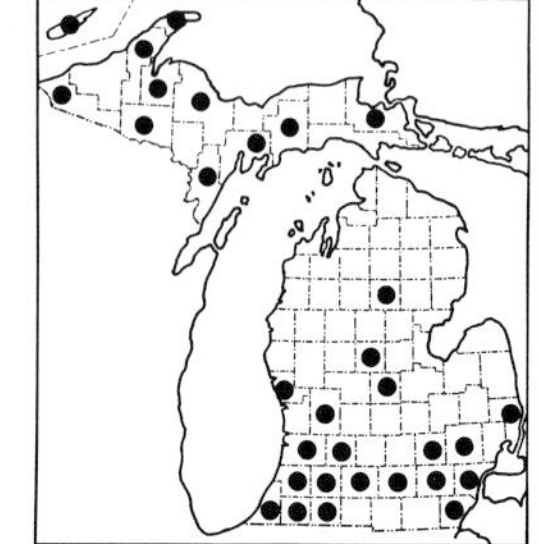

108. Polygonum sagittatum

22. Achene with sides granular or cellular in appearance; ocreae often swollen by included flowers or fruit; tepals usually rose-colored at tip . . . . 22. **P. hydropiper**

21. Tepals not gland-dotted; plant not peppery to taste

23. Inflorescences (at least the main ones) over 3 cm long, slender and loosely flowered; styles 3 and achenes 3-sided; plants perennial (often with stoloniferous base rooting at nodes); peduncles and stems above glabrous to ± densely strigose . . . . 23. **P. hydropiperoides**

23. Inflorescences (0.5) 1–3 cm long, densely flowered; styles 2 and achenes 2-sided (a few or very rarely all 3-sided); plants annual, from a small taproot; peduncles and stems glabrous or at most sparsely strigose

24. Bristles of ocreolae, if present, much shorter than perianth; achenes usually 2-sided (rarely all 3-sided) . . . . 24. **P. persicaria**

24. Bristles of ocreolae about equaling or exceeding perianth; achenes all 3-sided . . . . 25. **P. cespitosum**

1. **P. arifolium** L. Fig. 46 Tear-thumb

Map 107. Wet ground along streams and lakes and in swamps.

2. **P. sagittatum** L. Fig. 44 Tear-thumb

Map 108. Marshes, streamsides, shores, wet bogs—often in a distinct zone; also in damp fields, meadows, and roadsides.

3. **P. tenue** Michaux

Map 109. Dry sandy open ground on hills and old fields, borders of oak woods, rarely in heavily disturbed sites as along roadsides.

4. **P. douglasii** Greene

Map 110. More or less disjunct from western North America (see Mich. Bot. 20: 68–69. 1981), in sandy soil of open places (sometimes with *Polygonella*) in the west-central Lower Peninsula and counties bordering Lake Michigan in the Upper Peninsula, but on dry rock outcrops and "mountains" in the northwestern Upper Peninsula and Isle Royale.

5. **P. ramosissimum** Michaux

Map 111. Sandy or gravelly shores of the Great Lakes (or similarly sandy disturbed sites slightly inland); plants from southern Michigan are from waste ground and are doubtless introduced [they may in fact represent the Eurasian *P. patulum* Bieb.].

In the typical form, the tepal borders are yellow-green; in many of our plants they are ± red [f. *atlanticum* Robinson]. Sometimes confused with the next three species, but more erect than any form of the variable *P. aviculare* and with the fruiting perianths 3.5–4.5 mm long, at least partly on exserted pedicels; in *P. aviculare* the perianth is smaller and the pedicels nearly or quite hidden in the ocreae. The leaves are more narrow and acute than in any plants of *P. erectum* and *P. achoreum,* which may be erect but have more broadly elliptic to obovate blades (as well as having the perianth prolonged into a beak beyond the achene, especially in *P. achoreum*). In a tendency for the upper leaves of flowering branches to be reduced, *P. ramosissimum* may be confused with the preceding two species, but it has the tepals rugose with prominent raised veins, while in *P. tenue* and *P. douglasii* the tepals are nearly or quite smooth.

6. **P. aviculare** L. Knotweed

Map 112. A common weed of both Old and New Worlds, known in Michigan since the First Survey (1838). Its low stature permits it to endure mowing, and it survives in highly calcareous as well as saline areas. Sometimes on sandy shores or rock outcrops, but usually a durable and obvious weed of cleared and disturbed sites such as roadsides, lawns and gardens, sidewalk cracks, and railroad banks.

About the only taxonomic aspect of this complex agreed upon by botanists is that, in a broad sense, it is extremely variable and that more study is needed to be certain of the significance of all the possible subdivisions that could be (and have been) made (see, e. g., Löve & Löve 1956; Styles 1962; Mitchell & Dean 1978). Among the variable features are leaf size and shape (including apex, whether acute or rounded), reduction (or not) of leaves on flowering branches, achene surface and shape, color of sepal margins, and chromosome number. There has been disagreement regarding the extent to which any sets of characters correlate with breeding systems or the extent to which apomixis is involved, as well as the identity of North American plants with taxa described from Europe. "Heterophyllous" plants with much-reduced leaves on flowering branches and sometimes ± erect habit are typical hexaploid *P. aviculare*. Plants with the leaves more uniform or only gradually changing in size are frequently distinguished as *P. arenastrum* Boreau, a tetraploid. An occasional plant with more boat-shaped tepals than usual and obtuse leaves may be *P. buxiforme* Small.

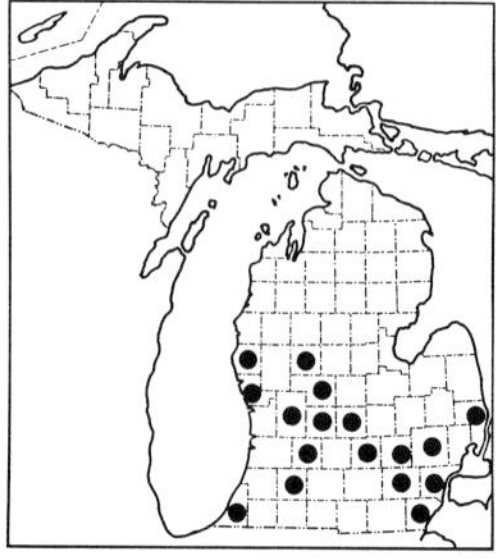

109. Polygonum tenue

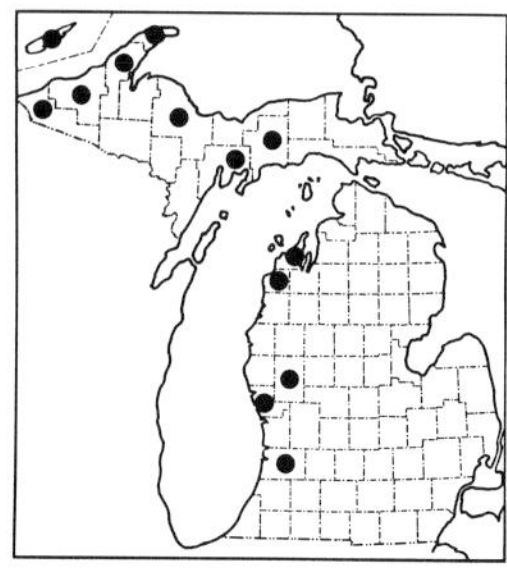

110. Polygonum douglasii

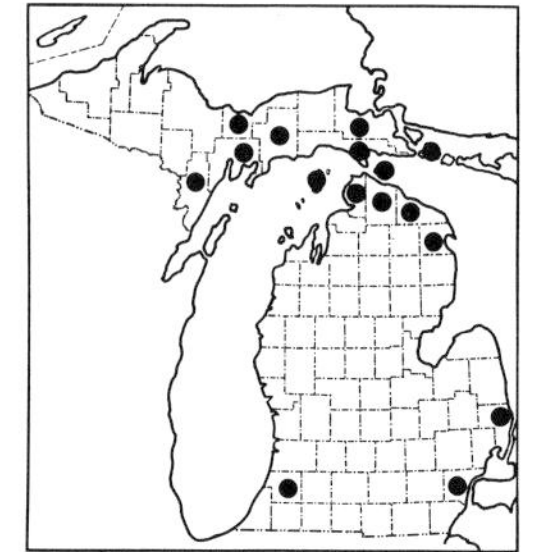

111. Polygonum ramosissimum

42. *Polygonella articulata* ×1/2
43. *Fagopyrum esculentum* ×1/2
44. *Polygonum sagittatum* ×1/2
45. *P. achoreum* ×1/2; fruiting calyx ×5

7. **P. erectum** L.

Map 113. Considered a native American plant, but not collected in Michigan before 1861 nor after 1939. During that 80-year period, recorded from farmyards, roadsides, and waste places.

Other species have frequently been misidentified as this one. The knotweeds are often troublesome to determine, as any of the characters cited is open to exception and one must too often make a judgment on a balance of characters. The leaves of *P. erectum* may be similar to the shapes in *P. aviculare,* but run distinctly larger. See comments on species 5, 6, and 8. Mature plants are rather strongly heterophyllous, with flowers (fruits) in axils of much-reduced leaves or bracts.

8. **P. achoreum** Blake Fig. 45

Map 114. A plant of unknown origin, now a weed across northern North America. Sandy and gravelly roadsides and clearings, railroads.

The prominent beak on the fruiting perianth, relatively short perianth lobes, and more rounded leaf tips make this species relatively distinctive in the group.

9. **P. virginianum** L. Fig. 47 Jumpseed

Map 115. Swamp forests, rich woods, and damp hollows.

When the fruit is mature, there is sufficient tension to propel it as much as 3 meters when it is disturbed, whence the common name. Often segregated in a separate genus, but the sectional name *Tovara* has been rejected at generic rank and if such segregation is made the correct name is *Antenoron virginianum* (L.) Roberty & Vautier.

10. **P. sachalinense** Friedr. Schmidt Giant Knotweed

Map 116. A native of the Far East, rarely escaped from cultivation to such sites as railroads, roadsides, and old homesites.

Also known as *Reynoutria sachalinensis* (F. Schmidt) Nakai.

11. **P. cuspidatum** Sieb. & Zucc. Japanese Knotweed; "Mexican Bamboo"

Map 117. A native of Japan, planted as an ornamental and occasionally found as an escape at gravel pits, filled ground, roadsides, dumps, and gullies.

Both this species and the preceding grow from sturdy rhizomes and are difficult to eradicate once established. They form dense thickets as tall as 2 m or more. Whenever there is a report of "bamboo" in Michigan, it is likely to be based on one of these stout species, which have hollow internodes. (True bamboos are grasses and do not grow in this region.) When segregated in the genus *Reynoutria,* the correct name is *R. japonica* Houtt. Some specimens, as those from Alcona County, approach the preceding species.

12. **P. cilinode** Michaux Fig. 48 Fringed False Buckwheat

Map 118. Characteristic of recently disturbed areas, especially in cleared forests, becoming abundant and persisting along logging (and other) roads, in dumps and gravel pits, on shores and dunes, and in open rocky ground.

The styles and stigmas closely resemble those of the preceding two species. Sometimes segregated as *Fallopia cilinodis* (Michaux) Holub or *Reynoutria cilinodis* (Michaux) Shinners.

13. **P. convolvulus** L. Black-bindweed; False Buckwheat

Map 119. A European weed usually in disturbed ground and waste places, including roadsides, railroads, gardens and cultivated (or recently cultivated) fields, gravel pits, railroads, clearings; sometimes on shores.

If segregate genera are recognized, this species may be called *Fallopia convolvulus* (L.) Á. Löve or *Reynoutria convolvulus* (L.) Shinners.

14. **P. scandens** L. False Buckwheat; Black-bindweed

Map 120. Disturbed ground (roadsides, dumps, railroads, etc.) and also shores, woods, thickets, and wetlands. Our plants are probably all or mostly native; some are doubtfully var. *dumetorum* (L.) Gleason, a native of Europe.

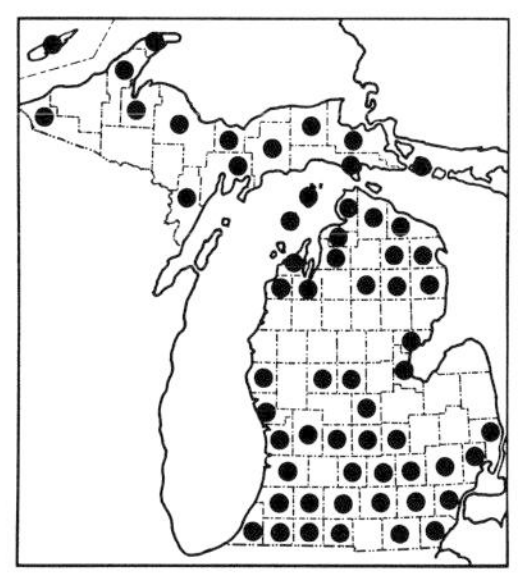

112. Polygonum aviculare

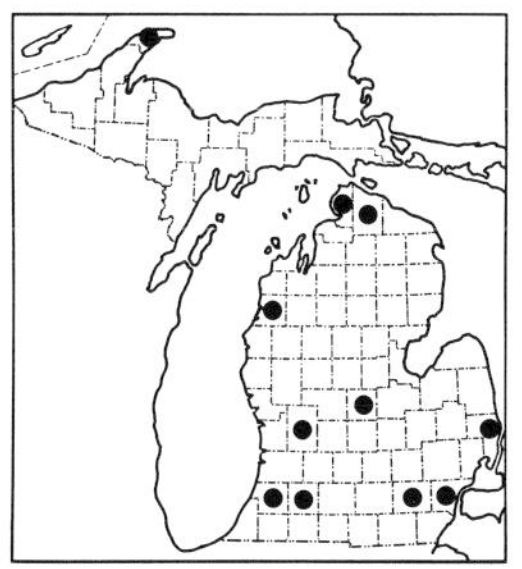

113. Polygonum erectum

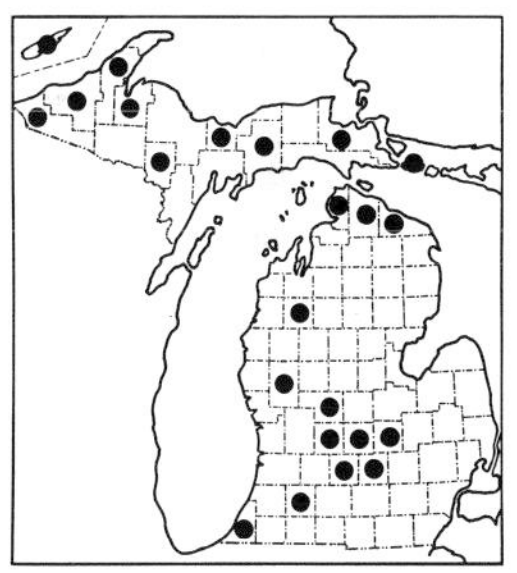

114. Polygonum achoreum

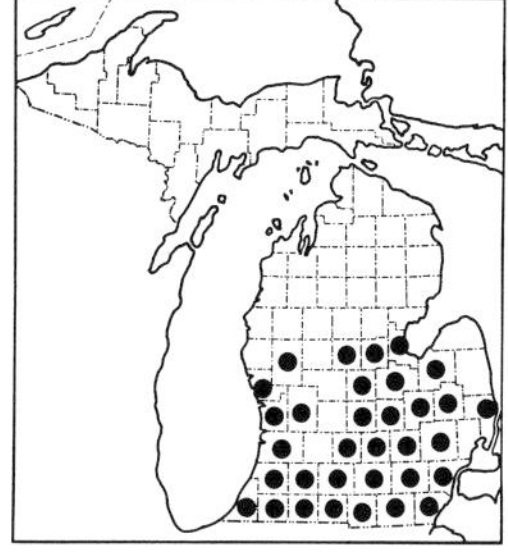

115. Polygonum virginianum

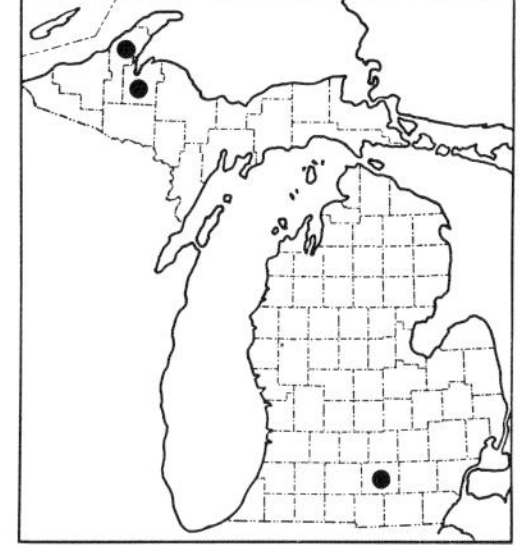

116. Polygonum sachalinense

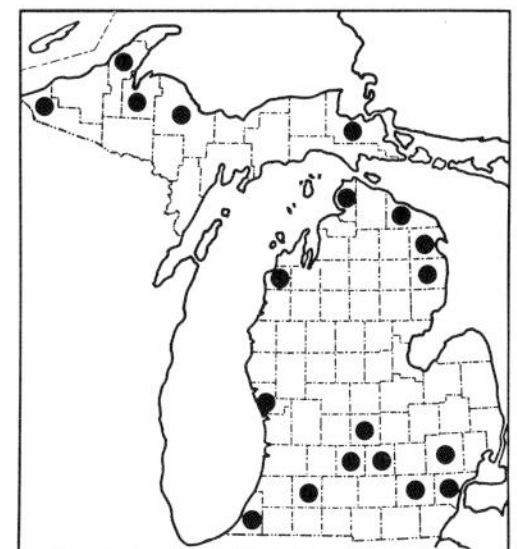

117. Polygonum cuspidatum

Variable in size of fruit and shape of wing on the outer tepals, here including plants sometimes referred to *P. dumetorum* L. and *P. cristatum* Engelm. & Gray. Also known as *Reynoutria scandens* (L.) Shinners (the combination in *Fallopia* has not been made).

Because of the ± prolonged extension of the tepal wings down the pedicel, the joint (which is always below the wings) generally appears to be on the lower half of the pedicel in this species, while in the preceding two it is on the upper half, near the perianth.

P. aubertii L. Henry, China fleece plant, a native of Asia, would key here, but differs from *P. scandens* in its perennial (even woody) habit and long, many-flowered inflorescences; it has been recently collected in Adrian, Lenawee County, perhaps planted or perhaps escaped from cultivation.

15. **P. viviparum** L. Alpine Bistort

Map 121. A circumpolar arctic-alpine species, ranging south in eastern North America to Lake Superior and the mountains of New England. In Michigan, it grows in rock crevices and associated sod along Lake Superior, as on the Canadian shore.

Sometimes segregated from *Polygonum* as *Bistorta vivipara* (L.) S. F. Gray.

16. **P. amphibium** L. Plate 2-A Water Smartweed

Map 122. Lakes and ponds, permanent or ephemeral; river margins and quiet backwaters, marshes; "terrestrial" forms in drier ground nearby (or after lowering of water levels).

The typical variety of this species is an Old World plant; we have two intergrading varieties—distinguished in "pure" form as follows:

(a) var. *stipulaceum* Coleman (TL: near Grand Rapids [Kent Co.]). Stems prostrate, in water (up to 2 m deep) producing flowering shoots with floating leaves (superficially resembling those of a *Potamogeton*). Young terrestrial shoots with a ± horizontal green flange at the summit of the ocrea, seldom flowering. (Both types of shoot may be found on a single stem which extends from a terrestrial to an aquatic habitat.) Inflorescence ± ovoid or conic, 1.5–3 (4) cm long. In some manuals, treated as *P. natans* Eaton. This variety occurs throughout the state.

(b) var. *emersum* Michaux. Stems ± erect, even when in the water. Floating leaves not produced. Ocreae without green flange. Inflorescence ± cylindrical, at least 3–8 cm long. These plants, as well as some floating-leaved plants now referred to var. *stipulaceum,* have long been known as *P. coccineum* Willd. This variety apparently occurs only very rarely north of the Straits of Mackinac in the state, though it is common in the southern Lower Peninsula.

Terrestrial shoots are usually very hairy, while aquatic ones are glabrous. Aquatic plants with the habit of var. *stipulaceum* and the elongate inflorescences of var. *emersum* (and often with ± cordate leaves) are intermediate and not, as long considered, merely aquatic forms of the generally erect var. *emersum* (*P. coccineum*). Over 100 names (specific and infraspecific) have been applied to this diverse complex; the simplified treatment of Mitchell (1968), based on considerable experimental work, is followed here. The species is sometimes segregated as *Persicaria amphibia* (L.) S. F. Gray.

17. **P. pensylvanicum** L. Pinkweed; Bigseed Smartweed

Map 123. Like the next species, with which it is often associated, may be abundant on recently exposed shores, river margins and banks, marshes, mucky hollows, old fields, damp or dry waste ground.

An annual, readily recognized by the large fruits and flowers, in thick, erect, pink spikes. Occasionally the upper branches of the plant, in addition to the peduncles, are ± densely covered with stalked glands; at the other extreme, an eglandular form is known from northern Ohio. The copious achenes are an important food for waterfowl. Also known as *Persicaria pensylvanica* (L.) Small.

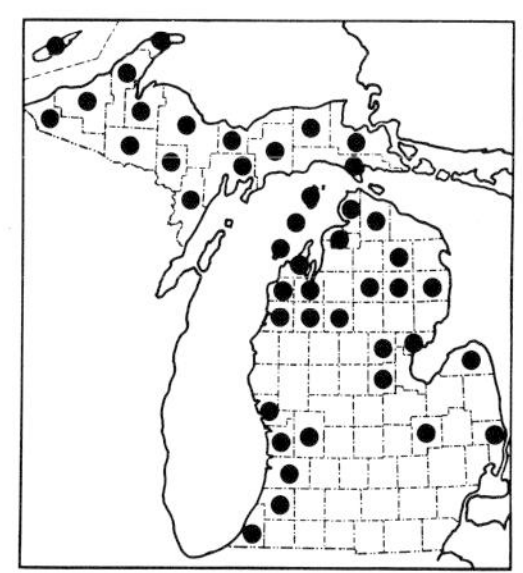
118. Polygonum cilinode

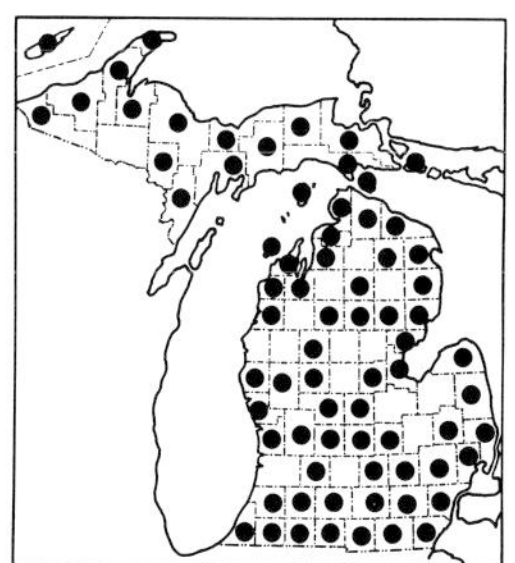
119. Polygonum convolvulus

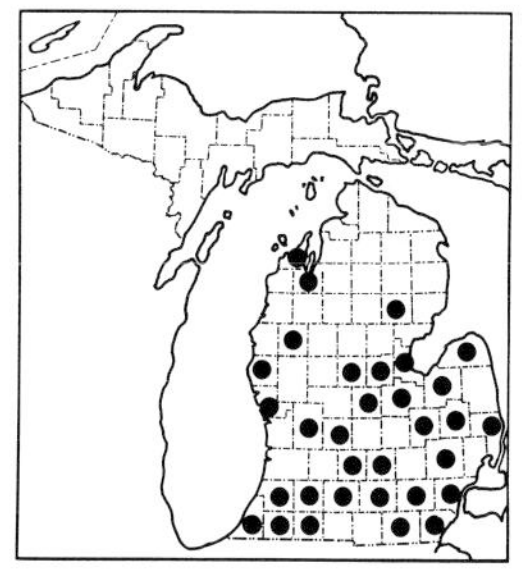
120. Polygonum scandens

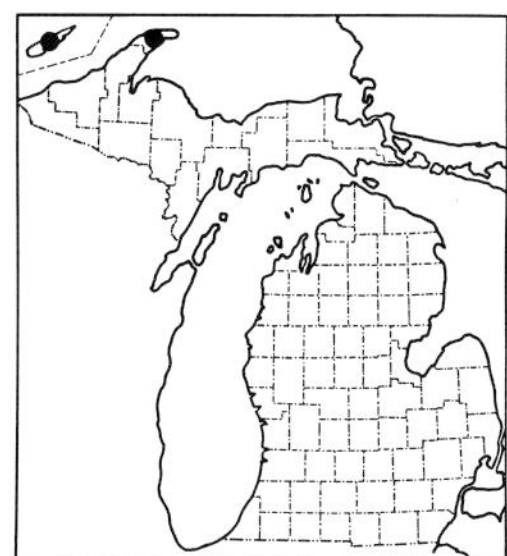
121. Polygonum viviparum

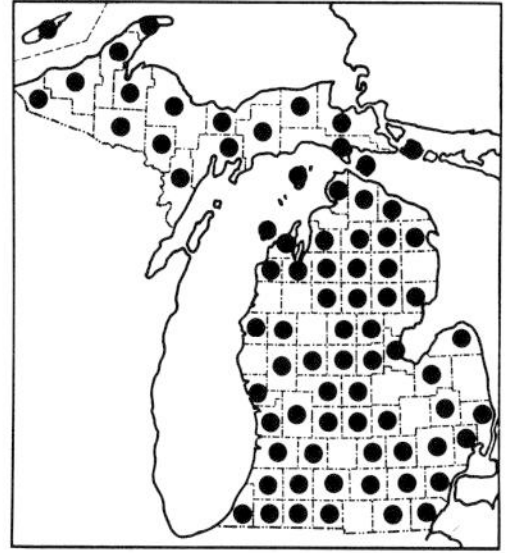
122. Polygonum amphibium

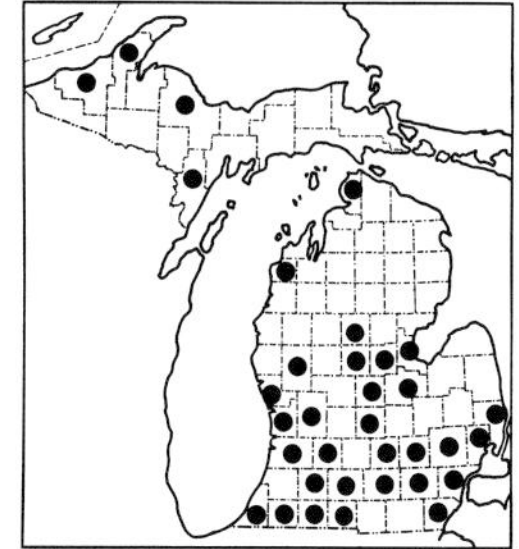
123. Polygonum pensylvanicum

46. *Polygonum arifolium*, leaf $\times^1/_2$
47. *P. virginianum* $\times^1/_2$
48. *P. cilinode* $\times^1/_2$; node $\times 2$
49. *P. punctatum* $\times^1/_2$; fruiting calyx $\times 8$

18. **P. lapathifolium** L. Fig. 50 Willow-weed; Nodding Smartweed

Map 124. Like the preceding, a species of moist shores, river margins, marshes, mucky hollows, and damp (rarely dryish) waste ground. Some of our plants may be introduced from Europe; the weedy habit has obscured any distinction between native and introduced sources.

Variable in pubescence, color, and stature. Some plants (especially when young) have leaves tomentose beneath. Plants range from several cm tall to as high as 2 m, and have very conspicuously swollen nodes and typically nodding inflorescences of pink to white flowers. Another annual often locally abundant and producing copious achenes. The smartweeds bloom late in the summer and into fall, and are readily laid low by the first frost. Also known as *Persicaria lapathifolia* (L.) S. F. Gray.

19. **P. careyi** Olney

Map 125. Damp sandy shores, banks, and made land, very local in Michigan.

If *Persicaria* is recognized as a distinct genus, this is *P. careyi* (Olney) Greene.

20. **P. orientale** L. Prince's Feather; Kiss-me-over-the-garden-gate

Map 126. A native of Asia with deep rose-purple flowers, cultivated as a garden annual and occasionally escaping to vacant lots, dumps, river banks, and waste places.

The summits of the ocreae often have green herbaceous lobes. Also known as *Persicaria orientalis* (L.) Spach.

21. **P. punctatum** Ell. Fig. 49

Map 127. Marshes, edges of lakes and rivers, shores, and mucky hollows, often in very wet places (even cold spring-fed areas) and seems frequently to grow in somewhat more shaded places than other smartweeds.

Another extremely variable species (see Fassett 1949). The inflorescences

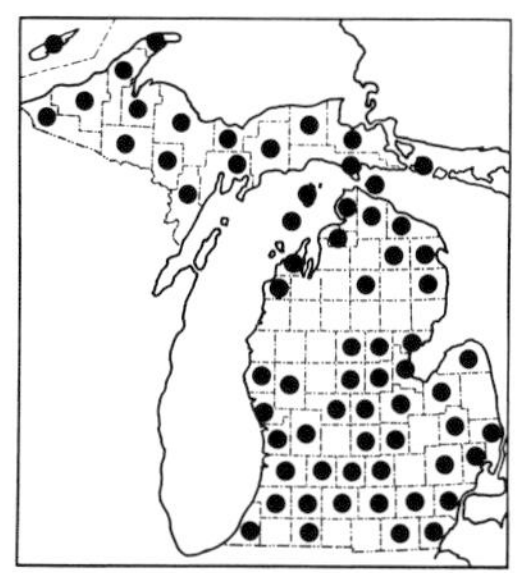

124. Polygonum lapathifolium

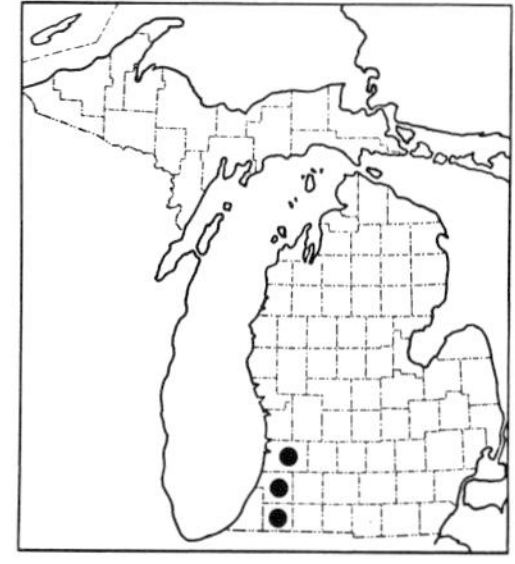

125. Polygonum careyi

126. Polygonum orientale

tend to be very straight, while in *P. hydropiper* they tend to nod. One collection from a river bank in St. Joseph County (*Rogers 12962* in 1963, WUD) may be the rather distinctive var. *majus* (Meissner) Fassett, often recognized as a separate species, *P. robustius* (Small) Fern., which occurs primarily along the Atlantic coast, south into northern South America and only very locally inland. It is much more robust in every way than other variants of *P. punctatum,* with leaves (in the Michigan specimen) mostly 2.5–3 cm broad, a stout stem, a denser inflorescence, and larger flowers; in this collection, however, the flowers are all infested with a smut and the ocreolae have a few cilia, unlike var. *majus*.

This species may be segregated as *Persicaria punctata* (Ell.) Small.

22. **P. hydropiper** L. Water-pepper

Map 128. Swamps, ditches, shores, marshes, streamsides, especially in slightly disturbed areas or even along roadsides. Usually considered naturalized from Europe, but sometimes though to be at least partly indigenous in North America.

Also called *Persicaria hydropiper* (L.) Opiz.

23. **P. hydropiperoides** Michaux Mild Water-pepper

Map 129. Swamps; wet bogs, fens, and ditches; borders of lakes, ponds, and rivers; often in shallow water.

A variable species or complex, including here at least Michigan plants sometimes referred to *P. setaceum* Baldwin and *P. opelousanum* Small. The ocreae in addition to bearing long, firm marginal bristles often are covered with similar strong, stiff, closely appressed bristles (unlike the slender usually soft ones of *P. persicaria*). Plants from throughout the state may bear on the leaves and/or perianth flat, plate-like, pale, easily dislodged, sparse and irregularly spaced "glands" resembling the farinosity of some Chenopodiaceae; but these plants seem not worthy of recognition as a distinct species (*P. opelousanum*). The flowers of this species often appear open in the field, while in our other smartweeds the tepals generally remain closed. Also known as *Persicaria hydropiperoides* (Michaux) Small.

24. **P. persicaria** L. Heart's-ease; Lady's-thumb

Map 130. A widespread weed, naturalized from Europe. Sometimes more weedy (gardens, roadsides) and in drier ground than customary for *P. pensylvanicum* and *P. lapathifolium,* but in many places grows with them: moist or dry waste ground, shores, fields, ditches, borders of marshes.

There is usually a dark blotch in the middle of the leaf, but such a mark is also found at times in other species, e. g., *P. lapathifolium* and *P. hydropiperoides*. If the genus *Persicaria* is recognized, this becomes *P. vulgaris* Web. & Moq.

25. **P. cespitosum** Blume

Map 131. A native of eastern Asia, becoming a weed in North America. First recognized in Michigan in 1978 as a weed in Ann Arbor.

Closely resembles the preceding species and may grow with it. The bristles at the summit of the ocreae are also long (about equaling the sheath, whereas they are shorter in *P. persicaria*). Besides the longer bristles and consistently 3-sided achenes, this smartweed has a deeper pink flower color and the inflorescence tends to be more slender and less dense than in most plants of *P. persicaria*. The latter usually has stronger, more prominent nerves toward the base of the tepals.

Plants such as ours—the common or only form becoming weedy in this country—have been called var. *longisetum* (De Bruyn) Steward. The typical variety has shorter bristles in the inflorescence.

CHENOPODIACEAE — Goosefoot Family

An important character is often whether the fruit is vertical (its longer axis being the axis of the flower) or horizontal (its longer axis at right angles to the axis of the flower). So far as possible, fruit characters are mentioned

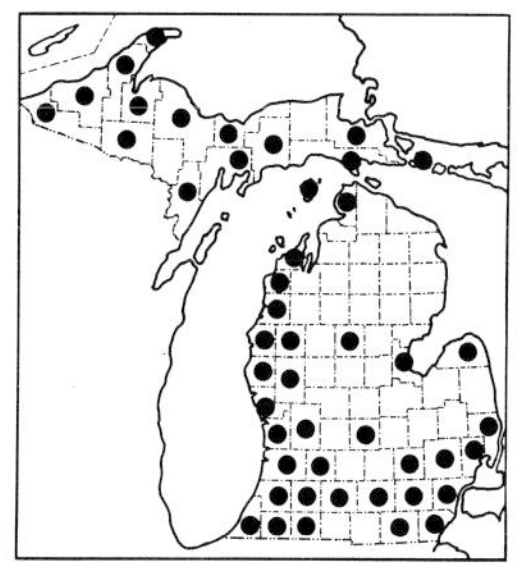

127. Polygonum punctatum

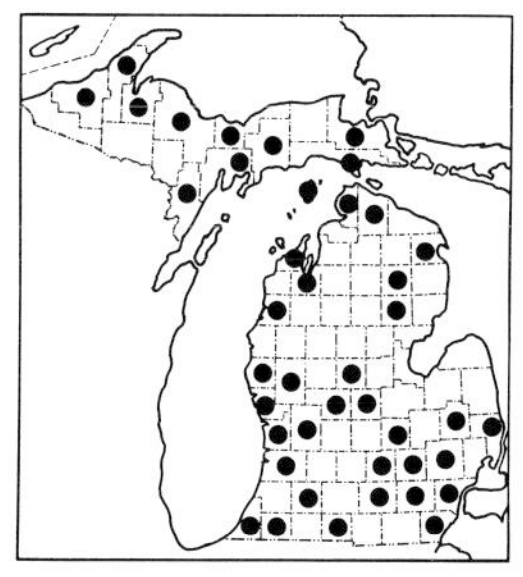

128. Polygonum hydropiper

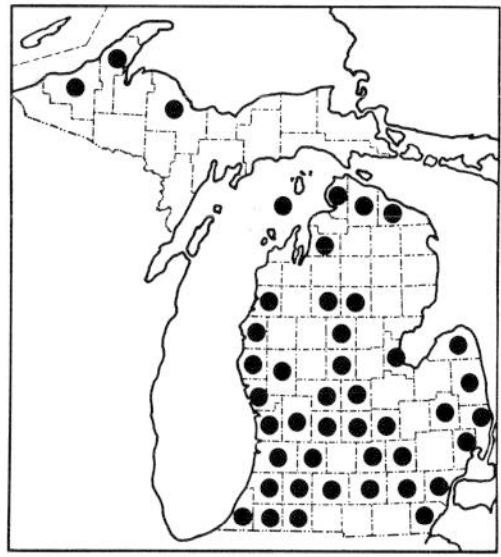

129. Polygonum hydropiperoides

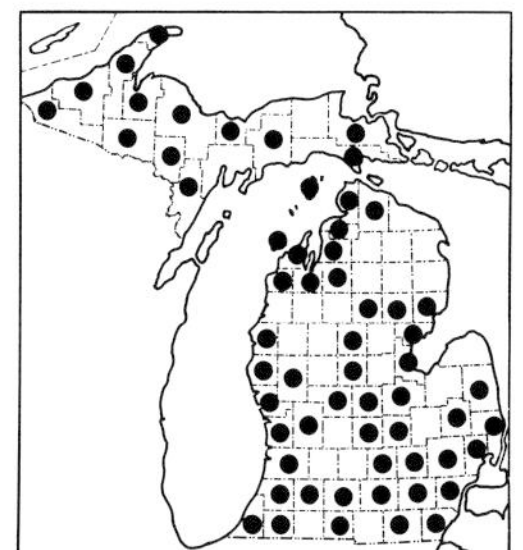

130. Polygonum persicaria

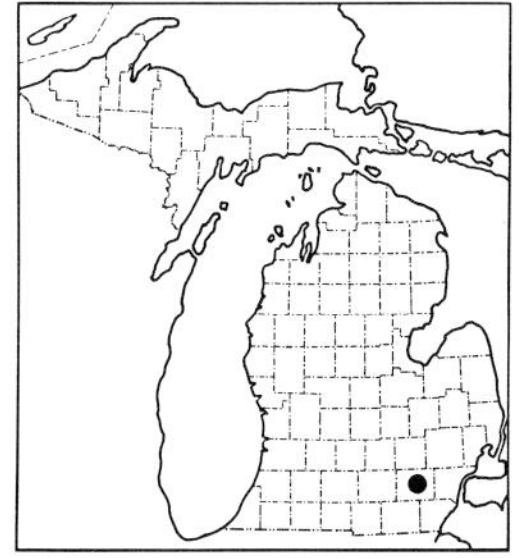

131. Polygonum cespitosum

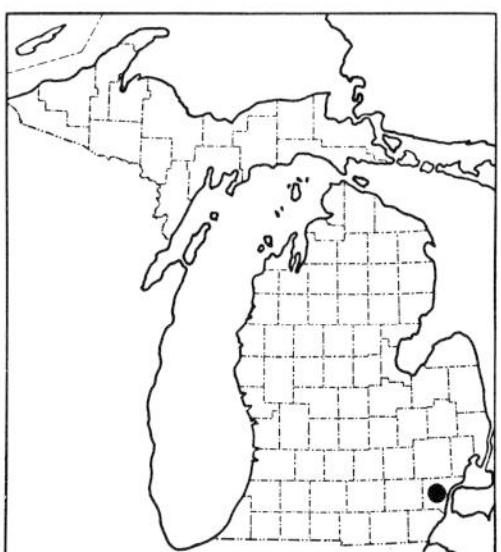

132. Salicornia europaea

secondarily in the keys, since maturity often comes in September and October, and many specimens are immature when collected. A number of species become a rich red-purple in the fall.

KEY TO THE GENERA

1. Leaves opposite, much reduced and scale-like, scarious, connate; stem branches succulent, appearing jointed, the flowers entirely sunk in the fleshy internodes . . . . 1. **Salicornia**
1. Leaves alternate (or the lower sometimes opposite), well developed; stem not succulent, the flowers pediceled or sessile but not embedded in the internodes
  2. Leaf tips with a sharp spine over 0.5 mm (usually ca. 1 mm, even longer on bracts subtending flowers); leaves filiform, ± terete; fruit horizontal, ca. 1–1.3 mm long, slightly broader, covered by the perianth; tepals with transverse keel or wing (fig. 54) sometimes longer than body of tepal . . . . 2. **Salsola**
  2. Leaf tips at most with mucro less than 0.5 mm long; leaves various in width, flat; fruit and perianth various
    3. Flowers unisexual (plants monoecious); fruit in most if not all flowers concealed by a pair of bracteoles (perianth absent) . . . . 3. **Atriplex**
    3. Flowers mostly perfect; fruit not concealed by bracts (perianth, however, may cover it)
      4. Leaves linear to narrowly lanceolate, less than 4 (6) mm broad, entire, 1 (–3)-nerved
        5. Plant with numerous delicate spines ca. 1–3 mm long in inflorescence (sterile pedicels and branch-tips); leaves without spines; fruit all horizontal . . . . 9. **Chenopodium (aristatum)**
        5. Plant without spines in inflorescence (or these less than 0.5 mm long and also on leaves); leaves and fruit various
          6. Inflorescence and leaves beneath farinose; flowers crowded on short branches which exceed their subtending bracts . . . . 9. **Chenopodium (subglabrum)**
          6. Inflorescence and leaves not farinose; flowers 1–3 in the axils of longer bracts
            7. Leaves and bracts green to the tips, not mucronate; bracts long-ciliate, especially basally; fruit horizontal, round, less than 1 mm long, enclosed by the perianth (each tepal with a transverse wing) . . . . 4. **Kochia**
            7. Leaves when intact tipped with a non-green sharp mucro less than 0.5 mm long (no longer, or even absent, on bracts subtending flowers); bracts glabrous to pubescent but not long-ciliate; fruit various
              8. Fruit vertical, flattened, usually narrowly wing-margined, ca. (2) 3–4.5 mm long, greatly exceeding the tiny scarious perianth; plant of sandy (not saline) habitats . . . . 5. **Corispermum**
              8. Fruit horizontal (or mostly so), less than 1 mm long and 1.5 mm broad, enclosed by the perianth; plant of saline habitats . . . . 6. **Suaeda**
      4. Leaves usually at least 4 mm broad, toothed to sinuate or crenulate on the margin (if entire, then pinnate- or triple-nerved and not linear)
        9. Plant from a stout swollen tap root, rarely persisting from cultivation or in waste ground; principal leaves at least 3 cm broad, the margins crenulate and ± crisped . . . . 7. **Beta**
        9. Plant with tap root, if present, no thicker than stem, not cultivated (though some species weedy); principal leaves narrower and/or prominently dentate with coarse teeth (or hastate)

10. Fruit horizontal, completely encircled by the connate wing of the perianth (fig. 52); styles 3 . . . . 8. **Cycloloma**

10. Fruit horizontal or vertical, but the perianth without connate wing; styles usually 2

11. Tepals not transversely winged (may be keeled); leaves and fruit various; bracts not ciliate . . . . 9. **Chenopodium**

11. Tepals with transverse (but separate) wings; leaves entire, not over 5 mm wide; fruit horizontal; bracts long-ciliate, especially basally . . . . 4. **Kochia**

1. Salicornia

1. **S. europaea** L. Fig. 51 Glasswort

Map 132. Plants at inland stations in North America are dubiously native there—apart from the broad sense in which this name has been applied to North American populations. Collected several times 1914–1930 in a salt marsh west of Woodmere Cemetery, Detroit (Rep. Mich. Acad. 19: 219. 1918. Rhodora 18: 243–244. 1916). Recollected in the state in 1984 in a saline expressway median (*Reznicek 7456,* MICH, MSC, GH).

2. Salsola

1. **S. kali** L. Fig. 54 Russian-thistle

Map 133. A Eurasian native (at least our variety), found on railroads, roadsides, dunes, and other sandy or cindery places. It was probably introduced independently at more than one point in the state, for it first appeared in Detroit in 1891; was collected at Caro, Charlevoix, and Grand Rapids in 1894; in 1895 it was collected at Muskegon and Port Huron—where within three years it became an abundant weed; in 1896 it was found in Iosco County—and it is now throughout the state.

The stems and branches are narrowly striped, usually in pink or reddish. This is a spiny, obnoxious weed. Our material is usually referred to *var. tenuifolia* Tausch or ssp. *ruthenica* (Iljin) Soó.

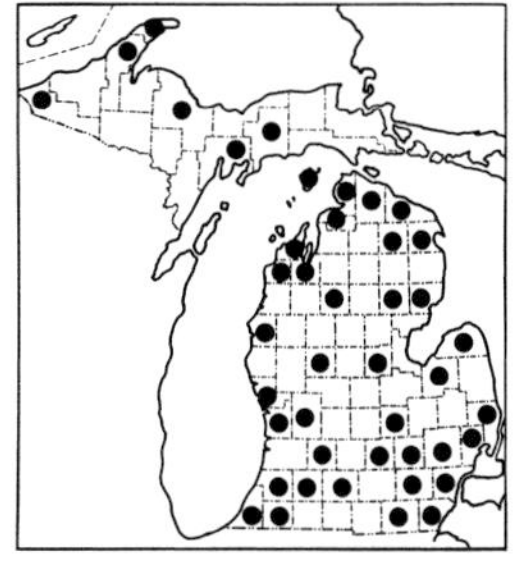

133. Salsola kali

134. Atriplex rosea

135. Atriplex hortensis

50. *Polygonum lapathifolium* $\times^2/_5$; fruiting calyx $\times 10$
51. *Salicornia europaea* $\times^1/_2$; flowering branch tip $\times 2$
52. *Cycloloma atriplicifolium* $\times^1/_2$; fruiting calyces $\times 1$, $\times 5$

3. **Atriplex** Orache

REFERENCES

Bassett, I. J., C. W. Crompton, J. McNeill, & P. M. Taschereau. 1983. The Genus Atriplex (Chenopodiaceae) in Canada. Agr. Canada Monogr. 31. 72 pp.

Taschereau, P. M. 1972. Taxonomy and Distribution of Atriplex Species in Nova Scotia. Canad. Jour. Bot. 50: 1571–1594.

KEY TO THE SPECIES

1. Bracteoles at maturity indurated, their margins fused ca. one-third their length; leaves with conspicuous fine network of dark green veinlets (seen when scraped lightly with a sharp blade) . . . . 1. **A. rosea**
1. Bracteoles thin at maturity, their margins fused less than a third of their length, if at all; leaves with normal, inconspicuous venation
   2. Bracteoles obtuse to rounded at apex, neither toothed nor tuberculate, the fruit appearing central; pistillate flowers of 2 kinds, most with bracteoles, vertical fruits, and no perianth, the others without bracteoles, the fruit horizontal, and with perianth . . . . 2. **A. hortensis**
   2. Bracteoles acute, usually toothed and/or tuberculate, the fruit nearly basal; pistillate flowers all alike, with bracteoles, vertical seeds (but these dimorphic: larger and pebbled brown or smaller and glossy black), and no perianth . . . . 3. **A. patula**

1. **A. rosea** L. Saltbush

Map 134. An Old World native, rarely a waif in our area but commonly established in western North America. Collected by Farwell along a railroad in Lapeer County in 1923, and in 1930 at River Rouge and Detroit.

Often misidentified as *A. argentea* Nutt., which has entire rather than sinuate leaves, bracteoles united about half their length, and the foliage densely covered with silvery-scurfy scales.

2. **A. hortensis** L. Garden Orache

Map 135. Grown as a potherb and very rarely collected as an escape from cultivation; native of Asia. Farwell gathered it on waste ground at Detroit in 1893 and 1912; and Mary B. Fallass, near a railroad at Grand Rapids in 1904.

3. **A. patula** L. Fig. 53 Spearscale

Map 136. Probably introduced in North America. A weed of roadsides, railroads, gardens, and waste ground generally, thriving where salting of streets and highways has killed the competition; also in marshy ground.

The characters used to distinguish taxa in this section of the genus do not correlate well in our material, which needs further study. As interpreted by Taschereau, true *A. patula* is a tetraploid species of variable leaf shape but the blades generally ± cuneate at the base, including narrow-leaved inland plants often erroneously referred to *A. littoralis* L., a strictly maritime diploid species. Mature brown seeds are 2.5 mm or more broad (in our material, ca. 2.3–2.7 mm). Taschereau separated *A. prostrata* DC. [*A. triangularis* Willd.] as a diploid species, always with broadly triangular-hastate leaves and with mature brown seeds less than 2.5 mm broad. It may grow with *A. patula* but apparently does not hybridize with it in nature. The name *A. hastata* has often been misapplied to this species. Only a few collections from Bay County (growing with *A. patula* sens. str.) appear to be *A. prostrata;* they have an erect central stem, opposite leaves on most of the plant, small seeds, and hastate blades. However, several collections (e. g., from Barry, Midland, and St. Clair counties) have hastate blades and large seeds (2.5–2.7 mm), as well as bracteoles with a thin spongy tissue inside, a character of *A. prostrata*. Some Kalamazoo County material with small seeds and hastate leaves lacks spongy tissue on the bracteoles. On the other hand, a Monroe County collection with narrow leaves has small seeds (1.7–2 mm). Immature specimens lacking fruit cannot in any event be placed confidently in anything but a broad *A. patula* complex.

4. **Kochia**

1. **K. scoparia** (L.) Schrader — Summer-cypress

Map 137. An Asian species, adventive especially along railroads and highways where salt applied in winter favors such plants of saline habitats. A very narrow-leaved form is frequently cultivated for its ornamental foliage which turns red in the fall; it occasionally escapes, and was first collected in Michigan in 1917. The typical form was not found in the state until even later.

5. **Corispermum** — Bugseed

This is a complex Eurasian genus and there is some doubt as to whether any taxa are native in North America. Characters used to distinguish the species are not all agreed upon and it is not clear what our specimens should

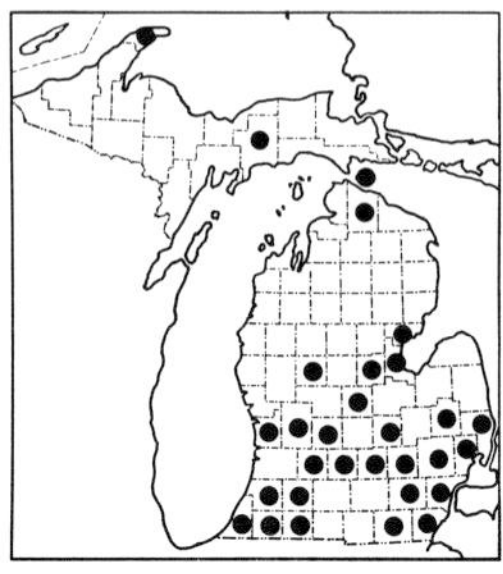
136. Atriplex patula

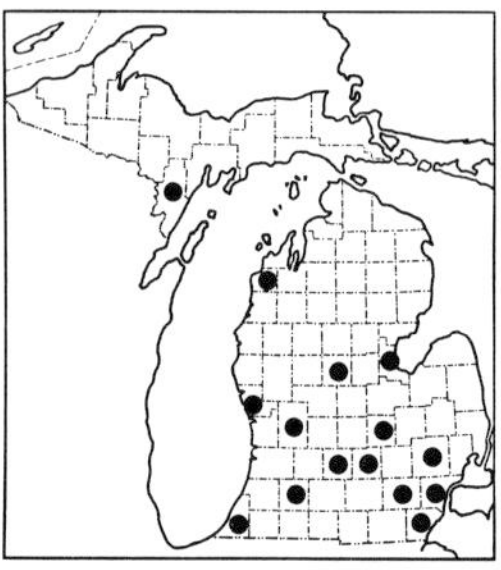
137. Kochia scoparia

138. Corispermum orientale

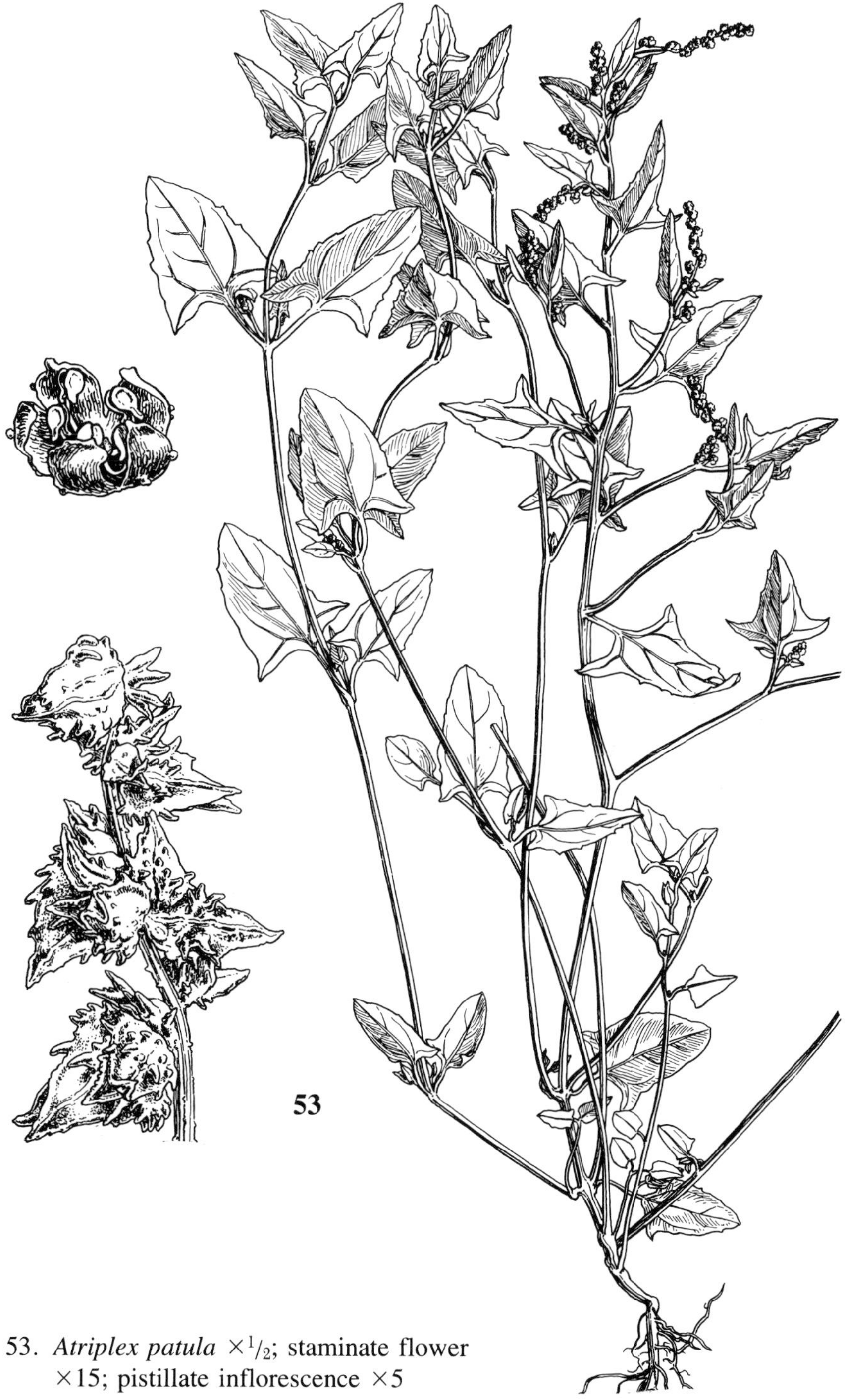

53. *Atriplex patula* $\times{}^{1}/_{2}$; staminate flower $\times 15$; pistillate inflorescence $\times 5$

be called. The winged character of the fruit is often used to separate taxa, and one must try not to confuse the true thin translucent wing with the pale border of the body of the achene. Some achenes in an inflorescence may be more strongly winged than others.

REFERENCE

Maihle, Nita J., & Will H. Blackwell, Jr. 1978. A Synopsis of North American Corispermum (Chenopodiaceae). Sida 7: 382–391.

KEY TO THE SPECIES

1. Fruit wingless, ca. 2–3 mm long . . . . 1. **C. orientale**
1. Fruit ± winged, ca. (3) 3.5–4.5 mm long . . . . 2. **C. hyssopifolium**

1. **C. orientale** Lam.

Map 138. Possibly adventive from western North America, collected on the shore of Lake Superior at Marquette in 1923 (*Sherff,* F).

2. **C. hyssopifolium** L. Fig. 55

Map 139. Not known from as far east as the shores of the Great Lakes until the 1860's and hence quite possibly not indigenous in our area even if so elsewhere on the continent. Here, occurs on dunes and sandy beaches of the Great Lakes; rarely inland on sand banks or in railroad yards and gravel pits.

This name is used here in a broad sense, pending the detailed attention which the taxonomy of the genus needs in North America. Plants referred to *C. nitidum* Schultes seem, first of all, not clearly distinguished from those referred to *C. hyssopifolium* and, furthermore, they do not match Old World material of *C. nitidum,* which has a very slender inflorescence (ca. 3–4 mm thick), the bracts remote (revealing almost the entire axis), the fruit less than

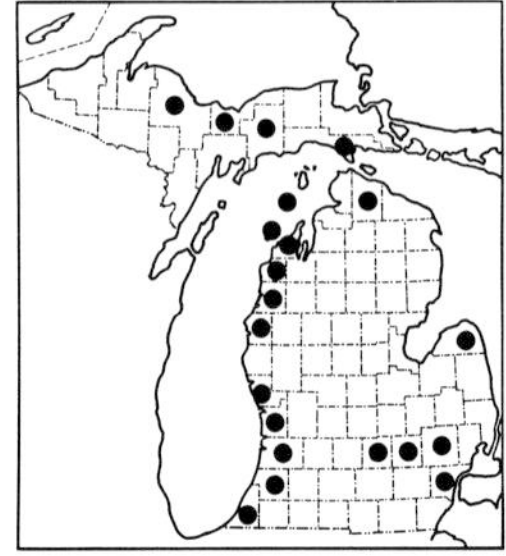

139. Corispermum hyssopifolium

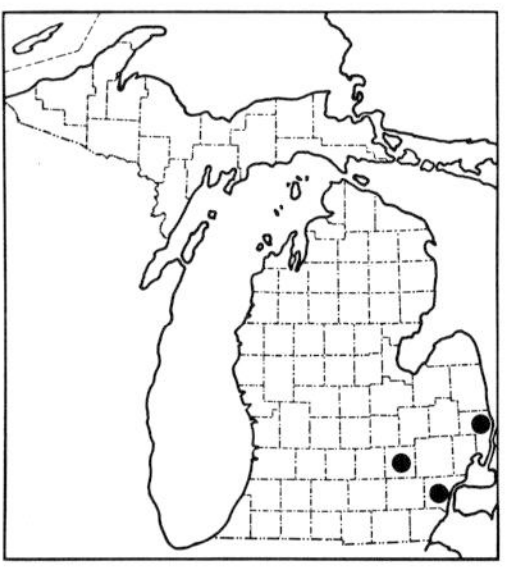

140. Suaeda calceoliformis

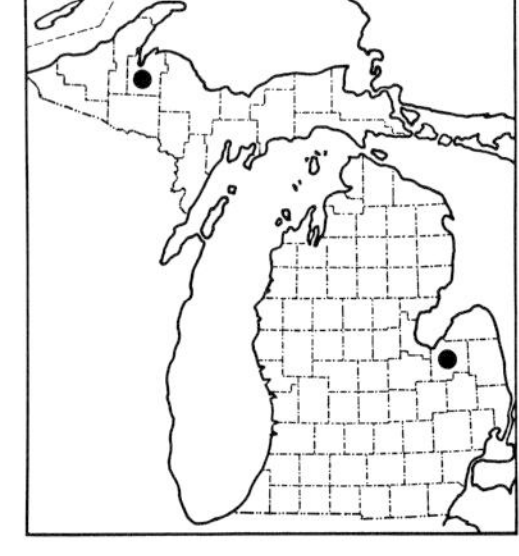

141. Beta vulgaris

3.5 mm long and nearly wingless, and the leaves filiform (strongly involute). *C. hyssopifolium* sensu stricto, as treated in *Flora Europaea* (1964), is a plant restricted to eastern Europe, with achenes somewhat larger than in our plants, the wing very narrow or absent; achenes of our plants vary from nearly wingless to conspicuously winged (the wing often grading into the pale margin) and usually have dense spikes ca. 5–8 mm thick. These plants may be what is treated as *C. leptopterum* (Asch.) Iljin in *Flora Europaea*—a name antedated by *C. americanum* (Nutt.) Nutt., which may apply to the same species if it is truly distinct from *C. hyssopifolium*.

6. **Suaeda**

REFERENCES

Bassett, I. J., & C. W. Crompton. 1978. The Genus Suaeda (Chenopodiaceae) in Canada. Canad. Jour. Bot. 56: 581–591.

Hopkins, Christine O., & Will H. Blackwell, Jr. 1977. Synopsis of Suaeda (Chenopodiaceae) in North America. Sida 7: 147–173.

1. **S. calceoliformis** (Hooker) Moq. Sea-blite

Map 140. A native of western North America, adventive eastward and discovered in 1978 in a saline expressway median in Wayne County (*Reznicek 4912 & 4913,* MICH, MSC, GH; see Mich. Bot. 19: 23–30. 1980). Later found elsewhere in similar situations.

This species has long been known as *S. depressa,* a name which applies to a different (Eurasian) species.

7. **Beta**

1. **B. vulgaris** L. Beet

Map 141. An Old World native, widely cultivated but hardly established. Both sugar beets, a major crop in Michigan, and red beets are forms of this species and have rarely been found in waste ground or persisting a few years after cultivation. Such occurrences are probably more frequent than the map suggests. Farwell's report of adventive beets in Detroit was based on a collection of *Amaranthus* sp. Swiss chard is another form of *B. vulgaris*.

8. **Cycloloma**

1. **C. atriplicifolium** (Sprengel) Coulter Fig. 52 Winged Pigweed

Map 142. Probably adventive in Michigan from farther west. Not collected in the state before 1895 (at Grand Rapids); most collections have been made in the last 50 years. Sandy disturbed ground, including dunes and

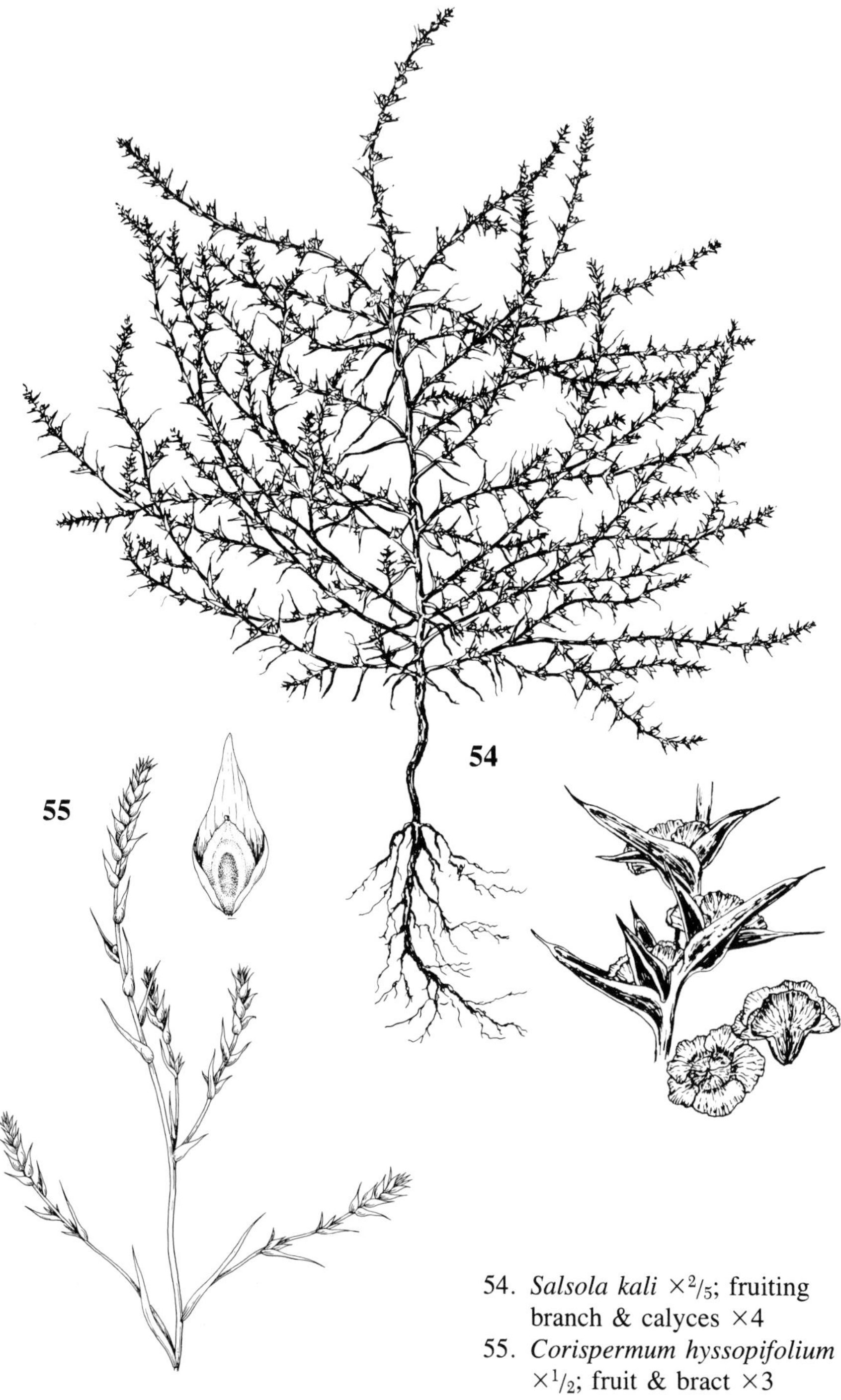

54. *Salsola kali* $\times^2/_5$; fruiting branch & calyces $\times 4$
55. *Corispermum hyssopifolium* $\times^1/_2$; fruit & bract $\times 3$

shores, newly made roadsides, dumps and gravel pits, railroad ballast, river banks; rarely an agricultural weed.

Very distinctive in its sessile, flat, saucer-like or wheel-like winged fruits 3–4 (5) mm broad and its sinuate almost holly-like leaves with sharp-tipped lobes. Young stems and leaves are ± loosely tomentose (sometimes glabrate). In the fall (September–October), the plant, especially the fruit, may become a rich purple-red, as in some other members of the family. The whole mature much-branched, dome-shaped plant is a fine tumbleweed.

9. **Chenopodium** Goosefoot

Statements about leaves refer to well developed foliage leaves, not necessarily the smaller, often less toothed bracts in the inflorescences. Statements about seeds, especially measurements, refer to fully mature ones.

REFERENCES

Bassett, I. J., & C. W. Crompton. 1978. The Biology of Canadian Weeds. 32. Chenopodium album L. Canad. Jour. Pl. Sci. 58: 1061–1072.

Bassett, I. J., & C. W. Crompton. 1982. The Genus Chenopodium in Canada. Canad. Jour. Bot. 60: 586–610.

Wahl, Herbert A. 1954. A Preliminary Study of the Genus Chenopodium in North America. Bartonia 27: 1–46.

KEY TO THE SPECIES

1. Leaves with orange resinous glands or gland-tipped hairs at least beneath, not farinose; bruised plant strongly aromatic
  2. Leaf blades with sessile glands, glabrous; stem (except youngest part) and perianth glabrous or glabrate, glandular or not . . . . 1. **C. ambrosioides**
  2. Leaf blades, stem, and perianth ± copiously covered with short spreading gland-tipped hairs, especially when young . . . . 2. **C. botrys**
1. Leaves (and rest of plant) neither glandular nor pubescent, but farinose in some species
  3. Leaves linear to linear-lanceolate, less than 4 (5) mm wide, 1 (–3)-nerved, entire
    4. Plant bushy-branched, at least the upper branches ending in subulate, delicate, spine-like points; flowers solitary, not farinose, in axils of branches . . 3. **C. aristatum**
    4. Plant erect in aspect, without spine-like branches; flowers in small clusters, at least lightly farinose . . . . 4. **C. subglabrum**
  3. Leaves usually broader, toothed or with irregular margin (if entire, then blades pinnate- or triple-nerved and not linear)
    5. Plants ± densely farinose on underside of leaves
      6. Perianth not farinose; blades of principal leaves mostly 1–2.5 (3) cm long, with margins sinuate and toothed at ± regular intervals; pericarp free from seed; fruit in some flowers often vertical . . . . 5. **C. glaucum**
      6. Perianth farinose; blades of principal leaves often longer, or entire to irregularly toothed, or both; pericarp closely adherent to seed; fruit in all flowers horizontal
        7. Principal leaves with blades broadly ovate to deltoid, less than 1.5 times as long as broad, entire; plant with fetid odor (like rotten fish) when bruised. . . . . 6. **C. vulvaria**

7. Principal leaves ovate and irregularly toothed to lanceolate and entire; plant not fetid .... 14. **C. album**

5. Plants with leaves not farinose (except sometimes when very young)

8. Fruit all or almost all vertical; tepals 3 or erose-toothed

9. Tepals usually 5, erose-toothed; styles conspicuous, stout, becoming 1 (1.2) mm long; inflorescence not becoming fleshy; leaves triangular-hastate, not toothed .... 7. **C. bonus-henricus**

9. Tepals usually 3, entire; styles less than 0.5 mm long; inflorescence in common species becoming fleshy; leaves various

10. Leaf blades broadly elliptical to rounded, ovate, or obovate, entire to sinuate; inflorescences (small dry heads less than 6 mm across) not becoming red and fleshy .... 8. **C. rubrum**

10. Leaf blades triangular-hastate and usually also coarsely dentate; inflorescences (often 1–2 cm across when fully developed) becoming bright red, fleshy and berry-like by coalescence of the perianths in each head .... 9. **C. capitatum**

8. Fruit all or almost all horizontal; tepals 5, entire

11. Largest leaf blades (2) 4–13 (15) cm long, truncate to subcordate at base when well developed, with 1–3 large acute teeth separated by broadly rounded sinuses on each side; seed 1.6–1.8 mm broad .... 10. **C. hybridum**

11. Largest leaf blades often shorter, narrowly to broadly cuneate at base, with teeth more numerous, obscure, or none; seed 1–1.5 (1.7) mm broad

12. Tepals scarcely if at all keeled and very slightly if at all farinose; mature fruit 1–1.5 mm broad, largely exposed (tepals too short to cover it)

13. Fruit ca. 1.3–1.5 mm broad; pericarp easily separable; lower leaf blades acute at base .... 11. **C. standleyanum**

13. Fruit ca. 1–1.1 mm broad; pericarp slightly adherent to seed; lower leaf blades truncate at base .... 12. **C. urbicum**

12. Tepals slightly to strongly keeled, moderately to densely farinose; fruit 1.2–1.5 (1.7) mm broad, completely covered by perianth (or sometimes slightly exposed in *C. murale*)

14. Styles very short (up to 0.2 mm) and stout; margin of seed ± sharply angled (fig. 58); principal leaf blades less than 1.5 times as long as broad, coarsely toothed .... 13. **C. murale**

14. Styles when intact usually longer, slender; margin of seed usually rounded; principal leaf blades usually at least 1.5 times as long as broad, entire (and lanceolate) to variously toothed .... 14. **C. album**

1. **C. ambrosioides** L. Wormseed; Mexican-tea

Map 143. Railroads, alleys, and waste ground, only very rarely collected in the state during the past half-century. Native of tropical America and grown as a vermifuge, collected as long ago in Michigan as the First Survey (Kalamazoo County, 1838).

The strong aroma of this plant may persist for at least 50 years in dry specimens. The leaves may be essentially entire or coarsely toothed.

2. **C. botrys** L. Jerusalem-oak

Map 144. A Eurasian native, found in sand and cinders along roads and railroads, in gravel pits, and on other disturbed sites.

3. **C. aristatum** L. Fig. 56

Map 145. Originally described from Siberia, and a weed from there across China to Korea and Japan. The species was found in Michigan in 1981 (see Mich. Bot. 23: 53–55. 1984)—the second verified occurrence in North America. It is now apparently an abundant annual weed around Alpena, turning beet-red in the fall. It was possibly introduced by foreign shipping at the port of Alpena.

4. **C. subglabrum** (S. Watson) A. Nelson

Map 146. Undoubtedly adventive in Michigan from farther west. Dry ground, railroads, bluffs, roadsides.

Referred here are lightly farinose plants with very narrow 1-nerved leaves and pericarp readily separable from the seed, which is well over 1 mm in diameter. Sometimes these have been included in a broad concept of the densely farinose *C. leptophyllum* (Moq.) S. Watson, which differs also in having an adherent pericarp and smaller seeds, according to Wahl (1954) and Crawford (Brittonia 27: 282. 1975). Young material lacking fruit from Bay and Kalamazoo counties with leaves 3-nerved at the base may be *C. desiccatum* A. Nelson [including *C. pratericola* Rydb.], which has small seeds and a non-adherent pericarp. One fruiting collection from Berrien County was referred to *C. pratericola* by Wahl.

5. **C. glaucum** L.

Map 147. Probably adventive in Michigan, perhaps in North America. Gravelly shores, dumps, roadsides, railroad tracks, river banks, and other naturally or artificially disturbed ground.

Most of our collections are var. *glaucum,* with rather obscure rounded or obtuse teeth on the leaves; some material is apparently var. *salinum* (Standley) Boivin, with distinct acute teeth with ± involute margins. Seeds of the latter variety, often recognized as a distinct species, are supposedly larger and the inflorescence more bracteate, but these characters are not clear in

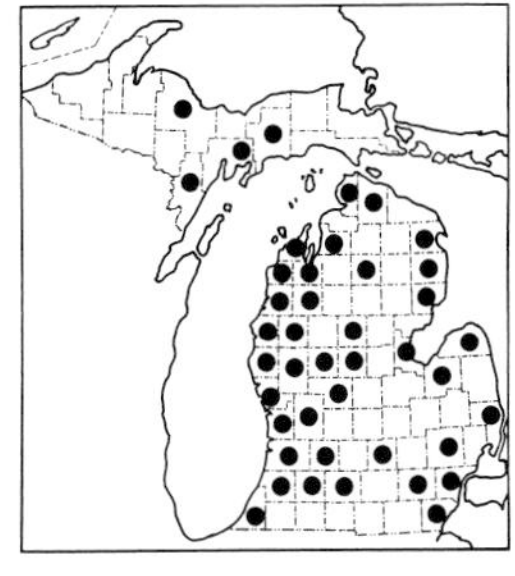

142. Cycloloma atriplicifolium

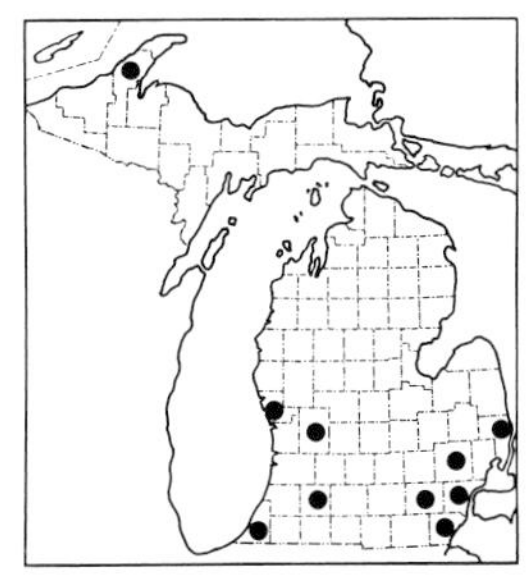

143. Chenopodium ambrosioides

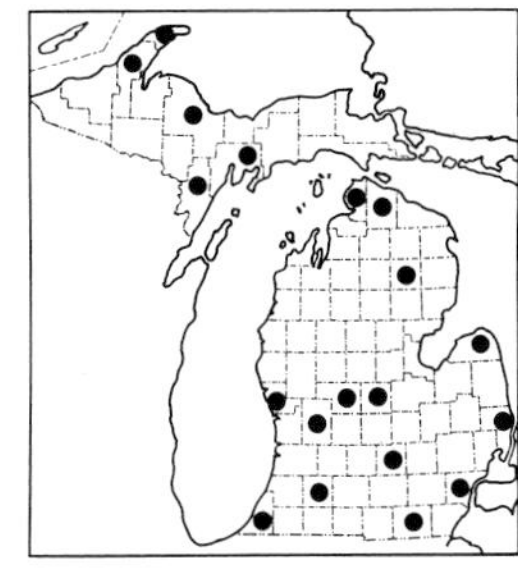

144. Chenopodium botrys

our limited material. This species is often confused with others if one does not remember the distinctive combination of leaves (which tend to be relatively small) farinose beneath but perianth *not* farinose.

6. **C. vulvaria** L. Stinking Goosefoot

Map 148. A native of Europe, collected by Farwell as a weed in Detroit in 1893 (*1432*, BLH, MSC; Rep. Mich. Acad. 20: 174. 1919).

The leaves are small on the Michigan specimen (blades less than 1 cm broad) and have retained their fetid odor when moistened and crushed.

7. **C. bonus-henricus** L. Good-King-Henry

Map 149. A locally occurring introduction from Europe, sometimes grown as a potherb and spreading to waste ground. Not collected in Michigan since 1937 (Houghton County, shores of Torch Lake, misidentified by Farwell as spinach).

The name is originally of German origin, without the "King." The small heads of flowers are ± closely crowded in a terminal spike, while in the next species the heads are more distinctly axillary or in axillary spikes.

8. **C. rubrum** L.

Map 150. Waste ground, shores, and river banks. Considered native in North America, but like other of our weedy species, perhaps not so in Michigan. I have seen no collections before 1894 (Bay City).

Our only collection of var. *humile* (Hooker) S. Watson, a prostrate form with smaller leaves (less than 2.5 cm) and heads, is from waste ground at Ann Arbor. In var. *humile* the seeds are 0.8–1 mm broad and the leaves entire or nearly so, while in the generally large and erect var. *rubrum* the seeds are only 0.6–0.8 mm broad and the leaves coarsely toothed or sinuate.

9. **C. capitatum** (L.) Asch. Plate 1-E Strawberry Blite

Map 151. Characteristic, along with *Corydalis aurea, C. sempervirens,* and *Leucophysalis grandiflora,* of sandy or gravelly (often very calcareous)

145. Chenopodium aristatum

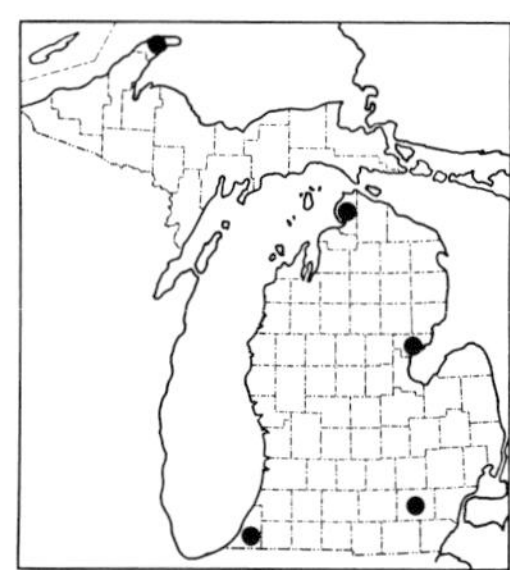

146. Chenopodium subglabrum

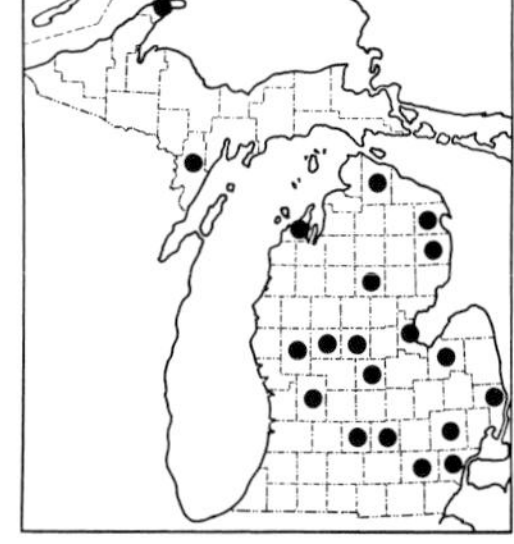

147. Chenopodium glaucum

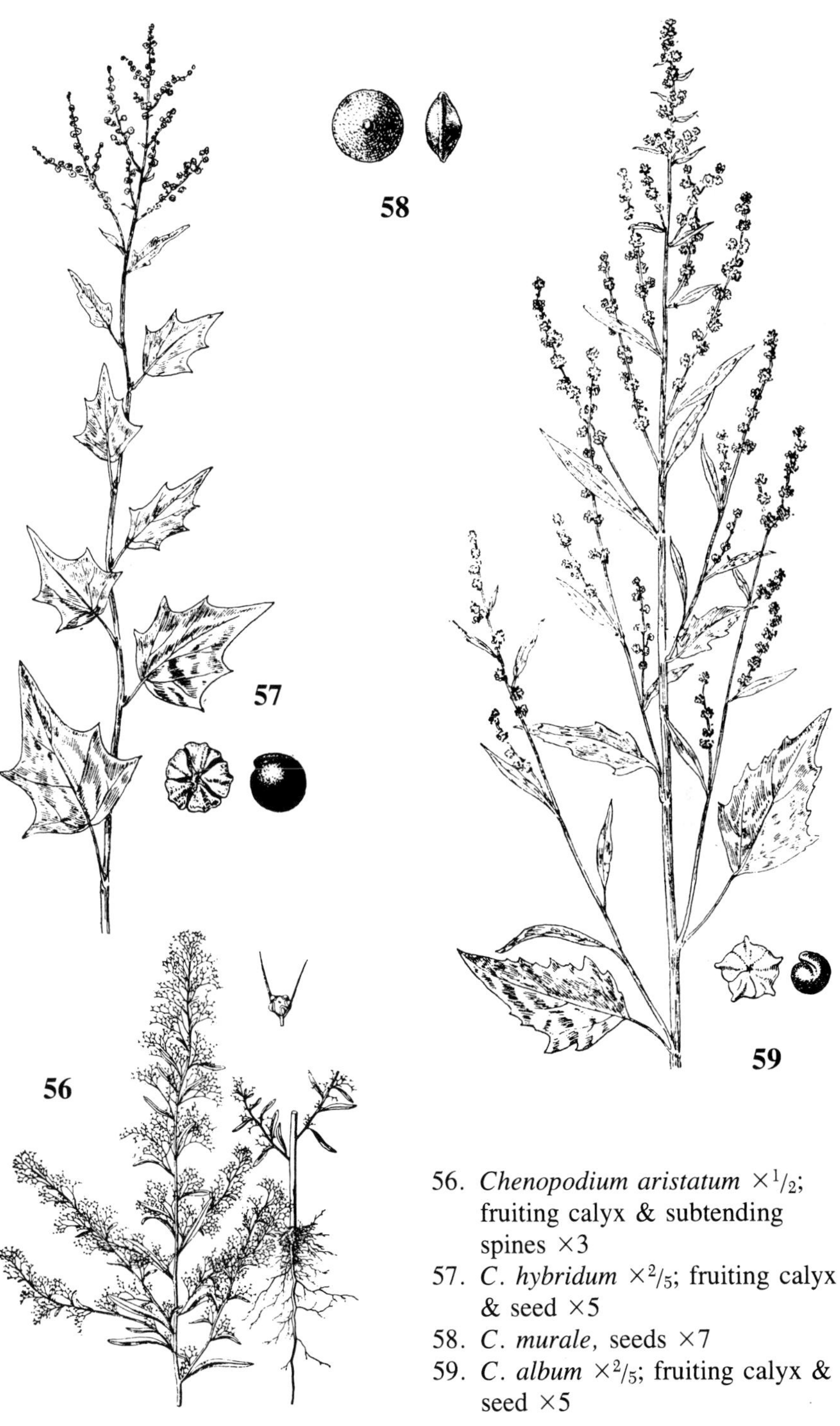

56. *Chenopodium aristatum* ×1/2; fruiting calyx & subtending spines ×3
57. *C. hybridum* ×2/5; fruiting calyx & seed ×5
58. *C. murale,* seeds ×7
59. *C. album* ×2/5; fruiting calyx & seed ×5

sites one or two years after disturbance (e. g., bulldozing) and dying out (if disturbance is not maintained) within another 2 or 3 years. Besides growing in dumps, gravel pits, clearings, and newly graded roadsides (all basically in wooded areas), strawberry blite is occasionally found on gravelly shores, limestone ledges, burned-over ridges, and other naturally disturbed raw sites.

Plants can be abundant and when in fruit are a striking sight, as brightly handsome as a poinsettia. Branches of the plant shown on the color plate were as long as 1 m. This conspicuous species often attracts attention along roadsides and such places, although it is ignored by most wildflower books. The strawberry-like fruiting heads are edible (at least they are nourishing and harmless), either raw or cooked. Sometimes maintained in a separate genus from *Chenopodium,* as *Blitum capitatum* L.

10. **C. hybridum** L. Fig. 57

Map 152. Not as weedy in habit as many species of the genus, but like others, grows along new roadsides and clearings in woods, on shores and river banks, in farms and gardens, and other ± disturbed places.

American plants have been separated from the European as var. *gigantospermum* (Aellen) Rouleau or *C. gigantospermum* Aellen. The pericarp varies, being ± easily separable from the seed or closely adherent. The mature fruit is largely exposed, as the tepals do not cover it above and their margins are well separated on the sides.

11. **C. standleyanum** Aellen

Map 153. A very few Michigan collections appear to be this supposedly widespread native species; all are without habitat data but are presumably from disturbed ground.

12. **C. urbicum** L.

Map 154. A native of Europe. Only three Michigan collections have been seen (1880–1920), presumably all from waste ground although only one (from Alma) is explicitly so labeled.

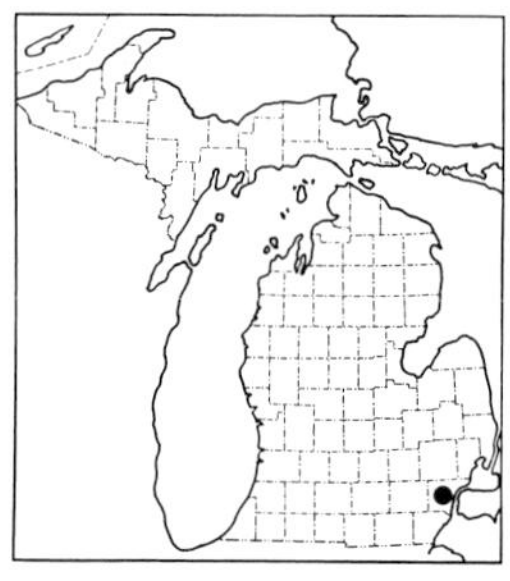
148. Chenopodium vulvaria

149. Chenopodium bonus-henricus

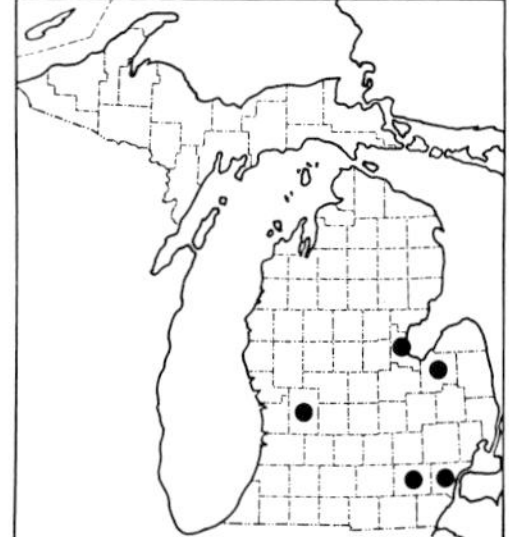
150. Chenopodium rubrum

Frequently confused with the next species, but differs (besides the characters in the key) in rounded margins of the smaller seeds (less than 1.2 mm broad). In *C. murale,* the seeds are angled on the margin and 1.3–1.5 mm broad. The stigmas are extremely short in both.

13. **C. murale** L. Fig. 58

Map 155. Like the preceding, to which it is closely related, a native of Europe. Collected by Mr. and Mrs. C. R. Hanes as a garden weed at Schoolcraft in 1935 and 1937. The species persisted there for at least a decade.

14. **C. album** L. Fig. 59 Lambs-quarters; "Pigweed"

Map 156. A cosmopolitan weed, probably at least in part native in North America and known in Michigan ("dry roadside") since the First Survey (1838); now in all kinds of waste places and disturbed sites, spreading into woods and marshy places.

An extremely variable species in leaf shape, amount and distribution of farinosity, perianth and fruit characters (pericarp variously colored and marked, seed rarely angled as in *C. murale,* style length variable). Many segregates have been proposed, and treatments are not in agreement on their disposition. Most Michigan specimens referred in the past to the principal segregates seem closer to the basic *C. album* (including *C. lanceolatum* with

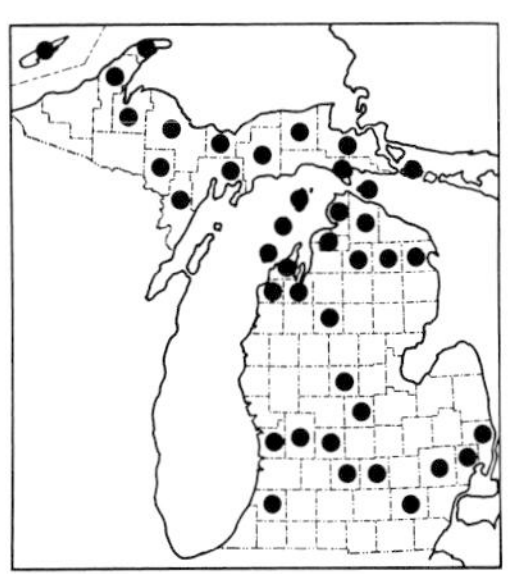

151. Chenopodium capitatum

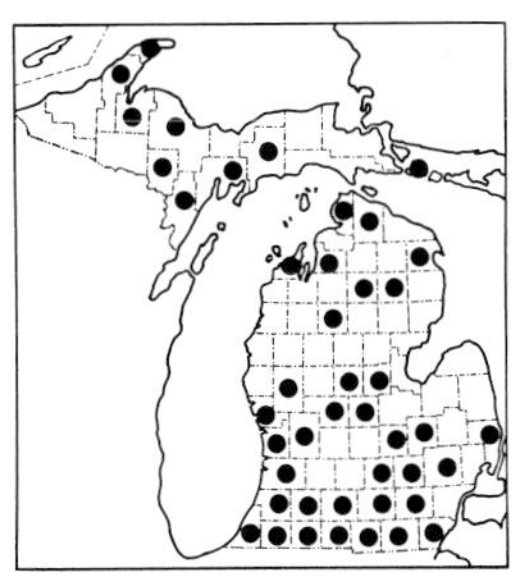

152. Chenopodium hybridum

153. Chenopodium standleyanum

154. Chenopodium urbicum

155. Chenopodium murale

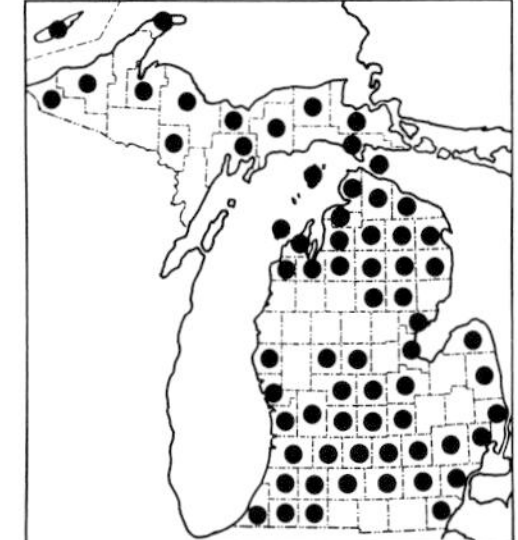

156. Chenopodium album

lanceolate often entire leaves and whitish-margined tepals). However, one collection from Livingston County could be referred to *C. berlandieri* Moq. on the basis of its very well developed keels (fully half as wide as the tepal) and distinctly pitted seed as well as pericarp. A few collections, mostly made around the turn of the century, have a large fruit (1.6–1.8 mm) with a very reticulate or honeycombed pericarp and unusually large leaves; these would be *C. bushianum* Aellen [*C. paganum* Reichb. of many manuals]. In a very few collections, the pericarp is loose and easily separable from the seed, but the specimens do not share the other characters of any species which normally has a loose pericarp. And of course many collections lack mature fruit and cannot safely be assigned except to a broad *C. album* complex.

While all of the species in the genus, except the first two with glandular parts, can be used as tasty potherbs, the abundant weed *C. album* is the one most frequently used for this purpose. The tender shoots and leaves can be cooked like spinach, to which they bear some resemblance. The seeds can be ground to obtain a flour and were used by various Indian tribes across North America, including Michigan.

AMARANTHACEAE — Amaranth Family

REFERENCES

Robertson, Kenneth R. 1981. The Genera of Amaranthaceae in the Southeastern United States. Jour. Arnold Arb. 62: 267–313.

Sauer, Jonathan, & Robert Davidson. 1962 ["1961"]. Preliminary Reports on the Flora of Wisconsin. No. 45. Amaranthaceae—Amaranth Family. Trans. Wisconsin Acad. 50: 75–87.

KEY TO THE GENERA

1. Leaves opposite, linear-lanceolate, less than 10 mm wide; stem, leaves (both surfaces), and perianth with white silky-woolly hairs; flowers perfect; stamens with the filaments united into an elongate membranous tube (with points prolonged between the anthers and exceeding them) . . . . 1. **Froelichia**
1. Leaves alternate, in most species usually more than 10 mm wide; stem, leaves, and perianth glabrous to somewhat pubescent but not white silky-woolly; flowers unisexual; stamens separate . . . . 2. **Amaranthus**

1. **Froelichia**

REFERENCE

Blake, S. F. 1956. Froelichia gracilis in Maryland. Rhodora 58: 35–38.

1. **F. gracilis** (Hooker) Moq. Cottonweed

Map 157. Generally considered native west of the Mississippi, this species has been spreading eastward, almost entirely along railroads. Blake (1956) discussed its spread but was unaware of the first Michigan collection, from a railroad in St. Joseph County in 1951 (*Rogers 7786*, WUD). In 1973, found to be abundant in sand at New Buffalo, Berrien County (*Schulenberg, J. & K. Kohout 73-418*, MOR).

2. **Amaranthus** Pigweed; Amaranth

The species in this genus are generally plants of weedy, disturbed habitats, and there is much doubt as to their "original" native range. All except *A. retroflexus, A. tuberculatus,* and possibly *A. albus* have presumably spread into Michigan from farther south or west (or from cultivation) in historic time. Several species have long been cultivated as grain crops, as ornamentals, or as potherbs.

Although Michigan collections of them have not been found, several additional species have been reported from the state, two of which, at least, might be expected in the southern Lower Peninsula: *A. spinosus* L. is readily recognized by the presence of sharp rigid spines in the axils of the leaves. *A. tamariscinus* Nutt. somewhat resembles *A. arenicola*; pistillate plants differ in having only 1 or 2 tepals (and these ± lanceolate), while staminate plants differ in the spiny-tipped tepals.

In most species, the fruit (a thin-walled, one-seeded structure called a *utricle*) is circumscissile—that is, it dehisces neatly at maturity along a circular belt (which is easily visible even on immature material), the top falling away like a lid (fig. 62). Sometimes younger specimens of monoecious species may falsely appear entirely staminate, or older ones appear falsely pistillate, and care must be taken not to assume that such a plant is really one of the truly dioecious species with characteristic slender elongate spikes. The key is based largely on specimens annotated 1952–1959 by Jonathan D. Sauer, and his nomenclature (updated) is followed.

REFERENCES

(National Research Council). 1984. Amaranth—Modern Prospects for an Ancient Crop. (BOSTID Report 47.) Natl. Acad. Press, Washington. 80 pp.

Sauer, Jonathan. 1955. Revision of the Dioecious Amaranths. Madroño 13: 5–46.

Sauer, Jonathan D. 1967. The Grain Amaranths and Their Relatives: A Revised Taxonomic and Geographic Survey. Ann. Missouri Bot. Gard. 54: 103–137.

Weaver, Susan E., & Edward L. McWilliams. 1980. The Biology of Canadian Weeds. 44. Amaranthus retroflexus L., A. powellii S. Wats. and A. hybridus L. Canad. Jour. Pl. Sci. 60: 1215–1234.

KEY TO THE SPECIES

1. Flowers in small axillary clusters; plants monoecious; utricles circumscissile
 2. Tepals (4–) 5, scarcely if at all exceeded by the bracts; seeds ca. 1.5 mm in diameter; stem and branches ± prostrate .... 1. **A. blitoides**
 2. Tepals 3, much surpassed by the slender bracts; seeds ca. 1 mm or slightly less in diameter; stem ± erect or ascending, stiffly branched .... 2. **A. albus**
1. Flowers in elongate simple or paniculate, chiefly terminal spikes; plants and utricles various (but dioecious with utricle not circumscissile in *A. tuberculatus,* of which some leafy-bracted pistillate plants might be said to have axillary clusters)
 3. Stem densely pubescent around the inflorescence; leaves with petioles and often also midrib beneath pubescent .... [go to couplet 8]
 3. Stem glabrous or sparsely pubescent, at least toward the summit; leaves glabrous or nearly so
  4. Plants dioecious
   5. Utricle circumscissile; pistillate flowers with 5 conspicuous spatulate tepals; staminate tepals ca. 2–2.5 mm long, often ± broadly rounded and apiculate at apex and half or slightly more as wide as long .... 3. **A. arenicola**
   5. Utricle indehiscent (irregularly breaking open); pistillate flowers with tepals absent or rudimentary; staminate tepals (at least the largest) ca. 2.3–3.5 mm long, less than half as wide, ± tapered or gradually rounded toward the apex, usually terminating in a short weak awn .... 4. **A. tuberculatus**
  4. Plants monoecious
   6. Utricle indehiscent, with a very rough or warty surface; bracts not over 1 mm long, at most minutely pointed (not spine-tipped); inflorescence of slender branches mostly less than 6 mm thick .... 5. **A. gracilis**
   6. Utricle circumscissile, with smoothish or rough surface; bracts longer and usually spine-topped; inflorescences often more robust
    7. Flower clusters not elongate (usually ± hemispherical or subglobose), present (though small) even in the lowermost leaf axils (those at top of plant more crowded into elongate spikes); tepals 3 ....6. **A. tricolor**
    7. Flower clusters ± elongate and crowded in middle and upper axils, absent from lowest axils; tepals ordinarily 5
     8. Larger tepals of at least some pistillate flowers ca. (2.5) 3–3.5 mm long
      9. Tepals of pistillate flowers ± truncate or broadly rounded (and spinulose or apiculate) at the tip; stamens 5; plants ± densely pubescent on stems (especially around the inflorescence), on the petioles, and on the base of the midrib of the blades beneath ....7. **A. retroflexus**
      9. Tepals of pistillate flowers ± acute or tapered to the weakly spinulose tip; stamens 3 or 5; plants glabrous or sparsely pubescent around the inflorescence and on leaves
       10. Seed obovate in outline, ca. 1 (–1.2) mm broad; stamens mostly 3 .... 8. **A. powellii**
       10. Seed nearly circular in outline, ca. 1.2–1.3 mm broad; stamens 5 .... 9. **A. hypochondriacus**
     8. Larger tepals of pistillate flowers not over 2.2 (2.4) mm long
      11. Tepals of pistillate flowers obovate or spatulate, conspicuously overlapping; terminal inflorescence thick and greatly elongate, drooping from near its base, much more strongly developed than lateral inflorescences...... .... 10. **A. caudatus**
      11. Tepals of pistillate flowers narrowly elliptic to oblong, scarcely if at all overlapping; terminal inflorescence not so disproportionately developed

12. Tepals of pistillate flowers ± obtuse or rounded at apex; bracts acute, narrowly acuminate, or short-spined, shorter than tepals or up to ca. 1.5 times as long; inflorescences normally reddish or purplish . . . . 11. **A. cruentus**

12. Tepals (at least the outer ones) of pistillate flowers acute; bracts generally long-spined, often twice as long as the tepals, inflorescences normally green . . . . 12. **A. hybridus**

1. **A. blitoides** S. Watson   Fig. 60

Map 158. This is apparently a native of western North America. It was first collected in Michigan at South Haven in 1880 by L. H. Bailey, and there are numerous collections from around the state in the 1880's and 1890's. Now a frequent weed of railroad embankments, roadsides, dumps, and other waste ground; also on sand dunes, in gardens, and on shores.

The anthers in this species tend to be a little larger (ca. 0.8–1.1 mm long) than in the next. Many recent works have called this species *A. graecizans* L., a name formerly applied to the next but actually referring to an Old World species. See also remarks under *A. tricolor.*

2. **A. albus** L.

Map 159. A native American tumbleweed, but not collected in Michigan before 1878 (Grand Rapids). Waste places of all kinds, especially sandy areas: roadsides, railroads, dumps, gravel pits, fields, beaches, and gardens.

3. **A. arenicola** I. M. Johnston

Map 160. Collected in railroad yards at Detroit in 1918 and at Tecumseh in 1920; the only other collection seen is from Palisade Park (Van Buren Co.) in 1941.

4. **A. tuberculatus** (Moq.) Sauer

Map 161. Wet meadows, shores, stream banks, and swamps.

An extremely variable plant in growth form and inflorescence, sometimes as tall as 2 m and sometimes nearly prostrate. Long placed in a separate genus, *Acnida,* sometimes as *A. altissima* (an illegitimate name). The tepals of the pistillate flowers are none or reduced, but those of staminate flowers

157. Froelichia gracilis

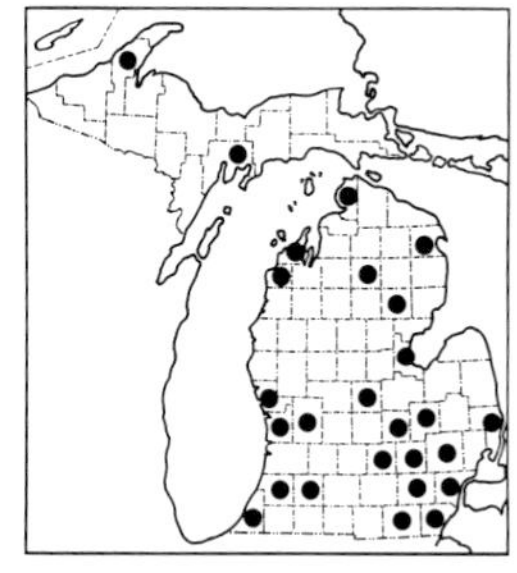

158. Amaranthus blitoides

159. Amaranthus albus

are 5 and well developed. Apparent hydrids with *A. retroflexus* occasionally occur.

5. **A. gracilis** Desf.

Map 162. Collected by Farwell at Detroit in 1932.

This species is included in *A. viridis* L. by some recent authors.

6. **A. tricolor** L.

Map 163. A species domesticated from the Far East and collected by Farwell at Detroit in 1912, presumably as an escape from cultivation. Strains with blotched and colored leaves are cultivated under the name of Joseph's Coat.

Specimens of *A. tricolor* might run in the key to the first two species because of the prominent axillary flower clusters. These are, however, more prominent than in *A. blitoides* and *A. albus,* and become contiguous and crowded terminally, though they may be leafy-bracted. The erect stature of *A. tricolor* immediately distinguishes it from *A. blitoides,* and the seeds are larger than in *A. albus*.

7. **A. retroflexus** L. Figs. 61, 62

Map 164. Probably native in the eastern United States. The earliest Michigan specimen was collected at Grand Prairie, Kalamazoo County, in 1838 by the First Survey. Now widespread in fields, shores, gardens, roadsides, and other disturbed areas.

8. **A. powellii** S. Watson

Map 165. A native of western North America. Although collected in Michigan as early as 1867 (at Detroit), becoming common only in recent years. Usually in disturbed ground, dumps, sandy fields, and roadsides.

The rigid spiny bracts of this species give it a superficial resemblance closer to the preceding species than the next.

160. Amaranthus arenicola

161. Amaranthus tuberculatus

162. Amaranthus gracilis

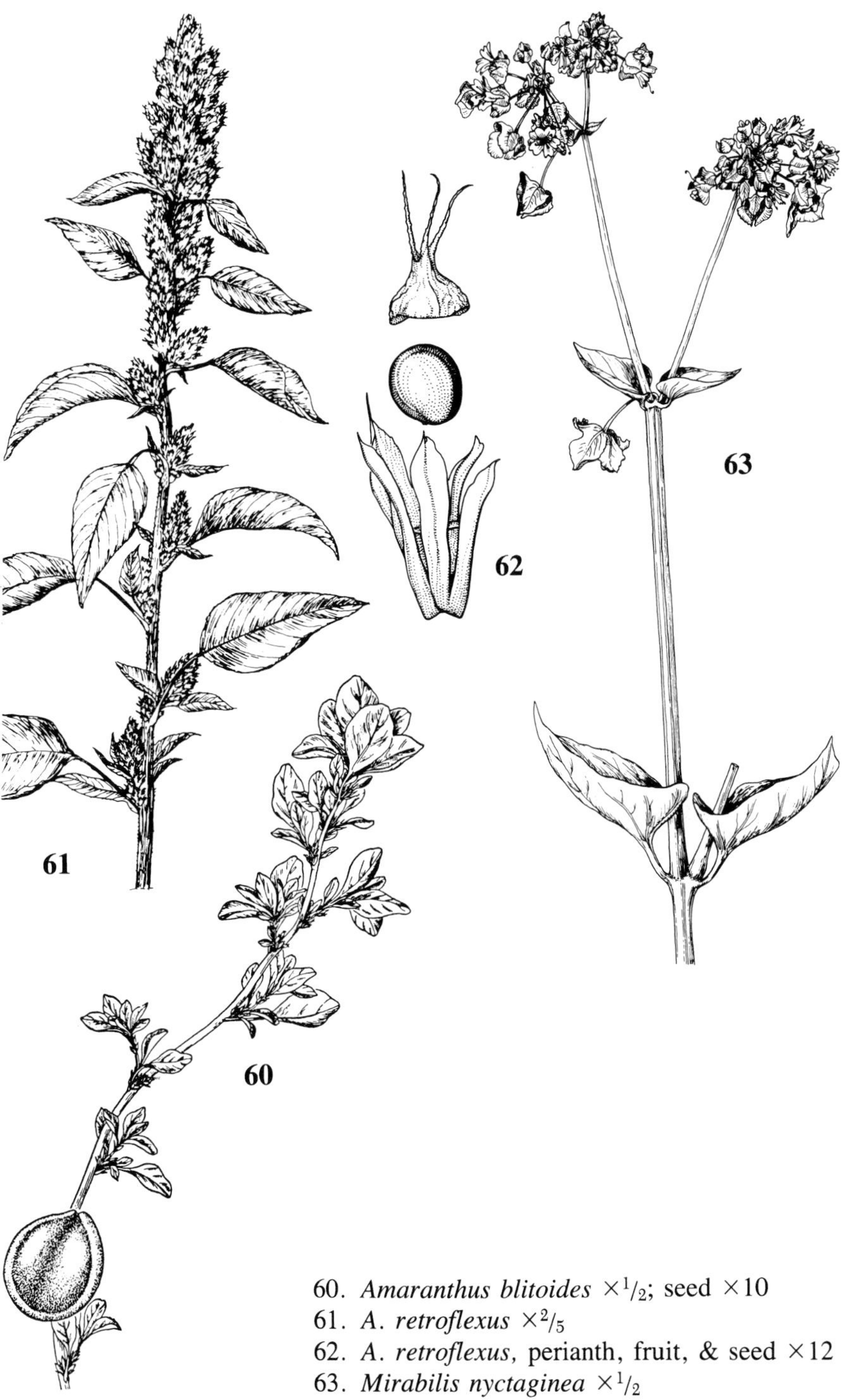

60. *Amaranthus blitoides* $\times ^1/_2$; seed $\times 10$
61. *A. retroflexus* $\times ^2/_5$
62. *A. retroflexus*, perianth, fruit, & seed $\times 12$
63. *Mirabilis nyctaginea* $\times ^1/_2$

9. **A. hypochondriacus** L.

Map 166. A species domesticated from Mexico. An ornamental form is grown under the common name of Prince's Feather. Plants from waste ground in Michigan are doubtless escaped from cultivation.

Our plants have sometimes been called *A. leucocarpus* S. Watson. In this species the seeds may be a distinctive pale cream color, especially in cultivated strains, but the few known Michigan plants have shiny dark reddish or almost black seeds. The leaf blades are large in *A. hypochondriacus,* mostly (4.5) 8–12 (20) cm long, acute or short-acuminate in general outline at the apex, the midrib excurrent or not as a short bristle. In the closely related *A. powellii,* which is sometimes distinguished with difficulty and from which this species was presumably derived, the blades tend to be smaller, the larger ones mostly 3–8 cm long with smaller ones usually present; the apex is broadly rounded in general outline and often slightly notched, the midrib excurrent as a distinct short bristle.

10. **A. caudatus** L. — Love-lies-bleeding

Map 167. A native of the Andes, collected once by Farwell, at Detroit in 1912, presumably as a garden escape. Another Detroit collection of Farwell's (in 1932) has been determined by Sauer as *A. caudatus* × *A. cruentus,* and another from the shores of Torch Lake, Houghton County (in 1940), as *A. caudatus* × *A. retroflexus* ?. Both are apparently sterile hybrids.

11. **A. cruentus** L. — Purple Amaranth

Map 168. Another garden species, originally from southern Mexico or Guatemala, occasionally escaped from cultivation or appearing on dump heaps.

The stem is pubescent, especially around the inflorescence.

12. **A. hybridus** L. — Green Amaranth

Map 169. Probably a native of tropical America and the southern United States, now a weed of dumps, farmyards and gardens, waste ground, and shores. The first Michigan collection seen is from Detroit in 1866.

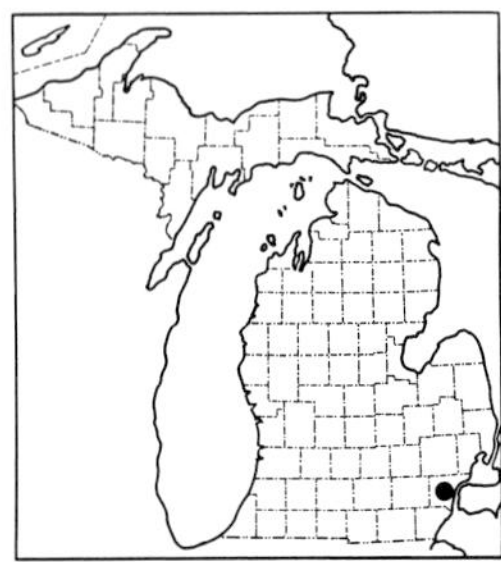

163. Amaranthus tricolor

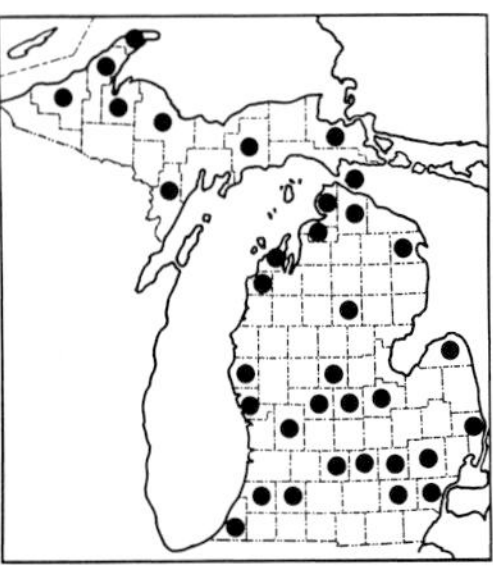

164. Amaranthus retroflexus

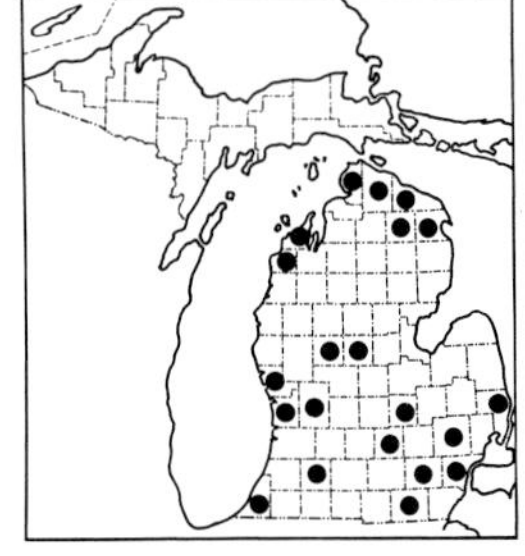

165. Amaranthus powellii

Our specimens of this species all have the stem quite pubescent, especially around the inflorescence, although generally less so than in *A. retroflexus*. Some collections from shores of Lake Erie in Monroe County are apparently sterile hybrids with *A. tuberculatus*.

NYCTAGINACEAE — Four-o'clock Family

1. **Mirabilis** — Umbrellawort

All of our species are sometimes placed in the segregate genus *Oxybaphus* (which, being a masculine name, means that the epithets become *nyctagineus, albidus, hirsutus*). The true garden four-o'clock, *Mirabilis jalapa* L., native to Latin America, rarely escapes from cultivation and apparently has not done so in Michigan. It has a perianth at least 2 cm long and 1-flowered, cup-shaped, herbaceous involucres that do not change in fruit. In our species, the perianth is less than 1.2 cm long and the cup-shaped involucres enlarge, broaden, and become papery as the fruit matures. Occasionally young plants will have only axillary involucres, on slender peduncles, giving them quite a different aspect from larger plants with a ± dense terminal cymose inflorescence.

REFERENCES

[Both of these references include mention of all our species.]

Bogle, A. Linn. 1974. The Genera of Nyctaginaceae in the Southeastern United States. Jour. Arnold Arb. 55: 1–37.

Shinners, Lloyd H. 1951. The North Texas Species of Mirabilis (Nyctaginaceae). Field & Lab. 19: 173–182.

KEY TO THE SPECIES

1. Leaf blades linear, less than 5 mm wide; middle and lower internodes (if not all) glabrous . . . . 1. **M. linearis**
1. Leaf blades broader, lance-ovate to deltoid; middle and lower internodes glabrous or pubescent

166. Amaranthus hypochondriacus

167. Amaranthus caudatus

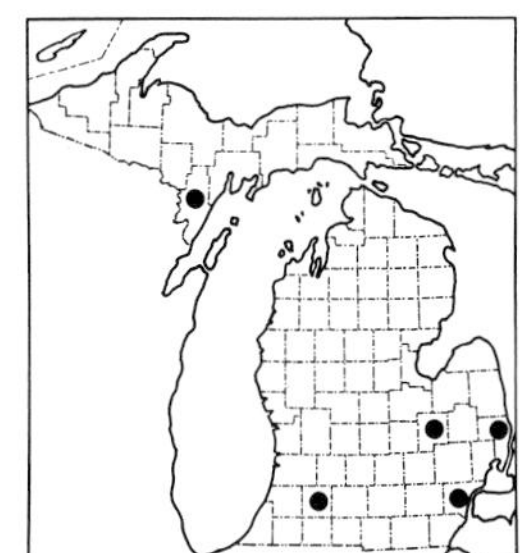
168. Amaranthus cruentus

2. Middle and lower internodes glabrous or nearly so; peduncles and involucres glabrate to pubescent with non-glandular hairs; blades of principal leaves broadly ovate to deltoid, on distinct petioles ± 1 cm or more long . . . . 2. **M. nyctaginea**

2. Middle and lower internodes pubescent, at least in strips; peduncles and involucres with some glandular hairs; blades of principal leaves lance-ovate to lance-oblong, nearly or quite sessile

3. Pubescence on middle and lower internodes of tiny upwardly incurved or appressed hairs up to ± 0.5 mm long . . . . 3. **M. albida**

3. Pubescence on middle and lower internodes including many spreading septate hairs at least 1–1.5 mm long . . . . 4. **M. hirsuta**

1. **M. linearis** (Pursh) Heimerl

Map 170. Native west of the Great Lakes, a rare adventive along railroads, first collected in Michigan in 1947 in Branch County by W. B. Drew.

2. **M. nyctaginea** (Michaux) MacM. Fig. 63 Wild Four-o'clock

Map 171. Native south and southwest of the Great Lakes, adventive here. Railroads, roadsides, dumps, shores, and other disturbed, usually dry ground. Collected at River Rouge, Wayne County, in 1890 by Farwell but not in the state again until 1919 at Ypsilanti.

3. **M. albida** (Walter) Heimerl

Map 172. Another species from south and southwest, collected twice in the 1930's near a cemetery in Midland, by Dreisbach.

4. **M. hirsuta** (Pursh) MacM.

Map 173. A species of the Great Plains, found primarily along railroads but also along roadsides, in fields, and elsewhere. First collected in Michigan in 1895 in Kent County.

PHYTOLACCACEAE — Pokeweed Family

We have only a single species of this family in this region, and it is not likely to be confused with anything else.

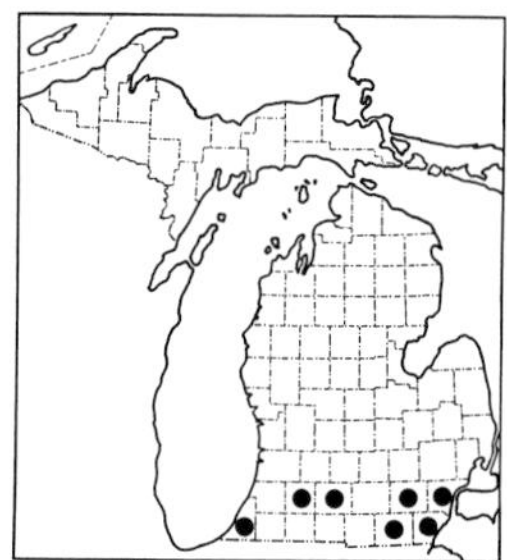

169. Amaranthus hybridus

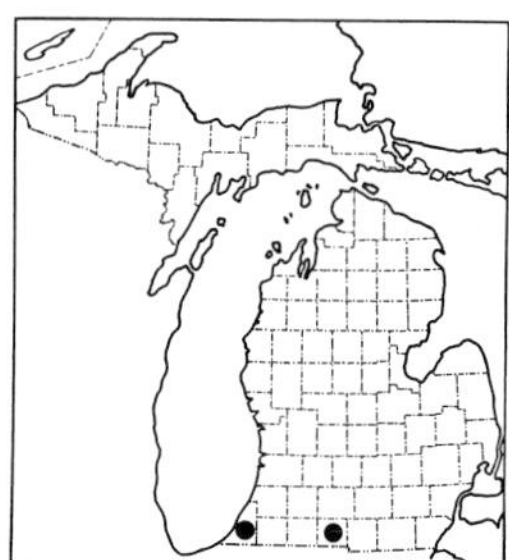

170. Mirabilis linearis

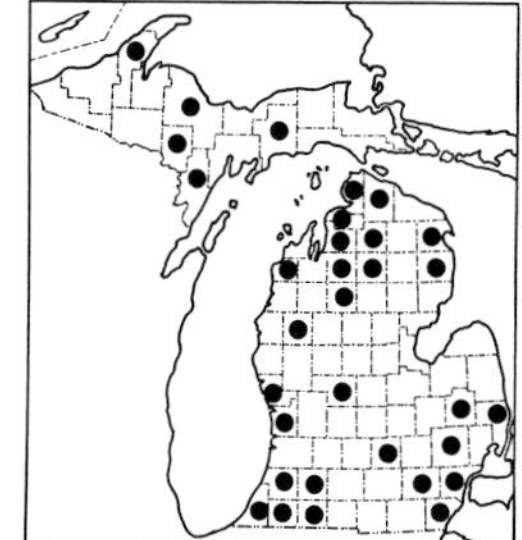

171. Mirabilis nyctaginea

1. Phytolacca

REFERENCE

Sauer, Jonathan D. 1952. A Geography of Pokeweed. Ann. Missouri Bot. Gard. 39: 113–125.

1. **P. americana** L. Fig. 64 Pokeweed; Poke

Map 174. A native species of somewhat disturbed soils: borders of woods and fields, roadsides, disturbed woods, fencerows, barnyards and edges of gardens, usually in fairly rich soils.

This is a large bushy herb, sometimes as much as 3 m tall, and it produces copious large racemes of attractive purple-black berries. These are a source of dye (though a difficult one to fix permanently) and of another common name, inkberry. The large perennial root is poisonous, but was of reputed medicinal value in early days. Very young shoots are often eaten as a palatable vegetable, and the berries, although sometimes considered poisonous (at least when raw), are used in pies. Use of any part of the plant for food should be done with caution, and with reference to experienced directions regarding preparation.

MOLLUGINACEAE Carpetweed Family

This family has often been united with the Aizoaceae.

REFERENCE

Bogle, A. Linn. 1970. The Genera of Molluginaceae and Aizoaceae in the Southeastern United States. Jour. Arnold Arb. 51: 431–462.

1. Mollugo

1. **M. verticillata** L. Fig. 65 Carpetweed

Map 175. Considered originally native in warmer climates of America, but known from Michigan as early as the 1837 collections of the First Survey. Dry sandy fields, trails, gravel pits, dumps, and roadsides; also in cindery railroad ballast and sometimes on shores and banks.

The whorled, ± narrowly spatulate leaves (broader at base of plant) and prostrate habit give this species the superficial aspect of a *Galium,* but it is readily distinguished by the 5 separate sepals, absence of petals, 3 (or 5) stamens, and clearly 3-carpellate superior ovary which ripens into a many-seeded capsule.

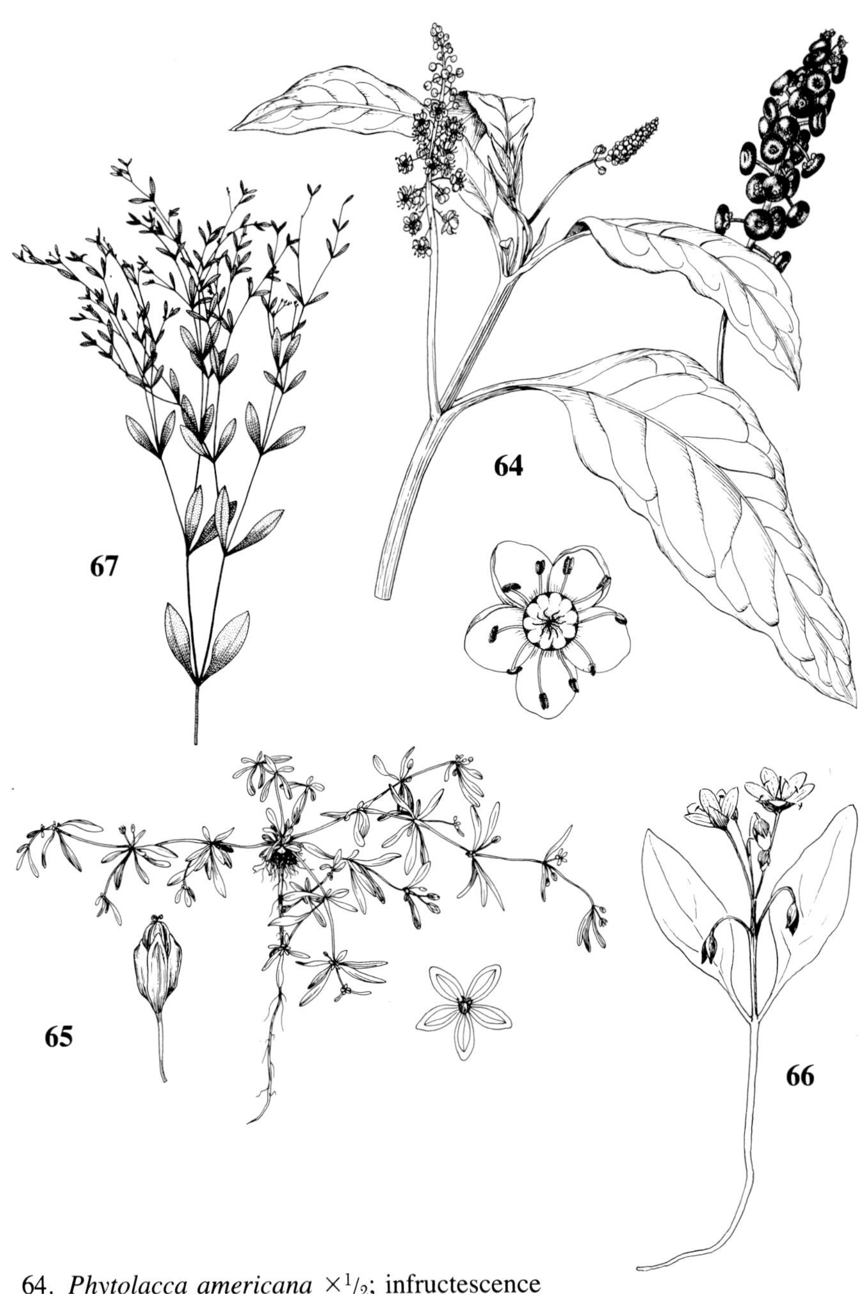

64. *Phytolacca americana* $\times 1/2$; infructescence $\times 1/2$; flower $\times 5$
65. *Mollugo verticillata* $\times 1/2$; fruit & flower $\times 3$
66. *Claytonia caroliniana* $\times 1/2$
67. *Paronychia canadensis* $\times 1/2$

PORTULACACEAE   Purslane Family

REFERENCE

Bogle, A. Linn. 1969. The Genera of Portulacaceae and Basellaceae in the Southeastern United States. Jour. Arnold Arb. 50: 566–598.

KEY TO THE GENERA

1. Cauline leaves numerous, the upper ones closely subtending the flowers; corolla yellow or white to red; capsule circumscissile; leaves very succulent; plant an annual in waste ground or a weed escaped from cultivation . . . . 1. **Portulaca**
1. Cauline leaves 2, the inflorescence and flowers conspicuously stalked; corolla pale to deep pink (or white with pink veins); capsule splitting longitudinally; leaves scarcely succulent; plant a perennial from a globose (but seldom collected) corm, common in woods . . . . 2. **Claytonia**

1. Portulaca   Purslane

KEY TO THE SPECIES

1. Leaves linear (scarcely 2 mm broad), hence nearly terete; stems with conspicuous tufts of long dense hairs at the nodes; flowers at least 2 cm broad, variously colored . . . . 1. **P. grandiflora**
1. Leaves spatulate to obovate, flat; stems glabrous or with a few short hairs at nodes; flowers not over 1 cm wide, yellow . . . . 2. **P. oleracea**

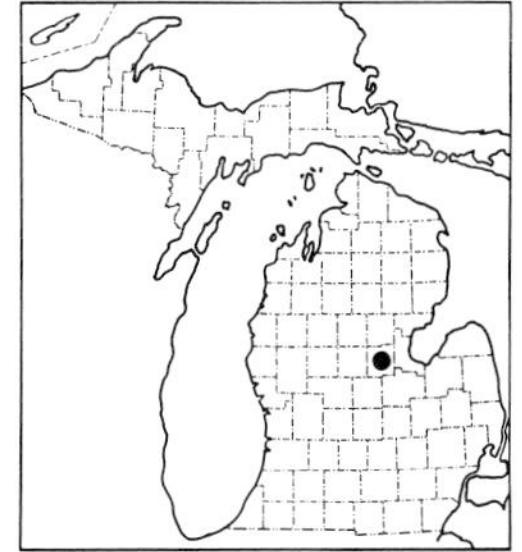

172. Mirabilis albida

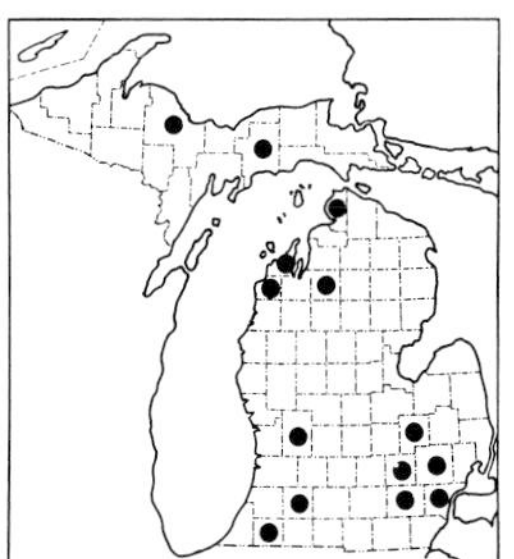

173. Mirabilis hirsuta

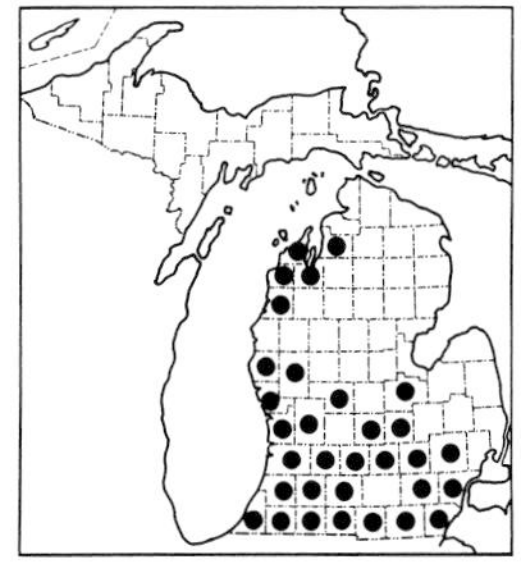

174. Phytolacca americana

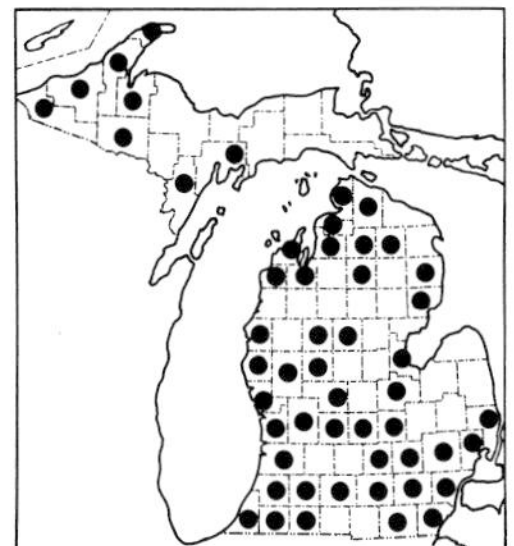

175. Mollugo verticillata

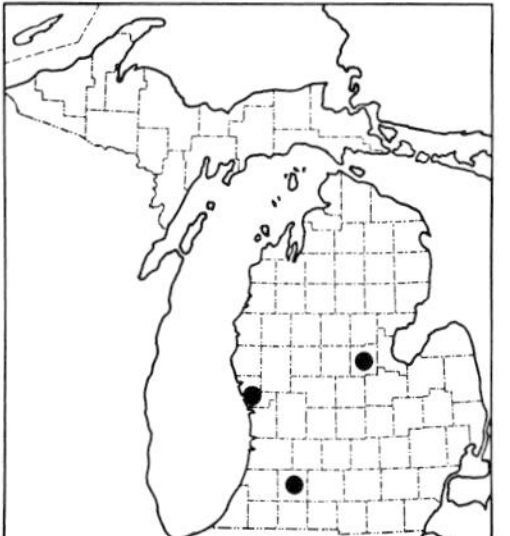

176. Portulaca grandiflora

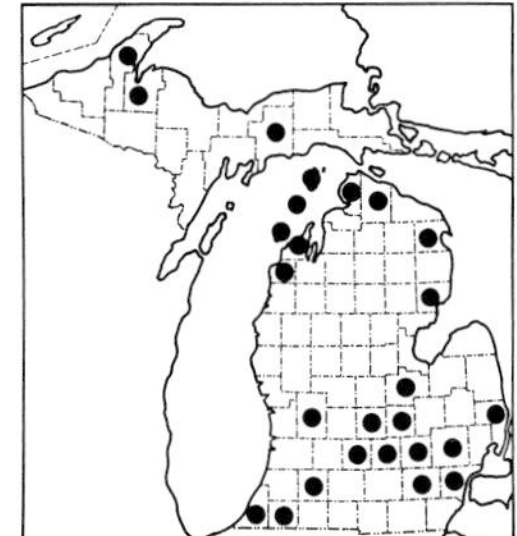

177. Portulaca oleracea

1. **P. grandiflora** Hooker Rose-moss

Map 176. A native of South America, occasionally escaped from gardens.

2. **P. oleracea** L. Fig. 68 Purslane; Pusley

Map 177. A cosmopolitan weed in temperate and warm climates, its true origin unknown. Much more widely distributed as an obnoxious weed of lawns, gardens, cultivated fields, alleys, and gravel pits than the map indicates, but neglected by collectors. (The succulent leaves tend to fall off as the plant shrivels and dries, so that herbarium specimens are as scorned as the live plant is in its habitat.)

2. Claytonia Spring-beauty

The two species of *Claytonia* very rarely grow side by side in the same woods in Michigan. Such colonies are known from Benzie, Clare, and Oceana counties and Round Island (Mackinac County). In such circumstances, under identical conditions, *C. virginica* reaches the peak of its flowering later by at most a few days than *C. caroliniana*. The vegetative parts of both species turn yellowish after a short flowering and fruiting season in the spring, and soon the plants are no longer seen above ground in woods which may have been carpeted (and perfumed) with them a month previously.

The petals of both species are usually pale pink with deeper colored veins, but the ground color ranges from white to very deep pink; the corolla may be as much as 27 mm broad. Usually several fertile stems as well as basal leaves arise from a globose corm deep in the ground. Both species are extremely variable in leaf shape and size as well as in other characters, such as the aberrant presence of extra leaves on the stem.

REFERENCES

Reznicek, T., & D. M. Britton. 1971. Chromosome Studies on the Spring Beauties, Claytonia, in Ontario. Mich. Bot. 10: 51–62.

Voss, Edward G. 1968. The Spring Beauties (Claytonia) in Michigan. Mich. Bot. 7: 77–93.

KEY TO THE SPECIES

1. Cauline leaves more than 8 times as long as broad, linear, without a distinct petiole . . . . 1. **C. virginica**
1. Cauline leaves less than 8 (usually 2.5–6.5) times as long as broad, including typically a ± diamond-shaped blade and a distinct petiole . . . . 2. **C. caroliniana**

1. **C. virginica** L. Plate 1-F

Map 178. Upland beech-maple and oak forests, persisting after clearing (consequently with as unlikely associates as *Opuntia* on Oceana Co. sands); also in lowland, even mucky forests.

An occasional robust plant may have large cauline leaves only 4–5 times as long as broad, but these are more elliptical or oblong than the characteristic shape of *C. caroliniana* leaves.

2. **C. caroliniana** Michaux Fig. 66
Map 179. Like the preceding species, typical of rich deciduous woods, but not so often in swamp forests; on the other hand, this northern species is more often associated with conifers and less often in oak forests than is *C. virginica*.

CARYOPHYLLACEAE Pink Family

Much aid in dealing with several of the genera in this family has generously been given me by Richard K. Rabeler.

REFERENCE

Schlising, Robert A., & Hugh H. Iltis. 1962. Preliminary Reports on the Flora of Wisconsin No. 46. Caryophyllaceae—Pink Family. Trans. Wisconsin Acad. 50: 89–139.

KEY TO THE GENERA

1. Stipules present, conspicuous (though small), scarious, sometimes fused and forming a membranous collar between the leaf bases
 2. Petals absent; fruit 1-seeded, indehiscent; styles 2-forked or none and stigmas 2; leaves elliptic to narrowly obovate
  3. Plant ± erect, with flowers on very slender terminal branches; leaves densely red-dotted, at least beneath; stipules eciliate; sepals ca. 0.7–1 mm long . . . 1. **Paronychia**
  3. Plant a prostrate, matted weed, with flowers crowded in axillary clusters; leaves not red-dotted; stipules ciliate; sepals ca. 0.5–0.6 mm long . . . . 2. **Herniaria**
 2. Petals present (but shorter than the sepals); fruit a several-seeded capsule; styles and stigmas 3 or 5; leaves linear-filiform
  4. Styles and valves of capsule normally 5; petals white; leaves in 2 principal fascicles at each node, appearing densely whorled . . . . 3. **Spergula**
  4. Styles and valves of capsule normally 3; petals usually pink; leaves more clearly opposite (with partly connate stipules), but usually with axillary fascicles . . . . 4. **Spergularia**
1. Stipules absent (though leaf bases may meet around the stem)
 5. Sepals (ca. 2–5 mm long) connate nearly or fully half their length to form a hard closely perigynous cup around the ovary and membranous fruit; fruit 1-seeded, indehiscent . . . . 5. **Scleranthus**
 5. Sepals connate much more than half their length, or separate nearly or quite to their base, or over 2.5 cm long; fruit a few–many-seeded capsule
  6. Calyx of sepals separate nearly or quite to their base
   7. Petals clearly notched at the apex or deeply cleft (the corolla then appearing 10-parted), or rarely absent
    8. Styles 3–4 (very rarely –6); capsules ± ovoid, 6 (–8)-cleft to the base . . . . 6. **Stellaria**
    8. Styles 5; capsules 10-cleft or 10-toothed, ovoid to cylindric

9. Leaves ovate, the principal ones at least (1.5) 2 cm broad; capsule ovoid . . . . 7. **Myosoton**

9. Leaves various, but less than 1 cm broad; capsule cylindric . . . . 8. **Cerastium**

7. Petals entire or toothed, but neither cleft nor merely notched at the tip

10. Inflorescence umbellate; styles 3; capsules 6-cleft; plant withering in May or soon afterwards . . . . 9. **Holosteum**

10. Inflorescence cymose, not umbellate, or flowers solitary; styles and capsules various; plant persisting beyond May

11. Styles (as well as sepals and petals) (4–) 5, the capsules splitting by an equal number of teeth . . . . 10. **Sagina**

11. Styles 3 (fewer than sepals and petals), the capsules splitting by 3 or 6 teeth . . . . 11. **Arenaria**

6. Calyx of sepals fused at least a fourth of their length

12. Sepals fused ca. one-fourth to half their length (hence with prominent free tips), ± densely pilose, (1.6) 3.5–5.5 (7) cm long . . . . 12. **Agrostemma**

12. Sepals fused half their length or more, glabrous or pubescent, less than 3 cm long

13. Calyx immediately subtended by closely appressed bracts (no naked pedicel evident between bracts and calyx)

14. Sepals each with 1 (–3) main rib(s), connate by membranous ribless margins . . . . 13. **Petrorhagia**

14. Sepals each with (3) 5–7 usually strong ribs, connate without membranous margins . . . . 14. **Dianthus**

13. Calyx subtended by no bracts, or with at least a short pedicel visible above spreading bracts

15. Styles 2, the capsule opening by 4 teeth; flowers perfect; calyx 5-nerved or very obscurely many-nerved

16. Calyx less than 5 mm long . . . . 15. **Gypsophila**

16. Calyx at least 7 mm long

17. Inflorescence open, the pedicels over 1 cm long; calyx with 5 green wings . . . . 16. **Vaccaria**

17. Inflorescence crowded, the pedicels less than 5 mm long; calyx wingless . . . . 17. **Saponaria**

15. Styles 3–5 (or more), or flowers entirely staminate; calyx 10–30-nerved (or nerves obscure)

18. Flowers unisexual (plants dioecious); styles (in pistillate flowers) 5–6 (sometimes more, rarely fewer) . . . . 18. **Silene (pratensis)**

178. Claytonia virginica

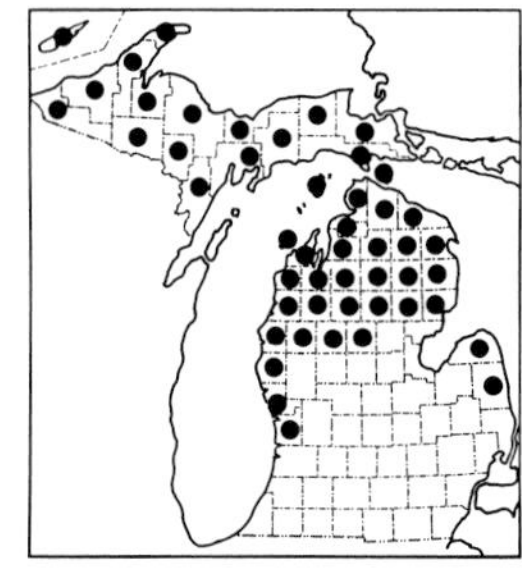

179. Claytonia caroliniana

180. Paronychia canadensis

18. Flowers perfect; styles various
 19. Styles 3, the capsule opening by 6 teeth; corolla often white (the stem in pink-flowered plants usually sticky) . . . . 18. **Silene**
 19. Styles 5, the capsule opening by 5 teeth; corolla red or purplish (stem not sticky) . . . . 19. **Lychnis**

1. Paronychia — Whitlow-wort; Forked Chickweed

REFERENCE

Core, Earl L. 1941. The North American Species of Paronychia. Am. Midl. Nat. 26: 369–397.

KEY TO THE SPECIES

1. Stems glabrous . . . . 1. **P. canadensis**
1. Stems finely pubescent with ± retrorsely incurved hairs . . . . 2. **P. fastigiata**

1. **P. canadensis** (L.) Wood Fig. 67

Map 180. Sandy woodlands and banks.

A plant with very slender branches, the leaves often alternate at the uppermost nodes.

2. **P. fastigiata** (Raf.) Fern.

Map 181. Barely enters Michigan from the south. The only collection seen from the state (*Farwell 2120$^1/_2$,* BLH) was made at Geddes (between Ann Arbor and Ypsilanti) in 1909. No habitat was stated, but this, too, is a species of dry places.

2. Herniaria

1. **H. glabra** L.

Map 182. An inconspicuous little Eurasian weed, first collected in Michigan at East Lansing in 1920 (*Walpole 979,* MSC, misidentified as *Scleranthus*), and not again until 1979 when it was discovered by Kenneth Dritz in cracks of an old pavement near a railroad in Berrien County. So easily overlooked that it may well occur elsewhere.

3. Spergula

1. **S. arvensis** L. Fig. 70 — Spurrey

Map 183. A European weed, local in North America and not widespread in Michigan, where it has been collected in various disturbed sites, including fields, gardens, roadsides, and railroads.

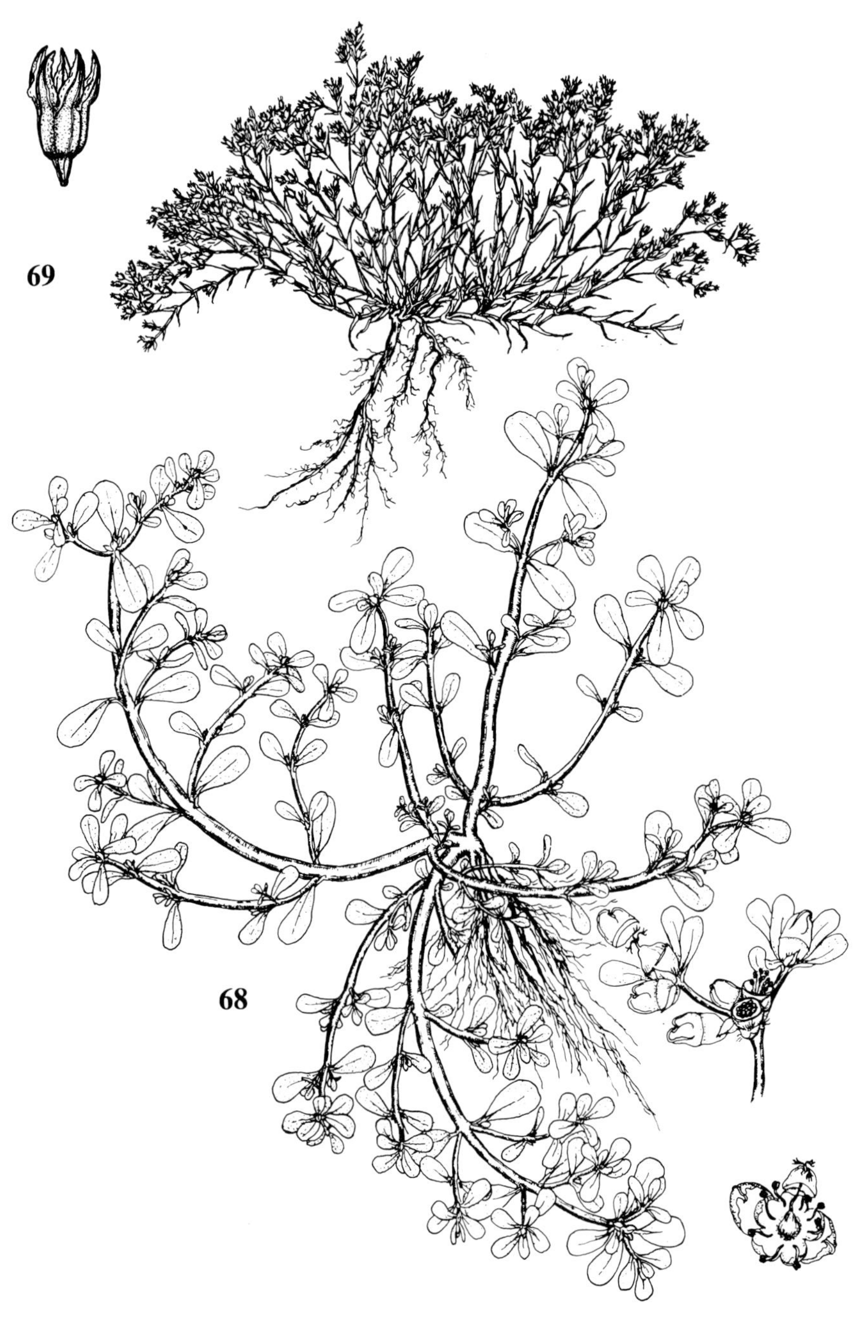

68. *Portulaca oleracea* $\times ^2/_5$; fruiting branch $\times 1^1/_4$; flower $\times 3$
69. *Scleranthus annuus* $\times ^2/_5$; fruiting calyx $\times 4$

4. **Spergularia** Sand-spurrey

REFERENCE

Cusick, Allison W. 1983. Spergularia (Caryophyllaceae) in Ohio. Mich. Bot. 22: 69–71.

KEY TO THE SPECIES

1. Sepals at least 4 mm long (usually 5); seed with a thin wing ca. 0.3 mm broad . . . . 1. **S. media**
1. Sepals mostly not over 4 mm (usually 2–3.5); seed usually wingless
   2. Stamens normally 10 (or some aborted); plant of non-saline habitats . . . . . 2. **S. rubra**
   2. Stamens 1–3 (rarely more); plant of saline habitats . . . . 3. **S. marina**

1. **S. media** (L.) Griseb.

Map 184. A species of European shores and inland saline areas, one of several halophytes collected by C. K. Dodge at freight yards in Port Huron (in 1910), where he reported it as well established (Univ. Mich. Mus. Zool. Misc. Publ. 4: 10. 1918). It has since been found at several places as a roadside halophyte.

The name is sometimes rejected as ambiguous, replaced by *S. maritima* (All.) Chiov.

2. **S. rubra** (L.) J. & C. Presl

Map 185. A European native, locally established in our area. Dry, usually sandy roadsides, driveways, lawns, and clearings. First collected in Michigan in 1896 in Hillsdale County, but only since 1970 becoming at all widely distributed in the state.

A low matted plant with attractive bright pink little flowers open only briefly in the sunshine.

3. **S. marina** (L.) Griseb. Fig. 71

Map 186. Like *S. media*, this is a species of saline habitats, probably introduced in North America from Eurasia. Discovered in a drainage ditch

181. Paronychia fastigiata

182. Herniaria glabra

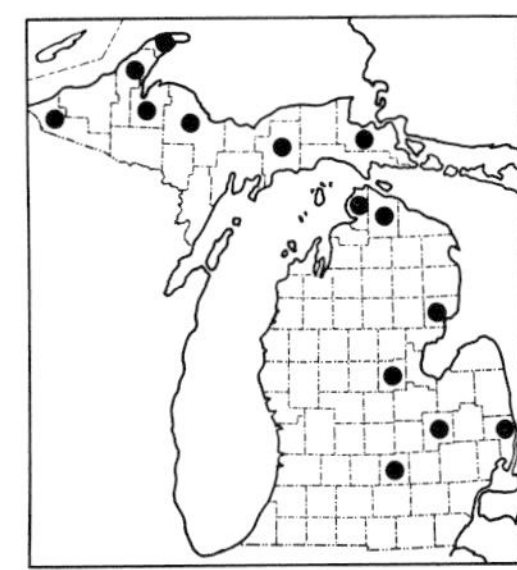
183. Spergula arvensis

in Washtenaw County in 1974 (*F. Omilian 992,* WUD), and subsequently found along a number of expressways where highway salt creates a favorable habitat (see Mich. Bot. 19: 24. 1980).

The plants look like a small version of *S. media,* with fewer stamens. Seeds on the Michigan material are only occasionally winged. The mature capsule exceeds the calyx. According to some interpretations of the complex nomenclatural history of this species, the correct name is *S. salina* J. & C. Presl.

5. **Scleranthus** Knawel

These are easily overlooked bushy little plants with linear-subulate leaves united around the slender slightly hairy stems. The flowers are numerous, sessile on the upper portions of the plant, perfect but lacking petals, the greenish calyx becoming brownish and indurated, readily shed while enclosing the fruit.

KEY TO THE SPECIES

1. Tips of sepals acute, barely if at all margined with white . . . . 1. **S. annuus**
1. Tips of sepals rounded, with conspicuous white border ca. 0.2–0.4 mm broad . . . . 2. **S. perennis**

1. **S. annuus** L. Fig. 69

Map 187. A native of Eurasia, locally found as a weed. Sandy trails, fields, roadsides, gravel pits, gardens. Known in Michigan only since the 1880's.

2. **S. perennis** L.

Map 188. Another Eurasian weed, apparently first recognized in the United States in Wisconsin (Schlising & Iltis 1961). Collections from the Lansing area all date from 1880–1893. In Marquette County, found in 1965 on an old railroad siding at Gentian (*J. S. Pringle 404,* MICH).

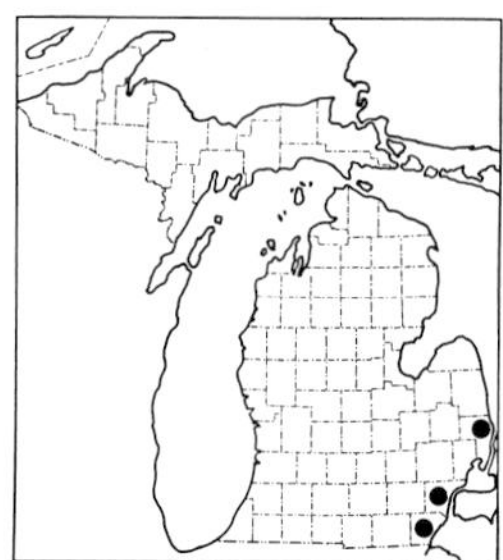
184. Spergularia media

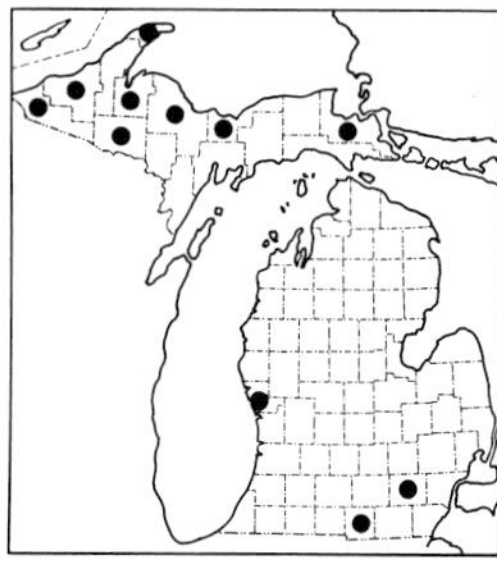
185. Spergularia rubra

186. Spergularia marina

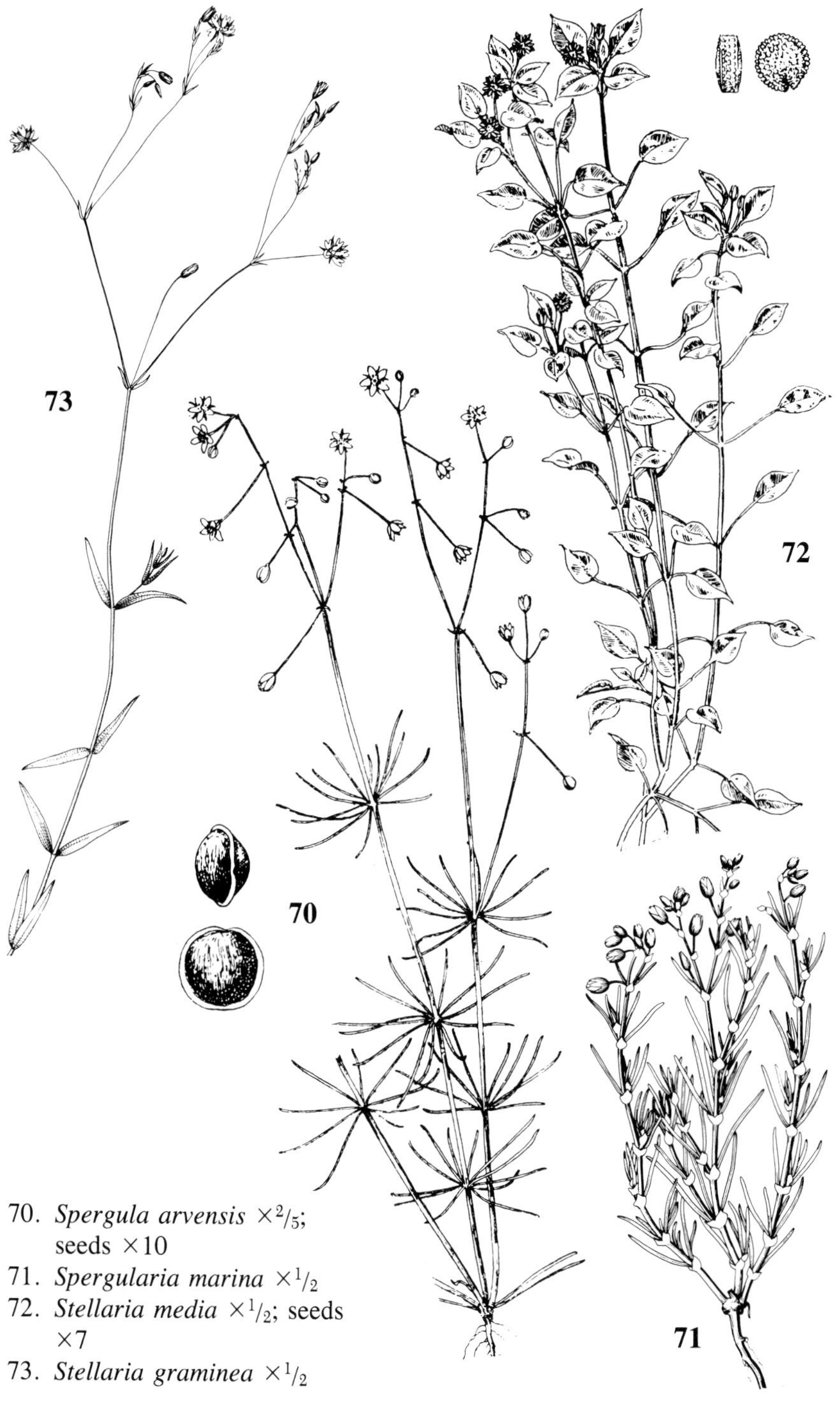

70. *Spergula arvensis* ×$^{2}/_{5}$; seeds ×10
71. *Spergularia marina* ×$^{1}/_{2}$
72. *Stellaria media* ×$^{1}/_{2}$; seeds ×7
73. *Stellaria graminea* ×$^{1}/_{2}$

6. Stellaria — Chickweed; Stitchwort

REFERENCES

Chinnappa, C. C., & J. K. Morton. 1976. Studies on the Stellaria longipes Goldie Complex—Variation in Wild Populations. Rhodora 78: 488–502.

McNeill, J., & Judy N. Findlay. 1972. Introduced Perennial Species of Stellaria in Québec. Nat. Canad. 99: 59–60.

Turkington, Roy, Norman C. Kenkel, & Gail D. Franko. 1980. The Biology of Canadian Weeds. 42. Stellaria media (L.) Vill. Canad. Jour. Pl. Sci. 60: 981–992.

KEY TO THE SPECIES

1. Leaves (at least the middle and lower ones) with broadly ovate blades and distinct petioles . . . . 1. **S. media**
1. Leaves linear to lanceolate, without petioles
  2. Bracts on middle (and lower) portion of inflorescence green, herbaceous; principal leaves ca. 4.5–7 times as long as broad; sepals ca. (1.5) 2–3.5 mm long, even in fruit
    3. Leaves completely glabrous, less than 12 (18) mm long; petals exceeding the sepals . . . . 2. **S. crassifolia**
    3. Leaves ± sparsely ciliate, at least toward the base, the principal ones 15–35 (40) mm long; petals absent or (rarely) nearly equaling the sepals . . . . 3. **S. calycantha**
  2. Bracts on middle (and usually lower) portion of inflorescence scarious (at most with green midrib); principal leaves often more than 8 times as long as broad; sepals often longer (especially in fruit)
    4. Plants short, 3–20 (25) cm tall, few-flowered (1–5 or rarely 7–10 flowers per stem axis); pedicels ascending; leaves narrowly triangular (tapering evenly from a base up to 2.3 mm broad), ± crowded and strongly ascending . . . . . 4. **S. longipes**
    4. Plants taller, few–many-flowered; pedicels and branches of inflorescence (at least when mature) strongly divaricate or reflexed; leaves linear to narrowly lanceolate, ± spreading
      5. Median (if not all) leaves linear or slightly broader near the middle; margins of at least upper leaves and angles of the internodes between them finely papillose [use 20× lens]; bracts and sepals usually eciliate; seeds slightly marbled but essentially smooth . . . . 5. **S. longifolia**
      5. Median (if not all) leaves broadest at or just above the base; margins of leaves and stems not papillose (rarely ciliate); bracts and usually sepals ciliate; seeds distinctly wrinkled with wavy sculpturing . . . . 6. **S. graminea**

187. Scleranthus annuus

188. Scleranthus perennis

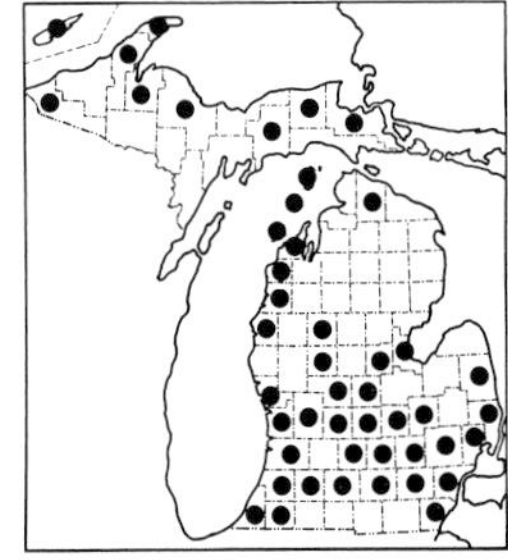

189. Stellaria media

1. **S. media** (L.) Vill. Fig. 72

Map 189. A too-familiar Eurasian weed, more widespread than the map suggests in lawns, gardens, and waste ground (roadsides, railroads, fields, etc.); also along trails and other disturbances in forests, both coniferous and deciduous. The earliest Michigan collections seen are from 1869 (Ann Arbor) and the 1880's.

2. **S. crassifolia** Ehrh.

Map 190. Brooksides and wet sandy or springy shores.

3. **S. calycantha** (Ledeb.) Bong.

Map 191. Most often in forests of spruce, fir, and birch, of tamarack, of cedar, or even of northern hemlock-hardwoods, especially in openings or hollows, along trails, etc.; also on gravelly shores and borders of ponds and marshes.

This is our only species in the genus characterized by ciliate leaf margins without any other pubescence; it is also our only one which regularly lacks petals (a very few of the specimens from Isle Royale have petals barely equaling the sepals). In our material, the styles are usually 4 (or even rarely 5 or 6) rather than 3. Most of our specimens appear to be var. *isophylla* (Fern.) Fern. with flowers few, arising from axils of herbaceous bracts or leaves. Some specimens are var. *floribunda* (Fern.) Fern., with numerous flowers, the upper bracts scarious. These varieties, however, are not always well marked.

4. **S. longipes** Goldie

Map 192. This is a northern circumpolar species which comes south in our area very locally to the upper sandy beaches and dunes of Lake Superior and the north end of Lake Michigan.

S. longipes represents a difficult taxonomic complex (see Chinnappa & Morton for a survey), but in Michigan it seems to be a quite distinctive plant, albeit difficult to discriminate in a key. The lower bracts of the relatively few-flowered inflorescence are generally leaf-like, but specimens can easily

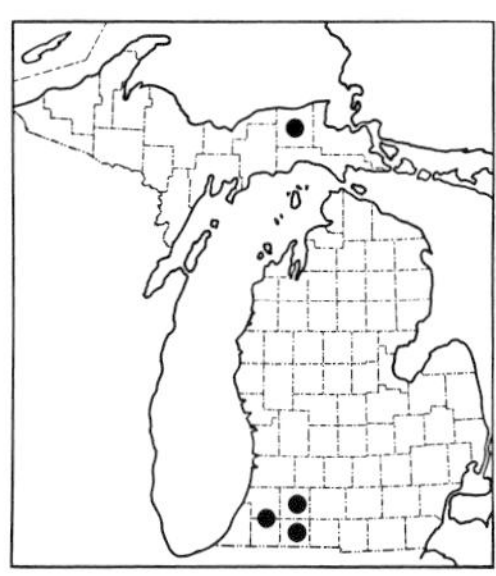

190. Stellaria crassifolia

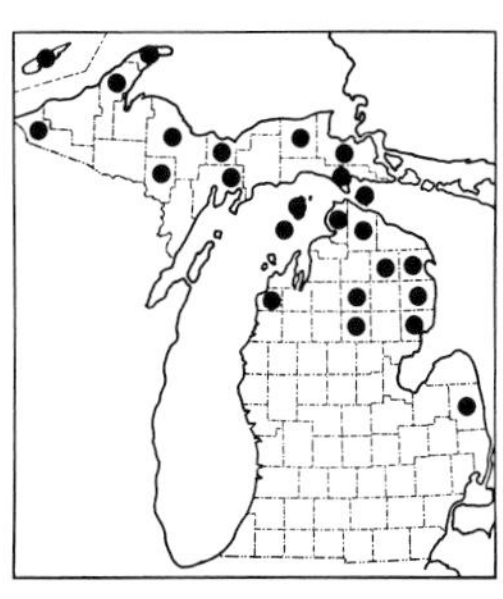

191. Stellaria calycantha

192. Stellaria longipes

be distinguished from the preceding flat-leaved species with herbaceous bracts; for the leaves are strongly ascending, very elongate-triangular, tapering from no more than 2 (2.3) mm wide at the base, with prominent midrib beneath and hence usually appearing ± folded or channeled rather than flat. The stem, leaves, and sepals are usually ± glaucous. The stems are glabrous or (in some Lake Superior plants) slightly pubescent, and the leaves of pubescent plants may be sparsely ciliate basally.

5. **S. longifolia** Willd.

Map 193. Nearly always in damp to wet places: swamps (both deciduous and coniferous) and marshes; shores and edges of ponds, creeks, and bogs.

Usually considered closely related to the preceding species and in some regions grading into it, but in Michigan the two are quite distinct in habit and morphology. Both have smooth seeds. Like the next species, *S. longifolia* is a rather weak-stemmed, lax plant with a widely spreading inflorescence at maturity. See also comments under *S. graminea*.

Reports of *S. palustris* Retz. [or *S. glauca* With.] from Michigan are apparently based on specimens of *S. longifolia* (cf. McNeill & Findlay 1972).

6. **S. graminea** L. Fig. 73

Map 194. A weed of European origin, locally well established. Fields, lawns, thickets and borders of woods, roadsides and other waste ground, shores; often in grassy places and usually in moist rather than very dry sites.

This species bears a strong superficial resemblance to the preceding, and the two are frequently confused—especially when one relies upon keys or descriptions that stress the nature of the inflorescence, the color of the mature fruit (far less reliable than the seed surface), or the character of the sepals (which tend in *S. graminea* to be more strongly 3-nerved and to run a little longer—up to 5.5 mm—than in *S. longifolia*). Only occasionally in *S. longifolia* are some sepals ciliate and even more rare are sparsely ciliate bracts; such plants can be distinguished from *S. graminea* by the shape and minutely papillose margins of the leaves. One truly ambiguous Michigan

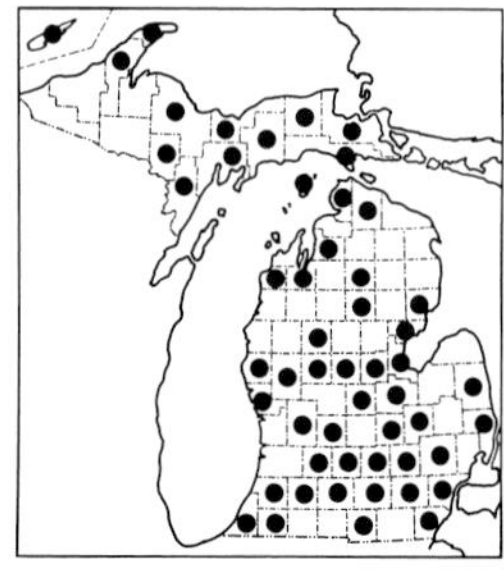

193. Stellaria longifolia

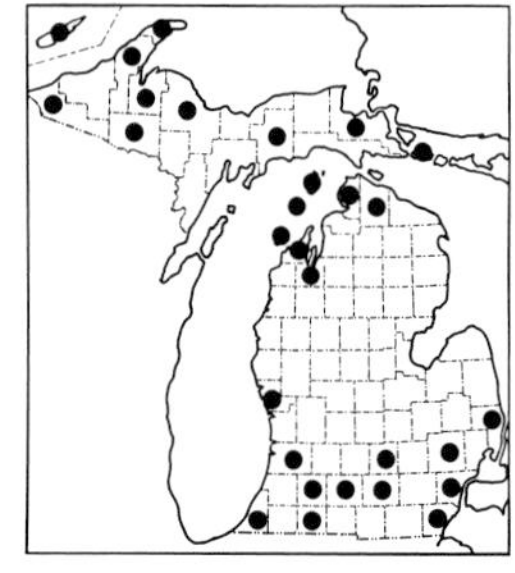

194. Stellaria graminea

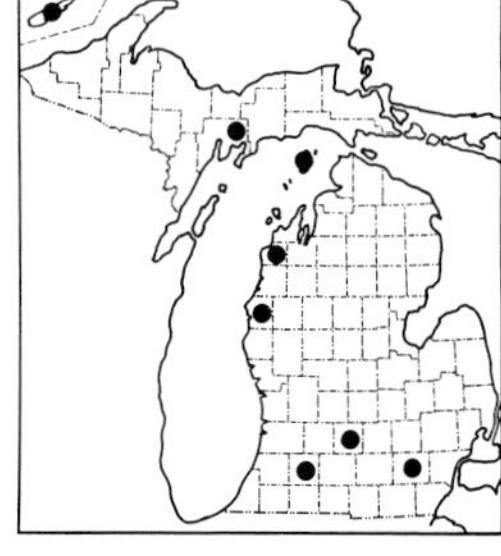

195. Myosoton aquaticum

specimen has been seen (from Beaver Island), with eciliate sepals and bracts *and* non-papillose leaves (the sepals are too short and the flowers too numerous for the ciliate and non-papillose *S. palustris*). The leaves (and bracts) of this species and the preceding one may occasionally have a few cilia at the very base, but both are readily distinguished from *S. calycantha* by the definitely scarious bracts and usually more elongate leaves.

6. **Myosoton**

1. **M. aquaticum** (L.) Moench — Giant Chickweed

Map 195. A native of Eurasia, found on banks of rivers and streams, ponds, and other damp ± shaded sites, generally growing in a tangled mass.

This is the only species in the genus and is sometimes included in *Stellaria* as *S. aquatica* (L.) Scop.

7. **Cerastium** — Chickweed

The 10 distinct short teeth by which the cylindrical, usually slightly curved capsule opens at its apex are a very characteristic feature of this genus and should allow confusion with no other. Except in the very earliest part of the long flowering season, some capsules are likely to be ripe on a plant (from the first flowers of a cyme to bloom). A tiny pentagonal "plug" to which the styles are attached falls free from the apex of the capsule.

KEY TO THE SPECIES

1. Leaves, stems, and sepals ± densely white-tomentose . . . . 1. **C. tomentosum**
1. Leaves, stems, and sepals with mostly straight or irregular hairs, not white, often sticky
   2. Axils of middle cauline leaves mostly with short sterile leafy tufts; leaves linear to narrowly oblong or lanceolate; corolla showy, exceeding the calyx by 2–5 mm or more; anthers ca. 0.7–1.1 mm long . . . . 2. **C. arvense**
   2. Axils usually without leafy tufts, or leaves narrowly to broadly ovate to elliptic; corolla shorter than the calyx or barely exceeding it; anthers less than 0.7 mm long

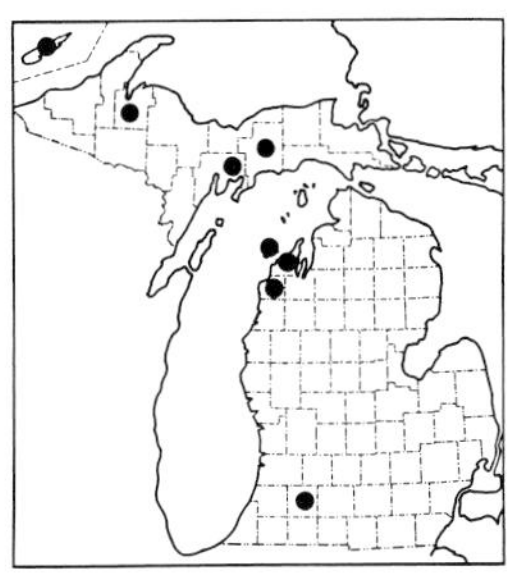

196. Cerastium tomentosum

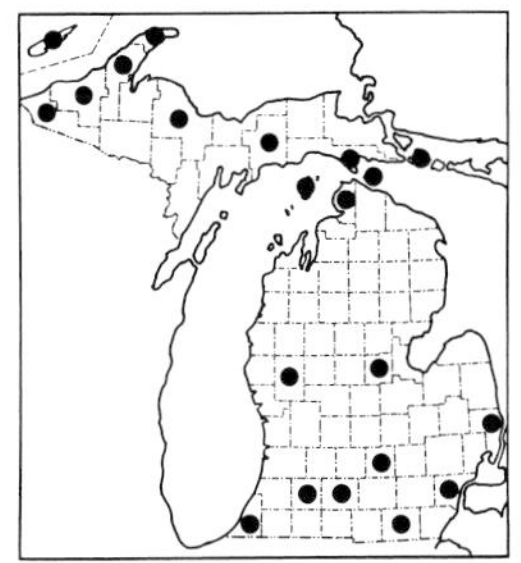

197. Cerastium arvense

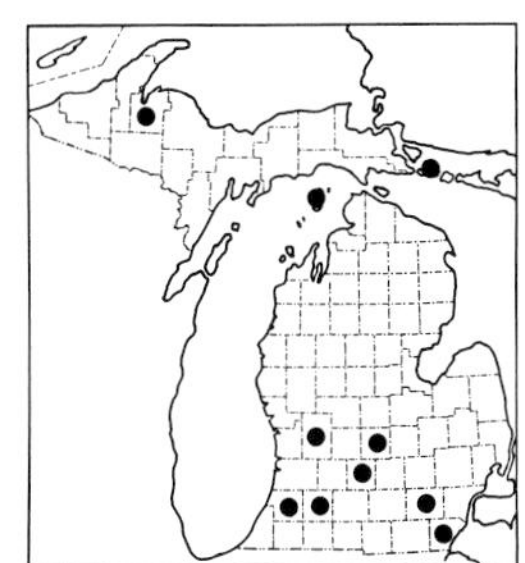

198. Cerastium nutans

3. Uppermost bracts of the inflorescence completely herbaceous, glandular-puberulent even at the tip; pedicels often ± strongly arched ("hooked") beneath the calyx

4. Pedicels mostly longer than the calyx; hairs of sepals not extending beyond tip of sepals . . . . 3. **C. nutans**

4. Pedicels mostly shorter than the calyx; hairs of sepals extending beyond tip of sepals . . . . 4. **C. glomeratum**

3. Uppermost bracts (often partly hidden in a congested inflorescence) with scarious margins at the tip, glabrous there or occasionally with a few stiff hairs; pedicels slightly if at all curved

5. Lowermost bracts of inflorescence with conspicuous scarious tips ca. 1–1.5 mm long; petals very shallowly notched (less than 0.6 mm); sepals ca. 3.5–4.5 mm long; seeds ca. 0.5 mm or less in longest dimension; stamens 5 . . . . 5. **C. semidecandrum**

5. Lowermost bracts herbaceous or with scarious margin less than 0.5 mm wide at tip; petals cleft 0.8–1.2 (1.4) mm; sepals (4) 4.5–7 mm long; seeds ca. 0.6–0.7 mm; stamens normally 10 . . . . 6. **C. fontanum**

1. **C. tomentosum** L. Snow-in-summer

Map 196. A showy perennial rock-garden plant, native to Italy, occasionally escaping from gardens and locally established in lawns, roadsides, and neighboring areas.

2. **C. arvense** L. Field Chickweed

Map 197. Rock outcrops in the Upper Peninsula from the limestones of Drummond Island to the granites and conglomerates farther west; also in dry open oak woods and sandy or grassy places (lawns, roadsides, dry prairies). This perennial species is widespread in temperate regions of the world and is sometimes cultivated; it is possible that some of our collections, especially from southern Michigan, represent escapes.

3. **C. nutans** Raf. Nodding Chickweed

Map 198. This species (an annual) and the preceding are the only native members of the genus in Michigan, and *C. nutans* is the only one not found in Europe. It grows on river banks, rocky ground, fields, and borders of woods.

The flowers are small, with sepals 2.5–4 (5) mm long and seeds ca. 0.5 (0.4–0.8) mm in their longest dimension. The mature capsules are 7–10 (12) mm long, thus running about the same size as those of *C. fontanum,* although the teeth may be a little shorter (as small as 0.4 mm but ranging as long as 1 mm).

4. **C. glomeratum** Thuill.

Map 199. A native of Europe, rather resembling the preceding species but with shorter (and often straighter) pedicels resulting in a more dense inflorescence. Collected in Michigan as a weed in Ann Arbor (*Reznicek 4952* in 1978, MICH, MSC), but probably more widespread in the state.

5. **C. semidecandrum** L.

Map 200. A European species collected in 1907 by Dodge in sandy ground in St. Clair County, where he said it was plentiful. Since 1970, it has been found elsewhere as a lawn weed or in disturbed gravelly ground.

This is an annual with small broadly elliptic-ovate leaves and conspicuous scarious tips on the bracts. The capsules in our material are about 4–7 mm long.

6. **C. fontanum** Baumg. Fig. 74 Mouse-ear Chickweed

Map 201. An all too common short-lived perennial weed, native to Eurasia. Collected in Michigan as early as 1838 by the First Survey and now thriving throughout in all kinds of dry waste places and disturbed ground, spreading into woods and onto shores and other damp ground.

The nomenclature of this species is unusually complicated, and *Flora Europaea* is here followed. Our plants have long been known as *C. vulgatum,* a name used by Linnaeus in different senses and hence rejected as ambiguous. Linnaeus also confused application of his own *Cerastium viscosum,* a name at one time applied to the present species and sometimes used for what is now called *C. glomeratum.*

The mature capsules in our specimens of *C. fontanum* are 7–10 mm long, including teeth 0.8–1.2 (1.4) mm.

8. Holosteum

REFERENCES

Piehl, Martin A. 1962. Holosteum umbellatum L., an Angiosperm New to Michigan. Rhodora 64: 222–225.

Shinners, Lloyd H. 1965. Holosteum umbellatum (Caryophyllaceae) in the United States: Population Explosion and Fractionated Suicide. Sida 2: 119–128.

1. **H. umbellatum** L. Jagged Chickweed

Map 202. A rather inconspicuous little Eurasian weed, very local in

199. Cerastium glomeratum

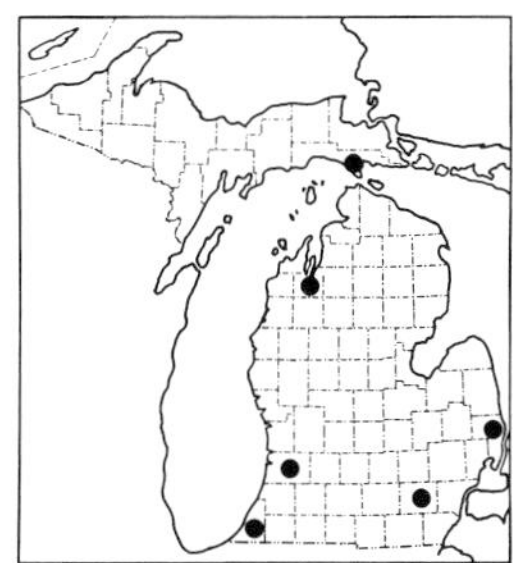

200. Cerastium semidecandrum

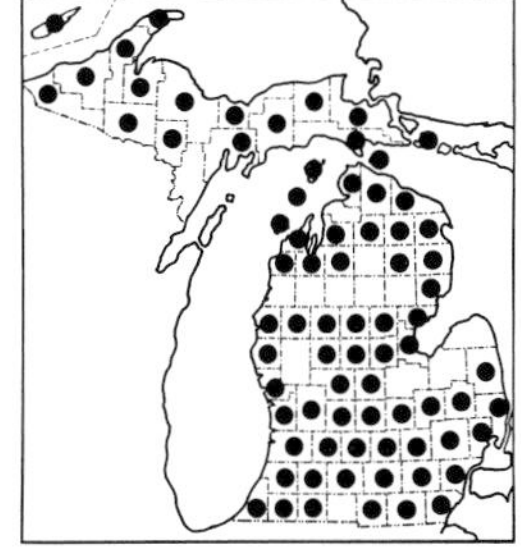

201. Cerastium fontanum

Michigan but perhaps spreading into the state from the south. A plant of roadsides, lawns, and cultivated ground.

Most of our specimens are slightly to densely glandular-pubescent on portions of the stem, but some are completely smooth. With us, the plant appears to be a winter annual, blooming as early as April, fruiting promptly, and ordinarily drying up in May; consequently, it is easily overlooked. Plants may flower when as short as 2 cm, although mature fruiting individuals may be as tall as 25 cm.

9. **Sagina** Pearlwort

REFERENCE

Crow, Garrett E. 1978. A Taxonomic Revision of Sagina (Caryophyllaceae) in North America. Rhodora 80: 1–91.

KEY TO THE SPECIES

1. Petals much shorter than the sepals, very inconspicuous; styles less than 0.5 mm long . . . . 1. **S. procumbens**
1. Petals conspicuously exceeding the sepals; styles ca. 1–1.5 mm long . . . . . 2. **S. nodosa**

1. **S. procumbens** L.

Map 203. A Eurasian native, sparingly found in lawns, sidewalk cracks, and meadows, as well as on wet rocks.

This species has much more conspicuous hyaline leaf bases than the next, connate around the stem.

2. **S. nodosa** (L.) Fenzl Plate 1-G

Map 204. Characteristically in crevices of rocks very near the edge of Lake Superior, where the spray from waves keeps the plants moist.

A northern species, the native North American plants distinguished as the glabrous ssp. *borealis* Crow, and ranging from Newfoundland and Labrador to Great Slave Lake, south to New England and Lake Superior.

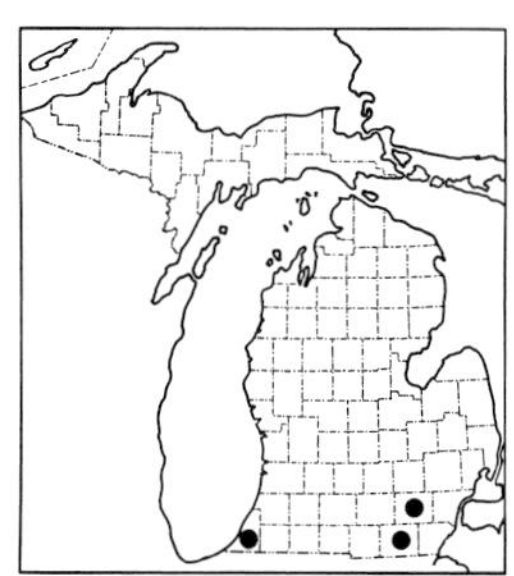

202. Holosteum umbellatum

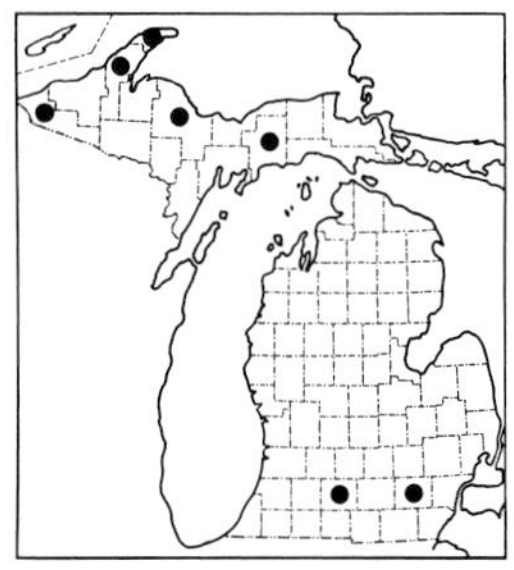

203. Sagina procumbens

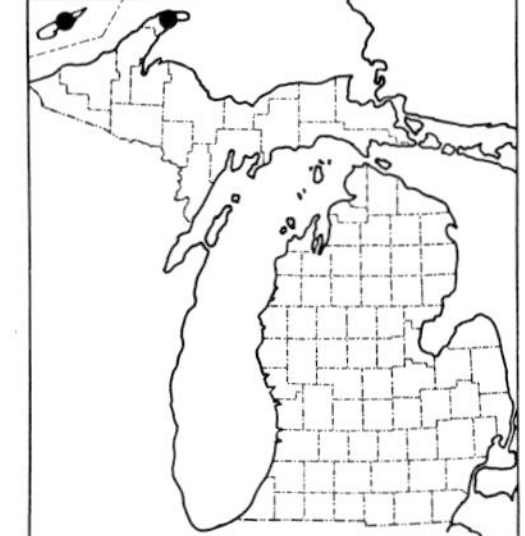

204. Sagina nodosa

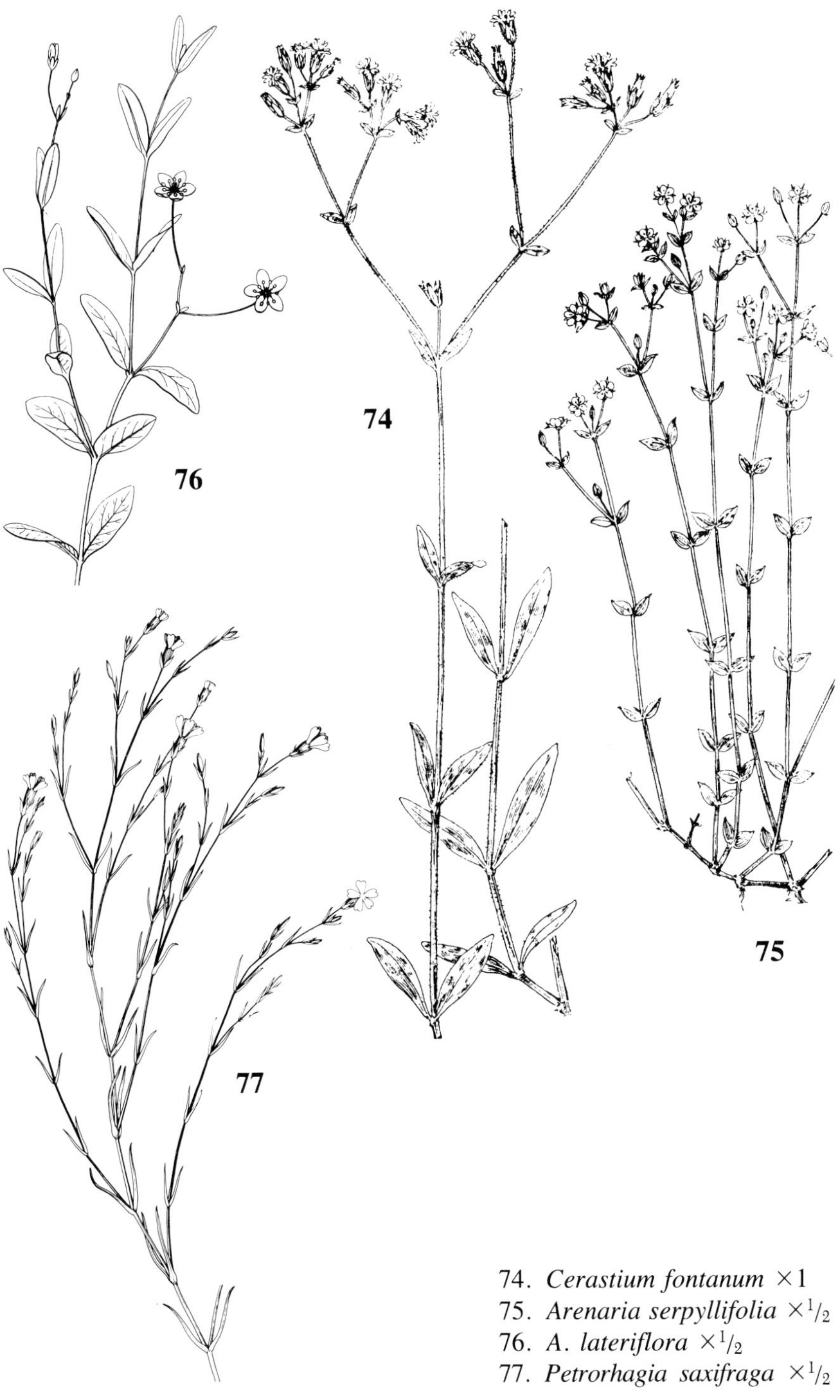

74. *Cerastium fontanum* ×1
75. *Arenaria serpyllifolia* ×$^1/_2$
76. *A. lateriflora* ×$^1/_2$
77. *Petrorhagia saxifraga* ×$^1/_2$

10. **Arenaria** Sandwort

The genus is here retained in a broad sense, following most American manuals. Most European authorities (and some North American) segregate at least two additional genera, as noted under the respective species below. The two positions are presented in the references cited.

REFERENCES

Maguire, Bassett. 1951. Studies in the Caryophyllaceae—V. Arenaria in America North of Mexico. Am. Midl. Nat. 46: 493–511.

McNeill, J. 1962. Taxonomic Studies in the Alsinoideae: I. Generic and Infra-generic Groups. Notes Roy. Bot. Gard. Edinburgh 24: 79–155.

KEY TO THE SPECIES

1. Leaves linear-subulate, the principal cauline ones subtending dense axillary fascicles; plant entirely glabrous; capsule dehiscing into 3 valves . . . . 1. **A. stricta**
1. Leaves ovate to elliptic or lanceolate, mostly without axillary tufts; plants puberulent at least on the stem; capsule with the 3 valves again split, resulting in a total of 6 teeth
   2. Ripe seeds minutely and regularly roughened (tuberculate) and unappendaged; leaves ovate-elliptic, acute to acuminate, but not over 7 (9) mm long; petals shorter than sepals; plants annual . . . . 2. **A. serpyllifolia**
   2. Ripe seeds smooth and shiny, with a pale appendage ("strophiole") at the point of attachment; leaves lanceolate to broadly or narrowly elliptic, mostly over 10 mm long; petals exceeding sepals; plants perennial
      3. Leaves lanceolate to narrowly elliptic, acute, eciliate (though margins may be slightly roughened); sepals acute . . . . 3. **A. macrophylla**
      3. Leaves broadly elliptic, the larger ones nearly oblong-elliptic and mostly rounded at apex, ciliate (and midrib pubescent beneath); sepals ± rounded or obtuse . . . . 4. **A. lateriflora**

1. **A. stricta** Michaux Rock Sandwort

Map 205. Dry open oak and jack pine woodlands, sandy ridges and dunes, limestone pavements (Drummond Island), rarely shores.

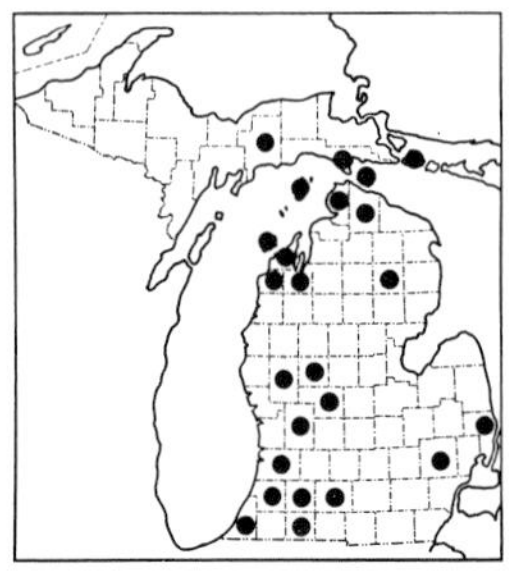
205. Arenaria stricta

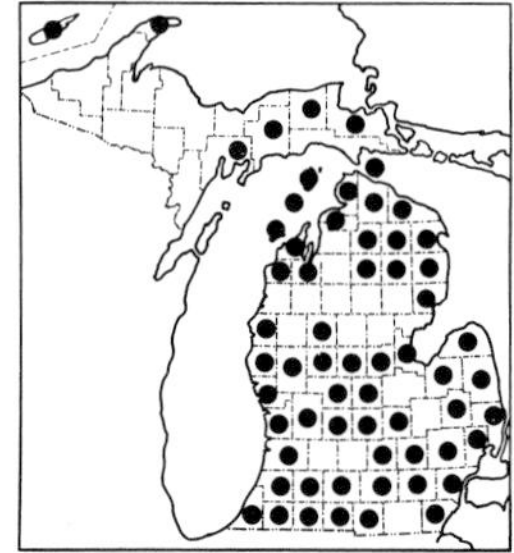
206. Arenaria serpyllifolia

207. Arenaria macrophylla

One collection (*C. & E. Erlanson 699* in 1924, MICH) from the shore of Scotty Bay, Mackinac Co., has the petals slightly shorter than the sepals, the two lateral ribs of the sepals slightly less strong than the midrib, and the leaves very crowded at the base of the plant; it may represent var. *litorea* (Fern.) Boivin [ssp. *dawsonensis* (Britton) Maguire], as may a very few other collections.

This species is often included in a segregate genus *Minuartia,* in which case the correct name becomes *M. michauxii* (Fenzl) Farw., as there is already an *M. stricta* referring to quite a different plant. Several similar species occur north and west of our area. *A. stricta* is readily distinguished from other "chickweed-like" plants by its stiff, needle-like leaves, acuminate prominently 3 (–5)-ribbed sepals, conspicuous petals, and totally glabrous foliage.

2. **A. serpyllifolia** L. Fig. 75

Map 206. A Eurasian weed, all too familiar, especially in dry sandy or gravelly ± disturbed places. The earliest Michigan collections seen date from the early 1860's.

A distinctive much-branched annual with very small leaves; the stem, leaves, and usually sepals are ± densely glandular-puberulent. Sometimes segregated from this species is *A. leptoclados* (Reichb.) Guss., a more delicate plant with capsule nearly cylindrical, sepals often smaller, and half as many chromosomes. I am not convinced that we have two species in Michigan. All our material has either ovoid capsules (definitely expanded below the middle) or, if very young, sepals at least 3 mm long. Overall, the sepals range 2.3–3.7 (4) mm long in our plants with ovoid capsules.

3. **A. macrophylla** Hooker

Map 207. Very rare, on rock outcrops.

Sometimes confused with *Stellaria calycantha,* but it should be remembered that the latter has glabrous stems, sparsely ciliate leaves, and petals usually absent; *A. macrophylla* has puberulent stems (and peduncles), eciliate leaves, and conspicuous petals. Although the petals are often said to be shorter than the sepals, in our freshest flowering material they are distinctly longer. This is included in a separate genus by many authors, as *Moehringia macrophylla* (Hooker) Fenzl.

4. **A. lateriflora** L. Fig. 76

Map 208. Usually in damp ± open woods and bluffs. The best known locality in the state is Grand Ledge, Eaton County.

Like the preceding, often placed in *Moehringia,* a genus separated from *Arenaria* by the strophiole on the seeds and a different base chromosome number. It is then called *M. lateriflora* (L.) Fenzl.

11. Agrostemma

1. **A. githago** L. Fig. 78 Corn-cockle

Map 209. Grain fields and other cultivated or dry disturbed areas, including roadsides, railroads, bluffs, clearings in woodlands. Known in Michigan at least since 1837, when it was collected by the First Survey in Lenawee County. Introduced from Eurasia, where it is also a weed of cultivated lands.

The prominent calyx lobes exceed the large pink-purple petals, making this a very distinctive as well as showy plant.

12. Petrorhagia

The illegitimate name *Tunica* has often been applied to this genus.

REFERENCES

Rabeler, Richard K. 1980. Petrorhagia prolifera, a Naturalized Species in Michigan. Mich. Bot. 19: 83–88.

Rabeler, Richard K. 1985. Petrorhagia (Caryophyllaceae) of North America. Sida 11: 6–44.

KEY TO THE SPECIES

1. Flowers in a head, closely enveloped by large scarious-margined bracts; calyx (mostly hidden by bracts) ca. 9–12 mm long . . . . 1. **P. prolifera**
1. Flowers solitary; calyx ca. 3–5 mm long (not hidden by bracts) . . . . 2. **P. saxifraga**

1. **P. prolifera** (L.) Ball & Heywood

Map 210. A Eurasian native, sparingly escaped from cultivation in sandy or gravelly places.

2. **P. saxifraga** (L.) Link Fig. 77

Map 211. A bushy little perennial, native to Eurasia and occasionally escaped from cultivation, especially in sandy and gravelly places, including roadsides, shores, and lawns.

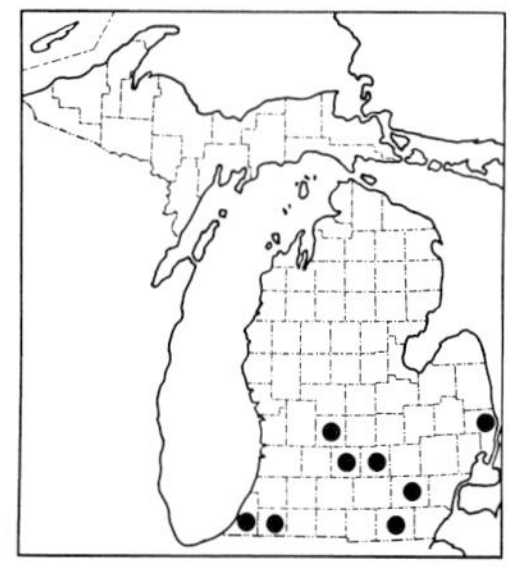
208. Arenaria lateriflora

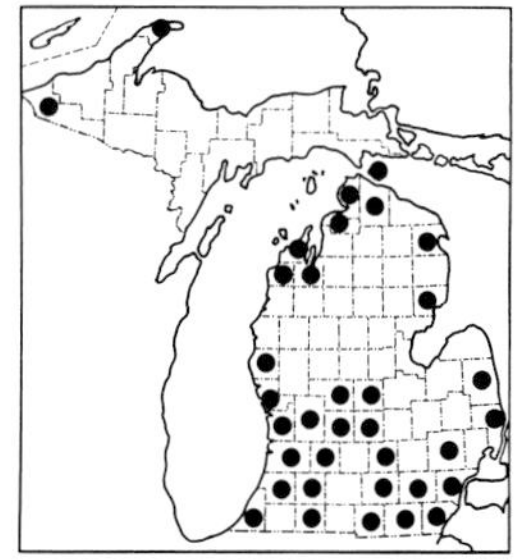
209. Agrostemma githago

210. Petrorhagia prolifera

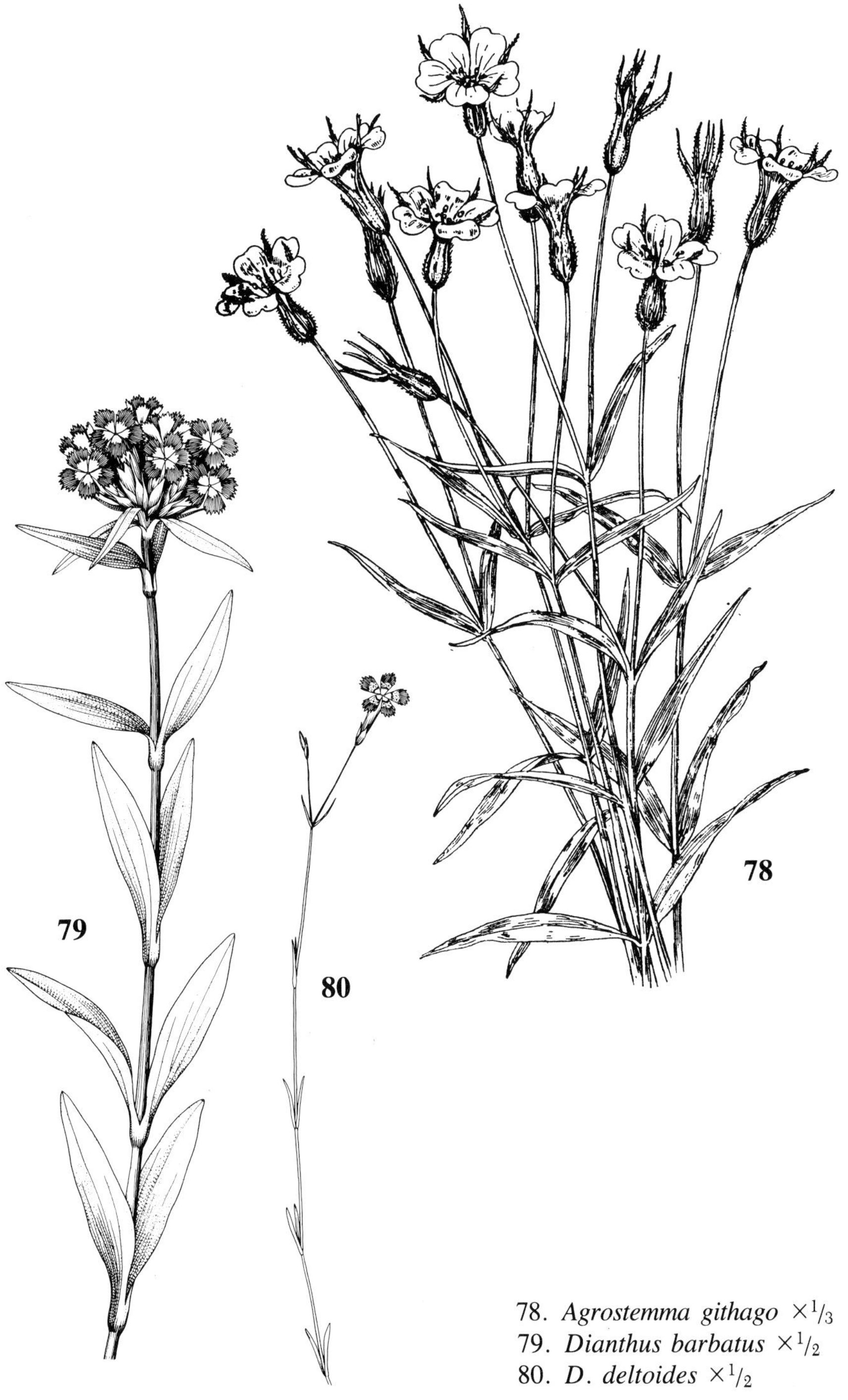

78. *Agrostemma githago* $\times {}^{1}/_{3}$
79. *Dianthus barbatus* $\times {}^{1}/_{2}$
80. *D. deltoides* $\times {}^{1}/_{2}$

13. **Dianthus** Pink

All of our species are natives of the Old World, escaped from cultivation. The corollas are various shades of red, pink, or white, sometimes variegated or spotted. The garden carnation is *D. caryophyllus* L., which does not escape.

KEY TO THE SPECIES

1. Flowers in a densely crowded inflorescence; bracts (including awns if present) at base of calyx mostly equaling or exceeding calyx tube
  2. Calyx pubescent . . . . 1. **D. armeria**
  2. Calyx (at least the tube) glabrous
    3. Largest cauline leaves ca. (1) 1.5–2.5 (3) cm broad, sheathing the stem for less than 3 times its diameter . . . . 2. **D. barbatus**
    3. Largest cauline leaves less than 5 mm broad, sheathing the stem for more than 3 times its diameter . . . . 3. **D. carthusianorum**
1. Flowers solitary or if few, on evident pedicels; bracts at base of calyx ca. half as long as calyx tube, or shorter
  4. Stems (especially lower internodes) minutely puberulent; calyx ca. 12–14 (18) mm long; lower leaves (on sterile shoots) less than 2 cm long and minutely ciliate; margins of leaves thin; petals toothed but not lacerate . . . . 4. **D. deltoides**
  4. Stems glabrous; calyx ca. 16–23 mm (or more) long; lower leaves elongate, those of the dense basal tufts mostly at least 3 cm long and smooth to scabrous-margined; margins of leaves usually ± thickened; petals various
    5. Petals strongly lacerate, with hairs on the darker lower portion; plant ± glaucous . . . . 5. **D. plumarius**
    5. Petals merely toothed to entire, glabrous; plant green, not glaucous . . 6. **D. sylvestris**

1. **D. armeria** L. Deptford Pink

Map 212. Sandy fields and roadsides, clearings in woods, ditches and banks.

2. **D. barbatus** L. Fig. 79 Sweet William

Map 213. Roadsides, clearings, ditches, and woods.

3. **D. carthusianorum** L. Cluster-head Pink

Map 214. Locally established in Hancock (see Rabeler & Gereau in Mich. Bot. 23: 39–42. 1984).

4. **D. deltoides** L. Fig. 80 Maiden Pink

Map 215. Roadsides, waste places, often near old homesites.

The calyx and bracts are glabrous to minutely puberulent. *D. chinensis* L., rainbow pink, would run near here in the key; it has a glabrous calyx subtended by more than 2 bracts, the basal leaves wither by flowering time, and the cauline leaves are broad. It was collected at Grand Rapids in 1896 but its status there is not clear—especially since Emma Cole did not list it in her *Grand Rapids Flora* (1901).

5. **D. plumarius** L. Garden or Grass Pink

Map 216. Roadsides, fields, open rocky woods, dunes, expected near homesites and cemeteries.

6. **D. sylvestris** Wulfen

Map 217. Found by Rabeler in 1982 in gravelly ditch at Hancock, with *D. carthusianorum*.

14. **Gypsophila** Baby's-breath

Our species are native from central Asia to eastern or central Europe, and are here escaped from cultivation. The petals are white to roseate.

REFERENCES

Darwent, A. L. 1975. The Biology of Canadian Weeds. 14. Gypsophila paniculata L. Canad. Jour. Pl. Sci. 55: 1049–1058.

Lawrence, G. H. M. 1953. Keys to Cultivated Plants 1. The Cultivated Species of Gypsophila. Baileya 1: 16–18.

Pringle, James S. 1976. Gypsophila scorzonerifolia (Caryophyllaceae), a Naturalized Species in the Great Lakes Region. Mich. Bot. 15: 215–219.

Rabeler, Richard K. 1981. Gypsophila muralis Is It Naturalized in Michigan? Mich. Bot. 20: 21–26.

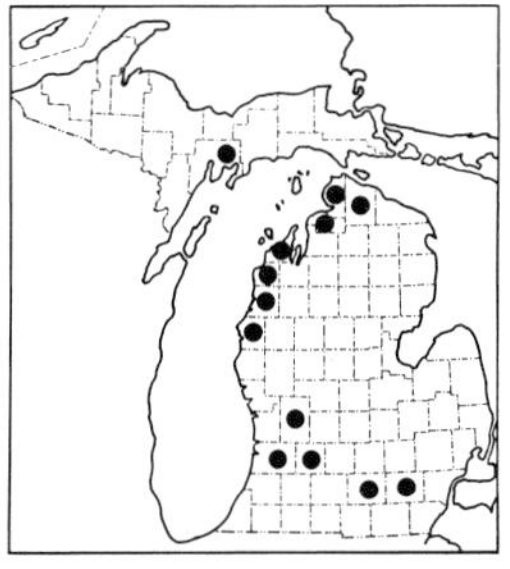

211. Petrorhagia saxifraga

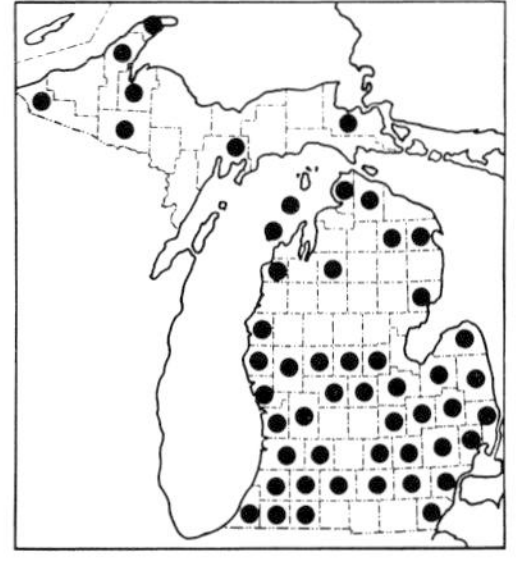

212. Dianthus armeria

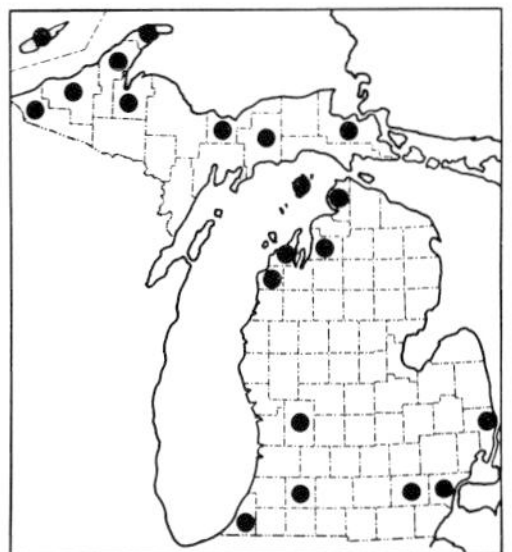

213. Dianthus barbatus

214. Dianthus carthusianorum

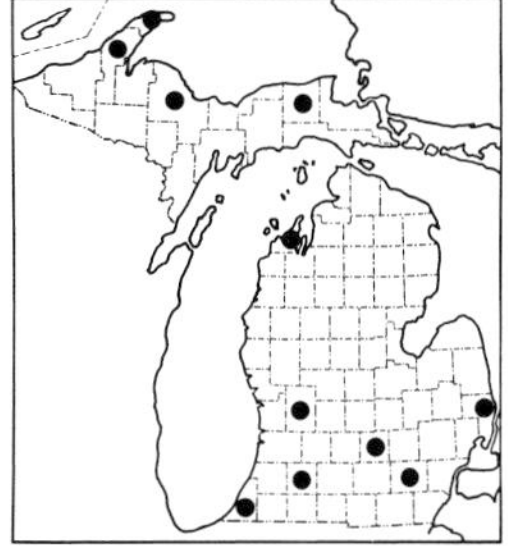

215. Dianthus deltoides

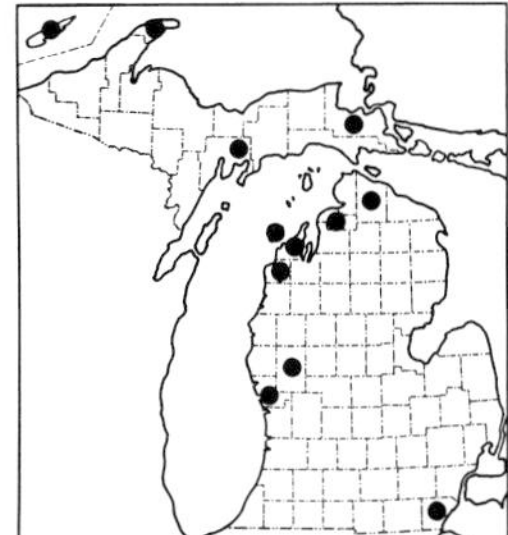

216. Dianthus plumarius

KEY TO THE SPECIES

1. Pedicels and calyces glandular-pubescent; largest leaves (except on depauperate stems) usually at least 1 cm broad . . . . 1. **G. scorzonerifolia**
1. Pedicels and calyces glabrous; largest leaves less than 1 cm broad
   2. Calyx 1–2.1 mm long; plant a perennial 5–8 dm or more tall, from a strong woody rhizome . . . . 2. **G. paniculata**
   2. Calyx larger; plant an annual up to 6 dm tall, from a slender tap root
      3. Leaves usually ca. 4–7 mm wide; lowermost internodes glabrous . . . . . . 3. **G. elegans**
      3. Leaves less than 3 mm wide; lowermost internodes minutely pubescent . . . . 4. **G. muralis**

1. **G. scorzonerifolia** Ser.

Map 218. Roadsides, shores, dunes, quarries, stamp sands (Houghton Co.), apparently thriving in very calcareous sites.

A tall sturdy plant, usually with several stems from a stout caudex, only recently recognized as established in this country (Pringle 1976).

2. **G. paniculata** L.

Map 219. Sandy roadsides, fields, and shores; ditches and railroad embankments; often abundant where found.

The large dome-shaped or tumbleweed-like plants, much-branched and bushy, covered with tiny flowers, are a familiar sight in many disturbed areas.

3. **G. elegans** Bieb.

Map 220. Very local in waste places such as vacant lots, roadsides, railroad embankments.

Depauperate little plants with narrower leaves than usual can be puzzling, but are stiffly erect in contrast with the habit of the next species.

4. **G. muralis** L.

Map 221. Roadsides and waste ground, not collected in Michigan since 1900 (Rabeler 1981).

217. Dianthus sylvestris

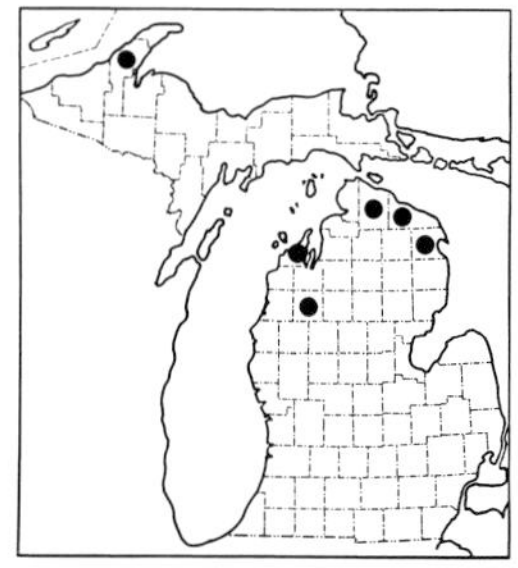

218. Gypsophila scorzonerifolia

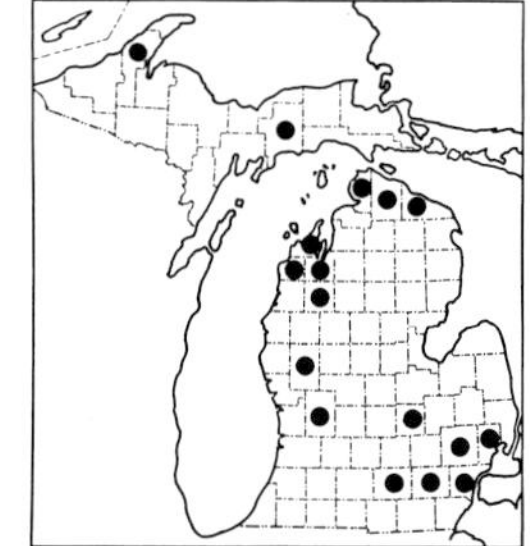

219. Gypsophila paniculata

A delicate diffuse little annual, not nearly so large or robust as usual for the preceding species, and easily mistaken for *Petrorhagia saxifraga*.

15. Vaccaria

1. **V. hispanica** (Miller) Rauschert Fig. 83 Cow Herb

Map 222. A Eurasian species found sporadically along railroads and in waste ground. Found by the First Survey (1838) but no Michigan collections since the 1950s have been seen.

Originally described by Linnaeus as *Saponaria vaccaria*. Also known as *V. segetalis* Garcke or *V. pyramidata* Medicus.

16. Saponaria

KEY TO THE SPECIES

1. Calyx ca. 7–9 mm long, glandular-pubescent (as are the stem and pedicels); plant ± spreading . . . . 1. **S. ocymoides**
1. Calyx (16) 18–21 (23) mm long, glabrous and eglandular (as is the rest of the plant); plant erect . . . . 2. **S. officinalis**

1. **S. ocymoides** L.

Map 223. A European species, locally established since at least 1967, presumably as an escape from cultivation, on the sandy grounds of the University of Michigan Biological Station (see Mich. Bot. 16: 133. 1977).

Distinctive in its sprawling habit, bright rose-magenta flowers, and conspicuously glandular-pubescent calyx.

2. **S. officinalis** L. Fig. 81 Bouncing Bet; Soapwort

Map 224. A familiar and thoroughly naturalized Eurasian species, often forming large beds along roadsides and railroad embankments, as well as occurring in fields and clearings, on dunes and shores, and in similarly open

220. Gypsophila elegans

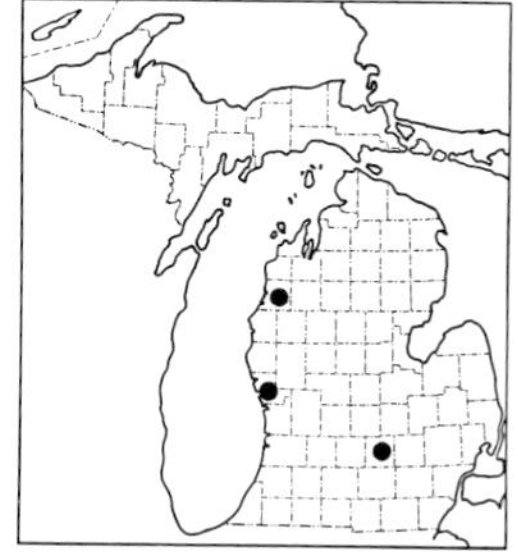
221. Gypsophila muralis

222. Vaccaria hispanica

places. Reported by the First Survey (1838) but specimens have not survived; collections before the 1890's in Michigan are very rare.

The corolla is usually pale pink to whitish, and often double. The calyx is usually ± flushed with reddish toward the apex.

17. Silene

The number of styles and hence of capsule teeth (the same number or, when teeth split, twice the number of the styles) may vary from the "typical," making absolute reliance on this key character less trustworthy than one would like. The nature of the inflorescence is also variable, as well as difficult to describe simply. As stated in *Flora Europaea*, "no key which depends on inflorescence-form can work with more than moderate success, and small specimens, whether of annual or perennial species, may be wrongly identified. In the present state of knowledge, there appears to be no remedy to this."

The western American *S. drummondii* Hooker has been reported (as *Lychnis drummondii*) from Michigan in manuals, but the basis for such reports seems to be completely unsubstantiated (see Mich. Bot. 6: 19–20. 1967). *S. gallica* L., a Eurasian weed, has also been reported from Michigan but no specimens have been found. It differs from *S. dichotoma*, with which it shares dichotomous one-sided inflorescences, in having broadly obtuse (rather than acute) principal leaves, calyx ca. 8–10 mm long, stipe of capsule ca. 1 mm long or shorter and puberulent (rather than 1.5–3 mm long and glabrous), and seeds less than 1 mm broad (rather than ca. 1.5 mm broad).

S. pendula L., a native of the Mediterranean region, was collected at Lansing in 1887 as "Escaped from flower garden" (MSC). It was not, however, listed in subsequent floras of the state nor mentioned in any other literature, so its status may have been in doubt. And it has not been collected since. It has a ± racemose inflorescence (as in *S. dichotoma* but not one-sided and with a larger calyx of distinctive shape). The calyx lobes are short-triangular as in *S. pratensis*, but the flowers are perfect with 3 styles. The calyx is distinctively narrowed toward the base (as in *S. armeria*) and the

223. Saponaria ocymoides

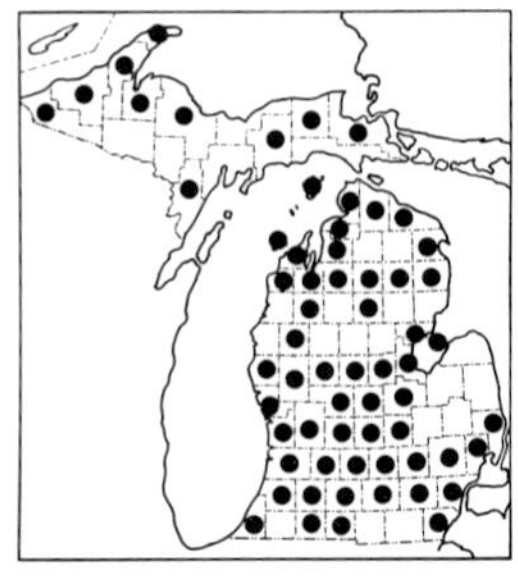

224. Saponaria officinalis

225. Silene antirrhina

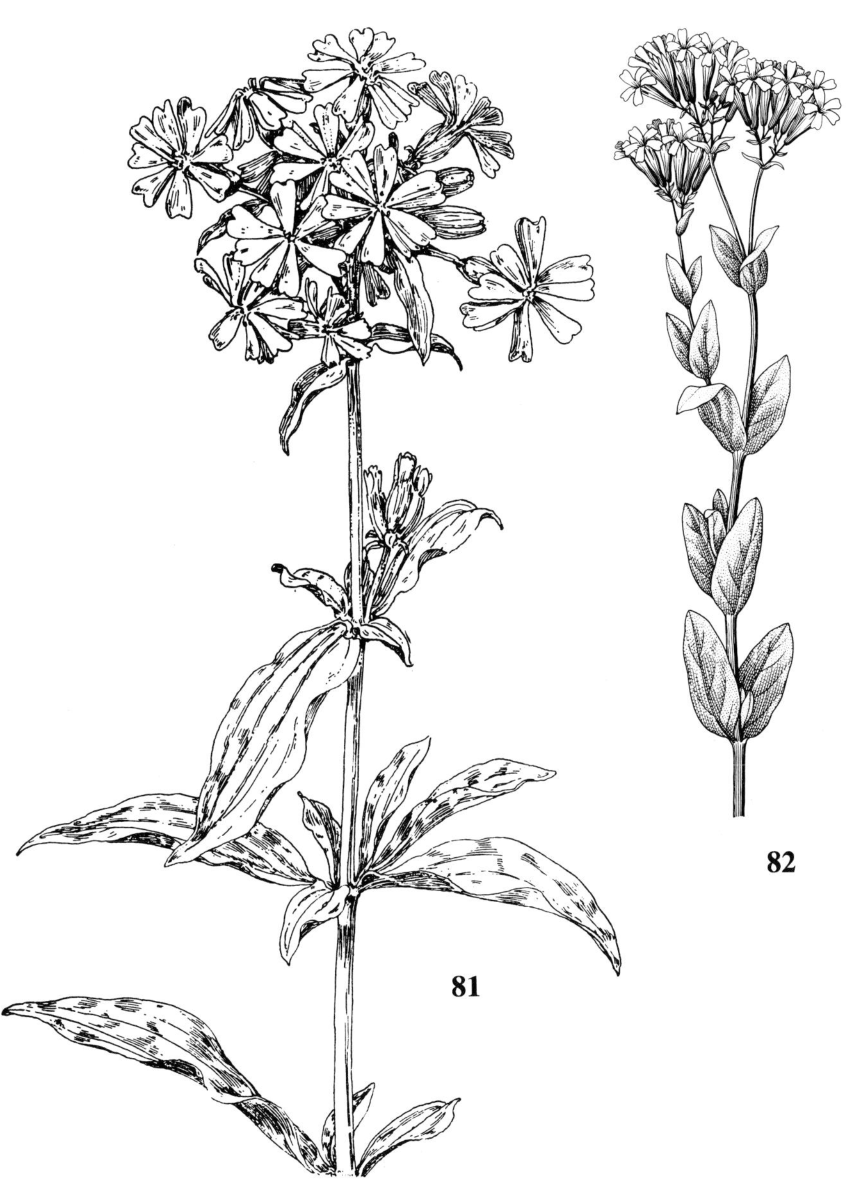

81. *Saponaria officinalis* ×1
82. *Silene armeria* ×1/2

fruiting pedicels are widely spreading or reflexed—quite unlike other species with which it might be confused, with ascending or erect fruit. The petals are pink.

REFERENCES

Bocquet, Gilbert, & Charles Baehni. 1961. Les Caryophyllacées-Silénoïdées de la Flore Suisse. Candollea 17: 191–202.

Hitchcock, C. Leo, & Bassett Maguire. 1947. A Revision of the North American Species of Silene. Univ. Washington Publ. Biol. 13. 73 pp.

McNeill, J. 1977. The Biology of Canadian Weeds. 25. Silene alba (Miller) E. H. L. Krause. Canad. Jour. Pl. Sci. 57: 1103–1114.

McNeill, J., & Honor C. Prentice. 1981. Silene pratensis (Rafn) Godron & Gren., the Correct Name for White Campion or White Cockle (Silene alba (Miller) E. H. L. Krause, nom. illeg.). Taxon 30: 27–32.

KEY TO THE SPECIES

1. Calyx glabrous
  2. Plant an annual, usually with glutinous zones on upper internodes; petals pink to rose (or absent); calyx with 10 prominent nerves
    3. Calyx (4.5) 6–9 mm long; capsule on a stipe (within calyx) scarcely 1 mm long; petals inconspicuous or lacking . . . . 1. **S. antirrhina**
    3. Calyx 12–16 mm long; capsule on a stipe ca. 7–8 mm long (the calyx therefore ± clavate); petals showy (fresh flowers ca. 1 cm or more across) . . . . 2. **S. armeria**
  2. Plant a perennial without glutinous zones on the stem; petals white; calyx with 20 nerves, at least at base, but these often obscure and anastomosing
    4. Mature calyx little if at all inflated around the capsule, without conspicuous anastomosing veins, ± tapered to base; upper bracts of inflorescence minutely ciliate; stipe of capsule (and ovary) minutely puberulent, ca. 1.5 mm long . . . . 3. **S. csereii**
    4. Mature calyx much inflated around the capsule, usually with conspicuous anastomosing veins, umbilicate at base (the pedicel attached in a dimple); upper bracts of inflorescence only very rarely ciliate; stipe glabrous, ca. 1.5–3.5 mm long . . . . 4. **S. vulgaris**
1. Calyx pubscent
  5. Leaves whorled (except at uppermost and lowermost nodes); petals deeply fringed (more than once cleft) . . . . 5. **S. stellata**
  5. Leaves all opposite; petals at most once cleft
    6. Calyx with ca. 30 prominent straight raised nerves . . . . 6. **S. conica**
    6. Calyx with at most ca. 20 nerves, these often obscure
      7. Petals bright red, cleft into two lance-acute lobes (± rounded at tips) . . . . 7. **S. virginica**
      7. Petals white to cream or at most pinkish, the lobes ± obtuse and broadly rounded
        8. Flowers in 1-sided spikes or racemes, the inflorescence simple or dichotomously once- or twice-forked and the flowers or flowering branches 1 per node; calyx ca. 10.5–13.5 (15) mm long . . . . 8. **S. dichotoma**
        8. Flowers in a panicle or compound cyme scarcely if at all 1-sided, the inflorescence typically bearing opposite branches, with flowers (or branches) thus 2–3 per node; calyx larger (except in rare perennial species with tufts of spatulate leaves at base)

9. Calyx ca. 10 mm long or shorter; plant a perennial with sterile tufts of narrowly spatulate leaves at base ....9. **S. nutans**

9. Calyx (11) 14–27 mm long; plant an annual, a biennial, or a short-lived perennial, without basal tufts of leaves (or these mostly elliptical)

10. Lobes of calyx 6–11 mm long, not over 1 mm wide at middle; total calyx length 20–27 mm; plants with perfect flowers; styles 3, the capsule opening by 6 teeth, these strongly recurved at maturity; upper internodes clammy-viscid in living plants, with ± dense glands sessile or on short hairs, the longer hairs mostly not gland-tipped .... 10. **S. noctiflora**

10. Lobes of calyx 2.5–6 (7) mm long, if as long as 6–7 mm then at least 1 mm wide at middle; total calyx length (11) 14–22 (27) mm; plants dioecious, the flowers of pistillate plants with 5–6 (often more, rarely fewer) styles, the capsule opening by twice as many teeth, these at most spreading; upper internodes not clammy-viscid to the touch when fresh, with glands as frequent on tips of longest hairs as on shorter ones (and usually less dense than in preceding species) ....11. **S. pratensis**

1. **S. antirrhina** L. Fig. 84 Sleepy Catchfly

Map 225. A native American plant, but characteristic of dry sandy weedy habitats, as well as on rock outcrops and in dry (especially somewhat disturbed) woods (jack pine, aspen, oak) and prairies.

This species is variable in stature, ranging from very small 1- or few-flowered plants to large much-branched ones, in size and shape of leaves, and in other characters. Plants lacking glutinous zones on the stem are scarce and are f. *deaneana* Fern. Those lacking petals are rather common and are f. *apetela* [sic] Farw. (TL: Washington [Macomb Co.]). Those with petals white beneath and pink above have been called f. *bicolor* Farw. (TL: Flat Rock [Wayne Co.]). The tips of the sepals are ± strongly reddened.

2. **S. armeria** L. Fig. 82 Sweet-William Catchfly

Map 226. A Eurasian species, locally established as an escape from cultivation in dry open fields and roadsides, borders of woods, and garden areas. Collected in Michigan as long ago as 1832 by Nathan Folwell in "oak openings" near Tecumseh, Lenawee County.

An attractive, usually small plant with glaucous foliage. The tips of the

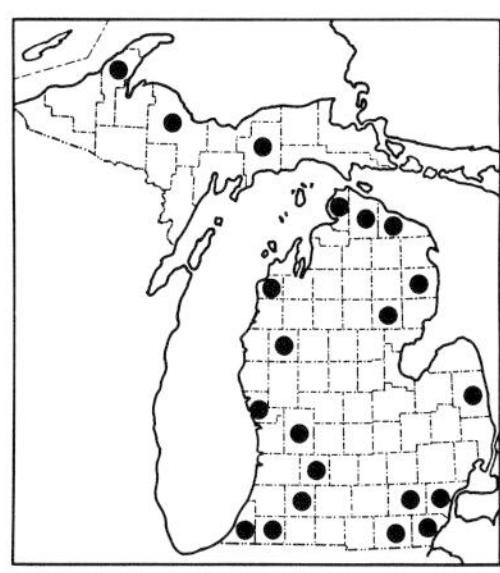

226. Silene armeria

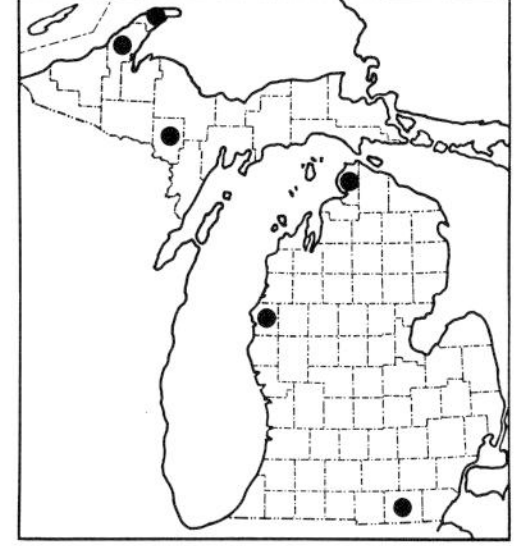

227. Silene csereii

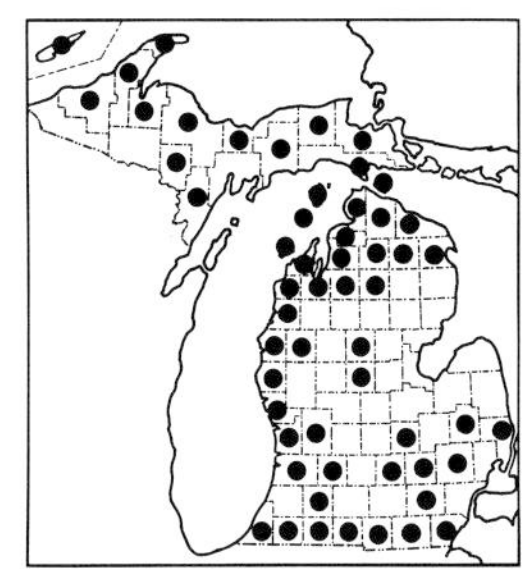

228. Silene vulgaris

sepals are usually not red. Some authors have repeated the statement by Hitchcock and Maguire that the stems are without glutinous bands; in all of our specimens, such bands are present, although an occasional specimen lacking them might be expected, as in the preceding species.

3. **S. csereii** Baumg.

Map 227. A native of southeastern Europe, found locally in Michigan, chiefly along railroads but also in other disturbed places. Apparently first collected in the state in 1932 in Mason County (see Brittonia 9: 91. 1957).

A strongly glaucous plant, very similar to the next species, but with a much stouter root, generally more elongate inflorescence, and capsules slightly exserted from the calyx.

4. **S. vulgaris** (Moench) Garcke — Bladder Campion

Map 228. A widespread weed of Eurasian origin. I have seen no Michigan specimens collected before the 1890's, but it is now common along roadsides and railroads, in clearings and gravel pits, on dunes and gravelly shores, and in other disturbed places; often thrives in calcareous sites.

The capsule is included in the inflated calyx. This species has been widely known under the illegitimate names *S. cucubalus* Wibel and *S. latifolia* (Miller) Britten & Rendle.

5. **S. stellata** (L.) Aiton f. — Starry Campion

Map 229. Dry oak woodland and banks.

Our plants, with densely puberulent calyx and pedicels, represent var. *scabrella* (Nieuwl.) Palmer & Steyerm.; the typical variety, with calyx and pedicels glabrous or nearly so, is more southern and eastern.

6. **S. conica** L.

Map 230. A Eurasian weed, found very sparingly in our area, in sandy or gravelly places.

7. **S. virginica** L. — Fire Pink

Map 231. The native range extends barely to southeastern Michigan, where the species was collected in open woods in the vicinity of Trenton and Grosse Ile (both Wayne Co.) as early as 1838 but no more recently than 1917. It has also been collected in waste ground (perhaps an escape from cultivation) at East Lansing (in 1938) and Bay City (1939).

8. **S. dichotoma** Ehrh. Fig. 85

Map 232. Roadsides, fields, railroads, and other weedy sites. Of European origin, naturalized in North America and not collected in Michigan before the first decade of this century; becoming conspicuous throughout the Lower Peninsula in the 1940's and 1950's.

9. **S. nutans** L. Nottingham Catchfly

Map 233. A Eurasian species, sometimes cultivated and presumably an escape when collected in 1913 by Dodge in dry open ground at a cemetery in Port Huron.

10. **S. noctiflora** L. Night-flowering Catchfly

Map 234. Another Eurasian species, naturalized as a weed in the usual sorts of disturbed sites as along roadsides and railroads, in fields and clearings. Collected in Michigan as long ago as 1832 by Folwell.

11. **S. pratensis** (Rafn) Godron & Gren. White Cockle; White Campion

Map 235. An Old World native, too familiar as a naturalized weed, but apparently not collected in Michigan before 1865. Common in all kinds of disturbed sites: roadsides, fields, shores, gravel pits, banks, and edges of woods; usually but not always on dry sites.

Long called *Lychnis alba* Miller in American manuals and also known as *Melandrium album* (Miller) Garcke, this species is now accepted as belonging to the same genus as the preceding one, with which it is often confused. The essential distinctions are given in the key. For some time it was also called *S. alba* (Miller) E. H. L. Krause, which is incorrect (see McNeill & Prentice 1981). More recently, it turns out that the correct name may be

229. Silene stellata

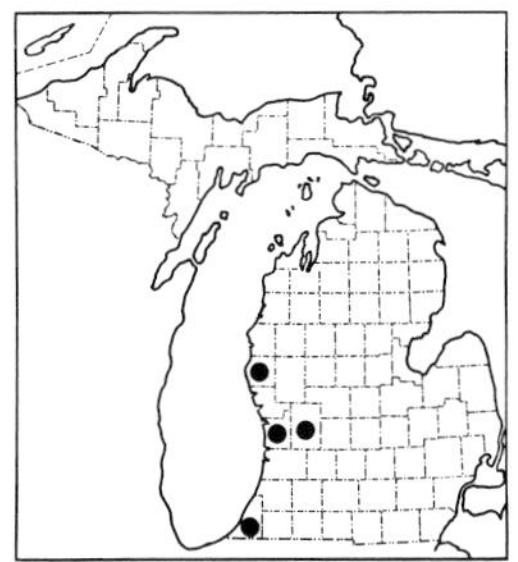
230. Silene conica

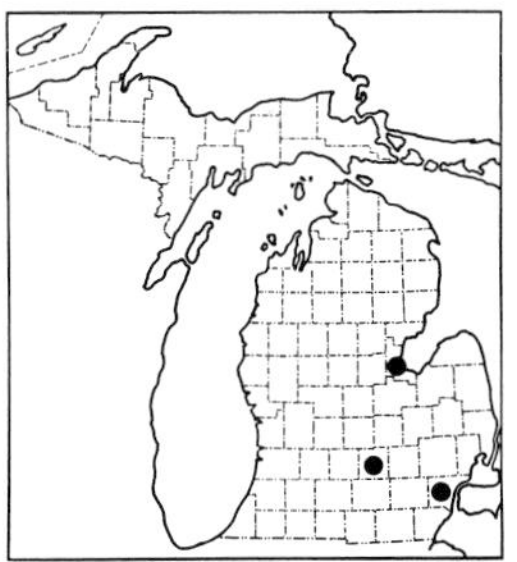
231. Silene virginica

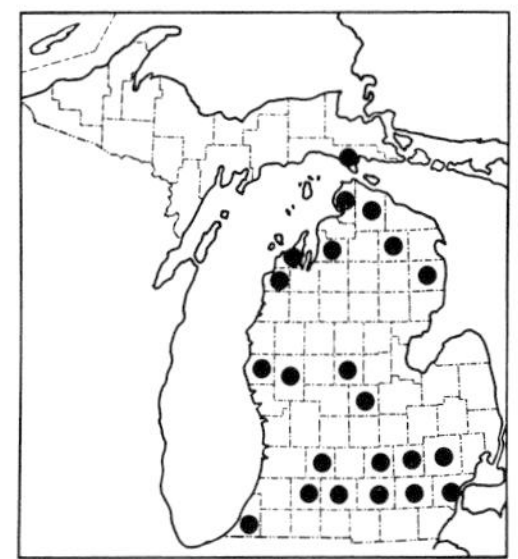
232. Silene dichotoma

233. Silene nutans

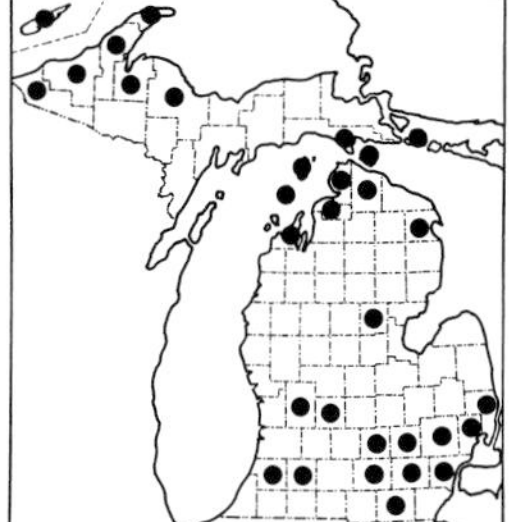
234. Silene noctiflora

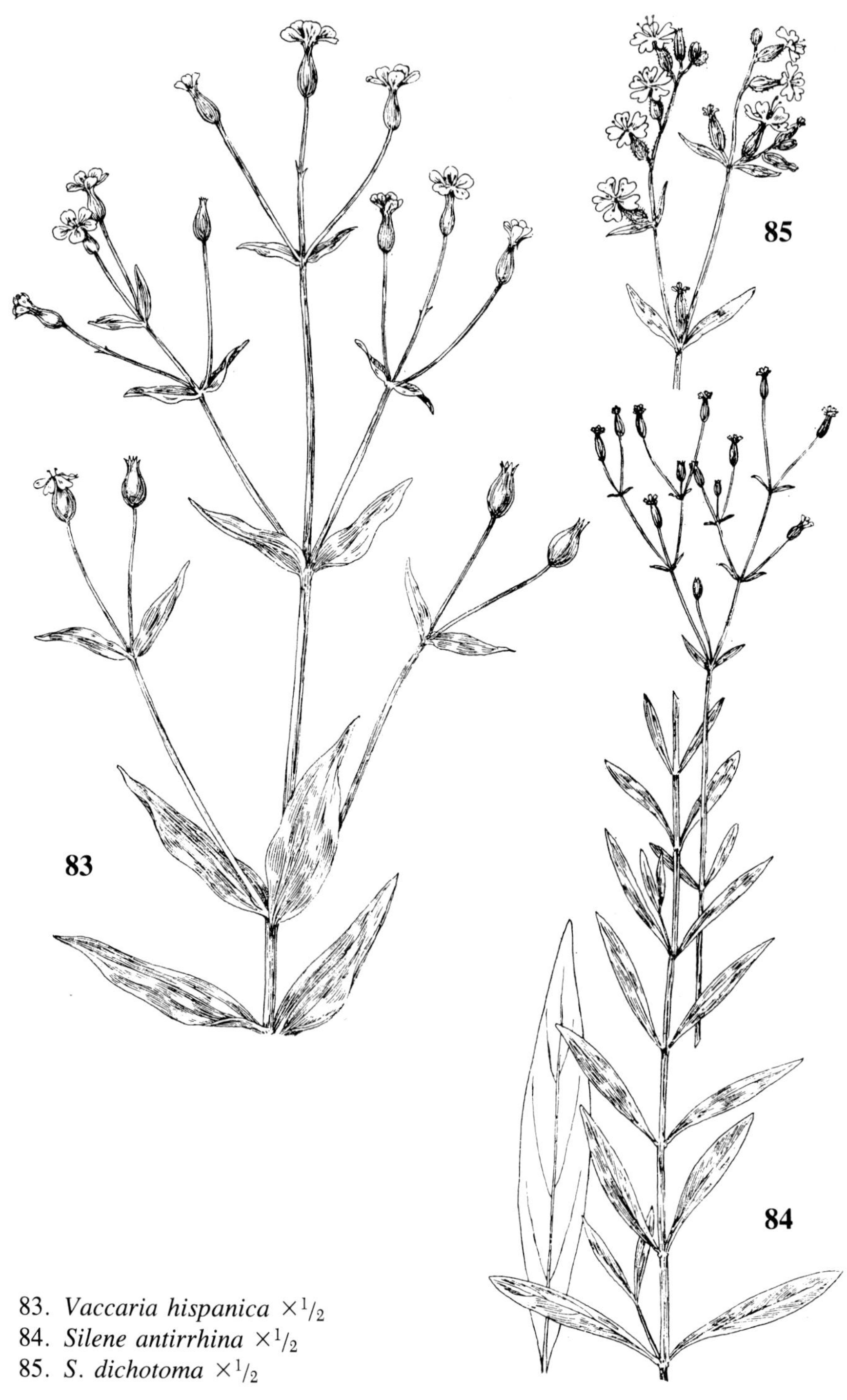

83. *Vaccaria hispanica* ×$^1/_2$
84. *Silene antirrhina* ×$^1/_2$
85. *S. dichotoma* ×$^1/_2$

S. latifolia Poiret, which would be quite confusing in view of the frequent use of the same binomial, with different authors, for *S. vulgaris*.

S. pratensis is an example of a dioecious plant in which sex determination results from X and Y chromosomes, as in the human species. The calyces of staminate plants are 10-nerved; those of pistillate plants, 20-nerved (with alternate nerves stronger than the others). (In *S. noctiflora*, the calyx is 10-nerved and there are prominent white glabrous areas between the nerves, especially in fruit.) Pink-flowered plants from Hillsdale County (*Voss 8782A* in 1959, MICH) have been determined by Prentice as a hybrid of this species with *S. dioica* (L.) Clairv.

18. **Lychnis**

Recent widely accepted generic realignments (see Bocquet & Baehni 1961, *Flora Europaea*, *Hortus Third*, etc.) have transferred from this genus to *Silene* several species, including the well known "*Lychnis alba*." A number of showy species are cultivated and might appear rarely as escapes; Lawrence (1953) will help in their identification.

REFERENCES

Bocquet, Gilbert, & Charles Baehni. 1961. Les Caryophyllacées-Silénoïdées de la Flore Suisse. Candollea 17: 191–202.

Lawrence, G. H. M. 1953. Keys to Cultivated Plants 2. The Cultivated Species of Lychnis. Baileya 1: 105–111; 114.

KEY TO THE SPECIES

1. Plant densely white-woolly; flowers few, on elongate pedicels; petals slightly if at all notched . . . . 1. **L. coronaria**
1. Plant glabrate to sparsely hairy; flowers numerous, crowded in a dense inflorescence; petals strongly 2-lobed . . . . 2. **L. chalcedonica**

1. **L. coronaria** (L.) Desr. Fig. 86 Mullein Pink

Map 236. An Old World native, often well established as an escape from

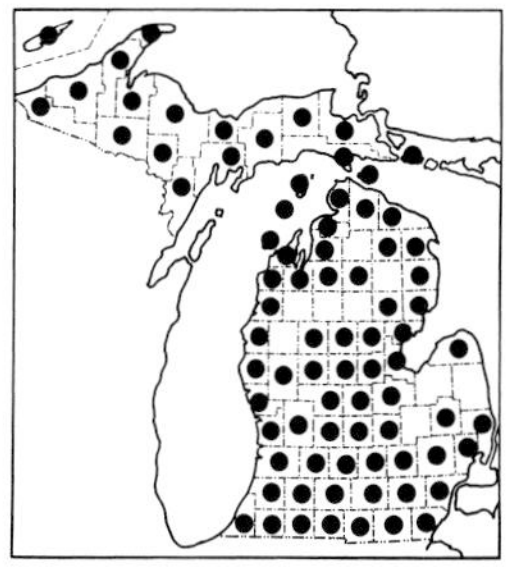

235. Silene pratensis

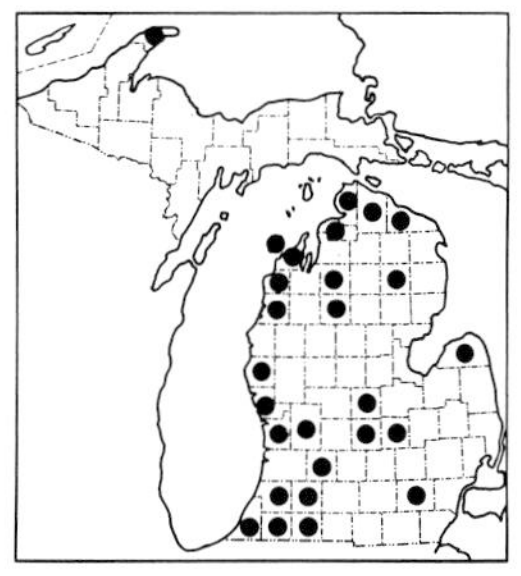

236. Lychnis coronaria

237. Lychnis chalcedonica

cultivation. Roadsides and banks, fields and clearings, open to dense deciduous woods (oak or beech-maple).

2. **L. chalcedonica** L. Maltese-cross; Scarlet Lychnis

Map 237. A native of the northern and central U.S.S.R., rarely established as an escape from cultivation along grassy roadsides and thickets.

The numerous large, ovate, clasping cauline leaves and crowded inflorescences, the numerous flowers with bilobed petals, should distinguish this species from other cultivated members of the genus.

CERATOPHYLLACEAE Hornwort Family

1. Ceratophyllum Coontail

These plants are strictly aquatic and cannot withstand emergence. The leaves are in dense whorls, varying in size, number, and other characters (but more than 4 at a node, unlike *Myriophyllum*); depending in part on water depth, internode length is also extremely variable. This is our only genus of aquatic vascular plants with whorled, dichotomously forked leaves. While the genus is easily recognized, the species have long been the subject of differing opinions.

Reproduction is usually by simple fragmentation of the rather brittle stem, on which the internodes are much shortened toward the tips, the more dense leaves giving the characteristic bushy "coontail" appearance. The flowers are inconspicuous, unisexual (plants monoecious), sessile or nearly so, lacking a perianth but subtended by a 10–12-parted involucre of cleft bracts. The method of pollination is unique among water plants: the stamens (numerous in each flower) are released at maturity and rise to the surface of the water, where the pollen is discharged before sifting down through the water and—with luck—landing on a pistillate flower. The fruit is a relatively large achene (body at least 4 mm long) and in our species bears conspicuous spines, including the long, hard, terminal style. Fruiting is often said to occur only when the water has been (for our climate) unusually warm, and achenes are rarely seen. However, both fruit and shoot fragments are known to be food for several species of waterfowl.

Although *Ceratophyllum* produces no structures that qualify anatomically as roots, the plants are usually anchored in the substrate by pale, whitish, modified leaves and are not necessarily free-floating as often described.

Donald H. Les has been very helpful in developing the treatment of this genus.

REFERENCES

Fassett, Norman C. 1953. North American Ceratophyllum. Com. Inst. Trop. Invest. Cient. 2: 25–45.

Lowden, Richard M. 1978. Studies on the Submerged Genus Ceratophyllum L. in the Neotropics. Aquat. Bot. 4: 127–142.

KEY TO THE SPECIES

1. Achene with only 2 spines (± basal) besides the style, not winged on sides; foliage usually ± stiff, and no leaves forked more than 2 times .... 1. **C. demersum**
1. Achene with several spines on each margin, their bases connected by a low but distinct wing; foliage usually very limp, and with at least some (larger) leaves of main axis forked 3 (–4) times .... 2. **C. echinatum**

1. **C. demersum** L. Figs. 87, 88

Map 238. Submersed in very shallow to deep (18 ft) water of ponds and lakes and quiet backwaters of streams and rivers.

A common species, typically with teeth along the margins of the ultimate segments of the leaves just visible to the naked eye, expanded and green basally (shaped like a rose thorn) and tipped with a pale spinule usually no longer than the green portion. Sometimes, especially on lateral branches and new growth, the leaves may be limper and less conspicuously toothed, and such specimens may resemble the next species. The work of Les supports leaf forking as the most reliable vegetative character, in the absence of fruit with which to identify the species.

2. **C. echinatum** A. Gray Fig. 89

Map 239. A species most often of acid, softwater lakes and ponds, in contrast to the more neutral or hardwater habitats of the preceding.

C. echinatum typically has very delicate and flaccid foliage, in this respect suggestive of two other species characteristic of softwater lakes, *Myriophyllum farwellii* and *Utricularia geminiscapa*. The lower segments of the leaves are sometimes slightly inflated, and the teeth of the ultimate segments are absent or sparse and inconspicuous, consisting almost solely of a pale spinule with little or no expanded green base (but see remarks under the preceding species, which may have similar leaves). Several other characters sometimes stated for distinguishing our two species of *Ceratophyllum* are even more unreliable, including the number of sides of the leaf bearing teeth and smoothness of the achene surface. The significance of the fruit differences in the genus is not agreed upon, and in the usual absence of fruit one is faced with leaves that can sometimes be distinguished only with difficulty. It is necessary to search very carefully for leaves that are forked 3 times (thus with 8 ultimate segments).

North American plants have sometimes been included in the Old World *C. muricatum* Cham. Some authors have treated them as, at best, *C. demersum* var. *echinatum* (A. Gray) A. Gray, considering that there is only one species here—and perhaps in the world.

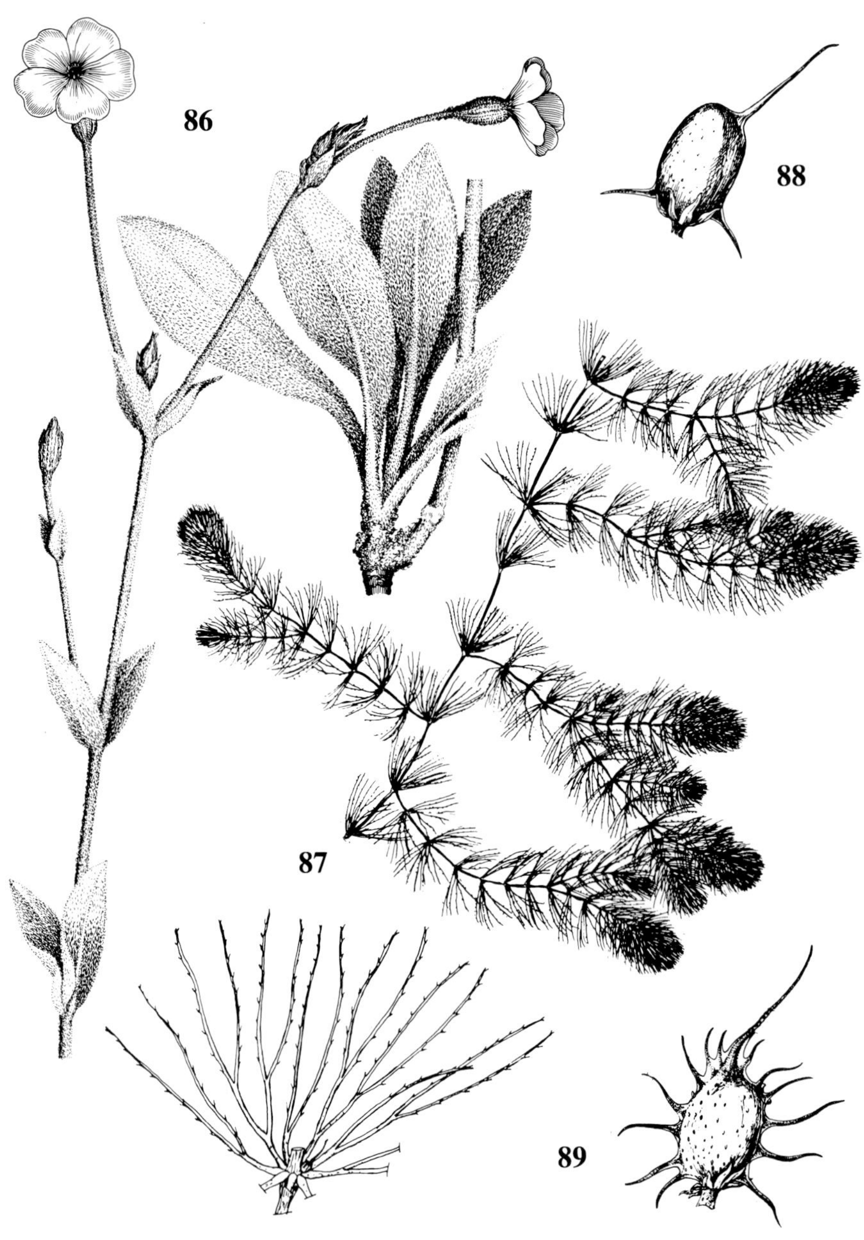

86. *Lychnis coronaria* ×$^{1}/_{2}$
87. *Ceratophyllum demersum* ×$^{2}/_{5}$; whorl of leaves with pistillate flower ×2$^{1}/_{2}$
88. *C. demersum*, fruit ×3
89. *C. echinatum*, fruit ×3

NYMPHAEACEAE Water-lily Family

This group of aquatic plants, mostly with showy flowers, has been variously classified, sometimes in at least three families in two orders. Of the segregates, the Nelumbonaceae (*Nelumbo*) are most often recognized, although Hutchinson retained *Nelumbo* in the Nymphaeaceae while recognizing the Cabombaceae (*Cabomba* and *Brasenia*) as distinct. Cronquist recognizes all these families. Thorne and Wood do not. Since maintaining the family in a broad sense (with three distinct subfamilies) is in accord with the standard manuals for our region, that is done here, recognizing that there are arguments both ways on the question of rank.

REFERENCES

Mitchell, Richard S., & Ernest O. Beal. 1979. Magnoliaceae through Ceratophyllaceae of New York State. New York St. Mus. Bull. 435. 62 pp.

Wood, Carroll E., Jr. 1959. The Genera of the Nymphaeaceae and Ceratophyllaceae in the Southeastern United States. Jour. Arnold Arb. 40: 94–112.

KEY TO THE GENERA

1. Principal leaves dissected, opposite, submersed (small, alternate, floating, peltate leaves may also be present); stamens 3–6; flowers whitish, less than 12 mm long .... 1. **Cabomba**
1. Principal leaves simple, alternate or basal, floating or emersed (entire basal submersed leaves also present in *Nuphar*); stamens 12 or more; flowers yellow, reddish, or white, 12 mm or more long
  2. Leaves peltate; carpels separate
    3. Leaf blades less than 1.5 dm long, ± elliptical, coated beneath (as are the petioles, stems, and buds) with a thick gelatinous coat; flowers maroon or purplish, less than 3 cm broad; stamens 12–18 (30); fruit 1–2-seeded, less than 1 cm long, not embedded in receptacle .... 2. **Brasenia**
    3. Leaf blades mostly more than 1.5 dm in diameter, ± circular, without gelatinous coat; flowers pale yellow, much broader (over 1 dm); stamens numerous; fruits 1-seeded, hard and nut-like, ca. 2 cm long, deeply embedded in the enlarged top-shaped receptacle .... 3. **Nelumbo**

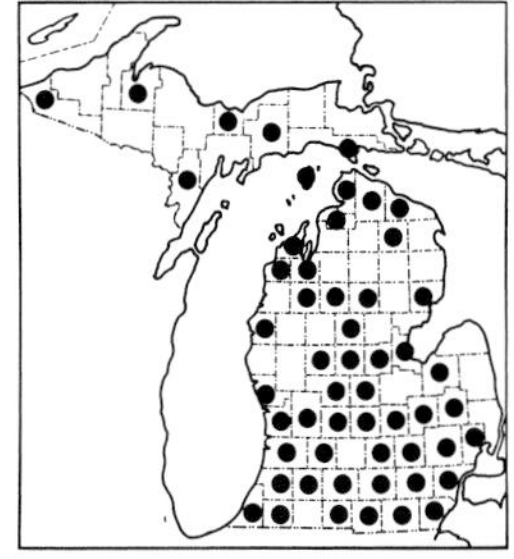

238. Ceratophyllum demersum

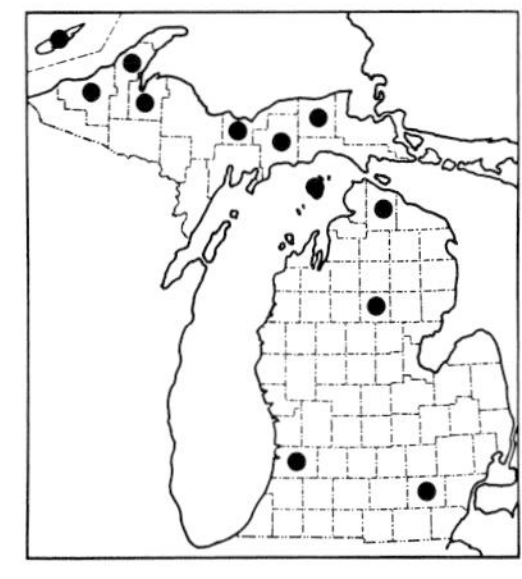

239. Ceratophyllum echinatum

240. Cabomba caroliniana

2. Leaves with marginal petiole inserted in a ± deep sinus; carpels united, at least basally
  4. Flowers with showy white petals, usually fragrant; sepals green or reddish (paler inside); blades of floating leaves circular (or elliptic in one rare species) in general outline, the lobes pointed, with more pairs of veins radiating from base of midrib than from along midrib; basal submersed leaves none . . . . 4. **Nymphaea**
  4. Flowers with small inconspicuous yellowish petals, not fragrant; sepals (at least the inner ones) bright yellow, showy; blades of floating or emersed leaves elliptic in general outline, the lobes rounded, with more pairs of veins arising from along the midrib than at its base; basal submersed leaves (often gone by late season) broad and membranous . . . . 5. **Nuphar**

1. Cabomba

REFERENCE

Fassett, Norman C. 1953. A Monograph of Cabomba. Castanea 18: 116–128.

1. **C. caroliniana** A. Gray Fig. 90 Fanwort

Map 240. A native of the southeastern states, from Texas to Florida northward, this species has become established as far north as New England and southern Michigan, perhaps escaped from cultivation as it is a popular aquarium plant. It was first discovered in Michigan in Kimble Lake, Kalamazoo County, by F. W. Rapp in 1935 (Pap. Mich. Acad. 23: 137. 1938). Soon afterwards, it was found to be abundant upstream, in Barton and Howard lakes, and down Portage Creek into St. Joseph County.

This is our only aquatic vascular plant with opposite (rarely whorled) much-dissected leaves on distinct petioles (5–15 mm long on well developed leaves). The palmate pattern of dissection results in a characteristic fan-shaped leaf.

2. Brasenia

The single species of this genus is widespread throughout the world, except for Europe. *Brasenia* and *Cabomba* are sometimes placed in a separate family, Cabombaceae.

1. **B. schreberi** J. F. Gmelin Fig. 91 Water-shield

Map 241. Quiet ponds and lakes, usually in acid and soft waters.

Mature leaf blades all float on the surface of the water. The maroon or dull purple flowers are held above the water on one day when the anthers discharge pollen, and on a second day when the stigmas are receptive, thus helping to ensure cross-pollination by the wind. Because the thick gelatinous coating of most parts of the plant will glue it effectively to any paper in which it is pressed, it is best to lay a specimen on a sheet of herbarium mounting paper and then to cover it with a layer of waxed paper (which

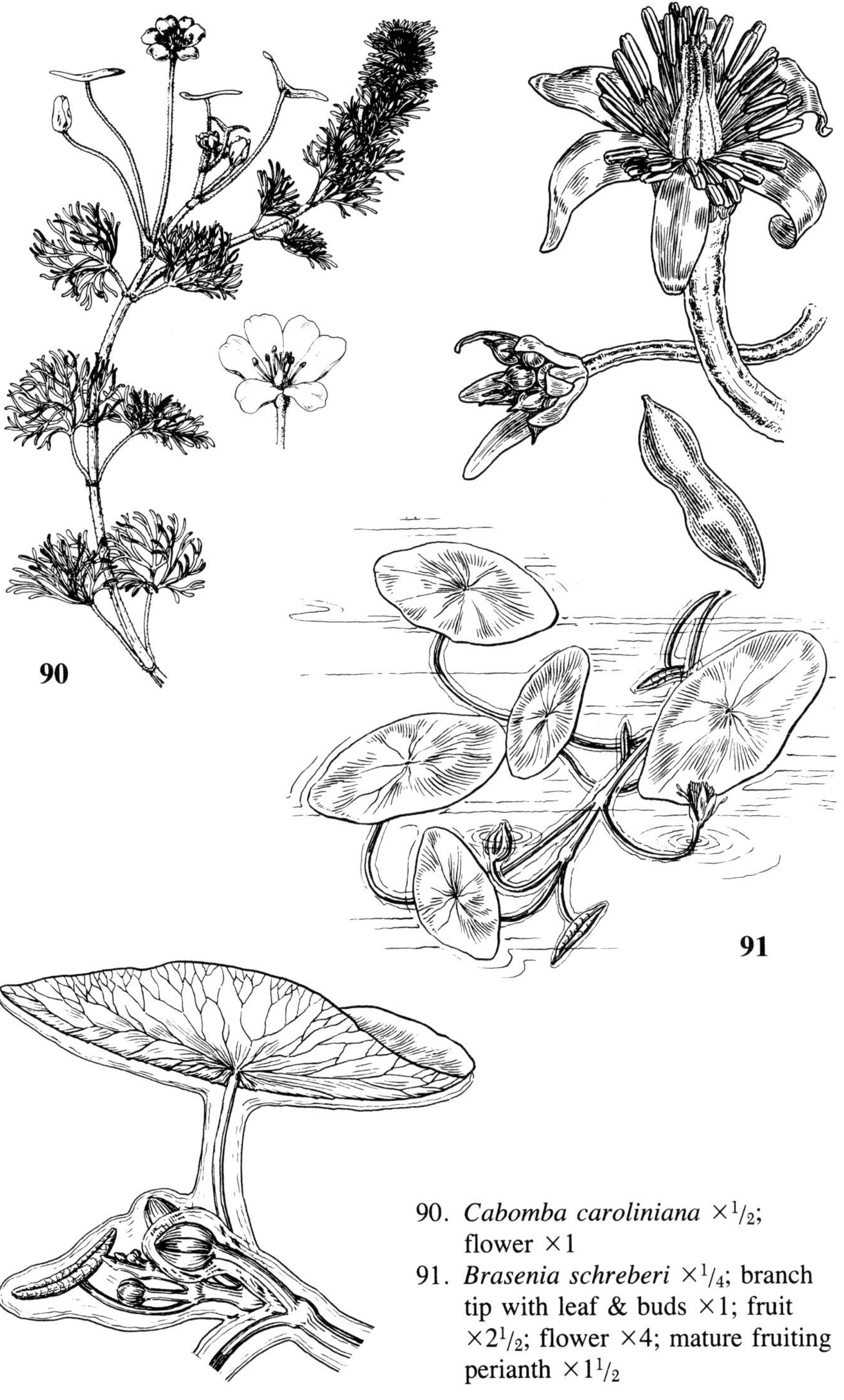

90. *Cabomba caroliniana* ×$^{1}/_{2}$; flower ×1
91. *Brasenia schreberi* ×$^{1}/_{4}$; branch tip with leaf & buds ×1; fruit ×2$^{1}/_{2}$; flower ×4; mature fruiting perianth ×1$^{1}/_{2}$

should peel away when the specimen is dry) before pressing. The leaves are about 1 dm or less in their longer dimension.

3. **Nelumbo** Lotus

The genus includes two species, one of which is *N. nucifera* Gaertner of Asia and Australia, the sacred lotus admired in various aspects of oriental religion, art, and architecture. That species has pink or white flowers, and is sometimes grown in botanical gardens; it has become established locally in warmer places in the United States. A single leaf collected at Grand Rapids, Michigan, in 1891 is labeled as this species, but no *Nelumbo* was mentioned by Emma Cole in her 1901 *Grand Rapids Flora*, so whatever species was collected it was presumably not considered established. Because of the very large showy flowers and mystic lore associated with *N. nucifera*, it has been the subject of much popular notice—and misconception. It was not introduced into Egypt until about 500 B.C. and is not the so-called "sacred lotus" (which was not sacred) of ancient Egypt; that was *Nymphaea lotus* L. or another water-lily. The seeds are unusually long-lived, although probably for only a few hundred years and not the much longer periods sometimes claimed. The seeds and rhizomes of both species have been used for many centuries as a source of food.

REFERENCES

Beal, W. J. 1878. Nelumbium luteum in Michigan. Bot. Gaz. 3: 13.

Farwell, Oliver A. 1936. The Color of the Flowers of Nelumbo pentapetala. Rhodora 38: 272. [Comments as well on history of beds in Monroe Co.]

Gillman, Henry. 1875. The Lotus in the Detroit River. Am. Nat. 9: 178–179.

Hall, Thomas F., & William T. Penfound. 1944. The Biology of the American Lotus, Nelumbo lutea (Willd.) Pers. Am. Midl. Nat. 31: 744–758.

Ward, Daniel B. 1977. Nelumbo lutea, the Correct Name for the American Lotus. Taxon 26: 227–234.

1. **N. lutea** (Willd.) Pers. Plate 2-B; fig. 92 American Lotus; "Lotus-lily"

Map 242. It was once suggested that perhaps lotus in Michigan was originally introduced by Indians as a food-plant, but there is no documentation one way or the other. Some of our colonies, at least, are assumed to be natural, at the northern edge of the range for the species. There were once enormous beds in the marshes at the west end of Lake Erie, but they became much reduced as a result of depredations by picking and by muskrats, according to Farwell (1936). The species has been reported from Berrien County, where plants would be close to present or former occurrences in the Chicago area. Plants in Sunset Lake (originally a millpond) in Vicksburg, Kalamazoo County, have reputedly been there since at least the middle of the 19th century, but whether they were introduced after construction of the pond in 1829 was debated in the 1870's (see Beal 1878). Henry Gillman attempted to introduce the species (source of seed not stated) in the Detroit River as early as 1868 and he (1875) mentioned other introductions there, so the status of lotus in the Detroit River is likewise debatable (Beal 1878)—except that it now appears to be extinct there, whatever the origin of old colonies. The origin of populations on the Ontario side in the Lake St. Clair region has similarly been questioned, sometimes thought possibly to be the result of Indian introduction, and sometimes to represent recent migration northward.

This is one of the species originally listed in Michigan's wildflower protection law (a 1943 amendment to the "Christmas Tree Law"), and it was later recognized as "threatened" in the state under the Michigan Endangered Species Act of 1974. Local conservation groups have been active and successful in protecting and restoring the remnants of lotus in Michigan. As attractive as this plant is, and regardless of uncertainty as to the status of any particular colony, it is clear that assertions to the effect that any site is the "only" place where it grows are in error. Our native yellow-flowered species ranges southward from New England, Michigan, and Minnesota to Florida and Texas.

While the flowers bloom in late midsummer, the very large peltate leaves are conspicuous throughout the season, their blades floating on the water like large flat plates or held above the water like umbrellas (but slightly funnel-shaped). The unique top-shaped receptacle enlarges tremendously as the fruits ripen, each embedded just below the broad flat surface, in a cavity with a circular opening to the exterior (fig. 92). The peduncle bends soon after the petals are shed, tilting the receptacle almost at right angles; it later returns to an erect position as the little fruits mature, and then bends downward before the receptacle breaks off and floats in the water, releasing the fruits.

At times it has been claimed that the correct name for our species is *N. pentapetala* (Walter) Fern., but such a change seems not to be necessary.

4. **Nymphaea** Water-lily

Plants of this genus are popular aquatics for cultivation, and many cultivars of species from all over the world are grown for colorful flowers and for the variegated leaves of some forms.

REFERENCES

Monson, Paul H. 1960. Variation in Nymphaea, the White Waterlily, in the Itasca State Park Region. Proc. Minnesota Acad. 25–26: 26–39.

Williams, Gary R. 1970. Investigations in the White Waterlilies (Nymphaea) of Michigan. Mich. Bot. 9: 72–86.

KEY TO THE SPECIES

1. Flowers ca. 5 cm or less across, opening in the afternoon; sepals ca. 3 cm or shorter at flowering time, becoming a little longer and stiffly erect over the fruit; stigmas 10 or fewer; leaf blades small (usually less than 7 cm broad), ca. 1.3–1.5 times as long as wide, with sinus extending half their length .... 1. **N. tetragona**
1. Flowers larger (except in dwarf bogpool form), opening early in the morning; sepals over 3 cm long, curving over the fruit; stigmas more than 10; leaf blades ordinarily larger, at most about 10% longer than wide, with sinus extending less than half their length .... 2. **N. odorata**

1. **N. tetragona** Georgi

Map 243. Thus far known in Michigan only from a stream at the head of Duncan Bay, Isle Royale (*Veirs* in 1963, MICH, IRP), but to be sought elsewhere in the Lake Superior region. A circumpolar species, described originally from Siberia, *N. tetragona* barely enters the northern United States.

I have not seen this species in the field and cannot confirm the supposed lack of fragrance of the flowers (but see remarks about "*N. tuberosa*" below) or their alleged afternoon opening. Herbarium specimens do not clearly reveal the four-angled receptacle (at insertion of the sepals) which is the source of the name and a key character in the field.

2. **N. odorata** Aiton   Plate 2-C

Map 244. Ponds and sheltered areas of lakes and rivers, the mature leaf blades all floating on the surface of the water. Occasionally, foliage and even flowers can be found amid sphagnum and shrubs in a bog mat, silent testimony to plant succession from the days before the water-lily was engulfed by the mat as it grew toward the center of the bog lake.

Our common white water-lilies have long been considered to represent two species, although their distribution and the characters used to distinguish them have been variously interpreted and some authorities have at least questioned recognition of the species. Extensive investigations by Monson (1960) and Williams (1970) in Minnesota and Michigan, respectively, led to in-

241. Brasenia schreberi

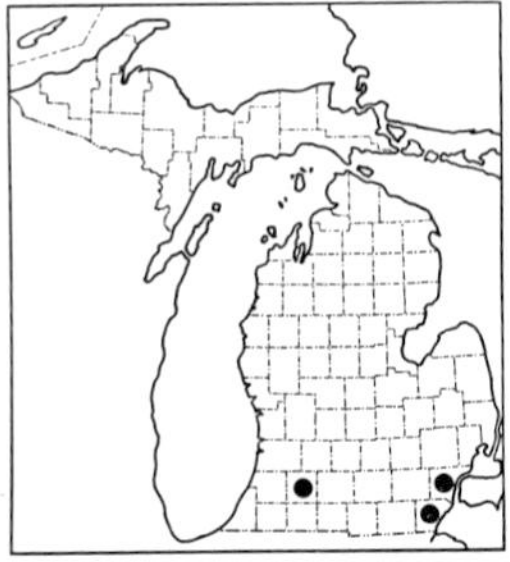
242. Nelumbo lutea

243. Nymphaea tetragona

dependent conclusions that there was only one taxon in the region studied. *N. tuberosa* Paine supposedly differs from *N. odorata* in lacking fragrance (certainly not true of fresh flowers in our area!); flowers closing later in the afternoon; leaf blades and flowers averaging larger; and other indistinct characters. More clearcut are green to dull purple undersides of the leaf blades in *N. tuberosa*, on green petioles prominently striped with purple; in *N. odorata*, the blades are deep purple or maroon beneath and the petioles (usually reddish) not striped. However, Williams' transplant studies revealed that a rhizome from a typical large plant of *N. tuberosa* when transplanted to a lake in which *N. odorata* grew could produce leaves and flowers resembling the latter in color and size—suggesting more importance than has usually been recognized of environmental influences on these characters.

The massive rhizomes of this species produce structures ranging from branches and barely constricted tubers to distinct tubers readily detaching at a strongly constricted base (later germinating new leaves and serving as a means of propagation)—the latter a character of typical *N. tuberosa*.

Some authors have said that *N. odorata* does not range south of northern Michigan and Minnesota in the Great Lakes region, referring all material from southern Michigan and adjacent states to *N. tuberosa*; while others recognize both species throughout this area. In view of the varying and intergrading nature of the characters of the plants, a single species is admitted here.

Plants with small leaves and flowers (as small as 5 cm) in pools of sphagnum bogs are a dwarf form which looks rather different from the usual ones but can be distinguished from *N. tetragona* by its shallower leaf sinus, more orbicular leaves, and fragrant flowers open in the morning (if indeed *N. tetragona* can be relied upon to lack fragrance and to open in the afternoon). At the other extreme are plants of the *N. tuberosa* sort, with leaves as broad as 3.5 dm or more and flowers 17 cm across. In some forms which may escape from cultivation, the petals are pink or rosy.

5. **Nuphar** Pond-lily; Cow-lily; Spatterdock

The generic name was feminine as originally published (see also Taxon 8: 270. 1959) and it is treated as such in the list of conserved generic names, but usage as neuter is persistent.

E. O. Beal (1956) recognized a single species in this genus, treating American plants as subspecies of the Eurasian *N. lutea*. He has been followed by many authors, but by no means all, exceptions including *Flora Europaea* and the works of Hultén, whose position regarding circumpolar complexes is often to treat such taxa as subspecies. Our plants are quite easily distinguished (much more so than the "species" usually recognized in *Nymphaea*) and they are treated here as closely related species.

The massive, starchy, spongy rhizomes (several cm in diameter) may be distinguished from those of *Nymphaea* by the more numerous closely spaced spiral rows of leaf scars, which are ± triangular, winged, or semicircular. In *Nymphaea*, the leaf scars are circular and more irregularly spaced (and there may also be tuberous branches or scars thereof). Peduncle scars are circular in both genera. The petioles and peduncles in both genera arise from the rhizome, and there is no branching in the water.

REFERENCE

Beal, E. O. 1956. Taxonomic Revision of the Genus Nuphar Sm. of North America and Europe. Jour. Elisha Mitchell Sci. Soc. 72: 317–346.

KEY TO THE SPECIES

1. Sepals 5, at most about 15 mm long; stigmatic rays 10 or fewer, on a red disc; pistil ca. 1 cm long or less at anthesis, ripening into a smooth fruit at most 2 cm long; anthers ca. 1–2 (2.5) mm long, shorter than filaments, the stamens deciduous from the base of the ripe fruit; blades of floating leaves up to 10 cm long, the sinus usually at least two-thirds as long as the midrib . . . . 1. **N. pumila**
1. Sepals usually 6 and longer; stigmatic rays (9) 10–21, on a green (rarely reddish) disc; pistil often larger, ripening into a ridged or striate fruit up to 4 cm long; anthers ca. (3) 4–6 (7) mm long, longer than filaments, the withered stamens persisting at the base of the fruit; blades of floating and emersed leaves usually larger, the sinus rarely more than half as long as the midrib
   2. Petioles flattened above, or even slightly winged; sepals with a prominent maroon (very rarely green) patch basally . . . . 2. **N. variegata**
   2. Petioles rounded in section; sepals with a green patch basally . . . . 3. **N. advena**

1. **N. pumila** (Timm) DC.

Map 245. In lakes, very local in Michigan (Clark Lake, Luce Co., and Mountain Lake, Marquette Co.).

While treating this taxon as a species distinct from our other two, I am

244. Nymphaea odorata

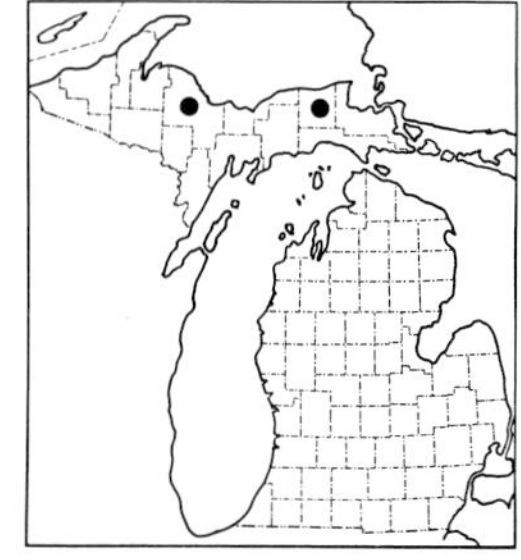
245. Nuphar pumila

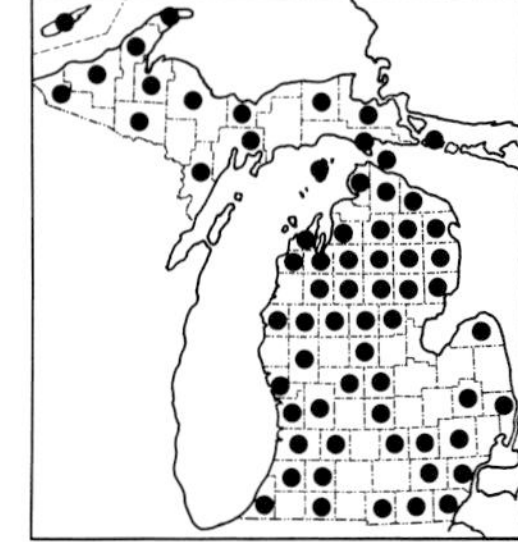
246. Nuphar variegata

following Beal (and Hultén) in not separating it from the Eurasian *N. pumila*—or *N. lutea* ssp. *pumila* (Timm) E. O. Beal. American plants have usually been called *N. microphylla* (Pers.) Fern. With its small flowers, small deeply lobed leaves, and extremely slender petioles rounded in section (peduncles similarly slender), this is a quite distinctive plant.

Individuals intermediate between this species and the next, the leaves and flowers larger than in *N. pumila* but stigmatic rays fewer than in *N. variegata* and a reddish disc, have been considered hybrids, *N.* ×*rubrodisca* Morong, by many authors. Beal has placed here specimens from Cheboygan and Otsego counties—far from known stations for *N. pumila*, but others have considered these specimens to be depauperate *N. variegata*. Similarly dubious specimens are known from Isle Royale. However, such plants growing with more typical *N. pumila* and *N. variegata* in Mountain Lake appear clearly to be hybrids.

2. **N. variegata** Durand

Map 246. Lakes, ponds, quiet rivers and streams.

In his treatment of *Nuphar* as a single polymorphic species, Beal called this taxon *N. lutea* ssp. *variegata* (Durand) E. O. Beal. The most reliable character for distinguishing it from the next (at whatever rank)—especially for use with herbarium specimens—is the flattened (or even narrowly winged) petiole. The leaves (except for submersed basal ones) of *N. variegata* usually have the blades floating, with basal lobes ± parallel, the sinus narrow, parallel-sided, or closed (the lobes slightly overlapping). In *N. advena*, the blades are often erect, held above the water (the plant thus perhaps better adapted for emergence on drying shores), with basal lobes rather divergent, the sinus wide-angled (often as much as 60° or even more). However, in *N. variegata* the lobes are frequently divergent, and in *N. advena* they are more rarely overlapping. Leaf aspect is thus more helpful in the field, when other characters can be considered, than in the herbarium, where the round or 2-edged petiole shape seems most helpful.

Michigan is sometimes said to be the type locality for *N. variegata*, but that is impossible; the original description mentioned only a specimen from New York, which must therefore be the type. (On the authorship and typification of this name, see Taxon 14: 159–160. 1965.) The rare plants with pure yellow sepals or with a green rather than red patch on them may be called *N. variegata* f. *lutescens* (Farw.) E. G. Voss (TL: Lakeville Lake [Oakland Co.]).

3. **N. advena** (Aiton) Aiton f. Fig. 93

Map 247. Lakes, ponds, river margins, creeks.

In the southern Lower Peninsula, this species is sometimes found in the same lakes as *N. variegata*, but it does not extend very far north in the state.

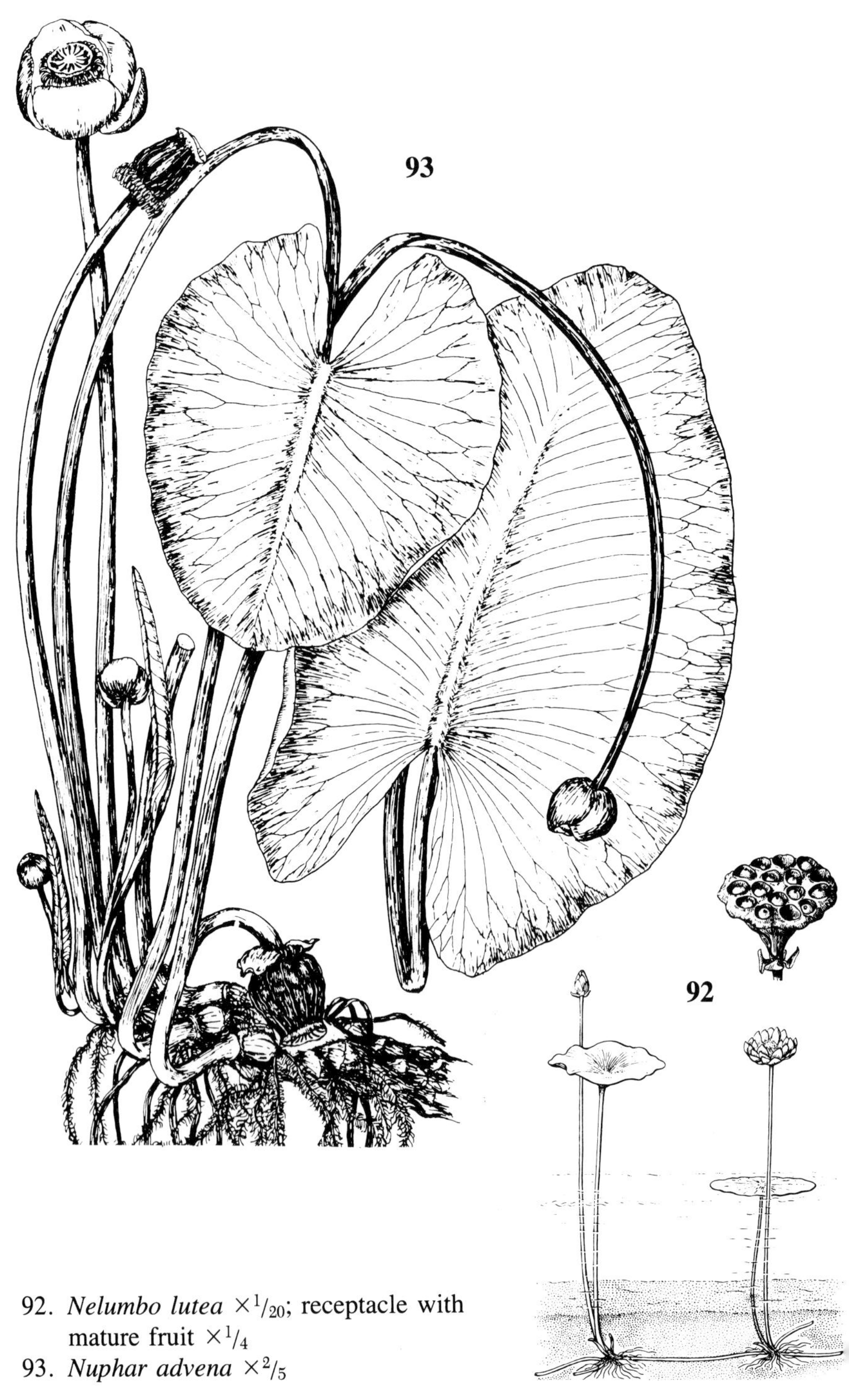

92. *Nelumbo lutea* $\times {}^1/_{20}$; receptacle with mature fruit $\times {}^1/_4$
93. *Nuphar advena* $\times {}^2/_5$

Although the two species are reported to intergrade where the ranges overlap, there is little if any intergradation in our area. See also comments under the preceding species. The broadest leaf among the herbarium material examined was 26 cm, but even larger leaves may occur in nature. This taxon has been called *N. lutea* ssp. *macrophylla* (Small) E. O. Beal, but under the Code as amended in 1981 would have to be called ssp. *advena* (a combination not yet published) if treated at that rank and position.

RANUNCULACEAE — Buttercup Family

Trollius laxus Salisb., globe-flower, a rare native plant of eastern North America, is often said to occur in Michigan but there are no specimens to substantiate such reports. One-flowered plants might run to *Anemone* in the key, but the alternate leaves and the fruit a follicle would quickly separate them from that genus. Other plants might key to *Helleborus*, but the flowers appear later in the spring than in *H. viridis* and are yellowish.

The garden monkshood, *Aconitum napellus* L., is a native of Europe. It is a highly poisonous plant if taken internally. A thriving colony of this (or an allied species) grows in the swampy interior of Passage Island, Isle Royale National Park, where it was seen and photographed by members of the Michigan Audubon Society in 1983. Presumably it is an escape from cultivation there, as it is at other scattered locations in eastern North America. The alliance of species associated with *A. napellus* is readily recognizable by the deeply divided leaves; erect stem with terminal raceme of large, bilateral, deep blue-violet flowers; and petaloid sepals, of which the upper forms a conspicuous hood.

REFERENCE

Mitchell, Richard S., & J. Kenneth Dean. 1982. Ranunculaceae (Crowfoot Family) of New York State. New York St. Mus. Bull. 446. 100 pp.

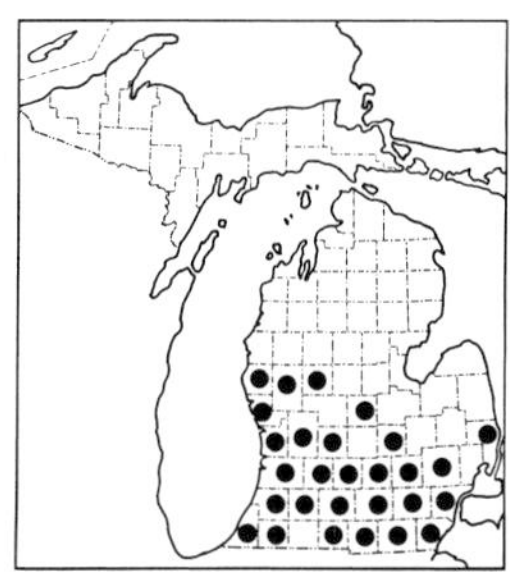

247. Nuphar advena

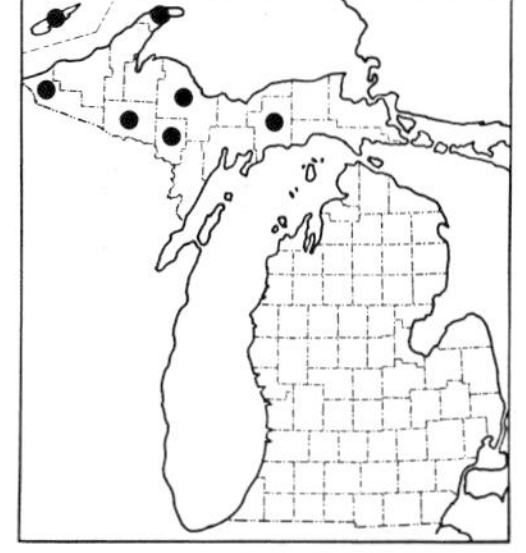

248. Clematis occidentalis

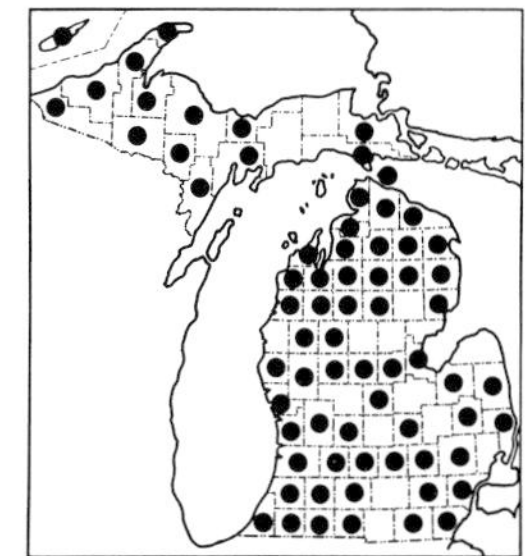

249. Clematis virginiana

KEY TO THE GENERA

1. Plant a vine (twining, if at all, by petioles); leaves opposite; fruit with a conspicuous persistent feathery style . . . . 1. **Clematis**
1. Plant an erect or stemless herb; leaves alternate or basal (sometimes a single whorl of "involucral" cauline leaves); fruit without a plumose style
  2. Perianth with 1 or more distinct spurs
    3. Flower regular, usually red, with all petals spurred; fruit a group of follicles (normally 5) . . . . .2. **Aquilegia**
    3. Flowers bilateral, blue, with 1 long sepal spur enclosing 1 or 2 petal spurs; fruit 1 or 3 follicles
      4. Follicles solitary; plants annual; leaves divided into narrowly linear segments . . . . 3. **Consolida**
      4. Follicles 3; plants perennial; leaves with broader segments or merely lobed . . . . .4. **Delphinium**
  2. Perianth without spurs
    5. Flowers numerous per inflorescence (6–many in a raceme or panicle), the perianth much less conspicuous than the stamens (or carpels), which offer the most showy portion of each flower; leaves clearly compound
      6. Inflorescence open, panicle-like; teeth or lobes of leaflets few, ± rounded or obtuse (occasionally acute); fruit an achene . . . . 5. **Thalictrum**
      6. Inflorescence a dense raceme; teeth or lobes of leaflets numerous, acute; fruit a follicle or fleshy berry
        7. Raceme elongate (over 10 cm), often branched at base; fruit a short plump follicle; plant very rare, in southern Lower Peninsula . . . . 6. **Cimicifuga**
        7. Raceme short (less than 10 cm), unbranched; fruit a fleshy berry; plants common throughout . . . . 7. **Actaea**
    5. Flowers solitary or few (not over 6) per inflorescence (though a plant may bear more than 1 inflorescence), the perianth more conspicuous than the stamens (except in one species with solitary flower and simple leaves); leaves simple or compound
      8. Flowers solitary, on a leafy stem or a scape; leaves simple (± lobed or cleft) or strictly once-compound
        9. Plant without a leafy stem, the flowers scapose (peduncles and leaves all basal); perianth white, pink, or blue (not yellow or yellowish)
          10. Leaves trifoliolate, toothed; fruit a follicle (ca. 3–7 per flower) on a distinct stipe; flowers white; plants essentially glabrous . . . . 8. **Coptis**
          10. Leaves simple, 3 (–5)-lobed, entire; fruit an achene (numerous per flower), sessile; flowers white, pink, or blue; plants with peduncles and new foliage long-pilose . . . . 9. **Hepatica**
        9. Plant with an erect, reclining, or stoloniferous leafy or leafy-bracted stem; perianth (or at least the petals) yellow, white, cream, or greenish white
          11. Perianth early deciduous (as the flower opens), the stamens comprising the conspicuous aspect to the flower; fruit a red berry . . . . 10. **Hydrastis**
          11. Perianth persistent, more conspicuous than the stamens; fruit an achene
            12. Flowers with petaloid sepals and no petals; cauline leaves (or bracts) opposite or whorled . . . . 17. **Anemone**
            12. Flowers with both petals and sepals; cauline leaves or bracts alternate . . . . .11. **Ranunculus**
      8. Flowers more than 1 on a leafy stem (never on a scape) [if flower solitary on small plants, the leaves dissected or at least twice-compound]; leaves simple or compound

13. Perianth with both sepals and petals .... 11. **Ranunculus**
13. Perianth of ± petaloid sepals and no petals (or these modified to staminodia or nectaries)
14. Leaves simple, toothed to almost entire but not cleft or lobed (except at cordate base of blade); perianth bright yellow .... 12. **Caltha**
14. Leaves deeply cleft, dissected, or compound; perianth not yellow (white, bluish, cream, or maroon)
15. Leaves pinnately dissected (including a conspicuous involucre beneath each flower), the segments less than 1.5 mm wide, glabrous; ovaries united; plant a rare escape from cultivation .... 13. **Nigella**
15. Leaves compound or palmately lobed, the leaflets or lobes at least 1.5 mm wide, pubescent or glabrous; ovaries separate; plants mostly native
16. Principal leaves (at least the basal ones) clearly twice-compound, with broad leaflets on slender petiolules; teeth or lobes of leaves broadly rounded; perianth white (rarely pinkish); foliage and fruit glabrous
17. Leaves all basal, plus an involucral whorl beneath the umbellate flowers; carpels essentially without styles, ripening into achenes .... 14. **Anemonella**
17. Leaves alternate and basal, the flowers terminal and often axillary; carpels with prolonged styles, ripening into beaked follicles .... 15. **Isopyrum**
16. Principal leaves palmately lobed, cleft, or barely once-compound (leaflets scarcely petioluled); teeth or lobes of leaves sharply acute; perianth white, cream, greenish, or maroon; foliage (including stems and petioles) and fruit ± pubescent (glabrous in *Helleborus*)
18. Cauline leaves alternate; flowers white to pinkish or ± green, ca. 4 cm or more in diameter; fruit a follicle; plant a rare escape from cultivation, blooming in earliest spring or late winter (often in snow) ... 16. **Helleborus**
18. Cauline leaves or bracts opposite or whorled; flowers white, cream, or maroon, usually smaller; fruit an achene; plants native, blooming spring–summer .... 17. **Anemone**

1. **Clematis** — Virgin's Bower; Clematis; Woodbine

REFERENCE

Pringle, James S. 1971. Taxonomy and Distribution of Clematis, Sect. Atragene (Ranunculaceae), in North America. Brittonia 23: 361–393.

KEY TO THE SPECIES

1. Perianth pink-purple, ca. 3–4.5 cm long; flowers perfect, solitary on elongate peduncles; mature achenes over 2 mm broad .... 1. **C. occidentalis**
1. Perianth white, ca. (5) 7–13 (15) mm long; flowers unisexual (plants dioecious, though pistillate flowers often with staminodia), in small to large inflorescences; mature achenes mostly less than 2 mm broad .... 2. **C. virginiana**

1. **C. occidentalis** (Hornem.) DC. Plate 2-D

Map 248. Rocky woods and thickets, on stream banks, and in burned or cleared areas.

A handsome large-flowered plant, long known by the later name of *C. verticillaris* DC.

2. **C. virginiana** L. Fig. 94

Map 249. Damp woods and thickets, river banks, borders of swamps and marshes.

The body of the mature achene (excluding narrowed beak and plumose style) is ca. 4 (4.5) mm or shorter, while in the preceding species it is about 5 mm long.

2. Aquilegia — Columbine

KEY TO THE SPECIES

1. Flowers red and yellow, not double; spurs essentially straight; plant a common native . . . . 1. **A. canadensis**
1. Flowers purple, white, or pink, without yellow, often "double"; spurs strongly incurved; plant an occasional escape from cultivation . . . . 2. **A. vulgaris**

1. **A. canadensis** L. Plate 2-E — Wild Columbine

Map 250. Generally associated with deciduous or mixed woods and thickets, but usually at borders or clearings, river banks, roadsides, or excavations; also frequent on gravelly shores, ridges, and banks; occasionally in swamp forests. Sometimes abundant for a few years on a recent excavation or sand bank.

This is one of the few plants in the local flora which is pollinated by hummingbirds—as one would expect from a red flower with long nectar-filled spurs.

2. **A. vulgaris** L. — Garden Columbine

Map 251. A European species, occasionally escaped from cultivation near buildings, along roadsides, on shores, or in woods.

A number of other species and hybrids of *Aquilegia* are cultivated. A collection from a roadside at Phoenix, Keweenaw County (*Bourdo* in 1959, MCTF), "apparently an escape," with mature spurs ± straight and ca. 3 cm long (longer than in the above 2 species), is probably a form or hybrid involving *A. chrysantha* A. Gray, a yellow-flowered species with spurs normally ca. 4–7.5 cm (native from Colorado south into Mexico, and popular in cultivation).

3. Consolida

The species of this genus have often been included in the genus *Delphinium*, from which they differ in having only 1 carpel and 1 petal spur (covered by the sepal spur).

1. **C. ambigua** (L.) Ball & Heywood — Larkspur

Map 252. Roadsides, borders of fields and woods, and other disturbed places. A native of the Mediterranean region, occasionally escaped from cultivation in North America.

This is the larkspur to which the name of *Delphinium ajacis* L. [or *C. ajacis* (L.) Schur] has long been misapplied. Much material of this species has also been misidentified in American gardens as *C. regalis*. And to further complicate the story, some material cultivated as *C. ambigua* (or *ajacis*) is actually *C. orientalis* (Gay) Schröd., which differs in having shorter spurs (not over 12 mm) on the flowers, and bracteoles on the pedicels of the lower flowers reaching to the base of the mature flower and the fruit; in *C. ambigua*, the tiny bracteoles are inserted lower on the pedicels and do not reach the flower and fruit. All of our material which appears to have escaped from cultivation seems to be *C. ambigua*, although some was originally identified as *D. consolida*, which in the genus *Consolida* is to be called *C. regalis* S. F. Gray. *C. ambigua* has the ovary and follicle pubescent, and the lower bracts of the inflorescence are dissected; in *C. regalis*, the ovary and follicle are glabrous, and the lower bracts are entire.

All of our specimens have blue flowers, although other colors occur in the species.

4. **Delphinium**

The native American *D. tricorne* Michaux ranges very close to Michigan in the Chicago area, and might be discovered in the southwestern part of the Lower Peninsula—or as an escape from cultivation. It is a fairly low plant (up to ca. 5 dm tall), few-flowered, with follicles spreading when mature and the leaves deeply lobed into ± narrow segments. *D. elatum* is a tall, many-flowered plant with erect follicles. Other escapes from cultivation are possible, as these plants, like the species of *Consolida*, seed readily in the garden.

1. **D. elatum** L. — Larkspur

Map 253. A native of Eurasia, this species (or at least plants belonging to this complex) has been collected once as established on a sandy shaded

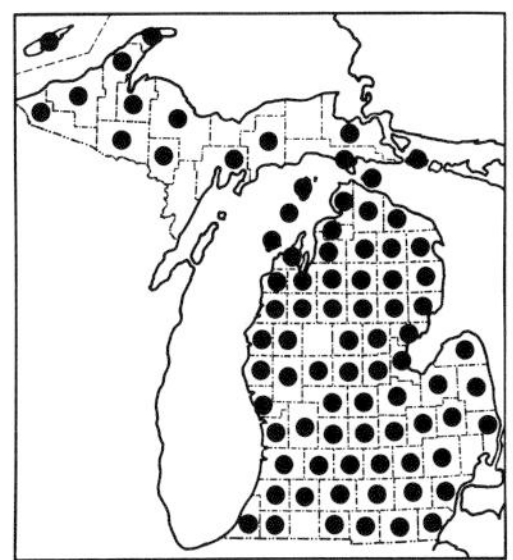

250. Aquilegia canadensis

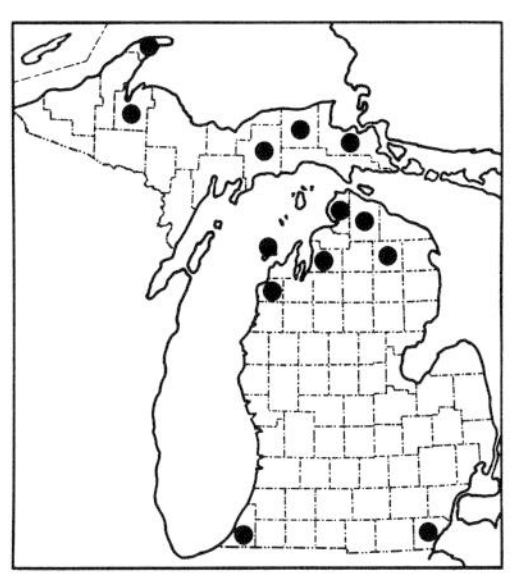

251. Aquilegia vulgaris

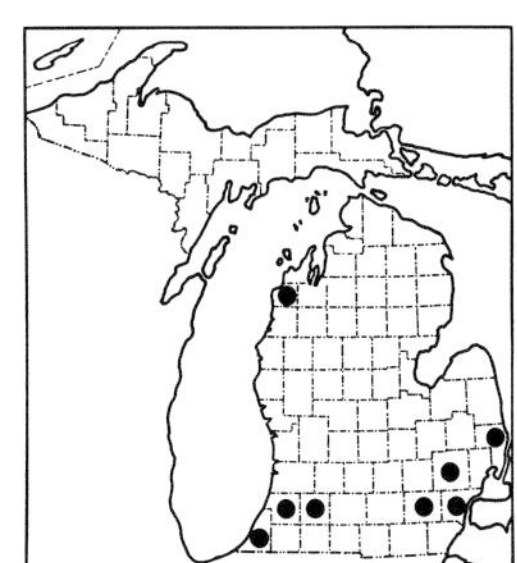

252. Consolida ambigua

knoll in Schoolcraft County (*Henson 479* in 1972, MICH). The leaves are palmately lobed (more deeply parted toward the base of the plant), somewhat resembling those of the wild geranium. The long, many-flowered raceme rising above the leaves has led to the common name of "candle larkspur." This species has been cultivated for centuries, and is involved in the production of many cultivars and hybrids; some of the plants in cultivation are frequently misidentified as other species. Reliable works on cultivated plants should be consulted for identification of such material.

5. **Thalictrum** Meadow-rue

KEY TO THE SPECIES

1. Upper cauline leaves long-petioled; plants blooming in April or May, before the leaves are fully expanded; leaflets glabrous and eglandular . . . . 1. **T. dioicum**
1. Upper cauline leaves sessile or nearly so (the three main divisions on stalks appearing to be 3 petioles at a node); plants blooming June–July (–September), after the leaves are fully expanded; leaflets glabrous or pubescent or with short-stalked glands
  2. Leaflets (especially on middle and lower leaves) with 2 or all 3 of the lobes again toothed or lobed, glabrous to sparsely glandular beneath . . . . 2. **T. venulosum**
  2. Leaflets mostly 3-lobed without further teeth, usually pubescent or glandular beneath
    3. Fruit and undersides of leaflets with sessile or short-stalked glands . . . . 3. **T. revolutum**
    3. Fruit and leaflets not glandular, the leaflets usually ± pubescent beneath with multicellular hairs . . . . .4. **T. dasycarpum**

1. **T. dioicum** L. Early Meadow-rue

Map 254. Typically in rich deciduous woods, often on slopes; also in oak-hickory woods and thickets along rivers.

The glabrous leaflets are thin in texture, with flat (not revolute) margins; the achenes are shorter, less prominently beaked, and more symmetrical than in our other species.

253. Delphinium elatum

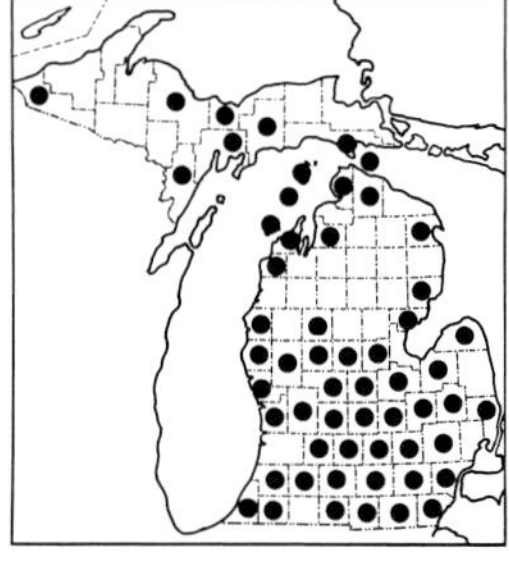

254. Thalictrum dioicum

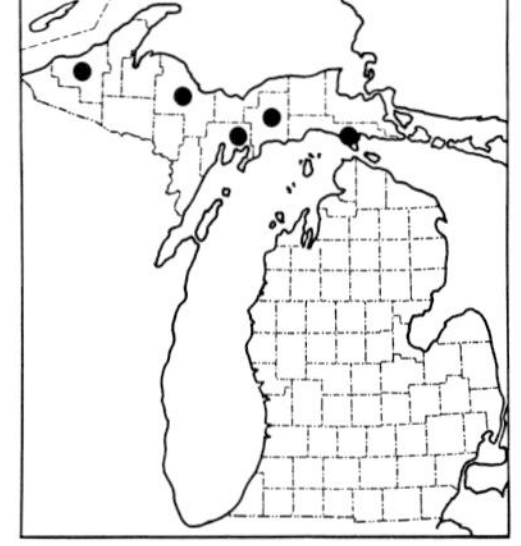

255. Thalictrum venulosum

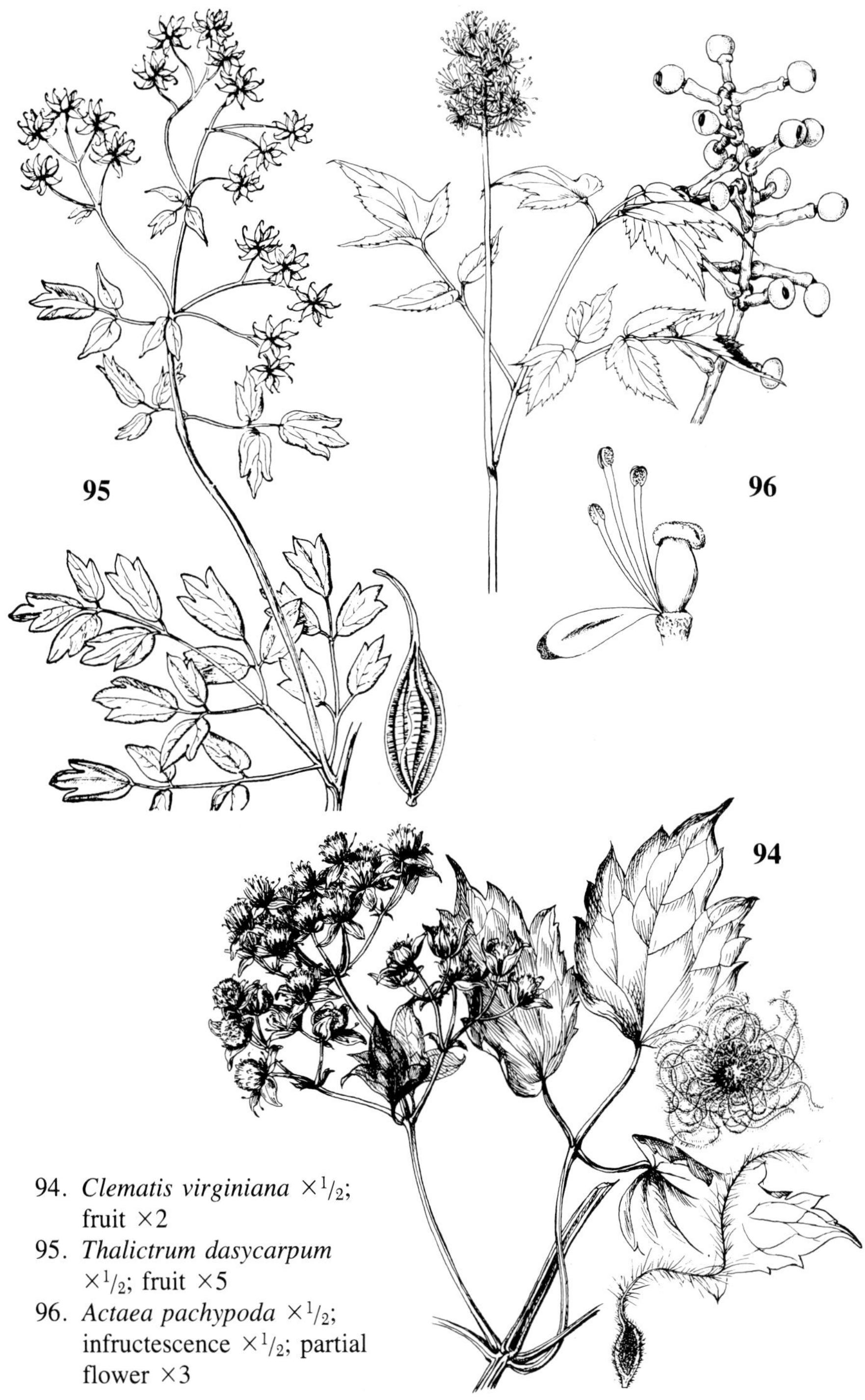

94. *Clematis virginiana* $\times ^1/_2$; fruit $\times 2$
95. *Thalictrum dasycarpum* $\times ^1/_2$; fruit $\times 5$
96. *Actaea pachypoda* $\times ^1/_2$; infructescence $\times ^1/_2$; partial flower $\times 3$

2. **T. venulosum** Trel.

Map 255. River-bank thickets and wet calcareous Great Lakes shores.

Our plants are referred to var. *confine* (Fern.) Boivin [*T. confine* Fern.]. Typical *T. venulosum* of farther north and west has thicker, more veiny leaflets. In var. *confine*, the leaflets are very similar to those of *T. dioicum*, and the upper leaves may have a relatively short petiole. But the fruit is longer and more prominently beaked than in *T. dioicum* and the undersides of the leaflets, many descriptions as glabrous notwithstanding, are usually at least sparsely glandular. Although less glandular than *T. revolutum*, such plants may be confused with that species. *T. revolutum* arises from a short erect underground base, while in *T. venulosum* there is an elongate horizontal rhizome.

When this species grows adjacent to *T. dasycarpum*, as at Horseshoe Bay in Mackinac County, the two apparently hybridize. Intermediate plants (*Brodowicz 208* in 1984, MICH) may have the leaflet shape of *T. venulosum* but without glands and with multicellular hairs as in *T. dasycarpum*.

3. **T. revolutum** DC.

Map 256. A southern species, barely ranging into Michigan in damp ground: meadows, thickets, near rivers.

The margins of the leaflets are slightly revolute, and the lobes tend to be more acute than in *T. venulosum*. See also remarks above.

4. **T. dasycarpum** Fisch. & Avé-Lall. Fig. 95 Purple Meadow-rue

Map 257. Marshy and open swampy ground generally: wet shores, stream and river margins, thickets, marshes, ditches, bog mats and tamarack swamps.

The leaflets are extremely variable in size, shape, acuteness of lobes, and pubescence. Only a very few of our specimens, however, have the leaflets completely glabrous beneath. The margins are usually slightly revolute, and the texture is thicker than in the preceding three species. The anthers are ca. (1) 1.5–2.5 (3) mm long and the stigmas (1.5) 2–3 (3.5) mm. Plants

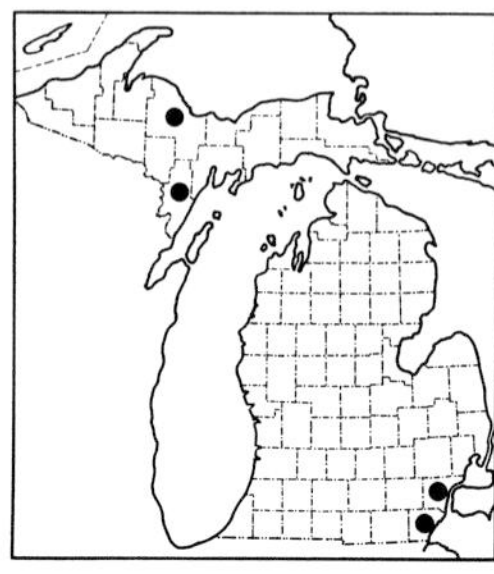
256. Thalictrum revolutum

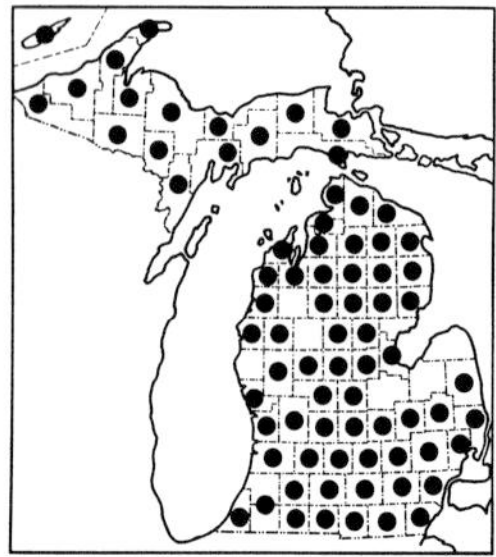
257. Thalictrum dasycarpum

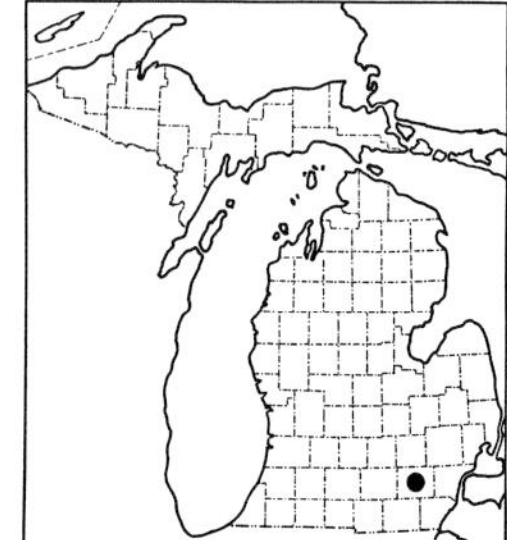
258. Cimicifuga racemosa

are nearly always dioecious, with unisexual flowers. In *T. pubescens* Pursh, a species occurring east of our area but often reported from Michigan under the later name of *T. polygamum* Barton, the anthers and stigmas run shorter and many of the flowers are perfect. The filaments are strongly clavate in *T. pubescens*, usually becoming broader than the anthers, while in *T. dasycarpum* they are filiform (often kinky and tangled at maturity) or occasionally slightly clavate (but narrower than the anthers).

6. **Cimicifuga**

1. **C. racemosa** (L.) Nutt. Fig. 97 Black Snakeroot; Black Cohosh

Map 258. Collected in 1918 from a farm woods in Superior Tp., Washtenaw County (*Walpole 730*, BLH). The species is at the northern edge of its range here.

The roots and rhizomes have long been reputed to have medicinal value. The foliage is quite similar to that of *Actaea*, but the elongate inflorescence, nodding at the tip when young, is very different, as are the fruits.

7. **Actaea** Baneberry

The colorful berries are reputed to be poisonous, although no fatalities from eating them are recorded.

REFERENCE

Fernald, M. L. 1940. What is Actaea alba? Rhodora 42: 260–265.

KEY TO THE SPECIES

1. Pedicels slender, not exceeding 0.5 (0.7) mm in diameter, even in fruit; fruit red or white, capped with a relatively inconspicuous stigma (less than 1.5 mm across); petals not modified at the tip as in *A. pachypoda* . . . . 1. **A. rubra**
1. Pedicels soon thickening, in fruit usually almost or quite as thick as the axis of the raceme and equally red; fruit white, capped by a conspicuous dark stigma ca. 1.5–2.5 mm across; petals ± truncate, withered in appearance or even anther-like at the tip (fig. 96) . . . . 2. **A. pachypoda**

1. **A. rubra** (Aiton) Willd. Red Baneberry

Map 259. Often in the same rich deciduous woods as the next species, but also in a broader range of mixed and coniferous forests and swamps, with aspen, oak and hickory, cedar, spruce and fir, hemlock, or tamarack.

Quite similar to the black-fruited European *A. spicata*, of which it is sometimes considered a variety. Plants with the fruit a pure white are frequent, sometimes locally more common than the red-fruited form, with which they are often growing. From the next species they differ in their slender, ± dull-colored pedicels and smaller "eye" formed by the persistent stigma.

White-fruited plants are f. *neglecta* (Gillman) Robinson and type locality (although none was stated in the original description) is probably Michigan, for at the time of publication (in 1885), Henry Gillman had served five years as librarian of the Detroit Public Library and had collected extensively in Michigan since the 1860's, when he was an engineer with the U.S. Lake Survey (see Voss 1978).

2. **A. pachypoda** Ell. Fig. 96 White Baneberry; Doll's-eyes

Map 260. Rich deciduous woods and northern hardwoods, less often under pine, cedar, or other conifers. Often growing with one or both color forms of the preceding species.

Readily distinguished from *A. rubra* in fruit, by its striking elongate raceme of porcelain-like berries on thick red stalks. In *A. rubra*, the raceme is more often compact and the pedicels remain very slender. Young flowering material, before the pedicels have begun to thicken, is more difficult to determine, although the peculiar tips of the petals are helpful (the sepals fall off very early). The broad stigma in *A. pachypoda* seems to be more clearly sessile (the style, if any, broad and thick beneath it) than in *A. rubra*, where the pistil tends to narrow somewhat beneath the stigma. Differences in leaf pubescence cannot be relied upon, although in *A. pachypoda* the leaves are more often glabrous or nearly so beneath while in *A. rubra* there is usually at least some small pubescence along the veins beneath. Specimens truly intermediate between the species are rare but some hybridization may occur. A red-fruited form of *A. pachypoda* is known [f. *rubrocarpa* (Killip) Fern.] but seems to be very rare if it occurs in Michigan at all.

This species has been called *A. alba* (L.) Miller in some manuals. The history and authorship of that name are complex. Fernald (1940) offered a convincing argument that it is based on an illustration of the European *A. spicata* and does not apply to either of our native American species.

8. Coptis

1. **C. trifolia** (L.) Salisb. Fig. 98 Goldthread

Map 261. Especially characteristic of cedar swamps, but occurs in a diversity of damp woods and banks, usually under conifers, often in moss.

A very easily recognized plant at any time of the year, by the trifoliolate evergreen leaves, absence of a leafy stem, and slender bright yellow-orange rhizome. The flowers, in early spring, are less than 2 cm broad, with conspicuous white sepals (sometimes a little pinkish on the back) and relatively inconspicuous spatulate, clavate, or knob-tipped nectariferous petals (or staminodia). The peculiar stipes of the carpels begin to elongate during anthesis, and the umbel-like cluster of stalked follicles is distinctive in summer, when even the mature scape is seldom much over 12 cm tall. The plants are glabrous except for a little puberulence at the juncture of the leaflets and sometimes on their main veins.

Plants of eastern North America are often separated from those of Alaska and Asia as *C. groenlandica* (Oeder) Fern., and may also be treated as *C. trifolia* var. *groenlandica* (Oeder) Fassett.

9. **Hepatica** Hepatica

Hepaticas are one of the first (and therefore most welcomed) wildflowers to bloom in the spring, at which time the new leaves are very small, densely pilose, and undeveloped. The distinctive 3-lobed leaves surviving from the previous year, however, may be found, sometimes buried among dry leaves shed by the trees of the forest canopy. The colored perianth in this genus consists of sepals; the three green structures beneath, resembling sepals, are actually bracts, inserted on the scape only a few mm below the flower. Occasionally abnormal (diseased?) forms with a green perianth are seen.

In our area only rarely do the two taxa appear to hybridize and they are treated here as the traditional two species. Steyermark and Steyermark (1960) treat each as a variety of the Old World *H. nobilis* Miller. *H. americana*, especially, is close to the Eurasian species, which may have either round or pointed leaf lobes. Yet another taxonomic opinion is to include this genus in *Anemone*, in which case *H. nobilis* becomes *A. hepatica* L.

Proportions given in the key for the middle lobe of the leaf are calculated by measuring the lobe (from the apex to a line connecting the bases of the sinuses on each side) and the total length of the blade (apex to the summit of the unexpanded petiole). In the case of ambiguous measurements, check more than one leaf on a plant.

REFERENCE

Steyermark, Julian A., & Cora S. Steyermark. 1960. Hepatica in North America. Rhodora 62: 223–232.

259. Actaea rubra

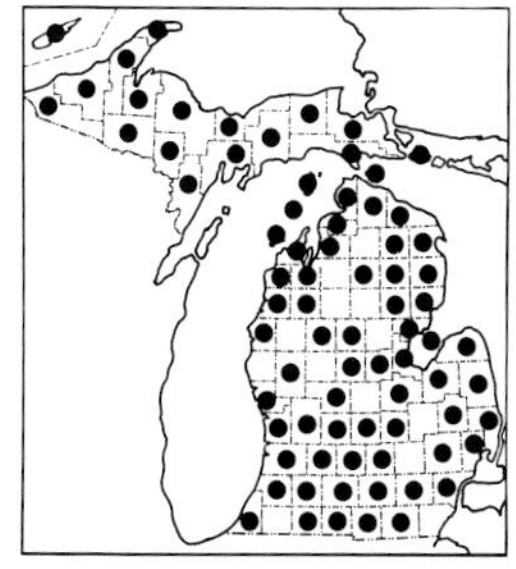

260. Actaea pachypoda

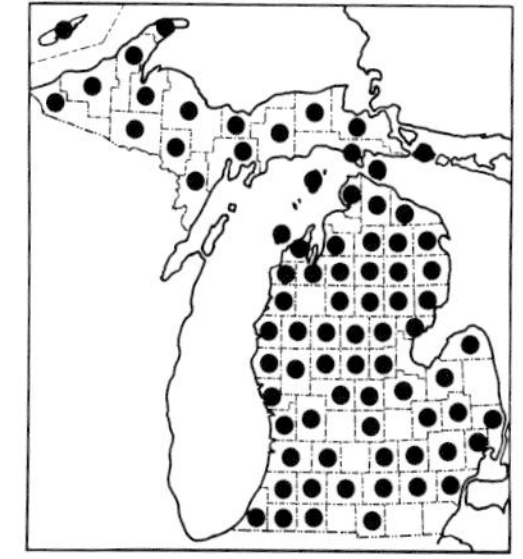

261. Coptis trifolia

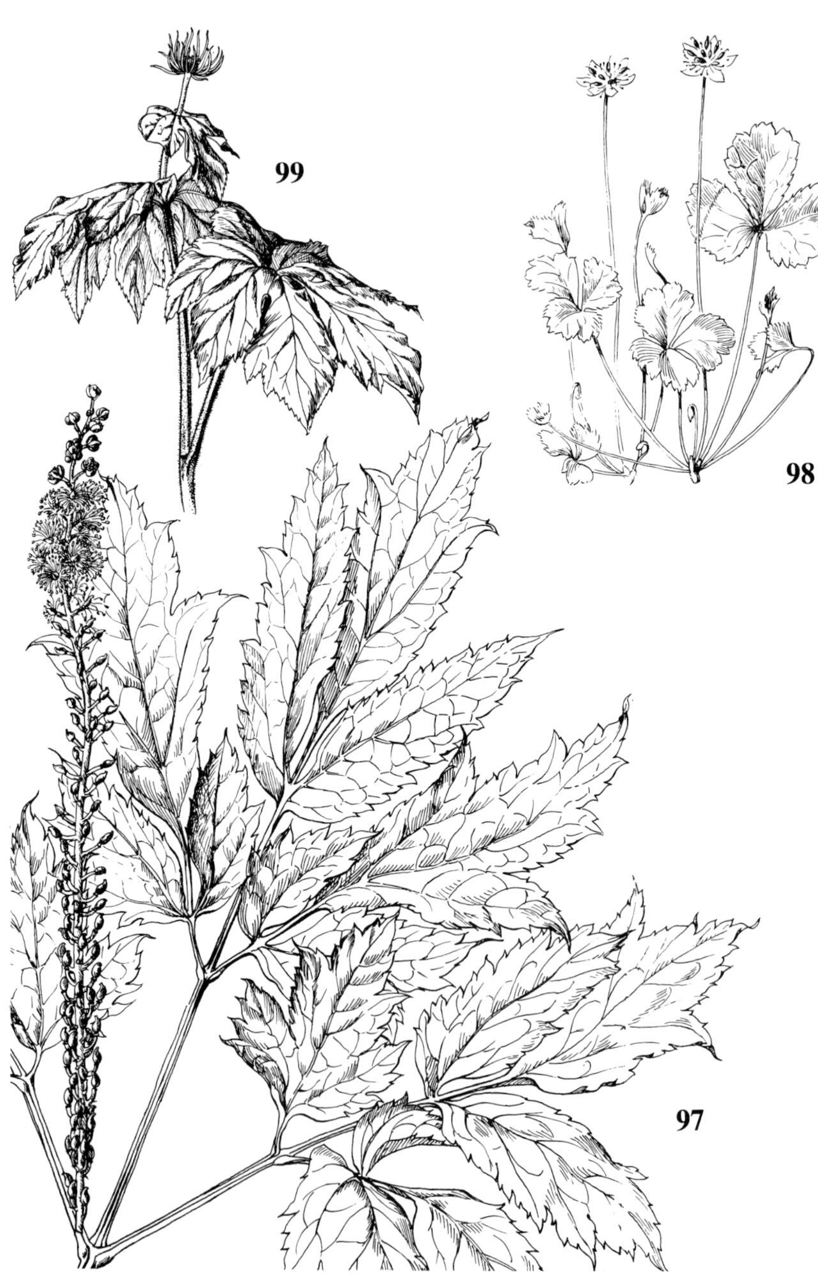

97. *Cimicifuga racemosa* $\times ^1/_2$
98. *Coptis trifolia* $\times ^1/_2$
99. *Hydrastis canadensis* $\times ^1/_2$

KEY TO THE SPECIES

1. Middle lobe of leaf ± acute (margins approaching apex at angle of 90° or, usually, less), 70–90% of the total blade length .... 1. **H. acutiloba**
1. Middle lobe of leaf rounded to obtuse, (45) 55–65% of the total blade length ... . 2. **H. americana**

1. **H. acutiloba** DC.   Fig. 100

Map 262. Almost entirely restricted to beech-maple woods (sometimes with additional trees dominant, as hemlock or oak), usually on rich soils. Only one collection has been seen from beyond the range of beech, from the lower Carp River valley in the Porcupine Mountains (*Darlington* in 1923, MSC).

Forms with white, pink, and deep blue flowers may grow together. There are frequently small additional lobes on the leaves of this species.

2. **H. americana** (DC.) Ker

Map 263. Rich beech-maple woods, as for *H. acutiloba*, but more often associated on drier sites with aspen, oak, hickory, pine, or even with spruce or cedar.

When the two hepaticas do grow together, intermediate leaf shapes are only rarely found, although what they mean has not been fully investigated. As in *H. acutiloba*, several forms occur within this species, and are hardly worth naming, although the type localities for some of them are in Michigan, including f. *purpurea* (Farw.) Farw. (TL: Rochester [Oakland Co.]), with deep purple flowers, and f. *cahniae* Farw. (TL: Loon Lake, Oakland Co.), with double flowers and short blue sepals.

10. **Hydrastis**

REFERENCE

Eichenberger, M. D., & G. R. Parker. 1976. Goldenseal (Hydrastis canadensis L.) Distribution, Phenology and Biomass in an Oak-Hickory Forest. Ohio Jour. Sci. 76: 204–210.

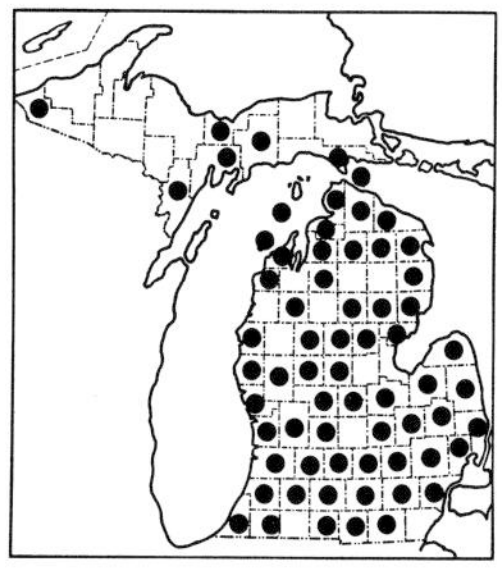

262. Hepatica acutiloba

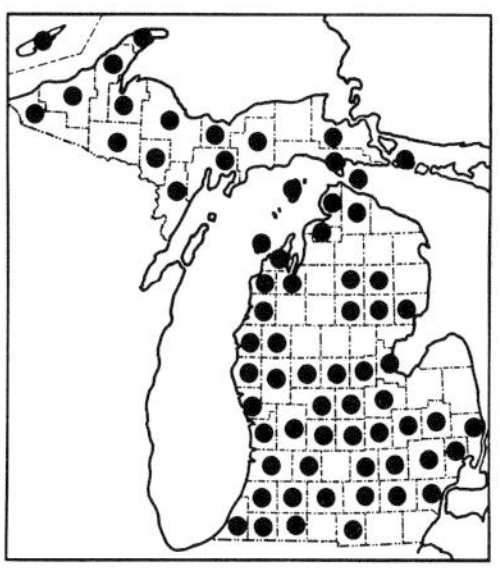

263. Hepatica americana

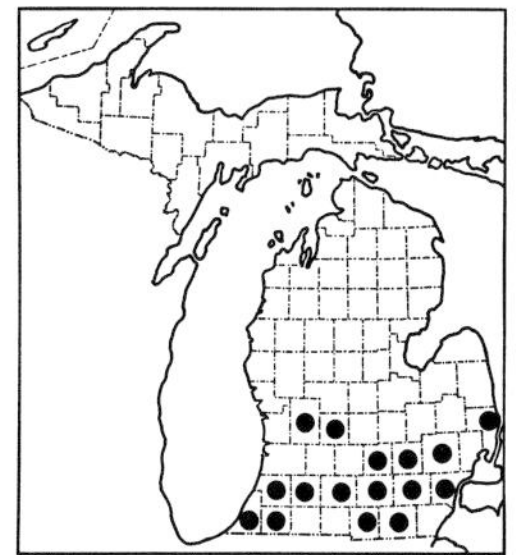

264. Hydrastis canadensis

1. **H. canadensis** L. Fig. 99 Goldenseal

Map 264. Rich deciduous woods, or less often, oak-hickory; quite local. The three petaloid sepals fall away as the flower opens, leaving the reproductive parts as the showy portion of the flower, conspicuous as in *Thalictrum* and *Actaea*. The two cauline leaves expand fully after the flower has opened. The raspberry-like fruit is considered inedible. The knotty yellow rhizomes, however, have long had a reputation for being of medicinal value, and as a result of ruthless exploitation the species has declined considerably in the United States.

11. **Ranunculus** Buttercup; Crowfoot

REFERENCES

Benson, Lyman. 1948. A Treatise on the North American Ranunculi. Am. Midl. Nat. 40: 1–261.

Cook, C. D. K. 1966. A Monographic Study of Ranunculus subgenus Batrachium (DC.) A. Gray. Mitt. Bot. Staatssam. München 6: 47–237.

Drew, W. B. 1936. The North American Representatives of Ranunculus § Batrachium. Rhodora 38: 1–47 (also Contr. Gray Herb. 110).

Duncan, Thomas. 1980. A Taxonomic Study of the Ranunculus hispidus Michaux Complex in the Western Hemisphere. Univ. California Publ. Bot. 77. 125 pp.

KEY TO THE SPECIES

1. Leaves dissected into capillary or very narrow segments, submersed
  2. Petals white; achenes transversely ridged; leaf segments not over 0.3 mm in diameter, rounded in section . . . . 1. **R. longirostris**
  2. Petals yellow; achenes smooth or warty but not transversely wrinkled; leaf segments (at least the broader ones) over 0.3 mm wide, definitely flattened
    3. Body of achene ca. 1.7–2.2 mm long, with a conspicuous corky keel; beak of achene ca. 1–1.5 (2) mm long; petals (6.5) 7–14 (16) mm long . . . 2. **R. flabellaris**
    3. Body of achene ca. 1.1–1.5 mm long, the sides ± thickened but without a corky keel; beak of achene ca. 0.5 mm long; petals 3.5–5 (6.5) mm long . . 3. **R. gmelinii**
1. Leaves simple, lobed, or compound but not finely dissected and usually not normally submersed
  4. Leaves all unlobed (except at cordate base in nos. 4 & 5), entire or at most denticulate or crenate
    5. Leaf blades crenate to entire, obovate to reniform, truncate to cordate at the base, mostly much shorter than the distinct petioles; achenes smooth and pubescent or longitudinally ridged
      6. Petals ca. 2.5–4 [5] mm long; sepals 5; achenes glabrous, less than 2 mm long (including beak), longitudinally ridged; roots not tuberous . . . . . . 4. **R. cymbalaria**
      6. Petals ca. 8–10 [14] mm long; sepals 3 [–4]; achenes pubescent, at least 2 mm long, not ridged; roots strongly tuberous-thickened . . . . 5. **R. ficaria**
    5. Leaf blades denticulate or entire, linear to lance-elliptic, tapering into the shorter petiole (or often indistinguishable from it in *R. reptans*); achenes smooth or finely reticulate, glabrous

7. Largest leaves at least (5) 7 mm wide and (5) 7 cm long (including petiole), clearly to obscurely denticulate with ± remote callus-like teeth; achenes with a distinct beak at least 0.5 mm long; petals usually at least 5 mm long .... 6. **R. ambigens**

7. Largest leaves less than 4 mm wide (usually 1 mm or less, often filiform), usually less than 7 cm long, entire; achenes with beak less than 0.3 mm long; petals ca. 2–4 (5) mm long .... 7. **R. reptans**

4. Leaves (at least the cauline ones) with blades deeply cleft or lobed more than halfway to the base

8. Plants with a very lax or creeping stem (growing in wet muddy or mossy places); leaf blades palmately lobed, the lobes crenate or with rounded teeth

9. Flowers solitary on long slender peduncles from the nodes; sepals 3; achenes with body ca. 3–4 mm long and a prominent beak recurved at the tip; principal leaf lobes usually at least 1 cm broad .... 8. **R. lapponicus**

9. Flowers usually at least 2 in a branched inflorescence; sepals 5; achenes with body less than 2.5 mm long and a beak straight at the tip; principal leaf lobes usually less than 1 cm broad

10. Body of achene ca. 1.7–2.2 mm long, with a conspicuous corky keel; beak of achene ca. 1–1.5 (2) mm long; petals (5) 7–14 (16) mm long .... 2. **R. flabellaris**

10. Body of achene ca. 1.1–1.5 mm long, the sides ± thickened but without a corky keel; beak of achene ca. 0.5 mm long; petals 3.5–5 (7) mm long .... 3. **R. gmelinii**

8. Plants with an erect leafy stem, or if trailing shoots or stolons present, the leaves clearly compound and sharply toothed; leaf blades various

11. Petals less than 5 mm long, shorter than the sepals; achenes either plump or flattened with a recurved beak; basal leaves often much less deeply lobed than cauline leaves

12. Stem and petioles with spreading hairs; achenes strongly flattened, with a prominent beak recurved (hooked) at the tip (evident on carpels at anthesis) .... 9. **R. recurvatus**

12. Stem and petioles glabrous or with a few delicate mostly curly hairs; achenes plump (with convex sides), the minute beak less than 0.3 mm long

13. Basal leaves ± lobed; fruiting heads usually cylindrical, 1.5–3 times as long as thick; achenes rather dull, ca. 1 mm long with terminal beak, ± corky margin, and slightly wrinkled or pebbled sides; plant of wet shores, ditches, etc. .... 10. **R. sceleratus**

13. Basal leaves (at least some of them) cordate to reniform, not at all lobed; fruiting heads less than 1.5 times as long as thick; achenes shiny, ca. 1.5–2 mm long with subterminal beak, the margins not corky, and the sides at most minutely puncticulate; plant of deciduous or mixed woods (often disturbed areas) .... 11. **R. abortivus**

11. Petals usually longer and exceeding the sepals (shorter in nos. 13 & 14); achenes flattened (except in *R. rhomboideus*), the beak straight or curved (or sometimes minutely hooked in species with large petals); basal leaves, if any, deeply lobed, cleft, or compound (unlobed only in *R. rhomboideus*)

14. Basal leaves simple, the blades crenate, unlobed; achenes plump (convex on sides), ca. 1.6–2.2 mm long .... 12. **R. rhomboideus**

14. Basal leaves deeply cleft, lobed, or compound (or absent); achenes flattened, the body at least 2 mm long and the beak at maturity at least 0.4 mm

15. Petals scarcely if at all longer than the sepals; anthers less than 1 mm long; achenes in a cylindrical or ovoid head; stems and petioles with spreading hairs

16. Petals 4.5–6.5 mm long, about equaling the sepals (often slightly longer); beak of achene ca. 0.7–1.5 mm long; fruiting head ovoid; plant a rare northern species at Lake Superior .... 13. **R. macounii**

16. Petals 2.5–4 (4.5) mm long, distinctly shorter than the sepals; beak of achene ca. 0.4–1 mm long; fruiting head ± cylindrical; plant frequent throughout .... 14. **R. pensylvanicus**

15. Petals much exceeding the sepals; anthers mostly (0.9) 1.1–2.8 mm long; achenes in a globose or subglobose head; stems often glabrous or with hairs mostly ± appressed (especially on upper portion) but sometimes spreading

17. Leaves simple, very deeply cleft but the terminal lobe not stalked; receptacle glabrous; plant a common introduced weed of fields and waste places (spread into damp ground everywhere) .... 15. **R. acris**

17. Leaves compound, at least the terminal primary lobe (leaflet) definitely stalked; receptacle ± bristly or hairy; plant a rare introduction (nos. 16 & 17) or native (nos. 18 & 19)

18. Style ± deltoid, short (less than 1 mm), stigmatic more than half its length; plants uncommon, introduced, usually in waste places

19. Plants usually strongly stoloniferous, without bulbous base; sepals spreading; beak of achenes ca. 0.8–1 mm long .... 16. **R. repens**

19. Plants erect, from a thick bulbous corm; sepals becoming reflexed; beak of achenes ca. 0.4–0.5 mm long .... 17. **R. bulbosus**

18. Style slender and elongate (rarely deltoid), soon becoming (1) 1.5–3 mm long, stigmatic only at the tip; plants native, widespread in wet or dry places

20. Plant erect, not stoloniferous, of dry sandy or rocky habitat, the roots often tuberous-thickened; petals linear-oblong, usually widest near or below the middle; ultimate major lobes of upper leaves linear-oblanceolate, frequently less than 4 mm wide; teeth or lobes of leaflets obtuse to rounded; rhizomes not surviving from year to year .... 18. **R. fascicularis**

20. Plant usually becoming ± stoloniferous, with repent stems, of swampy habitats (one scarce variety in uplands), the roots long and fibrous (sometimes thick but not tuberous); petals rounded-obovate, widest above the middle; ultimate leaf lobes broad (mostly over 5 mm); teeth or lobes of leaflets (at least upper ones) ± acute; rhizome only partly dying back each year .... 19. **R. hispidus**

1. **R. longirostris** Godron Fig. 101 White Water Crowfoot

Map 265. In a great diversity of lakes, ponds, slow to rapid streams, rivers, ditches, and pools.

The status of our white-flowered aquatic buttercups in *Ranunculus* subgenus *Batrachium* is by no means clear. As noted by Morton and Venn (1984) in connection with the Manitoulin Island flora, application of any published treatment for them is a frustrating experience. I here tentatively follow the interpretation of Boivin (Nat. Canad. 93: 590. 1966) that we have a single variable species, but I do not employ for it the name *R. aquatilis* L., which Cook (1966) maintains strictly as a heterophyllous species, capable of producing expanded, simple floating leaf blades as well as dissected submersed ones. It is an Old World species also found in western North America and the name has been often misapplied to our plants, which always produce only dissected leaves, even when stranded on wet flats. According to Cook,

R. longirostris appears to be a North American vicariant of *R. circinatus* Sibth., a name also often misapplied to our plants, but oddly enough he saw no living material of our common species and acknowledged loans from no North American herbaria. Nevertheless, the name *R. longirostris* is clearly typified by American material and has been widely enough used that the least ambiguity results from maintaining it until any identity with a European species can be firmly established.

Cook recognized only one other species of this subgenus as occurring in eastern North America, *R. trichophyllus* Chaix, characterized principally by having a very short style and achene beak, while in *R. longirostris* the persistent style forms a beak 0.7–1.1 mm long, or at least a third as long as the body of the achene. Supposed vegetative and floral characters which have been used to distinguish species in our area are simply not correlated, and Cook failed to account at all for some important names applicable to some of our material, such as *R. subrigidus* W. B. Drew (close to *R. longirostris*) and *R. trichophyllus* var. *calvescens* W. B. Drew, characterized by a glabrous rather than hairy receptacle (both of our species merely said by Cook to have the receptacle hairy). Several of our specimens have a short style or beak and have been referred in the past to what would now be called *R. trichophyllus*; however, most of these have little or no petiole beyond the adnate portion of a broad and hairy stipule—a condition most authors ascribe to *R. longirostris*, with *R. trichophyllus* supposedly having a further petiole nearly or quite as long as the adnate portion of a glabrous stipule. Sepals, when present, on these plants have a purplish center or tip on the back, a feature said by Cook to be "a useful and reliable character" for *R. longirostris*. The petals may be as long as 9 mm—larger than indicated for *R. trichophyllus*, but the maximum for *R. longirostris*. So the latter name is here applied in a broad sense to all of our Batrachian buttercups—permitting mapping of many fruitless specimens which otherwise would remain unnamed if the sole reliable distinction between two species were the length of the beak on the achene.

Some of our specimens have the stipule and petiole characters of *R. trichophyllus*, and the lax leaves often attributed to that species—but long styles. Even if we do have two species, *R. longirostris* is surely the commoner one, almost all of the supposed *R. trichophyllus* (fertile plants, at least) being from the Upper Peninsula. The size and degree of limpness or delicacy of the leaves are conspicuously variable, as is the length of internodes, and such characters probably depend in part on the flow, depth, and temperature of the water. The size, shape, and number of achenes, as pointed out by Cook, may depend on environmental factors including completeness of pollination and hence crowding on the receptacle.

I have found the flowers—which are sometimes abundant—to be quite fragrant. They may open at or barely above the surface of the water or as deep as 0.75 m beneath.

Sterile plants (e. g., from deep or rapidly flowing water) may be distinguished from the next two species as indicated under the first lead in the key, and these three species may be distinguished from all other submersed aquatics with dissected cauline leaves by the combination of alternate leaf arrangement (unlike *Myriophyllum*, *Ceratophyllum*, and some others), absense of bladders (unlike *Utricularia*), and absence of a definite central axis in the leaves (unlike *Armoracia*, *Proserpinaca*, *Myriophyllum*). The tips of the leaf segments in *R. longirostris* bear tiny translucent spicules.

2. **R. flabellaris** Raf. Fig. 102 Yellow Water Crowfoot

Map 266. Standing water of swamp forests, woodland pools, ponds, ditches, marshes; very local northwards.

Even in the absence of the distinctive corky-winged achenes, this species is usually easily distinguished from the next by its larger size, especially in the flowers but in general also in foliage (although the submersed leaves of the next may be quite large, usually with segments broader on the average than in the finer submersed leaves of *R. flabellaris*). Plants of both species produce smaller, less limp, less extensively lobed leaves when stranded on damp shores than they do when submersed.

3. **R. gmelinii** DC. Yellow Water Crowfoot

Map 267. Edges of rivers, streams, lakes, and ponds (in water and on banks); pools in bogs and cedar swamps.

See comments under the preceding species. The submersed parts are sometimes said to be more pubescent than the aerial parts, which would be a remarkable reversal of the usual situation in aquatic plants if true. Certainly in our material, submersed parts are glabrous, while terrestrial parts may sometimes be ± strigose. Plants of the more southern (and especially eastern) part of the North American range of this circumpolar species have usually been referred to var. *hookeri* (D. Don) Benson or ssp. *purshii* (Richardson) Hultén.

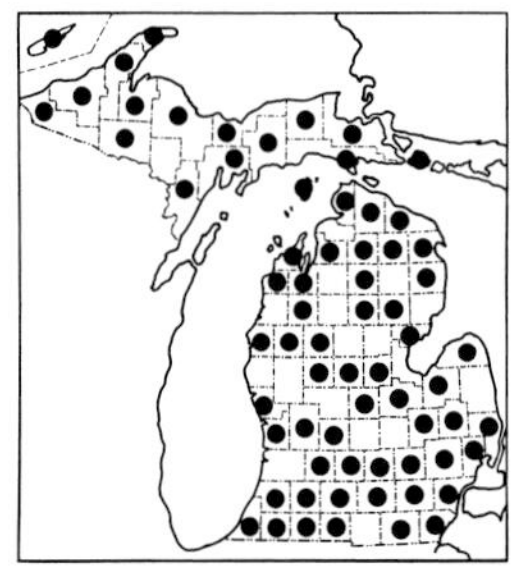

265. Ranunculus longirostris

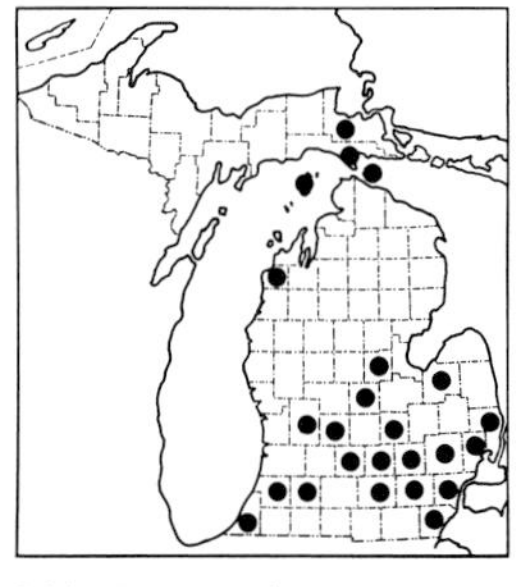

266. Ranunculus flabellaris

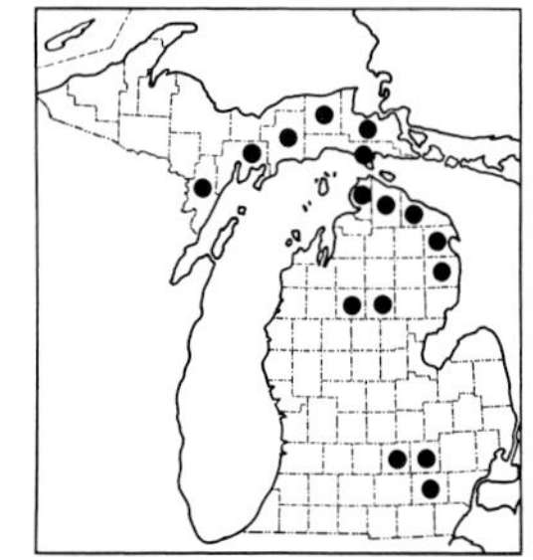

267. Ranunculus gmelinii

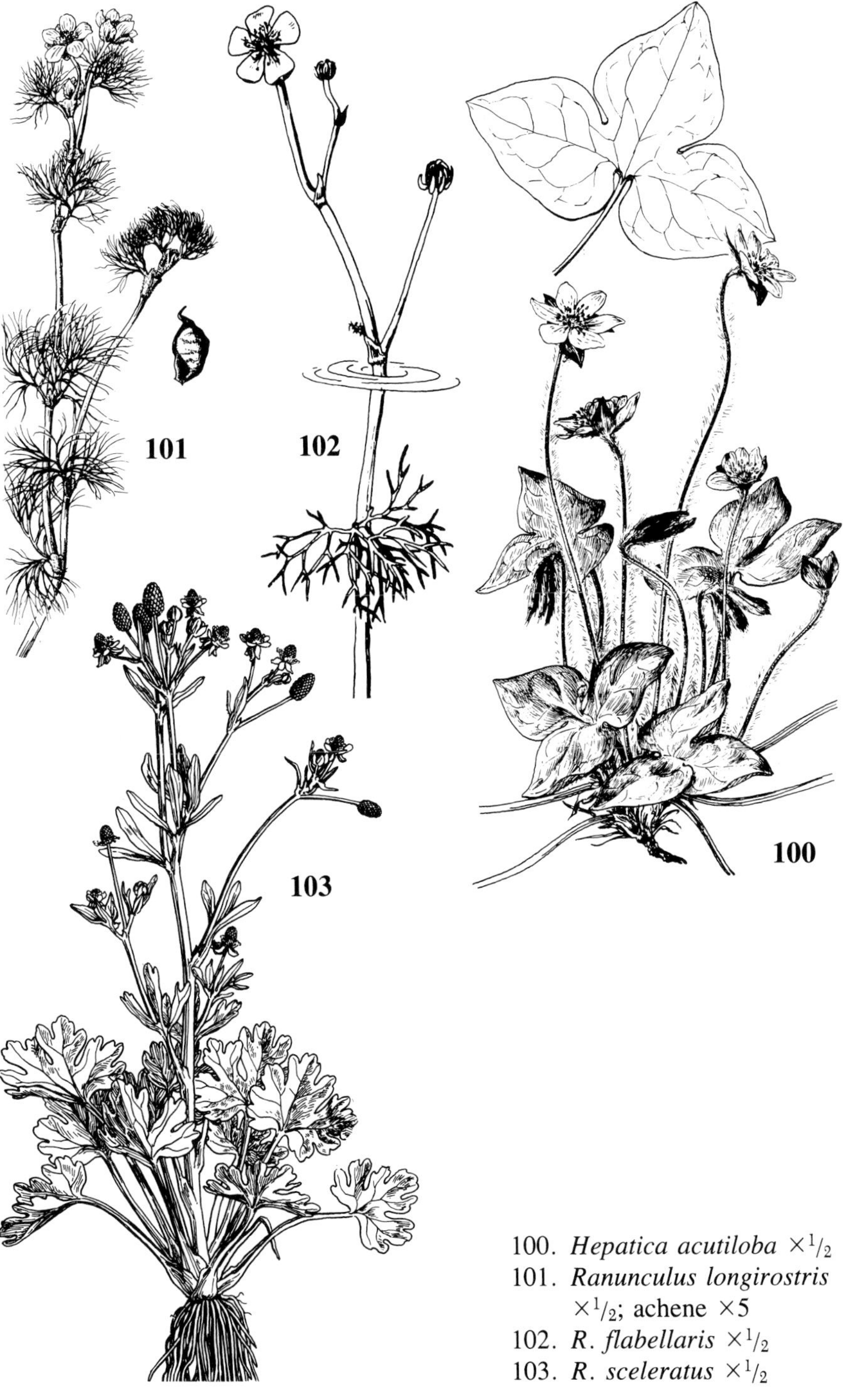

100. *Hepatica acutiloba* $\times ^1/_2$
101. *Ranunculus longirostris* $\times ^1/_2$; achene $\times 5$
102. *R. flabellaris* $\times ^1/_2$
103. *R. sceleratus* $\times ^1/_2$

4. **R. cymbalaria** Pursh

Map 268. A species of damp shores and ditches, apparently very rare in the state. On a moist rather boggy shore (Barney's Lake) on Beaver Island (*Voss 4929* in 1957, MICH, UMBS) and near Wakefield (*Rogers 12449* in 1960, WUD).

5. **R. ficaria** L. Lesser-celandine

Map 269. A Eurasian species, locally naturalized in North America, presumably as an escape from cultivation. First collected in Michigan in 1982 (*Gereau 966*, MICH, MSC, BLH; see Mich. Bot. 23: 51–52. 1984) in a damp wooded ravine along the Grand River in Clinton County. Two years later, discovered by Ralph Blouch in a swamp forest along the same river in Eaton County (Mich. Bot. 24: 125. 1985).

6. **R. ambigens** S. Watson

Map 270. Along streams and ditches in St. Clair County (collected 1903–1904) and in 1940 found by Bazuin (*1617*, MICH, MSC) on a springy hillside in Ottawa County.

7. **R. reptans** L. Creeping Spearwort

Map 271. Damp sandy or gravelly shores, especially of softwater lakes; less often on muddy or mucky shores or on rock. Sterile plants are often locally abundant in shallow to deeper water of many northern lakes, and can be easily recognized by the arching green stolons, with filiform leaves truncate at the tip. Both aquatic and terrestrial forms root at every node of the creeping stem.

Terrestrial plants with the broadest leaves (to ca. 5 mm in our specimens) have been called var. *ovalis* (Bigelow) T. & G. This is a circumpolar species, sometimes united by American authors with *R. flammula* L., almost entirely a species of Europe, where authors usually maintain the two as distinct.

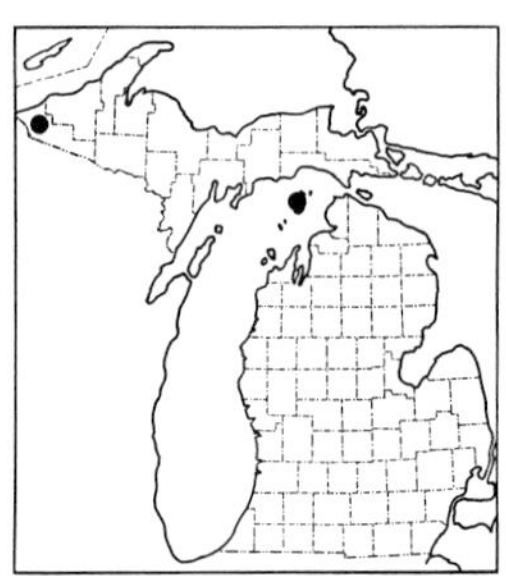

268. Ranunculus cymbalaria

269. Ranunculus ficaria

270. Ranunculus ambigens

8. **R. lapponicus** L. Lapland Buttercup

Map 272. Known in Michigan only from a cold, wet, mossy cedar swamp near the St. Mary's River. Northern Michigan and northeastern Minnesota are at the southern limit of this circumpolar species in North America.

A very distinctive little species, the foliage superficially resembling that of *Coptis*, but the leaves paler (not evergreen) and only deeply lobed (not actually compound).

9. **R. recurvatus** Poiret Hooked Crowfoot

Map 273. Usually in rich moist deciduous woods, especially in open or ± disturbed spots as along trails, banks, and borders; also in cedar swamps and deciduous swamp forests, and less often in oak-hickory woods.

Distinctive in the prominent, strongly hooked styles and beaks; no other short-petaled species has such hooks, although a hooked tip may occur in some of the large-petaled buttercups.

10. **R. sceleratus** L. Fig. 103 Cursed Crowfoot

Map 274. Shores, muddy banks, and pools, thriving on recently exposed surfaces; ditches, puddles and ruts in roads and trails, marshes, and wet meadows. A circumpolar species, and some of our plants may represent introductions from the Old World.

This species may be truly aquatic, with some floating leaves when growing in shallow water. It may bloom into November. It is usually readily distinguished from the next species by its cylindrical fruiting head, more succulent leaves, slightly paler yellow petals, lobed lower leaves, and damper habitat.

11. **R. abortivus** L. Small-flowered Buttercup

Map 275. Chiefly in upland woods—beech-maple, mixed hardwoods, even oak-hickory, especially in slightly disturbed places as along trails, by upturned tree roots, etc.; sometimes in swampy woods, meadows and fields (especially bordering woods), and shrubby thickets.

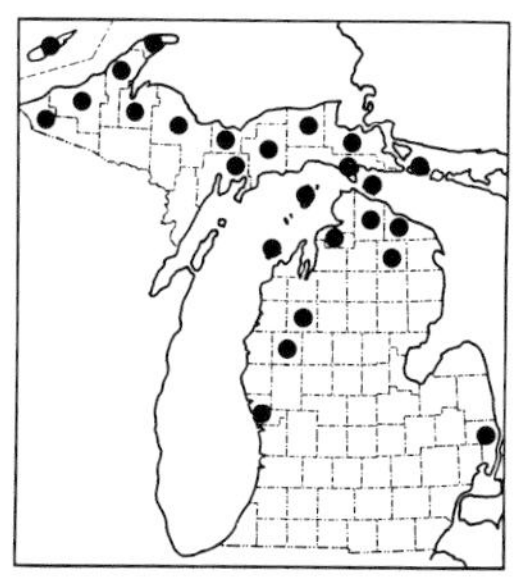

271. Ranunculus reptans

272. Ranunculus lapponicus

273. Ranunculus recurvatus

Throughout the state, sometimes growing with perfectly glabrous plants, are ± pubescent individuals bearing a few fine hairs on the stem. Most noticeable on some of these plants is a more dense pubescence of short incurved hairs on the pedicels. Pubescent plants have been called var. *acrolasius* Fern., but do not seem to represent a significant variant in our area. Occasionally the petals may appear to be as long as the sepals, but if the cupped sepals are flattened out they will exceed the petals.

12. **R. rhomboideus** Goldie — Prairie Buttercup

Map 276. Sandy banks and grasslands, very local in southern Michigan; also on rocky ridges (generally south-facing) on Isle Royale. A prairie species, originally described from near the eastern edge of its range, Lake Simcoe, Ontario.

Relationship to the preceding species is shown by the simple basal leaves, although the blades are ± truncate at the base, not cordate. But this species is easily distinguished from *R. abortivus* by the larger petals and conspicuously villous stems and petioles.

13. **R. macounii** Britton

Map 277. Known in Michigan only from wet ground (marsh, streamside, muddy hollow) at Isle Royale.

Occasionally plants of the next species with larger flowers than usual might appear to be close to this species, but their cylindrical heads of achenes in contrast to the thick-ovoid ones of *R. macounii* are quite different. The achene beak in *R. macounii* may be strongly turned toward one edge, but sometimes it is as short and straight as in some specimens of *R. pensylvanicus*. The sepals in both species usually become strongly reflexed.

14. **R. pensylvanicus** L. f. — Bristly Crowfoot

Map 278. Marshes and swampy places, river banks and streamside thickets, ditches and swales, damp fields and shores; often in muck but sometimes in sandy and disturbed sites.

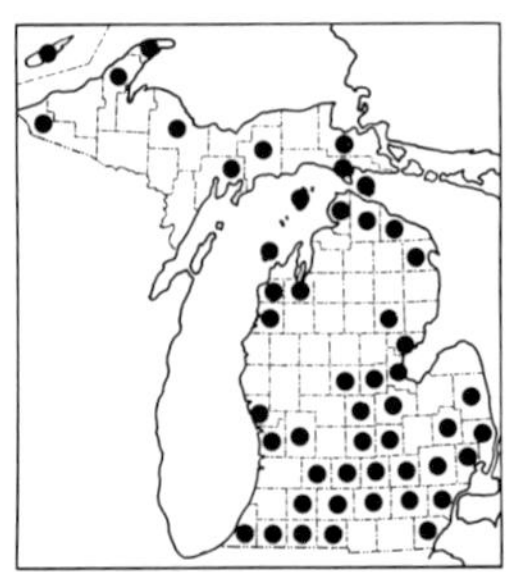

274. Ranunculus sceleratus

275. Ranunculus abortivus

276. Ranunculus rhomboideus

15. **R. acris** L. Fig. 104 Tall or Common Buttercup

Map 279. Roadsides, fields, clearings, shores, damp thickets—everywhere. A native of Europe, where it is also weedy. Reported by the First Survey (1838) but the oldest Michigan specimen I have seen was made by Dennis Cooley, who recorded that he first saw this species in Macomb County on June 11, 1845. Known from several places, from southern Michigan to Isle Royale, by the 1860's.

A double-flowered form occasionally escapes from cultivation. In our specimens of *R. acris*, the beak of the achene is ca. 0.5 mm long and the petals are 6–13 (15) mm long.

16. **R. repens** L. Creeping Buttercup

Map 280. A variable European species, occasionally found as a weed of roadsides, lawns, vacant lots, fields, springy shores and marshes.

The species is sometimes cultivated, especially in a double-flowered form which may escape.

17. **R. bulbosus** L. Fig. 105 Bulbous Buttercup

Map 281. Fields, roadsides, and waste ground. Another European species, sometimes cultivated for its showy flowers, which may be double (our few records are almost all old, of uncertain status, and the plants are single-flowered).

277. Ranunculus macounii

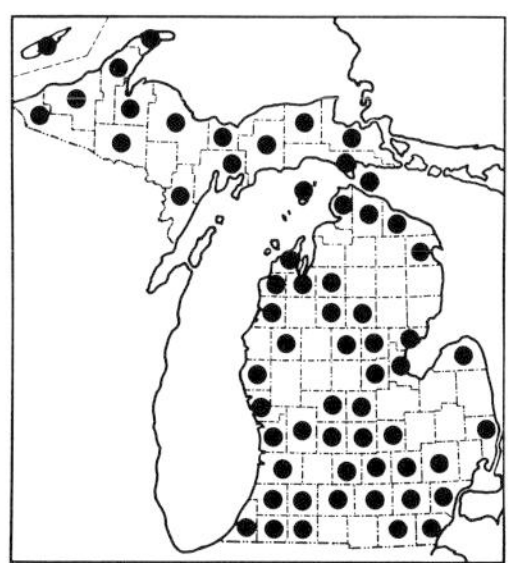

278. Ranunculus pensylvanicus

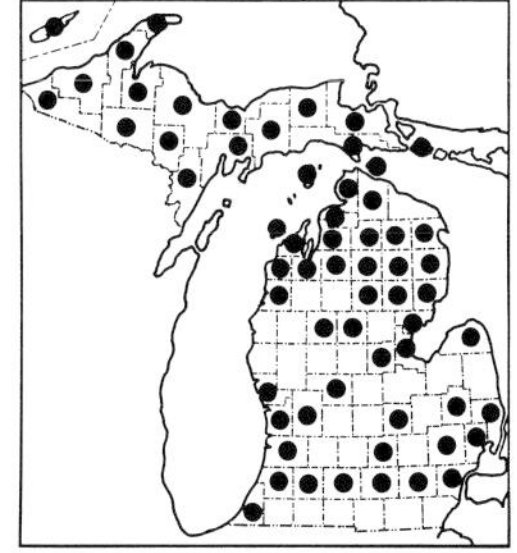

279. Ranunculus acris

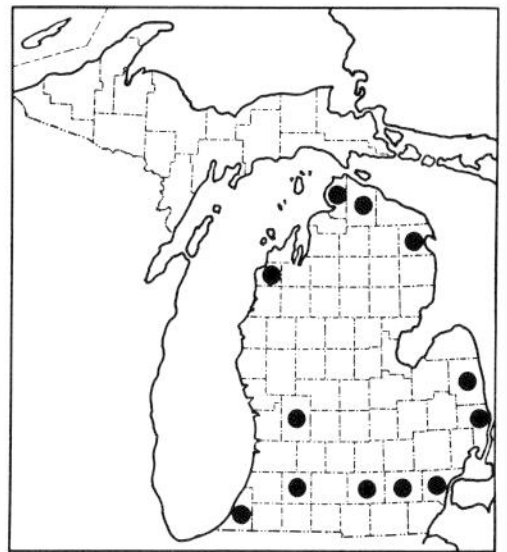

280. Ranunculus repens

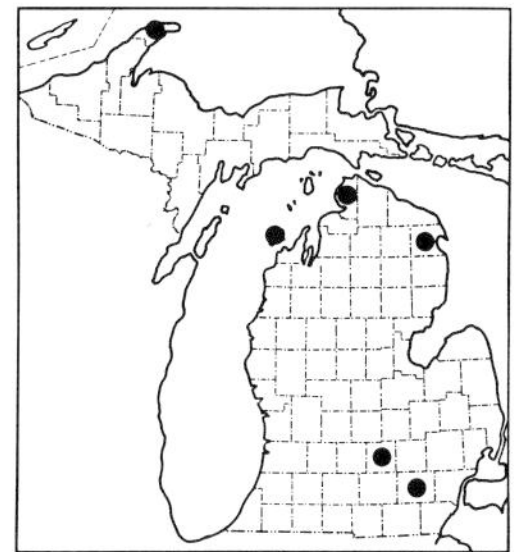

281. Ranunculus bulbosus

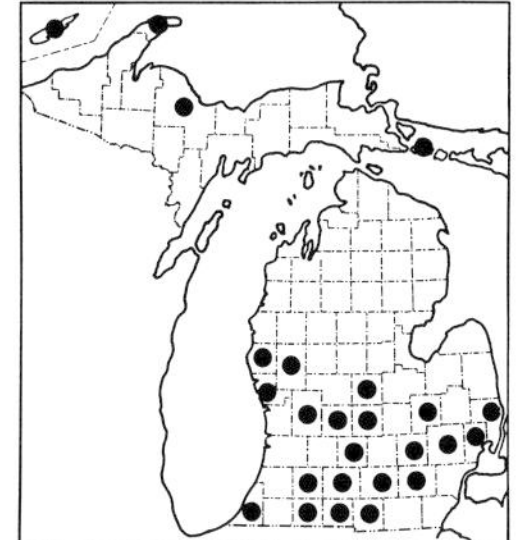

282. Ranunculus fascicularis

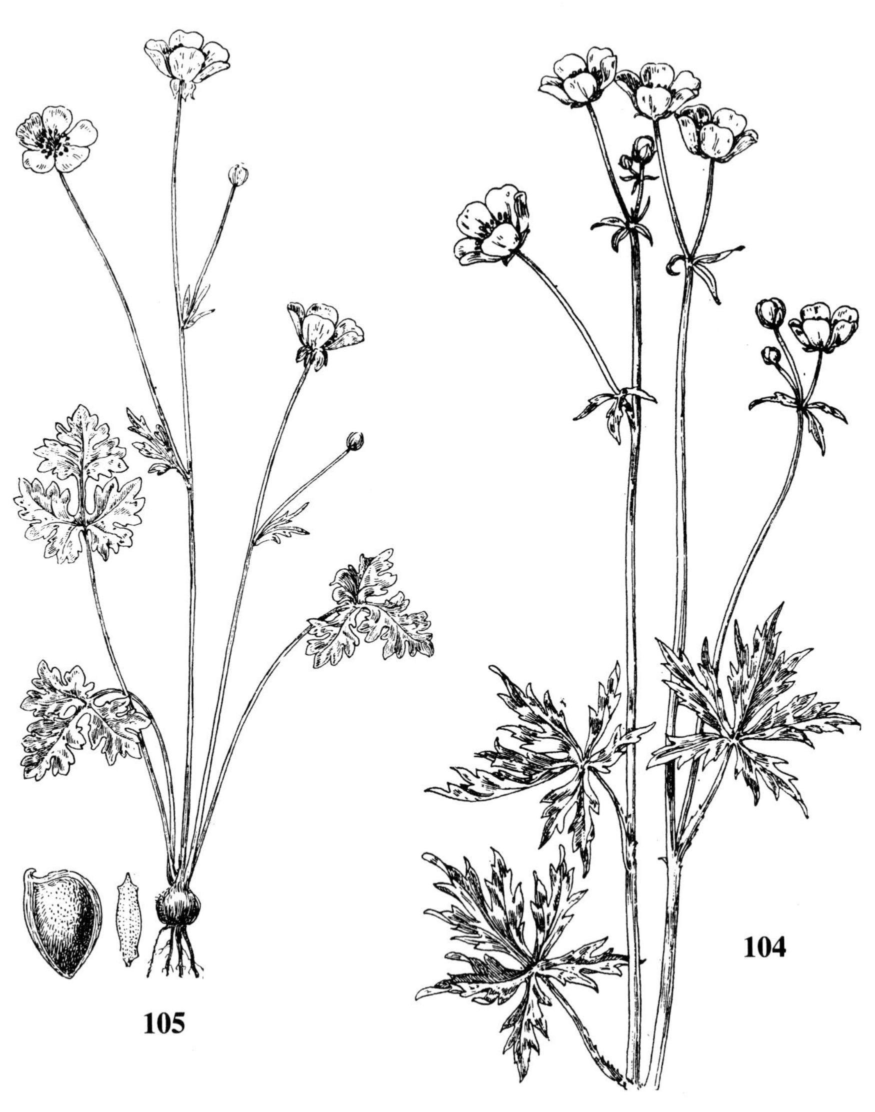

104. *Ranunculus acris* ×3/4
105. *R. bulbosus* ×1/2; achene ×5

18. **R. fascicularis** Bigelow Early Buttercup

Map 282. Sandy oak woodland, dry grassy places (and prairies), shallow soil over limestone pavement (Drummond Island), rock ledges (Keweenaw Co., Isle Royale, and Huron Mountains).

This is a very early-flowering species, usually readily distinguished from small plants of the next by the less sharply toothed leaves and the short tuberous-swollen roots.

19. **R. hispidus** Michaux Plate 3-A; fig. 106 Swamp Buttercup

Map 283. Typical var. *hispidus* is very local in the southernmost Lower Peninsula, from Monroe County to Kalamazoo County; it grows in moist to dry upland places (oak woods, sandy knolls, etc.). The other two, more widespread, varieties have practically the same habitats and may indeed grow together; they are characteristic of low woods and thickets along streams, ponds, and lakes, as well as ravines and wet spots in deciduous or cedar swamps.

This is an extremely variable species, with the lateral leaflets (as well as the terminal one) definitely petiolulate, at least the upper ones ± sharply toothed, and the roots even if slightly thickened long and not tuberous. Duncan (1980) notes that the leaf and pubescence characters often used to distinguish taxa in (or from) *R. hispidus* are not consistent within populations. He recognizes three varieties: (1) var. *hispidus* barely enters southernmost Michigan and is a small erect plant, not producing trailing shoots; the petals are less than 1 cm long and the achenes are only narrowly margined. (2) var. *caricetorum* (Greene) T. Duncan occurs throughout the state and is a large coarse often very pubescent plant (with twice the chromosomes of the other varieties), producing trailing shoots which often root (and which can be seen developing from the base of the plant early in the season); the flowers are the largest in this variety, the petals usually over 1 cm long and sometimes nearly 2 cm. (3) var. *nitidus* (Ell.) T. Duncan, found in the southern half or two-thirds of the Lower Peninsula, somewhat similar to var. *hispidus* in small stature and often confused with it, but characteristic of damp habitats and with trailing shoots as in var. *caricetorum*, from which it differs in more broadly winged achenes, petals seldom much over 1 cm long, and often less pubescence. *R. septentrionalis* Poiret is a name widely used for var. *caricetorum* but actually applying to var. *nitidus*, which is sometimes known as *R. carolinianus* DC.

12. **Caltha**

A second species, *C. natans* Pallas, summer-blooming and with white flowers, was attributed to Michigan in old manuals as a result of misinterpretation of locality data from Minnesota. *C. natans* ranges in North America from Alaska to northeastern Minnesota and adjacent Wisconsin, and might conceivably turn up in Michigan some day.

1. **C. palustris** L. Plate 3-B Marsh-marigold; Cowslip

Map 284. Edges of streams, marshy hollows, wet swamps (deciduous or coniferous), thriving best in open or only partly shaded sites.

A familiar plant to most people, sometimes carpeting wet places with brilliant yellow flowers early in the spring.

13. **Nigella**

1. **N. damascena** L. Love-in-a-mist

Map 285. A garden plant, native to the Mediterranean region, rarely escaped along roadsides, etc.

The white to pale blue flowers are large (up to ca. 4 cm across) and the fruit is inflated, of united follicles.

14. **Anemonella**

This genus, which contains a single species, is included by some authors in *Thalictrum*, the achenes being very similar to those of that genus.

1. **A. thalictroides** (L.) Spach Fig. 107 Rue-anemone

Map 286. Rich or sometimes dry deciduous woods.

The involucre is normally composed of sessile compound leaves, the petiolules appearing to be petioles of small simple leaves. Double-flowered and pink-flowered forms are known.

15. **Isopyrum**

1. **I. biternatum** (Raf.) T. & G. Fig. 108 False Rue-anemone

Map 287. Rich deciduous woods, often on shaded banks of streams.

There seems to be a widespread impression that this is a less common species than the preceding (and it is omitted in many wildflower books), but that is certainly not its status in Michigan. Superficially, it does closely resemble the preceding in its white flowers and delicate, clearly compound,

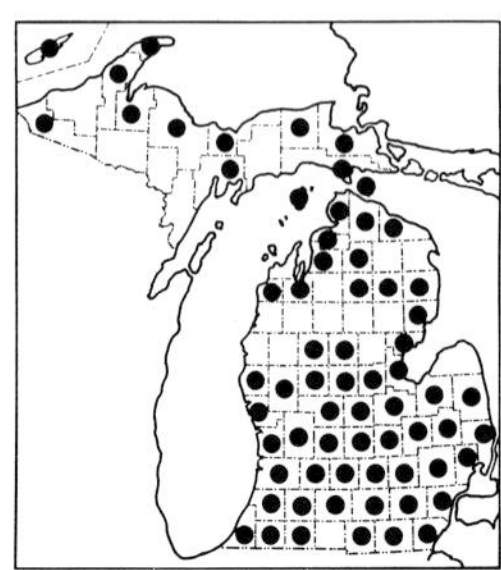

283. Ranunculus hispidus

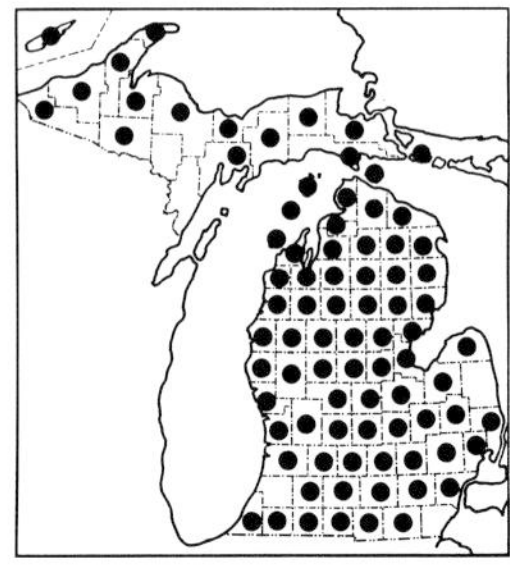

284. Caltha palustris

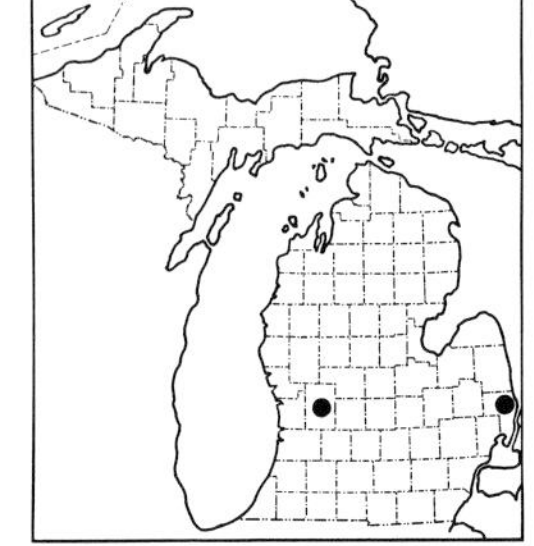

285. Nigella damascena

Thalictrum-like leaves, although the leaves are alternate as well as basal and the fruit is a few-seeded follicle rather than an achene. The leaflets of *Isopyrum* consistently look different from those of *Anemonella*, which are ± crenate; in *Isopyrum*, the larger leaflets are clearly lobed, one or more sinuses extending at least a third of the way to the base of the leaflet. Furthermore, the tips of the lobes in *Isopyrum* bear a minute projecting callus, while in *Anemonella*, on the contrary, the tips are usually minutely notched. The lobes in *Isopyrum* give a ± oblong impression, in contrast to the rounded leaflets of *Anemonella*.

16. **Helleborus** — Hellebore

These plants are not to be confused (because of name, not appearance) with the false hellebores (*Veratrum* spp.) but, like them, are poisonous if ingested.

KEY TO THE SPECIES

1. Leaves deciduous; bracts at inflorescence divided and serrate; flowers greenish . . . . 1. **H. viridis**
1. Leaves evergreen (overwintering); bracts simple, entire; flowers white to pinkish . . . . 2. **H. niger**

1. **H. viridis** L. — Green Hellebore

Map 288. A native of Europe, found as an escape from cultivation blooming April 18, 1896, on sandy shores of Belle Isle (*Farwell 860*, BLH).

2. **H. niger** L. — Christmas-rose

Map 289. Another garden plant of European origin, found blooming in woods November 5, 1919 (*Walpole 677*, BLH; reported by him as *H. viridis*).

286. Anemonella thalictroides

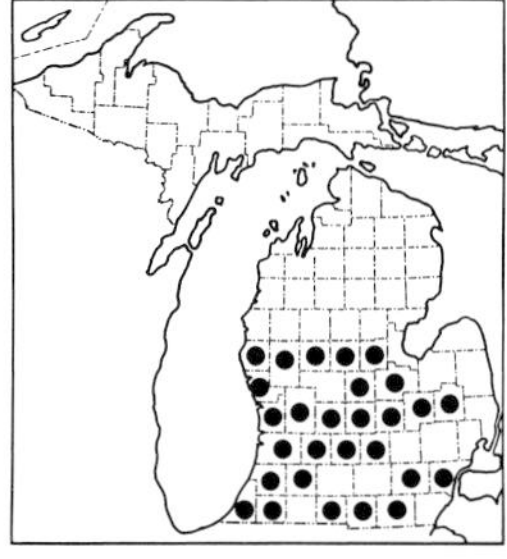

287. Isopyrum biternatum

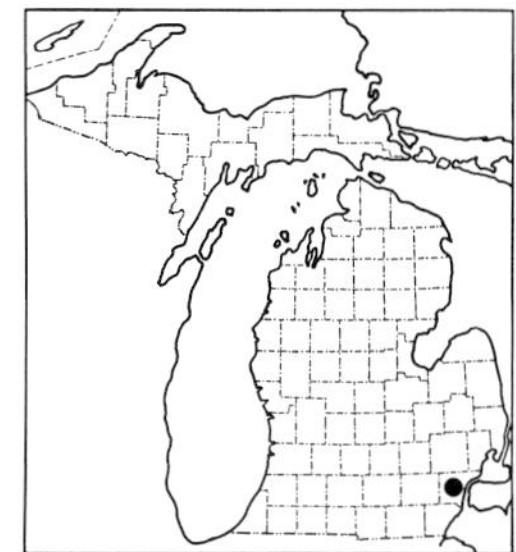

288. Helleborus viridis

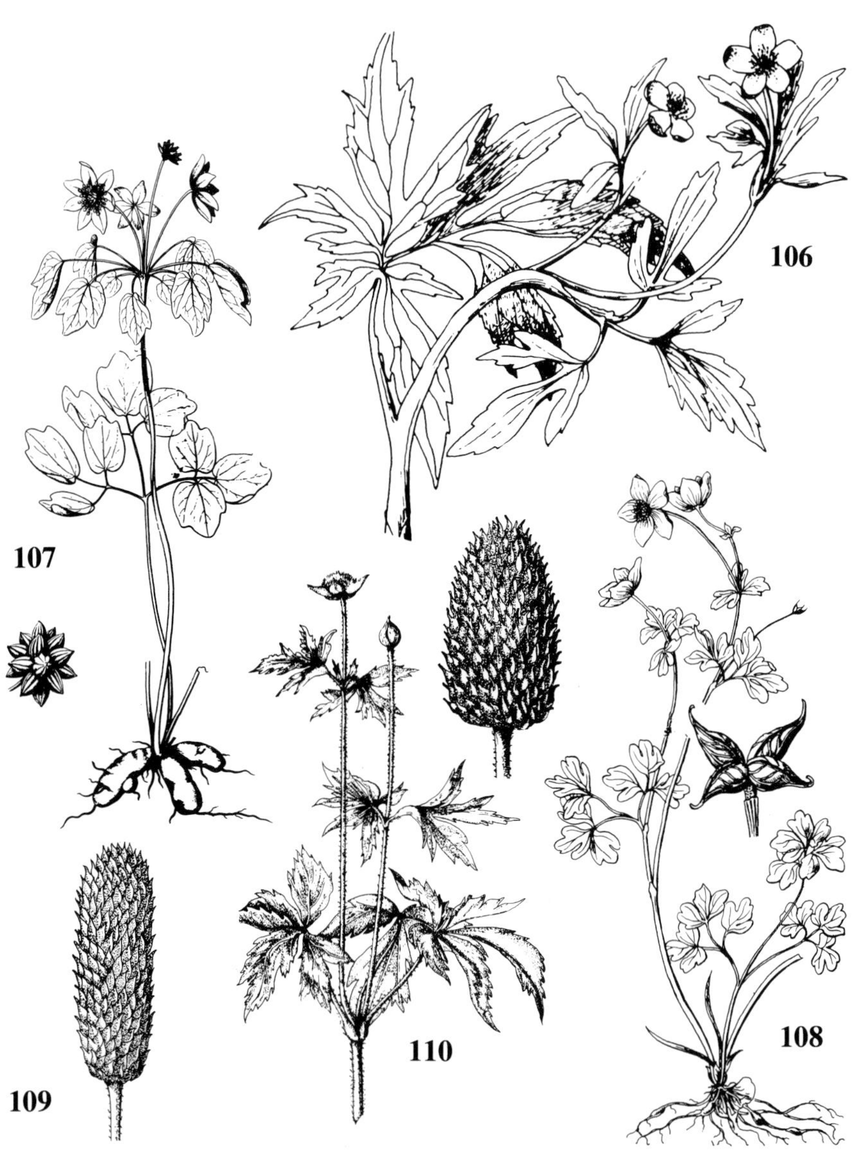

106. *Ranunculus hispidus* $\times {}^{2}/_{5}$
107. *Anemonella thalictroides* $\times {}^{1}/_{2}$; head of achenes $\times 1$
108. *Isopyrum biternatum* $\times {}^{1}/_{2}$; head of follicles $\times 1$
109. *Anemone cylindrica,* head of achenes $\times 1$
110. *Anemone virginiana* $\times {}^{1}/_{2}$; head of achenes $\times 1$

17. **Anemone** Anemone; Windflower

Several species of this genus are cultivated as garden plants. One small, very early-blooming one (from the Mediterranean region), with deep blue flowers, *A. blanda* Schott & Kotschy, has been collected as persisting at an old garden site in Schoolcraft County. The pasque-flower, *A. patens* L., which ranges east in prairies and other dry places as far as Lake Michigan, has been attributed to Michigan many times, but no specimens are known; it is a silky-pubescent plant with plumose styles in fruit (as in *Clematis*). *A. parviflora* Michaux is a northern species which ranges south to Lake Superior (Slate Islands) in Ontario and might some day be discovered in Michigan; it has one-flowered peduncles bearing a pair of lobed bracts and rounded or blunt-lobed basal leaves very similar to those of *Ranunculus lapponicus*; the sepals are whitish.

KEY TO THE SPECIES

1. Lobes of leaf linear, entire, ca. 1.5–4 (5) mm wide; perianth usually maroon (rarely cream); achenes densely long-woolly . . . . 1. **A. multifida**
1. Lobes of leaf ± cuneate to convex, sharply toothed, often broader; perianth never maroon (white, cream, or greenish); achenes various
  2. Achenes pubescent with ± straight hairs (silky when young but not woolly or sinuous and not so dense as to conceal the surface); fruiting head subglobose; cauline leaves sessile or basal leaves absent; mature perianth white
    3. Cauline (involucral) leaves sessile, cleft more than half their length but not compound; basal leaves long-petioled; achenes flat . . . . 2. **A. canadensis**
    3. Cauline leaves on distinct slender petioles, cleft so deeply as to be compound; basal leaves absent at flowering time (separate sterile leaves frequent); achenes plump . . . . 3. **A. quinquefolia**
  2. Achenes densely woolly with long sinuous hairs nearly or quite obscuring the surface; fruiting head ovoid to long-cylindrical; cauline leaves petioled and basal leaves also present; mature perianth white to cream or greenish
    4. Peduncles all (or usually at least 2 of them) bractless, the bracts and involucral (cauline) leaves crowded in one cluster of 4–9 at the base of the peduncles; divisions of leaf narrowly cuneate with straight to slightly concave sides toothed only well beyond the middle of the blade; fruiting head cylindrical, ca. 2–4 times as long as thick (not over 1 cm thick); pubescence (of peduncles, upper part of stem, and lower surfaces of leaves) silky and ± appressed or ascending . . . . 4. **A. cylindrica**
    4. Peduncles (all except the leading one) with a pair of bracts, the involucral leaves only 3 at the base of the peduncles; divisions of leaf broadly cuneate to (usually) ± convex on the sides, toothed to or below the middle of the blade; fruiting head ovoid to thick-cylindric, up to 2.5 times as long as thick (as thick as 1.7 cm); pubescence (especially of stems and leaves) sparser and more spreading (though usually silky on peduncles) . . . . 5. **A. virginiana**

1. **A. multifida** Poiret Plate 2-F Red Anemone

Map 290. Sandy (or gravelly) shores and dunes along the Great Lakes and nearby old shore ridges; on rocks at Isle Royale.

Our plants are almost all the showy red- or maroon-flowered f. *sanguinea* (Pursh) Fern. However, at the only known station on the south shore of Lake Superior, on the Grand Sable dunes, the typical f. *multifida* with yellowish or cream perianth is found. Plants of dunes on the north shore of Lake Superior, in Ontario, are similar, but plants of rocks at Isle Royale are red-colored (though the red is often somewhat paler).

2. **A. canadensis** L. Canada Anemone

Map 291. Open moist ground: shores and meadows, even marshes; roadsides and railroad banks; clearings and borders of woods; river banks and thickets, borders of streams. Often forms large colonies from spreading horizontal roots.

The sessile cauline leaves of this species will distinguish it from the next three, even if the carpels are too young to be sure of the pubescence; the long-beaked, flattened, and sparsely pubescent achenes are distinctive. The attractive white flowers are (2) 2.5–4.8 (5) cm across.

3. **A. quinquefolia** L. Wood Anemone

Map 292. In a diversity of dry to swampy deciduous and mixed woods—even under conifers and in cedar swamps; thickets and banks along streams.

Similar to the Old World *A. nemorosa* L., with which many older authors associated it. This is an eastern American species, and in the western portion of its range, including all of Michigan, the stems and petioles are spreading-pubescent. These plants were named *A. quinquefolia* var. *interior* by Fernald, in admitted, deliberate violation of the International Code of Botanical Nomenclature; for an older epithet was available, albeit an inappropriate one: var. *bifolia* Farwell (TL: Rochester [Oakland Co.]). Names may not be rejected on grounds of inappropriateness, however, and Farwell's name, based on plants with only 2 leaves instead of the usual 3, has clear priority. Fernald's later name may be used only if one believes that 2-leaved pubescent plants and 3-leaved pubescent plants constitute two different varieties worthy of names—a quite untenable position, as 2-leaved plants may turn up in almost any population and are of little significance. There is even

289. Helleborus niger

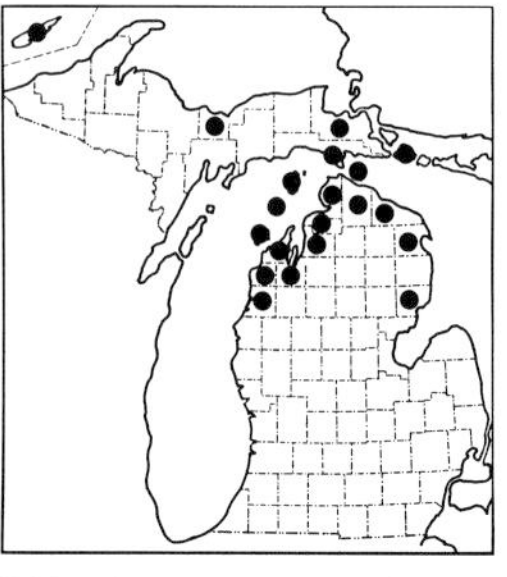

290. Anemone multifida

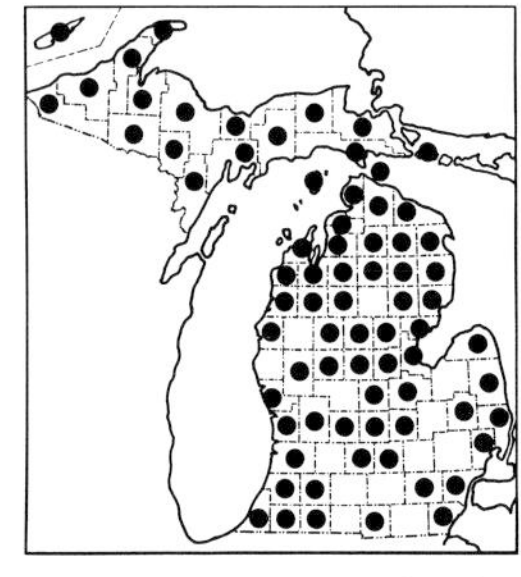

291. Anemone canadensis

considerable doubt whether pubescent plants are worthy of any taxonomic recognition.

This is a relatively short plant, compared to our other species, less than 2.5 dm tall (often only 1 dm), never with more than a single flower, which is 1.5–2.5 (3.2) cm across and blooms in the spring.

4. **A. cylindrica** A. Gray Fig. 109 Thimbleweed

Map 293. Dry sandy barrens, woodlands (jack pine, oak, aspen), dunes; fields, roadsides, shores; borders of woods.

A "typical" plant of this variable species has 3 larger involucral cauline leaves and 2 smaller ones, all subtending 2 peduncles; the 2 smaller leaves represent the bracts of one of the peduncles. Plants with more peduncles have even more leafy bracts in the cluster. But occasionally some of the peduncles bear bracts ("involucels") as in the next species. Such plants, and those with only 1 naked peduncle and therefore 3 involucral leaves, may be distinguished from *A. virginiana* by their narrow leaf segments and pubescence, if they are too immature to reveal the cylindrical head. The flowers of *A. cylindrica* are rarely if ever over 2.5 cm across.

5. **A. virginiana** L. Fig. 110 Thimbleweed

Map 294. Woods (often ± open), fields, meadows, river banks, conifer swamps.

Ordinarily quite distinct from the preceding species, although there may be an exception to any of the characters in the key. Less clear is the meaning of variation within this species, which is here treated in a broad sense. Occasional plants may have white flowers as large as 3–3.5 cm across, resembling those of *A. canadensis* (from which this species is easily separated by its petioled cauline leaves). Such plants were named as *A. virginiana* var. *alba* Wood and were included in *A. riparia* Fern., a name often applied (but frequently with some misgivings) to certain plants in our area. Distinctions in anther length, sometimes cited, appear useless, and the angle of the style is scarcely better. Typical var. *virginiana* tends to have fatter, more ovoid

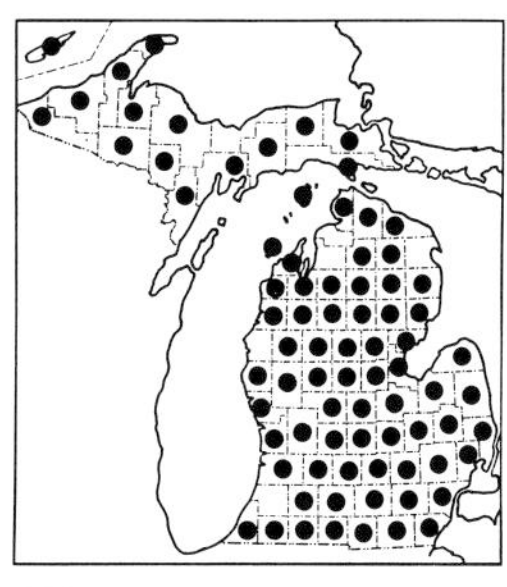
292. Anemone quinquefolia

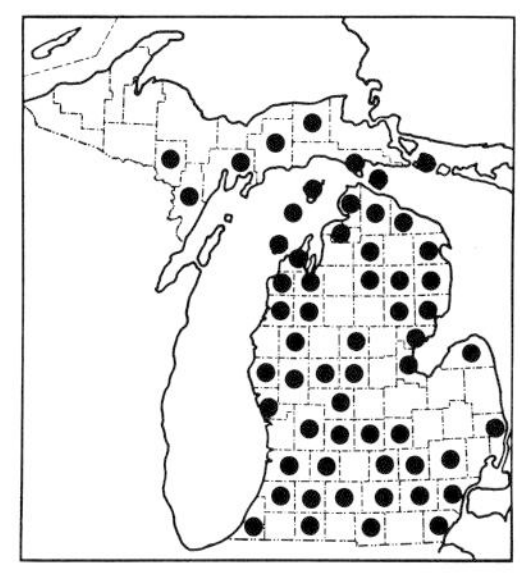
293. Anemone cylindrica

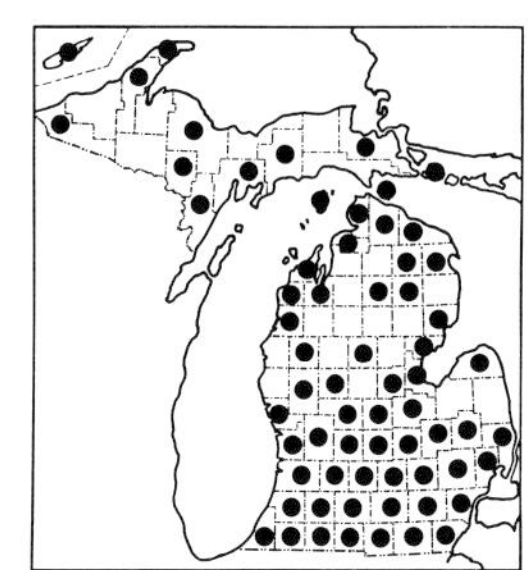
294. Anemone virginiana

fruiting heads (to 17 mm thick) and to be larger-leaved, the leaf segments more ovate, than var. *alba*, which in some ways approaches *A. cylindrica*. All three taxa have the same chromosome number and some hybridization and introgression might be expected.

Plants determined as *A. riparia* by M. Heimburger, who has studied this complex for many years, range throughout the Upper Peninsula and into the northern third of the Lower Peninsula. Typical *A. virginiana* apparently ranges from the southern border of the state throughout the Lower Peninsula but scarcely enters the Upper Peninsula (Menominee and Delta counties).

BERBERIDACEAE — Barberry Family

A diverse family, sometimes divided into two or more, and of considerable phytogeographic interest because so many of the herbaceous species of eastern North America have close counterparts in eastern Asia (e. g., *Caulophyllum*, *Jeffersonia*, *Podophyllum*).

REFERENCE

Ernst, Wallace R. 1964. The Genera of Berberidaceae, Lardizabalaceae, and Menispermaceae in the Southeastern United States. Jour. Arnold Arb. 45: 1–35.

KEY TO THE GENERA

1. Plant a woody shrub with the twigs or leaves spiny; leaves simple and unlobed or pinnately compound with at least 3 leaflets
  2. Leaves simple; stem spiny at nodes .... 1. **Berberis**
  2. Leaves compound; stem without spines .... 2. **Mahonia**
1. Plants herbaceous, spineless; leaves deeply lobed or, if compound, not pinnate
  3. Leaves thrice-compound, the leaflets petioluled; flowers yellowish green to purplish, less than 1.5 cm across, several in an inflorescence; "fruit" ca. 1 cm long or less, a blue, fleshy seed (the ovary wall withering away) ....3. **Caulophyllum**
  3. Leaves of only 2 leaflets or deeply lobed but simple; flowers white, broader, solitary; fruit larger, a many-seeded yellowish berry or a brownish capsule
    4. Leaves all basal, deeply 2-lobed or of 2 sessile leaflets; fruit ca. 1.5–2.5 cm long, a brownish capsule opening (incompletely) transversely near the end, as if by a lid .... 4. **Jeffersonia**
    4. Leaves cauline, deeply several-lobed; fruit larger, a yellowish (to purplish) berry .... 5. **Podophyllum**

1. Berberis — Barberry

KEY TO THE SPECIES

1. Leaves entire (at most with an apiculus or small terminal spine); flowers solitary or in small clusters .... 1. **B. thunbergii**
1. Leaves with spiny-toothed margins; flowers in a raceme .... 2. **B. vulgaris**

1. **B. thunbergii** DC. Japanese Barberry

Map 295. A native of Asia, widely cultivated, especially for hedges, and occasionally escaping into woods, swamps, fields, and dunes. Doubtless dispersed largely by birds that eat the bright red fruit; may become quite well established.

A few sterile collections have 3-forked rather than simple spines at the nodes, with larger leaves than expected in *B. thunbergii*, and might be some other species.

2. **B. vulgaris** L. Common Barberry

Map 296. Woods, thickets, meadows, river banks. A European species, formerly widely cultivated and sometimes escaped, but largely eradicated as an alternate host for wheat rust.

2. Mahonia

1. **M. aquifolium** (Pursh) Nutt. Oregon-grape

Map 297. A native of the Pacific Northwest, apparently escaped from cultivation in woods as well as closer to plantings.

Often included in the preceding genus as *B. aquifolium* Pursh.

3. Caulophyllum

REFERENCES

Dore, William G. 1964. Two Kinds of Blue Cohosh. Ontario Nat. 2(1): 5–9.

Loconte, Henry, & Will H. Blackwell. 1981. A New Species of Blue Cohosh (Caulophyllum, Berberidaceae) in Eastern North America. Phytologia 48: 483.

1. **C. thalictroides** (L.) Michaux Plate 3-C Blue Cohosh

Map 298. A species of rich deciduous woods.

There are two rather well marked varieties of this plant: var. *thalictroides* usually has yellow-green flowers (at least in our region) and a short style,

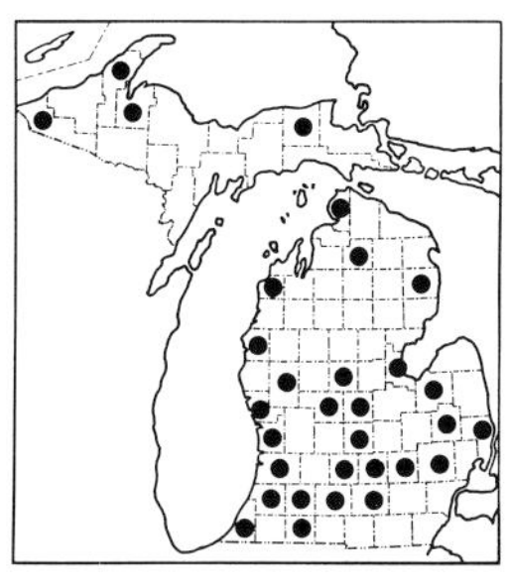

295. Berberis thunbergii

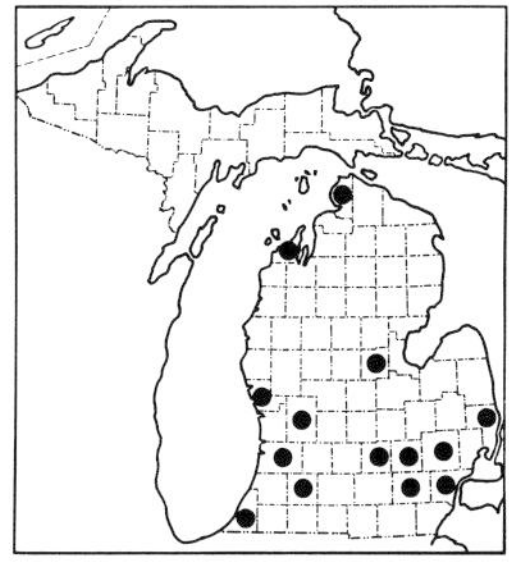

296. Berberis vulgaris

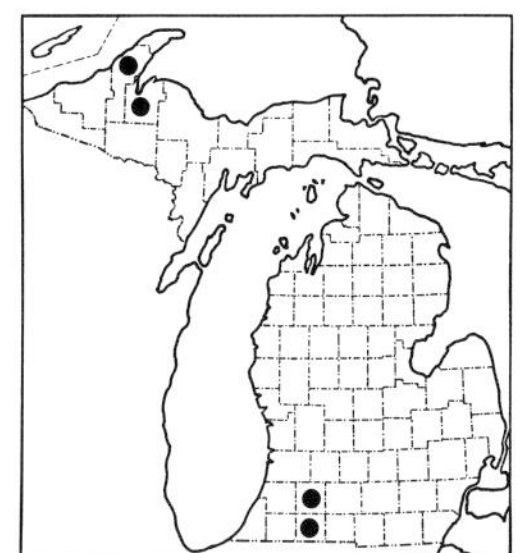

297. Mahonia aquifolium

distinctly less than 1 mm long (ca. 0.1–0.7 mm); var. *giganteum* Farw. (TL: Farmington [Oakland Co.]) has purplish flowers and a longer style (ca. 1–1.5 mm). The purple-flowered variety has purplish green foliage when young and blooms earlier in the spring; it also has slightly larger flowers. While var. *thalictroides* is found throughout the state, except apparently for the Keweenaw Peninsula and Isle Royale, var. *giganteum* seems to be found only in the Lower Peninsula, where it has not been collected in the central and southwestern portions. The two varieties may grow in the same woods, but intermediate plants are very rare. Loconte has found good separation of the taxa on the basis of total carpel length at flowering time, as well as other characters, and even recognizes *C. giganteum* (Farw.) Loconte & Blackwell as a distinct species—as was earlier suggested by Dore.

The bright glaucous blue "fruit" is more conspicuous and attractive than the flowers, and may remain on the plant through the winter; it is actually the fleshy seed, the ovary wall shriveling and falling away as the seed ripens. The style, necessary to distinguish the varieties, may usually be found even in mature plants, on aborted ovaries in which the seed has not enlarged. The foliage of this species, as the name suggests, resembles that of *Thalictrum*; it is always completely glabrous and ± glaucous.

4. Jeffersonia

1. **J. diphylla** (L.) Pers. Plate 3-D Twinleaf

Map 299. Rich deciduous woods, including floodplains and well-drained slopes.

An attractive, unusual plant, blooming very early (usually April), and not at all common in the state.

5. Podophyllum

1. **P. peltatum** L. Fig. 111 May-apple; Mandrake

Map 300. Deciduous woods and borders of woods, rarely in cedar swamps.

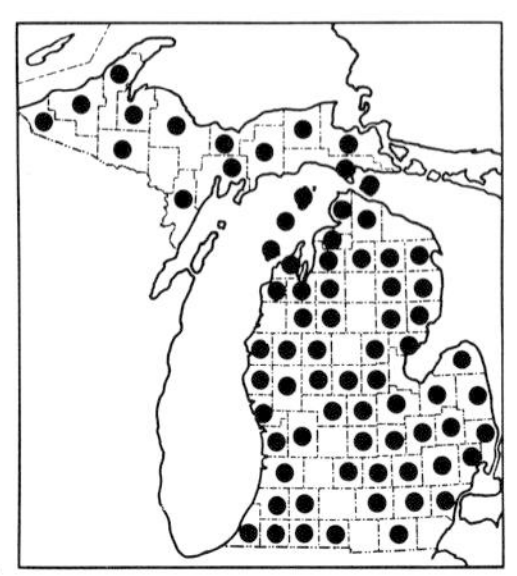

298. Caulophyllum thalictroides

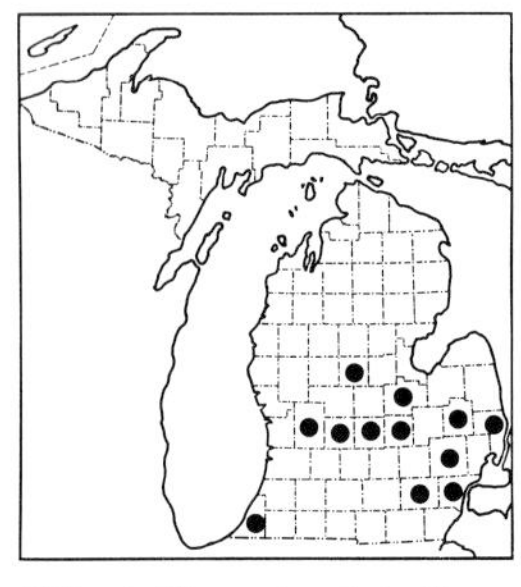

299. Jeffersonia diphylla

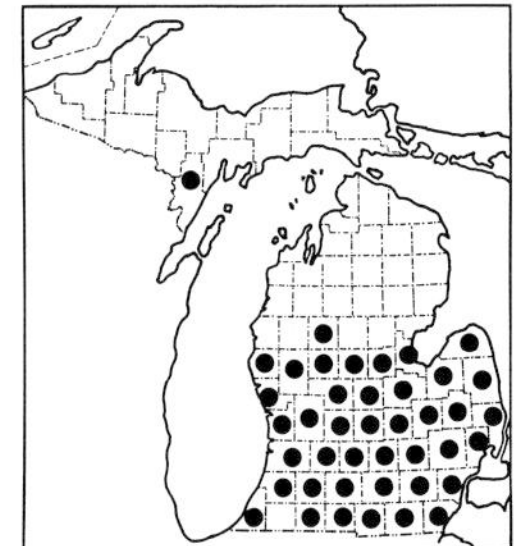

300. Podophyllum peltatum

A familiar wildflower, forming large colonies of deeply lobed, peltate leaves, whose umbrella-like aspect is emphasized as they emerge from the ground, neatly furled. Fertile plants bear a pair of cauline leaves, and similar sterile leaves are common. Rarely a leafless flowering plant is found. The flower may be as broad as 7 cm. The large berry is generally considered edible, but other parts, especially the roots, contain a poisonous resin which has some medicinal value. All Berberidaceae should be treated with caution because of their poisonous principles.

LARDIZABALACEAE — Lardizabala Family

1. Akebia

REFERENCE

Levenson, Burton E. 1975. Akebia quinata Established in Michigan. Mich. Bot. 14: 105–107.

1. **A. quinata** (Houtt.) Dcne. Fig. 112 — Akebia; Chocolate Vine

Map 301. A native vine of eastern Asia, the generic name being derived from the Japanese name of the plant. Collected at one locality in Ann Arbor where established in thickets near the Huron River, presumably an escape from cultivation.

MENISPERMACEAE — Moonseed Family

1. Menispermum

As in the Berberidaceae, this genus is represented by a species in eastern North America and another in eastern Asia.

1. **M. canadense** L. Fig. 115 — Moonseed

Map 302. A vine climbing on various trees and shrubs in swamp forests, rich woods, and thickets, especially along rivers. Very local northward, along rivers such as the Au Sable, Boardman, Manistee, Menominee, Muskegon, and Pentwater.

This dioecious vine is readily recognized by the peltate leaves, with petiole attached very near the margin of the ± cordate base of the blade, which is slightly pubescent along the veins beneath and usually ± shallowly (rarely deeply) lobed or with a few large teeth. The fruit is a glaucous drupe, looking rather like a wild grape, with a ± crescent-shaped pit.

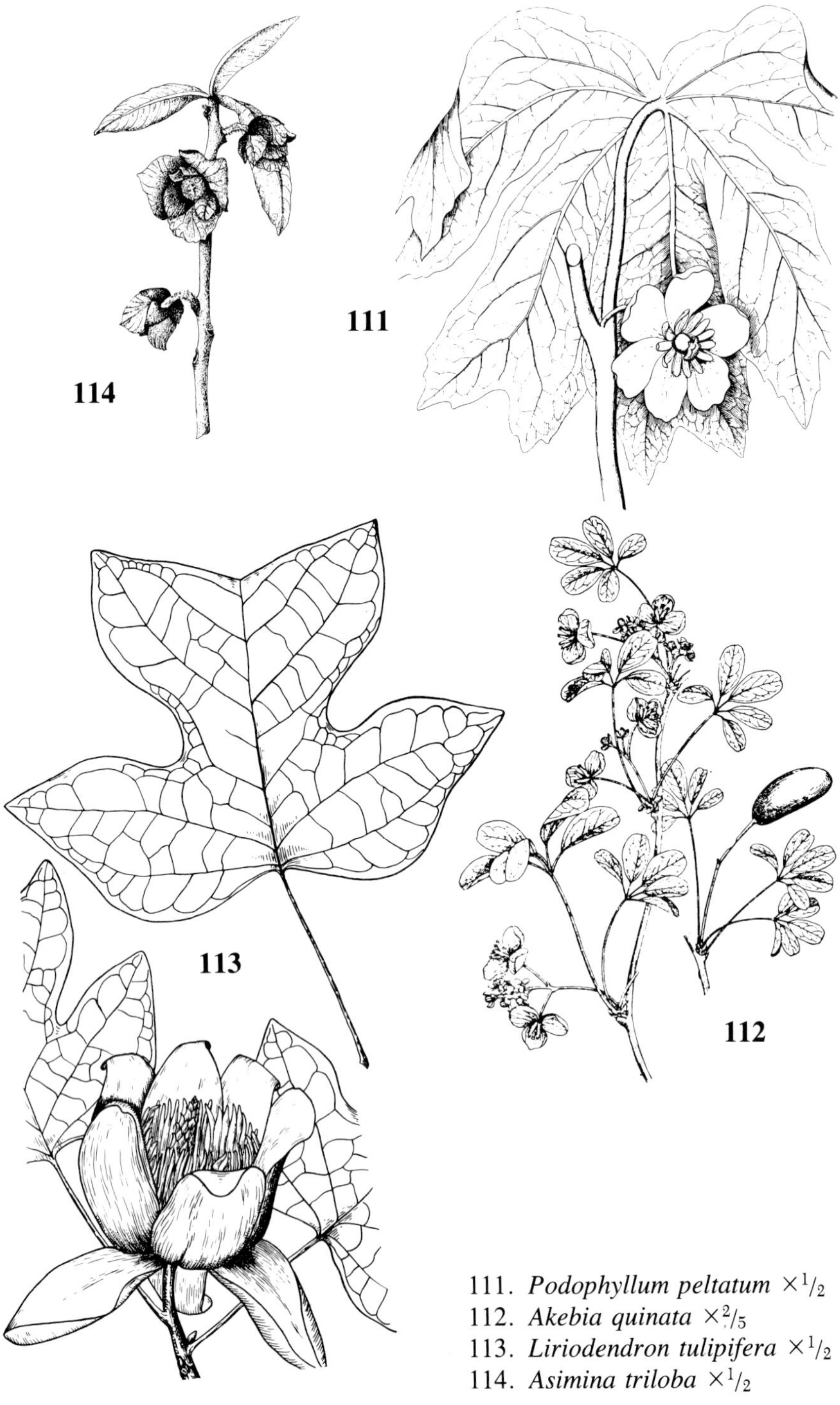

111. *Podophyllum peltatum* $\times ^1/_2$
112. *Akebia quinata* $\times ^2/_5$
113. *Liriodendron tulipifera* $\times ^1/_2$
114. *Asimina triloba* $\times ^1/_2$

MAGNOLIACEAE — Magnolia Family

Various species or hybrids of *Magnolia* are planted in southern Michigan, but there are no records of any spreading from cultivation, nor of any native species occurring in the state, although *M. acuminata* (L.) L., "cucumber-tree," grows as close as northeastern Ohio and southern Ontario.

1. Liriodendron

1. **L. tulipifera** L. Fig. 113 — Tulip-tree; Tulip-poplar; Yellow-poplar

Map 303. Rich deciduous woods. The "national champion," a handsome specimen in Cass County, 200 feet tall and 7.5 feet in diameter, was toppled by a windstorm in May of 1984.

A fine tree, easily distinguished by its leaves (basically 4-angled or with 4 acute lobes, the apex with an obtuse angular sinus), dry cone-like fruit, and large flowers in early summer. It is often planted as a shade tree. A second species of *Liriodendron* grows in central China.

ANNONACEAE — Custard-apple Family

This is a tropical family of which only a single genus occurs in temperate North America. Tropical fruits in the genus *Annona* include custard-apple, sweetsop, soursop, and cherimoya. Our only species in the family has the largest fruit of any native plant in the state; shaped rather like a chubby banana, it is considered by many to be not merely edible, but choice.

1. Asimina

REFERENCES

Bowden, Wray M., & Bert Miller. 1951. Distribution of the Papaw, Asimina triloba (L.) Dunal, in Southern Ontario. Canad. Field-Nat. 65: 27–31.

Willson, Mary F., & Douglas W. Schemske. 1980. Pollinator Limitation, Fruit Production, and Floral Display in Pawpaw (Asimina triloba). Bull. Torrey Bot. Club 107: 401–408.

1. **A. triloba** (L.) Dunal Fig. 114 — Pawpaw

Map 304. Deciduous woods, swamp forests, thickets along streams.

A town and a river in Van Buren County are named for this well known plant; the river's mouth at the St. Joseph River is in Benton Harbor, Berrien County. A distinctive shrub or small tree with very large ± obovate entire leaves, pawpaw has a large fleshy big-seeded fruit. The flowers appear before the leaves are expanded in late spring; as might be surmised from their dark color and strong odor, they are pollinated mostly by flies.

This is the family of the true laurel, *Laurus nobilis* L., a Mediterranean native, used not only to create wreaths for victorious athletes in ancient times but also as the source of bay leaves used in cooking. Besides laurel and our native species, several other well known aromatic plants are in the family, such as cinnamon and camphor, and the avocado also belongs here. Most laurels are not hardy in our climate. Both of our native species bloom in May before the leaves are mature, tend to be dioecious, and are spicy-aromatic when the bark is bruised.

REFERENCE

Wood, Carroll E., Jr. 1958. The Genera of the Woody Ranales in the Southeastern United States. Jour. Arnold Arb. 39: 296–346. [Includes also Annonaceae and Magnoliaceae.]

KEY TO THE GENERA

1. Leaves mostly (but not all) lobed; flowers in peduncled racemes, the pedicels longer than the perianth; fruit blue, on a conspicuous enlarged red pedicel; plant a tree . . . . 1. **Sassafras**
1. Leaves all unlobed; flowers in nearly sessile small umbels, the pedicels barely as long as the perianth; fruit bright red, on slender inconspicuous pedicel; plant a much-branched shrub . . . . 2. **Lindera**

1. Sassafras

Two additional species occur in eastern Asia, representing the same kind of distribution pattern so well displayed in several related families.

1. **S. albidum** (Nutt.) Nees Fig. 116 Sassafras

Map 305. Dry sandy woods (especially oak), often on old dunes; fence-rows; mixed deciduous woods and low wet woods.

301. Akebia quinata

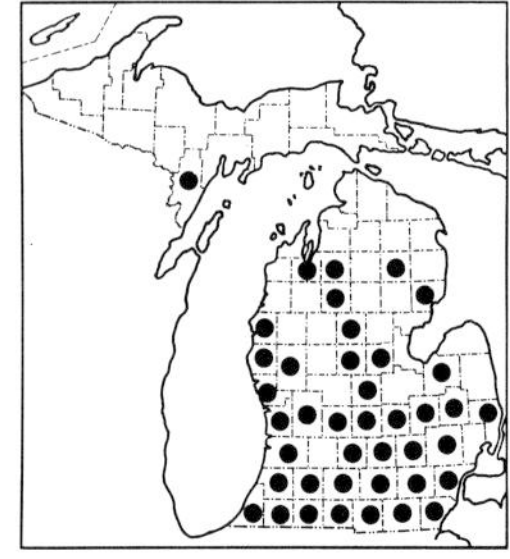

302. Menispermum canadense

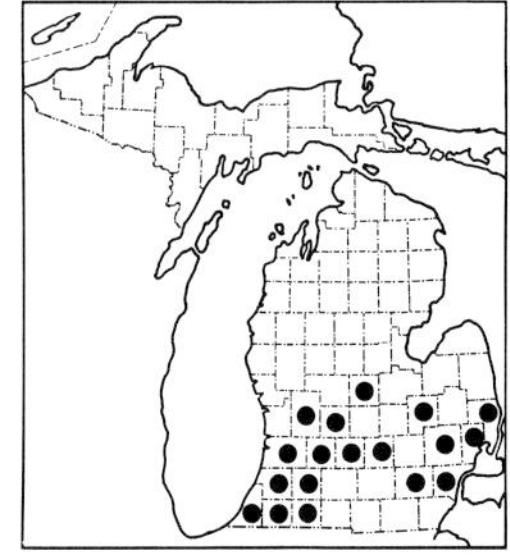

303. Liriodendron tulipifera

Some leaves have a characteristic "mitten" shape with one lobe as a "thumb," others have two "thumbs," and others are unlobed; all three shapes occur on the same plant, often even on the same branch. The twigs and undersides of the leaves vary from glabrous to pubescent. Plants often grow in colonies, from root suckers, and it is the bark of the root which is especially esteemed in folk medicine for its oil. Sassafras tea (made from the bark, not the leaves) is a long-time home remedy, the safety of which has recently been questioned.

2. **Lindera**

A large genus in eastern Asia, *Lindera* is represented by two species in North America, only one of which ranges as far north as Michigan.

1. **L. benzoin** (L.) Blume Fig. 117 Spicebush

Map 306. Low rich deciduous woods and swamp forests, rarely under cedar northward.

Like most aromatic plants, spicebush has reputed medicinal value as well as having served as a source of tea, but it is not the source of benzoin of the drug trade, which is derived from species of *Styrax* (Styracaceae).

The small but attractive yellow flowers (perhaps more conspicuous on staminate plants) appear early in the spring, before the leaves, calling the shrub to the attention of anyone who passes by.

PAPAVERACEAE Poppy Family

The Fumariaceae are often included as a subfamily of Papaveraceae, differing in bilateral symmetry of the flowers and watery sap. All of our Papaveraceae have a colored (yellow to red-orange or milky) sap.

REFERENCES

Ernst, Wallace R. 1962. The Genera of Papaveraceae and Fumariaceae in the Southeastern United States. Jour. Arnold Arb. 43: 315–343.

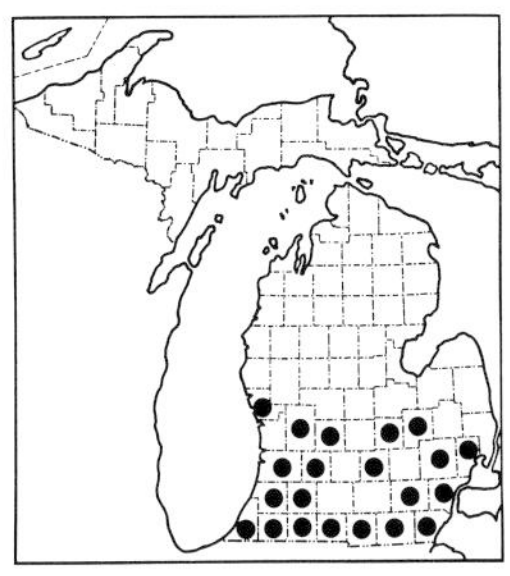

304. Asimina triloba

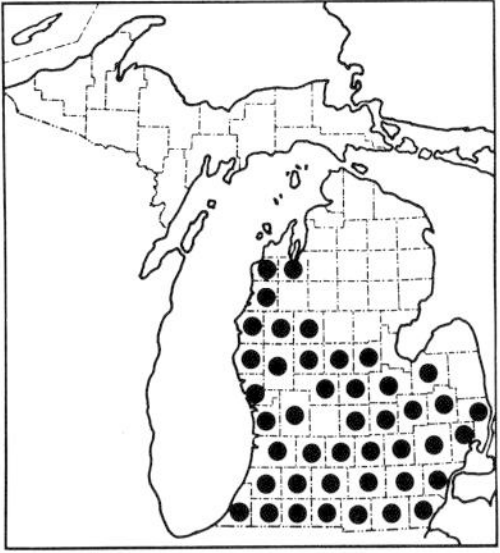

305. Sassafras albidum

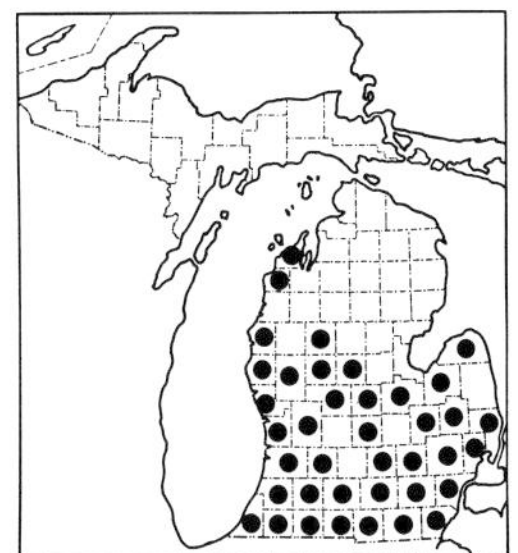

306. Lindera benzoin

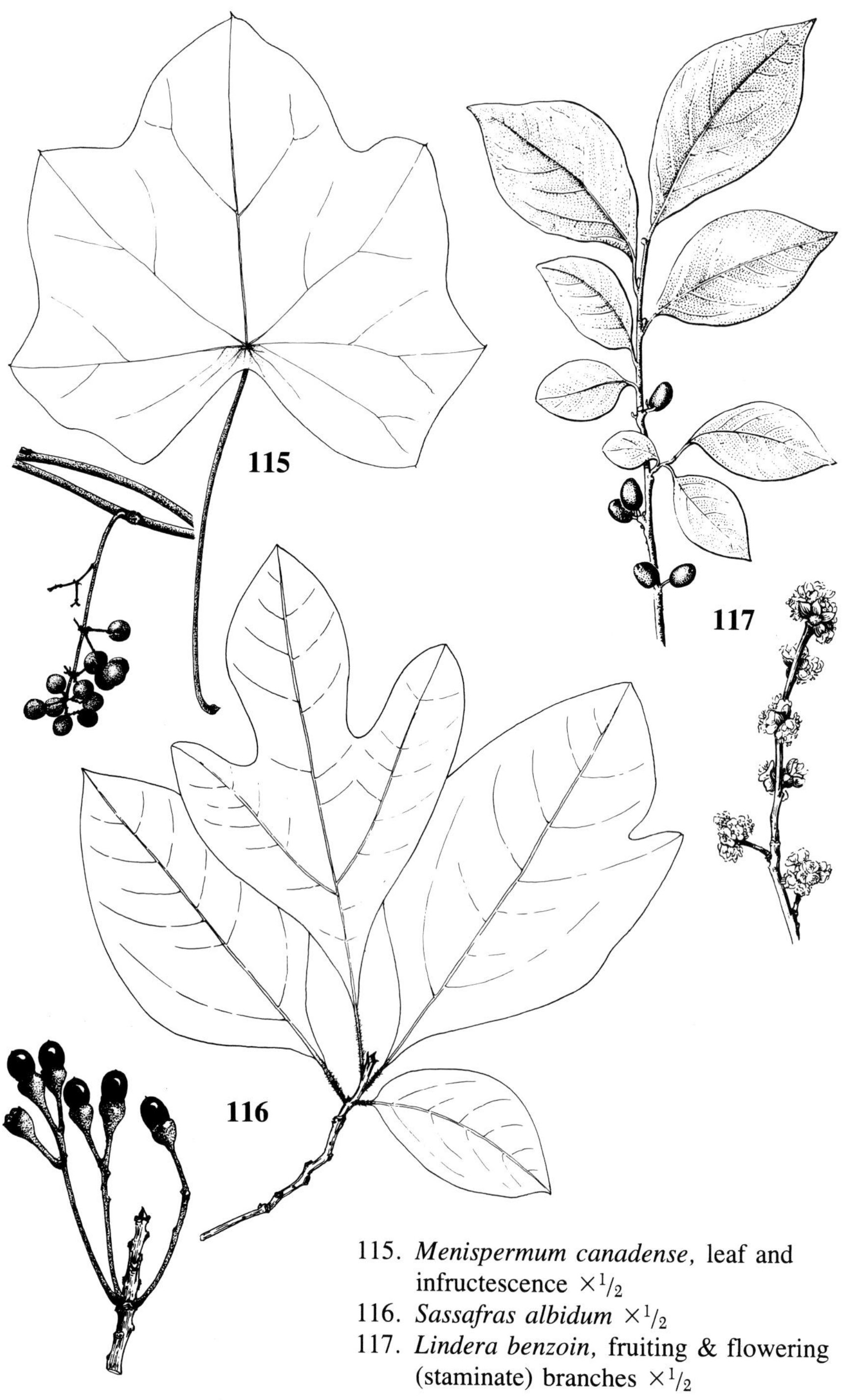

115. *Menispermum canadense,* leaf and infructescence $\times ^{1}/_{2}$
116. *Sassafras albidum* $\times ^{1}/_{2}$
117. *Lindera benzoin,* fruiting & flowering (staminate) branches $\times ^{1}/_{2}$

Mitchell, Richard S. 1983. Berberidaceae through Fumariaceae of New York State. New York St. Mus. Bull. 451. 66 pp.

KEY TO THE GENERA

1. Flowers apetalous (and the sepals early deciduous), less than 1 cm long or broad, in a large, branched, many-flowered panicle-like inflorescence .... 1. **Macleaya**
1. Flowers with showy petals, larger, solitary or in a few-flowered inflorescence
 2. Leaves ternately dissected into numerous very narrowly linear lobes, glabrous to puberulent .... 2. **Eschscholzia**
 2. Leaves simple to pinnatisect, but segments (if any) broader and often setose (or even spiny)
  3. Petals (7–) 8 (rarely more), over twice as long as wide, not wrinkled or crumpled in the bud; leaves basal (flowers scapose), palmately veined and lobed .... 3. **Sanguinaria**
  3. Petals 4 (more if flower "double"), less than twice as long as wide, wrinkled or crumpled in the bud; leaves cauline as well as basal, pinnately veined and usually lobed
   4. Leaves with distinct stiff sharp spines on teeth and lobes; spines also on buds, fruit, and (sparsely) stem .... 4. **Argemone**
   4. Leaves without spines (or only obscure, weak ones in *Papaver*); spines absent on fruit
    5. Petals white, pink, red, or purple; fruit opening by a ring of pores or chinks beneath the stigmatic disc; sap milky white .... 5. **Papaver**
    5. Petals yellow; fruit dehiscent to the base; sap yellow-orange
     6. Pistil with a definite style persisting and elongating (to ca. 8 mm) on the fruit; ovary and fruit long-pubescent, ellipsoid, splitting from apex to base; cauline leaves 2 (–3), opposite .... 6. **Stylophorum**
     6. Pistil with little or no definite style (less than 2 mm); ovary and fruit linear, splitting from base to apex (as in a silique), glabrous or at most (*Glaucium*) very scabrous or tuberculate; cauline leaves several, alternate
      7. Upper leaves cordate-clasping; fruit becoming over 10 cm long; petals at least 3 cm long .... 7. **Glaucium**
      7. Upper leaves sessile (not clasping) or petiolate; fruit ca. 3–5 cm long; petals not over 1.5 cm long .... 8. **Chelidonium**

1. Macleaya

1. **M. cordata** (Willd.) R. Br. — Plume Poppy

Map 307. A native of Asia, formerly often included in the genus *Bocconia*, sometimes cultivated and seldom escaping. Collected along a riverside (Benzie Co.) and in waste ground (roadside dump, ditch, old field near railroad, weedy area by an old building site).

2. Eschscholzia

1. **E. californica** Cham. — California Poppy

Map 308. This native of the west coast is cultivated and was collected by Farwell as an escape at Lake Linden in 1934. It will doubtless be found as an escape again.

3. **Sanguinaria**

1. **S. canadensis** L. Plate 4-A Bloodroot

Map 309. Usually in rich deciduous woods and floodplain forests, surviving considerable disturbance and clearing; rather local northward.

Double-flowered plants (stamens and carpels converted to petals) were found in 1907 near Whitmore Lake and transplanted to the garden of the Kempf House in Ann Arbor, where they thrived. Many gardens have progeny from this colony (or others).

The common and scientific names derive from the bright red-orange sap which "bleeds" from the broken rhizome and other parts. This familiar wildflower blooms in April (or early May in the north); the petals are normally white, though rosy-flowered forms are reported.

4. **Argemone** Prickly Poppy

These are very glaucous plants with spiny-margined leaves, spiny fruits and buds (sepals), and large showy flowers. Besides the two species recorded here, others are in cultivation.

REFERENCE

Ownbey, Gerald B. 1958. Monograph of the Genus Argemone for North America and the West Indies. Mem. Torrey Bot. Club 21: 1–159.

KEY TO THE SPECIES

1. Corolla white, over (5) 6 cm in diameter; stamens over 100; leaves without pale markings . . . . 1. **A. albiflora**
1. Corolla yellow, less than 5 [6] cm in diameter; stamens ca. 50 or fewer; leaves with conspicuous pale green or whitish markings along the main veins . . . . . 2. **A. mexicana**

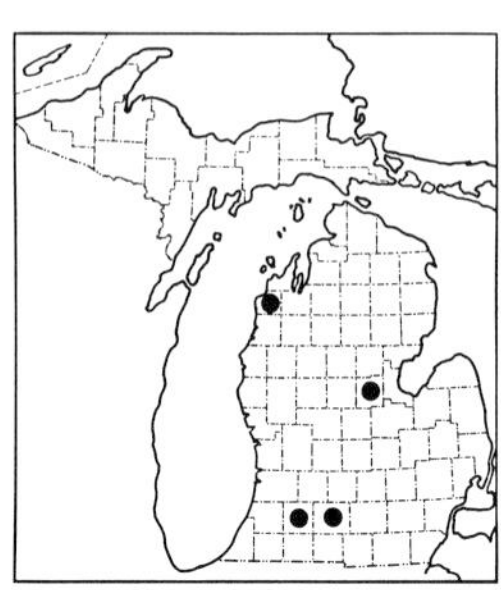

307. Macleaya cordata

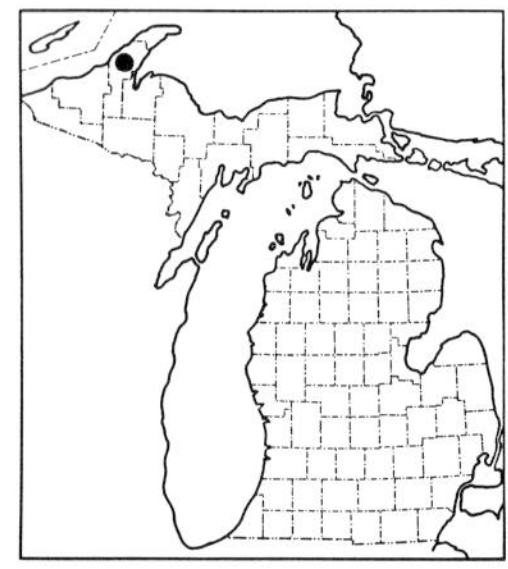

308. Eschscholzia californica

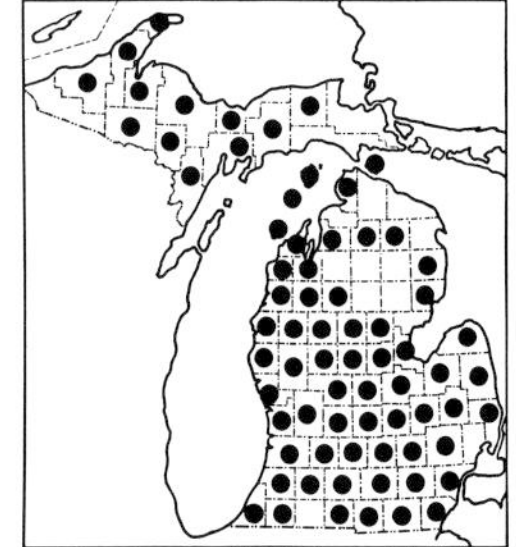

309. Sanguinaria canadensis

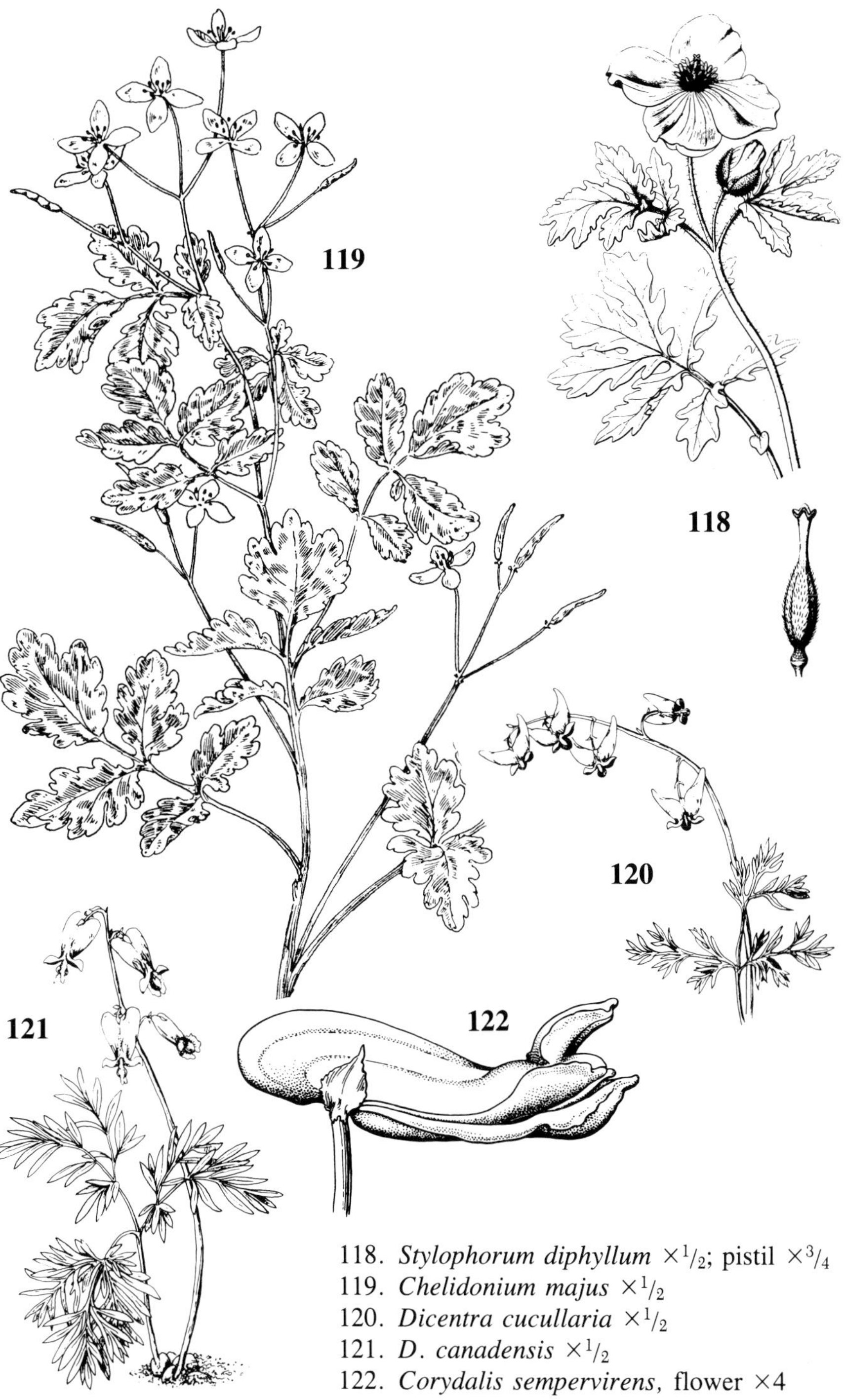

118. *Stylophorum diphyllum* ×1/2; pistil ×3/4
119. *Chelidonium majus* ×1/2
120. *Dicentra cucullaria* ×1/2
121. *D. canadensis* ×1/2
122. *Corydalis sempervirens*, flower ×4

1. **A. albiflora** Hornem.

Map 310. A native of southernmost United States, sporadically escaped from cultivation as far north as southern Michigan. Apparently became well established along a roadside in Jackson County.

The names *A. alba* and *A. intermedia* (attributed to various authors) have sometimes been applied, or misapplied, to this species.

2. **A. mexicana** L.

Map 311. A native of tropical America, now a widespread weed in warm regions of both hemispheres and occasionally reported as escaped from cultivation as far north as our region, but not long persisting. The status of specimens from Kent and Wayne counties is not stated, and no specimens have been found to support the reports of Allmendinger and Dodge from Washtenaw and St. Clair counties, respectively.

5. **Papaver** — Poppy

Several cultivated annual species, including those treated here, are ± weedy in habit and are occasionally found escaped in dumps and other waste places, though they seldom persist for long. Mitchell (1983) deals thoroughly with a number of species. The species thus far collected out of cultivation in Michigan have doubtless occurred at many places besides those mapped here on the basis of extant collections, and additional ones may be found. *P. argemone* L. is a weedy little annual from Europe with deeply pinnatifid leaves and with bristles on the broadly club-shaped fruit. There is one specimen from a field of crimson clover in Grand Traverse county (*no coll.* in 1896, MSC)—rather a flimsy basis for including it in our flora, especially since it was not listed by Beal (1905).

REFERENCE

Kiger, Robert W. 1975. Papaver in North America North of Mexico. Rhodora 77: 410–422.

KEY TO THE SPECIES

1. Petals mostly at least 5 cm long; plant perennial (non-flowering stems present) . . . . 1. **P. orientale**
1. Petals less than 5 cm long; plant annual (all stems bearing flowers)
   2. Plant usually glabrous; upper leaves strongly clasping the stem, very glaucous . . . . 2. **P. somniferum**
   2. Plant coarsely hairy or bristly; upper leaves sessile but not clasping, scarcely if at all glaucous . . . . 3. **P. rhoeas**

1. **P. orientale** L. — Oriental Poppy

Map 312. Originally native to southwestern Asia, commonly cultivated and rarely escaped as a conspicuous plant of roadsides and waste places.

Not distinguished on the map is the very closely related *P. pseudoörientale* (Fedde) Medv., differing in chromosome number and in having a large dark ± rectangular blotch at the base of each petal.

2. **P. somniferum** L. Opium Poppy

Map 313. A Eurasian species, rarely found as a garden weed or waif on roadsides, dumps, or fields; now generally replaced in cultivation by species that can legally be grown.

This plant is the source of poppy seed used in baking and for making a drying oil. The hardened sap of the unripe fruit, on the other hand, is the source of opium and its derivative drugs. The flowers of ornamental plants are often double. Some plants have stiff hairs on the peduncles.

3. **P. rhoeas** L. Fig. 123 Corn, Field, or Flanders Poppy

Map 314. Another Eurasian species of roadsides, dumps, railroads, and near gardens.

The petals are large, generally 2.5–4 cm long and broader than long. A similar species is *P. dubium* L., not yet collected as a weed in Michigan. It has smaller flowers, elongate, narrowly obovate fruit (at least twice as long as broad), and appressed hairs on the peduncles. In *P. rhoeas*, the fruit is only slightly if at all longer than broad and the stiff hairs of the peduncles

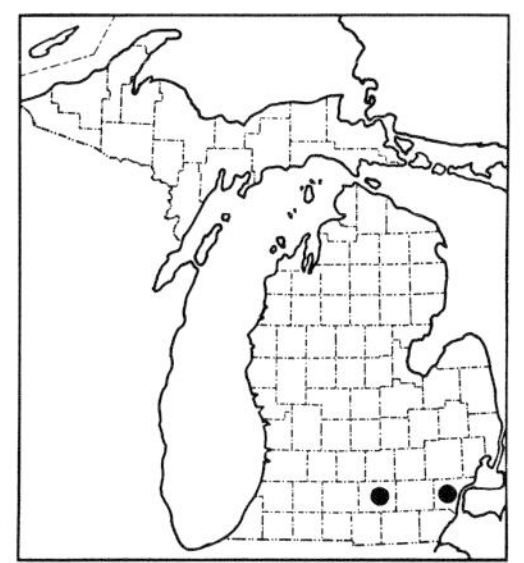
310. Argemone albiflora

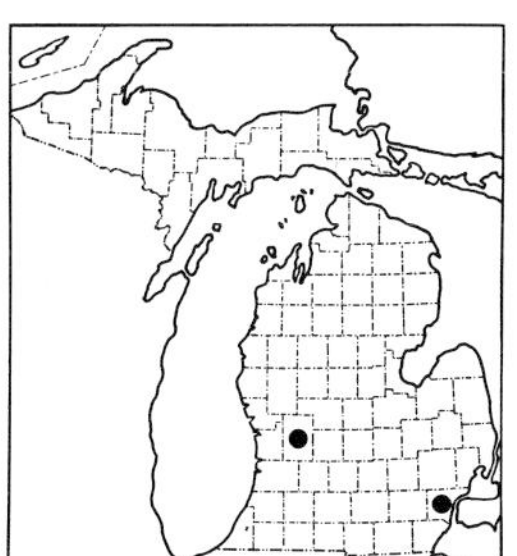
311. Argemone mexicana

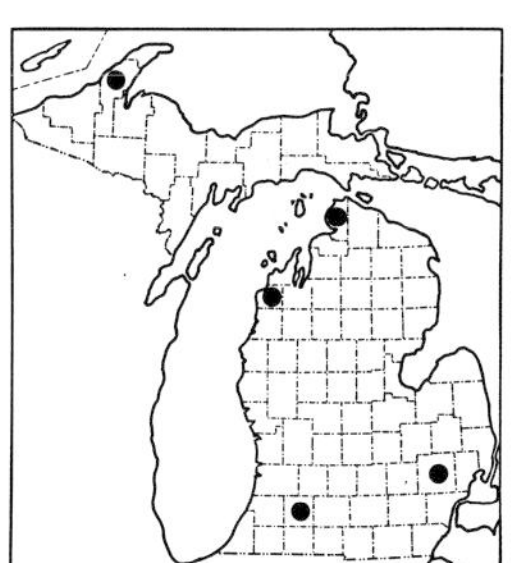
312. Papaver orientale

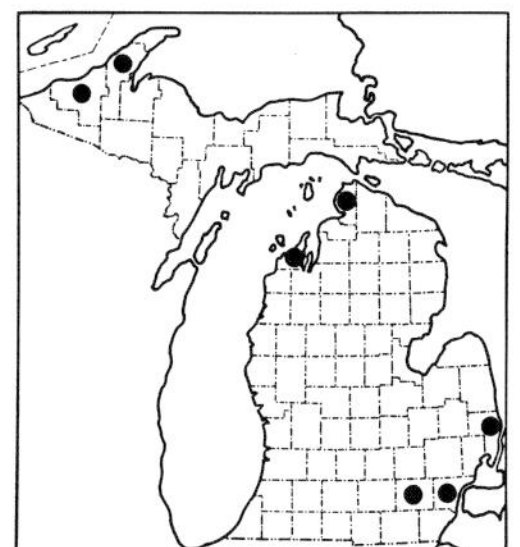
313. Papaver somniferum

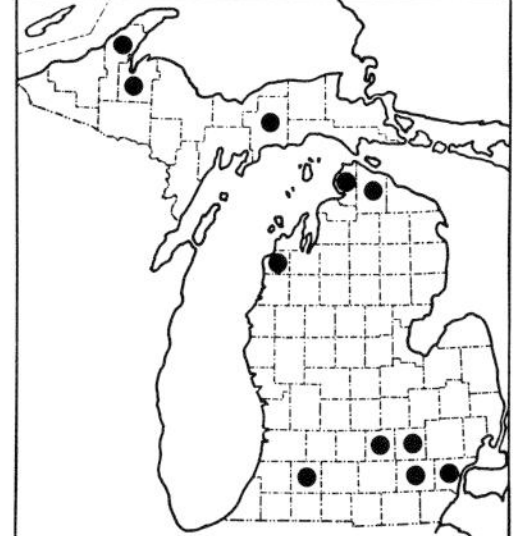
314. Papaver rhoeas

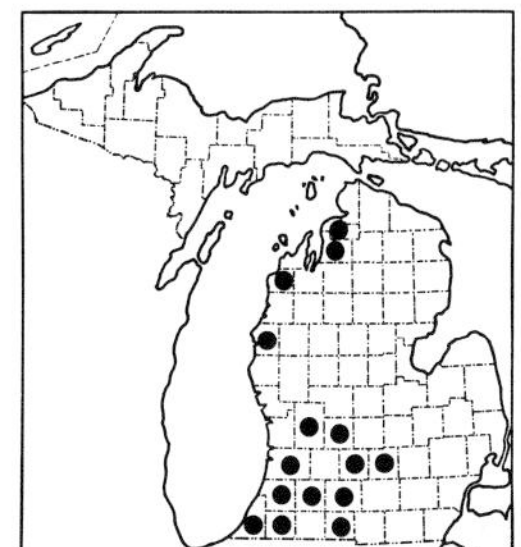
315. Stylophorum diphyllum

spread at right angles. Intermediate plants (e. g., with very large flowers but appressed hairs on peduncles), as from Berrien, Emmet, and Gratiot counties, are presumably hybrids [*P.* ×*strigosum* (Boenn.) Schur] though they may be only forms of *P. rhoeas*; such plants would add Berrien and Gratiot counties to the map.

6. **Stylophorum**

Besides our eastern North American woodland species, the genus includes a few others in eastern Asia.

1. **S. diphyllum** (Michaux) Nutt. Fig. 118 Celandine or Wood Poppy

Map 315. Deciduous woods, rather local.

This is our only native yellow-flowered poppy. The distinctive fat-ellipsoid fruit with pubescent body and long beak well distinguishes it from all other species we have in the family, native or not. The leaves are very pale beneath. The petals are 17–30 mm long. *Chelidonium majus* has very similar foliage but smaller flowers and slender fruit.

7. **Glaucium**

1. **G. flavum** Crantz Horned Poppy

Map 316. Collected in the 1890's at Les Cheneaux by W. C. Coryell, who noted that it was probably introduced from Ohio. A native of Europe and Africa, the species occasionally escapes from cultivation in the United States and can be found in waste places.

A very glaucous plant, this species has lower leaves deeply pinnatifid, the upper ones merely shallowly lobed and toothed with strongly clasping base. The common name refers to the very long fruit (said to attain as much as 40 cm).

8. **Chelidonium**

REFERENCE

Jakobsons, R. O. 1963. Strutenes Celandine (Chelidonium majus L.). [Published by author]. 55 pp.

1. **C. majus** L. Fig. 119 Celandine

Map 317. A native of Eurasia, sometimes a locally common weed of roadsides, dumps, railroads, woods, thickets, and gardens.

The foliage is very similar to that of *Stylophorum* but the cauline leaves are clearly alternate and the petals are distinctly smaller (ca. 8–14 mm long). The bitter sap of this plant, as of many other members of the family, has long been thought to be of medicinal value although fundamentally poisonous. The bilingual reference cited above, by a chemist with the Michigan Department of Health, reviews its cultivation and use in Latvian folk medicine, and its chemical content.

FUMARIACEAE — Fumitory Family

Often included in the Papaveraceae, and the same references as listed for that family may be helpful.

KEY TO THE GENERA

1. Flowers with the 2 outer petals identical, thus with 2 obscure to strong lobes or spurs at base, slightly compressed at right angles to the plane of symmetry (figs. 120, 121); leaves basal or, if cauline, plant a vine climbing by the petioles
 2. Leaves all basal; corollas white (occasionally pink-tinged), deciduous in the spring .... 1. **Dicentra**
 2. Leaves cauline (plant a delicate vine); corolla white to purplish, persistent till fall .... 2. **Adlumia**
1. Flowers with only 1 petal spurred, slightly compressed parallel to the plane of symmetry (fig. 122); leaves cauline (as well as basal) and plant not a vine
 3. Ovary and fruit subglobose; fruit indehiscent, 1-seeded; corolla purplish, the tip deep red .... 3. **Fumaria**
 3. Ovary and fruit linear; fruit dehiscent from the base upward (as in a silique), several-seeded; corolla yellow or at least yellow-tipped .... 4. **Corydalis**

1. Dicentra

The common bleeding-heart, *D. spectabilis* (L.) Lem., has red flowers (white in albinos) in a raceme resembling that of our two native species, but unlike any of the species keyed below, which all have only basal leaves, it has a leafy stem. It seems not to escape from gardens, but may last after cultivation. Plants of both our native species bloom in the early spring, soon set fruit, and die back to the ground by early summer.

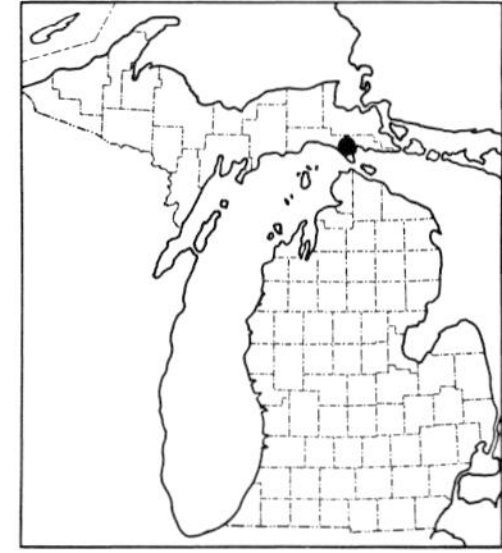
316. Glaucium flavum

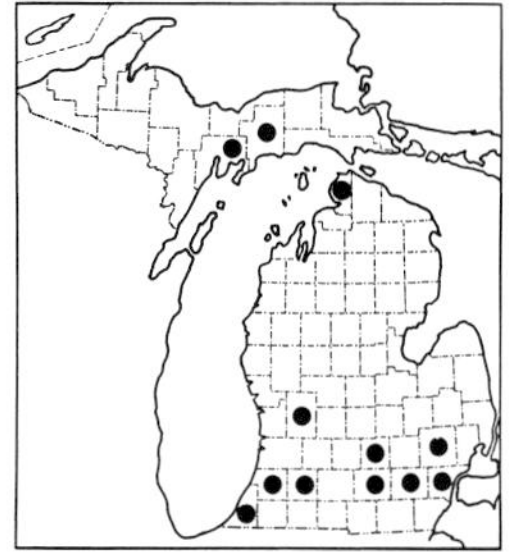
317. Chelidonium majus

318. Dicentra eximia

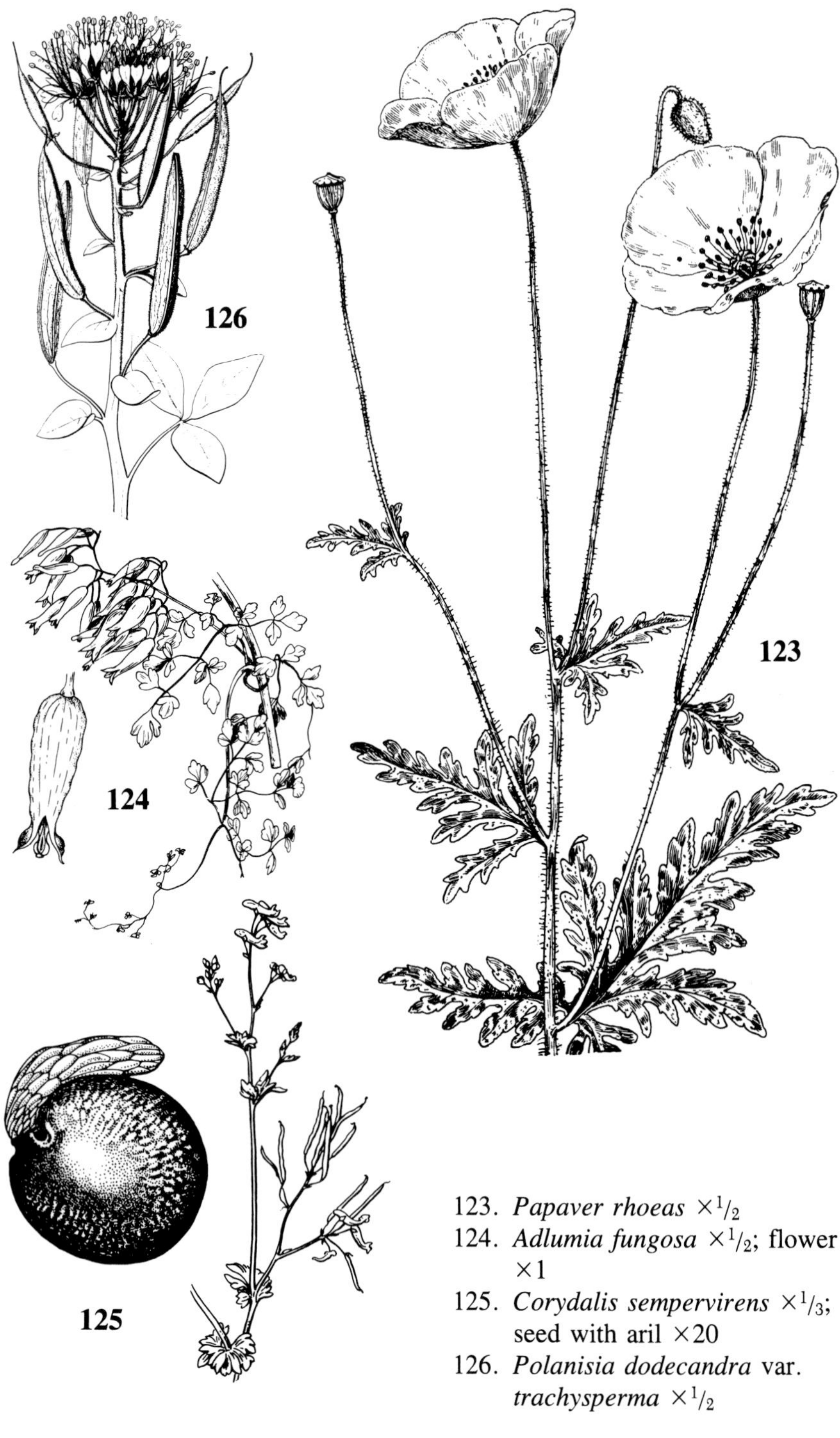

123. *Papaver rhoeas* $\times{}^{1}/_{2}$
124. *Adlumia fungosa* $\times{}^{1}/_{2}$; flower $\times 1$
125. *Corydalis sempervirens* $\times{}^{1}/_{3}$; seed with aril $\times 20$
126. *Polanisia dodecandra* var. *trachysperma* $\times{}^{1}/_{2}$

REFERENCE

Stern, Kingsley R. 1961. Revision of Dicentra (Fumariaceae). Brittonia 13: 1–57.

KEY TO THE SPECIES

1. Inflorescence compound; flowers pink to purple; plant a rare escape from cultivation . . . . 1. **D. eximia**
1. Inflorescence a simple raceme; flowers white; plant a common native forest species
   2. Spurs of corolla ± triangular, the axes quite divergent at maturity; plant with numerous closely packed, ± flattened (on at least 1 face), fleshy pink bulblets (leaf bases) at ground level . . . . 2. **D. cucullaria**
   2. Spurs of corolla rounded, the axes nearly or quite parallel (base of flower thus cordate); plant with ± spherical yellow bulblets on a short rhizome . 3. **D. canadensis**

1. **D. eximia** (Ker) Torrey — Wild Bleeding-heart

Map 318. Native southeast of our area; sometimes cultivated, and apparently escaped to rocky woods and hillsides in the vicinity of Marquette.

2. **D. cucullaria** (L.) Bernh. Fig. 120 — Dutchman's-breeches

Map 319. Rich deciduous woods, usually mesic beech-maple stands, occasionally swampy or relatively dry woods.

A familiar wildflower, reaching the peak of its blooming a week or more earlier than the next species when the two grow together in the same or nearby woods—as they often do. The distinctive basal bulblets, crowded in a large bulb-like assemblage, owe their pink color to copious minute red dots. These dots are rarely absent, leaving the bulblets white.

3. **D. canadensis** (Goldie) Walp. Fig. 121 — Squirrel-corn

Map 320. In just the same kind of deciduous woods as the preceding; rarely in swampy or dry oak woods.

This species has twice the chromosomes of the preceding. The two are not known to hybridize, although they often grow together and the flowering periods may overlap.

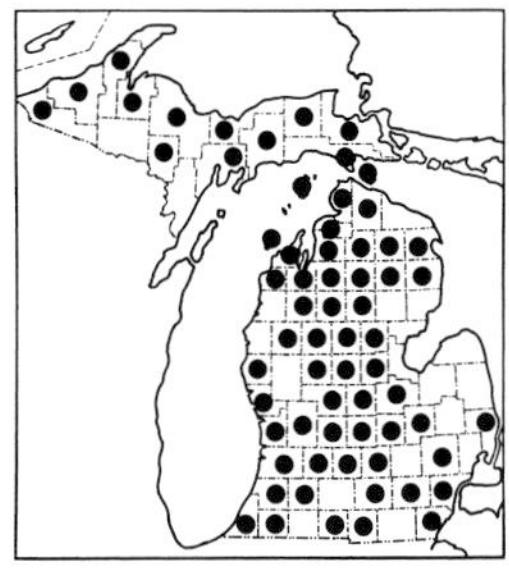

319. Dicentra cucullaria

320. Dicentra canadensis

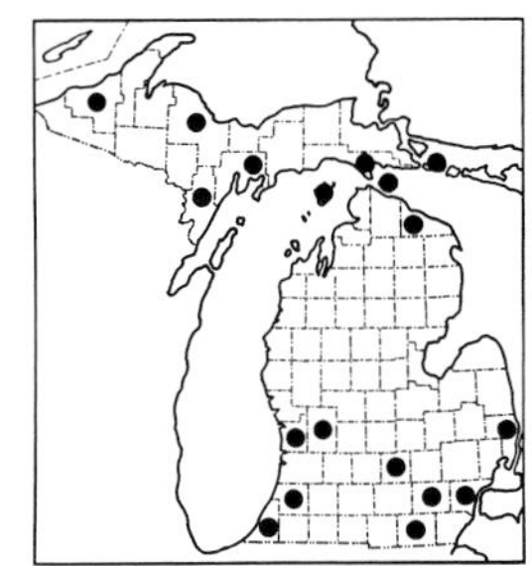

321. Adlumia fungosa

2. Adlumia

1. **A. fungosa** (Aiton) BSP. Fig. 124 Climbing Fumitory; Alleghany Vine

Map 321. Woods, rocky shores, and thickets; seems to do well on islands in the Great Lakes (some small ones as well as those mapped).

Our only vine in this family, distinctive in the persistent spongy corolla of definitely united petals. Sometimes cultivated as an ornamental, as well as native in woods.

3. Fumaria

1. **F. officinalis** L. Fumitory

Map 322. A European annual, rarely found as a weed in waste ground.

Even the largest flowers are less than 9 mm long, whereas in our species of *Corydalis* they are over 1 cm (except in *C. flavula*).

4. Corydalis

Our two commoner species share an interesting ecological situation with a few other plants, such as *Chenopodium capitatum* and *Leucophysalis grandiflora*. These are most typical of disturbed ground (especially sand and calcareous gravels) at previously wooded sites about 1–3 years after disturbance (clearing, bulldozing, fire, pasturing, etc.). Unless the area continues to be disturbed, these species then die out. On rock outcrops, shores, and places where soil is disturbed by erosion or overthrown tree roots, there are more natural niches for these plants than the easily noticed human-disturbed ground. The prominent appendage (aril) on the seed suggests dispersal by ants.

REFERENCE

Ownbey, Gerald Bruce. 1947. Monograph of the North American Species of Corydalis. Ann. Missouri Bot. Gard. 34: 187–259.

KEY TO THE SPECIES

1. Flowers pink, yellow-tipped; seeds ca. 1–1.5 mm wide; plants ± erect, the terminal inflorescences definitely surpassing the leaves . . . . 1. **C. sempervirens**
1. Flowers yellow; seeds ca. 1.8–2.2 mm wide; plants ± spreading or sprawling, the terminal inflorescences barely if at all surpassing the leaves
    2. Mature flowers 7–10 (11) mm long, including a spur ca. 1–1.5 (2) mm long; seeds very smooth and shiny . . . . 2. **C. flavula**
    2. Mature flowers (10) 11–14 mm long, including a spur ca. (2.5) 3.5–4 mm long; seeds shiny but reticulate . . . . 3. **C. aurea**

1. **C. sempervirens** (L.) Pers. Figs. 122, 125 Pink or Pale Corydalis; Rock Harlequin

Map 323. At home on rock ledges and summits, gravelly shores, and piney woodland, but more often to be expected 1–2 years after disturbance along roadsides, clearings, trails, gravel or sand pits, etc., dying out after about another two years if conditions are stable.

A very attractive plant; the leaves and stems are very glaucous. Fertile plants may vary in size from a few centimeters, unbranched, to over 1 meter tall with many bushy branches. The presence of fresh plants late in the summer suggests that seeds from early-blooming plants may germinate and mature in the same season.

2. **C. flavula** (Raf.) DC. Yellow Harlequin

Map 324. Thickets, wooded river banks, swampy ground.

3. **C. aurea** Willd. Plate 3-E Golden Corydalis

Map 325. Usually found with *C. sempervirens* and of similar ecological position. A less glaucous, more sprawling plant with more finely divided leaves than in that species.

CAPPARACEAE Caper Family

The family name is conserved as spelled here, although some botanists have preferred the spelling Capparidaceae. The pickled flower buds of *Capparis spinosa* L., a shrub of the Mediterranean region, are the capers used as a condiment in sauces, salads, etc.

REFERENCE

Ernst, Wallace R. 1963. The Genera of Capparaceae and Moringaceae in the Southeastern United States. Jour. Arnold Arb. 44: 81–95.

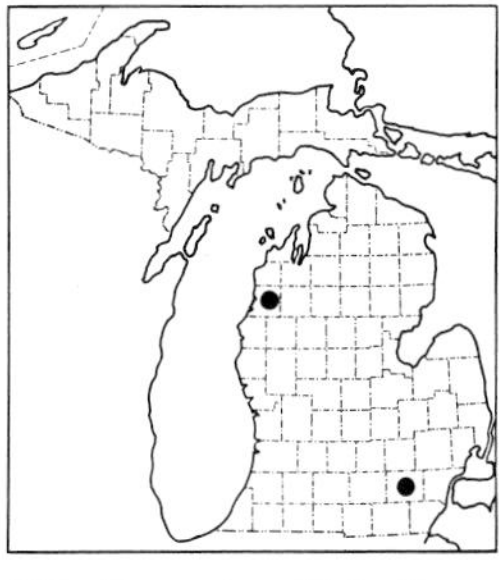

322. Fumaria officinalis

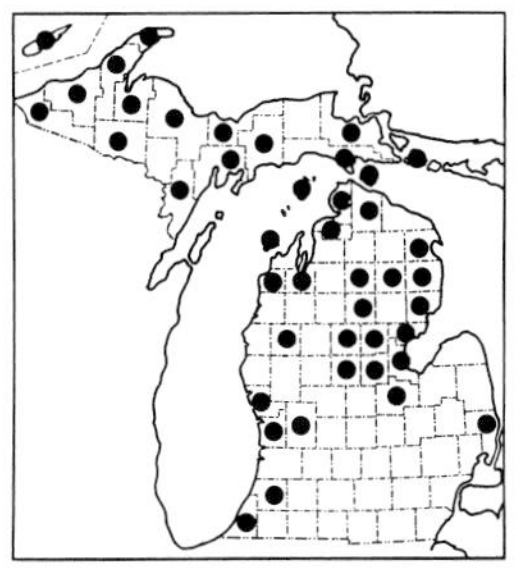

323. Corydalis sempervirens

324. Corydalis flavula

KEY TO THE GENERA

1. Petals definitely notched (hence ± 2-lobed) at apex; stamens more than 6; stipe of fruit (not to be confused with long pedicel) very short or absent .... 1. **Polanisia**
1. Petals not notched or lobed; stamens 6; stipe of fruit very long and slender . 2. **Cleome**

1. Polanisia

This genus is very closely related to the next and is sometimes combined with it (see Ernst 1963), in which case our species is called *Cleome dodecandra* L.

REFERENCES

Iltis, Hugh H. 1958. Studies in the Capparidaceae—IV. Polanisia Raf. Brittonia 10: 33–58.
Iltis, Hugh H. 1966. Studies in the Capparidaceae VIII. Polanisia dodecandra (L.) DC. Rhodora 68: 41–47.

1. **P. dodecandra** (L.) DC. Fig. 126 Clammy-weed

Map 326. Sandy shores of the Great Lakes and nearby disturbed sites; roadsides, railroads, gravelly bluffs, especially inland.

Often called *P. graveolens* Raf. The typical var. *dodecandra* is considered native in our area and elsewhere in eastern North America, and has petals that do not exceed 8 mm in length (usually ca. 4–6 mm). In var. *trachysperma* (T. & G.) Iltis—sometimes treated as a distinct species—the petals are longer (ca. 10–12 (15) mm in our material); this variety is considered to be adventive from western North America and is known from several places in the state. In both varieties, the leaves, stems, and fruit are ± densely glandular-pubescent and the clammy plant has a rank odor.

2. Cleome Spider plant

KEY TO THE SPECIES

1. Stem glandular-pubescent and with stipular spines; leaflets mostly 5–7 .... 1. **C. hassleriana**
1. Stem glabrous or nearly so, spineless; leaflets 3 .... 2. **C. serrulata**

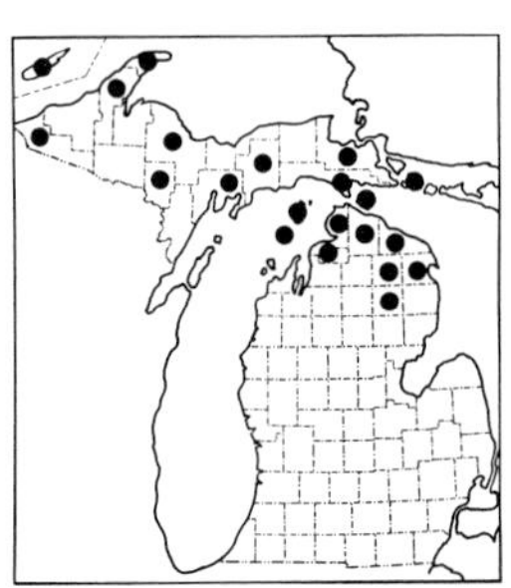

325. Corydalis aurea

326. Polanisia dodecandra

327. Cleome hassleriana

1. **C. hassleriana** Chodat

Map 327. A native of South America, widely cultivated and occasionally escaped. It occurred on exposed river-bottom and mud-flats at Ann Arbor late in 1968, after a dam had broken in June; the seeds had presumably lain dormant in the ponded area.

Long identified as *C. spinosa* Jacq., a species of Central and northern South America, not known to be in cultivation in the United States.

2. **C. serrulata** Pursh — Rocky Mountain Bee Plant

Map 328. A native of western North America, sparingly adventive or escaped from cultivation eastward; locally established in fields and waste ground, but not collected in Michigan since 1902.

CRUCIFERAE (BRASSICACEAE) — Mustard Family

This is a large and difficult family, the separation between genera (and often species) not always clear. It includes many common weeds as well as familiar ornamental and food plants. Unlike some families which include important food plants, however, such as Umbelliferae and Solanaceae, the Cruciferae include no species which are likely to cause poisoning under normal circumstances. (One does not get a toxic dose of mustard oils from anointing a hamburger, and most children are in no danger of overeating brussels sprouts or broccoli!) The pungent flavors and odors of the family are sometimes a helpful character in placing an unknown plant.

The fruit is composed of 2 carpels, which typically separate at maturity from the *base upwards*, leaving a membranous partition or septum in a little frame attached to the pedicel. An elongate linear fruit is called a *silique*; a short stubby one, a *silicle*. In a very few species, the fruit is indehiscent, or breaks transversely. The distinctive flower, typically with 4 sepals, 4 petals, and 6 stamens (4 long and 2 short), together with the unique fruit type, make this an easily recognized family even if the species are sometimes troublesome to distinguish.

Most keys to Cruciferae stress important characters of the ripe fruit, even of the embryo. If one has mature fruit, flowers with the color known, *and* an intact plant with cauline and (if any) basal leaves, identification is facilitated. Fortunately, most plants will still bear a few flowers while fruit is ripening on the lower part of the inflorescence. Even very young fruit at the base of the inflorescence will suggest the ultimate shape (silique or silicle); and with a little intuition one may also be able to predict the size to which partly developed fruits might grow.

The keys here avoid characters of ripe fruit (and seeds) as much as possible. If you have specimens with such parts, they can be identified by keys in other works. Often, characters of both flowers and fruit, as well as vegetative parts, are included, so that with some experimentation it should be possible to find where even an incomplete specimen belongs.

In part because of easy shriveling of the claw (or narrowed basal part of a petal), dry petals may appear to be shorter than they really are. Intact, flat (well-pressed), unshriveled petals should be used to determine length. The style persists as an indehiscent beak on the fruits of some species. It is thus a part of the fruit and fruit measurements *include* the beak if any, unless only the body or valves are mentioned. The length of the beak is measured from the end of the valves, and includes any stigma.

The common names "mustard" and "cress" are so widely used in the family that it is impossible to restrict them to single genera.

REFERENCE

Patman, Jacqueline-P., & Hugh H. Iltis. 1962. Preliminary Reports on the Flora of Wisconsin No. 44 Cruciferae—Mustard Family. Trans. Wisconsin Acad. 50: 17–72. [This is one of the most useful reports in its series, and is especially helpful with *Cakile* and *Arabis*.]

KEY TO THE GENERA

1. Petals pale to deep yellow
 2. Leaves (at least the middle or lower ones) ± deeply lobed (sinuses at least halfway to midrib), pinnatifid, or dissected
  3. Cauline leaves mostly deeply pinnatifid, appearing dissected (almost parsley-like), the ultimate segments mostly less than 2 mm broad; stem pubescence in our common species largely of forked, stellate, and/or glandular hairs
   4. Fruit a linear silique; all leaves ± dissected; stem pubescence largely of forked, stellate, and/or glandular hairs . . . . 1. **Descurainia**
   4. Fruit a round silicle; upper leaves simple, entire, strongly auriculate-clasping; stem glabrous . . . . 32. **Lepidium (perfoliatum)**
  3. Cauline leaves at most once-pinnatifid, at least the terminal segment more than 2 mm broad (except sometimes in *Sisymbrium altissimum*); stem pubescence absent or of simple hairs, not glandular (except in *Bunias*)
   5. Pedicels (at least lower ones) subtended by pinnatifid bracts . . . . . . 2. **Erucastrum**
   5. Pedicels all bractless
    6. Cauline leaves strongly clasping the stem with well developed auricles
     7. Petals less than 3 mm long; fruit less than 5 times as long as broad . . 8. **Rorippa**
     7. Petals 4–14 (16) mm long; fruit becoming over 10 times as long as broad
      8. Upper leaves mostly ± pinnatifid (with at least 1 pair of narrow lobes) or with angular teeth; fruit with beak not over 3.1 mm long . . . . 3 **Barbarea**
      8. Upper leaves entire to obscurely toothed or scalloped; fruit with beak 5–15 mm long . . . . 6. **Brassica** (couplet 2)
    6. Cauline leaves sessile to petioled but not clasping
     9. Petals 10–20 mm long; style at least 1.5 mm long, soon elongating into a prominent beak on fruit
      10. Ovary and fruit on a distinct short stipe ca. 0.5–1 mm long (easily seen as a zone of different shade or texture above the receptacle just after flowering); fruits indehiscent, very strongly constricted (and ultimately breaking) between the ripe seeds . . . . 4. **Raphanus (raphanistrum)**

10. Ovary and fruit sessile; fruits dehiscent at maturity, slightly if at all swollen around the seeds

11. Petals ca. 15–20 mm long, veined with purple; ripe fruit with body ca. 4–5 mm thick, conspicuously 1-nerved on each valve, with very flat beak .... 5. **Eruca**

11. Petals 10–15 mm long, without purple veins; ripe fruit less than 4 mm thick .... 6. **Brassica**

9. Petals less than 10 mm long; style usually shorter than 1.5 mm or absent, elongating at most into a beak ca. 2 mm long (longer in *Brassica*)

12. Fruit (and maturing ovary) subglobose to broadly ovoid or short-oblong, less than 5 times as long as wide

13. Fruit indehiscent, 1–4 (usually 2)-seeded, asymmetrically ovoid, the surface with irregular warts or ridges; principal leaves with terminal lobe much larger than lateral ones, 2–7 cm broad; stem with scattered sessile or short-stalked glands .... 7. **Bunias**

13. Fruit readily dehiscent, many-seeded, subglobose to oblong or cylindrical (often slightly curved), smooth; principal leaves with terminal lobe less than 2.5 cm broad (very rarely to 3.5 cm); stem without glands .. 8. **Rorippa**

12. Fruit (and maturing ovary) slender, ± linear, attaining a length at least 10 times as great as the width

14. Petals less than 5.5 mm long; fruit in common species less than 2 cm long

15. Plant usually in damp ground, the stem usually ± lax or prostrate, reproducing vegetatively (ripe seeds not formed); terminal lobe of most leaves scarcely if at all broader than lateral lobes ... 8. **Rorippa (sylvestris)**

15. Plant usually in dry ground, the stem erect, reproducing by seed; terminal lobe of most leaves much broader than lateral lobes .... 9. **Sisymbrium**

14. Petals (5) 5.5–9.5 mm long; fruit often longer than 2 cm

16. Beak of fruit 7–14 mm long at maturity, the style soon exceeding 4 mm after flowering .... 6. **Brassica**

16. Beak of fruit less than 4 mm long, or none

17. Buds overtopping open flowers; pedicels and fruit becoming closely appressed to axis of inflorescence .... 6. **Brassica (nigra)**

17. Buds mostly overtopped by open flowers; pedicels and fruit not spreading

18. Ovules and seeds in 2 rows in each locule (easily seen on pressed dried immature fruit or as 2 rows of seeds or depressions in septum of mature fruit) .... 10. **Diplotaxis**

18. Ovules and seeds in 1 row in each locule .... 8. **Sisymbrium**

2. Leaves all unlobed, entire or toothed

19. Cauline leaves sagittate- or auriculate-clasping at the base

20. Body of fruit ± globose or obovoid; upper leaves (like the lower) with forked or stellate and/or simple hairs (at least a few on margins)

21. Axis of inflorescence at least sparsely pubescent; petals up to 2.5 mm long; fruit indehiscent, 1–2-seeded, less than 2.5 mm broad, the body subglobose-compressed, strongly reticulate .... 11. **Neslia**

21. Axis of inflorescence glabrous; petals 3.5–4.5 (5) mm long; fruit dehiscent, several-seeded, ca. (3) 4–5 mm broad, the body obovoid, sometimes slightly compressed, weakly if at all reticulate .... 12. **Camelina**

20. Body of fruit elongate, linear (the shape evident as ovary matures); upper leaves (not always the lower) glabrous

22. Stem and leaves glabrous; petals ca. (7) 8–12 mm long; leaves broadly rounded at apex .... 13. **Conringia**

22. Stem and leaves pubescent at very base of plant; petals less than 7 mm long; leaves acute .... 21. **Arabis**

19. Cauline leaves merely sessile, not clasping

23. Plant glabrous or with simple hairs; fruit a linear silique ca. 1–4.5 cm long

24. Petals (5) 5.5–7.5 mm long; ovules and seeds in 2 rows in each locule (easily seen on pressed dried immature fruit or as 2 rows of seeds or depressions in septum of mature fruit) .... 10. **Diplotaxis**

24. Petals 7–15 mm long; ovules and seeds in 1 row in each locule .... 6. **Brassica** (couplet 4)

23. Plant pubescent with stellate or forked hairs; fruit various

25. Fruit 4-sided or ± terete, at least 15 mm long; petals 3.5–10 mm long .... 14. **Erysimum**

25. Fruit strongly flattened (parallel to the septum), less than 10 mm long; petals less than 4 [6] mm long

26. Fruit elongate (ca. 3–5 times as long as broad), ca. 5–9 mm long; ovules and seeds numerous in each locule; leaves slightly toothed, ovate to elliptic .... 29. **Draba**

26. Fruit round or nearly so, ca. 4 [5] mm or less long; ovules and seeds 1–2 in each locule; leaves entire, linear to oblanceolate .... 15. **Alyssum**

1. Petals white to purple, or none

27. Principal cauline leaves deeply lobed (e. g., lyrate or pinnatifid) or compound (uppermost leaves or bracts—at the inflorescence, or above water in aquatics—may be simple)

28. Leaves palmately compound or deeply palmately divided, the cauline only 2 or 3 (4) .... 16. **Dentaria**

28. Leaves pinnately lobed or divided, the cauline often more than 3

29. Plant truly aquatic, the submersed leaves dissected in a bipinnate pattern into filiform segments (midvein present, the lateral segments again dissected; fig. 141), frequently detaching readily from the stem ...... 17. **Armoracia (aquatica)**

29. Plant terrestrial or aquatic but even if in water the leaves with definite flat lobes (not bipinnately dissected) and not falling from the stem

30. Petals ca. 15–20 mm long ....5. **Eruca**

30. Petals less than 15 mm long

31. Leaf blades or leaflets with short stiff hairs ca. 0.3–0.5 mm long on margins or undersides; petals at least 10 mm long; mature fruit elongate, ca. 3–12 mm thick, indehiscent .... 4. **Raphanus**

31. Leaf blades or leaflets glabrous or with finer smaller hairs, or if blades ciliate, the petals much less than 10 mm long; mature fruit narrower or round, dehiscent

32. Fruits less than twice as long as broad, with 1 seed in each locule (or few seeds never maturing); plant a stiff erect weed of usually dry places

33. Petals less than 4 mm long; fruit strongly flattened; habitat dry .... 32. **Lepidium**

33. Petals at least 5 mm long; fruit ± globose; habitat dry to wet......... .... 17. **Armoracia (rusticana)**

32. Fruits soon becoming more than twice as long as broad, many-seeded; plant often ± lax, usually in wet ground or water

34. Petals at least 7 mm long .... 18. **Cardamine (pratensis)**

34. Petals less than 6 mm long

35. Fruits straight, ± erect, on ascending pedicels; seeds smooth, in a single row in each locule; plants often with a basal rosette (at least when young and if not submersed), the stems rooting only at the base; petals 0–3.5 (4.5) mm long .... 18. **Cardamine** (couplet 3)

35. Fruits often ± curved, spreading (the pedicels soon divergent after flowering); seeds reticulate (fig. 144), usually in 2 rows in each locule; plants without basal rosette, the stems usually rooting at the nodes (in water); petals (2.5) 3.5–5.7 mm long . . . . 19. **Nasturtium**

27. Principal cauline leaves (if any) simple, unlobed (except for clasping base in some species), entire or toothed

36. Stems and/or leaves (especially beneath and along margins) pubescent with many or all of the hairs stellate or forked

37. Cauline leaves sagittate- or auriculate-clasping

38. Fruit strongly flattened at right angles to the septum, ± triangular to obcordate, less than twice as long as broad; petals less than 3 mm long . 20. **Capsella**

38. Fruit plump or flattened parallel to the septum, linear (straight or curved), becoming at least 10 times as long as broad; petals usually at least 4 mm long (less than 3 mm in *A. perstellata*) . . . . 21. **Arabis**

37. Cauline leaves (if any) petioled or sessile but not clasping the stem

39. Body of fruit (and maturing ovary) slender, ± linear, attaining a length at least 10 times its width

40. Petals ca. (13) 15–20 (25) mm long, usually ± deep purple (occasionally white) . . . . 22. **Hesperis**

40. Petals less than 10 mm long, white or slightly tinged with color

41. Silique (8) 10–14 (16) mm long, less than 0.8 mm broad, with a style barely 0.2 mm long; petals less than 4 mm long; plant an introduced weedy annual (slender root, no old leaves at base) of waste ground . 23. **Arabidopsis**

41. Silique usually longer, sometimes broader, with a style ca. 0.5 (–1) mm long; petals various; plant a biennial or perennial (sturdier root, persistent old leaves), native (or rarely introduced)

42. Axis of inflorescence lightly pubescent; plant a rare native of Lake Superior region, less than 2 dm tall, with fruit less than 1 mm broad and at most 2.2 (2.4) cm long and petals less than 5 mm long . . . . .24. **Braya**

42. Axis of inflorescence glabrous or nearly so; plants of various range and habit, usually taller than the preceding, with larger fruit and/or petals . . . . 21. **Arabis**

39. Body of fruit (and maturing ovary) never linear, but short-oblong, ovate to narrowly elliptic, or round, less than 6 (8) times as long as wide

43. Petals deeply bilobed (fig. 152)

44. Stems leafy, without a basal rosette; petals 4–7 mm long; fruit pubescent, with ± convex sides and persistent slender style; plant perennial (usually), blooming in early to late summer . . . . 25. **Berteroa**

44. Stems leafless, all leaves in a basal rosette; petals 1.5–3 mm long; fruit glabrous, with flat sides and no evident style; plant annual, blooming in early spring (April) . . . . 26. **Erophila**

43. Petals entire or slightly notched at apex (or absent)

45. Pubescence mostly (or entirely) of appressed medifixed hairs (evident on leaves, stems, and inflorescences; fig. 153); seed 1 in each locule . . . . 27. **Lobularia**

45. Pubescence mostly of branched (or stellate) or simple hairs, without appressed medifixed ones; seeds 2–many in each locule

46. Fruit round or nearly so, ca. 4 mm or less long; ovules and seeds 2 in each locule; leaves all entire, cauline (or on short basal shoots), without basal rosette . . . . 15. **Alyssum**

46. Fruit elongate, elliptical or narrowly oblong to ovate (2.5–8 times as long as wide), at least 5 mm long; ovules and seeds numerous in each locule; leaves mostly in a basal rosette or toothed (or both) . . . . 28. **Draba**

36. Stems and leaves glabrous or with only simple hairs (occasionally a few forked hairs may be intermixed)

47. Cauline leaves sagittate- or auriculate-clasping

48. Fruit (and maturing ovary) linear, becoming at least (1.5) 3 cm long; petals ca. 3–12 mm long

49. Stem and leaves completely glabrous; leaves all cauline, entire and broadly rounded at the apex .... 13. **Conringia**

49. Stem and leaves not as above: pubescent (at least at base), some leaves basal, leaves toothed, and/or leaves acute .... 21. **Arabis**

48. Fruit (and maturing ovary) round to obovate (flattened to subglobose), less than twice as long as broad and not over 1.5 (2) cm long; petals not over 4 mm long (or absent)

50. Style ca. 0.7–1.2 (1.5) mm long on fruit; fruit not notched at apex, ± plump, wingless, indehiscent .... 29. **Cardaria**

50. Style less than 0.5 (0.7) mm long, usually barely if at all exceeding a notch at apex of fruit; fruit strongly flattened or at least winged, dehiscent

51. Plant glabrous; ovules and seeds 5–7 in each locule; fruit ca. (7) 9–12 (20) mm broad at maturity .... 30. **Thlaspi**

51. Plant slightly to densely pubescent; ovules and seeds 1 per locule; fruit less than 5.5 mm broad at maturity .... 31. **Lepidium**

47. Cauline leaves (if any) sessile or petioled, but not clasping the stem

52. Petals 15–25 mm long, usually purple (rarely white); fruit very flat, at least 15 mm broad; leaves opposite, at least at middle and lower nodes .. 32. **Lunaria**

52. Petals up to 14 mm long (to 16 mm in 2 species), white to pink (or absent); fruit even if flat less than 10 mm broad; leaves all alternate or basal

53. Plant a small aquatic with all leaves basal and awl-shaped, usually flowering and fruiting under water .... 33. **Subularia**

53. Plant with leafy stem, of damp or dry ground, not flowering under water

54. Middle cauline leaves with toothed, triangular-ovate blades little if at all longer than broad, on slender petioles; bruised plant with odor of onion or garlic .... 34. **Alliaria**

54. Middle cauline leaves with blades toothed or entire, distinctly longer than broad, sessile or tapered into petiole; plant without onion-garlic odor

55. Leaves succulent, the margin ± irregularly sinuate-toothed; plant an annual of sandy shores of the Great Lakes, without basal leaves; petals ± pink; fruit not longitudinally dehiscent, but breaking transversely between the two (usually 0–1-seeded) segments, the terminal segment larger, usually tapering into a prominent beak .... 35. **Cakile**

55. Leaves not succulent, the margin various (usually entire or regularly, even if remotely, toothed); plant of various habitat, most species with basal leaves (if somewhat succulent and on sandy shores, a small basal rosette of lyrate leaves usually present); fruit longitudinally dehiscent, with or without a beak

56. Fruit (and maturing ovary) linear, becoming at least 10 times as long as broad

57. Petals ca. 4–8 (9) mm long; basal leaves (if any) not distinctly petioled, arising from a slender root; pubescence (if any) usually including a few stellate hairs [check basal leaves carefully] .... 21. **Arabis** (couplet 3)

57. Petals 8–14 (16) mm long; basal leaves distinctly with a rounded blade and long slender petiole, arising from a tuberous root; pubescence (if any) of strictly simple hairs [check lower part of stem carefully] .... 18. **Cardamine** (couplet 2)

56. Fruit (and maturing ovary) round or globose, or nearly so, less than twice as long as broad

58. Largest cauline and basal leaves usually 3.5–30 cm broad, with ± rounded or obtuse teeth; basal leaves long-petioled; fruit ± ellipsoid-obovoid (not flattened), not notched at the apex, the several ovules not ripening into seeds . . . . 17. **Armoracia (rusticana)**

58. Largest leaves less than 2.5 (3.5) cm broad, entire or sharply toothed; basal long-petioled leaves none; fruit round or nearly so, strongly flattened, notched at apex, with 1 seed in each locule

59. Petals distinctly unequal (2 large and 2 small), the larger ca. 5–9 mm long; style at least 1 mm long . . . . 36. **Iberis**

59. Petals equal (if present), 0–2.2 [3] mm long; style at most ca. 0.5 mm long, usually shorter . . . . 31. **Lepidium**

1. Descurainia

KEY TO THE SPECIES

1. Plants green, with sparse stellate but numerous short gland-tipped hairs; mature fruit ca. (0.4) 0.7–1 cm long . . . . 1. **D. pinnata**

1. Plants ± whitened with dense mostly stellate hairs, but without glandular hairs; mature fruit over 1 cm long . . . . 2. **D. sophia**

1. **D. pinnata** (Walter) Britton Fig. 127 Tansy Mustard

Map 329. Considered a native species, but usually found in waste ground and the earliest Michigan collection dates from 1879 (Kent Co.), with several records from the 1890's. Roadsides, railroads, fields, gravel pits; also on disturbed shores and in open rocky (calcareous) woods.

The typical var. *pinnata* is southern, with denser stellate pubescence and fewer glands (only on stem) than in our var. *brachycarpa* (Richardson) Fern.

2. **D. sophia** (L.) Prantl

Map 330. A species of weedy places in Eurasia, naturalized in North America. The earliest Michigan collection seen is from Keweenaw County in 1895. Roadsides, railroad banks, and other waste ground.

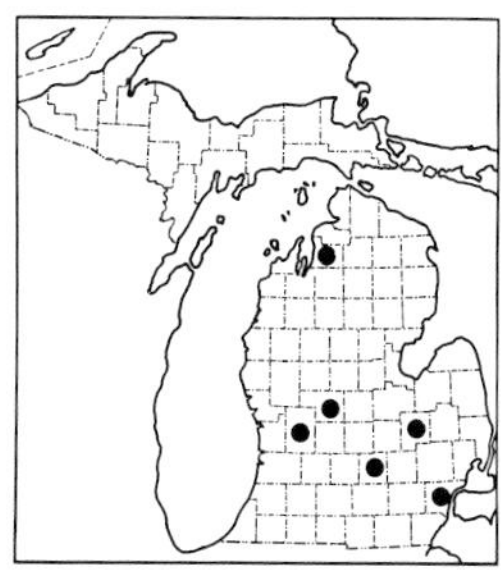

328. Cleome serrulata

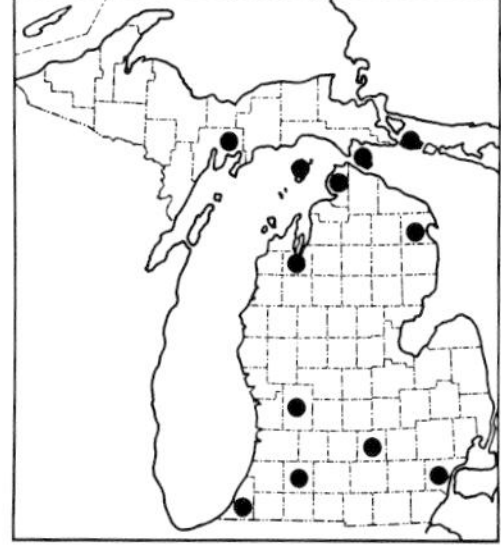

329. Descurainia pinnata

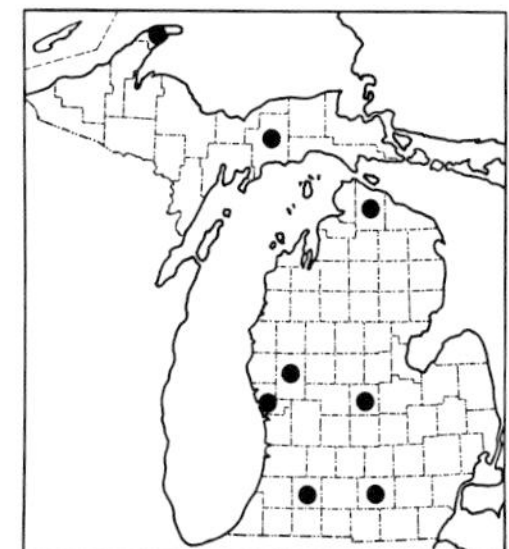

330. Descurainia sophia

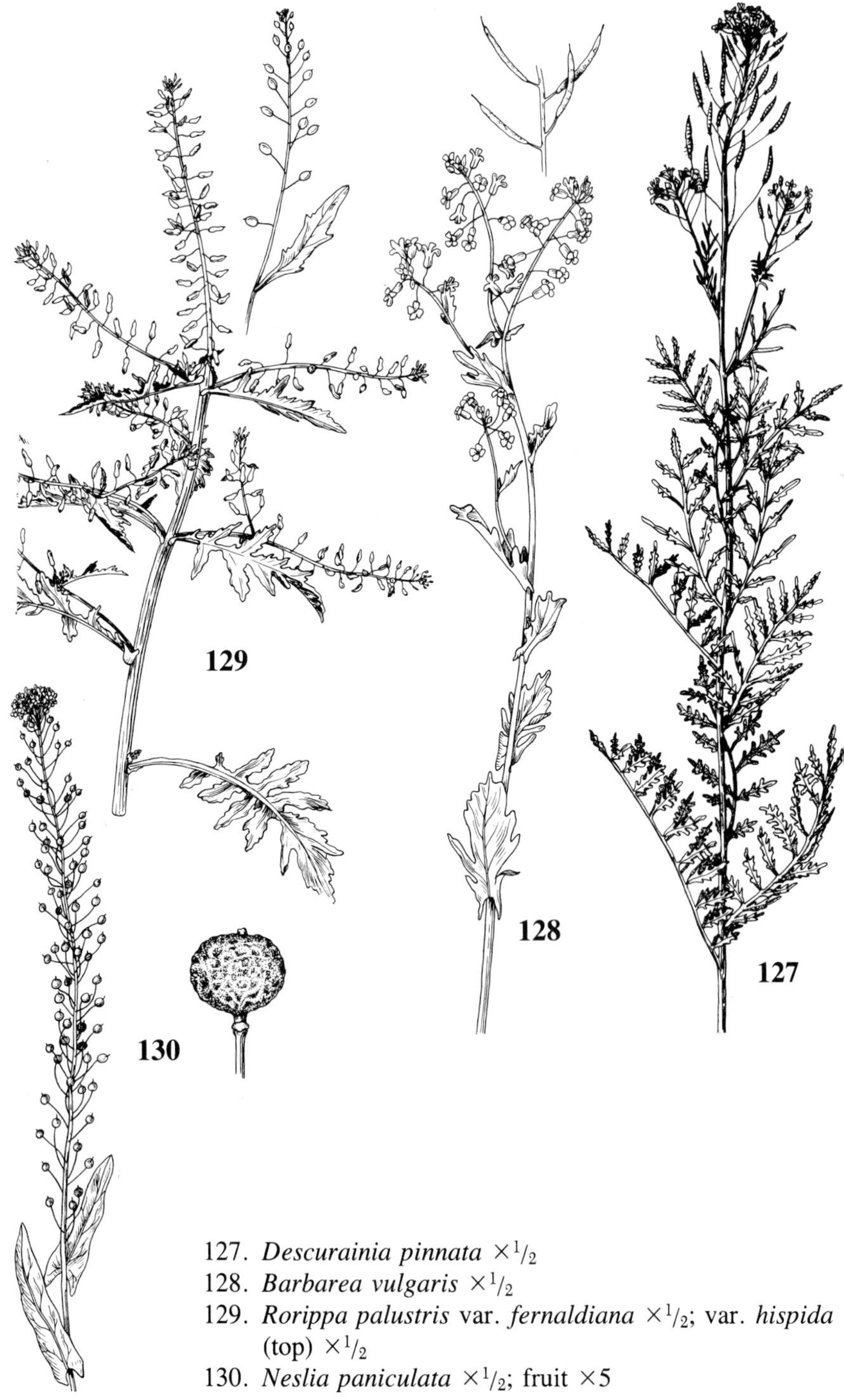

127. *Descurainia pinnata* ×$^1/_2$
128. *Barbarea vulgaris* ×$^1/_2$
129. *Rorippa palustris* var. *fernaldiana* ×$^1/_2$; var. *hispida* (top) ×$^1/_2$
130. *Neslia paniculata* ×$^1/_2$; fruit ×5

2. Erucastrum

1. **E. gallicum** (Willd.) Schultz   Fig. 131   Dog Mustard

Map 331. Roadsides, railroads, dumps, gravel pits, gardens, vacant lots, limestone pavements and quarries, shores, occasionally spreading into woods. A European native, first collected in Michigan in 1922 along railroad tracks at Ypsilanti and now a widespread weed, locally common in some places, such as the calcareous gravels in the vicinity of the Straits of Mackinac.

The principal hairs of the stem are simple and ± strongly retrorse; the lower leaves are often strongly flushed with purple.

3. Barbarea   Winter Cress

This is a difficult genus and the species are not easy to distinguish. Fortunately the nature of the beak (which develops from the style) can usually be predicted from immature fruit. Our plants are nearly always completely glabrous, but pubescent forms are known.

KEY TO THE SPECIES

1. Uppermost leaves (and bracts at base of peduncles) with angular teeth but no lobes (other than basal auricles); fruit with beak ca. (1.3) 1.5–3.1 mm long .. 1. **B. vulgaris**
1. Uppermost leaves ± pinnatifid with 1 or usually 2 pairs of narrow lateral segments; fruit with beak rarely as long as 1.5 mm
   2. Principal middle cauline leaves with at least 4–5 pairs of lateral lobes; mature fruit at least 3.5 cm long; petals ca. 5–8 mm long .... 2. **B. verna**
   2. Principal middle cauline leaves with 2–3 pairs of lateral lobes; mature fruit less than 3 (3.4) cm long; petals ca. 4–5 mm long .... 3. **B. orthoceras**

1. **B. vulgaris** R. Br.   Fig. 128   Yellow Rocket

Map 332. A common and noxious Eurasian weed, often called "wild mustard." Ubiquitous in dry to wet disturbed places of all kinds, spreading into woods (especially along roads and trails) and onto shores. The earliest Michigan collection seen was made by Dennis Cooley in 1842 in Macomb

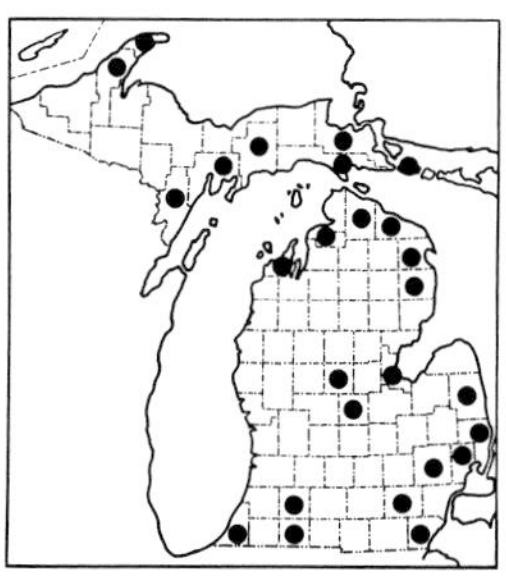

331. Erucastrum gallicum

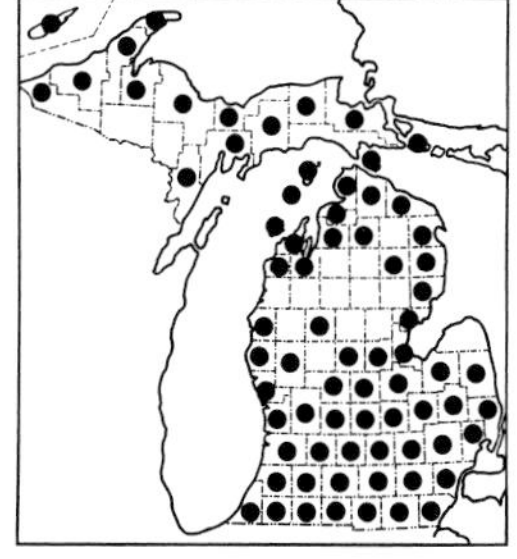

332. Barbarea vulgaris

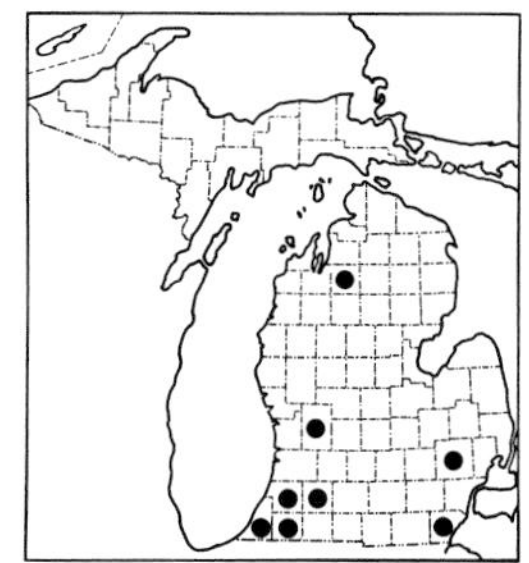

333. Barbarea verna

County. By the 1890's the species was apparently widespread about the state but not yet as aggressive as it has now become.

This is a variable species in regard to fruit and foliage. The slender pedicels and slender or narrowly tapered beak are in contrast with those of the next two species, especially *B. verna*, and help to give the plant a distinctive aspect once one is familiar with this rather subjective contrast. The fruit in our material does not exceed 2.5 cm long. The foliage of fresh plants has a characteristic glossy dark green appearance.

2. **B. verna** (Miller) Asch.

Map 333. A native of Europe, local in fields and open waste ground.

The pedicel and short beak are almost as thick as the body of the fruit in this species, in contrast to the more slender structures of the preceding. Furthermore, the silique is much longer. The basal leaves have more lateral lobes than either of our other species.

3. **B. orthoceras** Ledeb.

Map 334. A native species, mostly of northern North America and Asia, originally described from Siberia, and found with us on gravelly and rocky shores.

Our plants are not always easily distinguished from *B. vulgaris*, and some authors would include them with that species. The lower leaves and stem are often strongly flushed with purple. The petals tend to be smaller (ca. 4–5 mm long) than those of the other two species (which are 5–8 mm). A few collections with upper leaves resembling those of this species are referred to *B. vulgaris* on the basis of their larger petals and longer styles.

The name *B. stricta* Andrz. has sometimes been misapplied to this species and sometimes to *B. vulgaris*, but it belongs to a European species with pubescent buds and small petals.

4. **Raphanus** Radish

KEY TO THE SPECIES

1. Petals yellow to white but usually with conspicuous dark veins when fresh; mature fruit at most ca. 4 mm broad, strongly constricted with prominent corky partitions between the seeds (which are often 4 or 5) . . . . 1. **R. raphanistrum**
1. Petals purple or sometimes white; mature fruit ca. 8–12 mm broad, not constricted, usually 2–3-seeded . . . . 2. **R. sativus**

1. **R. raphanistrum** L. Wild Radish

Map 335. A Eurasian weed of dry roadsides, fields, and waste ground.

There is a distinct stipe (gynophore) ca. 0.5–1 (1.5) mm long at the base of the maturing ovary and fruit—a helpful character for specimens otherwise incomplete or ambiguous.

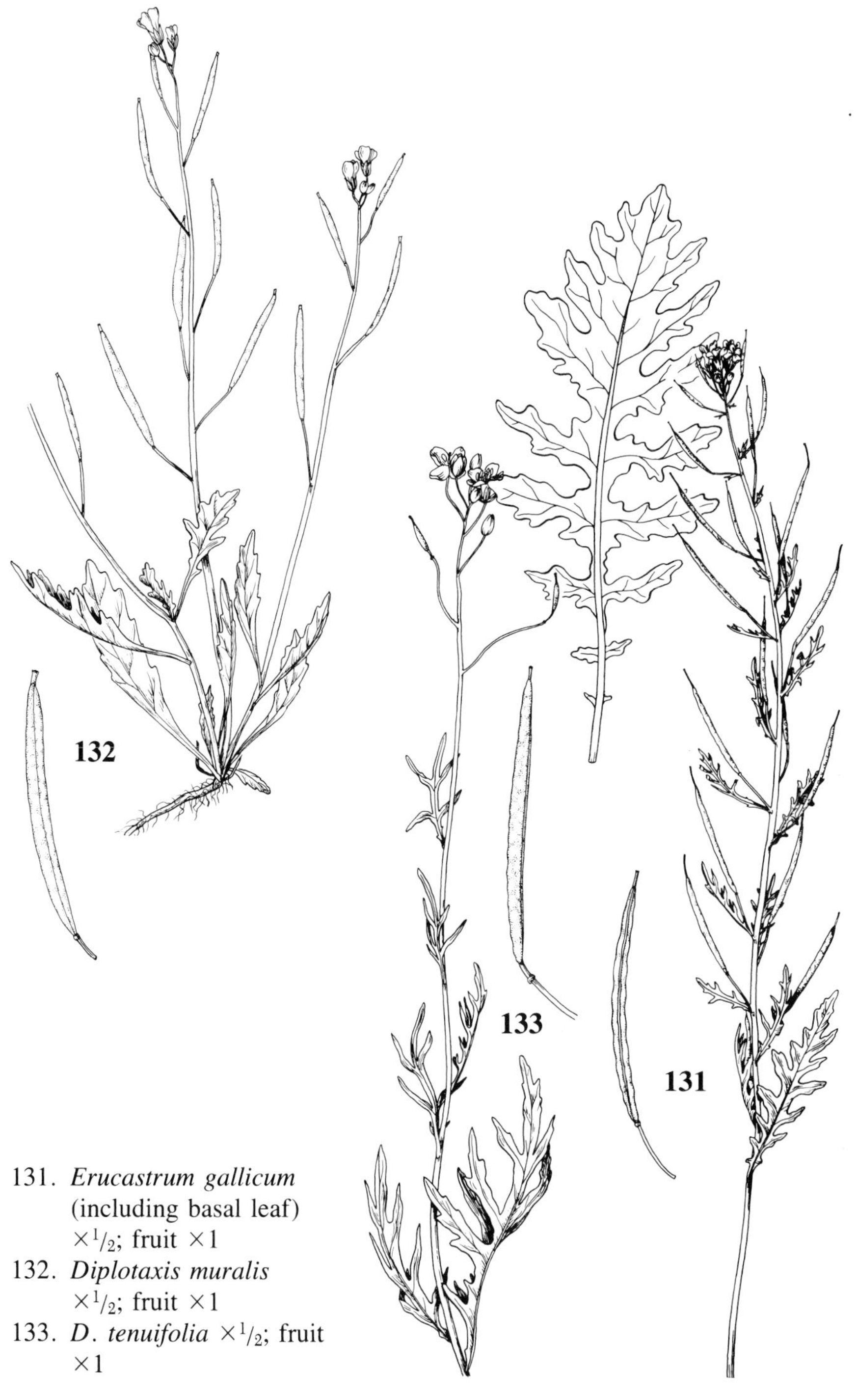

131. *Erucastrum gallicum* (including basal leaf) $\times ^1/_2$; fruit $\times 1$
132. *Diplotaxis muralis* $\times ^1/_2$; fruit $\times 1$
133. *D. tenuifolia* $\times ^1/_2$; fruit $\times 1$

2. **R. sativus** L. Radish

Map 336. An Old World native of unknown origin, long cultivated for its crisp, thick, edible roots; sporadic in waste ground such as roadsides, fields, disturbed shores.

Specimens of ambiguous flower color and lacking fruit mature enough to determine its distinctive fat shape may be distinguished from the preceding by the absence or obscurity of the stipe at the base of the young fruit.

5. Eruca

1. **E. vesicaria** (L.) Cav. Rocket-salad; Garden Rocket

Map 337. A native of the Mediterranean region, long cultivated in parts of the Old World for use of its tender new growth as a salad plant; more often in this country, a weed. All Michigan collections seen were made 1902–1920, from roadsides, fields, and railroads—insofar as any habitat is stated.

Often known under the name *E. sativa* Miller, now generally considered only a subspecies of the variable *E. vesicaria*.

6. Brassica

The taxonomy, as well as application of both scientific and common names,

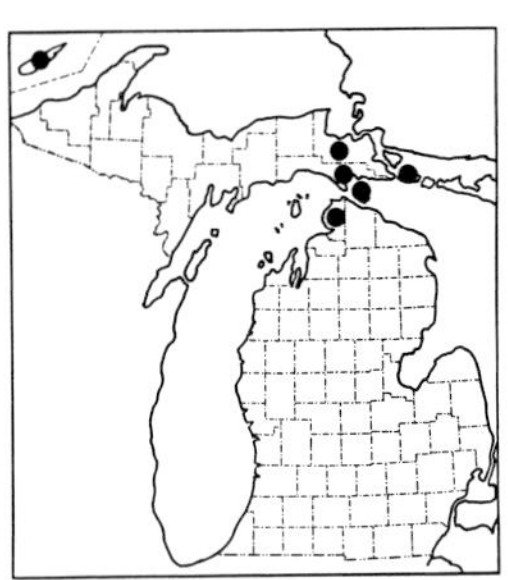

334. Barbarea orthoceras

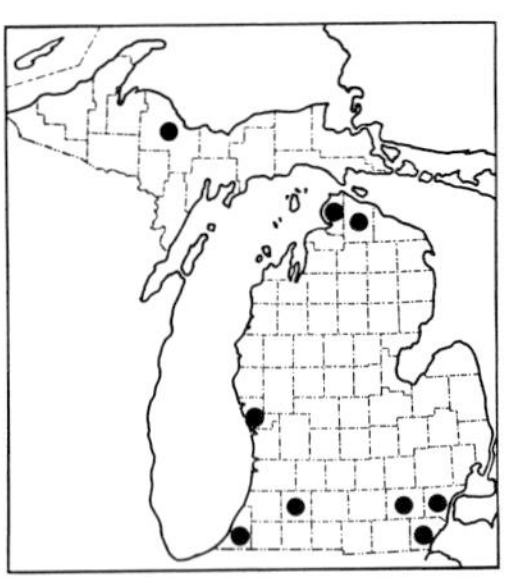

335. Raphanus raphanistrum

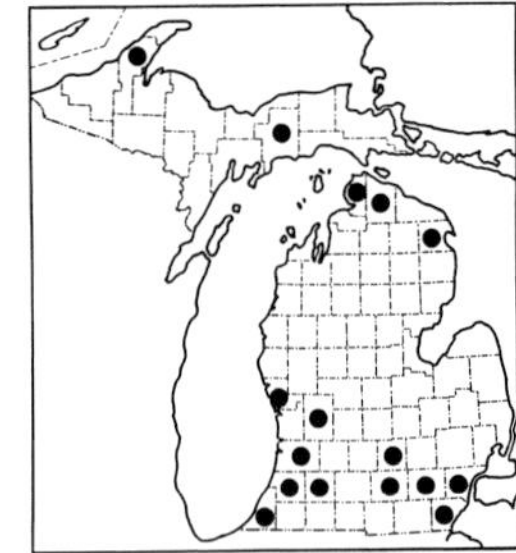

336. Raphanus sativus

337. Eruca vesicaria

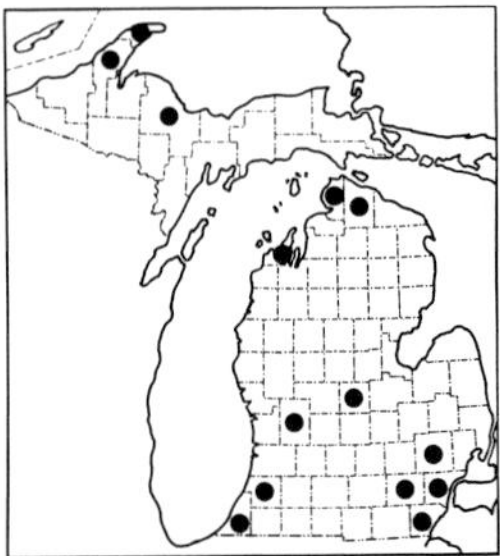

338. Brassica rapa

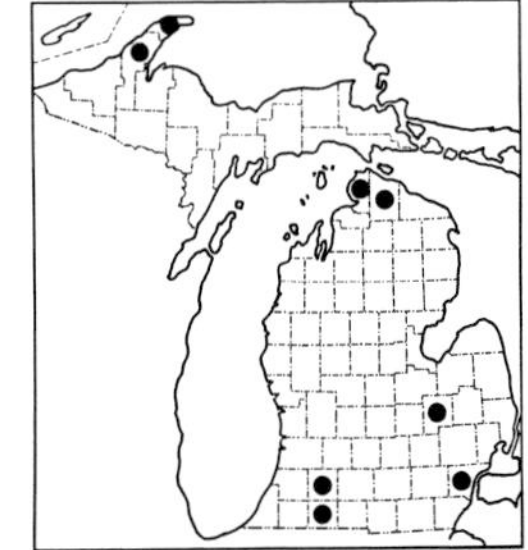

339. Brassica napus

has long been confused in this genus, especially among the species with strains cultivated for vegetables. Insofar as possible, *Hortus Third* is here followed in these matters, with results consistent with *Flora Europaea* (except that *Sinapis* is not here segregated). The characters used to distinguish species, and in their description, vary tremendously in different references, so that one may name available material in several ways depending on the source relied upon.

REFERENCES

Bailey, L. H. 1922. The Cultivated Brassicas. Gentes Herb. 1: 51–108.

Bailey, L. H. 1930. The Cultivated Brassicas Second Paper. Gentes Herb. 2: 209–267.

KEY TO THE SPECIES

1. Leaves (especially upper ones) strongly clasping with auriculate bases
  2. Sepals 3.4–6.5 mm long; petals 6.5–9 (10) mm long; beak of fruit ca. 8–15 (23) mm long, ca. 30–50% as long as valves; open flowers overtopping the buds of the inflorescence; foliage green or glaucous . . . . 1. **B. rapa**
  2. Sepals 6.5–9 mm long; petals 9–14 (16) mm long; beak of fruit ca. 5–8 (12) mm long, ca. 10–25% as long as valves; open flowers not clearly overtopping the unopened buds; foliage glaucous . . . . 2. **B. napus**
1. Leaves sessile or petioled, not clasping
  3. Fruit and even mature ovary densely hispid with spreading to ascending hairs; beak of fruit very flat, as long as the body when ripe . . . . 3. **B. alba**
  3. Fruit and ovary glabrous (or sometimes with a few reflexed hairs); beak rarely over half as long as the body, 4-angled to terete
    4. Uppermost leaves (e. g., at base of main branches of inflorescence) coarsely toothed to pinnatifid, ca. 2–4 times as long as broad; valves of fruit each with 2 or 4 distinct parallel nerves besides the midnerves and margins; petals (8) 10–15 mm long . . . . 4. **B. kaber**
    4. Uppermost leaves entire or nearly so, narrow (including petiole if any, ca. 5–10 times as long as wide); fruit without parallel nerves besides margins and midnerves (additional nerves, if any, ± looped or anastomosing); petals 7–10 mm long
      5. Fruit 0.9–1.7 (2) cm long, including beak less than 4 mm, on pedicels 3–5 (6) mm long and closely appressed to the axis of the inflorescence; young tips of inflorescences with unopened buds overtopping open flowers . . . . 5. **B. nigra**
      5. Fruit (2.5) 3.5–4.5 cm long at maturity (even when immature soon exceeding 1.5 cm), including beak 7–9 (10) mm long, on pedicels 7–10 mm long and diverging from axis; young tips of inflorescences ± corymbose, with flowers overtopping the unopened buds . . . . 6. **B. juncea**

1. **B. rapa** L. Field Mustard; Turnip

Map 338. A field weed, some strains of which are important vegetables and fodder crops, the Old World origins lost in antiquity. Found in disturbed ground, such as railroad ballast, fields, and roadsides.

Following most (but not all) recent authors, *B. campestris* L., field mustard or bird rape, is included here in *B. rapa,* the basic species not only of turnip but also of Chinese or celery cabbage and bok choy.

In all of the collections referred here (insofar as they are flowering), the open flowers overtop the buds at the end of the inflorescence, as usually stated for *B. rapa* (though not in *Flora USSR*). The foliage is usually completely glabrous, although sometimes there are stiff translucent hairs on the lower leaves, especially on the midrib beneath and the margins.

2. **B. napus** L. Rape; Rutabaga

Map 339. Like the preceding, long cultivated in several forms, its native origins unknown. And also found along railroads and roadsides, in fields and waste places.

Identification of this species and the similar *B. rapa* is not helped by the divergent diagnoses in various treatments. As interpreted by Fernald (1950), followed by Patman and Iltis (1962), the relative petal size is reversed from that given here; as interpreted in *Flora USSR,* the relative bud position is reversed. The basal leaves of *B. napus* are supposed to be glaucous, while in *B. rapa* only the upper leaves may be glaucous.

Another important cultivated *Brassica* is *B. oleracea* L., rather similar to *B. napus* in characters of flowers and fruit, but with sepals closely appressed to the corolla rather than slightly spreading. *B. oleracea* is apparently native to Europe but long grown in various forms including cabbage, broccoli, cauliflower, kale, kohlrabi, Brussels sprouts, etc. One roadside collection from Emmet County might be referred here, although it is probably a form of *B. napus*; it has a tough woody taproot and glaucous foliage, with petals shorter than is often the case in *B. oleracea.* Farwell claimed that kale was escaped in Houghton County.

3. **B. alba** (L.) Rabenh. White Mustard

Map 340. Railroad ballast, dumps, vacant lots, and similar waste places. A native of the Mediterranean region, widely cultivated for its large seeds, which are a commercial source of mustard; now naturalized as a weed in many parts of the world.

This species is often segregated as *Sinapis alba* L. The same epithet should be retained in *Brassica,* although the name *B. hirta* Moench has been used by many authors.

4. **B. kaber** (DC.) Wheeler Fig. 134 Charlock; Wild Mustard

Map 341. Probably originally native in the Mediterranean region, seldom cultivated but now a widespread weed. Next to *Barbarea vulgaris,* our commonest "yellow mustard" in Michigan. First noted by Dennis Cooley, apparently in Macomb County, about 1841. By the 1890's, widespread in the state, Dodge calling it a "vile weed" in St. Clair County. Now thoroughly established throughout in all kinds of waste places and disturbed ground, including roadsides, fields, railroads, dumps, shores, yards and gardens, fencerows; invading woods, both dry and moist.

Often segregated, with the preceding species, in the genus *Sinapis,* in which case the correct name is *S. arvensis* L.

A variable species usually easily distinguished, even in the absence of the distinctive fruit, from the next two species by the slightly larger flowers, broad toothed upper leaves (frequently all the leaves merely toothed and not pinnatifid), and tendency for the stem to be hispid with translucent hairs even into the inflorescence. The fruit is (2.2) 2.5–4.2 cm long in our material, including a beak 7–14 mm long and often 1-seeded. (In the next two species the beak has no seed.) The beak is rather sharply 2-edged along the continuation of the suture, but is basically 4-angled, there being a prominent midnerve on each side, with additional nerves as well. (These additional nerves are absent in the next two species, in which the 4-angled beak is not so clearly 2-edged.) The fruit may be so closely appressed as to resemble that of *B. nigra,* although usually it is more spreading.

In typical var. *kaber,* which we do not have, the fruit is supposed to be less than 2.5 cm long and nearly sessile. Plants with longer fruit are said to include three varieties, all of which we have although their significance is dubious: Plants with ± sparsely hispid fruits are var. *orientalis* (L.) Scoggan; those with very slender (less than 2 mm thick) fruits strongly constricted between the seeds are var. *schkuhriana* (Reichb.) Wheeler—and some plants combine the features of these varieties. Plants with glabrous fruit over 2 mm thick and not strongly constricted are var. *pinnatifida* (Stokes) Wheeler and are by far the commonest.

5. **B. nigra** (L.) Koch Fig. 135 Black Mustard

Map 342. A Eurasian species, widely cultivated for its seed, used as a condiment, and naturalized as a weed though less common than often thought. Collected in Michigan as early as 1838, when the First Survey found it along the margin of the Grand River in Jackson County. Roadsides, fields, vacant lots, and other waste or cultivated places; borders of rivers and other damp ground.

340. Brassica alba

341. Brassica kaber

342. Brassica nigra

The stem is usually described as hispid, especially basally, in contrast with a glabrous stem in the next species, but most of our specimens are nearly or quite glabrous. The shorter appressed fruit on very short pedicels will readily distinguish this species from the next when mature; *B. kaber* also has short pedicels, but they are thicker than in *B. nigra* and the fruit is longer—in addition to the differences in the leaves.

6. **B. juncea** (L.) Czern. Indian, Chinese, or Brown Mustard

Map 343. Shores, railroads, dumps, fields, and waste ground. Another Old World species sometimes cultivated for leafy greens or for the seeds as a source of oil or condiment, but also known as a weed. Like the preceding, much less common than often supposed (or misidentified!), perhaps declining as a result of stricter control of agricultural seed quality (reducing contaminants) and improved cultivation methods.

7. **Bunias**

1. **B. orientalis** L. Turkish Rocket

Map 344. Native from eastern Europe to Siberia, found by F. J. Hermann in 1936 as well established in a low meadow in the Nichols Arboretum in Ann Arbor. Also collected by L. H. Bailey in 1887 as adventive near Lansing.

8. **Rorippa** Yellow Cress

Our species have had a tangled taxonomic and nomenclatural history. They have sometimes been placed in the genus *Nasturtium* (although *Rorippa* is the older name and to be used if the two genera are combined) and sometimes in *Radicula* (a name not validly published when first used and when later validated, lacking priority). Our common species has often been combined with another, sometimes split into two, and usually divided into several varieties distinguished in various ways. The generic name has often been unjustifiably altered to *Roripa*.

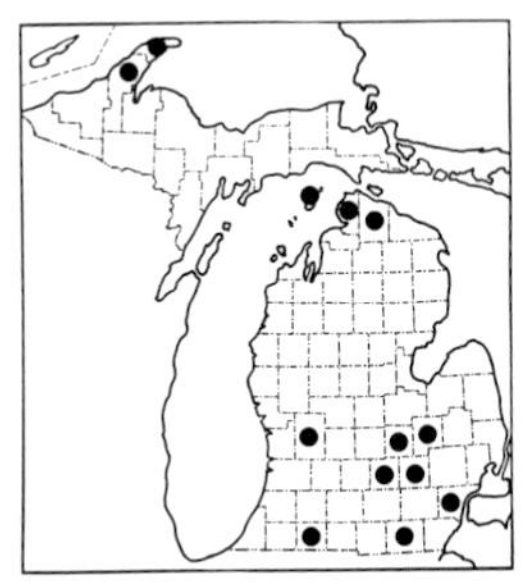

343. Brassica juncea

344. Bunias orientalis

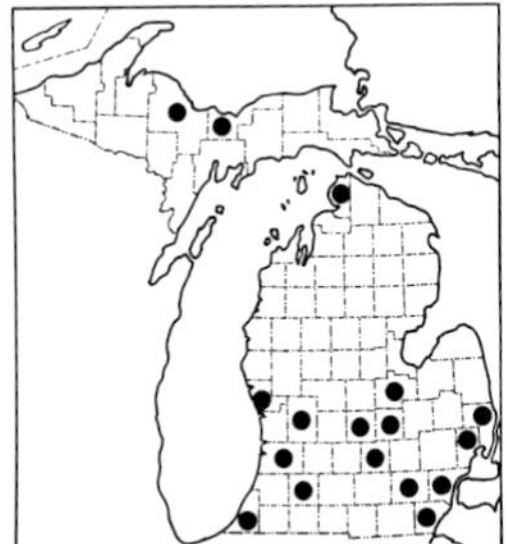

345. Rorippa sylvestris

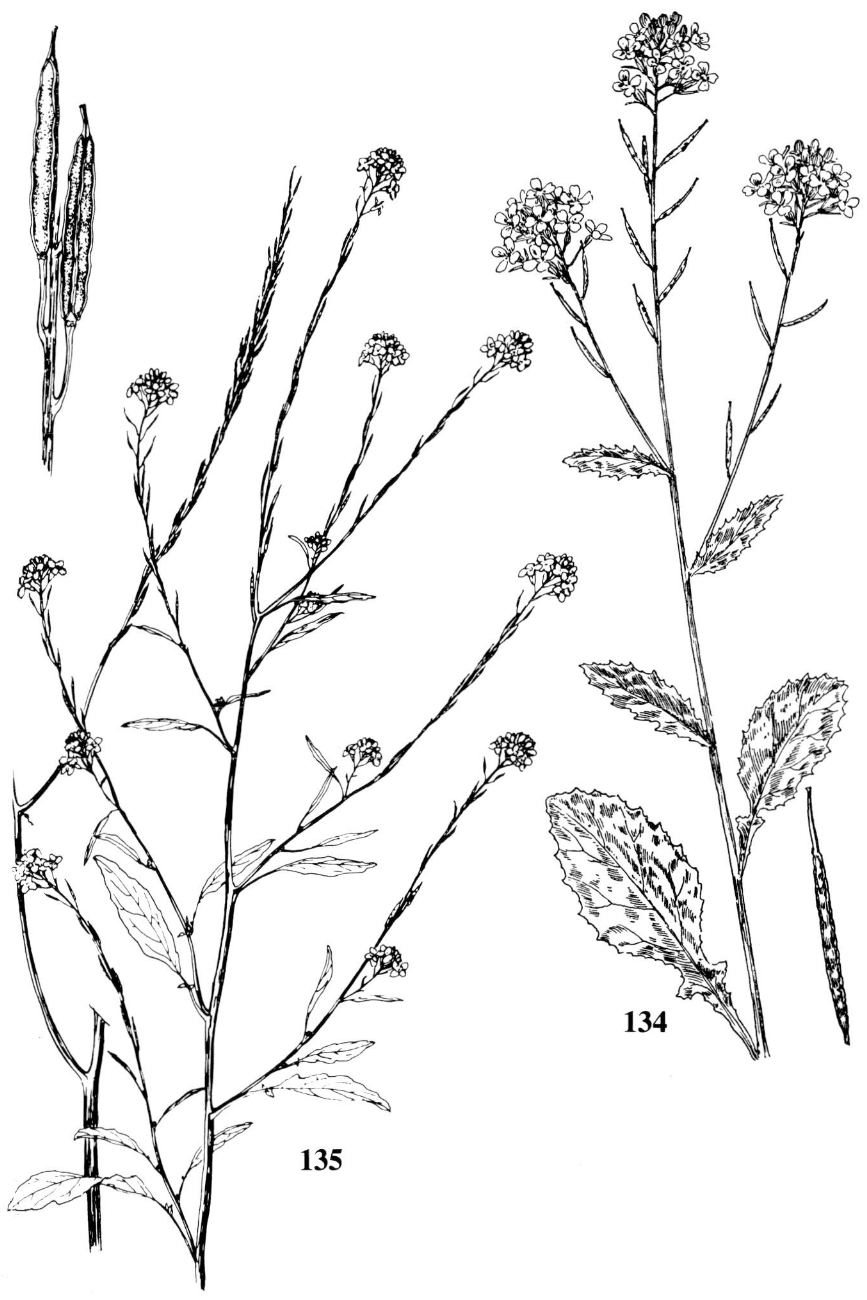

134. *Brassica kaber* $\times{}^{1}/_{3}$; fruit $\times 1$
135. *B. nigra* $\times{}^{2}/_{5}$; fruit $\times 2$

REFERENCES

Stuckey, Ronald L. 1966. The Distribution of Rorippa sylvestris (Cruciferae) in North America. Sida 2: 361–376.

Stuckey, Ronald L. 1972. Taxonomy and Distribution of the Genus Rorippa (Cruciferae) in North America. Sida 4: 279–430.

KEY TO THE SPECIES

1. Uppermost leaves (including those at base of branches of the inflorescence) deeply pinnatifid (lobed nearly or quite to the midrib); fruit linear, the ovary soon becoming more than 5 times as long as wide, but not (with us) forming ripe seed; petals 2.5–3.7 mm long, distinctly exceeding the sepals .... 1. **R. sylvestris**

1. Uppermost leaves often merely toothed or shallowly lobed; fruit subglobose to short-oblong or cylindrical, no more than 5 times as long as wide, normally setting copious seed; petals 0.5–2.3 mm long, barely if at all longer than the sepals (often shorter)

  2. Pedicels of mature fruit somewhat curved or arching downwards; petals ca. 1 mm long or shorter; leaves all glabrous beneath; mature fruit ca. 2–5.2 mm long, up to 1.7 mm thick, mostly broadest well below the middle, strongly tapering to ± acute apex .... 2. **R. curvipes**

  2. Pedicels of mature fruit usually ± straight, ascending or spreading; petals ca. (1) 1.5–2.3 mm long; leaves glabrous or hispid beneath; mature fruit thicker, the largest on a plant (1.6) 2–2.7 mm thick at or near the middle, not strongly tapering, the apex blunter .... 3. **R. palustris**

1. **R. sylvestris** (L.) Besser

Map 345. Swamps and muddy to rocky shores along rivers and lakes; also weedy in lawns, cultivated ground, ditches, and city streets. Native of Europe and western Asia, now widespread in the northeastern United States and adjacent Canada, more sporadically across the continent.

This is a perennial species, spreading by creeping roots (often erroneously termed rhizomes) and apparently distributed by fragments along streams and transportation routes as well as with transplanted nursery stock (see Stuckey 1966). Our other two species are annuals or biennials. The petals are usually described as being 4–5 mm long, but ours are almost always less than 3.5 mm.

2. **R. curvipes** Greene

Map 346. A native of the Rocky Mountain region, occasionally found eastward as presumably an introduction. Collected in 1949 as a weed along the edge of a cultivated field south of Houghton (*Richards 2656,* MICH, WUD, NY).

The pedicels are also sometimes recurved in *R. palustris* var. *fernaldiana,* which is also glabrous, but which has fruit broader and often longer than in *R. curvipes* and less tapered to the tip.

3. **R. palustris** (L.) Besser Fig. 129

Map 347. Wet shores and meadows, marshy ground, pools (sometimes in water a foot or more deep), mudflats, recently exposed lake- and stream-beds, ditches, swampy thickets; somewhat weedy along roadsides and in cultivated ground.

American plants have long been included in *R. islandica* (Murray) Borbás by many authors, but the two species are now considered, by botanists who have studied the genus, to be different, with *R. islandica* not known from North America.

Michigan plants represent three varieties according to Stuckey (1972). *R. palustris* var. *hispida* (Desv.) Rydb. has the stem and leaves beneath ± hispid. In var. *fernaldiana* (Butters & Abbe) Stuckey and var. *palustris* the leaves are glabrous beneath and the stems are at most hispid near the base. There may be some question whether var. *palustris* is native in our area; it was collected in wet ground once on the Keweenaw Peninsula (*F. E. Wood* in 1884, PH), and is native in western North America and in Europe. It is a small, delicate plant (stems under 3 mm in diameter), the leaves thinner textured than in our other varieties, the stigma (or extreme summit of the style) expanded in fruit. In var. *fernaldiana,* as in var. *hispida,* the plants may be stouter and the leaves are thicker textured; the stigma is not expanded in fruit (except usually in var. *fernaldiana*)—but this is a very difficult character to interpret. The fruit of var. *hispida* is distinctly more plump and subglobose than the slightly narrower more oblong fruit of var. *fernaldiana,* in which the fruit is slightly constricted at the middle (as it may be in *R. curvipes*).

Both of our common varieties occur throughout the state and apparent hybrids between them are occasionally found (with hispid leaves and stem of var. *hispida* but with fruit characters of var. *fernaldiana*). This is an extremely variable species, and a full classification places our three varieties, along with others, in subspecies *hispida* (Desv.) Jonsell, ssp. *fernaldiana* (Butters & Abbe) Jonsell, and ssp. *palustris*.

346. Rorippa curvipes

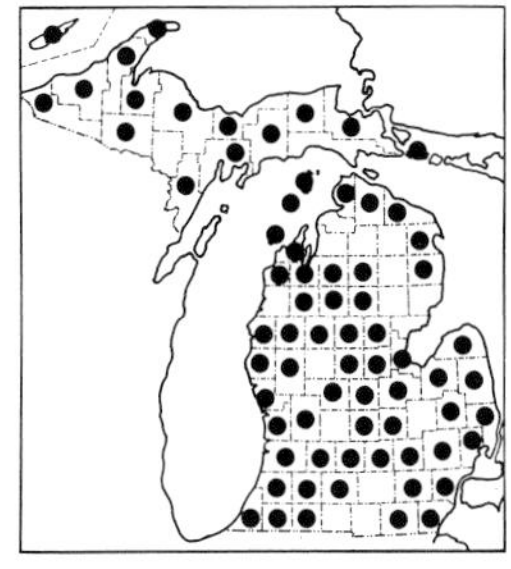

347. Rorippa palustris

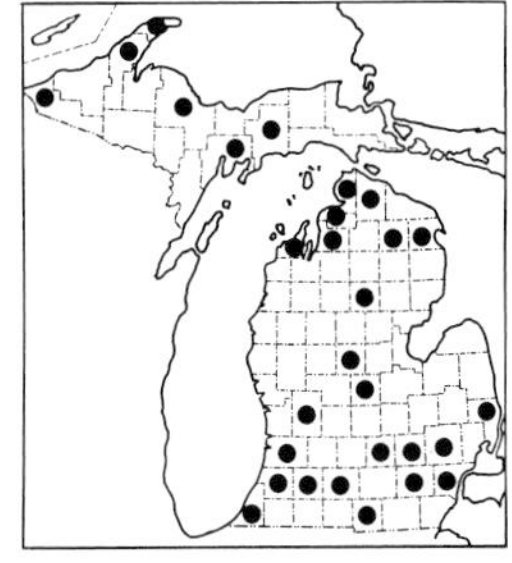

348. Sisymbrium officinale

9. Sisymbrium

KEY TO THE SPECIES

1. Fruit less than 1.8 cm long, on pedicels less than 4 mm, even when young closely appressed to axis of inflorescence—but branches, if present, of inflorescence often divergent nearly or quite at right angles to main axis; leaves with broad lobes; petals 3–4 mm long . . . . 1. **S. officinale**
1. Fruit ca. 2–5.9 cm long, on pedicels at least 6 mm, spreading; branches of inflorescence ascending; leaves in common species with very narrow lobes; petals 2–9.5 mm long
   2. Middle and upper leaves deeply pinnatifid, the narrow lobes less than 3 (5) mm wide (often less than 1 mm), the terminal lobes no broader than the lateral ones; petals (5.5) 7–9.5 mm long; mature fruit ca. 6–9 cm long, scarcely if at all thicker than the pedicels (the latter over 0.5 mm thick) . . . . 2. **S. altissimum**
   2. Middle and upper leaves entire to pinnatifid, with at least the terminal lobe much broader than 5 mm (and broader than the lateral lobes); petals ca. 3–6 [7] mm long; mature fruit ca. 2.5–4 cm long, thicker than the pedicels (the latter less than 0.5 mm thick, at least in the middle)
      3. Petals less than 4 mm long; flowers and buds overtopped by ends of uppermost young fruit; seeds ca. 1.2 mm long . . . . 3. **S. irio**
      3. Petals ca. 4.5–6 [7] mm long; flowers and buds barely if at all surpassed by tips of fruit; seeds less than 1 mm long . . . . 4. **S. loeselii**

1. **S. officinale** (L.) Scop. Hedge Mustard

Map 348. A Eurasian weed, naturalized in North America. For a plant known in Michigan as long ago as 1837 (collected by the First Survey on a dry roadside at Ann Arbor) and as widespread in the state, it has actually been collected remarkably few times. Roadsides, fields, railroads, vacant lots, farmyards, disturbed woods and banks.

Plants with the fruit and axis of the inflorescence glabrous [var. *leiocarpum* DC.] are at least as common as the typical var. *officinale,* which is pubescent in these parts.

2. **S. altissimum** L. Fig. 136 Tumble Mustard

Map 349. A native of Eurasia, widespread (as are so many other crucifers) as a weed. The earliest Michigan specimen seen is from Berrien County in 1896, but collections were made in Kent and St. Clair counties the following year and the species is now throughout on sandy or gravelly roadsides, shores, fields, railroads, gravel pits; spreading into woods.

A few of our specimens have sparsely hispid fruit [f. *ucrainicum* (Blonski) Thell.]. This is an easily recognized species, with its dissected-pinnatifid leaves (the upper ones often resembling a *Myriophyllum* leaf) and elongate linear siliques on thick pedicels, so that the fruit and pedicel appear to be one at first glance. The petals are a paler yellow than the bright ones of our other species.

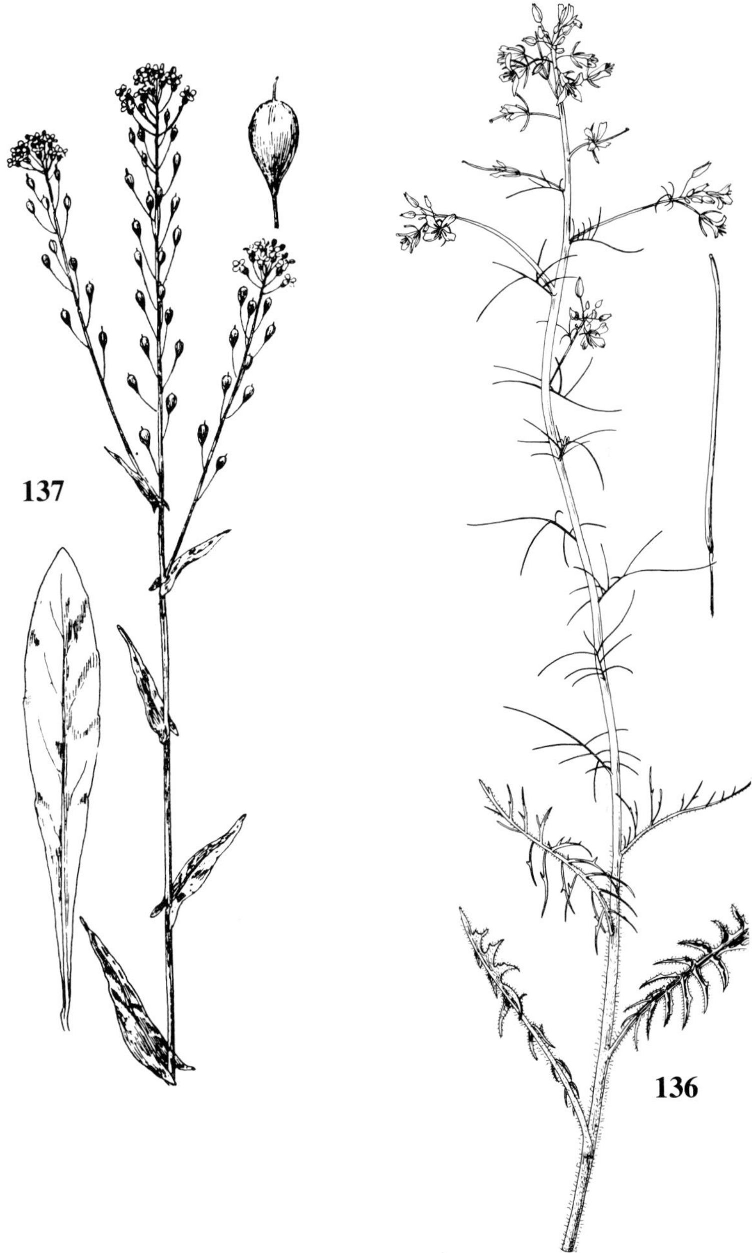

136. *Sisymbrium altissimum* $\times {}^{1}/_{2}$;
fruit $\times {}^{1}/_{2}$
137. *Camelina microcarpa* $\times {}^{1}/_{3}$;
fruit $\times 2$

3. **S. irio** L. London Rocket

Map 350. Another European native, but only rarely found as a weed with us. Collected by Farwell (*7730,* MICH) in waste ground at Detroit in 1926 (Am. Midl. Nat. 10: 212. 1927).

Supposed differences in anther length between this species and the next do not seem to hold up in our material.

4. **S. loeselii** L.

Map 351. A Eurasian species resembling the preceding and found with it by Farwell (*7731,* MICH, BLH) at Detroit in 1926. Also collected in 1900 by Emma Cole (MICH, GH) in a meadow along a road near Grand Rapids (and reported by her as *S. irio*).

10. **Diplotaxis** Wall Rocket

KEY TO THE SPECIES

1. Sepals (2.5) 3–4 mm long; petals 5–7.5 mm long; stem ± soft at base (plant annual or perennial), with most or all leaves in a basal rosette; leaves usually lyrate, sinuate-dentate, or sparsely toothed . . . . 1. **D. muralis**
1. Sepals (4) 5–7.5 mm long; petals 6.5–9.5 (10.5) mm long; stem hard and woody at base (perennial), leafy to the inflorescence, without a basal rosette; leaves all deeply pinnatifid . . . . 2. **D. tenuifolia**

1. **D. muralis** (L.) DC. Fig. 132

Map 352. A European native, found as a weed of railroads, roadsides, gravel pits, sidewalk cracks, lawns, and waste places. First collected in Michigan in the 1890's, but only quite recently becoming frequent in sandy or gravelly disturbed ground, especially in the northern Lower Peninsula.

Normally a slender annual with basal rosette, but sometimes close to the next species in habit. The stem extends at an angle from the ground, rather than being strictly erect. Supposed distinctions from *D. tenuifolia* based on the stipe of the ovary and fruit (ca. 0.5–1.5 mm, supposedly only in *D. tenuifolia*) do not hold up with our material—as also noted on Manitoulin Island by Morton and Venn (1984).

2. **D. tenuifolia** (L.) DC. Fig. 133

Map 353. Like the preceding, a native of Europe, found locally as a weed of roadsides, railroads, disturbed ground, sandy shores; apparently especially well established in Huron County. First collected in Michigan in 1916 at Port Huron.

A perennial, several-branched from the woody base, usually distinguished easily from the preceding by its larger, paler flowers, leafy stem, and deeply pinnatifid leaves with the narrow lateral segments about the same width as the rachis.

11. Neslia

1. **N. paniculata** (L.) Desv. Fig. 130. Ball Mustard

Map 354. A Eurasian species, widespread as a weed and known from scattered locations in North America. Collected in Washtenaw County in 1918 as a weed of grain fields, and on banks at Eloise in Wayne County in 1931. Specimens to support other reports from the state have not been found.

12. Camelina

False Flax

KEY TO THE SPECIES

1. Lower internodes and leaves glabrous or with small stellate hairs (rarely an occasional long hair surpassing them); body of mature fruit ca. (6.5) 7–9 (11) mm long, ca. 4–7 (9) times as long as the beak . . . . 1. **C. sativa**
1. Lower internodes and leaves with long simple or somewhat branched hairs surpassing the small stellate hairs; body of mature fruit 5–7 (7.5) mm long, ca. 2–3.7 times as long as the beak . . . . 2. **C. microcarpa**

1. **C. sativa** (L.) Crantz

Map 355. A Eurasian weed, rarely found in our area. Collected by Dodge in 1912 near a grain elevator at Point Edward, Ontario (across the river from

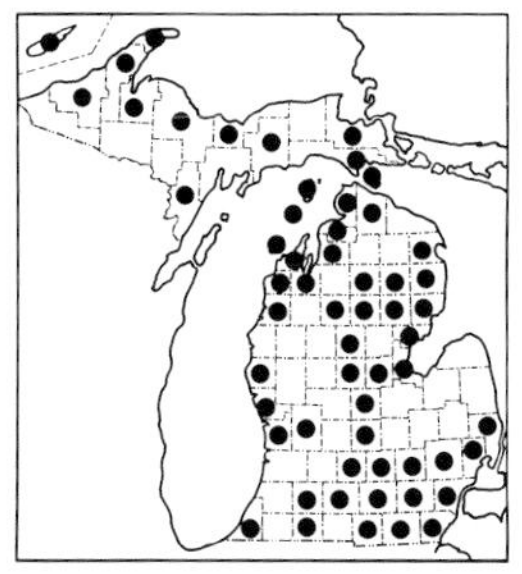
349. Sisymbrium altissimum

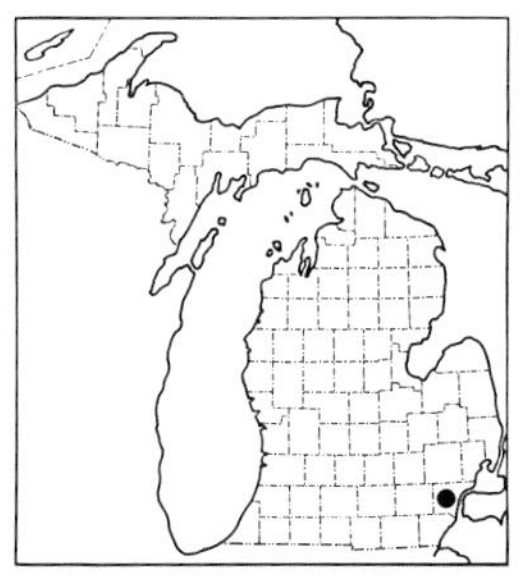
350. Sisymbrium irio

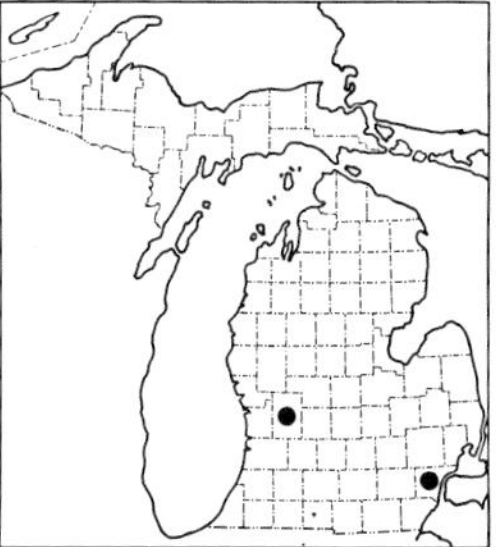
351. Sisymbrium loeselii

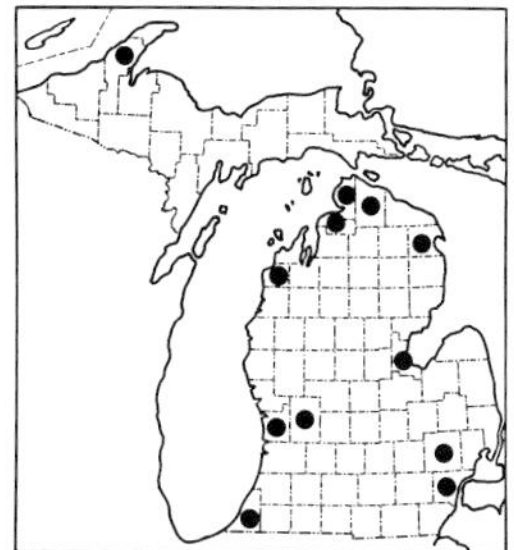
352. Diplotaxis muralis

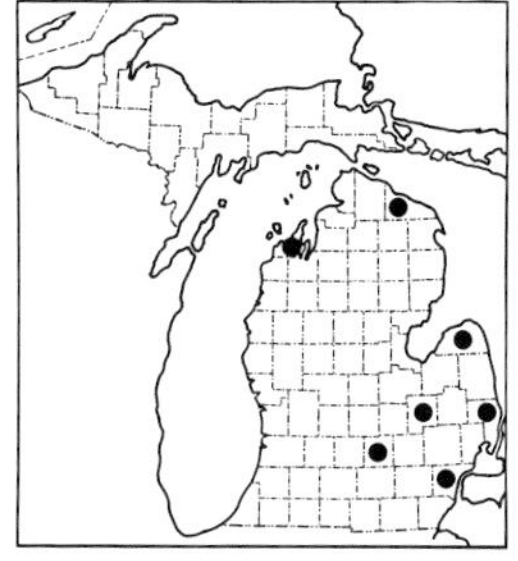
353. Diplotaxis tenuifolia

354. Neslia paniculata

Port Huron), but old Michigan collections labeled as this species are almost invariably the next. *C. sativa* was collected by Dennis Cooley in 1843, locality not stated but presumably in Macomb or Oakland county; by Farwell in 1884 in Keweenaw County; by Rogers in 1951 in an old cultivated field in Monroe County; and (identification perhaps doubtful) by Voss in 1957 along a roadside in Antrim County.

2. **C. microcarpa** DC. Fig. 137

Map 356. Another Eurasian weed, now a frequent plant of roadsides, fields, railroads, and other disturbed, usually sandy ground. Evidently in the state only since about 1890.

Manuals published before 1900 did not mention this species, and most old collections and reports of *C. sativa* in Michigan should be referred to *C. microcarpa* (see above). Some specimens are difficult to place; those lacking both basal parts and mature fruits can usually not be safely identified.

Both species have a very distinctive obovoid or pear-shaped fruit, often slightly compressed, with a prominent narrow-winged margin along the suture, the style persisting as a slender beak.

13. **Conringia**

1. **C. orientalis** (L.) Dumort. Fig. 138 Hare's-ear Mustard

Map 357. A Eurasian species, occasionally found in waste places and disturbed ground in North America. Michigan collections are from railroads, cultivated fields and gardens, vacant lots.

The petals may be so pale as to appear white, although they are usually said to be yellow.

14. **Erysimum**

REFERENCE

Rossbach, George B. 1958. The Genus Erysimum (Cruciferae) in North America North of Mexico—A Key to the Species. Madroño 14: 261–267.

KEY TO THE SPECIES

1. Petals ca. 18–24 mm long, (4) 6–9 mm broad .... 1. **E. capitatum**
1. Petals much smaller (less than 11 mm long and 4 mm broad)
   2. Sepals 2.2–3.2 mm long; petals 3.5–5 (5.5) mm long; fruit 1.5–2.5 cm long, ca. 2–3 times as long as the distinctly more slender pedicels; hairs on upper surface of leaves mostly 3- and 4-pronged .... 2. **E. cheiranthoides**
   2. Sepals (4) 4.5–6.5 (7.5) mm long; petals (6) 7–9 (10) mm long; fruit 2.5–8 cm long, ca. (5) 6–20 or more times as long as the thick pedicels (nearly as thick as the fruit); hairs on upper surface of leaves various

3. Fruit ca. 5–8 cm long, widely spreading; leaves remotely denticulate or shallowly sinuate, without 4-pronged hairs (all 2- and 3-pronged); plant annual, usually many-branched . . . . 3. **E. repandum**

3. Fruit 2.5–5 cm long, strongly ascending or erect; leaves entire and without 4-pronged hairs *or* toothed and with 4-pronged hairs above; plant biennial or perennial, usually stiffly erect and unbranched

4. Cauline leaves entire, essentially linear, up to 5 mm broad, without 4-pronged hairs (all 2- and 3-pronged) . . . . 4. **E. inconspicuum**

4. Cauline leaves remotely toothed or denticulate, elliptic-lanceolate, often more than 5 mm broad, with both 3- and 4-pronged hairs above . . . . 5. **E. hieraciifolium**

1. **E. capitatum** (Douglas) Greene Western Wallflower

Map 358. Collected in 1900 by Emma Cole in a meadow along a road near Grand Rapids. Also collected as an "accidental" in a nursery on Mackinac Island (*W. H. Manning* n.d. [1915?], GH). In both instances, undoubtedly introduced or a waif from the native range south and west of Michigan.

Although the Grand Rapids collection is not fruiting, it is apparently this species, sometimes called *E. arkansanum* Nutt. and sometimes misidentified as *E. asperum* (Nutt.) DC., under which name Miss Cole reported it. The foliage is densely covered with 2-pronged hairs.

2. **E. cheiranthoides** L. Fig. 139 Wormseed Mustard

Map 359. A species usually, though not always, considered introduced

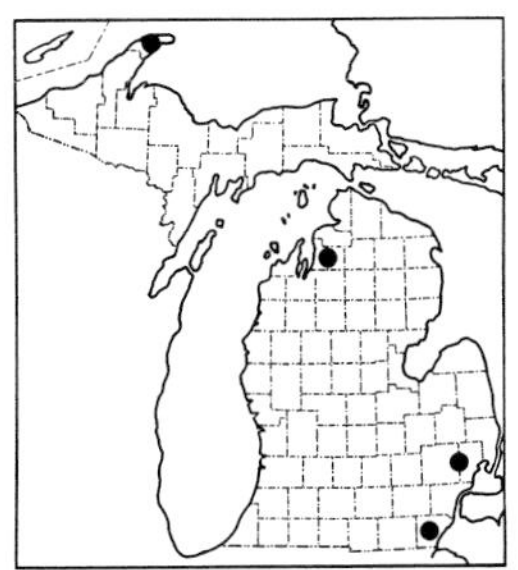
355. Camelina sativa

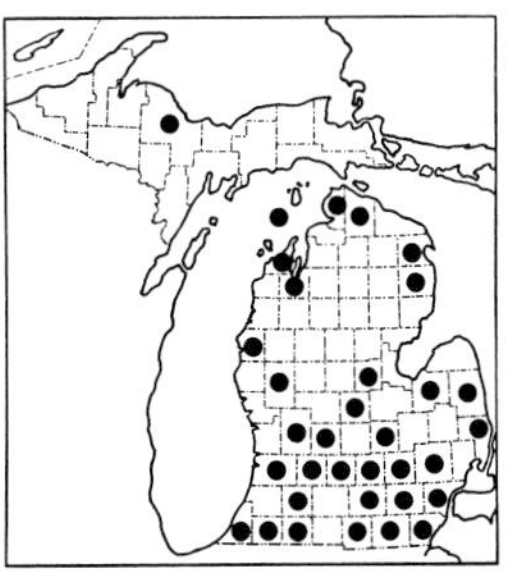
356. Camelina microcarpa

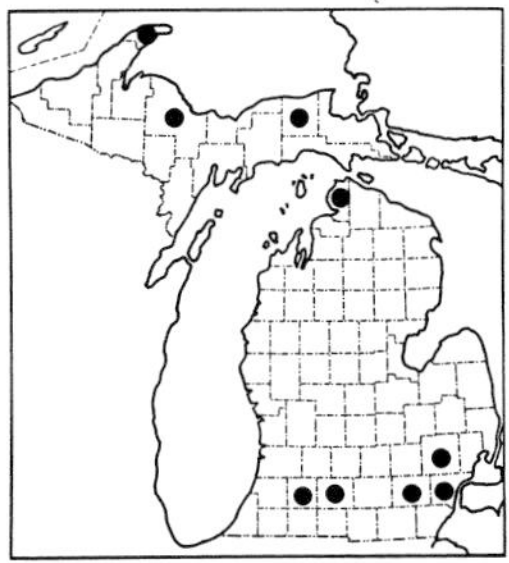
357. Conringia orientalis

358. Erysimum capitatum

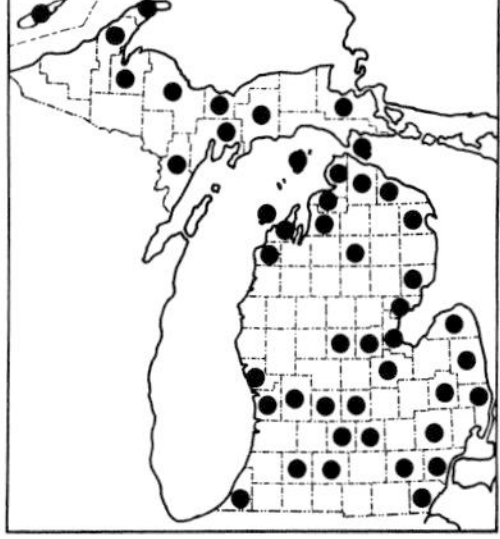
359. Erysimum cheiranthoides

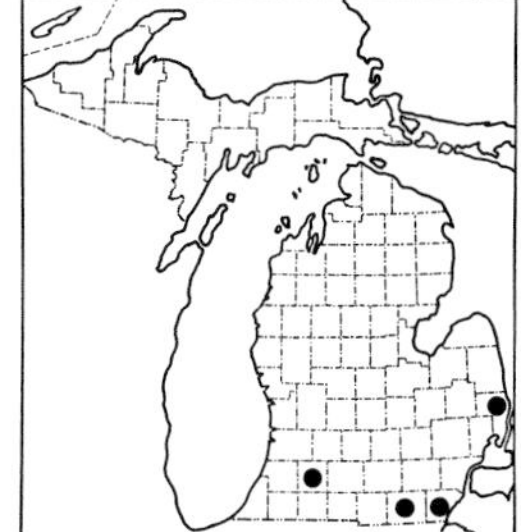
360. Erysimum repandum

in this part of North America, found in natural as well as human-disturbed ground in both hemispheres. The earliest collections seen from the state date from the 1880's (Crawford, Ingham, and Keweenaw counties). Roadsides, railroads, fields, gardens, and waste places; river banks, thickets, shores, marshy and muddy ground.

A species of quite distinctive aspect, even the young fruit generally more erect (at times like little candles) than the strongly divergent and very slender pedicels. This is our only common yellow-flowered crucifer with entire leaves and the fruit a silique.

3. **E. repandum** L. Treacle Mustard

Map 360. A native of the Mediterranean region, widely distributed as a weed but rather local in North America. The earliest collections seen from the state are from St. Clair County in 1900 and from Lenawee County in 1901—where it continues to be a weed. Fields, railroads, and other waste places.

4. **E. inconspicuum** (Watson) MacM.

Map 361. Native in western North America, occasionally adventive eastward. No earlier Michigan collections have been seen than from Keweenaw County in 1895. Roadsides, railroads, dry fields, borders of woods, river banks, rock crevices (Isle Royale).

5. **E. hieraciifolium** L.

Map 362. A European species, spreading in Ontario, but the only Michigan collections thus far seen are from Kalamazoo County in 1937 and 1939 (reported by Hanes as *E. inconspicuum*) and from Keweenaw County (*Farwell 844* in 1895, BLH).

15. **Alyssum**

KEY TO THE SPECIES

1. Plant a stiffly erect little taprooted annual, a common weed in waste ground, without basal leaves; petals pale yellow (or even white) .... 1. **A. alyssoides**
1. Plant a matted to erect perennial woody at base, rarely escaped from gardens, with basal leaves; petals bright yellow
   2. Basal leaves smaller than cauline leaves, but usually withering before flowering; silicles stellate-pubescent, with 1 seed per locule; petals entire or barely notched .... 2. **A. murale**
   2. Basal leaves much larger than the cauline leaves, persistent; silicles glabrous, with 2 seeds per locule; petals notched or bilobed .... 3. **A. saxatile**

1. **A. alyssoides** (L.) L. Fig. 143 Pale Alyssum

Map 363. A plant of Europe and Asia Minor, with us characteristically a weed of railroad embankments, but also common on roadsides, parking lots, gravel pits, and other sandy bare places.

This little densely stellate-pubescent annual is likely to be confused only with *Lepidium,* in which, however, the fruit is flattened at right angles to the septum rather than parallel to it. The petals are mostly ca. 3–4 mm long and pale yellow. Plants with white flowers hardly merit recognition, but were named f. *albineum* by Farwell (TL: Shelbyville [Allegan Co.]).

2. **A. murale** Waldst. & Kit. Yellowtuft

Map 364. A European species rarely escaped from cultivation to roadsides and such places.

3. **A. saxatile** L. Goldentuft

Map 365. Another garden plant of European origin, cultivated in rock gardens and rarely escaped. Collected on foredunes near Glen Arbor in 1982 (*S. Reznicek,* MICH).

The flowers are brighter yellow and slightly larger than in *A. murale.* Now usually separated from *Alyssum* as *Aurinia saxatilis* (L.) Desv.

16. Dentaria Toothwort

Some authors include this genus in *Cardamine,* as a distinct subgenus, but its habit is quite different and tradition is here followed in keeping the genus separate. This is our only genus of Cruciferae with the leaves *palmately* compound (or deeply cleft). The relatively large white flowers and few leaves usually opposite or whorled or nearly so also help recognition. The fruit frequently does not mature.

REFERENCE

Montgomery, F. H. 1955. Preliminary Studies in the Genus Dentaria in Eastern North America. Rhodora 57: 161–173.

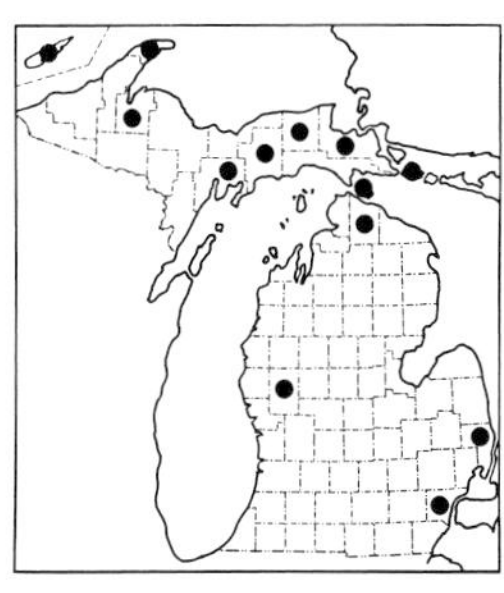

361. Erysimum inconspicuum

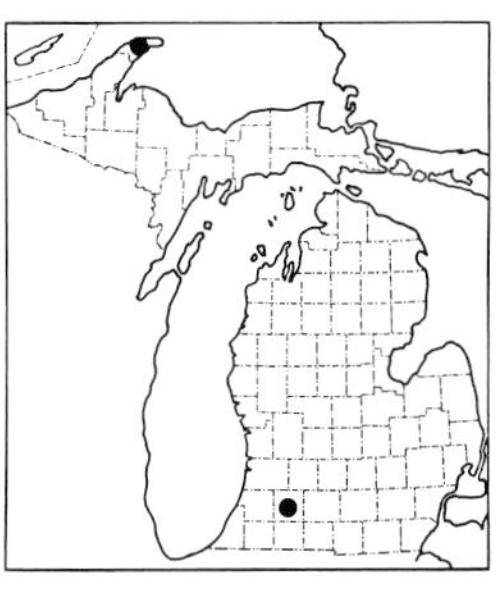

362. Erysimum hieraciifolium

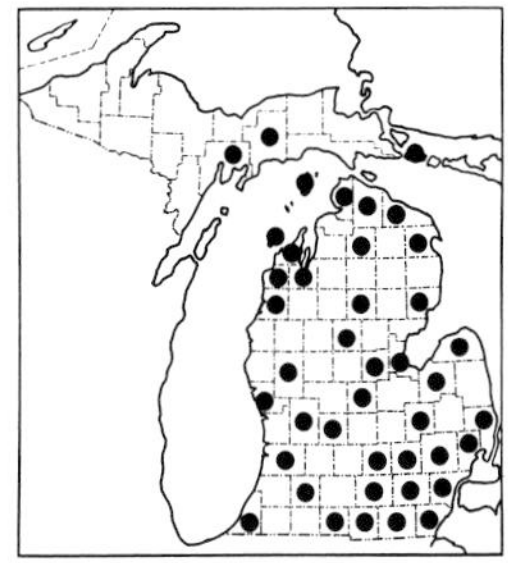

363. Alyssum alyssoides

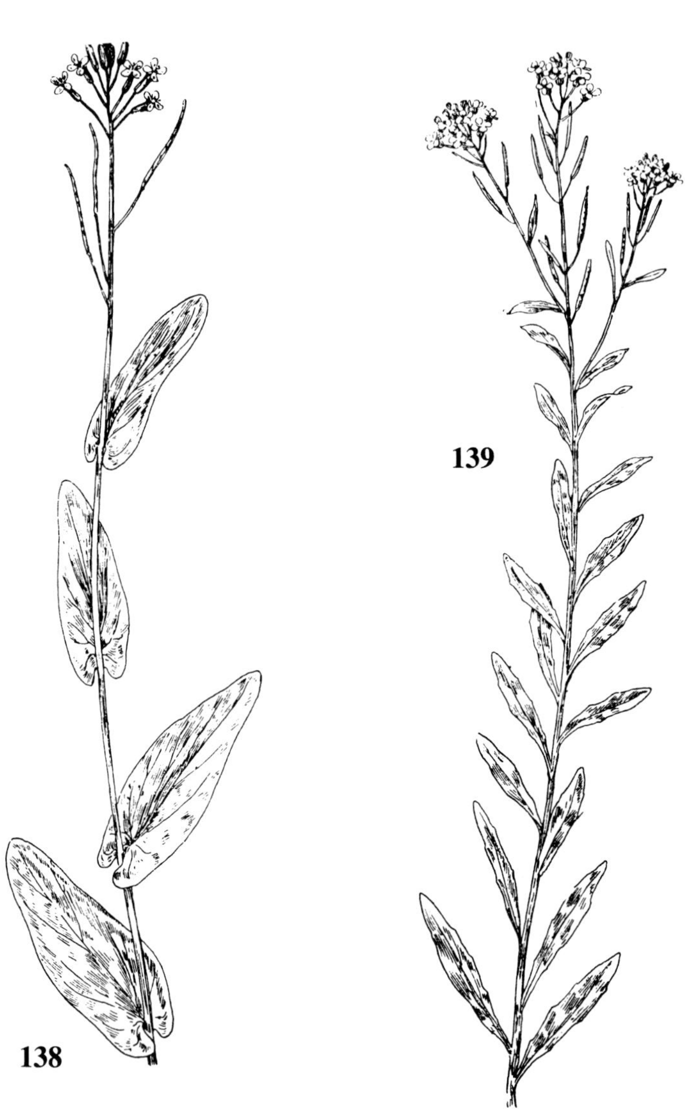

138. *Conringia orientalis* $\times {}^{1}/_{2}$

139. *Erysimum cheiranthoides* $\times {}^{1}/_{2}$

KEY TO THE SPECIES

1. Each leaf completely or deeply cleft into (3) 4–7 lanceolate or narrowly elliptic lobes; rhizome constricted at frequent intervals into easily separable fusiform obscurely toothed segments; peduncle and rachis of inflorescence minutely pubescent . . . . 1. **D. laciniata**
1. Each leaf consisting of 3 ovate leaflets; rhizome continuous and strongly toothed (somewhat constricted in the very rare *D. maxima*); peduncle and rachis glabrous or nearly so (except in *D. maxima*)
   2. Margins of leaves distinctly ciliolate with stiff hairs ca. 0.2 mm long; rhizome with constrictions (but not the series of fusiform tubers of *D. laciniata*); peduncle and rachis usually at least sparsely pubescent; cauline leaves usually 3–4, alternate . . . . 2. **D. maxima**
   2. Margins of leaves scarcely more than scabrous ("hairs" less than 0.2 mm long); rhizome of essentially uniform diameter (except for the prominent teeth); peduncle and rachis glabrous; cauline leaves usually 2, opposite or nearly so . . . . 3. **D. diphylla**

1. **D. laciniata** Willd. — Cut-leaved Toothwort

Map 366. Usually in rich beech-maple forests and northern hardwoods, very rarely in drier oak woods. Quite local (and often sterile) in the northern Lower Peninsula and eastern Upper Peninsula.

Blooms earlier than *D. diphylla,* although there may be some overlap. The segments of the leaves are usually ± sharply toothed, but occasionally entire or nearly so. Ordinarily there are 3 leaves, whorled or nearly so. Names (TL: Plymouth [Wayne Co.]) were bestowed by Farwell on some of the trivial variants in this species: var. *latifolia* with broader leaflets than usual; var. *opposita* with 2 leaves, opposite; and var. *alterna* with the leaves alternate. A colony of striking plants with widely alternate leaves (usually 4), broader than usual and jagged-toothed, was found in Ontonagon County (*Voss 6197* in 1958, MICH); the rhizome is typical of *D. laciniata*.

The rhizome segments in this species may be as thick as 1–2 cm although usually they are much smaller; they are whitish or yellowish when fresh (more brown when dry). They are edible, with characteristic radish-like taste; carbonized remains have been recovered from a prehistoric Indian archeological site on Bois Blanc Island.

364. Alyssum murale

365. Alyssum saxatile

366. Dentaria laciniata

2. **D. maxima** Nutt.

Map 367. Collected at the Iron River on Lake Superior, Ontonagon County, in 1868 (*Gillman,* NY, GH); in rich woods along the Black River in St. Clair County (*Dodge* in 1894 and 1913, MICH); and in rich deciduous woods in Clayton Tp., Arenac County (*Voss 3507* in 1957, MICH; *15173* in 1980, MICH, MSC, BLH, UMBS, GH, NY).

Our other two species of *Dentaria* are common and variable, although quite distinct in their normal appearance. *D. maxima* has an aspect strongly suggesting that it is of hybrid origin. In New York state, hybrids of *D. maxima* with both the other species have been reported, and some authors have suggested that a spectrum of intermediate forms may all represent hybrids (including backcrosses). On May 16, 1980, the Arenac County population was at the peak of blooming, the flowers distinctly pink (color intensifying upon drying), slightly smaller than the pure white flowers of adjacent *D. diphylla,* although most of the latter was still in bud. No *D. laciniata* was seen in the area.

3. **D. diphylla** Michaux Fig. 140 Two-leaved Toothwort

Map 368. Deciduous and mixed woods, especially in rather moist or seepy areas; cedar swamps.

The cauline leaves are usually nearly opposite (at most 1–2 cm distant), rarely more evenly alternate (as in the preceding species). The distinctive white strongly toothed rhizome is ample by itself for recognition of this species.

17. **Armoracia**

The fruit in both of our species is plump, ellipsoid, slightly longer than broad.

REFERENCE

La Rue, Carl D. 1943. Regeneration in Radicula aquatica. Pap. Mich. Acad. 28: 51–61.

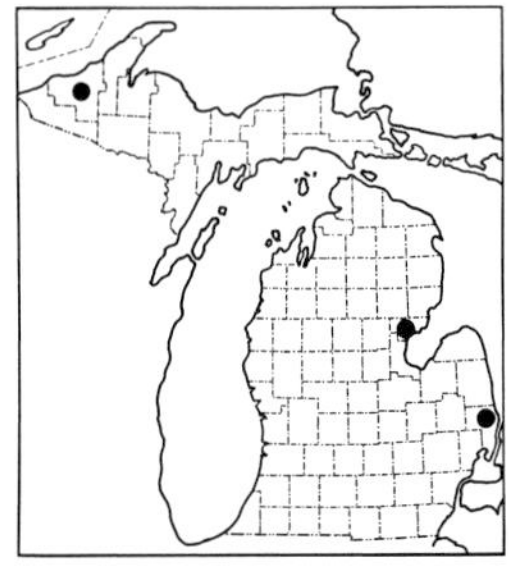
367. Dentaria maxima

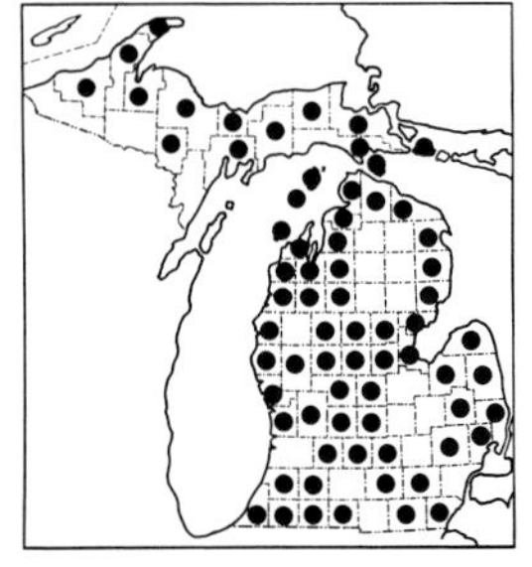
368. Dentaria diphylla

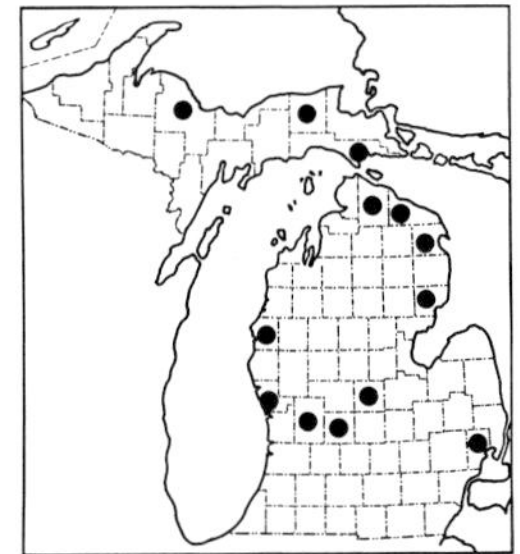
369. Armoracia aquatica

140. *Dentaria diphylla* (including rhizome) $\times ^1/_2$
141. *Armoracia aquatica*, submersed leaves $\times 2$
142. *Cardamine pensylvanica* $\times ^1/_2$

KEY TO THE SPECIES

1. Plant a lax native aquatic with ± finely pinnately dissected underwater leaves (simple toothed above-water leaves beneath the flowers), from a rhizome; style forming a beak ca. half as long as the body of the fruit . . . . 1. **A. aquatica**
1. Plant a coarse introduced escape from cultivation, with leaves mostly simple (a few lower ones sometimes once-pinnatifid), from a thick taproot; style forming a very short beak (ca. 0.5 mm or less) . . . . 2. **A. rusticana**

1. **A. aquatica** (Eaton) Wieg. Fig. 141 Lake Cress

Map 369. Rivers and lakes, especially in cold spring-fed waters.

A remarkable aquatic plant, rarely seen except by those who know where to seek it or who recognize sterile drifting fragments. The dissected submersed leaves fall off readily when mature; it is a unique experience to attempt to remove a nice-looking specimen of this species from the water and have the surface film strip off all the leaves as the stem is lifted! These leaves serve as vegetative propagules (although of course most of them never are lodged in a favorable site), growing leaves and roots for new plants from their base. The first leaves formed at the base of detached petioles are simple. The plant can also regenerate from stem fragments. The alternate leaves will readily distinguish this species from *Ceratophyllum, Megalodonta,* and our common species of *Myriophyllum*; the presence of a central axis will distinguish the leaves from those of aquatic *Ranunculus* and *Utricularia. Proserpinaca* and some rare Myriophyllums have alternate leaves with a central axis, but in these the lateral segments of the leaf are not again divided as they are in *Armoracia aquatica.*

This species has been variously placed in *Radicula* (an illegitimate synonym of *Rorippa*), *Rorippa,* and *Neobeckia.* It was also once known as *Nasturtium lacustre* A. Gray.

2. **A. rusticana** Gaertn., Mey., & Scherb. Horseradish

Map 370. A native of Europe, escaped from cultivation to roadsides and waste ground, especially wet ditches and shores.

Cultivated for a well known condiment prepared from the roots. The large lower leaves may be 2.5–3 dm broad and resemble those of *Silphium terebinthinaceum.* Often called *A. lapathifolia* Gilib., a name not validly published.

18. **Cardamine** Bitter Cress

Dentaria is sometimes treated as a subgenus of *Cardamine.*

REFERENCES

Hart, Thomas W., & W. Hardy Eshbaugh. 1976. The Biosystematics of Cardamine bulbosa (Muhl.) B.S.P. and C. douglassii Britt. Rhodora 78: 329–419.

Stuckey, Ronald L. 1962. Characteristics and Distribution of the Spring Cresses, Cardamine bulbosa and C. douglassii, in Michigan. Mich. Bot. 1: 27–34.

KEY TO THE SPECIES

1. Cauline leaves all simple, unlobed (margins entire to irregularly toothed or sinuate); plant from a prominent basal tuber

2. Upper part of stem and inflorescence with sparse to dense spreading hairs at least 0.25 mm long (occasionally glabrous); lower part of stem glabrous or with some kind of hairs (rarely entire stem glabrous); petals pink to purple (rarely white); sepals ± flushed with purple (becoming brown after long dry) . . . . 1. **C. douglassii**

2. Upper part of stem and inflorescence glabrous; lower part of stem glabrous or usually with minute appressed to incurved hairs; petals white; sepals bright green when fresh, becoming yellowish, rarely with a slight flush of purple at tips only . . . . 2. **C. bulbosa**

1. Cauline leaves deeply pinnately lobed or compound; plant without a basal tuber

3. Leaves with prominent lanceolate ciliate auricles at the base; petals none [or shorter than sepals] . . . . 3. **C. impatiens**

3. Leaves without auricles; petals longer than sepals

4. Petals (8) 9–14 mm long; leaflets of basal leaves nearly round; stems glabrous, unbranched . . . . 4. **C. pratensis**

4. Petals ca. 2–3.5 (4.5) mm long; leaflets of basal leaves usually distinctly longer than broad; stems glabrous or sometimes pubescent, usually ± branched

5. Petioles of cauline leaves pubescent; leaf blades and often stem up to the inflorescence also ± hispidulous

6. Stems ± hispidulous up to the inflorescence, usually flexuous and branched above; stamens 6 . . . . 5. **C. flexuosa**

6. Stem glabrous nearly or quite to the base, straight and unbranched above; stamens 4 . . . . 6. **C. hirsuta**

5. Petioles glabrous; leaves and stem usually glabrous (stem sometimes hispidulous, especially toward base)

7. Cauline leaves with narrowly linear or oblanceolate segments forming distinct leaflets not confluent with the rachis, the terminal leaflet scarcely if at all broader than lateral leaflets; stem and leaves glabrous . . . . 7. **C. parviflora**

7. Cauline leaves with broader (elliptical to obovate, often somewhat toothed) segments confluent with the winged rachis (the leaf thus deeply pinnatifid rather than compound), the terminal segment usually distinctly broader than the lateral segments; stem (especially toward the base) and margins of leaves occasionally hispidulous . . . . 8. **C. pensylvanica**

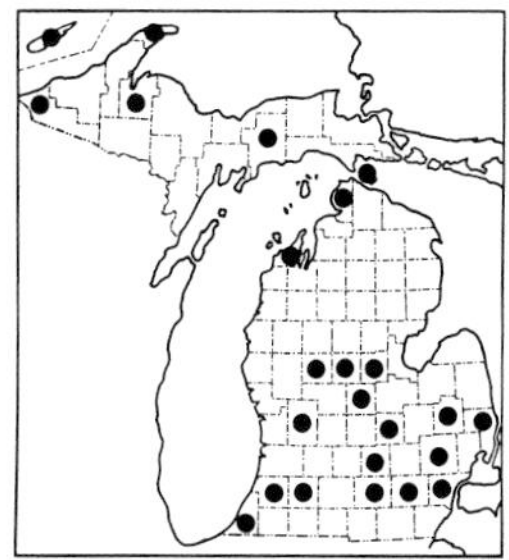

370. Armoracia rusticana

371. Cardamine douglassii

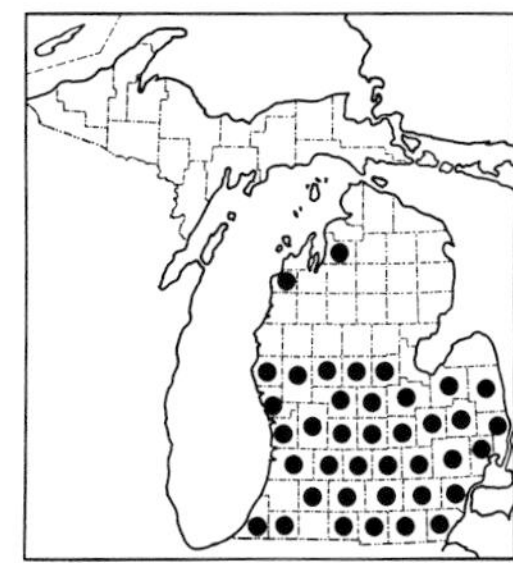

372. Cardamine bulbosa

1. **C. douglassii** Britton — Pink Spring Cress

Map 371. Low rich deciduous woods, floodplains, often in more dense shade than the next species and in not quite as wet spots.

This species is named for David Bates Douglass, who accompanied the exploring expedition under Lewis Cass in 1820 to the Upper Great Lakes. (The type material apparently came from Buffalo, however.)

2. **C. bulbosa** (Muhl.) BSP. — Spring Cress

Map 372. Wet hollows and streamsides in deciduous woods (occasionally even with cedar or tamarack), swamps, springy places, sometimes in wet ditches.

This species can generally be distinguished rather easily from the preceding by the characters in the key, although no single one of them is fully reliable and occasional ambiguous specimens occur. Hybridization between the two appears to be rare. In the same place and season, *C. douglassii* blooms ca. 1–4 weeks earlier than *C. bulbosa,* usually beginning in mid- or late April, whereas *C. bulbosa* usually does not begin until mid- or even late May. *C. douglassii* tends to be a shorter plant (especially measured to the lowest pedicel), is less often branched, and grows on the average in slightly drier places although the two may grow together.

3. **C. impatiens** L.

Map 373. A native of Eurasia, recently found established in Ontario and Michigan (longer known from farther east). Thus far found only on banks and thicket margins at several places around Ann Arbor, where first collected in 1978 (*Reznicek 4948,* MICH, MSC), but doubtless becoming more widespread.

4. **C. pratensis** L. — Cuckoo-flower

Map 374. Wet bogs and marshy ground, borders of woodland pools and creeks, tamarack and cedar swamps.

An attractive spring-flowering plant found around the world in northern latitudes. Native plants of Canada and the northeastern United States have been placed in var. *palustris* Wimmer & Grab. It is possible that a few of our records represent escapes from cultivation.

5. **C. flexuosa** With.

Map 375. A European species, sparingly introduced in North America. First collected in Michigan in 1892, on Belle Isle, Wayne County (*Farwell 1318,* BLH); not much found again until the late 1940's and 1950's. Moist roadsides and sandy shores, muddy creek banks and shores, flowering as late as early December (in 1981) along a sidewalk next to a building in Ann Arbor.

6. **C. hirsuta** L.

Map 376. Another little Eurasian introduction, first collected in Michigan in 1976 in Berrien County, and now a lawn, driveway, trailside, and garden weed in that county and Saginaw County—and doubtless overlooked elsewhere.

The petals in our material of both this species and the preceding one clearly exceed the sepals, unlike some descriptions of the species.

7. **C. parviflora** L.

Map 377. A variable circumpolar species, American plants sometimes separated as var. *arenicola* (Britton) Schulz. Rarely found except on rocky shores and ridges, limestone pavements, and "mountain" summits.

Except for one collection from a disturbed roadside on Drummond Island, which is 4.5 dm tall, our plants are not over 2.5 dm tall and often are as short as 1 dm; in the next species, they are often taller.

8. **C. pensylvanica** Willd. Fig. 142

Map 378. Low wet areas: swamps, creek margins, low wet woods, seepy shores and banks, cedar swamps, rocky to sandy shores, often in somewhat disturbed areas such as ditches or trails in damp woods, rarely in dry ground.

Although the stems are often described as pubescent, at least on plants

373. Cardamine impatiens

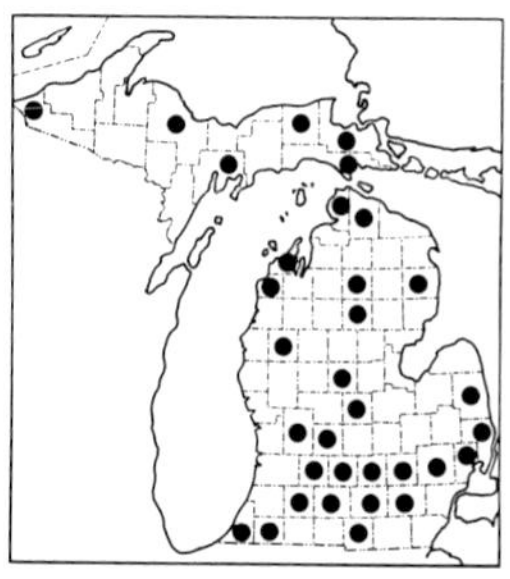

374. Cardamine pratensis

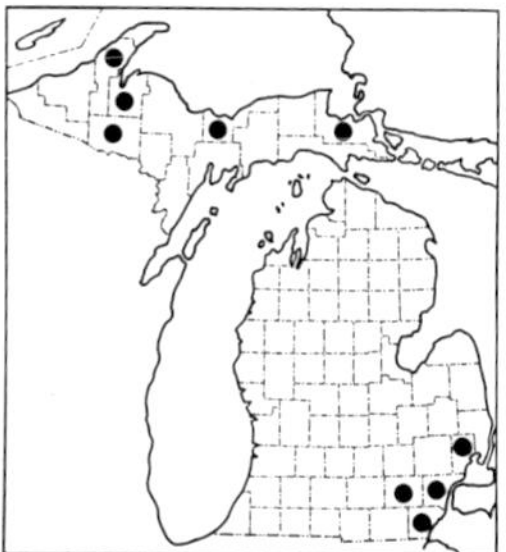

375. Cardamine flexuosa

376. Cardamine hirsuta

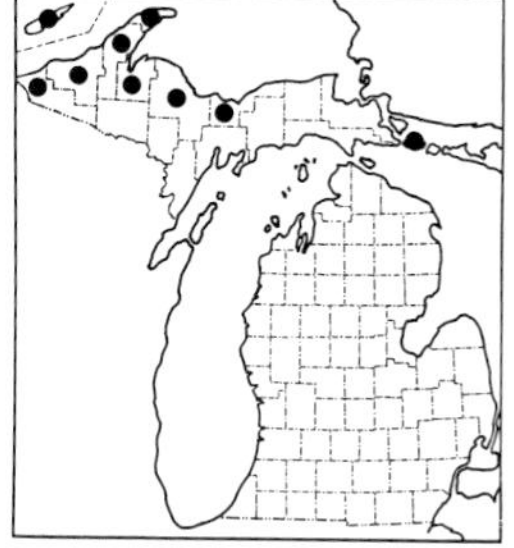

377. Cardamine parviflora

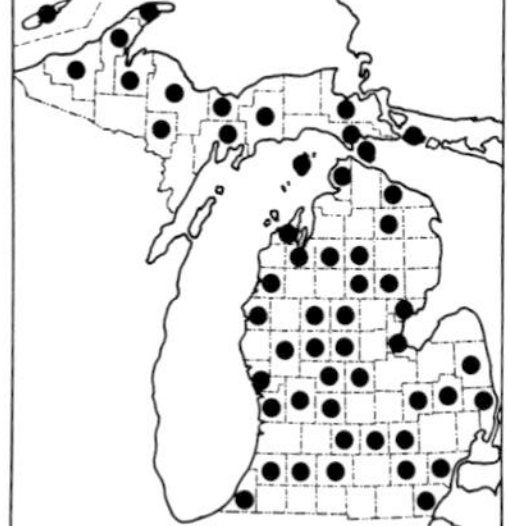

378. Cardamine pensylvanica

not growing in water, the great majority of our specimens have glabrous stems. Small plants in drier places are not always easily distinguished from the preceding species. A very few specimens of *C. pensylvanica* have been seen with sparsely ciliate petioles on the cauline leaves, but the plants are so tall and the leaf segments so broad that they can be readily separated from *C. flexuosa* or *C. hirsuta*. Specimens are occasionally misidentified as *Nasturtium officinale* (see next genus).

19. **Nasturtium**

This genus is often united with *Rorippa,* in which case the name of our common species becomes *Rorippa nasturtium-aquaticum* (L.) Hayek.

REFERENCES

Green, Peter S. 1962. Watercress in the New World. Rhodora 64: 32–43.
Rollins, Reed C. 1978. Watercress in Florida. Rhodora 80: 147–153.

1. **N. officinale** R. Br. Fig. 144 Watercress

Map 379. Margins of rivers and streams; ditches; seepy places and brooks in woods and cedar swamps—especially in cold spring-fed waters; also at home in sedgy tamarack swamps.

This is usually said to be a species native to the Old World, but thoroughly naturalized in North America. However, some authors have thought it native here, on the grounds that it is found in remote places associated with native plants (so are hawkweeds!). The earliest botanical explorers in our area did not collect or list watercress, there apparently being no published reports from the state before the 1870's. The oldest collection seen was from Ann Arbor in 1859. I believe we would therefore be unjustified in assuming that merely because the species is so well established it is necessarily indigenous. It seems unlikely that explorers by canoe, such as Douglass Houghton in the 1830's, would have overlooked so notable an aquatic plant. Furthermore, old manuals note that it was just being cultivated in the early and mid-19th century.

Watercress is a well known edible species, the foliage used in salads, mixed juices, soups, casseroles, and as a garnish.

In addition to the questions raised about its generic assignment and naturalized status, this poor plant also suffers a split personality taxonomically. It consists of two taxa, for besides typical *N. officinale* there can be recognized var. *longisiliquum* (Irmisch) Thell., often also treated as *N. microphyllum* Reichb. Typical *N. officinale* is a diploid ($2n = 32$) and *N. microphyllum* a tetraploid ($2n = 64$). A sterile triploid hybrid between them is common, readily propagated, as are the parents, vegetatively. Apart from the microscopical characters, the diploid has slightly shorter (often less than

15 mm) and thicker fruits with seeds in 2 rows in each locule, while in the tetraploid the fruits tend to be longer and more slender, with seeds in a single row. Almost all our material lacks ripe seeds, although in some they are mature enough to see a useful character of the surface: in the diploid the reticulations form fewer (less than 50) polygonal depressions on a side (fig. 144), while in the tetraploid there are about 100 or more such pits. Both taxa are extremely variable vegetatively. According to Green (1962) we have both (as well as the hybrid) in the state, and it appears from specimens that both may be widespread. However, since the great majority of our specimens cannot be assigned with confidence to one or the other, a single taxon is here mapped—following the example of several other local floras.

Specimens of *Cardamine pensylvanica* growing in water are sometimes misidentified as *Nasturtium,* for the foliage is quite similar, although the former does not make the large massed beds so characteristic of the latter. Unless the plants are completely sterile, such specimens of *Cardamine* can usually be identified by the absence of adventitious roots along the stem (at most they are at the base) and their ascending pedicels bearing straight slender fruit with smooth (not reticulated) seeds clearly in a single row; the valves of the fruit are flatter than in *Nasturtium,* and curl after discharge of the seeds.

20. **Capsella**

1. **C. bursa-pastoris** (L.) Medicus Fig. 145 Shepherd's-purse

Map 380. A widespread Eurasian weed, apparently established in Michigan by the middle of the 19th century, and reported as early as 1838 (First Survey). Roadsides, fields, gardens, lawns, and waste ground generally. In sheltered or warm places in lawns or along buildings, this is one of the first plants to bloom in the spring.

This is a variable species, but none of our specimens appear to belong to any of the other generally recognized species in the genus.

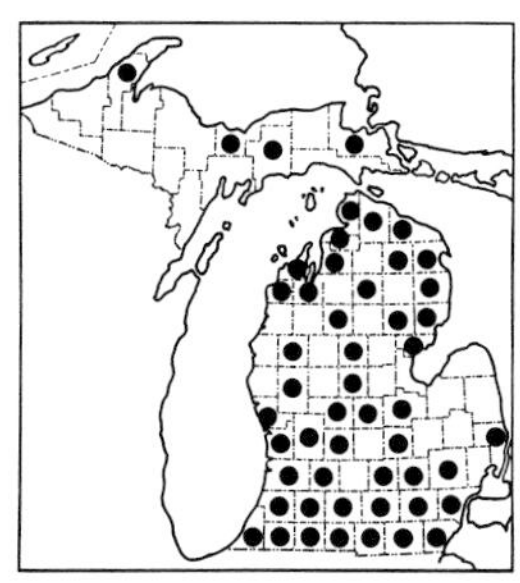

379. Nasturtium officinale

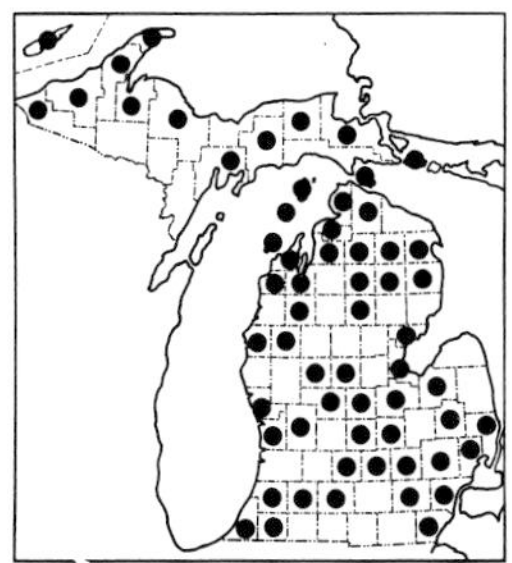

380. Capsella bursa-pastoris

381. Arabis procurrens

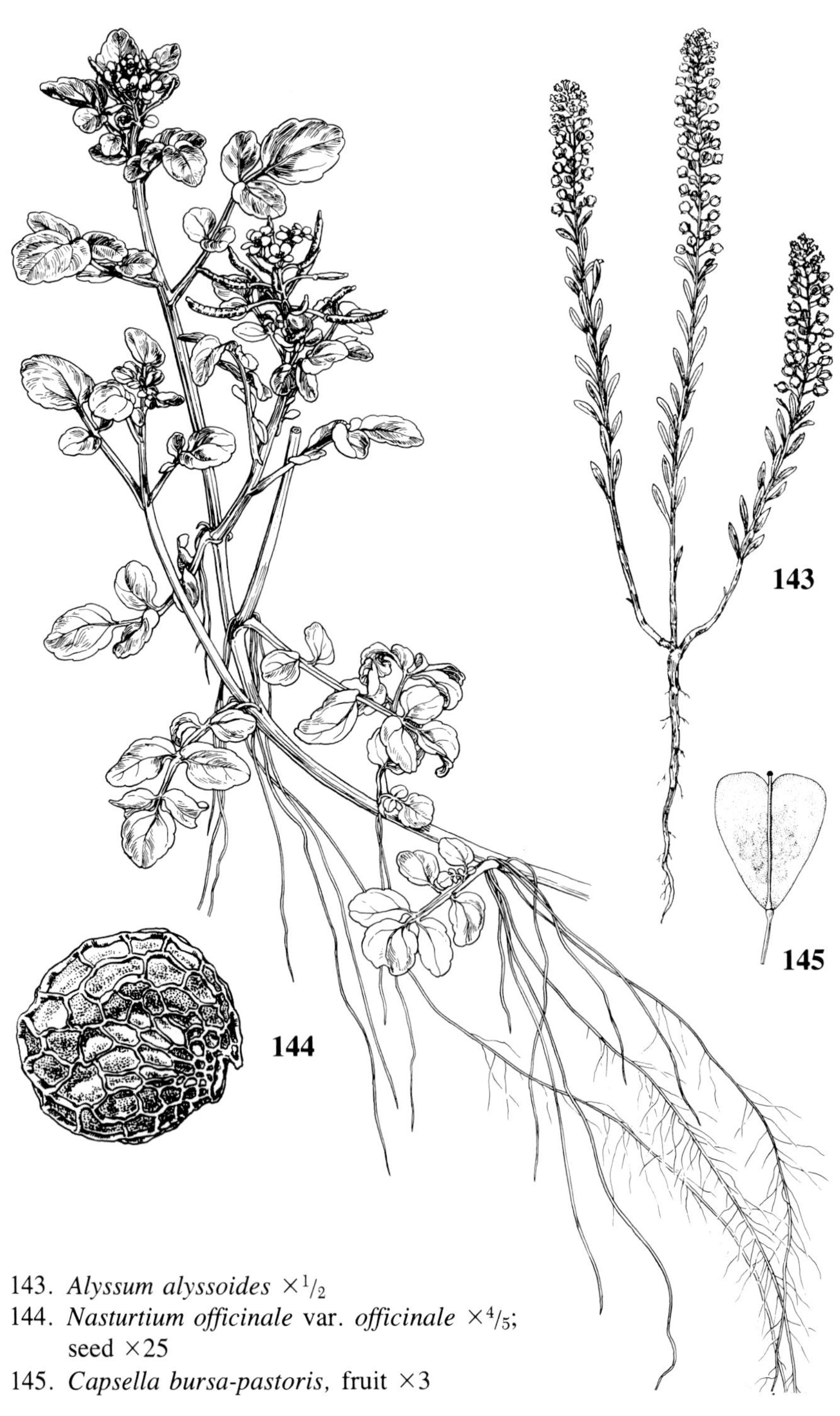

143. *Alyssum alyssoides* $\times ^1/_2$
144. *Nasturtium officinale* var. *officinale* $\times ^4/_5$; seed $\times 25$
145. *Capsella bursa-pastoris*, fruit $\times 3$

21. **Arabis** Rock Cress

Plants without mature or nearly mature fruit are often puzzling to identify. If one is confronted with an immature specimen, it may be necessary to try both sides of the key where fruit is required and to continue until further descriptions fail to apply on one side to the plant at hand, thus justifying the inference that the other choice is the correct one.

REFERENCES

Hopkins, Milton. 1937. Arabis in Eastern and Central North America. Rhodora 39: 63–98; 106–148; 155–196 (also Contr. Gray Herb. 116).

Rollins, Reed C. 1941. A Monographic Study of Arabis in Western North America. Rhodora 43: 289–325; 348–411; 425–481 (also Contr. Gray Herb. 138).

Rollins, Reed C. 1983. Interspecific Hybridization and Taxon Uniformity in Arabis (Cruciferae). Am. Jour. Bot. 70: 625–634.

KEY TO THE SPECIES

1. Cauline leaves not clasping at the base
 2. Plant with well developed stolons; leaves all entire, at least the basal ones with appressed medifixed hairs on margins and midrib beneath; petals ca. 8–9 (10) mm long . . . . 1. **A. procurrens**
 2. Plant without stolons; leaves ± toothed or lobed, pubescent or glabrous but without medifixed hairs; petals ca. 4–8 (9) mm long
  3. Fruit straight, mostly (0.7) 2–4 cm long, less than 1.3 mm broad, on ascending pedicels; leaves less than 3.5 (4.5) cm long, including those in a basal rosette of usually toothed or lyrate-lobed blades normally present at flowering time; stem often branched at base . . . . 2. **A. lyrata**
  3. Fruit curved, 4.5–7 (8) cm long, (2) 2.4–3.2 mm broad, pendent on reflexed pedicels; leaves (at least lower ones) (4) 6–12 (16) cm long, with no basal rosette evident at flowering time; stem simple . . . . 3. **A. canadensis**
1. Cauline leaves clasping with auriculate or sagittate bases
 4. Pedicels becoming distinctly reflexed even before shriveling of the petals, the fruit definitely pendent; sepals ca. half as long as mature petals or a little shorter
  5. Stem with stellate pubescence at least on lower half; sepals at least sparsely stellate-pubescent; mature fruit strongly pendent, the pedicels sharply reflexed . . . . 4. **A. holboellii**
  5. Stem with stellate pubsecence only at the very base, otherwise glabrous; sepals glabrous; mature fruit spreading to loosely pendent, the pedicels more arched than reflexed . . . . 12. **A. divaricarpa**
 4. Pedicels spreading or ascending to strongly appressed, even after anthesis, the fruit spreading to erect; sepals various
  6. Fruiting pedicels strongly ascending to appressed, the fruits straight, erect and closely appressed to the stem
   7. Stem and leaves entirely glabrous or with a very few scattered simple and/or appressed medifixed hairs at the *very base* of the plant (especially on leaf margins and petioles); sepals ca. half as long as the petals; mature fruit 1.4–2.5 [3.3] mm broad, with seeds in 2 rows in each locule . . . . 5. **A. drummondii**
   7. Stem and leaves pubescent, at least at the base, with spreading simple or stellate hairs (sparse appressed medifixed hairs in an uncommon variety of *A. hirsuta*);

sepals ca. 65–75% as long as the petals; mature fruit less than 1.3 (very rarely 1.5) mm broad, with seeds crowded into 1 row in each locule

8. Fruit ± terete or 4-angled, slightly if at all flattened at maturity; style-beak nearly or quite as wide as the fruit; stem pubescence only on the lowermost 1–3 full-grown internodes, and only the lowermost leaves pubescent; petals usually pale yellow [rarely pink] . . . . 6. **A. glabra**

8. Fruit rather strongly flattened; style-beak clearly narrower than mature fruit; stem pubescent with simple and/or forked (or stellate) hairs on at least the lower half or third, and leaves on the same portion ± pubescent (often stellate); petals white [or pink] . . . . 7. **A. hirsuta**

6. Fruiting pedicels ± spreading or divaricate, the fruits straight or somewhat curved and clearly spreading from the axis

9. Upper cauline leaves (i. e., below the lowermost pedicels or branches) ± dentate and pubescent on both surfaces

10. Petals at least (9) 10–14 mm long; plant a procumbent tufted escape from cultivation; fruit at least 3 cm long, glabrous . . . . 8. **A. caucasica**

10. Petals less than 3 mm long; plant a tall native; fruit usually less than 3 cm long, finely stellate-pubescent . . . . 9. **A. perstellata**

9. Upper cauline leaves entire or nearly so, glabrous

11. Basal leaves ± lyrate-pinnatifid, with at least a few simple hairs at the tips of the teeth or lobes; cauline leaves below the inflorescence numerous, ca. 30–40; sepals ca. half as long as the petals . . . . 10. **A. missouriensis**

11. Basal leaves entire or merely serrate (or absent at anthesis), completely glabrous or stellate-pubescent on both surfaces; cauline leaves various; sepals various

12. Cauline leaves ca. 10–15 (20) below the inflorescence, the longest (8) 9–15 cm; stem and leaves completely glabrous at base of plant (and elsewhere); petals white; sepals much more than half as long as petals . . . . 11. **A. laevigata**

12. Cauline leaves ca. (11) 25–35 or more below the inflorescence, the longest (2) 2.5–4 (6) cm; stem at the base and both surfaces of basal leaves ± stellate-pubescent; petals pink or pale purple; sepals at most barely more than half as long as the petals . . . . 12. **A. divaricarpa**

1. **A. procurrens** Waldst. & Kit. Hungarian Rock Cress

Map 381. A native of southeastern Europe, sometimes cultivated, and collected in 1968 in a marshy area near Intermediate Lake (*J. LaRue 2*, MICH).

2. **A. lyrata** L. Fig. 146 Sand Cress

Map 382. Dry open sandy ground, including woodlands but chiefly on dunes along the Great Lakes; on rock at Porcupine Mountains and Isle Royale and occasionally elsewhere; tending to spread into disturbed sandy areas.

Sometimes confused with *Arabidopsis thaliana*, but with larger petals, (4) 5–8 (9) mm long, and larger seeds, ca. 1 mm long.

3. **A. canadensis** L. Sickle-pod

Map 383. Dry woods, especially oak-hickory, banks, and bluffs; wooded dunes.

The broad, flat, curved, pendent siliques make this an easily recognized species in a difficult genus.

4. **A. holboellii** Hornem. Fig. 147

Map 384. Sandy dune ridges; rock ledges and summits (northwestern Upper Peninsula and Isle Royale).

The basal leaves are ± densely stellate-pubescent, in tight rosettes. Sterile rosettes at the ends of basal shoots are frequent, and the main stem often bears remnants of earlier rosettes at a lower level. On sand dunes, plants sometimes have a distinctive habit in that the rosettes are above the ground level.

Our plants are apparently all to be placed in the var. *retrofracta* (Graham) Rydberg. This is a northern species, more or less disjunct from the west to the Great Lakes region and occurring again in Quebec, but the disjunctions are not as great as once thought (see Marquis & Voss, 1981).

5. **A. drummondii** A. Gray

Map 385. Woods (oak, aspen, jack pine, or beech-maple), often in slightly disturbed areas; thickets on dunes; rocky openings (Isle Royale).

When the distinctive appressed fruiting habit is not yet developed, this species may be confused with *A. laevigata,* which also has a glabrous stem at the base, but in the latter the leaves are longer, the sepals distinctly more

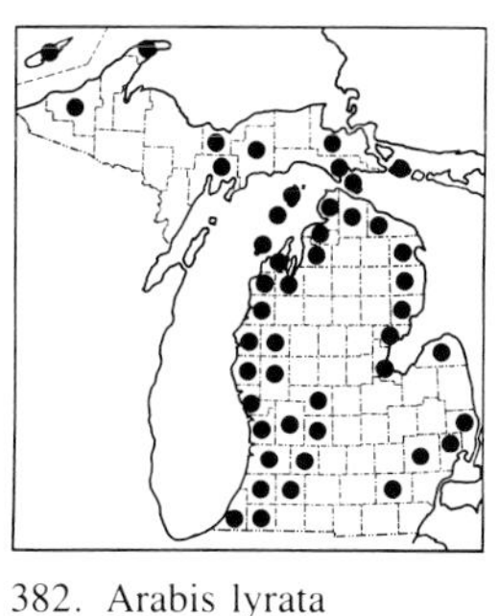
382. Arabis lyrata

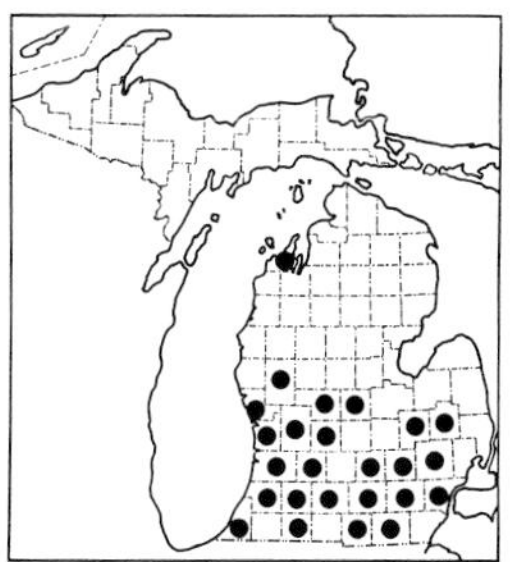
383. Arabis canadensis

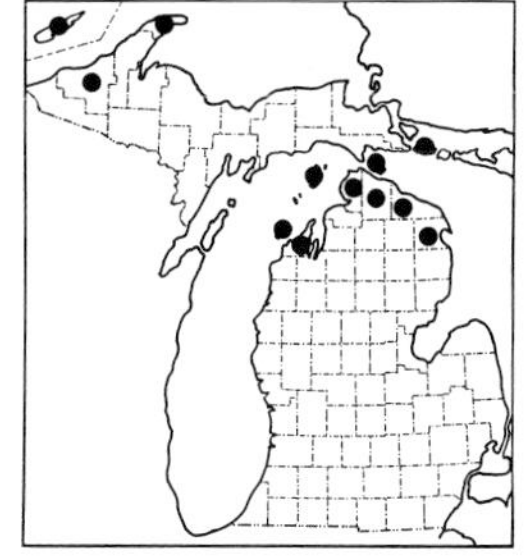
384. Arabis holboellii

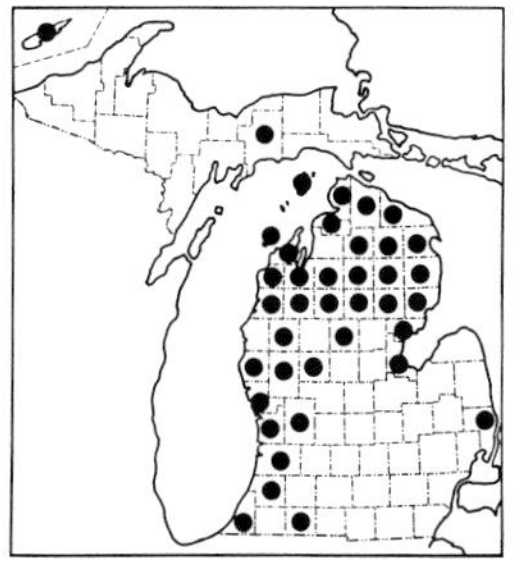
385. Arabis drummondii

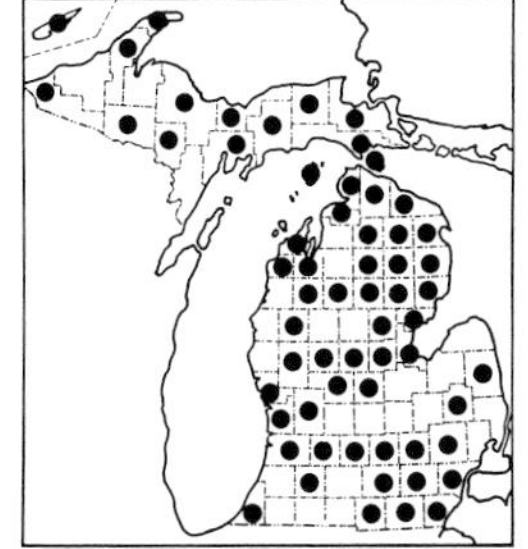
386. Arabis glabra

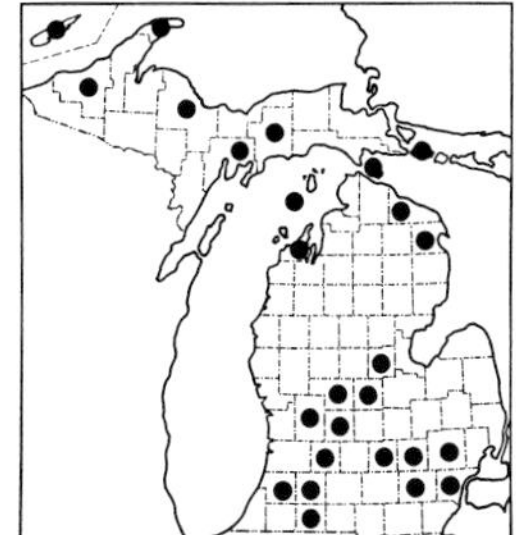
387. Arabis hirsuta

than half as long as the petals, and the cauline leaves usually less numerous. In *A. drummondii* the cauline leaves are (13) 18–25 below the lowest branch or pedicel of the inflorescence. The aspect of the fruiting plant is like that of *A. glabra,* which is more pubescent at the base (especially stellate-pubescent on basal leaves) and has sepals much more than half as long as the petals. There is also some habitat difference, *A. glabra* being characteristic of dry open areas with bracken, blueberry, sweet-fern, and scattered aspen or jack pine. *A. drummondii* is more characteristic of openings or slightly disturbed ground in woods.

6. **A. glabra** (L.) Bernh. Tower Mustard

Map 386. Sandy fields and barrens, gravel pits, open woodland (oak, aspen, jack pine) especially along roadsides and clearings, rocky openings and summits, rarely on gravelly shores and moist ground.

Rarely the very base of the stem is nearly or quite glabrous, but even so the basal leaves are more densely pubescent than in *A. drummondii,* with which this species is frequently confused, especially when young. The seeds are wingless or have a partial wing at most ca. 0.2 mm broad at the apex. *A. hirsuta* is usually said to have broader-winged seeds but most of our mature specimens are difficult to separate from some specimens of *A. glabra* on those grounds, having a wing perhaps 0.3 mm broad at the apex.

7. **A. hirsuta** (L.) Scop. Fig. 148

Map 387. Grows in a diversity of habitats, including sandy banks and hillsides, moist shores (especially calcareous rocks and gravels), river banks, and especially rock outcrops and associated grassy areas.

Most of our plants are var. *pycnocarpa* (Hopkins) Rollins, although var. *adpressipilis* (Hopkins) Rollins occurs rarely. The latter is much less pubescent, the hairs of the stem mostly medifixed and appressed, and the cauline leaves nearly or quite glabrous.

8. **A. caucasica** Schlecht. Wall Rock Cress

Map 388. A native of southern Europe, widely cultivated (often under the name *A. alpina* L., which applies to a similar species) and occasionally escaping to dry or moist areas.

The foliage is remarkably similar to that of the next species, but the petals are showy and the habit more that of a rock-garden plant.

9. **A. perstellata** E. L. Braun

Map 389. Floodplain forests and banks along the Grand River.

Plants from our region are var. *shortii* Fern., typically with minutely pubescent siliques and smaller flowers, compared to other varieties. Long known under the illegitimate name *A. dentata* (Torrey) T. & G.

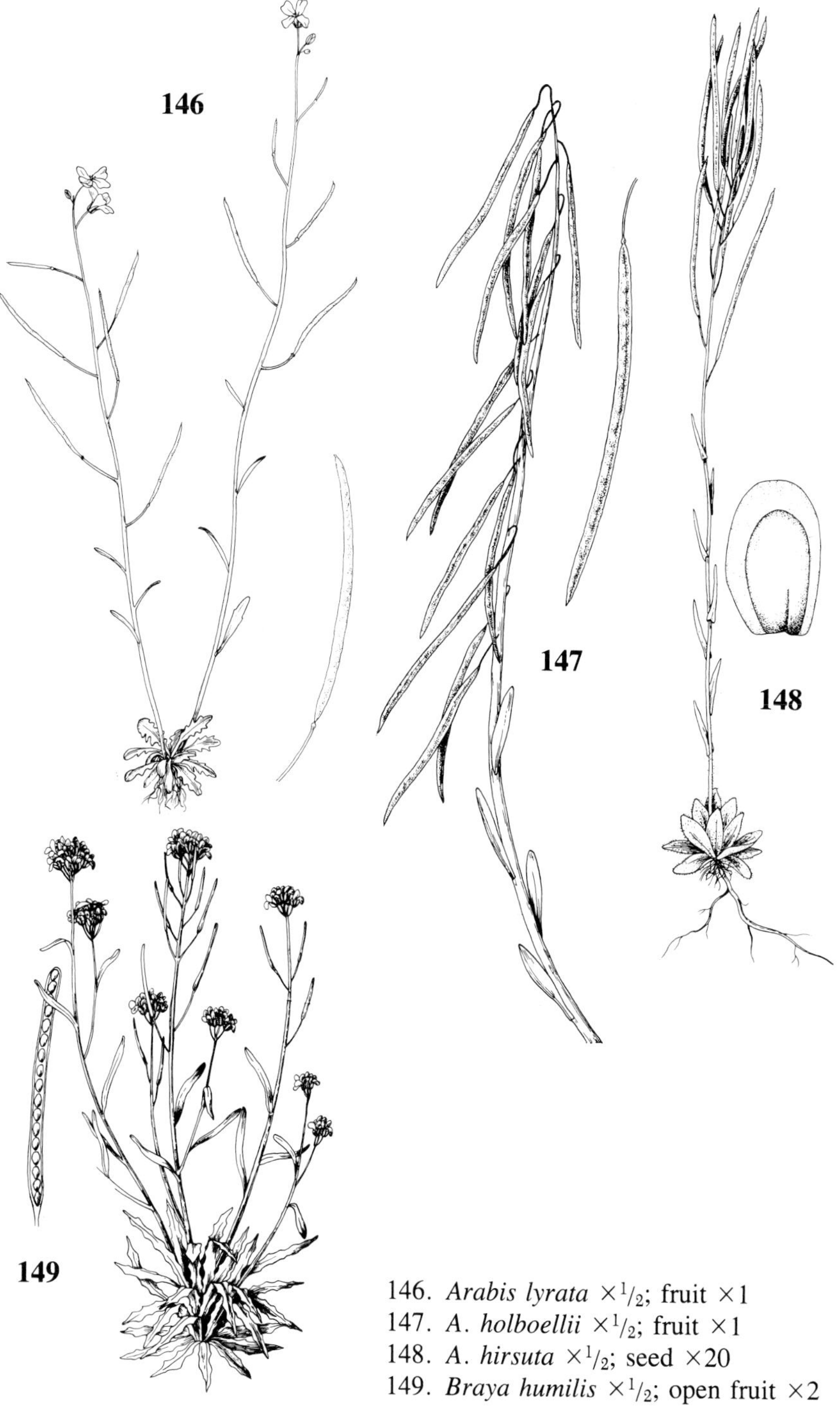

146. *Arabis lyrata* $\times ^1/_2$; fruit $\times 1$
147. *A. holboellii* $\times ^1/_2$; fruit $\times 1$
148. *A. hirsuta* $\times ^1/_2$; seed $\times 20$
149. *Braya humilis* $\times ^1/_2$; open fruit $\times 2$

10. **A. missouriensis** Greene

Map 390. Sandy open woodlands and fields.

Our plants are presumably all var. *deamii* (Hopkins) Hopkins, with the base of the stem and the basal leaves at least sparsely pubescent; the typical variety is entirely glabrous. Fruiting plants resemble *A. laevigata* except for a greener and less glaucous color and the more numerous, shorter leaves.

11. **A. laevigata** (Willd.) Poiret

Map 391. Floodplain, rich hardwood, and oak forests; river-bank thickets.

Distinctive in its long and relatively few cauline leaves (cf. previous and next species), which are usually entire and nearly linear on the upper part of the stem and strongly toothed toward the base of the stem. The rosette leaves of the second year, at the base of the fertile stem, are glabrous but usually soon wither; the rosettes of the first year, sometimes seen offsetting from the base of the fertile stem, are pubescent.

12. **A. divaricarpa** A. Nelson

Map 392. Sandy or gravelly clearings and borders of woods (especially aspen) and shores, rock outcrops and summits of "mountains."

A notably variable species, tempting the suggestion that if a plant fits "none of the above" in the key, it must be *A. divaricarpa*. The cauline leaves are usually numerous, short, and crowded, as in *A. missouriensis,* but the basal leaves are entire or slightly toothed and ± stellate-pubescent on both surfaces. The fruit is ordinarily widely spreading or even somewhat down-curved, up to 9.5 cm long; but on some plants the siliques are very short (ca. 2 cm), ascending on divaricate pedicels (aspect like *Erysimum cheiranthoides*)—and the cauline leaves on small plants may be relatively few (11–20) and/or short (the longest barely 2 cm). Plants without mature fruit displaying the strongly spreading aspect compared to the compact infructescence of *A. glabra* may be confused with the latter species, but differ in the shorter sepals and pink petals; in *A. glabra,* the pubescence at the

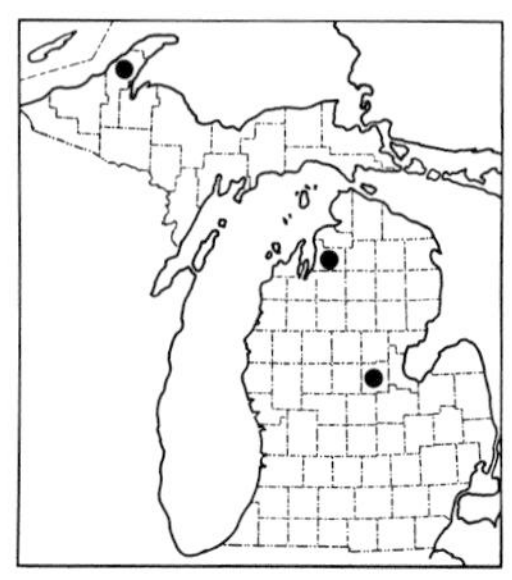

388. Arabis caucasica

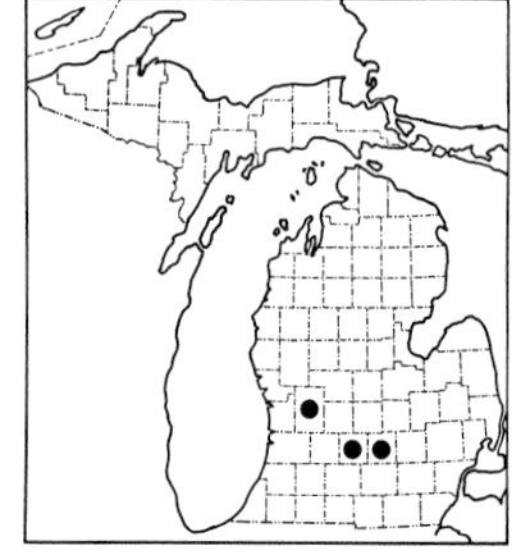

389. Arabis perstellata

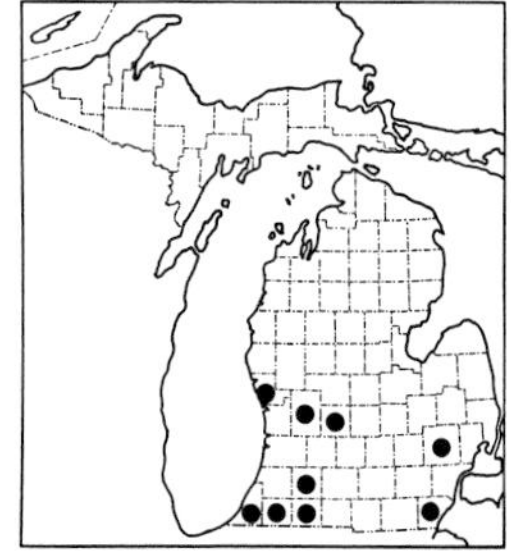

390. Arabis missouriensis

base of the stem is generally more dense, with more straight, spreading hairs.

Rollins (1983) has suggested that the great variability in *A. divaricarpa* results from hybridization with its neighbors (such as *A. drummondii* and *A. holboellii,* the former with strongly ascending fruit and the latter with strongly reflexed fruit); in fact, *A. divaricarpa* may be a mixture of old stable (presumably apomictic) populations of hybrid origin as well as newly produced hybrids.

22. **Hesperis**

1. **H. matronalis** L. Dame's Rocket

Map 393. A native of Europe, escaped from cultivation. Roadsides, overgrown dumps and weedy thickets, river borders, sometimes well established in damp woods (deciduous or coniferous).

No specimens have been found to document the occurrence of any species of *Matthiola,* stock, as escaped from cultivation in Michigan. They would key here to *Hesperis,* but differ in having rather dense white stellate hairs, especially on the pedicels, fruit, and calyx, giving the plants an ashy-gray aspect. The summit of the fruit has ± triangular projections ("horns"). In the commonly cultivated *M. incana* (L.) R. Br., these horns are inconspicuous (less than 1.5 mm long) and the leaves are entire or nearly so. In *M. longipetala* (Vent.) DC. [including *M. bicornis* (Sibth. & Sm.) DC.], reported from Michigan, the horns become 5 mm or more long and the leaves are rather coarsely toothed.

23. **Arabidopsis**

1. **A. thaliana** (L.) Heynh. Fig. 150 Mouse-ear Cress

Map 394. A Eurasian species found with other pioneer annuals in recently disturbed, usually sandy ground, including cultivated land, fields, and trails; spreading to sandy oak woods and even to damp sites.

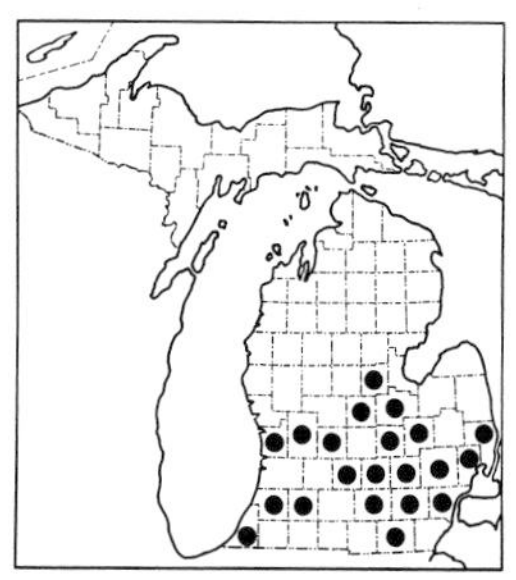

391. Arabis laevigata

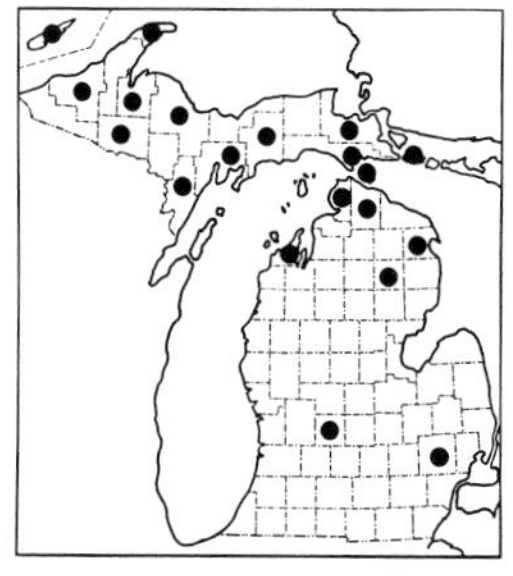

392. Arabis divaricarpa

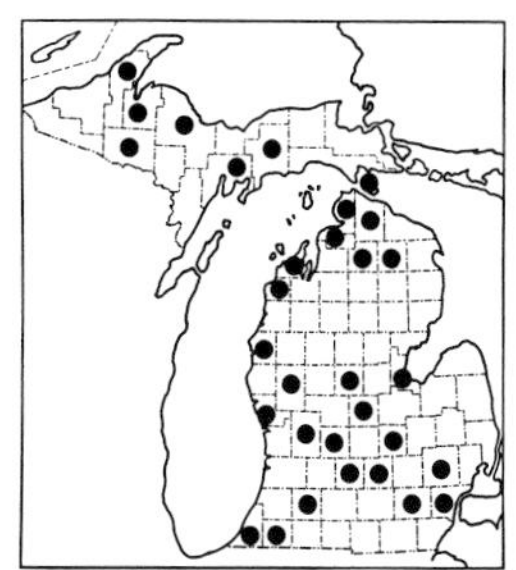

393. Hesperis matronalis

The petals are less than 4 (4.2) mm long and the rosette leaves are entire or somewhat toothed. Sometimes confused with *Arabis lyrata,* which has longer petals, usually slightly larger fruit, and often ± lyrate rosette leaves. Both species flower very early (April–May) in southern Michigan. These species, as well as *Braya humilis,* have most of their leaves in a basal rosette, from which one to several stems arise. *Erophila verna* differs from all three in its strictly scapose habit (*no* cauline leaves) and bifid petals, as well as in its broader fruit.

The very tiny seeds of *Arabidopsis*—barely 0.5 mm long—weigh about 50,000 to a gram.

24. **Braya**

1. **B. humilis** (C. A. Meyer) Robinson Fig. 149

Map 395. Our only representative of a boreal and arctic genus, collected at Copper Harbor in 1895 (*Farwell 850,* BLH) and on rocks at Isle Royale in 1933 (*Hebert,* ND).

The fruit is more terete (or 4-angled) than in those species of *Arabis* and *Arabidopsis* with which this might be confused, as well as being somewhat constricted between the seeds.

25. **Berteroa**

1. **B. incana** (L.) DC. Fig. 151 Hoary Alyssum

Map 396. A native of Europe, now a weed there as well as here. First found in Michigan about 1900, and subsequently spread throughout the state along roadsides and railroads, in sandy fields, on shores and banks; invading woodlands.

The rather dense pubescence gives the plant a distinctive gray-green aspect.

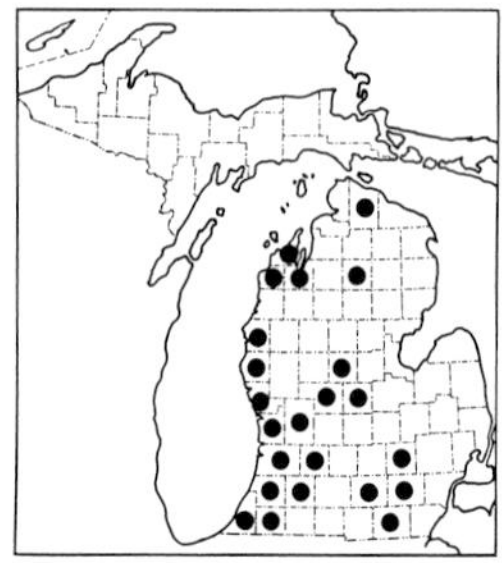
394. Arabidopsis thaliana

395. Braya humilis

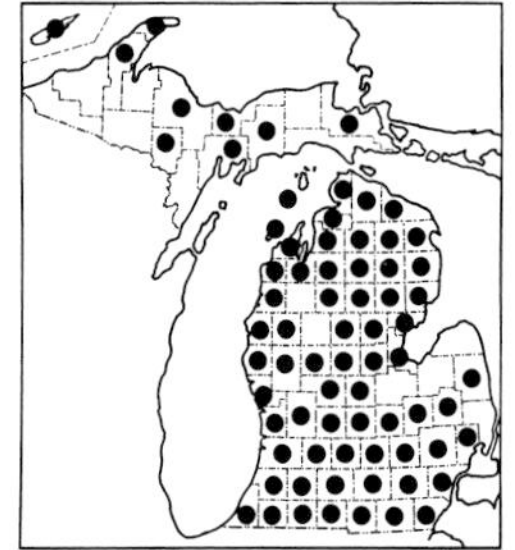
396. Berteroa incana

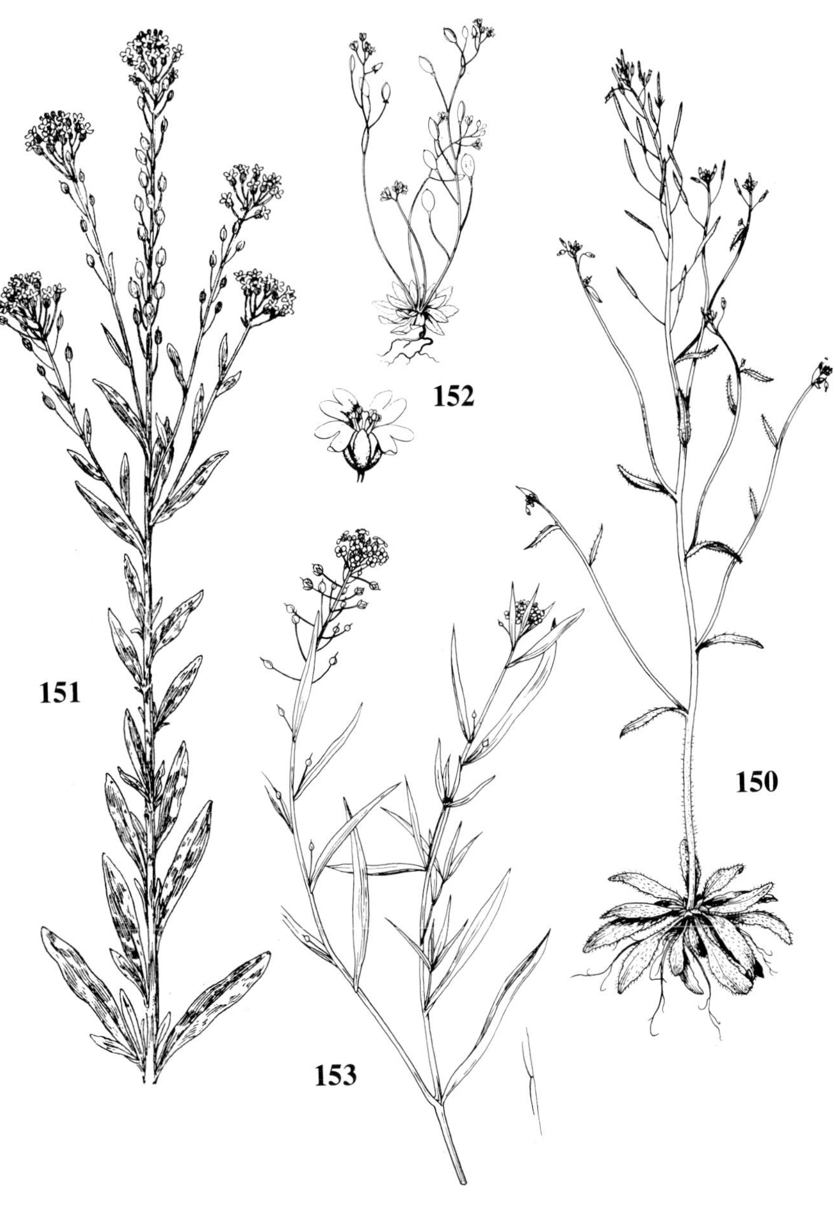

150. *Arabidopsis thaliana* ×1/2
151. *Berteroa incana* ×1/2
152. *Erophila verna* ×1/2; flower ×5
153. *Lobularia maritima* ×1/2; detail of hair ×200

26. **Erophila**

1. **E. verna** (L.) Besser   Fig. 152   Whitlow-grass

Map 397. A small annual or winter-annual, naturalized from Europe and found locally in disturbed ground: cultivated areas, fields, roadsides, graded areas, lawns. The oldest Michigan collection seen is from Lansing in 1886. The species appears to have spread considerably within the past few years.

Originally described as a species of *Draba* and often retained in that genus, especially in American manuals, but distinguished from *Draba*—and other small Cruciferae with which it might be confused—by the combination of scapose habit and bifid petals.

27. **Lobularia**

1. **L. maritima** (L.) Desv.   Fig. 153   Sweet Alyssum

Map 398. Vacant lots, dumps, near old gardens, roadsides, and waste places. A native of Europe, commonly grown as a rock-garden plant, much-branched and spreading; probably escaped from cultivation more widely than collections indicate.

28. **Draba**

Erophila verna is often included in this genus as *D. verna* L. All of our true Drabas, however, have at least a few cauline leaves (only near the base in *D. reptans*) and the petals are not bifid. *Draba* is a large and complex genus, mostly of boreal and arctic regions, represented by only a few relatively distinctive species in our area. The siliques are strongly flattened and in many species are usually ± twisted.

REFERENCES

Fernald, M. L. 1934. Draba in Temperate Northeastern America. Rhodora 36: 241–261; 285–305; 314–344; 353–371; 392–404 (also Contr. Gray Herb. 105).

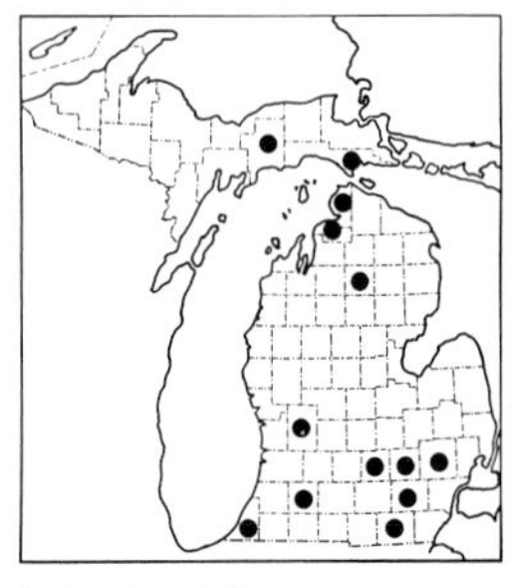

397. Erophila verna

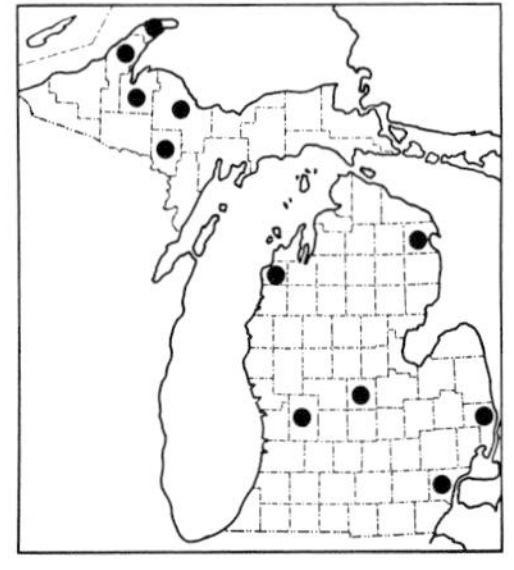

398. Lobularia maritima

399. Draba nemorosa

Mulligan, Gerald A. 1976. The Genus Draba in Canada and Alaska: Key and Summary. Canad. Jour. Bot. 54: 1386–1393.

KEY TO THE SPECIES

1. Plant a delicate slender-stemmed annual (or winter-annual); fruit narrowly elliptic or oblong, tapering no more toward the apex than the base and no broader below the middle than at it; axis of the inflorescence glabrous
  2. Petals pale yellow, less than 2.2 mm long; cauline leaves scattered nearly to the open, elongate raceme . . . . 1. **D. nemorosa**
  2. Petals white, ca. 3–4 mm long (or absent); cauline leaves near the base (sometimes scarcely distinguishable from the basal rosette), remote from the crowded inflorescence . . . . 2. **D. reptans**
1. Plant a ± strong perennial (or sturdy biennial); fruit mostly slightly broader below the middle, tapering more toward the apex than the base; axis of inflorescence usually sparsely to densely pubescent
  3. Stem very leafy (at least 20 cauline leaves between basal rosette and lowest pedicel); axis of inflorescence densely pubescent; fruit not twisted; style obsolete, ca. 0.3 mm long or shorter on fruit and no longer than broad . . . . 3. **D. incana**
  3. Stem with few cauline leaves but many basal leaves (often with short leafy basal shoots); axis of inflorescence sparsely pubescent or glabrate; fruit often twisted; style usually ca. 0.3–0.5 mm long on fruit and longer than broad
    4. Fruit pubescent; lower pedicels subtended by leaves or bracts . . . . 4. **D. cana**
    4. Fruit glabrous; lower pedicels without leaves or bracts (rarely one subtending lowermost pedicel) . . . . 5. **D. arabisans**

1. **D. nemorosa** L.

Map 399. The only Michigan collection located (NY) was made by Dr. Zina Pitcher, presumably in 1829, at Fort Gratiot (Port Huron), when he served there briefly as an Army surgeon. This is a circumpolar species primarily, in North America, of dry open ground west of the Great Lakes. Reported from Michigan on the basis of Pitcher's collection by Torrey and Gray (Fl. N. Am. 1: 108. 1838) and subsequently attributed to the state in manuals though apparently never again collected. Reported by Morton and Venn (1984) as local on Manitoulin Island—not far from Michigan.

The fruit is glabrous in the Michigan collection, although it may be pubescent in this species, unlike *D. reptans*.

2. **D. reptans** (Lam.) Fern.

Map 400. Sandy banks and fields.

The fruits are narrowly oblong, mostly (7) 10–14 mm long and 5–8 times as long as broad in our material. In the preceding species, they tend to be shorter and more elliptical.

3. **D. incana** L.

Map 401. A northern species in eastern North America, isolated in Michigan on Passage Island and Gull Islands of Isle Royale National Park, where it grows in exposed rock crevices above Lake Superior.

4. **D. cana** Rydb.

Map 402. A boreal species, isolated in Michigan on the limestone of Burnt Bluff, Delta County, and a very few other places.

Until recently, misidentified as the Himalayan *D. lanceolata* Royle, under which name it appears in most manuals.

5. **D. arabisans** Michaux Fig. 158

Map 403. Another northern species, the only one likely to be encountered in the state by most botanists. On large boulders and outcrops of limestone at Mackinac Island and the mainland of Mackinac County; on gravelly slopes and outcrops (conglomerates and igneous rocks) at Isle Royale and the mainland of Keweenaw County.

The stellate hairs of the leaves are more nearly sessile than in the preceding species. The petals appear yellow in bud, but are white when the flower is open.

29. Cardaria

REFERENCES

Mulligan, Gerald A., & Clarence Frankton. 1962. Taxonomy of the Genus Cardaria with Particular Reference to the Species Introduced into North America. Canad. Jour. Bot. 40: 1411–1425.

Mulligan, Gerald A., & Judy N. Findlay. 1974. The Biology of Canadian Weeds. 3. Cardaria draba, C. chalepensis, and C. pubescens. Canad. Jour. Pl. Sci. 54: 149–160.

Rollins, Reed C. 1940. On Two Weedy Crucifers. Rhodora 42: 302–306.

KEY TO THE SPECIES

1. Pedicels, ovary, and fruit pubescent, the latter ± globose to obovoid, at least as long as broad, not at all cordate . . . . 1. **C. pubescens**
1. Pedicels, ovary, and fruit glabrous, the latter slightly cordate at base, often broader than long . . . . 2. **C. draba**

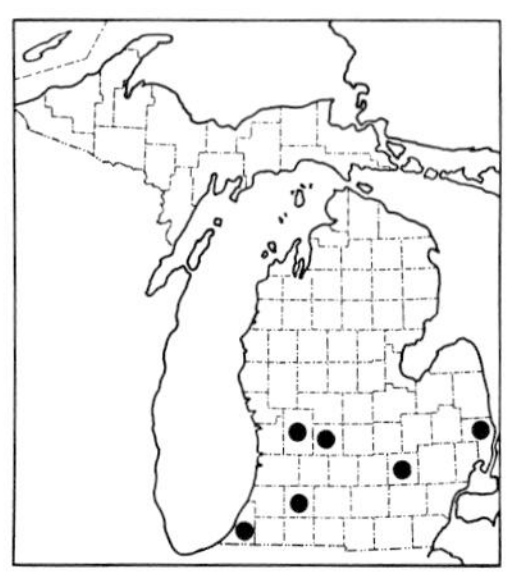
400. Draba reptans

401. Draba incana

402. Draba cana

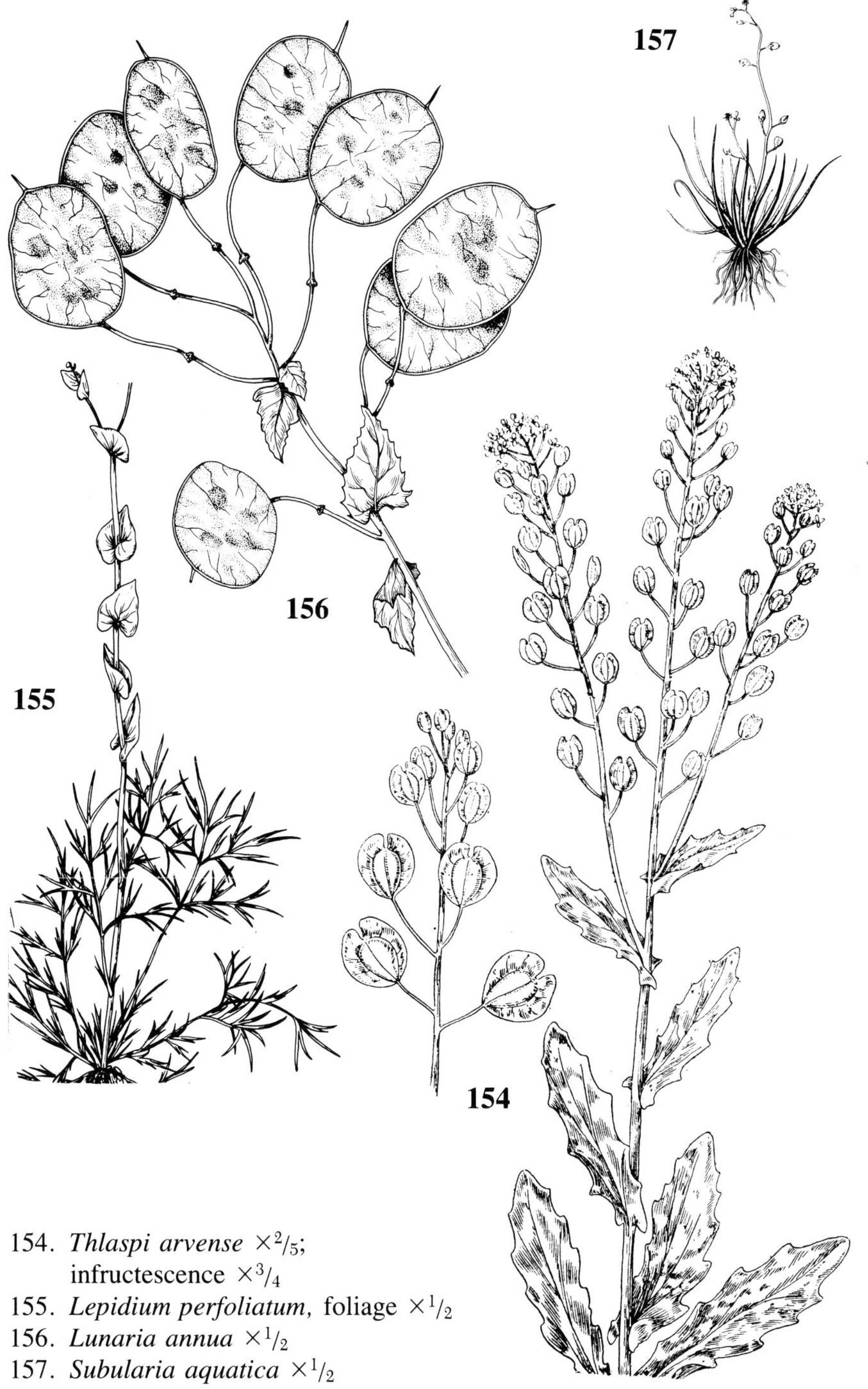

154. *Thlaspi arvense* $\times^2/_5$;
infructescence $\times^3/_4$
155. *Lepidium perfoliatum,* foliage $\times^1/_2$
156. *Lunaria annua* $\times^1/_2$
157. *Subularia aquatica* $\times^1/_2$

1. **C. pubescens** (C. A. Meyer) Jarm. White-top

Map 404. A native of central Asia, locally adventive in eastern North America, sometimes an aggressive weed farther west. The first North American collection was apparently from railroad tracks at Ypsilanti (*Walpole* and *Billington* in 1919, BLH; *Walpole* and *Farwell* in 1919 and 1920, BLH, ALBC, GH, MO, MSC; see Rep. Mich. Acad. 22: 183. 1921). Not since collected in the state.

Long kept in the genus *Hymenophysa,* in which it was originally described.

2. **C. draba** (L.) Desv. Fig. 159 Hoary Cress

Map 405. Another central Asian species, occasionally found as a weed. First collected in Michigan in 1920 in waste ground at Ypsilanti; found along a roadside in Berrien County in 1970; no other localities known in the state.

Originally placed by Linnaeus in the genus *Lepidium*. Both of our species of *Cardaria* have a single seed in each locule, as does *Lepidium*.

30. **Thlaspi**

REFERENCE

Best, K. F., & G. I. McIntyre. 1975. The Biology of Canadian Weeds 9. Thlaspi arvense L. Canad. Jour. Pl. Sci. 55: 279–292.

403. Draba arabisans

404. Cardaria pubescens

405. Cardaria draba

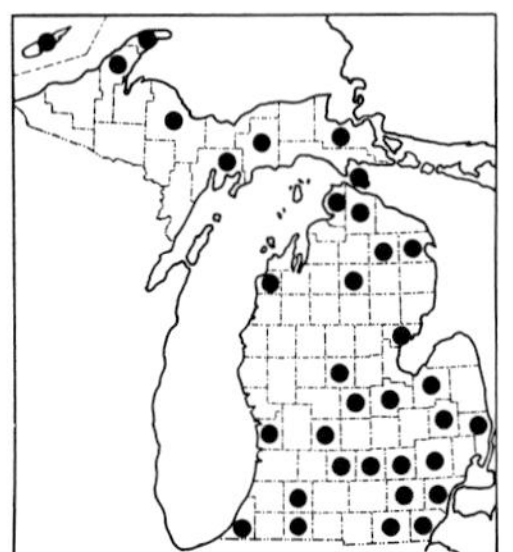
406. Thlaspi arvense

407. Lepidium perfoliatum

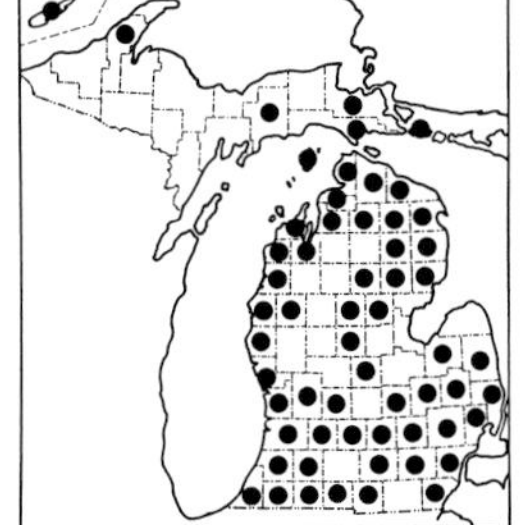
408. Lepidium campestre

1. **T. arvense** L. Fig. 154 Penny Cress

Map 406. The earliest Michigan collection (CU) I have seen is from Mackinac Island, by Horace Mann, Jr. (and Henry David Thoreau), in 1861. In 1863, it was collected (*L. Foote,* MICH) at Fort Wayne (Detroit). By 1885 it had been collected at both Ann Arbor and Lansing. It is still not as abundant as some of our other widespread Eurasian weeds.

31. **Lepidium** Pepper-grass

REFERENCES

Hitchcock, C. Leo. 1936. The Genus Lepidium in the United States. Madroño 3: 265–320.
Mulligan, Gerald A. 1961. The Genus Lepidium in Canada. Madroño 16: 77–90.

KEY TO THE SPECIES

1. Upper cauline leaves sagittate- or auriculate-clasping
 2. Inflorescence and upper stem glabrous; petals pale yellow; lower leaves ± finely divided; upper leaves so strongly auriculate-clasping as to appear perfoliate .... 1. **L. perfoliatum**
 2. Inflorescence and stem ± densely short-pubescent; petals white; lower leaves not divided; upper leaves sagittate-clasping .... 2. **L. campestre**
1. Upper cauline leaves narrowed and sessile but not clasping at the base
 3. Fruit ca. 5–6 mm long, often at least as long as the pedicel; upper cauline leaves deeply lobed or pinnatifid .... 3. **L. sativum**
 3. Fruit ca. 2.5–3.2 mm long, shorter than the pedicel; upper cauline leaves entire or merely toothed
  4. Petals absent or occasionally present and inconspicuous (shorter than sepals)
   5. Apex of fruit with ± acute lobes beside the style (fig. 160); upper leaves and bracts obtuse to rounded at tip, basal and lower cauline leaves pinnatifid or bipinnatifid; plant a rare waif, fetid when fresh.......... 4. **L. ruderale**
   5. Apex of fruit with rounded lobes; upper leaves and bracts sharply acute at tip, basal and lower cauline leaves entire to deeply toothed or (basal) once-pinnatifid; plant a common weed, nearly odorless.......... 5. **L. densiflorum**
  4. Petals present, conspicuous (equaling or surpassing sepals)
   6. Style nearly absent, much shorter than notch at apex of fruit; basal leaves toothed to pinnatifid; plant annual .... 6. **L. virginicum**
   6. Style ca. 0.5 mm long, exceeding notch of fruit; basal leaves deeply pinnatifid or bipinnatifid; plant perennial, nearly shrubby .... 7. **L. montanum**

1. **L. perfoliatum** L. Fig. 155

Map 407. A native of Eurasia, widely distributed as a weed but not found in Michigan until 1920, when it was collected along a railroad near Ypsilanti; collected 1935–1937 in an alfalfa field and waste places in Kalamazoo County.

Although, like our other species, this is a plant of dry waste ground, it has the aspect of many aquatics in its strong heterophylly, with dissected lower leaves and entire upper leaves.

2. **L. campestre** (L.) R. Br.

Map 408. Another introduced Eurasian species, now quite widespread in the state. Roadsides, railroads, sandy fields, sandy and gravelly shores, and other disturbed sites, sometimes spreading into dry woods. Collected at several places across the state in the 1890's. Although no earlier collections have been seen, it must have been well established by then.

Distinctive in its dense pubescence and sagittate-clasping leaves, yet surprisingly often other species of the genus and even plants of other genera are misidentified as *L. campestre*. The fruit is unique in being somewhat oblong (rather than round in the other species), and inflated in the middle (rather than flat), though still bordered with a distinct flat wing; furthermore, it is covered with minute pale papillae.

3. **L. sativum** L. Garden Cress

Map 409. An Old World native, collected by Dodge in the 1890's as a weed along the streets of Port Huron. The only other Michigan collection is as a probable garden escape along a gravel driveway in Birmingham in 1977. The species may turn up elsewhere as a casual escape.

The stamens are 6 in this species, while in the next three they are usually only 2 (or 4).

4. **L. ruderale** L. Fig. 160

Map 410. Another weedy species of Eurasian origin, probably to be found in the state more often than the single collection seen, made in 1877 in Detroit (*J. Blake,* WMU).

Although the flowers are similar to those of the common *L. densiflorum,* the fruit is distinctive: narrower, tapering more toward the apex, which has 2 ± sharply angular lobes. The 2 lobes and the entire apex in outline are more rounded in *L. densiflorum*.

5. **L. densiflorum** Schrader Fig. 161

Map 411. Sandy and gravelly disturbed ground (roadsides, clearings, fields,

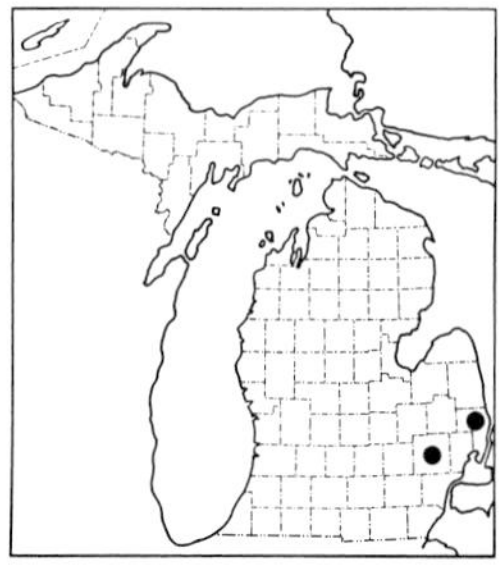
409. Lepidium sativum

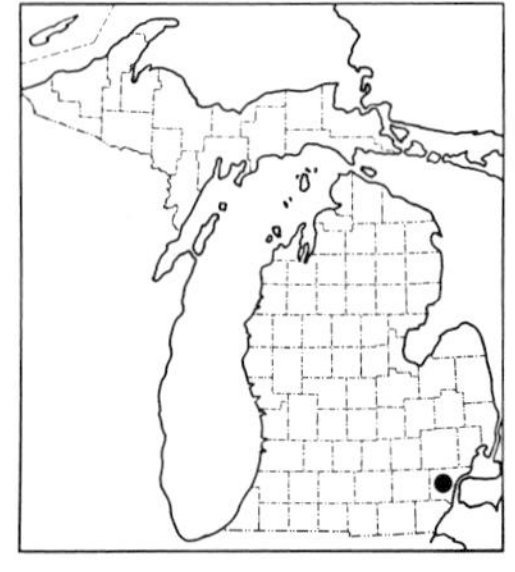
410. Lepidium ruderale

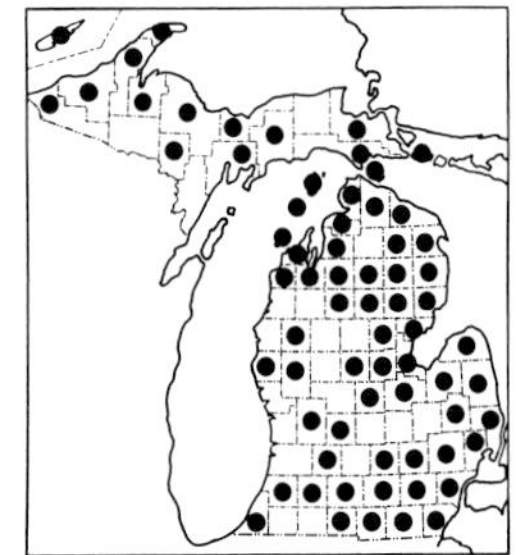
411. Lepidium densiflorum

gravel pits, dunes, railroads); shores; spreading to dry woods (oak, aspen, jack pine). By some authors considered native in eastern North America but introduced in the west; by others thought to be native primarily in the west; by some, to be introduced from Eurasia, and by others to be introduced into Europe. Whatever its history, now a widespread weed. The earliest Michigan collections date from the 1860's and 1870's, by which time it was noted as "common"; but the species was apparently not found by the First Survey (1837–1841).

Long known as *L. apetalum* Willd, and often misidentified as *L. virginicum*. The mature fruit of the latter is scarcely if at all longer than broad and tends to be broadest just below the middle. The fruit of *L. densiflorum* tends to be slightly longer than broad (± oblong-elliptic to obovate), often very slightly broader just above the middle—but the shape differences are subtle. A better character is in the seed: in *L. virginicum* (at least in our region), the cotyledons are accumbent, so that in a cross section one sees the round embryo at the end of the flattened, parallel cotyledons, diagrammatically shown as =o; in *L. densiflorum,* the cotyledons are incumbent, so that a cross section of the seed appears to have 3 structures in a row: ooo.

6. **L. virginicum** L. Fig. 162

Map 412. Like the preceding, a widespread weed of dry soil, often considered native in North America and introduced in Europe. Sandy and gravelly roadsides, fields, dumps, old dunes, disturbed areas in dry woods. Not as common as *L. densiflorum,* but collected as early as the First Survey in Monroe County in 1837. By the 1870's, likewise known from several counties in the state.

This species is especially variable in the western part of the continent. Plants in our area, var. *virginicum,* have seeds with accumbent cotyledons (see above) and this is a helpful character for specimens with mature fruit and no flowers which would show petals. This is the only species of *Lepidium* in the state with accumbent cotyledons (which can be seen easily under 20× magnification or more). In both this species and the preceding the stems and axis of the inflorescence are often minutely puberulent or glandular-puberulent. These two species are reported to hybridize in Wisconsin and may be expected to do so in Michigan as well.

7. **L. montanum** Nutt.

Map 413. A variable species of the western United States. Collected in Michigan only in 1928 by Farwell (*8218 & 8221,* BLH, GH; see Rhodora 39: 280. 1937) along the railroad beween Ann Arbor and Ypsilanti.

Some of the forms have all the leaves pinnatifid, but the plants collected in Michigan have narrow entire upper leaves. The fruit is ± ovate, narrowed to the apex, at which the style protrudes beyond 2 acute teeth.

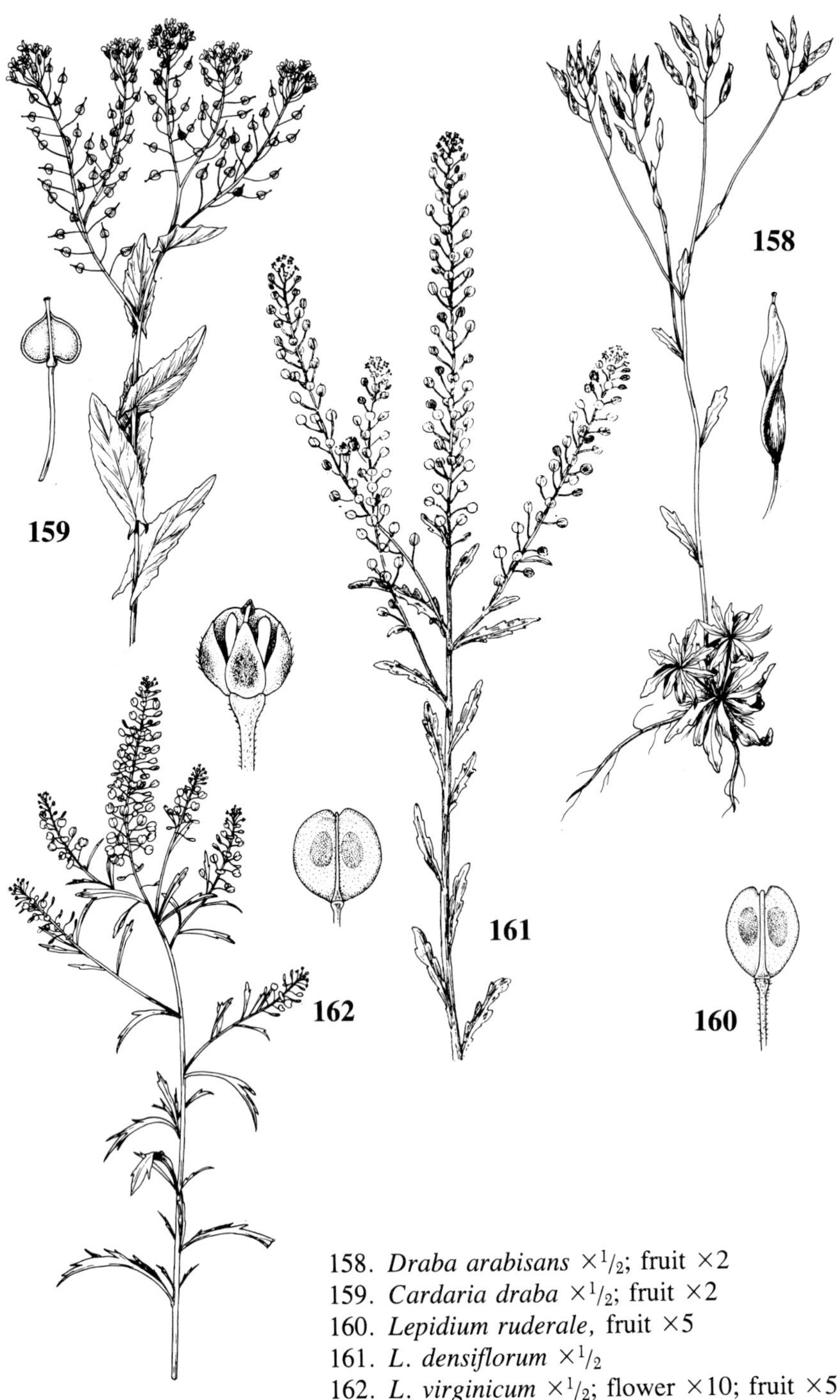

158. *Draba arabisans* ×1/2; fruit ×2
159. *Cardaria draba* ×1/2; fruit ×2
160. *Lepidium ruderale,* fruit ×5
161. *L. densiflorum* ×1/2
162. *L. virginicum* ×1/2; flower ×10; fruit ×5

32. **Lunaria**

Two species of this genus, both native to Europe, are cultivated in North America for their showy, usually purple flowers and especially for the large flat fruit with a satiny septum which is attractive in dry bouquets and arrangements.

1. **L. annua** L. Fig. 156 Money-plant; Honesty

Map 414. Locally escaped from cultivation and a well established biennial some places. Weedy thickets, roadsides, and banks; dumps. A long-persistent colony grows in sand under a grove of *Populus* on the shores of Lake Michigan at Cross Village, Emmet County.

The very flat fruit is ca. 1.5–3.5 cm broad at maturity, rounded at both ends, and usually not much longer than broad. The uppermost leaves are sessile or nearly so. In the perennial *L. rediviva* L., not yet collected as an escape in Michigan, the fruit is angled (nearly acute) at both ends, somewhat longer in relation to the width; and the uppermost leaves are distinctly petioled.

33. **Subularia**

REFERENCE

Mulligan, Gerald A., & James A. Calder. 1964. The Genus Subularia. Rhodora 66: 127–135.

1. **S. aquatica** L. Fig. 157 Awlwort

Map 415. A species of northern North America, Europe, and Eastern Asia. Although the species is known to flower on damp shores, Michigan collections are all from ca. 1–3 feet of water, where they flower and fruit on a sandy-gravelly bottom: Rock Harbor, Isle Royale (*Lowe & Brown 3647* in 1930, MICH); St. Mary's River at north end of Sugar Island (*Hiltunen 1854* in 1958, WUD; in 1965, MICH, UMBS).

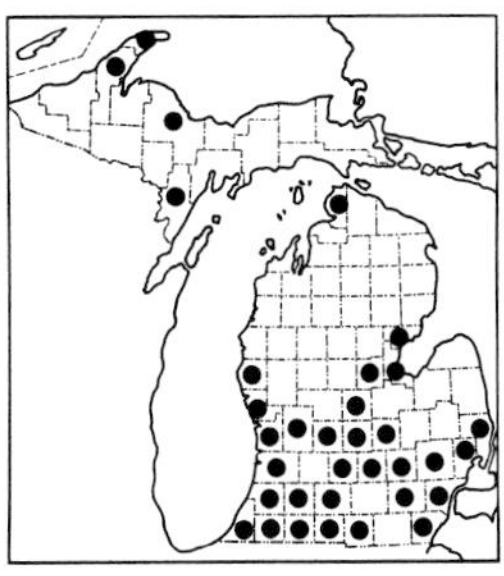

412. Lepidium virginicum

413. Lepidium montanum

414. Lunaria annua

The usual associates of *Subularia* include such species as *Myriophyllum tenellum* and *Eleocharis acicularis,* as well as other somewhat similar rosette-formers such as *Littorella uniflora, Isoëtes* spp., *Ranunculus reptans,* and *Juncus pelocarpus*. These are all distinguished on vegetative features in the aquatic key in Part I of this Flora (pp. 52–55). The leaves of *Subularia* are less than 4 cm long, somewhat flattened dorsoventrally (especially toward the base), with numerous small hollow areas of irregular size.

34. **Alliaria**

1. **A. petiolata** (Bieb.) Cavara & Grande Fig. 163 Garlic Mustard

Map 416. A native of Europe and Asia, naturalized locally in North America, but often abundant where found, in disturbed ground such as roadsides and in moist woods, even swamp forest.

The *Allium*-like odor of the bruised plant is unusual in a family where the odor, if any, is normally of a turnip, cabbage, or horseradish nature. *Alliaria* is also distinctive in its slenderly petioled leaves with broad ± deltoid often cordate blades. The white petals are ca. 4–7 mm long and the fruit is a slender silique.

Long known as *A. officinalis* Andrz., a later name.

35. **Cakile**

REFERENCE

Rodman, James E. 1974. Systematics and Evolution of the Genus Cakile (Cruciferae). Contr. Gray Herb. 205: 3–146.

1. **C. edentula** (Bigelow) Hooker Fig. 164 Sea-rocket

Map 417. A characteristic succulent strand plant seldom found far from sites where waves and wind have deposited fruit segments along the sandy beaches and low dunes of the Great Lakes. *Cakile* is often found closer to the water's edge than other beach and dune species, including the similarly

415. Subularia aquatica

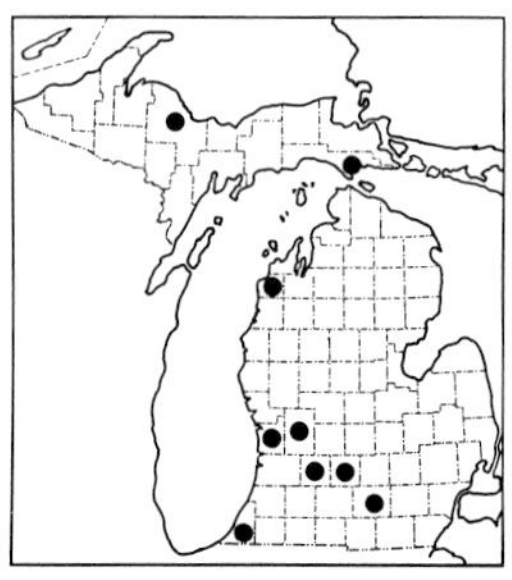

416. Alliaria petiolata

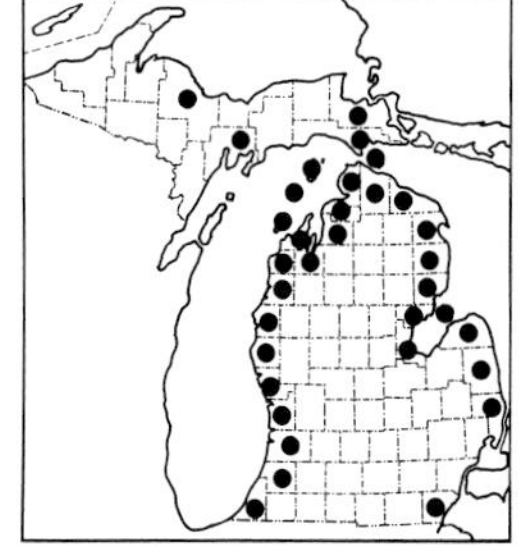

417. Cakile edentula

annual *Euphorbia polygonifolia*. Although I have explored the Lake Superior shore closely for this species (and other shoreline plants) from Whitefish Point to Grand Marais, I have been unable to find it there. The only known Michigan localities on Lake Superior are beaches in the vicinity of the Huron Mountains, Marquette County, where it has been collected several times and seems to be thriving. *Cakile* was collected at Sault Ste. Marie [Ontario or Michigan?] by Dr. Zina Pitcher, presumably in the 1820's (Rodman 1974; specimen not relocated); this may even represent a collection site farther to the west, somewhere on the Lake Superior shore. Macoun (Cat. Canad Pl., 1883) implied the occurrence of this species on the Canadian shore of Lake Superior as had Agassiz in 1850. In any event, the only collections yielded up by Lake Superior in the 20th century are from Marquette County, Michigan. I collected the species once in disturbed sand of a dump slightly inland at Mackinaw City.

Also found along the Atlantic coast, as var. *edentula*, with a very broad ovoid terminal segment in the fruit; this variety is very uncommon on the Great Lakes shores, where it is considered by Rodman to be introduced (also introduced on the Pacific coast). Most of our plants are var. *lacustris* Fern., with a lanceolate terminal segment in the fruit, only slightly broader than the basal segment. The European *C. maritima* Scop. is of similar appearance and habitat.

36. **Iberis** — Candytuft

KEY TO THE SPECIES

1. Fruit ca. 3–5 mm long; flowers usually white, fragrant; leaves ± irregularly toothed; inflorescence elongating in fruit . . . . 1. **I. amara**
1. Fruit ca. 5.5–6.5 [10] mm long; flowers usually pink, red, or purple, not fragrant; leaves entire; inflorescence remaining densely compact in fruit . . . . 2. **I. umbellata**

1. **I. amara** L. — Rocket Candytuft

Map 418. A native of Europe, widely cultivated and sometimes persisting for a while as an escape near gardens.

2. **I. umbellata** L. — Globe Candytuft

Map 419. A Mediterranean species, like the preceding an annual which occasionally spreads from cultivation for a time; our collections are from shores and waste places.

Other species of the genus, including some perennials such as *I. sempervirens* L., are also cultivated. *I. sempervirens* also differs from *I. umbellata* in its elongate fruiting racemes, and from *I. amara* in having larger fruit; it may spread and persist (as collected on the shore of Crystal Lake, Benzie County).

RESEDACEAE — Mignonette Family

This small family, native chiefly in the Mediterranean region, is peculiar in that the carpels are usually open apically, exposing the ovules and seeds. One genus is represented in our area by one or two European species, rarely escaped from cultivation, with pinnatifid leaves.

1. **Reseda** — Mignonette

1. **R. lutea** L.

Map 420. Collected at Jackson in 1892 (*S. H. Camp,* MSC) and said to spreading there (*Wheeler,* GH). Found by Farwell in a field at Livonia in 1929 and 1930. Not recognized in the state again until 1981 when it was found by H. Crum to be well established among grasses and weeds along Interstate 75 in Hebron Township, Cheboygan County.

Farwell reported collecting *R. alba* L. in 1897 on waste ground in Detroit, but no specimens are extant. The status of an 1895 Grand Rapids collection is not clear; it is not explicitly said to be escaped and the species was not included in Emma Cole's 1901 *Grand Rapids Flora*. *R. alba* differs from *R. lutea* in having greenish white (rather than greenish yellow) flowers, stamens persistent until fruit is ripe (rather than falling long before), and principal leaves with 4 or more (rather than 1–3) pairs of narrow lateral segments.

SARRACENIACEAE — Pitcher-plant Family

This is a small family, but one of immense scientific and popular interest. *Darlingtonia californica* Torrey is the California pitcher-plant, endemic to northern California and southern Oregon. The only other genus in North America is *Sarracenia,* of which all species are native in the southeastern United States. Only one of them ranges as far north as Labrador and as far west as Saskatchewan and the Mackenzie District.

418. Iberis amara

419. Iberis umbellata

420. Reseda lutea

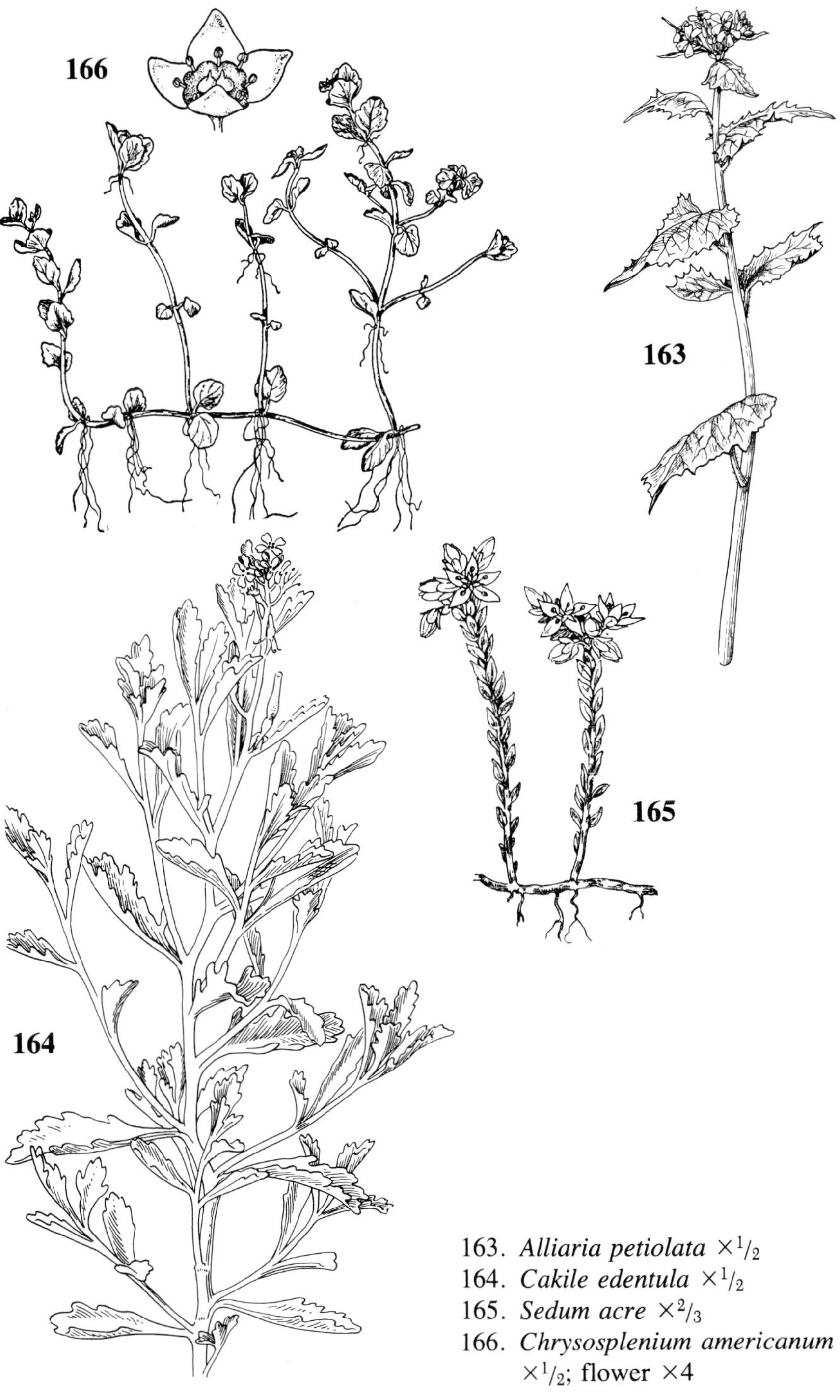

163. *Alliaria petiolata* ×1/2
164. *Cakile edentula* ×1/2
165. *Sedum acre* ×2/3
166. *Chrysosplenium americanum* ×1/2; flower ×4

1. **Sarracenia**

We have a single, absolutely distinctive species, the leaves modified into a pitcher-like shape in which water accumulates, drowning insects and other small organisms that slip into it, foiled in their escape by downward-pointing hairs on the inside of the pitcher. Apparently the plants do derive some nutrients from this source of protein. On the other hand, larvae of several species of moths feed on, or burrow in, our pitcher-plant and the larvae of a small, dark, non-biting mosquito, *Wyeomyia smithii* Coq., live only in the water held by pitcher-plant leaves.

The generic name honors Dr. Michel Sarrazin, who, as Royal Physician sent to Canada from France in 1697, was the first to make herbarium specimens of Canadian plants. His earliest collecting was in Newfoundland, for which the pitcher-plant is appropriately now the official flower.

In addition to a large popular literature on carnivorous plants, the following references are especially relevant locally.

REFERENCES

Case, Frederick W., Jr. 1956. Some Michigan Records for Sarracenia purpurea forma heterophylla. Rhodora 58: 203–207.

Mandossian, Adrienne J. 1965. Plant Associates of Sarracenia purpurea (Pitcher Plant) in Acid and Alkaline Habitats. Mich. Bot. 4: 107–114.

Mandossian, Adrienne J. 1966. Variations in the Leaf of Sarracenia purpurea (Pitcher Plant). Mich. Bot. 5: 26–35.

Mandossian, Adrienne J. 1966. Germination of Seeds in Sarracenia purpurea (Pitcher Plant). Mich. Bot. 5: 67–79.

1. **S. purpurea** L. Plate 3-F Pitcher-plant

Map 421. Sphagnum bogs and tamarack swamps, surviving a good deal of shade as succession advances; fens and boggy interdunal flats and pools—thriving in both acid and alkaline habitats.

The petals are normally a rich maroon, the calyx of similar red-purple shade, and the leaves at least slightly veined with red. Plants completely lacking this pigment (flowers yellow, leaves green to yellowish) have been called f. *heterophylla* (Eaton) Fern. This form was reported from Michigan by Farwell (Asa Gray Bull. [2](7): 45–46. 1894) and explicitly from the Marquette area by Gillman (Am. Nat. 4: 43. 1870), but the only material known is recent, from Montmorency County (Case 1956). This color form (*not* the whole species) is listed as "threatened" under Michigan's Endangered Species Act of 1974.

DROSERACEAE Sundew Family

The famous Venus' fly-trap, *Dionaea muscipula* Ellis, endemic to the Carolinas, is usually included in this family of insectivorous plants. The leaves of *Drosera,* fringed with gland-tipped "tentacles," are less dramatic than the bear-trap leaves of *Dionaea*. Nevertheless, our plants are able to hold not only very small animals, chiefly insects, but also larger ones. I once saw a monarch butterfly struggling unsuccessfully to free its legs and one forewing from entanglement in the leaves of *D. linearis* with their mucilaginous secretions.

1. **Drosera** Sundew

Very small or young leaves may be atypical in shape, and fully grown mature leaves should be examined whenever possible.

For many years a U.S. Coast Guard cutter named "Sundew" has operated on the Great Lakes. For over 20 years it was based in Charlevoix, whence it participated in many rescue missions on Michigan waters. It had previously been based in Wisconsin, and more recently, has been in Minnesota.

In addition to the references cited below and the annotations of C. E. Wood on certain critical material, I have received help from studies (unpublished) by L. J. Davenport and Allan Bornstein, largely on living material of all four taxa in the vicinity of the University of Michigan Biological Station.

REFERENCES

Cruise, James E., & Paul M. Catling. 1974. The Sundews (Drosera spp.) in Ontario. Ontario Field Biol. 28: 1–6.

Wood, Carroll E., Jr. 1955. Evidence for the Hybrid Origin of Drosera anglica. Rhodora 57: 105–130.

Wood, Carroll E., Jr. 1960. The Genera of Sarraceniaceae and Droseraceae in the Southeastern United States. Jour. Arnold Arb. 41: 152–163.

Wynne, Frances E. 1944. Drosera in Eastern North America. Bull. Torrey Bot. Club 71: 166–174.

KEY TO THE SPECIES

1. Leaves usually spreading (± prostrate or horizontal), with distinctly round to subreniform blades (at least as broad as long—excluding tentacles and slightly expanded summit of petiole); pollen (and hence mature fresh anthers) white; seeds fusiform, at least 4 times as long as broad, finely striate but otherwise smooth . . . . 1. **D. rotundifolia**

1. Leaves ± ascending or erect, with linear to obovate blades longer than broad; pollen (and hence fresh mature anthers) orange; seeds various (fusiform only in *D.* ×*anglica,* with rough surface)

  2. Blades of leaves linear (parallel-sided), 7–20 (25) times as long as wide; fresh petals pink-tinged . . . . 2. **D. linearis**

  2. Blades spatulate-obovate or cuneate, 2–6.5 times as long as wide; fresh petals pure white

    3. Scape arising laterally from base of plant before curving upward; petioles glabrous; blades 2–4 mm wide (excluding tentacles) . . . . 3. **D. intermedia**

    3. Scape strictly erect, arising centrally from base of plant; petioles at least sparsely pubescent or glandular; blades 3–8 mm wide . . . . 4. **D. ×anglica**

1. **D. rotundifolia** L.

Map 422. Bogs and cedar swamps, usually on sphagnum hummocks or mossy logs; also on damp sands (including excavations) and mossy crevices in rocks.

The petioles are ± pubescent with loose crooked hairs, sometimes a helpful character with depauperate sterile specimens that one might otherwise confuse with *D. intermedia,* which has glabrous petioles. Some pubescence, derived from this parent, is characteristic of the hybrid with *D. linearis,* and very rarely a hybrid with *D. intermedia* can be found.

2. **D. linearis** Goldie   Plate 4-B

Map 423. Marly bogs (or parts of bogs/fens) and shores, often abundant (when the water level is right) in interdunal hollows along the shores of Lakes Michigan and Huron and in open peaty hollows of bogs. Very rarely growing in sphagnum. This is the only one of our species not also found in Europe.

Normally found in more alkaline situations than the preceding or the next, but sometimes found growing side by side with *D. rotundifolia.* When these two are close, hybrids may be expected (see *D.×anglica*). The seeds of *D. linearis* are short and stubby (less than 4 times as long as wide), the surface densely and finely pebbled.

3. **D. intermedia** Hayne

Map 424. Usually in the wettest parts of bogs and on sandy shores subject to periodic inundation. It is almost never found with *D. linearis,* but all four taxa were found in 1984 within 1 meter of each other in a "string bog" in north-central Luce County.

Although in general this is a less robust plant than the next, with shorter scapes, measurements overlap a bit. The petioles tend to be a little longer, usually 2.5–3.5 times as long as the blades, whereas in the next they are usually 1–3 times as long as the blades. On unusually healthy plants, an occasional blade may be as broad as 5 mm. In water or very wet sphagnum, the internodes often elongate, so the plant bears alternate leaves at least

421. Sarracenia purpurea

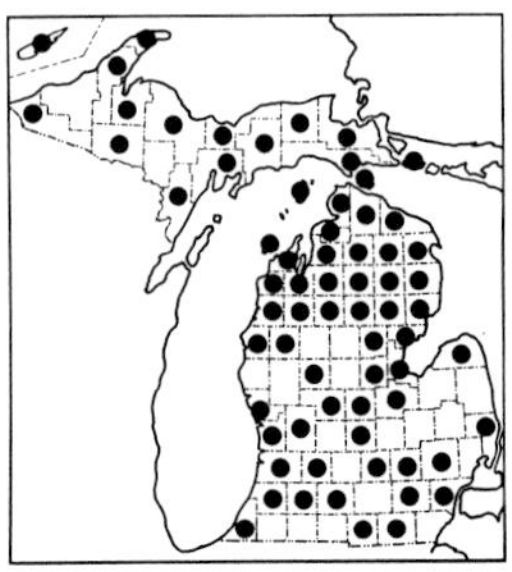

422. Drosera rotundifolia

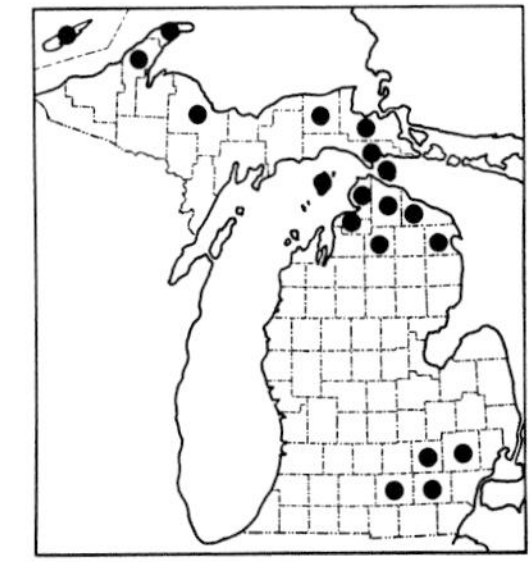

423. Drosera linearis

below a rosette. Such stem elongation apparently occurs only very rarely in *D. ×anglica*. Occasional specimens with pubescent petioles and spatulate leaf blades wider than in *D. intermedia* may be hybrids with *D. rotundifolia*.

4. **D. ×anglica** Hudson

Map 425. Interdunal calcareous flats, bogs (especially alkaline fens), rock pools (Isle Royale), marly shores.

One may expect F_1 hybrids of *D. linearis* with *D. rotundifolia* wherever the species grow near each other (e. g., *D. rotundifolia* on acid hummocks of sphagnum, the other on wet flats between the hummocks). These hybrids are sterile, the ovules and seeds aborting and no ripe fruit developing. Like the parents (and *D. intermedia*) they are diploid ($2n = 20$). Otherwise, they are morphologically indistingishable from the circumpolar *D. anglica* ($2n = 40$), a fertile species apparently derived from the hybrid by doubling of chromosome number. Such fertile plants are sometimes growing with the sterile hybrid in Michigan (see Wood 1955). All these plants are included here under the one name, in view of their identical vegetative morphology.

Some specimens have fairly narrow elongate blades and could conceivably be backcrosses with *D. linearis*. The petioles are ordinarily shorter, proportionately, than in *D. intermedia,* but are sometimes elongate, especially in very wet sites—where *D. intermedia* tends to produce elongate internodes. *D. ×anglica* is often confused with *D. intermedia,* but rarely occurs in the same bogs. The seeds of *D. intermedia* are less than 4 times as long as broad and are densely covered with prominent papillae. Seeds are not developed in the sterile hybrid; however, in the fertile amphiploid they are the same shape as in *D. rotundifolia* (fusiform, at least 4 times as long as broad) but with the surface roughened by dense elongate low knobs.

The name *D. longifolia* L. has sometimes been applied to *D. anglica* and sometimes to *D. intermedia,* and is now generally rejected.

CRASSULACEAE — Orpine Family

Species of several genera in this family of succulent plants are grown, especially in rock gardens, and some, such as *Sempervivum tectorum* L., hen-and-chickens, may turn up in waste heaps or other places where "transplanted." Additional species in *Sedum* are also likely to be found.

The genus *Penthorum,* here treated in the Penthoraceae, is sometimes included as a non-succulent member of the Crassulaceae.

1. **Sedum** — Stonecrop; Sedum; Orpine

Besides the species included here, others may spread from cultivation, for the plants propagate easily from detached fragments, often being distributed by water or other means. For identification one should consult works on garden plants or on the flora of the region whence the cultivated material came (if known). Clausen (1975) includes brief accounts (with keys) for species cultivated in North America as known as the time.

REFERENCES

Clausen, Robert T. 1975. Sedum of North America North of the Mexican Plateau. Cornell Univ. Press, Ithaca. 742 pp.

Cody, W. J. 1967. Sedum in the Ottawa District. Canad. Field-Nat. 81: 273–274.

Spongberg, Stephen A. 1978. The Genera of Crassulaceae in the Southeastern United States. Jour. Arnold Arb. 59: 197–248. [*Sedum,* pp. 206–226.]

KEY TO THE SPECIES

1. Leaves definitely flat (though succulent), the largest over 3 mm wide, toothed or entire, opposite, whorled, or alternate
  2. Leaves entire, some or all of them in whorls of 3; flowers white or yellow
    3. Flowers white, 4-merous; larger leaves 7–14 mm wide, obovate, on the sterile shoots crowded in terminal rosettes . . . . 1. **S. ternatum**
    3. Flowers yellow, 5-merous; larger leaves 3–4 mm wide, narrowly elliptic to lanceolate, not crowded in rosettes (internodes elongate) . . . . 2. **S. sarmentosum**
  2. Leaves irregularly toothed, all opposite or alternate; flowers white to (usually) pink or purple
    4. Leaves opposite, with prominent papillose-ciliate margins . . . . 3. **S. spurium**
    4. Leaves alternate, with smooth glabrous margins . . . . 4. **S. telephium**
1. Leaves terete to elliptical in cross-section, less than 3 mm wide, entire, alternate
  5. Flowers white to pink; leaves of flowering stem not overlapping, the internodes conspicuous
    6. Stem and branches of the inflorescence at least sparsely and minutely glandular-pubescent; petals usually 6, acuminate . . . . 5. **S. hispanicum**
    6. Stem (at least above) and inflorescence glabrous; petals 5, acute . . . . 6. **S. album**
  5. Flowers bright yellow; leaves of flowering stems ± densely overlapping, ordinarily obscuring the stem
    7. Leaves ± ovoid, broadest toward the base, not clearly ranked; petals ca. 5–7 mm long . . . . 7. **S. acre**
    7. Leaves linear-cylindric (like minute sausages), generally 6-ranked on sterile shoots; petals ca. 4–5 mm long . . . . 8. **S. sexangulare**

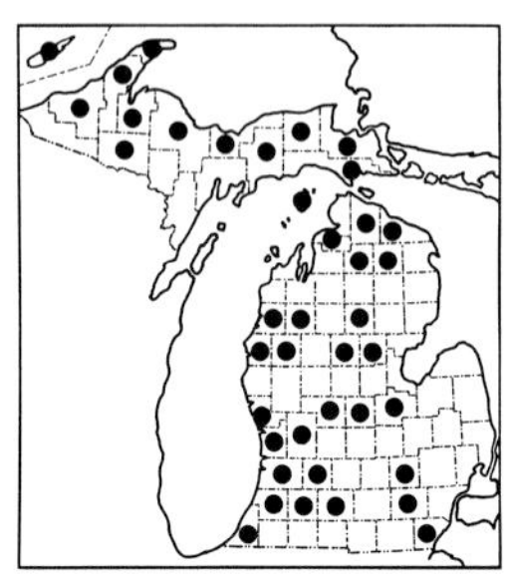

424. Drosera intermedia

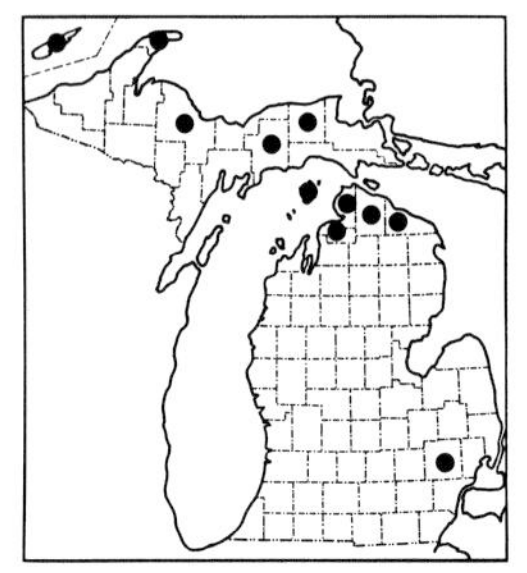

425. Drosera ×anglica

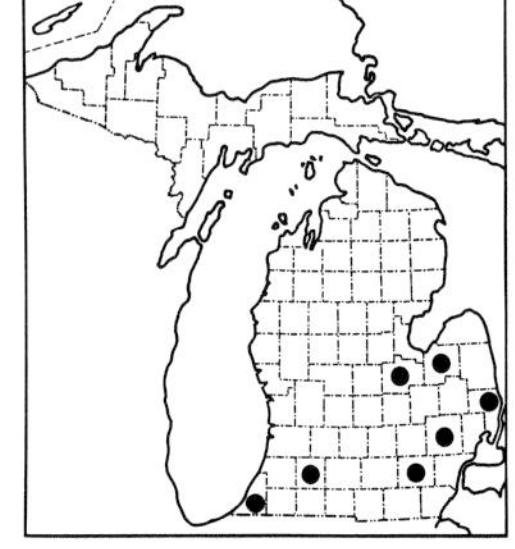

426. Sedum ternatum

1. **S. ternatum** Michaux

Map 426. This is the only one of our species of *Sedum* which is considered native in North America, though its range is south and east of Michigan, where it is naturalized in wooded areas or only a waif near cultivation, especially on rocky sites.

2. **S. sarmentosum** Bunge

Map 427. A native of eastern Asia, widely cultivated and occasionally spreading to waste places.

Easily recognized by the elongate prostrate stems with whorls of 3 narrowly elliptical pale green leaves, forming large mats.

3. **S. spurium** Bieb.

Map 428. A native of the Caucasus region, often cultivated and sometimes escaping onto shores, rock outcrops, and disturbed ground.

4. **S. telephium** L. Live-forever

Map 429. A highly variable and hardy Eurasian species, commonly cultivated and long-persisting around old homesites, sometimes spreading to shores, roadsides, railroads, clearings, and other disturbed ground, even invading woods and swamps.

Included in this species is *S. purpureum* (L.) Schultes [or *S. telephium* var. *purpureum* L.]. The tuberous perennial roots are thick, sometimes the size of a carrot (but white!). On some shoots the leaves may be opposite. The flowers are usually red-purple or at least rosy.

5. **S. hispanicum** L.

Map 430. A collection of what appears to be this species, a native of southeastern Europe and Asia, was made on a sunny roadside bank in Oakland Tp., Oakland County (*Churchill 747612* in 1974, MSC), where it was "well escaped and established."

The petals are distinctly papillose within, and the plants of the Michigan collection are somewhat taller than the usual low stature of this species.

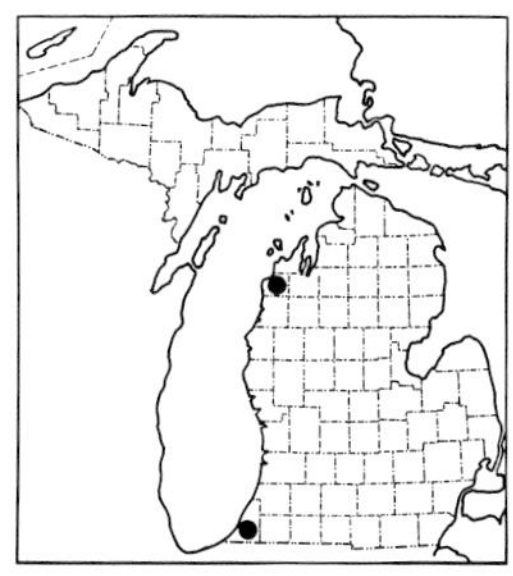

427. Sedum sarmentosum

428. Sedum spurium

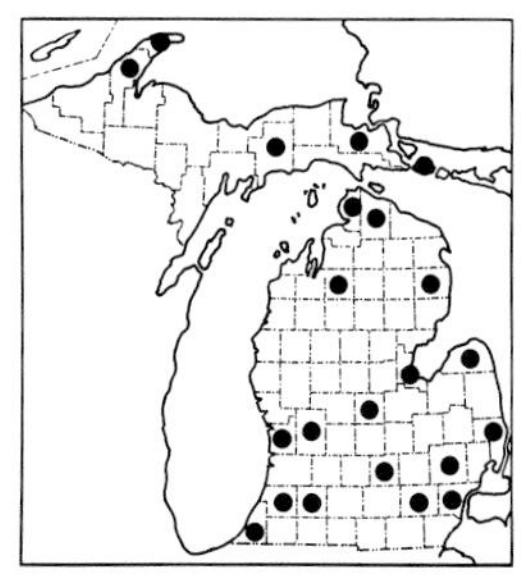

429. Sedum telephium

6. **S. album** L.

Map 431. An Old World species, occasionally found established outside of cultivation. Roadsides, shores, fields, and disturbed ground, especially near yards and cemeteries.

7. **S. acre** L. Fig. 165 Mossy Stonecrop

Map 432. This Old World species is widely naturalized in northern North America and is the only one commonly found in Michigan, where it grows on shores (where fragments have been washed) and dunes, along roadsides, near cemeteries, in lawns and fields, etc.

Forms dense moss-like mats which, because of the low stature, withstand mowing. When in bloom, the bright yellow flowers are often conspicuous along roadsides.

8. **S. sexangulare** L.

Map 433. A plant of Europe and adjacent Asia, superficially resembling *S. acre* in habit, but with smaller flowers and distinctive leaves. Locally established as an escape on the grounds of the University of Michigan Biological Station (*Thieret 48960* in 1976, MICH, UMBS) and to be expected elsewhere (see Mich. Bot. 16: 134. 1977).

PENTHORACEAE Ditch Stonecrop Family

This family consists of two species of *Penthorum,* one in eastern Asia and the other in eastern and midwestern North America. It is often placed as a subfamily of the Saxifragaceae and sometimes in the Crassulaceae. Its 5 nearly separate carpels, spreading and ripening rather like a group of follicles, closely resemble the latter family, but the thin non-succulent leaves and basal fusion of carpels and of sepals are closer to the Saxifragaceae. In its numerous alternate cauline leaves, *Penthorum* is readily distinguished from all of our Saxifragaceae, which have mostly basal or opposite leaves.

430. Sedum hispanicum

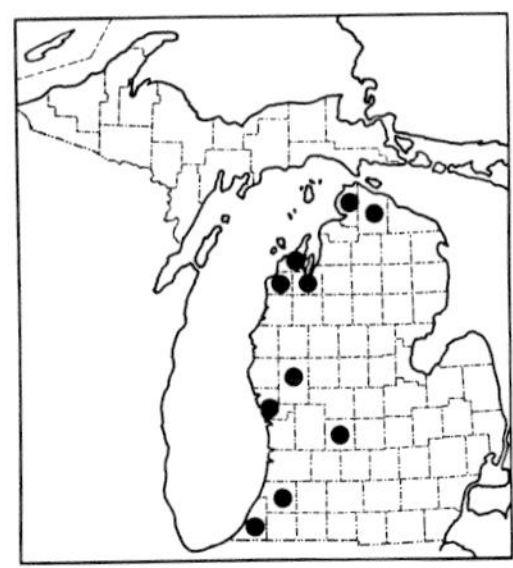

431. Sedum album

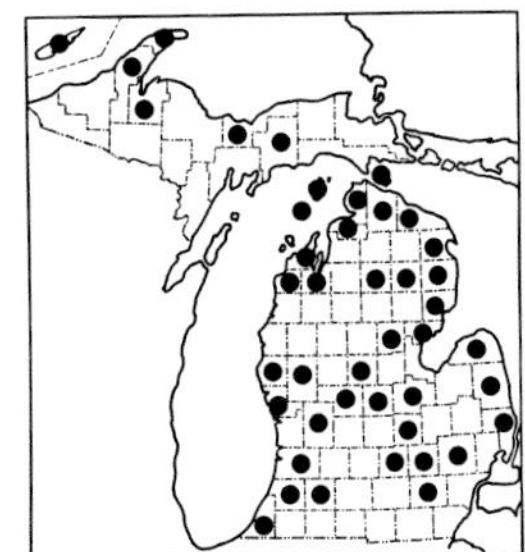

432. Sedum acre

1. Penthorum

REFERENCES

Baldwin, J. T., Jr., & Bernice M. Speese. 1951. Penthorum: Its Chromosomes. Rhodora 53: 89–91.

Spongberg, Stephen A. 1972. The Genera of Saxifragaceae in the Southeastern United States. Jour. Arnold Arb. 53: 409–498. [*Penthorum*, pp. 419–422.]

1. **P. sedoides** L. Plate 4-C Ditch Stonecrop

Map 434. Low ground, ditches, marshes, borders of streams and ponds.

Quite variable in stature, ranging from unbranched stems only a few cm tall to bushy branching plants over half a meter tall; but distinctly unlike any other species in our flora: the apetalous flowers are cream-colored, but as the carpels ripen into a star-like fruit (a cluster of follicles united at the base) they become quite red. The branches of the inflorescence are glandular-pubescent, ± curved, with flowers on the upper sides.

SAXIFRAGACEAE Saxifrage Family

There will probably never be full agreement on the delimitation of this family. Many authors define it very broadly, including a large and quite heterogeneous assortment of genera, placed in a number of subfamilies—as many as 17. Others recognize the subfamilies, or other aggregations of genera, as families. For convenience, in this work *Penthorum,* intermediate with the Crassulaceae and sometimes included in that family, is placed in the Penthoraceae; the woody plants with inferior ovaries and fleshy fruit—the genus *Ribes*—in the Grossulariaceae; and the remaining genera of our flora, all herbaceous with superior or mostly superior ovaries and dehiscent fruit, in the Saxifragaceae. Some authors would further segregate *Parnassia* in the Parnassiaceae. As Cronquist has aptly stated (Evol. Class. Fl. Pl. 230. 1968): "These are all matters of opinion, on which botanists may reasonably differ even when they have access to the same set of facts."

Several genera of cultivated shrubs with inferior ovaries are included in Saxifragaceae subfamily Hydrangeoidea or Hydrangeaceae, notably species and hybrids of *Philadelphus, Hydrangea,* and *Deutzia*. Some of these may spread (largely by suckers) in the garden but it is doubtful whether any are truly escaped or naturalized in our area. *Philadelphus,* known as mock-orange or "syringa" (also the generic name of the lilac, in the Oleaceae), of which some species are native in the southeastern United States and others in Europe and eastern Asia, has large, white, often fragrant flowers and simple, estipulate, opposite leaves; of the genera mentioned above, it is the most likely to be found persisting in thickets or slightly spreading from cultivation, but naming the species or hybrids is often extremely difficult.

REFERENCE

Spongberg, Stephen A. 1972. The Genera of Saxifragaceae in the Southeastern United States. Jour. Arnold Arb. 53: 409–498

KEY TO THE GENERA

1. Leaves all cauline (opposite), 4–11 (24) mm wide, on prostrate creeping stems; petals absent; flowers 4-merous . . . . 1. **Chrysosplenium**
1. Leaves all or mostly basal (no more than 2 cauline leaves, or several small alternate cauline bracts), often larger, the stems or scapes erect; petals present; flowers 5-merous
 2. Flowers solitary, with petals 5–16 (19) mm long and with branched staminodia, blooming in summer or fall; carpels (at least stigmas) 4; leaves entire . . . 2. **Parnassia**
 2. Flowers in racemes or panicles (or cymes), with petals less than 5 (5.5) mm long and no staminodia, blooming in spring or early summer; carpels 2; leaves toothed or crenulate
  3. Stamens 5 . . . . 3. **Heuchera**
  3. Stamens 10
   4. Petals deeply pinnatisect; carpels in fruit spreading widely, exposing the seeds in a shallow flattish cup . . . . 4. **Mitella**
   4. Petals entire; carpels in fruit not spreading but forming an elongate capsule or pair of follicles separate nearly to the base
    5. Leaf blades both lobed and irregularly toothed their entire length, strongly cordate, shorter than their petioles; carpels very unequal; inflorescence a simple raceme . . . . 5. **Tiarella**
    5. Leaf blades 3-toothed at apex, or at most with teeth or crenulations of equal size (or obscure), tapered at base (petiole usually shorter than the blade, obscure, or none); carpels equal; inflorescence branched . . . . 6. **Saxifraga**

1. Chrysosplenium

1. **C. americanum** Hooker Fig. 166 Golden Saxifrage

Map 435. Wet places, often forming large mats along streams and in cold springy spots or muddy hollows, especially in deciduous woods and cedar swamps.

A very inconspicuous little plant, with greenish flowers, only the 8 brick-red anthers offering any color when it blooms in May.

433. Sedum sexangulare

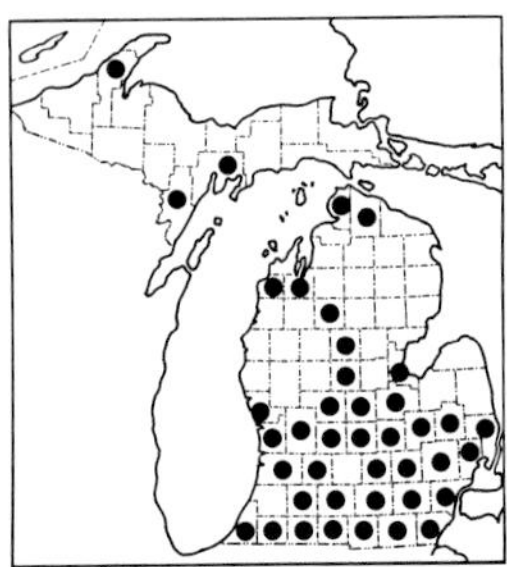

434. Penthorum sedoides

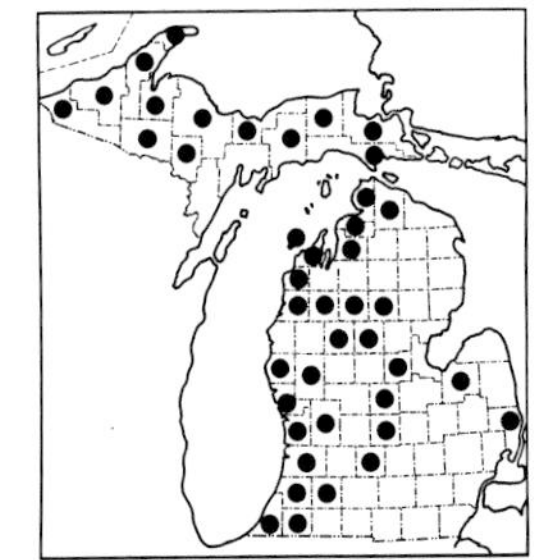

435. Chrysosplenium americanum

2. **Parnassia** Grass-of-Parnassus

An easily recognized genus, the conspicuous white or creamy petals strongly veined with greenish or yellowish. The solitary flowers are on scapes bearing at most one leaf-like bract, and appear late in the season (*P. parviflora* earlier than the others). Alternating with 5 anther-bearing stamens there are 5 staminodia, each cleft into 3 or more slender filaments bearing a glandular tip.

KEY TO THE SPECIES

1. Staminodia cleft nearly to the base into 3 segments; leaves coriaceous (tough, almost succulent); petals (11) 12–16 (19) mm long . . . . 1. **P. glauca**
1. Staminodia cleft 40–60 (75)% to the base into 5 or more segments; leaves thin and membranous; petals 5–13 mm long
   2. Blades of larger basal leaves 11–25 mm broad, at least slightly cordate; staminodia with 9 or more filaments; petals 8–13 mm long . . . . 2. **P. palustris**
   2. Blades of larger basal leaves 4–13 (15) mm broad, tapered and not at all cordate at base; staminodia with 5–7 (8) filaments; petals 5–9 mm long . . . . . . 3. **P. parviflora**

1. **P. glauca** Raf. Fig. 167

Map 436. Calcareous sandy and gravelly shores, interdunal flats, meadows, stream banks; ± alkaline bog mats and marshes.

This is our commonest and most widely distributed species in the state, and the latest to reach the peak of blooming. It also has the largest flowers and the largest leaves, although overlapping *P. palustris* in these parts. The leaf blades are as much as 4–5 (5.8) cm wide, although more often ca. 2–3 cm (and sometimes all the leaves even smaller), and they are at most subcordate at the base.

2. **P. palustris** L.

Map 437. Damp rocks, shores, meadows, and bogs (especially sedgy mats).

The single cauline leaf is usually nearly or quite as large as the basal leaves (sometimes even larger), whereas in the other two species it is smaller and usually inconspicuous—if present at all. Plants of this circumpolar species from northern North America have been called var. *neogaea* Fern. (other varieties occur in the west).

3. **P. parviflora** DC.

Map 438. Damp calcareous sandy shores, interdunal flats, excavations, bluffs, and rocks along the Great Lakes. Often with *P. glauca*.

3. **Heuchera** Alum-root

A red-flowered species of the southwest and Mexico, *H. sanguinea* Engelm. (coral-bells), is hardy in the north and often cultivated. Our native species have greenish white or dull yellowish flowers. In determining flower measurements as given in the key, mature, but not over-mature, flowers should be used; by the time the anthers are shed, the flowers are generally larger.

REFERENCES

Rosendahl, Carl Otto, Frederic K. Butters, & Olga Lakela. 1936. A Monograph on the Genus Heuchera. Minnesota Stud. Pl. Sci. 2. 180 pp.

Wells, Elizabeth Fortson. 1984. A Revision of the Genus Heuchera (Saxifragaceae) in Eastern North America. Syst. Bot. Monogr. 3: 45–121.

KEY TO THE SPECIES

1. Perianth at anthesis (5) 6–9 mm long, the deepest sinus ca. half the length of the calyx . . . . 1. **H. richardsonii**
1. Perianth at anthesis 3–5 (5.5) mm long, the deepest sinus usually less than half the length of the calyx . . . . 2. **H. americana**

1. **H. richardsonii** R. Br.

Map 439. Dry prairies and sandy river banks, rocky ground, oak woodland, rarely in lowland woods.

The flowers are quite bilaterally symmetrical, the upper lobe of the perianth nearly twice as long as the lower side; and the stamens (including anthers) are exserted less than 2 mm beyond the upper side of the perianth—or less than half its length. Considering the larger and more bilateral flowers, longer calyx lobes, and less exserted stamens, one can usually separate this species from the next.

2. **H. americana** L.

Map 440. Deciduous woods, often on dry to damp banks and ravines.

Typical plants of this species (in both the nomenclatural and the variational sense) have the flowers regular or nearly so and the petioles glabrous to very sparsely hirsute. Most of our plants unfortunately approach the pre-

436. Parnassia glauca

437. Parnassia palustris

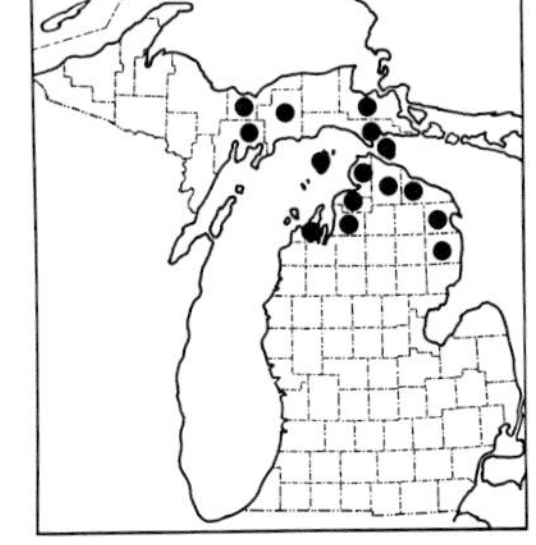

438. Parnassia parviflora

ceding species in a tendency to bilateral flowers and very hirsute petioles, and are referred to var. *hirsuticaulis* (Wheelock) Rosendahl, Butters, & Lakela. They presumably reflect the influence of hybridization in the past, and the lines between varieties, as well as species, are difficult to draw. Wells (1984) distinguishes var. *hirsuticaulis* from *H. richardsonii* "by its shorter calyces and free hypanthia [1.1–1.9 mm long]"; in *H. richardsonii,* the free hypanthium (floral tube above the half-inferior ovary) is "2–6.4 (–7) mm long." The stamens are strongly exserted in *H. americana,* usually protruding beyond the longest calyx lobe by 60% or more the length of the perianth (by 3–5 mm, according to Wells).

4. **Mitella** — Miterwort

KEY TO THE SPECIES

1. Flowers white or cream-colored, ca. 2.5–4.5 mm long; plants at full anthesis (1.4) 2–4 dm tall; flowering stem normally with 2 opposite leaves (other leaves basal); all leaves acute . . . . 1. **M. diphylla**
1. Flowers greenish yellow or even reddish, ca. 4.5–6.5 mm long; plants less than 1.5 (2.3) dm tall; flowering stem naked or with at most 1 small leaf; all leaves obtuse or rounded . . . . 2. **M. nuda**

1. **M. diphylla** L. Plate 4-D — Bishop's-cap

Map 441. Typically in rich deciduous woods, often in wet hollows or slopes; also in swampy woods, sometimes even cedar swamps.

Ordinarily easily recognized by the pair of opposite, horizontally oriented sessile leaves above the middle of the stem; very rarely plants occur with the cauline leaves 3 or only 1, or aborted, or (more often) with short petioles. In the prominent acute terminal lobe, the leaves are more like those of *Tiarella* than *M. nuda*. The rhizome is stout and the whole plant taller than *M. nuda,* though the flowers are a little smaller and the pedicels usually a little shorter (ca. 1.5 mm or less at anthesis).

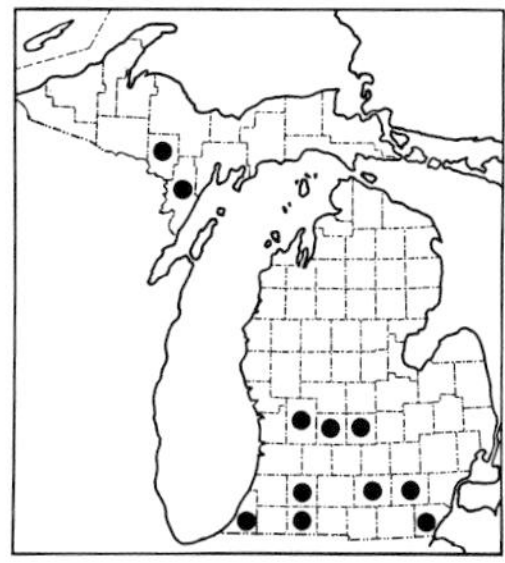
439. Heuchera richardsonii

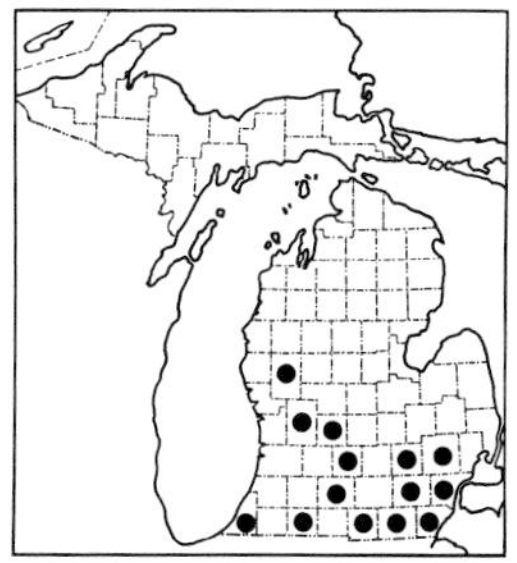
440. Heuchera americana

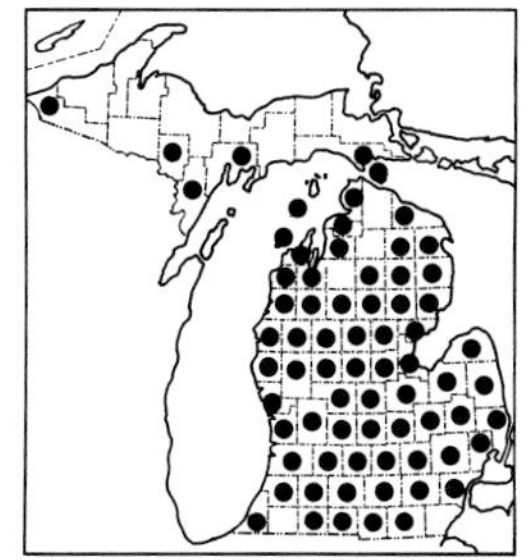
441. Mitella diphylla

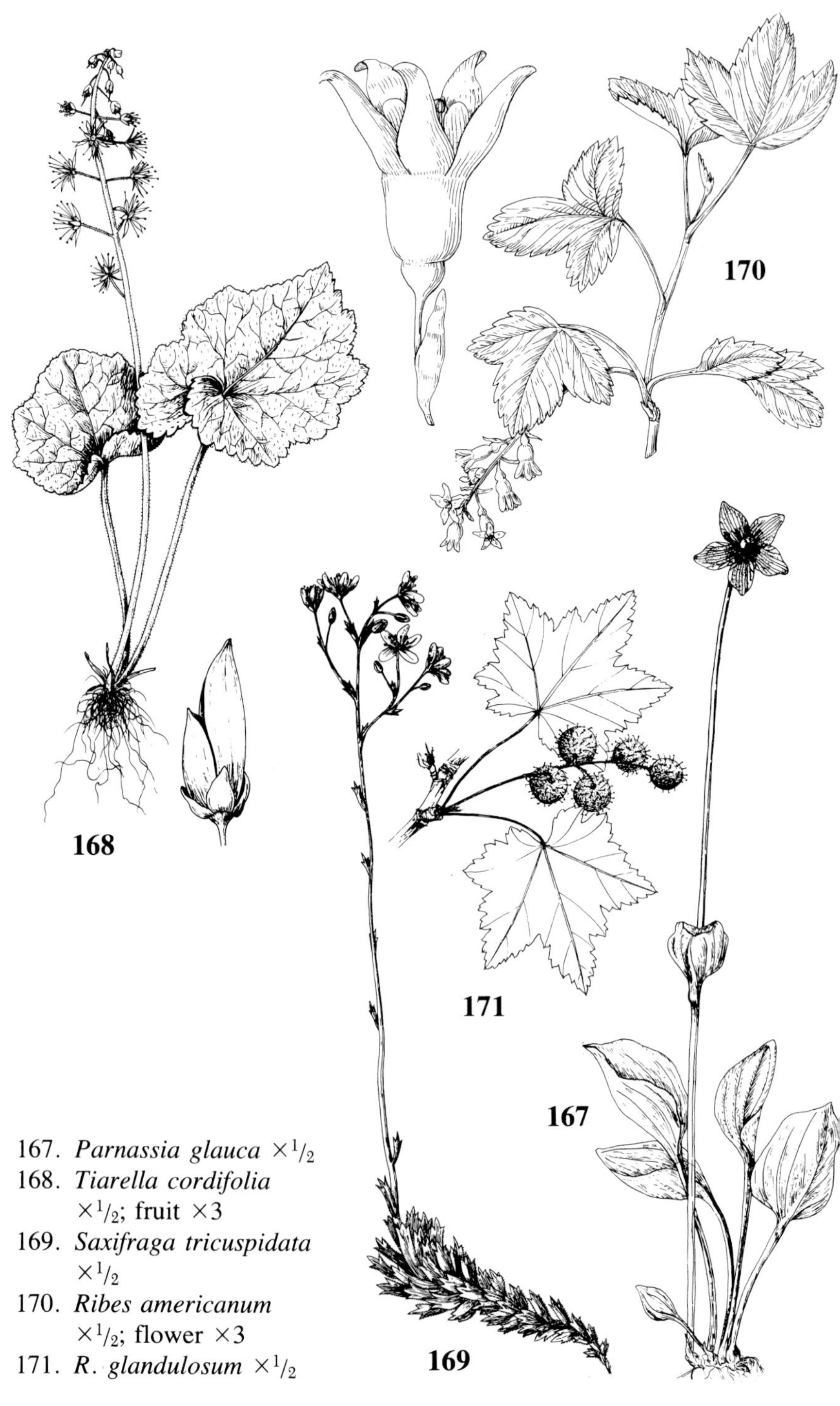

167. *Parnassia glauca* $\times{}^{1}/_{2}$
168. *Tiarella cordifolia* $\times{}^{1}/_{2}$; fruit $\times 3$
169. *Saxifraga tricuspidata* $\times{}^{1}/_{2}$
170. *Ribes americanum* $\times{}^{1}/_{2}$; flower $\times 3$
171. *R. glandulosum* $\times{}^{1}/_{2}$

2. **M. nuda** L. Plate 4-E Naked Miterwort

Map 442. Most often in cedar swamps and thickets, but also in moist mixed woods and spruce-fir stands; often on mossy logs and hummocks.

A plant with very slender rhizomes and stolons, abundant and distinctive even when sterile. The *erect* hairs on the upper side of the crenate leaf blades will tell it from everything similar except perhaps the rare *Dalibarda repens*. However, the leaf of *Dalibarda* has more regular small crenations (over 14 on a side) and the petioles are covered with fine appressed hairs. In *M. nuda*, the petioles have ± spreading stiff hairs, usually gland-tipped.

5. Tiarella

REFERENCE

Lakela, Olga. 1937. A Monograph of the Genus Tiarella L. in North America. Am. Jour. Bot. 24: 344–351.

1. **T. cordifolia** L. Fig. 168 Foamflower; False Miterwort

Map 443. Deciduous and mixed woods, often in wet hollows or springy places; swamp forests; sometimes associated with cedar and hemlock, especially at southern stations. Michigan is at the western edge of the range of this species in North America; it is known from Wisconsin only in two northeastern counties (Florence, adjacent to Michigan, and Door). The species seems to be absent from the southwestern Lower Peninsula except for Berrien County, where it was collected in 1917 (*E. Spaulding*, WIS) and again in 1977 (*M. Medley*, MICH, MSC, MOR).

The foliage is remarkably similar to that of *Mitella diphylla*, but the petals are entire, the flowers on longer pedicels (ca. 4–10 mm), the carpels nearly separate and quite unequal, and the scape naked (rarely with a small leaf or bract).

6. Saxifraga Saxifrage

KEY TO THE SPECIES

1. Basal leaves with cartilaginous sharp teeth (either 3 spines at the end, or regular flat teeth with lime encrustations); petals usually dotted with red or purplish; flowering stem with several bracts or small leaves
 2. Margins of basal leaves with crowded teeth and white lime-encrusted pores . . . . 1. **S. paniculata**
 2. Margins of basal leaves entire except for (2) 3 prominent spine-tipped apical teeth . . . . 2. **S. tricuspidata**
1. Basal leaves minutely crenulate, nearly entire, or with obtuse green teeth; petals not dotted; flowering stem at most with 1–2 small bracts between the inflorescence and basal leaves

3. Larger leaves 3–7.5 cm long or even shorter, distinctly toothed, especially around the apex; petals (3) 3.5–5 (5.5) mm long; plant of rocky, mostly dry habitat . . . . 3. **S. virginiensis**

3. Larger leaves (10) 14–35 cm long, obscurely crenulate-toothed to remotely denticulate, entire around the apex; petals (2) 2.5–3 (3.5) mm long; plants of damp, swampy habitat . . . . 4. **S. pensylvanica**

1. **S. paniculata** Miller — Lime-encrusted Saxifrage

Map 444. Crevices in basic rock along Lake Superior at Isle Royale, not yet found on the Keweenaw County mainland (but occurs farther south, east of Lake Huron in Ontario). A boreal species which ranges from Iceland, Greenland, and the eastern Canadian arctic south in North America to the northern Great Lakes, New England, and northern New York. In central and southern Europe as well as the Caucasus, it is a mountain plant, and it also grows locally in Norway. The range is thus quite different from the usual circumpolar arctic-alpine pattern of many of our Lake Superior specialties.

This is the species long known as *S. aizoön* Jacq., a later name.

2. **S. tricuspidata** Rottb. Fig. 169 — Prickly Saxifrage

Map 445. This is a North American arctic species, coming as far south as the United States only in Isle Royale National Park, where it grows on rock on the main island and on several of the offshore islands at the northeastern end of the archipelago. It is apparently very local on the nearby Canadian shore. I know of no evidence for the indication in *Gray's Manual* that the species is also Eurasian.

3. **S. virginiensis** Michaux Plate 4-F — Early Saxifrage

Map 446. Crevices and shallow soil on limestone (Drummond Island) and other rock, including the Porcupine and Huron mountains. Although very northern in this state, ranges south to Georgia and Arkansas.

4. **S. pensylvanica** L. — Swamp Saxifrage

Map 447. Swampy and marshy places, often calcareous, especially damp

442. Mitella nuda

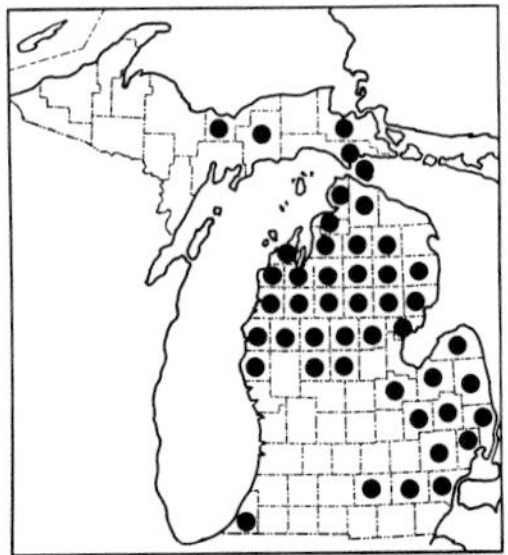

443. Tiarella cordifolia

444. Saxifraga paniculata

deciduous woods and grassy borders, tamarack swamps, and springy rocky ground.

A taller, larger-leaved, thicker-stemmed plant than our other species, differing also in its habitat. The flowers are greenish, while in *S. virginiensis* they are white (though without the dots of the first 2 species).

GROSSULARIACEAE — Gooseberry Family

As noted under the Saxifragaceae, this family is often included in that one. Even when it is not, its circumscription varies. Our only genus is readily distinguished by its inferior ovary ripening into a berry and its woody habit. Even this genus, *Ribes,* is sometimes split into two (or more) genera, but present opinion is largely uniform in recognizing the currants and gooseberries in a single genus.

1. **Ribes** — Currant; Gooseberry

The gooseberries have sometimes been separated into *Grossularia,* but the distinction is based almost solely on whether the pedicel is jointed, other characteristics not being well correlated.

Above the inferior ovary is a ± well developed floral tube, either cylindrical or flaring, with prominent calyx lobes. The petals are inserted on the tube and are rather small, never exceeding the calyx lobes, which are the more conspicuous part of the perianth. All species bloom in May or early June.

Ribes has achieved some notoriety as the alternate host for the white pine blister rust, *Cronartium ribicola,* a serious fungal disease of an important timber tree—the state tree of Michigan. The rust was introduced into North America from Europe about 1900 and spread rapidly. In the 1930's and early 1940's, aggressive campaigns were waged against both wild and cultivated species of *Ribes,* attempting to eradicate them within a certain distance of

445. Saxifraga tricuspidata

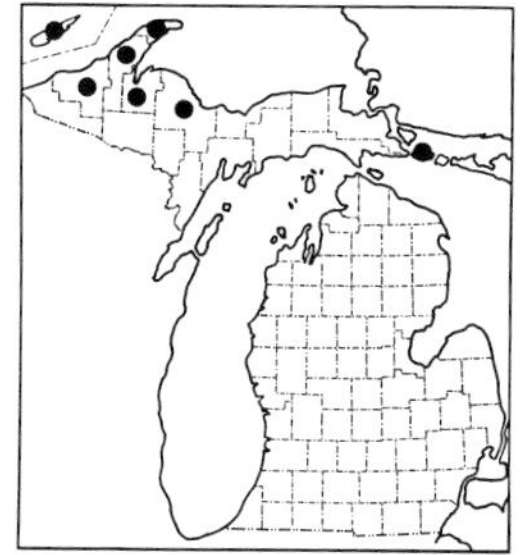
446. Saxifraga virginiensis

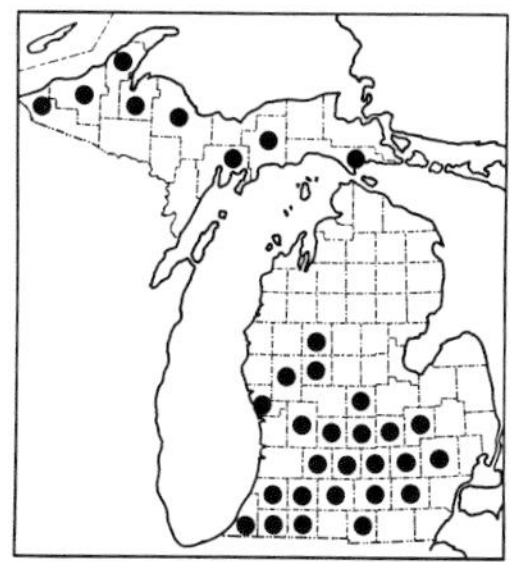
447. Saxifraga pensylvanica

white pine. The rust fungus does not spread from one pine to another but certain spores must grow on *Ribes* to complete the life cycle; other spores are then blown from *Ribes* to pine needles, where new infections occur. Control measures depend in part on local climatic conditions, which are important in governing spread of the airborne spores; pine stands can be managed to take maximum advantage of conditions unfavorable to the spread and establishment of the infecting spores.

Ribes alpinum L. (mountain or alpine currant), a European species, may persist from cultivation as an ornamental. It is functionally dioecious (though flowers have rudimentary organs of the opposite sex), with small greenish flowers (the fruit red) in strongly ascending racemes, rather small leaves, the ovaries and leaves essentially glabrous and glandless. It has been collected as "long persistent" in northwestern Baraga County.

REFERENCES

Berger, Alwin. 1924. A Taxonomic Review of Currants and Gooseberries. New York Agr. Exp. Sta. Tech. Bull. 109. 118 pp.

Darlington, Henry T., & Lawson B. Culver. 1939. Keys to the Species of Ribes Occurring in the Great Lakes Region. Mich. Agr. Exp. Sta. Circ. Bull. 170. 24 pp.

Fassett, Norman C. 1932. Preliminary Reports on the Flora of Wisconsin. XIX. Saxifragaceae. Trans. Wisconsin Acad. 27: 237–246.

Miller, Douglas R., James W. Kimmey, & Marvin E. Fowler. 1959. White Pine Blister Rust. U.S. Dep. Agr. For. Pest Leafl. 36. 8 pp.

Van Arsdel, Eugene P. 1961. Growing White Pine in the Lake States to Avoid Blister Rust. U.S. Dep. Agr. Lake States For. Exp. Sta. Sta. Pap. 92. 11 pp.

KEY TO THE SPECIES

1. Flowers solitary or in corymb-like clusters of 2–3 (4); pedicels not jointed at the summit; stems usually with bristles or spines at least at the nodes (sometimes completely unarmed, especially in *R. hirtellum*)
 2. Flowering peduncles (from base to lowest bract) mostly 7–18 (20) mm long; bracts fringed with gland-tipped hairs; calyx lobes of most flowers shorter than the tube; ovary and fruit at least sparsely prickly with stiff bristles (occasionally smooth, rarely pubescent) . . . . 1. **R. cynosbati**
 2. Flowering peduncles ca. 2–3 (5) mm long; bracts glandless (in common species) or ± glandular; calyx lobes about equaling or exceeding the tube; ovary and fruit smooth or (in rare species) ± prickly
  3. Leaves without glands, the blade cuneate to subcordate; bracts also glandless (or sparsely glandular-margined); stamens at maturity equaling or slightly exceeding calyx lobes, distinctly exceeding the petals; fruit smooth . . . . 2. **R. hirtellum**
  3. Leaves with sessile or stalked glands (at least on veins beneath, sometimes on both surfaces), the blades truncate to subcordate at base; bracts glandular-margined; stamens about equaling the petals (ends of anthers at most barely exceeding them); fruit smooth to glandular-setose or bristly . . . . 3. **R. oxyacanthoides**

1. Flowers in racemes of usually 5 or more; pedicels jointed at the summit (i. e., fruit articulated at base); stems without bristles or spines (except in the very prickly *R. lacustre*)

4. Stems ± densely prickly; ovary and fruit with gland-tipped hairs or bristles (berries black, in drooping racemes) . . . . 4. **R. lacustre**
4. Stems without bristles or spines of any kind; ovary and fruit smooth (except in *R. glandulosum,* with red berries in ascending racemes)
5. Floral tube 9–15 mm long; calyx lobes ca. half as long as tube or shorter; leaf blades less (usually much less) than 3.5 cm long, even when full grown, 3 (5)-lobed, each lobe with at most 2–3 (7) teeth . . . . 5. **R. odoratum**
5. Floral tube less than 4.5 mm long; calyx lobes about equaling the tube or longer; leaf blades often more than 3.5 cm long, at least in summer, and with many teeth
6. Leaves with resinous dots, at least on under side of blade; ripe fruit black
7. Bracts of inflorescence longer than the pedicels; calyx glabrous or with a few scattered hairs; ovary without resinous dots; flowers yellow (or cream) to greenish, in pendent (or rarely only spreading) racemes . . . . 6. **R. americanum**
7. Bracts of inflorescence much shorter than the pedicels; calyx ± pubescent; ovary usually with a few resinous dots; flowers white or whitish (greenish white to purplish in *R. nigrum*), in pendent to erect racemes
8. Flowers and fruit in erect or ascending racemes; calyx lobes ca. 3 times as long as the tube; plant native, in northern Michigan . . . . 7. **R. hudsonianum**
8. Flowers and fruit in spreading to pendent racemes; calyx lobes about equaling the tube or slightly longer; plant an escape from cultivation near roads and gardens . . . . 8. **R. nigrum**
6. Leaves without resinous dots (sometimes short-stalked glands on veins beneath); ripe fruit red
9. Ovary and fruit with gland-tipped hairs or bristles; bruised foliage and fruit with skunk-like odor; flowering and fruiting racemes ascending . . . . 9. **R. glandulosum**
9. Ovary and fruit smooth; plant without skunk-like odor; flowering and fruiting racemes spreading to pendent
10. Pedicels essentially glabrous and eglandular; anthers ± dumbbell-shaped, the lobes well separated by the connective; terminal lobe of leaf ± ovate (sides rounded in outline) . . . . 10. **R. rubrum**
10. Pedicels (or most of them) with a few very short-stalked glands and usually also ± loosely pubescent; anthers ± cordate, the lobes contiguous; terminal lobe of leaf usually broadly deltoid (sides straightish) . . . . 11. **R. triste**

1. **R. cynosbati** L. Wild or Prickly Gooseberry

Map 448. Especially characteristic of beech-maple woods and northern hardwoods (where other gooseberries are not expected), but also in more swampy woods and thickets, and even occasionally in oak-hickory stands, tamarack and poison sumac or cedar swamps, or dry woodlands on rock outcrops.

Sometimes the internodes are densely prickly but usually at least the younger ones are smooth or nearly so and the only spines are 1–3 at the nodes. Rarely even these are lacking, and on some plants the fruits are smooth [f. *inerme* Rehder]. The ovary and fruit (even in natural habitats far from any source of cultivated stock) are rarely slightly to densely pubescent, ordinarily a character of the garden gooseberry. The berry is reddish when ripe, and edible, but the tough prickly skin is not appealing.

Specimens with smooth fruit or with unusually stout spines (as on shoots) have sometimes been identified as *R. missouriense* Nutt., a species found south and west of our area; however, it has very long-exserted stamens, while in *R. cynosbati* the stamens do not exceed the calyx lobes. A collection from Van Buren County (*Pepoon 830* in 1906, MSC) bearing a single fruit with filaments apparently much exserted from the withered corolla, as well as stout spines, may indeed be *R. missouriense*.

The garden gooseberry, *Ribes uva-crispa* L. [also known as *R. reclinatum* L. and *R. grossularia* L.], a native of Europe, sometimes escapes from cultivation in North America and it or its hybrids might be expected in Michigan. The stem has well developed nodal spines. The fruit is usually pubescent and the calyx lobes are longer than the tube. If as usual there are long prickles on the fruit, they are prominently gland-tipped, while in *R. cynosbati* there are no glands (or only very tiny ones) terminating the prickles on the fruit. The peduncles are very short in *R. uva-crispa*.

Ordinarily *R. cynosbati* is easily distinguished by the long peduncles (and pedicels), the flowers and fruit thus held well away from the stems, and the short calyx lobes. The leaf blades are slightly pubescent above, truncate to subcordate, thus quite different from those of *R. hirtellum* var. *hirtellum* (but similar to other plants of *R. hirtellum*). In both this species and the next there are no glands on the leaves, but in *R. cynosbati* the bracts are ± prominently glandular-ciliate. See also comments below.

2. **R. hirtellum** Michaux — Swamp Gooseberry

Map 449. Cedar and tamarack swamps, rocky openings in mixed woods, gravelly shores and edges of woods, shrubby thickets along streams and lakes.

In typical var. *hirtellum,* the leaf blades are ± strongly cuneate at the base and glabrous or almost so above. However, rather commonly the blades are nearly truncate to subcordate, pubescent on both surfaces (± densely so beneath), representing var. *calcicola* (Fern.) Fern. The two varieties are not always clearly separated and may be of relatively little significance. The

448. Ribes cynosbati

449. Ribes hirtellum

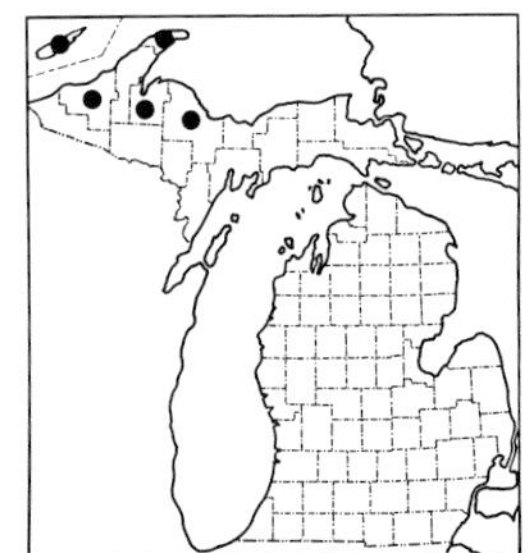
450. Ribes oxyacanthoides

bracts of the inflorescence are occasionally very sparingly glandular, at least on some inflorescences of a plant, but the glands are usually overtopped by the copious white cilia typical of the bracts in this species.

The gooseberries are the most difficult group of *Ribes* and one suspects that there is some hybridization, especially in the northern part of the state. Any one character in the key is open to exception.

3. **Ribes oxyacanthoides** L. Northern Gooseberry

Map 450. Gravelly and rocky clearings, rock cliffs and summits including the Huron and Porcupine mountains.

In many older works, *R. hirtellum* was not clearly distinguished from this species, now understood to be a subarctic American one, barely ranging south into the United States, chiefly near Lake Superior. Reports of *R. setosum* Lindley from this region seem to be based on the smaller-flowered *R. oxyacanthoides* (or even in some instances on *R. cynosbati*). In *R. setosum,* which is similar in the glandular bracts and leaves, the flowers have a conspicuous cylindrical floral tube and the perianth is usually at least 10 mm long. In our three wild gooseberries, the perianth (measured with extended calyx lobes) is ca. (4) 5–8 mm long.

A few plants vary in having leaves with extremely obscure glands, ovary pubescent, or stamens clearly exceeding the petals, and could presumably represent the result of hybridization rather than merely variation in *R. oxyacanthoides*. The rare specimen of *R. hirtellum* with glandular bracts may be distinguished from this species by eglandular leaves and longer stamens (and from *R. cynosbati* by the short peduncles and in some cases cuneate leaf bases and smooth fruit).

4. **R. lacustre** (Pers.) Poiret Swamp Black Currant

Map 451. In coniferous, mixed, or rarely deciduous woods, swamps, and thickets, often in dense shade, sometimes associated with rock outcrops.

Usually easily recognized by the densely spiny glossy light brown young twigs, as well as the relatively deeply lobed leaves (sinuses ca. two-thirds to three-fourths of the way to the base of the blade). The flowers are reddish

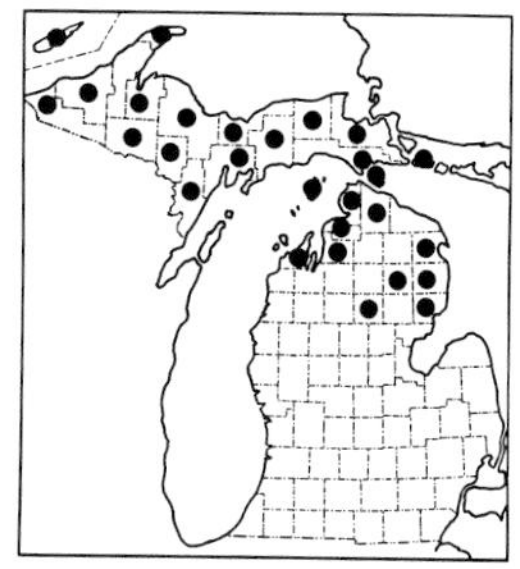
451. Ribes lacustre

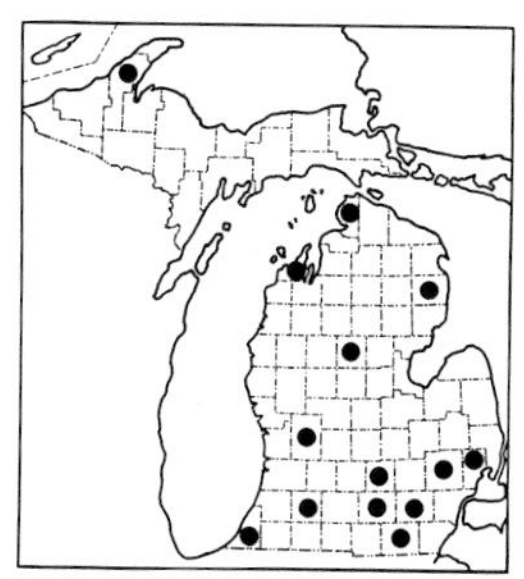
452. Ribes odoratum

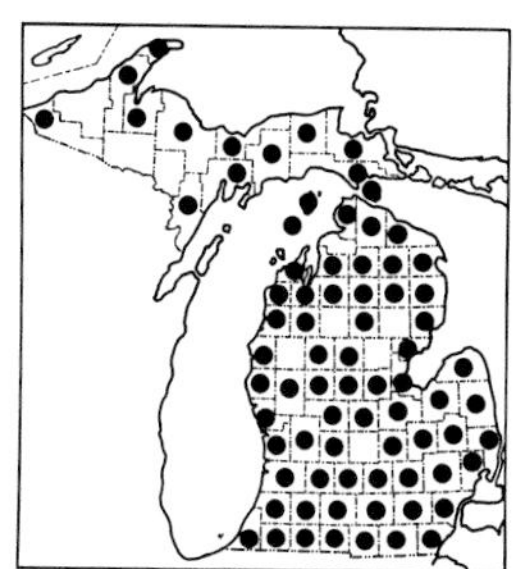
453. Ribes americanum

or yellowish and are quite flat or saucer-like—thus readily distinguishing them, even when fewer than 5 are present in a raceme, from the gooseberries (which have a clearly cylindrical or campanulate floral tube).

5. **R. odoratum** Wendl. f. Buffalo or Golden Currant

Map 452. Roadsides, thickets, fencerows, borders of fields, woods (especially near old yards and cemeteries). A native of mid-America, west of the Great Lakes, but widely cultivated for its yellow, spicy-fragrant flowers, and occasionally escaping.

Sometimes not distinguished from *R. aureum* Pursh, a native species centered in the Pacific Northwest with similar tubular but smaller flowers and the calyx lobes as long as the tube.

The young leaves and twigs are sometimes dotted with resinous atoms, but these are apparently largely or entirely lost later in the season. The current year's twigs at maturity are densely and finely pubescent. The fruit of at least some cultivated forms is considered edible.

6. **R. americanum** Miller Fig. 170 Wild Black Currant

Map 453. In a diversity of open woods and thickets, mostly damp, including beech-maple forests, swamp forests (deciduous or cedar), and old tamarack stands, as well as marshy sites and banks of streams and lakes.

The flowers are fairly large and conspicuous, and numerous in many racemes, so this is a frequently noticed *Ribes*. The fruit is considered quite palatable when cooked.

Most *Ribes* can be distinguished by vegetative characters alone, although these are not always easily expressed in a key. The leaves in *R. americanum* are usually glandular-dotted above as well as beneath, whereas in *R. hudsonianum* the leaves are gland-dotted only beneath. The leaf shapes are indistinguishable. Sterile plants from the range of *R. hudsonianum* without glands on the upper surfaces of the leaf blades cannot safely be named when dry, although fresh foliage of *R. americanum* lacks the strong scent of *R. hudsonianum* and *R. nigrum*.

7. **R. hudsonianum** Richardson Northern Black Currant

Map 454. Usually in swamps mostly or entirely of cedar, sometimes in other moist, especially coniferous, woods.

The year-old twigs of this species are ± heavily glandular with resinous atoms, while in the next these are relatively few if any.

8. **R. nigrum** L. Black Currant

Map 455. A Eurasian species, cultivated and sometimes escaped to damp places along rivers, swamps, and ponds.

This is the black currant of gardens—the true currant of black currant preserves and not the small raisin called a currant in commerce.

9. **R. glandulosum** Grauer Fig. 171 Skunk Currant

Map 456. Moist or boggy woods and thickets, cedar (and tamarack) swamps, ravines and banks in deciduous woods, with spruce and fir northward.

The fruit very rarely lacks the characteristic bristles.

10. **R. rubrum** L. Red Currant

Map 457. A native of western Europe, but widely cultivated and sometimes escaped to fields, fencerows, vacant (and not so vacant) lots, thickets, and woods.

Following *Flora Europaea,* here including *R. sativum* Syme and plants called *R. vulgare* Lam., an illegitimate name. The difference in leaf shape from *R. triste* is subtle, and the species are not always easy to distinguish, especially in the herbarium. *R. rubrum* is more of an erect shrub and the flowers are greenish. The anthers are generally said to have the lobes separated by a connective as broad as themselves, but this is not always well displayed, especially in young anthers. Old anthers may often be found persisting in withered flowers at the end of partly matured fruit. In general, only plants with both smooth pedicels and well developed anther connectives have been accepted here as *R. rubrum*.

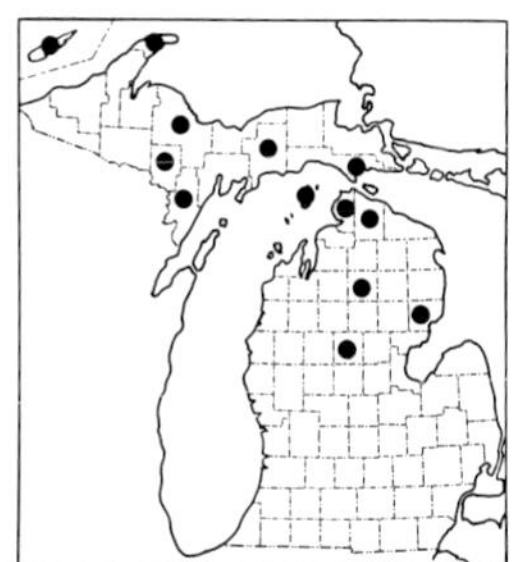

454. Ribes hudsonianum

455. Ribes nigrum

456. Ribes glandulosum

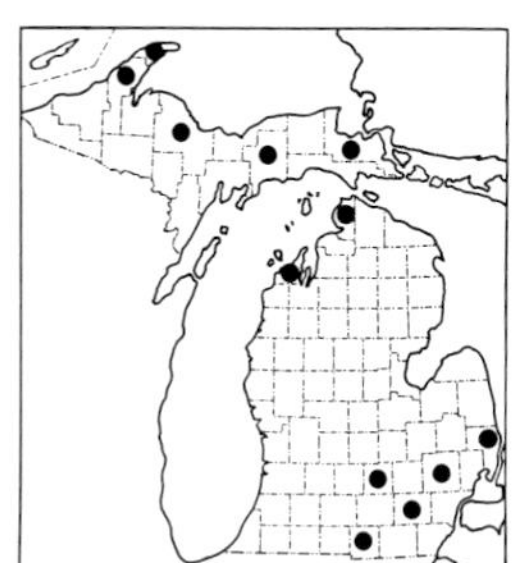

457. Ribes rubrum

458. Ribes triste

459. Hamamelis virginiana

11. **R. triste** Pallas Plate 5-A Swamp Red Currant

Map 458. Usually in swampy woods (deciduous or coniferous) or wet hollows and seepy slopes in upland rich deciduous woods.

Like *R. glandulosum,* with which it often grows, this is a low shrub with reclining stems, from which fertile leafy shoots ascend. The flowers are pinkish to red, very flat or saucer-like. Sometimes considered a variety of the preceding, as *R. rubrum* var. *propinquum* (Turcz.) Trautv. & Meyer. It is a good source of acid fruit for jams and jellies, like its garden cousin.

HAMAMELIDACEAE Witch-hazel Family

A single species is native in our area. Another genus is represented by *Liquidambar styraciflua* L., sweet gum, native north into southern Illinois, Indiana, and Ohio, but well known as an ornamental planted as far north as southern Michigan.

REFERENCES

De Steven, Diane. 1983. Floral Ecology of Witch-hazel (Hamamelis virginiana). Mich. Bot. 22: 163–171.

Ernst, Wallace R. 1963. The Genera of Hamamelidaceae and Platanaceae in the Southeastern United States. Jour. Arnold Arb. 44: 193–210.

1. Hamamelis

In addition to our late-blooming species, in the southern states is an early spring-blooming one, *H. vernalis* Sarg., which is sometimes cultivated. Two Asian species (as well as their hybrids) are also widely cultivated, *H. japonica* Sieb. & Zucc. and *H. mollis* Oliver, both also blooming in the late winter or early spring. *H. vernalis* has duller yellow flowers than the others (which range from bright yellow to red), but they are more fragrant.

1. **H. virginiana** L. Fig. 172 Witch-hazel

Map 459. May be found in rich deciduous woods, but more often in ± sandy dry woods and woodlands with oak, hickory, aspen, or pine.

A very distinctive shrub, with asymmetrical crenate leaves and very late-blooming flowers as the leaves turn yellow and fall (late September to November). The 4 petals are yellow and narrowly linear (at most ca. 1 mm wide, 10–15 times as long). A form with pink flowers [f. *rubescens* Rehder] was found in oak-hickory woods near South Lyon, Oakland County, in 1921.

A distilled extract from the leaves, twigs, and bark has long been used in lotions and linaments, e. g., for sore muscles. American Indians, including tribes in the Great Lakes region, used the plant extensively for its reputed medicinal properties.

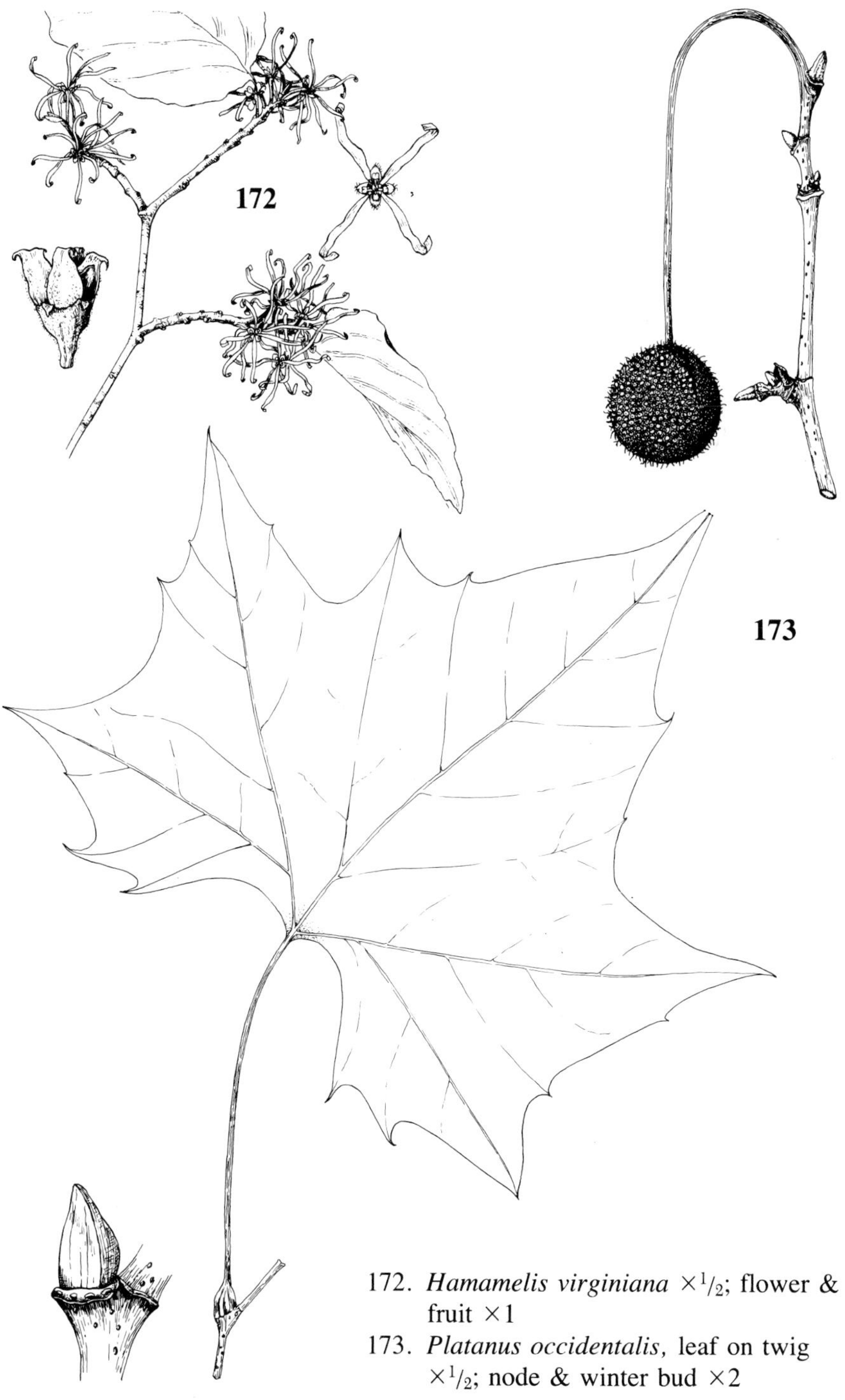

172. *Hamamelis virginiana* $\times{}^{1}/_{2}$; flower & fruit $\times 1$

173. *Platanus occidentalis*, leaf on twig $\times{}^{1}/_{2}$; node & winter bud $\times 2$

PLATANACEAE — Plane-tree Family

In this family there is a single genus of North American and Eurasian trees.

REFERENCE

Ernst, Wallace R. 1963. The Genera of Hamamelidaceae and Platanaceae in the Southeastern United States. Jour. Arnold Arb. 44: 193–210.

1. **Platanus**

In addition to our eastern species, there are others in western North America, from California south to Guatemala. The oriental plane-tree, *P. orientalis* L., has 2 or more inflorescences on a peduncle. It and our native species have perhaps produced a hardy hybrid, the London plane-tree, *P.* ×*hybrida* Brot. [*P.* ×*acerifolia* (Aiton) Willd.], which is widely planted as a shade tree; its origin, however, is unknown and it may be merely a cultivar of *P. orientalis* and not a hybrid. All species have a distinctive mottled bark, exfoliating in ± large, dark, irregular, brittle plates, leaving paler green to white areas. The flowers are unisexual, in spherical, peduncled staminate or pistillate heads. The large, firm, alternate, palmately veined leaves roughly resemble in shape those of a maple (or grape); the enlarged base of the petiole covers a conical axillary bud.

1. **P. occidentalis** L. Fig. 173 — Sycamore

Map 460. Low woods, as on floodplains, river banks, and borders of lakes.

Our native species differs from the planted ones in having the spherical inflorescences only 1 (2) on a peduncle. This is not the "sycamore" of the Bible, which is the sycomore fig, *Ficus sycomorus* L., nor is it the English "sycamore," which is a maple, *Acer pseudoplatanus* L.

ROSACEAE — Rose Family

This is a diverse family, worldwide in occurrence but especially well developed in North Temperate regions. Although the sepals are 5, often there are additional large or small bractlets in the calyx, sometimes giving the impression of 10 sepals altogether. Except for *Dalibarda,* all of our herbaceous species have deeply lobed or compound leaves; many of the woody species have simple leaves.

Several genera (see discussions of *Amelanchier, Crataegus,* and *Rubus*) are well known sources of great taxonomic difficulty, thanks to hybridization, polyploidy, asexual reproduction, and other behavioral problems, so that the only practical course in a local flora is to recognize species complexes. Some genera, on the other hand, are among our favorite ornamentals or the source of many edible fruits.

Four subfamilies are all well represented in our area:

Maloideae ("Pomoideae"): Ovary inferior, developing into a *pome,* a distinctive fruit in which papery (or even bony) carpels (e. g., forming a "core" in apples) are surrounded by fleshy tissue. The carpels are 1–5. These are woody plants, including *Amelanchier* (serviceberry), *Aronia* (chokeberry), *Crataegus* (hawthorn), *Malus* (apple), *Pyrus* (pear), and *Sorbus* (mountain-ash).

In the other subfamilies, the ovaries are superior, and there is a perigynous structure, often called a hypanthium, ranging from ± flat to deeply cup-like, representing some degree of fusion of the other floral parts (sepals, petals, stamens). The petals generally fall free from this floral tube, while the stamens are ± persistent, and this habit helps to distinguish Rosaceae from Ranunculaceae for many beginners who tend to confuse the two families with 5-merous perianth, numerous stamens, and often numerous carpels.

Amygdaloideae (Prunoideae): Carpel 1, superior, ripening into a *drupe,* a fruit with the seed enclosed by a stony "pit" surrounded by fleshy tissue. These are woody plants, with only a single genus in eastern North America, *Prunus* (cherry, plum, peach, etc.).

Spiraeoideae: Carpels few (usually 5), ripening into follicles. Our genera include *Aruncus* (goatsbeard), *Physocarpus* (ninebark), *Porteranthus* (bowman's root), *Sorbaria* (false spiraea), and *Spiraea* (spiraea).

Rosoideae: Includes all of the other genera established in our flora, with indehiscent fruits (achenes or drupelets), usually from numerous carpels (10 or more). In *Rosa* the floral tube or hypanthium becomes fleshy, enclosing the achenes in its deep flask-like cup.

REFERENCES

Mason, Harriet Gale, & Hugh H. Iltis. 1959. Preliminary Reports on the Flora of Wisconsin. No. 42—Rosaceae I—Rose Family I. Trans. Wisconsin Acad. 47: 65–97.

Robertson, Kenneth R. 1974. The Genera of Rosaceae in the Southeastern United States. Jour. Arnold Arb. 55: 303–332; 344–401; 611–662.

KEY TO THE GENERA

1. Plant a woody tree or shrub, or a bristly to thorny erect or trailing bramble
2. Leaves (at least principal ones) compound; plants of many species armed with numerous thorns or prickles

3. Stems biennial, ± thorny, bristly, or prickly (very rarely completely smooth); leaves (at least on flowering stems) trifoliolate or palmately compound (pinnate in the jagged-toothed and cut *R. laciniatus*); fruit an aggregate of juicy drupelets; flowers normally white . . . . 1. **Rubus**

3. Stems perennial, smooth or armed; leaves pinnately compound (often trifoliolate in the pink-flowered *Rosa setigera*); fruit various but not juicy nor composed of drupelets; flowers white, pink, or yellow

4. Corolla pink (rarely white or, in escapes from cultivation, yellow), 2 cm or more broad; stems usually ± thorny or bristly; stipules adnate to petiole; fruit a fleshy red to orange, globose to urn-shaped, floral tube enclosing the achenes (and thus resembling an inferior ovary except for the narrow open throat at the summit) . . . . 2. **Rosa**

4. Corolla white or yellow, less than 2 cm broad (to 2.5 cm in *Potentilla*); stems unarmed; stipules free; fruit a pome from a truly inferior ovary or a cluster of follicles or hairy achenes

5. Flowers solitary or few in an inflorescence, the petals yellow and 6–11 (12) mm long; leaflets 5–7, entire, less than 6 (8) mm broad; fruit a hairy achene . . . . 18. **Potentilla (fruticosa)**

5. Flowers numerous in a dense inflorescence, the petals white and less than 5 mm long; leaflets 12–23, toothed, over 10 mm broad; fruit a pome or follicle

6. Inflorescence an elongate panicle (much longer than broad); ovaries 5, superior, forming 5 follicles; stamens ca. twice as long as petals; leaflets doubly serrate (each primary tooth with several fine teeth); plant a colonial shrub, occasionally escaped from cultivation . . . . 3. **Sorbaria**

6. Inflorescence a broad flat or dome-shaped corymb (much broader than long); ovary 1, inferior, forming a pome; stamens no longer than the petals; leaflets singly serrate (or a few teeth with a secondary tooth); plant a small erect tree, native (or rarely escaped) . . . . 4. **Sorbus**

2. Leaves all simple; plants all unarmed or with only strong elongate spines

7. Style and ovary 1; ovary superior, ± surrounded by floral tube, ripening into a drupe; leaves unlobed . . . . 5. **Prunus**

7. Styles more than 1 (solitary in *Crataegus monogyna*); ovary inferior and ripening into a pome or more than 1 and superior ripening into a group of drupelets or dry dehiscent fruits; leaves lobed or not

8. Ovaries superior

9. Petals 18–25 mm long; leaves mostly at least (7) 9 cm broad; ovaries numerous, ripening into an aggregate of dryish or juicy red drupelets . . . . 1. **Rubus** (couplet 3)

9. Petals 1–4.5 mm long; leaves less than 5 (8) cm broad; ovaries 3–5, ripening into dry dehiscent fruits

10. Leaf blades ± lobed as well as toothed; stipules (or their scars on an elevated ridge) present; fruits ca. 5–9 (12) mm long (not including persistent style), red (drying light brown), inflated; bark conspicuously peeling into longitudinal shreds . . . . 6. **Physocarpus**

10. Leaf blades toothed but not lobed; stipules none; fruits ca. 2.5–4 mm long, yellow-brown to blackish, scarcely inflated; bark at most slightly shredding. . . . . 7. **Spiraea**

8. Ovary inferior

11. Leaves with elongate red ± appressed glands on the midrib above, at least basally, not at all lobed; plants without thorns or spines; petals (less than 7 mm long) less than twice as long as wide . . . . 8. **Aronia**

11. Leaves without glands on midrib (or leaves not yet open at flowering time), lobed or not; plants armed or not; petals various

12. Petals more than twice as long as broad; leaves (if open) regularly toothed but not lobed . . . . 9. **Amelanchier**

12. Petals less than twice as long as broad (or plants with only leaves and fruit); leaves various

13. Leaves regularly to coarsely toothed and usually also at least slightly lobed or notched; branches almost always armed with distinct sturdy spines ca. 1.5–11.5 cm long; inflorescence usually a compound (branched) corymb (simply and obscurely corymbose in *Malus*)

14. Seeds enclosed in very hard bony nutlets; inflorescence (at least the larger ones) usually a compound corymb; styles separate to the base and glabrous; petals white; stamens ca. 20 or fewer; spines usually shiny; bud scales glabrous . . . . 10. **Crataegus**

14. Seeds only in papery-cartilaginous carpels; inflorescence a simple cluster; styles connate and very pubescent at the base, petals pink; stamens ca. 20; spines dull; bud scales pubescent . . . . 11. **Malus** (couplet 2)

13. Leaves regularly toothed or crenulate but not at all lobed; branches without definite spines; inflorescence simple (unbranched)

15. Fruit berry-like, less than 15 mm thick, apparently 10-locular with 1 seed per locule; flowers and fruit usually in racemes (flowers 1–3 in one species); plants common native shrubs . . . . 9. **Amelanchier**

15. Fruit a pear or apple, much larger, 5-locular with 2 seeds normally in each locule; flowers crowded with the leaves on short spur-branches; plants trees occasionally escaped from cultivation

16. Young branchlets densely pubescent; petals pink, at least outside when young; styles densely pubescent basally and connate; fruit an apple, without grit cells . . . . 11. **Malus**

16. Young branchlets glabrous or nearly so; petals white; styles glabrous and separate to the base; fruit a pear, with grit cells . . . . 12. **Pyrus**

1. Plant herbaceous, unarmed (stem dying to the ground each year—or slightly woody at the ground)

17. Leaves all simple, cordate and crenulate . . . . 13. **Dalibarda**

17. Leaves all or mostly compound or pinnatifid

18. Leaves all trifoliolate or palmately compound

19. Style with a prominent joint near or above the middle, the lower portion hooked at apex and persistent as a long beak on the achene, the terminal portion hooked at its base (fig. 205) and deciduous; plant with an erect leafy stem . . . . . 21. **Geum**

19. Style not jointed, entirely deciduous from fruit; plant erect or with only basal leaves (stemless or stoloniferous)

20. Calyx with no bractlets between the sepals (or rarely an incomplete set of minute ones)

21. Plant with the largest leaves sessile or nearly so (petiole, if any, less than a fourth as long as leaflets); carpels 5, ripening into follicles . . . 14. **Porteranthus**

21. Plant with the largest leaves long-petioled; carpels 2–many, ripening into achenes or fleshy drupelets

22. Petals yellow; carpels 2–6, ripening into achenes, ± concealed within the floral tube (long styles protruding); leaves all basal (at least at flowering time) . . . . 15. **Waldsteinia**

22. Petals white or pink; carpels normally more than 6, ripening into fleshy drupelets conspicuous on receptacle; leaves cauline . . . . . . 1. **Rubus** (couplet 4)

20. Calyx with bractlets nearly or quite as large as the sepals, hence appearing 10-lobed

23. Bractlets of calyx with 3 (–5) teeth or lobes at summit (fig. 198) . . . . 16. **Duchesnea**

23. Bractlets entire (very rarely 2-toothed at tip)

24. Receptacle greatly enlarging in fruit, becoming fleshy and red; plants with elongate stolons flat on the ground; petals white, the flowers several on a peduncle; leaflets 3 . . . . 17. **Fragaria**

24. Receptacle not enlarging, dry; plants without conspicuous stolons (or these ± arching and the flowers solitary); petals yellow (white in 1 species with nearly entire leaves); leaflets 3, 5, or 7 . . . . 18. **Potentilla**

18. Leaves mostly pinnately compound or divided

25. Leaflets dissected into linear segments; stamens 5, opposite the white petals . . . . 19. **Chamaerhodos**

25. Leaflets definite, with broad flat blades; stamens and petals various but not as above (stamens 4 and petals absent, or numerous, or petals yellow)

26. Leaves twice (or thrice) compound, estipulate; flowers unisexual (plants dioecious or with polygamous tendency); inflorescence a large (ca. 1–4 dm long) panicle of numerous spike-like racemes; carpels 3 (–4), ripening into strongly reflexed follicles . . . . 20. **Aruncus**

26. Leaves once-pinnate, stipulate; flowers perfect (except in *Sanguisorba*); inflorescence various; carpels 2–many, ripening into achenes (or indehiscent 1-seeded "follicles")

27. Carpels numerous on a hemispherical or conical receptacle; calyx with bractlets alternating with the true sepals, thus appearing 10-parted (bractlets absent in *Geum vernum*)

28. Styles short, slightly if at all exceeding the achene and deciduous from it, neither jointed nor pilose nor greatly elongating in fruit. . . . . 18. **Potentilla**

28. Styles elongating and much exceeding the achene at maturity, and at least partly persistent on it, in most species jointed near or above the middle and usually at least partly pilose on lower and/or terminal segment . . . . . 21. **Geum**

27. Carpels ca. 5 or fewer, on a flat to concave receptacle; calyx 5-lobed without additional bractlets

29. Petals pink; carpels ca. 5, evident on a flattish receptacle, ripening into fruits resembling follicles (but indehiscent); inflorescence a diffuse much-branched broad panicle; leaflets deeply lobed or cleft . . . . 22. **Filipendula**

29. Petals none or yellow; carpels 1–2, set in a deeply concave floral tube, the ovaries and achenes largely or entirely covered but the styles protruding; inflorescence a compact short to narrow and elongate spike-like raceme; leaflets not lobed or cleft

30. Floral tube angled or winged but not bristly; petals none; sepals 4, white to greenish or brown; inflorescence crowded and compact . . . 23. **Sanguisorba**

30. Floral tube with hooked bristles at its summit but not winged; petals yellow; sepals 5, green; inflorescence an elongate short-pediceled raceme . . . . 24. **Agrimonia**

1. **Rubus** Bramble; Raspberries, Dewberries, and Blackberries

The first six species treated here are normally all diploid and are fairly clearcut, although hybrids apparently exist between the members of each of the three pairs of species. The remainder of our taxa are in *Rubus* subgenus *Rubus*—often still called subgenus or section *Eubatus* although those combinations have been incorrect under the Code since 1950 and 1954 respectively. These are the dewberries and blackberries, a challenge to the botanist in more than one way.

Rubus subgenus *Rubus* includes diploids and polyploids. Some taxa are sexually reproduced and others are apomicts; the role of hybridization has been considered variously as important or unimportant, a recent study (Steele & Hodgdon 1970) concluding that hybrids are more frequent than supposed by followers of Bailey. Liberty Hyde Bailey, who spent most of his long life studying the brambles, declared in his last major contribution on the subject (1949): "*The problem in Rubus is to recognize the lines of evolution*" (italics his). Hardly anyone can question this statement, although many have questioned Bailey's recognition of so many of what might be termed "microspecies" or minor variants—as he rather impatiently acknowledged in the same paper. This subgenus has therefore long been a notoriously thorny problem—a problem not to be solved in a local flora. Over 400 so-called species have been described from eastern North America. These opportunisistic plants parallel in several ways the explosion into disturbed ground, laid bare by the onslaughts of "civilization," illustrated by *Amelanchier* and *Crataegus,* which also appear to be actively evolving. The special problem with *Rubus* is the biennial habit and the presence of characters, thought to be important in distinguishing taxa, in the sterile first-year stems, which are rarely collected by casual botanists.

Entirely apart, then, from the evolutionary problems, which may be amenable to cytological and genetic investigations, there are practical taxonomic problems of what the evolutionary lines *mean*—not to mention how we can deal with plants when most herbarium specimens are inadequate. Good specimens, to show all characters thought to be helpful in identification, should include portions of the first-year unbranched sterile stem, or *primocane* (generous sections from the tip as well as the middle), and of the second-year stem, or *floricane,* with inflorescences (or, better, infructescences) which are on short leafy lateral shoots branching from the old primocane—now the floricane. Notes on the habit of the plant are important: whether the stems are erect, ascending, arched, immediately prostrate, or trailing (starting upward but soon back on the ground), and whether they root at the tips. Further discussion of habit and other characters can be found in Bailey (1941) or Hodgdon and Steele (1966). The latter also give a good history of *Rubus* study in North America.

The fruit in *Rubus* is an aggregate one, of numerous small ± juicy drupelets (each from a separate carpel of the flower); the receptacle on which these are inserted is fleshy and falls with the fruit in the dewberries and blackberries, but is firm and remains on the plant in the thimbleberries and raspberries. Vegetative parts exhibit a full range of pubescence, glandular or not, and of armature, ranging from rather soft bristles or setae through slender but firm prickles at most slightly expanded at the base, to stout broad-based thorns or prickles like those in many roses. The "broad" base, incidentally, is elongated in the same axis as the stem or cane, not transversely. In *R. canadensis* the canes are rarely completely unarmed, but no alert student could fail to recognize such plants as blackberries. Leaves of primocanes in some taxa usually have 5 leaflets; floricane leaves have 3 leaflets (or some of them are simple). The shape of primocane and floricane leaflets may also differ.

The treatment which follows is a practical one which should allow the user to make a preliminary sorting of most specimens, even if they lack the primocane with its characters of habit, armature, and leaflet shape so often used in defining segregate species. Those who wish to go further with a detailed study of their local flora will need to use the full descriptions and illustrations in the work of Bailey, updated taxonomically in the outline of Davis, Fuller, and Davis. Hodgdon and Steele treat only the New England plants, but their work is also helpful. Overall, the broad approach here taken parallels closely the treatment by E. Lucy Braun in her very practical *The Woody Plants of Ohio* (1961). The "complexes" here suggested in *Rubus* subgenus *Rubus* follow in one way or another the apparently natural groupings recognized by students of *Rubus* in the past. They are essentially equivalent to the sections (found in this region) as remodeled by Davis et al. but are not identical to any other set of "sections" or "collective species." And there is no claim that they are of equal significance or diversity. The *R. flagellaris* and *R. setosus* complexes are undoubtedly the most heterogeneous and unsatisfactory.

I have tried to list all names for which the types are from Michigan or which have been otherwise applied in published reports for Michigan by Bailey (including also his determinations for the assiduous collectors of Kalamazoo County, C. R. and F. N. Hanes and F. W. Rapp) or by Davis et al. And I have tried to indicate the disposition of all such names as aligned in the full treatment of subgenus *Rubus* for eastern North America by Davis et al., especially since their work (with a few further alterations) is the basis for the genus in the checklist by Kartesz and Kartesz (1980), and represents a modern synthesis in the Bailey tradition.

Since the broad treatment here is not in the Bailey tradition, however, it has not seemed necessary to examine all the relevant material collected or cited by him and his followers. Davis et al. note more than one case in which type collections of Bailey are evidently mixed. Because their disposition of the named species is the most thorough one for eastern North America, not likely to be soon superseded, it has been the principal source for the listing of taxa reported from the state. There is, of course, no guarantee that all specimens referred by me to a particular species complex would be similarly aligned by anyone else—or that if I saw a specimen reported by others, I would place it in the same way as they did. As always, the maps are based only on herbarium specimens I have seen.

The characters used in the key are not necessarily—any more than in other keys—the ones considered to be of most biological significance in the evolution of the genus. They are simply the ones which seem easiest to use and which correlate best with the natural groupings generally recognized on the basis of assemblages of characters, some of which are less easily described and interpreted.

The edible qualities of *Rubus* fruits are well known. Some taxa are better than others, even allowing for different tastes. The boysenberry and loganberry are cultivars derived from *R. ursinus* Cham. & Schlect., one of the few species of *Rubus* native only to the West Coast. The cultivars of blackberries are often not traceable to their original wild sources.

REFERENCES

Bailey, L. H. 1941–1945. Species Batorum. The Genus Rubus in North America (North of Mexico). Gentes Herb. 5: 1–932. [Fasc. I, pp. 1–64, 1941; fasc. II, Hispidi, pp. 65–125, 1941; fasc. III, Setosi, pp. 127–198, 1941; fasc. IV, Verotriviales, pp. 199–228, 1941; fasc. V, Flagellares, pp. 229–422, 1943; fasc. VI, Cuneifolii, pp. 423–461, 1943; fasc. VI, Canadenses, pp. 463–503, 1944; fasc. VIII, Alleghenienses, pp. 505–588, 1944; fasc. IX, Arguti, pp. 589–856, 1945; fasc. X, subg. Idaeobatus, subg. Anoplobatus, pp. 857–932, 1945. See next two titles for Addenda.]

Bailey, L. H. 1947. Species Studies in Rubus. Gentes Herb. 7: 193–349.

Bailey, L. H. 1949. Rubus Studies—Review and Additions. Gentes Herb. 7: 479–526.

Davis, H. A., Albert M. Fuller, & Tyreeca Davis. 1967. Contributions Toward the Revision of the Eubati of Eastern North America. Castanea 32: 20–37. II, Setosi, Castanea 33: 50–76, 1968. III, Flagellares, Castanea 33: 206–241, 1968. IV, [Verotriviales, Canadenses, Alleghenienses], Castanea 34: 157–179, 1969. V, Arguti, Castanea 34: 235–266, 1969. VI, Cuneifolii, Castanea 35: 176–194, 1970.

Fassett, Norman C. 1941. Mass Collections: Rubus odoratus and R. parviflorus. Ann. Missouri Bot. Gard. 28: 299–374.

Ho[d]gdon, A. R., & Frederic Steele. 1966. Rubus subgenus Eubatus in New England a Conspectus. Rhodora 68: 474–513.

Steele, Frederic L., & A. R. Hodgdon. 1970. Hybrids in Rubus subgenus Eubatus in New England. Rhodora 72: 240–250.

460. Platanus occidentalis

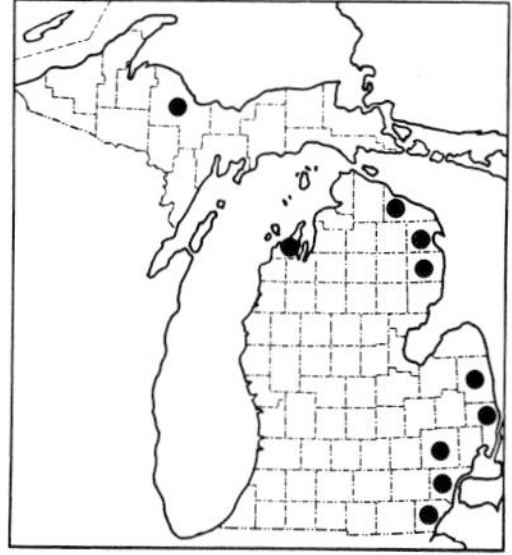

461. Rubus odoratus

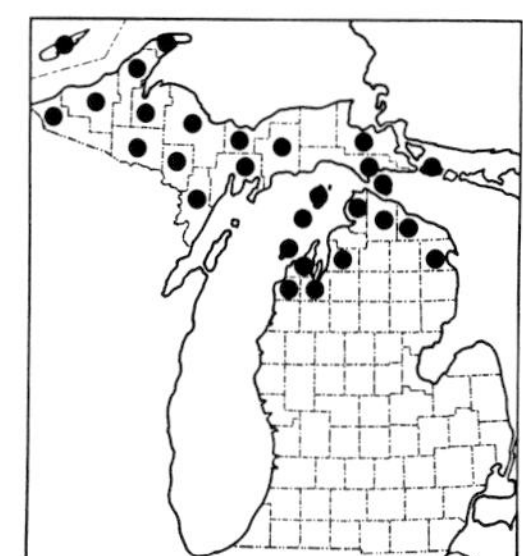

462. Rubus parviflorus

KEY TO THE SPECIES

1. Stems without thorns or prickles (new stems may be glandular-bristly if leaves are simple)
   2. Leaves all simple (palmately lobed), mostly (7) 9–22 cm or more broad; petals 18–25 mm long; fruit readily separating from receptacle; plant with erect woody stems and shreddy bark
      3. Petals pink to purple (rarely white); calyx lobes with ± dense purple elongate gland-tipped hairs or bristles; range from southeastern Lower Peninsula as far as southeastern Presque Isle Co. .... 1. **R. odoratus**
      3. Petals white; calyx lobes with yellow-orange very short gland-tipped hairs; range from Upper Peninsula to Alpena Co. .... 2. **R. parviflorus**
   2. Leaves trifoliolate, the leaflets less than 5 (7) cm broad; petals 4–14 [18] mm long; fruit not separating from receptacle; plant with trailing woody stems at or below ground level and with tight bark, the erect shoots herbaceous
      4. Petals deep rose-pink, ca. 10–14 [18] mm long; flowers solitary on glandless pedicels; leaflets nearly or quite obtuse to rounded at apex .... 3. **R. acaulis**
      4. Petals white (sometimes drying pinkish or turning with age), 4–7 (8) mm long; flowers 1–3 (4), the pedicels usually with a few stipitate glands; leaflets sharply acute to acuminate .... 4. **R. pubescens**
1. Stems with thorns or prickles, or at least bristles (rarely completely smooth in the erect *R. canadensis*); leaves all or mostly compound
   5. Floricane leaves densely white-pubescent beneath; fruit red or black at maturity, readily separating from the receptacle
      6. Pedicels and peduncles with glandless mostly broad-based slightly recurved prickles; mature fruit purple-black with narrow belts of white tomentum between the drupelets .... 5. **R. occidentalis**
      6. Pedicels and peduncles with gland-tipped ± straight bristles; mature fruit red, without tomentum between the drupelets .... 6. **R. strigosus**
   5. Floricane leaves green on both sides (pubescence if any not both dense and conspicuously white); fruit black at full maturity, separating from the floral tube *with* the fleshy receptacle included
      7. Leaflets (all or many of them) pinnately lobed or ± incised nearly or quite to the midrib, besides the usual teeth; calyx lobes armed with firm prickles .... 7. **R. laciniatus**
      7. Leaflets merely toothed, at most shallowly lobed; calyx lobes unarmed (may be glandular-bristly)
         8. Floricane prostrate or trailing, slender, the flowering shoots arising ± perpendicularly from the essentially horizontal cane; primocanes ± ascending or trailing, rooting at the tips
            9. Floricanes (flowering shoots and/or trailing old canes) armed only (or almost entirely) with slender prickles or bristles, these slightly if at all expanded at the base; leaflets usually ± obovate, rounded or obtuse at apex (especially lower on flowering shoots), firm (even persistent or winter-green); flowers small, the petals 3.5–6 (9) mm long .... 8. **R. hispidus** complex
            9. Floricanes armed mostly or entirely with strong broad-based prickles or thorns; leaflets mostly acute to acuminate, thin; flowers larger, the petals 8–15 (20) mm long .... 9. **R. flagellaris** complex
         8. Floricane as well as primocane erect or arching, usually not rooting at tip (plants depressed by snow or their own weight might be confusing if they appear prostrate, and *R. setosus* is often trailing—see text)

10. Canes armed with bristles or slender narrow-based prickles; plants relatively low, even well developed ones scarcely 1 m tall . . . . 10. **R. setosus** complex
10. Canes armed with stout broad-based thorns or prickles (these sparse and small in *R. canadensis*); plants when well developed stout and at least 1 m tall
11. Pedicels and axis of inflorescence (also petioles and young growth) ± copiously stipitate-glandular; leaflets pubescent beneath . . . . 11. **R. allegheniensis** complex
11. Pedicels and other parts glandless or nearly so; leaflets glabrous or pubescent beneath
12. Leaflets glabrous beneath (or a few hairs only on principal veins); old canes with few or no prickles (though new growth usually has a few) . . . . 12. **R. canadensis** complex
12. Leaflets pubescent beneath (between the principal veins) or canes strongly armed (or both conditions present) . . . . 13. **R. pensilvanicus** complex

1. **R. odoratus** L. Flowering Raspberry

Map 461. Clearings and borders of coniferous or deciduous woods. This is an eastern species, at the edge of its range in Michigan. Collections from Marquette (some along a railroad) undoubtedly represent an escape from cultivation; Oakland and Leelanau county collections may also be of dubious origin.

White-flowered plants rarely occur [f. *albiflorus* House] and are superficially confused with the next species. This form was originally described from the state of New York, where the flower color could hardly have resulted from any influence of *R. parviflorus*. However, our white-flowered plants occur in the only region where the ranges of the two species overlap: the Grand Lake area of Presque Isle County and adjacent Alpena County (particularly Thunder Bay Island), Michigan. Here, the species are evidently hybridizing. A hybrid was artificially produced early in this century and named *R.* ×*fraseri* Rehder. Suspected of being this hybrid in the wild are plants with pale or partly pink flowers and less glandularity than usual for *R. odoratus,* as well as those with white flowers but more glandularity than usual for *R. parviflorus*.

In *R. odoratus,* the fresh young branches and peduncles are very sticky with clammy purple pubescence and the lobes of the leaves tend to be more acuminate at the tip, with basically straight sides. Such plants with pure white petals are referred to f. *albiflorus*. In *R. parviflorus,* the branches and peduncles are not so sticky and the sides of the leaf lobes are slightly convex in general outline. The best discrimination seems to be made on the calyx lobes (not necessarily the broad tube), which in *R. parviflorus* are sparsely to densely covered with glands sessile or nearly so or on stalks at most ca. 5 times as long as the glands; these are generally yellow to orange, occasionally with a suggestion of purple. The much longer-stalked glands on the calyx lobes of *R. odoratus* are dense and rich purple—as also on the peduncles and new growth. Other pubescence characters are variable in both species, leading to the naming of numerous forms now rarely thought to be of any importance. (Discussions may be found in Fassett 1941, Bailey 1945, and by Fernald in Rhodora 37: 273–284. 1935.)

2. **R. parviflorus** Nutt. Frontispiece Thimbleberry

Map 462. Forming large thickets in northern hardwoods and moist mixed (or aspen-birch) woods, especially along borders and clearings and not far from the Great Lakes; abundant in thin woods of the northwestern Upper Peninsula and Isle Royale.

This is one of the classic species common in western North America, disjunct in the Black Hills and northern Great Lakes region (see Mich. Bot. 20: 63–64. 1981). It is one of the commonest and most widespread of the "western disjuncts" in our region. The type locality is Mackinac Island, where it was found in 1810 by Thomas Nuttall, who gave it the peculiarly inept epithet meaning "small-flowered"—when in fact the flowers may be 5 cm broad.

As discussed above, hybrids occur in the very limited area where this species and the preceding grow together. Elsewhere in Michigan these two species can be easily distinguished by flower color and range. A variant with the leaf blades cleft to the base and hence essentially compound was discovered in Keweenaw County in 1928 by Hermann, amid normal plants, and was named by him f. *pedatifidus* (Rhodora 37: 61 + pl. 326. 1935). However, a less lacerate form named with an older epithet is f. *lacera* (Kuntze) Fern., discovered on Vancouver Island in 1858. It is doubtful whether such a sport needs even one name.

The bright red fruit is produced abundantly where this species is common, as on Isle Royale and the Keweenaw Peninsula and in the Porcupine Mountains area. It is extensively used in the making of tarts and jams, and can be enjoyed by the handful raw. Although there are those who question its merits, I find it delicious, a bit more tart than a red raspberry and a most welcome treat when one is hiking along a trail bordered by bushes laden with the easily picked fruit. The fruit separates very readily from the receptacle at maturity, like a broad flattish thimble.

3. **R. acaulis** Michaux Plate 5-B

Map 463. A boreal species coming south very locally to the Lake Superior region of Ontario and bogs or conifer swamps in northern Schoolcraft County.

Sometimes treated as a subspecies or variety of *R. arcticus* L. In Michigan, as at many places in its range, it apparently hybridizes with *R. pubescens,* producing *R.* ×*paracaulis* Bailey. Such plants are presumably the basis for reports of true *R. arcticus* from eastern North America, as the latter occurs only in the western part of this continent and the Old World. Hybrid plants generally have pale pink petals but more acute tips on the leaflets than does *R. acaulis*.

4. **R. pubescens** Raf. Dwarf Raspberry

Map 464. Low woods and swamps (often with cedar or tamarack, sometimes only deciduous trees); rocky shores and openings; cliffs and ledges of sandstone and igneous rock.

The bright red fruit has the flavor of a wild raspberry, but the few drupelets do not readily separate from the receptacle. (The practical hiker merely picks the pedicel and bites off the tasty portion!)

5. **R. occidentalis** L. Black Raspberry

Map 465. Fields, thickets, woods (oak, disturbed beech-maple), clearings and borders of woods, fencerows, boggy ground.

Where this species and the next grow together, intermediate plants, presumably hybrids, are sometimes found. These and other intermediate plants have been called *R.* ×*neglectus* Peck, but there is disagreement whether that name truly applies to a hybrid or to a form of *R. strigosus*. Intermediate plants may have the characteristic purple, heavily glaucous, strongly prickly old canes of *R. occidentalis,* but also the glandular bristles of *R. strigosus* (these mixed with, and generally shorter than, the stouter and slightly recurved prickles in the inflorescence). The fruit is ± intermediate. Such plants have been collected in Gratiot, Ingham, Isabella, Kent, Monroe, and Wayne counties.

R. michiganus (Greene) Fedde (TL: near the Agricultural College [Ingham Co.]) is included by Bailey in the synonymy of *R. occidentalis*.

6. **R. strigosus** Michaux Wild Red Raspberry

Map 466. In all sorts of dry to moist open or slightly shaded ground: thickets, woods, shores, stream banks, rocky openings; especially common following clearing, burning, erosion, or other disturbance in damp woods.

For plants intermediate with *R. occidentalis,* see discussion above. The canes in *R. strigosus* may be somewhat glaucous, but not conspicuously so, and they lack the stout broad-based prickles or thorns of *R. occidentalis*. There may be some slender prickles in the inflorescence, even without glan-

463. Rubus acaulis

464. Rubus pubescens

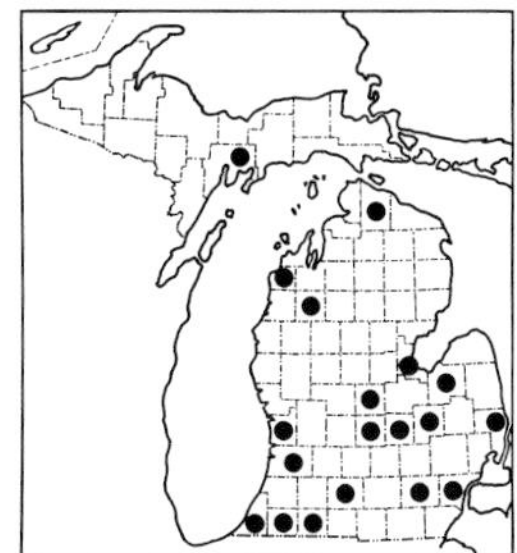
465. Rubus occidentalis

dular tips, but they are ± straight and only slightly if at all broad-based. Pale-fruited forms are known of both raspberry species.

The Old World *R. idaeus* L. is similar, but glandless. Both it and *R. strigosus,* which is often treated as a variety or subspecies of it (though according to Bailey they apparently do not hybridize in nature), are the sources of the numerous cultivated red raspberries. *R. phoenicolasius* Maxim., wineberry, an Asian species, is sometimes cultivated and was collected once in Berrien County, of unknown status; it has very dense, long, gland-tipped, red or purple bristles.

7. **R. laciniatus** Willd. Cut-leaf Blackberry

Map 467. An Old World plant of unknown origins, sometimes cultivated; occasionally escaped and naturalized along roadsides, railroads, fields, and shores.

8. **R. hispidus** L. Fig. 174 Swamp Dewberry

Map 468. Usually in ± low, damp, shaded places: moist thickets, swamps (coniferous or deciduous), borders of bogs and marshes; occasionally in dry woods.

Once the prostrate or trailing habit is known, this complex is easily distinguished from the next, which shares the habit but has strong, broad-based prickles (sometimes sparse), thinner more acute or acuminate leaflets, and larger flowers. In *R. hispidus* the firm texture of the mostly obovate leaflets, usually ± glossy when fresh but duller and darkish green when dry, the veins distinctly impressed above, is distinctive. The inflorescence in both complexes is few-flowered, the pedicels usually ± strongly ascending. The longest prickles in *R. hispidus* are often ca. 1.5 times as long as the diameter of the cane where they arise, so even if slightly broad-based they are noticeably slender and elongate; in the *R. flagellaris* complex, the prickles have broad bases and are at most ca. 1.2 times as long as the diameter of the cane.

The inflorescence in *R. hispidus* is only rarely glandular, and then only

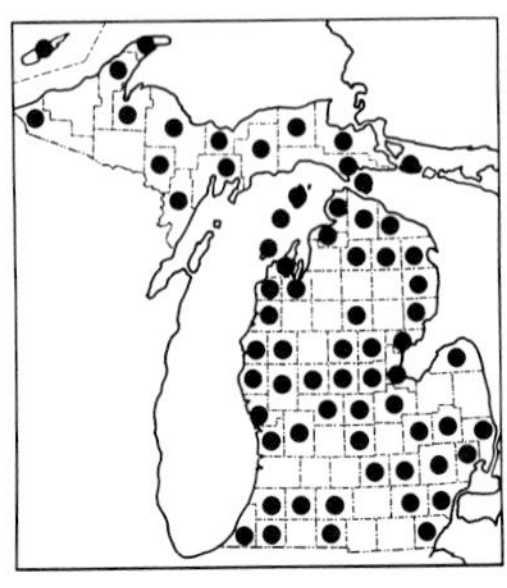

466. Rubus strigosus

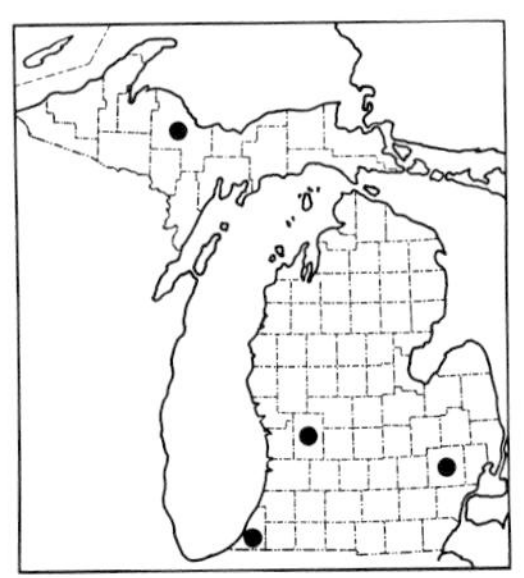

467. Rubus laciniatus

468. Rubus hispidus

sparsely so. In the *R. flagellaris* complex the inflorescence is frequently glandular and in the *R. setosus* complex it is usually distinctly glandular-setose. The latter has a broader, more cymose inflorescence than *R. hispidus*. However, these two complexes with slender prickles or bristles and no stubby broad-based ones are rather close, and apparently hybridize.

Among the segregates that have been named in *Rubus* sect. *Hispidi* are the following which have been reported from Michigan (and all of which are in some respects intermediate with the *R. setosus* or other complexes and could represent hybrids):

R. distinctus Bailey [TL: Alamo Tp., Kalamazoo Co.; included by Davis et al. in *R. permixtus* Blanchard, which is treated by Steele & Hodgdon as a hybrid with *R. allegheniensis*. Specimens of the type number for *R. distinctus* certainly have the aspect of such a hybrid, with numerous gland-tipped setae among the small but broad-based prickles on the rather stout canes (compared to those of *R. hispidus*).]

R. kalamazoensis Bailey [TL: Schoolcraft Tp., Kalamazoo Co.; included by Davis et al. in *R. plus*.]

R. plus Bailey [TL: Brady Tp., Kalamazoo Co.]

R. signatus Bailey

9. **R. flagellaris** Willd. Fig. 177 Northern Dewberry

Map 469. Sandy plains, prairie-like areas, fields, bluffs, and shores; meadows and swamp borders; oak-hickory woods, mixed hardwoods, cedar thickets; more often in dry habitats than the preceding complex, which is characteristic of moist sites.

See comments under *R. hispidus*. When the prostrate or trailing habit is not known, plants of this very diverse complex may be difficult to place, since the broad-based prickles resemble those of the blackberries. The few-flowered inflorescence, with strongly ascending pedicels, is helpful when well displayed. The southern dewberry, *R. enslenii* Tratt., is said to range into southern Michigan, and such plants are included here, as are those placed by Gleason in his collective species *R. arundelanus* Blanchard. Variants based on leaflet shape, inflorescence, armature, glandularity, pubescence, and other characters are nearly unlimited.

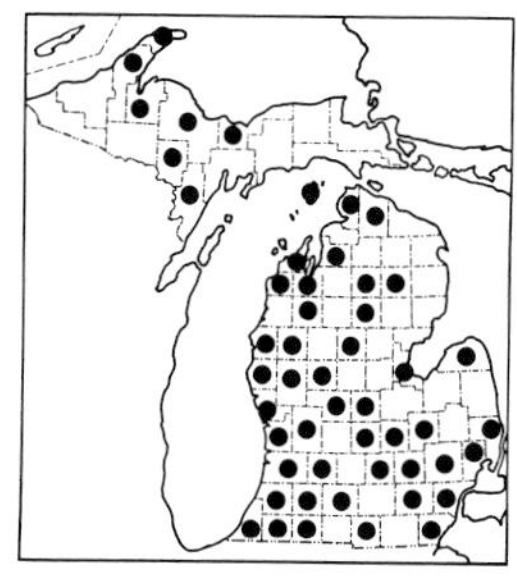

469. Rubus flagellaris

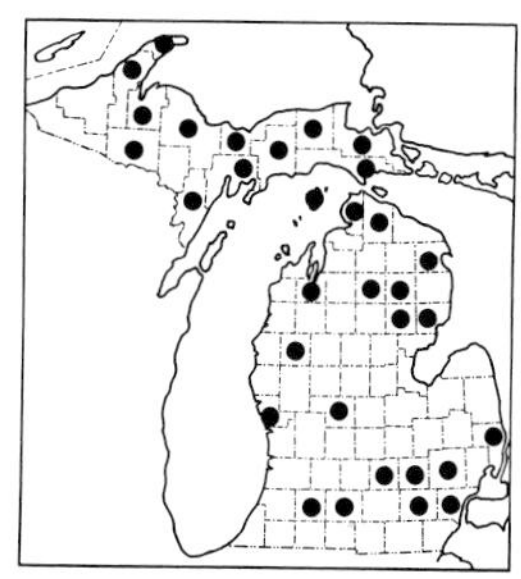

470. Rubus setosus

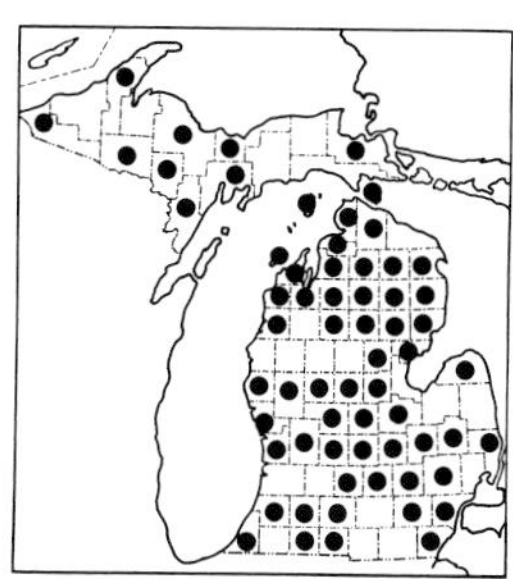

471. Rubus allegheniensis

Among the segregates named in this complex, others reported from Michigan include the following:

R. baileyanus Britton
R. centralis Bailey
R. complex Bailey [TL: South Haven, Van Buren Co.; included by Davis et al. in *R. michiganensis*.]
R. exutus Bailey [TL: Vicksburg, Kalamazoo Co.; included by Davis et al. in *R. plicatifolius*.]
R. florenceae Bailey [TL: Brady Tp., Kalamazoo Co.; included by Davis et al. in *R. michiganensis*.]
R. geophilus Blanchard [= *R. flagellaris* according to Davis et al.]
R. ithacanus Bailey [Originally placed by Bailey in his section Canadenses.]
R. meracus Bailey [TL: vicinity of Alamo, Kalamazoo Co.]
R. michiganensis (Bailey) Bailey [TL: South Haven, Van Buren Co. Some of the Kalamazoo Co. material identified as this species looks more like the *R. pensilvanicus* complex except for tip-rooting primocanes.]
R. pauper Bailey [Davis et al. have included this in *R. plicatifolius*, but the Kalamazoo Co. collections referred here by Bailey look more like the *R. pensilvanicus* complex and, indeed, Bailey had originally placed *R. pauper* in the Arguti rather than the Flagellares.]
R. peracer Bailey [TL: Cheboygan Co.; included by Davis et al. in *R. multiformis* Blanchard. Originally placed by Bailey in Canadenses.]
R. plicatifolius Blanchard
R. roribaccus (Bailey) Rydberg
R. schoolcraftianus Bailey [TL: Pavilion Tp., Kalamazoo Co.]
R. tantulus Bailey [TL: near Vicksburg, Kalamazoo Co.; a stunted plant, included by Davis et al. (Castanea 47: 219. 1982) in *R. meracus*.]
R. tenuicaulis Bailey [TL: Austin Lake, Kalamazoo Co.; included by Davis et al. in *R. baileyanus*.]
R. vagus Bailey [TL: Alamo Tp., Kalamazoo Co.]

10. **R. setosus** Bigelow

Map 470. Sandy fields, barrens, pine or aspen woodland, shores, and wooded dunes; also in moist places: meadows, borders of marshes and swamps, even bog mats.

Plants of this complex are in some ways often quite similar to those of the *R. hispidus* complex, especially in the setose bristles or elongate slender prickles; and the leaflets may also have the veins impressed above. However, the habit is generally a little stouter, with erect, ascending, or arching (but sometimes trailing) canes of medium size and not ordinarily rooting at the tip. The leaflets tend to be larger and acute to acuminate. The flowers, too, run larger, and the inflorescence tends to be a broad but relatively few-flowered cyme, often ± glandular. Plants exhibiting all the expected characters are easy to place; those lacking some of them, or of ambiguous expression, are what make *Rubus* an exciting genus. Some plants, such as those named *R. jejunus*, are rather delicate in stature, like typical *R. hispidus*, but are distinguishable by their ± erect habit, canes not rooting at the tip, and acute though small leaflets. I fear that I have treated this complex somewhat like a wastebasket, for plants with the armature of *R. hispidus* but differing in enough other characters so as not to fit comfortably in that complex. Furthermore, hybridization between *R. hispidus* and *R. setosus* is said to be rather frequent in New England (Steele & Hodgdon 1970), in which case they doubtless behave similarly here.

This is a group of essentially northern range, well represented in Michigan. Segregates named in this complex and reported from the state include the following:

R. angustifolius Bailey
R. compos Bailey [TL: near Kinross, Chippewa Co.; referred by Davis et al. to *R. wheeleri*, but close to *R. hispidus*.]
R. conabilis Bailey [TL: Alamo Tp., Kalamazoo Co. As stated in the original description, there are no setose prickles, only broad-based ones. I do not see why Davis et al. have placed this in the Setosi, though without further alignment. It appears to belong in the Flagellares, where originally placed by Bailey, although some collections from the type colony seem to represent a more erect habit than other apparently prostrate ones.]
R. dissensus Bailey [TL: Texas Tp., Kalamazoo Co.; included in *R. stipulatus* by Davis et al. (Castanea 47: 218. 1982).]
R. glandicaulis Blanchard
R. jejunus Bailey [TL: Pavilion Tp., Kalamazoo Co.; included by Davis et al. in *R. missouricus*.]
R. junceus Blanchard
R. mediocris Bailey [TL: Portage Tp., Kalamazoo Co.; included by Davis et al. in *R. missouricus*.]
R. notatus var. *boreus* Bailey [TL: Kalkaska, Kalkaska Co.; not placed by Davis et al. (1968) but included in *R. notatus* Bailey by Kartesz & Kartesz (1980).]
R. perdebilis Bailey [TL: Porcupine Mountains, Ontonagon Co.; included by Davis et al. in *R. superioris*.]
R. perspicuus Bailey [TL: Pavilion Tp., Kalamazoo Co. Although this species was placed in the Setosi when described and was retained there by Davis et al., the glandular inflorescence and stout prickles would seem to place it better in the *R. allegheniensis* complex—where Bailey indeed considered it to belong until very shortly before the time of publication. A tendency to have some long slender prickles could suggest a little influence from the *R. setosus* complex.]
R. potis Bailey [TL: Schoolcraft Tp., Kalamazoo Co.; referred by Davis et al. to *R. wheeleri*, but resembles a large-flowered *R. hispidus*.]
R. regionalis Bailey
R. spectatus Bailey
R. stipulatus Bailey
R. superioris Bailey [TL: Rock River, Alger Co.; included in *R. vermontanus* by Davis et al. (Castanea 47: 218. 1982).]
R. variispinus Bailey [TL: northeast of Vicksburg, Kalamazoo Co.; material considered by Davis et al. to be inadequate for evaluation but belonging to this group. Although prickles on Kalamazoo Co. specimens referred here are relatively long and slender they are strong and broad-based; I would be inclined to leave this in the *R. pensilvanicus* complex (Arguti) where originally described.]
R. vermontanus Blanchard
R. wheeleri (Bailey) Bailey [TL: East Lansing, Ingham Co.]
R. wisconsinensis Bailey

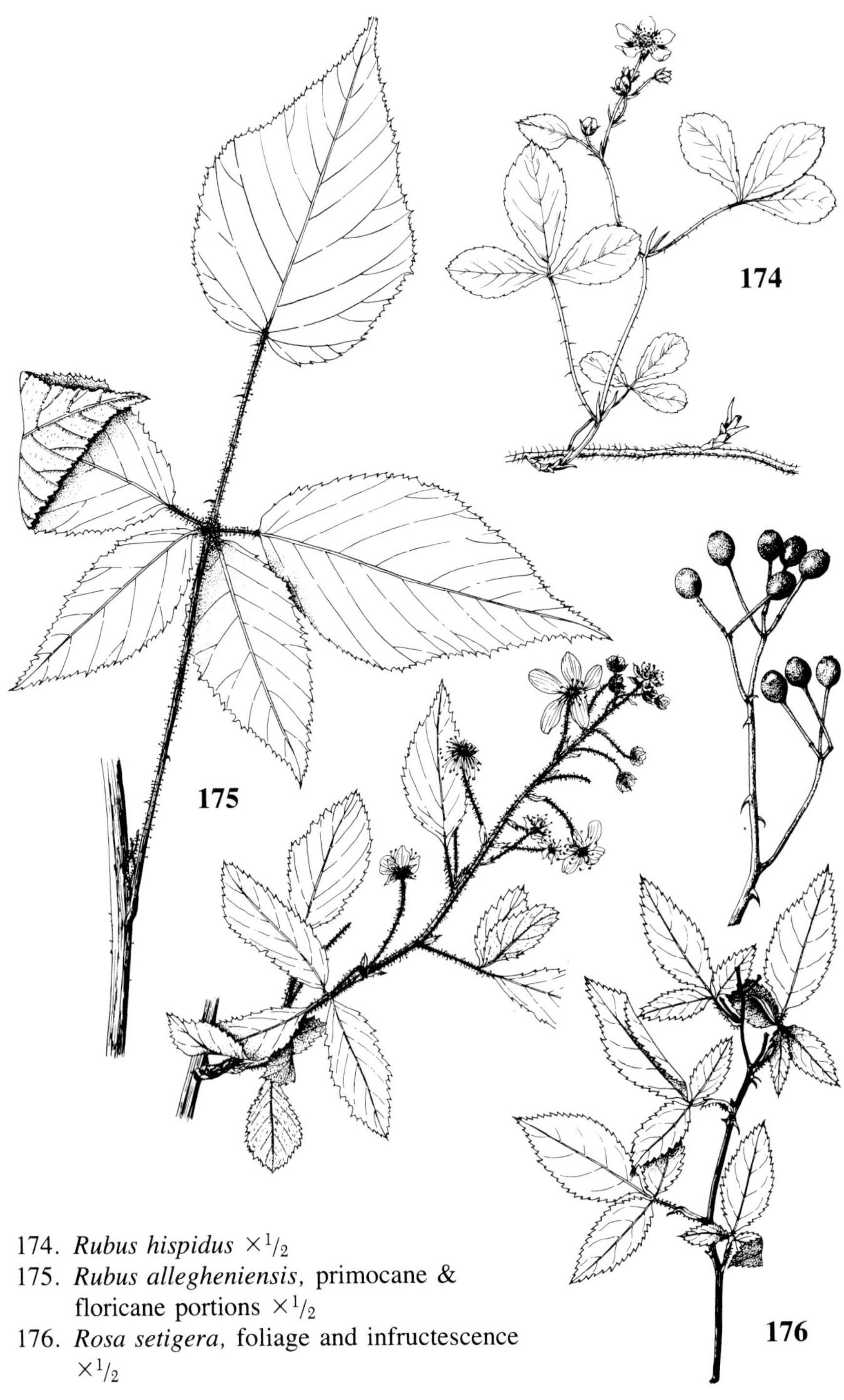

174. *Rubus hispidus* ×1/2
175. *Rubus allegheniensis*, primocane & floricane portions ×1/2
176. *Rosa setigera*, foliage and infructescence ×1/2

11. **R. allegheniensis** Porter Fig. 175 Common Blackberry

Map 471. Woodlands, clearings, old fields, roadsides; usually on dry uplands, occasional in marshy or swampy ground.

This complex includes the familiar tall blackberries with rather densely stipitate-glandular inflorescences, which when well developed appear to be elongate racemes. Depauperate plants may occur in shade or other unfavorable conditions but can often be recognized by the numerous glands and soft-pubescent undersides of the leaflets.

Yellow-fruited forms [f. *albinus* (Bailey) Fern., said to have been based on Michigan plants] are known; Michigan specimens have been seen from Branch County.

Most of our specimens are apparently true *R. allegheniensis*, including var. *plausus* Bailey (TL: near Lapeer, Lapeer Co.) and *R. rappii* Bailey (TL: near Vicksburg, Kalamazoo Co.). Other segregates attributed to Michigan include the following:

R. attractus Bailey [TL: south of Holland, in Allegan Co.; not able to be placed by Davis et al.]

R. rosa Bailey [*R. orarius* Blanchard is treated by Gleason as a different collective species (including *R. rosa*) from *R. allegheniensis*; however, this name is interpreted by Davis et al. as applicable to a glandless plant known from Maine and belonging in the section Arguti or perhaps Canadenses.]

12. **R. canadensis** L.

Map 472. Woodlands, clearings, fields, roadsides; occasionally in low ground.

Plants with a very few stalked glands on some pedicels may demonstrate influence from *R. allegheniensis*. Following the lead of Davis et al., this complex is here restricted to plants similar to *R. allegheniensis* but unarmed or nearly so, glandless, with foliage nearly or quite glabrous. Plants once associated here but with pubescent leaves have been shifted to the *R. pensilvanicus* complex (the section Arguti of some authors). Plants with a tendency for the canes to trail and root at the tip have been shifted to the *R. flagellaris* complex. What remains, in our area, may be only *R. canadensis* itself, although some other segregates are known from Wisconsin and Minnesota. *R. darlingtonii* Bailey (TL: Porcupine Mountains, Ontonagon Co.), originally described in the Canadenses, may belong in this complex or in the *R. flagellaris* complex; Davis et al. were unable to evaluate it but did include two other segregates in the synonymy of *R. canadensis*:

R. besseyi Bailey [TL: Porcupine Mountains, Ontonagon Co.; a distinctive form with the terminal primocane leaflets broadly rounded with abruptly acuminate tips.]

R. laetabilis Bailey [TL: Wakeshma Tp., Kalamazoo Co.; the specimens I have seen of the type number have leaves ± pubescent beneath and the pedicels somewhat glandular, thus not appearing like good *R. canadensis*. However, other Kalamazoo Co. material does appear to be better for this species, unusually far south in the state.]

13. **R. pensilvanicus** Poiret

Map 473. Roadsides and fields; thickets, woods, and borders of woods; often in low ground, such as borders of marshes and swamps.

This is a more variable complex than the two rather narrowly defined ones immediately preceding. The inflorescence is more often loose and ± corymbose, compared to the distinct racemose shape expected in *R. allegheniensis* and *R. canadensis*. As in *R. canadensis*, rare specimens with a few scattered glands may show some influence of *R. allegheniensis*.

This complex consists of the plants often placed in *Rubus* sect. *Arguti*, as remodeled by Davis et al. In the narrowest sense, *R. pensilvanicus* itself occurs only east of our region. Most of our plants are presumably *R. frondosus* Bigelow. More recent segregates attributed to Michigan are as follows:

R. abactus Bailey [Specimens assigned by Bailey to this species are apparently included in *R. frondosus* by Davis et al., though not in Kartesz & Kartesz.]

R. associus Hanes [TL: West Lake, Kalamazoo Co.; included by Davis et al. in *R. uvidus*. Trailing floricanes, which may occur in this taxon, resemble the *R. flagellaris* complex.]

R. avipes Bailey [Included by Davis et al. in *R. pergratus*.]

R. bellobatus Bailey [Perhaps escaped from early cultivation as the "Kittatinny Blackberry."]

R. cauliflorus Bailey [TL: Schoolcraft Tp., Kalamazoo Co.; included by Davis et al. in *R. recurvans*.]

R. hanesii Bailey [TL: north of Vicksburg, Kalamazoo Co.; as noted by Davis et al., this "odd plant" looks close to the *R. setosus* complex.]

R. licens Bailey [TL: Alamo Tp., Kalamazoo Co.; originally placed in the Flagellares, but included in *R. uvidus* by Davis et al. Material I have seen from the type colony and other numbers cited by Bailey certainly looks close to *R. flagellaris* in some respects, including long-trailing canes, the primocanes tip-rooting.]

R. limulus Bailey [TL: Kalamazoo, Kalamazoo Co.; included by Davis et al. in *R. recurvans*.]

R. localis Bailey [TL: Schoolcraft (actually Prairie Ronde Tp.), Kalamazoo Co.]

R. pergratus Blanchard

R. recurvans Blanchard

R. uvidus Bailey [TL: near Vicksburg, Kalamazoo Co.; Davis et al. considered the type to be mixed and referred the primocane material to *R. baileyanus* in the Flagellares, typifying the name with the floricane material, which belongs here. Material from the type number (WMU, MICH) appears to support this separation.]

2. **Rosa** — Rose

Most of the cultivated roses, which seldom escape, are "double," i. e., some or all of the stamens are converted into extra petals. Our native wild roses are "single," with 5 petals, usually pink but sometimes a deep purplish and rarely albino. Following widespread usage, the floral tube, which ripens into the fruit—a bright red or orange ± fleshy "hip," is here termed the *hypanthium*; it closely resembles an inferior ovary from a superficial external view. The calyx lobes are then simply termed sepals; these are ± pinnately lobed or pinnatifid in many of the cultivated species and also frequently in two or three of our native ones.

Much of the pioneer work studying the wild roses and reducing the excessive number of described "species" (including many of her own) was done at the University of Michigan some 50 years ago by Eileen W. Erlanson. More recently, Walter H. Lewis has studied the roses extensively, and his annotations on many specimens have also been of great help. Even our native species are quite variable in such features as pubescence, fruit shape, armature, and glands. It seems best to be quite conservative in recognizing named infraspecific taxa or in assuming hybridization—which does occur—to be an explanation for variability (see discussion under *R. blanda*). The present keys and descriptive notes will require a certain amount of judgment, experience, and intuition to use for some ambiguous specimens. In general, I have followed Erlanson's advice (1928) to "treat intermediate forms under the species they most nearly resemble rather than to disregard them under the category of unnamed crosses."

The garden species which have been collected as established escapes in Michigan are here treated quite superficially, based on very limited material, and some specimens are dubiously named (while others are quite unidentifiable). Hybridization is involved in many of the garden roses, and identification is difficult even in works on cultivated plants. While roses are known primarily as ornamentals, an important component of the florist and nursery industries, they are also the source of essential oils ("attar of roses") used in perfume. A major source of flowers from which the oils are distilled is damask rose [*R. damascena* Miller or *R. ×bifera* (Poiret) Pers.], extensively grown in Bulgaria for this purpose. This species (or hybrid), *R. ×alba* L., and dog rose [*R. canina* L.] have all been reported as persisting (and perhaps spreading a little) after cultivation in Michigan—as do others, undoubtedly—but do not seem to have spread to new locations as have those included in the key.

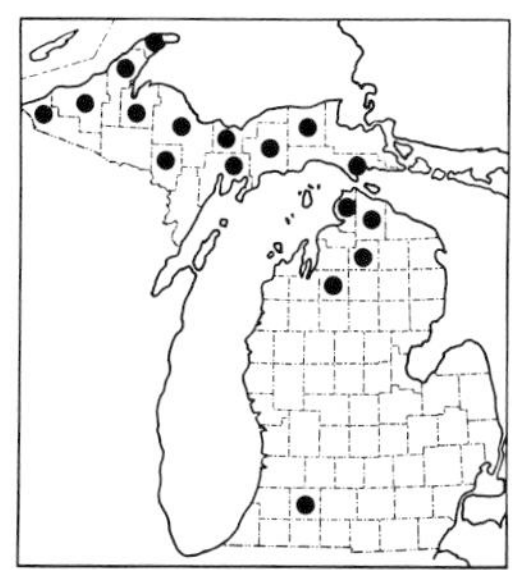
472. Rubus canadensis

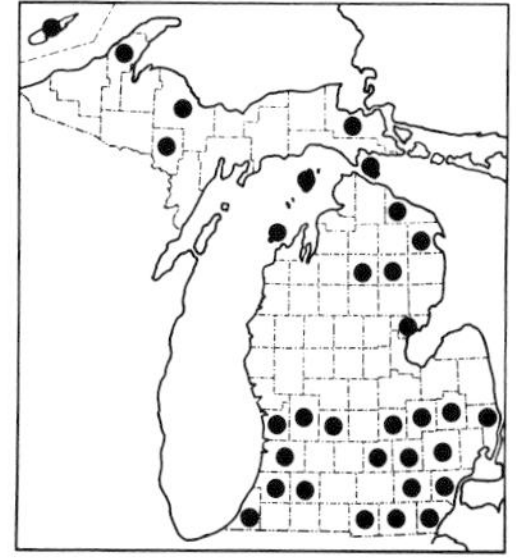
473. Rubus pensilvanicus

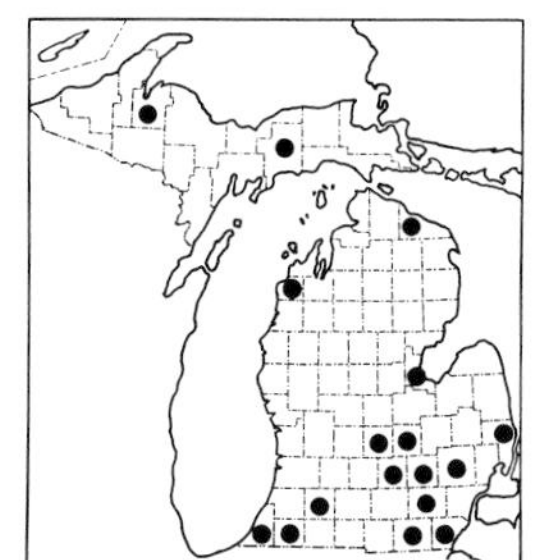
474. Rosa multiflora

REFERENCES

Erlanson, Eileen Whitehead. 1926. The Wild Roses of the Mackinac Region of Michigan. Pap. Mich. Acad. 5: 77–94.

Erlanson, Eileen Whitehead. 1928. Ten New American Species and Varieties of Rosa. Rhodora 30: 109–121.

Erlanson, Eileen Whitehead. 1934. Experimental Data for a Revision of the North American Wild Roses. Bot. Gaz. 96: 197–259.

Krüssman, Gerd. 1981. The Complete Book of Roses. Timber Press, Portland, Oregon. 436 pp. [Translated from 1974 German edition; deals little with our native roses, but contains a wealth of rose lore and information on cultivated roses.]

Lewis, Walter H. 1969. A Monograph of the Genus Rosa in North America. I. R. acicularis. Brittonia 11: 1–24.

KEY TO THE SPECIES

1. Sepals (3 outer ones) pinnatifid or at least with narrow lateral lobes; plant a local escape from cultivation (except for 3 native species keyed here as well as under the alternative because they often have lobed sepals)
  2. Inflorescence several–many-flowered; styles glabrous; stipules deeply pinnatifid . . . . . 1. **R. multiflora**
  2. Inflorescence 1–4-flowered (rarely more); styles villous; stipules entire (or with stipitate-glandular margins)
    3. Leaflets on most if not all leaves 9–11; pedicels and hypanthium smooth and glabrous . . . . . 12. **R. arkansana**
    3. Leaflets 5–7; pedicels and usually hypanthium ± glandular-hispid
      4. Flowers densely double, ± nodding . . . . . 2. **R. centifolia**
      4. Flowers usually single, not nodding
        5. Petals less than 2 cm long; largest leaflets less than 2.5 (very rarely 3) cm long; plants with stipitate glands more conspicuous than the pubescence on lower surface of leaflets and ± dense on pedicels, sepals, and stipules . 3. **R. eglanteria**
        5. Petals and usually largest leaflets longer; plants with stipitate glands absent or less conspicuous than dense pubescence on lower surface of leaflets and absent to dense elsewhere
          6. Plant a rare escape from cultivation with spines or thorns (if any) below the nodes no more prominent than others; leaflets doubly serrate
            7. Leaflets densely pubescent beneath and also glandular; mature fruit ca. 12–18 mm in diameter; petals less than 3 cm long . . . . . 4. **R. villosa**
            7. Leaflets (except on midvein) at most rather sparsely pubescent and glandular beneath; mature fruit smaller; petals often larger . . . . . 5. **R. gallica**
          6. Plant native, with spines or thorns below the nodes and these often more prominent than any others; leaflets usually singly serrate
            8. Nodal thorns stout, down-curved; leaflets with ± 20 or more fine teeth on a side; plant of wet habitats . . . . . 9. **R. palustris**
            8. Nodal thorns straight and slender, similar to internodal ones in form if not size; leaflets with ± 15 or fewer primary teeth on a side (sometimes doubly serrate); plant of dry habitats . . . . . 11. **R. carolina**
1. Sepals entire; plant an escape or native (all native species are included here)
  9. Styles glabrous, united into a single column (stigmas ± separate) protruding from the hypanthium nearly or quite as far as the stamens; leaflets 3 (–5) . . . 6. **R. setigera**
  9. Styles villous, separate but forming a dense head in the throat of the hypanthium, much shorter than the stamens; leaflets mostly 5–11

10. Leaflets ± rugose, densely soft-pubescent beneath; young branchlets, thorns, leaf rachises, and pedicels densely pubescent .... 7. **R. rugosa**
10. Leaflets not rugose, glabrous or lightly pubescent beneath; young branchlets and thorns, and often other parts, glabrous or stipitate-glandular
11. Stems with strong, slightly down-curved thorns immediately below most branchlets and petioles but few if any other thorns or prickles (except sometimes at base of plant)
12. Pedicels and hypanthia smooth, glabrous; sepals erect on the mature fruit; plant a rare escape from cultivation along roadsides and fields .... 8. **R. cinnamomea**
12. Pedicels and hypanthia glandular-hispid; sepals widely spreading on mature fruit; plant native in wet ground ....9. **R. palustris**
11. Stems with thorns, if any, below the nodes straight and internodal spines often of similar size and shape
13. Leaflets mostly 6–12 mm long, glabrous beneath; flowers solitary; stems ± well armed with slender broad-based thorns and more slender needle-like prickles; plant an uncommon escape from cultivation ...... 10. **R. spinosissima**
13. Leaflets mostly ca. (15) 18 mm or more long, usually pubescent at least on midvein beneath; flowers 1–several in an inflorescence; stems variously smooth or armed; plant a native shrub
14. Pedicels and hypanthia glandular-hispid; sepals ± widely spreading and usually deciduous from mature fruit; plant native in dry ground in southern Lower Peninsula, with straight narrowly tapered thorns or prickles, especially below the nodes [hybrids of *R. blanda* × *R. palustris* with glandular hypanthia would also run here—see text] .... 11. **R. carolina**
14. Pedicels and hypanthia smooth and glabrous; sepals ± erect and persistent on fruit; plant of various distribution and habitat
15. Leaflets (7) 9–11, usually ± soft-pubescent beneath and obovate; flowers mostly clustered at the ends of long, ± strongly glaucous, at least sparsely prickly new shoots; stipule margins mostly glandless or irregularly glandular .... 12. **R. arkansana**
15. Leaflets 5–7 (rarely 9 on vegetative shoots), variously pubescent and shaped (mostly ovate-elliptic if new shoots prickly); flowers mostly on prickly short lateral shoots or on smooth (sometimes glaucous) shoots; stipule margins various
16. Lateral (floral) branches bristly; leaflets usually doubly serrate, especially toward apex, the primary teeth ovate (biconvex); margins of floral bracts and stipules usually ± copiously glandular .... 13. **R. acicularis**
16. Lateral branches smooth; leaflets singly serrate, the teeth sharp, often straight or concave on inner side; margins of floral bracts and stipules glandless or sparsely glandular or sometimes copiously glandular ...... 14. **R. blanda**

1. **R. multiflora** Murray Fig. 178 Multiflora or Japanese Rose

Map 474. Roadsides, disturbed woods and borders, fencerows and fields, thickets and untended yards; sometimes in low ground although usually in dry places. A native of eastern Asia, long inadvisedly recommended for "living fences" and now an obnoxious aggressive weed in some parts of the country.

The flowers are usually white, and the styles are united into a column that protrudes more conspicuously from the hypanthium than in most species. The clusters of numerous small red fruits are distinctive late in the season.

2. **R. centifolia** L. Cabbage Rose

Map 475. A widely cultivated plant, interpreted as either a native of the Caucasus region or a hybrid involving as many as four species. Rarely spreading to roadsides, as in Kalamazoo County. Specimens from St. Clair County in 1903, referred here by Dodge, appear somewhat doubtful, as the flowers do not appear to have been nodding and the stems are very sparsely armed; very similar material has been collected more recently in Berrien County.

3. **R. eglanteria** L. Sweetbrier

Map 476. A Eurasian species, locally naturalized as an escape from cultivation on roadsides, fields, railroads, old dunes, shores and river banks, borders of woods. Its very thorny stems are avoided by grazing animals, so that it can dominate the vegetation when well established.

This name is rejected by some authors, especially European, as ambiguous; *R. rubiginosa* L. is then the name employed.

The leaflets, stipules, and sepals are all provided with abundant short-stipitate glands and the foliage is ± fragrant. Some other species in the genus will rarely have a narrow lobe on a sepal, but such plants differ from the species in the first half of the key in other characters (e. g., *R. setigera* has mostly trifoliolate leaves and glabrous styles).

R. micrantha J. E. Smith differs from *R. eglanteria* in having glabrous styles; it has been reported from Michigan, but all specimens seen seem closer to *R. eglanteria*. *R. micrantha* differs from *R. multiflora*, which also has glabrous styles, in its pink flowers, separate styles, and entire stipules.

4. **R. villosa** L.

Map 477. Another garden species, primarily of European range, grown for its large and showy fruit, and rarely escaped to roadsides.

475. Rosa centifolia

476. Rosa eglanteria

477. Rosa villosa

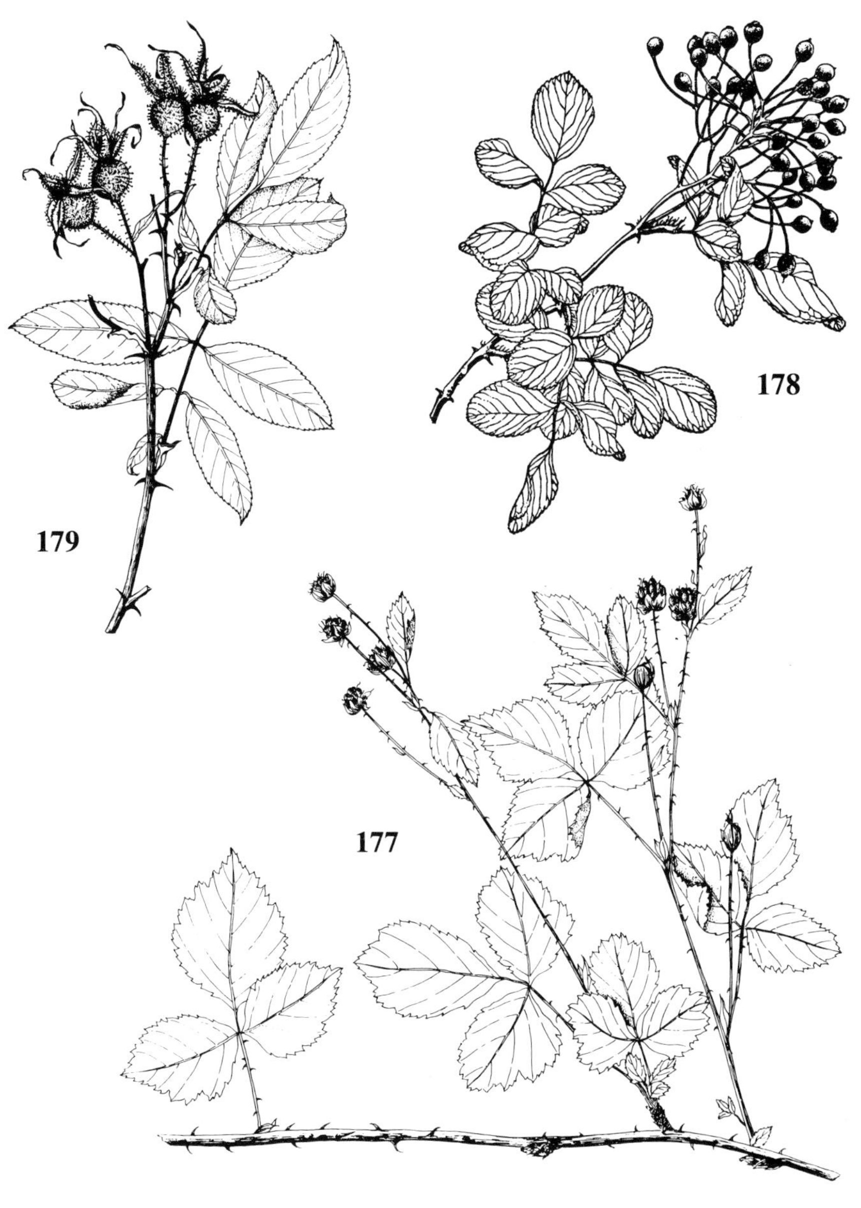

177. *Rubus flagellaris* $\times{}^1/_2$
178. *Rosa multiflora* $\times{}^1/_3$
179. *Rosa palustris* $\times{}^1/_2$

5. **R. gallica** L. French Rose

Map 478. Originally a Eurasian species, occasionally escaped from cultivation to roadsides and fields.

6. **R. setigera** Michaux Fig. 176 Prairie Rose

Map 479. Woods and thickets, the long stems often "climbing" over other shrubs.

The leaflets on most of our specimens are ± softly pubescent beneath, but sometimes they are completely glabrous.

7. **R. rugosa** Thunb. Japanese Rose

Map 480. An Asian species grown for ornament and hedges, rarely escaping to roadsides, fields, shores, and disturbed ground.

A densely thorny plant, with large, thick, rugose leaflets and large red-purple (or white) flowers. The thick pubescence tends to obscure whatever stipitate glands may also be present.

8. **R. cinnamomea** L. Cinnamon Rose

Map 481. A Eurasian species, occasionally escaped from cultivation to roadsides, fields, river banks, and sites near dwellings.

9. **R. palustris** Marsh. Fig. 179 Swamp Rose

Map 482. Bogs, wet conifer swamps, thickets, and swales; margins of ponds, lakes, and streams.

Usually easily recognized by its wet habitat, curved infranodal thorns, and very finely and evenly serrate margins of the leaflets (ca. 20 or more teeth on a side of the larger leaflets). The flowers are very fragrant. The sepals (as in *R. carolina* and *R. arkansana* among our native roses) are often ± pinnately lobed, as in most of our garden species. In *R. blanda* and *R. acicularis*, only very rarely is there a hint of such lobing.

R. palustris f. *inermis* (Regel) W. H. Lewis lacks the characteristic stout curved infranodal thorns. Such plants can be recognized by the leaf margins

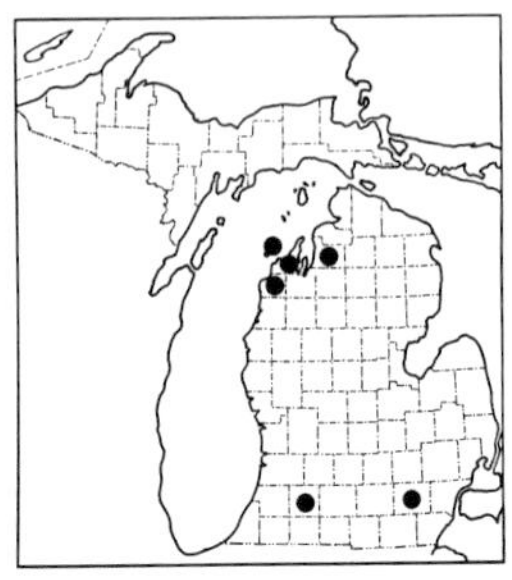

478. Rosa gallica

479. Rosa setigera

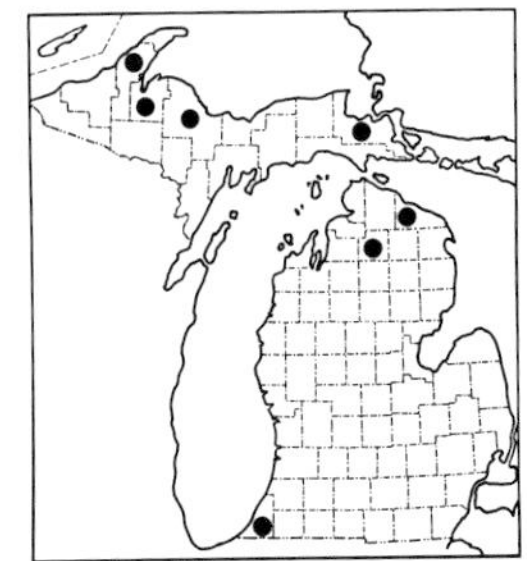

480. Rosa rugosa

and glandular-hispid hypanthium. Hybrids with *R. blanda* are discussed under that species.

10. **R. spinosissima** L. Scotch Rose

Map 483. Another cultivated species of Eurasian origin, occasionally established as an escape. Roadsides, railroads, near old dwellings, dry woodland.

The flowers are solitary and pink, white, or yellow. This name is sometimes rejected as ambiguous, and the species is then called *R. pimpinellifolia* L.

11. **R. carolina** L. Pasture Rose

Map 484. Dry sandy woods, oak (and jack pine) woodland; sandy banks, fields, dunes, and fencerows; roadsides and railroad embankments; borders of woods; occasionally in moist ground.

A variable species. In f. *glandulosa* (Crépin) Fern. there are stipitate glands on the rachis of the leaf. A few plants lack the gland-tipped bristles on hypanthium and pedicel, but can be placed here by the slender, straight, usually terete prickles. From *R. acicularis* they also differ in range and in usually singly serrate margins of the leaflets, and from *R. palustris* f. *inermis* in the more coarsely serrate leaflets as well as habitat.

In old literature this name is applied to *R. palustris*, and the present species is called *R. humilis* Marsh.; it is also sometimes misidentified as *R. virginiana* Miller, a more eastern species.

12. **R. arkansana** Porter Prairie Rose

Map 485. Roadsides and fields, along railroads, river banks and lake shores.

Our plants are mostly if not all the eastern representative of this prairie species, sometimes designated as var. *suffulta* (Greene) Cockerell; there may be some doubt as to whether they are truly native or only adventive here. This species hybridizes rather freely with *R. carolina*, also tetraploid, farther

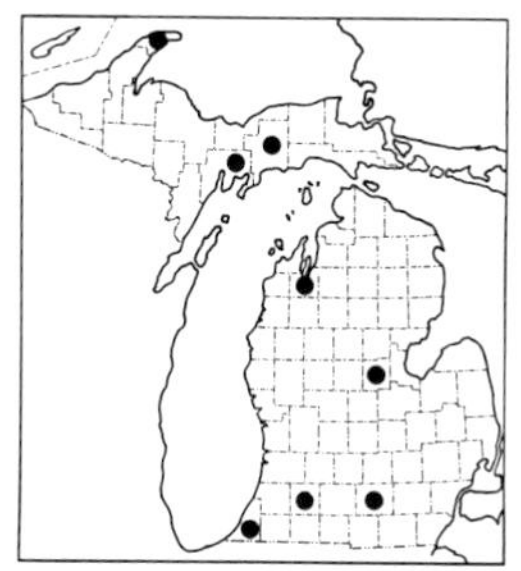
481. Rosa cinnamomea

482. Rosa palustris

483. Rosa spinosissima

west, but it is so local in our area that hybrids seem not to be a problem with us. The Chippewa County collection has sparsely setose hypanthia and may represent a hybrid.

13. **R. acicularis** Lindley   Plate 5-C   Wild Rose

Map 486. Sandy and gravelly shores and dunes; rocky ridges, outcrops, and shores; dry sandy woodlands with jack pine and oak; moist thickets and openings in coniferous woods and swamps.

This is a circumpolar species, the only one of our native roses that is also indigenous in the Old World—the largest total range of any species in the genus. Typical *R. acicularis* is Eurasian, ranging into Alaska, and is octoploid (2n = 56). The American plant is hexaploid (2n = 42) and differs in various morphological tendencies; it may be called ssp. *sayi* (Schw.) W. H. Lewis or var. *bourgeauana* Crépin.

This species sometimes hybridizes with the diploid *R. blanda;* the tetraploid progeny are partly sterile, especially if *R. blanda* is the pollen parent, according to Erlanson, whose later work (1934) placed less emphasis on hybridization as a source of variation in our common wild roses (and also less stress on naming variants than in earlier papers, e. g. 1926 & 1928). Plants of this species are extremely variable, including characters of pubescence, glandularity, and fruit shape.

14. **R. blanda** Aiton   Wild Rose

Map 487. Dunes, sandy bluffs, and shores; jack pine woodland, river banks, and borders of woods and thickets; rocky openings and outcrops; fields and fencerows.

Plants with white flowers [f. *alba* (Erl.) Fern.] are rarely found. Plants with bristly floral branches were named var. *hispida* by Farwell (TL: Belle Isle [Wayne Co.]). The type is very similar to the more northern *R. acicularis*, but the leaf serrations and eglandular stipules are appropriate for *R. blanda*. It could be of hybrid origin, perhaps involving *R. arkansana*. Plants from northern Michigan referred to this variety by Erlanson have glandular-

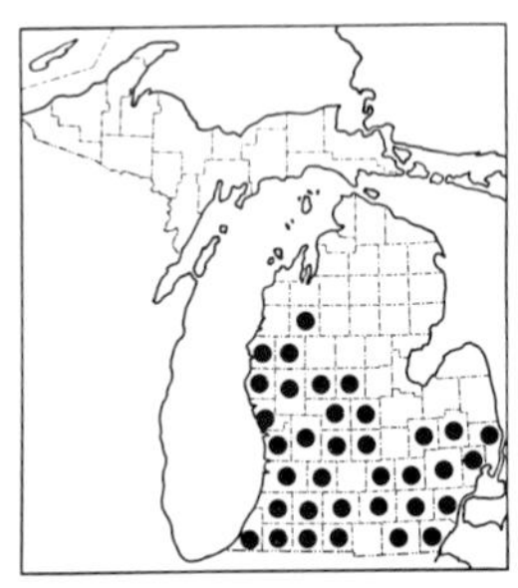

484. Rosa carolina

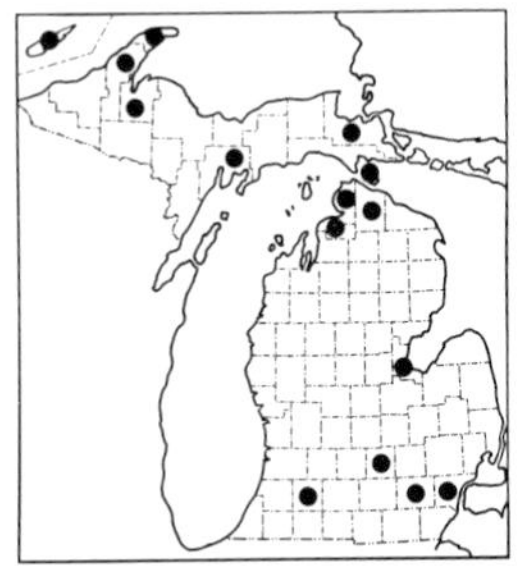

485. Rosa arkansana

486. Rosa acicularis

margined stipules and long lateral shoots (as in *R. blanda*, but sparsely prickly); these could be hybrids with *R. acicularis*.

Hybrids with the hexaploid *R. acicularis* are further mentioned under that species. Hybrids with *R. palustris*, also a diploid, are more frequent and may be called *R. ×palustriformis* Rydb. [including *R. ×michiganensis* Erl. and *R. ×schuetteana* Erl. (TL for both: Douglas Lake, Cheboygan Co.)]. In this hybrid, the influence of *R. palustris* is usually seen in finely serrate leaflets and, sometimes, in a glandular-hispid hypanthium although the pedicel and hypanthium are usually smooth; the influence of *R. blanda* is seen in the smooth lateral shoots and the usual absence of stout curved infranodal thorns, although there are small internodal prickles. The hybrid flowers between the early-blooming *R. blanda* (peaking in June) and the late-blooming *R. palustris* (peaking in late July). (*R. acicularis*, incidentally, tends to flower in late May or early June—some days before *R. blanda*.)

3. Sorbaria

1. **S. sorbifolia** (L.) A. Br. False Spiraea

Map 488. A native of Asia, locally well established as an escape from cultivation, spreading readily by suckers to form large colonies on roadsides and banks, at edges of fields, and near old homesites.

The long new shoots are herbaceous, but older stems are woody though somewhat soft.

4. Sorbus Mountain-ash

Often included with other segregates in the comprehensive genus *Pyrus*. Although distinctive in our area with its compound leaves, the genus does include simple-leaved species, and even some compound-leaved ones do hybridize with *Aronia*. (But they can also hybridize with *Amelanchier*!)

The bright red fruits of all our species are attractive, and persist on the trees until birds get to them—which is usually not long.

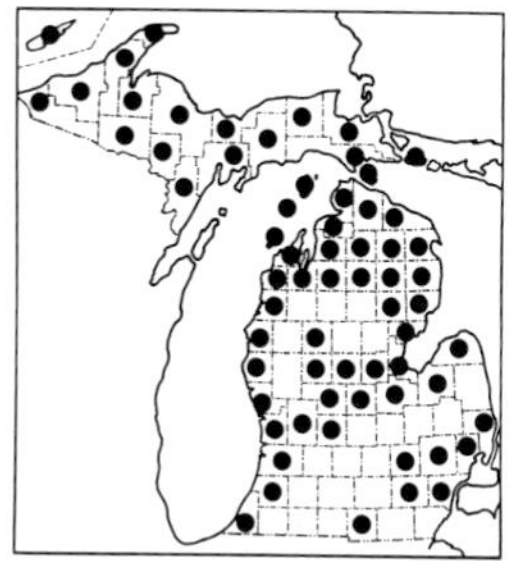
487. Rosa blanda

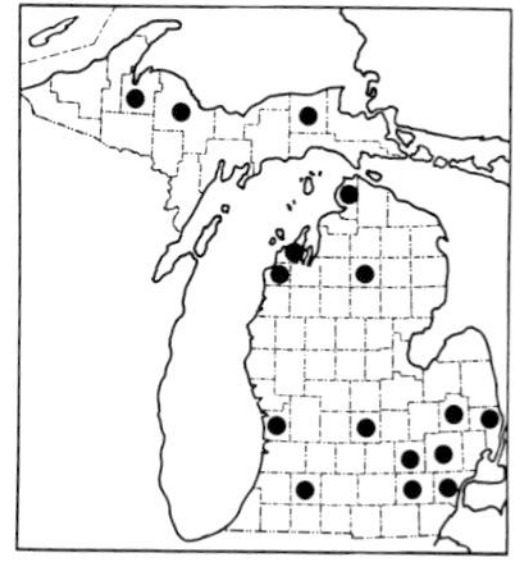
488. Sorbaria sorbifolia

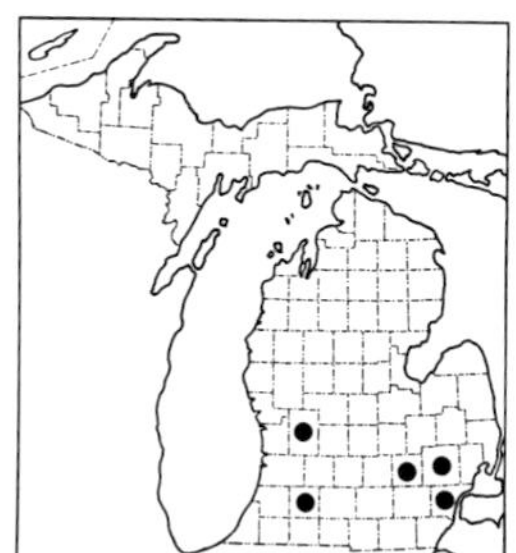
489. Sorbus aucuparia

REFERENCES

Jones, George Neville. 1939. A Synopsis of the North American Species of Sorbus. Jour. Arnold Arb. 20: 1–43.

Jones, George Neville. 1953. Nomenclature of American Mountain-ash. Rhodora 55: 358–360.

KEY TO THE SPECIES

1. Sepals, ovary, and winter-buds ± densely white-pubescent; leaflets ± soft-pubescent beneath; plant a rare escape from cultivation . . . . 1. **S. aucuparia**
1. Sepals, ovary, and winter buds (outer scales) glabrous or sparsely pubescent; leaflets glabrous or glabrate over much or all of the surface beneath; plant native
   2. Lateral leaflets 2.4–3 (3.2) times as long as broad, ± abruptly acute to obtuse at apex, often retaining some pubescence at least along midrib beneath at maturity; petals (2.7) 3–4.5 mm long; pedicels and branches of inflorescence ± hairy; mature fruit ca. [6] 8–10 mm in diameter, often becoming glaucous especially in drying . . . . 2. **S. decora**
   2. Lateral leaflets (3.2) 3.4–4.5 (6.4) times as long as broad, ± tapered or acuminate to sharp apex, usually completely glabrous at maturity; petals (2) 2.4–3 mm long; pedicels and branches of inflorescence glabrous or nearly so; mature fruit ca. 5–6.5 mm in diameter, not glaucous even when dried . . . . 3. **S. americana**

1. **S. aucuparia** L. European Mountain-ash; Rowan

Map 489. This European species is widely cultivated and rarely escapes, the seeds dispersed by birds, to woods and swamps as well as sites nearer habitations.

A common cultivar has orange rather than red fruit. In the genus *Pyrus*, this species is called *P. aucuparia* (L.) Gaertner.

2. **S. decora** (Sarg.) Schneider Plate 5-D

Map 490. Wooded dunes and bluffs, especially frequent at edges of woods along Lake Superior; deciduous, mixed, and coniferous woods, often with fir, cedar, and pine, but sometimes in beech-maple stands; seems to do particularly well along shores, perhaps because of the moister climate and perhaps because released from competition of larger trees in the forest.

A few unusually pubescent flowering specimens closely resemble the preceding species. There is a subtle distinction usually evident in the tips of the leaflets, where the terminal tooth is prolonged in *S. decora* (as in *S. americana*) while in *S. aucuparia* it is no more conspicuous than the lateral teeth (which, furthermore, are frequently partly doubly serrate). *S. decora* varies considerably in overall shape of the apex of the leaflet and in pubescence. The dried fruit in herbarium specimens is often strikingly blue-glaucous.

This is our commonest species of *Sorbus*, a handsome small tree when in flower (or fruit), often reported in earlier years under a variety of misidentifications and misapplied names, including *S. dumosa, S. sambucifolia,*

S. scopulina, S. sitchensis, and *S. subvestita*. In the genus *Pyrus*, the name becomes *P. decora* (Sarg.) Hylander.

3. **S. americana** Marsh. Fig. 180

Map 491. Swamps (both cedar and deciduous) and stream banks; mixed woods, especially at borders, as on bluffs along Lake Superior.

Specimens of the preceding species with more prolonged acute apices on the leaflets than is often the case are frequently misidentified as *S. americana* if one is not familiar with the overall lanceolate, longer leaflets of this species (usually ca. 4 times as long as broad). Checking 2 or 3 representative leaflets in case of doubt should be adequate to decide on which side of the length-width ratio a specimen belongs (3.2 apparently being a nearly certain dividing line, as indicated in the key). (Compare the leaves in the color plate of *S. decora* with the leaf of Fig. 180.)

Jones (1953) argues that in the genus *Pyrus* this species must be called *P. microcarpa* (Pursh) DC. and not *P. americana* (Marsh.) DC., as is usually done. [The combination *P. microcarpa* was actually first published by Sprengel.]

5. **Prunus** Cherry; Plum

A useful vegetative feature often mentioned for this genus is the presence of a pair of glands (sometimes 1 or 3) near the summit of the petiole or toward the base of the blade. These are obscure or absent in some species, and of course they will not in themselves separate *Prunus* from some quite different genera, such as *Salix*, in which certain species may have similar glands. In flower or fruit, however, *Prunus* is an easily recognized genus, with deep perigynous cup and the fruit a drupe with a single large stone or pit enclosing the seed. Many species flower before the leaves are fully grown, and the key is designed for use with or without flowers.

A considerable number of species are cultivated for flowers or fruit, and some may persist in old yards or appear as occasional waifs besides those recorded here as escaped to roadsides or elsewhere. One of the flowering almonds, *P. glandulosa* Thunb., a native of China, has for example been

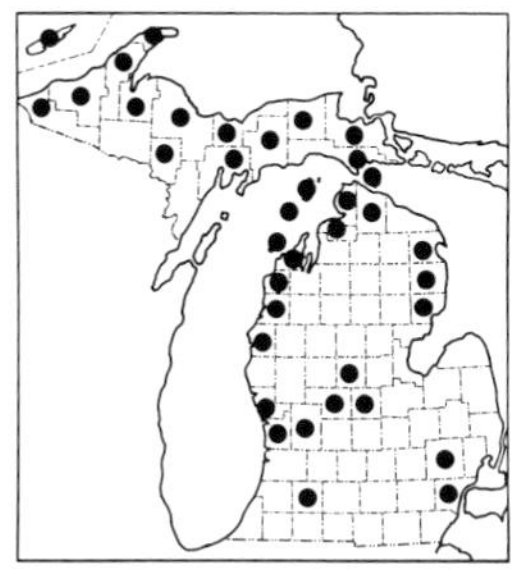
490. Sorbus decora

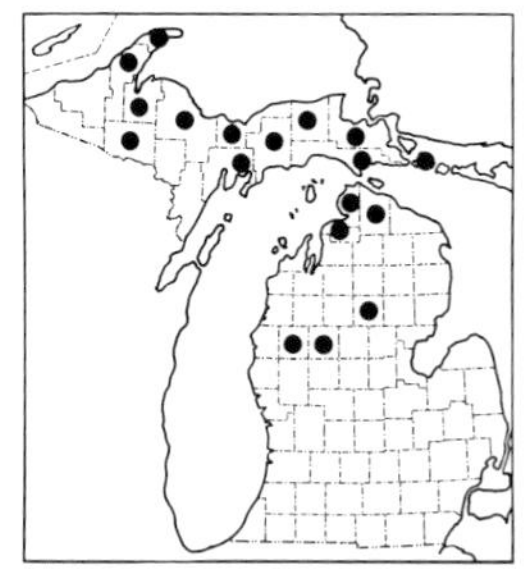
491. Sorbus americana

492. Prunus serotina

"spreading somewhat" from an old planting along a roadside in southern Benzie County. It has double flowers and glandular-serrate margins on the leaves and sepals. There are numerous horticultural forms of the cultivated species, including double-flowered ones for the ornamentals, and they may also hybridize. The treatment of these species here is superficial and of necessity draws more on literature and less on specimens than would be desirable for accurate and comprehensive keys.

REFERENCES

Groh, Herbert, & Harold A. Senn. 1940. Prunus in Eastern Canada. Canad. Jour. Res. C 18: 318–346.

Hedrick, U. P. 1911. The Plums of New York. Rep. New York Agr. Exp. Sta. 18 (for 1910), II. 616 pp.

Hedrick. U. P. 1915. The Cherries of New York. Rep. New York Agr. Exp. Sta. 22 (for 1914), II. 371 pp.

Wight, W. F. 1915. Native American Species of Prunus. U.S. Dep. Agr. Bull. 179. 75 pp.

KEY TO THE SPECIES

1. Inflorescence an elongate raceme of many (at least 12, usually at least 20) flowers, terminating a new leafy branchlet of the current year
  2. Leaves with blades ovate or elliptical to lanceolate, mostly broadest at or below the middle, ± glossy above, the teeth ± incurved, giving the margin a crenulate aspect; calyx lobes entire or nearly so (at most 5 glandular teeth), persistent with the floral tube beneath the maturing fruit . . . . 1. **P. serotina**
  2. Leaves with blades elliptical to obovate, mostly broadest at or above the middle, dull above, the teeth narrowly acuminate and ascending to spreading, giving the margin a sharply and finely serrate aspect; calyx lobes with numerous irregular gland-tipped teeth, deciduous with the floral tube, leaving only a smooth disc beneath the maturing fruit . . . . 2. **P. virginiana**
1. Inflorescence a corymb or umbel of fewer than 12 flowers (or flowers solitary), sessile or on short lateral shoots (leafy or not)
  3. Flowers and fruit sessile or nearly so, mostly solitary (or 2); ovary and fruit densely pubescent; flowers pink to white
    4. Petals white (pink in bud), ca. 1.3 cm long or shorter; leaf blades less than 5 (7) cm long; fruit less than 1.5 cm in diameter (a cherry) . . . . 3. **P. tomentosa**
    4. Petals pink, usually larger (often 1.5–2 cm long); leaf blades longer; fruit much larger (peach) . . . . 4. **P. persica**
  3. Flowers and fruit on distinct slender pedicels, often not solitary; ovary and fruit glabrous; flowers in most species white
    5. Calyx lobes glabrous throughout (at most, glandular-margined); fruit (cherry) ± globose, not grooved, not glaucous, the pit rounded rather than 2-edged
      6. Plant a low spreading shrub (native), with mostly decumbent or ascending elongate branches; leaf blades oblanceolate (to obovate-elliptic), the teeth obscure or absent on lower third or half; calyx lobes with irregularly glandular-toothed margins; petals less than 7.5 (9) mm long . . . . 5. **P. pumila**
      6. Plant an erect small to large tree or tall bushy shrub; leaf blades ovate to obovate, regularly toothed to the base; calyx lobes without glands (or a few glandular teeth in the large-flowered *P. cerasus*); petals various

7. Petals 4–7.5 mm long; fruit less than 1 cm in diameter
8. Inflorescence a few-flowered corymbose raceme (i. e., with a definite central axis); petals glabrous; leaf blades round-ovate, less than 1.5 times as long as wide; fruit nearly black . . . . 6. **P. mahaleb**
8. Inflorescence usually umbellate (occasionally ± corymbose); petals hairy on the outside at the base; leaf blades usually at least twice as long as wide; fruit bright red . . . . 7. **P. pensylvanica**
7. Petals ca. 9–15 mm long; fruit ca. 1.5–2.5 cm in diameter
9. Calyx with entire lobes, constricted below them; bud scales at base of umbel not leaf-like, the inner ones divergent or reflexed; leaves retaining some pubescence, especially along the midrib, beneath, the blades ca. 7–15 cm long at maturity and the petioles with conspicuous glands near the summit; fruit sweet . . . . 8. **P. avium**
9. Calyx with glandular-toothed lobes, not constricted below them; bud scales often with leaf-like tips, the inner ones erect; leaves becoming glabrous beneath, the blades mostly 4–8 cm long at maturity with glands toward the base (rather than on the petiole); fruit sour . . . . 9. **P. cerasus**
5. Calyx lobes pubescent, at least sparsely at the base above (rarely glabrous in *P. nigra*); fruit (plum) with a longitudinal shallow furrow or groove, usually glaucous, the pit 2-edged, often somewhat flattened
10. Teeth of leaf sharp (almost bristle-tipped), glandless; calyx lobes without marginal glands (rarely a few teeth at end in *P. americana*)
11. Petals (at least the largest) (6.5) 7–11 mm long; calyx lobes usually glabrous on the lower (outer) side; leaf blades mostly oblong-elliptic to obovate in general outline but abruptly acuminate to a conspicuous prolonged tip; fruit ca. 10–25 mm in diameter when dry (larger fresh), the pit ca. 10–15 mm broad; plant usually of mesic woodland, thickets, and stream borders . . . . 10. **P. americana**
11. Petals 4–6 mm long; calyx lobes usually slightly pubescent on the lower side; leaf blades mostly ovate in outline, acute or barely acuminate; fruit less than 10 (12) mm in diameter when dry [said to be ca. 15 mm when fresh], the pit ca. 5–8 mm broad; plant of dry jack pine plains and sandy oak woodland (rarely mesic sites) . . . . 11. **P. alleghaniensis**
10. Teeth of leaf gland-tipped (evident on young leaves unfolding as the flowers open), the rounded tip with a callous scar if gland is shed; calyx lobes with glandular margins (except in *P. cerasifera*)
12. Margins of calyx lobes pubescent but not glandular; petals ca. 5–7 mm long, white; young twigs shiny reddish brown; leaves small, the blades mostly less than 5 cm long . . . . 12. **P. cerasifera**
12. Margins of calyx lobes glandular; petals ca. 5–12 mm long, white or pink; twigs and leaves various
13. Flowers in clusters of 2–4; petals 8–12 mm long, often pink; leaves with blades acuminate at tip and petioles with glands at summit; plant a native shrub of thickets and woodland borders . . . . 13. **P. nigra**
13. Flowers solitary or in pairs; petals ca. 5–12 mm long, white; leaves with obtuse to rounded blades and petioles usually without glands; plant a rarely escaped shrub or tree
14. Twigs spiny, the young ones finely pubescent; petals ca. 5–8 mm long; fruit usually less than 1 (1.5) cm in diameter, erect at maturity . 14. **P. spinosa**
14. Twigs not spiny, often glabrous; petals ca. 8–12 mm long; fruit over 2.5 cm in diameter, pendent at maturity . . . . 15. **P. domestica**

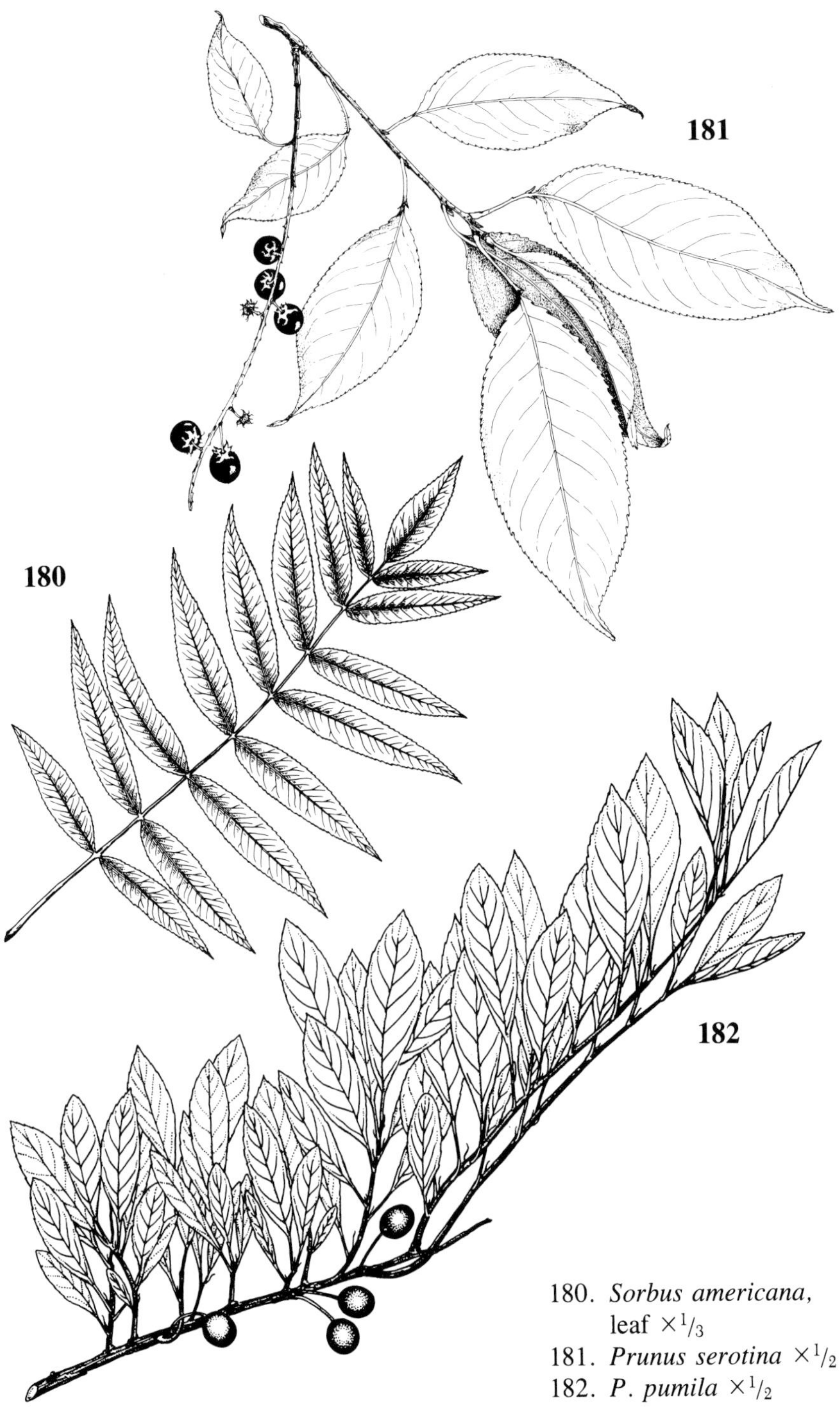

180. *Sorbus americana*, leaf $\times^1/_3$
181. *Prunus serotina* $\times^1/_2$
182. *P. pumila* $\times^1/_2$

1. **P. serotina** Ehrh. Fig. 181 Wild Black Cherry

Map 492. A common tree of fencerows and borders of fields and forests—almost anywhere that birds have deposited the seeds; can be very scrubby in rocky ground or dry open jack pine or aspen woodland, a fine tree in deciduous forests (oak, beech-maple, or others), attaining considerable size in rich hardwoods. The "national champion" is a tree in Washtenaw County nearly 4.5 feet in diameter.

There is often a strip of pubescence, usually rust-colored, along the midrib on the underside of the leaves, although sometimes there are no hairs at all. See also comments under the next species.

The purple-black fruit makes an excellent jelly. The aromatic bark of this species (and the non-aromatic bark of the next) has long had reputed medicinal value, especially when brewed in a beverage. On the other hand, several species of *Prunus*, especially *P. serotina*, are known as stock-poisoning plants because of their cyanogenic glycosides that produce dangerous amounts of hydrogen cyanide, particularly in the succulent young leaves and slightly wilted ones and in the pits, which may fatally poison children. The wood is valuable in cabinet-work and fine furniture (often now as a veneer).

2. **P. virginiana** L. Choke Cherry

Map 493. Almost everywhere except wet ground. Even more commonly than the preceding, a species of fencerows and roadsides, dry open rocky and sandy ground, shores and openings, as well as dune thickets, jack pine plains, and river banks; often at borders of woods and in thin woods, but not an important forest species.

The leaves are usually glabrous beneath. If there is pubescence along the midrib or in tufts in axils of the lateral veins, it extends out along the lateral veins instead of being restricted to the midrib as in the preceding species. The veinlets form a characteristic fine network of very small areas. Choke cherry is a shrub or at best a small tree. The fruit, deep red-purple when ripe, nearly or quite as dark as in *P. serotina*, is extremely astringent, puckering the mouth of anyone who experiments with it; but the juice—mixed with apple or other source of pectin—makes a fine jelly.

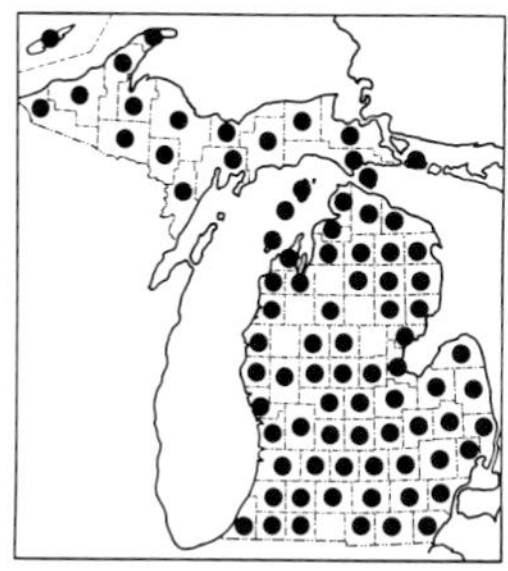
493. Prunus virginiana

494. Prunus tomentosa

495. Prunus persica

The European bird cherry, *P. padus* L., is very similar to *P. virginiana*, but grows to a larger size. The petals in *P. virginiana* are ca. (2) 3–4.5 (5) mm long, in the largest extreme approaching those of *P. padus*, which is usually distinguished by its longer, more elliptical petals and in having the floral tube pubescent within. However, many of our specimens of *P. virginiana* have a pubescent floral tube, especially basally. Like the other cherries popular with birds, *P. padus* may spread from cultivation, although it has apparently not been recorded as doing so in Michigan.

3. **P. tomentosa** Thunb. Nanking Cherry

Map 494. Native to eastern Asia, spreading readily from seed in cultivation to fencerows, borders of woods, and vacant or untended lots.

The petals of this small bushy shrub are white, appearing just before the leaves in early spring; the calyx and filaments are flushed with red.

4. **P. persica** (L.) Batsch Peach

Map 495. A well known cultivated tree, originating in China, rarely spontaneous where pits have been discarded or otherwise spread, as along roadsides and rivers, in thickets, or near yards.

The nectarine is a glabrous variety of this species. The apricot, *P. armeniaca* L., also has solitary sessile flowers but they are paler (or white) and the pit is smooth rather than deeply sculptured as in the peach.

5. **P. pumila** L. Fig. 182 Sand Cherry

Map 496. Sandy or gravelly shores, dunes, and old beach ridges around the Great Lakes; inland, on sandy plains and fields, often with jack pine and/or oak, in both Upper Peninsula and Lower Peninsula; also on rocky ledges, pavements, and summits.

A variable species, the extremes distinctive but thoroughly intergrading. The broad-leaved form, with oblong-elliptic or obovate leaf blades quite obtuse or rounded, and pale beneath, is found on jack pine plains and other dry wooded areas. It has been called *P. pumila* var. *susquehanae* (Willd.) Jaeger [much later named var. *cuneata* (Raf.) Bailey] or *P. susquehanae* Willd. In var. *pumila* the leaves are more narrowly oblanceolate, more acute, and less pale beneath. A few collections are close to *P. pumila* var. *besseyi* (Bailey) Gl., a western plant sometimes recognized as *P. besseyi* Bailey, with broader, shinier leaves, more acutely toothed.

The fruit is dark purple-black when ripe (red when green) and is the largest of all our native cherries. It varies in taste, as do individuals in their evaluation of it, but some persons find at least some of the fruit very tasty; others find it slightly bitter. But it makes a good jam.

The low branches of this shrub illustrate well the microclimatic effect of lying on the warm sand of dunes, for flowers and fruit in contact with the sand tend to develop distinctly earlier than those on branches of the same plant slightly elevated. The leaves in fall are a colorful red. Sterile plants are sometimes misidentified as a *Salix*, to which the leaves have a remote resemblance. The leaf glands typical of *Prunus* are sometimes obscure; when evident they tend to be at the base of the ± decurrent blade rather than at the summit of the petiole. On young shoots, the stipules are narrowly linear with elongate glandular teeth—quite unlike any of our species of *Salix*.

6. **P. mahaleb** L. Perfumed Cherry

Map 497. A Eurasian species, occasionally well established as an escape from cultivation. Recorded from low woods and river banks, roadsides, fields and thickets, dump sites and railroads—doubtless established in other sites as well.

This cherry is grown not for the small inedible fruit but for its fragrant flowers and wood.

7. **P. pensylvanica** L. f. Pin or Fire Cherry

Map 498. Sandy clearings, shores, and plains; borders of woods and fields and in thin (rarely swampy) woods, usually with aspen, white birch, and/ or jack pine. A small tree especially characteristic of recently burned or cleared areas; frequently clonal from root suckering. The seeds are very long-lived in the soil, germinating after fire or other disturbance.

Attractive and conspicuous in bloom late in the spring, especially along highways in the northern part of the state a few years after construction or remodeling of roadsides. The umbel of a few small bright red fruits is absolutely distinctive in midsummer. The cherries are sour, but tasty to many people (as well as birds). They make an excellent jelly, although the high percentage of pit to flesh requires a considerable amount of fruit to obtain much juice. The bark is reddish, a helpful character for quick distinction from other species which may be superficially similar, such as *P. mahaleb* (contrasted in the key) and *P. americana* (with calyx lobes pubescent within, slightly larger petals, and no gland-tips on the leaf teeth). The leaves are

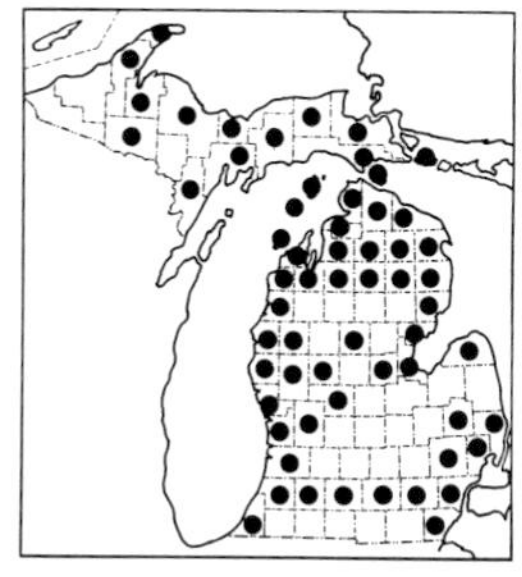
496. Prunus pumila

497. Prunus mahaleb

498. Prunus pensylvanica

usually glabrous or nearly so, but a few specimens are densely pubescent with ± straight hairs on the midrib beneath, the petiole, and the new branchlets.

8. **P. avium** (L.) L. Sweet Cherry

Map 499. Dumps, fencerows, thickets, and forests. A species of Eurasian origin, widely cultivated for the fruit, including the large and popular "Bing" cherries (named for a Chinese workman) sold along so many Michigan roadsides. A large tree at maturity, not suckering, rarely escaped from cultivation. Michigan ranks third in the nation in sweet cherry production.

9. **P. cerasus** L. Sour or Pie Cherry

Map 500. Roadsides, fencerows, borders of woods, locally well established. Another Eurasian species, grown for both flowers (including double-flowered forms) and fruit. Michigan's sour cherry crop is a major one in the fruit-growing region on the west side of the Lower Peninsula. The state ranks first in this crop, well over half of the nation's red tart cherries being Michigan-grown. Unlike the sweet cherry, this one suckers freely from the roots and is often more of a large shrub.

10. **P. americana** Marsh. Wild Plum

Map 501. A small tree or tall shrub often forming thickets on sandy open to wooded sites; along streams, ponds, lakes, and borders of woods; fencerows and roadsides. A native species, but also cultivated in various forms. Some collections may represent escapes.

The ripe fruit is yellow or red, often ± glaucous. The fruit of all three of our wild plums has been used by some people for jams and conserves. See comments under *P. nigra*.

11. **P. alleghaniensis** Porter Alleghany Plum

Map 502. A shrub ca. 2 m tall or less in woods of jack pine and oak, open woodlands, or sandy plains; occasionally at borders of mesic woods.

Plants from Michigan have been named as var. *davisii* (Wight) Sarg. (TL:

499. Prunus avium

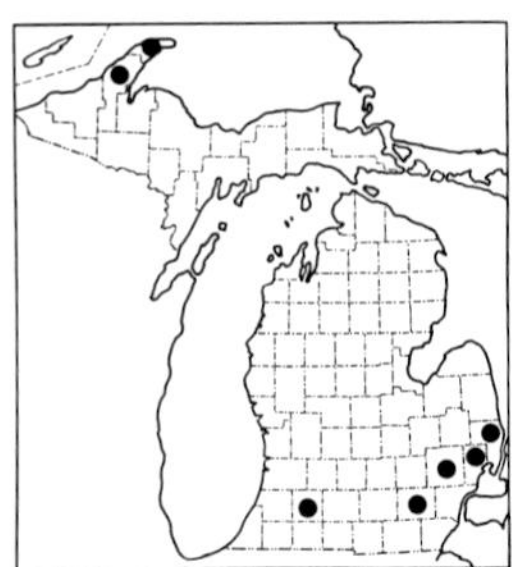

500. Prunus cerasus

501. Prunus americana

[Crawford and Roscommon cos.]), named in honor of C. A. Davis, prolific collector of Michigan plants around the turn of the century. While the variety may be ill-defined, even the presence of the species in any form in Michigan has been overlooked by most botanists. Our plants are disjunct from the main range of the species from the mountains of West Virginia north through central Pennsylvania and into southern New England. The Livingston County record is based on a specimen from a hillside in an old field at the Island Lake Recreation Area; it has leaves as in this species but lacks flowers or fruit, so can only tentatively be placed here.

According to Wight, the mature fruit is blue-glaucous, ca. 15 mm in diameter; all herbarium specimens (perhaps immature) appear to have had smaller fruit (as does typical *P. alleghaniensis*).

12. **P. cerasifera** Ehrh. — Cherry Plum

Map 503. A Eurasian species, which has been an important non-suckering stock onto which species given to suckering have been grafted. Certain collections from fencerows or near old homesteads, with characters as stated in the key, appear to be a form of this variable species.

The flowers are mostly in pairs (or solitary) on our specimens, but several buds, each giving rise to a pair of flowers, in addition to others originating leaves, may be crowded together on very short spur branches.

13. **P. nigra** Aiton — Canada Plum

Map 504. Hardwoods and borders of woods, especially on banks along rivers and streams; thickets and fencerows.

Sometimes confused with *P. americana*, but rather easily distinguished not only by the glandular margins on the leaves and calyx but also by the tendency for the flowers to be pink (or to become pink in age) and the conspicuous glands toward the summit of the petiole (in *P. americana* the glands are generally on the lower margin of the blade). The upper surfaces of the calyx lobes are rather densely pubescent in *P. americana,* while in *P. nigra* the pubescence is restricted to the very base of the lobes or is

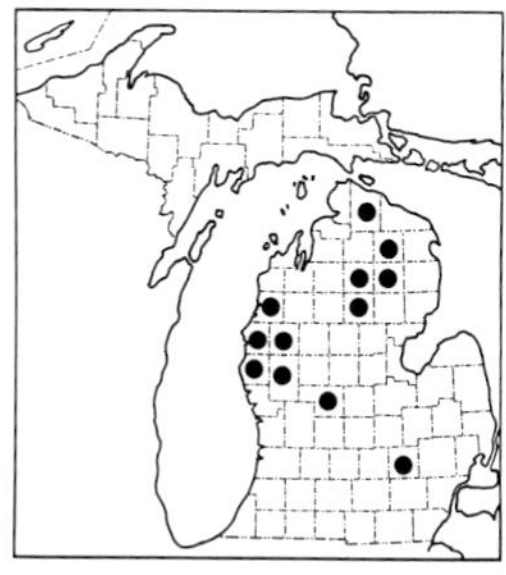

502. Prunus alleghaniensis

503. Prunus cerasifera

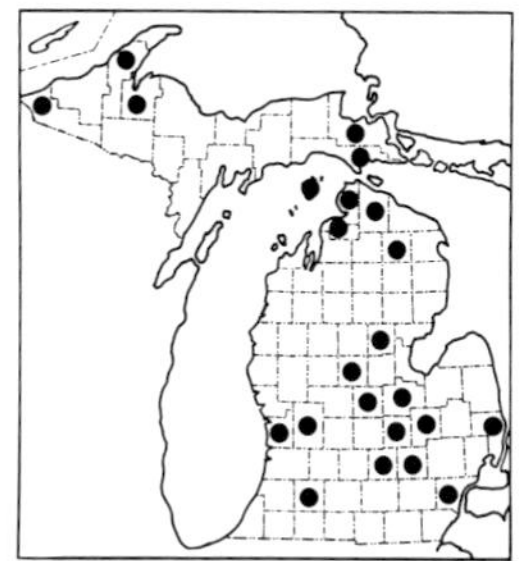

504. Prunus nigra

sometimes completely absent. In our specimens of *P. americana* (and also *P. alleghaniensis*) the petioles are ± densely pubescent above, while in *P. nigra* they are almost always sparsely pubescent or glabrous. When *P. nigra* and *P. americana* grow together, the former is reported to bloom about a week earlier.

14. **P. spinosa** L. Sloe; Blackthorn

Map 505. Spreading into thickets in Texas Tp., Kalamazoo County, and apparently also in the Island Lake Recreation Area, Livingston County. A Eurasian species, much branched and suckering, grown for its flowers, rarely escaping from cultivation. In Europe the hard wood is used for walking sticks and tool handles.

15. **P. domestica** L. Common Plum

Map 506. Roadsides, fencerows, clearings, and shores. A species so long grown in various forms for its fruit that its precise origin in the Old World is obscure.

The characters of this taxon and the preceding are unclear, and contradictory in reliable sources. This is often thought to have arisen as a hybrid. European specimens seem often to lack the conspicuous glands on the teeth of the leaf that characterize ours. Characters of pubescence and leaf size appear unreliable.

6. Physocarpus

1. **P. opulifolius** (L.) Maxim. Fig. 183 Ninebark

Map 507. Banks of rivers and streams, shores, often in very calcareous sites; occasionally in swamps or swales farther from borders of water; also on rock outcrops in the northwestern Upper Peninsula.

A handsome shrub in flower (resembling a large spiraea) or fruit. The pedicels (crowded in a dense umbellate or corymbose raceme) and calyx are usually lightly to heavily pubescent with stellate hairs. The fruit is often

505. Prunus spinosa

506. Prunus domestica

507. Physocarpus opulifolius

somewhat stellate-pubescent also; such plants are found from southern Michigan to Lake Superior and do not seem to warrant varietal recognition.

7. **Spiraea** Spiraea

Several taxa are cultivated in addition to those recorded here as rarely escaped in Michigan; suspicious specimens should be checked in works on garden plants.

REFERENCE

Kugel, Agnes R. 1958. Variation in the Spiraea alba—latifolia Complex. Ph.D. thesis, Univ. Mich. 124 pp.

KEY TO THE SPECIES

1. Inflorescence a dense (almost umbellate) raceme; flowering in midspring; petals white; plant commonly cultivated, very rarely escaped . . . . 1. **S. ×vanhouttei**
1. Inflorescence a branched panicle or corymb; flowering in midsummer; petals pink or (only in common native species) white
   2. Flowers in a broad corymb wider than long; floral tube pubescent; leaves glabrous beneath; flowers bright pink . . . . 2. **S. japonica**
   2. Flowers in an elongate cylindrical to conical panicle; floral tube, leaves, and flowers not combined as above
      3. Leaves beneath and inflorescence densely tomentose . . . . 3. **S. tomentosa**
      3. Leaves and often inflorescence nearly or quite glabrous
         4. Petals white (stamens may give a pinkish cast to the flowers); plant a common native of usually wet ground . . . . 4. **S. alba**
         4. Petals bright pink or rose; plant a rare escape from cultivation . . . . . 5. **S. salicifolia**

1. **S. ×vanhouttei** (Briot) Carr. Bridal-wreath

Map 508. A very well known, commonly cultivated taxon of garden origin as a hybrid between two Asian species, and seldom established where not planted. Our few specimens are from an abandoned quarry, lake shores, and along a railroad.

2. **S. japonica** L. f. Japanese Spiraea

Map 509. A native of eastern Asia, occasionally naturalized in North America; collected once along a railroad in Schoolcraft County (*Henson 485* in 1972, MICH).

3. **S. tomentosa** L. Fig. 184 Hardhack; Steeplebush

Map 510. Bogs, tamarack swamps, meadows, sandy-peaty shores and dried lake-beds, marshes, borders of ponds; often with *S. alba*.

The flowers are normally an attractive pink. A white-flowered form is known but has apparently not yet been found in Michigan; the tomentose leaves would readily distinguish it from *S. alba*. The tomentum on the undersides of the leaves is yellowish or brownish. In *S. douglasii* Hooker, a native of the Pacific coast of North America, the tomentum is white; this species is sometimes cultivated and has apparently been planted (along with *S. salicifolia* and many other cultivated plants) at Ives Lake in Marquette County, where it is conspicuous.

4. **S. alba** Duroi — Meadowsweet

Map 511. Wet shores, marshes, sedge meadows, tamarack swamps, peatlands, edges of streams, interdunal swales; moist borders of woods and shallow soil over rock; when occurring in apparently dry places they are often periodically flooded.

There are two thoroughly intergrading varieties, one of them, primarily eastern in range, often recognized as a distinct species, *S. latifolia* (Aiton) Borkh. The more midwestern var. *alba* is said to have a ± densely pubescent inflorescence, yellow-brown stems, and leaves 4–8 times as long as broad, finely and sharply serrate. In var. *latifolia* (Aiton) Dippel, the inflorescence is glabrous or nearly so, the stem red- to purple-brown, and the leaves less than 3 times as long as broad, coarsely and bluntly serrate. (The combination in varietal rank was independently—and unnecessarily—made at least three times in the 1960's, suggesting anyway some present consensus on rank!) In an extensive analysis of variation, primarily in Michigan, Kugel (1958) noted that it is frequently possible to find populations intermediate in all characters. The rampant intermediacy in the Great Lakes region is presumably the result of long hybridization between eastern and western (or southern) populations which are more distinct—or the whole complex can be looked upon as a cline. The zone of intermediacy centers in northern Michigan, and it is quite unprofitable to attempt to distinguish two taxa here. The few specimens referred by Kugel to *latifolia* are all from the northeastern Lower Peninsula (Tawas City to the Straits of Mackinac) and the Upper Peninsula. "Pure" *alba* is widespread throughout.

5. **S. salicifolia** L.

Map 512. A Eurasian species, sometimes cultivated for its handsome pink

508. Spiraea ×vanhouttei

509. Spiraea japonica

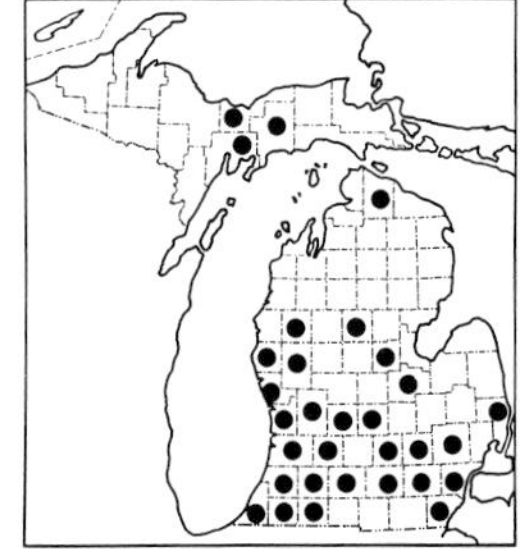

510. Spiraea tomentosa

flowers and rarely escaping. Apparently spreading on Ives Lake shores in Marquette County, and in a field by an abandoned homesite in Schoolcraft County.

If a specimen of *S. alba* with truly pink petals were to be found, it might be difficult to distinguish from this species, which generally has a narrower more cylindrical inflorescence and a tendency for the leaves to be broadest at or below the middle, while in *S. alba* they tend to be broadest above the middle.

This name was long applied in older manuals to include our *S. alba*, and until general recognition of the latter as a distinct species, published reports of *S. salicifolia* from the state almost certainly refer to *S. alba*.

8. **Aronia**

Frequently treated as a subgenus of *Pyrus* (see comments under *Sorbus*), but I agree with Hardin (and others) that without any unanimity of opinion (or scientific evidence) on the subject, maintenance of separate genera, as is almost universal in horticulture, is useful for communication.

REFERENCES

Farwell, O. A. 1918. Rare or Interesting Plants in Michigan. Rep. Mich. Acad. 19 (for 1917): 251–261.

Hardin, James W. 1973. The Enigmatic Chokeberries (Aronia, Rosaceae). Bull. Torrey Bot. Club 100: 178–184.

Uttal, Leonard J. 1984. Nomenclatorial Changes, Lectotypification, and Comments in Aronia Medikus (Rosaceae). Sida 10: 199–202.

1. **A. prunifolia** (Marsh.) Rehder Fig. 185 Chokeberry

Map 513. Although not nearly so abundant when found as is usual for, e. g., *Chamaedaphne*, this must surely be about as widespread a wetland shrub as any in the state, occurring throughout in bogs, tamarack swamps, boggy swales, marshy and swampy thickets and shores, and low ground generally; occasionally in moist to dry sandy oak-pine woods.

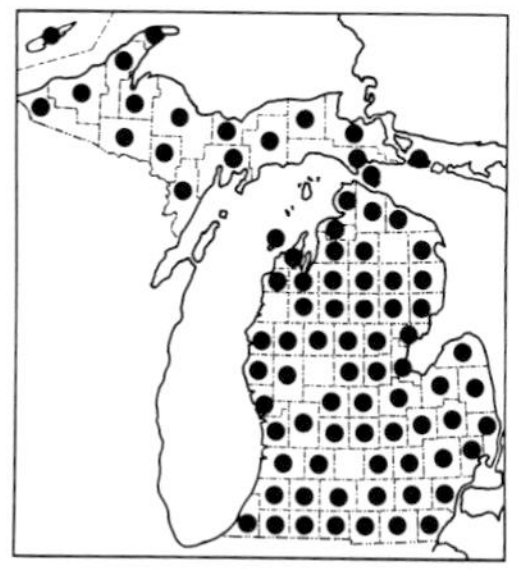

511. Spiraea alba

512. Spiraea salicifolia

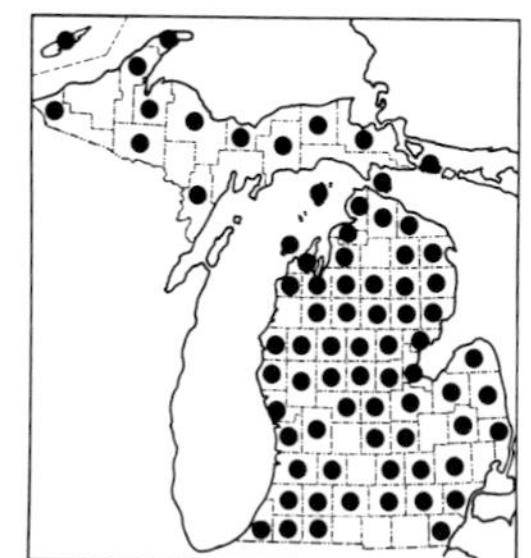

513. Aronia prunifolia

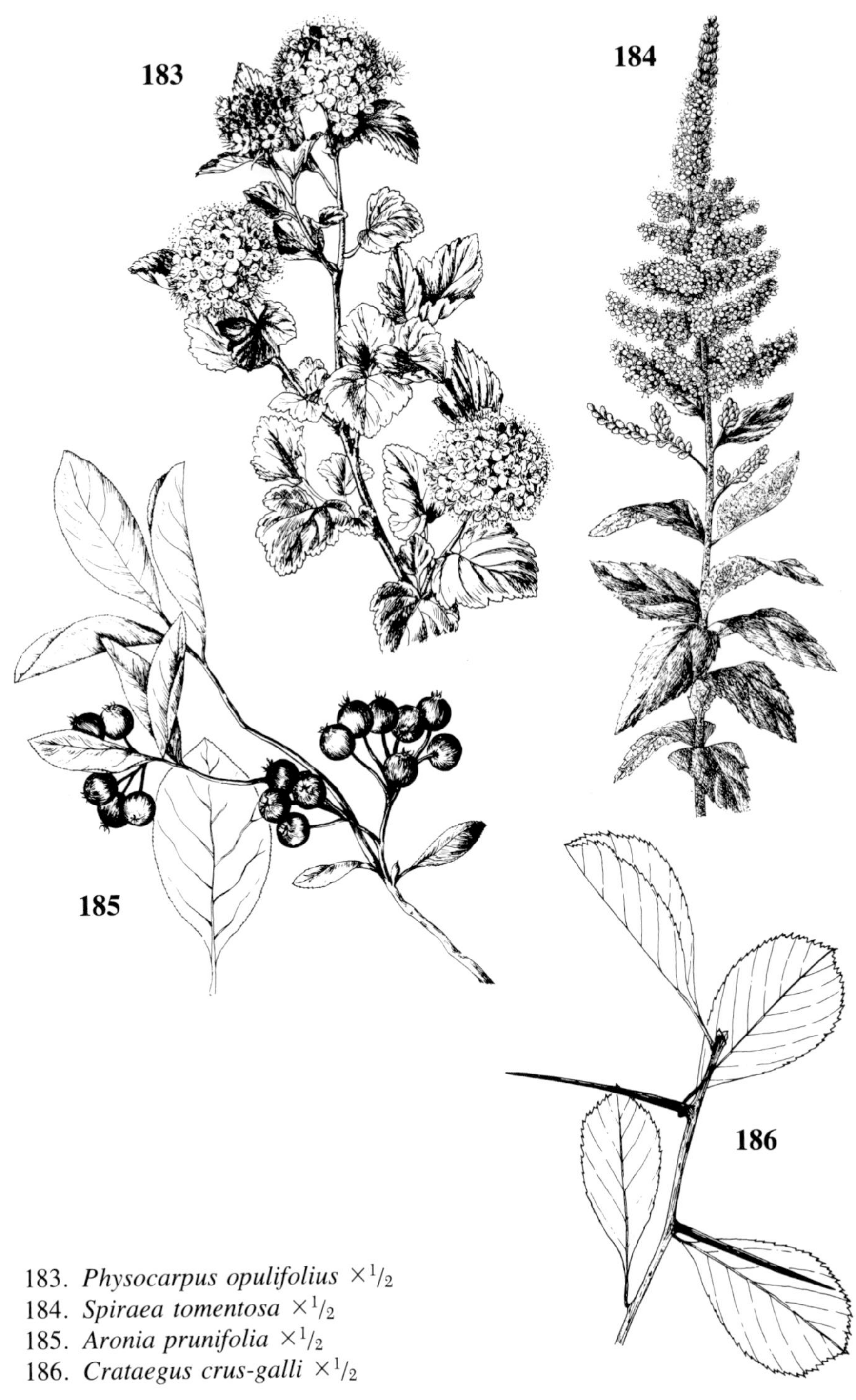

183. *Physocarpus opulifolius* $\times^1/_2$
184. *Spiraea tomentosa* $\times^1/_2$
185. *Aronia prunifolia* $\times^1/_2$
186. *Crataegus crus-galli* $\times^1/_2$

Plants unusually pubescent in the inflorescence and on the young branchlets have sometimes been aroniously identified as *A. arbutifolia* (L.) Ell., a red-fruited species of the Coastal Plain and southern Appalachians, with leaves densely pubescent beneath. Completely glabrous plants, on the other hand, have usually been recognized as *A. melanocarpa* (Michaux) Ell., which in our area seems quite impossible to separate consistently from *A. prunifolia*. Farwell (1918) discussed these species (using the later name of *A. atropurpurea* for *A. prunifolia*) and concluded that we have only one species in Michigan. Mason and Iltis, 40 years later, likewise concluded in Wisconsin that "*A. prunifolia*" was merely a pubescent form of *A. melanocarpa*. Hardin (1973), on the basis of more detailed studies, felt that the evidence suggested ancient as well as current hybridization and introgression as the origin of *A. prunifolia* and he concluded that under his treatment *A. prunifolia* would be a synonym of *A. melanocarpa*. In all of these papers, it was evidently overlooked that *A. prunifolia* is the name based on the oldest epithet at the rank of species, so that *A. melanocarpa* is a later synonym if the two are not treated as different taxa. Uttal (1984), working in the Southeast, formally designated *A.* ×*prunifolia* as a hybrid which "introgresses continually" with its presumed parents. In the genus *Pyrus*, the name *P. floribunda* Lindley has been applied to this plant (the epithet *prunifolia* having already been used in *Pyrus* for a crab apple).

The fruit is black or dark purple, less than 12 mm in diameter, ripens fairly early (compared to the late-fruiting *A. arbutifolia* of the east), and puckers the mouth when raw (like the choke cherry, *Prunus virginiana*).

9. **Amelanchier** — Serviceberry; Shadbush; Shadblow; Juneberry; Sugarplum

The species of this genus are said to have more than 80 common names. "Service" has been explained as a corruption and misapplication of the "Sarviss" of Elizabethan England, in itself a corruption of the old Roman "Sorbus," a name applied to some fruit (not necessarily the genus now called *Sorbus*—and not *Amelanchier*, of which no species are native in Great Britain and only one in the Mediterranean region).

The fruits of some species, at least in some seasons, are juicy and tasty; they are good mixed with something tart (like rhubarb or citrus) in jams and preserves. One common name, saskatoon, refers to a species (*A. alnifolia* of central North America) of which the fruit was used by Blackfoot Indians, fresh or dried or added to meat to make pemmican; the name is more widely known as that of a major city in the southern part of the Canadian province of Saskatchewan. Even the dry relatively tasteless fruits are popular with birds and mammals. Like those of some related Rosaceae, they are frequently infected with rust fungi of the genus *Gymnosporangium*, which at another point in its life cycle forms colorful orange galls on *Juniperus*.

It seems strange that *Amelanchier* is not more used in horticulture in this country, since the species are winter-hardy and the numerous white (very rarely rosy) flowers are among the first to open in the spring. In the wild, they brighten the drab woodlands, especially in dry ground such as the barren jack pine plains, and may be confused by the passing motorist with pin cherry, which begins to bloom a little later (and on close examination can be seen to have shorter petals). Next to bloom in dryish places is choke cherry, and in damp ones, chokeberry—followed by most hawthorns and (on better sites) mountain-ash and wild black cherry. Truly the native white- and small-flowered rosaceous shrubs fill the spring with beauty—just as do the larger-flowered and mostly cultivated plums, apples, and pears.

Taxonomically, *Amelanchier* has long been a trial to botanists, resembling in this respect the larger genera in the family, *Rubus* and *Crataegus*. As can be seen in the references cited below (as well as manuals), no two authors fully agree on classification; every large collection includes a high percentage of intermediate plants (usually dismissed as hybrids), and no existing treatment works perfectly in our region. Almost every presentation of the genus is cautiously proclaimed by its author to be "provisional," "tentative," or "preliminary," and the need for further study is stressed. The treatment below can only be described as "secondary," drawing from existing treatments (by those who have been able to devote more time to the investigation) but in no way solving the problem of when to blame hybridization for the failure of so many specimens to exhibit the characters of leaf toothing, pubescence, venation, petal length, habit, and inflorescence that are supposed to correlate in defining certain species. Quite possibly we should recognize "aggregate species" (not necessarily the "complexes" below) composed of imperfectly distinguished "microspecies." The only virtue *Amelanchier* has over *Crataegus* and *Rubus* is that, by being smaller, it lures us to the hope that it may be more manageable.

The first three species below are diploid, and the remainder are a polyploid mixture including many triploid and tetraploid plants—the "Sanguinea complex" of Robinson and Partanen (1980). This would include the three complexes recognized in the compromise treatment offered by Morley in his *Spring Flora of Minnesota* (1966, revised 1969) and largely followed here. Hybridization, polyploidy, and asexual vegetative reproduction have created a vast array of forms, which thrive (as does *Crataegus*) in cleared, disturbed, or burned-over areas, where even the growth form may differ from ancestral plants (clumped stems, for example, replacing single destroyed trunks—as in red maple, paper birch, and some other species).

It is usually desirable to check several representative leaves before deciding the nature of the preponderant vein and tooth pattern. Likewise, it may be necessary to check the summit of several fruits before concluding that the top of the ovary is not to be considered pubescent. Some of the specimens mapped may well be hybrids which resemble one parent so closely (or are so incomplete) that their origin was not recognized.

REFERENCES

Cinq-Mars, Lionel. 1971. Le Genre Amélanchier au Québec. Nat. Canad. 98: 329–345 (also Ludoviciana 9).

Jones, George Neville. 1946. American Species of Amelanchier. Illinois Biol. Monogr. 20(2). 126 pp. [Note extended reviews and remarks by Fernald in Rhodora 48: 125–134 (1946) and by McVaugh in Madroño 8: 237–240 (1946).]

McKay, Sheila Mary. 1973. A Biosystematic Study of the Genus Amelanchier in Ontario. M. S. thesis, Univ. Toronto. 255 pp.

Nielsen, Etlar L. 1939. A Taxonomic Study of the Genus Amelanchier in Minnesota. Am. Midl. Nat. 22: 160–206.

Robinson, W. Ann, & Carl R. Partanen. 1980. Experimental Taxonomy in the Genus Amelanchier I: A New Look at the Chromosome Numbers of the Amelanchier Species Growing in the Northeastern United States. Rhodora 82: 483–493.

Robinson, W. Ann. 1982. Experimental Taxonomy in the Genus Amelanchier. II: Do the Taxa in the Genus Amelanchier Form an Agamic Complex? Rhodora 84: 85–100.

Schroeder, F.-G. 1970. Exotic Amelanchier Species Naturalised in Europe and Their Occurrence in Great Britain. Watsonia 8: 155–162.

KEY TO THE SPECIES

1. Pedicels 1–3 in axils of leaves; petals less than twice as long as broad; leaves at least partly open and essentially glabrous (except margins and petioles) at flowering time, the blade tapering trough-like into raised petiole margins and the petioles less than 8 [15] mm long . . . . 1. **A. bartramiana**

1. Pedicels more numerous (at least scars if some have fallen with fruit), the inflorescence a raceme (or corymbose); petals at least twice as long as broad; leaves various (glabrous to tomentose) but the blade rounded or truncate to subcordate, not tapered at base, and petioles usually longer than 8 mm

 2. Summit of ovary glabrous, even in flower (or with a few hairs at base of style only); leaf blades short-acuminate, finely and closely serrate with 22–45 (50) teeth per side

  3. Leaves just beginning to unfold at flowering time, densely white-tomentose beneath, otherwise green, retaining some of the pubescence on petioles and along midrib beneath into maturity . . . . 2. **A. arborea**

  3. Leaves mostly half-grown at flowering time, usually bronze-red, glabrous or nearly so, completely glabrous at maturity . . . . 3. **A. laevis**

 2. Summit of ovary tomentose (± densely so in flower, sometimes more sparsely, but evenly, in fruit); leaf blades variously shaped and toothed

  4. Larger leaves with ca. (22) 25–50 (55) fine teeth on a side (more than twice as many teeth as lateral veins), acute to short-acuminate, at flowering time open though not fully grown and often glabrous or soon becoming so . . . . 4. **A. interior** complex

  4. Larger leaves with fewer than 20 (25) teeth on a side (no more than twice as many teeth as lateral veins), the blades at flowering time ± folded and white-tomentose beneath, when mature acute to rounded

5. Most leaves finely toothed at least toward apex (5–8 teeth per cm when mature), the veins anastomosing and becoming indistinct near the margin, at most with weak veinlets ending in the teeth; petals ca. 5–9 (10) mm long; plants typically spreading underground forming colonies of low shrubs ..... 5. **A. spicata** complex
5. Most leaves more coarsely toothed (2–5 teeth per cm toward apex when mature), the veins prominent and running to tips of the teeth (or a principal fork into the teeth) at least toward apex of blade; petals (10) 11–18 (20) mm long; plants typically solitary or in tall many-stemmed clumps (though sometimes colonial) .... 6. **A. sanguinea** complex

1. **A. bartramiana** (Tausch) M. J. Roemer — Mountain or Northern Juneberry

Map 514. Bogs and wet conifer swamps (tamarack, spruce); thickets and old dune or rock ridges; borders of hardwoods; may be low and sprawling on bare rock shores and ledges, otherwise a tall shrub. Presumed hybrids from Cheboygan and Crawford counties are very close to good *A. bartramiana*. The disjunct southernmost station in Kent County was east of Grand Rapids (*Cole* in 1894, MICH), but the material looks quite typical. The species ranges from Labrador to the Lake Superior region, eastward and southward into the Lower Peninsula (very rarely), through Ontario to the mountains of New York, northeastern Pennsylvania, and New England.

Fortunately, the range of *A. bartramiana* in Michigan largely avoids the concentration of other species in the Lower Peninsula, for it has the bad habit (as Cinq-Mars expressed it) of hybridizing with almost all other species. It is ordinarily our most distinct species, with short petioles, blades ± pointed at both ends, and petals shaped like those of a *Prunus*. The summit of the ovary is densely tomentose and tapers into the base of the style, while in the other species the ovary is flat or rounded at the top. When *A. bartramiana* grows with another species, intermediates may be expected. In hybrids with *A. laevis,* which appear to be rather common, the pubescence on the ovary may be sparser and the flowers more numerous, although the leaf blades are shaped more like those of *A. bartramiana* than the subcordate ones of *A. laevis*. Hybrids with other species possessing a tomentose ovary are less easily recognized, but in addition to the combination of leaf shape and several-flowered racemes, the leaves of hybrids may be more pubescent than in the essentially glabrous-leaved *A. bartramiana*.

2. **A. arborea** (Michaux f.) Fern.

Map 515. In rich or swampy to dry woods and borders, but most often noticed on dry sandy open woodland with red maple, aspen, oaks, and/or jack pine.

Where this species grows with the next, intermediates can be expected. These presumed hybrids are much more recognizable at flowering time, when the parent species are ordinarily quite distinct. The intermediates generally combine ± tomentose leaves (as in *A. arborea*) with red flush and sometimes more open condition (as in *A. laevis*)—or they may be closely folded at anthesis, but red and glabrous. Hybrids of either of these two species, our only ones with glabrous summits on the ovaries, with the several species having tomentose ovaries are not easy to assign to parentage.

In dry aspen woods and adjacent open areas, as at the University of Michigan Biological Station east of Pellston, *A. arborea* and *A. laevis* may be found with plants of the *A. sanguinea* and *A. spicata* complexes along with a full array of intermediates of all kinds, including those referable to the *A. interior* complex.

3. **A. laevis** Wieg. Fig. 187

Map 516. Like the preceding, ranges from a shrub to a small tree, most often in dry sandy open woodland, rocky sites, sandy bluffs and shores; also on river banks, at borders of coniferous and deciduous woods, even bog borders.

Referred here are fruiting (and a few sterile) specimens with short-acuminate, finely toothed, and completely glabrous leaves, although it is possible that some of them could be hybrids, the *interior* complex, or unusually glabrous *A. arborea*—to which have been referred similar specimens retaining at least sparse pubescence on the petiole and midrib beneath.

The sepals of this species are often a little longer and more narrowly lanceolate than in the preceding, but there appears to be no clear distinction in our material. Although very different at anthesis, this is sometimes considered a variety of the preceding [var. *laevis* (Wieg.) Ahles]. Both species were long misidentified as *A. canadensis* (L.) Medicus, a species as now understood ranging entirely east of the upper Great Lakes.

Plants with glabrous summit on the ovary but obtuse leaves (and perhaps a low colonial habit) are presumably hybrids with the *A. spicata* complex expressing their characters in a different combination from the plants included in the *A. interior* complex. Such plants seem to be quite scarce. Presumed hybrids with *A. bartramiana* and *A. arborea* are mentioned under those species.

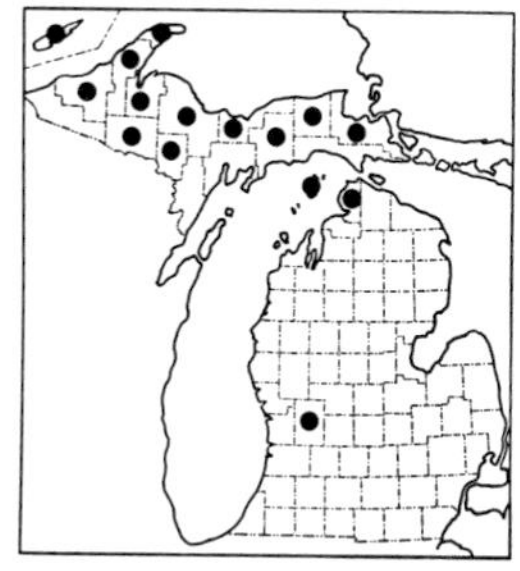

514. Amelanchier bartramiana

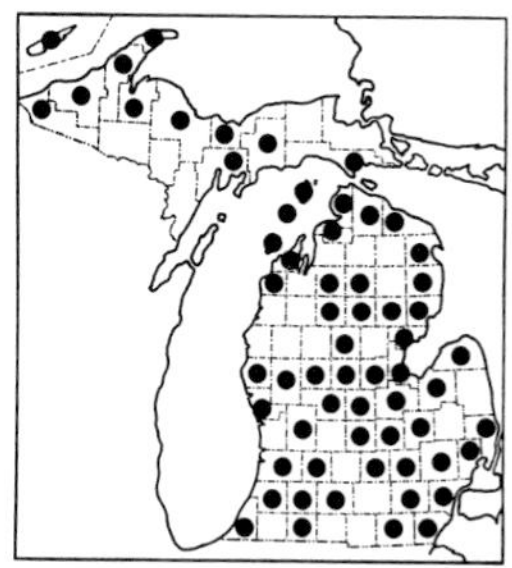

515. Amelanchier arborea

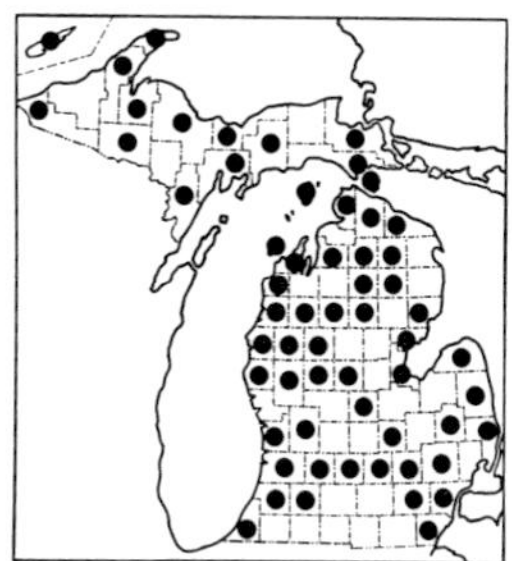

516. Amelanchier laevis

4. **A. interior** Nielsen

Map 517. Often a tall, clumped shrub, typical of sandy open woodland and dunes (oak, jack pine, aspen), shallow soil on rock outcrops, shores; less often in or at borders of hardwoods and conifer swamps or in other wetlands.

Here treated as a complex of uncertain origin, characterized by finely toothed acute to short-acuminate leaves. The commonest and most easily recognized expression has leaves nearly glabrous (or soon glabrate) and bronze-red at anthesis—similar to *A. laevis* but the summit of the ovary densely to sparsely tomentose. There is gradation, however, to plants with some tomentum on at least young leaves and this may persist on petioles or midribs. The suggestion that *A. interior* represents a hybrid swarm involving *A. laevis* (or sometimes *A. arborea*) and plants of the *A. spicata* and/or *A. sanguinea* complexes has been made and is a likely explanation for the range of plants included here. Plants with veins extending clearly into the leaf teeth suggest *A. sanguinea* as a parent; those with ± tomentose young leaves suggest that *A. arborea* rather than *A. laevis* could be a parent (though such a character could come from either *A. sanguinea* or *A. spicata*).

If the application of the name *A. intermedia* Spach were to be settled in this complex, rather than for a plant with glabrous summit on the ovary (as often stated), it would of course antedate *A. interior*.

5. **A. spicata** (Lam.) K. Koch

Map 518. Dry, sandy plains, dunes, and woodlands, usually with jack pine and/or oaks, often little if at all taller than the *Comptonia* and *Vaccinium* with which it is frequently associated; very rarely on rock outcrops or in moist places.

This is perhaps the most heterogeneous complex recognized here. Many of our specimens are undoubtedly the weakly distinguished *A. stolonifera* Wieg., often (but not always) now considered a variety (for which other epithets are available) of *A. spicata*. The latter name is alleged by some to be applicable to a hybrid that arose in European gardens, but this opinion is flatly rejected by European authors (see Schroeder 1970).

With us, *A. spicata* (sensu latissimo) is most characteristically encountered as a low (under 1 m) colonial shrub of jack pine plains and similar dry places. Separation from the *A. sanguinea* complex is not easy (unless perhaps one were arbitrarily to select some single character to distinguish them and apply it rigorously). Plants with prominent veins into the leaf teeth but with short petals are, for instance, problematic and rather frequent. Specimens with leaves having more teeth than usual and perhaps more acute apices grade into the *A. interior* complex.

Certain ± intermediate plants approach *A. alnifolia* Nutt., often thought to range to the west of the Great Lakes but by some (see Cinq-Mars 1971; McKay 1973) recognized east to Quebec and Ontario. This taxon has very compact dense racemes—less than 2.5 cm long in flower, up to 3.5 cm in fruit—with very short pedicels; in our specimens which resemble this, the general aspect is that of *A. spicata* but the leaf veins enter the teeth as in *A. sanguinea* and the leaves are mostly obtuse to acute, while in *A. alnifolia* they are strikingly truncate or very broadly rounded at the apex. The similar (identical?) *A. humilis* Wieg. has been variously recognized as associated with *A. sanguinea* or *A. spicata* by authors, and specimens referred to it often combine characters of both complexes; the teeth on the leaf are sometimes very few or nearly absent.

Specimens referred to the overall *sanguinea-spicata* group either because they seem truly intermediate or because they are inadequate for assignment to one or the other are known from the following counties in addition to those from which one or both are otherwise mapped: Chippewa (Drummond I.), Lake, and Mecosta. Some such intermediates in northern Michigan may in fact be hybrids involving *A. bartramiana*.

6. **A. sanguinea** (Pursh) DC. Fig. 188

Map 519. Perhaps somewhat more often than our other species along shores, on low dunes, and on calcareous gravels, but often in essentially similar dry open sandy woodlands (aspen, jack pine, oak, red maple) and clearings, sandy thickets, borders of woods (coniferous or deciduous) and bogs.

A complex fairly well recognizable as defined in the key, especially when one is familiar with the admittedly subjective quality of prominent veins running into the teeth and well displayed toward the end of the leaf blade. Characteristically a tall solitary or clumped shrub with longer petals than the *A. spicata* complex. The inflorescence is often more open and lax (even corymbose) than in *A. spicata*, though sometimes the raceme is compact and stiff (including plants referred to *A. gaspensis* Wieg.). Also included here is *A. huronensis* Wieg. (TL: Sand Point, Huron Co.), originally described as a large-flowered calciphile of the upper Great Lakes region; in

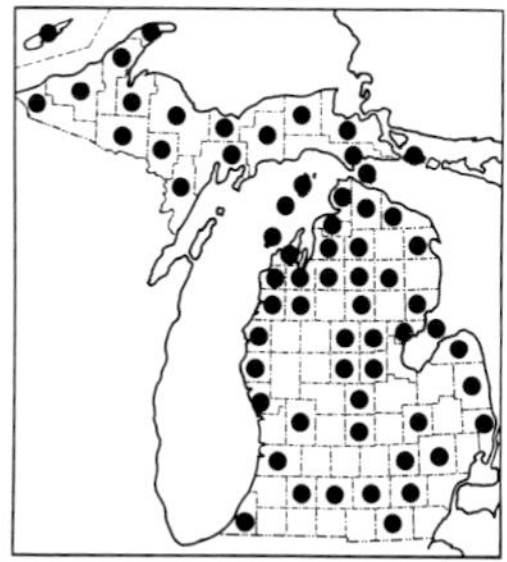

517. Amelanchier interior

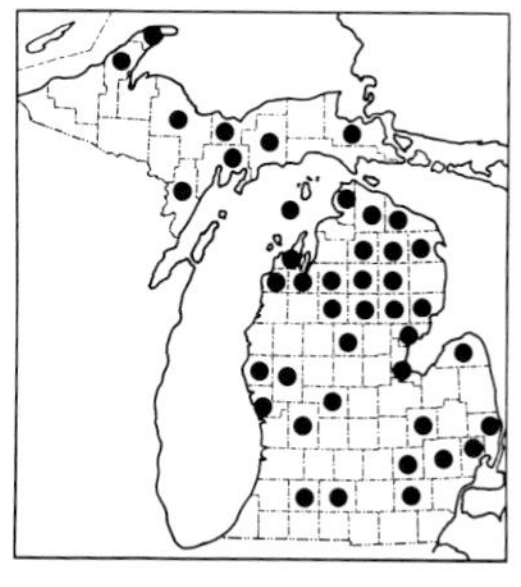

518. Amelanchier spicata

519. Amelanchier sanguinea

it, the veins are said to fork so that the principal branches, not just the main lateral veins, extend straight into the teeth.

Intermediates with the *A. spicata* complex are rather frequent and are discussed above.

10. **Crataegus** Hawthorn; Thornapple

Like *Rubus* and *Amelanchier, Crataegus* is a genus easy to recognize, but the species (or other taxa) of which drive most botanists to distraction. Hybridization, polyploidy, and apomixis presumably account for much of the complexity, asexual populations breeding true and acting as species but better thought of as individuals. Concise summaries of the problem, with history and references, can be found in Robertson (1974) and in Little's revised *Checklist of United States Trees* (1979).

Before 1900, regional manuals in the United States listed scarcely one or two dozen species, varieties, and forms in *Crataegus*. Between 1900 and 1925, well over 1000 supposedly new species were described for North America, over 700 of them by C. S. Sargent (Director of the Arnold Arboretum at Harvard and outstanding authority on North American trees). The remainder were described by W. W. Ashe and C. D. Beadle, both of North Carolina, and a few others. Most of these "species" were based on trivial characters of little biological significance. In 1921, late in Sargent's life, E. J. Palmer joined the staff of the Arnold Arboretum. There he had (until his retirement in 1948) extraordinary facilities for the study of *Crataegus*, including a herbarium which attained some 30,000 specimens in this genus and outdoor plantings, largely of seed from type plants, of 1400 trees and shrubs. Much of the available herbarium material in the United States passed through Palmer's hands.

Palmer, as early as 1925, published a useful index to *Crataegus* names, assigning them to groups (now usually termed series). In 1946, he published a discussion of the species problem in the genus, with a listing of species in northeastern North America, reducing many of the names to synonymy. This was the basis for his essentially identical treatments of *Crataegus* contributed to Fernald's edition of *Gray's Manual* (1950) and Gleason's *Illustrated Flora* (1952). Fernald introduced Palmer's work on *Crataegus* as by "the only student who professes to understand it." Palmer spent over half a century in active field and herbarium studies of the hawthorns, until his death in 1962, in his 87th year.

In an effort to tackle promptly some of the major worries in a state flora, I sent to Mr. Palmer in 1957 the approximately 1200 Michigan specimens of *Crataegus* then belonging to herbaria in the state and which had not already been examined by him. He promptly and generously annotated all of these (in addition to another hundred or so over the next four years), and I was able to study them, plus the material previously checked by him, en masse. A draft manuscript on the genus in Michigan was sent to him the following year, along with a number of questionable specimens for rechecking (including a few apparent duplicates which had received differing annotations). He graciously offered his comments on all these. Because of Palmer's vast experience with the genus and the fact that practically all treatments in manuals and local floras had been contributed by him (including those in the excellent state floras for Indiana, Ohio [woody plants], and Missouri), it was intended from the beginning to offer in *Michigan Flora* an account reflecting his identifications, as well as an alternative outline with broader concepts of collective or aggregate species. In the meantime, others have been wrestling with the *Crataegus* problem.

In 1965, E. P. Kruschke published a synoptic outline of *Crataegus* in the northern United States and adjacent Canada, based in large part on his own extensive and detailed field studies in Wisconsin. This treatment offered some welcome improvements in aligning synonyms, but in the absence of keys and descriptions cannot be used for identification. One of the major premises in Kruschke's work is that no species may have both 20- (15–20-) stamened and 10- (5–10-) stamened flowers—a concept not so strictly accepted by others. Although there was some reduction in the number of species, Kruschke's work was in the Palmer tradition (and I had, in 1958, sent him also a copy of my draft manuscript, on which he offered very few comments).* Kruschke's list is evidently the major basis for this genus in the checklist by Kartesz and Kartesz (1980), and it is also largely followed, but with further refinements, by Phipps and Muniyamma (1980) in an excellent treatment for Ontario which includes distribution maps, illustrations (some in color), and keys (unfortunately with several misleading errors in numbering).

In quite a different vein, Cronquist presented a "drastic" reduction in number of species of *Crataegus* in his 1963 compression of Gleason's *Illustrated Flora* into a one-volume manual. He listed a large number of names apparently applicable to hybrids (as often suggested, indeed, more tentatively by Palmer). Cronquist's alignments were the principal source for the "practical" classification in Little's revised checklist of 1979. Whether termed "drastic" or "practical," as done by their respective authors, these treatments (because of their broad geographic as well as taxonomic scope) are bound to influence American manuals and lists for some time.

The account of *Crataegus* here presented attempts (perhaps ambiguously) to be useful to workers of various taxonomic persuasions. The basic treatment permits users to know what species Palmer thought we have and there are keys to most of these except for those names that he applied only with question marks to our plants. Where I have had difficulty following his dispo-

*Users are cautioned that names for Kruschke's new species and infraspecific taxa were not validly published because of failure to designate a single type; and many of his new combinations were illegitimate because not based on the oldest available epithet in the rank adopted.

sition of our material, I have frankly stated so and have had to quote his keys at these points without feeling able to apply the distinctions therein drawn. It did not seem practicable to reexamine all Michigan *Crataegus* when Kruschke's 1965 list was published, but his opinions are generally noted and several of his alignments adopted. Finally, the condensed alternative treatment drafted in 1958 has been reworked, suggesting "species complexes" (similar to those in *Amelanchier* and *Rubus*) which are close in most instances to those of Cronquist, but with a few more species recognized, especially in the Rotundifoliae. It is, of course, not to be expected that all groups to which binomials are applied are of equal distinctness or taxonomic merit—any more than in other genera. The often more broadly defined species or species complexes are the ones for which maps are provided, sometimes under a name appropriate only for such broader concepts. (A treatment at this level less provincial geographically might ultimately result in use of different names, based on specimens from beyond our region, for some of the other species complexes as well.) Altogether, then, the present account was developed over a period of 25 years, and is admittedly based largely on herbarium specimens and literature.

Michigan has not had a modern collecting and research program for *Crataegus* as have neighboring areas in Wisconsin and Ontario. However, this state was one of the active sites in the early "expansionist" days of hawthorn study. Emma J. Cole, who did such admirable work in the Grand Rapids area at the turn of the century, sent a large amount of material to Sargent. C. K. Dodge likewise sent many specimens from the Port Huron region to Ashe and to Sargent. O. A. Farwell sent some specimens to Sargent, also. Ashe visited Dodge in the falls of 1901 and 1902 (according to Dodge 1908; in 1902 and 1903 according to Ashe 1904). Sargent was on Belle Isle with Farwell in September of 1901 and visited Miss Cole the same month; he also collected on Belle Isle in May of 1899 and visited Dodge in the falls of 1904 and 1906 (Dodge 1908). As a consequence of all the activity by these persons, the type localities for 52 described species of *Crataegus* lie in Michigan; most of the names were reduced to synonymy by Palmer, but some, including those named for two collectors (*C. coleae* and *C. dodgei*) he recognized as quite widespread in the northeastern states.

In the text below, all of the species originally described from Michigan are accounted for in synonymy or otherwise, with two exceptions: *C. borealis* Ashe and *C. fallax* Ashe. No material has been located of these two species, and it is impossible to tell from their descriptions what their proper disposition might be. In Dodge's notes for one of his trees (No. 81) on which *C. prona* Ashe (= *C. macrosperma*) is based, there is this statement: "April 24-1910, W. W. Eggleston says this is the same as Ashe's C. borealis." Dodge's notes on his No. 84, the type tree of *C. incerta,* have this statement: "W. W. Eggleston tells me this is C. fallax." The basis for these observations by Eggleston (who prepared the treatment of *Crataegus* for the 7th edition of *Gray's Manual*) is not known. (*C. incerta* is near *C. coleae*; Kruschke included *C. fallax* in the synonymy of *C. macracantha*; *C. borealis* was placed by Kruschke in the synonymy of *C. macrosperma* on the basis of a specimen in the herbarium of the University of Minnesota.) Except for these two names and four others (*C. albicans, C. michiganensis, C. sitiens,* and *C. superata*), which are more easily placed, I have seen type or paratype material for names of all species described from Michigan (cultivated specimens from seed from types in the cases of *C. incerta* and *C. comparata*). Although names for which the types came from Michigan are thus accounted for, other names applied to Michigan specimens, even with types from adjacent areas (or paratypes cited from Michigan) are usually not mentioned if they are not accepted, for the line has to be drawn somewhere. Names based on Ontario plants are largely covered by Phipps and Muniyamma (1980).

During the period 1902–1909, C. K. Dodge studied this genus most extensively in St. Clair County and across the St. Clair River in the neighborhood of Sarnia, Ontario. He permanently marked approximately 175 individual trees and shrubs, and made regular observations and collections from them. Dodge's notebooks on these plants are in the University of Michigan Herbarium. Among the most useful kinds of information included are tracings evidently made from longitudinal sections of mature fresh fruit of each plant. Emma Cole, around Grand Rapids in Kent County, made many collections of flowering and fruiting specimens from the same plants. Later, Virginia Angell made a study of *Crataegus* in the Grand Rapids area, observing (1933): "We in Grand Rapids have been totally unable to classify satisfactorily any local individuals according to Sargent's key." O. A. Farwell made many collections, some of them with flowers and fruits from the same plants, and a few other collectors have done similarly. Thus, a considerable percentage of our herbarium specimens include both flowers and fruit and in some cases, additional notes. More recently, in the 1940's, C. W. Bazuin made very extensive collections in the southwestern and west-central portions of the state.

Since Miss Angell did not, for the most part, employ botanical names for the entities which she described in some detail, the following chart is offered as an aid in associating her descriptions and illustrations with currently recognized species:

Name as given by Angell in her Table I	*Determination by Palmer*
C. mollis "broad, not lobed"	C. mollis (mostly var. sera)
C. mollis "grape-leaf"	C. mollis
C. mollis "elliptical"	C. mollis
C. mollis "elliptical, lobed"	C. mollis

C. "Bronze"	C. pedicellata var. albicans (5A, the cited number; 3 uncited numbers are C. holmesiana)
C. "cut-leaf"	C. jesupii
C. "red-calyx"	C. filipes
C. "claw-spine"	C. macrosperma
C. "Woodcliffe"	C. pruinosa (the cited 10A, labeled as "coccinoides"; 4 uncited numbers also labeled "coccinoides" are C. filipes, C. leiophylla, C. pruinosa, & C. rugosa)
C. "dainty"	C. gravis? (9A, the only number cited by Angell, was thus determined although it has 20 stamens; 3 other numbers are C. compacta)
C. "right-angle"	C. pruinosa var. dissona
C. "burnt-orange"	C. intricata
C. "flabellate"	C. dodgei (an uncited number is C. margaretta)
C. crus-galli	C. crus-galli (2 uncited numbers are C. fontanesiana)
C. punctata	C. punctata
C. "yellow-punctata" and "brown-punctata"	(no specimens so labeled)
C. succulenta "big-bud"	C. succulenta
C. succulenta "nearly smooth"	C. coleae
C. succulenta "hispid"	C. pedicellata var. albicans
C. succulenta "slow"	C. calpodendron
C. succulenta "large-fruit"	C. coleae
C. succulenta "shrubby"	C. succulenta
C. succulenta "smooth" [not in table]	C. compacta

In any serious study of *Crataegus,* it is essential to mark individual plants with permanent and unmistakable tags or blazes, collecting both flowering and fruiting specimens from the *same* plant (not from another which "looks the same" nor from a branch of another which may have penetrated into the crown of one). The color of the fresh anthers should be noted in the flowers; and in fruiting specimens, the color, texture, and size of fully ripe fruit recorded. The principles of making good *Crataegus* collections were stated concisely by Dodge (1908) and much more amply by Kruschke (1955). Unfortunately, the keys presented here will not work well without both flowers and fruit (or at least experience with certain species). *C. succulenta* is one of our most widespread as well as most variable species, so that an early choice in the key must refer to the nutlets, which are distinctive in this species.

The number of stamens can be determined in most cases by a careful study of the apex of the fresh fruit, where remnants of the filaments are usually present. Likewise, the margins of the calyx lobes can usually be examined in fruit which is not too old. The number of styles and nutlets is usually the same, although in some cases the former could be more, due to failure of nutlets to develop. Often, old fruit may be found on a flowering plant, or beneath it, and can be studied if there is no chance of mixture with fruit from another plant. The necessity of having characters of both flowers and fruit can thus often be met in part at a single time. Anther color is safely determined only from fresh, unopened anthers. Corymbs and leaves at flowering time tend to be more pubescent than the same organs at fruiting time; a few scattered hairs, therefore, in a late-season collection are more significant than a similar amount of pubescence in the spring. (Although the inflorescence is widely called a corymb, technically it is cymose.)

Incidentally, a "lobed" leaf in *Crataegus* has lobes relatively small (or shallow) compared to most other genera (e. g., *Quercus*). The more one becomes used to the variability of the genus and to the meanings of terms, the easier the recognition of species becomes. The statements in the keys are based largely on the branches—the portions ordinarily collected and studied. The field student will note that the trunks often bear larger, compound spines and will, in general, note differences in foliage, growth habit, etc., about which little can be determined from inadequate herbarium specimens scantily labeled. The leaves of flowering branchlets, or *floreal leaves,* often differ in shape from the *vegetative leaves* of sterile branches and shoots, the latter being more variable. In cases of trouble, try an unknown plant under both leads of a couplet in the key. Soon it will be found that under one of them no species has the particular combination of stamen number, anther color, pubescence of leaves and corymbs, etc., demonstrated by the specimen. These combinations are distinctive, although separation in a key is difficult on the basis of any single character. Stamen number, as stressed by Kruschke, is at least objectively determinable and easy to employ in keys. But both Palmer and, later, Phipps and Muniyamma question its absolute importance. I have not accepted Kruschke's separations based solely on stamen number unless this is correlated, albeit less precisely, with additional characters, e. g., in leaves or other organs.

Even when narrowly defined, 19 species, mostly easily recognizable, constitute the vast majority of all specimens thus far made in the state, and are known from a dozen to many collections each. *Crataegus succulenta* and *C. macrosperma* are apparently common throughout Michigan. Others which are probably fairly common in the Upper Peninsula and northernmost Lower Peninsula are *C. chrysocarpa* (including *C. brunetiana*), *C. irrasa,* and *C. douglasii*; of these, *C. chrysocarpa* ranges occasionally farther south. The commonest additional species in the Lower Peninsula appear to be *C. pruinosa, C. crus-galli, C. punctata, C. calpodendron, C. pedicellata,* and *C. mollis*. Others fairly common are *C. coleae, C. brainerdii, C. holmesiana, C. fontanesiana, C. intricata, C. jesupii, C. dodgei,* and *C. margaretta*. The remaining 30 or so species admitted by Palmer to our flora are known from very few collections each.

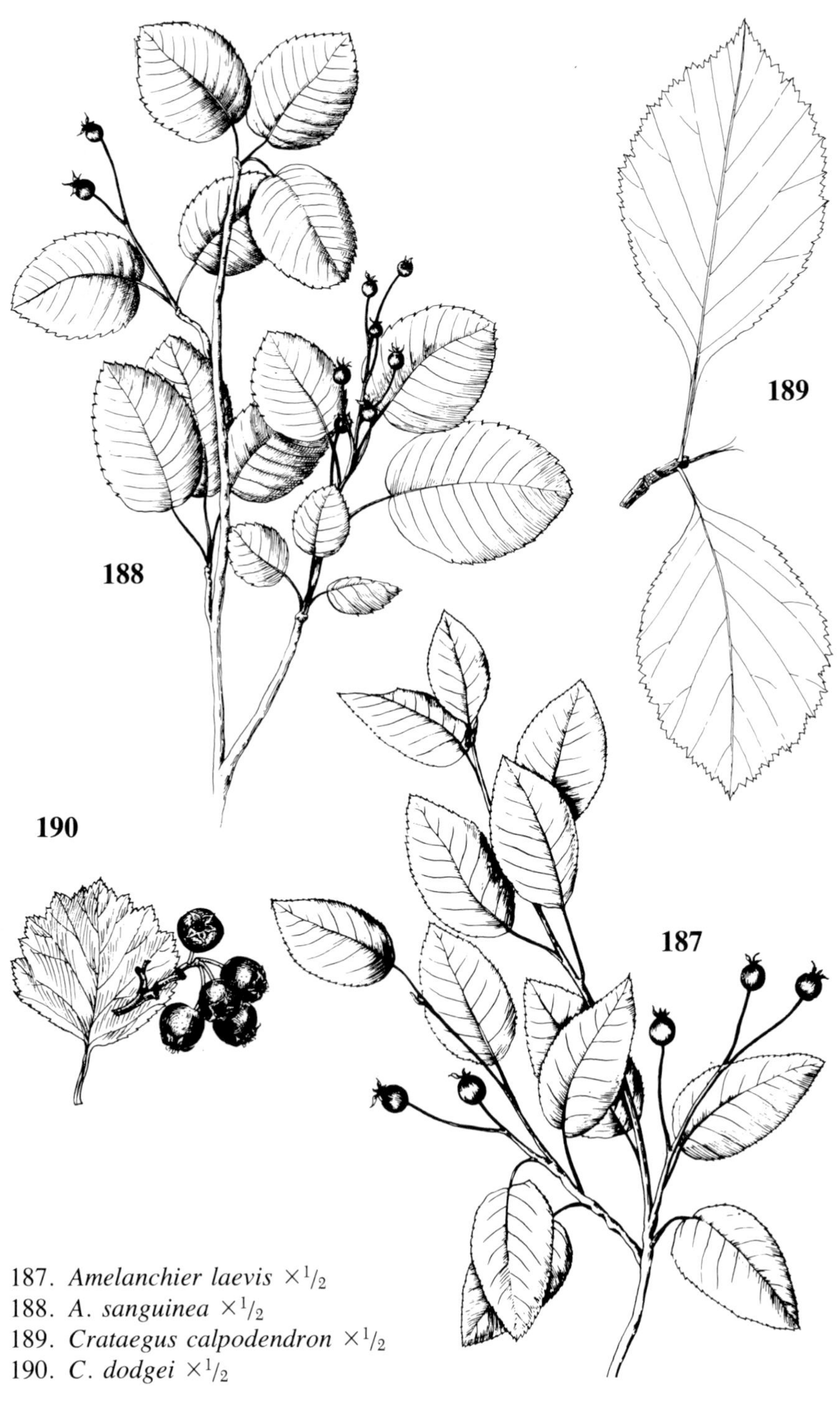

187. *Amelanchier laevis* ×1/2
188. *A. sanguinea* ×1/2
189. *Crataegus calpodendron* ×1/2
190. *C. dodgei* ×1/2

REFERENCES

Note: Practically all the species originally described from Michigan were published in the four papers cited by Ashe and the two works by Sargent.

Angell, Virginia C. 1933. The Crataegi of Grand Rapids, Michigan, and Vicinity. Pap. Mich. Acad. 17: 1–50 + 9 pl.

Ashe, W. W. 1900. New North American Plants—Some New Species of Crataegus. Bull. North Carolina Agr. Exp. Sta. 175: 109–114.

Ashe, W. W. 1902. New East American Thorns. Jour. Elisha Mitchell Sci. Soc. 18: 17–28.

Ashe, W. W. 1903. New North American Thorns. Jour. Elisha Mitchell Sci. Soc. 19: 10–31.

Ashe, W. W. 1904. East American Thorns. Jour. Elisha Mitchell Sci. Soc. 20: 47–56.

Dodge, C. K. 1908. Observations on the Collection and Study of Crataegi in the Vicinity of Port Huron, Michigan. Rep. Mich. Acad. 9 (for 1907): 123–125.

Kruschke, Emil P. 1955. The Hawthorns of Wisconsin Part I Status, Objectives, and Methods of Collecting and Preparing Specimens. Milwaukee Public Mus. Publ. Bot. 2. 124 pp.

Kruschke, Emil P. 1965. Contributions to the Taxonomy of Crataegus. Milwaukee Public Mus. Publ. Bot. 3. 273 pp. [For review, see Voss 1965.]

Palmer, Ernest J. 1925. Synopsis of North American Crataegi. Jour. Arnold Arb. 6: 5–128.

Palmer, Ernest J. 1946. Crataegus in the Northeastern and Central United States and Adjacent Canada. Brittonia 5: 471–490.

Phipps, J. B., & M. Muniyamma. 1980. A Taxonomic Revision of Crataegus (Rosaceae) in Ontario. Canad. Jour. Bot. 58: 1621–1699 + 4 col. pl.

Phipps, J. B. 1983. Crataegus—A Nomenclator for Sectional and Serial Names. Taxon 32: 598–604.

Sargent, C. S. 1907. Crataegus in Southern Michigan. Rep. State Board Geol. Surv. Mich. 1906: 509–570.

Sargent, Charles Sprague, ed. 1902–1913. Trees and Shrubs. Houghton Mifflin, Boston. 2 vols.

Sutton, S. B. 1970. Charles Sprague Sargent and the Arnold Arboretum. Harvard Univ. Press, Cambridge. 382 pp. [Ch. 11, pp. 279–298, is "Crataegus: A Thorny Problem" and puts matters in a readable perspective.]

Voss, E. G. 1965. A Review on Crataegus. Mich. Bot. 4: 93–96.

". . . *you can form no idea how delightful it is to travel in a country where there is no* Crataegus."

—C. S. Sargent, upon returning from South America in 1906 (Sutton 1970, p. 290).

520. Crataegus phaenopyrum

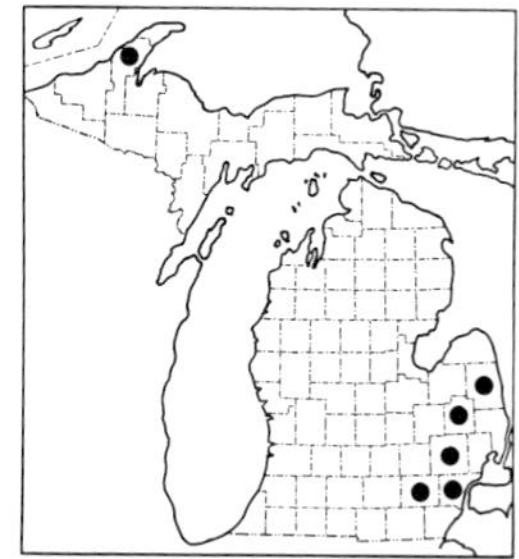

521. Crataegus monogyna

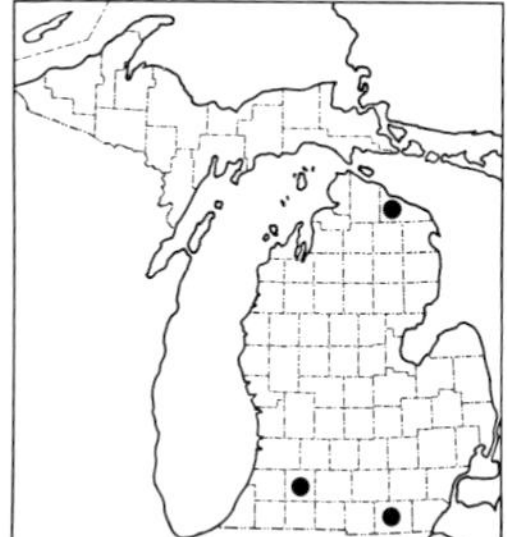

522. Crataegus laevigata

KEY TO THE GROUPS

Note: as was done for *Carex,* the familiar names for these groups are employed for convenience, without implying formal nomenclatural acceptance of these as epithets in legitimate names at the rank of section or series, though they have all been validly published at both ranks (Phipps 1983).

1. Leaves with some of the primary lateral veins running to (or toward, forking just before) the sinuses as well as to the points of the lobes; blades ± deltoid in general outline (Cordatae) or small and deeply lobed (Crataegus); thorns under 5 cm long; stamens ca. 20; plants escaped from cultivation
  2. Blades of leaves 3-lobed in general outline, ± deltoid, the apex sharply acute, the sinuses not more than halfway to the midrib; calyx deciduous in fruit, the ends of the (3–) 5 nutlets clearly exposed—almost slightly exserted; anthers yellow; thorns up to 4.5 (5.5) cm long . . . . 1. CORDATAE
  2. Blades of leaves mostly rather deeply 3–7-lobed, the apex ± rounded and toothed, the sinuses in many leaves extending at least halfway to the midrib; calyx persistent on fruit, the ends of the 1–2 nutlets ± hidden in depression; anthers pink; thorns (not sharp-tipped spur-branches) up to 2.5 cm long—usually much shorter . . . . 2. CRATAEGUS
1. Leaves with the primary lateral veins running only to (or toward) the points of the lobes (if any); blades, thorns, and stamens various; plants native, mostly not cultivated
  3. Nutlets with deep to shallow pits or depressions on their ventral surfaces; plants flowering in June or latter half of May
    4. Mature fruit purplish black, glaucous; thorns mostly 1.5–2.5 cm long; corymb glabrous or very sparsely villous; stamens 10 or fewer; nutlets 3–4 (5), rounded at the ends. . . . . 3. DOUGLASII
    4. Mature fruit red or orange; thorns mostly 2.5–9.5 cm long; corymb glabrous to densely villous; stamens ca. 10 or 20; nutlets 2–3 or if more, acute at the ends
      5. Leaves acute to rather broadly rounded or obtuse in outline at the apex, at least the midrib beneath often pubescent; nutlets 2–3, with a definite pit occupying most of each half of the ventral face, smoothly rounded or nearly so dorsally; thorns mostly 2.5–9.5 cm long; corymbs often ± villous . . . . 4. MACRACANTHAE
      5. Leaves ± narrowly acute or short-acuminate in general outline at the apex, completely glabrous beneath (the rare exceptions hybrids?); nutlets 2–5, with rather shallow and irregular depressions on the ventral face, usually very strongly ridged and grooved dorsally; thorns mostly 3.5–4.5 (5.5) cm long; corymbs mostly glabrous . . . . 8. BRAINERDIANAE
  3. Nutlets plane, not pitted, ventrally; plants flowering in April, May, or early June
    6. Blades of at least the floreal leaves (in many species also the vegetative leaves) ± acute to broadly or (more commonly) narrowly tapered or cuneate at their bases (figs. 189, 191, 192)
      7. Bracts of the corymb numerous, conspicuously and copiously stipitate-glandular on the margins; petioles glandular; stamens ca. 10; corymbs and leaves essentially glabrous (or a few hairs on the veins above); corymb few- (rarely more than 8-) flowered, ± simple or umbelliform . . . . 5. INTRICATAE
      7. Bracts of corymb usually few and/or the glands sparse or absent; petioles eglandular or the corymbs villous; stamens and leaves various; corymbs usually many-flowered
        8. Blades (especially of floreal leaves) mostly obovate to oblong-elliptic, broadest

above or rarely at the middle, unlobed or very obscurely lobed near the apex, mostly 1.5–3 or more times as long as broad, usually thickish to coriaceous (except sometimes in shade) (fig. 189)

9. Leaves glossy above (less so in *C. disperma*), the veins not (or only slightly) impressed; petioles mostly less than 1 cm long; styles and nutlets 1–3; corymb glabrous (except in a var. of *C. crus-galli*); stamens ca. 10 or 20 . . . . 6. CRUS-GALLI

9. Leaves dull above, the veins rather conspicuously impressed; petioles mostly 1–2 cm long; styles and nutlets 3–5; corymb ± villous (except in rare and questionable *C. nitidula*); stamens ca. 20. 7. PUNCTATAE

8. Blades (at least of floreal leaves) mostly elliptic to ovate, broadest at or below the middle, often ± definitely lobed, usually 1–1.5 times as long as broad, often thin

10. Corymbs and under surfaces of leaves glabrous or nearly so (the rare exceptions hybrids?); leaves strigose with short appressed hairs above, at least when young; nutlets often slightly pitted ventrally, especially before full maturity; stamens usually 20 with pink anthers (occasionally 5–15 with either pink or yellow anthers); vegetative leaves usually ovate or broadly elliptic . . . . 8. BRAINERDIANAE

10. Corymbs and under surfaces of leaves villous to glabrous; leaves glabrate or with short appressed hairs above when young; nutlets plane ventrally; stamens various but anthers usually yellow; vegetative leaves often broad, ± rotund . . . . 9. ROTUNDIFOLIAE

6. Blades of both floreal and vegetative leaves mostly broadly rounded, truncate, or subcordate at their bases (figs. 193, 194)

11. Corymb, calyx, and leaves (at least along main veins) beneath ± densely villous-tomentose; fruit short-villous at least at the ends; stamens in common species 20, the anthers white or yellow (rarely pink?) . . . . 10. MOLLES

11. Corymb, calyx, and leaves glabrous or pubescent; fruit glabrous; stamens ca. 10 or 20 but anthers in most species pink to purple

12. Stamens ca. 15–20 (10 in certain taxa that may not belong here anyway); anthers pink to purple or white to yellowish; young leaves glabrous to pubescent above

13. Leaves glabrous or nearly so on both sides when young; corymbs glabrous; calyx lobes entire or weakly and sparsely serrate; ripe fruit with rather thin dry flesh . . . . 11. PRUINOSAE

13. Leaves strigose above, at least when young; corymbs glabrous to villous; calyx lobes usually glandular-serrate; ripe fruit mellow or succulent

14. Flowers 2–2.5 cm wide . . . . 12. DILATATAE

14. Flowers less than 2 cm wide . . . . 14. TENUIFOLIAE (*C. lucorum* complex)

12. Stamens 10 or fewer; anthers pink to purple; young leaves strigose above

15. Corymbs villous or glabrous; leaves often with a few hairs, at least along veins or in axils, beneath; calyx lobes rather prominently glandular-serrate; ripe fruit with mellow or succulent flesh . . . . 13. COCCINEAE

15. Corymbs glabrous (villous in 2–3 rare taxa); leaves completely glabrous beneath; calyx lobes usually entire or weakly serrate; ripe fruit various

16. Fruit bright red, soft and succulent when ripe; leaves thin at maturity, the tips of the lobes generally spreading or reflexed (note out-curving of tips of lower main lateral veins) (fig. 195) . . . . 14. TENUIFOLIAE

16. Fruit bright or dull red or greenish with thin dry firm or mealy flesh when ripe; leaves firmer, the tips of the lobes usually not spreading . . . . 15. SILVICOLAE

1. CORDATAE

1. **C. phaenopyrum** (L. f.) Medicus — Washington Thorn

Map 520. Native south of our area, but with us presumably an escape from cultivation to fields and thickets.

2. CRATAEGUS (OXYACANTHAE)

1. Style and nutlet 1; leaves deeply divided (most sinuses extending more than halfway to the midrib) . . . . 2. **C. monogyna**
1. Styles and nutlets usually 2; leaves less deeply divided (most sinuses scarcely halfway to the midrib) . . . . 3. **C. laevigata**

2. **C. monogyna** Jacq. — English Hawthorn

Map 521. An Old World species, occasionally escaped from cultivation to roadsides, shores, pastures, hillsides, meadows.

Very distinctive in its deeply lobed leaves, which are smaller than in other species in the genus (blades up to 6.5 cm broad but usually—except on sprouts—under 3 cm broad).

3. **C. laevigata** (Poiret) DC. — English Hawthorn

Map 522. Another Old World species, of which the red-flowered cultivar 'Rosea' and double-flowered cultivars are grown, as is the typical form, of which we have very few records as apparent escapes: on a sandy knoll in a marsh in Kalamazoo County, in an abandoned field in Lenawee County, and along a roadside in Presque Isle County.

Quite similar to the preceding vegetatively, but the leaf lobes and stipules tend to be more abundantly toothed. Often called *C. oxyacantha* L., a name correctly applied to the preceding species and hence rejected as ambiguous although it questionably provides the type of the generic name (hence the epithet for the group if recognized in formal rank of series or section would be *Crataegus* and not *Oxyacanthae*).

3. DOUGLASII

4. **C. douglasii** Lindley — Black Hawthorn

Map 523. Borders of woods, sometimes quite common locally, as in the vicinity of Delaware, Keweenaw County; often in rocky woodlands and on rock summits; also in thickets on sand dunes and shores. This is one of the species which is strikingly disjunct from the West (see Mich. Bot. 20: 56–57. 1981).

A tall shrub, our only *Crataegus* with dark blue-black fruit and hence our easiest to recognize when mature. Michigan and Ontario plants were separated by Sargent from typical material of the Rocky Mountains and Pacific Coast as *C. brockwayae* (TL: near Clifton, Keweenaw Co.).

4. MACRACANTHAE

1. Mature leaf blades ± coriaceous, thickened at margins, the veins usually rather deeply impressed above, glabrous to pubescent on both surfaces; corymbs, new branchlets, and petioles glabrous to sparsely villous (if corymb somewhat villous, at least the young branchlets nearly always glabrous, the veins deeply impressed, and/or the stamens ca. 10); thorns ca. 2.5–9.5 cm long, usually numerous; stamens ca. 20 or 10 .... 5. **C. succulenta**
1. Mature leaf blades thin, the veins (except sometimes for midrib) scarcely if at all impressed above, strigose above and usually pubescent beneath; corymbs, branchlets of current year, and petioles all usually villous or lightly tomentose; thorns ca. 2.5–5 cm long, often sparse or even absent; stamens ca. 20 .... 6. **C. calpodendron**

5. **C. succulenta** Link

Map 524. Woods of diverse sorts, borders of woods, river banks and floodplains, lake shores, ravines, thickets, hillsides, rocky openings; roadsides, fencerows, fields. The Ontonagon County record is discussed under *C. suborbiculata* at species no. 13 below.

Most of our material was placed by Palmer in either typical var. *succulenta* (stamens ca. 20, anthers pink) or var. *macracantha* (Loudon) Eggleston (stamens ca. 10, anthers usually yellow). These are recognized as distinct species by Kruschke and by Phipps and Muniyamma. Both occur throughout the state, although var. *macracantha* appears to be more widespread in the Upper Peninsula, where var. *succulenta* has been collected only rarely and in the more southern counties.

Associated by Kruschke with *C. succulenta* (20 stamens with tiny anthers) are var. *michiganensis* (Ashe) Palmer (anthers usually white; TL: "Michigan," presumably St. Clair Co.), known from several counties in the southern half of the Lower Peninsula; var. *gemmosa* (Sarg.) Kruschke; var. *neofluvialis* (Ashe) Palmer, known from Kent and Wayne counties; and var. *laxiflora* (Sarg.) Kruschke. One collection of the latter, determined by Palmer as questionably *C. laxiflora*, was made by C. F. Wheeler on the river bank in East Lansing in 1902; it is supposedly distinguished from typical

523. Crataegus douglasii

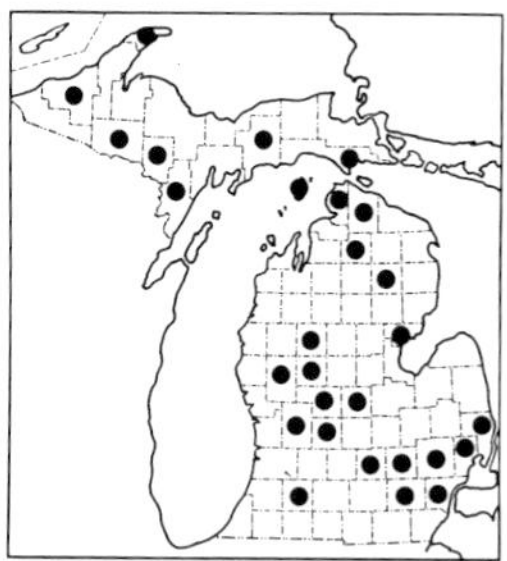
524. Crataegus succulenta

525. Crataegus calpodendron

succulenta by a more lax corymb and smaller flowers (which are not demonstrated by the East Lansing specimen, which has large flowers). The type plant of *C. gemmosa* was an unusually large tree from which Emma Cole and C. S. Sargent collected specimens in Grand Rapids; at the turn of the century, the tree was ca. 30 feet tall and 1 foot in diameter, its age estimated by Sargent as at least 100.

With *C. macracantha* Loudon (stamens 5–10 with larger anthers), Kruschke included var. *occidentalis* (Britton) Eggleston [*C. succulenta* var. *occidentalis* (Britton) Palmer]; Palmer referred a Menominee County collection here.

6. **C. calpodendron** (Ehrh.) Medicus Fig. 189

Map 525. Lowland as well as upland woods and thickets, river banks, roadsides and fencerows, pastures.

The anthers are ordinarily pink. Plants with pale yellow anthers have been segregated as *C. structilis* Sarg., and have been collected in the southeastern portion of the Lower Peninsula. The leaves of *C. calpodendron* tend to be acute more often than those of *C. succulenta,* but the more abundant pubescence of the plants will distinguish them from the Brainerdianae. *C. calpodendron* is quite a late-flowering hawthorn, typically blooming into mid-June in our area.

5. INTRICATAE

1. Leaves usually with 4–5 pairs of acute, spreading lobes with their tips slightly recurved (sometimes seen more easily in the outcurving of the main lateral veins); fruit mostly obovoid or pyriform; anthers white or pale yellow (or pink in f. *straminea,* with yellowish fruit) . . . . 7. **C. intricata**
1. Leaves usually with 3–4 pairs of broad, shallow lobes not recurved at the tips; fruit subglobose; anthers white, yellow, or pink . . . . 8. **C. foetida**

7. **C. intricata** Lange

Map 526 (the complex, including sp. 8). Thickets and borders of woods, wooded ravines, bluffs and sandy hillsides; roadsides, fencerows, fields, pastures.

Included here is *C. diversifolia* Sarg. [= *C. wheeleri* Sarg., not A. Nelson] (TL: near Grand Rapids [Kent Co.]). Occasional plants with the fruit short-oblong or subglobose rather than oblong or pyriform have been called *C. pusilla* Sarg. (TL: near Grand Rapids [Kent Co.]). More generally recognized is f. *straminea* (Beadle) Kruschke, with the mature fruit yellowish and the anthers pink; this form is known from Kent, Livingston, Oakland, Osceola, and St. Clair counties.

8. **C. foetida** Ashe

Included in this segregate are *C. bealii* Sarg. (TL: near Grand Rapids [Kent Co.]), the form with pale pink anthers; and *C. meticulosa* Sarg. (TL:

near Grand Rapids, Kent Co.). Although Palmer listed the latter in the synonymy of *C. intricata* (as did Kruschke), he placed a collection from the type tree as *C. foetida,* and the fruit is ± subglobose.

Specimens collected by H. H. Bartlett in 1951 in a small sedge bog east of Cavanaugh Lake, Washtenaw County, have been identified by Palmer as probably *C. rubella* Beadle. The fruit, in a few-flowered inflorescence, is characteristic of the Intricatae; the leaves are more broadly oblong-ovate than in the other species, and more shallowly lobed. Following Kruschke's treatment, this would be *C. intricata* var. *rubella* (Beadle) Kruschke. Kruschke also included *C. foetida* as a synonym under *C. intricata* var. *boyntonii* (Beadle) Kruschke. Phipps and Muniyamma are uncertain of the status of *C. foetida,* which they view as differing from *C. intricata* in having larger flowers and leaves. Altogether, recognition of *C. intricata* as a species complex seems to be an appropriate course.

6. CRUS-GALLI

1. Stamens ca. 10 with either pink or yellow anthers; young branchlets usually reddish or brownish; veins (except sometimes midrib) not at all impressed above .... 9. **C. crus-galli**
1. Stamens ca. 10 with pink anthers or ca. 20 with either pink or yellow anthers; young branchlets olive or yellowish green; veins sometimes slightly impressed on upper surfaces of leaves
   2. Leaves tending to be yellowish green and shiny above at maturity, strictly glabrous beneath .... 10. **C. fontanesiana**
   2. Leaves tending to be dark green and somewhat dullish above at maturity, sometimes slightly pubescent toward the bases of the veins beneath ...... 11. **C. disperma**

9. **C. crus-galli** L. Fig. 186 Cockspur Thorn

Map 527 (the complex, including sp. 10). Along creeks and river banks, borders of woods, shores, often on sandy hillsides, sometimes in wet ground; roadsides, fields, pastures. A collection from a yard in Menominee is presumably from a cultivated plant, as this native species is sometimes grown.

The thorns are often 5–7 cm long and may be as long as 11.5 cm, while in the next two species they are nearly always under 5 cm, at least on the younger branches (as long as 6.5 cm in a very few collections). Most Michigan plants are var. *crus-galli* (including *C. attenuata* Ashe, TL: Port Huron

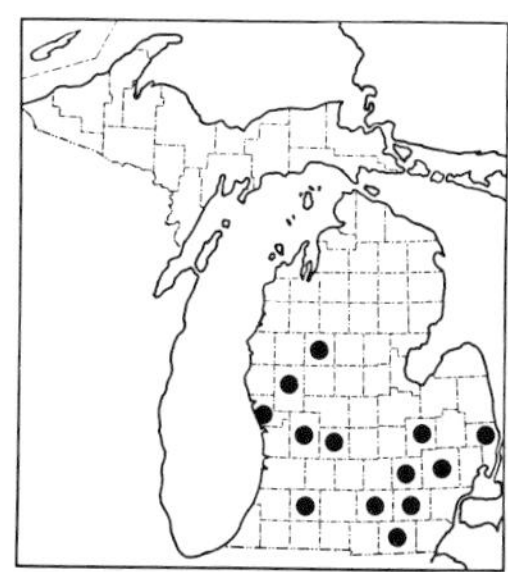
526. Crataegus intricata

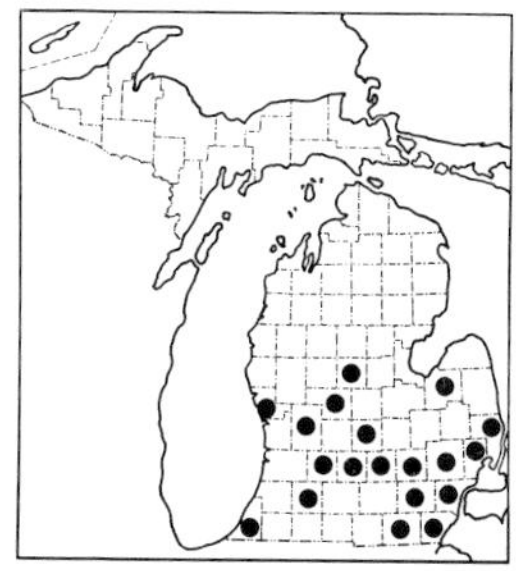
527. Crataegus crus-galli

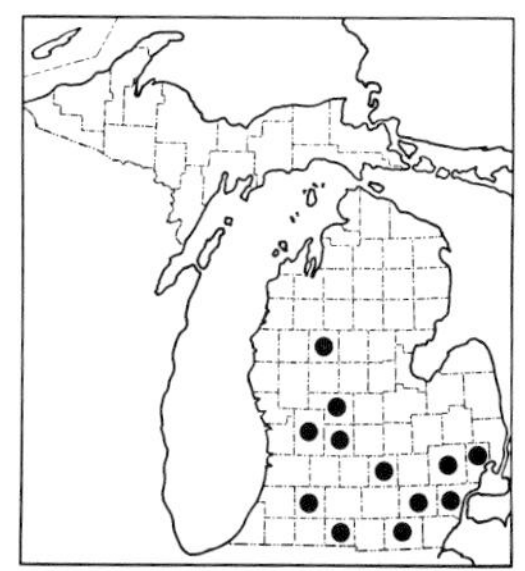
528. Crataegus disperma

[St. Clair Co.]). In addition, Palmer referred several specimens to other varieties: var. *barrettiana* (Sarg.) Palmer, with many of the petioles 10–14 mm long, collected in Kent, Lenawee, Tuscola, and Washtenaw counties; var. *capillata* Sarg. (*C. farwellii* Sarg., TL: Belle Isle [Wayne Co.]), with the corymb slightly villous, known from a few sites in Wayne County; var. *pyracanthifolia* Aiton, with relatively narrow leaves, collected once in Oakland County. Of these, Kruschke recognized only var. *capillata,* considering the others indistinguishable from typical *crus-galli.*

10. **C. fontanesiana** (Spach) Steudel

The stamens are ordinarily 10–20 with pink anthers in this species (including *C. tenax* Ashe, TL: vicinity of Port Huron [St. Clair Co.]). Plants with 20 stamens and yellow anthers appear to represent a variant placed with this species by the olive or yellowish green branchlets and yellow-green leaves. The veins of this variant are not at all impressed above, although they are usually so in typical plants. Kruschke apparently considered this species to have 20 stamens, which means that plants with 10 stamens must be referred elsewhere, possibly to *C. crus-galli.* This species is very close to *C. crus-galli,* the characteristics of leaves and branchlets being not entirely consistent, and I am not convinced that they warrant specific recognition. Both are included on the map. The range as indicated by specimens identified by Palmer is essentially the same for both. Older branchlets, incidentally, are gray in all three species of this group.

11. **C. disperma** Ashe

Map 528 (the complex, with *C. disperma* as determined by Palmer only from Ingham, Kalamazoo, Kent, Macomb, and Oakland counties). Woods and thickets—insofar as any habitat is suggested on the few specimens.

Intermediate in appearance between *C. crus-galli* and *C. punctata,* and quite possibly of hybrid origin; usually placed in the Punctatae, but I include it here because of the smaller number of stamens, styles, and nutlets, and the somewhat shiny leaves. Kruschke logically included *C. punctata* var. *pausiaca* (Ashe) Palmer as a synonym of *C. disperma,* for it also seems intermediate. Michigan specimens referred by Palmer to var. *pausiaca* are included on the map for the *C. disperma* complex.

7. PUNCTATAE

1. Corymb, young branchlets, and veins on underside of leaves ± villous; leaf blades mostly 1.5–2 or more times as long as broad; petioles winged nearly to their bases . . . . 12. **C. punctata**
1. Corymb, young branchlets, and veins glabrous or nearly so; leaf blades mostly 1–1.5 times as long as broad; petioles scarcely winged below the middle . . . . 13. **C. nitidula**

12. **C. punctata** Jacq. Dotted Hawthorn

Map 529 (this sp., sens. str.). Woods (beech-maple, oak, aspen, jack pine) and borders, swamp forests, river banks and creek bottoms; roadsides, fencerows, fields. Can grow to considerable size on rich bottomlands, but survives also on dry hillsides.

The leaves of vegetative shoots usually have distinct, narrow, acute lobes, although those of the flowering branchlets are at most indistinctly lobed near the apex. Our plants are largely typical var. *punctata,* including f. *aurea* (Aiton) Rehder, with the fruit and anthers yellow; and f. *canescens* (Britton) Kruschke, with the corymb very densely pubescent. In addition, the very narrow-leaved var. *microphylla* Sarg. was collected by Beal at East Lansing in 1899. *C. punctata* var. *pausiaca* (Ashe) Palmer, with the leaves tending to be slightly shiny above, the corymb glabrate, and often fewer stamens, appears to be intermediate with *C. crus-galli* and is included with *C. disperma* (no. 11).

13. **C. nitidula** Sarg.

Known with certainty in Michigan only from the colony of over 20 trees in a "thornapple orchard" from which the type came (TL: near St. Clair, St. Clair Co.).

Three puzzling specimens were identified by Palmer as questionably *C. suborbiculata* Sarg. This is a species he placed in the Punctatae, but the identity of these particular specimens is very doubtful. The yellow-green leaves are slightly longer than wide, rather small, with a small lobe at the tip of each deeply impressed lateral vein. A collection from the Porcupine Mountains, Ontonagon County (*C. Messner* in 1949, BLH), and another from an old field in Emmet County (*Voss 3288* in 1956, MICH, UMBS, BLH), both have 10 stamens; the former has 2 styles and the latter, 3. (*C. suborbiculata* has 20 stamens and the styles normally 5.) A sterile branch (*Bingham* in 1937, BLH) from edge of a beech-maple forest on Bois Blanc Island, Mackinac County, resembles the other two collections vegetatively—especially the Porcupine Mountain one. The petioles of the latter are scarcely 5–7 mm long (whereas they are described as 1–2 cm long in *C. suborbiculata*); Kruschke examined this specimen thoroughly and referred it to *C. macracantha* var. *divida* (Sarg.) Kruschke.

A river-bank collection from Menominee County (*Grassl 3294* in 1933, MICH) was determined by Palmer as questionably *C. celsa* Sarg., which differs from *C. punctata* in having the leaves rather broadly elliptic, widest about the middle (rather than obovate). The specimen differs from both *C. punctata* and *C. celsa* in that the fruit has remains of apparently only 10 stamens; the leaves are very dull above with impressed veins, but perhaps this is in the *C. disperma* complex.

Kruschke transferred a number of species from the Punctatae to a new subseries Suborbiculatae of the Brainerdianae. Among these species are *C. suborbiculata, C. celsa,* and *C. nitidula*. He also described here two new entities with 10 or fewer stamens and 2–4 styles: *C. wisconsinensis* and *C. desueta* var. *wausaukiensis* [neither name validly published]. Both came from northern Wisconsin, and it is quite possible that the puzzling Michigan specimens cited above belong somewhere in Kruschke's subseries Suborbiculatae. In 1974 he determined a Mackinac County specimen (*Bourdo* in 1963, MCTF5959) as *C. wisconsinensis*. Phipps and Muniyamma raise this group to the rank of a series, separating it from the Brainerdianae, and also include in it *C. compacta* (of the Pruinosae or Rotundifoliae) as well as *C. compta* (including *C. levis*) and *C. beata* (both of the Silvicolae). They consider the species of this group to be intermediate between the Tenuifoliae and the Pruinosae, but I would suggest Rotundifoliae and Pruinosae.

8. BRAINERDIANAE

The Brainerdianae seem in some ways to be a rather artificial series, too variable to include easily in a key. On one side they approach the Macracanthae and on the other the Tenuifoliae or other series. Such intermediacy suggests (but by no means proves) a hybrid origin, and this point has also been suggested by Phipps and Muniyamma, who do not include in this group the taxa transferred to it by Kruschke from the Punctatae (see note at end of the preceding series).

1. Calyx tube of mature fruit 4–5 mm broad, 0.5–1 mm long, elevating the prominent calyx lobes . . . . 14. **C. coleae**
1. Calyx tube of fruit at most 3–4 mm broad, the lobes sessile or nearly so, not conspicuously elevated
   2. Stamens ca. 20 (5–10 in two vars.); corymb glabrous (rarely very sparsely villous); nutlets 2–5 (usually 3) . . . . 15. **C. brainerdii**
   2. Stamens ca. 10 or fewer; corymb slightly villous; nutlets 2–3 . . . . 16. **C. pinguis**

14. **C. coleae** Sarg.

Named for Miss Emma J. Cole (1845–1910), author of the excellent *Grand Rapids Flora* and an outstanding high school teacher, who inspired a large number of people to do careful collecting and study of Michigan plants. (TL: near Grant Rapids [Kent Co.]) The stamens are ca. 20 in this species.

C. incerta Sarg. (TL: [near Ruby, Clyde Tp.], northwest of Port Huron [St. Clair Co.]) was included in the synonymy of *C. coleae* by Palmer. The calyx is not so prominent as in the latter, and the stamens are only ca. 10. It is known only from the type plants.

15. **C. brainerdii** Sarg.

Map 530 (the complex, including the entire group). Open woodlands, sandy roadsides and bluffs, river banks, fields and pastures; usually in dry ground, rarely in moist places.

529. Crataegus punctata

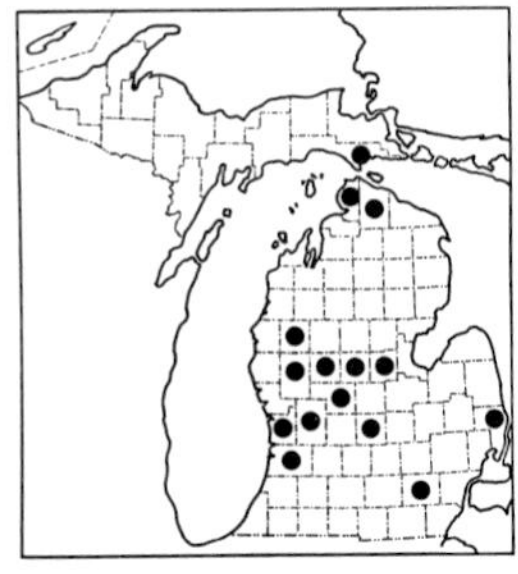

530. Crataegus brainerdii

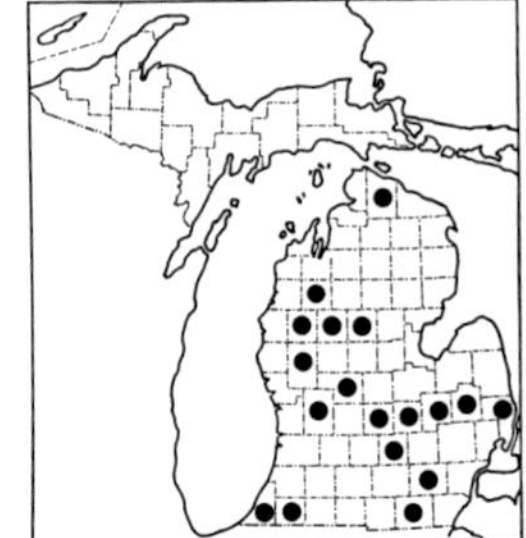

531. Crataegus dodgei

Included here is *C. urbana* Sarg. (TL: near Grand Rapids [Kent Co.]). According to Palmer, most Michigan plants are typical var. *brainerdii,* although we also have var. *scabrida* (Sarg.) Eggleston (stamens 5–10, floreal leaves mostly elliptic to obovate) and var. *egglestonii* (Sarg.) Robinson (similar, but floreal leaves oval, more obscurely lobed). On the basis of stamen number, Kruschke recognized *C. scabrida* Sarg. as a distinct species, including in it var. *egglestonii* and the questionable *C. pinguis* and *C. honesta* (see below). Phipps and Muniyamma also keep *C. scabrida* separate. On the other hand, the entire group (or series) can be recognized as a species complex, as here mapped. Varieties could be sorted out simply: var. *coleae* having a prominent calyx, the others not; var. *brainerdii* with 20 stamens and pink anthers; var. *scabrida* with glabrous corymbs and yellow anthers and var. *pinguis* with slightly pubescent corymbs and pink anthers, these two with 5–10 stamens.

16. **C. pinguis** Sarg.

A little-known plant, found nowhere except in southern Michigan (TL: near Lowell [Kent Co.]).

C. honesta Sarg. (TL: near Grand Rapids [Kent Co.]) is a questionable plant known only from the type locality. Palmer suggested that it may not be distinct from *C. brainerdii* var. *scabrida* or may be a hybrid between *C. brainerdii* and *C. macrosperma*. The flowers have 10 or fewer stamens and are in villous corymbs; the leaves are less cuneate at the base than in *C. pinguis*.

9. ROTUNDIFOLIAE

1. Stamens ca. 10 or fewer
   2. Blades of floreal leaves with lobes shallow or obsolete, mostly developed only beyond the middle; corymbs glabrous (loosely villous in one uncommon variety); styles and nutlets mostly 2–3; calyx lobes essentially entire (glands few and small or none); vegetative leaves with the apex often more acute than the base; leaves glabrous beneath, with teeth obscurely if at all gland-tipped .... 17. **C. dodgei**
   2. Blades of floreal leaves with lobes rather sharp and spreading, usually developed below the middle; corymbs glabrous to densely villous; styles and nutlets mostly 3–5; calyx lobes ± conspicuously glandular-serrate; vegetative leaves with the base about as acute as, or more acute than, the apex; leaves glabrous to sparsely villous beneath, the teeth rather conspicuously gland-tipped .... 18. **C. chrysocarpa**
1. Stamens ca. 15–20
   3. Corymb and leaves on both surfaces ± villous; range in Upper Peninsula .... 19. **C. irrasa**
   3. Corymbs at most slightly villous; leaves completely glabrous beneath (and usually quite or nearly so above); range in southern half of Lower Peninsula
      4. Leaf serrations rather crenate; apex and lobes of blades ± rounded in outline; styles 2–4; anthers yellow .... 20. **C. margaretta**
      4. Leaf serrations sharp; apex and lobes acute; styles 4–5; anthers pink .... 21. **C. immanis**

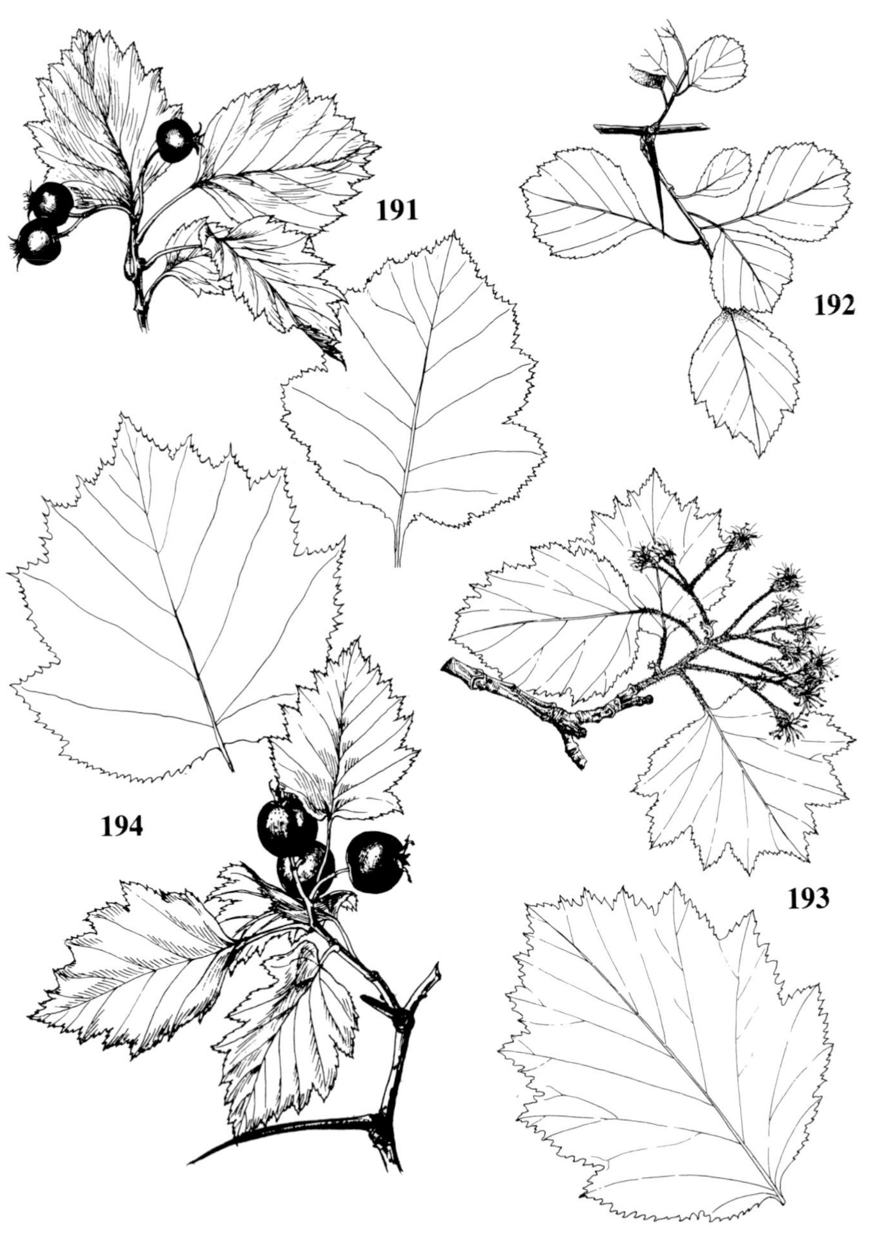

191. *Crataegus irrasa* $\times 1/2$
192. *C. margaretta* $\times 1/2$
193. *C. mollis* $\times 1/2$
194. *C. pedicellata* $\times 1/2$

17. **C. dodgei** Ashe Fig. 190

Map 531. Thickets, borders of woods and swamps; dry prairie-like ground and jack pine plains; roadsides, fencerows, fields.

Most plants are the glabrous var. *dodgei* (TL: "Port Huron" [St. Clair, St. Clair Co.]), including those with the fruit yellowish rather than red, described as *C. flavida* Sarg. (TL: London, Ontario, but paratypes from Port Huron). Plants with the corymbs and to some extent the leaves loosely villous are var. *lumaria* (Ashe) Sarg. This species could be treated as a variety of *C. margaretta*.

18. **C. chrysocarpa** Ashe

Map 532. Forming thickets at edges of woods and along shores (especially the Great Lakes); sandy bluffs and old dunes; aspen, oak, jack pine woodlands; roadsides, clearings, fields, pastures; occasionally on river banks and floodplains.

Phipps and Muniyamma treat *C. brunetiana* Sarg. as a synonym of *C. chrysocarpa,* and I readily concur that these taxa run completely together and cannot be distinguished, although Palmer maintained them and Kruschke allied *C. brunetiana* with *C. laurentiana* Sarg. Typical *C. chrysocarpa* has the corymbs, petioles, and young branchlets slightly villous to glabrous or nearly so, white or pale yellow anthers, flowers 1.3–1.6 mm broad, petioles sparsely glandular and seldom half as long as the blades. In *C. brunetiana* the corymbs, and usually the petioles and young branchlets, are villous; the anthers white, yellow, or pink [the latter in var. *fernaldii* (Sarg.) Palmer]; flowers 1.6–2 cm broad; petioles often rather prominently glandular and often at least half as long as their blades. There are also subtle differences in leaf shape. But too many intermediates occur to allow recognition of two species even in a narrow treatment. The range in Michigan is essentially the same, and specimens identified as either species by Palmer are included on the map.

One could sort out as varieties in this species var. *fernaldii* with pink anthers, the others having white to yellow anthers: var. *phoenicea* with glabrous corymbs and var. *brunetiana* and var. *chrysocarpa* with ± villous corymbs.

A collection by Kruschke (K-69-41, MICH) from L'Anse, Baraga County, was referred by him to *C. faxonii* var. *praetermissa* (Sarg.) Palmer. This species is supposedly similar to *C. chrysocarpa* but has a more densely villous inflorescence; the collection, however, has the large flowers, long petioles, ± elliptical leaves, and other features of "*C. brunetiana.*"

19. **C. irrasa** Sarg. Fig. 191

Map 533. Ravines, hillsides, lake shores, often in rocky places.

Some of the specimens from Baraga, Houghton, and Marquette counties said by Palmer to "appear to be typical C. irrasa" have ca. 10 stamens and therefore are either an undescribed form of this 20-stamened species or actually *C. chrysocarpa*. The leaves of the two species can be very similar.

20. **C. margaretta** Ashe   Fig. 192

Map 534. Thickets, fencerows, roadsides, fields, pastures; borders of woods, stream valleys, river bluffs.

Occasional plants occur with the fruit bright yellow rather than red or orange-red [f. *xanthocarpa* Sarg.]. A form with the leaves very narrow is var. *angustifolia* Palmer. This species closely resembles *C. dodgei,* and sometimes the two can be safely distinguished only by the stamen number. Although the thorns are usually relatively short, if present at all, on a shoot of one specimen seen they are as long as 8 cm.

21. **C. immanis** Ashe

Map 535 (the complex as here described). The few ± ambiguous specimens included here are from floodplains, woods, fencerows, and old fields.

Strictly speaking, this species is known only from a fencerow at the type locality ("Port Huron" [actually Clyde Tp., St. Clair Co.]). However, I am associating certain other plants in the species complex.

C. compacta Sarg. (TL: London, Ontario, but paratypes from St. Clair Co.) was placed by Palmer in the Pruinosae, and he referred a number of collections additional to the types to it. The taxon is a puzzling one (as evidenced by the fact that the various names cited as synonyms had previously been placed in three different series). Sargent originally placed *C. compacta* in the Punctatae. Kruschke recognized it as a species in the Silvicolae, while Phipps and Muniyamma transferred it as a "very distinctive" species to their new series Suborbiculatae near the Brainerdianae—stating that it "also has similarities to" the Pruinosae and is probably an apomict. Specimens from the trees which furnished the Michigan paratypes, as well as several others, appear practically indistinguishable from *C. immanis,* as the floreal leaf blades taper to the base and the leaves are glabrous or nearly so even when young. The tendency for the veins of mature leaves to be

532. Crataegus chrysocarpa

533. Crataegus irrasa

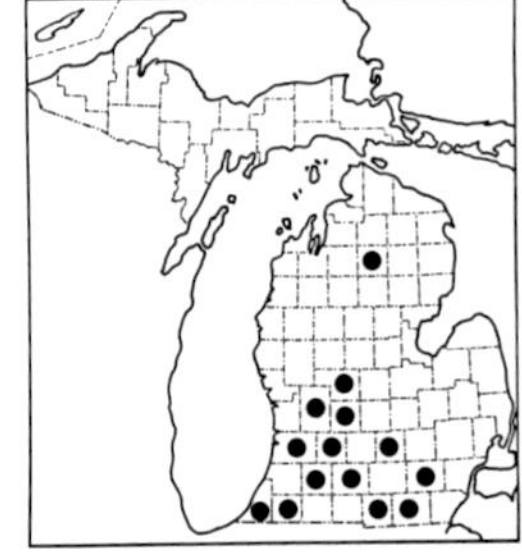

534. Crataegus margaretta

impressed above is a character of the Rotundifoliae, although the 20 stamens with pink to purple anthers are typical of the Pruinosae. (Specimens referred to *C. compacta* seem to have the lobes more restricted to the apical half of the blades and to have the bases of the vegetative leaf blades more tapered to the base than do those referred to *C. immanis*. The calyx lobes are rather obscurely glandular-serrate, particularly near the middle, in both taxa.)

C. parvula Sarg. (TL: near Grand Rapids [Kent Co.]) and *C. ater* Ashe (TL: Port Huron and St. Clair [St. Clair Co.]) differ from *C. pruinosa,* in the synonymy of which they are often cited, in having the floreal leaves ± tapered or rounded at the base, with a very few hairs above when young. While they may nevertheless belong in that series, they seem similar to the *C. immanis* complex; Kruschke cited both names in the synonymy of typical *C. pruinosa*; Phipps and Muniyamma made a separate variety of *C. parvula*. At least one of Dodge's trees referred to *C. ater* by Ashe was cited as a paratype of *C. compacta* by Sargent.

A collection from the Cadillac airport, Wexford County (*Bazuin 7249,* BLH, WUD), which Palmer could only say was a *Crataegus* "sp. or hyb." seems associated here also.

This whole complex of 20-stamened forms appears to include either extremes of *C. pruinosa* or hybrids of that species with several others (such as *C. margaretta* [or *C. chrysocarpa*?] and *C. punctata*). They quite probably are not closely related to the Rotundifoliae, but happen to key here because of their intermediate nature. Kruschke included *C. immanis* in the synonymy of the variable *C. pruinosa* var. *virella* [an illegitimate name] and has been followed by Phipps and Muniyamma. Cronquist, followed by Little, adopted a hybrid interpretation, as had been suggested by Palmer. It is quite possible that an older name in the rank of species would replace *C. immanis* when names applied to taxa beyond our area have, along with *C. immanis,* been more thoroughly evaluated.

10. MOLLES

1. Stamens ca. 20; fruit subglobose (occasionally oblong-obovoid) . . . . 22. **C. mollis**
1. Stamens ca. 10; fruit pyriform or obovoid . . . . 23. **C. submollis**

22. **C. mollis** Scheele   Fig. 193

Map 536 (the complex, including sp. 23). Woods and woodland borders, floodplains and river bluffs, sandy hillsides, fields, meadows, roadsides. The "national champion" is a tree 2.8 feet in diameter in Wayne County.

This is a species with large flowers (up to 2.5 cm or sometimes more in width), large fruit (up to 2.2 cm in diameter), and large leaves, sometimes laciniate on vigorous vegetative shoots (often as wide as 8 cm). Included here are var. *sera* (Sarg.) Eggleston, differing in its extremes by thinner leaves rounded at the base and later fruiting, but grading into typical plants; *C. nutans* Sarg. (TL: [Roberts Landing] near Algonac, St. Clair Co.), with thin leaves and very late-ripening fruit (according to Dodge's notes, the type tree was in a hedge with trees of *C. mollis*!); and *C. mollipes* Sarg. (TL: near Grand Rapids [Kent Co.]), with obovoid fruit and small flowers (not over 1 cm broad).

23. **C. submollis** Sarg.

A flowering collection made at Ontonagon in 1860 was questionably placed here by Palmer. Kruschke had previously annotated it as *C. arnoldiana* Sarg. In any event, it appears to be a 10-stamened Molles, with 3–4 styles, quite beyond the range of the rest of our *C. mollis* complex, which is otherwise only in the Lower Peninsula. *C. submollis* should be more common in the Upper Peninsula, judging from Ontario and Wisconsin distributions.

C. hillii, listed in the Coccineae, might key here and is probably a hybrid, resembling a somewhat less pubescent *C. mollis* but with 20 pink anthers.

11. PRUINOSAE

1. Stamens 10 or fewer; vegetative leaves deeply lobed . . . . 24. **C. jesupii**
1. Stamens ca. 20 (or 10 in one variety of *C. pruinosa* with shallowly lobed leaves)
   2. Mature leaves yellowish green, the blades of the vegetative ones often as broad as long, or broader, and not deeply lobed . . . . 25. **C. rugosa**
   2. Mature leaves bluish or dark green, the blades of the vegetative ones seldom as broad as long, or if broader, deeply divided
      3. Floreal leaves with ± triangular, sharp, acuminate, spreading lobes; nutlets 2–3 . . . . 26. **C. leiophylla**
      3. Floreal leaves with ± shallow, obscure lobes; nutlets 4–5 . . . . 27. **C. pruinosa**

24. **C. jesupii** Sarg.

Palmer included here *C. bellula* Sarg. (TL: Grand Rapids [Kent Co.]), which Kruschke placed as a variety of his *C. dissona* (cf. *C. pruinosa* below). *C. jesupii* was transferred from this series to the Silvicolae by Kruschke; the species is often quite difficult to distinguish from *C. stolonifera* Sarg.

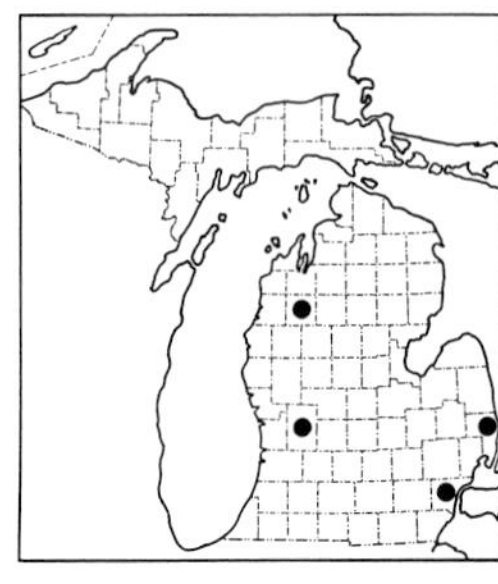
535. Crataegus immanis

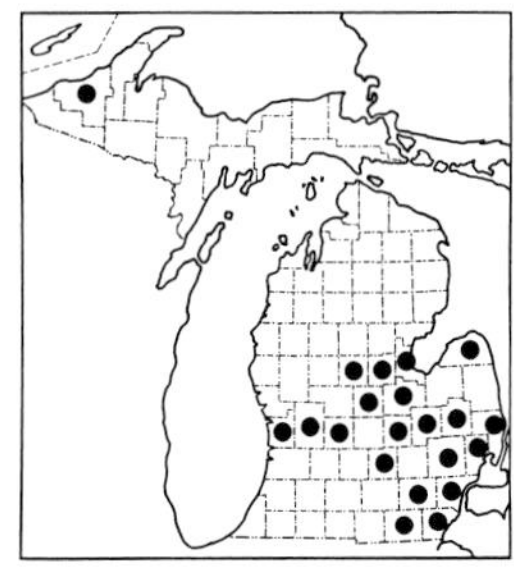
536. Crataegus mollis

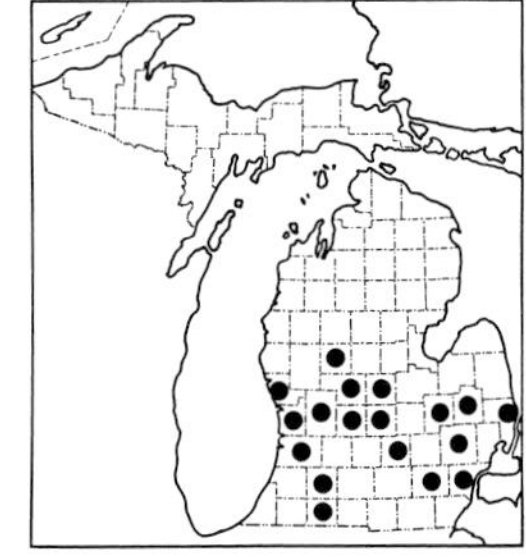
537. Crataegus pruinosa

of that group. Two collections from Oakland County identified as *C. jesupii* by Palmer distinctly have ca. 20 stamens and must be *C. pruinosa*.

25. **C. rugosa** Ashe

Palmer indicated in *Gray's Manual* that this species has 2–3 nutlets, but in the *Illustrated Flora* that it has 3–5 nutlets. Our specimens referred here by him are all flowering and have 4–5 or rarely 3 styles.

26. **C. leiophylla** Sarg.

This species is said to have 2–3 nutlets, in contrast to 4–5 in *C. pruinosa*. However, the original description said "usually 4" and all of our specimens referred here by Palmer have 3–4 (mostly 4–5) styles. I am not convinced of the distinctness of this and the preceding species from *C. pruinosa;* Kruschke included them both as *C. pruinosa* var. *rugosa* (Ashe) Kruschke, and Phipps considered them large-leaved forms of *C. pruinosa*.

27. **C. pruinosa** (Wendl.) K. Koch

Map 537 (the complex, including spp. 25, 26, & 27 but not *C. pruinosa* var. *dissona*). Sandy open upland woods as well as richer wooded banks, stream borders, wooded hills, river bluffs; roadsides, fencerows, fields, meadows, pastures.

Plants with unusually large flowers may generally be distinguished from *C. dilatata* by the complete absence of hairs on the upper surfaces of the young leaves. In the typical variety, the stamens are ca. 20 with pink anthers (including *C. horridula* Sarg., TL: near Grand Rapids [Kent Co.]; and *C. perampla* Sarg., TL: near Grand Rapids [Kent Co.]); Palmer also referred here *C. sitiens* Ashe (TL: [Clyde Tp.], St. Clair Co.), which Kruschke referred to var. *rugosa*. In var. *latisepala* (Ashe) Eggleston (TL: "eastern Michigan" [St. Clair, St. Clair Co.]), the stamens are 20 with white or yellowish anthers and the sepals are ± deltoid and broad-based. In var. *dissona* (Sarg.) Eggleston, the stamens are 10 with pink anthers (including *C. superata* Sarg., TL: south of Grant Rapids, Kent Co.). This variety is scarcely distinguishable from *C. macrosperma* and the Silvicolae, as the sepals often have a few teeth on each side. The closeness of this resemblance is illustrated by the fact that a collection of flowering and fruiting material from the paratype tree (*Cole 44*) of *C. remota,* listed by Palmer as a synonym of *C. gravis,* was identified by him as *C. pruinosa* var. *dissona* (the young leaves are completely glabrous above). Kruschke kept *C. dissona* as a good species, based on stamen number, and made *C. bellula* a variety of it. Phipps and Muniyamma also retained *C. dissona* (but did not mention *C. bellula*).

C. pruinosa var. *dissona* is not included on Map 537 of the *C. pruinosa* complex, but specimens referred to it are known from several additional counties: Arenac, Berrien, Eaton, and Livingston. Including *C. jesupii* on the map would add only Barry County.

Collections of flowering and fruiting material from a shrub near Ann Arbor were placed by Palmer as questionably *C. porteri* Britton. That species ordinarily differs from *C. pruinosa* in having pyriform to obovoid fruit, thick leaves with prominent spreading lateral lobes, and white or pale yellow anthers. However, our specimens are described by the collector as having dark crimson anthers, and the fruit appears to have been subglobose. A scanty fruiting specimen from Barry County was determined by Palmer as questionably *C. gattingeri* Ashe, a southern species which is supposed to differ from *C. pruinosa* in thinner (but firm) leaves, smaller flowers, and more prominently lobed leaves. I find the specimen inadequate for any evaluation.

C. glareosa Ashe (TL: "Lake St. Clair" [actually Port Huron, St. Clair Co.]) differs from *C. pruinosa* in its white to yellowish anthers and large elevated calyx with abruptly narrowed lobes. The application of the name is questionable, as the original description mentions 10 stamens and pubescent corymbs. The trees said by Dodge to have provided the types have 20 stamens and glabrous or very sparsely villous corymbs. The general appearance of the leaves suggests *C. succulenta,* of which this may be a hybrid. Kruschke expressed doubt as to the status of *C. glareosa* and tentatively transferred it to the Dilatatae, where it was listed among "Doubtful Taxa" by Phipps and Muniyamma.

There is a small complex of so-called species which appear intermediate between the Pruinosae and other series. They often resemble the Pruinosae in general aspect, in fruit, and in having the leaves nearly glabrous above when young. Blades of the floreal leaves, however, tend to be more tapered to the base. Hence, they will run in the key to the Rotundifoliae, where they are discussed under No. 21, *C. immanis*. Sinnott and Phipps (Syst. Bot. 8: 66. 1983) are convinced that *C. parvula* and its allies belong in the Pruinosae.

12. DILATATAE

28. **C. dilatata** Sarg.

Map 538. Referred here by Palmer as "very good typical material" are collections in both flower and fruit from Livonia (*Farwell 8334* in 1929, MICH, BLH) with no habitat indicated.

The dry flowers on the Michigan collection seem scarcely 2 cm broad, but the leaves are relatively broad with prominent lower lobes, as in *C. dilatata*. Phipps and Muniyamma state that the Dilatatae could "easily be included in the Coccineae" but might be hybrids with the Pruinosae.

13. COCCINEAE

If one prefers to recognize a broader species complex, I suggest with some hesitancy following Cronquist (and Little) in taking up the disputed Linnaean name *C. coccinea*. If the epithet "Coccineae" is adopted for a section (as was originally done by Loudon) or a series (as later recognized by Rehder) or any other rank of subdivision of a genus, the type of the subdivisional name is, under the Code, the same as for the species from which "Coccineae" is derived. So one cannot use this name for a series (or section) without having *C. coccinea* L. in it. Long-standing controversy over typification of the Linnaean name is described by Kruschke (1965, pp. 202–204), although by retaining the series epithet Coccineae he automatically (though surely unintentionally) indicated that *C. coccinea* belongs in this series—a judgment he declared he was unwilling to make in view of the mixed basis of the Linnaean name. At present, it seems in the interests of stability to retain the respective epithets for both the series (or informal group) and the major species complex placed here.

1. Stamens 15–20; corymb and leaves (at least when young) villous (see also variety of *C. holmesiana*) . . . . 29. **C. hillii**
1. Stamens ca. 10 or fewer; corymb and leaves glabrous or pubescent
  2. Calyx tube and corymb villous; fruit dull dark red . . . . 30. **C. pringlei**
  2. Calyx tube and corymb glabrous or slightly villous; fruit bright red
    3. Fruit distinctly obovoid or oblong, noticeably longer than thick; floreal leaves mostly oblong-ovate to elliptic, often slightly narrowed at the base . . . . 31. **C. holmesiana**
    3. Fruit at most slightly longer than thick; floreal leaves ovate, broadest at the base . . . . 32. **C. pedicellata**

29. **C. hillii** Sarg.

Referred here by Palmer are a fruiting collection from Detroit and material with both flowers and fruit from Eberwhite Woods, Ann Arbor. The latter has flowers up to 2.5 cm broad, as well as quite large leaves. The mature leaves and fruit on these specimens are only slightly less pubescent than in the more glabrate extremes of *C. mollis,* with which—at least as one parent of a hybrid—they might well be related. This was included by Kruschke in *C. corusca* Sarg.—which was suggested by Palmer to be a hybrid of *C. mollis* with one of the Coccineae, a position accepted by Cronquist but not by Phipps.

30. **C. pringlei** Sarg.

A single specimen is referred here, from a swamp at East Lansing (*Beal* in 1899, MSC). Sargent also cited a Wheeler collection from East Lansing (as well as "Port Huron" collections actually from Ontario and furthermore later reidentified).

31. **C. holmesiana** Ashe

In the typical variety, the corymbs and calyx are essentially glabrous (including Dodge's no. 34, *C. caesa,* but not the type). In var. *villipes* Ashe, they are villous (including *C. lenta* Ashe, TL: Port Huron [St. Clair Co.]). This species sometimes superficially resembles *C. macrosperma,* but may usually be distinguished by the glandular-serrate sepals. However, the species in the Coccineae are extremely difficult to distinguish from each other (except for *C. hillii,* which as suggested above may belong elsewhere). Supposedly duplicate collections have sometimes been referred to one species and sometimes to another.

A fruiting collection from Houghton County (*Hyypio 365,* MSC) was determined by Palmer as questionably this species, although it looks to me more like *C. chrysocarpa*—which would be more reasonable geographically as *C. holmesiana* is otherwise a southern species in the state. It has not been mapped. Besides more clearcut material from Washtenaw County, one collection (*Farwell 5454,* MICH, BLH) with 20 stamens and large flowers (over 2 cm broad) was referred by Palmer questionably to *C. holmesiana* var. *magniflora* (Sarg.) Palmer. Because of its 20 stamens, Kruschke did not consider *magniflora* to be associated with *holmesiana,* but made it a variety of *C. fulleriana* Sarg., with which he also associated *C. miranda* Sarg. [under an illegitimate name] (see note after species 34, *C. lucorum*).

32. **C. pedicellata** Sarg. Fig. 194

Map 539 (the *C. coccinea* complex, excluding *C. hillii*). Oak, aspen, and beech-maple woods, especially along borders and stream banks; low moist ground including floodplains and riverside thickets; and the usual field and fencerow habitats of other hawthorns.

The corymb is villous in the typical variety (including *C. pura* Sarg., TL: near Grand Rapids [Kent Co.]; and *C. caesa* Ashe, TL: Port Huron [St. Clair Co.]; each placed in a different variety by Kruschke). It is glabrous in var. *albicans* (Ashe) Palmer (TL: "Eastern Michigan," presumably St. Clair Co.). Less common varieties which we have, according to Palmer (but not distinguished from each other by Kruschke) are var. *robesoniana* (Sarg.) Palmer, with large deeply serrate leaves, these and the corymbs slightly villous; and var. *ellwangeriana* (Sarg.) Eggleston, with the leaves beneath and the corymbs ± villous and the fruit often slightly pubescent at the ends (suggesting hybridization with *C. mollis*). The latter variety apparently includes at least the Michigan plants of *C. pascens* Ashe, as Dodge's no. 22 (collected in Port Huron near the type of *C. lenta*) was referred to *C. pascens* by Ashe and to *C. ellwangeriana* by Sargent, the latter concept confirmed by Palmer. The plants and Dodge's notes, however, do not fit Ashe's description in every respect (e. g., the latter calls for very late flowers and fruit, while Dodge's 22 has flowers in May and fruits in September). On the basis of the original description, assuming these plants not to represent the type, Kruschke placed *C. pascens* in the synonymy of var. *albicans*.

14. TENUIFOLIAE

The leaves of species in this group are completely glabrous beneath, though short appressed pubescence occurs above when young. In the Pruinosae, the leaves are essentially glabrous or, in the rare instances of pubescence, it is of longer or scattered hairs. All of C. K. Dodge's extensive field notes on plants of this group state that the ripe fruit is edible and palatable, "much sought after by children." Anthers in all species are pink to red or purple.

1. Flowering corymbs ± villous
  2. Stamens mostly 5–7 (often fewer, rarely as many as 10) .... 33. **C. apiomorpha**
  2. Stamens ca. 15–20 .... 34. **C. lucorum**
1. Flowering corymbs completely glabrous
  3. Stamens ca. 15–20 .... 35. **C. basilica**
  3. Stamens 10 or fewer
    4. Leaves usually with 4–5 pairs of lobes; fruit ripening in September .... 36. **C. macrosperma**
    4. Leaves with 3–4 pairs of lobes; fruit ripening in October .... 37. **C. merita**

33. **C. apiomorpha** Sarg.

In Michigan known only from St. Clair, St. Clair County, the type locality (as "Summerville") of the indistinguishable *C. uber* Ashe. The species seems scarcely distinguishable from the Coccineae, and Cronquist (following a suggestion of Palmer) indicated a hybrid origin (see next species). Phipps and Muniyamma suggest that *C. apiomorpha* may be a mutant of *C. macrosperma*.

34. **C. lucorum** Sarg.

Map 540 (the complex, including sp. 35). Woods, thickets, sandy hillsides, and stream banks; roadsides, fields, and pastures.

Included here is *C. asperata* Sarg. (TL: Port Huron [St. Clair Co.]). Our specimens are from St. Clair and Washtenaw counties and all have ca. 20 stamens. Mapped as part of this complex are *C. basilica,* with the same number of stamens, and the specimen cited under *C. tortilis*. Although the species is currently described as usually having 5–10 stamens, the original descriptions of *C. lucorum* and all synonyms listed by Palmer state that the stamens are ca. 20. Palmer suggested a hybrid origin for *C. lucorum,* and Cronquist included with it *C. apiomorpha, C. fretalis,* and *C. merita*—all of which have fewer stamens.

C. tortilis Ashe differs from *C. basilica* in having larger flowers (18–20 mm across) and glandular-serrate calyx lobes. One collection from a creek bank in Midland County (*Dreisbach 7119,* MICH) was determined by Palmer as questionably this species. It differs, however, in having the calyx lobes essentially entire. According to Kruschke, *C. schuettei* Ashe is the same species, and an older name. Phipps and Muniyamma include *C. basilica* in *C. schuettei* but do not deal with *C. tortilis*. Our specimen, at least, seems to belong to the *C. lucorum* complex of possibly hybrid forms.

C. fretalis Sarg. is likewise doubtfully admitted to our flora on the basis of a specimen with very few mature fruits from the East Lansing area, collected in 1898 (no collector, MSC). This species is normally distinguished by large flowers and glandular-serrate calyx lobes, as in the preceding, but

538. Crataegus dilatata

539. Crataegus coccinea

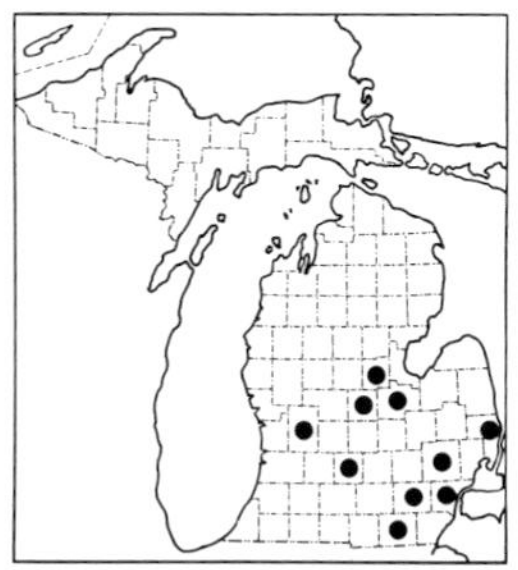

540. Crataegus lucorum

differs in having 10 or fewer stamens and usually 3 styles (*C. tortilis* usually has 5 styles). Kruschke listed *C. fretalis* as a synonym of *C. macrosperma* var. *matura*.

C. miranda Sarg. (TL: Sarnia, Ontario, but paratypes from Port Huron) has the flowering corymbs slightly villous to glabrate, but hardly differs from *C. lucorum*. (Specimens with more glabrous corymbs might run to *C. basilica*.) According to Dodge's notes, Sargent originally thought the type trees of *C. miranda* to be the same as his *C. asperata*; the resemblance is indeed very close. Kruschke referred *miranda* to *C. fulleriana* in the Coccineae and called attention to the inconspicuous glands on the fine leaf teeth, in contrast to the larger glands and coarser teeth of *C. lucorum*.

35. **C. basilica** Beadle

Included here is *C. taetrica* Sarg. (TL: north of Port Huron [Clyde Tp., St. Clair Co.]). A collection from Houghton County (*Richards 3040,* MICH) determined as this by Palmer is not mapped with this complex, as the leaf blades are ± cuneate-based and the calyx lobes nearly entire, suggesting the *C. immanis* complex.

36. **C. macrosperma** Ashe Fig. 195

Map 541 (as the *C. flabellata* complex; see also note under the Silvicolae below). Frequently in dry sandy ground, with jack pine, oak, and/or aspen; rocky ground and summits; deciduous woods and river banks, even with conifers; roadsides, fencerows, clearings, fields, and pastures.

In addition to the typical variety (including *C. prona* Ashe, TL: Port Huron [St. Clair Co.]; and *C. otiosa* Ashe, TL: "Summerville" [St. Clair], St. Clair Co.), Palmer referred some of the Michigan specimens to the following: var. *acutiloba* (Sarg.) Eggleston, var. *demissa* (Sarg.) Eggleston, var. *matura* (Sarg.) Eggleston (including *C. retrusa* Ashe, TL: "Summerville" [St. Clair], St. Clair Co.), and var. *roanensis* (Ashe) Palmer (including *C. multifida* Ashe, TL: Port Huron [St. Clair Co.], the type considered by Sargent to be the same as his *C. streeterae*). The calyx lobes are only slightly or not at all glandular-serrate in this species, except sometimes in var. *acutiloba* and in var. *eganii* (Ashe) Kruschke (to which Kruschke has referred some Michigan specimens).

These entities were considerably rearranged by Kruschke, who transferred *prona* and *demissa* to the Silvicolae (associated with *gravis* and *stolonifera,* respectively) and recognized *C. roanensis* as a distinct species (still including *multifida,* although a collection from the type tree was determined by him as *C. macrosperma* var. *matura*). The other varieties here mentioned he left with *C. macrosperma*—with which I would propose to include all of our Silvicolae anyway, under the older name *C. flabellata* (Spach) Kirchner, which in a narrower sense applies to a species ranging from Georgian Bay to southern Quebec, New England, and New York. This common and variable species complex, then, might also include *C. jesupii* and *C. pruinosa* var. *dissona,* 10-stamened members of the Pruinosae (although these are not included on the map). Some of the taxa in this large assemblage may represent hybrids or other ambiguous states, such as *C. fretalis, C. compta,* and *C. macrosperma* var. *acutiloba,* all with glandular-serrate calyx lobes; and *C. merita* and *C. pruinosa* var. *dissona,* with young leaves nearly glabrous above.

37. **C. merita** Sarg.

Known only from the type locality (near Grand Rapids [Kent Co.]). Palmer suggested that this may be a hybrid of *C. macrosperma* (from which it is scarcely distinguishable, although the leaves are nearly glabrous above when young) and one of the Coccineae; Kruschke included it in the synonymy of *C. apiomorpha.*

15. SILVICOLAE

The Silvicolae constitute a group of exceptional difficulty, even in this difficult genus. The problems of placing species in the Silvicolae and the Tenuifoliae are well illustrated by the divergent treatments cited for many of the taxa. Most post-Palmer authors, no matter how narrowly or broadly they define species, do not treat the Silvicolae as a recognizable series or section.

The key below indicates the tendencies as given by Palmer which supposedly distinguish the species, but I have not been able to detect any consistent differences among our few specimens (which do not, of course, always show the characters desired, seldom having both flowers and fruit). The tips of the lateral lobes on the leaves are slightly to distinctly spreading or recurved on most of these specimens, although only *C. beata* of this group is ordinarily said to have this character, which supposedly is typical of the Tenuifoliae. Thus only the texture of the ripe fruit (and a tendency in many species for it to be definitely oblong or pyriform in shape) appears to remain as a good criterion for the group. If this characteristic is considered of sufficient importance, all of these species might be grouped under a single name (*C. populnea* and *C. brumalis* being the oldest names in the group for such a complex). However, one might with reason include all of our material (except for a few apparent misidentifications) with the very variable *C. macrosperma* in one grand species complex characterized by 10 stamens with pink to red or purple anthers, essentially eglandular sepals, glabrous corymbs, young leaves strigose above and glabrous beneath, and leaf bases rounded to truncate or subcordate. Incidentally, under *C. macrosperma,* including all varieties, Palmer already cited 70 synonyms! For such a complex, following Cronquist, the name *C. flabellata* can be used (see species 36).

1. Fruit obovoid or oblong, longer than broad
  2. Vegetative leaves sharply lobed, often deeply incised (sometimes halfway to middle) . . . . 38. **C. filipes**
  2. Vegetative leaves ± lobed, usually rather shallowly rather than deeply incised
    3. Fruit obovoid to pyriform, gradually narrowed below the middle, up to 1 cm thick; lateral lobes of leaves small and ± acuminate . . . . 39. **C. levis**
    3. Fruit oblong-obovoid, rounded or slightly narrowed at the base, up to 1.8 cm thick
      4. Sepals ± glandular-serrate, at least at the middle; lateral lobes of leaves often acute and prominent . . . . 40. **C. compta**
      4. Sepals entire; lateral lobes of leaves rather broad and shallow, ± convex-sided . . . . 41. **C. gravis**
1. Fruit subglobose or short-oblong, usually as broad as long or broader

5. Flowers 1.2–1.5 cm wide; mature leaves thin
6. Leaves very finely serrate, with narrow acuminate teeth; corymbs mostly 3–7-flowered .... 42. **C. iracunda**
6. Leaves with coarser acute broad-based teeth; corymbs loose, mostly 6–12-flowered .... 43. **C. brumalis**
5. Flowers 1.5–2 cm wide; mature leaves thick or firm
7. Fruit 1.3–1.6 cm in diameter .... 44. **C. beata**
7. Fruit 0.9–1.2 cm in diameter .... 45. **C. populnea**

38. **C. filipes** Ashe

Yet another taxon originally described from Michigan (TL: "Port Huron" [Clyde Tp., St. Clair Co.]). Two collections from Menominee County were referred here by Palmer, but Kruschke called them *C. chrysocarpa* and I concur, as the floreal leaves are narrowed at the base, the corymbs and under surfaces of the leaves are ± villous, and the (immature) fruits are subglobose with glandular-serrate calyx lobes. Palmer also placed here a collection from Houghton County, which appears to be *C. irrasa* (corymbs and bases of veins sparsely villous, calyx lobes glandular-serrate, stamens ca. 15, and styles 5). These reidentifications are also supported by the distribution patterns of these species in the state. However, even the remaining specimens of *C. filipes*—all from the southern half of the Lower Peninsula—are quite varied in appearance. Kruschke made *C. filipes* a synonym of *C. jesupii,* which he transferred to the Silvicolae from the Pruinosae (see species 24).

39. **C. levis** Sarg.

The original description states that the calyx lobes are serrate; in our two collections (Emmet and Newaygo cos.), they are scarcely if at all so. Kruschke included *C. levis* in the synonymy of the next species.

40. **C. compta** Sarg.

Included here by Palmer are *C. allecta* Sarg. (TL: Grand Rapids [Kent Co.]), which differs in having slightly smaller fruit and calyx lobes frequently entire; *C. ambitiosa* Sarg. (TL: east of Grand Rapids, Kent Co.); and *C. comparata* Sarg. (TL: east of Grand Rapids, Kent Co.). The latter two have 20 stamens, rather than 10 as in *C. compta,* and so were maintained by Kruschke (as *C. ambitiosa*). Phipps and Muniyamma also include *C. levis* in *C. compta,* which they place in their new series Suborbiculatae (see comments following species 13).

41. **C. gravis** Ashe

The original description, followed by later authors, states that the fruit is 7–9 mm in diameter. However, Dodge's notes on type plants (TL: Port Huron [St. Clair Co.]) include traced outlines of sections of subglobose fruit 1–1.5 cm thick. A collection from Grand Rapids is referred here with a question by Palmer; it has flowers ca. 23 mm broad (quite large!) and 20 stamens, but otherwise resembles this species. Palmer included here *C. remota* Sarg. (TL: east of Grand Rapids, Kent Co.) although a collection from the paratype tree (*Cole 44*) was determined by him as *C. pruinosa* var. *dissona* (young leaves completely glabrous above, unlike *C. gravis*); specimens from plants grown at the Arnold Arboretum from seed of the type tree of *C. remota* (*Cole 86*) also have leaves glabrous, as noted in the original description of the species. *C. remota* was originally placed in the Pruinosae. Kruschke adopted the name *C. prona* Ashe (see under *C. macrosperma*) for this species and included in the synonymy both *C. allecta* (see under *C. compta* above) and *C. remota*.

42. **C. iracunda** Beadle

In our only collection determined as this by Palmer, from hills by Penoyer Pond, Newaygo (*Bazuin 6987,* MSC), the teeth are hardly fine or acuminate, as described in the key, but the fruiting inflorescence is apparently very few-flowered. The leaves are rather thick and firm. The specimen was placed in var. *silvicola* (Beadle) Palmer, which Kruschke included with typical var. *iracunda*. Altogether, Kruschke offered numerous realignments in this species—and in the Silvicolae as a whole. *C. stolonifera* (see under *C. populnea*) he treated as a variety of *C. iracunda,* with *C. macrosperma* var. *demissa* as a synonym; *C. populnea* and *C. brumalis* also as varieties of *C. iracunda* (but the new combinations nomenclaturally illegitimate).

43. **C. brumalis** Ashe

Credited to Michigan in the manual treatments. Our only specimen, a flowering one from East Lansing (*Wheeler* in 1902, MSC), has been determined as questionably this by Palmer.

44. **C. beata** Sarg.

A Dodge collection with flowers and fruit from Port Huron represents var. *opulens* (Sarg.) Palmer (distinguished from the typical variety by stamens 10 or fewer rather than 15–20 and hence maintained as a distinct species by Kruschke). Even the dry flowers are over 1.5 cm wide, though they are described as smaller in this variety. The dry fruits are small, but Dodge's notes indicate a diameter of 1.4–1.7 cm. (Dodge's notes also show that Ashe first referred this tree to *C. filipes* and that Sargent first referred it to *C. gravis* before concluding that it was his *C. opulens*. This suggests well the ambiguity characteristic of the Silvicolae!) Also referred here by Palmer is a collection with old dry fruit from Oakland County (*Farwell 3270,* MICH, BLH) and a sterile collection from Allegan County (*Bazuin 7717,* BLH) without designation of variety. Phipps and Muniyamma include *C. beata* in their new series Suborbiculatae but do not treat *C. opulens*.

45. **C. populnea** Ashe

As suggested by the name, the leaves of this plant (as of some others in this group) bear some resemblance to those of *Populus* spp. Included here is *C. perlaeta* Sarg. (TL: near Grand Rapids [Kent Co.]), with young leaves nearly glabrous above and the calyx lobes sometimes with a few glandular teeth toward the apex. Our only collections are from Mecosta and St. Clair counties.

C. stolonifera Sarg. differs from *C. populnea* in having vegetative leaves with the lowest pair of lobes separated by relatively deep sinuses. Two fruiting collections from Ottawa County referred here are probably not this species. One of them, with 20 stamens, Kruschke has called *C. pruinosa,* and the other he has called *C. jesupii,* which it matches closely except for a few small hairs on the leaves above.

11. **Malus** — Apple

This genus is included in *Pyrus* perhaps more often than the other segregates (*Sorbus, Aronia*). There are several cultivated taxa which brighten gardens and roadsides in the spring (especially the "flowering crabs"). In addition, the closely related flowering quinces (*Chaenomeles* spp.) and the true quince (*Cydonia oblonga* Miller) are widely grown. These may persist after cultivation but seem not to be spreading.

The apple blossom was made the official state flower of Michigan in 1897, by a resolution of the State Legislature that explicitly mentioned the native crab apple (as *Pyrus coronaria*) as "one of the most fragrant and beautiful flowered species of apples." So evidently *any* apple, not merely the commercially important crop, represents the state flower.

The petals in our species are ca. 13–24 mm long.

KEY TO THE SPECIES

1. Leaves all of similar shape, unlobed, closely and regularly singly or doubly serrate (or crenate-serrate); anthers yellow; fruit often over 4 cm in diameter . . . . 1. **M. pumila**
1. Leaves, especially on sterile shoots, usually with at least shallow acute lobes and sinuses, as well as coarse ± irregular teeth; anthers pink; fruit less than 3 [4] cm in diameter
    2. Calyx lobes (outside), pedicels, petioles, and leaf blades beneath retaining tomentum well past flowering time—even into fruit . . . . 2. **M. ioënsis**
    2. Calyx lobes (outside), pedicels, petioles, and leaf blades beneath glabrous or if tomentose becoming glabrate toward end of anthesis . . . . 3. **M. coronaria**

1. **M. pumila** Miller — Apple

Map 542. This is the common garden apple, a major fruit crop in Michigan (which stands third in the nation in apple production) and frequently spread to roadsides, shores, railroad grades, fields, waste places, and even wooded areas. Human beings discarding cores are doubtless a major agent of dispersal.

Originally native in Eurasia, the domestic apple has been greatly modified—and diversified—by culture and hybridization, so that its taxonomy is obscure. The name *M. pumila* is used here in accordance with *Hortus Third* and many manuals in a sense that includes *M. domestica* Borkh., which *Flora Europaea* segregates as the common apple. In the genus *Pyrus,* the apple is called *P. malus* L. in most works, but *Hortus Third* treats this as a synonym of *Malus sylvestris* Miller, one of the crab apples.

The leaves are usually ± tomentose beneath even when mature.

2. **M. ioënsis** (Wood) Britton Prairie Crab

Map 543. The range of this species is usually thought to be west of Michigan, and the Emmet County station, especially, at the edge of woods not far from the old Indian center of Cross Village, may represent an introduction by native Americans. Also collected along a small stream in Berrien County (*Dodge* in 1917, MICH, NY).

A double-flowered cultivar is known as the Bechtel crab.

3. **M. coronaria** (L.) Miller Wild, American, or Sweet Crab

Map 544. Rich deciduous woods and borders to dry oak-hickory or (Oscoda Co.) jack pine woodland; stream banks, roadsides and fencerows, sandy prairie-like ground.

A variable species, sometimes with the parts nearly as pubescent as in the preceding, but the loose flocculent tomentum very quickly shed. When there are "spines," as there usually are, they reveal much more than in *Crataegus* their morphological nature as modified branchlets. In *Malus,* the spines often bear full-sized leaves or leaf scars, and have a more knobby appearance than normally found in *Crataegus,* which this species (and the preceding) might superficially resemble. Occasionally the leaves may have a few red glands on the midrib above, as in *Aronia,* but the much larger, pink

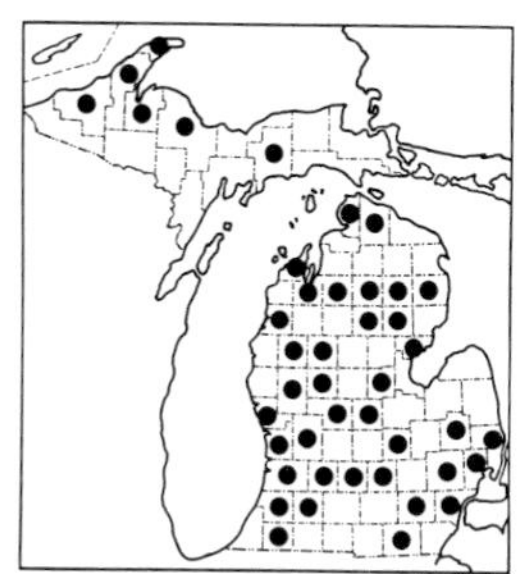

541. Crataegus flabellata

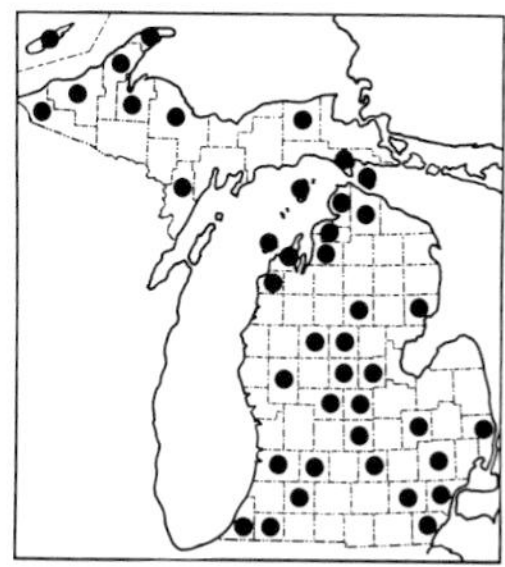

542. Malus pumila

543. Malus ioënsis

flowers and shallowly lobed, coarsely toothed leaves readily distinguish the *Malus*.

A hybrid with *M. pumila,* different from the latter in slightly lobed leaves, is known as *M. ×platycarpa* Rehder and might be expected in the state.

12. **Pyrus**

For genera often included in this genus, see *Sorbus, Aronia,* and *Malus.*

1. **P. communis** L. Pear

Map 545. The origins of the orchard pear are lost in antiquity and it is not certain which species of wild pear are its Old World ancestors. Countless cultivars are grown and occasionally a discarded core or other means of dispersal will result in a tree strayed from cultivation to fencerows, clearings, and other disturbed sites. Pears also spread by suckering, forming dense thickets at old orchards or farmyards.

Pear leaves are shinier than those of apple, with a much closer more distinct pinnate venation, and are glabrous at maturity.

13. **Dalibarda**

The single species in this genus was sometimes included in *Rubus,* as *R. repens* (L.) Kuntze. Linnaeus himself thought the plant was closely allied to the circumpolar *Rubus chamaemorus* (cloudberry), an essentially herbaceous, functionally dioecious species with ± 5-lobed simple leaves, solitary flowers, and yellowish orange very juicy fruit composed of large drupelets; it is known as far south as the north side of Lake Superior but is not known or reported from Michigan. Otherwise, any Rosaceous plant with simple leaves should be sought on the woody side of the key to genera, for *Dalibarda* is our only simple-leaved herbaceous species.

REFERENCE

Gorton, Peter B., Daniel & Karyn Townsend. 1977. Dalibarda repens Confirmed in Michigan. Mich. Bot. 16: 189–190.

1. **D. repens** L. Plate 5-E False Violet; Dewdrop

Map 546. There are several old published reports from Michigan, but none of them (from Cheboygan, Emmet, Macomb, and Washtenaw counties) are supported by extant specimens; however, there *is* material from Antrim County collected in 1894 (*O. E. Close,* MSC). The species was apparently not collected in the state again until 1975, in Hartwick Pines State Park, where it is locally frequent in moist coniferous woods (pines, spruces, and fir).

The plants are strongly stoloniferous, with leaves somewhat like those of many violets—or *Mitella nuda* (even to the stiff hairs on the upper side of the blade), but the margin is more finely crenulate and the petioles are densely hairy with prominent stipules at their bases. In addition to the white petaliferous flowers borne on long ascending peduncles, there are apetalous cleistogamous flowers on peduncles shorter than the petioles; these are usually the only ones that set fruit, which consist of dryish drupelets enclosed by the calyx (see Robertson 1974, p. 361),

14. **Porteranthus**

This genus is often called *Gillenia,* which may be treated as a later homonym of a different name.

REFERENCE

Ballard, Harvey E., Jr. 1985. Porteranthus trifoliatus, Bowman's Root, Verified in the Michigan Flora. Mich. Bot. 24: 14–18.

KEY TO THE SPECIES

1. Stipules linear, entire or nearly so, very inconspicuous; fruit pubescent .... 1. **P. trifoliatus**
1. Stipules broadly ovate, deeply toothed, conspicuous; fruit glabrous or nearly so .... 2. **P. stipulatus**

1. **P. trifoliatus** (L.) Britton   Fig. 196   Bowman's Root

Map 547. Although earlier reported from Michigan, no authentic specimens seem to exist prior to discovery of the species by Ballard in 1978 on the border of a clearing in oak woods in Schoolcraft Tp., Kalamazoo County.

Distinctive in several ways among our trifoliolate Rosaceae. The leaves are nearly sessile, the leaflets acuminate, double serrate with acuminate gland-tipped teeth.

2. **P. stipulatus** (Willd.) Britton   American Ipecac

Map 548. The only Michigan records are two collections (BLH) by Far-

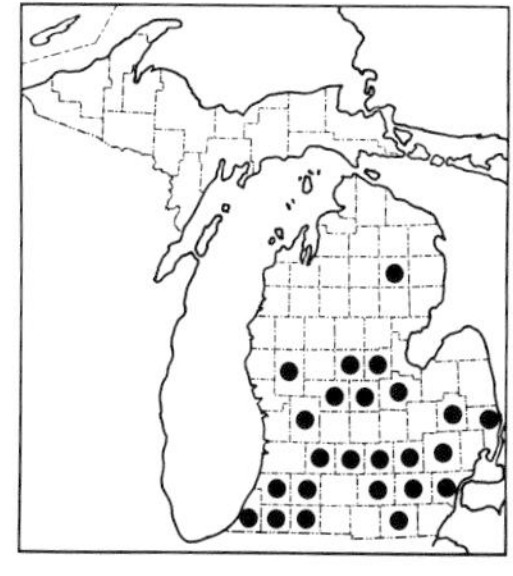
544. Malus coronaria

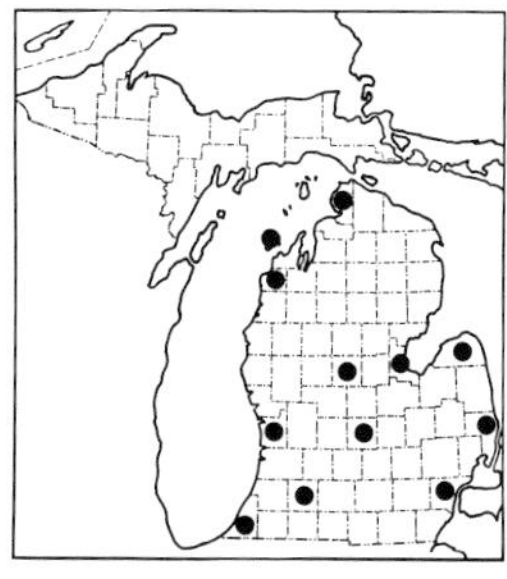
545. Pyrus communis

546. Dalibarda repens

well at Detroit in 1900 (Belle Isle) and 1902; the latter was reported by him from "fields" and misidentified as *P. trifoliatus*.

Very similar to the preceding except for the stipules, which almost suggest additional leaflets.

15. **Waldsteinia**

1. **W. fragarioides** (Michaux) Tratt. Fig. 197 Barren-strawberry

Map 549. Rather local, but often abundant when found. In northern Michigan, generally in open thickets or woodland, including thin soil over rock outcrops, with conifers and/or aspen and birch; in the southern part of the state, mostly in low deciduous woods.

Although this plant somewhat resembles a wild strawberry in habit, the leaflets are obtuse or rounded (not acute), the achenes and receptacle are dry, and the petals are yellow; bractlets in the calyx, if present at all, are minute and not all developed. The leaves are ± winter-green, old ones being evident with the new growth at flowering time (May–June).

16. **Duchesnea**

1. **D. indica** (Andrews) Focke Fig. 198 Indian-strawberry

Map 550. A native of Asia, sometimes planted in the U.S. as a ground cover, and locally established as a weed at the edges of gardens, walks, and streets.

The flowers are yellow but this species otherwise resembles a strawberry, with extensively developed runners or stolons. However, the enlarged receptacle, dotted with red achenes, lacks flavor or aroma.

17. **Fragaria** Strawberry

At fruiting time, the enlarged red juicy receptacle of these plants, dotted with tiny hard achenes ("seeds"), is a favorite with many people. Our common wild strawberry makes up in flavor what it lacks in size, and is justly

547. Porteranthus trifoliatus

548. Porteranthus stipulatus

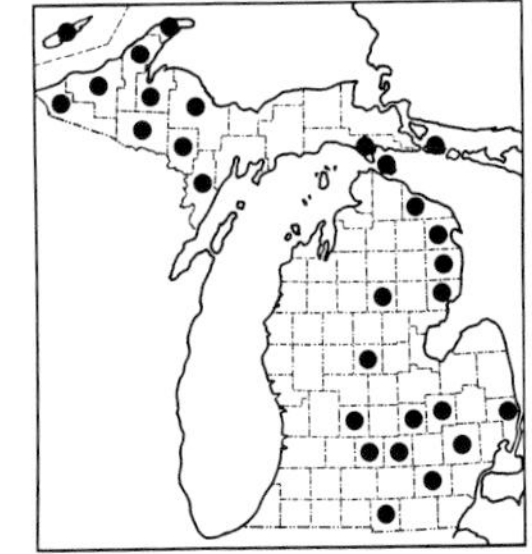

549. Waldsteinia fragarioides

rated as the best. Most cultivated strawberries are derived from the hybrid [*F.* ×*ananassa* Duchesne (1788)] between *F. virginiana* and *F. chiloënsis* (L.) Miller, a western American (and also octoploid) species. Garden strawberries may persist after cultivation, reported as such by Hanes to be flourishing in Kalamazoo County. They were also found once by Farwell (*12457* in 1940, BLH) as a presumed escape in Keweenaw County; and once by B. Hitt in Wayne County near Plymouth (in 1961, MCTF). They differ from *F. virginiana* in having achenes barely if at all sunken on the ripe receptacle, which is much larger; in coarser foliage; and in consistently perfect as well as larger flowers (those of *F. virginiana* sometimes being unisexual). Michigan's commercial strawberry crop ranks fourth in the nation. Its wild crop is first in the hearts of many of us.

REFERENCES

Guédès, M. 1981. No Species Names for Strawberries Ever Published by A. N. Duchesne. Taxon 30: 299. [See also Taxon 33: 724–726. 1984.]

Robertson, Kenneth R. 1974. Fragaria, pp. 362–371 in Robertson (1974).

KEY TO THE SPECIES

1. Mature fruiting receptacle ± conical, the achenes not sunken but projecting from the surface; calyx lobes widely spreading to reflexed from the developing fruit; terminal tooth of leaflets equaling or exceeding the adjacent pair of lateral teeth; hairs on peduncles and petioles usually ± appressed. . . . . 1. **F. vesca**
1. Mature fruiting receptacle ± ovoid to globose, the achenes sunken in depressions; calyx lobes appressed to the developing fruit; terminal tooth of leaflets reduced, usually surpassed by the adjacent pair of lateral teeth and only half (or less) as broad; hairs on petioles (and, often, peduncles) ± strongly spreading . . . . 2. **F. virginiana**

1. **F. vesca** L. Woodland Strawberry

Map 551. In rich hardwoods, mixed woods, and swamps; edges of woods and shores, openings in coniferous woods, cedar and tamarack swamps; rocky woodland and damp ledges. Also grows in Eurasia.

550. Duchesnea indica

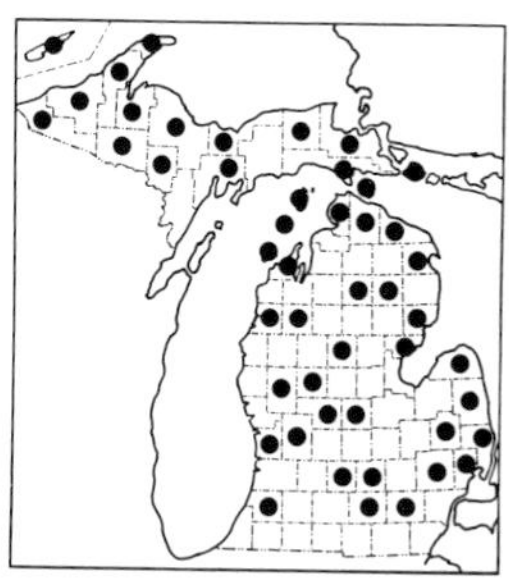
551. Fragaria vesca

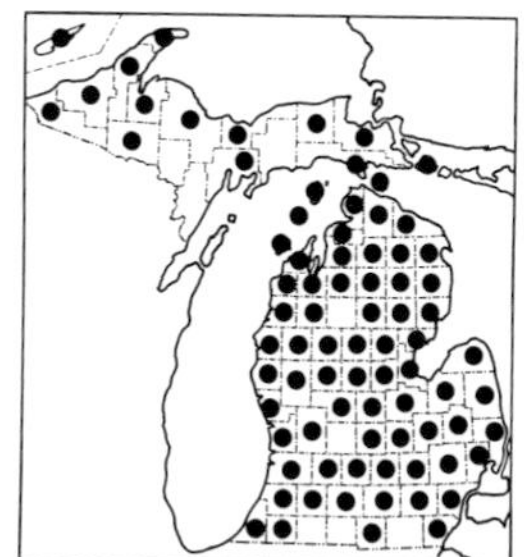
552. Fragaria virginiana

Plants of eastern North America may be recognized as var. *americana* Porter or ssp. *americana* (Porter) Staudt. As this strawberry is sometimes difficult to distinguish from the next if fresh mature fruit (or a chromosome count) is not available, one must balance several characters none of which alone is adequate for discrimination. Ambiguous specimens cannot readily be assumed to be hybrids, however, for *F. vesca* is diploid ($2n = 14$) and *F. virginiana* is octoploid ($2n = 56$).

2. **F. virginiana** Miller Fig. 199 Wild Strawberry

Map 552. Nearly ubiquitous: like *F. vesca,* in a diversity of deciduous, mixed, and coniferous woods and (not-too-wet) swamps, clearings, and shores; but also common in dry sandy woods (jack pine, oak, etc.), roadsides, and fields; along railroads, on dry rocky summits and bluffs, and in grassy places. While more often in dry open sunny places than the preceding, may grow with it.

Flowering specimens, especially, of the two species are not always easy to separate. Various authors have stressed different characters, but all are subject to more exception than one would like. Even more than for most other genera, a judgment based on an assemblage of characters is necessary. The mature leaflets in *F. virginiana* are ± dark or bluish green when fresh and (especially the terminal one) usually distinctly petiolulate; the peduncle at maturity is generally shorter than the longest petioles; the flowers tend to be larger (petals up to 12 mm long). In *F. vesca* the leaflets are a brighter green and sessile; the peduncle at maturity usually exceeds even the longest petioles; and the petals are less than 7 mm long (often as short as 4 mm or even smaller, but in *F. virginiana* they may also be less than 7 mm). After the petals have fallen, the attitude of the calyx lobes seems to be as dependable a character as any for distinguishing the species. It does not seem worthwhile here to consider the varieties (or subspecies) sometimes recognized.

18. **Potentilla** Cinquefoil; Five-finger

This is a complex and difficult genus, including diploids and polyploids, complicated by hybridization and apomixis. The genus is here treated in a broad sense, but various segregates have often been recognized and may be noted below. *Duchesnea* is closely allied to *Potentilla* and is sometimes included in it.

REFERENCES

Kohli, B., & John G. Packer. 1976. A Contribution to the Taxonomy of the Potentilla pensylvanica Complex in North America. Canad. Jour. Bot. 54: 706–719.

Werner, Patricia A., & Judith D. Soule. 1976. The Biology of Canadian Weeds. 18. Potentilla recta L., P. norvegica L., and P. argentea L. Canad. Jour. Pl. Sci. 56: 591–603. [Includes data from Michigan populations.]

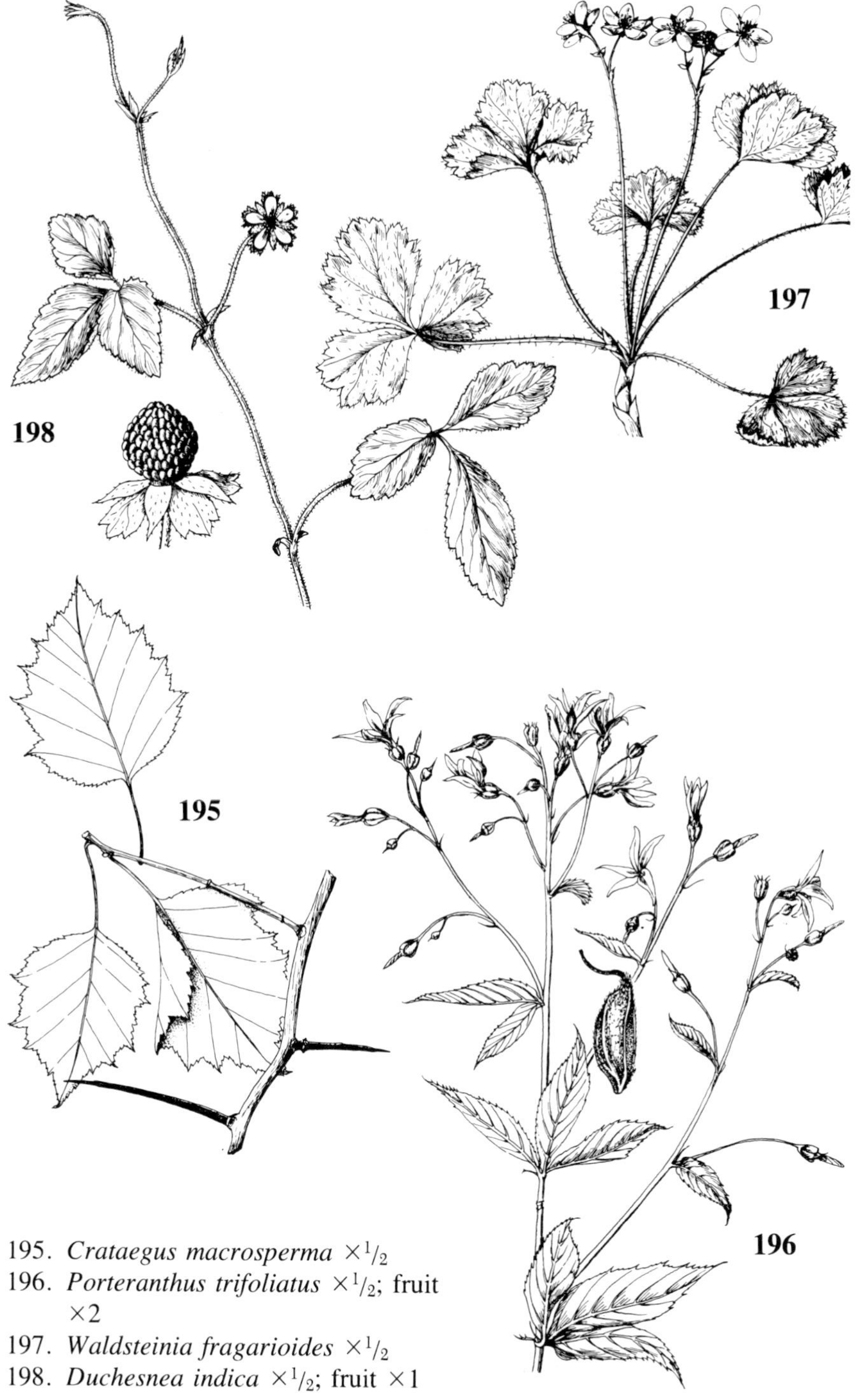

195. *Crataegus macrosperma* ×1/2
196. *Porteranthus trifoliatus* ×1/2; fruit ×2
197. *Waldsteinia fragarioides* ×1/2
198. *Duchesnea indica* ×1/2; fruit ×1

KEY TO THE SPECIES

1. Plant a bushy shrub; ovaries and achenes hairy; leaves pinnately compound, the leaflets entire and less than 6 (very rarely 7.5) mm broad . . . . 1. **P. fruticosa**
1. Plant herbaceous (at most, woody at base); achenes glabrous (except in *P. tridentata*); leaves variously compound but the leaflets toothed or lobed
  2. Leaves (at least basal and lower cauline ones) pinnately compound
    3. Petals deep maroon to purple, ca. half or less as long as the red-tinged sepals; stem usually decumbent, the lower portion in water or wet ground, rooting at nodes but without slender stolons (runners); leaves all cauline, with 5 (–7) leaflets . . . . 2. **P. palustris**
    3. Petals pale to deep yellow; stem usually erect (if decumbent in common species, stolons slender and elongate), not normally in water; leaves various
      4. Leaves all in basal tufts, the plant spreading by slender elongate stolons; flowers solitary on basal peduncles; leaflets mostly 13–21 (plus some alternating tiny ones), strongly whitened beneath with silvery-silky hairs . . . . 3. **P. anserina**
      4. Leaves all or mostly cauline, the plants without stolons; flowers in cymose inflorescences; leaflets 5–11, if whitened beneath then at least partly tomentose
        5. Lower surface of leaflets ± strongly whitened or silvery gray with mixture of dense tomentum and longer straight hairs
          6. Calyx lobes with ± dense sessile to short-stipitate glands (sometimes obscured by dense long hairs); style prominently glandular-warty on at least its basal half; stipules glandular (like the sepals), lanceolate to ovate, usually coarsely few-toothed or lobed . . . . 4. **P. pensylvanica**
          6. Calyx lobes without glands; style at most glandular only at very base; stipules eglandular, lanceolate, entire or nearly so . . . . 5. **P. hippiana**
        5. Lower surface of leaflets with green not obscured by pubescence, which is whitish to yellowish but not tomentose
          7. Plant much-branched, the stem often decumbent; petals ca. 3–5 mm long; pubescence eglandular; achenes at maturity bearing a corky appendage nearly as large as the body . . . . 6. **P. paradoxa**
          7. Plant simple, erect; petals ca. (5.5) 7–10 mm long; pubescence glandular-viscid; achenes without corky appendage . . . . 7. **P. arguta**
  2. Leaves all (except sometimes the uppermost bracts) palmately compound
    8. Principal leaves with only 3 leaflets
      9. Leaflets entire except for 3- (–5) toothed apex, evergreen; petals white, exceeding the sepals; ovaries and achenes pubescent, the achenes otherwise smooth . . . . 8. **P. tridentata**
      9. Leaflets toothed along their margins, not evergreen; petals yellow, about equaling the sepals or shorter; ovaries and achenes glabrous, the mature achenes usually rugose . . . . 9. **P. norvegica**
    8. Principal leaves (lower cauline and basal) with 5 (–7) leaflets
      10. Flowers on solitary peduncles; stem at first erect, becoming arching (or stoloniferous) and rooting at tip . . . . 10. **P. simplex**
      10. Flowers in cymes; stems various, erect or prostrate to ascending but without stolons and not rooting at tips or nodes
        11. Leaflets very *white*-tomentose beneath (with any straight silky hairs also white and almost entirely restricted to the main veins); achenes smooth (or rarely with obscure pattern)
          12. Stems erect; petals ca. 5[6–10] mm long, distinctly exceeding the sepals; leaflets deeply pinnatifid with numerous segments . . . . 11. **P. gracilis**

12. Stems usually prostrate to ascending; petals (2) 3–4.5 (5) mm long, scarcely if at all exceeding the sepals; leaflets sparsely but coarsely toothed to pinnatifid with (1) 2–3 (5) teeth per side .... 12. **P. argentea**

11. Leaflets not tomentose (though often glandular) or with fine to dense dull grayish tomentum besides long hairs beneath; achenes at maturity ± strongly rugose, with irregular curved ridges on surface

13. Leaflets with only ± straight hairs; petals (6.5) 8–11 mm long, pale yellow; anthers (at least the largest) 1–1.4 mm long .... 13. **P. recta**

13. Leaflets with gray tomentum beneath in addition to straight hairs; petals (3) 3.5–5.5 (6) mm long, bright yellow; anthers ca. 0.5–0.8 (1) mm long ... . 14. **P. inclinata**

1. **P. fruticosa** L. Shrubby Cinquefoil

Map 553. Wet open ground and thickets around lakes and streams, especially calcareous interdunal flats and rocky shores; bogs and open sedgy conifer swamps (especially with tamarack), an indicator of alkaline conditions (fens), meadows.

The numerous bright yellow flowers make this an attractive shrub for cultivation, and it is sometimes grown as an ornamental. Tetraploid plants of northern Europe and Asia represent typical ssp. *fruticosa*; diploid plants from North America and southern Europe east to Kamchatka may then be called ssp. *floribunda* (Pursh) Elkington. Or the shrubby cinquefoils may be segregated into a separate genus, *Pentaphylloides,* and our plant treated as *P. floribunda* (Pursh) Löve or *P. fruticosa* (L.) O. Schwarz sens. lat.

2. **P. palustris** (L.) Scop. Marsh Cinquefoil

Map 554. Bogs and conifer swamps, swales and marshes, shores and stream borders, usually in quite wet situations.

Sometimes segregated as *Comarum palustre* L.

3. **P. anserina** L. Fig. 200 Silverweed

Map 555. Sandy and gravelly shores of lakes and ponds or occasionally rivers, sometimes on rock outcrops, extending to marshy ground, ditches, and roadsides; especially in damp places, often calcareous or even marly, and most frequent around the Great Lakes shores.

Sometimes segregated as *Argentina anserina* (L.) Rydb. Rarely the leaves are ± silky pubescent above, but it is the distinctive aspect of the lower surface which provides the common name.

4. **P. pensylvanica** L.

Map 556. Rock outcrops and adjacent clearings.

A variable species complex, mostly occurring to the west and north of the Great Lakes. Michigan plants seem more easily recognized by the dense substipitate glands than by the technical characters usually given for the style. Some if not all of our specimens are presumably *P. bipinnatifida* Hooker, recognized by Kohli and Packer (1976) as a distinct octoploid species (*P. pensylvanica* sens. str. is tetraploid); but the morphological characters, such as length of petals and bractlets of the calyx, are ambiguous.

5. **P. hippiana** Lehm.

Map 557. Another variable western American species, presumably adventive here. Collected at a mine site near Calumet (*Farwell 11713* in 1937, BLH, GH). It is also known from the end of Sibley Peninsula, Ontario, north of Isle Royale. (See Rhodora 40: 135. 1938; Mich. Bot. 20: 60. 1981.)

6. **P. paradoxa** Nutt.

Map 558. The only Michigan collection is from loose sand of a barrier beach at Pte. Mouillée (*M. McDonald 5159* in 1949, MICH, MSC).

7. **P. arguta** Pursh — Tall or Prairie Cinquefoil

Map 559. Dry sandy old fields, river banks, prairies, sandy hillsides, shores, and roadsides, often in barren or grassy places; on thin rocky soil and rock summits. Collected in the 1830's from the southern Lower Peninsula, but almost all other collections are from the 20th century and there may be some question whether the plant is native in most of its range in the state.

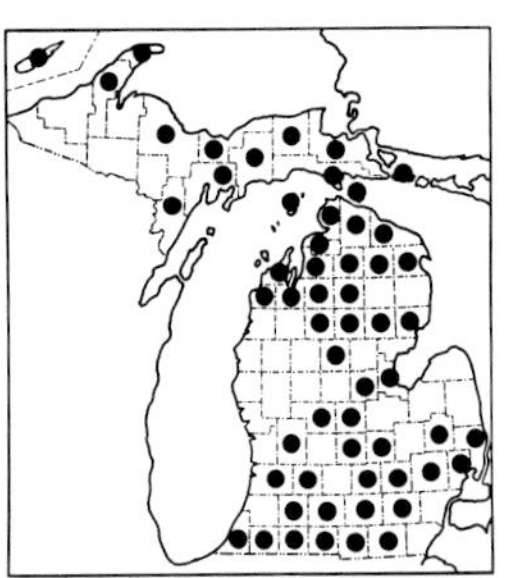

553. Potentilla fruticosa

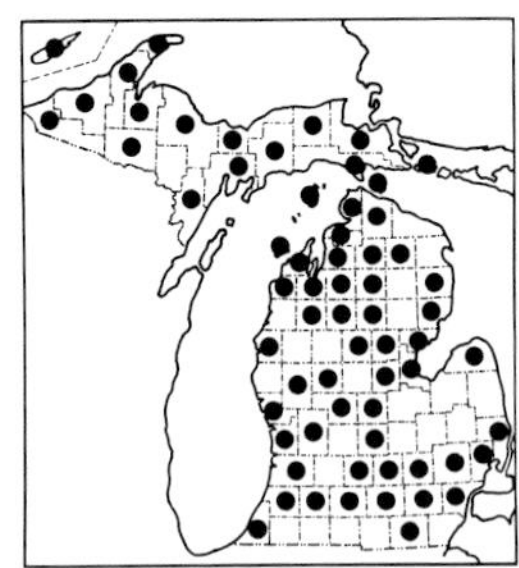

554. Potentilla palustris

555. Potentilla anserina

556. Potentilla pensylvanica

557. Potentilla hippiana

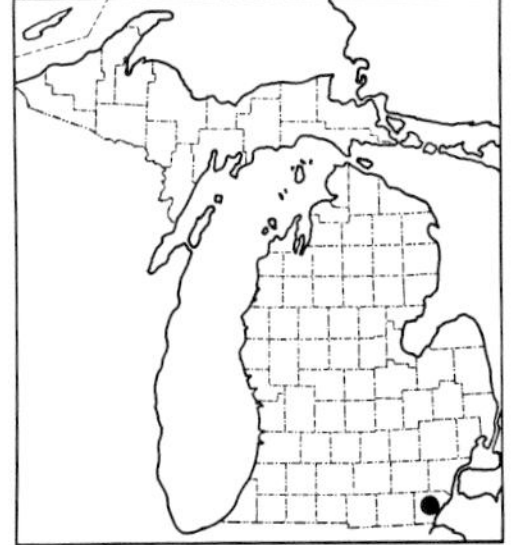

558. Potentilla paradoxa

The petals are pale yellow, cream, or even white. Sometimes segregated as *Drymocallis arguta* (Pursh) Rydb.

8. **P. tridentata** Aiton Fig. 201 Three-toothed Cinquefoil

Map 560. Open sandy ground, such as old dune ridges and glacial outwash plains; dry woodland of jack pine and oak; rock shores, outcrops, and summits. Locally abundant along Lake Superior from Whitefish Point westward, typically with scattered jack pine, blueberries and other ericads, *Deschampsia flexuosa,* and *Hudsonia tomentosa* but thriving in mowed lawns (as at Coast Guard installations); reappearing in the jack pine plains of the northern Lower Peninsula. The Emmet County record is based on plants presumably adventive along a railroad in 1920.

Sometimes segregated as *Sibbaldiopsis tridentata* (Aiton) Rydb.

9. **P. norvegica** L. Rough Cinquefoil

Map 561. Moist or dry, usually ± disturbed ground, including roadsides, railroads, fields, shores, meadows, rock outcrops, gardens; sometimes in quite marshy places or in aspen (or other) woods.

Native North American plants have been segregated from the European ones as *P. monspeliensis* L. [or as *P. norvegica* ssp. *monspeliensis* (L.) Asch. & Graebner or var. *hirsuta* (Michaux) Lehm.]. It is quite likely that both native and European forms now occur here. Plants vary a great deal in their stature and depth of teeth on the leaflets. Slender stems only a few cm tall may flower, while other plants are tall and coarse.

10. **P. simplex** Michaux Common or Old-field Cinquefoil

Map 562. Usually in dry open sandy woods, with oak, hickory, aspen, sassafras, and/or jack pine; fields, roadsides, and sandy barren ground; also in moist thickets and deciduous woods, and on rocky ledges.

Readily recognized by the combination of solitary flowers and palmately compound leaves. Very similar to *P. canadensis* L., which grows east and

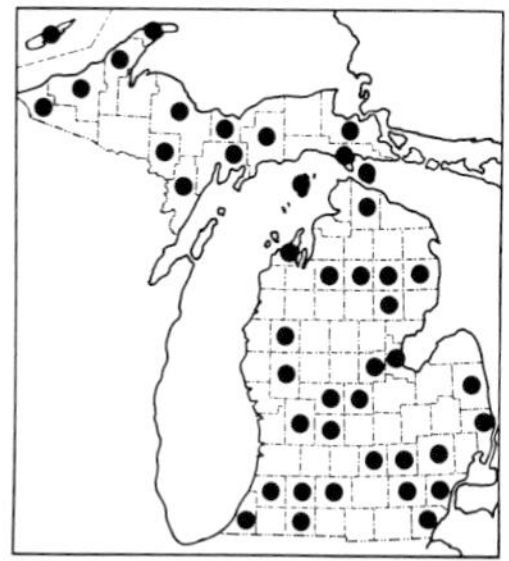

559. Potentilla arguta

560. Potentilla tridentata

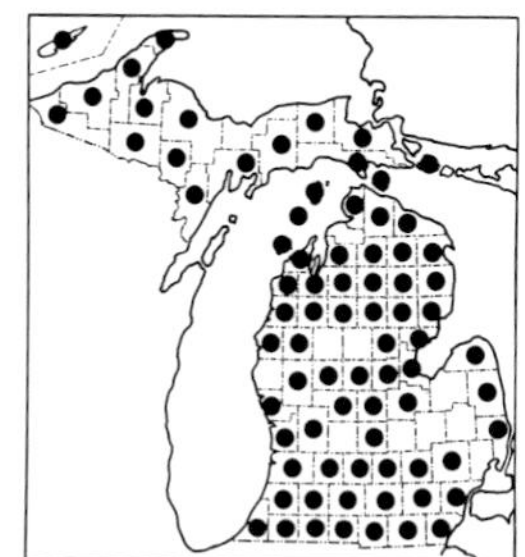

561. Potentilla norvegica

south of our area, and sometimes included in it. Further complicating records is the fact that the name *P. canadensis* was long applied to what is now called *P. simplex,* while the true *P. canadensis* was called *P. pumila* Poiret. (See Fernald in Rhodora 33: 180–191. 1931.) Typically in *P. simplex* the stem is initially ± erect, ca. 1 mm thick or thicker, and the peduncle of the lowest flower arises in the axil of the *second* developed leaf. The stem as it elongates becomes ± arching and essentially stoloniferous. *P. canadensis* is a more delicate plant, the stems less than 1 mm thick (often ca. 0.5 mm) and ± prostrate from the beginning; the peduncle of the lowest flower is in the axil of the lowermost developed leaf. The cauline (stolon) leaves are less than half grown at flowering time and there are said to be subtle differences in shape and toothing from those of *P. simplex*. A very few of our plants approach *P. canadensis* but are here treated as depauperate forms of *P. simplex,* which varies a great deal in habit depending on habitat and season of collection. One collection from Van Buren County (*Pepoon 746* in 1906, MSC) may be good *P. canadensis*.

11. **P. gracilis** Hooker

Map 563. Referred here is the plant reported by Farwell (Rhodora 37: 164. 1935) as *P. blaschkeana* Turcz. from Houghton County, collected by him (*9723,* MICH, BLH, WUD, GH) at Gregoryville (Lake Linden) in 1934 and 1935. The deeply pinnatifid leaflets (on at least some of the material) with narrowly revolute margins indicate *P. gracilis* var. *flabelliformis* (Lehm.) T. & G.—sometimes treated as a distinct species. F. J. Hermann collected from presumably the same colony in 1936 (*7926,* GH), citing the habitat as a dry pasture. Very similar material was found along a railroad in Schoolcraft County in 1981 (*Henson 1326,* MICH).

P. gracilis is overall an extremely variable species. Other variants, the leaflets not pinnatifid although strongly toothed, and sometimes with more straight hairs mixed with the white tomentum, have been collected on the north side of Lake Superior in Thunder Bay District, Ontario. Probably all are adventive from the west in the Great Lakes region.

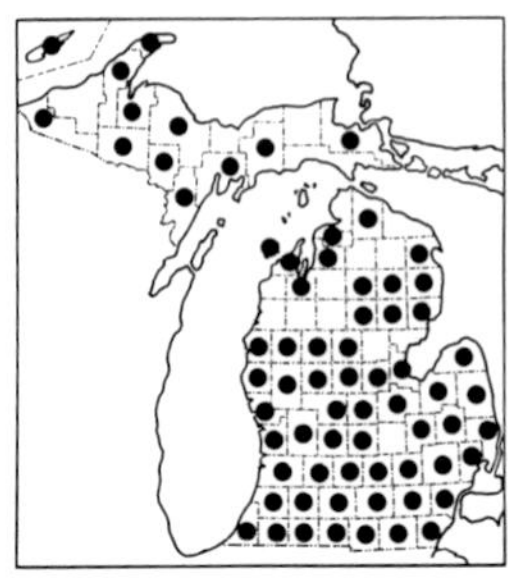

562. Potentilla simplex

563. Potentilla gracilis

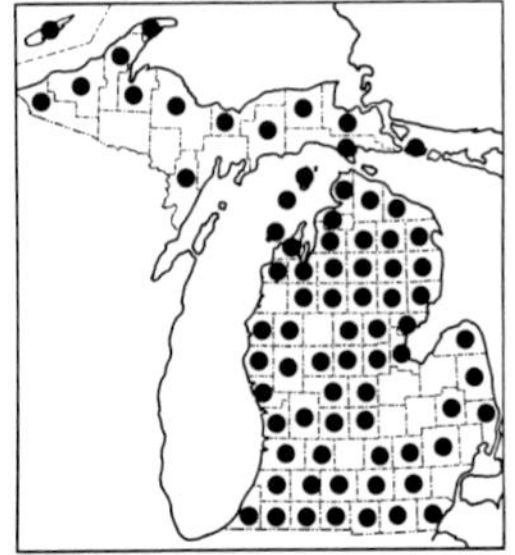

564. Potentilla argentea

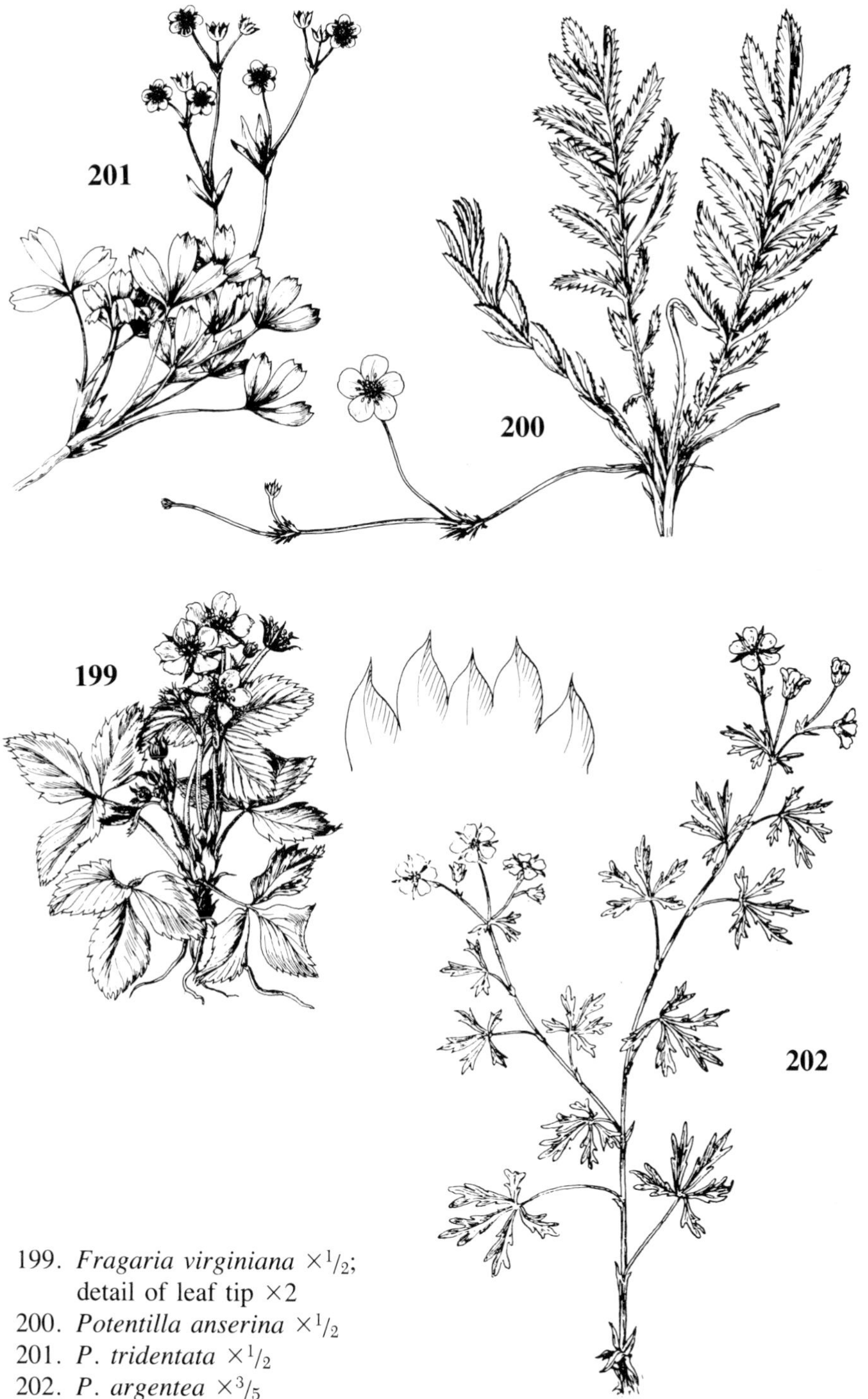

199. *Fragaria virginiana* $\times^1/_2$; detail of leaf tip $\times 2$
200. *Potentilla anserina* $\times^1/_2$
201. *P. tridentata* $\times^1/_2$
202. *P. argentea* $\times^3/_5$

12. **P. argentea** L. Fig. 202 Silvery Cinquefoil

Map 564. Except for an 1866 Lenawee County specimen, the earliest Michigan collections date from the late 1880's, by which time it was evidently well established across the state. Now a common weed of dry roadsides, trails, barnyards, lawns, railroads, fields, gravel pits, and waste ground generally; invading dry sandy upland prairies and woodlands; because of its low stature, withstands mowing well. A polymorphic European species (or complex).

13. **P. recta** L. Fig. 203 Rough-fruited Cinquefoil

Map 565. Originally from Europe, now a common weed of roadsides, fields, railroads, clearings, gravel pits, and dry waste ground; invading dry open woods such as aspen. More recently established and spreading in Michigan than many of our common weeds; first collected in the state in Washtenaw County in 1894.

Numerous sessile or short-stipitate glands can nearly always be found on the lower surface of the leaflets.

14. **P. inclinata** Vill.

Map 566. Another weed of European origin, often known as *P. canescens* Besser, first collected in Michigan in 1896 at Whitmore Lake (Livingston or Washtenaw co.) and apparently well established by the 1920's and 1930's as another weed of sandy roadsides and fields, railroads, and exposed rocks.

The leaflets are very similar in shape to those of *P. recta* (rather narrowly cuneate at the base, rather than the more rounded shape of *P. norvegica*), although averaging a little smaller; but they are gray-tomentose, with intermixed long hairs, on the lower surface. The flowers are smaller than in *P. recta,* with deeper yellow petals (as in *P. argentea*). The intermediate nature of these plants has led to the suggestion that they may be of hybrid origin. Referred here are also most Michigan plants identified in the past as *P. intermedia* L., often considered to be intermediate between *P. argentea* and *P. norvegica*. Although some of these specimens have the small flowers attributed to *P. intermedia* (those of *P. inclinata* supposed to be a little larger—but petals not over 7 mm long), the aspect of the foliage on them is uniform (and like that of *P. recta*) and the flowers are so similar, that they seem best referred to *P. inclinata*. These plants may be found with *P. recta* or *P. argentea* or with both of them, and while sometimes nearly erect are usually intermediate in habit between the other two: ± bushy but not as low as *P. argentea* usually is. It is entirely possible that the plants we have include a mixture of F_1 hybrids and stabilized apomicts. A few relatively recent collections from Ingham and Washtenaw counties have leaves more closely resembling *P. norvegica* than *P. recta* and may indeed represent hybrids between that species and *P. argentea,* as "*P. intermedia*."

19. **Chamaerhodos**

1. **C. nuttallii** Rydb. Fig. 204

Map 567. Known in Michigan only from West Bluff (Brockway Mountain), where it is a relatively inconspicuous and rare low plant on disintegrating conglomerate.

Our plants were named by Fernald as an endemic var. *keweenawensis*, but there is doubt as to their distinctness from plants of the main range of the species (Alaska to Colorado, eastward to northwestern Minnesota). On the other hand, this species is often included in *C. erecta* (L.) Bunge of eastern Asia. This is one of the rarest of our species disjunct from the west (see Mich. Bot. 20: 56. 1981).

20. **Aruncus**

REFERENCE

Mellichamp, Thomas Lawrence. 1976. A Comparative Study of Aruncus (Rosaceae) and Astilbe (Saxifragaceae), and the Problem of their Relationships. Ph. D. thesis, Univ. Mich. 180 pp.

1. **A. dioicus** (Walter) Fern. Goatsbeard

Map 568. Near homesites and borders of woods. Our plants are the Eurasian (and western North American) variant occasionally escaped from cultivation and sometimes called *A. sylvester* Kostel., but better treated as *A. dioicus* var. *acuminatus* (Rydb.) Hara. The typical variety is native mostly in the Appalachian and Ozark regions of eastern North America.

21. **Geum** Avens

A very distinctive genus, especially in young or mature fruit, with a globose to ovoid head of achenes, the basal portion of the style persistent as a long firm beak hooked at the apex, and the terminal portion (hooked at the base) eventually deciduous (fig. 205). Only *G. triflorum* (with especially

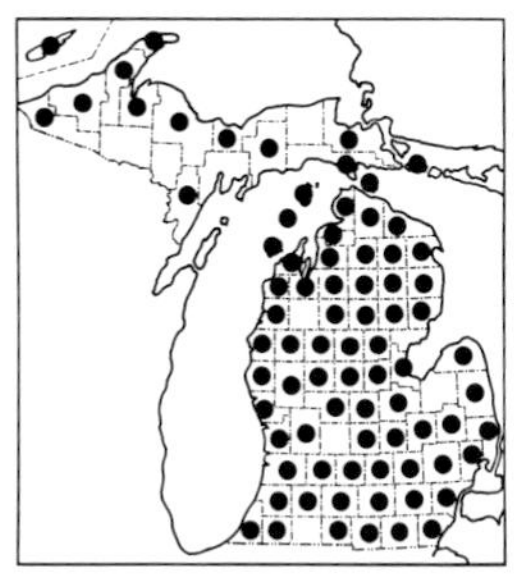
565. Potentilla recta

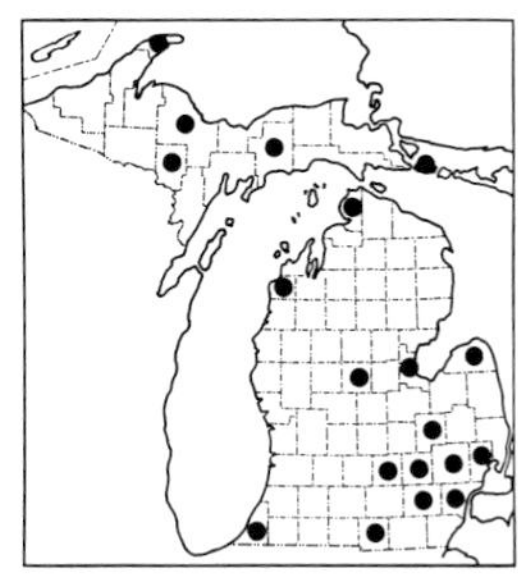
566. Potentilla inclinata

567. Chamaerhodos nuttallii

long plumose styles) among our species lacks this unique double kink in the style. Specimens with flowers and young (or mature) fruit are most easily identified. Specimens with only flowers (especially if dry and the collector neglected to record petal color on the label) are often difficult. A number of characters are useful in determining species, but they do not correlate in a way designed for use in a key employing several characters in a lead. If both flowers and at least immature fruit are present, it is best to try both leads at couplet 4 below.

REFERENCE

Gajewski, Wacław. 1957. A Cytogenetic Study on the Genus Geum. Monogr. Bot. 4. 416 pp.

KEY TO THE SPECIES

1. Calyx campanulate, red or purplish, the lobes ± erect (sometimes spreading in fully mature fruit); flowers nodding (becoming erect in fruit); petals yellow suffused with purple or purple-veined
  2. Style without a joint, elongating to 2–6 times as long as the perianth; receptacle sessile; bractlets of the calyx longer than the sepals; terminal leaflet scarcely larger than principal lateral leaflets (all ± narrowly cuneate); plant of dry habitats .... 1. **G. triflorum**
  2. Style with a prominent joint near the middle, barely twice as long as the perianth at maturity; receptacle stipitate, elevated above the perianth in fruit; bractlets of calyx much shorter than the sepals; terminal leaflet ± rotund or broadly cuneate, much larger than the lateral leaflets; plant of wet habitats .... 2. **G. rivale**
1. Calyx top-shaped to saucer-shaped, green when fresh, the lobes promptly reflexing; flowers ± erect; petals white or yellow
  3. Calyx without bractlets, ca. 3–4 mm long (including lobes before reflexing); plant ripening fruit by the first week of June; petals ca. 1.2–2 mm long; mature head of fruit ca. 1–1.2 cm (or less) in diameter (including beaks), elevated on a stipe well above the perianth; both segments of style completely glabrous ... 3. **G. vernum**
  3. Calyx with small bractlets between the sepals, at least 4 mm long; plant only beginning to bloom in June (or later); petals 2.5–8.5 mm long (sometimes smaller in *G. virginianum*); mature head of fruit 1–2.2 cm in diameter, essentially sessile (but cf. *G. aleppicum*); one or both segments of style usually ± pubescent
    4. Key to flowering material
      5. Petals white (often drying pale yellow), cream, or pale yellow
        6. Petals (3) 4–7 (7.5) mm long, equaling or exceeding the sepals; stem glabrous or with a few scattered appressed to spreading hairs; pedicels closely (sometimes glandular-) puberulent but at most with scattered long hairs.......... .... 4. **G. canadense**
        6. Petals [2] 2.5–4 (5.5) mm long, distinctly shorter than the sepals; stem ± densely pubescent with spreading hairs (many 2 mm or longer); pedicels various
          7. Pedicels conspicuously hirsute with spreading to reflexed hairs (also ± puberulent); largest stipules less than 10 mm broad .... 5. **G. laciniatum**
          7. Pedicels puberulent with at most scattered long hairs; largest stipules over 10 mm broad .... 6. **G. virginianum**

5. Petals bright yellow

8. Lower section of style with scattered short-stalked glands, especially toward base; terminal leaflet of basal leaves cordate to reniform, much larger than lateral segments . . . . 7. **G. macrophyllum**

8. Lower section of style without glands; terminal leaflet of basal leaves various, usually ± cuneate

9. Terminal segment of style glabrous or nearly so; petals ca. 3–4 [6] mm long, about equaling or shorter than the sepals; plant a weed of waste places . . . . 8. **G. urbanum**

9. Terminal segment of style conspicuously pilose (hairs much longer than thickness of style); petals 5–8 (8.5) mm long, equaling or usually exceeding the sepals; plant a widespread native . . . . 9. **G. aleppicum**

4. Key to fruiting material

10. Receptacle glabrous or only sparsely hairy; plants with either glandular-beaked achenes or ± dense long hairs overtopping puberulence of the pedicels

11. Pedicels ± densely hirsute with spreading to reflexed hairs; beak of achene eglandular . . . . 5. **G. laciniatum**

11. Pedicels closely puberulent, at most with scattered long hairs; beak with short-stalked glands especially toward the base . . . . 7. **G. macrophyllum**

10. Receptacle ± densely pilose; plants with neither glands on the beaks nor (usually) dense long hairs on pedicels

12. Beak of achene with a few long hairs at base; cauline leaves pinnately compound (often including very small leaflets); achenes more than 150 in a head. . . . . 9. **G. aleppicum**

12. Beak glabrous (or only minutely pubescent); cauline leaves mostly 3-lobed or trifoliolate; achenes fewer than 100 (150) in a head

13. Terminal segment of style glabrous or only minutely pubescent; stipules of cauline leaves mostly 12–35 (40) mm broad; plant a local introduction in waste places . . . . 8. **G. urbanum**

13. Terminal (deciduous) segment of style pilose at base (hairs much longer than thickness of style); stipules various; plant native, in natural habitats

14. Stems glabrous or slightly (usually appressed-) pubescent; stipules of cauline leaves less than 5 (10) mm broad . . . . 4. **G. canadense**

14. Stem ± densely pubescent, especially below, with many hairs 2 mm long; stipules, at least the largest, over 10 mm broad . . . . 6. **G. virginianum**

1. **G. triflorum** Pursh

Map 569. Sandy prairies, bluffs, and oak woodland; thin soil over limestone (Drummond Island).

The beautiful head of wind-dispersed fruits with very long plumose styles reminds one of *Dryas,* an arctic-alpine genus of which two species are known from islands along the north shore of Lake Superior but not from Michigan. (They differ from *Geum* in having simple leaves and ca. 8 petals and sepals, the latter without associated bractlets.)

2. **G. rivale** L. Fig. 205

Map 570. Bogs (or fens), conifer swamps (cedar, tamarack, spruce-fir), along streams and lakes, swampy woodland and rich deciduous woods, wet meadows; often in marly or other calcareous places.

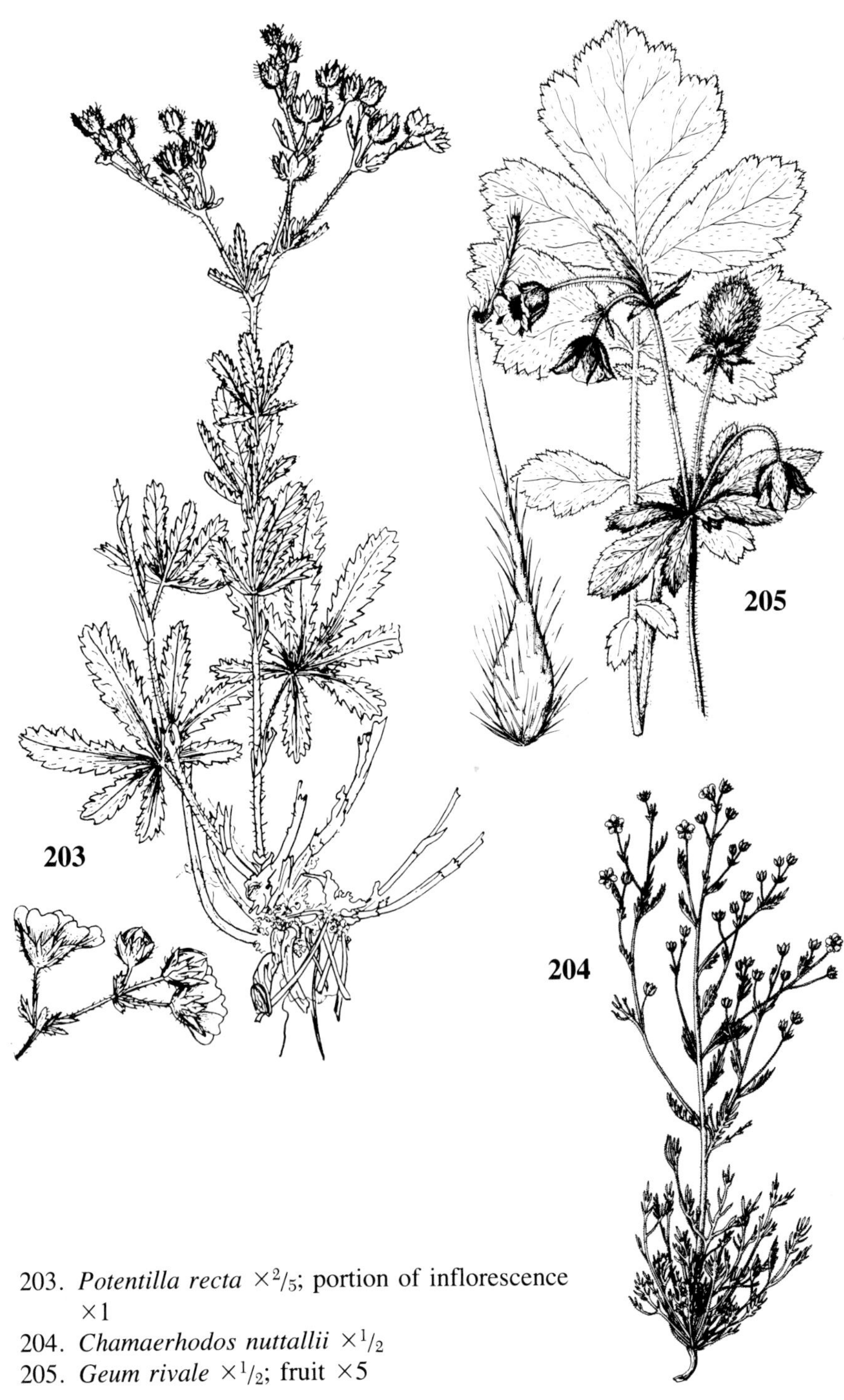

203. *Potentilla recta* $\times^{2}/_{5}$; portion of inflorescence $\times 1$
204. *Chamaerhodos nuttallii* $\times^{1}/_{2}$
205. *Geum rivale* $\times^{1}/_{2}$; fruit $\times 5$

3. **G. vernum** (Raf.) T. & G.

Map 571. Local in rich damp woods.

Distinctive in its very small, early-blooming flowers, basal leaves with usually simple cordate blades, and strongly elevated head of achenes.

4. **G. canadense** Jacq.

Map 572. Rich deciduous woods (and northern hardwoods), especially in damp places, but occasionally in drier oak-hickory woods; swamps and streamsides.

This is a variable species—perhaps even more so than others in the genus—but it does not seem worthwhile to apply names to the variants. It is a much commoner and more widespread white-flowered species than the next, though the petals in both usually dry ± yellowish. The small heads, prominent white petals, and tendency to trifoliolate leaves will help in identification. The slender puberulent pedicels have at most a few long hairs intermixed, in contrast with *G. laciniatum*. One collection from Delta County has a glandular beak on the fruit, as in *G. macrophyllum*.

5. **G. laciniatum** Murray

Map 573. Low woods and floodplains, ponds and wet areas in hardwoods, damp fields and fencerows, ditches.

Distinctive in its rather stout, densely spreading-hairy pedicels; small white

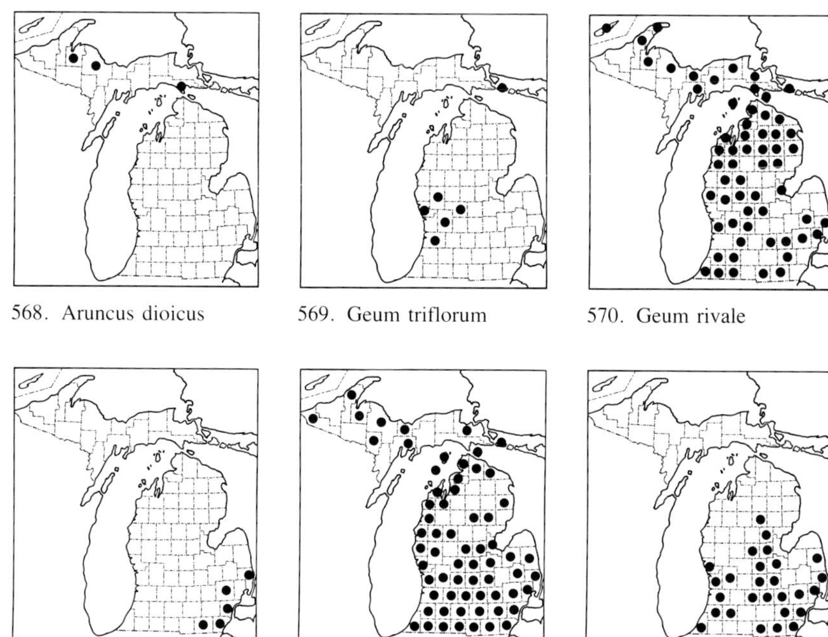

568. Aruncus dioicus

569. Geum triflorum

570. Geum rivale

571. Geum vernum

572. Geum canadense

573. Geum laciniatum

petals; and receptacle glabrous or nearly so. Some specimens of the yellow-flowered *G. aleppicum* may approach it in denseness of hairs on the pedicels.

6. **G. virginianum** L.

Map 574. Openings and banks in woods (oak etc.).

This species ranges mostly south and east of Michigan, but has often been reported from the state, usually because the name was once applied to what is now called *G. laciniatum*. Most other reports appear to represent misidentifications of *G. rivale* or *G. aleppicum*. This plant is sometimes thought to have originated as a fertile hybrid of *G. aleppicum* × *G. canadense*; specimens referred here may also include sterile F_1 hybrids. *G. virginianum* has the aspect of *G. canadense* in trifoliolate leaves and fewer-carpeled heads (than *G. aleppicum* and *G. laciniatum*); but the coarsely toothed terminal leaflet is usually much larger than the lateral ones, the stipules are much longer than in *G. canadense* (the largest 2 cm or more long), and the short petals are yellowish rather than pure white when fresh. The stem (but not the pedicels) is if anything even hairier than in *G. laciniatum*—which has more achenes in a head and a glabrous receptacle.

7. **G. macrophyllum** Willd.

Map 575. Northern hardwoods, especially along borders, trails, clearings, and rivers; meadows and ditches.

The characteristic large cordate to reniform terminal segment of the basal leaves is a feature also sometimes occurring in *G. aleppicum* and *G. rivale*, both of which are therefore sometimes confused with *G. macrophyllum*. However, the head of achenes in *G. rivale* is elevated on a stalk and the beak is conspicuously hairy though it may also be glandular; in *G. macrophyllum* the receptacle is essentially sessile and the beak is only glandular-puberulent. The beak of *G. aleppicum* has only a few long hairs at its base. Both may have pilose achene bodies. Plants with the leaflets of the basal

574. Geum virginianum

575. Geum macrophyllum

576. Geum urbanum

(and other) leaves incised, with ± cuneate segments, have been called *G. macrophyllum* var. *perincisum* (Rydb.) Raup.

8. **G. urbanum** Jacq.

Map 576. A Eurasian weed, locally established in Ann Arbor and doubtless overlooked elsewhere. First recognized in the state by A. A. Reznicek in 1978 at edges of thickets and scrappy woods; a garden weed.

9. **G. aleppicum** Jacq.

Map 577. Usually in moist places: meadows, marshy ground, along creeks, thickets and swamps (coniferous and deciduous); deciduous, mixed, and coniferous forests, especially along trails and in clearings; ditches and roadsides.

Quite variable in leaf shape and other characters, but recognized by the achene beak sparsely hairy at the very base (otherwise smooth), densely short-pilose receptacle, and yellow petals. Descriptions of the receptacle as sessile notwithstanding, there is often a short thick stipe (up to 4 mm long) in the fruiting head—not so pronounced as in *G. rivale* and *G. vernum*. This is our only species in which the pedicels may be almost as densely long-hairy as in *G. laciniatum,* which differs in having white petals shorter than the sepals, essentially glabrous receptacle, and achenes with glabrous beak (and sometimes also body, though this is usually pilose at the summit). American plants have been separated from Eurasian ones as var. *strictum* (Aiton) Fern., but the differences are obscure.

22. Filipendula

A white-flowered Eurasian species with leaves usually tomentose beneath and fruit twisted, *F. ulmaria* (L.) Maxim., Queen-of-the-meadow, is sometimes cultivated and occasionally escapes, although there are no reports of its doing so in Michigan. In our native species, the petals are pink, the leaves green beneath, and the fruits straight.

Sorbaria might be keyed here if not recognized as woody. It has white flowers and differs from *Filipendula* in the 5 dehiscent several-seeded follicles and serrate but unlobed leaflets, as well as the woody older stems.

1. **F. rubra** (Hill) Robinson — Queen-of-the-prairie

Map 578. Wet prairies, meadows, and shores, usually calcareous; hillside and perhaps escaped at Marquette. This species is sometimes cultivated, and some collections may represent escapes rather than natural populations, although except for the Marquette record there is little evidence on labels or in local floras to suggest this.

23. **Sanguisorba** Burnet

KEY TO THE SPECIES

1. Inflorescence ovoid, less than twice as long as thick; leaflets less than 2 cm long; flowers unisexual or perfect, the carpels 2 and the stamens numerous ..... 1. **S. minor**
1. Inflorescence cylindrical, ca. 3–15 (or more) times as long as thick; leaflets mostly ca. 3–7 cm long; flowers perfect, the carpel 1 and the stamens 4..... 2. **S. canadensis**

1. **S. minor** Scop. Fig. 206 Garden or Salad Burnet

Map 579. Gravelly or rocky calcareous banks, shores, openings, and roadsides; sandy open ground, railroad banks. Despite the common names, this Old World species is apparently not cultivated to any extent in this country, but is a very local weed.

2. **S. canadensis** L. American Burnet

Map 580. Wet prairies and open damp calcareous marshy sites.

24. **Agrimonia** Agrimony

The outer rim of the floral tube ("calyx tube") bears 2 or more rows of hooked bristles which elongate and stiffen in fruit, closely resembling the beaks on achenes of *Geum*. In *Agrimonia,* however, it is the entire floral tube which breaks free from the plant, the 2 achenes enclosed, and is well adapted for animal dispersal. The floral tube is usually top-shaped and ± grooved and furrowed. The roots in some species are fibrous and in others (nos. 2 & 4) become fusiform-thickened.

REFERENCE

Skalický, Vladimír. 1973. Amerikanische Arten der Gattung Agrimonia L. ser. Tuberosae ser. nova. Folia Geobot. Phytotax. 8: 95–104.

577. Geum aleppicum

578. Filipendula rubra

579. Sanguisorba minor

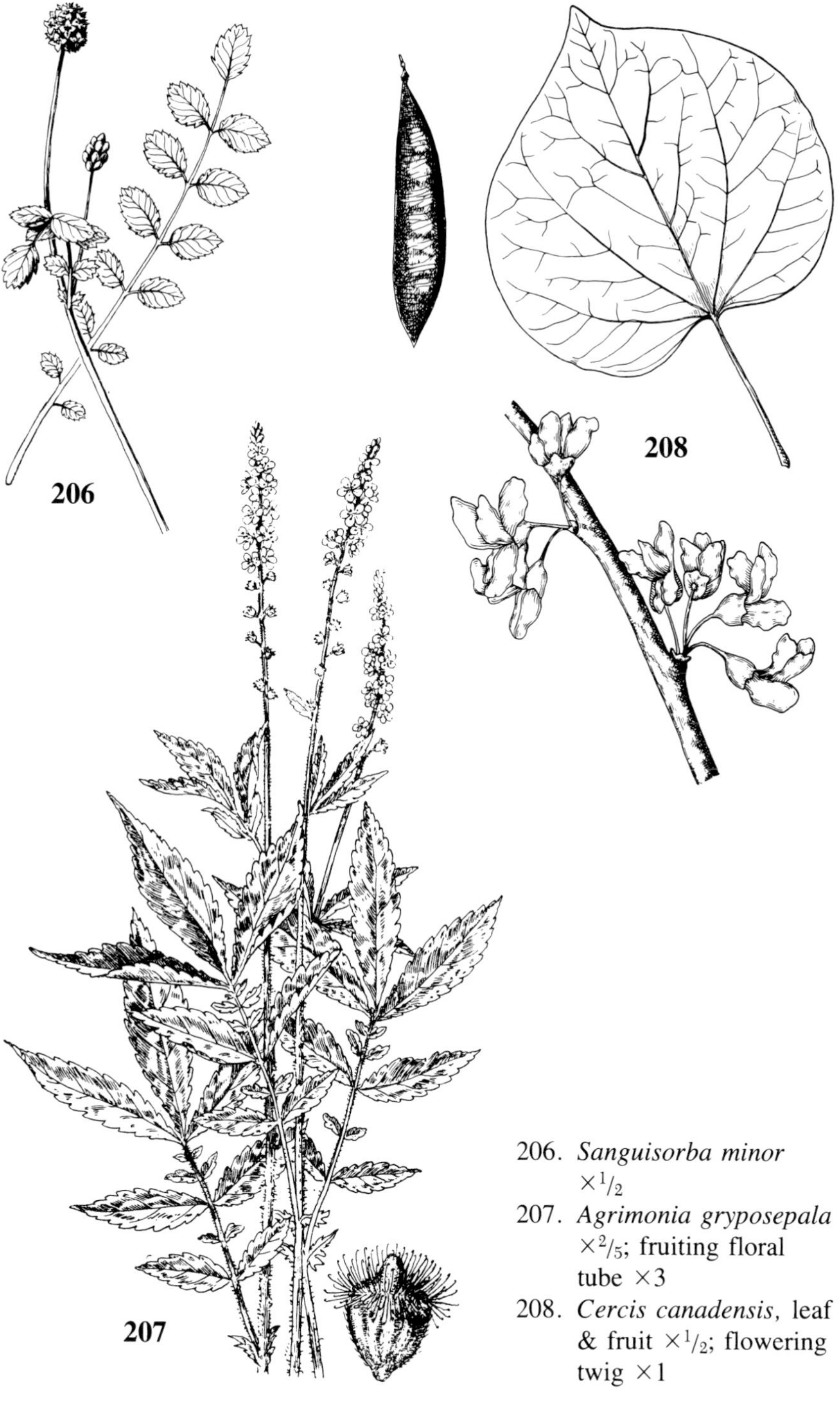

206. *Sanguisorba minor* ×1/2
207. *Agrimonia gryposepala* ×2/5; fruiting floral tube ×3
208. *Cercis canadensis,* leaf & fruit ×1/2; flowering twig ×1

KEY TO THE SPECIES

1. Principal leaflets (excluding tiny intermediate ones) (11) 13–17, narrowly elliptical (mostly ca. 3–4 times as long as wide); axis of inflorescence with glands largely obscured by ± dense short hairs (besides scattered long hairs) .... 1. **A. parviflora**
1. Principal leaflets 5–9 (11), narrowly to broadly elliptic or slightly obovate (but usually 1.5–2 (3) times as long as wide); axis of inflorescence with glands conspicuous, obscured, or absent
 2. Axis of inflorescence conspicuously glandular; grooves of floral tube without appressed hairs; bristles of floral tube in common species widely spreading or reflexed (except for inner ones)
  3. Bristles of floral tube elongating at most to 2 mm, ascending or slightly spreading; axis of inflorescence with very few if any long hairs; sepals (calyx lobes) ca. 1.5–1.8 mm long; fruiting floral tube ± hemispherical with rounded sides, slightly if at all grooved, ca. 2–2.5 mm wide; plant very rare, in southern Lower Peninsula .... 2. **A. rostellata**
  3. Bristles of calyx tube elongating to as much as 3.5 (4) mm, the outer ones widely spreading to reflexed; axis of inflorescence with conspicuous widely spreading scattered long hairs; sepals ca. 2–2.7 (3) mm long; fruiting floral tube ± top-shaped, often conspicuously grooved, ca. 3–4.5 mm wide (excluding bristles); plant common throughout..................................3. **A. gryposepala**
 2. Axis of inflorescence without glands, or these sparse and ± hidden by pubescence; grooves of floral tube with a strip of white appressed hairs (strigose); bristles of floral tube ± strongly ascending or erect
  4. Lower surface of leaflets velvety to the touch, the hairs ± strongly spreading; stipules of middle cauline leaves ovate-reniform, ± coarsely but regularly toothed; plant of the southern half of the Lower Peninsula .... 4. **A. pubescens**
  4. Lower surface of leaflets smooth or scabrous to the touch, the hairs usually ± appressed; stipules of middle cauline leaves mostly with a prolonged lanceolate terminal tooth or lobe; plant of the Upper Peninsula and northernmost Lower Peninsula .... 5. **A. striata**

1. **A. parviflora** Aiton

Map 581. Dry or moist fields and sandy openings, wet prairies and meadows, moist deciduous woods and low ground along streams.

Easily recognized by the relatively narrow leaflets, more numerous than

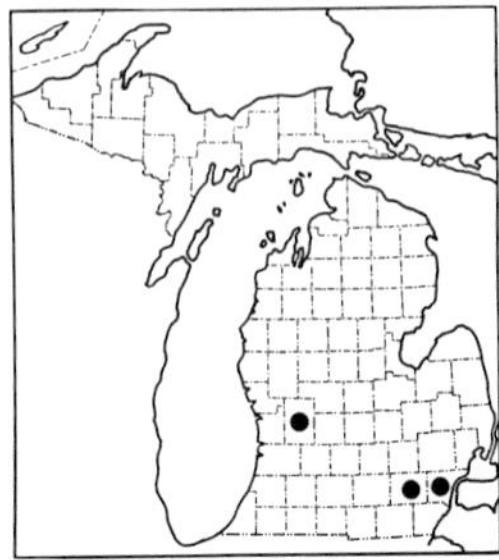
580. Sanguisorba canadensis

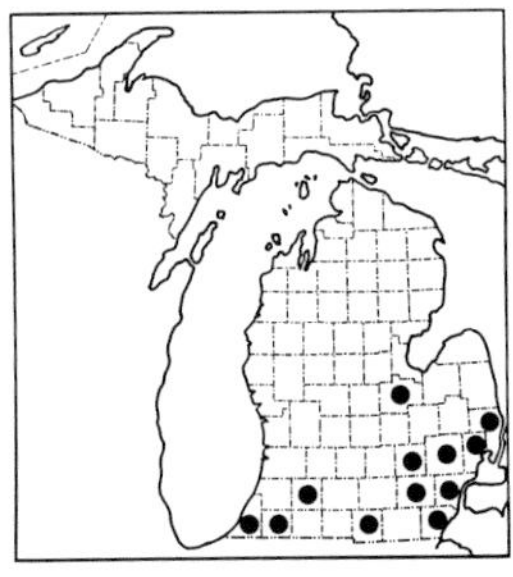
581. Agrimonia parviflora

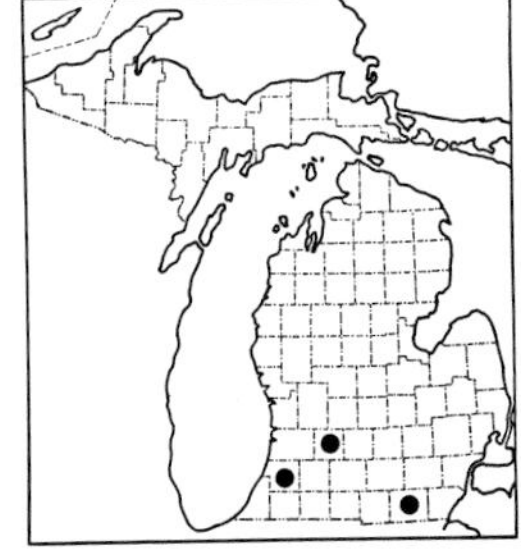
582. Agrimonia rostellata

in our other species, with unusually numerous tiny leaflets interspersed (often 3–4 pairs between adjacent pairs of large leaflets).

2. **A. rostellata** Wallr.

Map 582. Oak-hickory woods, sandy clearings, and thickets—insofar as our few specimens have any habitat data. A southern species barely entering southern Michigan.

A more delicate plant than the next, with smaller flowers and fruit, the latter with distinctly fewer hooked bristles.

3. **A. gryposepala** Wallr. Fig. 207

Map 583. Deciduous (oak-hickory or beech-maple) or mixed woods, especially in ± disturbed sites as along roads, trails, banks, and clearings; moist thickets and shores; fields; marshy and swampy ground, sometimes under cedar or tamarack.

Easily recognized by the ± densely and conspicuously glandular axis of the inflorescence (with only very long spreading hairs in addition), pedicels, floral tubes, and lower leaf surfaces, together with the widely spreading or reflexed outer rows of hooked bristles on the floral tube.

4. **A. pubescens** Wallr.

Map 584. Dry to mesic upland woods, swamps and floodplains, dry sandy open ground.

Some plants included in this species by most recent authors have again been recognized as a closely related species, *A. bicknellii* (Kearney) Rydb., by Skalický on the basis of more narrowly lance-elliptic leaflets, 7–13 per leaf, and spreading outer bristles on the floral tube. He included southeastern Michigan in its range. Some of our specimens do have one or more of the stated characters, but do not seem consistently or significantly different from more typical *A. pubescens* with 5–7 (9) leaflets tending to be obovate and the calyx bristles all erect or ascending.

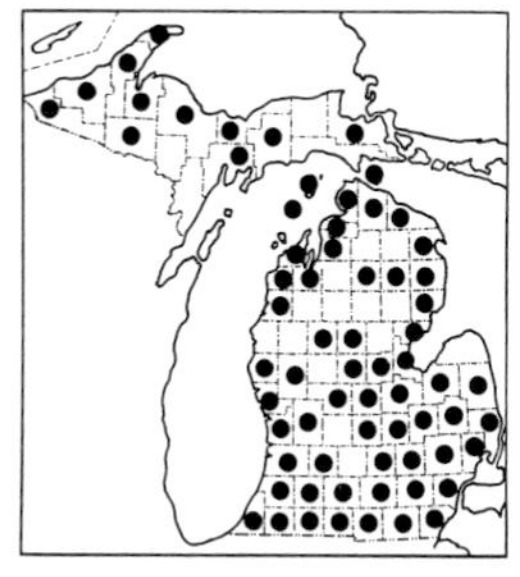

583. Agrimonia gryposepala

584. Agrimonia pubescens

585. Agrimonia striata

5. **A. striata** Michaux

Map 585. Thickets, shores, disturbed areas and borders of moist to dry deciduous or mixed woods, cedar swamps.

The leaflets are often more conspicuously glandular beneath than in the preceding species.

LEGUMINOSAE (FABACEAE) — Pea Family

This is one of the largest and most important families of flowering plants in the world, including major food and forage crops as well as some ornamentals and timber trees. Most species have root nodules in which certain bacteria (*Rhizobium* spp.) are able to convert free atmospheric nitrogen into compounds useful to the plant in forming amino acids. Many legume crops are grown to enrich the soil with nitrogen, and members of the family play a similar role in natural succession on sterile soils. The seeds are rich in food values, especially protein, and are used in many parts of the world as a major portion of the diet. Many of our most conspicuous roadside "weeds" are legumes originally escaped from cultivation and now thoroughly naturalized.

The family, in the broad sense, consists of three subfamilies, often recognized as families, as follows:

Mimosoideae (Mimosaceae): Mainly tropical and subtropical. The flowers are small, regular, in dense heads with prominent (often numerous) exserted stamens; leaves are typically twice-pinnately compound. Includes the very large genus *Acacia*. Our only representative, as a waif, is *Schrankia*.

Caesalpinioideae (Caesalpiniaceae): Also well represented in tropical and subtropical regions, and mainly woody. The flowers vary from nearly regular to papilionaceous, with no more than 10 stamens (usually distinct). Our representatives are *Cassia, Cercis, Gleditsia,* and *Gymnocladus*.

Papilionoideae (Faboideae) (Papilionaceae or Fabaceae—not Leguminosae if defined in this way): Worldwide, mostly herbaceous, and including all of our other genera. The flowers (except in a few odd genera, such as those with a single petal) are *papilionaceous,* i. e., bilaterally symmetrical with an upper petal or *standard* usually larger than the others and enclosing them in the bud; 2 lateral petals or *wings*; and the 2 lower petals ± fused to form a *keel,* which encloses the stamens. The 10 stamens are sometimes all connate (monadelphous), but usually diadelphous, i. e., in two groups, generally 9 + 1, occasionally 5 + 5.

The leaves of legumes are often sensitive to environmental stimuli, sometimes even to touch (as in the well known sensitive-plant) but especially to physiological conditions causing them to "close up" during the night or under stress such as on sand dunes on a hot sunny day. Such movements are the direct result of changes in turgor in specialized swellings or *pulvini* at the bases of the leaflets and the petiole, and which act like hinges. Sometimes the petiolule of a leaflet consists of little more than the pulvinus. Just as there are stipules at the base of the leaf in most legumes, in many species there are little *stipels* at the base of the leaflets.

Because of the difficulty in interpreting technical characters involving stamens and other parts when flowers are pressed and dry, the keys are very artificial and stress vegetative characters—which can be employed to excellent advantage in this family, as demonstrated by Fassett (1939).

REFERENCES

Fassett, Norman C. 1939. The Leguminous Plants of Wisconsin. Univ. Wisconsin Press, Madison. 157 pp.

Gambill, William G., Jr. 1953. The Leguminosae of Illinois. Illinois Biol. Monogr. 23(4). 117 pp.

Isely, Duane. 1973– . Leguminosae of the United States. Mem. New York Bot. Gard. 25. [Nos. 1 (152 pp.), 2 (228 pp.), & 3 (264 pp.) thus far published.]

Wilbur, Robert L. 1963. The Leguminous Plants of North Carolina. North Carolina Agr. Exp. Sta. Tech. Bull. 151. 294 pp.

KEY TO THE GENERA

1. Leaves all simple or apparently so (or absent at flowering time)
 2. Plant woody, the vegetative parts glabrous; petals pink, conspicuously exceeding the sepals; flowers in small sessile umbelliform racemes borne on old wood and opening in May (mostly before the leaves); leaf blades ovate to rotund, ± cordate; fruit flat, ca. 5–9 cm long . . . . 1. **Cercis**
 2. Plant herbaceous, the vegetative parts hairy; petals yellow, shorter than the sepals; flowers few, in small peduncled racemes, opening in mid- or late summer (after the leaves); leaf blades narrowly elliptic to lanceolate, tapering to both ends; fruit inflated, ca. 1.3–2.5 [4] cm long . . . . 2. **Crotalaria**

1. Leaves all or mostly compound
 3. Leaflets 3 (except for the rare "4-leaved clover")
  4. Margins of leaflets strongly to minutely toothed, at least at the apex (or with unusually prominent vein tips)
   5. Inflorescences ca. (2) 4–15 times as long as wide; stipules setaceous, entire, 1-veined, glabrous or nearly so, the free portion over 8 times as long as wide . . . . 3. **Melilotus**
   5. Inflorescences ca. 2 (rarely 3) times as long as wide, or shorter; stipules with distinct flat blades at least 2–3-veined, sometimes hairy and sometimes toothed, usually less than 8 times as long as wide (a long setaceous very hairy tip in *Trifolium arvense*)
    6. Leaflets all sessile or with petiolules (the pulvini) of uniform length . . . . 4. **Trifolium**
    6. Leaflets with terminal one on distinctly longer petiolule than the others
     7. Calyx with glabrous tube and the teeth very unequal (the longest often ca. twice as long as the shortest); stipules entire; fruit ovate-oblong, straight, enclosed in the persistent corolla . . . . 4. **Trifolium** (couplet 9)
     7. Calyx with hairy tube and the teeth often ± equal; stipules usually somewhat toothed, at least toward base; fruit reniform or elongate, ± curved, the corolla deciduous . . . . 5. **Medicago**

4. Margins of leaflets entire
8. Terminal leaflet with petiolule (if any) no longer than those of the lateral leaflets; leaflets not stipellate
9. Leaflets dotted with dark glands, linear, ca. 0.5–1.7 mm broad . . . . 27. **Dalea**
9. Leaflets not dotted, broader
10. Flowers in long-peduncled umbellate inflorescences; leaves glabrous or nearly so (actually 5-foliolate but appearing 3-foliolate with the lowest pair of the sessile leaf suggesting stipules about as large as the other leaflets) . . . . 29. **Lotus**
10. Flowers in racemes, in small axillary clusters, or solitary; leaves pubescent or glabrous (3-foliolate with small stipules except in one pubescent species)
11. Leaves of upper branches all or mostly simple; flowers bright yellow; stems strongly ridged or angled . . . . 6. **Cytisus**
11. Leaves all trifoliolate; flowers white to cream, blue, or purple; stems ± smooth
12. Flowers in a peduncled raceme terminating the stem (and often branches); fruit many-seeded, inflated; stamens distinct . . . . 7. **Baptisia**
12. Flowers solitary or few in leaf axils; fruit 1-seeded, flattened; stamens diadelphous . . . . 13. **Lespedeza** (spp. 1 & 2)
8. Terminal leaflet with longer petiolule than the lateral leaflets (at least a short extension of the rachis, and of similar color and texture, in addition to the pulvinus); leaflets stipellate or not
13. Stems ± vine-like, twining or trailing
14. Leaflets less than 2.5 cm long, without stipels; fruit 1-seeded . . . . 13. **Lespedeza (procumbens)**
14. Leaflets mostly more than (2.5) 3 cm long, stipellate; fruit with more than 1 seed
15. Leaflets suborbicular, at least as broad as long, very broadly rounded at apex; fruit of 1-seeded indehiscent segments covered with tiny hooked hairs . . . . 11. **Desmodium (rotundifolium)**
15. Leaflets longer than broad, acute; fruit neither segmented nor covered with hooked hairs, usually longitudinally dehiscent
16. Midvein of leaflets not excurrent; flowers without bractlets beneath (not to be confused with bracts at base of pedicel); plants with cleistogamous apetalous flowers at base (often setting 1-seeded fruit underground); calyx ± equally 4-toothed . . . . 8. **Amphicarpaea**
16. Midvein of each leaflet excurrent as a minute non-green bristle; flowers subtended by a pair of bractlets; plants without cleistogamous flowers or underground fruit; calyx 5-lobed (or ± 2-lipped)
17. Keel of corolla twisted or coiled at the tip; calyx lobes all shorter than the tube; seeds glabrous; leaflets not lobed . . . . 9. **Phaseolus**
17. Keel of corolla strongly arched but not twisted; calyx lobes (at least the longest) longer than the tube; seeds densely woolly; leaflets often broadly 2–3-lobed . . . . 10. **Strophostyles**
13. Stems erect, ± straight
18. Calyx lobes ± deltoid or rounded, much shorter than the tube; stipels none (or obsolete); fruit long-stalked above the calyx . . . . 11. **Desmodium** (couplet 2)
18. Calyx lobes triangular to lanceolate, equaling or longer than the tube; stipels present or not; fruit sessile or stalked only slightly above the calyx
19. Leaflets without stipels; calyx not bilabiate, all 5 lobes definite (the lower sometimes longer); fruit 1-seeded, glabrous or variously pubescent
20. Flowers in terminal spike-like racemes, on peduncles longer than subtending leaves; leaflets lance-elliptic, mostly 3 or more times as long as broad, gland-dotted (as are calyx and bracts) . . . . 12. **Psoralea**

20. Flowers few or crowded in dense inflorescences axillary as well as terminal, on peduncles often shorter than subtending leaves; leaflets various, but broadly rounded or oblong-elliptic in species with long-peduncled inflorescences and not gland-dotted . . . . 13. **Lespedeza**

19. Leaflets stipellate; calyx appearing somewhat 2-lipped, the upper 2 calyx teeth ± united, the lower 3 more deeply divided and longer; fruit 2–several-seeded, with spreading pubescence

21. Fruit composed of (1) 2 or more 1-seeded, flat, indehiscent segments covered with tiny hooked hairs; stem and axis of inflorescence (unless glabrous) also with minute hooked hairs; plant a native perennial . . . . 11. **Desmodium**

21. Fruit not segmented, ultimately dehiscent, covered with straight (or curly) hairs; stem and axis of inflorescence without hooked hairs or these long (many times as long as thick); plant a cultivated annual, rarely spread from fields

22. Stems, petioles, and fruit glabrate or with some scattered long (sometimes hooked) hairs . . . . 9. **Phaseolus**

22. Stems, petioles, and fruit densely pubescent with long sharp-tipped hairs . . . . 14. **Glycine**

3. Leaflets (i. e., flat blades, not necessarily including tendrils) 2, or 4 or more

23. Plant a tree, shrub, or woody vine

24. Leaves even-pinnate (if leaflets not opposite, appearing falsely odd-pinnate)—and sometimes twice-pinnate

25. Flowers yellow, papilionaceous; fruit straight, slender, ca. 3–5.5 cm long, dehiscent into twisted valves; leaves strictly once-pinnate . . . . 15. **Caragana**

25. Flowers greenish white, regular or nearly so; fruit often somewhat curved, ca. 2–4 cm wide and [6] 10–35 [45] cm long, tardily if at all dehiscent; leaves all or partly (or not at all) twice-pinnate

26. Leaflets entire, ovate with rounded sides and short-acuminate tip, mostly 2–4 cm broad when mature; fruit ca. 3–4.5 cm broad and [6] 10–15 cm long; flowers all perfect . . . . 16. **Gymnocladus**

26. Leaflets obscurely crenulate with dark glands, lance-oblong, less than 1.5 cm broad; fruit ca. 2–3 cm broad and 18–35 [45] cm long; flowers both perfect and unisexual . . . . 17. **Gleditsia**

24. Leaves clearly odd-pinnate (and leaflets—except for the odd terminal one—nearly or quite in opposite pairs)

27. Plant a high-climbing vine; leaflets often as broad as 3–4 cm, short-acuminate . . . . 18. **Wisteria**

27. Plant a shrub or tree; leaflets less than 3 cm broad, rounded at apex

28. Flowers less than 1 cm long, in narrow, elongate, erect, spike-like racemes; petal 1, blue to purple; fruit less than 1 cm long, indehiscent, 1–2-seeded; plant a low shrub or if tall, with leaflets gland-dotted beneath . . . . . 19. **Amorpha**

28. Flowers ca. 1.5–2.8 mm long, in broad ± pendent or spreading racemes; petals 5 in typical papilionaceous flower, pink (or nearly cream) to rose-red; fruit ca. (3.5) 4.5–10 cm long, 3–several-seeded; plant becoming a tall shrub or tree, the leaflets not gland-dotted . . . . 20. **Robinia**

23. Plant herbaceous (at most somewhat woody at the ground)

29. Leaves palmately or twice-pinnately compound

30. Flowers pink, regular, in globose axillary heads; leaves twice pinnately compound; plant armed with stiff prickles . . . . 21. **Schrankia**

30. Flowers blue (rarely rose or white), papilionaceous, in terminal racemes; leaves palmately compound; plant unarmed . . . . 22. **Lupinus**

29. Leaves once-pinnately compound
31. Leaves with an even number of leaflets, the terminal one at most represented by a bristle or tendril
32. Terminal "leaflet" a bristle or none; flowers yellow, slightly irregular but not papilionaceous; stamens 5–10, separate; petioles with a prominent gland near the base (or on the rachis at the lowest pair of leaflets) .... 23. **Cassia**
32. Terminal leaflet replaced by a well developed tendril; flowers various in color but not yellow in most species and clearly papilionaceous; stamens united (diadelphous, 9 + 1 or 5 + 5); petioles without glands
33. Stipules larger (both longer and broader) than the lowest leaflets ... 24. **Pisum**
33. Stipules smaller than lowest leaflets (narrower or shorter—usually both)
34. Leaflets 2 (not including tendrils). .... 25. **Lathyrus** (couplet 5)
34. Leaflets 4 or more
35. Larger stipules at least (7) 10 mm broad (hastate or semi-sagittate); principal leaflets at least 1.2 cm broad .... 25. **Lathyrus**
35. Larger stipules less than 7 mm broad (semi-sagittate or lanceolate); principal leaflets in most species all less than 1 cm broad
36. Leaflets with 10 or more pairs of lateral veins running from the midrib nearly or quite to the margins .... 26. **Vicia**
36. Leaflets with 6 or fewer pairs of lateral veins
37. Leaflets mostly 10 or more, less than 8 (9) mm broad, less than 3 cm long; stem wingless.................................... 26. **Vicia**
37. Leaflets mostly 4–8, or at least 8 mm broad (or both), over 2.5 cm long; stem in some forms narrowly winged ..... 25. **Lathyrus** (couplet 4)
31. Leaves odd-pinnate, the terminal leaflet developed
38. Leaflets linear to oblanceolate-elliptic, mostly less than 3 mm wide, covered with prominent glandular dots; inflorescence a dense cylindrical spike of tiny (less than 9 mm long) flowers .... 27. **Dalea**
38. Leaflets broader (mostly at least 4 mm wide), glandless; inflorescence various, of larger flowers
39. Inflorescence an umbel or involucrate head
40. Stem and calyx pubescent; flowers in a head subtended by 3–4-cleft bracts; terminal leaflet often distinctly larger than the lateral ones...... 28. **Anthyllis**
40. Stem and calyx glabrous or nearly so; flowers in an umbel, essentially bractless or subtended by a trifoliolate leaf; terminal leaflet about equaling the lateral ones
41. Flowers yellow (to orange); leaflets 5 (the lower pair resembling stipules); fruit dehiscent .... 29. **Lotus**
41. Flowers pink (to purple); leaflets numerous; fruit breaking transversely into 1-seeded indehiscent segments .... 30. **Coronilla**
39. Inflorescence a simple spike or raceme
42. Stem vine-like, twining; principal leaflets ca. 1.5–4 cm wide, acuminate; inflorescences all axillary .... 31. **Apios**
42. Stem erect or ascending, not twining; principal leaflets less than 1 (1.5) cm wide, obtuse, rounded or notched at apex (except for excurrent midvein); inflorescences terminal or axillary
43. Racemes all or mostly terminal; flowers 14–20 mm long, bicolored with yellow standard and pink to purple wings; stem, rachis of leaves, calyx, and fruit densely villous with simple mostly spreading hairs; calyx lobes longer than the tube .... 32. **Tephrosia**
43. Racemes all or mostly axillary; flowers 10–14 (15) mm long, uniformly white or cream to purplish; stem, rachis of leaves, calyx, and fruit glabrous or nearly so or strigose with straight or forked hairs

44. Fruit not segmented, 4.5–15 mm broad; flowers white to cream . . . . 33. **Astragalus**

44. Fruit segmented (as in *Desmodium*, but glabrous) with 2 or more distinct and very narrow constrictions, less than 6 mm broad; flowers pink or magenta . . . . 34. **Hedysarum**

1. Cercis

1. **C. canadensis** L. Fig. 208 Redbud; Judas Tree

Map 586. Rich woods, especially along rivers and streams. A handsome flowering shrub or small tree reaching the northern edge of its range in southern Michigan, where it brightens low woods locally in early spring. Also widely cultivated.

2. Crotalaria

Genista tinctoria L., dyer's greenweed, is an Old World plant grown for its yellow flowers, which yield a dye. It has been reported from Michigan (Wayne Co.), presumably as a waif escaped from cultivation, but no specimens have been found and the reputed basis (*Farwell 1987*$^{1}/_{2}$ in 1906) is suspicious because of the '1/2' number. It would run in the key close to *Crotalaria,* but differs most conspicuously in its larger corolla, nearly or quite glabrous parts, and flat fruits.

1. **C. sagittalis** L. Rattlebox

Map 587. Native south of our latitude, but an occasional waif along railroads and other dry waste places.

3. Melilotus Sweet-clover

REFERENCES

Isely, Duane. 1954. Keys to Sweet Clovers (Melilotus). Proc. Iowa Acad. 61: 119–131.

Turkington, Roy A., Paul B. Cavers, & Erika Rempel. 1978. The Biology of Canadian Weeds. 29. Melilotus alba Desr. and M. officinalis (L.) Lam. Canad. Jour. Pl. Sci. 58: 523–537.

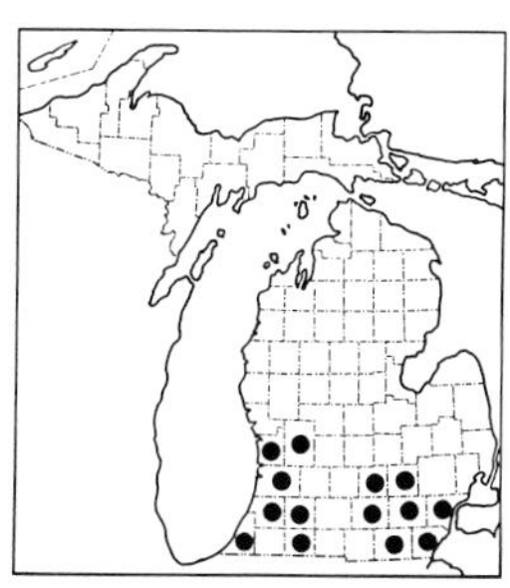

586. Cercis canadensis

587. Crotalaria sagittalis

588. Melilotus alba

KEY TO THE SPECIES

1. Corolla white .... 1. **M. alba**
1. Corolla yellow
 2. Ovary and young fruit glabrous .... 2. **M. officinalis**
 2. Ovary and young fruit short-hairy .... 3. **M. altissima**

1. **M. alba** Medicus White Sweet-clover

Map 588. An Old World species, widely grown as a forage plant (and also attractive to bees), thoroughly naturalized. Established in southern Michigan, at least, by the 1880's. Characteristic of recently disturbed places in dry, open, often calcareous ground, such as sand dunes, prairies, and roadsides, as well as fields, railroads, and shores.

2. **M. officinalis** (L.) Pallas Fig. 209 Yellow Sweet-clover

Map 589. Of similar status to the preceding, becoming common somewhat more recently although apparently also established by the 1880's. Roadsides, fields, railroads, disturbed areas in woods, shores, waste places generally.

This species begins to bloom several days earlier than *M. alba*. The only consistently reliable way to distinguish the two is by flower color—which careless collectors too often fail to record on their labels. Some yellow is usually detectable even in old dry corollas of *M. officinalis*. The racemes in *M. alba* at peak of flowering are often much longer (8–15 times as long as wide) than in *M. officinalis,* where they are rarely more than 6 times as long as wide. In *M. alba,* the standard often distinctly exceeds the wings, while in *M. officinalis* it is usually about the same length, but there are many exceptions. The leaflets of *M. officinalis* tend to be broader, usually no more than about twice as long as broad, while in *M. alba* they are often narrower (ca. 2.5–3.5 times as long as broad).

3. **M. altissima** Thuill.

Map 590. A European species, collected once in Michigan (*Rusby* in 1876,

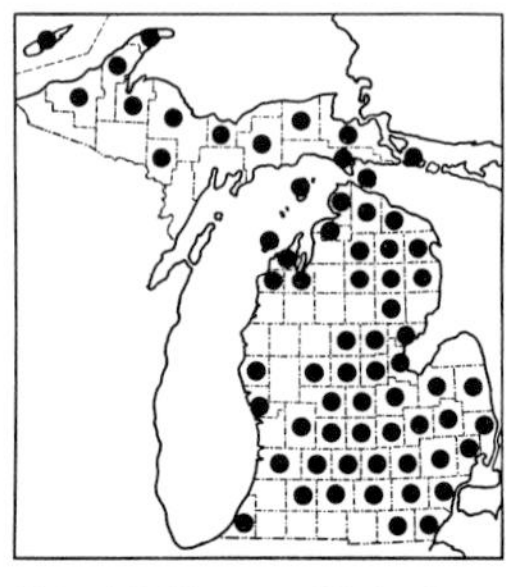

589. Melilotus officinalis

590. Melilotus altissima

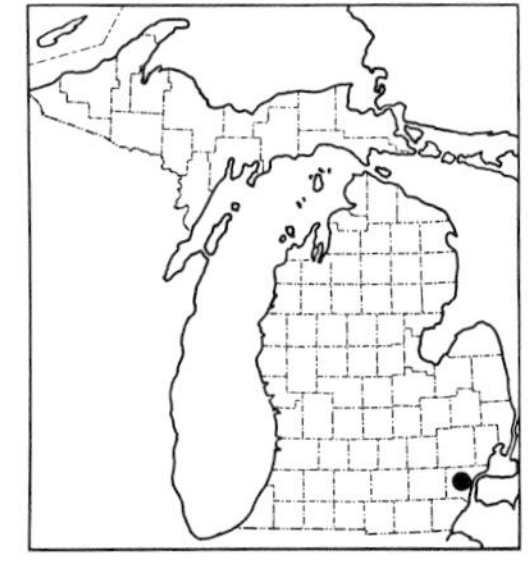

591. Trifolium depauperatum

MICH), at Byron. There are also only old collections in Wisconsin and Illinois, so one suspects that the species is no longer to be found in this part of the country.

Very similar to the preceding in almost every respect, but distinctive in its pubescent fruit; the leaflets are also much more narrow and elongate than is usual for *M. officinalis*. Fortunately most specimens in this genus have at least a few immature fruits at the base of one or more racemes and the pubescence character is easily checked.

4. **Trifolium** Clover

The first two species were collected in Michigan long ago as waifs from the West Coast (see Asa Gray Bull. 4: 46. 1896). The other species are all natives of Europe and the Mediterranean region and have long been grown as forage crops or for "green manure" in crop rotation—as are other legumes. The nomenclature of the European species follows *Flora Europaea*, which rejects the names *T. procumbens* and *T. agrarium* as ambiguous, for they have been commonly applied in senses that exclude their respective types.

The identity (if any) of the Irish "shamrock" is not clearly established, as the name has been used for several plants over the past 500 years, including species of *Oxalis*, although it appears to have been most often applied to a *Trifolium*, such as *T. dubium* or *T. repens*. (See A. R. Vickery in Plant Lore 5: 19–28. 1981.)

REFERENCES

Gillett, John M., & Theodore S. Cochrane. 1973. Preliminary Reports on the Flora of Wisconsin No. 63. The Genus Trifolium—the Clovers. Trans. Wisconsin Acad. 61: 59–74.

Hermann, F. J. 1953. A Botanical Synopsis of the Cultivated Clovers (Trifolium). U.S. Dep. Agr. Agr. Monogr. 22. 45 pp.

KEY TO THE SPECIES

1. Inflorescence subtended by an involucre consisting of simple bracts (at most 2-toothed at apex) connate at the base or reduced to a narrow ring of tissue; corolla becoming distinctly inflated after anthesis; plant an annual waif, probably no longer found in the state

2. Flowers ca. 5–8 mm long; involucre often reduced to a mere ring . . . . 1. **T. depauperatum**

2. Flowers ca. 11–22 mm long; involucre with well developed lobes . . . . . 2. **T. fucatum**

1. Inflorescence without an involucre, although in some species the stipules of the uppermost trifoliolate leaves may resemble one; corolla not inflated; plant an annual or perennial, most species common

3. Flowers sessile (or pedicels distinctly less than 0.5 mm) in a dense head, never yellow, the middle and upper ones not reflexed in maturity; calyx ± pubescent (rarely glabrous)

4. Head (or a short peduncle) subtended by a pair of opposite leaves with expanded involucre-like stipules; calyx teeth glabrous or (usually) with scattered, irregular, long hairs; plant a short-lived perennial, abundant . . . . 3. **T. pratense**

4. Head pedunculate with no subtending leaves or bracts and all leaves alternate; calyx teeth densely plumose with straight hairs; plant an annual, local

5. Corolla white or pink, inconspicuous, much exceeded by the calyx teeth; stipules with prolonged setaceous tips; heads ca. 8–14 mm broad . . . . 4. **T. arvense**

5. Corolla bright or deep red, equaling or exceeding the calyx teeth; stipules without setaceous tip; heads often broader . . . . 5. **T. incarnatum**

3. Flowers on pedicels ca. 0.5 mm or more long, yellow in some species, the middle and upper ones as well as lower becoming reflexed after anthesis; calyx glabrous (at most a few hairs at tips of teeth)

6. Flowers white to pink; petioles of lower leaves all or mostly much longer than the leaflets (usually at least twice as long); heads ca. 1.5–2.5 (3) cm in diameter

7. Peduncles arising from prostrate stems; stipules ± abruptly truncate or obtuse with setaceous mucronate tip up to 2 (3) mm long; calyx teeth shorter than the tube (or the longest of the unequal teeth equaling the tube) . . . . 6. **T. repens**

7. Peduncles arising from erect or ascending stems; stipules gradually tapering to tip; calyx teeth all or mostly exceeding the tube . . . . 7. **T. hybridum**

6. Flowers yellow; petioles of all leaves about equaling the leaflets or shorter; heads ca. 0.3–1.5 cm in diameter

8. Mature leaves with stipules (except on lower leaves) about equaling or exceeding petioles and the leaflets all sessile or nearly so; heads ca. 12–15 mm in diameter . . . . 8. **T. aureum**

8. Mature leaves with stipules (except sometimes on the uppermost leaves and in the small-headed *T. dubium*) much shorter than petioles and the terminal leaflet usually on petiolule distinctly longer than those of the lateral leaflets (which consist of only the pulvinus); heads ca. 3–12 mm in diameter

9. Heads mostly ca. 9–12 (13) mm in diameter, often ca. 20-flowered (or more); corolla 4.5–6 mm long, very strongly veined . . . . 9. **T. campestre**

9. Heads 3–7 (9) mm in diameter, ca. 4–15 (20)-flowered; corolla 3–4 mm long, only weakly veined . . . . 10. **T. dubium**

1. **T. depauperatum** Desv.

Map 591. Collected by Farwell (*1459,* BLH) in 1894 as a waif at Detroit. Native to the Pacific coast of the United States and so far as known not found in Michigan since.

The corolla is said to be white to purple. Plants with developed lobes on the involucre are var. *amplectens* (T. & G.) McDer.

2. **T. fucatum** Lindley

Map 592. A plant of identical status with us as the preceding, not found since 1894 (*Farwell 1458,* BLH).

The corolla is said to be cream or light yellow, becoming (or fading to) pink.

3. **T. pratense** L. Red Clover

Map 593. Fields, roadsides, gravel pits, and other waste ground; invading shores, dunes, open woods, rocky openings, and damp habitats. One of the most familiar of all our thoroughly naturalized "weeds," originally escaped from agricultural use. Collected at Ann Arbor in the 1860's; otherwise the earliest collections I have seen from the state date from the 1880's although the species was reported by the First Survey (1838). Now doubtless well established in every county, despite the incomplete representation on the map.

A variable species in stature, pubescence, and color of flowers. The latter range from deep reddish purple to rarely white. The heads are the largest of any of our clovers, sometimes as broad as 3 cm, and ± globose; in the next two species they are ± cylindrical.

4. **T. arvense** L. Rabbitfoot Clover

Map 594. Roadsides, sandy fields, weedy lots, plantations, and gardens; found more often in loose dry sand than any of our other species.

5. **T. incarnatum** L. Fig. 210 Crimson Clover

Map 595. Escaped to river banks, roadsides, and other disturbed sites—but not so found in Michigan since 1939.

6. **T. repens** L. Fig. 211 White Clover

Map 596. Roadsides, fields, lawns, trails, waste ground; invading dunes, rock openings, and damp habitats. The map underestimates its thorough establishment in the state. The earliest collection from Michigan was by the First Survey, at White Pigeon, St. Joseph County, in 1837.

7. **T. hybridum** L. Alsike Clover

Map 597. Roadsides, fields, clearings, disturbed ground, shores; frequently in damp habitats of various sorts. Already widespread across the state in the 1890's.

8. **T. aureum** Poll. Fig. 212 Hop Clover

Map 598. Roadsides, fields, railroads, meadows, gravel pits, clearings,

592. Trifolium fucatum

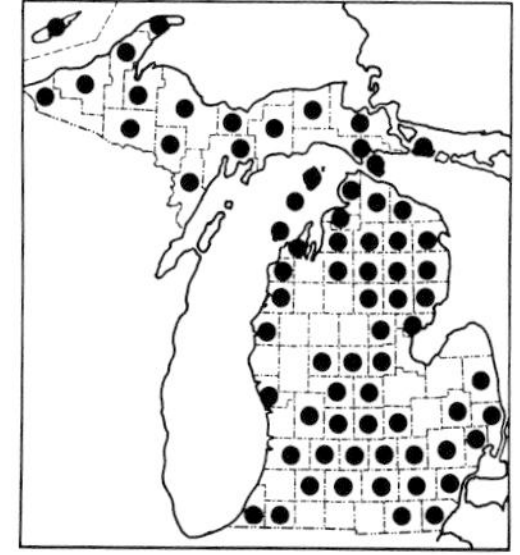

593. Trifolium pratense

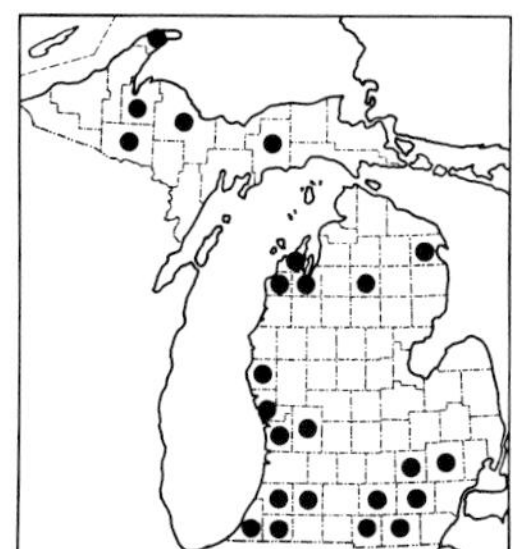

594. Trifolium arvense

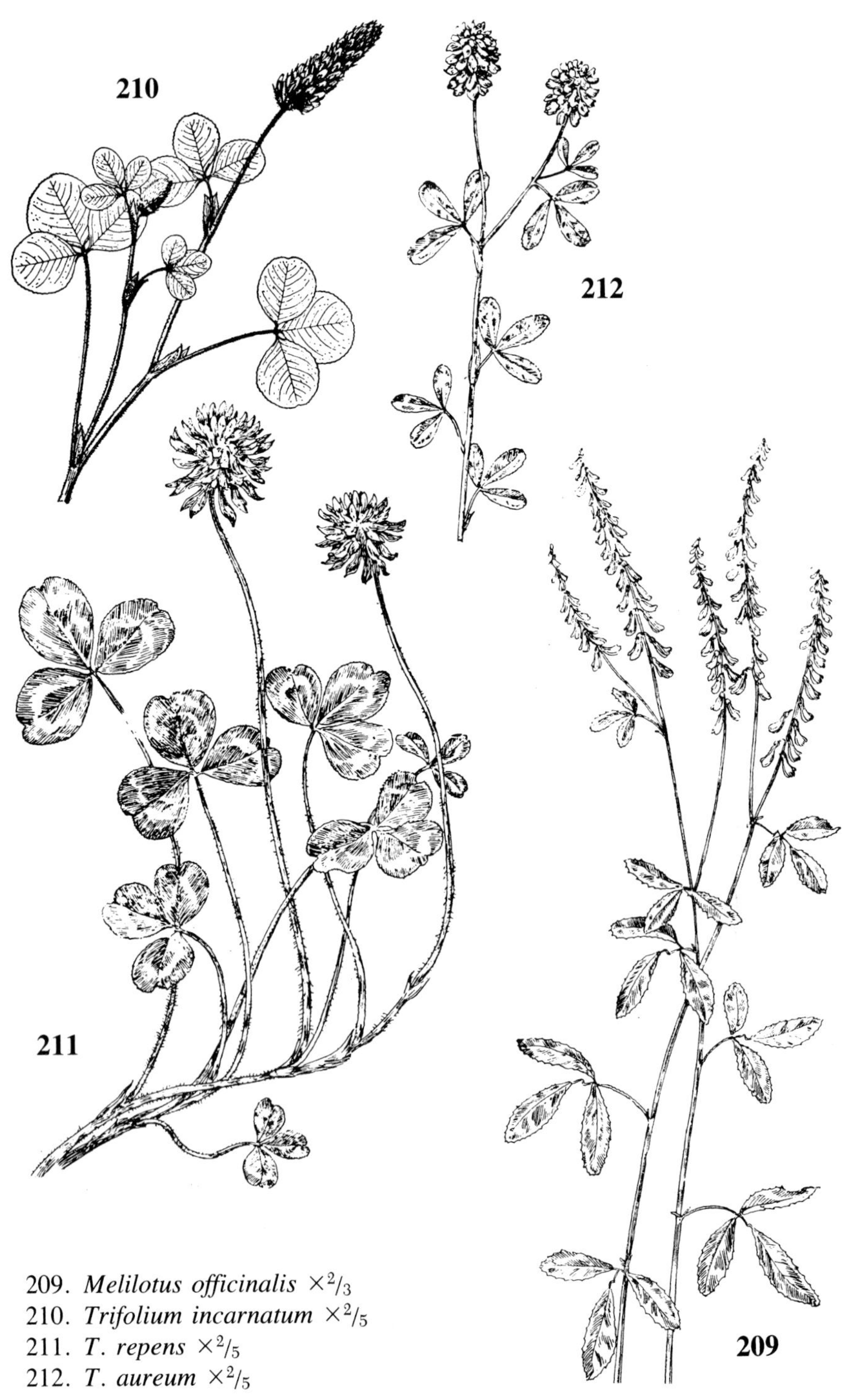

209. *Melilotus officinalis* $\times^2/_3$
210. *Trifolium incarnatum* $\times^2/_5$
211. *T. repens* $\times^2/_5$
212. *T. aureum* $\times^2/_5$

and other disturbed ground; invading damp habitats and rocky or sandy woods. Evidently this species did not appear out of cultivation in Michigan until the 1880's and 1890's.

Long known as *T. agrarium* L., a name which has also been applied to *T. campestre*. The broad ± ovate (but acute) stipules of the next two species will usually distinguish them readily from *T. aureum* with its prolonged lanceolate-ovate stipules.

9. **T. campestre** Schreber — Low Hop Clover

Map 599. Roadsides, fields, railroads, lawns, waste places; oak woods and clearings in woods. Apparently became a weed in the state about the same time as the preceding; the earliest collections seen are from the 1870's.

Often called *T. procumbens* L., a name which originally applied largely to the next species. The inflorescence closely resembles that of the preceding species, but runs a little smaller.

10. **T. dubium** Sibth. — Little Hop Clover

Map 600. Fields, roadsides, and turf; can be a serious weed of lawns. The oldest Michigan collections seen are from the 1890's.

Resembles *Medicago lupulina,* but distinguished as indicated in the key to genera (couplet 7) and discussed under that species below. The corolla

595. Trifolium incarnatum

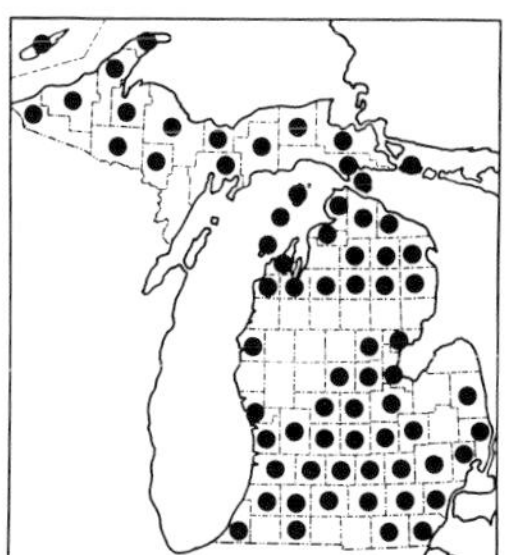
596. Trifolium repens

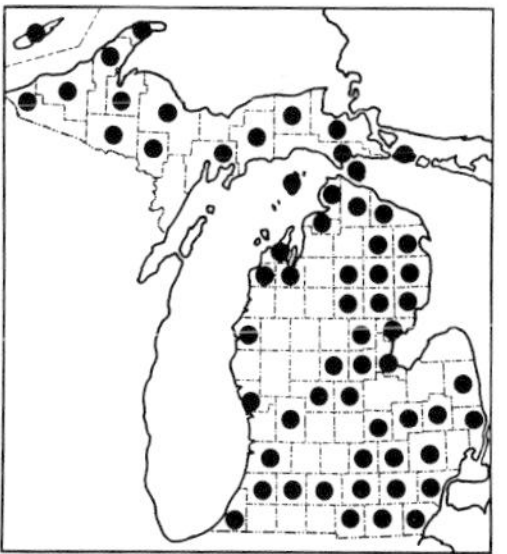
597. Trifolium hybridum

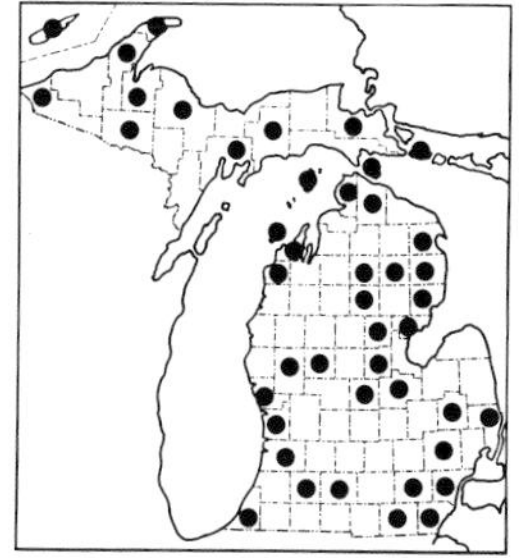
598. Trifolium aureum

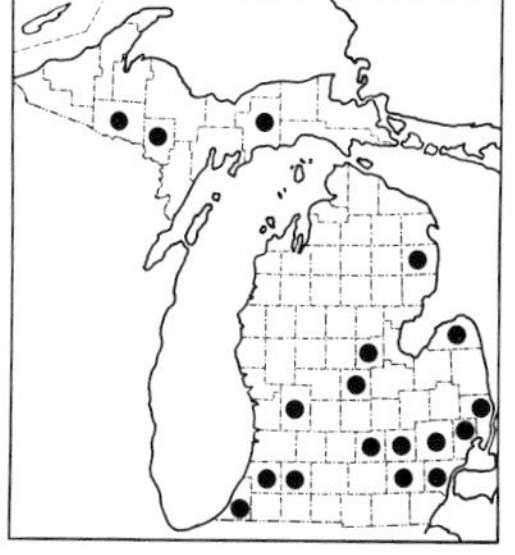
599. Trifolium campestre

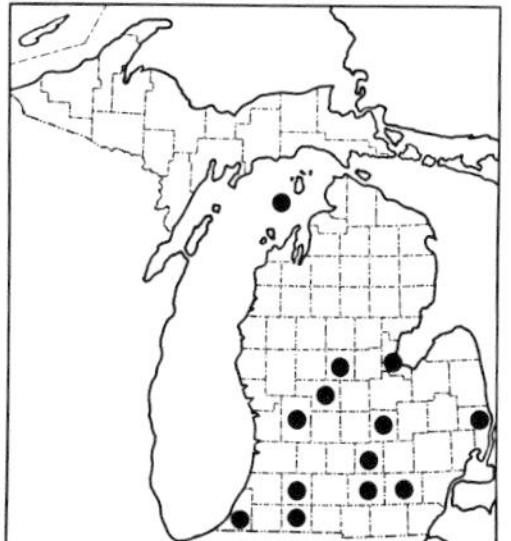
600. Trifolium dubium

is less strongly veined than in *T. campestre* but more strongly than in *M. lupulina,* in which the veins are scarcely if at all evident. The terminal leaflet is sometimes not distinctly stalked.

5. **Medicago**

REFERENCES

Gunn, Charles R., W. H. Skrdla, & H. C. Spencer. 1978. Classification of Medicago sativa L. Using Legume Characters and Flower Colors. U.S. Dep. Agr. Tech. Bull. 1574. 84 pp.

Isely, Duane. 1951. The Leguminosae of the North-central United States: I. Loteae and Trifolieae. Iowa St. Coll. Jour. Sci. 25: 439–482. [*Medicago,* pp. 451–459.]

Isely, Duane. 1983. Classification of Alfalfa (Medicago sativa L.) and Relatives. Iowa St. Jour. Res. 57: 207–220.

Lesins, Karlis Adolfs, & Irma Lesins. 1979. Genus Medicago (Leguminosae) a Taxogenetic Study. Junk, The Hague. 228 pp. [Contains much information but can hardly be considered a taxonomic monograph; the authors explicitly decline to follow the International Code of Botanical Nomenclature.]

Small, Ernest, & Brenda S. Brookes. 1984. Taxonomic Circumscription and Identification in the Medicago sativa—falcata (Alfalfa) Continuum. Econ. Bot. 38: 83–96.

KEY TO THE SPECIES

1. Fruit, even when immature, covered with hooked spines; stipules deeply toothed or lacerate . . . . 1. **M. polymorpha**
1. Fruit without spines; stipules at most shallowly toothed
  2. Corolla 1.8–2.2 (2.5) mm long, yellow; plant low and ± prostrate or slightly ascending; fruit ± reniform, 1-seeded, nearly black when mature . . . . . . 2. **M. lupulina**
  2. Corolla (7) 8–10 (11) mm long, blue-purple or (less often) yellow, cream, or variegated (white in albinos); plant erect; fruit curved to spirally coiled, several-seeded, green when mature . . . . 3. **M. sativa**

1. **M. polymorpha** L. Bur-clover

Map 601. A Eurasian species, the basis for early reports of *M. hispida* Gaertner [nom. illeg. = *M. nigra* (L.) Krocker] and *M. denticulata* Willd. from waste ground in Michigan. Quite probably no longer in the state. Collected by Robbins on the Keweenaw Peninsula in 1863 (GH, MICH), and by Farwell at Detroit in 1894 (*1460,* BLH).

Even in a narrower sense than the original breadth assigned to it by Linnaeus, this is an extremely variable species. The fruit is glabrous, strongly spiraled. A collection from a farm at Grand Rapids in 1902 (*Cole,* MICH, ALBC) appears to be *M. minima* (L.) L., with fruit ± pubescent between the spines and weakly glandular on the surface; unlike *M. polymorpha,* the stipules are at most shallowly toothed at the base; the stem is pubescent. Its status in 1902 is not clear, whether cultivated or a waif from contaminated seed. The corolla in both these species is slightly longer than in *M. lupulina,* but distinctly smaller than in *M. sativa.*

2. **M. lupulina** L. Fig. 213 Black Medick

Map 602. A Eurasian species, first collected in Michigan in 1870 (at Ann Arbor) and now a pernicious weed of roadsides, lawns, fields, railroads, grassy banks, and waste places throughout; invading disturbed woods and rock outcrops.

Often confused with *Trifolium campestre* and *T. dubium,* but differing in its even smaller flowers, black reniform fruit not enclosed in the old perianth, square stems, and other features (see key to genera, couplet 7). Michigan specimens almost invariably have ± glandular peduncles and calyces.

3. **M. sativa** L. Alfalfa

Map 603. Roadsides, fields, and waste ground everywhere. A very important and variable crop plant. Apparently a man-selected tetraploid hybrid of uncertain origin, widely naturalized. It is not clear when alfalfa became established as an escape from cultivation in Michigan, but it was at least by the 1890's.

The corolla of typical alfalfa is blue-purple (white in albinos), and the fruit is coiled at least 1.5–2 times at maturity. A yellow-flowered alfalfa with straight to C-shaped fruits has been called *M. falcata* L. or treated as a subspecies of *M. sativa*. It has not been collected in Michigan (although one cannot be certain of yellow-flowered plants lacking fruit). Our yellow-flowered plants as occasionally found outside of cultivation were discussed by Rabeler and Gereau (Mich. Bot. 23: 43–46. 1984), who concluded they were hybrids, which have been called *M.* ×*varia* Martyn or *M. sativa* ssp. ×*varia* (Martyn) Arcang.; insofar as they have fruit, they are indeed intermediate, and the corollas may be blue-green or variegated. Another approach is taken by Isely (1983), who simply included these intermediate plants in *M. sativa,* with which they are confluent and which is hardly "pure" anyway, having been so hybridized and selected for hardiness and forage yield that the original "natural" species is thoroughly obscured. Small and Brookes (1984), in yet another recent study of alfalfa, also note that ssp. ×*varia* is closer to *sativa* than to *falcata*.

601. Medicago polymorpha

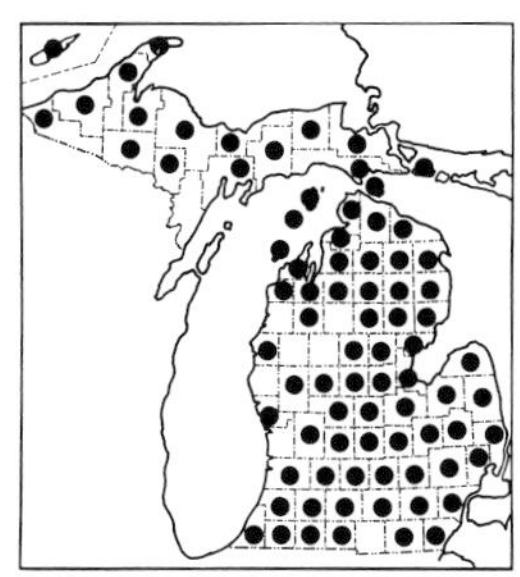
602. Medicago lupulina

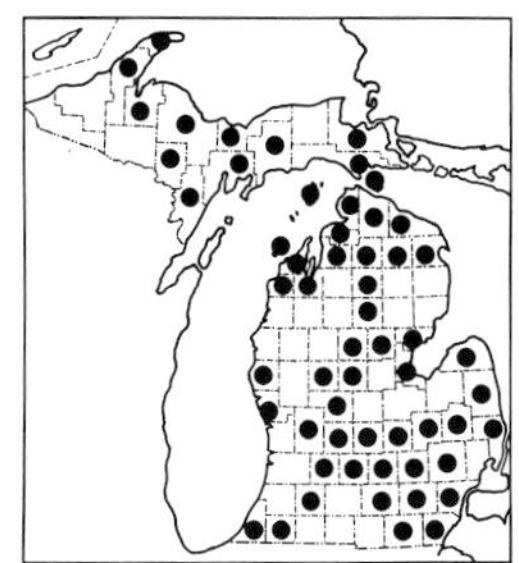
603. Medicago sativa

Yellow-flowered alfalfa has been collected in Michigan, mostly within the past 20 years, in the following counties: Alpena, Bay, Berrien, Cass, Chippewa, Clinton, Delta, Eaton, Emmet, Houghton, Ingham, Jackson, Kalamazoo, Kent, Leelanau (South Manitou I.), Livingston, Oakland, Schoolcraft, and Shiawassee. It will doubtless be found in many more, as it seems to be rapidly spreading (see Rabeler & Crowder in Mich. Bot. 24: 125. 1985).

6. **Cytisus**

1. **C. scoparius** (L.) Link — Scotch Broom

Map 604. A European species, locally naturalized in North America as an escape from cultivation. Evidently established in a disturbed area near Lake Michigan (*F. G. Goff 1223* in 1979, MSC).

7. **Baptisia** — False Indigo

B. australis (L.) R. Br., a more southern species with large blue flowers and glabrous foliage similar to that of *B. lactea* except for slightly larger stipules, is native no closer to Michigan than western Pennsylvania and the Ohio River valley of Indiana and Ohio. It has been collected on the campus of Western Michigan University in Kalamazoo, presumably planted but possibly escaped from cultivation.

KEY TO THE SPECIES

1. Flowers bright yellow, 1.1–1.5 cm long; body of fruit (excluding stipe and beak) ca. (0.6) 0.8–1 cm long; leaflets 0.6–2.3 (3.3) cm long . . . . 1. **B. tinctoria**
1. Flowers white or cream, (1.8) 2.1–2.5 cm long; body of fruit ca. 2–3 cm long; leaflets mostly (2.1) 2.5–6 cm long
   2. Stem and leaves pubescent; stipules sometimes almost as large as the leaflets . . . . 2. **B. leucophaea**
   2. Stem and leaves glabrous; stipules usually very small . . . . 3. **B. lactea**

604. Cytisus scoparius

605. Baptisia tinctoria

606. Baptisia leucophaea

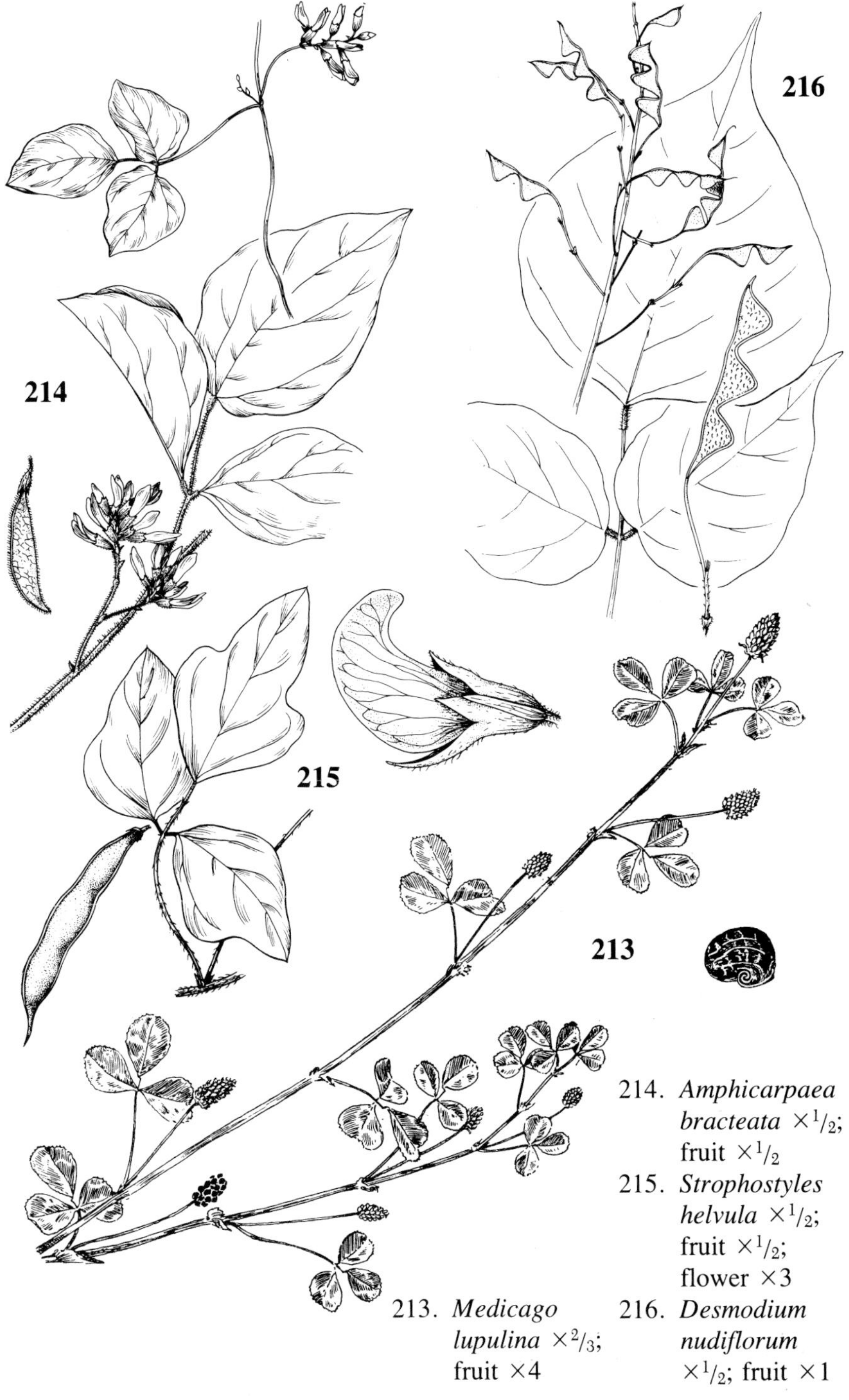

214. *Amphicarpaea bracteata* ×1/2; fruit ×1/2
215. *Strophostyles helvula* ×1/2; fruit ×1/2; flower ×3
213. *Medicago lupulina* ×2/3; fruit ×4
216. *Desmodium nudiflorum* ×1/2; fruit ×1

1. **B. tinctoria** (L.) R. Br.

Map 605. Open woods (oak and hickory), sandy openings and borders, fencerows.

2. **B. leucophaea** Nutt.

Map 606. A conspicuous plant, at least as large in all its parts as the next, but pubescent, with large stipules, persistent bracts, and cream-colored flowers. Collected in Michigan, however, only by Mr. and Mrs. Hanes on prairie roadsides and railroads in Schoolcraft and Texas tps., Kalamazoo County.

Perhaps not a distinct species from *B. bracteata* Ell. (see Isely in Brittonia 30: 470. 1978) but the combination as var. *leucophaea*, correct under the Code since 1981, has not yet been published.

3. **B. lactea** (Raf.) Thieret

Map 607. Prairies and associated or similar dry open roadsides, railroads, and fencerows.

A conspicuous glaucous bushy herb a meter or more tall, with showy white flowers and a large black pod. The leaves usually blacken in drying. Long known as *B. leucantha* T. & G.

8. Amphicarpaea

This generic name has been conserved as spelled here, although many authors continue to use the variant form *Amphicarpa*.

1. **A. bracteata** (L.) Fern. Fig. 214 Hog-peanut

Map 608. A low vine generally of open woods and thickets, ranging from sandy oak and oak-hickory woods to lowland swamps; shores, river banks, damp areas in deciduous forests.

Extremely variable in pubescence, stature, and size of parts, but recognizable varieties cannot be clearly distinguished. The petioles are usually retrorsely pubescent, rarely glabrous or with a few antrorse hairs; the stems vary from glabrate to heavily pubescent with reflexed brownish hairs.

9. Phaseolus Bean

KEY TO THE SPECIES

1. Raceme many-flowered, slender, elongate at maturity, exceeding the subtending petiole (or even the entire leaf); pedicels much longer than the tiny bractlets (less than 1 mm) at base of calyx . . . . 1. **P. polystachios**
1. Raceme few-flowered, scarcely if at all exceeding the subtending petiole; pedicels shorter than the pair of large bractlets which subtend and conceal the calyx at anthesis . . . . 2. **P. vulgaris**

1. **P. polystachios** (L.) BSP. Wild Bean

Map 609. Collected in thickets on Belle Isle in 1896 (*Farwell 1550*, BLH; see Rep. Mich. Acad. 2: 53. 1902), where its status is uncertain. If native, this was surely at the northern edge of the range for this species.

2. **P. vulgaris** L. Green, String, Snap, Kidney, Wax, or Common Bean

Map 610. This bean was apparently domesticated in Mexico over 7000 years ago, and numerous cultivars are now grown throughout the world, some of them erect "bush" beans and some vine "pole" beans. Rarely one persists as a relic from cultivation or spreads in a dump or along a watercourse or roadside. Beans are one of the most important agricultural crops in the state, especially in the "Thumb" area. Michigan produces more dry edible beans than any other state in the nation.

10. Strophostyles

1. **S. helvula** (L.) Ell. Fig. 215 Wild Bean

Map 611. Thickets on disturbed ground, roadsides, ditch banks, beaches, and dunes.

The specific epithet is usually misspelled *helvola*.

607. Baptisia lactea

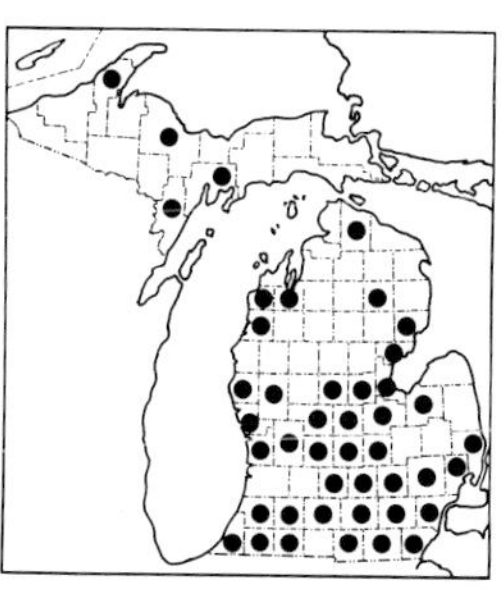

608. Amphicarpaea bracteata

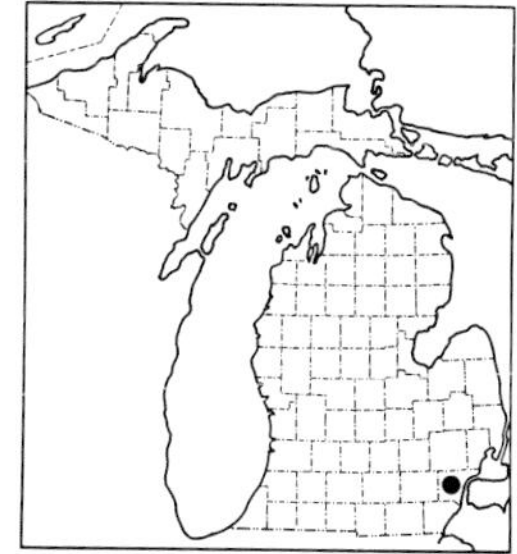

609. Phaseolus polystachios

610. Phaseolus vulgaris

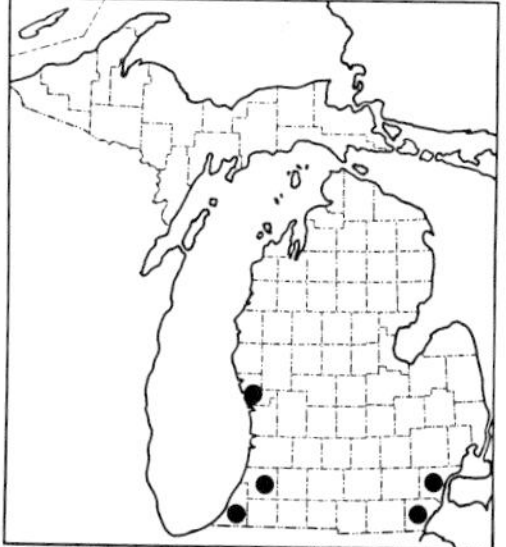

611. Strophostyles helvula

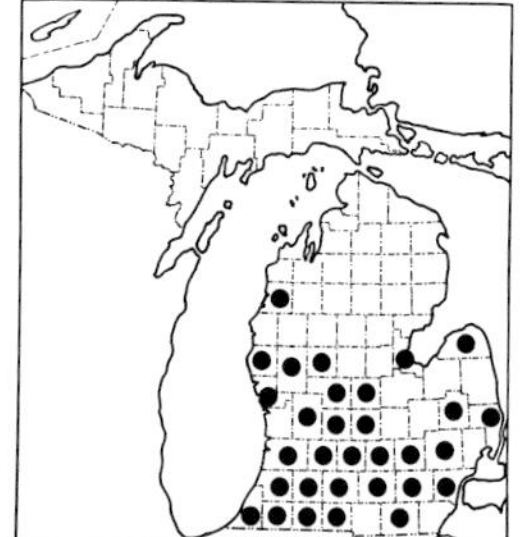

612. Desmodium nudiflorum

11. **Desmodium** Beggars-tick; Tick-trefoil

The characteristic fruit of this genus immediately sets mature material apart in our flora. The modified legume is a *loment,* consisting of (1) 2 or more 1-seeded, flat, indehiscent segments which are ultimately dispersed as separate units. As these are covered with tiny hooked hairs in *Desmodium,* they readily attach to fur or clothes.

The stipules in many species are deciduous, but they and the flowers may provide helpful characters when the fruit is not yet ripe. The flowers of all species are some shade of pink or purple, or whitish; they sometimes fade (or dry) to a greenish shade. This is a difficult genus, with a number of characters often having to be considered in order to make a sure identification of species. The names employed here can be easily correlated with those used in current manuals, but little effort is made to explain synonymy or names so frequently misapplied in old manuals or lists.

The generic name *Desmodium* is conserved against the older *Meibomia,* which was often used in the past. Some systems divide the genus into two, in which case all our species except the first two go into *Meibomia.* Those remaining in *Desmodium,* then, differ from the others, in addition to the characters mentioned in the key, in having the narrowest neck connecting the segments of the loment, which has conspicuous sinuses extending almost all the way to the opposite suture (fig. 216).

REFERENCES

Isely, Duane. 1953. Desmodium paniculatum (L.) DC. and D. viridiflorum (L.) DC. Am. Midl. Nat. 49: 920–933.

Isely, Duane. 1955. The Leguminosae of the North-central United States II. Hedysareae. Iowa St. Coll. Jour. Sci. 30: 33–118. [*Desmodium,* pp. 38–72.]

Isely, Duane. 1983. The Desmodium paniculatum (L.) DC. (Fabaceae) Complex Revisited. Sida 10: 142–158.

KEY TO THE SPECIES

1. Calyx lobes ± deltoid or rounded, much shorter than the tube; leaflets with stipels none (or obsolete); stipe of fruit at least 3 times as long as the calyx
   2. Pedicels, at least after anthesis, becoming 1 cm or more long; inflorescence on a long naked (very rarely leafy) peduncle arising from ground level, the stem terminated by a cluster of leaves; blade of terminal leaflet at least 20% longer than broad . . . . 1. **D. nudiflorum**
   2. Pedicels less than 1 cm long, even in fruit; inflorescence borne on the leafy stem (leaves usually crowded at base of peduncle); blade of terminal leaflet often about as broad as long or broader . . . . 2. **D. glutinosum**
1. Calyx lobes triangular to lanceolate, about equaling or longer than the tube; leaflets with stipels ± persistent; stipe of fruit at most ca. 1.5 times as long as the calyx, usually shorter
   3. Plant a prostrate, trailing vine; leaflets suborbicular; stipules ovate, persisting and conspicuous, becoming ± reflexed . . . . 3. **D. rotundifolium**

3. Plant erect; leaflets longer than broad; stipules various
4. Leaflets suborbicular to ovate-oblong, rounded-obtuse at the apex, the terminal one about the same size as the lateral ones, up to 2.5 (3) cm long
5. Stem and leaves lightly hairy; mature pedicels ca. 4–9 (10) mm long; petioles (3) 6–9 (12) mm long . . . . 4. **D. ciliare**
5. Stems and leaves essentially glabrous; mature pedicels mostly 7–15 mm long; petioles (8) 12–22 (24) mm long . . . . 5. **D. marilandicum**
4. Leaflets ± ovate to lanceolate and acute at the apex, or narrowly linear-oblong, the terminal one usually larger than the lateral ones and over 3.5 cm long
6. Leaves sessile or nearly so, the petioles less than 3 mm long; leaflets narrowly oblong or linear . . . . 6. **D. sessilifolium**
6. Leaves with petioles of middle (or all) leaves at least 4 mm long; leaflets ± ovate to lanceolate
7. Lower surface of leaflets (at least on midvein and other main veins) with tiny hooked hairs (besides any other kinds); stipules persistent, ovate (short- or long-acuminate)
8. Axis of inflorescence with numerous spreading eglandular straight-tipped hairs longer than its diameter; stem or inflorescence branched, the plant with more than 1 raceme; leaflets weakly reticulate-veiny beneath; segments of mature fruit ca. 7–11 mm long . . . . 7. **D. canescens**
8. Axis of inflorescence with all hairs shorter than its diameter, the longest ones glandular or hooked; stem often simple, the plant then with a single raceme; leaflets very strongly reticulate-veiny beneath; segments of mature fruit ca. 4–6.5 mm long . . . . 8. **D. illinoense**
7. Lower surface of leaflets without hooked hairs (or a very few scattered ones present); stipules deciduous to somewhat persistent, lanceolate to linear
9. Calyx less than 3 mm long (including the longest tooth); flowers less than 6 mm long; segments of fruit (1) 2–3, rounded (not at all angled) along both sutures (more strongly on one than on the other). . . . . 9. **D. obtusum**
9. Calyx or flowers (or both) usually longer; segments of fruit mostly 2–4, rounded or triangular in aspect (triangular in species with small flowers)
10. Pubescence on lower surface of leaflets denser on the midvein than on the branch veins, and denser on these than on the smaller veins; petiole shorter than the width of the terminal leaflet; segments of fruit rounded in aspect, not triangular (fig. 217) . . . . 10. **D. canadense**
10. Pubescence (if any) on lower surface of leaflets quite uniform in density on all veins (and surface); petiole (at least on middle leaves) equaling or exceeding the width of the terminal leaflet (sometimes shorter in *D. paniculatum*); segments of fruit distinctly triangular in aspect (or even unequally diamond-shaped; fig. 218)
11. Stipules lanceolate, broadest just above the base, (9) 11–18 (21) mm long; terminal leaflet broadly acuminate, (9) 10–15 cm long; segments of fruit 7–11 mm long; flowers ca. 10–13 mm long . . . . 11. **D. cuspidatum**
11. Stipules narrowly triangular, broadest at the very base, 3–8 mm long; terminal leaflet elliptic-lanceolate, or less than 10 cm long, or both, not acuminate; segments of fruit (5.5) 6–7.5 mm long; flowers (5) 6–7.5 (9.5) mm long . . . . 12. **D. paniculatum**

1. **D. nudiflorum** (L.) DC. Fig. 216

Map 612. A species of forests, ranging from rich hardwoods to oak and oak-hickory. Frequently in the same woods as *D. glutinosum*, and apparently blooms when the latter is in fruit.

Plants with scattered leaves on the flowering stem have been called f. *foliatum* (Farwell) Fassett (TL: Royal Oak [Oakland Co.] & Dearborn [Wayne Co.]). Such plants may be the basis for early reports of *D. pauciflorum* (Nutt.) DC. from the state, a species of more southern range, similar to *D. glutinosum* but with flowers averaging smaller, white rather than pink, the stipe of the fruit minutely hooked-puberulent, and the leaves scattered.

2. **D. glutinosum** (Willd.) Wood

Map 613. Rich deciduous woods and river-bank thickets to dry upland oak-hickory woods, including disturbed borders.

In material too young (or fragmentary) to display the stipitate fruit and mature pedicel, the linear stipules may be present, aiding in identification; these are 6.5–10 (12) mm long (rarely shorter), and the bracts in the inflorescence are similar although a bit shorter. In *D. nudiflorum* the stipules tend to be deciduous earlier, but are shorter (up to 3.5 mm) and the bracts in the inflorescence are likewise very small.

3. **D. rotundifolium** DC.

Map 614. Oak woods (sometimes with hickory or pine); dry thickets and openings.

4. **D. ciliare** (Willd.) DC.

Map 615. Dry open ground and oak woodlands.

5. **D. marilandicum** (L.) DC.

Map 616. Oak woods, borders and openings in woods; open (usually sandy) ground.

6. **D. sessilifolium** (Torrey) T. & G.

Map 617. Sandy fields, openings, and prairies; borders of oak woods.

613. Desmodium glutinosum

614. Desmodium rotundifolium

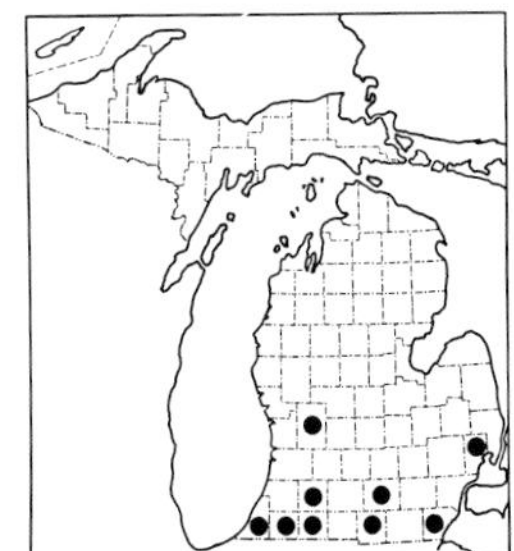

615. Desmodium ciliare

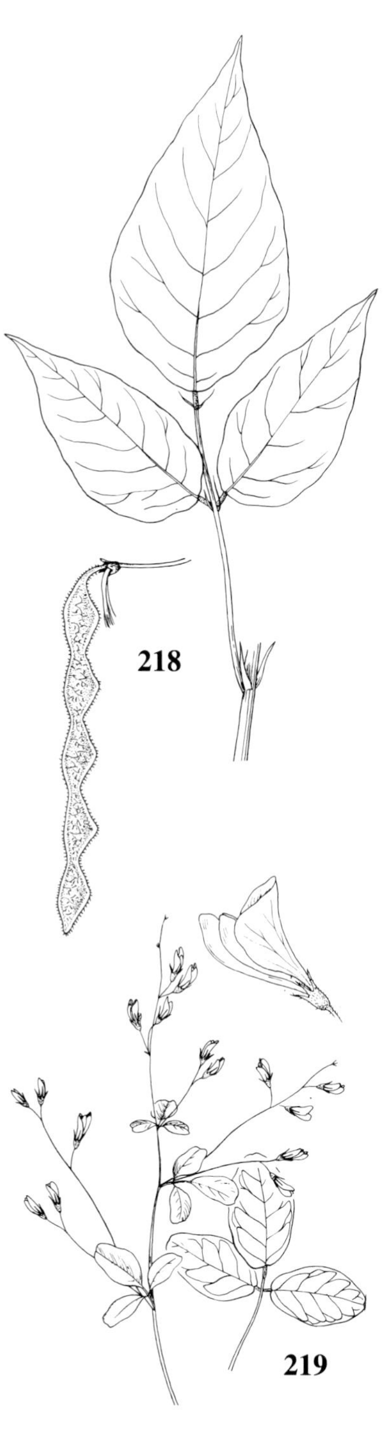

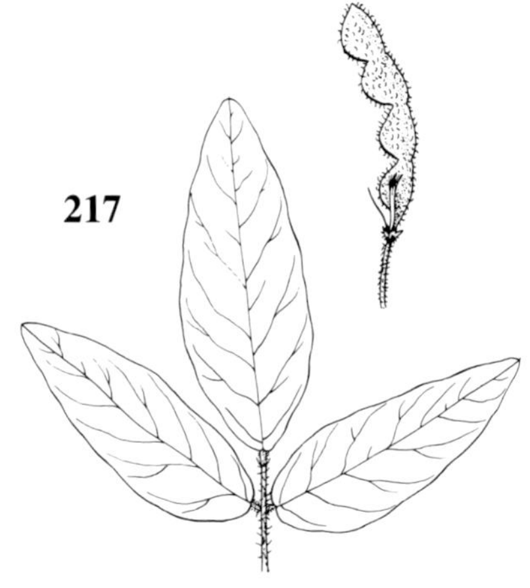

217. *Desmodium canadense,* leaf $\times {}^{1}/_{2}$; fruit $\times 1$
218. *D. cuspidatum* $\times {}^{1}/_{2}$; fruit $\times 1$
219. *Lespedeza violacea* $\times {}^{1}/_{2}$; flower $\times 2$
220. *Caragana arborescens* $\times {}^{2}/_{5}$; fruit $\times {}^{2}/_{5}$

The leaflets usually have some hooked hairs beneath, especially along the midveins.

7. **D. canescens** (L.) DC.

Map 618. Moist to dry sandy open ground.

The segments of the fruit are more angled on the central margin than in the next species, although the apex of the triangle is rounded.

8. **D. illinoense** A. Gray

Map 619. Prairies and associated or similar roadsides, fields, railroads; borders and openings in oak woods.

The segments of the fruit are more broadly rounded (on both margins) than in the preceding species, with no aspect of being triangular.

9. **D. obtusum** (Willd.) DC.

Map 620. Dry open, usually sandy, ground, as on hillsides or bluffs; clearings and openings in oak woods.

This species was long called *D. rigidum* (Ell.) DC.

10. **D. canadense** (L.) DC. Fig. 217

Map 621. In both dry and moist ground: river banks and shores, wet

616. Desmodium marilandicum

617. Desmodium sessilifolium

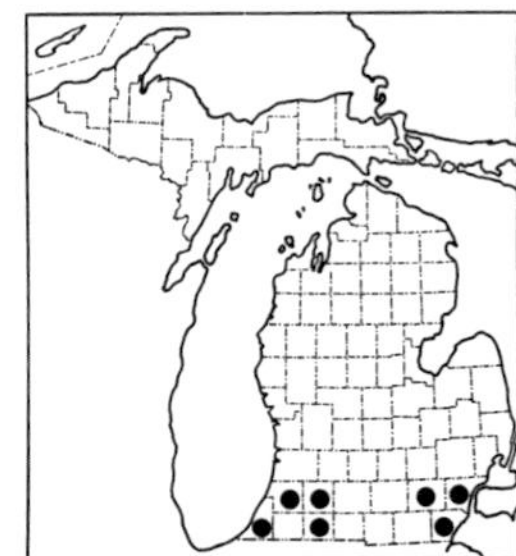

618. Desmodium canescens

619. Desmodium illinoense

620. Desmodium obtusum

621. Desmodium canadense

prairies and fens, borders of woods and thickets, sandy open ground (fields, clearings, prairie relics along roadsides and railroads).

This is both our most widespread and our showiest species in the genus, with red-purple flowers ca. 8–12 mm long.

11. **D. cuspidatum** (Willd.) Loudon   Fig. 218

Map 622. Oak and oak-hickory woods, borders and clearings, thickets and river banks.

The leaflets almost invariably are glabrous or essentially so beneath in our specimens.

12. **D. paniculatum** (L.) DC.

Map 623. Moist to dry (chiefly oak) woods, clearings, shores, ravines; prairies, sandy hillsides and banks.

A variable complex, as here treated following Isely. Most of our specimens can be placed in one of two varieties: var. *paniculatum,* with leaflets narrowly elliptic to lanceolate, generally 3–5 times as long as wide, typically glabrate beneath; or var. *dillenii* (Darl.) Isely, with leaflets more broadly ovate-elliptic, less than 3 times as long as wide, often more heavily pubescent beneath with appressed or ascending hairs. (The latter variety may be treated at specific rank as *D. dillenii* Darl. or, if that name is rejected, as *D. perplexum* Schubert.) Intermediates occur, however, and the varieties are troublesome throughout the range of the species. The distributions of the two in Michigan are essentially the same, and they may grow in the same woods. Some material from Allegan and Muskegon counties has the small flowers and uniform pubescence on the leaflets characteristic of *D. paniculatum,* but rounded fruit segments as in *D. canadense*.

D. viridiflorum (L.) DC. has been reported from Michigan in old literature, but as now known occurs only south of our area, as does *D. nuttallii* (Schindler) Schubert, a somewhat questionable segregate from it. This complex differs from the *D. paniculatum* complex in having the leaflets velvety

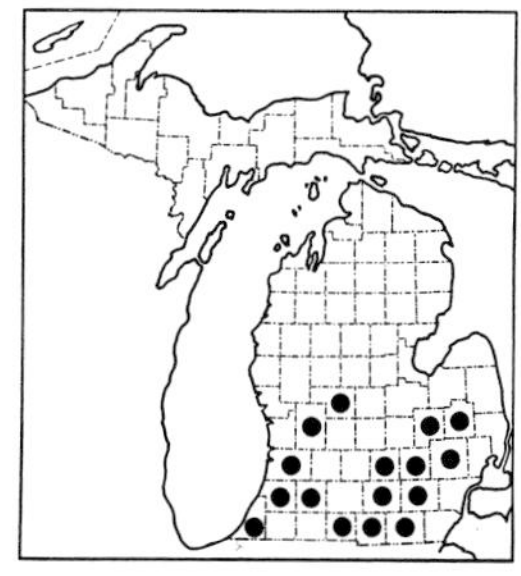

622. Desmodium cuspidatum

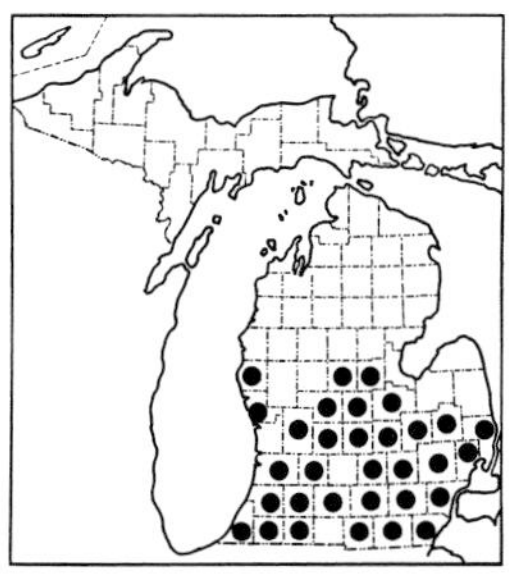

623. Desmodium paniculatum

624. Psoralea psoralioides

tomentose beneath and the stipules broadest just above the base. Michigan reports seem to have been based on collections of *D. paniculatum* var. *dillenii*—if not even more seriously misidentified.

12. Psoralea

1. **P. psoralioides** (Walter) Cory — Sampson's Snakeroot

Map 624. Native south of our region; collected once near a railroad at Royal Oak (*Farwell 5260,* BLH; see Rep. Mich. Acad. 22: 184. 1921), presumably a waif.

13. Lespedeza — Bush-clover

Sometimes considered a difficult genus, but our species are rather easily recognized, the major problems arising from hybridization (or from efforts to name recognizable varieties in some of the highly variable species). Hybrids are fairly frequent in the genus, and combine parental characters in various ways, including intermediate sizes and shapes. Plants in the shade may have larger leaflets, longer petioles, and longer internodes than sun forms and hence appear quite different.

The fruit is suggestive of a one-segmented *Desmodium* and indeed the two genera are traditionally (if not necessarily accurately) treated as fairly closely related. In addition to conspicuous petaliferous flowers, species of *Lespedeza* generally produce cleistogamous flowers in inconspicuous axillary clusters lower on the stem. The fruits from these flowers are easily recognized by the sharply recurved little style, which bends back on itself 180 degrees.

A shrubby species with showy racemes of relatively large flowers (and no cleistogamous flowers), probably the Japanese *L. thunbergii* (DC.) Nakai or a relative, has been collected in northwestern Washtenaw County. Its status, whether planted or established as an escape, is not known. Many species, including exotic ones, are planted for wildlife food and cover, agricultural use, or erosion control.

REFERENCES

Clewell, Andre F. 1966a. Natural History, Cytology, and Isolating Mechanisms of the Native American Lespedezas. Bull. Tall Timbers Res. Sta. 6. 39 pp.

Clewell, Andre F. 1966b. I. Identification of the Lespedezas in North America II. A Selected Bibliography on Lespedeza. Bull. Tall Timbers Res. Sta. 7. 29 pp. [Includes key & photos of specimens of native and introduced species.]

Clewell, Andre F. 1966c. Native North American Species of Lespedeza (Leguminosae). Rhodora 68: 359–405. [Supplemented by mimeographed list of "Voucher Specimens for Distribution Maps of Lespedeza."]

Isely, Duane. 1955. The Leguminosae of the North-central United States II. Hedysareae. Iowa St. Coll. Jour. Sci. 30: 33–118. [*Lespedeza*, pp. 76–110.]

KEY TO THE SPECIES

1. Stipules conspicuous, ovate, brownish, strongly striate, persistent, much exceeding the petioles; calyx lobes rounded; lateral veins of leaflets strongly parallel, running to the margins; plant an annual .... 1. **L. stipulacea**
1. Stipules inconspicuous, narrowly triangular to subulate, at most 3-veined, deciduous, the length various; calyx lobes elongate, sharp-pointed; lateral veins of leaflets ± branched and anastomosing before reaching the margins; plant perennial
  2. Corolla yellow or cream (often drying orange-brown), often with a purplish spot at the base of the standard; calyx nearly equaling to exceeding the fruit
    3. Flowers solitary or 2–3 in the axils of numerous cauline leaves (which exceed them); leaflets mostly 2–4.5 mm wide, broadest near the apex .... 2. **L. cuneata**
    3. Flowers crowded into dense head-like clusters at and near the end of the stem, these often exceeding the subtending leaves; leaflets (at least the terminal ones) mostly 5 mm or more wide, broadest near or slightly above the middle
      4. Leaflets less than twice as long as wide, elliptic-oblong to obovate; peduncles usually ± equaling or longer than the inflorescences and longer than subtending leaves; calyx lobes (3.5) 4–6.5 mm long .... 3. **L. hirta**
      4. Leaflets over 2 (usually ca. 3) times as long as wide, narrowly elliptic-oblong; peduncles usually shorter than inflorescences and shorter than subtending leaves; calyx lobes (4) 5–8 (10) mm long .... 4. **L. capitata**
  2. Corolla purple; calyx at most about half as long as the fruit
    5. Stems trailing, downy with spreading pubescence. .... 5. **L. procumbens**
    5. Stems erect (or ascending), glabrous or with appressed hairs [if pubescence spreading, suspect a hybrid]
      6. Petaliferous flowers in inflorescences most or all of which are on very slender peduncles exceeding the leaves (cleistogamous flowers in axillary clusters); calyx teeth of cleistogamous fruit less than a fourth as long as the fruit; keel of corolla longer than the wings; leaflets less than twice as long as broad .... 6. **L. violacea**
      6. Petaliferous flowers in inflorescences on peduncles not exceeding the leaves; calyx teeth of cleistogamous fruit usually about a fourth as long as the fruit or slightly longer; keel of corolla shorter than the wings; leaflets ca. 2–6 times as long as broad
        7. Leaflets glabrous (or rarely with a few hairs near midvein) above, at least the longer ones 2–3 times as long as wide, oblong-elliptic .... 7. **L. intermedia**
        7. Leaflets finely strigose above, at least the longer ones (3) 4–6 (8) times as long as wide, quite narrowly oblong- or linear-elliptic .... 8. **L. virginica**

1. **L. stipulacea** Maxim. — Korean Bush-clover

Map 625. Introduced from eastern Asia to the United States in 1919 as a potential crop plant, and spreading rather rapidly in some areas as an escape. In Michigan, however, collected only by the Haneses in fields and on a lakeshore bank in Kalamazoo County, 1937–1938.

2. **L. cuneata** (Dumont) G. Don — Sericea

Map 626. Another eastern Asian species, grown (from Japanese seed) at Michigan State University as early as 1894. However, only in the last 40 or 50 years has it become locally conspicuous along roadsides and fields, well established as an escape from cultivation or planting.

This is a very late bloomer, flowering with us from September into October.

3. **L. hirta** (L.) Hornem.

Map 627. Dry open usually sandy ground, including fields, roadsides, and river banks; oak and oak-hickory woods.

Even if flower color and fruit are unknown, this species can be easily recognized by the ± dense pubescence on the stems and both surfaces of the leaves. No other species in our area with oblong-elliptic leaves has so much pubescence or the flowers in such dense heads. Both this species and the next can be distinguished from our common purple-flowered species both by the dense heads and the spreading pubescence of the stems. Pubescence, if any, is appressed on the stems of species 6–8, only the rare and prostrate *L. procumbens* having spreading pubescence on the stem. (See also comments on *L.* ×*nuttallii* under *L. intermedia*.)

4. **L. capitata** Michaux

Map 628. Dry open usually sandy ground, including prairies, fields, bluffs, and roadsides. The Cheboygan County record doubtless represents a roadside waif well beyond the normal range (found only in 1924).

The leaves are nearly sessile or on very short petioles—shorter than the petiolules of the terminal leaflets. Clewell has pointed out that in all other native American Lespedezas, the petiole, at least on the largest (middle) leaves, is at least as long as the petiolule. This character is helpful for material without fruit or known flower color. In many respects this is an extremely variable species. The leaflets vary from glabrous to silky pubescent above, but are always ± silky pubescent beneath.

L. capitata hybridizes with *L. hirta,* and this hybrid has been identified by Clewell from Berrien, Kent, St. Clair, and Shiawassee counties. Additional collections from Lapeer, Oakland, and Washtenaw counties appear to be the same. Such plants are intermediate in leaf shape, and often (less noticeably) other characters. *L. capitata* × *L. intermedia* has been identified

625. Lespedeza stipulacea

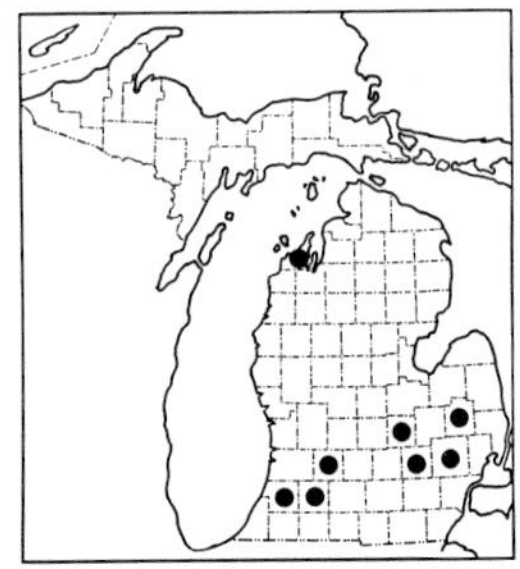

626. Lespedeza cuneata

627. Lespedeza hirta

by Clewell from Berrien and Cass counties; and a collection from Livingston County appears to be *L. capitata* × *L. virginica*.

5. **L. procumbens** Michaux

Map 629. Collected in Michigan only on the sandy shore of Dewey Lake 1902–1906.

6. **L. violacea** (L.) Pers. Fig. 219

Map 630. Dry usually ± open woods (especially oak), thickets, banks, and prairies.

An apparent hybrid with *L. virginica* has been identified by Clewell from Washtenaw County.

7. **L. intermedia** (Watson) Britton

Map 631. Oak-hickory woods; dry woodlands, plains, and bluffs.

The distinction between this species and the preceding is not clearcut on the basis of leaflet shape alone, although the separation from the next species is quite clear.

The hybrid with *L. hirta* [*L.* ×*nuttallii* Darl.] has spreading hairs on the stem (as in *L. procumbens*) but otherwise may resemble *L. intermedia*. It has been identified by Clewell from Kalamazoo, Kent, Oakland, St. Joseph, Van Buren, and Washtenaw counties; collections from Berrien, Calhoun, Ingham, and Newaygo counties appear to belong here also. One collection from Oakland County has the aspect of *L. violacea* but the keel is distinctly shorter than the wings and standard, and it may be *L. intermedia* × *L. violacea*.

8. **L. virginica** (L.) Britton

Map 632. Dry open woods (especially oak), shores, fields, railroad banks, hills.

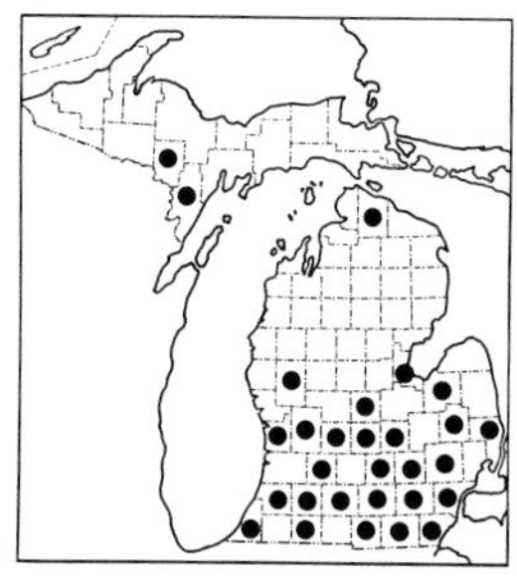

628. Lespedeza capitata

629. Lespedeza procumbens

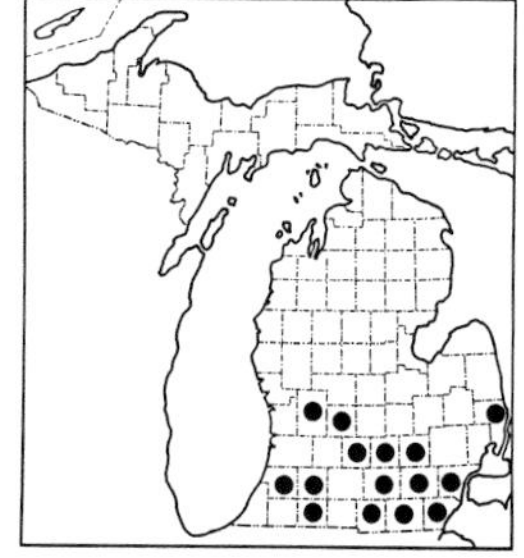

630. Lespedeza violacea

14. **Glycine**

REFERENCE

Hermann, F. J. 1962. A Revision of the Genus Glycine and Its Immediate Allies. U.S. Dep. Agr. Tech. Bull. 1268. 82 pp.

1. **G. max** (L.) Merr. Soybean

Map 633. An important world crop plant (for oil, seeds, fodder), its origins lost in antiquity. An annual, widely grown in southern Michigan and rarely spontaneous near cultivation as a temporary waif.

15. **Caragana**

1. **C. arborescens** Lam. Fig. 220 Pea-tree

Map 634. A native of northern Asia, sometimes planted, especially for windbreaks or hedges in the north-central states. Rarely found established as an escape in Michigan (see Mich. Bot. 16: 135. 1977). Open woods along a highway in Dickinson County and deciduous woods in Emmet County. Labels on collections from areas of human occupation in Baraga, Houghton, Kalamazoo, and Marquette counties do not state whether the plants were considered to be escaped.

16. **Gymnocladus**

1. **G. dioicus** (L.) K. Koch Fig. 221 Kentucky Coffee-tree

Map 635. Rich mesic and floodplain forests. Native as far north as southern Michigan. Sometimes also planted as a shade tree, but unlike *Gleditsia* does not spread from seeds or sprouts.

The heavy, woody, large-seeded pods, as much as 15 cm long and a good 4 cm wide, remain on the tree until early spring.

17. **Gleditsia**

1. **G. triacanthos** L. Fig. 222 Honey Locust

Map 636. Apparently indigenous on river banks and floodplains as far north as the Detroit and Kalamazoo rivers, but not the Grand River (see Cole 1901 for discussion of spread from cultivation in Kent County). Escaped to natural habitats as well as filled land, vacant lots, roadsides (seeding from planted trees), shores. Records mapped north of the southern three tiers of counties represent clear escapes, but a few of those farther south may be based only on planted trees when label data with specimens are not clear. Hardy as a planted tree north at least to the Straits of Mackinac.

The long, often ± sickle-shaped, rather thin pods are as impressive in their way as the shorter massive ones of the preceding species. The stems are often armed with stout spines—compound on older branches and trunk. Unarmed plants [f. *inermis* Schneider] are preferred for planting as an ornamental and shade tree.

18. **Wisteria** Wisteria

REFERENCES

Bowden, Wray M. 1976. A Survey of Wisterias in Southern Ontario Gardens. Royal Bot. Gard. (Hamilton) Tech. Bull. 8. 15 pp.

Gillis, William T. 1980. Wisteria in the Great Lakes Region. Mich. Bot. 19: 79–83.

KEY TO THE SPECIES

1. Ovary and fruit glabrous; pedicels and calyx with club-shaped glands .... 1. **W. frutescens**
1. Ovary and fruit densely pubescent; pedicels and calyx without glands ... 2. **W. sinensis**

1. **W. frutescens** (L.) Poiret

Map 637. A native plant south of our region [including *W. macrostachya* (T. & G.) Robinson & Fern.]. Its status in Cass County in an undisturbed floodplain forest, where it was collected in 1978 (*Gillis 14720, 14894,* MSC,

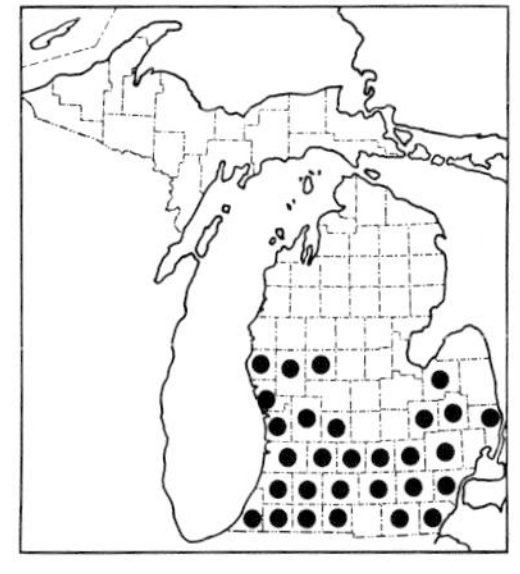
631. Lespedeza intermedia

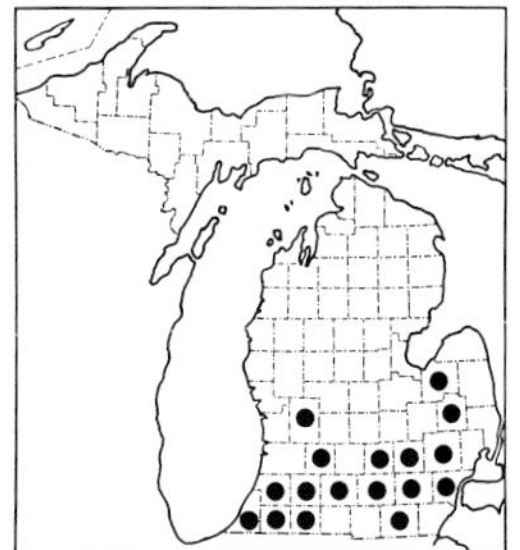
632. Lespedeza virginica

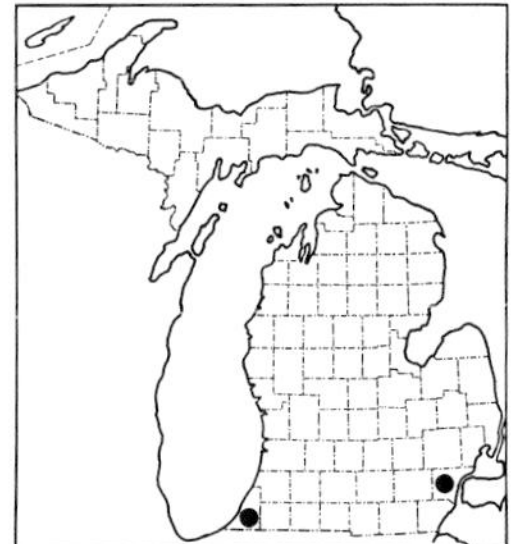
633. Glycine max

634. Caragana arborescens

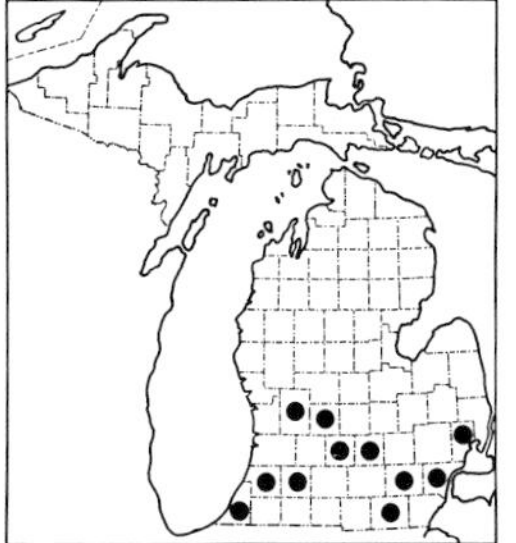
635. Gymnocladus dioicus

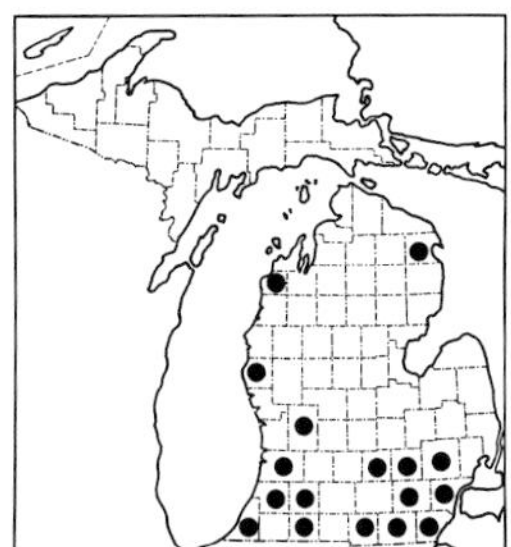
636. Gleditsia triacanthos

MICH, GH), is perhaps questionable, but Gillis (1980) suspected it was native there.

This is a showy species, sometimes cultivated, with racemes (up to 20 cm long) of large blue to purple flowers.

2. **W. sinensis** (Sims) Sweet — Chinese Wisteria

Map 638. A popular cultivated vine, introduced from China, occasionally escaping in North America. Apparently established in a thicket near Harbor Springs (*Voss 14860,* MICH, UMBS; see Mich. Bot. 16: 135. 1977. Also *15634* in 1983, MICH, MSC, UMBS, ISC).

Like the preceding, this species has large pendent racemes—although the Michigan plants appear to be sterile. The petioles and petiolules average a little longer than in *W. frutescens*.

19. **Amorpha**

KEY TO THE SPECIES

1. Leaflets strongly gland-dotted beneath but nearly or quite glabrous; calyx glabrous except on the lobes; fruit glabrous, conspicuously glandular; plant a much-branched shrub . . . . 1. **A. fruticosa**
1. Leaflets with glands very inconspicuous if present beneath, but usually densely pubescent; calyx ± densely pubescent; fruit densely pubescent, the large glands thus concealed; plant simple or branched from the base . . . . 2. **A. canescens**

1. **A. fruticosa** L. Fig. 223 — False Indigo

Map 639. Roadsides, fencerows, banks, fields, sandy knolls, river banks. Probably not originally native in Michigan, but recently spread into the state from farther south and/or (particularly in the Upper Peninsula) from cultivation.

2. **A. canescens** Pursh — Lead-plant

Map 640. Prairies, dry bluffs and hills, sandy roadsides and clearings.

637. Wisteria frutescens

638. Wisteria sinensis

639. Amorpha fruticosa

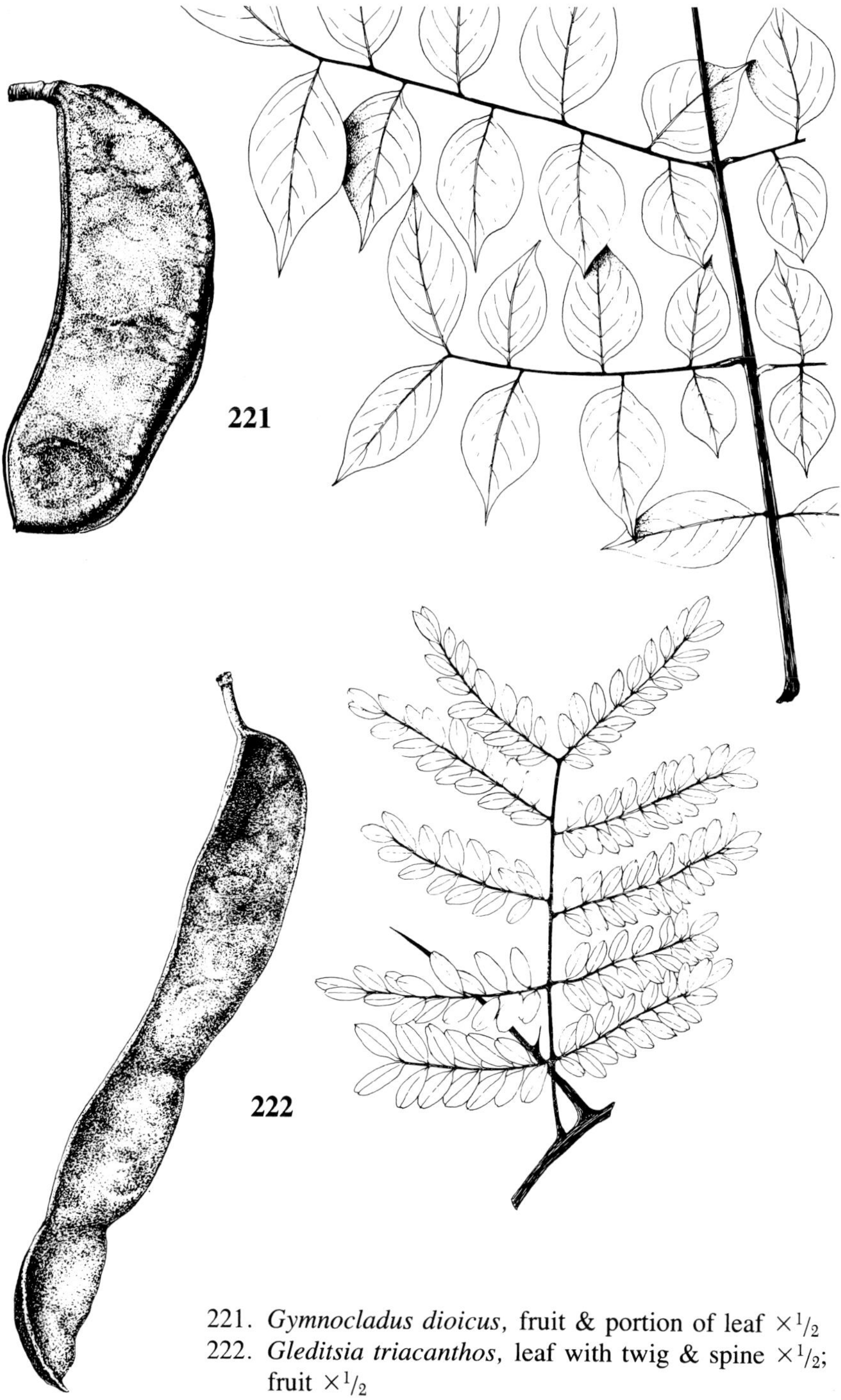

221. *Gymnocladus dioicus*, fruit & portion of leaf $\times ^1/_2$
222. *Gleditsia triacanthos*, leaf with twig & spine $\times ^1/_2$; fruit $\times ^1/_2$

20. **Robinia** Locust

The flowers of all our species have a yellow spot at the base of the standard, whatever the color of the rest of the corolla. All are shrubs or small trees here, escaped from cultivation or spread by the vigorous suckering habit.

An attractive rose-flowered species from the southwest, *R. neomexicana* A. Gray, is planted in Houghton (and perhaps elsewhere) and spreads by suckers though perhaps not as truly escaped as our other species. It has smooth (neither viscid nor pubescent) branchlets, but pubescent and lightly glandular-hispid peduncles and inflorescences; the stipels are small and the leaflets lightly strigose or puberulent on both surfaces.

Shrubs resembling a *Robinia* but with yellow flowers and inflated fruit (over a third as broad as long) are likely to be *Colutea arborescens* L., bladder-senna, a native of southern Europe and northern Africa, which might barely spread from cultivation as in Muskegon Heights.

REFERENCES

Hanes, Clarence R. 1956. Viability of Seed of the Black Locust. Rhodora 58: 26–27.

Isely, Duane, & F. J. Peabody. 1984. Robinia (Leguminosae: Papilionoidea). Castanea 49: 187–202.

Welsh, Stanley Larson. 1960. Legumes of the North-central States: Galegeae. Iowa St. Jour. Sci. 35: 111–249. [*Robinia,* pp. 210–217.]

KEY TO THE SPECIES

1. Corolla white; branchlets smooth, the year-old ones glabrous . . . . . . 1. **R. pseudoacacia**
1. Corolla pink to rose-purple; branchlets hispid or glandular-viscid
   2. Branchlets hispid with stiff spreading hairs (at most with tiny gland-tips) . . . . 2. **R. hispida**
   2. Branchlets viscid with conspicuous sessile or subsessile warty glands, but not hispid . . . . 3. **R. viscosa**

1. **R. pseudoacacia** L. Fig. 224 Black Locust

Map 641. Native south of our latitude, but widely planted in Michigan

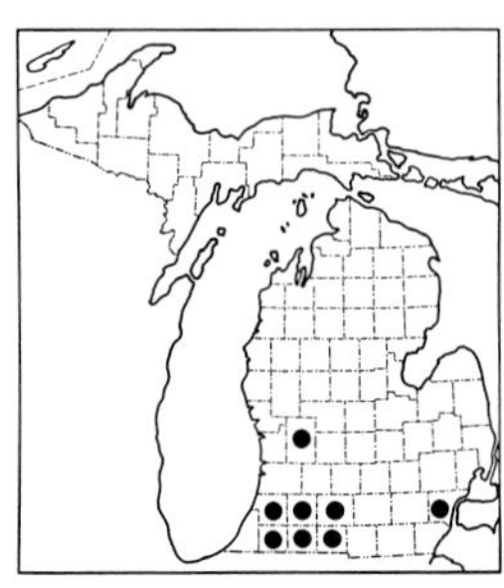

640. Amorpha canescens

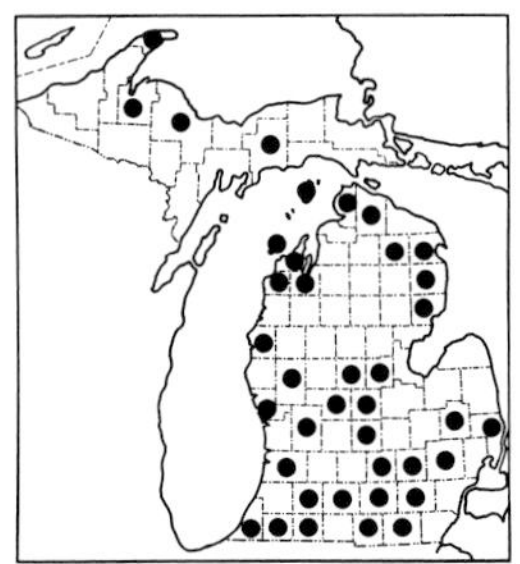

641. Robinia pseudoacacia

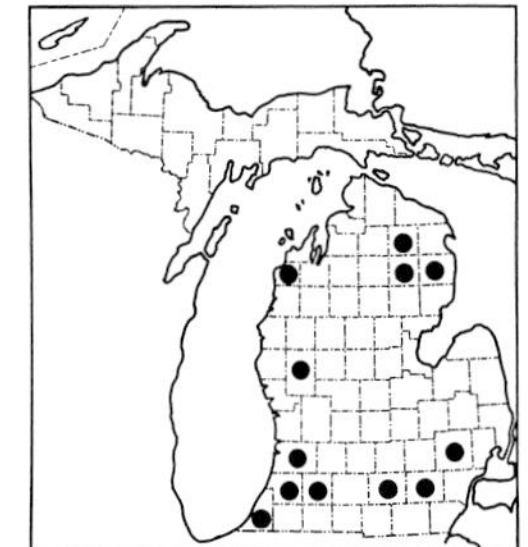

642. Robinia hispida

and locally established as a weed difficult to eradicate. Roadsides, fencerows, fields, farmyards, made land and waste ground; invading woods (beech-maple, aspen) and lightly wooded dunes, river banks.

The seeds may survive in the soil at least 88 years (Hanes 1956). Trees have been reported in Michigan as large as 3–5 feet in diameter, although rather sterile thickets of much smaller stems are common. The flowers are somewhat smaller than in the other species.

2. **R. hispida** L. Bristly Locust; Rose-acacia

Map 642. Native in the southeastern United States. Occasionally spreading from cultivation (usually in sterile clones) to roadsides, wooded dunes, disturbed ground, and sites near habitations.

The stipels are as long as the petiolules or slightly longer—more conspicuous in this species than in our other two.

3. **R. viscosa** Vent. Clammy Locust

Map 643. Native from Pennsylvania to the southeast—or perhaps only in North Carolina. Spreading from cultivation to roadsides, bluffs, old farm sites, and such places.

21. Schrankia

1. **S. nuttallii** (Britton & Rose) Standley Sensitive Brier; Cat-claw

Map 644. Native in the central states, from North Dakota south to Texas and Louisiana, rarely adventive eastward. Collected from a well established colony in the Island Lake Recreation Area in 1947 (*D. M. Lynch,* MSC).

This is the only representative of the subfamily Mimosoideae found outside of cultivation in Michigan. The leaflets are sensitive to touch as in the true sensitive-plants (*Mimosa*).

22. Lupinus Lupine

A genus well developed in western North America, whence (perhaps via cultivation) additional species are likely to spread, as has *L. polyphyllus.*

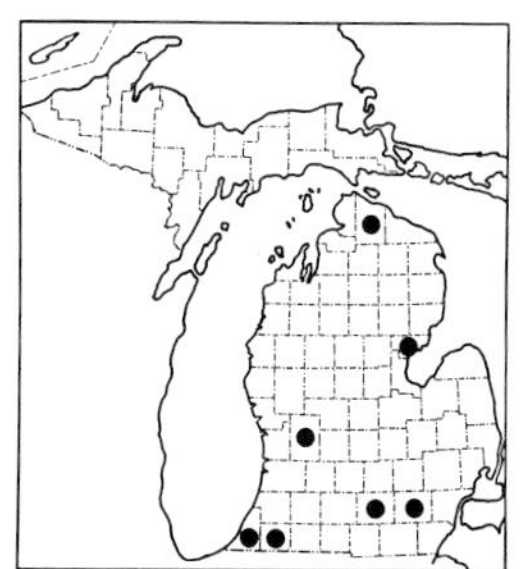

643. Robinia viscosa

644. Schrankia nuttallii

645. Lupinus polycarpus

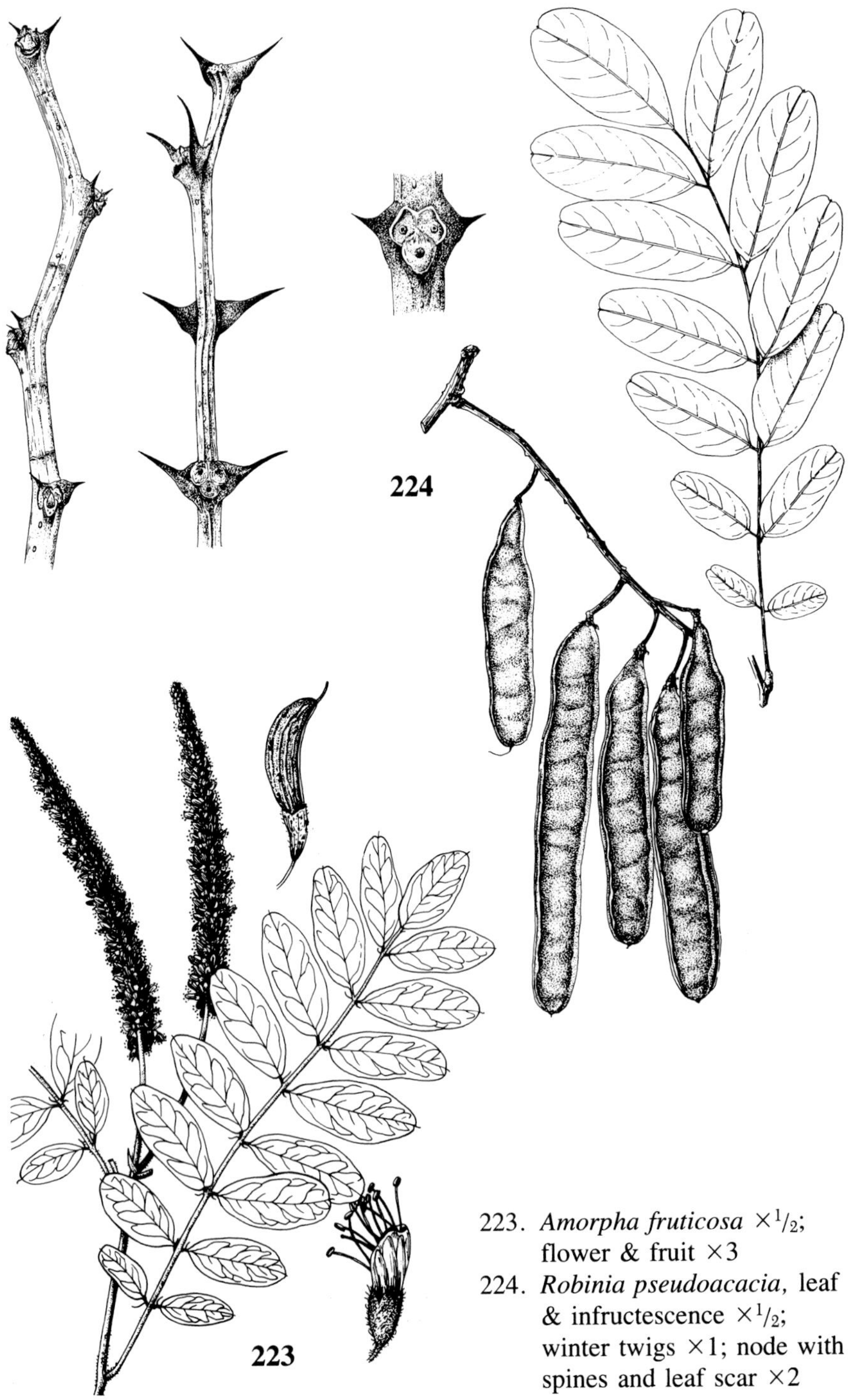

223. *Amorpha fruticosa* ×1/2; flower & fruit ×3

224. *Robinia pseudoacacia,* leaf & infructescence ×1/2; winter twigs ×1; node with spines and leaf scar ×2

REFERENCE

Dunn, David B., & John M. Gillett. 1966. The Lupines of Canada and Alaska. Canada Dep. Agr. Monogr. 2. 89 pp.

KEY TO THE SPECIES

1. Flowers less than 8 mm long, on pedicels ca. 1–2.5 mm long; plant an annual waif, probably no longer found in the state . . . . 1. **L. polycarpus**
1. Flowers 10–15 mm long, on pedicels (4) 5–9 mm long; plant a perennial (including common native species)
   2. Leaflets of larger leaves 8–11, ca. (2) 2.5–4.5 (4.8) cm long, obtuse to rounded . . . . 2. **L. perennis**
   2. Leaflets of larger leaves 12–14, ca. 6–11 cm long, acute . . . . 3. **L. polyphyllus**

1. **L. polycarpus** Greene

Map 645. This is another Pacific coast species (cf. *Trifolium*) collected as a ballast waif at Detroit in 1894 (*Farwell 1454*, BLH; see Asa Gray Bull. 4: 46. 1896).

2. **L. perennis** L. Fig. 225 — Wild Lupine

Map 646. Dry usually sandy ground, ranging from prairies and open barrens or clearings to woodlands of oak, jack pine, and/or aspen.

Quite variable in regard to pubescence on stems and petioles, the most hirsute extreme (tending to be more western in range and well represented in Michigan) having been named var. *occidentalis* Watson. The flowers occasionally exhibit assorted combinations of blue, purple, pink, and white, including entirely white corollas [f. *albiracemus* A. H. Moore], the standard purple and the rest rose-pink [f. *roseus* Britton], the standard violet and the rest blue [f. *bicolor* Farwell (TL: Ortonville [Oakland Co.])]. Emma Cole collected specimens in Kent County with the keel blue, the standard purple, and the wings white.

3. **L. polyphyllus** Lindley

Map 647. A native of western North America, locally escaped from cultivation and established, especially along roadsides, in the Lake Superior region of Michigan and Ontario.

23. **Cassia** — Senna

In the world as a whole, this (as broadly circumscribed) is one of the largest genera of flowering plants—and certainly of the legumes. Woody species are well represented in the tropics; both woody and herbaceous species are cultivated for their showy yellow flowers. Some species have been of medicinal importance. Our species are all herbaceous, and all but one are probably adventive in the state, there being no old collections despite the normal habitat of open disturbed places where they would be conspicuous.

Irwin and Barneby recognize three genera, of which the true *Cassia* is strictly tropical.

REFERENCE

Irwin, Howard S., & Rupert C. Barneby. 1982. The American Cassiinae a Synoptical Revision of Leguminosae tribe Cassieae subtribe Cassiinae in the New World. Mem. New York Bot. Gard. 35. 918 pp.

KEY TO THE SPECIES

1. Leaflets of best-developed leaves 1–5 mm broad, in 10–21 or more pairs; stamens 5 or 10
   2. Largest petals less than 8 mm long; stamens 5; pedicels ca. 2–3 mm long . . . . 1. **C. nictitans**
   2. Largest petals ca. 10 mm or more long; stamens 10; pedicels ca. 7–15 mm long . . . . 2. **C. chamaecrista**
1. Leaflets of best-developed leaves 9–35 [40] mm broad, in 3–8 pairs; stamens ca. 7
   3. Flowers solitary or few in a cluster; mature fruit ± 4-sided; leaflets mostly in 3 pairs; glands ± appressed to the rachis, between each of the two lowest pairs of leaflets . . . . 3. **C. tora**
   3. Flowers numerous in well developed racemes; mature fruit flat; leaflets in (6) 7–8 pairs; gland divergent, near base of petiole . . . . 4. **C. hebecarpa**

1. **C. nictitans** L. Wild Sensitive-plant

Map 648. Sandy, open, ± disturbed ground, possibly adventive in our area from the south.

Treated as *Chamaecrista nictitans* (L.) Moench when the broad genus *Cassia* is divided.

2. **C. chamaecrista** L. Partridge-pea

Map 649. Sandy, disturbed ground, as along roads and railroads. Apparently not collected in Michigan before 1906 (Berrien Co.), so it, too, may be a relatively recent invader of our area from the immediate south or west.

This is the plant often called *Cassia fasciculata* Michaux (see Brittonia 28: 437–439. 1977). And when *Cassia* is divided, it is called *Chamaecrista fasciculata* (Michaux) Greene. The leaflets of this species and the preceding are more sensitive to touch than those of the next two.

3. **C. tora** L. Sickle-pod

Map 650. Native in the Old World, collected in Michigan only in 1919 near a railroad in Detroit (*Farwell 5402*, BLH; see Rep. Mich. Acad. 22: 184. 1921).

Most recent authors have followed Linnaeus, who originally distinguished from this species the New World *C. obtusifolia,* although for many years they were considered to be the same. The latter grows in the southern states, and differs from true *C. tora* in longer pedicels (over 1 cm) and a gland only between the lowest leaflets (see Brenan in Kew Bull. 13: 248–252. 1958). When *Cassia* is divided, this becomes *Senna tora* (L.) Roxb.

4. **C. hebecarpa** Fern. Fig. 226 — Wild Senna

Map 651. River banks and floodplains, shores; sandy or dry upland sites. Our only unquestionably native *Cassia*.

Long called *C. marilandica* L., a name now restricted to a slightly more southerly species. If *Senna* is recognized, our species becomes *S. hebecarpa* (Fern.) Irwin & Barneby.

24. **Pisum**

1. **P. sativum** L. — Common, Garden, Green, or Field Pea

Map 652. An Old World species, cultivated since prehistoric times; rarely spontaneous. Shores, waste ground, fields.

The edible-podded snow or sugar pea is a variety of the same species.

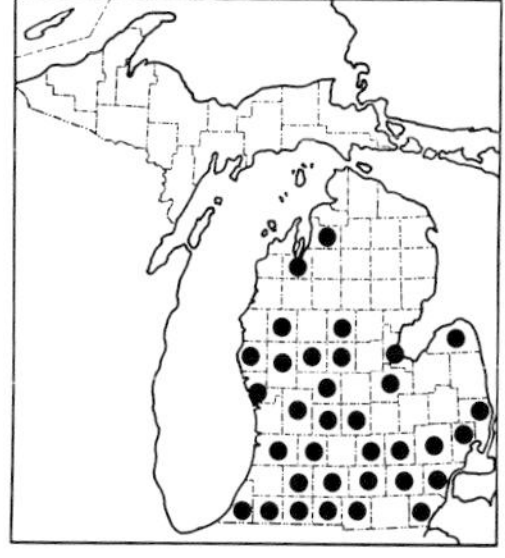
646. Lupinus perennis

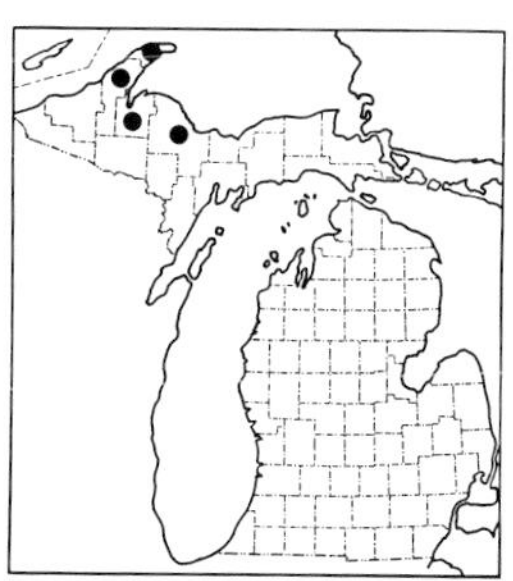
647. Lupinus polyphyllus

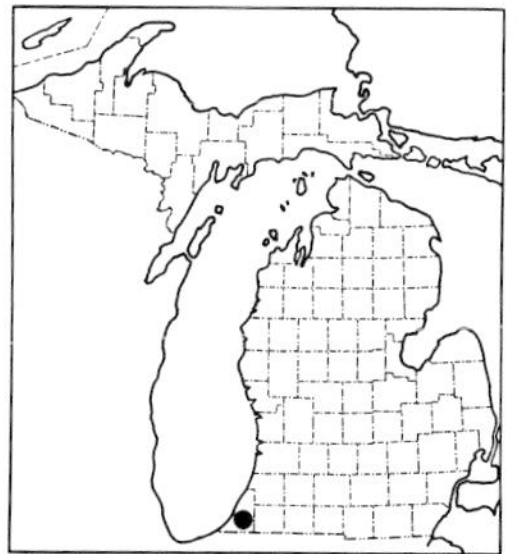
648. Cassia nictitans

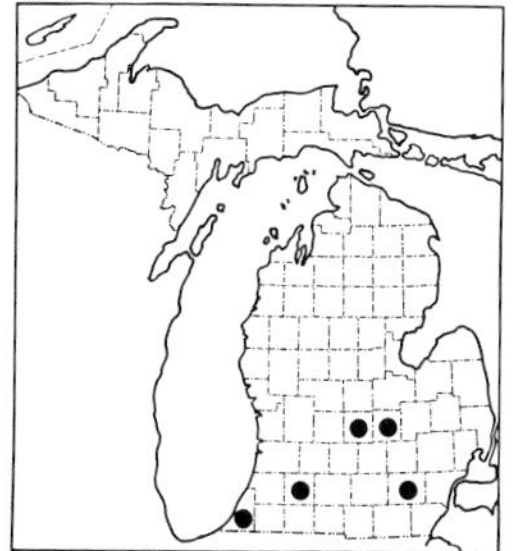
649. Cassia chamaecrista

650. Cassia tora

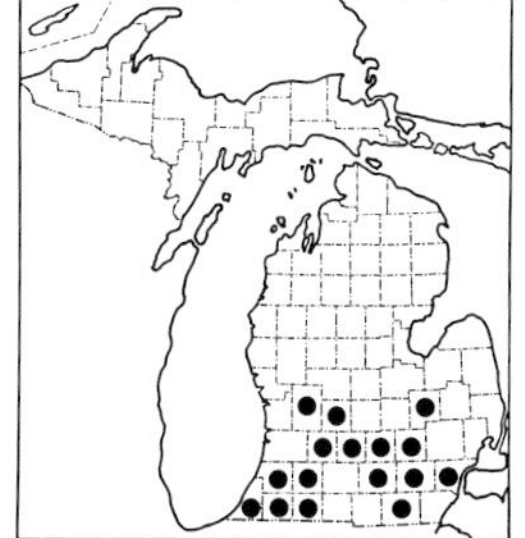
651. Cassia hebecarpa

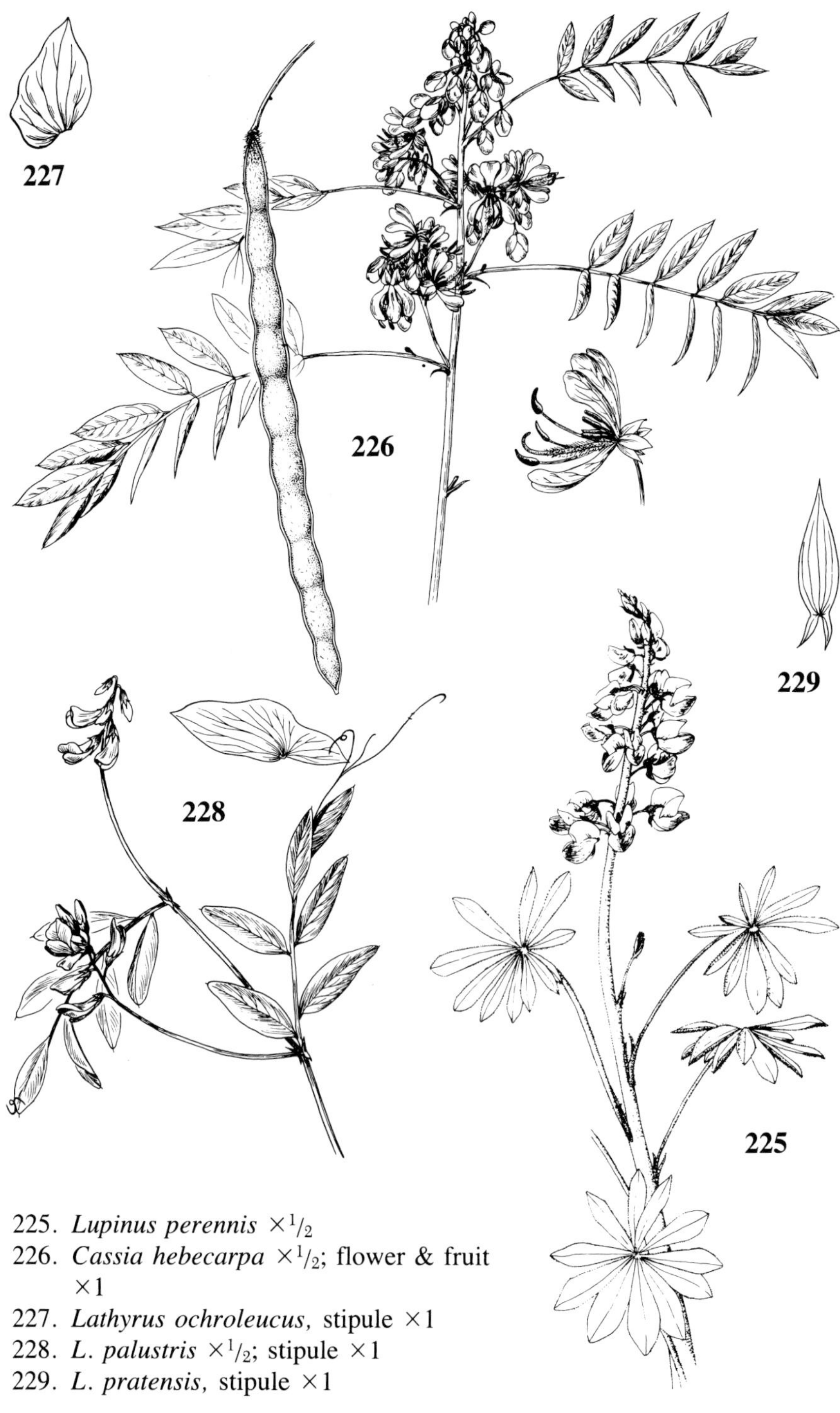

225. *Lupinus perennis* $\times^1/_2$
226. *Cassia hebecarpa* $\times^1/_2$; flower & fruit $\times 1$
227. *Lathyrus ochroleucus,* stipule $\times 1$
228. *L. palustris* $\times^1/_2$; stipule $\times 1$
229. *L. pratensis,* stipule $\times 1$

25. **Lathyrus**

REFERENCE

Hitchcock, C. Leo. 1952. A Revision of the North American Species of Lathyrus. Univ. Washington Publ. Biol. 15. 104 pp.

KEY TO THE SPECIES

1. Leaflets mostly 4–12 (in depauperate plants, some leaves with only 2 leaflets); plants native
  2. Stipules (at least the larger ones, on upper part of plant) (7) 10–20 (25) mm broad; larger leaflets (1.2) 1.5–3 (3.5) cm wide; stems angled but not winged
    3. Flowers (except in rare albinos) purple to blue (or pinkish); stipules with 2 basal lobes (± hastate); plant of sandy shores, beaches, and dunes........ 1. **L. japonicus**
    3. Flowers cream; stipules with 1 basal lobe (semi-sagittate or semi-cordate); plant of thickets and woodlands .... 2. **L. ochroleucus**
  2. Stipules all less than 7 mm broad (very rarely broader, usually on plants with winged stems); larger leaflets 0.4–2.4 cm wide; stems winged or not
    4. Calyx (and usually to some extent other parts) ± densely pubescent; principal leaves with 8–12 (14) leaflets; racemes 6–26-flowered; stem not winged...... .... 3. **L. venosus**
    4. Calyx glabrous or sparsely pubescent; principal leaves with 4–6 (10) leaflets; racemes 2–8-flowered; stems often narrowly winged ....4. **L. palustris**
1. Leaflets 2; plants escaped from cultivation, usually near settlements or along roadsides
  5. Stems not winged (at most angled); flowers ca. 11–19 mm long
    6. Stipules (at least on main stem) with 2 basal lobes (hastate or sagittate); flowers bright yellow; leaflets narrowly acute at the tip .... 5. **L. pratensis**
    6. Stipules with 1 basal lobe (semi-sagittate); flowers rose-red; leaflets broadly acute to rounded at the tip (except for excurrent midvein) ....6. **L. tuberosus**
  5. Stems distinctly winged with strips of green leaf-like tissue; flowers of various size
    7. Ovary and fruit pubescent with pustular-based hairs; plant annual
      8. Flowers over 2 cm long; leaflets elliptic ....7. **L. odoratus**
      8. Flowers less than 1.5 cm long; leaflets linear-lanceolate .... 8. **L. hirsutus**
    7. Ovary and fruit glabrous; plant perennial
      9. Main lobe of stipules up to half as wide as the winged stem; flowers ca. 12–18 mm long .... 9. **L. sylvestris**
      9. Main lobe of stipules more than half as wide as the winged stem; flowers ca. 16–26 mm long ....10. **L. latifolius**

1. **L. japonicus** Willd. Beach Pea

Map 653. Dunes, sandy or even gravelly beaches (rarely rock crevices), and adjacent ± disturbed ground (or relic on old dune ridges) along the Great Lakes. Our only inland collection is from Higgins Lake, presumably the north end in Crawford County, by one of the Government surveyors in 1849 (*G. H. Cannon,* MSC). [Collections of *Potentilla anserina* and *Polygonum lapathifolium* by Cannon are from the Crawford Co. shore of the lake, although the *Lathyrus* label is less precise.] Also occasionally found off the lake shores in sandy dumps (where I have found it at a Mackinaw City dump and, a bit farther from the shore, the Paradise dump). This is a circumpolar species, found on shores of the oceans and some of the largest inland lakes, including Lake Champlain and Lake Winnipeg as well as all five Great Lakes.

Some plants [var. *pellitus* Fern.] are rather densely pubescent on the stems, calyces, and other parts. All gradations occur to completely glabrous plants [var. *glaber* (Ser.) Fern.]. White-flowered plants have been found on Lake Superior at Whitefish Point and the Grand Sable Dunes, and on Lake Michigan at Saugatuck, and may be expected anywhere. Sometimes called *L. maritimus* Bigelow, a later name which has been incorrectly cited as *L. maritimus* (L.) Bigelow.

2. **L. ochroleucus** Hooker Fig. 227 Pale Vetchling

Map 654. Diverse woods and thickets: oak-hickory forests and hemlock-hardwoods; open woods, sand or rock ridges and openings in coniferous or mixed woods, as along trails; hillsides and river banks.

3. **L. venosus** Willd.

Map 655. Sandy open ground, prairies, shores, hillsides and banks (often shaded), oak-hickory woods, ridges and rock summits, roadside thickets.

4. **L. palustris** L. Fig. 228 Marsh Pea

Map 656. Damp thickets, shores, beach flats, marshes, boggy meadows, conifer swamps (cedar or tamarack), stream and river banks, ditches and swales, borders of woods.

Variable in size and shape of leaflets and stipules, in pubescence, and in other characters. The several named varieties all run into each other. One of the clearest is var. *myrtifolius* (Willd.) A. Gray, with the stem wingless. Other forms have the stem narrowly winged, and it is mostly among these plants that rare ones occur with stipules more than 7 mm broad. These can usually also be distinguished from the wingless *L. ochroleucus* by their bluish to pink or purple flowers (except in albinos) and the often ± sharply acute

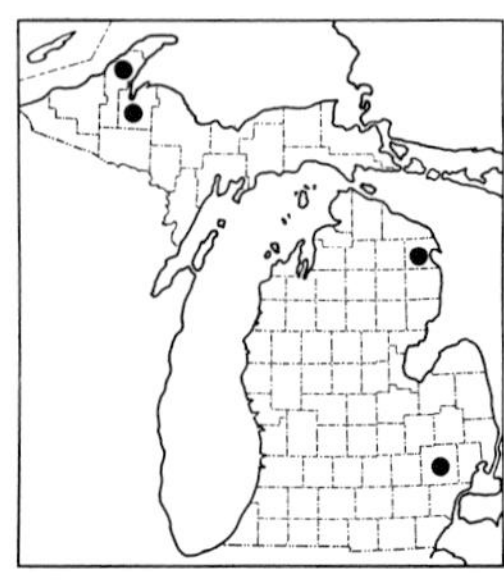

652. Pisum sativum

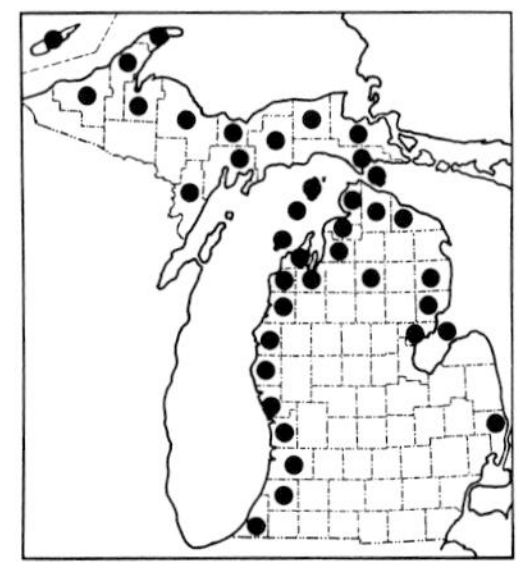

653. Lathyrus japonicus

654. Lathyrus ochroleucus

basal lobes on the stipules (compared to rather rounded, though toothed, lobes in *L. ochroleucus*). The lateral calyx lobes in *L. ochroleucus* tend to be more ovate-deltoid than the triangular ones of *L. palustris*.

5. **L. pratensis** L. Fig. 229 Yellow Vetchling

Map 657. Roadsides, fields, open rocky ground. A Eurasian species, locally established as an escape from cultivation.

The pods are sometimes slightly pubescent. The pedicels are pubescent with ± appressed hairs, while in the next species they are nearly or quite glabrous.

6. **L. tuberosus** L. Tuberous Vetchling

Map 658. Roadsides, fields, railroads, open rocky ground. Another Eurasian species, apparently a little more often found as an escape.

As the name suggests, underground tubers are produced in this species.

7. **L. odoratus** L.

Map 659. The only Michigan collection is as an escape from an old cemetery at Ypsilanti (*Farwell 1211* in 1891, BLH). Native to the Mediterranean region, cultivated for its showy and fragrant flowers; rarely (or barely?) escaped.

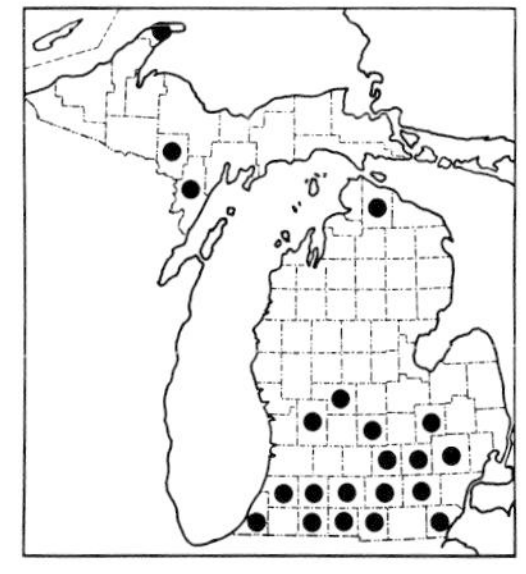
655. Lathyrus venosus

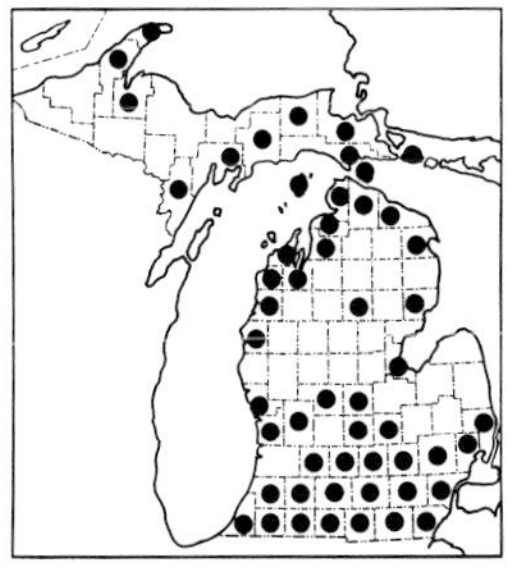
656. Lathyrus palustris

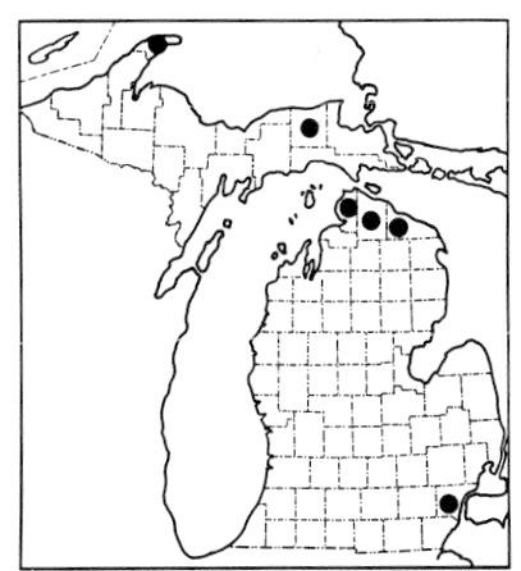
657. Lathyrus pratensis

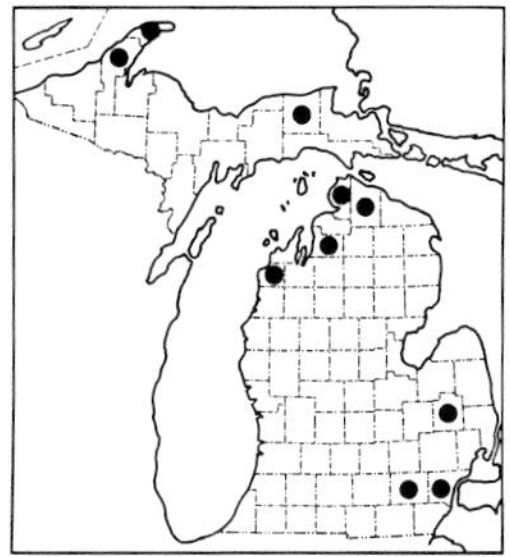
658. Lathyrus tuberosus

659. Lathyrus odoratus

660. Lathyrus hirsutus

8. **L. hirsutus** L.

Map 660. Collected in 1949 in waste ground at East Lansing (*R. S. Hauser 219,* MSC). A European native, like the preceding a rarely escaped annual.

9. **L. sylvestris** L. Perennial or Everlasting Pea

Map 661. Roadsides, fields, fencerows, and waste ground; occasionally invading woods. A European species, locally well established.

The leaflets average much longer and narrower than in the next species, being ± lanceolate to narrowly elliptic-linear.

10. **L. latifolius** L. Perennial or Everlasting Pea

Map 662. Roadsides, vacant lots, hillsides, railroad beds, and other waste places, especially near areas of human activity. Another European species, like *L. sylvestris* locally well established as an escape from cultivation.

The leaflets are ± elliptic to ovate-lanceolate.

26. Vicia — Vetch

The technical character separating this genus from *Lathyrus* is often obscure, especially in small flowers and more so when they are pressed and dried. Hence the key to genera is based on vegetative characters. In *Vicia,* the style is glabrous except for a tuft or fringe of hairs at the summit, just below the stigma. In *Lathyrus,* the style has a longer strip of pubescence. Only two of our species are native in North America. The others are all European, doubtless originally grown for forage or as a cover crop. The species delimitations of *Flora Europaea* are followed here for these.

REFERENCE

Hermann, F. J. 1960. Vetches in the United States—Native, Naturalized, and Cultivated. U.S. Dep. Agr. Agr. Handb. 168. 84 pp.

KEY TO THE SPECIES

1. Inflorescence sessile or on peduncle shorter than the leaflets, 1–2-flowered
   2. Calyx oblique at apex, the teeth unequal (upper ones smaller) and all less than half as long as the tube and less than 2 mm long; plant perennial .... 1. **V. sepium**
   2. Calyx not oblique, the teeth ± equal, ca. half as long as the tube or longer and more than 2 mm long; plant annual
      3. Corolla yellow (sometimes marked or suffused with purplish), ca. 25–35 mm long; calyx teeth distinctly shorter than the tube .... 2. **V. grandiflora**
      3. Corolla blue to purple (or white), ca. [10] 13–16 [30] mm long; calyx teeth nearly or quite as long as the tube .... 3. **V. sativa**
1. Inflorescences on peduncles longer than the leaflets, with usually more than 2 flowers

4. Leaflets with ca. 10 or more pairs of firm lateral veins running nearly or quite to the margins; stipules with several teeth or lobes; inflorescences slightly exceeded by subtending leaves, 3–8-flowered . . . . 4. **V. americana**

4. Leaflets with ca. 6 or fewer pairs of lateral veins running nearly or quite to the margins; stipules at most with 1 basal lobe; inflorescences shorter or longer than the leaves

5. Flowers less than 6 [8] mm long, 1–6 (7) in an inflorescence; calyx teeth (at least the longer ones) about as long as or longer than the tube

6. Calyx teeth ± equal; fruit pubescent, [6] 8–9 [10] mm long, 2-seeded . . . . 5. **V. hirsuta**

6. Calyx teeth unequal, the calyx clearly oblique; fruit glabrous, 10–13 mm long, mostly 4-seeded . . . . 6. **V. tetrasperma**

5. Flowers at least 8 mm long, (8) 12–40 or more in an inflorescence; calyx teeth often shorter than the tube

7. Calyx ca. 3 mm long or shorter, the teeth deltoid (about as long as broad), ± equal; corolla white (except for blue-tipped keel) or occasionally tinged with bluish . . . . 7. **V. caroliniana**

7. Calyx ca. 3.5–8 (9) mm long, the teeth elongate (at least the lower ones) and distinctly unequal; corolla (except in albinos) blue to purple (or rarely pink)

8. Calyx projecting at its base beyond the attachment of the pedicel, the latter thus appearing subterminal or ventral (fig. 232); blade of standard about half as long as the claw or shorter; stem and pedicels usually covered with fine spreading hairs (rarely glabrate) . . . . 8. **V. villosa**

8. Calyx not extended beyond the pedicel, which appears basal (though off-center; fig. 231); blade of standard about equaling the claw; stem and pedicels with mostly appressed or incurved pubescence . . . . 9. **V. cracca**

1. **V. sepium** L. Hedge Vetch

Map 663. A European species, locally established in North America. Said by Dodge to be becoming common (1909) in pastures in St. Clair County; but the only collection made in the state since 1915 (when Farwell found the species at Detroit) is from a trail on Isle Royale in 1959.

2. **V. grandiflora** Scop.

Map 664. Another European species, still very local in the United States. Collected in waste ground at East Lansing in 1949 (*R. S. Hauser 217,* MSC).

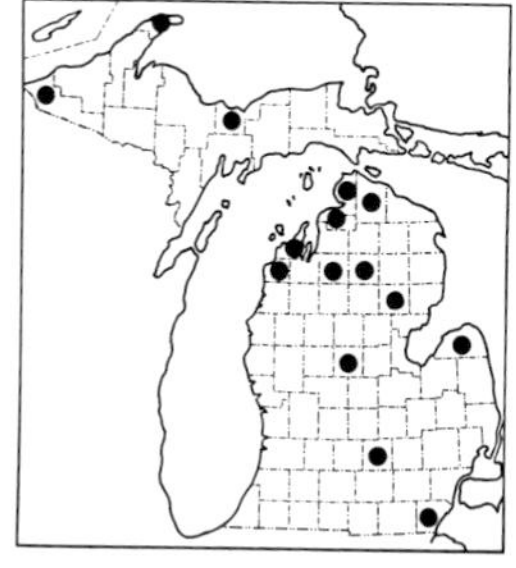

661. Lathyrus sylvestris

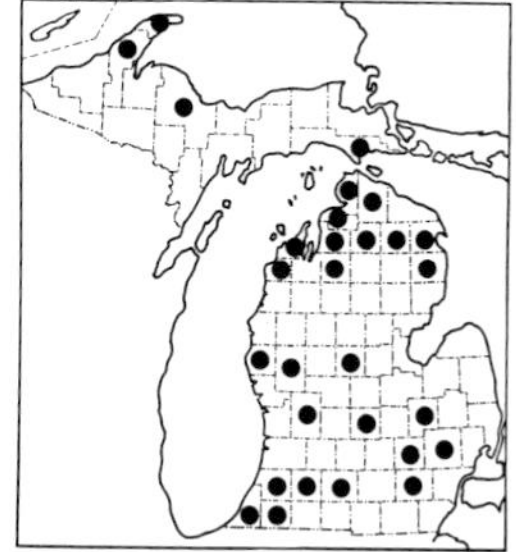

662. Lathyrus latifolius

663. Vicia sepium

3. **V. sativa** L. Common or Spring Vetch

Map 665. A forage crop, originally from Europe and only quite sporadically found in Michigan. Roadsides, railroads, fields, clearings, waste places.

All or most of our specimens are the segregate often recognized as *V. angustifolia* L., or as *V. sativa* var. *nigra* L. *Flora Europaea* follows recent work concluding that this is not a distinct species. True *V. sativa* has large flowers (to 3 cm long) and large fruit constricted between the seeds.

4. **V. americana** Willd. American Vetch

Map 666. In a diversity of moist to dry, open to shaded habitats: swampy woods and borders; oak, pine, or mixed forests and clearings; rocky thickets, bluffs, and thin woods; river banks; roadsides and railroads.

Usually glabrous or nearly so, but variable in pubescence and leaflet shape. The preceding three species may also have the numerous pairs of prominent lateral veins which characterize the leaflets of this native one.

5. **V. hirsuta** (L.) S. F. Gray

Map 667. A European species. Established for over 50 years in sandy disturbed ground at the University of Michigan Biological Station, Cheboygan County; collected in 1945 by Bazuin on bluffs along the Grand River.

The flowers are even smaller than in the next species, running ca. 2–4 mm long.

6. **V. tetrasperma** (L.) Schreber Sparrow Vetch

Map 668. A Eurasian species, locally established along roadsides and railroads, in fields and damp open ground, and frequently in thin woods on limestone.

7. **V. caroliniana** Walter Pale or Wood Vetch

Map 669. Oak and oak-hickory woods, borders of woods, dry open ground and clearings; less often in moist places, banks of streams and lakes.

8. **V. villosa** Roth Fig. 232 Hairy Vetch

Map 670. A Eurasian species, thoroughly naturalized and our commonest vetch in Michigan, but no collections have been seen before 1902, when it was recorded as rare in Bay County. Roadsides, fields, railroads, fencerows, dumps, and other waste places; invading shores, open woods, and dunes.

Pink- and white-flowered forms are occasionally found. The calyx teeth are much more elongate and even hair-like in this species compared to the next (cf. figs. 231 & 232). In a very few collections, the stems and other parts are glabrous or nearly so and the calyx teeth shorter. These have been called *V. dasycarpa* Ten., but such plants are now treated as *V. villosa* ssp. *varia* (Host) Corb.

9. **V. cracca** L. Figs. 230, 231 Bird Vetch

Map 671. Roadsides, fields, railroads, often forming large tangled patches. No Michigan collections have been seen before the 1890's. Although the species is sometimes thought to be native in northern North America as well as Eurasia, our plants appear surely to be recently introduced.

The leaflets often have a more prominent little "spine" at the tip than the minute tip in *V. villosa*.

27. Dalea

A specimen of *D. villosa* (Nutt.) Sprengel is attributed to Ann Arbor (*M. H. Clark* n. d., NY, not labeled in Miss Clark's hand). Since the species was not listed in the Ann Arbor Flora of 1876, prepared by E. C. Allmendinger with the help of Miss Clark (who died in 1875), it was apparently not considered a member of the flora—even if labeled correctly. This is a villous plant, the leaflets 13 or more and linear-involute.

REFERENCE

Barneby, Rupert C. 1977. Daleae Imagines. Mem. New York Bot. Gard. 27. 891 pp.

664. Vicia grandiflora

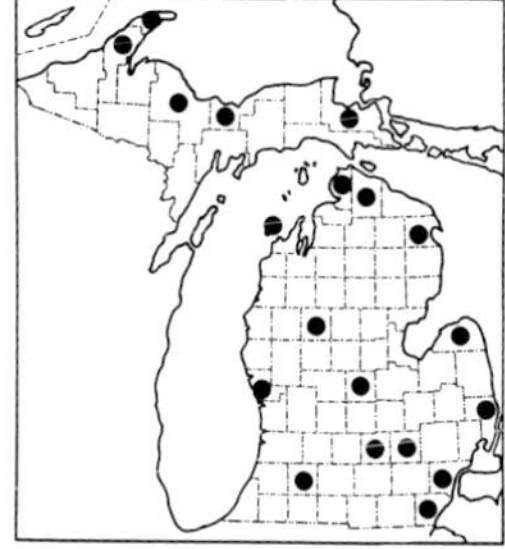
665. Vicia sativa

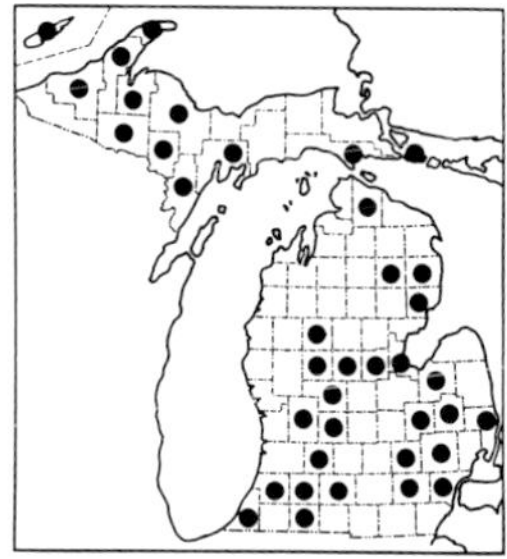
666. Vicia americana

667. Vicia hirsuta

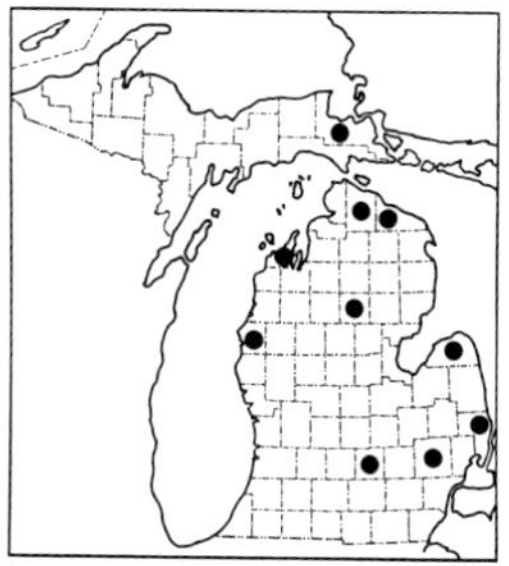
668. Vicia tetrasperma

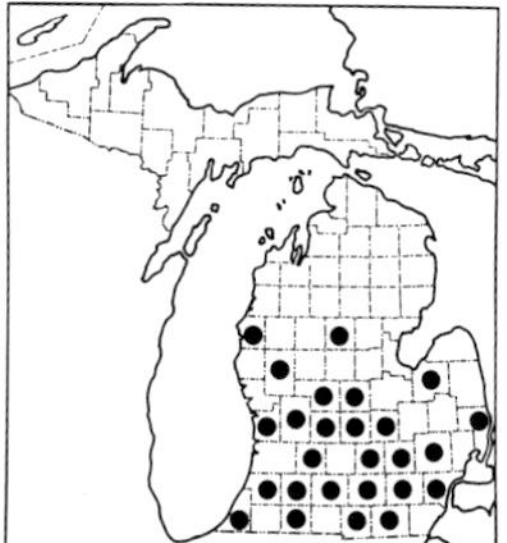
669. Vicia caroliniana

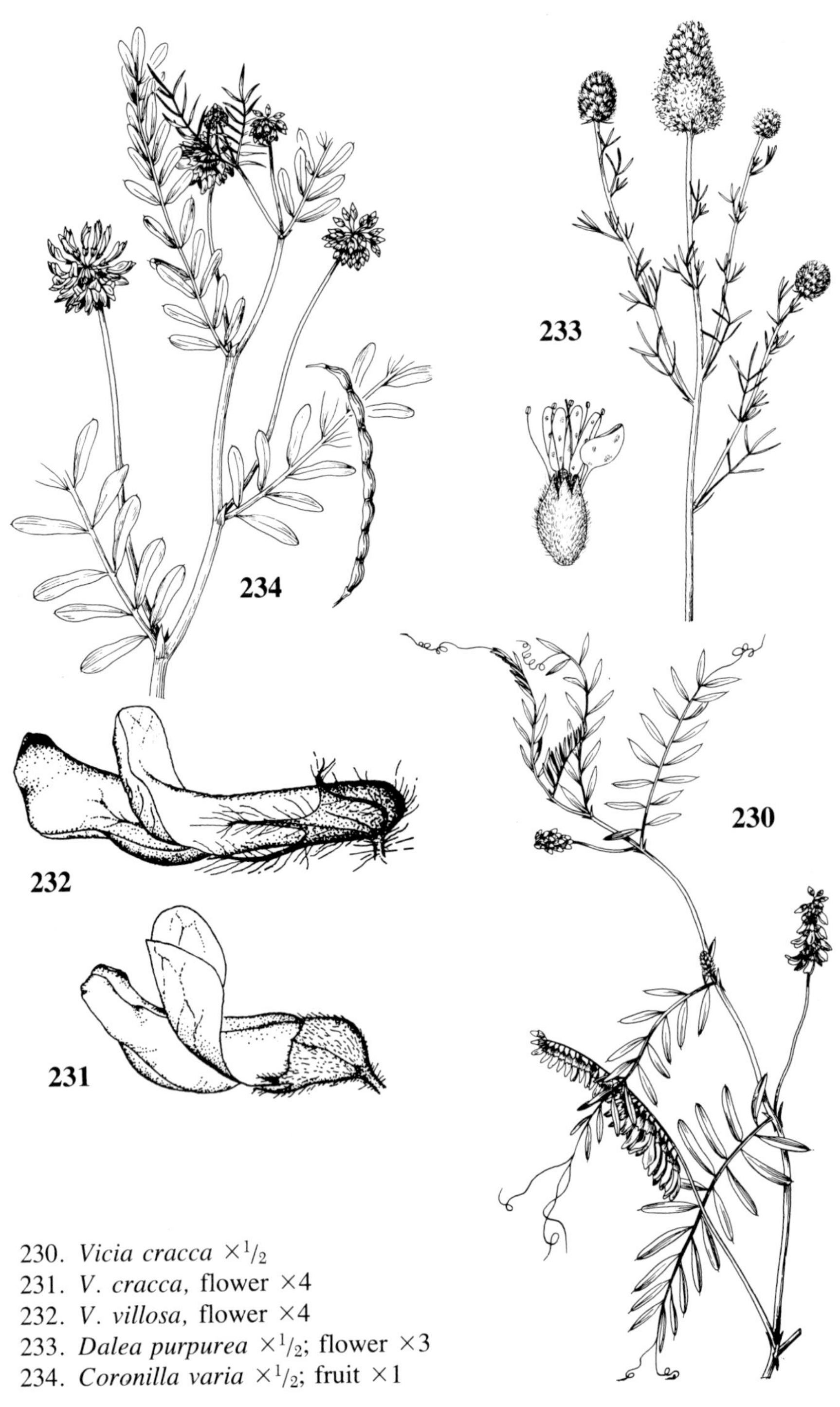

230. *Vicia cracca* $\times^1/_2$
231. *V. cracca,* flower $\times 4$
232. *V. villosa,* flower $\times 4$
233. *Dalea purpurea* $\times^1/_2$; flower $\times 3$
234. *Coronilla varia* $\times^1/_2$; fruit $\times 1$

KEY TO THE SPECIES

1. Leaflets 3–5, involute, linear, ca. 0.9–2 cm long, ca. 0.5–1.7 mm broad .... 1. **D. purpurea**
1. Leaflets more than 5 (15–31 on many of the larger leaves), flat, oblanceolate-elliptic, less than 1 cm long, mostly ca. 2–3 mm broad .... 2. **D. leporina**

1. **D. purpurea** Vent. Fig. 233 Purple Prairie-clover

Map 672. Dry open sandy ground; very rare and local.

Sometimes placed in a segregate genus as *Petalostemon purpureum* (Vent.) Rydb. (the generic name sometimes spelled *Petalostemum*, an orthography rejected against the conserved one).

2. **D. leporina** (Aiton) Bullock

Map 673. Native from the prairie regions southwest into Mexico, occasionally adventive eastward. Collected in 1924 (*Walpole 1642*, BLH) at Buckley, presumably as a waif.

Long known as *D. alopecuroides* Willd. The numerous glabrous leaflets suggest the very rare *D. foliosa* (A. Gray) Barneby—which is how the Michigan collection was originally identified (as *Petalostemon foliosum*)—but, among other differences, that species has a glabrous calyx and *D. leporina* a villous calyx.

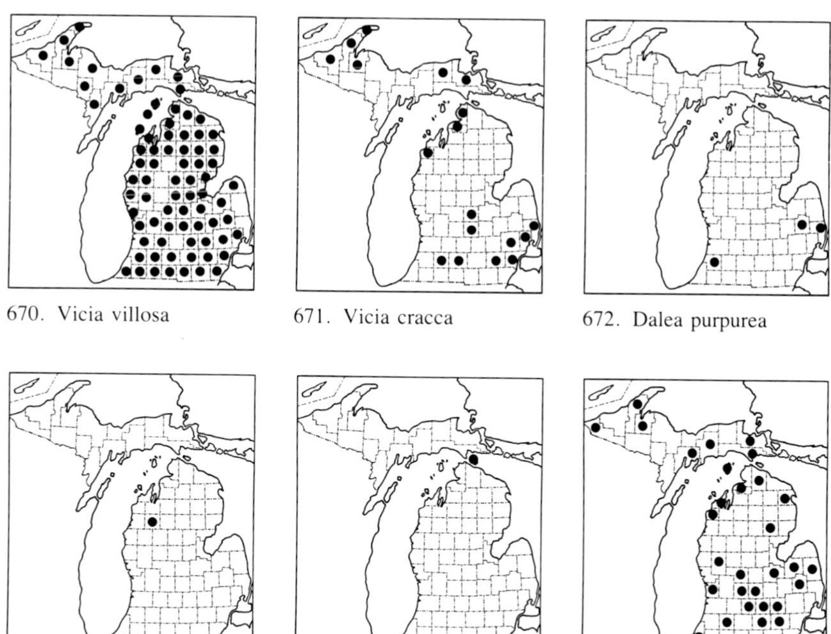

670. Vicia villosa
671. Vicia cracca
672. Dalea purpurea
673. Dalea leporina
674. Anthyllis vulneraria
675. Lotus corniculata

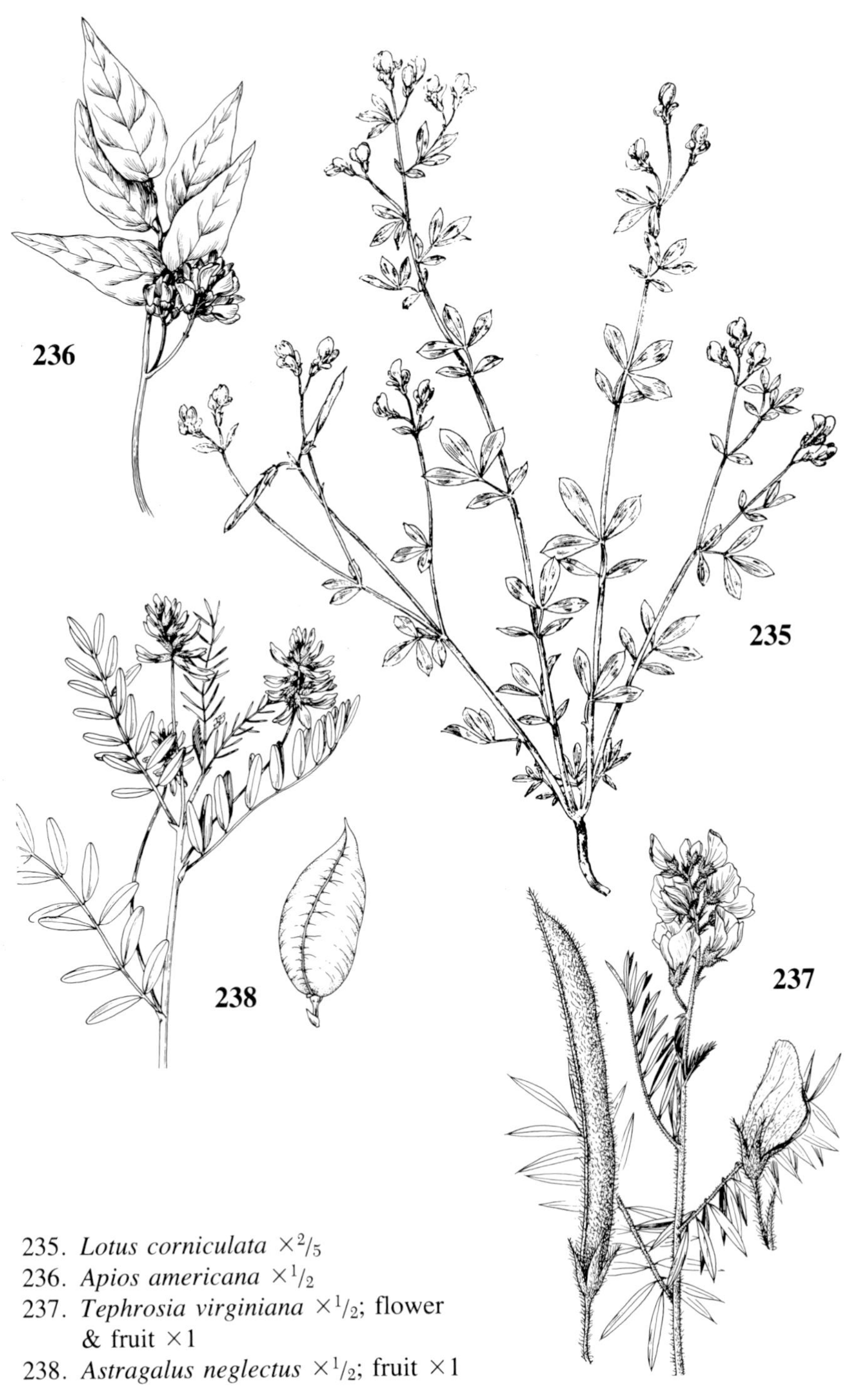

235. *Lotus corniculata* $\times{}^{2}/_{5}$
236. *Apios americana* $\times{}^{1}/_{2}$
237. *Tephrosia virginiana* $\times{}^{1}/_{2}$; flower & fruit $\times 1$
238. *Astragalus neglectus* $\times{}^{1}/_{2}$; fruit $\times 1$

28. Anthyllis

1. **A. vulneraria** L. Woundwort

Map 674. A European species collected at Mackinac Island by Potzger (*4559*, ND) in 1934 on an "open waste hillside." An Ingham County collection made at the Agricultural College in 1897 from "Crimson Clover field E. of Grain Barn" was presumably not considered established as the species was not listed in the subsequent Michigan Flora (Beal 1905).

29. Lotus

1. **L. corniculata** L. Fig. 235 Birdfoot Trefoil

Map 675. Roadsides, fields, clearings, grassy turf. A Eurasian species grown for forage and erosion control, apparently first collected as a weed in Michigan in 1941 (Charlevoix Co.) and now locally well established. Doubtless much more common than the relatively few scattered collections indicate.

The classical gender of the generic name, when used for a legume (rather than a water-lily), is feminine.

30. Coronilla

1. **C. varia** L. Fig. 234 Crown-vetch

Map 676. A European species, heavily promoted for erosion control and a rapid groundcover, becoming too well established sometimes and spreading extensively on roadsides, fields, and banks.

31. Apios

1. **A. americana** Medicus Fig. 236 Groundnut; Wild-bean; Indian-potato

Map 677. A low vine of moist woods, shores, river-bank thickets, marshes and meadows, wet prairies, streamsides.

676. Coronilla varia

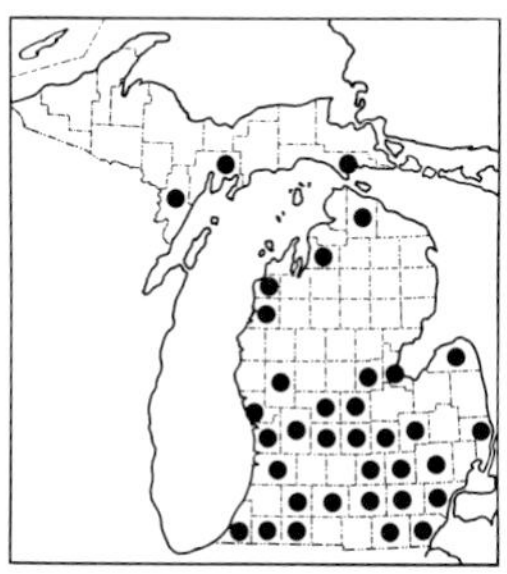

677. Apios americana

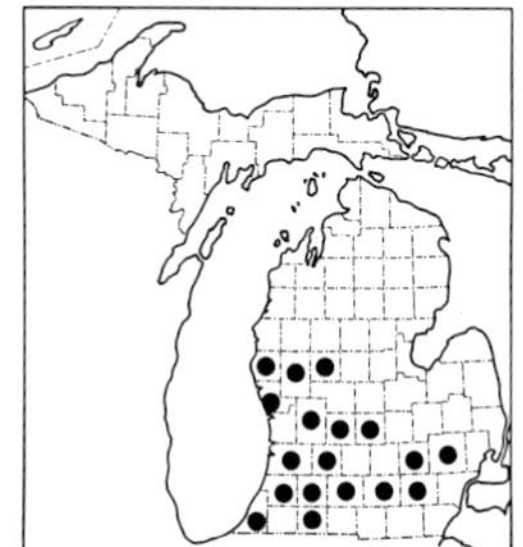

678. Tephrosia virginiana

Strings of edible tubers are produced underground, and can be prepared like potatoes. The flowers are normally a very deep red- or purple-brown, with a strongly curved, even coiled, keel.

32. **Tephrosia**

1. **T. virginiana** (L.) Pers. Fig. 237 Goats-rue; Rabbit-pea

Map 678. Sandy barrens, fields, prairie-like areas, and oak-pine woodlands.

One of the handsomest wildflowers of dry woodlands in southern Michigan with its large bicolored flowers.

33. **Astragalus** Milk-vetch

This is a very large genus, well developed west of the Great Lakes, represented in Michigan by only two distinctive native species. These are concisely compared by Barneby (Mem. New York Bot. Gard. 13: 596. 1964).

Galega officinalis L. (catgut or goats-rue) was reported from Michigan by Fogg (Morris Arb. Bull. 15: 17–18. 1964), but the material cited is clearly labeled as cultivated on Experiment Station plots. It is a Eurasian species and would key here to *Astragalus,* from which it differs in having calyx teeth nearly filiform, ± equal, and about as long as the tube; semi-sagittate stipules with a prolonged basal lobe; and fruit strongly striate. Our species of *Astragalus* have calyx teeth much shorter than the tube, deltoid to narrowly triangular stipules with neither teeth nor a basal lobe, and smooth or somewhat wrinkled fruit.

REFERENCES

Isely, Duane. 1983. Astragalus L. (Leguminosae: Papilionoideae) I: Keys to United States Species. Iowa St. Jour. Res. 58: 3–172.

Isely, Duane. 1984. Astragalus L. (Leguminosae: Papilionoideae) II: Species Summary A–E. Iowa St. Jour. Res. 59: 99–209.

679. Astragalus cicer

680. Astragalus canadensis

681. Astragalus neglectus

KEY TO THE SPECIES

1. Ovary and fruit ± densely pubescent with long hairs; plant an escape from cultivation in waste places . . . . 1. **A. cicer**
1. Ovary and fruit glabrous or nearly so; plant native
   2. Pubescence of stem and leaves including many 2-pronged hairs; stipules (at least the lower) connate with a belt of tissue around the stem; fruit 4.5–6.5 mm thick, 2-locular; plants rhizomatous . . . . 2. **A. canadensis**
   2. Pubescence of stem and leaves all of simple hairs; stipules not connate; fruit 10–15 mm thick, 1-locular; plant from a taproot . . . . 3. **A. neglectus**

1. **A. cicer** L. — Chick-pea Milk-vetch

Map 679. A European species, well established since at least 1974 along a railroad on the Michigan State University campus, doubtless escaped from cultivation (see Rabeler & Crowder in Mich. Bot. 24: 126. 1985).

The distinctive fruits are nearly globose, covered with long hairs, both black and white. As is often the case also in *A. neglectus,* the calyx has a mixture of short black and white hairs. The stipules are connate around the stem and the fruit is 2-locular. However, the fruit is thicker than in *A. canadensis* and all the hairs are simple (none 2-pronged).

2. **A. canadensis** L. — Canada Milk-vetch

Map 680. Dry prairies, moist shores, river banks, marshy ground, and other open or partly shaded ground.

3. **A. neglectus** (T. & G.) Sheldon Fig. 238 — Cooper's Milk-vetch

Map 681. Marshy to dry open, sometimes rocky, clearings, shores, thickets, and river banks; often in calcareous sites.

Also widely known as *A. cooperi* A. Gray. A yellow-flowered form was collected near Hadley by Farwell (*6179* in 1922, BLH, GH), f. *limonia* (Farw.) Fern. (TL: southwest corner of Lapeer Co.).

682. Hedysarum alpinum

683. Linum catharticum

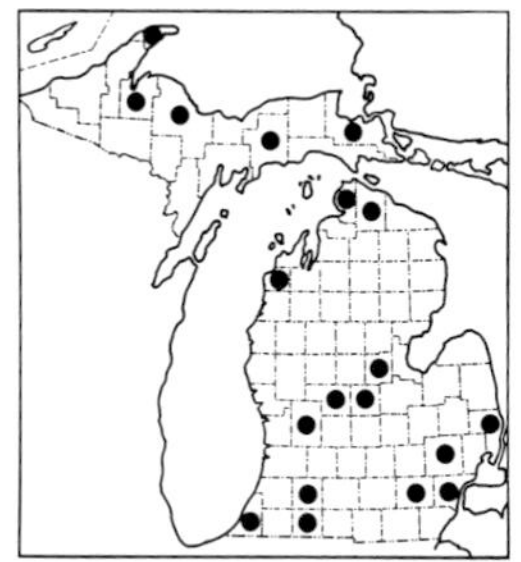

684. Linum usitatissimum

34. **Hedysarum**

1. **H. alpinum** L.

Map 682. A circumpolar boreal species long known as far south as the north shore area of Lake Superior, but first discovered in Michigan in 1984 in a partly shaded meadow area on limestone along the Escanaba River (*K. A. Chapman,* MSC).

LINACEAE Flax Family

1. **Linum** Flax

REFERENCES

Mosquin, Theodore. 1971. Biosystematic Studies in the North American Species of Linum, section Adenolinum (Linaceae). Canad. Jour. Bot. 49: 1379–1388.

Robertson, Kenneth R. 1971. The Linaceae in the Southeastern United States. Jour. Arnold Arb. 52: 649–665.

Rogers, C. M. 1957. Linaceae in Michigan. Asa Gray Bull., N. S. 3: 199–204.

Rogers, C. Marvin. 1963. Yellow Flowered Species of Linum in Eastern North America. Brittonia 15: 97–122.

Rogers, C. M. 1971. Changing Abundance of Two Species of Linum in Eastern North America. Mich. Bot. 10: 113–116.

KEY TO THE SPECIES

1. Petals blue or white (yellow-based); pedicels mostly (4) 6–15 (20) mm long (at least in fruit)
 2. Cauline leaves opposite (except uppermost bracts); petals ca. 3–4 mm long, white with yellow claw . . . . 1. **L. catharticum**
 2. Cauline leaves all alternate; petals ca. (8) 9–15 mm long, blue (except in albinos)
  3. Inner sepals ciliate; plant annual, homostylous . . . . 2. **L. usitatissimum**
  3. Inner sepals eciliate; plant perennial, heterostylous . . . . 3. **L. perenne**
1. Petals yellow throughout; pedicels (even in fruit) mostly less than 3.5 mm long
 4. Outer sepals glandular-ciliate (as are the inner ones), (3.5) 4–4.5 (5.5) mm long; leaves with a pair of dark stipular glands at the base; plant annual . . . . 4. **L. sulcatum**
 4. Outer sepals (and often inner ones) entire, less than 3.5 mm long; leaves without stipular glands; plant perennial
  5. Inner sepals glandular-ciliate; leaves all alternate (except sometimes at the lowermost 1–2 nodes); inflorescence with stiffly ascending branches . . . . . 5. **L. medium**
  5. Inner sepals glandless (or with a few glands on the distal half); leaves usually opposite at several or more of the lower nodes; inflorescence with branches usually ± divaricate
   6. Upper part of stem (below the inflorescence) terete or nearly so (at most, some slightly raised lines) . . . . 6. **L. virginianum**
   6. Upper part of stem distinctly angled (narrowly winged) with 3 ridges decurrent from each leaf base for 1–2 internodes . . . . 7. **L. striatum**

1. **L. catharticum** L. Fairy Flax

Map 683. A European species, locally naturalized usually in calcareous habitats where it is associated with many interesting native calciphiles such as (depending on locality) *Iris lacustris, Carex crawei, C. garberi, Sisyrinchium mucronatum, Scirpus cespitosus,* and *Primula mistassinica*. First found in Michigan in 1940 along a grassy roadside in Keweenaw County (*Farwell 12555,* BLH, MCTF; see Rhodora 43: 634. 1941).

This is a delicate little annual, rarely if ever over 2 dm tall. Its small white (yellow-eyed) flowers and sparse opposite leaves (dense on basal offshoots) readily distinguish it.

2. **L. usitatissimum** L. Common Flax

Map 684. An Old World plant of uncertain origin, long cultivated for the fiber of the stem (from which linen is derived) and the oil of its seed (linseed). Rarely found as a weed or garden escape to roadsides, railroads, shores, fields, and other disturbed or waste places.

3. **L. perenne** L. Perennial Flax

Map 685. Sandy or gravelly disturbed ground. Native in Europe, rarely escaped from cultivation in our area, where it is locally established.

The flowers are heterostylous, some plants in a population having long stamens and short styles, while others have short stamens and long styles. Some old collections, presumably of garden plants, from Wayne and Washtenaw counties are apparently homostylous (stamens and styles about the same length) and hence the western North American *L. lewisii* Pursh, which is also sometimes cultivated. Although some authors consider *L. lewisii* to be a variety or subspecies of *L. perenne,* there are arguments for keeping it distinct (Mosquin 1971).

The sepals have the 2 lateral nerves nearly or quite as strong as the midnerve, and the inner sepals are more obtuse or rounded than in *L. usitatissimum,* in which the midnerve is distinctly stronger than the lateral nerves. The larger leaves, however, of *L. usitatissimum* tend to be 3-nerved, while those of *L. perenne* are 1-nerved (except perhaps at the base).

685. Linum perenne

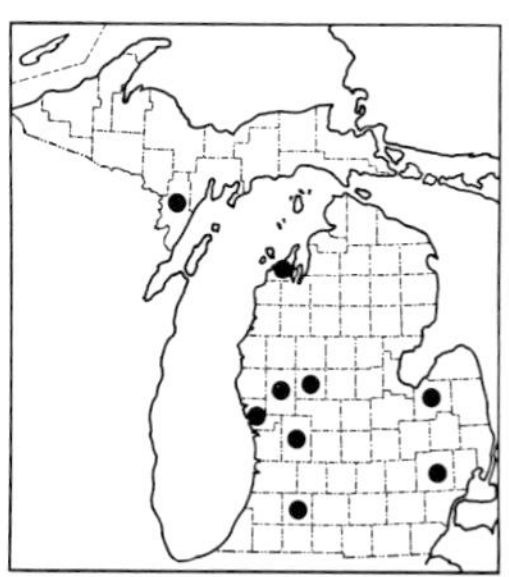
686. Linum sulcatum

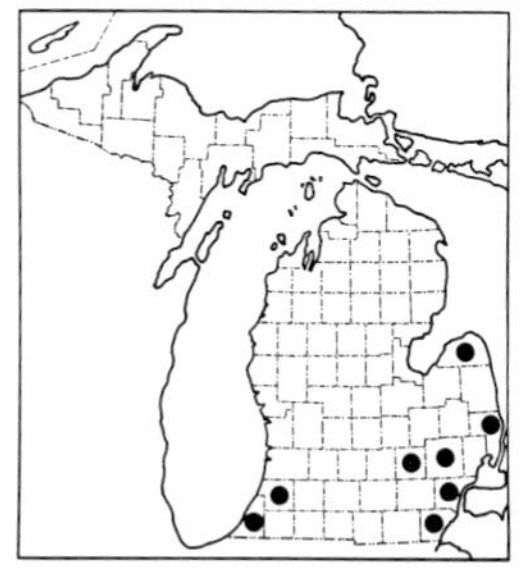
687. Linum medium

4. **L. sulcatum** Riddell Fig. 239
Map 686. Prairies and dry sandy open ground.

5. **L. medium** (Planchon) Britton
Map 687. Sandy fields and meadows, including interdunal flats, shore meadows, and oak openings.

Our plants are var. *texanum* (Planchon) Fernald, by far the commoner and more widespread variety, and a diploid; the tetraploid var. *medium* differs in having the inner sepals sparsely if at all glandular-ciliate (as well as in fruit characters). The type locality of *L. medium* is on the east side of Lake Huron, and the typical variety ranges only from that area south to western Lake Ontario and Lake Erie.

6. **L. virginianum** L.
Map 688. Dry woods, hillsides, and sandy banks; also moist shaded ground, shores, and river banks. This species seems to have become much less common during the past half-century or so; in fact, no collections appear to have been made in Michigan since 1938 (see Rogers 1971).

The inflorescence is more corymbose than in the next species, sometimes quite similar to that of *L. medium*.

7. **L. striatum** Walter
Map 689. Dry to (usually) damp sandy shores, even marshy places and sometimes with jack pine.

This species tends to have opposite leaves at more nodes than the preceding one—usually at more than half (and sometimes all) the nodes below the inflorescence (counting the lowermost nodes, from which leaves may have fallen). Furthermore, the inflorescence is more open and elongate ("paniculate") than in *L. virginianum*.

OXALIDACEAE — Oxalis or Wood-sorrel Family

1. Oxalis

Our common weedy yellow-flowered species are quite variable in pubescence, but fortunately not all variants occur in our area. They have also had a tortured nomenclatural history. The sour taste of the foliage makes it a pleasant nibble. The distinctive trifoliolate leaves with obcordate leaflets have led some to consider the true "shamrock" to be an *Oxalis* (cf. notes under *Trifolium*).

REFERENCES

Eiten, George. 1963. Taxonomy and Regional Variation of Oxalis section Corniculatae. I. Introduction, Keys and Synopsis of the Species. Am. Midl. Nat. 69: 257–309.

Lourteig, Alicia. 1979. Oxalidaceae Extra-austroamericanae II. Oxalis L. Sectio Corniculatae DC. Phytologia 29: 57–198.
Robertson, Kenneth R. 1975. The Oxalidaceae in the Southeastern United States. Jour. Arnold Arb. 56: 223–239.
Rogers, C. M. 1953. Oxalidaceae in Michigan. Asa Gray Bull., N.S. 2: 267–272.

KEY TO THE SPECIES

1. Plant stemless (leaves and scapes all basal); petals white to pink or purple
  2. Flowers in an umbel; petals purple; sepals glabrous, with thickened orange tips; leaves glabrous; plant bulbous at the base, in dry ground in southern Michigan .... 1. **O. violacea**
  2. Flowers solitary on each scape; petals white to pale pink with deeper pink veins; sepals ciliate, with normal tips; leaves with sparse hairs; plant rhizomatous but not bulbous, in damp woods of northern Michigan .... 2. **O. acetosella**
1. Plant with leafy stem; petals yellow
  3. Pubescence largely of septate hairs, mostly spreading on the stem (or stem glabrate) and pedicels, the capsules glabrous or with only septate ± spreading hairs; pedicels remaining erect or ascending in fruit; stipules absent .... 3. **O. fontana**
  3. Pubescence with very few or no septate hairs, antrorse-appressed at least on pedicels, the capsules with minute retrorse non-septate hairs; pedicels usually becoming ± strongly deflexed in fruit (but the capsules erect); stipules often evident, ± oblong, adnate to base of petiole
    4. Stems ± erect (or decumbent at base or in age), with whorled or fascicled leaves and ± dense antrorse-appressed pubescence; fully ripe seeds with the transverse ridges white-edged .... 4. **O. stricta**
    4. Stems prostrate (runners rooting at many nodes), with clearly alternate leaves and scattered spreading pubescence; fully ripe seeds uniform brown ... 5. **O. corniculata**

1. **O. violacea** L.

Map 690. Michigan is at the northern edge of the range of this species, which was found in 1837 by the First Survey in damp sandy soil (presumably a prairie remnant) at Monroe (*Sager?*, NY, as '1838'). The only other Michigan collection seen (despite other published reports) was made in 1964 at Lemon Creek (presumably at Andrews University), Berrien County (*E. Zollinger*, AUB).

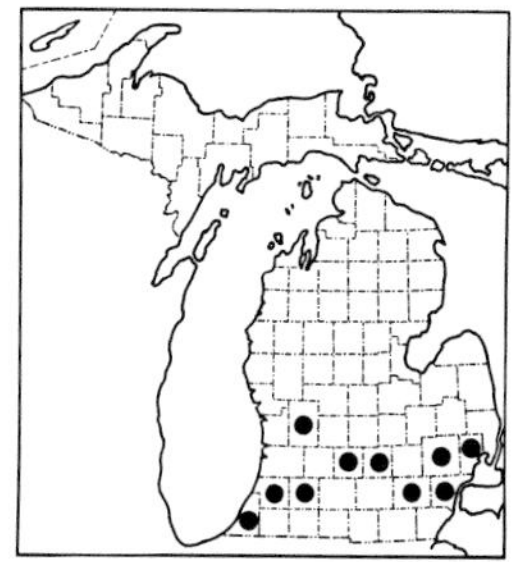
688. Linum virginianum

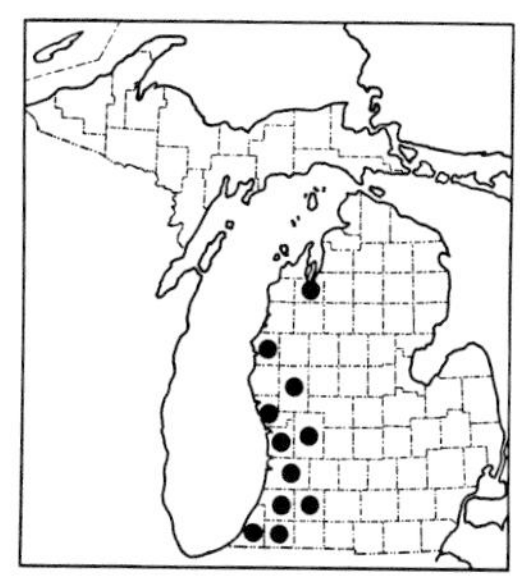
689. Linum striatum

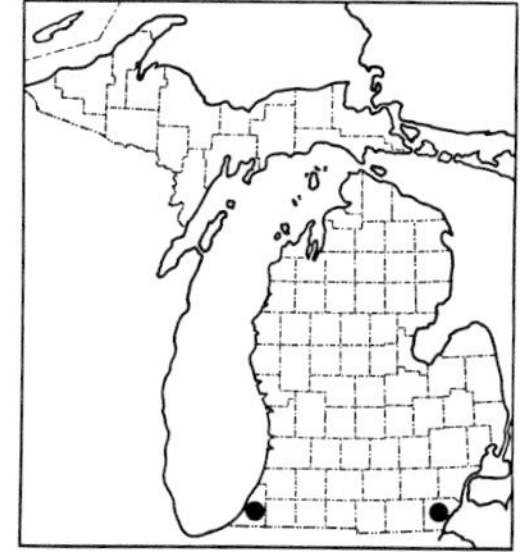
690. Oxalis violacea

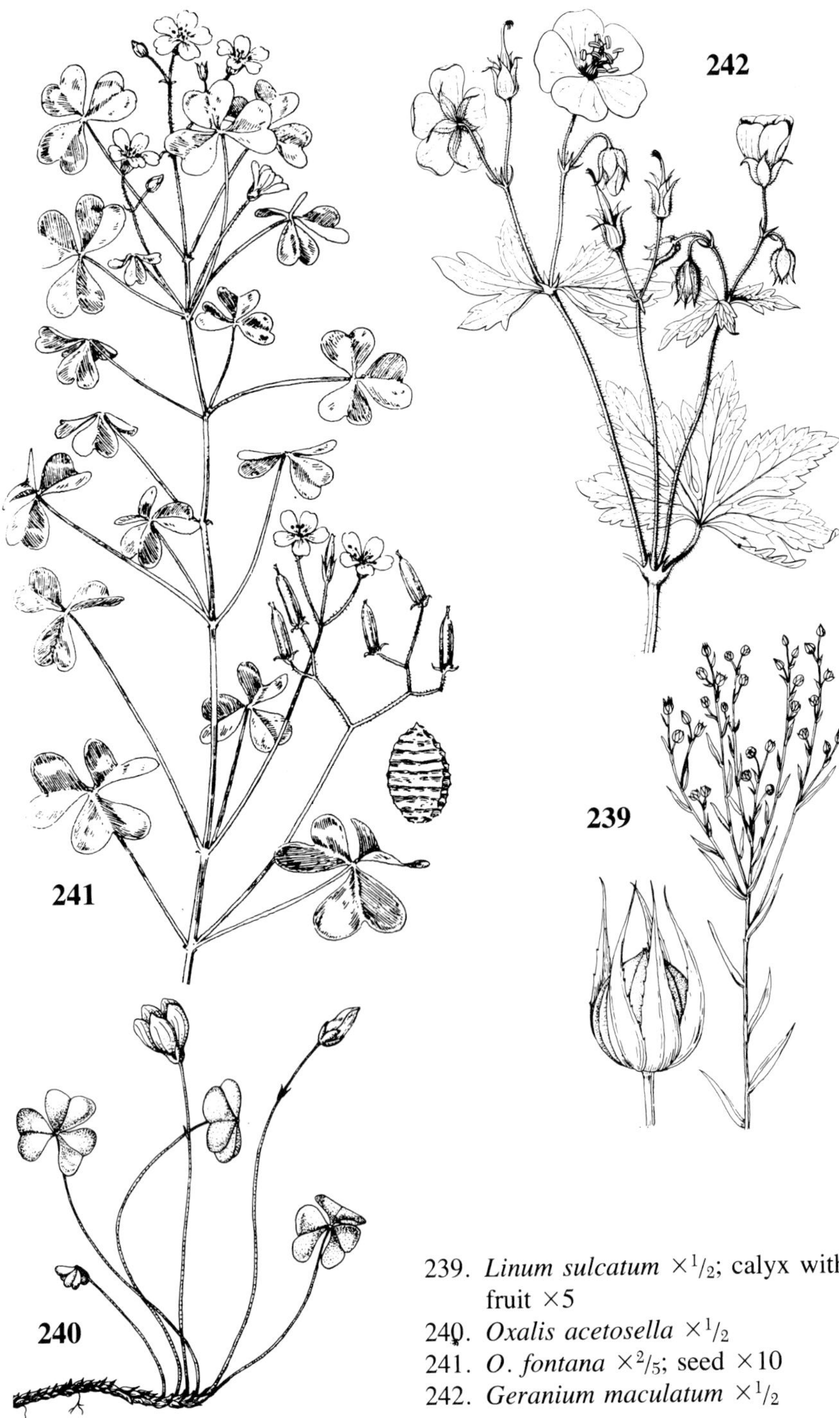

239. *Linum sulcatum* $\times^1/_2$; calyx with fruit $\times 5$
240. *Oxalis acetosella* $\times^1/_2$
241. *O. fontana* $\times^2/_5$; seed $\times 10$
242. *Geranium maculatum* $\times^1/_2$

2. **O. acetosella** L.   Fig. 240

Map 691. Especially characteristic of rich hemlock-hardwoods, but also in various sorts of deciduous, mixed, and coniferous woods, even on hummocks in cedar swamps.

In the broad sense this is a circumpolar species. North American plants have been called ssp. *montana* (Raf.) D. Löve or *O. montana* Raf. The attractive flowers are distylous, with the style either exceeding the anthers or slightly shorter. Small cleistogamous flowers are often produced later in the season.

3. **O. fontana** Bunge   Fig. 241

Map 692. Usually said to be native to North America as well as eastern Asia (whence it was originally described), and collected in Michigan as early as 1832 on a river bank. Now, mostly a common weed of roadsides, railroads, gardens, lawns, fields, and waste places generally; also in woods, especially along trails and other disturbed sites.

This is the species long known as *O. europaea* Jordan, and to which Eiten applied *O. stricta* L. Plants with the leaflets appressed-pubescent above have been called var. *bushii* (Small) Hara. Among the several other variants described is f. *villicaulis* (Wiegand) Hara (TL: Port Huron [St. Clair Co.]); it is ± densely villous with spreading hairs on pedicels and stems. Forms with appressed pubescence have apparently not been collected in Michigan.

4. **O. stricta** L.

Map 693. A native of North America but not collected in Michigan before the 1890's. Often in drier situations than the preceding but sometimes growing with it in gardens and lawns, along roadsides and railroads, in fields and gravel pits, and at other disturbed sites; invading woods, especially dry sandy areas.

Those who have followed Eiten in applying this name to the preceding species, as well as those who reject it in view of its various typifications, use the name *O. dillenii* Jacquin for the present species.

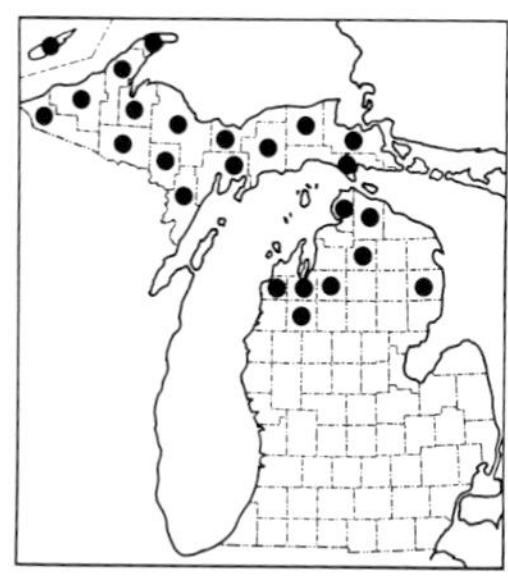

691. Oxalis acetosella

692. Oxalis fontana

693. Oxalis stricta

5. **O. corniculata** L.

Map 694. Native in southern Europe, but spread as a weed elsewhere. Apparently scarce in Michigan, unlike the preceding two species, but local in lawns and gardens (also in greenhouses).

GERANIACEAE — Geranium Family

Although some species of the genus *Geranium* are cultivated, the "geraniums" of the florist, including those commonly grown as house plants, are various species and hybrids in the genus *Pelargonium*—plants native mostly to South Africa.

The complex fruit in this family is a schizocarp, consisting of five 1-seeded mericarps (which may split open) at the base of an elongate beak which terminates in the more slender style. (Measurements of this stylar tip do not include the 5-fid apex.) At maturity, the mericarps separate, each with an "awn" that peels away upwardly from the beak and often remains attached to the apex of the beak. The awns are hygroscopic and curve outward or spiral with changes in humidity.

REFERENCE

Robertson, Kenneth R. 1972. The Genera of Geraniaceae in the Southeastern United States. Jour. Arnold Arb. 53: 182–201.

KEY TO THE GENERA

1. Leaves pinnately lobed or dissected; awn of mericarp ± spirally twisted at full maturity; anther-bearing stamens 5 (the other 5 represented by staminodia) . . . 1. **Erodium**
1. Leaves palmately lobed or compound; awn of mericarp curving in a simple arc or coil at maturity; anther-bearing stamens typically 10 (except in one species or—as is usually the case—if some of the readily deciduous anthers have fallen) . . . . 2. **Geranium**

1. Erodium

1. **E. cicutarium** L'Hér. Fig. 243 — Stork's-bill; Alfileria

Map 695. Native to the Mediterranean region, but a naturalized weed in most temperate areas of the world. No Michigan collections have been seen from before the 1870's. Usually in dry sandy (or rocky) places, including roadsides, fields, cultivated ground (gardens etc.), clearings.

2. Geranium — Wild Geranium; Crane's-bill

Reports of *G. rotundifolium* L. (a small-flowered plant with leaves cleft ca. 50–60% the length of the blade) from Michigan appear to have been based on misidentifications of *G. sanguineum* and *G. molle*.

KEY TO THE SPECIES

1. Petals (10) 12–18 (22) mm long, much exceeding the sepals, the margin densely ciliate at the base; anthers 2–3 mm long; plant perennial from a strong rhizome
   2. Stem with 1–2 pairs of petioled leaves; basal leaves conspicuous, with largest blades (6) 8–12 (18) cm broad; petals entire or obscurely emarginate; pedicels glandless or with sessile glands overtopped by glandless hairs; plant a common native of woods . . . . 1. **G. maculatum**
   2. Stem with several pairs of petioled leaves; basal leaves none or few with blades no larger than the cauline leaves (ca. 4–6 cm broad); petals notched at apex; pedicels with glands inconspicuous, sessile or nearly so . . . . 2. **G. sanguineum**
1. Petals (except in *G. robertianum,* with long claws) less than 10 mm long, scarcely if at all exceeding the sepals, glabrous or sparsely ciliate; anthers less than 1.5 mm long; plant annual or with an easily uprooted taproot or short rhizome
   3. Leaf blades cleft to the very base (hence compound with usually 3 deeply bipinnatifid leaflets); petals 9–13 (15) mm long, narrowed basally to a distinct long claw . . . . 3. **G. robertianum**
   3. Leaf blades cleft two-thirds or more of their length (sometimes close but not completely to the base); petals less than 8 (9) mm long, nearly or quite without a claw
      4. Sepals with at most a minute callous tip
         5. Mature mericarps (excluding awns) glabrous, smooth or with strong transverse ridges; style tip on beak of fruit ca. 0.5–1.5 mm long; stems with ± dense long spreading eglandular hairs and short often gland-tipped hairs . . . . 4. **G. molle**
         5. Mature mericarps pubescent (± strigose), smooth (not ridged); style tips obsolete; stem pubescence mostly of short gland-tipped hairs
            6. Petals less than 5 mm long; stems without long hairs; anthers ca. 0.5 mm long or shorter . . . . 5. **G. pusillum**
            6. Petals ca. 5–8 [10] mm long; stem with sparse long eglandular hairs overtopping dense short glandular hairs; anthers ca. 1–1.2 mm long . . . 6. **G. pyrenaicum**
      4. Sepals with a distinct subulate or awn-like tip (0.7) 1–2 mm long
         7. Mature pedicels about equaling or shorter than the calyx, at most becoming ca. 1.5 times as long in fruit, usually with few if any of their longest hairs gland-tipped; style tip on beak of fruit 1–2 mm long . . . . 7. **G. carolinianum**
         7. Mature pedicels mostly distinctly longer than the calyx, becoming more than twice as long in fruit, in commonest species with most of their longest hairs gland-tipped; style tip on beak of fruit (2.2) 2.5–4 (4.5) mm long
            8. Pedicels with dense spreading hairs of various length, the longest mostly glandular; mericarps with long hairs . . . . 8. **G. bicknellii**
            8. Pedicels with only tiny retrorse non-glandular hairs; mericarps glabrous or nearly so . . . . 9. **G. columbinum**

694. Oxalis corniculata

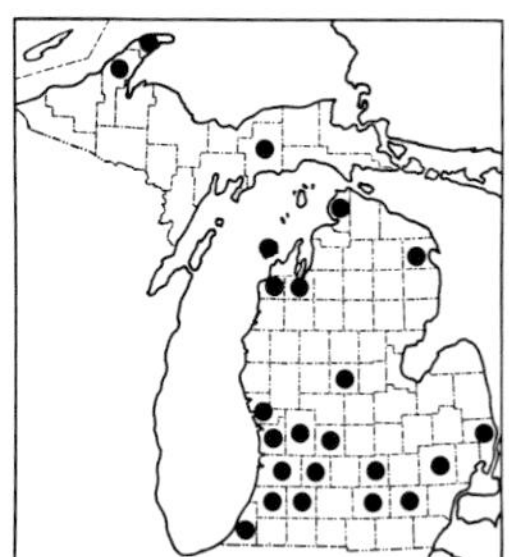
695. Erodium cicutarium

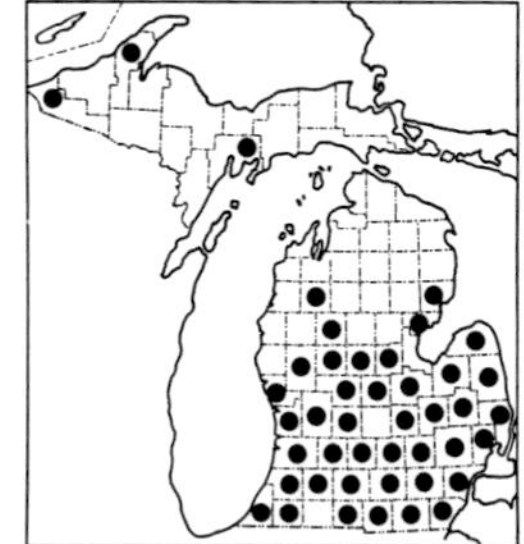
696. Geranium maculatum

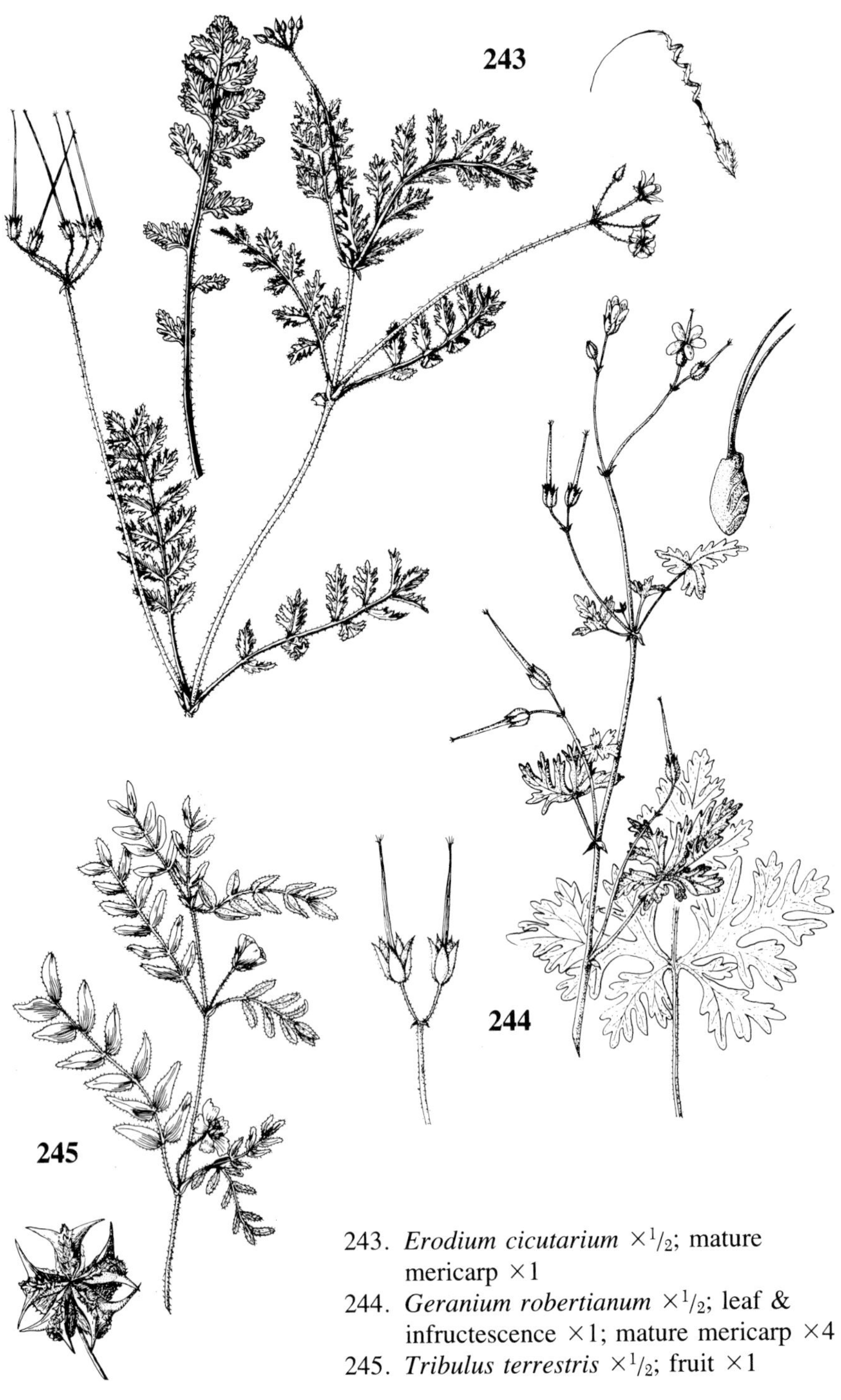

243. *Erodium cicutarium* ×1/2; mature mericarp ×1
244. *Geranium robertianum* ×1/2; leaf & infructescence ×1; mature mericarp ×4
245. *Tribulus terrestris* ×1/2; fruit ×1

1. **G. maculatum** L. Fig. 242

Map 696. Usually in rich deciduous woods, especially in moist sites such as streamsides, pond borders, wet hollows, and swamp forest; less often in upland oak-hickory woods.

Plants with white flowers are rare [f. *albiflorum* (Raf.) House]. A white-flowered form of a cultivated Eurasian species, *G. pratense* L., was collected along a driveway in Ann Arbor in 1954, but its status, whether truly escaped or not, is doubtful. It resembles *G. maculatum,* but the pedicels (and outer sepals) bear dense gland-tipped hairs.

2. **G. sanguineum** L.

Map 697. A Eurasian species, sometimes cultivated, locally established in dry soil, vacant lots, roadsides, and lawns.

When first reporting this species from the state, Dodge (Univ. Mich. Mus. Zool. Misc. Publ. 4: 11. 1918) described the rhizomes as "fully as red" as those of the common bloodroot. The petals are reddish purple (except in white forms).

3. **G. robertianum** L. Fig. 244 Herb Robert

Map 698. Rich deciduous woods (oak-hickory or more often beech-maple), especially in clearings, along roads and trails, and at borders; rocky openings, gravelly shores, and rubble—generally a calciphile; rarely associated with cedar. A circumpolar species, presumably native in our area although characteristic of disturbed places, but sometimes considered introduced in western North America.

The whole plant has a strong and rather unpleasant scent, quite evident especially after specimens have been shut in a confined space, not entirely unlike that of *Solanum dulcamara*.

4. **G. molle** L.

Map 699. A widespread weed, originally native to Europe, first collected in Michigan at Harbor Springs in 1903 (see Pap. Mich. Acad. 42: 24. 1957) and occasionally found in lawns and disturbed areas.

The petals are up to 7 mm long, deeply notched.

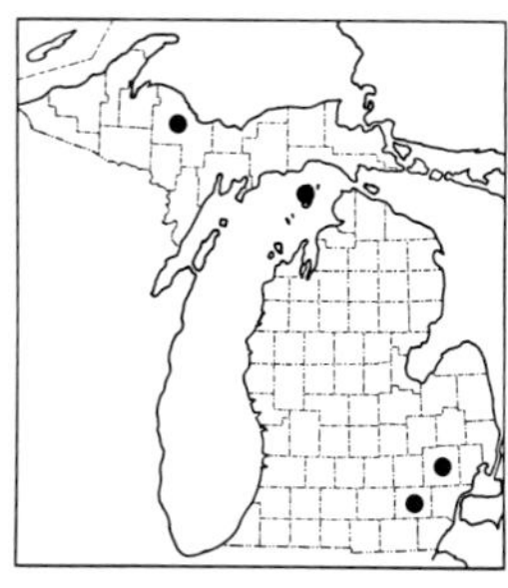

697. Geranium sanguineum

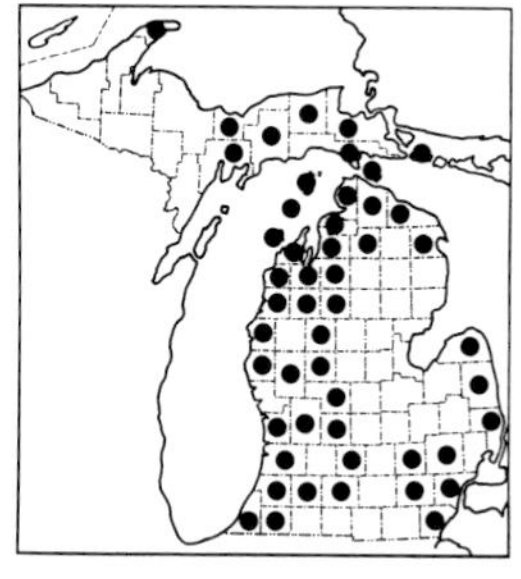

698. Geranium robertianum

699. Geranium molle

5. **G. pusillum** L.

Map 700. Another Eurasian species, naturalized as a weed. The earliest Michigan collection seen is from Flint in 1865 [*D. Clarke,* MSC]. Now in lawns, gardens, and cultivated ground; sandy disturbed places, clearings, gravel pits, railroads.

A rather inconspicuous prostrate or erect several-stemmed plant. The petals are more shallowly notched than in the preceding species, and smaller, scarcely if at all exceeding the sepals. Some of the stamens are reduced to staminodia, but since anthers are so readily deciduous in this genus, that character must be used with caution.

6. **G. pyrenaicum** N. L. Burman

Map 701. A weedy European species, collected at Rochester, Oakland County, in 1896 (*Farwell,* BLH) and on a roadside west of Sutton's Bay in 1954 (*Thompson L-1760,* BLH).

Generally more erect and larger-flowered than the preceding two species, and the lobes of the leaf are more broadly cuneate in outline, not so narrowly cleft.

7. **G. carolinianum** L.

Map 702. Sandy and rocky fields, clearings, and bluffs; thin soil over limestone pavement (Drummond Island); shores and grasslands; waste places.

Not nearly so common as the next species, and often confused with it. In our area, *G. carolinianum* seems quite consistently to have a rather compact, rounded inflorescence, giving the impression of a tight umbel, while in *G. bicknellii* the inflorescence is more diffuse and open. The mericarps in both species are smooth, with long ascending hairs (the longest ca. 1–1.5 mm, and darkish). The seeds in both are weakly reticulate with usually ± elongate cells. At least some of our plants may be the weakly distinguished var. *sphaerospermum* (Fern.) Breitung, with broader sepals (at least 5 mm) and subglobose (rather than oblong) seeds.

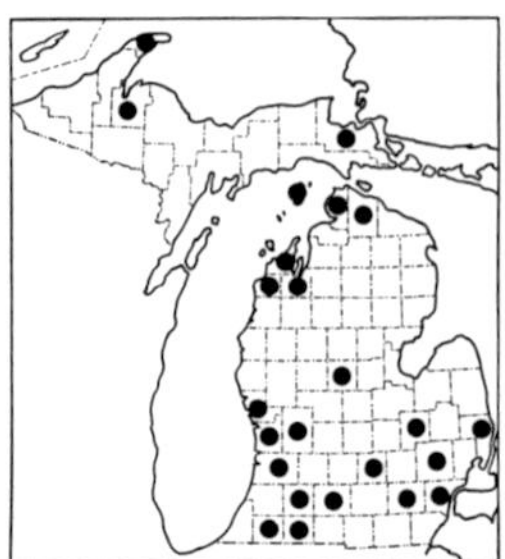

700. Geranium pusillum

701. Geranium pyrenaicum

702. Geranium carolinianum

G. dissectum L., an adventive from Europe, resembles this species and the next in its subulate-tipped sepals, and especially *G. carolinianum* in the short pedicels (ca. twice or less the length of the sepals) and stylar tips. However, the mericarps are pubescent with short spreading hairs (not over ca. 0.5 mm, white). The pedicels bear long gland-tipped hairs (as well as short hairs) and the petals are sometimes said to be a much deeper pink than the pale ones of *G. bicknellii* and *G. carolinianum*. Michigan has been included in its range by authors who copy unverified reports from one another, but no authentic specimens have been encountered from the state. A piece of fruiting *G. dissectum* is mounted with a flowering specimen of *G. bicknellii* from Delta County (*Wheeler* in 1892, MSC), originally labeled as the latter; it appears that some mixture has occurred, and that the *G. dissectum* does not belong.

8. **G. bicknellii** Britton Plate 5-F

Map 703. Rock outcrops (pavement, summits, ridges); clearings and burns, gravel pits, and trails in woods; open, usually dry, sandy or gravelly ground.

This is the commonest of our small-flowered geraniums (excluding *G. robertianum*), characteristic of recently disturbed ground, especially in the northern part of the state. It is often associated with *Corydalis* spp., but tends to be more widespread and to persist longer in dry open ground, in this respect associated with such species as *Carex houghtoniana, Polygonum cilinode,* and *Aralia hispida*. Flowering plants may be as small as a few cm tall, or large and bushy, close to a meter in height.

9. **G. columbinum** L.

Map 704. A native of Europe, collected as a lawn weed on the Andrews University campus in 1974 (*T. Rule 57,* AUB). Earlier reports from the state are not supported by specimens.

A distinctive-looking little geranium, with very long slender pedicels and deeply cleft leaves with narrow lobes.

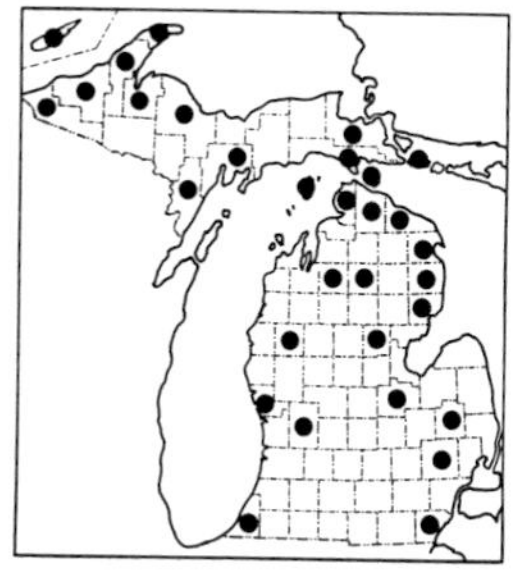

703. Geranium bicknellii

704. Geranium columbinum

705. Tribulus terrestris

ZYGOPHYLLACEAE — Caltrop Family

A family of uncertain affinity. Cronquist includes it in the Sapindales, Takhtajan in the Rutales, and several authors in the Geraniales—all closely allied orders, not uniformly defined. Our only representative is a seldom-collected weed.

REFERENCE

Porter, Duncan M. 1972. The Genera of Zygophyllaceae in the Southeastern United States. Jour. Arnold Arb. 53: 531–552.

1. **Tribulus**

1. **T. terrestris** L. Fig. 245 — Caltrop; Puncture Vine

Map 705. Originally native, apparently, to the Mediterranean area and now a noxious weed in warm climates. Rarely found as far north as Michigan, where it was first collected in 1930 along a roadside at Grand Rapids. It was collected in several other counties 1931–1937, so probably was dispersed before 1930. Still not common, but found also in poor lawns, gardens, parking lots, and such waste ground.

Easily recognized by its prostrate habit; hairy stems and leaves, the latter consisting usually of 6–7 pairs of small leaflets; and spiny nutlets—the two principal spines on each nutlet stouter than those of the equally unpleasant *Cenchrus*.

RUTACEAE — Rue Family

Plants of this family have characteristic oil glands, often evidenced by translucent dots on the leaves and an aromatic fragrance when crushed or wounded. The genus most familiar to many people is *Citrus*. The family is best represented in tropical regions and the southern hemisphere. We have only two easily recognized native species.

REFERENCE

Brizicky, George K. 1962. The Genera of Rutaceae in the Southeastern United States. Jour. Arnold Arb. 43: 1–22.

KEY TO THE GENERA

1. Flowers showy, ca. 2 cm long, bilaterally symmetrical; fruit a very deeply 5-lobed, several-seeded, woody capsule; plant herbaceous (except at base); leaflets closely but distinctly denticulate . . . . 1. **Dictamnus**

1. Flowers greenish or yellowish, much less than 1 cm long, radially symmetrical; fruit a 1-seeded samara or follicle; plant woody; leaflets entire or nearly so
  2. Leaves trifoliolate; stem unarmed; flowers perfect and/or unisexual (plants usually polygamous) with both calyx and corolla; inflorescences appearing with the leaves, terminal on short branches; fruit a flat 1-seeded samara, ca. 15–30 mm in diameter . . . . 2. **Ptelea**
  2. Leaves pinnately compound with 5–11 leaflets; stem with strong broad-based prickles; flowers with 1 series of perianth parts, unisexual (plants dioecious); inflorescences appearing before the leaves, sessile on old wood; fruit a plump wingless ellipsoid 1-seeded follicle less than 6 mm long . . . . 3. **Zanthoxylum**

1. Dictamnus

1. D. albus L. Gas-plant

Map 706. A European species, grown as a long-lived ornamental and rarely escaping, presumably by seed. Well established in an open area near the highway west of Conway (*H. Crum* in 1978, MICH).

The terminal raceme of white to pink or bluish flowers is densely glandular-pubescent. The volatile, inflammable oil produced by the plant may be ignited (it is said) in very calm weather by holding a match to the base of an inflorescence.

2. Ptelea

REFERENCES

Bailey, Virginia Long. 1960. Historical Review of Ptelea trifoliata in Botanical and Medical Literature. Econ. Bot. 14: 180–188.

Bailey, Virginia Long. 1962. Revision of the Genus Ptelea (Rutaceae). Brittonia 14: 1–45.

1. P. trifoliata L. Fig. 246 Wafer-ash; Hop-tree

Map 707. Wooded to ± open dunes along Lake Michigan; sandy fields and knolls; fencerows and dry bluffs or banks; rarely in swampy places.

706. Dictamnus albus

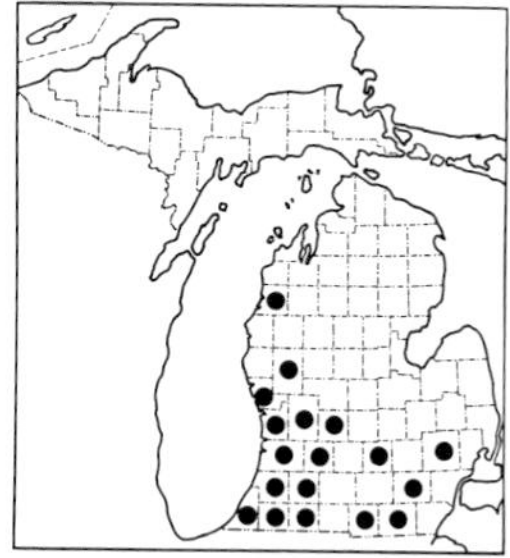

707. Ptelea trifoliata

708. Zanthoxylum americanum

The bitter and aromatic properties of this plant led to various medicinal uses and the fruits were tried as a substitute for hops in brewing. It was early cultivated in European botanical gardens, once it was discovered in colonial America. It is especially attractive in fruit, and apparently thrives when planted considerably north of its native range; a good stand is along a private road by Ives Lake in the Huron Mountains, Marquette County. This species grows as a shrub or small tree; the largest in the nation is in a park in Kent County, with a trunk over 10 inches in diameter.

The species is variable, but almost all Michigan plants have been placed by Bailey in var. *trifoliata*. A few from the Lake Michigan area have been referred to var. *mollis* T. & G., which differs in having much more dense pubescence on twigs, leaves, and inflorescences.

3. **Zanthoxylum**

REFERENCE

Porter, Duncan M. 1976. Zanthoxylum (Rutaceae) in North America North of Mexico. Brittonia 28: 443–447.

1. **Z. americanum** Miller Figs. 247, 248 Prickly-ash

Map 708. Sometimes in upland deciduous woods, but usually in low woods and thickets, even swamp forests, along streams and river banks.

Some persons have held that the correct name for this plant is *Z. fraxineum* Willd., but that is an illegitimate name and hence cannot be used under any circumstances. The generic name is often "corrected" to *Xanthoxylum,* but nothing authorizes such corrections.

Like the preceding species, this one has oils which have given it reputed medicinal properties. The plant is usually a tall shrub, and spreads readily by suckers, forming large, thorny thickets. Sterile specimens might superficially resemble *Robinia pseudoacacia* in the prickly stems and pinnately compound usually sparsely pubescent leaves. In *Zanthoxylum americanum,* the leaflets are ovate, ± acute but notched at the tip with a gland but no mucro, sessile or nearly so. In *Robinia pseudoacacia,* the leaflets are more elliptical, rounded at the tip but with a small mucro (even if somewhat notched), each with a distinct petiolule-pulvinus. The leaf scars on winter twigs are different in the two genera (cf. figs. 224 & 248).

SIMAROUBACEAE Quassia Family

REFERENCE

Brizicky, George K. 1962. The Genera of Simaroubaceae and Burseraceae in the Southeastern United States. Jour. Arnold Arb. 43: 173–186.

246. *Ptelea trifoliata* ×$^1/_2$; flower ×2
247. *Zanthoxylum americanum* ×$^1/_2$; staminate flower ×4; fruit ×3
248. *Z. americanum,* winter twig ×1; node with spines, buds, & leaf scar ×2

1. **Ailanthus**

1. **A. altissima** (Miller) Swingle   Fig. 249   Tree-of-Heaven

Map 709. Native of China, introduced into North America in 1784 as an ornamental, now about as loved as some other introductions such as starlings and gypsy moths. It is a fast-growing weed tree, sprouting from seeds and suckers, especially in alleys, beside walls and buildings, and in similar narrow places; also in fields, meadows, dumpsites; on shores and river banks; and along railroads. Doubtless far more widespread than the map indicates, but by-passed by collectors.

This robust smooth-barked tree, with large pinnately compound leaves and large inflorescences of small flowers, is polygamo-dioecious, some trees producing copious elongate samaras with a single central seed. The foliage and especially the staminate flowers are strong-smelling.

POLYGALACEAE   Milkwort Family

1. **Polygala**   Milkwort; Polygala

The bilaterally symmetrical flowers superficially resemble those of the Leguminosae, but the structures are not fully homologous. Of the 5 sepals, the 2 lateral ones are large and petaloid—usually as large as the petals; the petals are 3, the lower one forming a keel which encloses the stamens and style and usually has a fringe or appendage near the end. The stamens are 6 (in *P. paucifolia*), 7, or 8. The seeds are pubescent and bear an aril (or aril-like structure) which apparently aids in dispersal by ants. All of our species (except one blue-flowered escape from cultivation) have the flowers pink-magenta or greenish white or pure white, but yellow-flowered species occur in the genus. Some species also produce reduced cleistogamous flowers, underground or above ground; except when explicitly stated, the key does not refer to these. The lowermost leaves are often reduced to scales.

REFERENCES

Gillett, John M. 1968. The Milkworts of Canada. Canada Dep. Agr. Monogr. 5. 24 pp.

Miller, Norton G. 1971. The Polygalaceae in the Southeastern United States. Jour. Arnold Arb. 52: 267–284.

KEY TO THE SPECIES

1. Flowers 1–4 (5) on each stem, (12) 15–20 (22) mm long, the wings readily deciduous; well developed leaves few, crowded toward summit of stem . . . . . . 1. **P. paucifolia**
1. Flowers numerous, ca. 3–7.5 mm long, the wings persistent; well developed leaves several to many (or often none at anthesis in *P. incarnata*)
   2. Leaves (at least the lower ones) whorled (or opposite—look for scars if necessary)

3. Well developed inflorescences ca. 10–15 mm broad, sessile or on peduncles at most ca. 5 mm long . . . . 2. **P. cruciata**

3. Well developed inflorescences ca. 3–5 mm broad, on distinct peduncles usually 1–5 cm long . . . . 3. **P. verticillata**

2. Leaves all alternate

4. Inflorescence a distinct loose raceme, the lower flowers well separated from each other and on pedicels mostly 1–2 (3) mm long; plant (in common species) producing subterranean cleistogamous flowers or (rare escape) with flowers at least 6 mm long

5. Flowers ca. 6–7.5 mm long, normally blue; cleistogamous flowers none . . . . 4. **P. vulgaris**

5. Flowers ca. 3.5–4.5 (5.5) mm long, normally pink-magenta; cleistogamous flowers developing underground . . . . 5. **P. polygama**

4. Inflorescence a dense spike-like raceme, the flowers crowded and on pedicels mostly no more than 1 mm long; plant producing no cleistogamous flowers, nor others (except in *P. incarnata*) exceeding 6 mm

6. Leaves subulate, less than 1 mm wide, soon deciduous; flowers ca. 5–7 mm long, the petals united into a tube (open on one side) much exceeding the wings and other sepals . . . . 6. **P. incarnata**

6. Leaves broader (mostly 1.2–20 mm), persistent; flowers ca. 3–5.5 (6) mm long, the petals only very slightly united and scarcely if at all exceeding the wings

7. Plant perennial, with many glandular-puberulent stems from a woody root; inflorescence ca. 4–8 (9) mm broad, normally at least 3–5 times as long; flowers ca. 3–4 mm long . . . . 7. **P. senega**

7. Plant annual, with glabrous stem from a poorly developed taproot; inflorescence ca. 8–12 mm broad, at most ca. twice as long when in good flower (rachis elongate after lower flowers are shed); flowers ca. 3.5–5.5 (6) mm long . . . . 8. **P. sanguinea**

1. **P. paucifolia** Willd.   Fig. 250   Fringed Polygala; Gay-wings; Flowering-wintergreen

Map 710. In all sorts of mixed and coniferous woods except the wettest and the driest, most often with fir, cedar, birch, aspen, hemlock, and pines; on old shoreline ridges of calcareous rubble and sandy beach meadows, usually ± shaded.

The flowers are bright magenta or pink-purple, varying from pale to deep (in shade) and rarely pure white. (The attractive albinos [f. *alba* Wheelock]

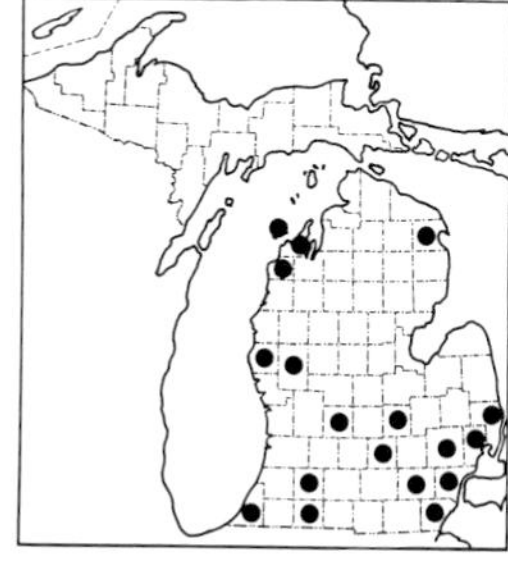

709. Ailanthus altissima

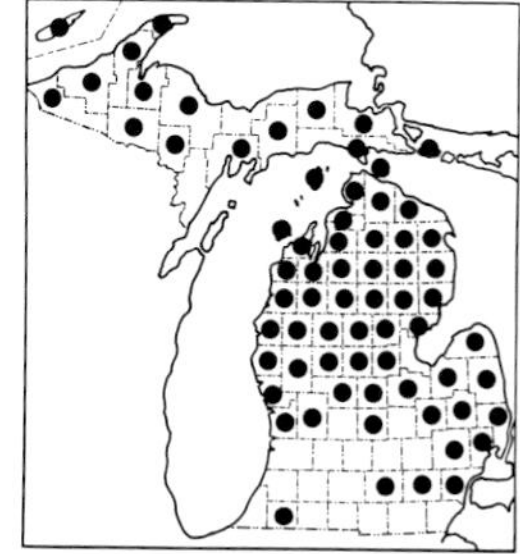

710. Polygala paucifolia

711. Polygala cruciata

are known from Emmet and Cheboygan counties and are reported from Keweenaw County.) The leaves on most of the stem are reduced to scales, the few flowers and leaves at the summit providing a striking touch of color, often abundant in northern damp woods in spring. An old beach ridge carpeted with polygala, calypso, and dwarf lake iris is a feast of beauty not soon forgotten. Cleistogamous flowers are produced at the base of a plant, often buried; their fruit is similar to those of the other flowers: a 2-seeded somewhat heart-shaped capsule.

2. **P. cruciata** L. Fig. 251

Map 711. Moist sandy or somewhat organic soil, especially shores of fluctuating lakes and old interdunal hollows.

See comments under *P. sanguinea*.

3. **P. verticillata** L.

Map 712. Moist to (usually) dry often sandy fields, shores, hillsides, woodlands.

Although a variable species, distinctive in its slender, ± conical, inflorescences and graceful aspect.

4. **P. vulgaris** L.

Map 713. A European species, rarely established and undoubtedly an escape from cultivation, as on Grand Island (where collected by Dodge in 1916) and the Quincy Mine site (*Bourdo* in 1974).

5. **P. polygama** Walter

Map 714. Dry sandy barren ground, often with oaks and jack pine; thin woods on granite (Marquette Co.); sandy meadows near ponds and lakes.

Plants are occasionally found with above-ground racemes or spikes of cleistogamous flowers, at least in axils of several of the leaves if not terminal. Such rather leafy-looking plants may be called (if they deserve a name) f. *ramulosa* (Farw.) E. G. Voss (TL: Algonac [St. Clair Co.]).

712. Polygala verticillata

713. Polygala vulgaris

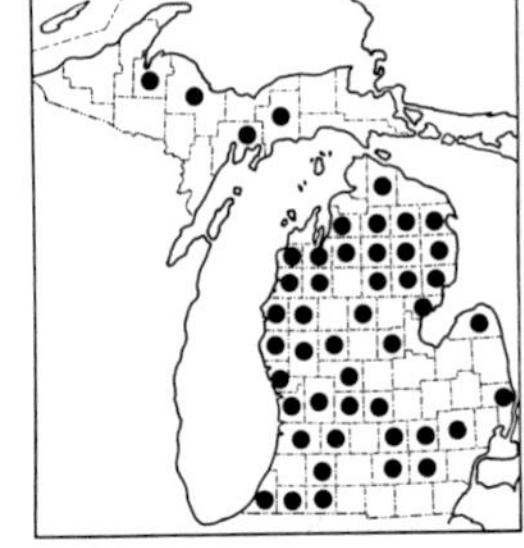

714. Polygala polygama

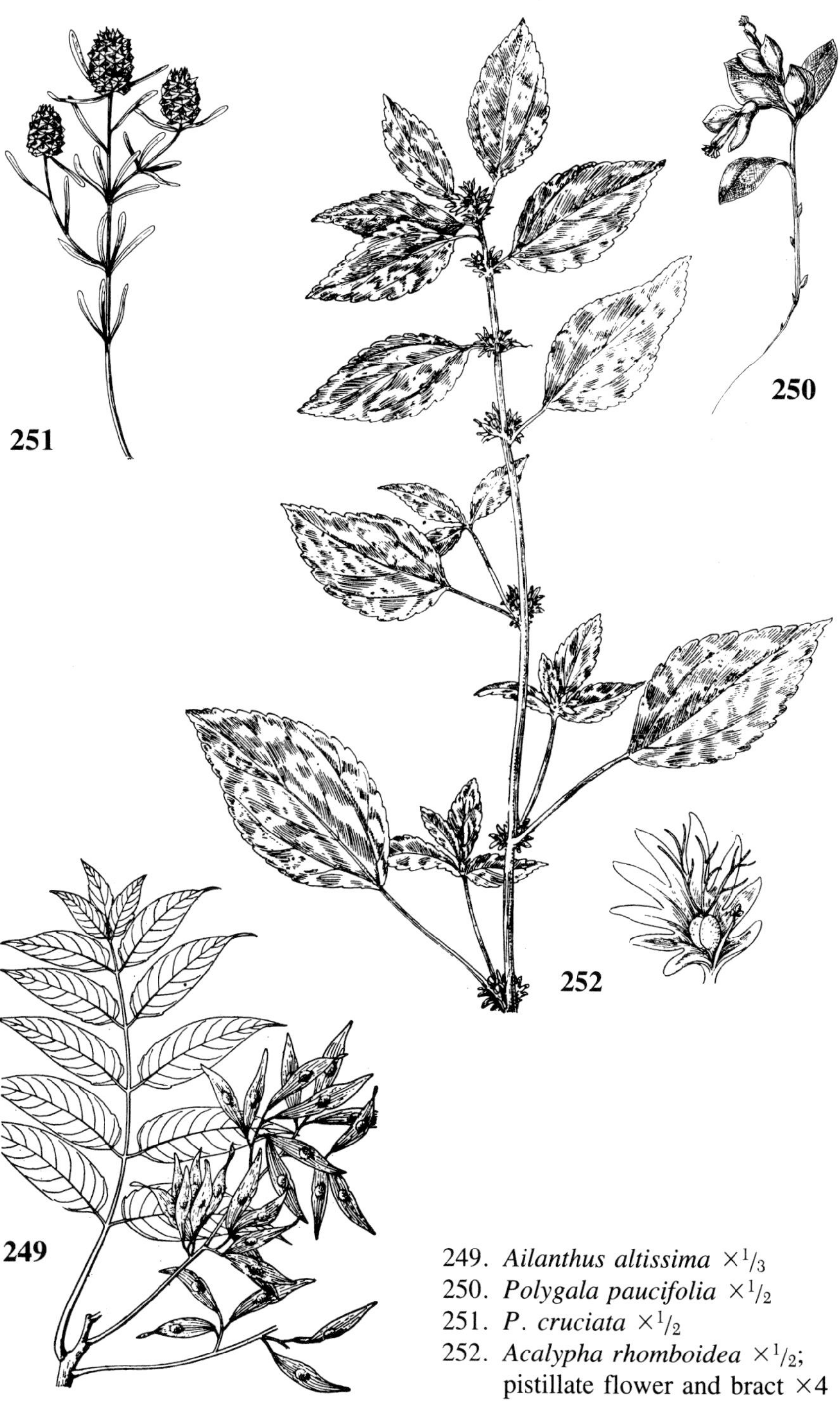

249. *Ailanthus altissima* $\times 1/3$
250. *Polygala paucifolia* $\times 1/2$
251. *P. cruciata* $\times 1/2$
252. *Acalypha rhomboidea* $\times 1/2$; pistillate flower and bract $\times 4$

6. **P. incarnata** L.

Map 715. Known from Squirrel and Walpole islands, on the Ontario side of the mouth of the St. Clair River (the only known Canadian localities); collected by Dodge in dry ground on nearby Harsen's Island in 1896.

7. **P. senega** L. Seneca Snakeroot; Senega Root

Map 716. In southern Michigan, occurs in a diversity of wooded, boggy, swampy, rocky, or even prairie-like habitats, shores and banks; in northern Michigan, on calcareous rocks and gravels in openings and at borders of coniferous woods. Surprisingly spotty in distribution, although locally frequent.

The stout knotty root has long had reputed medicinal properties for a wide variety of disorders, including snakebite. The root and the minutely glandular-puberulent stem readily distinguish this species, which has greenish white or nearly pure white flowers.

8. **P. sanguinea** L.

Map 717. Dry to moist, often sandy fields, excavations, and borders of marshes. The Keweenaw County collection (MICH), which looks so out of range, is by O. B. Wheeler, from Mount Houghton, undated but presumably collected during the 1866–1870 period when Wheeler was on Lake Superior as an astronomer with the U.S. Lake Survey (Voss 1978, pp. 52–53); the label is in the hand of Lewis Foote, who was also with the Lake Survey.

The flower color varies from bright pink-magenta to white [f. *albiflora* Wheelock]. The broad compact inflorescence is rather similar to that of *P. cruciata,* which differs not only in its tendency to whorled leaves but also in its ± deltoid, almost caudate wings.

EUPHORBIACEAE Spurge Family

This is one of the largest families of flowering plants in the world, with the number of species estimated as exceeding 5,000, perhaps 7,000. These are predominantly tropical. In some genera a milky (or colored) latex is produced, as in the largest genus, *Euphorbia,* and in the chief source of natural rubber, *Hevea brasiliensis* (Willd.) Muell. Arg. There are many poisonous plants in the family, and some species produce useful oils and waxes. *Manihot esculenta* Crantz is the source of tapioca, cassava meal, and other edible products (once the cyanide is removed)—one of the most important sources of starch in the tropics. The most familiar ornamental species in the family, unless certain South African succulents are considered, is probably *Euphorbia pulcherrima* Klotzsch, the poinsettia (often carelessly misspelled—and pronounced "poinsetta"). This very diverse and interesting family is poorly represented in our flora.

REFERENCE

Webster, Grady L. 1967. The Genera of Euphorbiaceae in the Southeastern United States. Jour. Arnold Arb. 48: 303–361; 363–430.

KEY TO THE GENERA

1. Leaves peltate, all but the smallest blades over 10 cm broad, palmately veined and ± deeply divided into acute lobes; inflorescence a large terminal raceme or panicle . . . . 1. **Ricinus**
1. Leaves not peltate, the blades less than 5 cm broad, pinnately veined (or 3-veined at the base), unlobed; inflorescence various, usually ± umbellate or axillary
  2. Stem and leaves with some (or all) hairs forked or stellate . . . . 2. **Croton**
  2. Stem and leaves glabrous or with only simple hairs
    3. Plant with watery sap; flowers not in cyathia, but pistillate flowers enclosed in a deeply lobed foliaceous bract; stem pubescent with incurved hairs . . . . . . 3. **Acalypha**
    3. Plant with milky juice; flowers grouped into cup-like cyathia with a central stipitate pistillate flower and several staminate flowers consisting of 1 stamen each (figs. 254, 256, 260); stems glabrous or variously pubescent . . . . 4. **Euphorbia**

1. Ricinus

1. **R. communis** L. Castor-bean

Map 718. A plant of tropical (probably African) origin, widely cultivated as an ornamental (for its impressive leaves) and locally escaped even as far north as our region. Collected in waste ground at Mount Clemens in 1924 (*Farwell 7028,* BLH; see Am. Midl. Nat. 9: 272. 1925; Farwell's earlier cited collection from Detroit has not been found).

The seed coats contain one of the most poisonous substances known. The large seeds, however, yield an oil (castor oil) useful as a lubricant (for machinery as well as the digestive tract) and in many industrial applications. The plant is glabrous and without spines, unlikely to be confused with any other species with large palmate leaves.

715. Polygala incarnata

716. Polygala senega

717. Polygala sanguinea

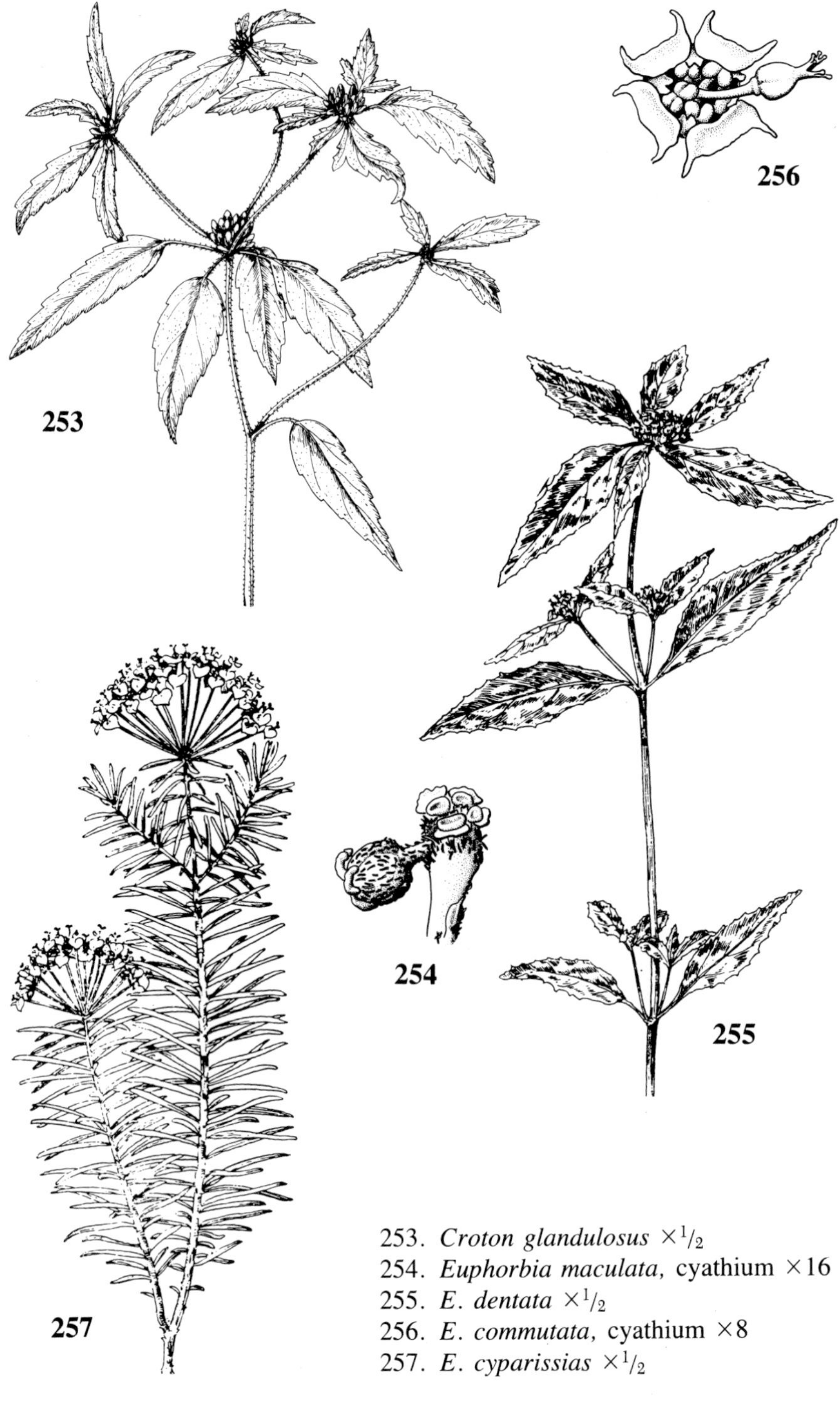

253. *Croton glandulosus* ×1/2
254. *Euphorbia maculata,* cyathium ×16
255. *E. dentata* ×1/2
256. *E. commutata,* cyathium ×8
257. *E. cyparissias* ×1/2

2. Croton

KEY TO THE SPECIES

1. Leaves with toothed blades and 1–2 prominent saucer-like glands at the summit of the petiole . . . . 1. **C. glandulosus**
1. Leaves with entire blades and no glands . . . . 2. **C. monanthogynus**

1. **C. glandulosus** L. Fig. 253

Map 719. Native farther south, adventive in our area along railroads and in disturbed ground.

2. **C. monanthogynus** Michaux Prairie-tea

Map 720. Native farther west, adventive in our area on a path to a lake shore (*C. R. & F. N. Hanes 7041* in 1941, WMU; see Pap. Mich. Acad. 28: 39. 1943).

3. Acalypha Three-seeded Mercury

KEY TO THE SPECIES

1. Bracts beneath pistillate flowers mostly with at least 10 narrowly lanceolate lobes bearing long straight hairs (and sometimes a few shorter gland-tipped ones); blades of well developed leaves at least twice as long as their petioles . . . . 1. **A. virginica**
1. Bracts mostly with 7–9 lanceolate to ovate lobes (fig. 252), glabrate or with a few gland-tipped hairs; blades of well developed leaves usually less than twice as long as their petioles . . . . 2. **A. rhomboidea**

1. **A. virginica** L.

Map 721. Disturbed ground; probably adventive in our area from states to the south.

This name was once misapplied to the next species, so old literature has to be used with caution. On the other hand, in view of inconstancy of dis-

718. Ricinus communis

719. Croton glandulosus

720. Croton monanthogynus

tinguishing characters, the next may be treated as *A. virginica* var. *rhomboidea* (Raf.) Cooperrider (Mich. Bot. 23: 165. 1984). *A. gracilens* A. Gray, with shallowly lobed bracts and narrower (but short-petioled) leaves, has sometimes been confused with *A. virginica.*

2. **A. rhomboidea** Raf. Fig. 252

Map 722. Considered to be native, and collected in Michigan as early as 1838, but generally a weed of gardens, hedgerows, waste places in cities, often far too common; also on river banks, shores, and floodplains; in fields, ditches, and disturbed ground.

4. **Euphorbia** Spurge

A very large and diverse genus. Following most recent authors (but not Webster), it is treated here in an inclusive sense. *Poinsettia* and *Chamaesyce* are the segregates most frequently recognized. Species 1–8 are in *Euphorbia* subg. *Chamaesyce* and while all of these are considered to be native in North America the status of some of them in Michigan is dubious. Reports of *E. humistrata* A. Gray by Beal (1905) are all based on collections of *E. vermiculata*; reports from Berrien and Emmet counties are based on specimens of *E. maculata.*

The staminate and pistillate flowers consist, respectively, of a single stamen and a single 3-carpellate pistil. These are arranged in a "false flower" or pseudanthium, in this genus called a *cyathium* and characterized by a cup-like structure or involucre in the center of which is a pistillate flower, on a stalk—the pedicel—that elongates as the fruit matures. Several staminate flowers surround the pistil, most easily being interpreted as not merely stamens if one notes the joint in the stalk of each, the portion below the joint corresponding to pedicel and that above, to filament. Around the margin of the cyathium are (1) 4–5 glands (fig. 256), and in some species (including most of subgenus *Chamaesyce*) these have petaloid appendages (figs. 254, 260) so that the whole cyathium mimics a single flower in a quite realistic way! The milky juice is ± poisonous, and for some people produces a dermatitis like that caused by poison-ivy.

721. Acalypha virginica

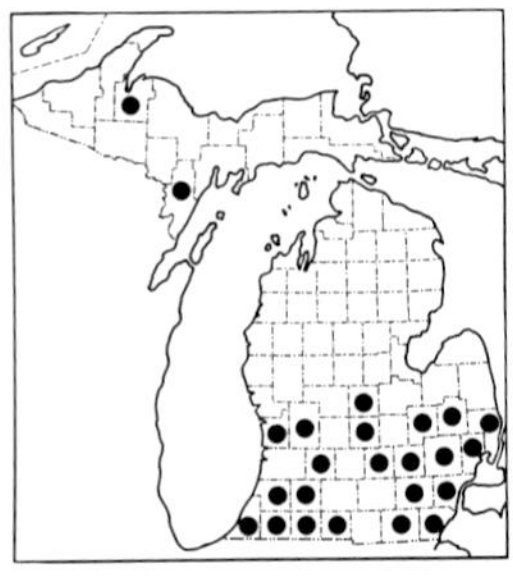

722. Acalypha rhomboidea

723. Euphorbia hirta

REFERENCES

Burch, Derek. 1966. The Application of the Linnaean Names of Some New World Species of Euphorbia subgenus Chamaesyce. Rhodora 68: 155–166.

Wheeler, Louis Cutter. 1941. Euphorbia subgenus Chamaesyce in Canada and the United States Exclusive of Southern Florida. Rhodora 43: 97–154; 168–205; 223–286 + pl. 654–668 (also Contr. Gray Herb. 136).

KEY TO THE SPECIES

1. Leaf blades distinctly asymmetrical at base; plants annual, with opposite leaves and prostrate or erect habit
 2. Stems (at least the young internodes) ± pubescent (ranging from dense spreading hairs to a fine line of puberulence); leaf blades usually at least sparsely pubescent beneath when young
  3. Capsules and ovaries pubescent; stems with ± conspicuous mostly spreading hairs
   4. Plant erect; cyathia in dense peduncled axillary and terminal heads; blades of leaves acute, mostly at least 15 mm long . . . . 1. **E. hirta**
   4. Plant prostrate; cyathia solitary in axils but often on short axillary shoots; blades mostly obtuse or rounded, less than 12 (14) mm long . . . . 2. **E. maculata**
  3. Capsules and ovaries glabrous; stems often with only short incurved hairs
   5. Plant branching from the base, widely spreading or prostrate; largest leaf blades less than 16 mm long . . . . 3. **E. vermiculata**
   5. Plant with erect stem and any main branches strongly ascending; largest leaf blades 16–27 (33) mm long . . . . 4. **E. nutans**
 2. Stem and leaves glabrous
  6. Leaves serrulate (sometimes weakly so) at least on apical third or half or along one side; mature seeds distinctly angled with shallowly pitted, rugulose, or ridged facets
   7. Seeds with irregularly pitted or rugulose surface (fig. 258); leaf blades usually ± oblong to obovate. . . . . 5. **E. serpyllifolia**
   7. Seeds with 3–4 strong transverse ridges and furrows on each facet (fig. 259); leaf blades usually linear-oblong to slightly broader below the middle . . . . 6. **E. glyptosperma**
  6. Leaves completely entire (at most mucronate); seeds plump, rounded, smooth
   8. Seeds ca. 1.3–1.5 mm long; capsule ca. 1.5–1.8 mm long; cyathium with small petaloid appendages; leaf blades less than 2.5 (3) times as long as wide . . . . 7. **E. geyeri**
   8. Seeds ca. 2.4–2.7 mm long; capsule ca. 3–3.5 (4) mm long; cyathium without petaloid appendages; leaf blades mostly 3.2–4.4 times as long as wide . . . . 8. **E. polygonifolia**
1. Leaf blades essentially symmetrical at base; plants annual or perennial, with mostly alternate leaves (except sometimes the uppermost, subtending cyathia or rays of the inflorescence, and in *E. dentata*) and erect habit
 9. Cyathia with conspicuous white petaloid appendages on the glands (fig. 260)
  10. Upper leaves and bracts conspicuously white-bordered, with distinct mucro; capsules and involucres of cyathia pubescent; plant annual . . . . 9. **E. marginata**
  10. Upper (and all) leaves without white borders, blunt; capsules and involucres glabrous or nearly so; plant perennial . . . . 10. **E. corollata**
 9. Cyathia without petaloid appendages

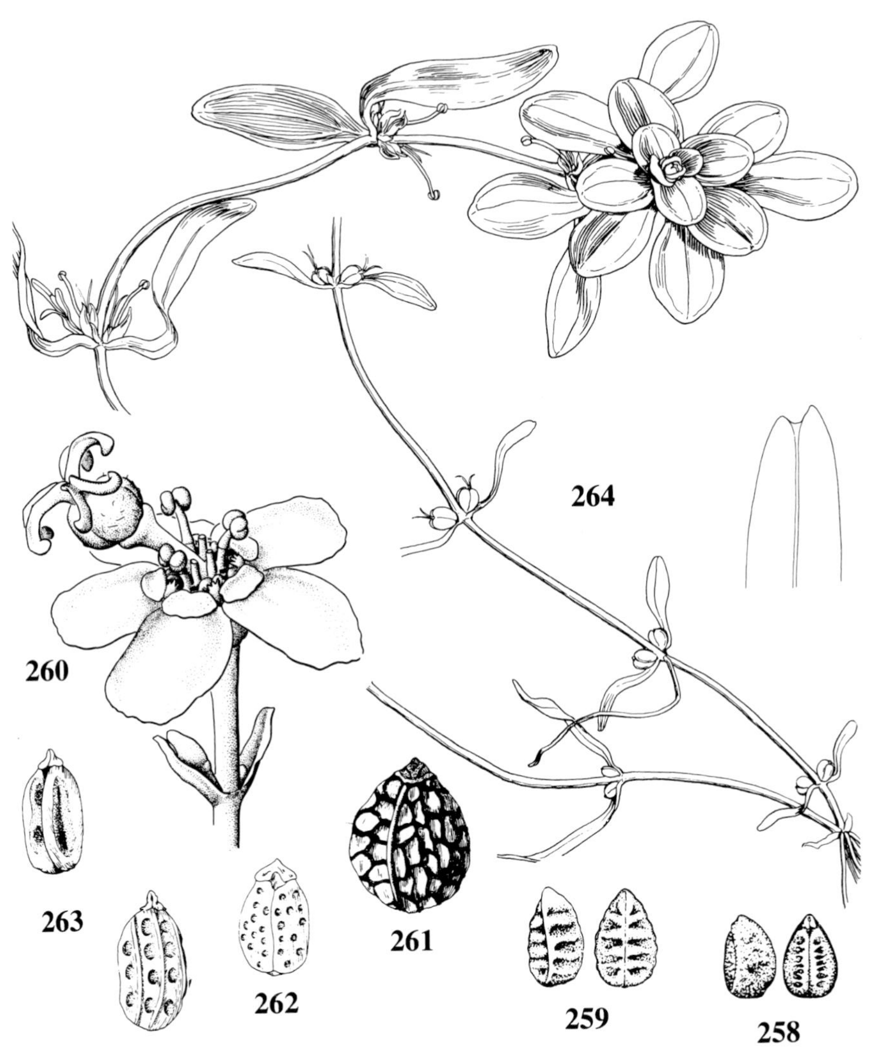

258. *Euphorbia serpyllifolia,* seeds ×10
259. *E. glyptosperma,* seeds ×10
260. *E. corollata,* cyathium ×8
261. *E. helioscopia,* seed ×10
262. *E. commutata,* seed ×10
263. *E. peplus,* seeds (ventral & dorsal face) ×10
264. *Callitriche verna,* submersed portion ×1 1/2; upper portion with floating rosette ×1 4/5; tip of submersed leaf ×12

11. Inflorescence consisting of clusters of cyathia subtended by opposite leaves; cyathium bearing 1 large gland; cauline leaves distinctly toothed, all or mostly opposite, ± antrorsely scabrous-pubescent; stem ± retrorsely scabrous-pubescent .... 11. **E. dentata**

11. Inflorescence consisting of umbellate rays subtended by a whorl of leafy bracts; cyathium bearing 4 glands; cauline leaves finely toothed or entire, all alternate, glabrous or at most loosely villous; stem glabrous or loosely villous

12. Leaves minutely toothed, at least on part of their margins; glands of cyathium rounded to obscurely reniform

13. Capsules smooth (except as dried over reticulate seed); seeds with strong raised reticulate pattern (fig. 261); principal inflorescences consisting of 5 umbellate rays and cyathia glabrous on the outside .... 12. **E. helioscopia**

13. Capsules tuberculate with prominent papillae; seeds smooth; principal inflorescences consisting of 3 rays *or* cyathia pubescent on the involucre

14. Leaves beneath (at least along midrib) and outside of cyathia ± loosely villous; rays of principal inflorescences usually 5; leaves ending in abrupt acute tip .... 13. **E. platyphylla**

14. Leaves and outside of cyathia glabrous; rays of principal inflorescences 3; leaves obtuse .... 14. **E. obtusata**

12. Leaves strictly entire; glands of cyathium crescent-shaped, with pointed tips or even prolonged horns (fig. 256)

15. Leaf blades ± rotund to obovate, less than 3 times as long as wide; seeds conspicuously pitted; plants annual, with usually 3 primary rays in the inflorescence

16. Seeds with numerous uniform small pits narrower than the ridges between them (fig. 262); bracteal leaves of inflorescence slightly broader than long .... 15. **E. commutata**

16. Seeds with few large pits broader than the ridges between them (1 on each ventral facet, 2–4 on each dorsal facet; fig. 263); bracteal leaves slightly longer than broad .... 16. **E. peplus**

15. Leaf blades linear to lanceolate, (6) 8–15 (18) times as long as wide; seeds smooth; plants perennial, with more than 5 rays in inflorescence

17. Main cauline leaf blades all less than 2.5 (3) mm broad; floral bracts (not those at base of rays) 3–6 (7) mm wide .... 17. **E. cyparissias**

17. Main cauline leaves 3.2–10 (15) mm broad; floral bracts (at least the largest) 8–16 (27) mm wide .... 18. **E. esula**

1. **E. hirta** L.

Map 723. Probably adventive from the southern states. Found by Farwell (*8756, 9352,* MICH, BLH, WUD, GH) on waste ground in Detroit 1930–1933. He identified his plants as the Old World *E. pilulifera* L. and assumed them to have come from bales of that species imported from India for the pharmaceutical industry (Rhodora 38: 331–332. 1936).

2. **E. maculata** L. Fig. 254

Map 724. Roadsides, railroads, shores, fields, dry barren places and waste ground; all too familiar to many people as a lawn and garden weed. Collected as early as 1837 by the First Survey, and at a number of places throughout the state in the 19th century, so either very early introduced or, as usually assumed, native here.

Wheeler and those who have followed him called this species *E. supina* Raf., but the controversial name *E. maculata* is now generally agreed to apply (see Burch 1966)—as it was interpreted for many years before 1941.

3. **E. vermiculata** Raf.

Map 725. Railroads, roadsides, shores, and waste ground. Considered a native species, but there are no Michigan collections before the 1880's and 1890's, at which time it was apparently locally common.

The ripe seeds are slightly concave on their ventral faces in this species, while in *E. nutans* they are slightly convex.

4. **E. nutans** Lag.

Map 726. Dry open ground, shores, prairie remnants; roadsides, railroads.

Wheeler applied the name *E. maculata* to this species, and for about 25 years it was thus widely (but not universally) known. It has also been called *E. preslii* Guss., and the name *E. hypericifolia* L. has sometimes been applied to it.

5. **E. serpyllifolia** Pers. Fig. 258

Map 727. Only two collections have been seen from Michigan since Farwell's from waste ground in Keweenaw County in the 1880's: under quaking aspen in Cheboygan County (1936) and in open woods along a highway in Otsego County (1952). Native in western North America, south into Mexico, but probably adventive here.

6. **E. glyptosperma** Engelm. Fig. 259

Map 728. Railroads, roadsides (sandy shoulders), gravel pits, and other dry raw sites. Probably our commonest prostrate mat-forming species of disturbed ground, usually considered native but the earliest Michigan collections seen are from 1892 (Dickinson Co.), 1893 (Ottawa Co.), and 1900 (Alger and Ingham cos.).

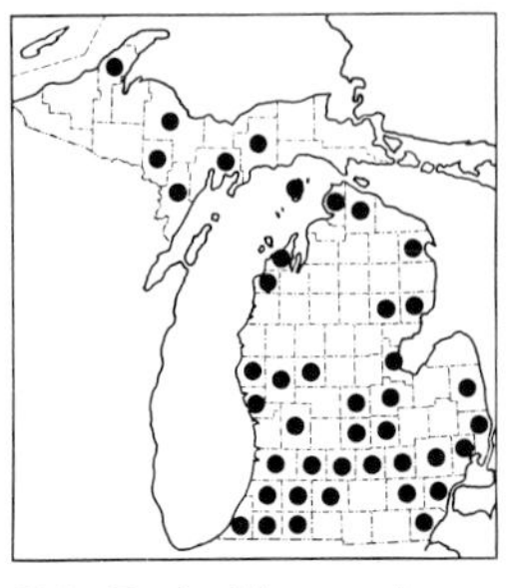
724. Euphorbia maculata

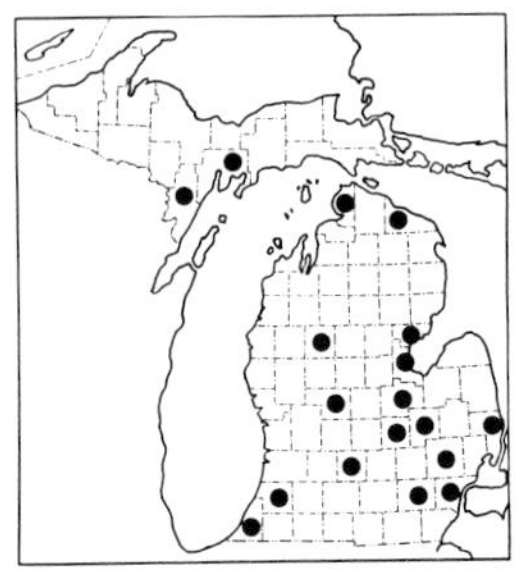
725. Euphorbia vermiculata

726. Euphorbia nutans

7. **E. geyeri** Engelm.

Map 729. Undoubtedly adventive from farther west. First found in the state along a railroad on the west side of Manistique in 1979 (*Henson 1112*, MICH, WIS).

8. **E. polygonifolia** L. Plate 5-G Seaside Spurge

Map 730. Restricted in North America to the sandy beaches and dunes of the Great Lakes and the Atlantic Ocean; one of the few native annuals (along with *Cakile*) in such habitats (see Mich. Bot. 2: 106, 108. 1963). Evidently absent from the Lake Superior shore. Also naturalized on seashores in France and Spain.

This species tends to develop and fruit a little later than other common prostrate ones, but even if its distinctive relatively large fruits and smooth large seeds are not developed, the slender entire leaf blades (a little less oblique than in other species) and lack of petaloid appendages on the cyathium should help in recognizing it—apart from the restricted habitat.

9. **E. marginata** Pursh Snow-on-the-mountain

Map 731. Native west of our area, but much cultivated and occasionally escaped to roadsides and waste ground.

727. Euphorbia serpyllifolia

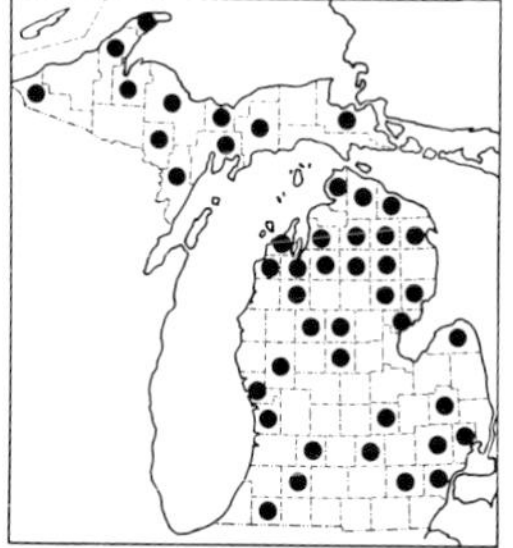
728. Euphorbia glyptosperma

729. Euphorbia geyeri

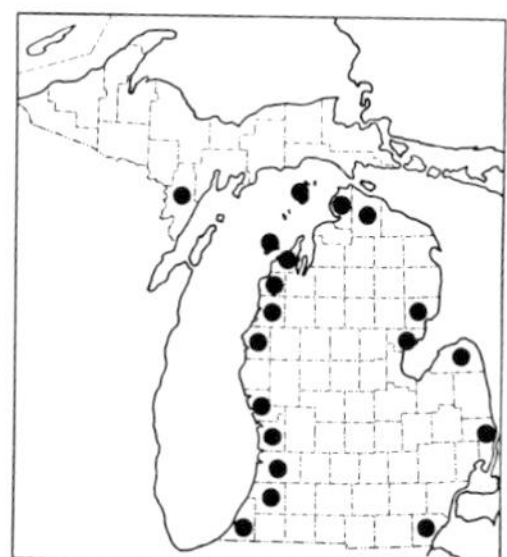
730. Euphorbia polygonifolia

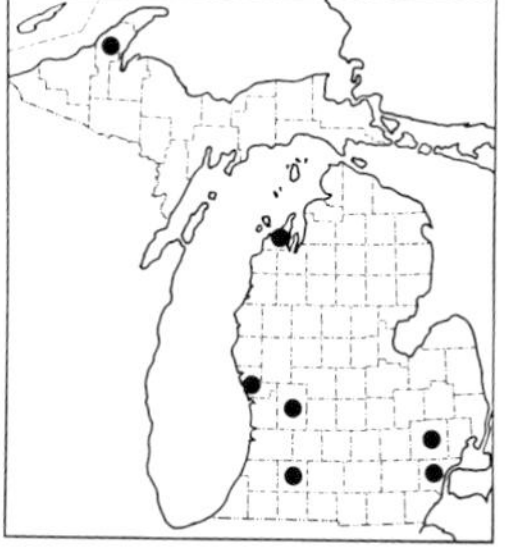
731. Euphorbia marginata

732. Euphorbia corollata

10. **E. corollata** L.   Fig. 260   Flowering Spurge

Map 732. Sandy, dry plains and woodland (oak, sassafras, jack pine); fields and prairies; railroads and other disturbed ground. The northernmost occurrences in the state presumably represent spread (along roadsides and railroads) from the native range, which reaches about the middle of the Lower Peninsula. Some stands may represent escapes from cultivation.

A form with green rather than white appendages was named var. *viridiflora* Farwell (TL: Oxford [Oakland Co.]).

11. **E. dentata** Michaux   Fig. 255

Map 733. Native to the southwest of Michigan, adventive here along railroads and roadsides; rarely on sandy shores.

This is our only species in *Euphorbia* subg. *Poinsettia,* sometimes recognized as a distinct genus.

12. **E. helioscopia** L.   Fig. 261

Map 734. A Eurasian species, locally established in our area in fields and waste ground and on shores. Collected most often at Mackinac Island, beginning with Douglass Houghton in 1839 and Horace Mann, Jr. (with Henry D. Thoreau) in 1861; a number of subsequent collectors gathered it on the island, including Thomas Morong, C. F. Wheeler, O. A. Farwell, Emma Cole, C. K. Dodge, and later botanists. It must have been very noticeable near the docks!

13. **E. platyphylla** L.

Map 735. A native of Europe, sparingly introduced in North America, especially near the lower Great Lakes. Collected in Michigan as early as 1850 by Dennis Cooley along the canal near Utica in Macomb County, and in 1869 by Henry Gillman near Fort Wayne, Detroit. Roadsides, shores, and vacant lots, but apparently not collected in the state since 1930.

733. Euphorbia dentata

734. Euphorbia helioscopia

735. Euphorbia platyphylla

14. **E. obtusata** Pursh

Map 736. A southern species, barely ranging north to Michigan, where it is probably adventive (Pap. Mich. Acad. 2: 28. 1924). Farwell's old Monroe County site was a "disused garden" and a more recent one is from the edge of a marsh.

15. **E. commutata** Engelm. Figs. 256, 262

Map 737. Sandy soil of plains, stream borders, and woods; not collected in Michigan since 1939.

16. **E. peplus** L. Fig. 263

Map 738. A weedy European species, locally established in North America and only rarely collected in Michigan, where it is a minor and inconspicuous weed of yards and gardens.

17. **E. cyparissias** L. Fig. 257 Cypress Spurge

Map 739. A European species, sometimes grown as a low ground cover (usually 15–30 cm tall), but too easily spreading like the next and becoming a weed of roadsides, banks, clearings, fields, railroads, old homesites, and lawns gone wild; often spreading from cemeteries.

The abundant, crowded, narrowly linear leaves with milky juice make this an easily recognized plant in the field, even when sterile. The leaves are especially dense on lateral shoots, which eventually overtop the inflorescence (which is not overtopped in *E. esula*).

18. **E. esula** L. Leafy Spurge

Map 740. Roadsides, railroads, fields, gravel pits, and such places. Another European species, more widely naturalized than the preceding, and an abundant weed in some areas of the northern Lower Peninsula, centering in Otsego County, where it may carpet open areas with a bright yellow-green color in early summer. Almost all Michigan collections have been made in

736. Euphorbia obtusata

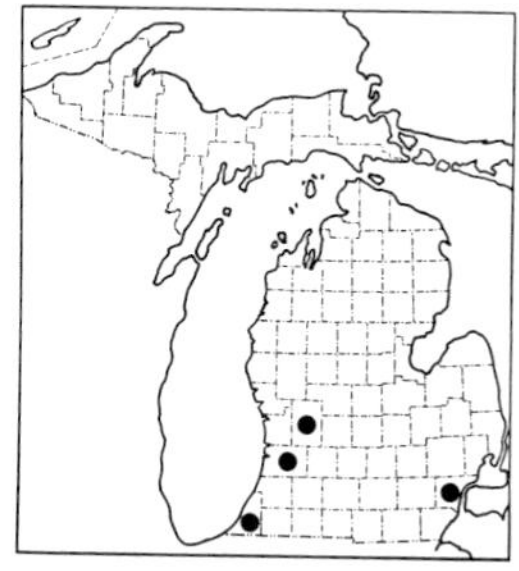

737. Euphorbia commutata

738. Euphorbia peplus

the past 50 years (and only in Ingham and St. Clair counties was this species collected before 1900).

The sturdy horizontal roots, by which it spreads, make this a difficult species to eradicate. It is almost as leafy as *E. cyparissias,* but the leaves and bracts are larger and the plant taller. Both species have reduced scale-like leaves at the base of the stems.

CALLITRICHACEAE — Water-starwort Family

1. **Callitriche** — Water-starwort

This is the only genus in this small family, whose affinities are not agreed upon. The flowers are unisexual, without a perianth, consisting of a single stamen or a single 2-carpellate pistil. The species have long been a source of taxonomic as well as nomenclatural uncertainty. Like many aquatics, the vegetative form may be quite variable, leaving little more than the fruits from which to draw reliable distinctions. *C. hermaphroditica* is reasonably distinctive in the genus. Insofar as fruiting or other characters are adequate on specimens examined, almost all other Michigan *Callitriche* is *C. verna.* Submersed leaves have a distinctive bidentate apex (fig. 264), which readily identifies specimens that might otherwise be confused with depauperate *Elodea* (if only opposite-leaved) and any similar aquatics.

REFERENCE

Fassett, Norman C. 1951. Callitriche in the New World. Rhodora 53: 138–155; 161–182; 186–194; 209–222 + pl. 1167–1175.

KEY TO THE SPECIES

1. Leaves all submersed, linear; fruit becoming fully ripe very late in the season (apparently September–November, if at all), and then conspicuously winged the full length of each margin .... 1. **C. hermaphroditica**
1. Leaves rarely (when young) all submersed and linear, some or all often terrestrial or crowded into a terminal floating rosette, and/or oblanceolate to spatulate; fruit ripening early to late summer and wingless or with a narrow wing noticeable toward the apex
   2. Fruit about as long as wide, wingless, with pits of punctate surface not in rows .... 2. **C. heterophylla**
   2. Fruit mostly longer (by 0.2 mm) than wide, winged across the apex, the wings abruptly narrowed or absent basally, with pits of punctate surface tending to run in longitudinal rows .... 3. **C. verna**

1. **C. hermaphroditica** L.

Map 741. A circumpolar species. Easily overlooked in shallow to deep water of lakes and streams.

The foliage tends to be deep green, and the bases of the leaves, although they may touch each other, are not confluent at the nodes. In the other two species, the foliage is a light green (especially when submersed) and the leaf bases are slightly confluent at each node, forming a narrow (not always easily detected) membranous wing across the sides of the stem.

2. **C. heterophylla** Pursh

Map 742. A North American species, evidently very local in Michigan, which is at the northern edge of its range. The two collections seen are from the Belle River near Capac and the Maple River east of Brutus (Emmet Co.)—where *C. verna* is also known. Fassett (1951) cited no Michigan material.

The fruit is 0.6–1 (1.1) mm long, while in the next species it is usually a bit longer.

3. **C. verna** L. Fig. 264

Map 743. In muddy or sandy substrates in shallow water of lakes, ponds, creeks, river margins, swamps, ditches, and (often) puddles in old trail roads; often in cold spring-fed waters, ± terrestrial on recently exposed shores and banks.

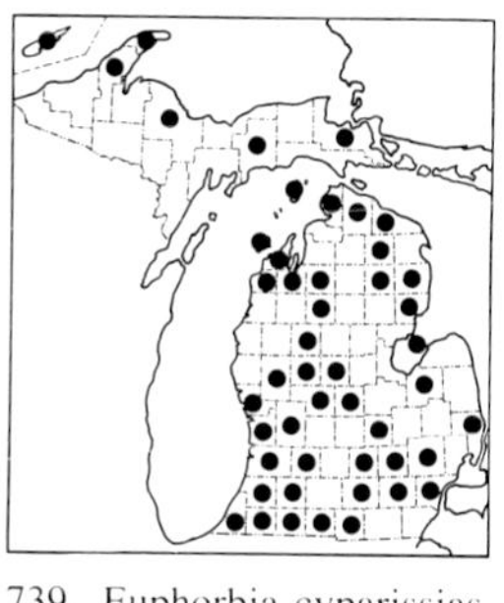

739. Euphorbia cyparissias

740. Euphorbia esula

741. Callitriche hermaphroditica

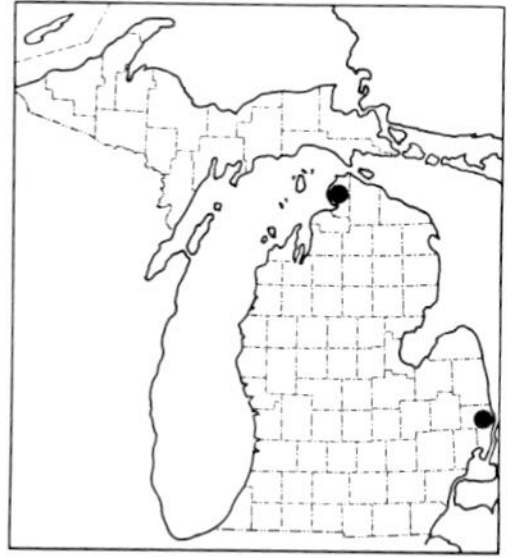

742. Callitriche heterophylla

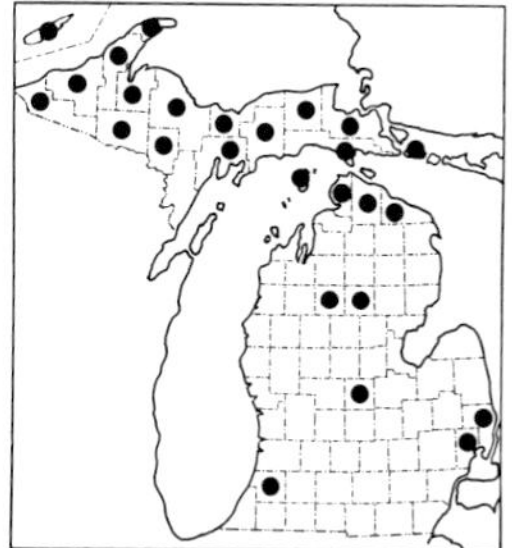

743. Callitriche verna

744. Empetrum nigrum

Since fruit starts ripening fairly early in the season, most collections have some far enough advanced to demonstrate the characteristic narrow wings at the apex, as well as the more elongate shape compared to *C. heterophylla*. There are arguments for using the name *C. palustris* L. for this circumpolar species, but that name has not been typified although it probably refers to the same species and is used by some authors (e. g., in *Flora Europaea*); recent checklists covering North America and the Soviet Union uniformly adopt *C. verna,* as did Fassett.

EMPETRACEAE — Crowberry Family

1. Empetrum

REFERENCE

Soper, James H., & Edward G. Voss. 1964. Black Crowberry in the Lake Superior Region. Mich. Bot. 3: 35–38.

1. **E. nigrum** L. Plate 6-A — Black Crowberry

Map 744. Bare rock outcrops, cedar or black spruce bogs, and exposed sandy bluffs and old dune ridges (under pines and with Ericaceae, *Linnaea, Maianthemum, Trientalis, Melampyrum, Deschampsia flexuosa,* etc.). Like a number of subarctic/boreal species which range south to Lake Superior, this interesting and seldom-seen species is found in the Isle Royale archipelago and the tip of the Keweenaw Peninsula. However, it has also been long known from the Pictured Rocks and from sandy banks in eastern Luce County. In the former area, it is found not only on sandstone cliffs (where Douglass Houghton collected it as early as 1831) but also on stabilized storm beaches near Big Beaver Lake (*Read 70* in 1973, MICH). In Luce County, its occurrence on sandy banks under conifers facing Lake Superior, where found by Davis in 1905, was confirmed in 1974 at two sites between Deer Park and the mouth of the Two-Hearted River (*Voss 14508,* MICH, MSC, UMBS, WIS, MCTF; *14510,* MICH, MSC, BLH, UMBS, LKHD, GH, NY). More recently, in 1977, crowberry was found south of the Lake Superior basin in an open marly cedar swamp with some spruce and tamarack near Moran, where it was discovered by R. and F. Case, P. Harley, and E. Voss (*14925,* MICH, MSC, UMBS, GH; see also Weitzman in Mich. Bot. 23: 11–18. 1984). Quite possibly there are additional outlying colonies of this plant relic in other microclimatically favorable sites in northern Michigan. It is frequent along the north shore of Lake Superior.

This is a very low shrub with short needle-like evergreen leaves, extremely inconspicuous flowers, and black (but not tasty) berries. The species is here treated broadly as a circumpolar "complicated complex, where different authors rarely, if ever, arrive at the same conclusion" (Hultén 1971). Most taxonomists who have considerable experience with the northern flora (e. g., Calder & Taylor, Hultén, Porsild & Cody, Scoggan, Soper) treat this species at least tentatively in an inclusive sense, at most assigning our Great Lakes plants to *E. nigrum* ssp. *hermaphroditum* (Hagerup) Böcher. This is sometimes recognized as a distinct species on the basis of perfect flowers, typical *E. nigrum* being dioecious. However, polygamous and monoecious plants also occur in North America. Characters of pubescence and glandularity are likewise extremely variable.

LIMNANTHACEAE — False Mermaid Family

1. Floerkea

There is only one species in this genus, and it is the only one in the family which occurs in eastern North America (as well as in the west, to which the rest of the family is native). Our species is distinct among our dicots in its completely 3-merous flowers: 3 sepals, becoming much larger than the 3 persistent tiny white petals; (3) 6 stamens; and (2–) 3 carpels. The very deeply pinnately lobed leaves will separate it readily from any monocot with which it might be confused.

1. **F. proserpinacoides** Willd. Fig. 265 — False Mermaid

Map 745. Rich deciduous woods, especially in moist, even springy, areas.

A weak little plant, blooming usually in late April and the first half of May. Even though it may be abundant when found, it is easily overlooked, thanks to its delicate habit, inconspicuous small flowers on short solitary axillary peduncles, and ephemeral nature, the foliage withering away by summer. The plant is at least as conspicuous when the foliage yellows and the dark tuberculate achenes (or mericarps) ripen.

ANACARDIACEAE — Cashew Family

A predominantly tropical family, including several edible products (mango, cashew, pistachio) as well as poisonous plants. *Cotinus coggygria* Scop., smoke-tree, is a widely cultivated ornamental shrub from Eurasia, with simple, entire leaves; it was collected once in a field in Leelanau County, possibly as an escape but perhaps only persistent from a planting.

REFERENCES

Barkley, Fred Alexander. 1937. A Monographic Study of Rhus and Its Immediate Allies in North and Central America, Including the West Indies. Ann. Missouri Bot. Gard. 24: 265–498 + pl. 10–26.

Brizicky, George K. 1962. The Genera of Anacardiaceae in the Southeastern United States. Jour. Arnold Arb. 43: 359–375.

Brizicky, George K. 1963. Taxonomic and Nomenclatural Notes on the Genus Rhus (Anacardiaceae). Jour. Arnold Arb. 44: 60–80.

KEY TO THE GENERA

1. Inflorescences axillary, at maturity spreading to pendulous with glabrous whitish fruit; leaves with leaflets entire or some irregularly notched or dentate, the rachis not winged . . . . 1. **Toxicodendron**
1. Inflorescences terminal, at maturity erect or ascending with red glandular-pubescent fruit; leaves with leaflets either entire and the rachis winged or ± regularly toothed along at least half the margin . . . . 2. **Rhus**

1. Toxicodendron

Often included as a subgenus of *Rhus,* but there is a growing tendency to maintain *Toxicodendron* as a separate genus—a practice dating from pre-Linnaean times.

The resin canals contain a poisonous oil to which many people are allergic on contact, some strongly so and others only mildly. Hundreds of thousands of cases of dermatitis occur in this country each year. The itching (and blistering) reaction is delayed, usually 1–2 days after exposure but sometimes in especially sensitive persons only a few hours, and in others sometimes a week or more. The oil is not volatile, and despite claims sometimes heard, one cannot get an allergic reaction from merely being near a plant. The plant must be bruised to break the resin canals, but the oil can then be carried on shoes or clothing, on an animal such as a cat or dog, as droplets in smoke from burning plants, or on the surface of water (as may happen with poison sumac in a bog). Only an extremely minute amount of oil is required to produce a reaction in a sensitive person.

The chief reason for the usual advice, upon exposure, to wash as quickly as possible with soap or preferably alcohol (in which the oil dissolves) is to remove excess oil beyond what has, almost immediately, reacted with proteins in the skin. It is this excess oil which can be transferred by clothing or scratching fingers to other sites on the body. The fluid in the blisters of an infection does not spread the rash. Poison-ivy and poison sumac are a cause of grief even to persons who know themselves to be sensitive and who try to avoid contact, because the long-active oil may persist on clothing and other objects and, in some cases, because of inability to recognize the offending plants. The delay in appearance of symptoms does not reinforce one's identification very effectively. Conversely, some people avoid harmless plants because of confusion in identification and unreliable folklore. A person's sensitivity may change, perhaps suddenly, so supposed immunity cannot be firmly relied upon. Medical treatment has changed considerably over the years, with improved knowledge of the nature of the dermatitis and with the development of steroids and other new drugs. Ordinarily, relief from itching and swelling is all that is necessary, as the allergy will in time run its course regardless of treatment (or lack of it).

In a flora of Michigan, especially, it is fitting to acknowledge with gratitude the aid and inspiration of William T. Gillis, for whom these plants held a special fascination during his doctoral studies on them at Michigan State University and beyond. He was always ready to share his thorough knowledge of their taxonomy and lore, and would surely have reviewed the present text of this genus had his life not prematurely ended in 1979. Those of us who were privileged to have been in the field with him found his enthusiasm for it to be more contagious than his subject itself.

REFERENCES

Crooks, Donald M., & Leonard W. Kephart. 1945. Poison-ivy, Poison-oak, and Poison Sumac: Identification, Precautions, Eradication. U. S. Dep. Agr. Farmers' Bull. 1972. 30 pp. [Frequently revised or reprinted; much other literature is available from agencies in many states.]

Gillis, William T. 1962. Poison-ivy in Northern Michigan. Mich. Bot. 1: 17–22.

Gillis, William T. 1971. The Systematics and Ecology of Poison-ivy and the Poison-oaks (Toxicodendron, Anacardiaceae). Rhodora 73: 72–159; 161–237; 370–443; 465–540.

Kligman, Albert M. 1958. Poison Ivy (Rhus) Dermatitis. A. M. A. Arch. Dermat. 77: 149–180.

Ulbrich, A. P. 1968. Contact Dermatitis Caused by Plants. Mich. Bot. 7: 265–268.

KEY TO THE SPECIES

1. Plant a large shrub with pinnately compound leaves of (5) 7–11 (13) leaflets, in wet habitats . . . . 1. **T. vernix**
1. Plant a climbing vine or low shrub with trifoliolate leaves, in dry, mesic, or flood-plain habitats . . . . 2. **T. radicans**

1. **T. vernix** (L.) Kuntze Fig. 267 Poison Sumac

Map 746. Often with tamarack in swamps and on floating mats of bogs (fens); swampy woods, lakeshore thickets, swales, borders of marshes.

Closely related to the Asian *T. vernicifluum* (Stokes) Lincz., whose similarly poisonous sap is the principal source of lacquer in the Orient. The flaming red foliage in the fall is tempting to some persons who risk more than wet feet to gather it. The alternate pinnately compound leaves with entire shiny leaflets, together with the habitat, should readily serve to identify even sterile plants.

2. **T. radicans** (L.) Kuntze Fig. 266 Poison-ivy

Map 747. Woods, thickets, and open ground, often in disturbed (or formerly disturbed) sites throughout the state. (Long thought to be absent from Isle Royale, but a very large colony was discovered in 1981 on a rocky slope near Lane Cove.) The northern shrubby variety grows along roadsides and railroads; on shores, dunes, and banks; in clearings and at forest borders; usually in sandy, gravelly, or rocky soil; even when (as rarely) in a swamp forest it does not climb. The southern vine variety is typical of swamp forests and floodplains, but also grows in upland woods such as beech-maple and oak-hickory.

This is the most widely distributed species in all the Anacardiaceae, and is the commonest source of allergic disease in the United States. For a plant so well known, at least by reputation, there is much misunderstanding about it, including aspects of its allergenic effects (see above). The popular ditty "leaflets three, let it be" is practically worthless by itself, as it would exclude from one's attention such plants as all clovers and alfalfa, strawberries and raspberries, goldthread, buckbean—and even jack-in-the-pulpit, not to mention some harmless *Rhus*. Another problem is that the only form of poison-ivy in half the state is *not a vine,* the form familiar in southern Michigan and southward. The best practice, as with mushrooms, is to *learn the poisonous species*.

Remember that poison-ivy is a low shrub or a vine, with alternate leaves, whitish fruit, and leaflets entire to sparsely and coarsely dentate and larger than those of most other trifoliolate plants. The closest "look alike," when sterile, is trifoliolate seedlings of box-elder, *Acer negundo*—which has opposite leaves less shiny than those of poison-ivy and without the tendency to cupped or trough-like leaflets that are characteristic of our northern non-viny poison-ivy.

The form of poison-ivy which ranges north in Michigan to Ogemaw and Manistee counties (just north of the latitude of Saginaw Bay) is an often vigorous climbing vine especially of lowland habitats. It may have spectacular horizontal branches a meter or two long; the leaflets are essentially flat when fresh, ovate to somewhat lance-ovate, and the petioles are ± pubescent. This is *T. radicans* in the narrow sense, usually included simply in var. *radicans* but treated by Gillis, who eventually split the species more

745. Floerkea proserpinacoides

746. Toxicodendron vernix

747. Toxicodendron radicans

265. *Floerkea proserpinacoides* $\times 1/2$; flower $\times 2\,1/2$; fruit $\times 2$
266. *Toxicodendron radicans* $\times 2/5$; flowers $\times 3$

finely, as ssp. *negundo* (Greene) Gillis. South to the northern edge of var. *radicans,* and southward along the lake shores to Bay County on Lake Huron and to the south end of Lake Michigan, is var. *rydbergii* (Rydb.) Erskine, treated by Gillis as *T. rydbergii* (Rydb.) Greene. This plant produces no aerial roots, so does not climb as a vine, and the leaflets are typically somewhat trough-like (so as often to be folded when pressed); the petioles are usually glabrous. Suspected hybrids between these varieties (or species) were identified by Gillis from Gladwin, Iosco, Lapeer, Oceana, Oscoda, and St. Clair counties.

Although admittedly there are no firm rules governing the application of common names, there is no plant in Michigan to which the name "poison-oak" should be applied. That name is sometimes used for the non-climbing form of poison-ivy (and I have even heard it applied to *Aralia nudicaulis*!), but properly belongs to species of the Pacific coast and the states well south of the Great Lakes.

2. **Rhus** — Sumac; Sumach

REFERENCES

Barkley, Fred A. 1938. Studies in the Anacardiaceae. III. A Note Concerning the Status of Rhus pulvinata Greene (R. glabra × typhina Koehne). Am. Midl. Nat. 19: 598–600.

Gilbert, Elizabeth F. 1966. Structure and Development of Sumac Clones. Am. Midl. Nat. 75: 432–445. [For later study of the same site, see Larch & Sakai in Mich. Bot. 22: 3–9. 1983.]

KEY TO THE SPECIES

1. Leaflets 3, with 3–6 large teeth per side; flowers in small (ca. 1 cm) dense inflorescences on short lateral shoots (± coalescing in fruit), opening before the leaves; foliage somewhat aromatic when bruised . . . . 1. **R. aromatica**
1. Leaflets 7–23, entire or with 6–15 teeth per side; flowers in large inflorescences, opening after the leaves; foliage not aromatic
   2. Leaves with entire leaflets, the rachis ± winged, especially between distal leaflets . . . . 2. **R. copallina**
   2. Leaves with toothed leaflets, the rachis not winged
      3. Petioles and new branches with ± dense pubescence including many long (mostly 1–2 mm) straight hairs; hairs on fruit mostly 1–2 mm long, tapering to a sharp point (very short obovoid hairs may also be present); leaflets pubescent beneath, at least on veins . . . . 3. **R. typhina**
      3. Petioles and new branches glabrous or nearly so; hairs on fruit less than 0.5 mm long, mostly clavate to obovoid; leaflets glabrous beneath (or with a few hairs) [see text for intermediate plants] . . . . 4. **R. glabra**

1. **R. aromatica** Aiton — Fragrant Sumac

Map 748. Sandy, gravelly soil, open or with oak, hickory, and/or pine; river banks; thin soil over limestone (Drummond Island).

A colorful shrub often less than 1 m tall, with yellow flowers before the leaves, and red fruit and foliage later in the season.

2. **R. copallina** L. Shining or Dwarf Sumac

Map 749. Sandy hillsides, fields and clearings, old lake shores, and open (often oak) woodlands.

3. **R. typhina** L. Fig. 268 Staghorn Sumac

Map 750. Fields, clearings, roadsides, hillsides, rocky openings and ledges, thickets, borders of woods and in thin woods.

Frequently hybridizes with the next. Both species (as well as the hybrid) form large, uniform colonies from root suckers. This clonal habit has been extensively studied in Michigan (Gilbert 1966). Both may have ± laciniate leaflets, often associated with a pathological inflorescence which includes many small curled and distorted leaves. The bright red inflorescences of these plants can provide a pleasant cold drink. One easy way to prepare "rhusade" is to place a clean fully mature cluster of fruit (no stem or peduncle) in a large jar of cold water and shake it vigorously for several minutes, to break the hairs and release their acid juice. Strain the water through a clean cloth.

4. **R. glabra** L. Fig. 269 Smooth Sumac

Map 751. Fields and clearings, roadsides, dry banks and hillsides, rocky slopes and summits, in open ground or thin woods.

Notorious for hybridizing with *R. typhina,* presumed hybrids being more common in the state than pure *R. glabra*. These have sometimes been called *R.* ×*borealis* Greene (a name *not* based on *R. glabra* var. *borealis* Britton), originally described from Michigan (TL: near Alma [Gratiot Co.], C. A. Davis in 1895, F, MICH, WIS). Type material of Greene's name, however, scarcely differs from *R. glabra*—chiefly in having sparse pubescence on the undersides of the leaflets. It seems better to follow recent authors who accept the later name *R.* ×*pulvinata* Greene for the hybrid. Several variants are here included in *R.* ×*pulvinata:*

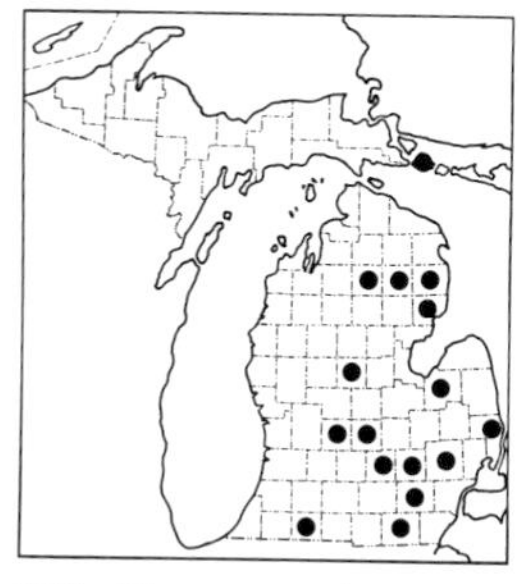

748. Rhus aromatica

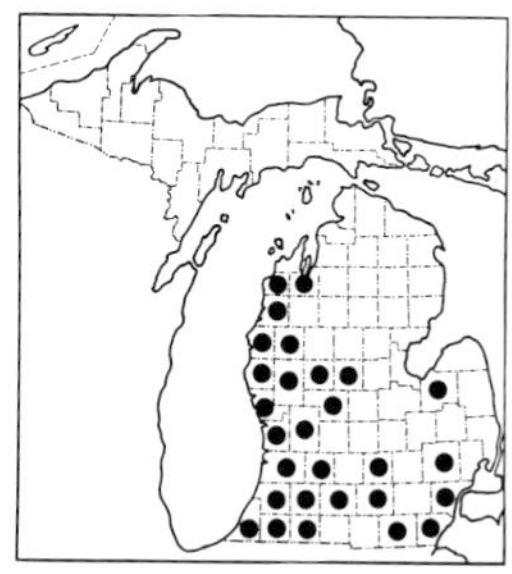

749. Rhus copallina

750. Rhus typhina

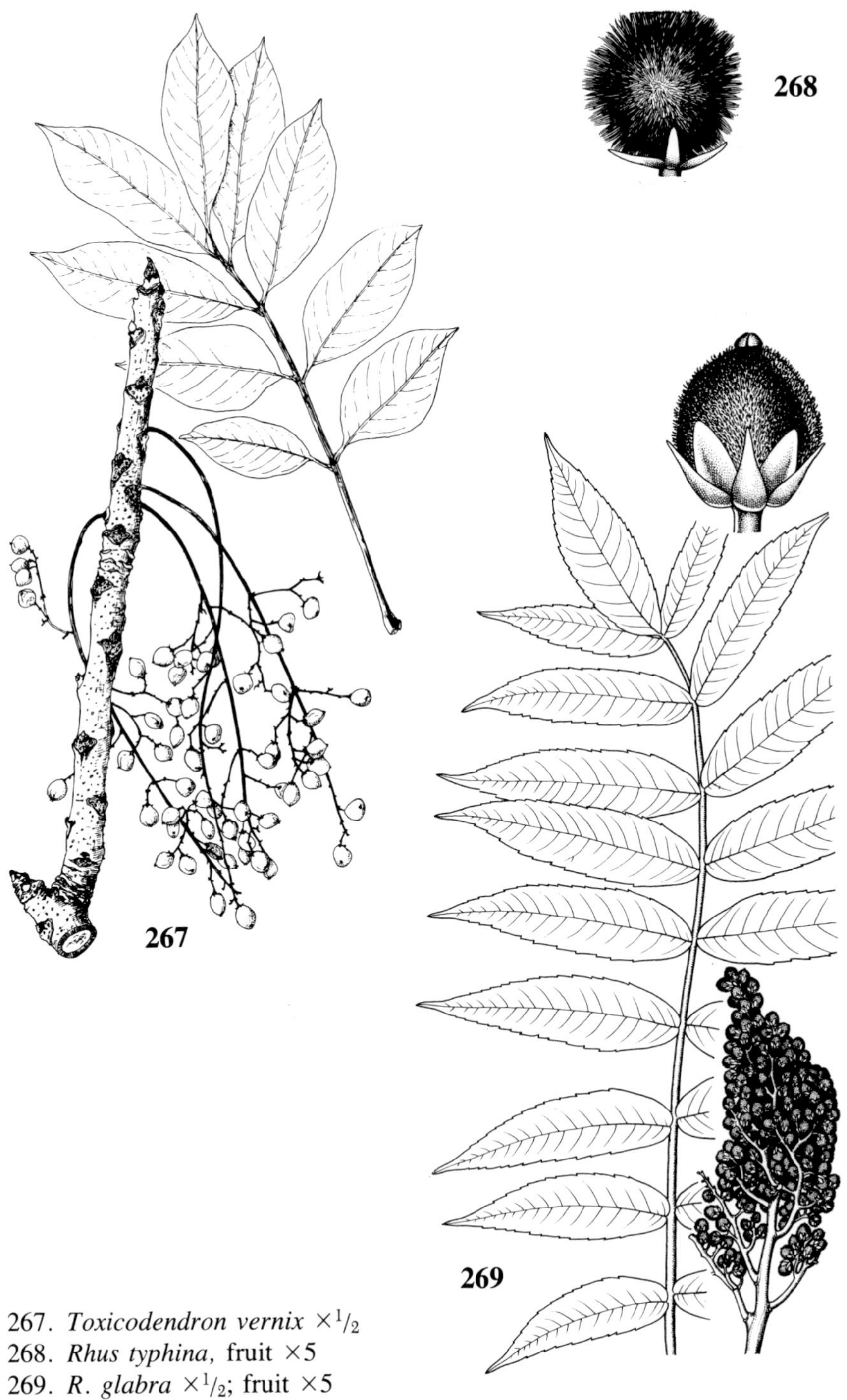

267. *Toxicodendron vernix* $\times^1/_2$
268. *Rhus typhina*, fruit $\times 5$
269. *R. glabra* $\times^1/_2$; fruit $\times 5$

(a) New branches and petioles with dense hairs but few if any as long as in *R. typhina*; fruit (if present) with hairs shorter than in *R. typhina*, ± blunt or some of them clavate. [These may be *R. glabra* var. *borealis* Britton.]
(b) New branches and petioles glabrous or nearly so but fruit with the characteristic long sharp-pointed hairs of *R. typhina*. [Sterile or staminate specimens of such a variant would probably be referred to *R. glabra* by most observers.]
(c) Hairs of fruit partly clavate and partly sharp-pointed, shorter than in *R. typhina* but often a bit long for *R. glabra*; pubescence of vegetative parts ranging from dense and long as in *R. typhina* to glabrous. These are among the commonest of the presumed hybrids, and the intermediate nature of the fruit pubescence is the most convincing evidence. [Some specimens lacking fruit may have been referred to the parent species if the pubescence (or its absence) is clear, but those with ambiguous pubescence on vegetative parts and annotated as hybrids by G. M. Baker (during an unpublished study of the sumacs) are included here.]
(d) Hairs of fruit as in *R. glabra* but pubescence of vegetative parts too abundant or long for that species. Such plants are very rare.

It is perhaps unnecessary to add that all these presumed hybrid variants are inclined to grade into one another. They are found in the same sorts of dry, usually sandy or rocky, open or lightly shaded ground as the parent species. In view of the abundance of *R.* ×*pulvinata* in the state, Map 752 is provided.

AQUIFOLIACEAE — Holly Family

Our two species both bear conspicuous red fruit on axillary pedicels, and have long been protected from exploitation by Michigan's "Christmas Tree Law." Since these species are dioecious (flowers rarely with rudiments of the opposite sex), fruit will be found only on female plants. Technically the fruit is a drupe (with more than one stone), not a berry. Our hollies are deciduous, but evergreen species of *Ilex* are familiar Christmas decoration; the traditional holiday colors reflect the red fruit and green leaves.

KEY TO THE GENERA

1. Pedicels several times as long as the petals, mostly twice as long as the fruit, or longer; petals yellowish, linear, separate, the flowers opening in May (or June near Lake Superior); leaves entire (very rarely with a few sharp teeth) . . . . . 1. **Nemopanthus**
1. Pedicels ca. twice as long as the petals, shorter than the fruit; petals white, broadly oblong-elliptic (less than twice as long as wide), connate at the base, the flowers opening in July (or June in southern Michigan); leaves with close, sharp teeth . . 2. **Ilex**

1. **Nemopanthus**

The generic name is often treated as feminine (*N. "mucronata"*), but the Code has long recommended that all names ending in *-anthus* should be masculine (rather than neuter, which would in fact be the correct classical gender).

1. **N. mucronatus** (L.) Loes. Plate 6-B Mountain Holly

Map 753. Bogs, especially in a characteristic zone of high shrubs near the outer margin; swamps and thickets; swales, interdunal hollows, and low places in woods; margins of lakes; damp coniferous woods on sandy banks facing Lake Superior.

Rare individuals with toothed leaves (stunted, juvenile, or dense shade forms?) may, if sterile, resemble *Ilex verticillata* but can be distinguished by the glabrous, purple petioles typical of *Nemopanthus*. In our *Ilex,* the petioles are green and often pubescent. A yellow-fruited form, f. *chrysocarpus* (Farw.) Fern., was described from Michigan (TL: Rice Lake, Houghton Co.).

2. **Ilex** Holly

1. **I. verticillata** (L.) A. Gray Fig. 270 Michigan Holly; Winterberry; Black-alder

Map 754. Bogs and swamps, damp shores and thickets, ditches and swales, margins of lakes and streams.

The fruit is often persistent well into the winter and is eaten by birds and mammals. The leaves of some other species, not in our range, contain caffeine and have been used for tea. Leaves vary somewhat in shape, texture, and pubescence, but recognition of named variants does not seem warranted. Most often the blades are ± obovate and resemble those of choke cherry (*Prunus virginiana*) in shape. The yellow-fruited form, f. *chrysocarpa* Robinson, has been collected in Wexford County.

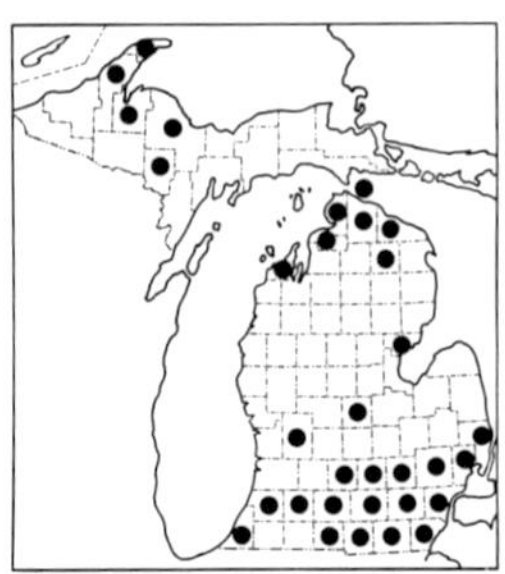

751. Rhus glabra

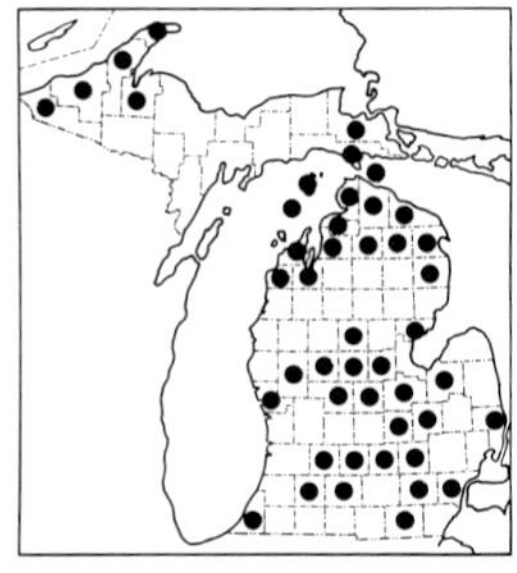

752. Rhus ×pulvinata

753. Nemopanthus mucronatus

REFERENCE

Brizicky, George K. 1964. The Genera of Celastrales in the Southeastern United States. Jour. Arnold Arb. 45: 206–234.

KEY TO THE GENERA

1. Leaves alternate; stems usually twining; flowers unisexual (or functionally so), the plants essentially dioecious . . . . 1. **Celastrus**
1. Leaves opposite; stems erect or trailing but not twining; flowers perfect . . 2. **Euonymus**

1. Celastrus

Bittersweet

Several species are cultivated, including our native *C. scandens,* for the fruit is showy when ripe. The seed is enclosed in a bright red appendage (the aril), and the valves of the capsule are orange. The seeds are sometimes said to be poisonous, and it would be well to avoid ingesting them, even though documented reports seem to be lacking. They are of limited use to birds. Both of our species are vines capable of climbing to considerable heights. Like the hollies, the American bittersweet is protected from exploitation under Michigan's "Christmas Tree Law."

The generic name is a Greek one for some tree and was adopted by Linnaeus, as was often the case, for a different plant. Nevertheless, since 1935 the Code has recommended that such names retain their classical gender, and *Celastrus* is accordingly here treated as feminine, although most authors cite the Oriental species as *C. orbiculatus*.

KEY TO THE SPECIES

1. Flowers in axillary cymes; leaf blades ± broadly obovate to nearly orbicular; plant a local escape from cultivation . . . . 1. **C. orbiculata**
1. Flowers in terminal racemes or panicles (at ends of branchlets); leaf blades ± ovate-lanceolate to elliptic; plant a common native . . . . 2. **C. scandens**

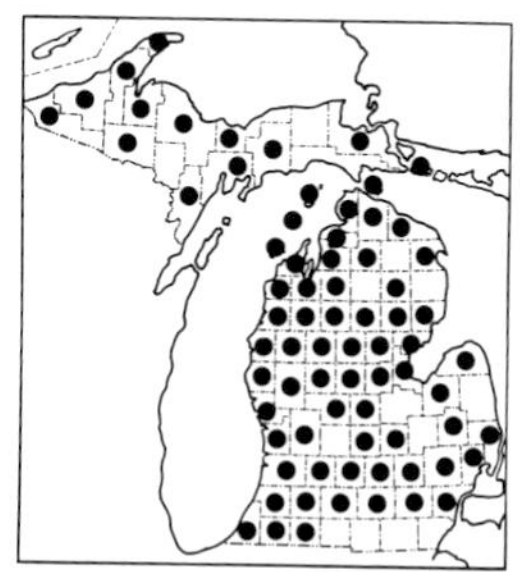

754. Ilex verticillata

755. Celastrus orbiculata

756. Celastrus scandens

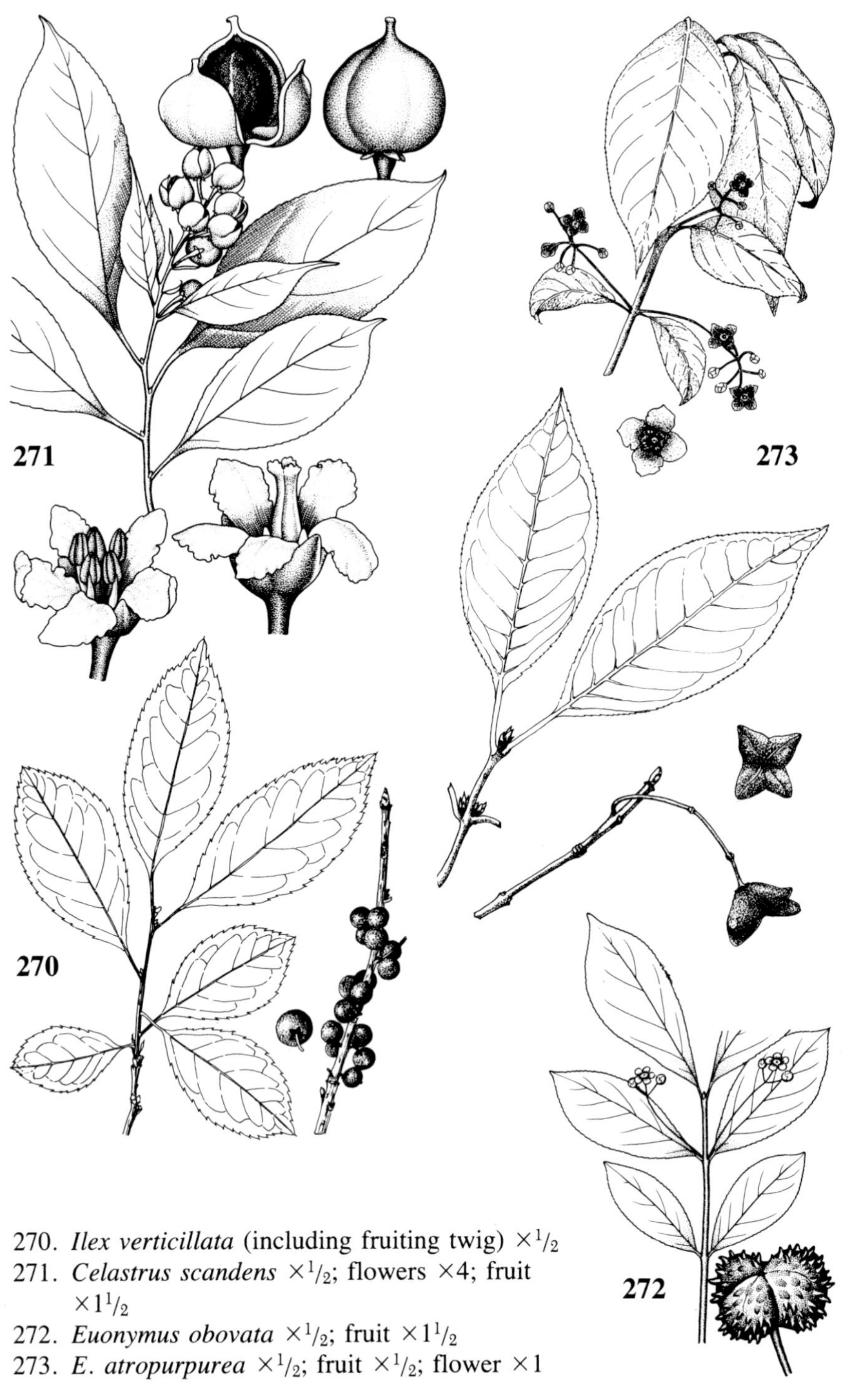

270. *Ilex verticillata* (including fruiting twig) $\times{}^1/_2$
271. *Celastrus scandens* $\times{}^1/_2$; flowers $\times 4$; fruit $\times 1{}^1/_2$
272. *Euonymus obovata* $\times{}^1/_2$; fruit $\times 1{}^1/_2$
273. *E. atropurpurea* $\times{}^1/_2$; fruit $\times{}^1/_2$; flower $\times 1$

1. **C. orbiculata** Thunb. Oriental Bittersweet

Map 755. A native of Asia, sometimes cultivated but it can be aggressive when escaped. Reported locally as a weed of woods and near habitations in southern Michigan, but actual collections are sparse.

2. **C. scandens** L. Fig. 271 Climbing or American Bittersweet

Map 756. Shores and dune thickets, stream and river banks, roadsides and fencerows, rock outcrops and talus slopes, dry to moist woods as well as open ground.

2. Euonymus

As in the preceding genus, the seed is enclosed in a colorful aril (red or orange); the color of the capsule is red to purple in our species. And here, too, the classical gender of the generic name is feminine, although some authors treat it as masculine.

KEY TO THE SPECIES

1. Leaves evergreen, pale along the principal veins; plant trailing or climbing by adventitious roots; young branchlets and midrib above papillose or warty . . 1. **E. fortunei**
1. Leaves deciduous, not pale along the veins; plant trailing or an erect shrub; young branchlets (unless becoming winged) and midrib smooth (at most slightly pubescent)
  2. Branches with corky wings; plant rarely escaped from cultivation . . . . 2. **E. alata**
  2. Branches without wings; plant native or not
    3. Plant a trailing, creeping shrub with erect or ascending flowering shoots; flowers 5-merous; fruit shallowly 3-lobed, strongly tuberculate, pink to red when ripe; leaf blades glabrous beneath (except sometimes on principal veins), usually obtuse . . . . 3. **E. obovata**
    3. Plant an erect bushy shrub or small tree; flowers 4-merous; fruit distinctly 4-lobed but nearly or quite smooth, red to purple when ripe; leaf blades pubescent or glabrous beneath, acute to short-acuminate
      4. Leaf blades finely pubescent over entire lower surface; flowers dark purple, usually numerous in inflorescence; aril red; plant a native species of woods . . . . 4. **E. atropurpurea**
      4. Leaf blades glabrous beneath; flowers yellow-green, rarely more than 5 in an inflorescence; aril orange; plant an uncommon escape (especially in cultivated areas or dumps) . . . . 5. **E. europaea**

1. **E. fortunei** (Turcz.) Handel-Mazz. Wintercreeper

Map 757. A Chinese species, of which there are several cultivars. Seldom flowering or escaping from cultivation, but collected in an oak woods west of Marshall in 1984 (*Freudenstein 1094*, MICH), where evidently growing from a seed (juvenile foliage); collected in a gravel pit on dumped roadside debris near Honor in Benzie County (mature foliage) in 1982 (*Overlease 2769*, MICH).

2. **E. alata** (Thunb.) Siebold — Winged Euonymus

Map 758. A species of eastern Asia, widely cultivated for its distinctive winged branches and brilliant autumn foliage. It spreads occasionally to woods, thickets, and untended yards.

3. **E. obovata** Nutt. Fig. 272 — Running Strawberry-bush

Map 759. Deciduous forests of all kinds, from low swampy woods to rich upland beech-maple and oak-hickory; at its northern limit, sometimes with cedar or hemlock.

Reports of *E. americana* L. from Michigan are apparently based on this species. *E. americana* is a much more southern species, an erect or ascending shrub with usually 5-lobed fruit and acute leaf blades which are ± obtuse to broadly rounded at the base. A few specimens referred to *E. obovata* have acute leaf blades, but they are glabrous and the bases taper to the short petioles, the habit is prostrate, and the fruit is 3-lobed.

4. **E. atropurpurea** Jacq. Fig. 273 — Burning-bush; Wahoo

Map 760. River banks and floodplain forests.

5. **E. europaea** L. — Spindle Tree

Map 761. Native to Eurasia, and adventive on shores, dumpsites, borders of woods, and near habitations.

757. Euonymus fortunei

758. Euonymus alata

759. Euonymus obovata

760. Euonymus atropurpurea

761. Euonymus europaea

762. Staphylea trifolia

STAPHYLEACEAE — Bladdernut Family

REFERENCE

Spongberg, Stephen. 1971. The Staphyleaceae in the Southeastern United States. Jour. Arnold Arb. 52: 196–203.

1. Staphylea

1. **S. trifolia** L. Fig. 274 — Bladdernut

Map 762. Deciduous woods and thickets, especially on floodplains and river banks. Sometimes cultivated, and possibly plants forming thickets at the edge of a deciduous woods in Charlevoix County are of such an origin rather than disjunct.

Our only species in this small family is a distinctive large shrub with opposite trifoliolate leaves and an inflated, usually 3-lobed indehiscent "capsule" with papery walls that pop when crushed. Confusion with *Ptelea* is surprisingly frequent, and if one is puzzled by a sterile specimen it is necessary only to remember that *Ptelea* has alternate leaves with leaflets entire or nearly so, whereas *Staphylea* has leaflets finely and closely toothed. The whitish flowers appear in rather showy drooping panicles with the leaves in the spring; the unique bladder-like fruit in late summer is ca. 3–5 cm long, at first green but becoming yellow-brown.

ACERACEAE — Maple Family

In most species, the flowers are both perfect and unisexual, the plants polygamo-dioecious or sometimes -monoecious. Just as acorns are distinctive for oaks, the characteristic fruits of maples readily identify them. These are samaras and grow in pairs, each with a seed at the base and a long broad wing beyond. Several additional species are planted, of which at least the Amur maple, *A. ginnala* Maxim., may seed in yards or ditches; it is usually only a shrub, with brilliant red foliage in the fall, the leaves 3-lobed with prolonged middle lobe.

1. Acer — Maple

REFERENCES

Boivin, Bernard. 1966. Les Variations d'Acer negundo au Canada. Nat. Canad. 93: 959–962 (also in Ludoviciana 4).

Brizicky, George K. 1963. The Genera of Sapindales in the Southeastern United States. Jour. Arnold Arb. 44: 462–501.

Desmarais, Yves. 1952. Dynamics of Leaf Variation in the Sugar Maples. Brittonia 7: 347–387 (also Contr. Dép. Biol. Univ. Laval 1).

Hanes, Clarence R. 1954 ["1953"]. Syrup from the Sap of Various Trees. Asa Gray Bull., N. S. 2: 210–211.

Kriebel, Howard B. 1957. Patterns of Genetic Variation in Sugar Maple. Ohio Agr. Exp. Sta. Res. Bull. 791. 56 pp.

Sutherland, Roger & Mary. 1975. Maple Syrup—A Family Project. Mich. Bot. 14: 57–61.

Wagner, W. H., Jr. 1975. Notes on the Floral Biology of Box-elder (Acer negundo). Mich. Bot. 14: 73–82.

KEY TO THE SPECIES

1. Leaves compound; leaf scars joined around the stem; flowers apetalous, unisexual (plants dioecious), on drooping pedicels becoming many times their length . . . . 1. **A. negundo**
1. Leaves simple; leaf scars not quite meeting around the stem; flowers various
 2. Sinuses between the principal leaf lobes broadly rounded or obtuse and entire at their base
  3. Flowers with conspicuous petals (ca. 4–7 mm long), in glabrous erect or ascending corymbs; samaras (3.5) 4–5 cm long, the wings of the 2 in each pair often diverging almost 180°; petioles with milky sap; plant cultivated, rarely escaped . . . . 2. **A. platanoides**
  3. Flowers without petals, on pubescent soon drooping pedicels; samaras ca. 2–3.5 cm long, the wings diverging at an acute angle (often becoming nearly parallel); petioles with watery sap; plant a common native . . . . 3. **A. saccharum**
 2. Sinuses between the principal leaf lobes acute at their base and/or the sinuses toothed all the way to the base
  4. Flowers (or at least the bud scales subtending them) red, in sessile umbel-like inflorescences (compact when first blooming, the pedicels later elongating), opening well before the leaves; terminal main lobe of leaf with sides parallel or angled inward near the base (a subordinate lobe on each side; figs. 275, 277); bud scales 6–10 (12)
   5. Terminal main lobe of leaf ca. two-thirds the length of the blade, the margins flaring from a narrow base, so that each adjacent sinus is usually narrower at its mouth than it is closer to the base (fig. 277); sepals mostly connate part or all their length; petals none; ovary and at least young fruit pubescent; mature samaras 3.5–5.5 (6) cm long . . . . 4. **A. saccharinum**
   5. Terminal main lobe of leaf ca. half the length of the blade, the margins parallel toward the base or nearly so, the adjacent sinuses much wider at their mouth than elsewhere (fig. 275); sepals separate; petals present, nearly or quite as long as the sepals; ovary and fruit glabrous; mature samaras 1.8–2.5 cm long . . . . 5. **A. rubrum**
  4. Flowers yellowish or greenish, in peduncled, elongate racemes or panicles, opening with or after the leaves; terminal main lobe of leaf (except in *A. pseudoplatanus*) broadest at base, tapered ± uniformly to apex (with no subordinate lobes); bud scales 2, valvate
   6. Leaf blades whitened beneath, mostly 5-lobed (including a pair of smaller basal lobes), the terminal lobe ± parallel-sided (or even broader above the middle); petals about equaling the sepals or slightly shorter; plant a large cultivated tree, rarely escaped . . . . 6. **A. pseudoplatanus**
   6. Leaf blades not whitened beneath, 3-lobed, the terminal lobe ± triangular (broadest at base); petals at maturity exceeding the sepals; plant a native large shrub or small tree

7. Leaves coarsely and simply toothed; inflorescence an ascending or erect puberulent panicle; petals linear-oblanceolate; fruit with strong raised veins over the seed; bark drab, not striped; year-old branchlets minutely pubescent .... 7. **A. spicatum**

7. Leaves finely doubly serrate; inflorescence a pendent glabrous raceme; petals obovate; fruit nearly or quite smooth over the seed; bark of young trunks and branches green with pale stripes; year-old branchlets glabrous .... 8. **A. pensylvanicum**

1. **A. negundo** L. Box-elder; Ash-leaved or Manitoba Maple

Map 763. Native at least in the southern half of the Lower Peninsula, and apparently northward on some river banks and floodplains as far as Baraga and Houghton counties of the northwestern Upper Peninsula. But widely cultivated and readily escaping throughout the state, almost a weed in some places. Swamp forests, shores, and banks; spreading aggressively so that seedlings and small (to large) trees are common along fencerows, sidewalks, railroads, ditches, and waste places generally; invading old fields and open woods.

The leaflets are usually 3–5, rarely 7. Small plants, such as seedlings, with 3 leaflets can easily be confused with poison-ivy until one notices that the leaves are opposite. Usually the branchlets are glabrous and glaucous (± purple beneath the waxy bloom); such plants have been called var. *violaceum* (Kirchner) Jaeger. Plants with puberulent branchlets have not yet been found in our area, but occur in Minnesota and westward [var. *interius* (Britton) Sarg.]. In any of these varieties, forms may occur with the wings of the fruit strikingly red or purple. The floral biology of this species in southern Michigan has been thoroughly studied by Wagner (1975).

2. **A. platanoides** L. Norway Maple

Map 764. A native of Europe and western Asia, very widely planted as a shade tree and hardy throughout Michigan. Occasionally becoming established, and seedlings or small trees large enough to fruit are sometimes found near cultivated plants along hedges, at edges of woods, by buildings, and in open waste land.

The seed end of the samara is very much flattened in this species, while in *A. saccharum* it is distinctly swollen. Numerous cultivars are grown, including some in which the leaves and normally yellow flowers are deeply suffused with red.

3. **A. saccharum** Marsh. Sugar or Hard Maple

Map 765. This is the maple of "beech-maple" forests, where it is also often associated with hemlock, yellow birch (northward), basswood, ironwood, and other species (beech itself being absent from the western Upper Peninsula); it is also a component of diverse mixed hardwood forests, especially in the southern part of the state; thrives on a diversity of soils,

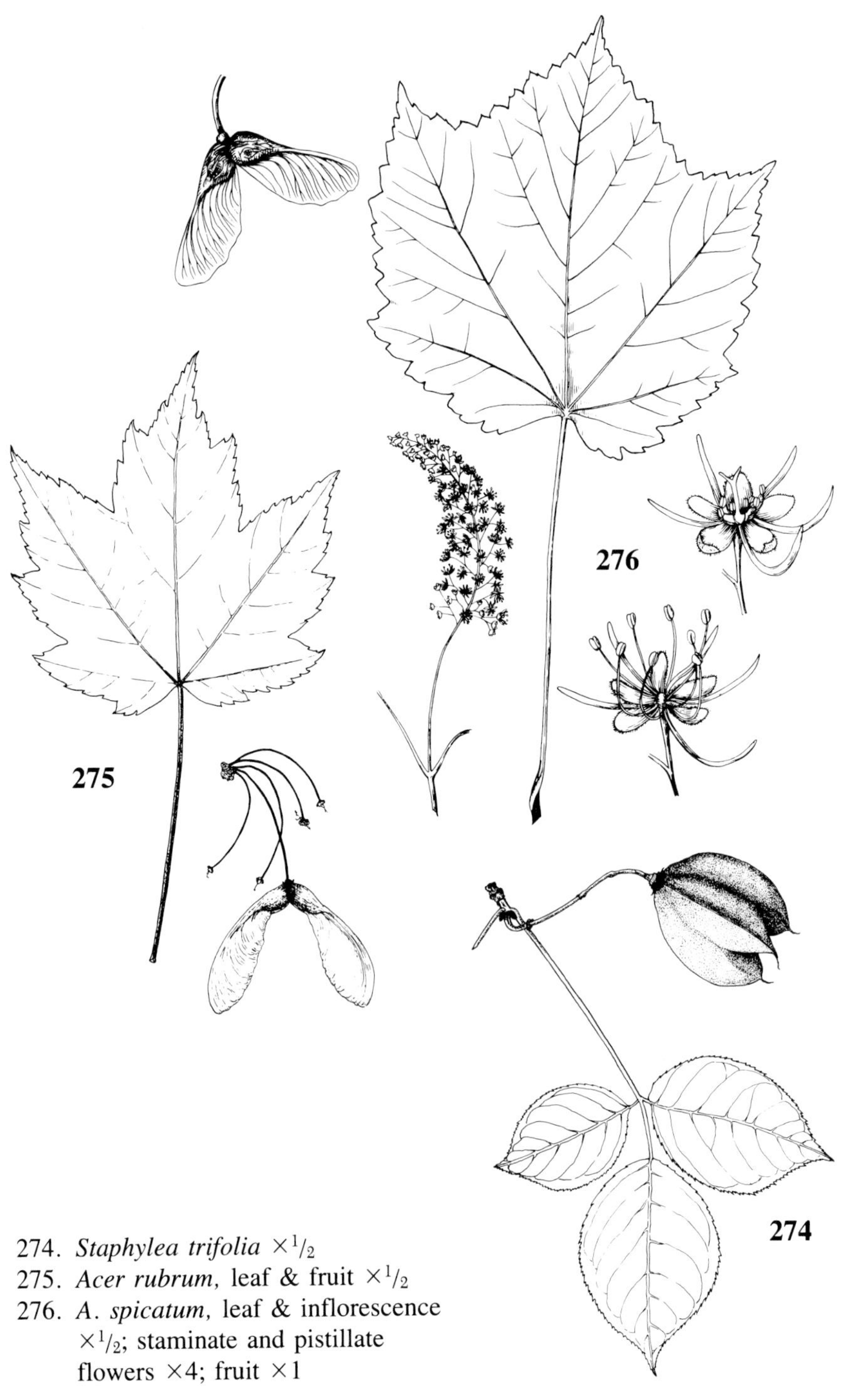

274. *Staphylea trifolia* $\times ^1/_2$
275. *Acer rubrum*, leaf & fruit $\times ^1/_2$
276. *A. spicatum*, leaf & inflorescence $\times ^1/_2$; staminate and pistillate flowers $\times 4$; fruit $\times 1$

including stabilized (forested) dunes receiving moisture-laden winds off the Great Lakes, but rarely in swamp forests.

Black maple is often recognized as a distinct species, *A. nigrum* Michaux f., said usually to differ in ± dense puberulence on petioles and undersides of blades, drooping or wilted look to the leaf lobes, expanded and often stipulate petiole bases, darker bark, and failure to develop red color in autumn foliage (which remains yellow). Quite possibly this was once a more distinct taxon than now, when supposed hybrids with sugar maple are so common as to make recognition of two good species questionable (see Desmarais 1952; Kriebel 1957). Stress on different characters leads to different conclusions as to where any line between taxa should be drawn. Black maple, which can be called *A. saccharum* ssp. *nigrum* (Michaux f.) Desm. or *A. saccharum* var. *viride* (Schmidt) E. G. Voss, ranges north in Michigan to about the middle of the Lower Peninsula. Trees can be found into the Upper Peninsula with leaf blades ± densely puberulent beneath but the petioles glabrous or nearly so—perhaps a sign of introgression from black maple. Leaf shape is extremely variable in both taxa, although black maple tends to have fewer major lobes. However, few-lobed leaves glabrous beneath are not rare.

The two taxa are not distinguished in commerce, and "hard maple" is a major timber tree in Michigan, the most important hardwood, for durable uses such as flooring, mallets and handles, and chopping blocks—not to mention furniture and veneer. Production of maple syrup and sugar from the sap has been important since before Europeans invaded the region, and indeed maple sugar was a major article of trade among the Indians. Michigan's maple syrup production is the fifth highest in the nation. The sap of some other species of *Acer* also produces good syrup, as does that of other trees as well (see Hanes 1954 and Sutherland 1975 for local references). Finally, this is a handsome shade tree, the foliage becoming various shades of yellow, orange, and red in the fall.

4. **A. saccharinum** L. Fig. 277 Silver Maple

Map 766. Swamp forests (including floodplains), river banks, lake shores (where samaras may wash up and germinate in dense rows); also sometimes in low hardwoods, as well as in fencerows along roadsides.

Often planted as a shade tree, for which "cut-leaf" cultivars, the leaves more deeply lobed than usual, are popular. Sometimes apparently hybridizes with red maple, with resultant intermediate samara size and leaf shape. Leaves of both species are whitened beneath, and the wood of both is called "soft maple" in commerce.

5. **A. rubrum** L. Fig. 275 Red Maple

Map 767. In almost all sorts of woods and woodlands: commonly in lowland forests and swamps (deciduous and coniferous) throughout the state;

northward, most often in upland forests and dry sandy woodland of plains, dunes, and hills (often with aspens, oaks, and/or pines). Sprouts readily after fire or cutting, and clumps of smooth gray young trunks (from old stumps) are often common in the dry cut- and burned-over woods of northern Michigan.

This is a well-named tree, for the flowers, young fruit, and autumn foliage are a brilliant red. This and silver maple are among the first plants to bloom in the spring. Red maple is also often planted as a shade tree.

6. **A. pseudoplatanus** L. Sycamore Maple

Map 768. Like *A. platanoides,* a native of Europe and western Asia, but less often planted in our area than that species and much less often seen as an escape. However, it has been collected as a self-established volunteer in vacant lots near Lake Superior in Marquette (*Bourdo* in 1961, MCTF, MSC; in 1962; MCTF, BLH).

The teeth on the leaves are quite obtuse or rounded, even callus-tipped, with no little point at all. These, together with the distinctive blades 5-lobed about halfway to the base, should help to identify even sterile plants.

7. **A. spicatum** Lam. Fig. 276 Mountain Maple

Map 769. Mixed and coniferous woods and thickets, including deciduous woods and wooded bluffs or banks, conifer swamps; sometimes growing with *A. pensylvanicum* but usually in somewhat moister woods.

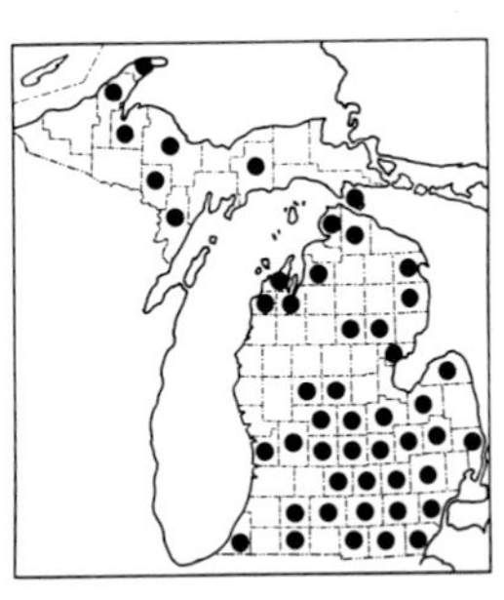

763. Acer negundo

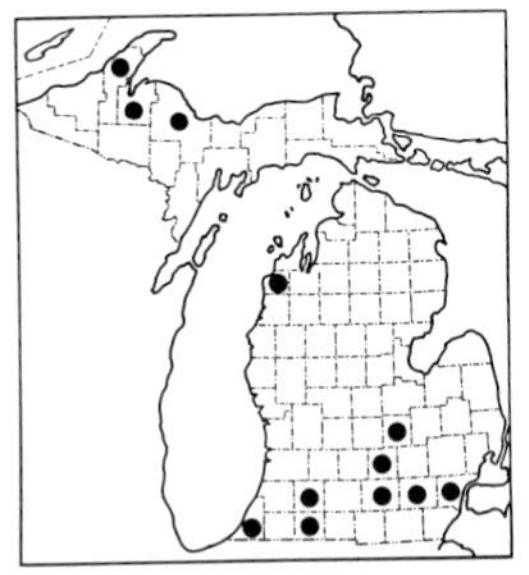

764. Acer platanoides

765. Acer saccharum

766. Acer saccharinum

767. Acer rubrum

768. Acer pseudoplatanus

This is a large shrub, at the very most the stems about 10 inches in diameter. The leaves may be orange or red in the fall. It is more pubescent than the next species in all its parts: inflorescence, young fruits, lower surface of leaves, twigs. Both retain their fruit into late summer or fall, whereas the early-flowering species (*A. rubrum* and *A. saccharinum*) have generally lost their fruit by early summer.

8. **A. pensylvanicum** L. Fig. 278 Striped or Goosefoot Maple; Moosewood

Map 770. Rich deciduous woods, conifer-hardwoods, sandy wooded bluffs, occasionally in cedar swamps. The westernmost point in the range of this species (which occurs east to Gaspé and south in the Appalachians) is apparently in hardwood forest of the Huron Mountains in northeastern Baraga County, except for a 1976 collection (MCTF) without habitat but said to be from one mile east of Eagle River (perhaps suspicious as the species was never previously reported from the well-explored Keweenaw County). It thrives especially well close to the Great Lakes and is less common inland. Fine stands grow along Lake Superior in the area of Muskellunge Lake State Park and elsewhere.

A tall shrub or small tree, this species has a trunk rarely over 8 inches in diameter, although trees as large as 12–14 inches occur in the Huron Mountains, including the "national champion." The bark of all but old trunks is a beautiful smooth green, striped with pale green or almost white lines. The distinctive leaf blades have 3 acuminate tips and are otherwise sharply and finely doubly serrate, often as wide as 15 cm or more; they turn a bright clear yellow in the fall.

HIPPOCASTANACEAE Buckeye Family

1. Aesculus Buckeye

A very easily recognized genus, with its opposite, palmately compound leaves and distinctive fruit. The latter is a capsule, the outside ± prickly (in our species) and only 1 or occasionally 2 (very rarely more) large seeds developing within. The seed has an unusually broad pale scar (*hilum*) representing the area of attachment, the rest of it being a shiny rich reddish brown when ripe. The raw nuts, like the young bark, are bitter and poisonous, at least to livestock; while they can reportedly be made edible, the effort hardly seems worthwhile.

REFERENCES

Brizicky, George K. 1963. The Genera of Sapindales in the Southeastern United States. Jour. Arnold Arb. 44: 462–501.

Hardin, James W. 1955. Studies in the Hippocastanaceae, I. Variation Within the Mature Fruit of Aesculus. Rhodora 57: 37–42.

Hardin, James W. 1957a. A Revision of the American Hippocastanaceae. Brittonia 9: 145–171.

Hardin, James W. 1957b. A Revision of the American Hippocastanaceae—II. Brittonia 9: 173–195.

Hardin, James W. 1957c. Studies in the Hippocastanaceae, IV. Hybridization in Aesculus. Rhodora 59: 185–203.

KEY TO THE SPECIES

1. Leaflets mostly 7, abruptly acuminate to a short tip; winter buds very glutinous; hilum about two-thirds or more as broad as the seed; flowers whitish (or pink) . . . . 1. **A. hippocastanum**
1. Leaflets 5, tapered ± gradually to tip; winter buds not glutinous; hilum not over about half as broad as the diameter of the seed; flowers yellowish green . . 2. **A. glabra**

1. **A. hippocastanum** L. Horse-chestnut

Map 771. A native of the eastern Mediterranean area, long cultivated in Europe and North America, occasionally seeding in waste ground, fence-rows, woods, rarely swamps—even surviving till old enough to fruit. Doubtless escaped more often than the few existing collections document. Hardy and often planted throughout the Lower Peninsula and into the Upper Peninsula.

Often called "chestnut," especially in Europe, but that name is very misleading as this plant is wholly different from the true chestnut, *Castanea*. Even Henry Wadsworth Longfellow, however, referred to *Aesculus* as a "spreading chestnut." Pink-flowered forms are sometimes cultivated, and there are also pink-flowered cultivars of a polyploid hybrid [*A.* ×*carnea* Hayne] derived from *A. hippocastanum* × *A. pavia,* the latter species being the red buckeye of the southern states. The coarser-looking, obovate leaflets and larger fruit than in buckeye will usually suffice to separate this species quickly from the next.

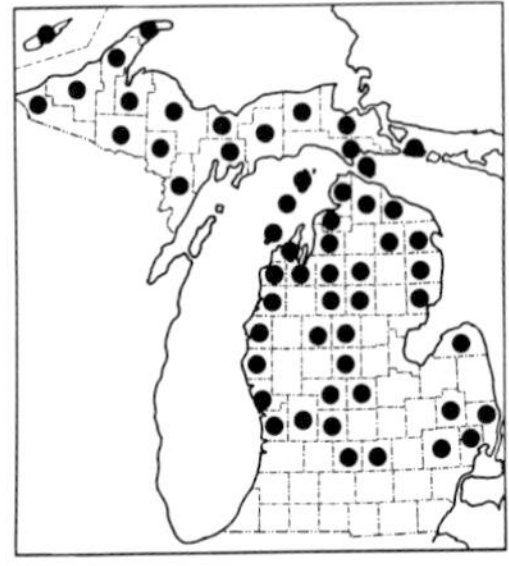
769. Acer spicatum

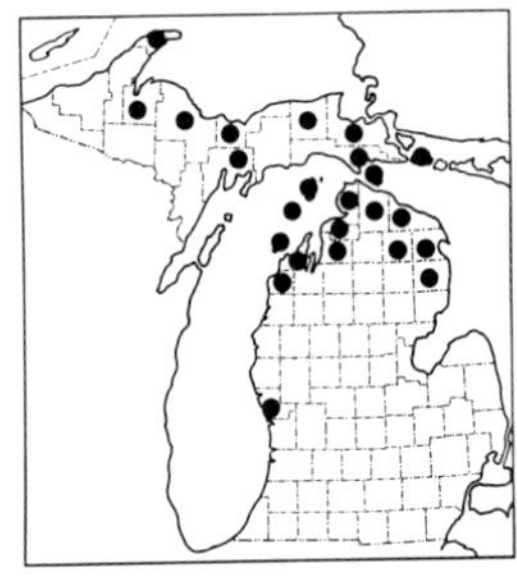
770. Acer pensylvanicum

771. Aesculus hippocastanum

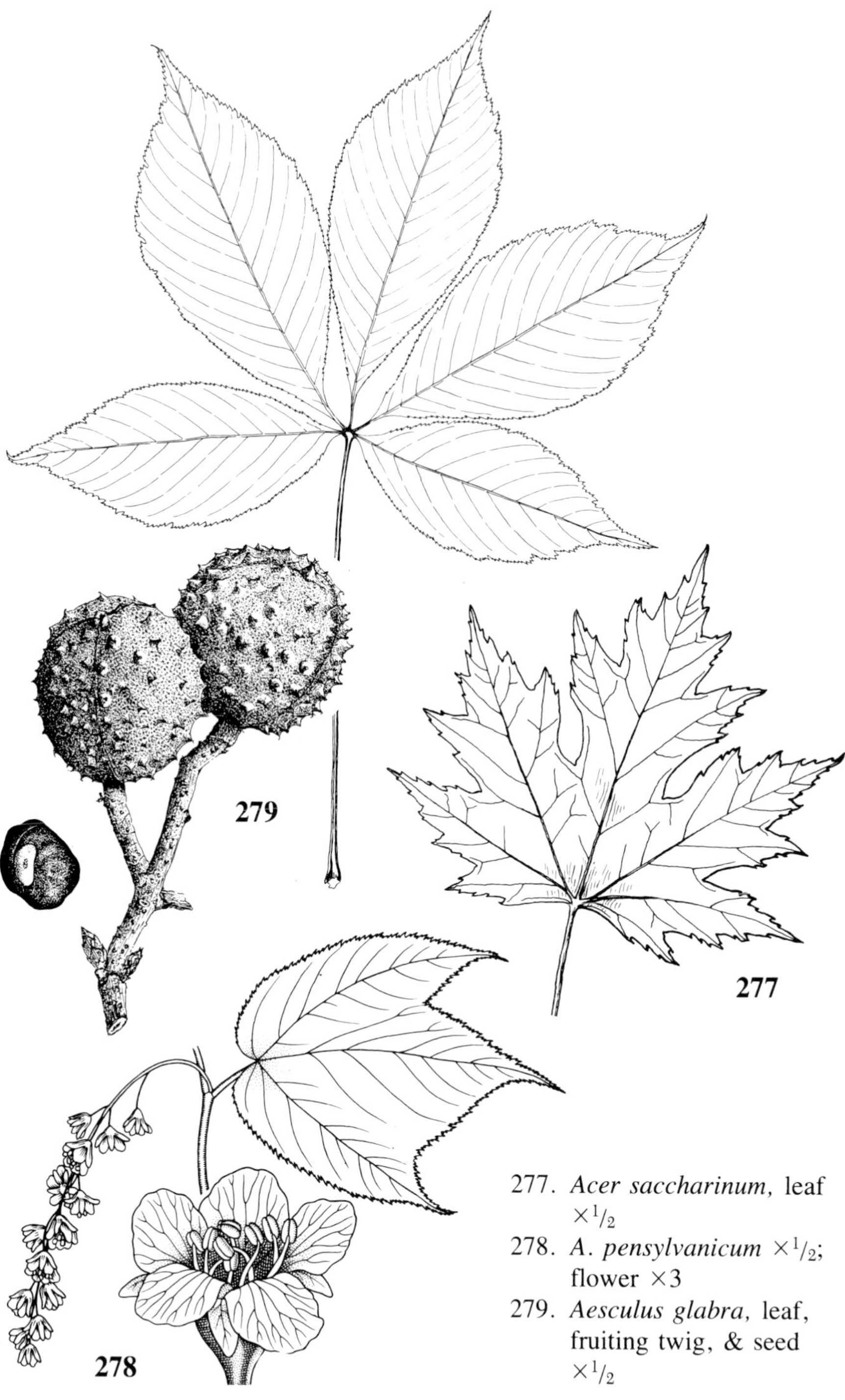

277. *Acer saccharinum,* leaf $\times 1/2$
278. *A. pensylvanicum* $\times 1/2$; flower $\times 3$
279. *Aesculus glabra,* leaf, fruiting twig, & seed $\times 1/2$

2. **A. glabra** Willd. Fig. 279 Ohio Buckeye

Map 772. Deciduous forests, swampy woods, and river-bank thickets. Southern Michigan is at the northern edge of the range of this species.

A. octandra Marsh. has often been reported from the state, but that species, the sweet or yellow buckeye, occurs only as far north as southern Ohio, Indiana, and Illinois. However, Hardin interpreted the presence of stipitate glands on the pedicels of some specimens of the normally eglandular *A. glabra* to be evidence of introgression long ago from *A. octandra*—a slight influence indicating gene flow as far north as Michigan but not evidence of *A. octandra* in the state. Although native only in the southernmost Lower Peninsula (and perhaps seeding from cultivated plants a little farther north), *A. glabra* is hardy northward, at least in sheltered places. For a number of years before construction of a boardwalk, a small buckeye thrived along the trail in woods on the bank of the Tahquamenon River, just below the Upper Falls—presumably where a tourist had stuck a seed.

SAPINDACEAE — Soapberry Family

1. Cardiospermum

One species of this tropical American genus is the only representative of this largely tropical family ever collected outside of cultivation in Michigan.

REFERENCE

Brizicky, George K. 1963. The Genera of Sapindales in the Southeastern United States. Jour. Arnold Arb. 44: 462–501.

1. **C. halicacabum** L. Balloon-vine

Map 773. Sometimes cultivated and escaped, especially in the southern United States. Collected once in Michigan, on "banks of marshes" at River Rouge (*Farwell 9320* in 1932, MICH, BLH).

This is a rather slender herbaceous vine climbing by tendrils, with alternate leaves usually twice-ternately compound, the leaflets ± coarsely toothed. The flowers are small (barely 5 mm), usually unisexual, and white; the fruit is an inflated, papery capsule.

BALSAMINACEAE — Touch-me-not Family

1. Impatiens

These are familiar plants to those who note the colorful flowers late in summer or who play with the fruit, a capsule that, at maturity, dehisces explosively when touched, scattering the seeds. The flower structure is highly modified, although it is clearly bilaterally symmetrical. There are 3 sepals, of which the two upper (lateral) ones are small but the lower one large, petaloid, sac-like, and slender-spurred. The petals appear to be 3, each of the two lateral ones with a lobe presumably representing a fused petal. White-flowered forms may occur in any of our species, and all are variable in color and marking.

Our two native species are impossible to distinguish in the absence of flowers. Seedlings are often abundant in damp places, looking pale and watery, with rather rounded or crenate teeth.

REFERENCE

Wood, Carroll E., Jr. 1975. The Balsaminaceae in the Southeastern United States. Jour. Arnold Arb. 56: 413–426.

KEY TO THE SPECIES

1. Leaf blades sharply and closely toothed; flowers usually red (pink) to purple; plant a rare waif, escaped from cultivation into waste places
   2. Stem ± pubescent; flowers axillary on main stem; petiole bearing prominent saucer-like glands below the blade; upper leaves (like the lower) alternate . . . . 1. **I. balsamina**
   2. Stem glabrous; flowers in terminal or axillary inflorescences; petioles eglandular or with ± glandular teeth at base of decurrent blade; upper leaves opposite or whorled . . . . . . 2. **I. glandulifera**
1. Leaf blades with rather rounded teeth (their tips obscure), crenate in aspect and only ca. 12 (15) or fewer per side; flowers orange or yellow; plant native, usually in damp places
   3. Flowers yellow; mature spur ca. 5 mm long or less; sac of spurred sepal about as broad as long or broader . . . . . . 3. **I. pallida**
   3. Flowers orange (rarely yellow); mature spur ca. (7) 9–10 (12) mm long; sac of spurred sepal usually distinctly longer than broad . . . . . . 4. **I. capensis**

1. **I. balsamina** L. Balsam; Garden Balsam

Map 774. A native of Asia, cultivated as an ornamental and collected by Farwell in waste ground at Farmington in 1927 (*8161*, BLH; Am. Midl. Nat. 11: 63. 1928; no specimen has been found to support his earlier report from Detroit).

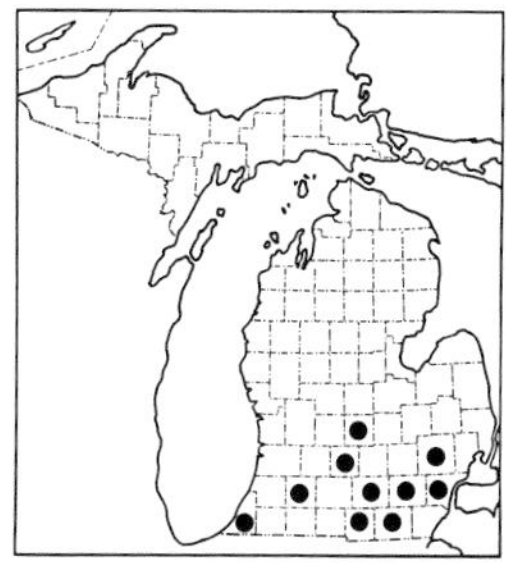
772. Aesculus glabra

773. Cardiospermum halicacabum

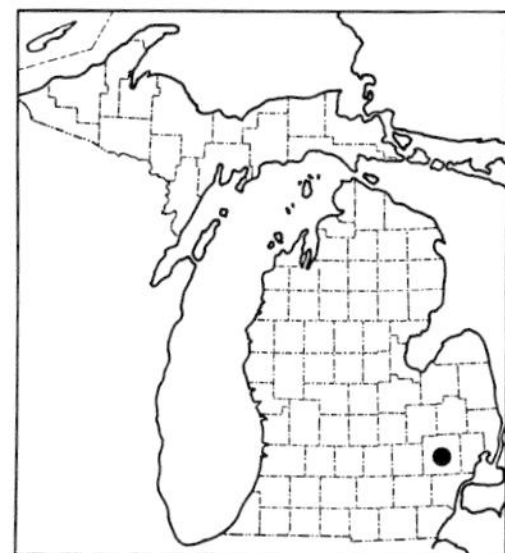
774. Impatiens balsamina

2. **I. glandulifera** Royle

Map 775. A native of the Himalayas, a common ornamental in gardens, collected rarely as a waif: by Dodge on the streets of Port Huron in 1912; by Hiltunen in disturbed areas on Sugar Island in 1956; and by Henson on a slope along Lake Superior at Grand Marais in 1984.

Both this species and the preceding have larger (especially broader) flowers than our native ones, as well as much more sharply acute and serrate leaves.

3. **I. pallida** Nutt. Pale Touch-me-not

Map 776. Woods (especially swampy), streamsides, ditches; much less common than the next and perhaps tending toward more mesic habitats, although the two may grow together.

The spur is less strongly bent than in the next species, sometimes nearly 180° but sometimes only about 90° from the axis of the saccate sepal.

4. **I. capensis** Meerb. Plate 6-C Spotted Touch-me-not

Map 777. Swamps, streamsides, ditches, lake shores, marshy areas, thickets, ravines and wet spots in woods; often in somewhat disturbed areas, including excavations.

Long known as *I. biflora* Willd. The epithet *capensis* results from an error in thinking that the plant came from the Cape of Good Hope, whereas it is in fact a native of North America (although established as a weed in Britain and France).

The flowers are usually more copiously and conspicuously spotted with deeper orange, rose, or brown dots than in the preceding species, and the long spur tends to be very strongly bent back below the sac, even a little more than 180°. Variants with yellow (sometimes very pale) flowers, without spots, or with the spots confluent, are rarely found.

RHAMNACEAE Buckthorn Family

REFERENCE

Brizicky, George K. 1964. The Genera of Rhamnaceae in the Southeastern United States. Jour. Arnold Arb. 45: 439–463.

KEY TO THE GENERA

1. Leaf blades with 3 prominent veins from the base; flowers white, in dense inflorescences at the ends of branches or on axillary peduncles; fruit dry, dehiscent . . . . 1. **Ceanothus**
1. Leaf blades clearly pinnately veined, not strongly 3-nerved from the base; flowers yellow-green, solitary or numerous but all axillary; fruit fleshy, indehiscent . . . . 2. **Rhamnus**

1. **Ceanothus**

REFERENCE

Brizicky, George K. 1964. A Further Note on Ceanothus herbaceus versus C. ovatus. Jour. Arnold Arb. 45: 471–473.

KEY TO THE SPECIES

1. Inflorescences on peduncles arising from branches of the previous year; shrub usually over 1 m tall, known only from Keweenaw County . . . . 1. **C. sanguineus**
1. Inflorescences on peduncles or branchlets of the current year; shrub less than 1 m tall, widespread
   2. Peduncles axillary, bractless or nearly so (sometimes bracts if inflorescence is branched), often equaling or surpassing the subtending leaves; principal leaf blades (1.8) 2.5–5 (6) cm broad . . . . 2. **C. americanus**
   2. Peduncles terminal on leafy branches, shorter than the leaves; principal leaf blades 0.8–2.2 (2.5 or very rarely 3) cm broad . . . . 3. **C. herbaceus**

1. **C. sanguineus** Pursh — Wild-lilac; Redstem Ceanothus

Map 778. Rocky bluffs and borders of woods. An attractive shrub blooming in late May and June in northern Keweenaw County, where it has been known for a century. Its principal range is in the Pacific Northwest, though it has also been reported [incorrectly?] from the Black Hills of South Dakota—as are some other western disjuncts (see Mich. Bot. 20: 62. 1981). Some western species of *Ceanothus* with blue flowers are more appropriately called "wild-lilac" than any of our white-flowered ones, although the true lilacs (*Syringa*) are in the olive family (Oleaceae).

2. **C. americanus** L. — New Jersey Tea

Map 779. Dry open sandy plains and prairie-like areas (spreading to roadsides), dry woodlands (with oak, aspen, pines); often in openings or transitional areas between lakes or marshes and woodland; river banks.

775. Impatiens glandulifera

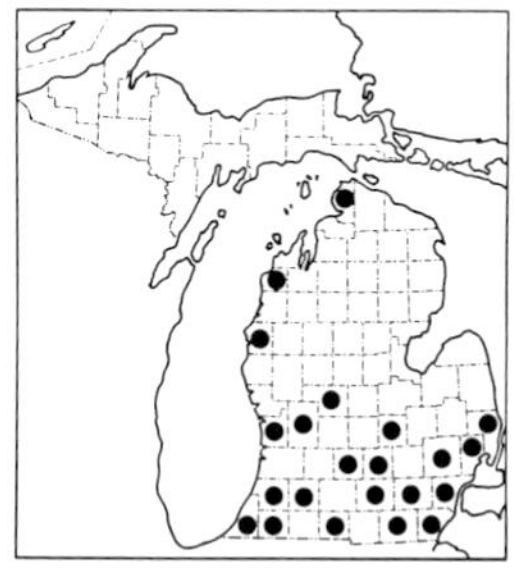

776. Impatiens pallida

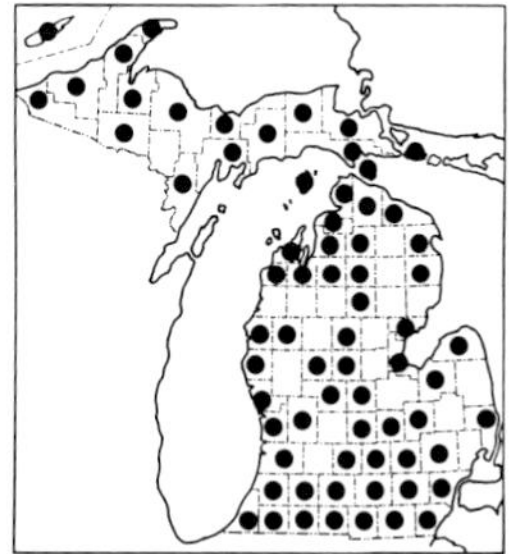

777. Impatiens capensis

This species blooms much later than the next, which will be in young fruit when *C. americanus* is flowering in any given area and season.

3. **C. herbaceus** Raf. Fig. 280 New Jersey Tea

Map 780. Jack pine plains, open rocky woods and outcrops, river and roadside banks, oak and other woodlands.

Often known as *C. ovatus* Desf., which is not only a later name but also applies to a different species and hence is doubly incorrect. It is also a misleading name, as the leaves of the preceding species are more ovate than the elliptical ones of *C. herbaceus*—and both species are woody, not herbaceous. The leaves vary from glabrous to densely pubescent beneath.

2. Rhamnus Buckthorn

The classical gender of the generic name was feminine and hence it is so used here, although Linnaeus treated it as masculine. The fruit is a small drupe with 2–4 stones or pits. The bark and fruit of some—perhaps all—species have laxative properties, and certain species have long been used medicinally. The fruit is popular with birds, and some species have become serious weeds thanks to their copious and easily dispersed fruits and ease of sprouting from stumps and roots. All of our species except one are escapes from cultivation.

KEY TO THE SPECIES

1. Flowers all perfect; sepals 5; petals 5; leaves entire (sometimes slightly crisped but at most very obscurely and irregularly crenulate) .... 1. **R. frangula**
1. Flowers unisexual or perfect (plants dioecious or polygamo-dioecious); sepals and petals 4, or sepals 5 and petals 0; leaves with close, small, often crenulate teeth
  2. Petals none; sepals and stamens 5; pits of drupe 3; leaves definitely alternate; plant a low native spineless shrub (less than 1 m tall) in wet habitats .... 2. **R. alnifolia**
  2. Petals, sepals, and stamens 4; pits of drupe 2 or 4; leaves (and branchlets) often appearing opposite or subopposite; plant introduced, tall, weedy, usually with spine-tipped branchlets
    3. Leaf blades twice as long as broad or shorter, less than 6 cm long, usually with 3 (4) pairs of lateral veins, dull above when fresh; drupes usually with 4 pits .... 3. **R. cathartica**
    3. Leaf blades distinctly more than twice as long as broad, mostly 8–12 cm long, usually with 5–6 pairs of lateral veins, shiny above when fresh; drupes with 2 pits .... 4. **R. utilis**

1. **R. frangula** L. Glossy Buckthorn

Map 781. A Eurasian species, locally aggressive and becoming a serious pest as a tall shrub in bogs, especially alkaline ones (fens), and other damp places, including tamarack and cedar swamps (particularly in disturbed areas as along new power lines and other clearings), thickets along rivers, lake shores, ditches, fencerows, and low woods. First collected in Michigan in

Delta County in 1934 (*Grassl 6781*, MICH). Doubtless much more widespread than the map would suggest.

Sometimes segregated from *Rhamnus*, as *Frangula alnus* Miller. The leaves are shinier than in our other species and the lateral veins somewhat more straight, though they do curve upward near the margin. The drupe has 2–3 pits.

2. **R. alnifolia** L'Hér. Fig. 284 Alder-leaved Buckthorn

Map 782. Tamarack and cedar (also sometimes spruce) bogs and swamps, boggy streamside thickets, interdunal swales, sedge marshes and mats; generally a calciphile and a good indicator of alkaline fens; low deciduous woods, rocky openings and outcrops.

The leaf blades are of similar proportions to those of the next species, but are more acute, may average a little larger, and have more pairs of lateral veins. These are strongly ascending and curve along the margins much as in leaves of *Cornus* (which, however, are strictly entire).

3. **R. cathartica** L. Common Buckthorn

Map 783. A Eurasian species, sometimes cutivated, and locally an obnoxious weed, forming thickets in vacant (or occupied!) lots; at borders (even interiors) of woods; along roadsides, fencerows, railroads, river banks, and

778. Ceanothus sanguineus

779. Ceanothus americanus

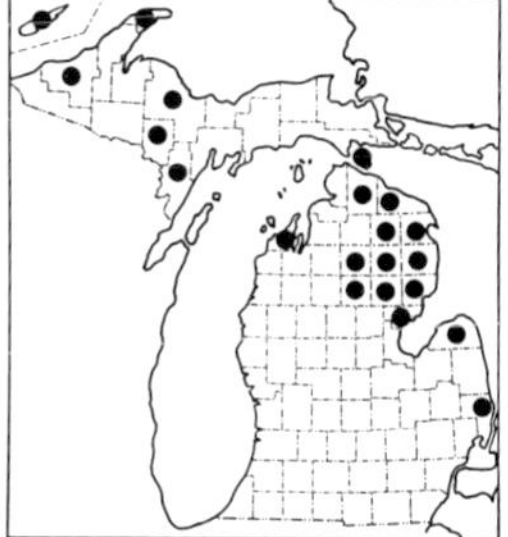

780. Ceanothus herbaceus

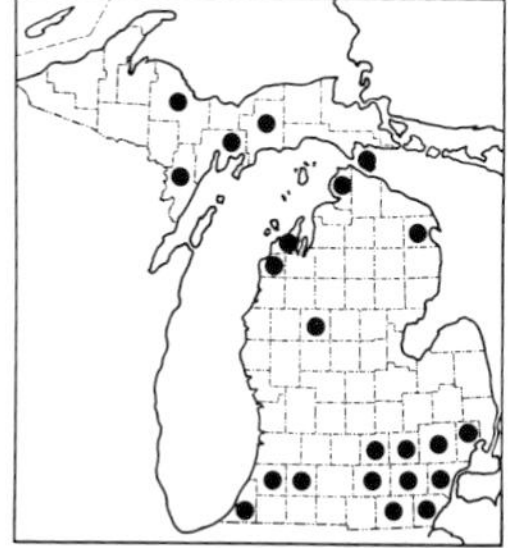

781. Rhamnus frangula

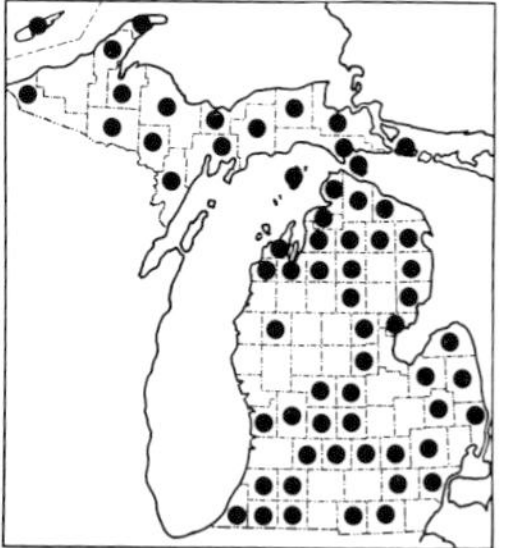

782. Rhamnus alnifolia

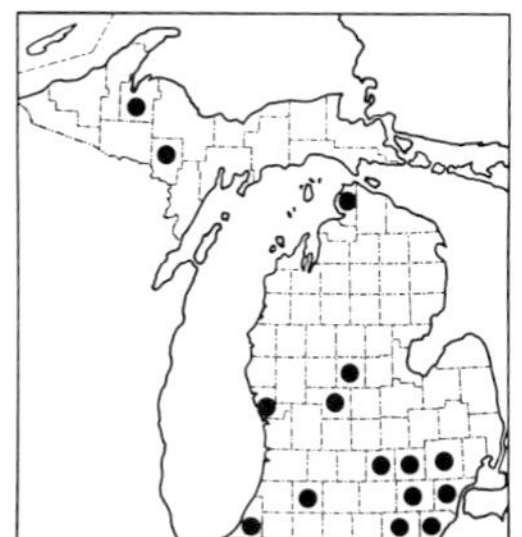

783. Rhamnus cathartica

clearings; occasionally in swampy sites. First collected in Michigan at Ann Arbor in 1914, but in the next two years recorded as plentiful and naturalized around Birmingham and Bloomfield Hills.

4. **R. utilis** Decne.

Map 784. Certain plants growing in old fields and thickets in Ann Arbor (e. g., *Reznicek 6165* in 1980, MICH, MSC; *7461* in 1984, MICH) are apparently this species, a native of China, although they are similar to forms of the closely related *R. davurica* Pallas (also Asian). The fruit is still green when that of neighboring *R. cathartica* is fully ripe. Plants with intermediate leaf shape may well represent hybrids.

VITACEAE — Grape Family

REFERENCE

Brizicky, George K. 1965. The Genera of Vitaceae in the Southeastern United States. Jour. Arnold Arb. 46: 48–67.

KEY TO THE GENERA

1. Leaves palmately compound; bark tight; petals distinct, falling separately .... 1. **Parthenocissus**
1. Leaves simple; bark shredding and peeling; petals coherent distally, falling as a unit .... 2. **Vitis**

1. Parthenocissus — Virginia Creeper; Woodbine

"Woodbine" is one of the dozen plants commemorated in the name of a Coast Guard cutter; like the "Sundew" (see *Drosera*), it operated for many years on the Great Lakes and was familiar in Michigan waters. The "Woodbine" was decommissioned in 1972.

Boston-ivy, *P. tricuspidata* (Sieb. & Zucc.) Planchon, a native of Asia,

784. Rhamnus utilis

785. Parthenocissus quinquefolia

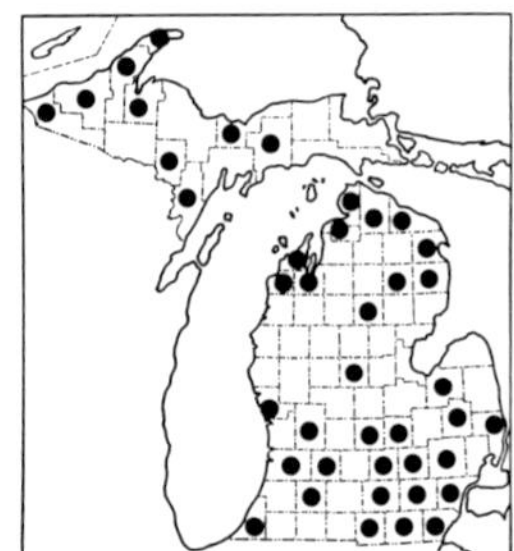

786. Parthenocissus inserta

is very commonly cultivated but does not seem to escape in our area. It has simple, 3-lobed, maple-like leaves except on basal shoots, where they are trifoliolate. Our two native species are not always easy to tell apart, especially from herbarium specimens which lack well developed tendrils, mature inflorescences, or notes on the label as to appearance of the fresh leaflets and tendrils. A good intuition, tempered by experience, is helpful in separating specimens. Both species are cultivated, *P. quinquefolia* preferred for climbing walls (as well as trees). Unlike *Vitis,* the fruits are not recommended for eating, although a reputation for being poisonous may be exaggerated.

REFERENCE

Webb, D. A. 1967. What is Parthenocissus quinquefolia (L.) Planchon? Feddes Repert. 74: 6–10. [Considers both species, including nomenclature.]

KEY TO THE SPECIES

1. Leaflets dull above when fresh, pale green and whitened or slightly glaucous beneath; tendrils with each branch forming an adhesive disc at the end if it comes in contact with a support; inflorescence with a definite central axis (i. e., branches forking unequally) . . . . 1. **P. quinquefolia**
1. Leaflets ± shiny above when fresh, at most slightly paler beneath but not whitened; tendrils not developing discs (though sometimes club-shaped at the ends when in a crevice); inflorescence dichotomous (or trichotomous) throughout (branches of each set ± equal in thickness) . . . . 2. **P. inserta**

1. **P. quinquefolia** (L.) Planchon Fig. 281

Map 785. Thickets, swamp forests, and upland deciduous forests (climbing not only rough-barked trees such as elm but also smooth-barked *Fagus,* by means of the adhesive discs on the tendrils).

The undersides of the leaflets, at least on the main veins, and the petioles are usually ± puberulent in our specimens of this species, although rarely they are completely glabrous. The fruit is not reported to have more than 3 seeds.

2. **P. inserta** (A. Kerner) Fritsch Fig. 282

Map 786. Rock outcrops, talus slopes, and rocky woodland; swamp forests and upland woods; fencerows, stream banks, and thickets.

In our specimens of this species, the undersides of the leaves and especially the petioles are usually completely glabrous, although occasionally puberulent. The berries average a little larger than in the preceding species, and are often 4-seeded. In some works, this is called *P. vitacea* (Knerr) Hitchc.

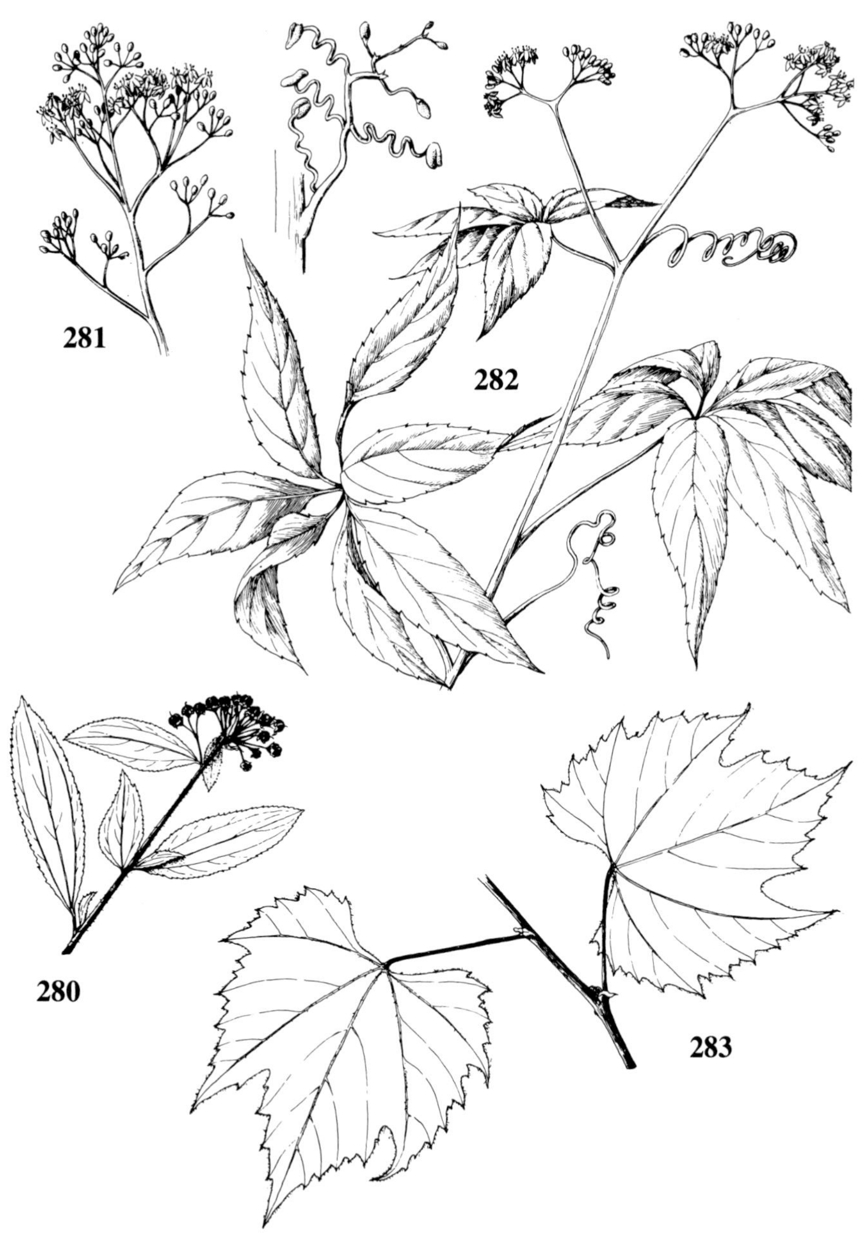

280. *Ceanothus herbaceus* $\times^1/_2$
281. *Parthenocissus quinquefolia,* inflorescence & tendrils $\times^1/_2$
282. *P. inserta* $\times^1/_2$
283. *Vitis riparia* $\times^1/_2$

2. **Vitis** Grape

A genus of vines very important historically and commercially, for grapes, raisins, and wine. Michigan is the fourth largest wine-producing state in the country. Grapes grown in eastern United States, including Michigan, such as 'Concord' and other cultivars, are mostly *V. labrusca* or hybrids of that species with others. *V. vinifera* L., the wine grape of the Old World, thrives much better in the Mediterranean climate of California. This species also produces the "table grapes," with the skin adherent to the flesh, not easily peeled or slipped off. Other species have contributed to the cultivated grapes, but in our wild ones the extent of hybridization is not known. Whatever the reasons for variability, the species are often difficult to distinguish. Flower and fruit characters are even less useful than vegetative ones.

REFERENCES

Bailey, L. H. 1934. The Species of Grapes Peculiar to North America. Gentes Herb. 3: 149–244.

Brown, David. 1984. Our Wine Industry Coming of Age. Mich. Natural Resources 53(5): 22–29.

Duncan, Wilbur H. 1967. Woody Vines of the Southeastern States. Sida 3: 1–76. [*Vitis*, pp. 22–32.]

KEY TO THE SPECIES

1. Underside of fully expanded leaf blades densely and evenly covered with a persistently adherent rust-colored tomentum concealing the entire surface between the veins; stems regularly with tendrils (or peduncles) at 3 or more consecutive nodes; ripe fruit at least 15 mm in diameter . . . . 1. **V. labrusca**
1. Underside of fully expanded leaf blades glabrous or with thin or very uneven (tufted) pubescence exposing much of the surface between the veins; stems with tendrils (or peduncles) at no more than 2 consecutive nodes; ripe fruit less than 10 (15) mm in diameter
 2. Mature leaf blades distinctly whitened or glaucous beneath, with scattered or tufted pale to rust-colored cobwebby hairs at least along the principal veins; leaf margins with shallow, obtuse teeth usually not ciliolate, the sinuses between lobes (if any) broadly angled or rounded . . . . 2. **V. aestivalis**
 2. Mature leaf blades green or yellowish green beneath, not glaucous, with sparse or no cobwebby hairs; leaf margins with teeth various but acute and ciliolate in common species, without lobes or (usually) with acute sinuses between lobes
  3. Lobes of leaf well developed, at least on later leaves of the season, pointed forwards; margins with acute teeth, ciliolate; ripe fruit glaucous; plant common, throughout the state . . . . 3. **V. riparia**
  3. Lobes of leaf none or weakly developed and pointing outwards; margins with obtuse teeth, usually not ciliolate; ripe fruit purple-black, not glaucous; plant rare, in southernmost Michigan . . . . 4. **V. vulpina**

1. **V. labrusca** L. Fox Grape

Map 787. Thickets, woods, fields, meadows, fencerows, sandy hills, railroad embankments. Barely ranges north into southern Michigan; some collections from the state surely represent escapes from cultivation.

2. **V. aestivalis** Michaux Fig. 285 Summer Grape

Map 788. Thickets, woods (usually rather dry), fencerows, sandy hillsides and dunes.

Included here is var. *argentifolia* (Munson) Fern., with less pubescence and a more glaucous underside to the leaf. Some herbarium specimens are certainly this species on characters of pubescence and leaf shape, including very low broad teeth, but are green beneath, perhaps the result of loss of glaucous bloom in drying. The overall shape of leaves of this species when deeply lobed (which is not often) is suggestive of *Humulus lupulus*. Most leaves are very shallowly lobed and shallowly toothed, compared to the jagged-toothed and often deeply lobed *V. riparia,* our commonest species.

3. **V. riparia** Michaux Fig. 283 River-bank Grape

Map 789. Lowland to upland woods, especially along borders; thickets, fencerows, river banks; shores and dunes.

Widespread and variable, some forms (especially on dunes) being more pubescent (e. g., on petioles) than others; but usually recognizable by the ciliolate sharp teeth and forward-pointing lobes on the leaf blades, which are longer than broad. The fruit is glaucous when ripe and sour until frost, but makes a good jelly. This species has sometimes been called *V. vulpina,* a name here applied to the next one.

Bushy plants of dry habitat, with tendrils few or none and short (less than 10 cm) leaf blades mostly broader than long, resemble *V. rupestris* Scheele, but our specimens are probably only depauperate forms of *V. riparia* (which, indeed, was noted nearby by collectors of "*V. rupestris*" in Livingston and Oakland counties).

4. **V. vulpina** L. Frost Grape

Map 790. Probably to be expected in Michigan in the same habitats as the preceding, but too scarce for much variety to be offered by existing

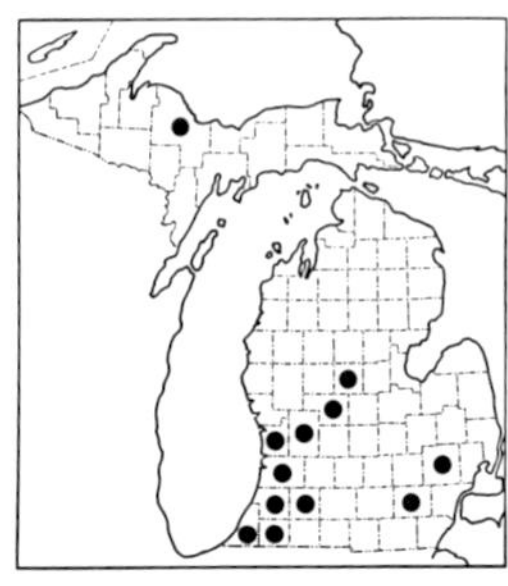

787. Vitis labrusca

788. Vitis aestivalis

789. Vitis riparia

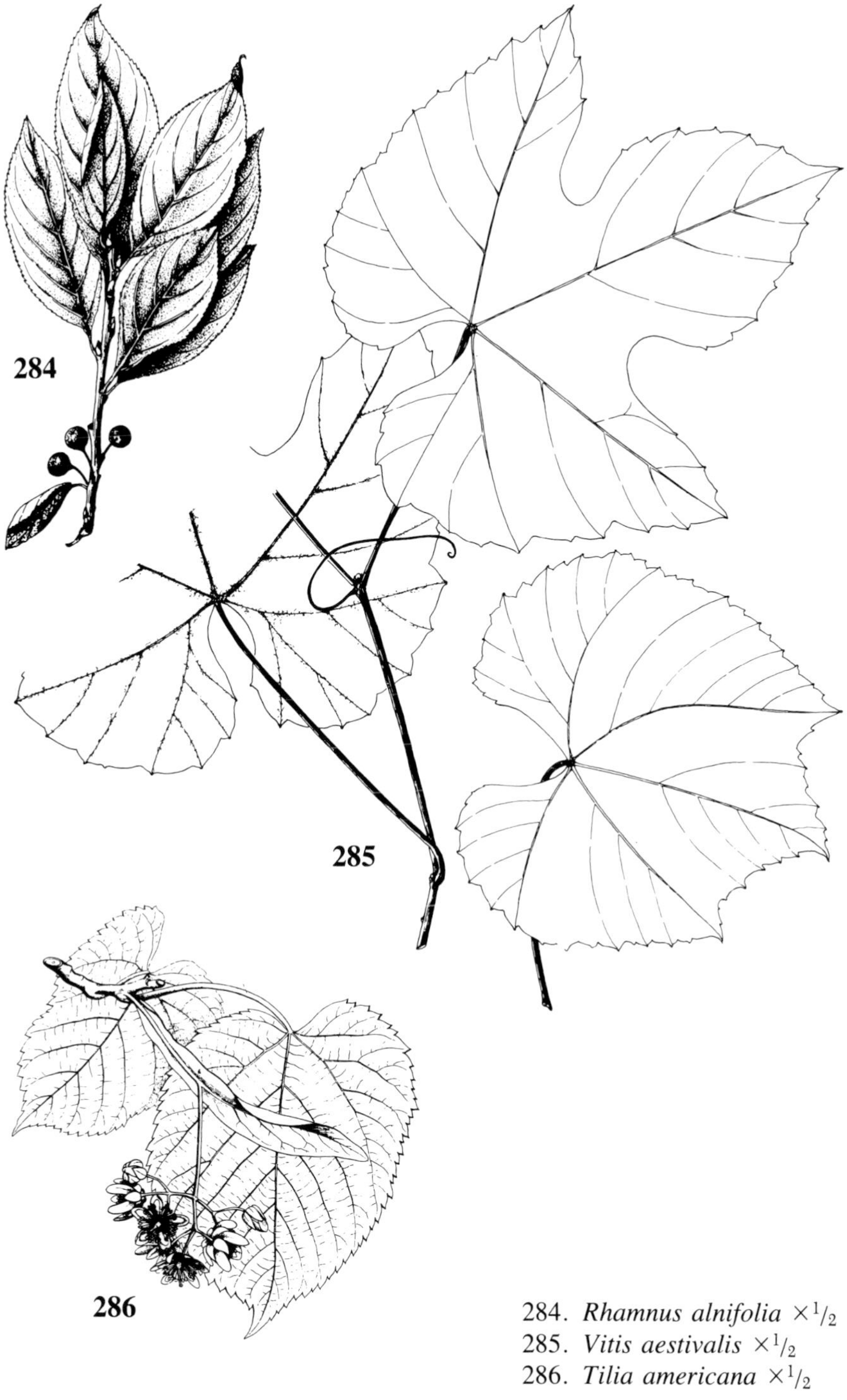

284. *Rhamnus alnifolia* $\times ^1/_2$
285. *Vitis aestivalis* $\times ^1/_2$
286. *Tilia americana* $\times ^1/_2$

specimens. Two of our three collections are from rich mesic forests and the third lacks habitat data.

This name is applied by some authors to the preceding species, in which case this one becomes *V. cordifolia* Lam.—an apt name, as the overall outline of the leaf blade is decidedly heart-shaped.

TILIACEAE — Linden Family

1. Tilia

An easily recognized genus of trees, the inflorescence with peduncle adnate about half its length to a subtending tongue-shaped bract which soon becomes twisted with the weight of the inflorescence so that the latter hangs beneath the bract. The cordate leaves in most species, including our only native one, are asymmetrical (oblique) at the base, one basal lobe being larger than the other. Several European species and hybrids, generally with smaller leaves than in *T. americana,* are planted as shade trees. *Tilia* is one of the few forest trees in our latitude to have flowers visited heavily by bees, and basswood honey is considered excellent. The wood has long been used for crates, boxes, pulp, and carving.

In England, *Tilia* is known as "lime" and in Sweden, as "lind" or "lin." In respect for a large tree on the family property, a certain 17th century Swede, son of a prosperous farmer and destined to become a Lutheran clergyman, took the name Nils Linnaeus when enrollment in the university obliged him to adopt a surname. In 1707 his first son was born, Carl Linnaeus—who, half a century later, had classified and cataloged the entire plant and animal kingdoms so far as known, and established the binomial system for naming of species. (Later he was ennobled and from 1762 had the title of "von Linné"—though he often did not use it.)

Tilia also provided the names for several communities in Michigan, including Linden, Lake Linden, and Basswood. To the French explorers, it was known as *bois blanc* (white wood)—a name given to a large island near Mackinac in northern Lake Huron and to a small island on the Ontario side

790. Vitis vulpina

791. Tilia americana

792. Abutilon theophrasti

of the Detroit River, both corrupted to "Boblo" by the barbarism that so often deals with French names. The common name "basswood" refers to the strong bast (or bass) fibers of the inner bark, extensively used by Indians for thread, ropes, mats, nets, snowshoes—and long used by invading European peoples for similar purposes.

There is some question whether *Tilia* in the United States consists of more than one or two species, but at least in this state there is only one. The amount of pubescence—if any—on the underside of the leaves is quite variable.

REFERENCES

Brizicky, George K. 1965. The Genera of Tiliaceae and Elaeocarpaceae in the Southeastern United States. Jour. Arnold Arb. 46: 286–307.

Jones, George Neville. 1968. Taxonomy of American Species of Linden (Tilia). Illinois Biol. Monogr. 39. 156 pp.

1. **T. americana** L. Fig. 286 Basswood; Linden

Map 791. Characteristic of rich upland deciduous forests, usually with sugar maple, beech, hemlock, and other trees; sometimes in swamps (even with cedar).

This species sprouts readily after fire or cutting, and clumps of large trunks are common, with their neatly furrowed bark (smoother and gray when young). The larger leaf blades are ordinarily ca. 8–18 cm long, although some may be even longer, especially on sprouts.

MALVACEAE Mallow Family

This is an easily recognized family, with palmately veined leaves and flowers with stamens united (monadelphous) into a column around the carpels. The flowers of *Hibiscus,* of which a number of species are familiar ornamentals in warm climates or greenhouses, are typical. There are other Malvaceae grown for their flowers, and cotton (*Gossypium* spp.) is useful both for the hairs on its seeds and for the oil in them. The mucilaginous juice often found in the family has led to the use of several species for assorted culinary purposes.

Only *Hibiscus* and the similar *Abelmoschus,* in our flora, have the fruit a many-seeded capsule. In our other genera, the carpels are in a single ring around a central axis, each carpel 1-seeded (except in *Abutilon*) and eventually separating as a mericarp.

REFERENCE

Utech, Fred H. 1970. Preliminary Reports on the Flora of Wisconsin No. 60 Tiliaceae and Malvaceae—Basswood and Mallow Families. Trans. Wisconsin Acad. 58: 301–323.

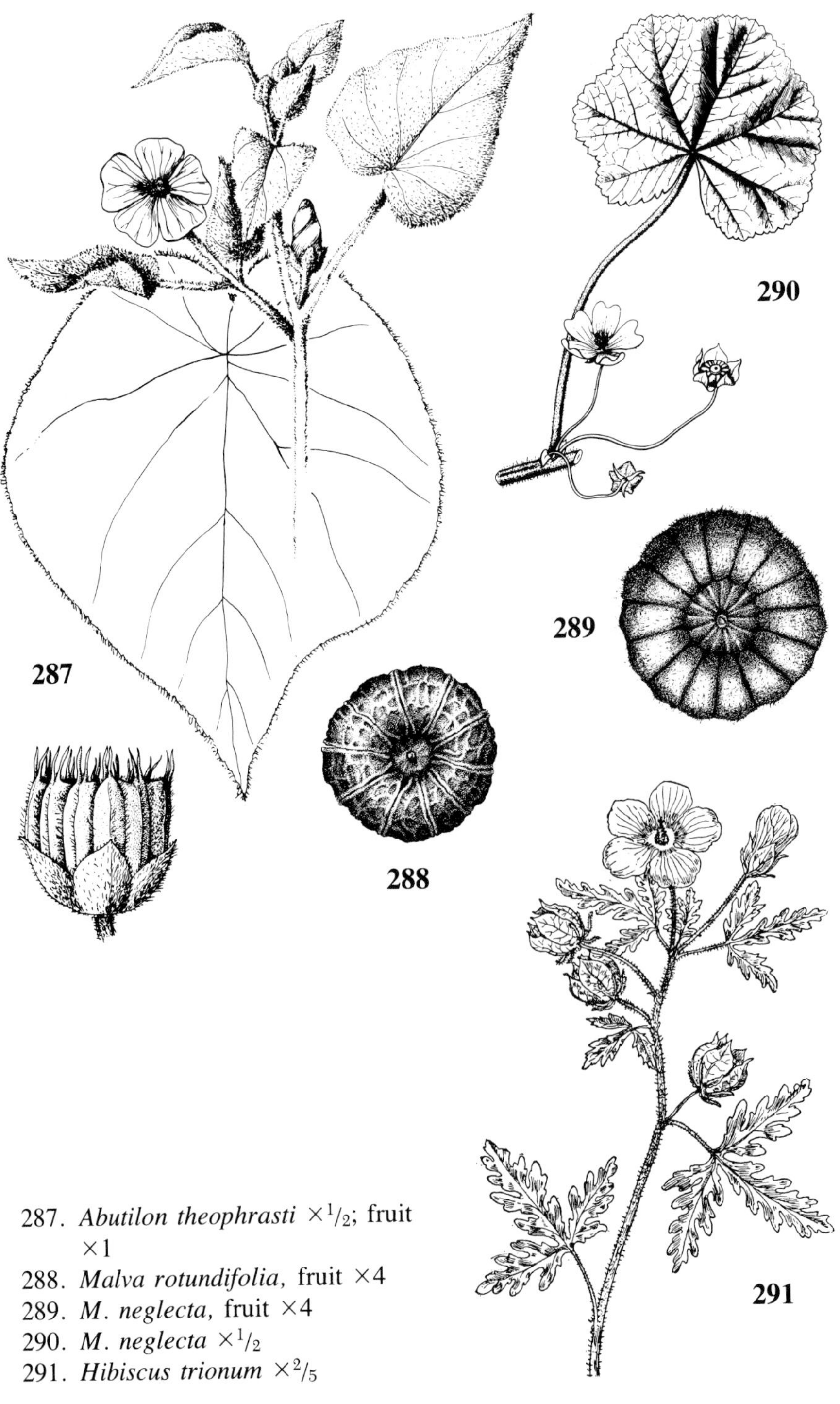

287. *Abutilon theophrasti* ×1/2; fruit ×1
288. *Malva rotundifolia,* fruit ×4
289. *M. neglecta,* fruit ×4
290. *M. neglecta* ×1/2
291. *Hibiscus trionum* ×2/5

KEY TO THE GENERA

1. Calyx without bractlets at its base
 2. Calyx with some unbranched hairs (especially on margins) over 0.5 mm long; leaf blades cordate, acuminate, densely velvety-pubescent beneath; carpels more than 10, with beaks becoming ca. 4–5 mm long in fruit .... 1. **Abutilon**
 2. Calyx finely (mostly stellate-) pubescent, all hairs shorter than 0.5 mm; leaf blades not cordate (ovate-elliptic, oblong, or deeply lobed), lightly velvety to glabrous beneath; carpels 5 or 10, with beaks ca. 1 mm long or less.................. 2. **Sida**
1. Calyx with bractlets at its base additional to the 5 sepals (figs. 290, 291)
 3. Bractlets subtending the calyx 3 (or fewer)
  4. Blades of all leaves shallowly (if at all) divided or lobed (less than halfway to their base) .... 3. **Malva** (couplet 3)
  4. Blades of at least middle and upper cauline leaves divided below the middle into coarsely toothed or pinnatifid lobes
   5. Plant erect, with linear-lanceolate stipules; petals obcordate (notched at tip, sometimes with additional irregular teeth); most if not all pedicels shorter than petals at anthesis.. 3. **Malva** (couplet 2)
   5. Plant prostrate or trailing, with broadly ovate stipules; petals truncate or rounded (not notched, but rather finely erose at tip); most if not all pedicels longer than petals at anthesis... 4. **Callirhoë**
 3. Bractlets subtending the calyx at least 6
  6. Carpels (and styles) numerous (more than 15), forming in fruit a ring of 1-seeded indehiscent mericarps which separate from a central axis; bractlets subtending calyx triangular
   7. Stem (especially above) covered with dense velvety gray pubescence; petals less than 2.5 [3] cm long; carpels ca. 20 or fewer .... 5. **Althaea**
   7. Stem with ± scattered, mostly setose hairs, the surface fully visible; petals over 3 cm long; carpels more than 20.................................... 6. **Alcea**
  6. Carpels (and styles) 5, forming in fruit a single capsule; bractlets subtending calyx narrowly linear
   8. Capsule more than twice as long as broad; calyx tube (splitting on one side) at least 3 times as long as the lobes; corolla yellow .... 7. **Abelmoschus**
   8. Capsule scarcely if at all longer than broad; calyx tube at most about equaling the lobes (or slightly longer if inflating in fruit); corolla, at least at base, purple or pink (or white) .... 8. **Hibiscus**

1. Abutilon

REFERENCE

Spencer, Neal R. 1984. Velvetleaf, Abutilon theophrasti (Malvaceae), History and Economic Impact in the United States. Econ. Bot. 38: 407–416.

1. **A. theophrasti** Medicus Fig. 287 Velvet-leaf

Map 792. Fields, roadsides, fencerows, gardens, vacant lots, and other disturbed or waste ground. A native of China, introduced into the New World before 1750 as a potential fiber crop but now a serious weed of row crops such as corn and soybeans. Frequent as a weed in southern Michigan, scarce northwards.

The petals are ca. 7–10 mm long, usually yellow-orange.

2. **Sida**

KEY TO THE SPECIES

1. Calyx divided less than halfway, with a terete base; petals white; leaf blades deeply palmately lobed; carpels (8–) 10 . . . . 1. **S. hermaphrodita**
1. Calyx divided more than halfway, with an angled base; petals yellow; leaf blades ovate-elliptic to oblong, not lobed; carpels ca. 5 . . . . 2. **S. spinosa**

1. **S. hermaphrodita** (L.) Rusby

Map 793. Native south and east of our area, but collected by Farwell in 1924 and 1931 in Wayne County, where apparently a well established adventive. (The 1924 site was "along the railway right of way near Wayne" and the 1931 site was "in low moist grounds" near Sheldon; it is conceivable that these represent the same introduction.) The status of an 1887 collection from Lansing is uncertain. Also reported in the 19th century from Kalamazoo County, but no specimens have been found.

2. **S. spinosa** L.

Map 794. A species of much more southern range, adventive northward and first collected in Michigan by Farwell in waste ground at Detroit in 1906 (*2005,* BLH); collected by him at Taylor in 1930 (*8835,* BLH). The only other Michigan record is from Britton in 1920 (*E. M. Shumac,* MSC).

3. **Malva** Mallow

KEY TO THE SPECIES

1. Blades of middle and upper (if not lower) cauline leaves divided below the middle into coarsely toothed to pinnatifid lobes
   2. Pedicels and outside of calyx with many short stellate hairs sometimes overtopped by longer simple hairs; bractlets subtending calyx at most 3 times as long as broad . . . . 1. **M. alcea**

793. Sida hermaphrodita

794. Sida spinosa

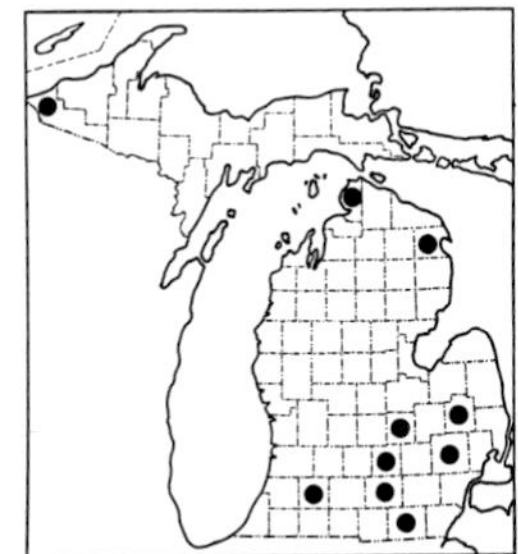
795. Malva alcea

2. Pedicels and outside of calyx with only simple hairs (rarely a few stellate ones); bractlets subtending calyx usually 3–5 times as long as broad . . . . 2. **M. moschata**

1. Blades of all leaves shallowly (if at all) divided or lobed (less than halfway to their base)

3. Petals ca. 15–24 mm long, deep pink to purple; bractlets subtending calyx oblong-ovate, less than 3 times as long as broad . . . . 3. **M. sylvestris**

3. Petals less than 15 mm long, white (sometimes with pinkish veins, tips, or shading); bractlets subtending calyx more than 3 times as long as broad

4. Leaf blades crisped or puckered along the margin; most flowers sessile or nearly so; stems erect . . . . 4. **M. crispa**

4. Leaf blades flat; most flowers at anthesis on pedicels longer than the calyx; stems ± prostrate or ascending

5. Petals barely if at all exceeding the sepals; carpels 9–11, at maturity strongly reticulate and glabrate on the outer surface (fig. 288) . . . . 5. **M. rotundifolia**

5. Petals at least 1.5–2 times as long as the sepals; carpels 12–14 (15), at maturity smooth (or very obscurely reticulate) and short-pubescent on the outer surface (fig. 289) . . . . 6. **M. neglecta**

1. **M. alcea** L. — Vervain Mallow

Map 795. Roadsides, railroads, fields, gardens, and waste ground. The earliest Michigan collection seen is from 1860 (Shiawassee County). A European species, locally adventive or escaped from cultivation. Especially well established in the vicinity of Middle Village, Emmet County.

The upper leaves are less narrowly incised than in the next species, and the lower leaves are only shallowly lobed. Basal leaves in both species tend to be merely cordate. *M. alcea* is generally characterized by having some stellate hairs on all vegetative parts; in *M. moschata,* there are rarely a few stellate hairs on the outside of the calyx. The mericarps in both species may bear straight white hairs at maturity.

2. **M. moschata** L. — Musk Mallow

Map 796. Another European species, established as a weed or garden escape. Locally common and colorful along roadsides and railroads, in fields and clearings, on dry or rocky ground. Apparently well established at a number of places throughout the state by the 1880's.

Both this species and the preceding are erect perennials with handsome pink or white flowers.

3. **M. sylvestris** L. — High Mallow

Map 797. Roadsides, fields, and waste ground. Michigan collections are mostly from the 19th century, beginning in 1860; none have been seen since 1936. A native of Europe, cultivated as an ornamental and (in our area) rarely escaping.

4. **M. crispa** (L.) L. — Curled Mallow

Map 798. A Eurasian species, sometimes cultivated as a salad plant and rarely found as a waif in waste ground. Collected by Farwell at Detroit (*1440* in 1893, BLH; see Asa Gray Bull. [2] (7): 48. 1894).

Often treated as *M. verticillata* var. *crispa* L. The petals at anthesis are distinctly longer than the sepals, but the latter enlarge considerably and are persistent around the nearly concealed fruit, while in the next two species the sepals scarcely enlarge and the fruit is ± exposed. *M. crispa* tends to be a more erect plant than the weedy *M. neglecta,* which also has relatively long petals.

5. **M. rotundifolia** L. Fig. 288

Map 799. Roadsides, filled land, railroads, and gardens—much less common than the next species. Evidently first collected in Michigan in 1901 in a barnyard in Marquette County (*B. Barlow,* MSC, GH). A weed of European origin.

This name was long applied to the next species, and is rejected by some as ambiguous, in which case *M. pusilla* J. E. Smith applies. *M. borealis* Wallman is a still later name for the same species.

6. **M. neglecta** Wallr. Figs. 289, 290 Common Mallow; Cheeses

Map 800. Another Eurasian weed, much more common here than the preceding. Roadsides, railroads, filled land, dumps, gravel pits, lawns, gar-

796. Malva moschata

797. Malva sylvestris

798. Malva crispa

799. Malva rotundifolia

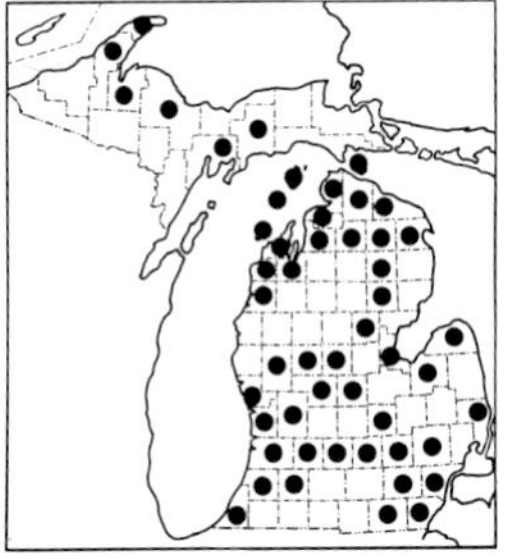
800. Malva neglecta

801. Callirhoë involucrata

dens, and waste places generally. The earliest collections seen from the state date from the 1860's, but this was listed (as *M. rotundifolia*) by the First Survey.

Published reports of this species from Michigan, especially before 1940, usually called it *M. rotundifolia*.

4. **Callirhoë**

1. **C. involucrata** (T. & G.) A. Gray — Poppy Mallow

Map 801. A showy-flowered native west of our area, occasionally escaped from cultivation eastward. Collected from a dry sandy roadside at Wayland in 1924 (*Fallass 3260,* ALBC).

5. **Althaea**

1. **A. officinalis** L. — Marsh Mallow

Map 802. A Eurasian species, rarely escaped from cultivation to low ground and waste places.

This plant has long been used medicinally, and its high mucilaginous content, especially in the roots, provided the original "marshmallow" before the days of synthetic substitutes.

6. **Alcea**

1. **A. rosea** L. — Hollyhock

Map 803. Roadsides, dumps, railroads, vacant lots; first collected in Michigan in the 1860's. A familiar cultivated ornamental of uncertain origin, perhaps from the Balkans (not from China, as sometimes stated). Doubtless in more waste places than the few collections mapped would indicate.

Alcea has often been included in *Althaea*. The technical distinctions include a 5-angled (rather than terete) staminal column in *Alcea,* as well as

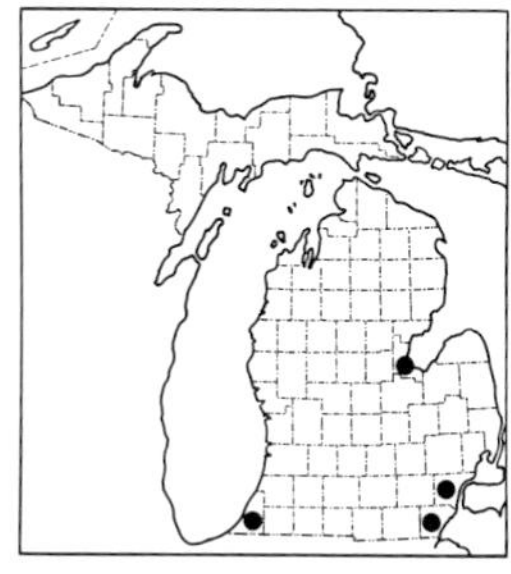

802. Althaea officinalis

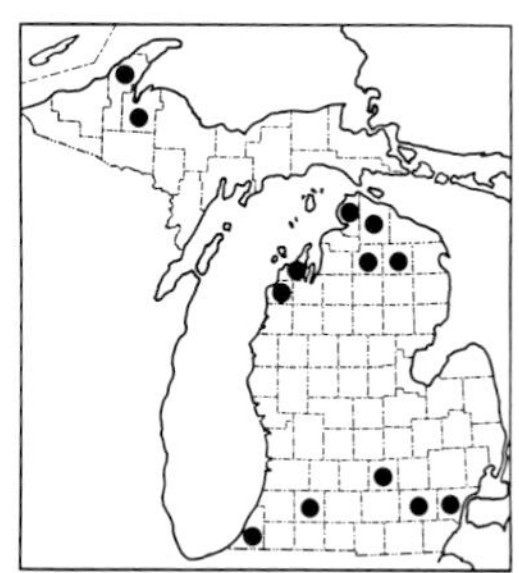

803. Alcea rosea

804. Abelmoschus esculentus

keeled mericarps containing 2 locules (although one of them is empty). The flowers are sometimes "double," and vary from white to pink or purplish.

7. **Abelmoschus**

1. **A. esculentus** (L.) Moench — Okra; Gumbo

Map 804. Collected once in Michigan as spontaneous on a dump site west of Buchanan (*Schulenberg* in 1970, MOR). A native of tropical Africa, widely cultivated in warm climates for its young mucilaginous fruits, which are used as a vegetable.

This genus has often been included in *Hibiscus,* this species then called *H. esculentus* L.

8. **Hibiscus** — Rose Mallow; Hibiscus

A popular genus of cultivated plants. We have one (or two) attractive native species and one introduced weed. *H. syriaca* L., rose-of-Sharon or "Althaea," is a commonly cultivated shrub as far north as Michigan.

KEY TO THE SPECIES

1. Leaf blades (at least the upper) deeply lobed to the base, glabrate or with scattered hairs beneath; calyx and bractlets with mostly long setose hairs (especially on nerves and margins); petals ca. 1.7–2.6 cm long, yellow with purple base; capsule hairy; plant a low branched annual weed of waste places .... 1. **H. trionum**
1. Leaf blades more shallowly if at all lobed, in common species velvety-pubescent beneath; calyx and bractlets minutely but densely stellate-pubescent, with no setose hairs; petals ca. 4.5–8 (9) cm long, pink [or white]; capsule glabrous; plant an erect perennial of damp ground
   2. Leaves velvety-pubescent on blades beneath .... 2. **H. moscheutos**
   2. Leaves completely glabrous on blades and petioles .... 3. **H. laevis**

1. **H. trionum** L. Fig. 291 — Flower-of-an-hour

Map 805. An Old World species, the original native range uncertain but apparently in southeastern Europe; now widely naturalized in mostly warm climates. Locally a weed of fields, gardens, roadsides, and waste places. Reported by the First Survey (1838).

The flowers are attractive but short-lived, lasting only a few hours.

2. **H. moscheutos** L. Plate 6-D — Rose or Swamp Mallow

Map 806. Marshes, river bottoms, and often adjacent disturbed ground. The species is more common in salt marshes of the Atlantic coast, but is locally frequent in southern Michigan and other inland sites. Some records, especially northward, may represent escapes from cultivation.

This species as now understood includes *H. palustris* L.

3. **H. laevis** All.

Map 807. Usually said to range south of Michigan, but a sterile plant was collected in 1910 along Lake Erie (*Farwell 2188*, BLH). The species has evidently been spreading northward (e. g., to the Chicago region and the islands of western Lake Erie) and is likely to be found again in Michigan, probably in the Erie marshes.

The leaf blades are often broadly hastate. Long known as *H. militaris* Cav.

GUTTIFERAE (CLUSIACEAE) — St. John's-wort Family

Opinion differs whether the Hypericaceae should be treated as a separate family from the Guttiferae (Clusiaceae), but most recent authors on these plants include them all in one family. (The Guttiferae sens. str. have been called the Mangosteen Family, and do not grow in our area.) The family is easily recognized by the opposite, entire, punctate leaves (translucent dots ± visible when held to the light, resulting from oil-filled cavities); the inflorescence is cymose, with the terminal flower in each cluster developing first.

REFERENCES

Gillett, John M., & Norman K. B. Robson. 1981. The St. John's-worts of Canada (Guttiferae). Natl. Mus. Canada Publ. Bot. 11. 40 pp.

Utech, Fred H., & Hugh H. Iltis. 1970. Preliminary Reports on the Flora of Wisconsin No. 61. Hypericaceae—St. John's-wort Family. Trans. Wisconsin Acad. 58: 325–351.

Wood, Carroll E., Jr., & Preston Adams. 1976. The Genera of Guttiferae (Clusiaceae) in the Southeastern United States. Jour. Arnold Arb. 57: 74–90.

KEY TO THE GENERA

1. Petals pink to purple, the flowers rarely open; stamens 9, in 3 distinct groups of 3 each, the groups of basally connate stamens alternating with conspicuous glands; capsules (young as well as mature) prominently striate longitudinally . . . . 1. **Triadenum**
1. Petals yellow (occasionally flushed with pinkish), the flowers ordinarily opening; stamens more than 9, or not in distinct groups, without alternating glands; capsules not striate . . . . 2. **Hypericum**

1. Triadenum — Marsh St. John's-wort

There is now fairly general agreement that this genus should be recognized as separate from *Hypericum,* although the technical characters are often difficult to see, especially in herbarium specimens. Once one is familiar, however, with the usually reddish aspect of these plants, with broadly elliptic-oblong subcordate leaves (larger than the somewhat similarly shaped leaves in *Hypericum mutilum* and *H. boreale*), there is no problem.

There is less agreement as to whether one or two species should be recognized, the rank at which the differences are treated being still a matter of taste and convenience. In presenting two species, here, the position of Utech and Iltis, Wood and Adams, and Gleason is followed, in contrast to that of Gillett and Robson, Fernald, and others.

REFERENCE

Gleason, H. A. 1947. Triadenum. Phytologia 2: 288–291.

KEY TO THE SPECIES

1. Sepals 5–7 mm long, acute; styles becoming 1.8–3 mm long (best seen in fruit) . . . . 1. **T. virginicum**
1. Sepals 2.5–4.3 (5) mm long, usually blunt or rounded; styles 0.5–1 (1.5) mm long . . . . 2. **T. fraseri**

1. **T. virginicum** (L.) Raf. Plate 6-E

Map 808. Sphagnum bogs, lake shores, old interdunal swales, wet meadows. A species of Coastal Plain affinity, disjunct to southern Michigan, northwestern Indiana, central Wisconsin, and other areas known for such plants.

The sepals are narrowly oblong to lanceolate, tapering to the acute apex, compared to the more elliptic, rounded sepals typical of *T. fraseri*. Plants intermediate between the two species are occasionally found.

2. **T. fraseri** (Spach) Gl.

Map 809. Bogs, marshes, swales, sedgy meadows, damp sandy (even marly) shores, conifer swamps and alder thickets.

This is the northern species long included in the preceding and usually known as *Hypericum virginicum* L.—sometimes as var. *fraseri* (Spach) Fern. Those who now recognize the segregate genus *Triadenum* but do not accept these as two good species may call this taxon *T. virginicum* ssp. *fraseri* (Spach) J. M. Gillett. The flowers are usually closed, especially in sunshine,

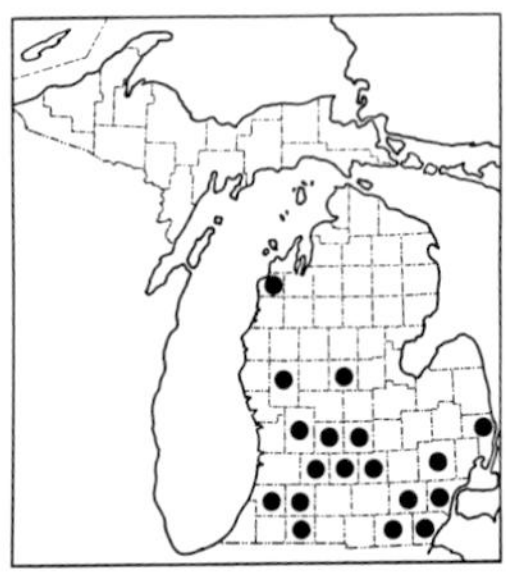
805. Hibiscus trionum

806. Hibiscus moscheutos

807. Hibiscus laevis

appearing to be perpetually in bud, but are occasionally seen open; some which were closed when collected opened most dramatically for me when shut up in a vasculum.

2. **Hypericum** St. John's-wort

KEY TO THE SPECIES

1. Plant woody, a bushy, often small, but much-branched shrub; styles coherent at anthesis, splitting in fruit; stamens very numerous (over 100)
  2. Carpels (styles, apparent locules) 3; inflorescences axillary as well as terminal; larger leaves 9–18 mm broad . . . . 1. **H. prolificum**
  2. Carpels 5 (rarely 4 or 6 in a few flowers); inflorescences all terminal; larger leaves 4–8 (9) mm wide . . . . 2. **H. kalmianum**
1. Plant herbaceous, with a central stem and in some species few if any lateral branches; styles distinct at anthesis (except in nos. 7 & 8); stamens (except in *H. ascyron*) fewer than 100
  3. Carpels (styles and locules) 5; flowers ca. 4–6.5 cm broad; stamens very numerous; capsules 12–19 mm long . . . . 3. **H. ascyron**
  3. Carpels 3 (styles 3; locules 1 or 3); flowers less than 3 cm broad; stamens fewer than 100; capsules less than 10 mm long
    4. Leaves minute, scale-like, linear-subulate, ± appressed, less than 3 mm long . . . . 4. **H. gentianoides**
    4. Leaves not so reduced, linear to elliptic or lanceolate
      5. Black spots (streaks and/or dots) clearly visible on petals (sometimes also on sepals, anthers, and/or leaves)
        6. Styles ca. 1–2 (2.5) mm long, shorter than the ovary; black spots on petals scattered all across them, on sepals usually similar (rarely absent), and on lower surface of leaves likewise throughout . . . . 5. **H. punctatum**
        6. Styles ca. 4–6.5 mm long, longer than the ovary; black spots on petals largely restricted to margins and apical portion, on sepals and leaves few or none . . . . 6. **H. perforatum**
      5. Black spots none
        7. Leaves definitely pinnately veined (one central longitudinal vein plus weaker lateral veins), elliptic-oblong; petals (5) 6.5–10 mm long, deciduous after anthesis; stamens more than 20; styles (1.5) 2–3.5 [4.5] mm long, coherent into a beak (until capsule dehisces—rarely separate)

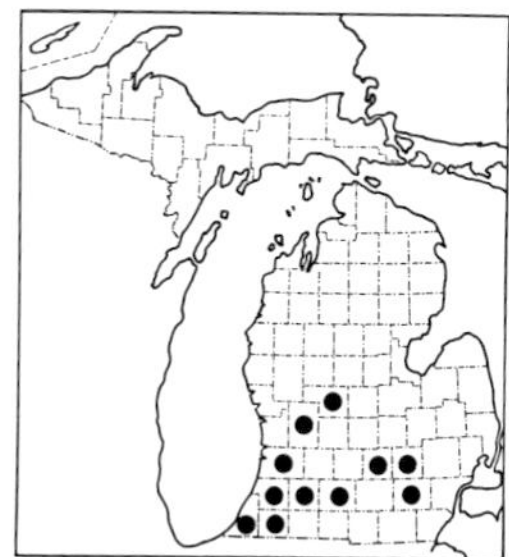
808. Triadenum virginicum

809. Triadenum fraseri

810. Hypericum prolificum

8. Plant 10–32 cm tall, strictly herbaceous, from slender easily collected superficial rhizomes, occurring in the Upper Peninsula; principal leaves 1.5–2.8 (3.4) cm long; seeds numerous, distinctly less than 1 mm long . . 7. **H. ellipticum**

8. Plant ca. (30) 40–75 cm tall, almost woody at base, from a deep rarely collected rhizome, occurring only in southernmost part of the state; principal leaves 4–7 cm long; seeds few, at least 2 mm long . . . . . . 8. **H. sphaerocarpum**

7. Leaves with 3–5 strong longitudinal veins from the base, or if only 1 then narrowly linear, the lateral veins obscure or not visible; petals less than 6.5 mm long, persistent (but withered) at base of fruit; stamens fewer than 20; styles not over 1 mm long, separate

9. Sepals and outline of fruit oblong to elliptic, broadest about the middle and tapering (if at all) scarcely more toward the apex than toward the base, their tips rounded or obtuse; branches of inflorescence ± divergent at maturity; leaves elliptic-oblong

10. Ultimate bractlets of inflorescence greatly reduced, ca. 0.2 mm wide (or less) . . . . 9. **H. mutilum**

10. Ultimate bractlets of inflorescence leaf-like, at least 0.5–2 mm wide . . . . 10. **H. boreale**

9. Sepals and often outline of young fruit lanceolate, broadest below the middle and more strongly tapering to acute tip; branches of inflorescence strongly ascending; leaves lanceolate or linear-oblanceolate

11. Cauline leaves with 1 (–3) nerves, linear or slightly oblanceolate, tapering to the base; sepals mostly 2.5–4 mm long . . . . 11. **H. canadense**

11. Cauline leaves with 3–5 (7) nerves, usually lanceolate, broadest below the middle; sepals (at least the largest, as around fruit) mostly (4) 4.5–6 (7) mm long . . . . 12. **H. majus**

1. **H. prolificum** L. Shrubby St. John's-wort

Map 810. Swamp borders, thickets, meadows, fields, roadsides, sandy open woods (oak). The North Manitou Island collection (*F. Wislizenus 467* in 1886, MO) seems rather out of range; it includes a fragment of *H. perforatum* (the original identification) and could represent an inadvertent mixture from some other locality.

H. spathulatum (Spach) Steudel is a later name for the same species, as now interpreted.

811. Hypericum kalmianum

812. Hypericum ascyron

813. Hypericum gentianoides

2. **H. kalmianum** L. Plate 6-F Kalm's St. John's-wort

Map 811. Largely restricted to the Great Lakes region, and often very well developed along lakes Michigan and Huron in sand dunes, especially calcareous interdunal hollows, and in moist limestone areas, but also inland on calcareous (even marly) shores, meadows, and river banks; occasionally in bogs (fens).

When specimens of this species are thought to have fewer than 5 styles (and thus may be misidentified as *H. prolificum*), it is usually because 2 or more of them are still cohering and have not yet separated.

3. **H. ascyron** L. Giant St. John's-wort

Map 812. Low ground along rivers and streams, ditches and wet meadows, borders of marshes and swampy woods.

This is the tallest of our species of *Hypericum,* attaining a height of ca. 2 m; it also has by far the largest flowers, capsules, and leaves. The plants of eastern North America have been called, by some recent authors, *H. pyramidatum* Aiton, but the distinction from plants of eastern Asia is not clear. *Hortus Third* and Gillett and Robson are here followed in reuniting the two taxa.

4. **H. gentianoides** (L.) BSP. Orange-grass

Map 813. Marshy or dry open ground; sandy clearings.

A bushy little plant with wiry, strongly ascending branches, appearing leafless.

5. **H. punctatum** Lam. Fig. 292 Spotted St. John's-wort

Map 814. Marshy or swampy ground, stream banks, borders of woods, dry to moist fields.

6. **H. perforatum** L. Fig. 293 Common St. John's-wort; Klamath Weed; Goatweed

Map 815. Abundant on fields, roadsides, and railroads; spreading to rock outcrops, shores, and elsewhere. A thoroughly naturalized weed, originally from Europe. First collected in Michigan in 1839 by Dennis Cooley in Macomb County.

Although other species of *Hypericum* are poisonous if ingested, this one because of its abundance has been an important stock-poisoning plant, especially in western North America. One of its properties is to make animals, especially white-skinned ones, hypersensitive to light.

This species is much less black-spotted than the preceding, but it does have prominent translucent dots in the leaves. The stem is angled (compared to terete in *H. punctatum*) for there is a ridge extending downward from the base of the midrib of each leaf.

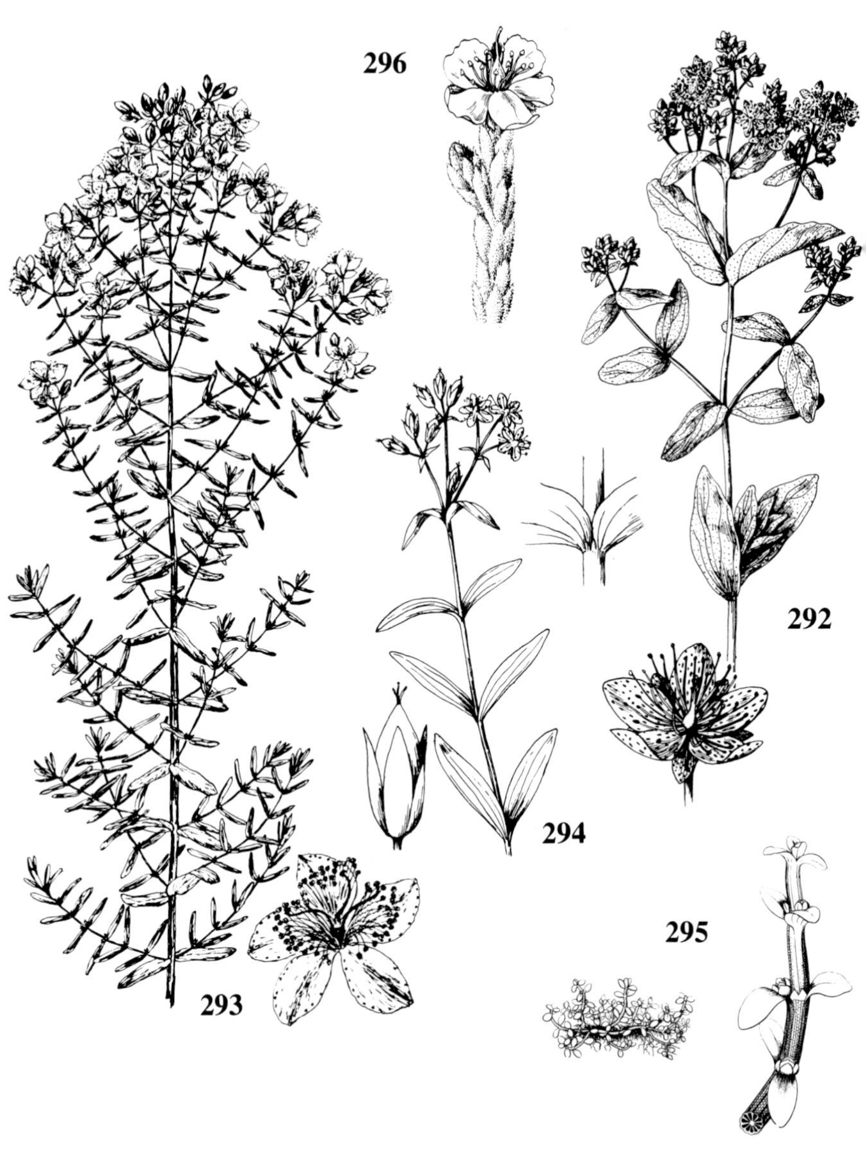

292. *Hypericum punctatum* ×$^1/_2$; flower ×2
293. *H. perforatum* ×$^2/_5$; flower ×$1^1/_2$
294. *H. majus* ×$^1/_2$; fruit ×3; node with leaf bases ×$1^1/_2$
295. *Elatine minima* ×$^1/_2$; flowering shoot ×$2^1/_2$
296. *Hudsonia tomentosa* ×2

7. **H. ellipticum** Hooker

Map 816. Damp sandy shores and clearings, open moist thickets, marshy beach flats, borders of rivers, bogs.

8. **H. sphaerocarpum** Michaux

Map 817. Michigan is at the northeastern extreme of the range of this species, which is also known from along the Maumee River in Wood County, Ohio. It was collected in openings of shrub thickets on the upper banks of a creek near Erie in 1983 (*Reznicek 7216,* MICH, MSC).

9. **H. mutilum** L.

Map 818. Damp sandy open ground, meadows and marshes, borders of streams.

10. **H. boreale** (Britton) Bickn.

Map 819. Bogs, marshy ground, damp sandy to mucky shores and interdunal hollows.

Recently treated by Gillett and Robson as a subspecies of the preceding species. Certainly the oft-stated characters of relative sepal and capsule length do not hold. However, the nature of the ultimate (not lower) bracts in the

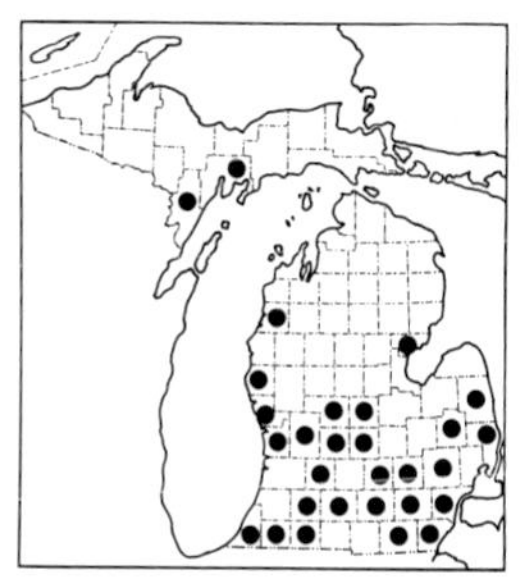

814. Hypericum punctatum

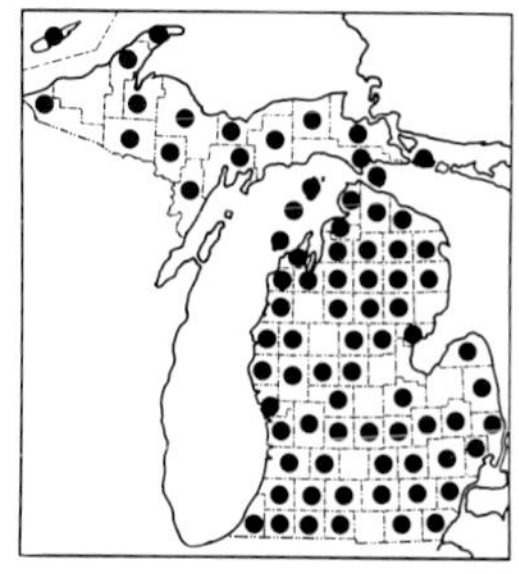

815. Hypericum perforatum

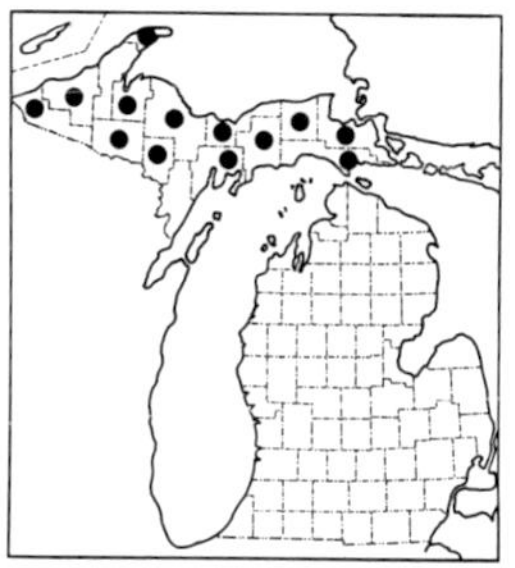

816. Hypericum ellipticum

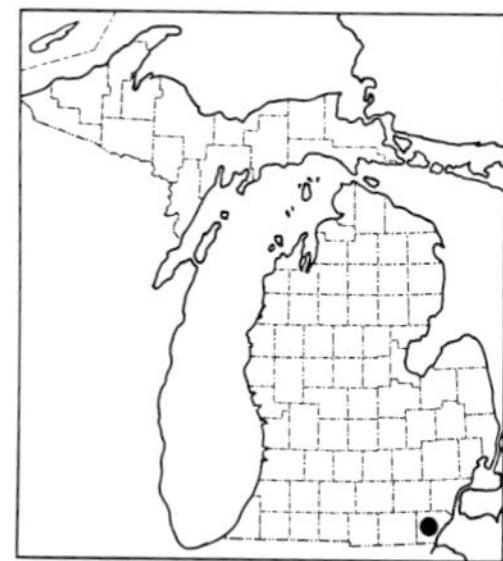

817. Hypericum sphaerocarpum

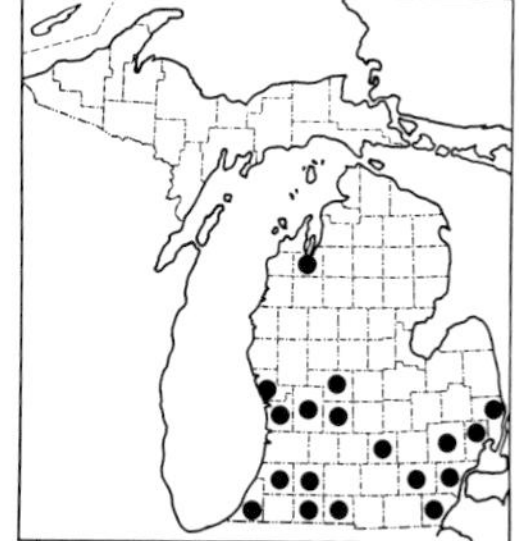

818. Hypericum mutilum

819. Hypericum boreale

inflorescence is quite consistent and no intermediates have been evident even in the area of overlapping range in Michigan.

The broadly elliptic-rounded, nearly clasping leaves of this species may suggest a small version of *Triadenum*. Submersed stems may exhibit long internodes and small 3-nerved (but not punctate) rounded leaves [f. *callitrichoides* Fassett]; identification can usually be confirmed by association with fertile plants on an adjacent shore.

11. **H. canadense** L.

Map 820. Damp sandy shores and excavations ("borrow pits"); peaty bog margins and disturbed trails.

Depauperate plants of *H. majus,* such as seedlings blooming the first year, may resemble this species and are often thus misidentified. The sepals of *H. majus* attain a greater length, especially in fruit. Both species have very reduced, subulate ultimate bractlets in the inflorescence, as in *H. mutilum*.

12. **H. majus** (A. Gray) Britton Fig. 294

Map 821. Damp sandy to mucky (or marly) shores, excavations, interdunal hollows; marshy or sedgy ground; borders of streams and cedar swamps; crevices in sandstone.

This species is distinctive when it has (ovate-) lanceolate 5-nerved leaves, but tends to run close to *H. canadense,* with which it is alleged to hybridize. Except for some depauperate individuals, however, only one truly intermediate specimen has been encountered, with short sepals (even in fruit) but leaves up to 5-nerved (*Davis* in 1905, Dickinson Co., MICH). Both these species also reportedly hybridize with *H. mutilum*; and *H. boreale* has been accused of hybridizing with *H. canadense*—but not with *H. mutilum*.

ELATINACEAE — Waterwort Family

1. Elatine

REFERENCE

Fassett, Norman C. 1939. Notes from the Herbarium of the University of Wisconsin—XVII. Elatine and Other Aquatics. Rhodora 41: 368–377.

1. **E. minima** (Nutt.) Fischer & C. Meyer Fig. 295 Waterwort

Map 822. All Michigan populations seen have been completely submersed or at the edge of the water in small lakes with sandy or sandy-mucky bottoms. Often locally abundant, forming small moss-like mats in shallow water (up to 2 ft.).

An easily overlooked little plant, at most a few cm tall, with tiny opposite rounded leaves, and likely to be mistaken for, perhaps, seedling *Hypericum*—until one notices the axillary fruit. The carpels are usually 2 in this species, the only member of this small family in Michigan, and the seeds have rows of round-ended (rather than angled) pits.

TAMARICACEAE

Tamarisk Family

1. **Tamarix**

REFERENCE

Baum, Bernard R. 1967. Introduced and Naturalized Tamarisks in the United States and Canada (Tamaricaceae). Baileya 15: 19–25.

1. **T. parviflora** DC. Tamarisk; Salt-cedar

Map 823. Native of the northeastern Mediterranean area, but one of the commonest cultivated species of the genus in North America and occasionally escaped. Collected in roadside thickets ca. 1 mile south of Clare (*Voss 3824* in 1957, MICH) and in Mason County (no loc.) (*C. W. Laskowski 259* in 1961, MICH).

Several species of these shrubs or small trees are cultivated, and this is

820. Hypericum canadense

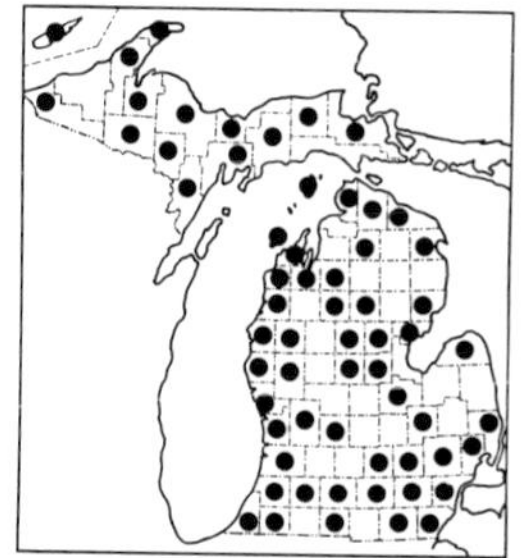

821. Hypericum majus

822. Elatine minima

823. Tamarix parviflora

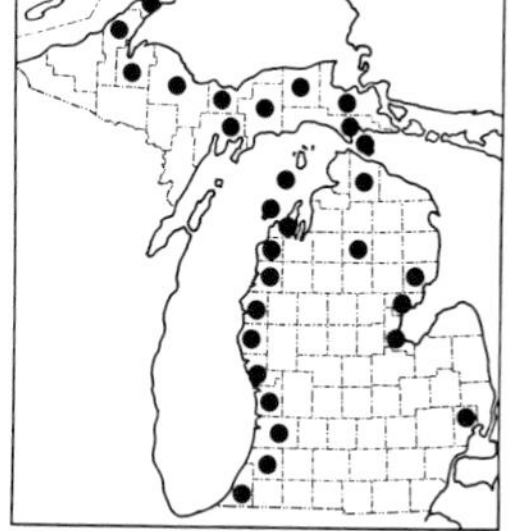

824. Hudsonia tomentosa

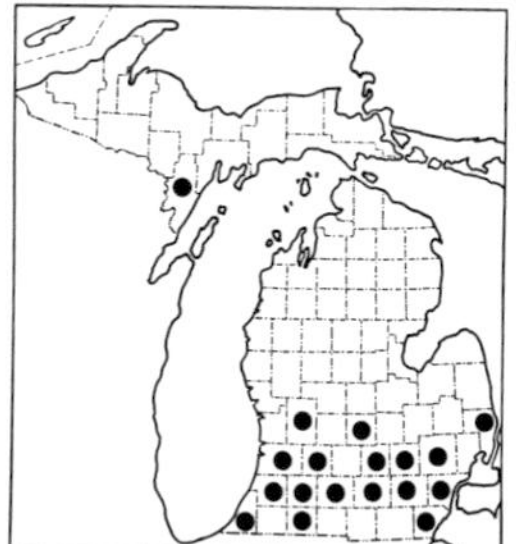

825. Helianthemum bicknellii

one of the easiest to recognize in a difficult genus, being our only introduced one with the sepals, petals, and stamens each 4 (rather than 5). The flowers are pink and appear in late spring in numerous compact racemes on the wood of previous years. The leaves are tiny and scale-like, appearing with or after the flowers. Many species (not *T. parviflora*) are halophytes and thrive in deserts (especially in Mediterranean regions) or in sites subject to salt spray. The Biblical "manna" is believed by some to be the sweet excretions of certain insects that live on *Tamarix*.

CISTACEAE — Rockrose Family

REFERENCE

Brizicky, George K. 1964. The Genera of Cistaceae in the Southeastern United States. Jour. Arnold Arb. 45: 346–357.

KEY TO THE GENERA

1. Plant a low shrub with scale-like, overlapping, closely appressed leaves; style distinctly longer than the ovary . . . . 1. **Hudsonia**
1. Plant an erect herb with flat linear to narrowly elliptic leaves that expose the stem; style (if any) shorter than the ovary
   2. Petals 5, yellow, showy but readily falling (or absent in cleistogamous flowers); stigma on a distinct style; leaves with stellate hairs . . . . 2. **Helianthemum**
   2. Petals 3, reddish, seldom expanded, mostly concealed by the sepals (cleistogamous flowers none); stigmas sessile or nearly so; leaves glabrate or with simple hairs . . . . 3. **Lechea**

1. **Hudsonia**

1. **H. tomentosa** Nutt. Plate 7-A; fig. 296 Beach-heath; False Heather
Map 824. Sandy shores and dunes, including old dune ridges, often with scattered pine, oak, or other trees; especially characteristic along Lake Superior from Whitefish Point to Grand Marais and along Lake Michigan; occasionally inland on sandy shores and plains, again with pines and oaks.

A bushy little shrub, ± white-pubescent, sometimes forming large mats on rather stabilized dunes and beaches. It might be mistaken for a form of juniper—except for the small but copious bright yellow flowers.

2. **Helianthemum** — Rockrose; Frostweed

Our species produce two kinds of flowers: showy (chasmogamous) ones that open in sunshine but shed their petals, which are (5) 8–17 mm long, late in the day; and tiny apetalous cleistogamous ones crowded and sessile or nearly so on the branches. The 2 outer sepals are rudimentary or linear, much narrower than the 3 inner sepals, and look like bractlets.

REFERENCE

Daoud, H. S., & Robert L. Wilbur. 1965. A Revision of the North American Species of Helianthemum (Cistaceae). Rhodora 67: 63–82; 201–216; 255–313.

KEY TO THE SPECIES

1. Petaliferous flowers in clusters of 2 or more, not overtopped by leafy branches; upper surface of cauline leaves with only stellate hairs; capsules of cleistogamous flowers rarely over 2 mm in diameter, 1–2 (3)-seeded; outer sepals of cleistogamous flowers with linear free portion ca. 0.6–1.6 mm long; ripe seeds with reticulate surface [use 20× lens] . . . . 1. **H. bicknellii**
1. Petaliferous flowers solitary, soon overtopped by leafy branches; upper surface of cauline leaves (at least when young) with scattered simple hairs much overtopping the stellate ones; capsules of cleistogamous flowers ca. 2.5 mm or more in diameter, with at least 5 seeds; outer sepals of cleistogamous flowers rudimentary, usually less than 0.5 mm long; ripe seeds with papillate surface . . . . 2. **H. canadense**

1. **H. bicknellii** Fern. Fig. 297

Map 825. Sandy soil, including clearings and disturbed places, usually open but sometimes under oaks. May grow with the next species (so that mixed collections are sometimes made).

2. **H. canadense** (L.) Michaux

Map 826. Dry sandy plains, hillsides, dunes, usually ± open or with thin tree cover (pines, oak, and/or aspen) or scattered junipers.

Older plants (collected later in the summer) may not display the scattered long hairs on the upper surface of the leaves, which are quite evident when early petaliferous flowers are present; but such late plants will usually have ample cleistogamous flowers or fruits (with their larger capsules and rudimentary outer sepals) to make identification possible. Furthermore, if they have the large fruits developing from early petaliferous flowers, these will be much overtopped by leafy shoots.

3. **Lechea** Pinweed

This genus is distinctive, but the species have long been puzzling. The flowers rarely open, but are reported to do so in early morning on bright days. The tiny flowers are pear-shaped to nearly globose, with 5 sepals (of which the 2 outer are linear), 3 reddish petals shorter than the sepals, a variable number of stamens, and essentially sessile dark red plumose stigmas. Sterile shoots with crowded leaves are typically developed at the base of the plant, especially late in the season.

REFERENCES

Hodgdon, Albion R. 1938. A Taxonomic Study of Lechea. Rhodora 40: 29–69; 87–131 + pl. 488–491 (also Contr. Gray Herb. 121).

Wilbur, Robert L., & Hazim S. Daoud. 1961. The Genus Lechea (Cistaceae) in the Southeastern United States. Rhodora 63: 103–118.

Wilbur, Robert L. 1966. Notes on Rafinesque's Species of Lechea (Cistaceae). Rhodora 68: 192–208.

KEY TO THE SPECIES

1. Pubescence of stems and leaves mostly divergent or spreading; inner sepals with ± prominent green keel (or warty midstrip), glabrous except sometimes on the keel .... 1. **L. villosa**
1. Pubescence of stems and leaves mostly appressed; inner sepals not keeled (but may have 3–5 inconspicuous veins), pubescent across their width
  2. Outer sepals distinctly longer than the inner ones .... 2. **L. minor**
  2. Outer sepals distinctly shorter than the inner ones
    3. Leaves with a hard shiny ± conical brownish tip ca. 0.2–0.3 mm long; fruiting calyx less than 1.5 mm broad (before dehiscence of capsule), ± pear-shaped; seeds 2–3; plant ± diffusely branched above with the branches spreading .... 3. **L. pulchella**
    3. Leaves pointed but without such a differentiated tip (flat and green to the margin); fruiting calyx mostly at least 1.5 mm broad, subglobose; seeds (3) 4–6; plant compact above, the branches ascending
      4. Seeds 4–6 (5–6 in largest capsules), when fully mature covered with a reticulate white membrane; plant frequent in northern Michigan .... 4. **L. intermedia**
      4. Seeds 3–4, without a white membrane; plant very rare .... 5. **L. stricta**

1. **L. villosa** Ell.

Map 827. Sandy shores, fields, and open woods.

It has been suggested (Wilbur 1966) that the older name *L. mucronata* Raf. applies to this species, but enough doubt seems to exist that others have been reluctant to agree. The outer sepals are about the same length as the inner ones.

2. **L. minor** L.

Map 828. Dry to moist sandy ground, especially shores.

3. **L. pulchella** Raf.

Map 829. Dry to moist sandy plains, ridges, shores, and open woods.

This is the species long known as *L. leggettii* Britton & Hollick, but Rafinesque's older name has been shown to apply (Wilbur 1966). Our plants are the distinctive form [= *L. leggettii* var. *moniliformis* (Bickn.) Hodgdon] which is quite different in aspect from the next two closely related species; it gives the first impression of a delicate *Panicum*. The longest slender spreading branches are usually 5–9 cm long, while in the next two species the branches are more ascending and usually (though not always) less than 5 cm long.

4. **L. intermedia** Britton   Fig. 298

Map 830. Sandy or rocky ground: plains (with jack pine, oak, and/or aspen), outcrops, shores.

Greener in aspect than the next species, but with numerous appressed white hairs.

5. **L. stricta** Britton

Map 831. A collection from oak woods at Menominee (*Grassl 3244* in 1933, MICH) appears to belong here (seeds 3), and one from a dry railroad right-of-way in Kalamazoo County (*W. B. Drew et al. M126* in 1947, MSC), although lacking fruit, also is tentatively referred here on the basis of habit and pubescence. Other reports from the western Upper Peninsula are unconfirmed or are apparently based on specimens of *L. intermedia*. *L. stricta* is known from Minnesota, Wisconsin, and northwestern Indiana westward, and should be expected in the southwestern Lower Peninsula as well.

Very similar to the preceding species, but more gray in aspect as a result of the denser white hairs on the young leaves, branchlets, and buds—but the differences defy any quantitative statement.

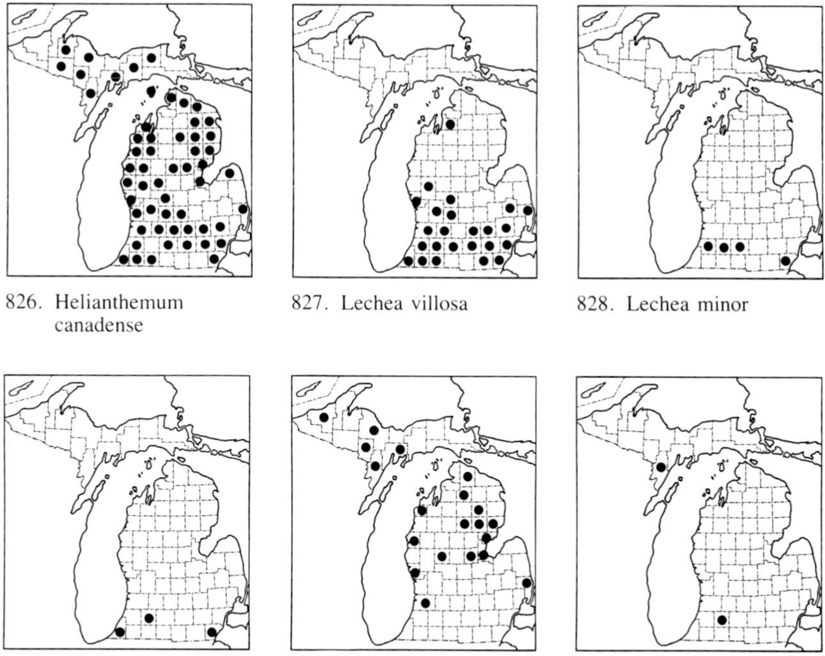

826. Helianthemum canadense

827. Lechea villosa

828. Lechea minor

829. Lechea pulchella

830. Lechea intermedia

831. Lechea stricta

297. *Helianthemum bicknellii* $\times ^1/_2$; fruiting calyx $\times 2$
298. *Lechea intermedia* $\times ^1/_2$; fruiting calyx $\times 10$
299. *Viola rostrata* $\times ^1/_2$
300. *Viola renifolia* $\times ^1/_2$

VIOLACEAE

Violet Family

REFERENCE

Brizicky, George K. 1961. The Genera of Violaceae in the Southeastern United States. Jour. Arnold Arb. 42: 321–333.

KEY TO THE GENERA

1. Leaves all cauline, more than 10 and usually entire; sepals without a basal auricle; lower petal notched or 2-lobed at apex and slightly saccate (but not spurred) at base . . . . 1. **Hybanthus**
1. Leaves all basal or fewer than 8 if cauline, toothed to crenulate; sepals with an appendage or auricle extending beyond their attachment at base of flower; lower petal not at all notched, but with a definite spur projecting backward at the base . . . . 2. **Viola**

1. Hybanthus

REFERENCE

Cole, Emma J. 1898. Cleistogamous Flowers on Solea concolor. Asa Gray Bull. 6: 50.

1. **H. concolor** (T. F. Forster) Sprengel Fig. 301 Green Violet

Map 832. Rich mesic to swampy woods, including floodplains and river banks.

The small greenish flowers grow 1–3 in the axils of the middle leaves, and are pendulous. Rarely, even smaller erect cleistogamous flowers appear in the axils of the upper leaves late in the summer.

2. Viola

Violet

The flowers of this genus, whether large as in the garden pansy or smaller as in our wild violets, are familiar to almost everyone. The species, however, are often difficult to distinguish, especially if one tries to accept the excess of species that have been named. I have been happy to follow personal advice from Arthur Cronquist (and others) to recognize fewer species than have been traditional. Harvey E. Ballard, Jr., has studied Michigan violets intensively in recent years, and I am very grateful to him for suggesting the alignments here adopted and most of the characters used in the key, as well as for his thorough annotation of Michigan specimens. I have relied thoroughly on his collaboration in dealing with this genus, and almost all dots on the maps are supported by specimens determined by him. Norman H. Russell kindly examined most specimens from Michigan herbaria in the 1950's and 1960's, and thus began the process of bringing some order into our collections, although he maintained a few more species than treated below. His studies documented much of the extensive variation and hybridization that occur in this genus.

The flowers are bilaterally symmetrical, with a spurred lower petal, 2 lateral petals, and 2 upper petals. The petals (especially the spurred one) of all our species have purple lines toward the base: nectar guides that offer a "road map" to visiting insects. These are omitted in the statements about flower color in the key. Below the middle of the inside of the lateral petals there may be a tuft of hairs, termed a *beard,* and in some species there is also a beard on the spurred petal. Unless otherwise indicated, statements about leaf length in the key refer to the entire length of the blade, including the basal lobes if the blade is cordate. The seeds usually have an appendage (caruncle) which is associated with dispersal by ants. Except in *V. pedata,* cleistogamous flowers are produced later in the season than the familiar spring-blooming petaliferous flowers, and on shorter peduncles (which may be erect or prostrate).

Especially in the stemless species there are many ambiguities, and exceptions to any character in the key mean that good judgment and intuition are required to base identifications on a judicious balance of characters. Habitats are often quite distinctive, and should be checked in addition to the key.

"My advice is just to detour this bunch of bastards—Viola as a whole. Many are doubtless interspecific hybrids and are hopeless."

—C. C. Deam to C. R. Hanes, May 20, 1942

REFERENCES

Cinq-Mars, Lionel. 1966. Mise au Point sur les Violettes (Viola spp.) du Québec. Nat. Canad. 93: 895–958 (also in Provancheria 1).

Lévesque, Lucien, & Pierre Dansereau. 1966. Études sur les Violettes Jaunes Caulescentes de l'Est de l'Amérique du Nord. I. Taxonomie, Nomenclature, Synonymie et Bibliographie. Nat. Canad. 93: 489–569.

Russell, Norman H. 1960. The Violets of Minnesota. Proc. Minnesota Acad. 25–26: 126–191.

832. Hybanthus concolor

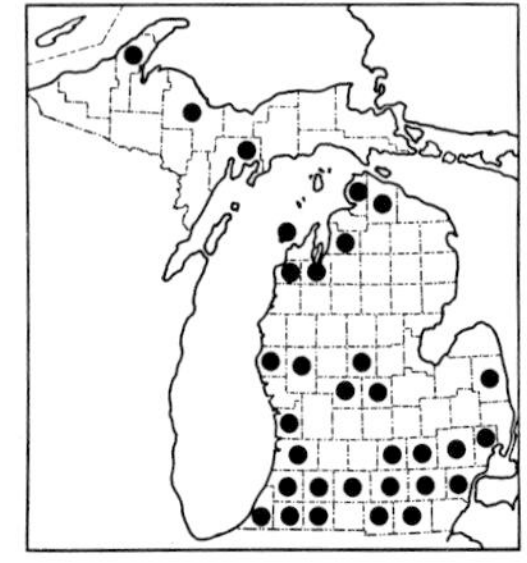

833. Viola arvensis

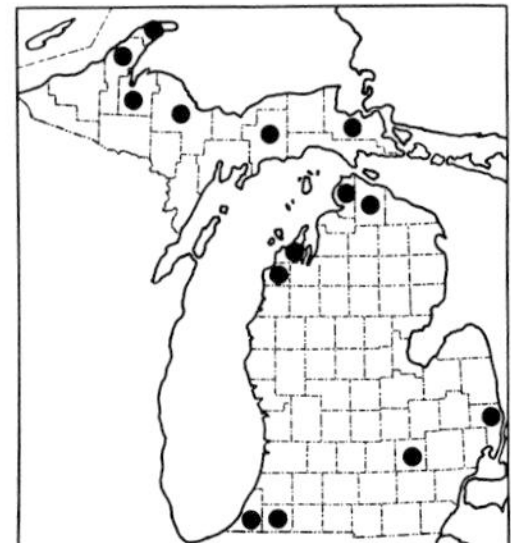

834. Viola tricolor

Russell, Norman H. 1965. Violets (Viola) of Central and Eastern United States: An Introductory Survey. Sida 2: 1–113. [Cites many previous papers.]

KEY TO THE SPECIES

1. Plant leafy-stemmed (i. e., with all petaliferous flowers and at least some leaves borne on above-ground stem)
 2. Stipules of cauline leaves well developed, leaf-like, deeply pinnatifid (or bipinnatifid) at the base with a large terminal lobe, the longest basal lobes ca. 30–50% as long as the whole stipule
  3. Petals all ± yellow or cream within (occasionally with purple spot), ± equal in length, shorter than the sepals or exceeding them by at most 2 mm . . . 1. **V. arvensis**
  3. Petals (at least the upper, which are longer) deep blue or purple distally, exceeding the sepals by at least 3 mm . . . . .2. **V. tricolor**
 2. Stipules of cauline leaves entire or fringed with lateral lobes less than 25% as long as the whole stipule
  4. Stipules entire or weakly toothed; petals yellow or white (never blue except outside)
   5. Petals white inside (yellow at base), usually ± bluish outside; upper stipules (like the lower) pale and scarious, narrowly tapered or acuminate, scarcely if at all broadened above the base; seeds 1.8–2.2 (2.4) mm long . . . . . . 3. **V. canadensis**
   5. Petals yellow; upper stipules green and herbaceous, acute to obtuse but without prolonged tapered or acuminate tip, often distinctly broader just above the base; seeds mostly 2.4–3.2 (3.5) mm long . . . . 4. **V. pubescens**
  4. Stipules prominently fringed along the margins, especially toward the base; petals cream or blue inside (and outside)
   6. Petals creamy-white on both surfaces; sepals ciliate (rarely nearly glabrous) . . . . 5. **V. striata**
   6. Petals blue (except in albinos); sepals eciliate
    7. Spur ca. (5) 8–13 (15) mm long (at least as long as the rest of the petal); flowers deeper blue toward the center, the lateral petals glabrous . . . . 6. **V. rostrata**
    7. Spur less than 5 (6) mm long (shorter than the rest of the petal); flowers paler blue (or white) toward the center, the lateral petals bearded
     8. Leaf blades glabrous (or with small flocculent tufts of knob-shaped hairs or scattered tiny hairs above toward the margins), varying from ovate (definitely cordate at base, acute at apex) to reniform (subcordate, obtuse to rounded in outline at apex); petals pale blue . . . . .7. **V. conspersa**
     8. Leaf blades usually finely hispidulous on one or both surfaces, subcordate to ± truncate at the base, ovate (usually narrowly so), acute to barely obtuse in outline but blunt at the apex, neither strongly cordate nor reniform and subcordate nor strongly acute; petals deep blue-violet . . . . 8. **V. adunca**
1. Plants with petioles and peduncles all basal (from rhizome or stolon), hence "stemless"
 9. Leaf blades tapered to the petiole, narrowly elliptic to lanceolate, neither cordate nor lobed; plant entirely glabrous . . . . 9. **V. lanceolata**
 9. Leaf blades cordate (or only truncate) at base, or lobed, or both; plant glabrous or pubescent
  10. Ovary and capsule pubescent; seeds ca. 3.5–4 mm long (including prominent appendage ca. 1.5 mm long); style scarcely enlarged but with a recurved conic-tapered tip; plant escaped from cultivation, spreading by extensive green stolons; leaves finely pubescent, at least above, slightly rugose . . . . 10. **V. odorata**

10. Ovary and capsule glabrous; seeds less than 2.3 mm long; style capitate or expanded (and concave or with tiny lateral beak) at tip; plant native, with stolons none or white (buried in litter)

11. Flowers white, (5) 7–12 (14) mm long; rhizome less than 2.5 mm thick (except sometimes at nodes); leaves all cordate to reniform, not otherwise lobed

12. Leaf blades ovate (longer than broad), glabrous, the crenulations on spring leaves ± straight most of their length with the minute sharp tooth in a notch ± equally deep on both sides; lateral petals glabrous or rarely with scanty short beard; capsules green, speckled with scattered orange dots; seeds ca. 1–1.3 mm long, black at full maturity . . . . 11. **V. macloskeyi**

12. Leaf blades ovate to reniform, often sparsely to densely pubescent, especially toward the base and margins or along veins, the crenulations on spring leaves rounded with the tooth usually in an asymmetrical notch deeper on one side than the other (the margin thus ± jagged or scalloped); lateral petals glabrous to heavily bearded; capsules mostly heavily spotted, flecked, or mottled with purple; seeds ca. 1.8–2.2 mm long, brown at full maturity

13. Plants strongly stoloniferous (with surficial or shallow runners), even at flowering time; largest leaf blades (not necessarily the first of the season) ± ovate, the midrib ca. (70) 75–95 (115) % as long as blade is broad; pubescence on upper surface of blade if present on only one surface (otherwise, blades glabrous or pubescent on both surfaces) . . . . 12. **V. blanda**

13. Plants not at all stoloniferous (but with short ± vertical, sometimes twisted rhizomes); largest leaf blades ± reniform, usually with midrib ca. 50–77 (82) % as long as blade is broad; pubescence on lower surface of blade if on only one surface (otherwise, blades glabrous or pubescent on both surfaces) . . . . 13. **V. renifolia**

11. Flowers (except in albinos) blue to violet, (12) 13–22 (25) mm long; rhizome various, at least 2 mm thick (except in *V. selkirkii* & *V. epipsila*), in most species much thicker; leaves various, in some species ± deeply lobed

14. Leaves basically ternate, the blade divided nearly to the base into 3 parts of which at least the lateral 2 are further deeply divided, all the segments linear (central portion scarcely if at all broader than the linear segments); rhizome erect, usually short

15. Lateral petals (and usually the spurred one) bearded; rhizome ca. 5 mm in diameter or less; stamens not protruding beyond throat of corolla; cleistogamous flowers produced . . . . 14. **V. pedatifida**

15. Lateral (and all other) petals glabrous; rhizome (except in juvenile plants) ca. 5–12 mm in diameter; stamens conspicuously protruding beyond throat of corolla; cleistogamous flowers never produced . . . . 15. **V. pedata**

14. Leaves more shallowly or broadly lobed (with central portion broader than the lobes) or not lobed at all (except for cordate base); rhizome usually elongate and ± angled or horizontal

16. Lateral petals (like the others) beardless; spur 4–6 mm long; rhizomes very slender (less than 2 mm); leaf blades with narrow sinus (the basal lobes often nearly or quite touching each other), glabrous beneath but pubescent with tiny ± erect hairs above (especially on veins and toward lobes and margins); petioles glabrous . . . . 16. **V. selkirkii**

16. Lateral (and often spurred) petal bearded; spur less than 4 mm long; rhizomes stouter (except in *V. epipsila*); leaf blades with broader sinus (lobes not touching)—or truncate to subcordate, and pubescence not usually distributed as above

17. Leaf blades all or mostly lobed or at least with basal teeth larger and coarser than the rest of the margin
18. Blades of leaves (including lobes) scarcely if at all longer than broad; sepals ± oblong but acute to rounded at tip, ciliate, their auricles in fruit less than 1.5 mm long; spurred petal glabrous within or nearly so; capsules of cleistogamous flowers heavily flecked with purple . . . . 17. **V. palmata** complex
18. Blades of leaves averaging at least 1.5 times as long as broad; sepals narrowly acute, ciliate or eciliate, their auricles in fruit ca. 1.5–4 mm or more long; spurred petal densely bearded within; capsules of cleistogamous flowers green . . . . 18. **V. sagittata**
17. Leaf blades all ± evenly toothed, without lobes (other than at a cordate base) or unusually large teeth
19. Sepals narrowly acute or tapered (though they may be blunt at the tip); leaf blades ovate-triangular (or sagittate), most of them distinctly longer than broad and strongly acute in outline at the apex
20. Leaves ± pubescent on blades and petioles, the blades truncate to subcordate at base . . . . 18. **V. sagittata**
20. Leaves often glabrous or glabrate on blades or petioles (or both), the blades strongly cordate at base
21. Spurred petal beardless; sepal auricles well developed, becoming 1–2.5 mm long (or 5 mm on fruit of cleistogamous flowers); beard of lateral petals often of short strongly clavate hairs; flowers usually much overtopping the leaves, which are completely glabrous except (usually) for a few tiny hairs only on upper surface of basal lobes and toward margins . . . . 19. **V. cucullata**
21. Spurred petal with at least sparse beard; sepal auricles poorly developed, less than 0.8 mm long; beard of at most slightly clavate hairs; flowers slightly if at all overtopping leaves, which are glabrous to somewhat pubescent throughout . . . . 20. **V. affinis**
19. Sepals ± oblong (or occasionally ovate), rather abruptly rounded at the tip; leaf blades usually broadly ovate-cordate to reniform, most of them little if any longer than broad and nearly or quite obtuse in outline at the apex
22. Rhizome at most scarcely 2 mm thick; foliage and spurred petal completely glabrous; capsules green; leaf blades ± crenulate . . . . 21. **V. epipsila**
22. Rhizome much stouter; foliage or spurred petal (or both) usually at least sparsely pubescent; capsules green or marked with purple; leaf blades serrate
23. Foliage glabrous (at most a few tiny hairs on petioles or upper surface of basal lobes of the blades); spurred petal at least sparsely bearded (very rarely glabrous); capsules of cleistogamous flowers green; sepals glabrous . . . . 22. **V. nephrophylla**
23. Foliage usually pubescent, at least on base of the blade beneath and summit of petiole (also on peduncle), though sometimes completely glabrous; spurred petal glabrous (rarely sparsely bearded); capsule of cleistogamous flowers flecked or mottled with purple; sepals often ciliate . . . . 23. **V. sororia**

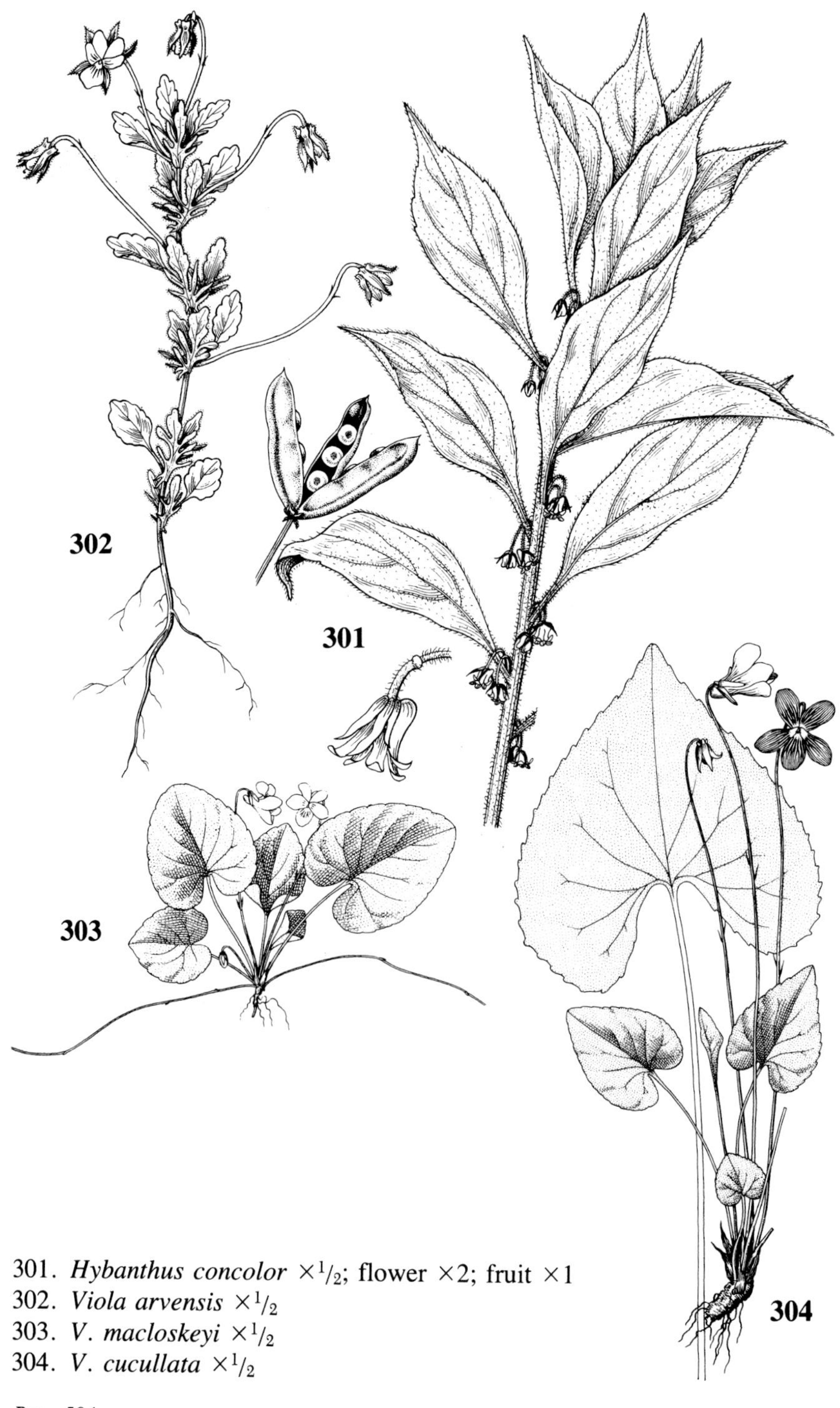

301. *Hybanthus concolor* $\times 1/2$; flower $\times 2$; fruit $\times 1$
302. *Viola arvensis* $\times 1/2$
303. *V. macloskeyi* $\times 1/2$
304. *V. cucullata* $\times 1/2$

1. **V. arvensis** Murray Fig. 302 Field Pansy

Map 833. A little European weed, usually in dry sandy disturbed ground: roadsides, fields, dumps, lawns.

2. **V. tricolor** L. Johnny-jump-up

Map 834. Another European species, occasionally escaped from cultivation to roadsides, fields, dumps and gravel pits, waste ground.

This species looks like a miniature pansy, and indeed it is one of the parents of the garden pansy, *V.* ×*wittrockiana* Gams, which may persist briefly after cultivation. The flowers are generally at least 1.5 cm long, whereas in the preceding species they are less than 1.5 cm. They are also small in *V. rafinesquii* Greene, but its petals are much longer than the sepals, unlike *V. arvensis,* and the upper leaves as well as the large terminal lobes of the stipules are entire or nearly so (rather than crenulate as in *V. arvensis* and *V. tricolor*). *V. rafinesquii* has sometimes been reported from Michigan, but no specimens have been found (at least not correctly identified).

3. **V. canadensis** L. Canada Violet

Map 835. Quite consistently a species of deciduous woods, usually beech-maple or hemlock-hardwoods, persisting after clearing.

Tardily blooming flowers may be found on individuals of this species—as rarely of some others—into the very late summer or even fall.

4. **V. pubescens** Aiton Yellow Violet

Map 836. In all kinds of deciduous and mixed woods (except the swampiest), persisting in clearings and fencerows.

Typical *V. pubescens* is densely pubescent, usually with only 2 cauline leaves (appearing opposite), and at most 1 basal leaf. Glabrous or glabrate plants with 3 or more cauline leaves and 1 or more basal leaves are more common and often have been treated as a separate species under the name *V. pensylvanica* Michaux or *V. eriocarpa* Schweinitz—with the admission of numerous intermediate plants as "hybrids." The differences traditionally employed to separate these taxa were thoroughly reviewed by Lévesque and Dansereau (1966). However, current students of the problem are largely agreed that there is only one extremely variable species here—our only native yellow-flowered violet in Michigan. The capsules of both glabrous and pubescent plants may be either glabrous or woolly. Those with the glabrous fruit may be called *V. pubescens* f. *leiocarpa* (Fern. & Wieg.) Farw. Plants with each fruit type may be found in the same woods.

5. **V. striata** Aiton Cream Violet

Map 837. Rich deciduous woods, thickets by streams, occasionally in swamp forests. Sometimes tends to be ± aggressive or weedy in habit.

A form lacking the purple lines on the petals has been named f. *albiflora* Farw. (TL: Eloise [Wayne Co.]). The species is easily distinguished from the more common and widespread white-flowered *V. canadensis* by the green, fringed stipules (rather than scarious, entire ones) and absence of blue color on the outside of the petals, which is normally conspicuous in *V. canadensis*.

6. **V. rostrata** Pursh   Fig. 299   Long-spurred Violet

Map 838. Upland or low deciduous woods.

The leaf blades are strongly cordate, ± acute in outline, and glabrous. The spur may be shorter in very young flowers than when the flowers are fully mature and the spur is longer than the petals.

Apparent hybrids with the next species [*V.* ×*malteana* House] have been found where they grow together and are intermediate in appearance—most noticeably in spur length, with the lateral petals bearded. They have been collected in Antrim, Clare, Emmet, Isabella, Kalamazoo, Midland, Osceola, Sanilac, Tuscola, and Wayne counties. Both of these blue-flowered species also hybridize with *V. striata,* with resulting intermediate flower color. The influence of the cream violet is also seen in a tendency to ciliate sepals. Plants referable to *V. rostrata* × *V. striata* have been collected in Berrien, Branch, Cass, Kalamazoo, Lenawee, and Saginaw counties. [This hybrid has been named *V.* ×*brauniae* Grover, but that name was not validly published.] With shorter spurs, specimens presumably *V. conspersa* × *V. striata* have been collected in Berrien, Cass, Ingham, Lenawee, Macomb, Midland, Monroe, Sanilac, Washtenaw, and Wayne counties.

7. **V. conspersa** Reichb.   Dog Violet

Map 839. Deciduous, mixed, and coniferous woods and swamps, usually in moist ground and often in slightly disturbed areas such as rocky ridges, trailsides, clearings; sometimes in open ground, especially at borders of woods and shores.

The leaves are quite variable on the same plant, the lower ones often ± reniform. The spurs may be a little longer than 5 mm if the other petals are

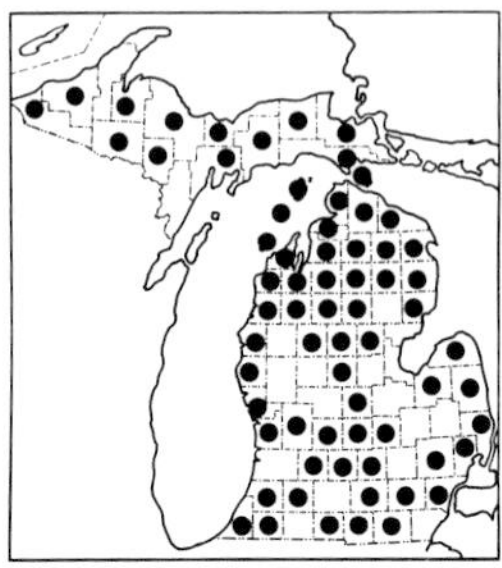

835. Viola canadensis

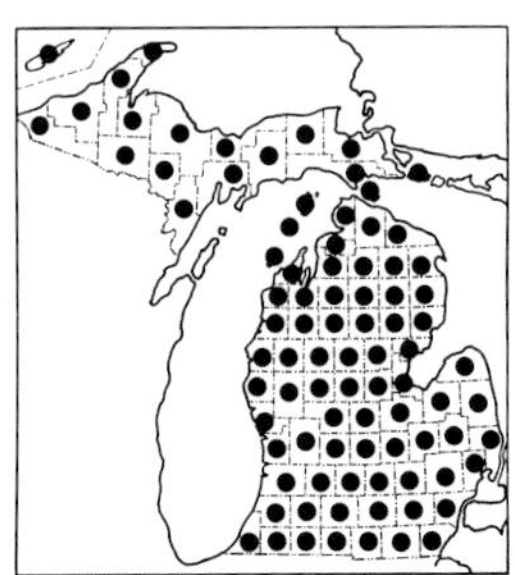

836. Viola pubescens

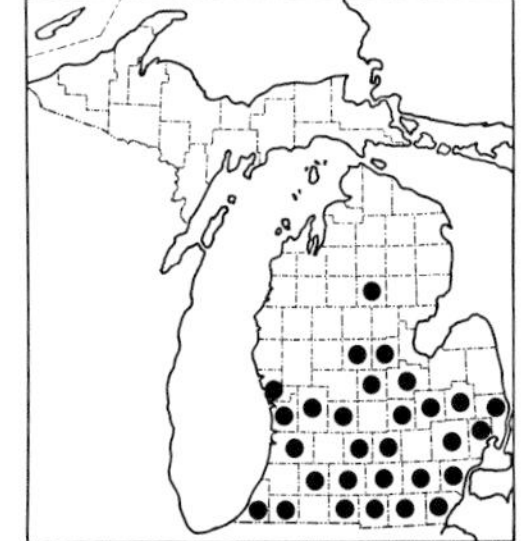

837. Viola striata

also unusually large, so that the spur is still no longer than the petals, unlike *V. rostrata*. The leaves are rather pale or even yellowish green and thin in texture compared to those of the next species. Rarely the flowers are pure white; such plants have been called f. *masonii* (Farwell) House (TL: near Utica [Macomb Co.]), but the type material is *V. striata* and there is no available name for those who think white-flowered *V. conspersa* deserves one. Apparent hybrids with *V. adunca* are known from Alpena County and Bois Blanc Island (Mackinac Co.) and perhaps elsewhere.

8. **V. adunca** J. E. Smith — Sand Violet

Map 840. Dry sandy ± open ground (plains, barren spots, ridges, banks), often with jack pine and oaks; crevices in rock outcrops on both summits and shores in the Lake Superior region.

The fresh leaves are darker (more of a blue-green) than in *V. conspersa,* and thicker in texture. Their shape is more easily used to distinguish the two species than it is to describe. In the jack pine plains and elsewhere in the Lower Peninsula, plants nearly always show at least some minute pubescence on peduncle, petioles, and blades (at least on the main veins and margins). In the Lake Superior region, glabrous plants are often found; these have sometimes been called *V. adunca* var. *minor* (Hooker) Fern. or *V. labradorica* Schrank, without agreement as to whether these names are indeed synonymous. Such plants supposedly also have entire stipules, but at least the upper stipules of both *V. conspersa* and *V. adunca* may occasionally be entire—and glabrous plants often have fringed stipules. Some plants identified in the past as *V. labradorica* seem better referred to *V. conspersa* by their leaf shape. Specimens from Keweenaw County referred by Fernald to *V. labradorica* appear to be *V. conspersa* or possible hybrids with *V. adunca,* according to Ballard. For the present, recognition of three species seems adequate to accommodate our stemmed blue violets.

9. **V. lanceolata** L. — Lance-leaved Violet

Map 841. Sandy to peaty shores of lakes and ponds; drying borders of marshes and bogs; damp sand of ditches, excavations, and fields.

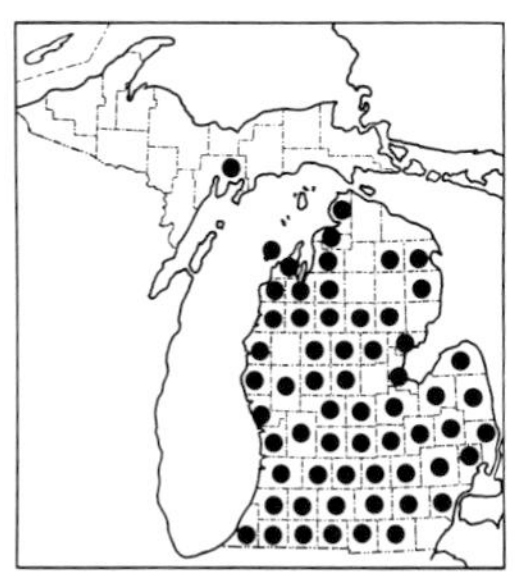
838. Viola rostrata

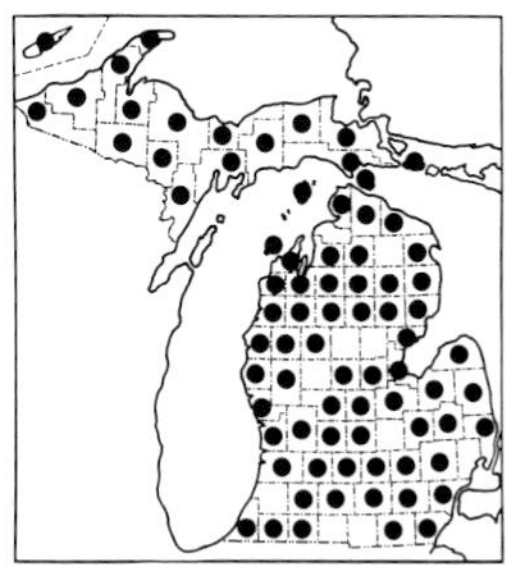
839. Viola conspersa

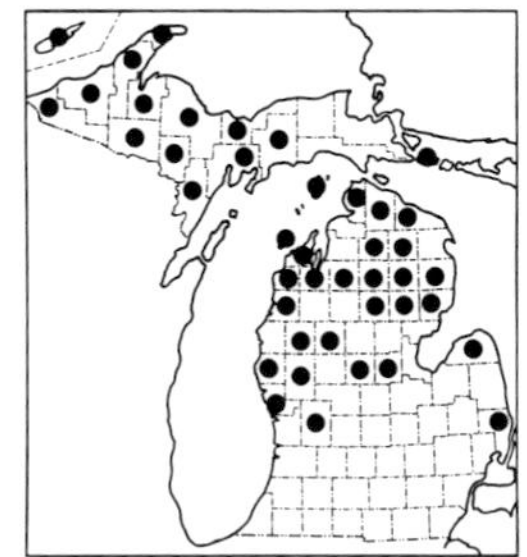
840. Viola adunca

Easily recognized by leaf shape alone (blades usually at least 4 times as long as wide), although the size is variable (blades ranging from 1 to 10 cm long). The petals are typically glabrous, although occasionally there are a few tiny hairs on the lateral ones—scarcely a beard.

Hybrids with *V. macloskeyi* or *V. blanda* have broader leaf blades ± truncate at the base but not really cordate. These have been the basis for Michigan reports of *V. primulifolia* L., which is evidently of such a hybrid origin. Except for an 1891 collection from Grand Rapids, our records of *V. ×primulifolia* (Map 842) are very recent (1983–1984), from bogs, lake shores, and swales, near one or both presumed parent species.

10. **V. odorata** English or Sweet Violet

Map 843. This Old World species is widely cultivated, and becomes weedy in lawns and vacant places in gardens, spreading even to woods. Doubtless much more widespread than indicated by the map.

The flowers are blue or white, and are often very fragrant. There are many horticultural forms, including double-flowered ones.

11. **V. macloskeyi** F. E. Lloyd Fig. 303 Smooth White Violet

Map 844. Wet ground: especially bogs and conifer swamps, often on sphagnum hummocks; swampy woods, alder thickets, wet hollows in rich deciduous woods, sometimes on mossy logs; mossy rock crevices, meadows.

Included here is *V. pallens* (DC.) Brainerd, now generally considered to be the same species as the western American one, at best distinguished as ssp. *pallens* (DC.) M. S. Baker or var. *pallens* (DC.) C. L. Hitchc. Many keys and descriptions notwithstanding, the lateral petals in var. *pallens* are glabrous or at most with a few short hairs (not strongly bearded). The flowers are usually somewhat smaller than in the next two species. Flowering plants may be as small as 1.5 cm tall—or 10 times that size; the flowers almost invariably overtop the leaves. In *V. blanda* and *V. renifolia,* the flowers are often about equal to the leaves or even shorter.

841. Viola lanceolata

842. Viola ×primulifolia

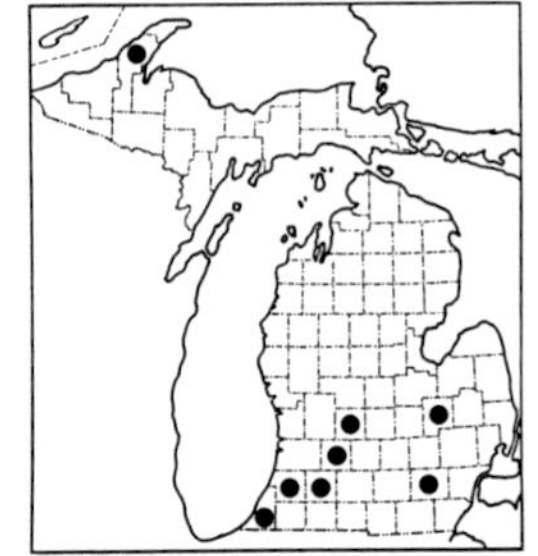

843. Viola odorata

12. **V. blanda** Willd. Sweet White Violet

Map 845. Mesic beech-maple woods or, more often, moister deciduous, mixed, or coniferous woods, swamps, and thickets, including northern hemlock-hardwoods, cedar (or tamarack) swamps, floodplain forests, creek borders; sometimes on mossy logs and hummocks in wet areas, occasionally in open ground.

True *V. blanda* is an eastern and Appalachian species which does not range into Michigan. More common on glaciated territory is var. *palustriformis* A. Gray, often treated as *V. incognita* Brainerd (and conceded to be very closely related if indeed distinct). *V. blanda* is much more widespread both geographically and ecologically in Michigan than *V. renifolia,* which is largely restricted to coniferous woods and swamps in the northern part of the state.

13. **V. renifolia** A. Gray Fig. 300 Kidney-leaved Violet

Map 846. Usually in cedar swamps; woods and thickets of other conifers as well.

Closely related to the preceding, but generally distinguishable as indicated in the key. Both species may have the lateral petals beardless or bearded, although the beard tends to be more pronounced in *V. blanda*. Leaf pubescence, if present, tends to be more dense in *V. renifolia*. Simplistic keys to the stemless white violets based on single characters, such as beard of petals or pubescence of leaves, are unreliable; careful observers will find many contradictions between different publications and even between keys and text in the same work.

Collections from a cedar swamp at Round Lake, Emmet County, have been placed by both Russell and Ballard as *V. blanda* × *V. renifolia* (*Fallass* in 1924, ALBC; *Voss 1885A* in 1954, UMBS). Such hybrids are surely to be expected elsewhere and a number of ambiguous specimens may be this.

14. **V. pedatifida** G. Don Prairie Violet

Map 847. A prairie species barely ranging into Michigan, where it is

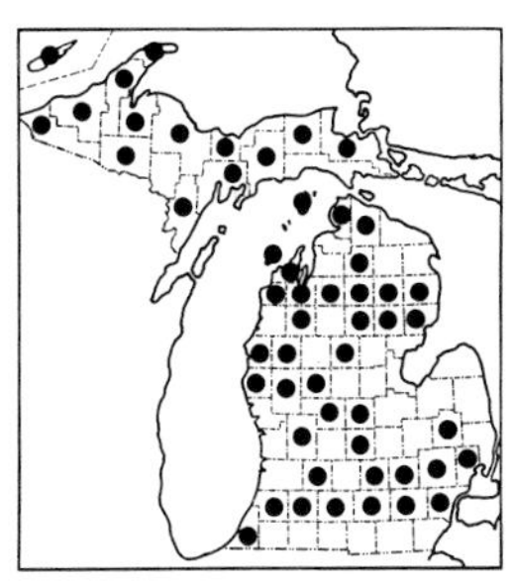

844. Viola macloskeyi

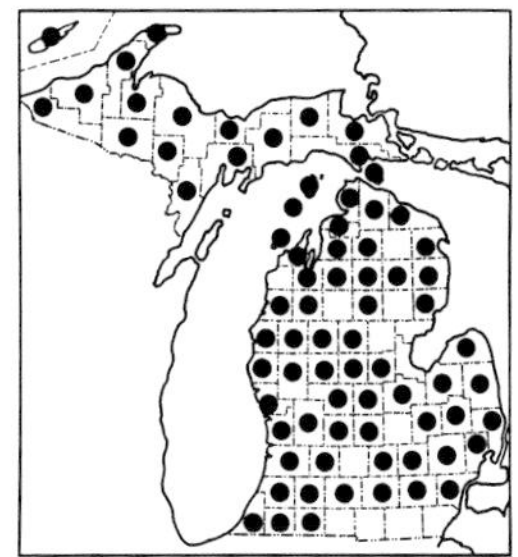

845. Viola blanda

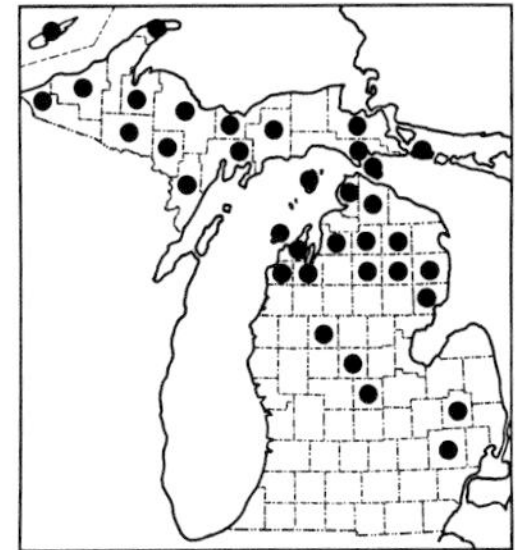

846. Viola renifolia

found on prairie remnants, including those along railroads; on a limestone bluff with *Andropogon* in Delta County. Some plants from Washtenaw County mapped as *V. palmata* are very close to *V. pedatifida*.

15. **V. pedata** L. Plate 7-B — Birdfoot Violet

Map 848. Sandy open plains, slopes, and woodlands, usually with jack pine and/or oaks.

The flowers have a flat, pansy-like aspect, unlike the preceding (and other) species, and may be as broad as 3 cm, making this an especially handsome plant. The upper petals may be deeper in color than the lower 3, or may be rosy with the lower ones blue or white. Or all the petals may rarely be white [f. *alba* (Thurber) Britton].

16. **V. selkirkii** Goldie — Great-spurred Violet

Map 849. Deciduous woods, especially beech-maple, frequently on rotting logs or mossy crevices in limestone; less often in coniferous woods and river bottomlands. This is a circumpolar violet, occurring also in northern Europe and Asia as well as Greenland, and is something of a calciphile.

As indicated by its position in the key, a very distinct species; and not known to hybridize with any other. Although small and easily overlooked, perhaps mistaken for the common leafy-stemmed *V. conspersa* with similarly colored light blue flowers, this species can be easily recognized by the long, thick spur or by the leaves with a tendency for the lobes to overlap and with minute pubescence only on the upper surface (very rarely a few hairs beneath).

17. **V. palmata** L. — Wood Violet

Map 850. Dry oak, oak-hickory, or even beech-maple woods, prairies, and thickets between woods and prairies; in shaded ground on a limestone bluff in Delta County, near *V. pedatifida*.

An extremely variable species, often with many leaf forms on the same individual (at the same time or as the season progresses), treated as a broad complex by Ballard (and others) to include Michigan plants previously re-

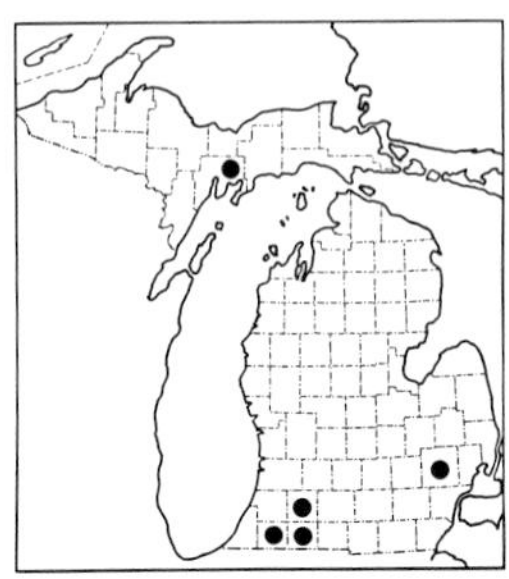
847. Viola pedatifida

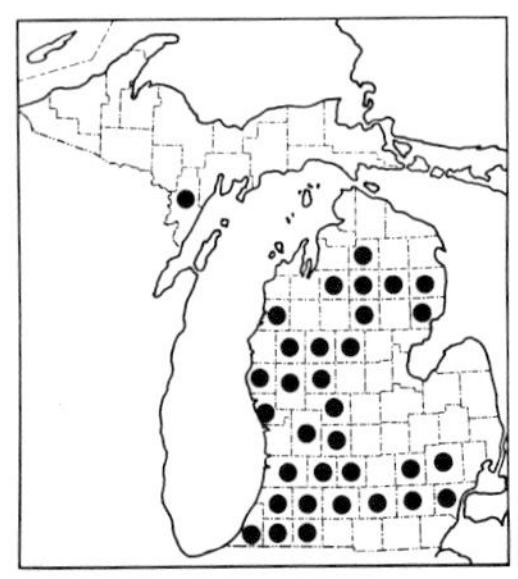
848. Viola pedata

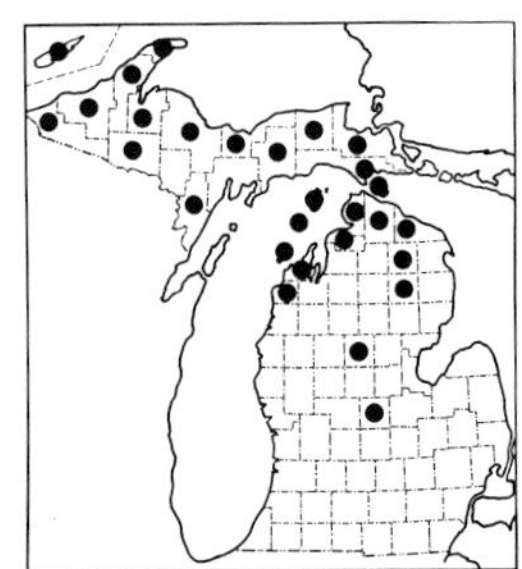
849. Viola selkirkii

ferred to *V. triloba* Schwein. as well as some probably of hybrid origin with *V. pedatifida* (× *V. sororia* and/or *V. sagittata*) or even hybrids of *V. sagittata* × *V. sororia*.

18. **V. sagittata** Aiton   Arrow-leaved Violet

Map 851. Dry to moist sandy open ground or woodlands; fields and meadows; often in transitional areas toward swamps and lakes.

The leaves vary from densely pubescent to glabrate, and the coarse teeth or short lobes at the base of the blade develop as the leaf matures, being much more prominent in summer. Included here are Michigan plants, at least, of *V. fimbriatula* J. E. Smith, sometimes distinguished as being more pubescent than *V. sagittata,* with leaves more obscurely toothed and petioles shorter, but perhaps only an environmental variant.

19. **V. cucullata** Aiton   Fig. 304   Marsh Violet

Map 852. Moist hardwoods (including low spots or trails in beech-maple woods), but more often in swamps (especially cedar) and bogs, often along streams, sometimes on mossy logs; alder thickets, wet meadows, shores.

The flowers may be quite large (to 25 mm long). Albinos are occasionally found [f. *albiflora* Britton]. The sepals are frequently ciliate and the auricles are often very prominent. Rarely there are a few hairs on the spurred petal, but large auricles will help to distinguish such plants from other species—if they are not hybrids. Some specimens approach *V. sororia* and others *V. affinis* or *V. nephrophylla*.

The name *V. obliqua* Hill has recently been resurrected for this plant, on rather unsubstantial grounds, by some European authors; but American students of the genus have been so unsure of the correct application of *V. obliqua* (see Rhodora 9: 93. 1907) that upsetting the name for one of our commonest wildflowers seems ill-advised.

20. **V. affinis** Le Conte

Map 853. Moist to swampy deciduous woods and thickets, river banks, meadows.

850. Viola palmata

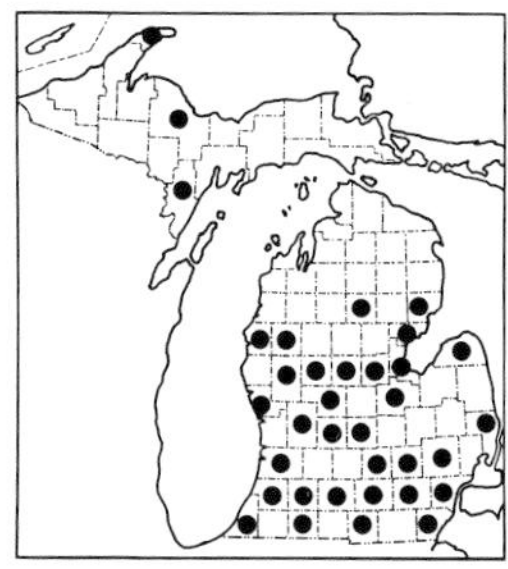

851. Viola sagittata

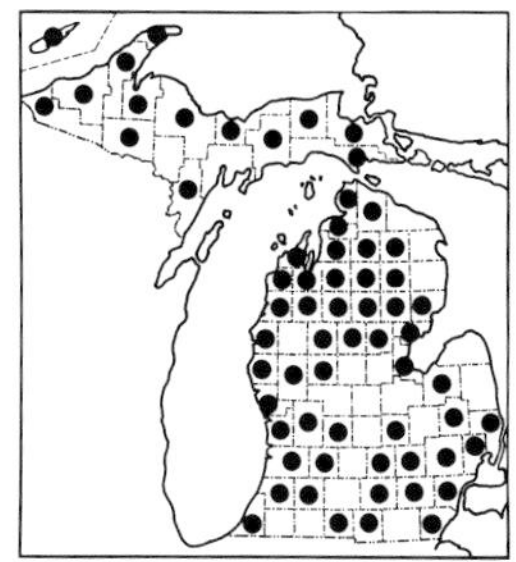

852. Viola cucullata

A rather difficult species to define, and perhaps dubiously distinct from *V. sororia,* with which it is apparently prone to hybridize, at least in this region. Plants with some pubescence or more rounded sepals or leaves than usual approach *V. sororia*. Those with ciliate sepals have been the basis for some reports of *V. septentrionalis* Greene from Michigan, and other pubescent plants have been referred to *V. novae-angliae* House. Hybrids with *V. cucullata* also apparently occur rarely.

21. **V. epipsila** Ledeb.

Map 854. A circumpolar violet, occurring, ± scattered, south to Lake Superior from more western and northern stations. In Michigan, known only from rich moist woods on Manitou Island, off the tip of the Keweenaw Peninsula (*Richards 3130* in 1950, BLH).

Great Lakes region material is referable to ssp. *repens* (Turcz.) Becker, and has sometimes been confused with *V. palustris* L. The small flowers, which are only light blue or lilac, very slender rhizomes (plus stolons), leaves very obtuse and more crenulate than serrate, and beardless spurred petal (with only a slight beard on lateral petals) will distinguish this violet from others in our region.

22. **V. nephrophylla** Greene

Map 855. Damp to wet places, including bogs (often marly), swamps, streamsides, borders of marshes, meadows, rock crevices, gravelly and rocky shores; a pronounced calciphile, often seen in interdunal hollows with *Primula mistassinica,* on damp limestone flats, and in similar habitats.

The large deep violet flowers are held well above the leaves. The blades are often only subcordate or nearly truncate at the base. Plants with white flowers have been named f. *albinea* Farwell (TL: near Erie [Monroe Co.]).

23. **V. sororia** Willd. Common Blue Violet

Map 856. Dry to moist, sometimes swampy, deciduous woods and thickets (including oak-hickory, beech-maple, very rarely pine).

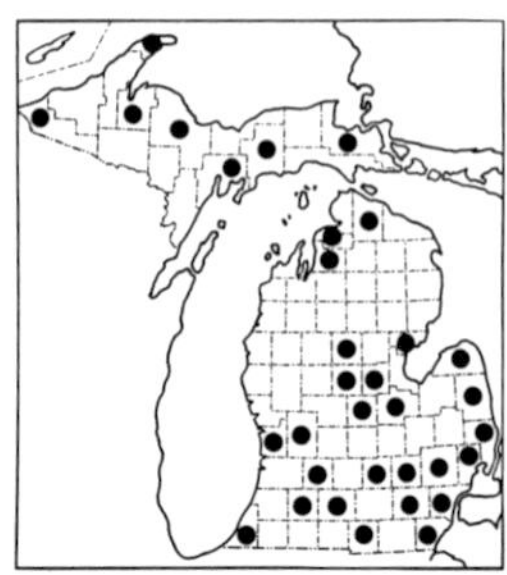
853. Viola affinis

854. Viola epipsila

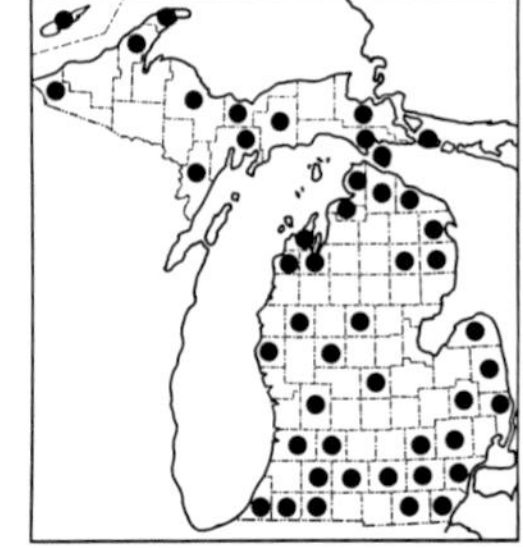
855. Viola nephrophylla

Included here are the completely glabrous plants long known under the name of *V. papilionacea* Pursh—a name apparently not clearly typified and hence uncertain as to correct synonymy, though probably belonging here. All gradations from glabrous through lightly to strongly pubescent leaves occur in *V. sororia*. Some reports of *V. septentrionalis* from Michigan have been based on plants now referred to *V. sororia*.

The spring leaves equal or surpass the flowers. The characters of leaf shape given in the key for these last five species may be rather subtle, and open to occasional exceptions (more so than the distinctions in sepal shape, although these too are ambiguous on some specimens). If the characteristic oblong-rounded sepals and ± obtuse short leaf blades of *V. sororia* are not clear enough to distinguish it from *V. affinis,* it may help to remember that pubescent plants of the latter (which may, of course, result from hybridization with *V. sororia*) should have the spurred petal at least lightly bearded. Occasional plants apparently *V. sororia* except for more acute sepals than usual may represent hybrids. Glabrous plants of *V. sororia* differ from *V. cucullata* in having shorter peduncles and poorly developed sepal auricles (which are relatively large in *V. cucullata,* especially as the fruit matures).

Intermediate or ambiguous specimens are frequent in the "stemless blues" but it is harder to recognize them as hybrids than it is to determine hybrids in the leafy-stemmed species. They may only represent variability within a species—or recognition of too many species!

CACTACEAE — Cactus Family

A well known family of the Western Hemisphere, but poorly represented in our area. Like most of the family, our two species are leafless, with fleshy green stems bearing spines. The strong spines arise from areoles containing numerous short barbed bristles, which are more irritating than the spines themselves. Modern systems of classification place the Cactaceae far from the families with which they were long associated, but rather in the same order as the Caryophyllaceae.

REFERENCES

Benson, Lyman. 1982. The Cacti of the United States and Canada. Stanford Univ. Press, Stanford. 1044 pp.

Reznicek, A. A. 1982. The Cactus in Southwestern Ontario. Ontario Field Biol. 36: 35–38. [Notes on both species in Ontario.]

1. **Opuntia** — Prickly-pear

An easily recognized genus, with jointed or segmented and branching stems, usually strongly flattened. The flowers are showy and (at least in our species) yellow. The plants are practically indestructible and if uprooted will survive many days of desiccation in hot sun before transplantation and recovery. Some of our northernmost records for *O. humifusa* probably represent colonies spread from ornamental planting. Published listings of an *Opuntia* from sandy ground "west of Alpena" have never been documented, but local reports indicate that it must have been *O. humifusa,* now exterminated by the advance of civilization.

KEY TO THE SPECIES

1. Segments of stem scarcely if at all flattened, readily detaching, less than 3 cm long (excluding spines); areoles mostly with 3 or more well developed spines . . 1. **O. fragilis**
1. Segments of stem strongly flattened, not detaching, usually more than 5 cm long; areoles mostly with 0–1 well developed spines . . . . 2. **O. humifusa**

1. **O. fragilis** (Nutt.) Haw. Plate 7-C

Map 857. A species of the western part of North America, Michigan being close to the eastern end of its range. Thriving on Huron Mountain in Marquette County on sunny rock surfaces. In Ogemaw County, of uncertain status on a hill east of St. Helen (where collected in 1967, MSC).

2. **O. humifusa** (Raf.) Raf.

Map 858. Sandy fields and plains, oak woods, sometimes in ± disturbed ground as along roadsides. The natural range in the state seems to be north to southernmost Oceana County (south of the South Branch of the White River, on soil classified as Sparta Loamy Sand); farther north, presumably escaped from cultivation.

Michigan is in an area of transition between this species and the more southwestern *O. macrorhiza* Engelm., with longer (over 3.8 cm) and more numerous spines, and some of our material has sometimes been referred to the latter. (Identical duplicates of one Muskegon County collection [*Voss 2861*] were cited as each species by Benson, but all five extant sheets of that number have spines less than 3 cm long, with at most a single well developed one at an areole.) Other names have been used for our plant but

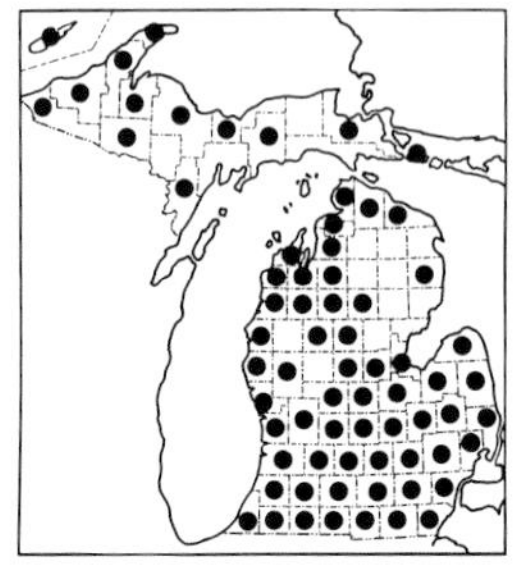

856. Viola sororia

857. Opuntia fragilis

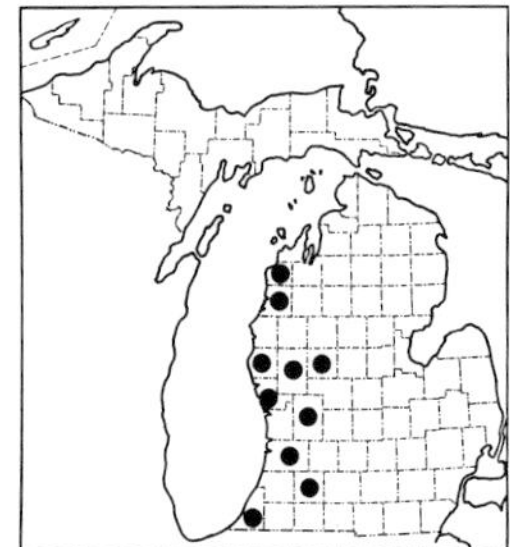

858. Opuntia humifusa

are nomenclaturally incorrect: *O. compressa* Macbride is a name often applied, under the impression that instead of dating from 1922 it is to be based on an older name which is, however, illegitimate. *O. rafinesquei* Engelm. is also a later name than either *O. humifusa* or *O. macrorhiza*.

THYMELAEACEAE — Mezereum Family

Besides our one native species, a cultivated shrub from Eurasia might be discovered as an escape, since it is locally established in southern Ontario. This is *Daphne mezereum* L., with fragrant, usually purplish flowers in which the floral tube ends in 4 spreading petal-like lobes. In *Dirca palustris,* which may also be fragrant, the flowers are pale yellow and the floral tube has no lobes (at most irregular at the margin). The flowers of both species appear before (or with) the leaves in the spring.

1. **Dirca**

REFERENCES

Nevling, Loren I., Jr. 1962. The Thymelaeaceae in the Southeastern United States. Jour. Arnold Arb. 43: 428–434.

Vogelmann, Hubert. 1953. A Comparison of Dirca palustris and Dirca occidentalis (Thymelaeaceae). Asa Gray Bull., N.S. 2: 77–82.

1. **D. palustris** L. Fig. 305 — Leatherwood

Map 859. Deciduous woods, especially rich moist sites.

An interesting shrub, attractive in its early-blooming yellow flowers with strongly exserted stamens. The bud scales form a sort of involucre with conspicuous dark hairs on the outside. The bark is very tough and pliable, the source of the common name. Some controversy has surrounded the color of the fruit, many sources saying (without original data) that it is reddish. In Michigan (as elsewhere in recent eye-witness accounts), the fruit is clearly

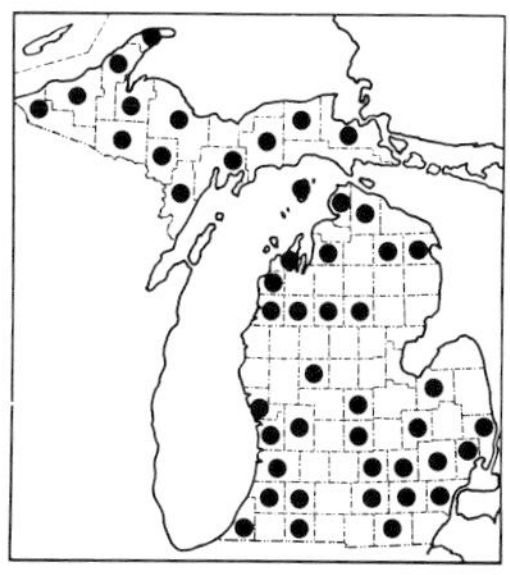

859. Dirca palustris

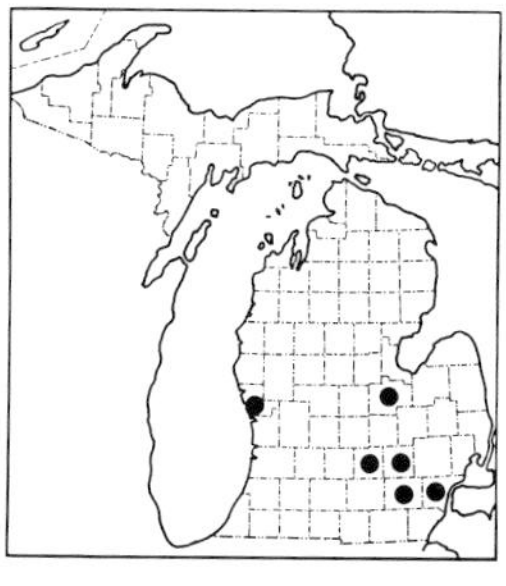

860. Elaeagnus angustifolia

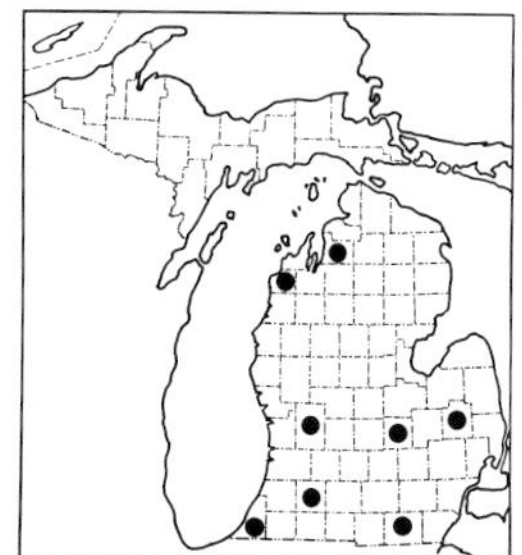

861. Elaeagnus umbellata

green when it is mature enough to fall; it may turn purplish when pressed and dried.

The alternate entire leaves give the foliage something of the look of the aromatic *Lindera,* but *Dirca* has peculiar angular swollen nodes with the petioles ± circular at the base, enclosing the developing axillary bud. A sharp line usually marks the contrast between the green petiole and the reddish brown twig.

ELAEAGNACEAE — Oleaster Family

This is a small family of shrubs and small trees, easily recognized by the dense peltate or stellate, silvery or rusty, scales or hairs on leaves, twigs, and buds. As in the Leguminosae, Myricaceae, *Alnus, Ceanothus,* and a few other plants, nodules on the roots are induced by certain soil bacteria (not the same as in the legumes, but Actinomycetes). The bacteria which inhabit these nodules in a symbiotic relationship are capable of fixing atmospheric nitrogen and thus enriching the soil.

REFERENCE

Graham, Shirley A. 1964. The Elaeagnaceae in the Southeastern United States. Jour. Arnold Arb. 45: 274–278.

KEY TO THE GENERA

1. Leaves alternate; stamens 4; flowers mostly perfect, appearing with (or after) the leaves . . . . 1. **Elaeagnus**
1. Leaves opposite; stamens 8; flowers unisexual (the plants dioecious), appearing before the leaves . . . . 2. **Shepherdia**

1. Elaeagnus — Oleaster

KEY TO THE SPECIES

1. Leaves and branchlets with only silvery scales; expanded floral tube (i. e., beyond the constriction around the ovary) less than twice as long as the lobes . . . . 1. **E. angustifolia**
1. Leaves and branchlets, at least when young, with some scattered rusty brown scales; expanded floral tube at least twice as long as the lobes . . . . 2. **E. umbellata**

1. **E. angustifolia** L. — Russian-olive

Map 860. A small tree, native to Europe and western Asia, frequently planted and very rarely escaping to fields, river banks, and other sites.

The fruit is silvery and drupe-like; the foliage is very silvery, the leaves narrowly elliptic.

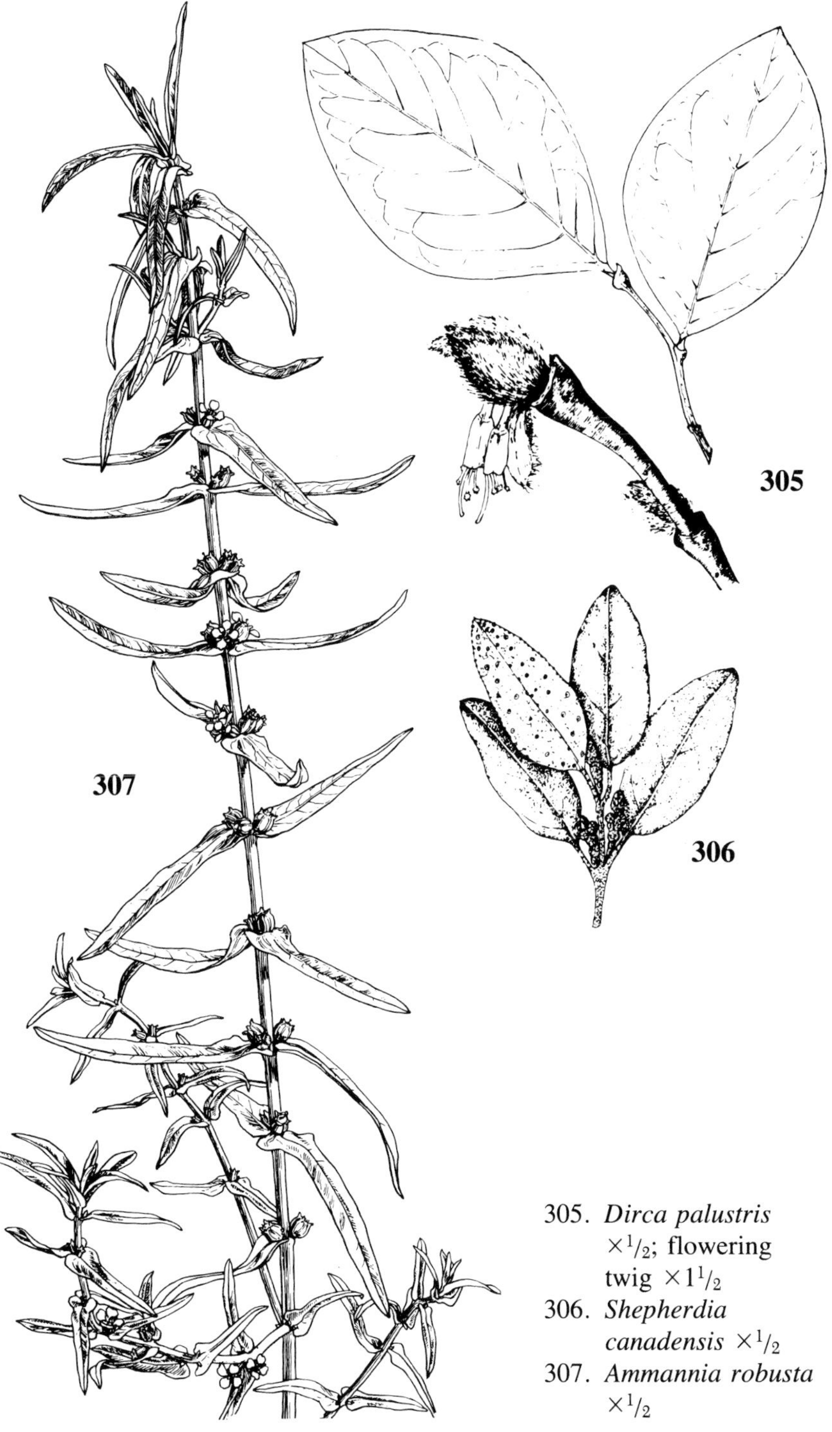

305. *Dirca palustris* ×1/2; flowering twig ×1 1/2
306. *Shepherdia canadensis* ×1/2
307. *Ammannia robusta* ×1/2

2. **E. umbellata** Thunb.

Map 861. An Asian species, planted for ornament or wildlife habitat and too freely escaping to roadsides, woods, fields, filled land, gravel pits—and almost anywhere. Thoroughly established as a weed in parts of Lapeer County, for example, in both dry and wet habitats.

The fruit is red and juicy when ripe. The leaves are green or nearly so above, especially when mature, and usually more broadly elliptic to ovate than in the preceding species.

E. commutata Rydb. is native in northeastern North America, south to Lake Superior, but has not yet been found in Michigan. It differs from *E. umbellata* in having dry mealy fruit, and leaves ± silvery above even when mature; it also has a shorter floral tube, but the rusty brown scales mixed with silvery ones and the habitat in natural communities should distinguish it from *E. angustifolia*.

2. Shepherdia — Buffalo-berry

1. **S. canadensis** (L.) Nutt. Plate 7-D; fig. 306 — Soapberry

Map 862. Dunes, gravelly and rocky shores, ridges, openings; rock outcrops (especially limestone); open woods (aspen, pine, birch, cedar, fir); clay and calcareous river banks; moist thickets, even fens; readily invading fields and clearings.

The flowers open just before the leaves in May or late April. They are very inconspicuous, with 4 tiny yellowish sepals (scurfy-pubescent outside), 8 nectariferous glands but no petals, and either 8 stamens or a single pistil. The fleshy red-orange fruit follows (on pistillate plants) in early summer and is not especially tasty. This is an easily recognized shrub with its opposite leaves, densely scurfy-pubescent beneath with mixed silvery and rusty brown scales and hairs, sparsely stellate-pubescent above.

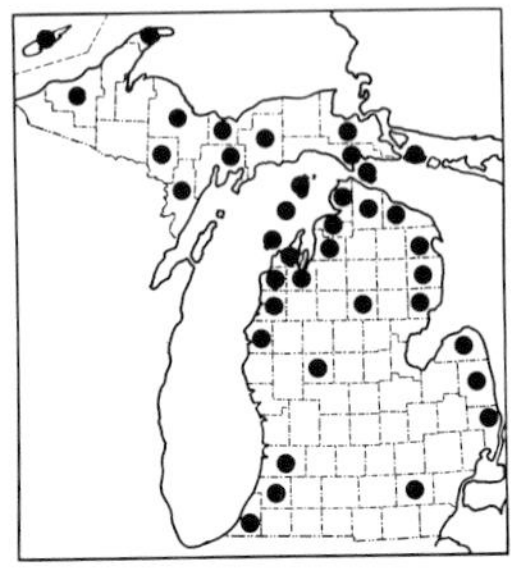

862. Shepherdia canadensis

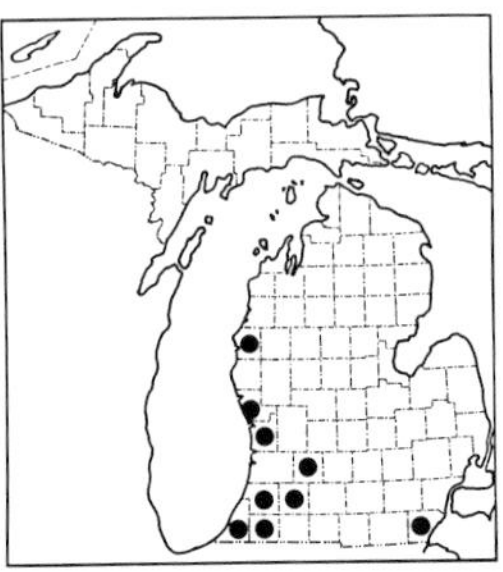

863. Rotala ramosior

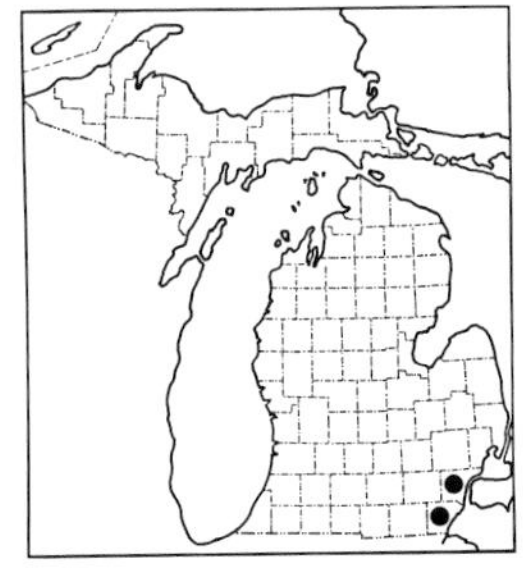

864. Ammannia robusta

LYTHRACEAE

Loosestrife Family

Many species in this family are noted for dimorphic or trimorphic flowers, that is, with styles and stamens of various lengths (e. g., long styles and short stamens or vice versa), thus helping to ensure cross-pollination. The calyx lobes (in our species) alternate with appendages—as in many Rosaceae. The ovary is free from the perigynous floral tube and hence is superior.

The references listed below may help not only in understanding our species but also in identifying additional species which might stray into our area (e. g., in *Didiplis* and *Cuphea*).

REFERENCES

Blackwell, Will H., Jr. 1970. The Lythraceae of Ohio. Ohio Jour. Sci. 70: 346–352.

Graham, Shirley A. 1964. The Genera of Lythraceae in the Southeastern United States. Jour. Arnold Arb. 45: 235–250.

Graham, Shirley A. 1975. Taxonomy of the Lythraceae in the Southeastern United States. Sida 6: 80–103.

Ugent, Donald. 1963 ["1962"]. Preliminary Reports on the Flora of Wisconsin. No. 47. The Orders Thymelaeales, Myrtales, and Cactales. Trans. Wisconsin Acad. 51: 83–134.

KEY TO THE GENERA

1. Calyx with 4 teeth or shallow lobes slightly exceeding the appendages; plant at maturity ± sprawling in habit, less than 4 dm tall
   2. Flowers and fruits solitary in the axils; leaves (at least middle and upper ones) tapered to the base . . . . 1. **Rotala**
   2. Flowers and fruits mostly clustered in the axils; leaves cordate-clasping at the base . . . . 2. **Ammannia**
1. Calyx with 5–6 (7) teeth or lobes distinctly shorter (and broader) than the appendages; plant at maturity erect or high-arching, only rarely less than 4 dm tall
   3. Floral tube campanulate (about as broad as long, excluding the calyx lobes); flowers mostly in distinct axillary clusters (not appearing crowded terminally); leaf blades tapered to a short petiole . . . . 3. **Decodon**
   3. Floral tube elongate (at least twice as long as broad); flowers (though axillary) often crowded in dense terminal inflorescences; leaf blades sessile or even clasping . . . . 4. **Lythrum**

1. Rotala

1. **R. ramosior** (L.) Koehne Fig. 308 Tooth-cup

Map 863. Sandy to mucky shores, interdunal swales, and clearings—usually periodically inundated; drying and burned marshes were found by the Haneses to be the preferred habitat in Kalamazoo County.

Under good magnification [20×] and illumination, the surface of the capsule appears covered with minute transverse striations, while in *Ammannia* it is smooth. This easily overlooked little plant is sometimes confused with

Ammannia and also with the common *Ludwigia palustris,* which usually has much broader leaves than the linear to narrowly elliptic ones of *Rotala* and which also has longer calyx lobes with no appendages between them.

2. Ammannia

REFERENCE

Graham, Shirley A. 1979. The Origin of *Ammannia* ×*coccinea* Rottboell. Taxon 28: 169–178.

1. **A. robusta** Heer & Regel Fig. 307

Map 864. Probably spread recently into southern Michigan. Except for a 1914 collection from Detroit (*Farwell 3929,* BLH), this species was not collected in the state before 1972. Now well established in marshy ground and muddy flats, especially disturbed areas.

This is the species previously known in this region as *A. coccinea* Rottb., a name which applies to an amphidiploid fertile hybrid derived from *A. auriculata* × *A. robusta*; *A. coccinea* as thus interpreted occurs to the south of our area.

Plants of this species and of *Rotala* are often strongly flushed with red. Petals in both are quickly deciduous, but are a little larger and showier in *Ammannia*.

3. Decodon

1. **D. verticillatus** (L.) Ell. Fig. 309 Whorled or Swamp Loosestrife

Map 865. Shallow water and mucky shores of ponds, lakes, marshes, and bogs, sometimes forming floating mats. Often grows in very soft substrates; a fringe of *Decodon* and *Cephalanthus* along the water's edge usually indicates a good place to go plunging to one's waist with the slightest misstep.

This is the only species in the genus. The stems form an extensive pale

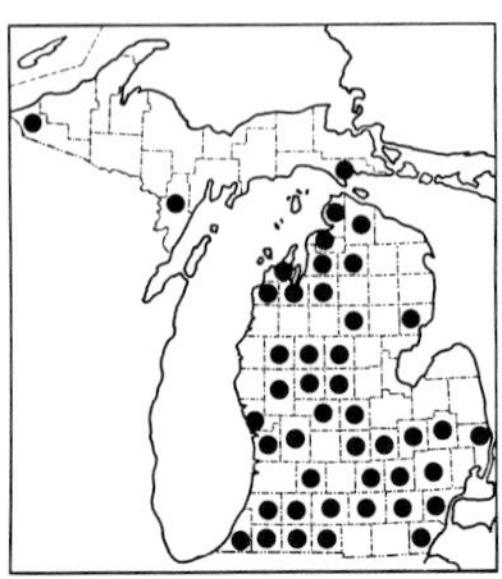
865. Decodon verticillatus

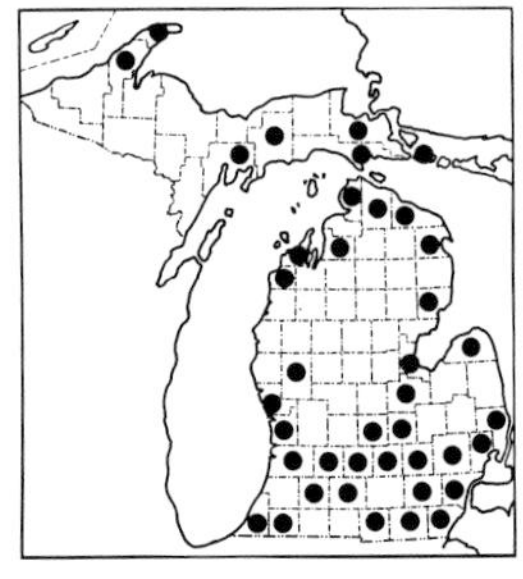
866. Lythrum salicaria

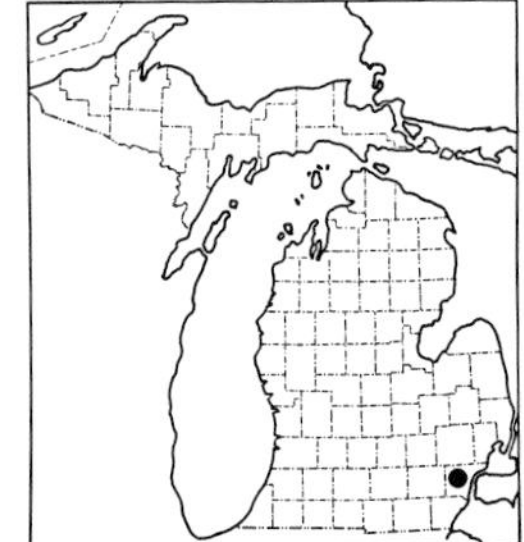
867. Lythrum hyssopifolia

spongy cork-like tissue (but containing cytoplasm) in the water. This "aerenchyma" forms at the base of the stem and at the tip of an arching stem when it strikes water and roots. The arching stems with enlarged floating tips—sometimes with another low arch developing beyond—are a unique aspect of this loosestrife.

4. **Lythrum** Loosestrife

REFERENCES

Cody, William J. 1978. The Status of Lythrum alatum (Lythraceae) in Canada. Canad. Field-Nat. 92: 74–75.

Shinners, Lloyd H. 1953. Synopsis of the United States Species of Lythrum (Lythraceae). Field & Lab 21: 80–89.

Stuckey, Ronald L. 1980. Distributional History of Lythrum salicaria (Purple Loosestrife) in North America. Bartonia 47: 3–20.

Teale, Edwin Way. 1982. Stems Beyond Counting, Flowers Unnumbered. Audubon 84(4): 38–43.

KEY TO THE SPECIES

1. Flowers in axillary clusters of 2–several, the ± ovate-acuminate bracts and floral tube pubescent; petals ca. 7–11 mm long . . . . 1. **L. salicaria**
1. Flowers solitary in the axils of linear-lanceolate leaves or bracts, these and the floral tube glabrous; petals ca. 2–6.5 (7) mm long
  2. Petals ca. 2–3 mm long; flowers borne all (or almost all) the way to the base of the plant in axils of small ± linear leaves . . . . 2. **L. hyssopifolia**
  2. Petals ca. 4–6.5 (7) mm long; flowers borne in the axils of small linear bracts on the upper half of the plant, which bears larger leaves below . . . . 3. **L. alatum**

1. **L. salicaria** L. Purple Loosestrife

Map 866. A Eurasian species, introduced into North America by the earliest 19th century for its colorful flowers, and now an attractive but persistent weed spreading vigorously in wet ground (see Stuckey 1980, esp. p. 9 for Michigan history). Marshes, shores, borders of rivers and streams, ditches; aggressively crowding out the native wetland flora. Reported by the First Survey (1838), but the oldest Michigan collection located by Stuckey dates from 1879 (Muskegon). Almost all other records from the state date from the 1890's onward.

The rather dense pubescence, especially on the upper part of the plant, will readily distinguish this species from the others, as will the larger and more numerous, hence much more showy, flowers in thick purple spike-like inflorescences. Variable in leaf shape and in pubescence, some plants being more glabrate and/or with narrower leaves than most. One collection from near Sheldon in western Wayne County (*Farwell 7139* in 1924, MICH, BLH) has not only relatively narrow leaves but also narrow bracts with only 1–2 flowers in the axils; the appendages are scarcely longer than the calyx lobes and the whole floral tube is only lightly pubescent. The thought of a hybrid with *L. alatum* is tempting—or with the entirely glabrous cultivated *L. virgatum* L. (which has, however, more flowers per axil).

2. **L. hyssopifolia** L.

Map 867. A European species, reported from Michigan as early as 1824 by Amos Eaton, but no supporting specimens have been found that old. The one Michigan record (see Stuckey in Mich. Bot. 17: 178–179. 1978) is a specimen (PH) apparently collected by N. W. Folwell in 1832, but with no data except "Detroit."

3. **L. alatum** Pursh Fig. 312

Map 868. Shores and wet meadows (especially frequent along Saginaw Bay), wet prairies, marshy ground, moist sandy openings.

Following Graham (1975), this name is applied in a broad sense, including *L. dacotanum* Nieuwl., which has sometimes been used for midwestern plants.

NYSSACEAE Tupelo Family

Modern authors are agreed that this family is very closely related to the Cornaceae. In the past, it has even been included in that family, although sometimes placed in a different order, near the Melastomataceae.

REFERENCE

Eyde, Richard H. 1966. The Nyssaceae in the Southeastern United States. Jour. Arnold Arb. 47: 117–125.

1. Nyssa Tupelo

1. **N. sylvatica** Marsh. Fig. 313 Sour-gum; Black-gum; Pepperidge

Map 869. Dry or usually moist woods or wet depressions in woods, borders of swamps (even with tamarack), shores.

A distinctive tree in silhouette, with often drooping branches becoming very crooked and "twiggy" toward their ends. The shining, entire leaves are alternate (but crowded when at ends of shoots), otherwise rather like dogwood leaves, and turn bright red in the fall—at least in part (for they tend to retain patches of green). Leaves on seedlings may be coarsely toothed. The petioles and base of the midrib beneath are usually ± pilose. The flowers, crowded at the ends of long peduncles, are often unisexual, the plants dioecious or polygamo-dioecious.

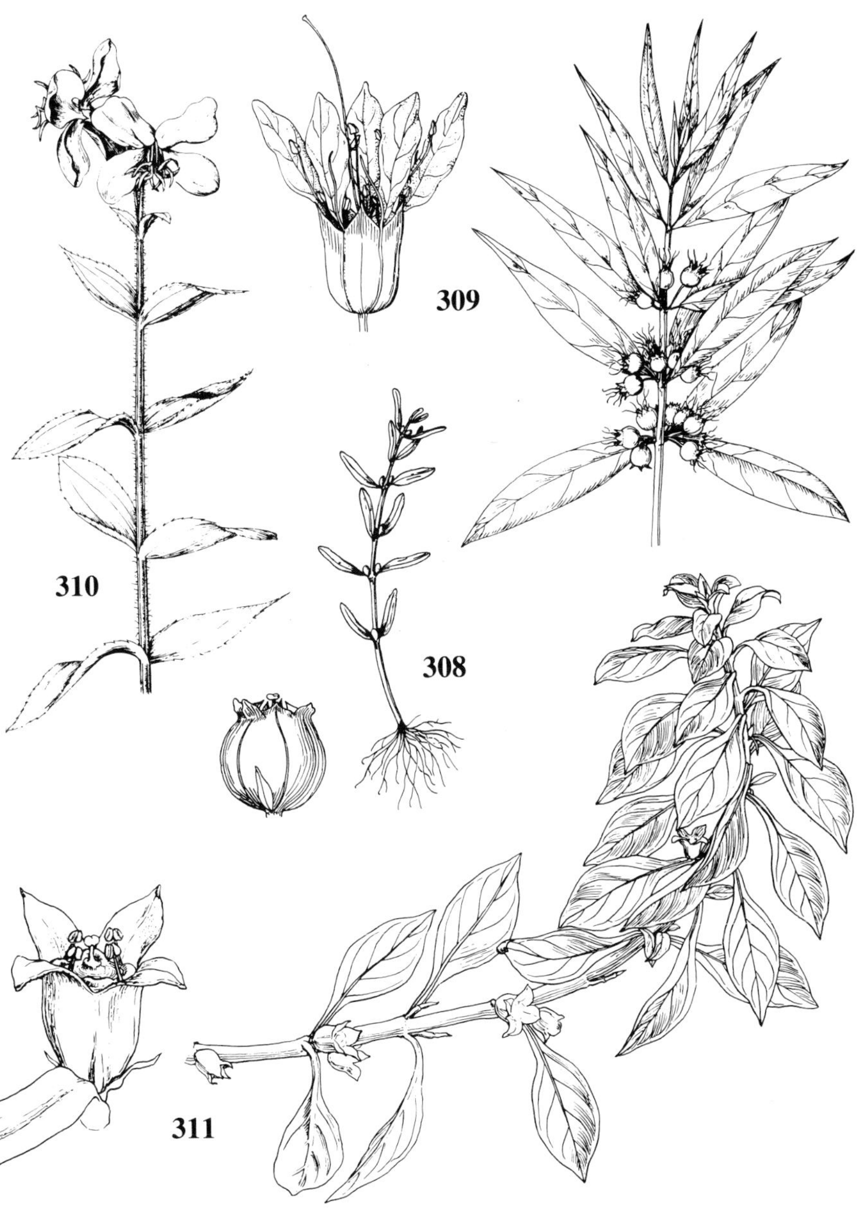

308. *Rotala ramosior* $\times{}^{1}/_{2}$; fruit $\times 5$
309. *Decodon verticillatus* $\times{}^{1}/_{2}$; long-styled flower $\times 2$
310. *Rhexia virginica* $\times{}^{1}/_{2}$
311. *Ludwigia palustris* $\times 1$; flower $\times 5$

MELASTOMATACEAE

Melastome Family

This is a very large, predominantly tropical family, of which a single genus ranges into temperate North America. The leaves in the entire family typically have 3 or more prominent longitudinal veins from the base, and are opposite; the anthers open by apical pores.

1. **Rhexia**

REFERENCE

Kral, R., & P. E. Bostick. 1969. The Genus Rhexia (Melastomataceae). Sida 3: 387–440.

1. **R. virginica** L. Fig. 310 Meadow-beauty

Map 870. While not itself sufficiently disjunct to be considered representative of the "Coastal Plain disjunct" element in our flora, this species is in a genus centered in that part of the continent and is often found associated with Coastal Plain species in open, damp, sandy to mucky marshy places and shores.

Meadow-beauty is locally common and a fine sight when in bloom, for it is one of our most attractive wildflowers, with pink-purple corolla ca. 2–3 cm broad and large bright yellow anthers. The flask-shaped floral tube is ± loosely glandular-bristly, and is adnate to the ovary, which is thus inferior.

ONAGRACEAE

Evening-primrose Family

The challenges offered in this family have been taken up by a number of workers in recent years. The treatment here owes much to Munz (1965), often modified in some genera by the research of Raven and his collaborators, to whom I am grateful for many helpful comments and suggestions.

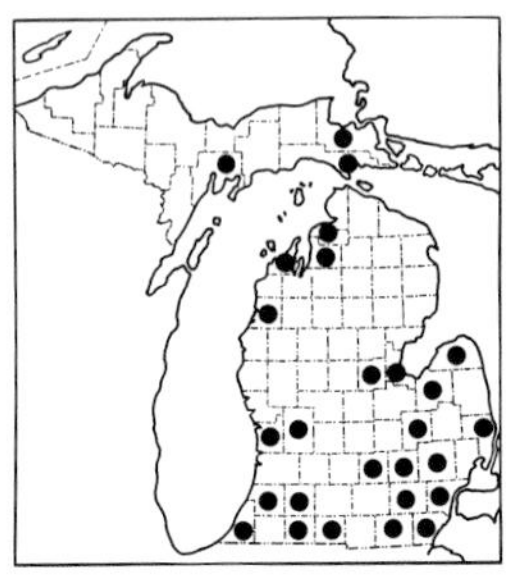

868. Lythrum alatum

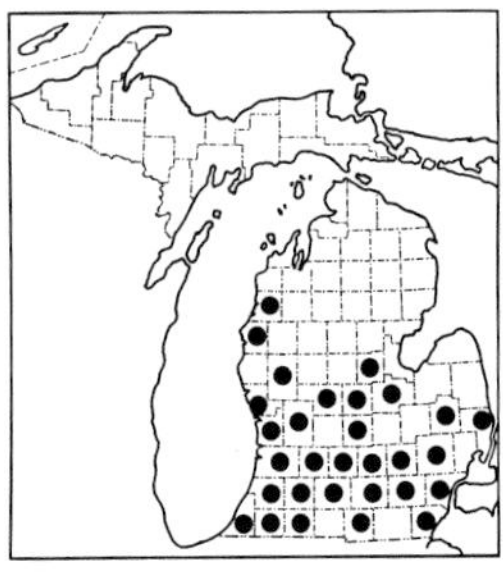

869. Nyssa sylvatica

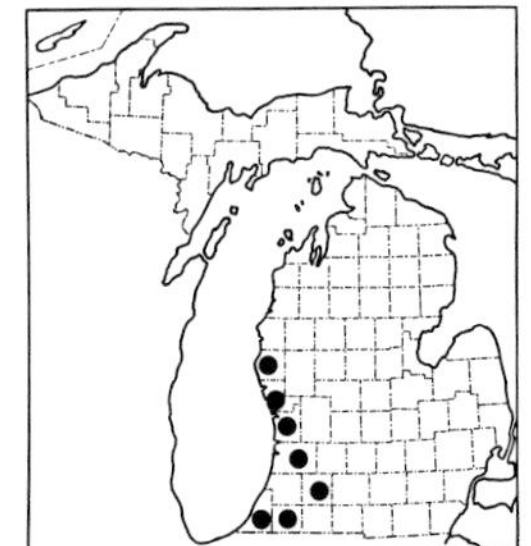

870. Rhexia virginica

REFERENCES

Munz, Philip A. 1965. Onagraceae. N. Am. Fl. II(5). 278 pp.
Ugent, Donald. 1963 ["1962"]. Preliminary Reports on the Flora of Wisconsin. No. 47. The Orders Thymelaeales, Myrtales, and Cactales. Trans. Wisconsin Acad. 51: 83–134.

KEY TO THE GENERA

1. Floral parts (sepals, petals, stamens) in 2's; fruit ± densely pubescent with hooked hairs, indehiscent; leaves opposite .... 1. **Circaea**
1. Floral parts in 4's; fruit glabrous or with hairs lacking hooked tip, dehiscent or indehiscent; leaves opposite or alternate
 2. Floral tube scarcely if at all prolonged beyond the end of the ovary; leaves opposite or alternate
  3. Petals none or yellow to greenish and minute (at most, in one species, equaling the sepals); calyx lobes persistent on the oblong to subglobose fruit; stamens 4; seeds without hairs .... 2. **Ludwigia**
  3. Petals white to magenta, exceeding the sepals; calyx lobes deciduous from the linear fruit; stamens 8; seeds with a tuft of silky hairs .... 3. **Epilobium**
 2. Floral tube prolonged beyond the ovary for about half as long as the ovary itself, or more; leaves alternate
  4. Petals white, turning pink, less than 1 cm long; fruit indehiscent, 2–6-seeded .... 4. **Gaura**
  4. Petals yellow (if white to pink, then over 1 cm long); fruit dehiscent, many-seeded
   5. Ovary 4-angled; fruit sharply 4-angled or -winged, tapered to the base .... 6. **Oenothera** (couplet 2)
   5. Ovary terete (or with several low ridges); fruit terete to obscurely 4-sided, abruptly rounded at the sessile base
    6. Floral tube ± funnel-shaped, about half as long as the ovary (or even shorter); stigma broad, with 4 rounded (not linear) shallow lobes; leaves linear-lanceolate, less than 7 mm broad, sharply but remotely toothed .... 5. **Calylophus**
    6. Floral tube linear, longer than the ovary; stigmas with 4 linear lobes; leaves various .... 6. **Oenothera** (couplet 4)

1. **Circaea** — Enchanter's-nightshade

REFERENCES

Boufford, David E. 1983. The Systematics and Evolution of Circaea (Onagraceae). Ann. Missouri Bot. Gard. 69: 804–994.
Haber, Erich. 1977. Circaea ×intermedia in Eastern North America with Particular Reference to Ontario. Canad. Jour. Bot. 55: 2919–2935.

KEY TO THE SPECIES

1. Flowers on glabrous, erect to ascending pedicels, open before elongation of the axis of the raceme (hence the open flowers appearing clustered at end of raceme); fruit 0.9–1.2 mm thick (excluding hairs), not ridged and grooved .... 1. **C. alpina**
1. Flowers on pubescent, widely spreading pedicels, open after elongation of the axis of the raceme; fruit 2–2.8 mm thick, when fully mature with several strong ridges and grooves .... 2. **C. lutetiana**

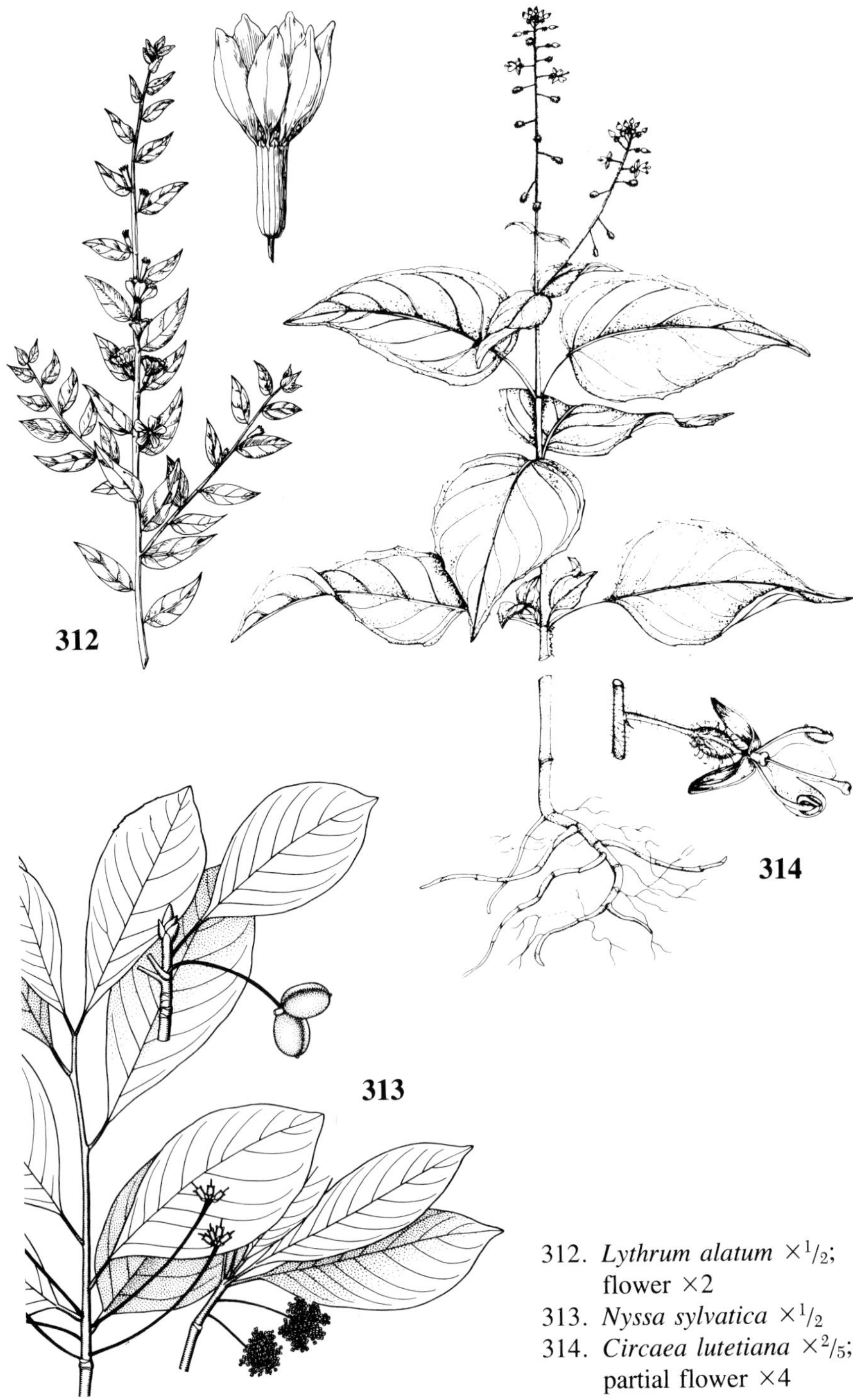

312. *Lythrum alatum* ×1/2; flower ×2
313. *Nyssa sylvatica* ×1/2
314. *Circaea lutetiana* ×2/5; partial flower ×4

1. **C. alpina** L.

Map 871. Swamps (especially cedar), frequently on rotting logs; hardwood forests, especially in damp depressions, ravines, and springy slopes.

Usually a small delicate-looking plant, occasionally as tall as 30 cm. The next species is only very rarely as short as 30 cm. *C. alpina* blooms earlier than *C. lutetiana,* with only a slight overlap in some seasons. The hybrid [*C.* ×*intermedia* Ehrh.], quite intermediate between the two (with ovaries aborting and highly sterile pollen), is likely to be mistaken for either parent, depending on size of the plant. It is very rare in the state, having been collected a few times in Marquette County, as identified by Raven, Haber, and Boufford; a Kent County collection appears to be the same.

2. **C. lutetiana** L. Fig. 314

Map 872. Dry (e. g., oak-hickory) or more often rich moist deciduous woods, frequently in ravines and wet spots; sometimes in swamps, even with tamarack, or along streams.

American plants may be designated as var. *canadensis* L. [or ssp. *canadensis* (L.) Asch. & Magnus] to distinguish them from the European ones. Our plants have also been called *C. quadrisulcata* (Maxim.) Franchet & Savat.

2. **Ludwigia** — False Loosestrife

REFERENCE

Munz, Philip A. 1944. Studies in Onagraceae—XIII. The American Species of Ludwigia. Bull. Torrey Bot. Club 71: 152–165.

KEY TO THE SPECIES

1. Leaves opposite; plants prostrate or floating in shallow water, usually rooting at the nodes . . . . . . 1. **L. palustris**
1. Leaves alternate; plants erect
  2. Flowers on distinct pedicels ca. 2–5 mm long; petals conspicuous, about equaling the sepals; roots usually tuberous-thickened . . . . . . 2. **L. alternifolia**
  2. Flowers sessile or nearly so; petals apparently none; roots not becoming thickened (but plant stoloniferous and the stem often with well developed spongy tissue at base)
    3. Bractlets at base of ovary 0–1 mm long; capsules, leaves, and upper part of stem at least sparsely strigose . . . . . . 3. **L. sphaerocarpa**
    3. Bractlets at or (usually) above base of ovary 2.5–6 (8) mm long; capsules, leaves, and stem glabrous . . . . . . 4. **L. polycarpa**

1. **L. palustris** (L.) Ell. Fig. 311 — Water-purslane

Map 873. Margins of rivers and lakes in shallow water, often fruiting heavily on recently exposed banks; ditches and marshy ground.

An easily recognized prostrate plant of wet places, the opposite leaves with ± diamond-shaped blades and bearing 4-merous flowers or fruit in the axils.

2. **L. alternifolia** L.

Map 874. Marshy ground, borders of swamps, wet thickets, shores, clearings.

3. **L. sphaerocarpa** Ell.

Map 875. Disjunct from the Coastal Plain to northwestern Indiana and southwestern Michigan. Although Michigan is mentioned in the range as stated in manuals, the only collection seen was made in 1981 (*Wells & Thompson 81316,* MICH, BLH), along the shore of Gilligan Lake.

4. **L. polycarpa** Short & Peter

Map 876. Marshy and swampy ground; ditches and sandy excavations; the Houghton County collection so far from the rest of the range is from wet places along a railroad.

3. Epilobium — Willow-herb

A collection from Washtenaw County (*Erlanson & Hermann 5177* in 1927, MICH) has been identified by Hoch as *E. obscurum* Schreber, a Eurasian and North African species locally naturalized in New Zealand, Australia, and South America, but not previously recorded from North America. Its status at "Gunther Gardens" in 1927 is obscure, though it was probably a weed at the site, which was a private garden (including greenhouse) that has now been abandoned for over half a century and reverted to the wild. Among other characters, the leaves of *E. obscurum* have flat (not revolute) margins and are denticulate, as in *E. ciliatum;* but the inflorescence and stem have ± dense whitish pubescence of minute incurved non-glandular hairs, as in *E. leptophyllum* (*E. ciliatum* has glandular puberulence).

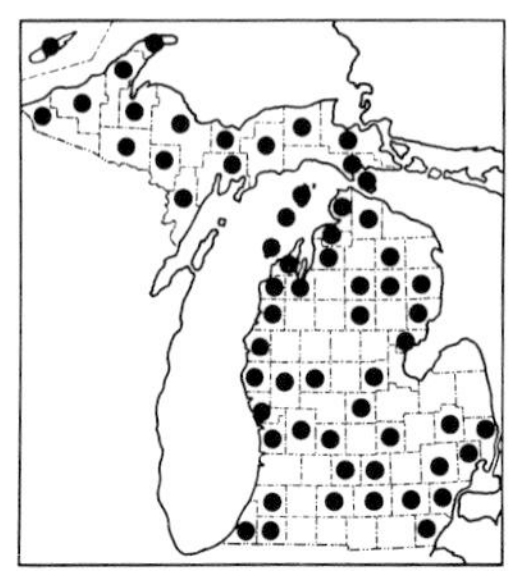

871. Circaea alpina

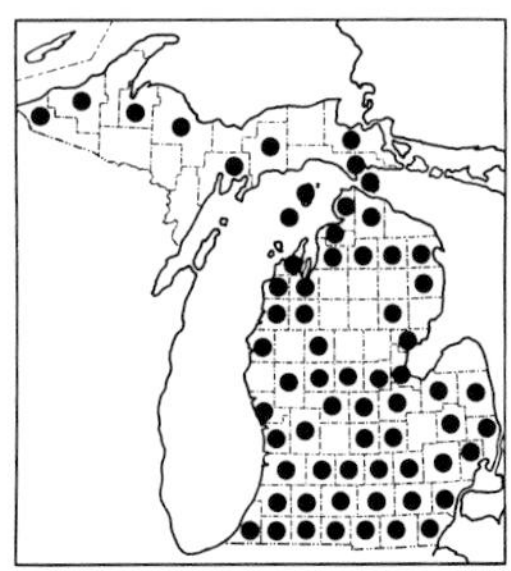

872. Circaea lutetiana

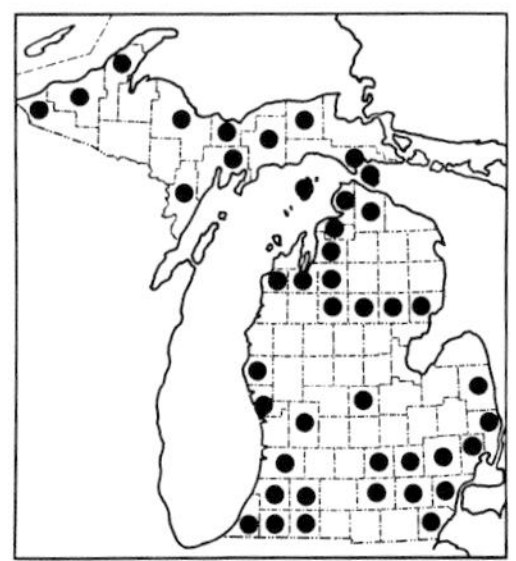

873. Ludwigia palustris

REFERENCES

Mosquin, Theodore. 1966. A New Taxonomy for Epilobium angustifolium L. (Onagraceae). Brittonia 18: 167–188.

Purcell, Nancy J. 1976. Epilobium parviflorum Schreb. (Onagraceae) Established in North America. Rhodora 78: 785–787.

Stuckey, Ronald L. 1970. Distributional History of Epilobium hirsutum (Great Hairy Willow-herb) in North America. Rhodora 72: 164–181.

KEY TO THE SPECIES

1. Leaves all alternate; flowers in long, terminal, showy racemes, the buds reflexed; floral tube none; stems and leaves glabrous or very minutely pubescent . . . . 1. **E. angustifolium**
1. Leaves, except sometimes the upper ones, opposite; flowers all or mostly in the axils of leaves or leafy bracts, the buds not reflexed; floral tube ca. 0.3–3.5 mm long beyond the ovary; stems glabrous to spreading-hairy (if upper leaves alternate and flowers showy, the plant with horizontally spreading pubescence)
  2. Stems pubescent with horizontally spreading (not incurved) hairs
    3. Stigma entire; leaves less than 8 (10) mm wide, entire or nearly so with slightly revolute margins . . . . 2. **E. strictum**
    3. Stigma deeply 4-lobed; leaves (at least the larger ones) over 8 mm wide, toothed or at least remotely but distinctly denticulate
      4. Leaves sessile or subsessile but the blades not at all clasping or decurrent; petals at most 10 mm long . . . . 3. **E. parviflorum**
      4. Leaves sessile and (except sometimes the smallest) the blades clasping up to halfway around the stem or often (the largest) slightly decurrent; petals (9) 11–17 mm long . . . . 4. **E. hirsutum**
  2. Stems glabrous or with pubescence all or in large part of minute incurved hairs (glandular puberulence may be present)
    5. Leaves linear to narrowly lanceolate (usually less than 4 mm, rarely as much as 1 cm, wide), with entire ± revolute margins or rarely (when broader) with flat but entire to obscurely undulate margins; internodes not angled (except sometimes immediately beneath the nodes); puberulence not glandular, appressed-incurved
      6. Leaves essentially glabrous except for minute incurved hairs on or near the midrib (above and/or below); tip of stem ± strongly nodding when flowers are still in bud . . . . 5. **E. palustre**
      6. Leaves (at least many of them) with ± evenly distributed minute incurved hairs on upper surface; tip of stem erect . . . . 6. **E. leptophyllum**
    5. Leaves lanceolate to lance-elliptic (the largest often as much as 1.5 cm wide), definitely toothed or denticulate; internodes often ± angled from slender ridges extending well below the nodes; puberulence glandular
      7. Teeth of leaf margin often scarcely more than dark (non-green) gland-like denticulations; petioles of midcauline leaves none or less than 2 (3) mm long, often doubled or tripled in width by broad wings; hairs of seeds whitish . . . 7. **E. ciliatum**
      7. Teeth of leaf definite (apart from gland-like tips) between quite irregular rounded sinuses; petioles of midcauline leaves mostly 2–5 (6) mm long, narrowly if at all winged; hairs of seeds copper-colored or brownish at maturity . 8. **E. coloratum**

1. **E. angustifolium** L. Fireweed; Great Willow-herb

Map 877. Dry woods (aspen, jack pine, etc.), fields, roadsides, rocky ground; clearings and borders of woods, upper shores; gravel pits and other disturbed ground; frequently in rather low wet places; as the common name suggests, thrives in burned-over areas, blooming in profusion as soon as three months after a spring fire. "One of the most completely circumpolar of all plants" (Hultén), this species grows all around the world in northern latitudes.

Our plants have been referred to ssp. *circumvagum* Mosquin, a tetraploid; typical diploid ssp. *angustifolium* occurs much farther to the north and at high elevations in the west. A similar-looking northern species which does not range south to the Great Lakes region is *E. latifolium* L. (with style glabrous, rather than pubescent toward its base). Those who segregate these plants into a separate genus call our species *Chamerion angustifolium* (L.) Holub. White-flowered plants are at least as handsome as the normal ones and may be expected anywhere; they have been collected in Emmet, Marquette, and Keweenaw counties [f. *albiflorum* (Dumort.) Haussk.].

2. **E. strictum** Sprengel

Map 878. Bogs (fens), tamarack and cedar swamps, marshes; usually in sedgy places.

Easily mistaken at a glance for other species until one notices the distinctive glistening horizontal pubescence (on stems, leaves, and capsules), which is shorter than in the next two, larger-leaved species. The entire leaves will distinguish it not only from these but also from specimens of *E. ciliatum* or *E. coloratum* with enough spreading pubescence that they might otherwise key here.

3. **E. parviflorum** Schreber Fig. 315

Map 879. A Eurasian species, first collected in Michigan in 1966 in Benzie County (*Overlease 493,* MICH), and now known from several counties where it grows in swamps and ditches and on shores. Also a recent intro-

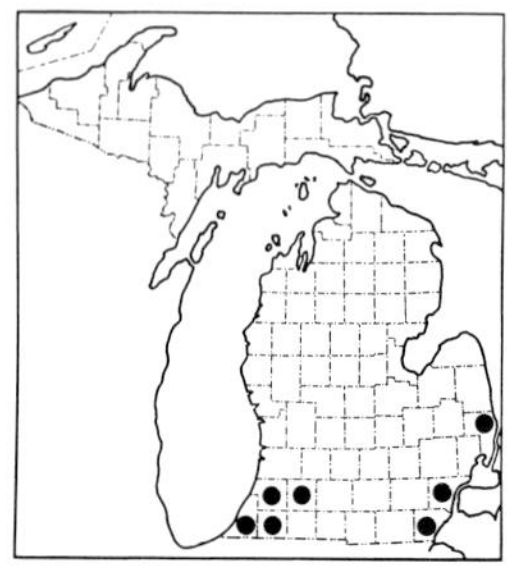

874. Ludwigia alternifolia

875. Ludwigia sphaerocarpa

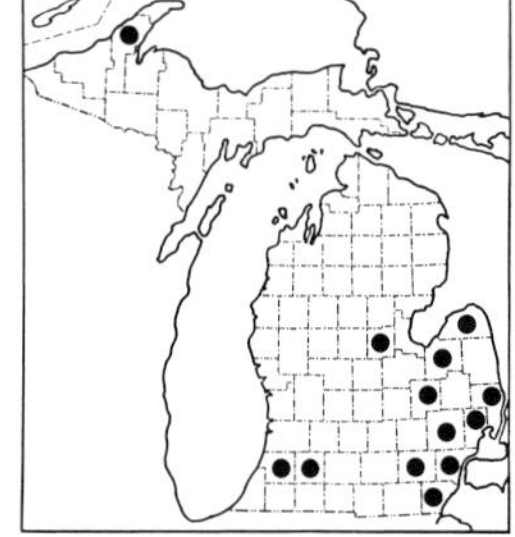

876. Ludwigia polycarpa

duction in Ontario (Purcell 1976) including Manitoulin Island (Morton & Venn 1984).

4. **E. hirsutum** L. Plate 8-A Great Hairy Willow-herb

Map 880. Another Eurasian species, now very well established from New England to the Great Lakes region (Stuckey 1970). It was first collected in Michigan in 1943 (Kalamazoo County). Now locally common in marshes, swamps, wet fields and thickets; at borders of rivers and moist woods; and on shores.

5. **E. palustre** L.

Map 881. Almost wholly restricted to sphagnum bogs; collected in an alder thicket on Isle Royale and in a boggy-sandy excavation on the Keweenaw County mainland.

Most Michigan plants are the linear-leaved, few-flowered form sometimes called var. *oliganthum* Fern.; others (var. *palustre*) have lanceolate leaves approaching those of *E. ciliatum*—but not clearly toothed. This is a variable species, however, and Hoch does not recognize distinct infraspecific taxa. Some authors include *E. leptophyllum* in this species.

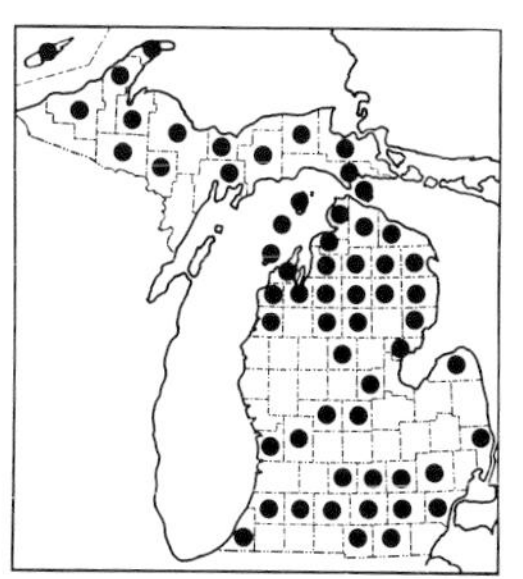

877. Epilobium angustifolium

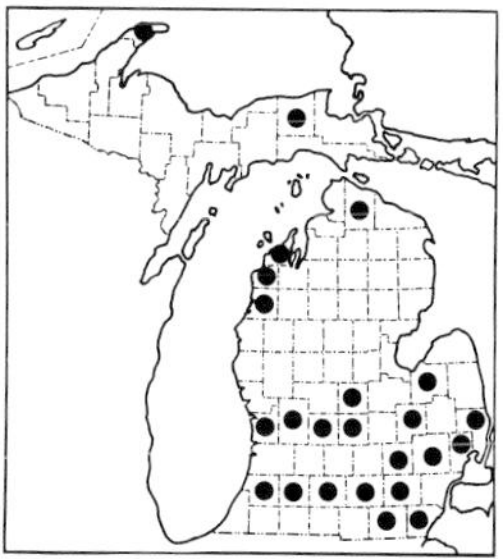

878. Epilobium strictum

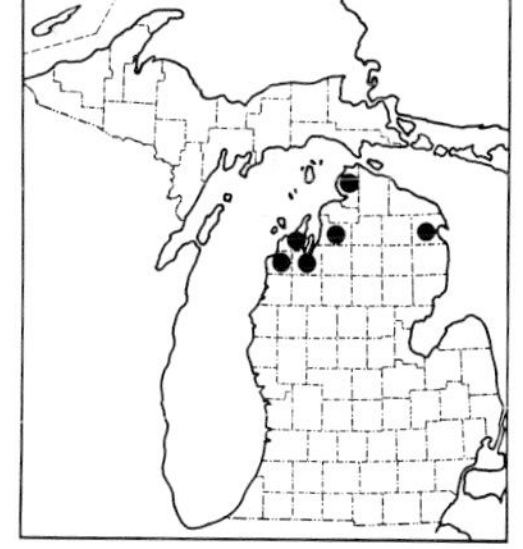

879. Epilobium parviflorum

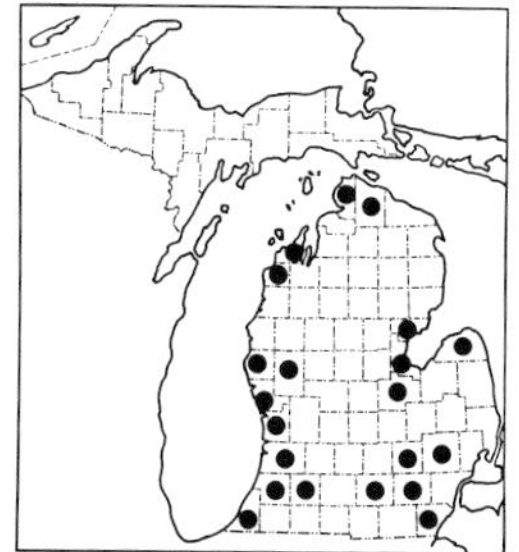

880. Epilobium hirsutum

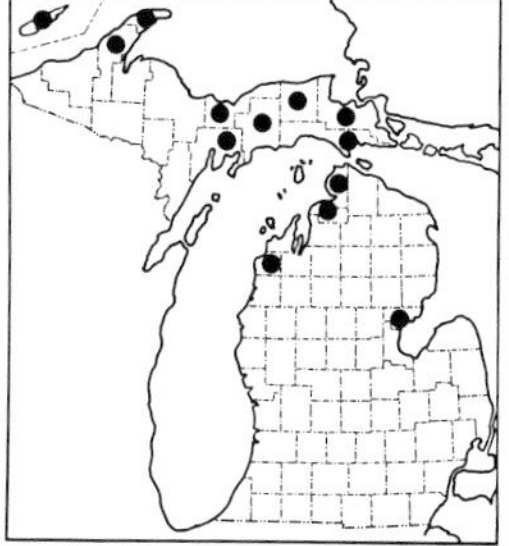

881. Epilobium palustre

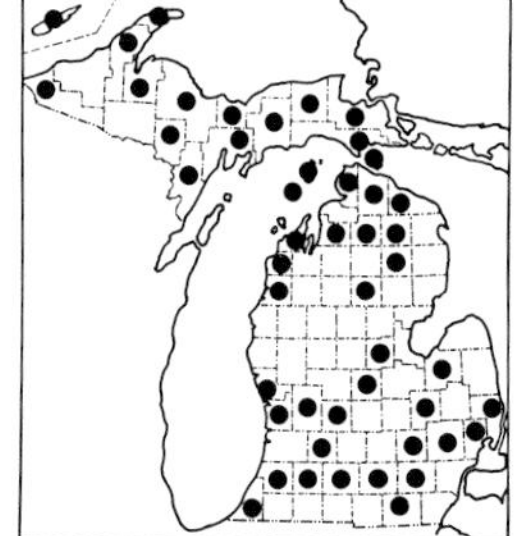

882. Epilobium leptophyllum

6. **E. leptophyllum** Raf.

Map 882. Low ground, including bogs (on mats) and marshes, swamps (cedar, tamarack, or mixed cover, especially on cleared, burned, or otherwise disturbed areas), wet shores and streamsides, sedgy meadows.

Sometimes resembling the preceding species rather closely, but more generally pubescent with minute incurved hairs, and usually with more branches, leaves, and flowers.

7. **E. ciliatum** Raf.

Map 883. Wet places generally: marshy, boggy, and swampy ground; ditches and swales; sandy and rocky shores; clearings and disturbed areas in cedar swamps and hardwood forests; seepy or periodically wet rock outcrops (including limestone crevices).

All Michigan specimens examined by Hoch have been referred by him to ssp. *ciliatum* [= *E. adenocaulon* Haussk.], and none to ssp. *glandulosum* (Lehm.) Hoch & Raven, a more northern taxon often reported from Michigan [as *E. glandulosum*]. Dwarf specimens, with a tendency to nod at the tip, from Drummond Island identified by Munz as *E. leptocarpum* var. *macounii* Trel. have been identified by Hoch as *E. ciliatum*.

This common species is not always easy to distinguish from the next. Hybrids [*E.* ×*wisconsinense* Ugent] have been reported from Wisconsin, and a few Michigan collections are apparently this.

8. **E. coloratum** Biehler Fig. 316

Map 884. Like the preceding species, grows in a diversity of wet places, such as shores, stream banks, swamps (including tamarack), meadows, low spots in forests, and ditches.

The seeds of *E. coloratum* are minutely papillose (as in most of its relatives), while in *E. ciliatum* the seeds are longitudinally striate, and this character may help with over-mature specimens when color of the hairs on the seeds may not be clear. The buds of *E. coloratum* are more abruptly contracted to distinct (sometimes spreading) sepal tips than is usual in *E. ciliatum*. This species also tends to flower and fruit later than the preceding, with capsules dehiscing from late August to October, depending on latitude, season, and individual variation. Capsules of *E. ciliatum* may be found dehiscing as early as late June. Still another helpful tendency in distinguishing the two is that *E. coloratum* is usually more bushy-branched.

4. **Gaura**

REFERENCE

Raven, Peter H., & David P. Gregory. 1972. A Revision of the Genus Gaura (Onagraceae). Mem. Torrey Bot. Club 23(1). 96 pp.

KEY TO THE SPECIES

1. Ovary and fruit with appressed to ± incurved hairs; fruit abruptly contracted just below the middle to a thick stipe-like, truncate base . . . . 1. **G. coccinea**
1. Ovary and fruit with mostly wide-spreading hairs; fruit ± equally tapered to both ends
  2. Main stem with numerous straight hairs (overtopping shorter or appressed hairs) . . . . 2. **G. biennis**
  2. Main stem with all hairs (whether long or short) ± curled or appressed . . . . 3. **G. longiflora**

1. **G. coccinea** Pursh

Map 885. Adventive from the west, collected along a railroad at White Pigeon (St. Joseph County) in 1903 (*Beal,* MSC) and in sandy ground at Rochester (Oakland County) in 1924 (*Farwell 7017,* BLH, GH).

2. **G. biennis** L. Fig. 317

Map 886. River banks, roadsides, fields, vacant lots. Some plants, at least, may be adventive in Michigan from a native range to the south. Not mapped is a very poor specimen (MICH) of unknown origin, labeled "Houghton Mich."; whether "Houghton" is intended to be the locality or the collector is not clear. This species was listed by the First Survey, under Douglass Houghton, although no specimens from that era have been found; possibly this specimen was received from Houghton by whoever first possessed it and labeled it so cryptically with a scrawled penciled slip. The earliest dated Michigan collections seen are from the 1890's.

This species in the restricted sense of Raven and Gregory is a self-compatible complex-heterozygote with pollen about half fertile, derived from the next species.

3. **G. longiflora** Spach

Map 887. Roadsides and waste ground, meadows, dry fields. No Michigan collections have been seen from before 1911, and the species is probably only adventive in Michigan from a range to the southwest.

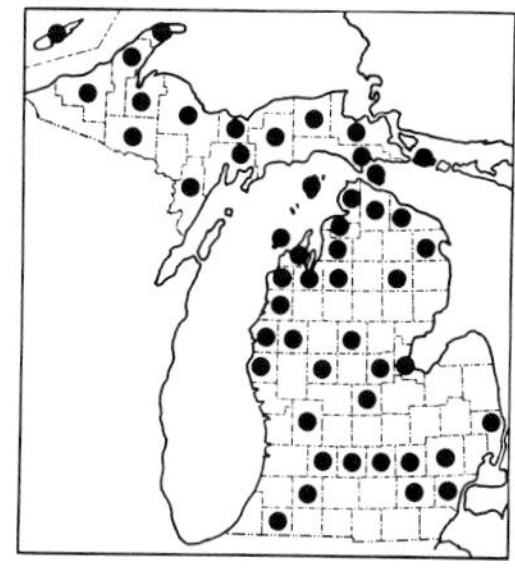

883. Epilobium ciliatum

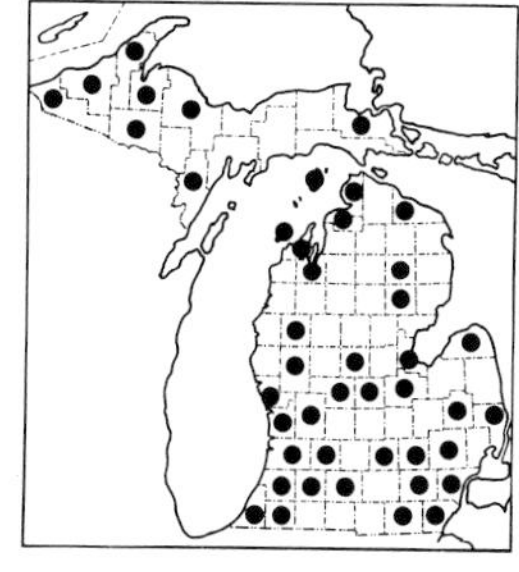

884. Epilobium coloratum

885. Gaura coccinea

This taxon, which could be called *G. biennis* var. *pitcheri* T. & G. and which was sometimes known as *G. filiformis* Small, was often not recognized as a distinct species until Raven and Gregory determined that it is a self-incompatible species with pollen almost fully fertile. The two taxa hybridize readily, at least experimentally. Our specimens of the two are very similar morphologically; all have very short, dense, weakly gland-tipped, straight hairs in the inflorescence and numerous short hairs on the leaves.

5. **Calylophus**

REFERENCE

Towner, Howard F. 1977. The Biosystematics of Calylophus (Onagraceae). Ann. Missouri Bot. Gard. 64: 48–120.

1. **C. serrulatus** (Nutt.) Raven

Map 888. A species presumably adventive in our area from the central United States. Collected in dry fields near Sleeping Bear Dunes (*Thompson L-1106* in 1948, MICH, BLH; *L-2376* in 1976, BLH; *Hazlett 2663* in 1984, MICH).

Calylophus was long included in *Oenothera,* and this species known as *O. serrulata* Nutt. In the short floral tube and short stigmatic lobes it might possibly be confused with *O. perennis,* but the pubescence is not glandular as in that species, nor is the ovary 4-angled.

6. **Oenothera** Evening-primrose; Sundrops

This genus is well known as a difficult one, with very extensively studied cytological features—especially in the section or subsection *Oenothera*: the last five species here listed. This treatment follows the work of Peter H. Raven and his associates, who have not yet published a full monograph. The comments of W. L. Wagner on an early draft are very deeply appreciated. Some taxa are of very minor or even dubious status in the state.

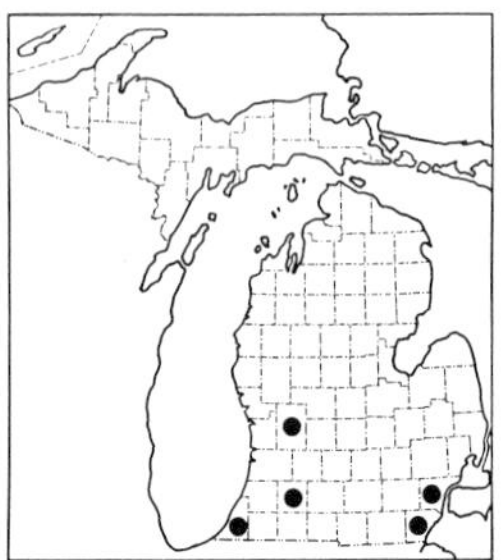

886. Gaura biennis

887. Gaura longiflora

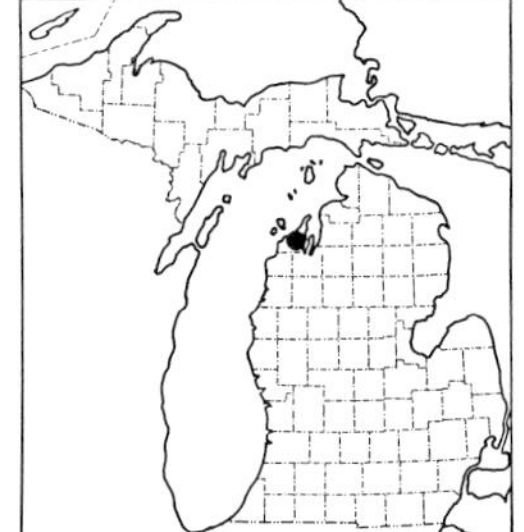

888. Calylophus serrulatus

Notes on genetic analysis and other details may be found in the references cited below and in the remarks on the species.

REFERENCES

Raven, Peter H., Werner Dietrich, & Wilfried Stubbe. 1980. An Outline of the Systematics of Oenothera subsect. Euoenothera (Onagraceae). Syst. Bot. 4: 242–252.

Straley, Gerald B. 1978. Systematics of Oenothera sect. Kneiffia (Onagraceae). Ann. Missouri Bot. Gard. 64: 381–424.

KEY TO THE SPECIES

1. Ovary 4-angled; fruit sharply 4-angled or -winged, tapered to the base
 2. Petals (3) 4.5–7 (9) mm long; tip of stem usually nodding when bearing buds; anthers ca. 1–2 (2.7) mm long; body of capsule ca. 8–11 mm long. . . . 1. **O. perennis**
 2. Petals (11) 13–27 (30) mm long; tip of stem erect or nearly so; anthers 4–7 mm long; body of capsule various
  3. Calyx copiously long-hirsute, at least on the free tips, which are (1.7) 2–3.5 mm long; capsules ca. 12–13 mm long . . . . 2. **O. pilosella**
  3. Calyx glabrous or nearly so, the free tips at most 2 mm long (often less than 1 mm); capsules ca. 5–10 mm long . . . . 3. **O. fruticosa**

1. Ovary terete (with several low ridges); fruit terete to obscurely 4-sided, abruptly rounded at the sessile base
 4. Petals white (turning pink); anthers ca. (7) 9–10 mm long; stems silvery white, glabrous (except for fine glandular pubescence toward the inflorescence) . . . . 4. **O. nuttallii**
 4. Petals yellow (sometimes turning reddish in age); anthers 3.5–8 mm long; stems green to brownish or reddish, usually pubescent
  5. Capsules nearly or quite linear, slender, 2–3.2 mm thick, 5–12 (14) times as long; seeds finely pitted but not strongly angled; cauline leaves either pinnately lobed or linear to narrowly lanceolate (less than 1 cm broad)
   6. Leaves pinnately lobed (even ± pinnatifid); flowers few, in the axils of upper and middle leaves (not a distinct bracted inflorescence); plant with long spreading hairs much overtopping the fine pubescence . . . . .5. **O. laciniata**
   6. Leaves (except in basal rosettes) linear to lanceolate and entire or nearly so; flowers numerous, crowded in the axils of reduced leaves (bracts) in a terminal spike; plant with only appressed hairs
    7. Petals 0.5–1.5 cm long; stigma surrounded by the anthers at anthesis . . . . .6. **O. clelandii**
    7. Petals 1.5–2.5 [3.5] cm long; stigma well elevated above the anthers at anthesis . . . . 7. **O. rhombipetala**
  5. Capsules tapered upward from near the base, ca. (4) 5–7 mm thick and 3.5–5 times as long; seeds sharply angled but not pitted; cauline leaves unlobed, elliptic-lanceolate, often at least 1 cm broad
   8. Petals ca. 3.5–4.5 cm long . . . . 8. **O. glazioviana**
   8. Petals ca. 1–2 cm long (rarely longer)
    9. Subulate sepal tips subterminal, their bases slightly separated in bud and with a distinct protuberance within (visible at anthesis); tip of stem often bent at early anthesis, with another bend "correcting" it to become erect

10. Calyx, ovary, capsule, and other parts glabrate to ± sparsely pubescent, often with some long spreading hairs as well as shorter glandular hairs; largest leaves various, typically at least 15 mm broad and nearly entire .... 9. **O. parviflora**

10. Calyx, ovary, capsule, and upper leaves or bracts ± densely pubescent with appressed whitish non-glandular hairs; largest leaves typically less than 15 mm broad and denticulate .... 10. **O. oakesiana**

9. Subulate sepal tips terminal, their bases contiguous in bud and with at most a mere transverse ridge within at anthesis; tip of stem straight at anthesis

11. Plant, especially the upper portion and inflorescence, gray in aspect, with rather dense appressed non-glandular pubescence .... 11. **O. villosa**

11. Plant green in aspect, with mostly spreading long hairs and often shorter glandular ones .... 12. **O. biennis**

1. **O. perennis** L. Fig. 318 Sundrops

Map 889. Moist sandy (or rocky) to boggy soil, especially of shores, borrow pits, ditches, and meadows; damp hollows in jack pine plains, along roadsides, in fields and gravel pits; rarely in truly dry soil.

This species and the next two are placed in *Oenothera* sect. *Kneiffia.*

2. **O. pilosella** Raf. Sundrops

Map 890. Dry to moist ground of fields, roadsides, woodlands, clearings, ditches. It is possible that some or even all of our collections of this showy species represent escapes from cultivation; none were made before the 20th century.

Readily distinguished in our area by the long, spreading pubescence on stems and inflorescence, compared to the glabrous or glabrate other species we have in this section (they are at most slightly glandular-pubescent).

3. **O. fruticosa** L. Sundrops

Map 891. Dry hillsides, fields, and open ground, especially disturbed; clearings, meadows, even marshy ground.

O. tetragona Roth is included here by Straley, who has referred almost all Michigan specimens seen by him to *O. fruticosa* ssp. *glauca* (Michaux)

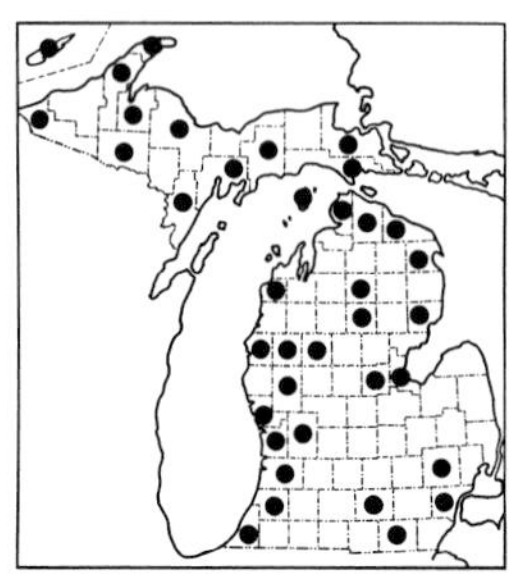

889. Oenothera perennis

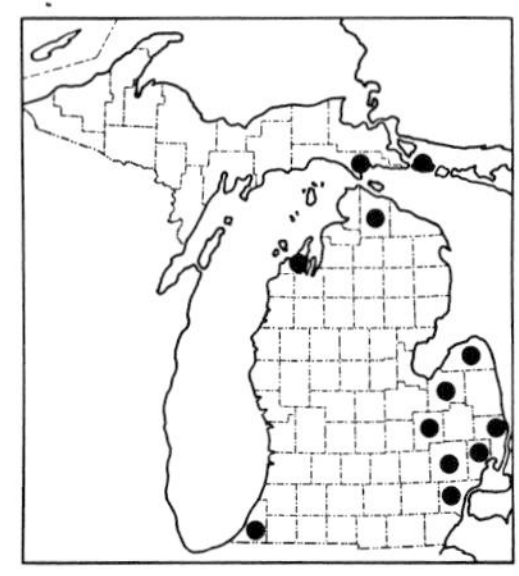

890. Oenothera pilosella

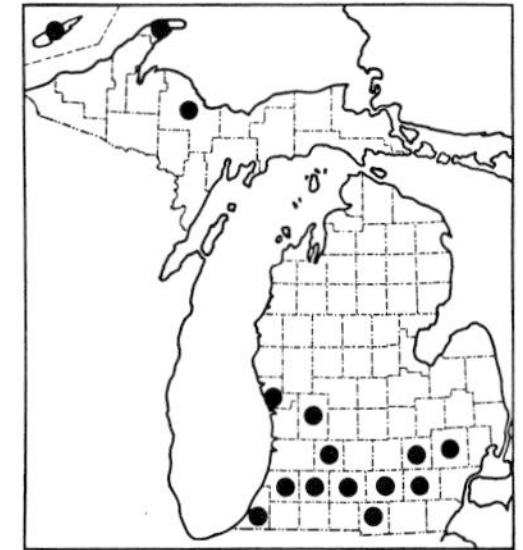

891. Oenothera fruticosa

Straley; they had been earlier identified by Munz as *O. tetragona* var. *longistipata* (Pennell) Munz. This is a variable complex and our specimens have smaller fruit than usually stated for this species.

4. **O. nuttallii** Sweet

Map 892. Another uncommon adventive from prairies and plains of the central United States and Canada. A small colony thrived between the railroad and the highway, near the airport, at Pellston in the early 1950's (see Brittonia 9: 94. 1957). A specimen (MSC) from Lansing ca. 1870 but without collector or date is perhaps dubious as to origin.

5. **O. laciniata** Hill

Map 893. Sandy roadsides and fields, fencerows and railroads, and other disturbed sites. Perhaps adventive in our region. The earliest Michigan collection seen is from waste ground at Detroit in 1894.

This is our only *Oenothera* with deeply lobed to pinnatifid leaves.

6. **O. clelandii** Dietrich, Raven, & W. L. Wagner

Map 894. Sandy roadsides, fields, and railroads; plains and dry woodland (oak, sassafras), generally in disturbed areas.

Included here are almost all specimens from the state previously identified as *O. rhombipetala,* which is now interpreted as a fully fertile outcrossing species with larger flowers and stigma well elevated above the anthers. *O. clelandii* was described in 1983 (Ann. Missouri Bot. Gard. 70: 196; TL: Amber Tp., Mason Co.) for complex-heterozygotes, with pollen half fertile, differing from *O. rhombipetala* in smaller flowers and the stigma at the same level as the anthers.

7. **O. rhombipetala** T. & G.

Map 895. Native farther west in North America. The only Michigan collection was made in 1948 in a disturbed sandy pasture in Grass Lake Tp., Jackson County (*Parmelee 764,* MSC).

892. Oenothera nuttallii

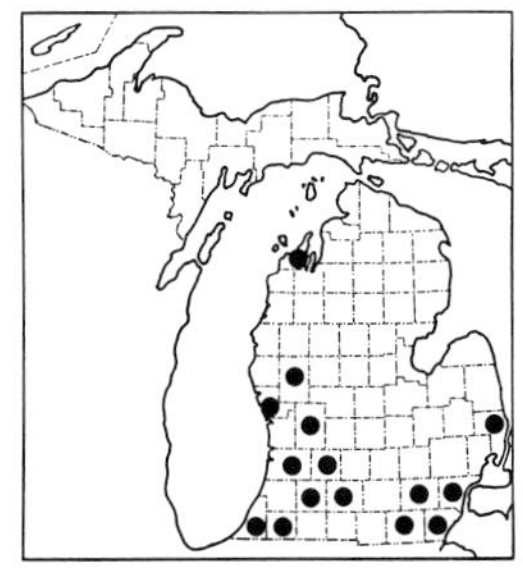

893. Oenothera laciniata

894. Oenothera clelandii

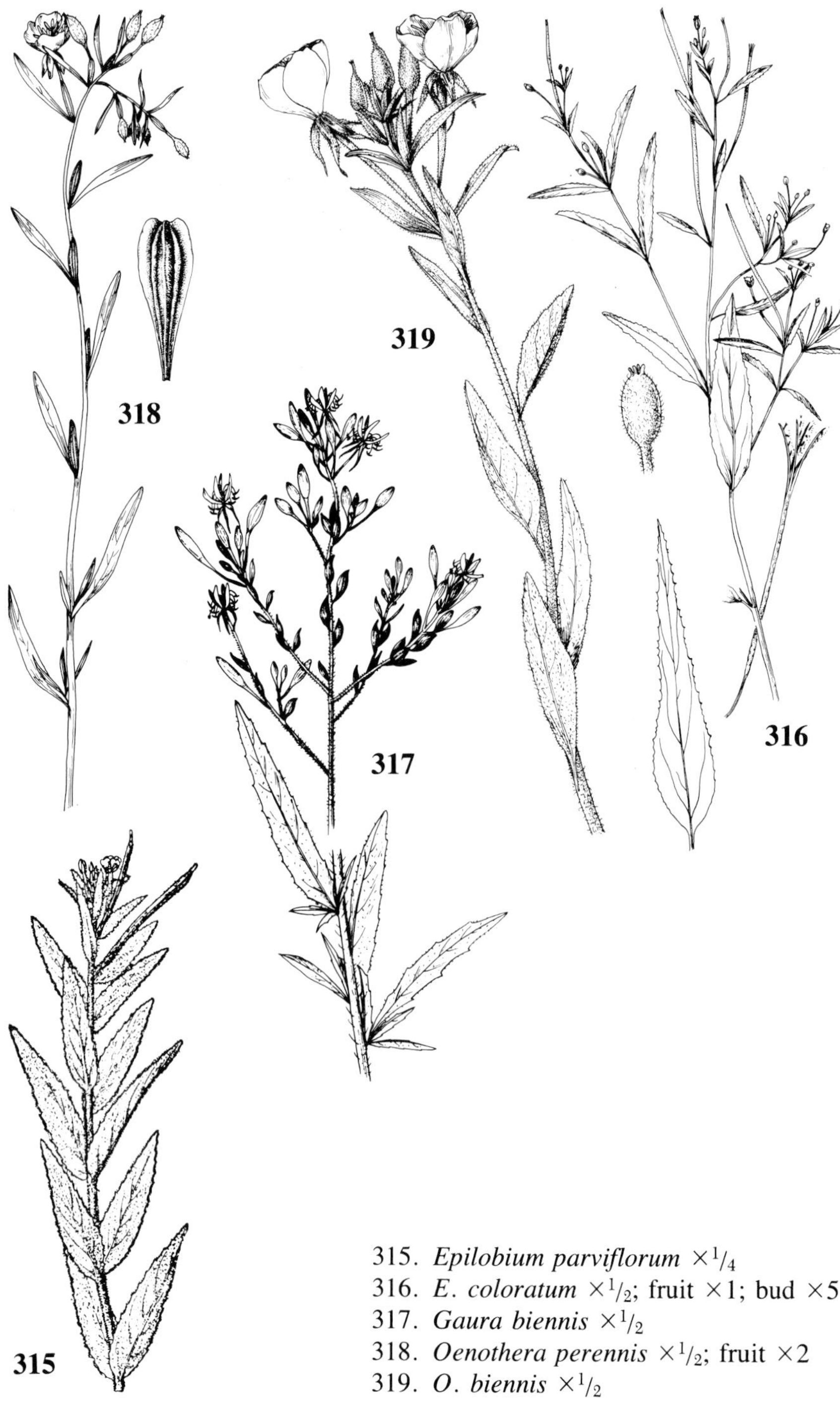

315. *Epilobium parviflorum* $\times 1/_4$
316. *E. coloratum* $\times 1/_2$; fruit $\times 1$; bud $\times 5$
317. *Gaura biennis* $\times 1/_2$
318. *Oenothera perennis* $\times 1/_2$; fruit $\times 2$
319. *O. biennis* $\times 1/_2$

8. **O. glazioviana** Micheli

Map 896. A taxon apparently of garden origin, collected in sod along a roadside at Cedar Springs, Kent County, in 1895 (*Fallass 813,* ALBC, MSC); the Clinton County record is from a yard at Bath in 1880 (*Bailey,* BH) and may have been a cultivated plant.

Easily recognized by its very large flowers (petals ca. 4–4.5 cm long on our specimens), the leaves crisped when fresh, and the stigmas elevated above the anthers at anthesis. Also known as *O. lamarckiana* De Vries (an illegitimate homonym) and *O. erythrosepala* Borbás (a later name).

9. **O. parviflora** L.

Map 897. On shores, but not as restricted to them as the next species; usually on roadsides, railroads, fields; in rocky or sandy openings in woods, gravel pits; on river banks and sometimes even in wet boggy or marshy places.

This species and the next three might justifiably be included, for the purposes of this Flora, in an "*O. biennis* complex"—as has been done by some cautious authors. The distinctions are not as clear as the key may suggest. Hybridization has been extensive, especially in the upper Great Lakes region. See further comments under *O. biennis* below.

10. **O. oakesiana** (A. Gray) Watson & Coulter

Map 898. Sandy or rocky shores, dunes, and clearings along the Great Lakes; occasionally inland along railroads, on sandy shores, in clearings or other disturbed places.

Although this was treated by Munz as a variety of the preceding species, Raven et al. maintain it as distinct. The two run close together morphologically, and doubtless hybridize.

11. **O. villosa** Thunb.

Map 899. Fields, roadsides, railroads, shores. Supposedly native in our

895. Oenothera rhombipetala

896. Oenothera glazioviana

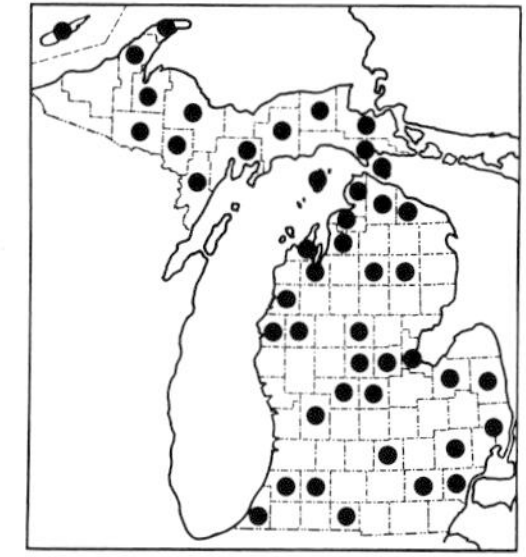

897. Oenothera parviflora

area; the earliest Michigan collections seen date from 1889 (Keweenaw Co.) and 1901 (Marquette Co.).

This species includes *O. strigosa* (Rydb.) Mack. & Bush, now considered a subspecies and very rare in Michigan (see Dietrich & Raven, Ann. Missouri Bot. Gard. 63: 382–383. 1976). Some individuals of *O. villosa* look very much like the preceding species.

11. **O. biennis** L. Fig. 319

Map 900. Usually on dry often sandy roadsides, fields, clearings, and disturbed ground; on stream banks and at borders of woods; rarely on beaches or dunes.

The genetic complexities of the true evening-primroses have been studied for many years. Raven, Dietrich, and Stubbe (1980) have presented a clear, up-to-date summary of the situation. Unfortunately, genetic analysis and study of chromosomes, while leading to a rational, natural classification, are not possible for the average field and herbarium botanist. The important taxonomic distinctions are not necessarily correlated with unequivocal morphological characters. It has long been lamented that only an expert can tell the Oenotheras apart. While this was more true when a great overabundance of "species" were recognized, there are still many specimens that do not "fit." Specimens which appear to me quite ambiguous (including immature or fruiting ones)—and that is a large number—have been mapped only when identified by Munz, Raven, or Dietrich. Many persons drawing up lists for local sites will be quite satisfied to include these last four species as merely the *O. biennis* complex.

Even the character distinguishing the *parviflora* group from the *biennis* group is not as clearcut as most keys imply. The sepals are connate in bud, separating at anthesis and becoming reflexed, when the narrow apical appendages are most easily seen. These are more clearly subterminal in the *parviflora* group, with a tiny projecting shelf-like apex of the sepal proper. In the *biennis* group, there is generally a transverse (convex) line, or even a boat-shaped tip, marking the end of the sepal proper; so experience (not

898. Oenothera oakesiana

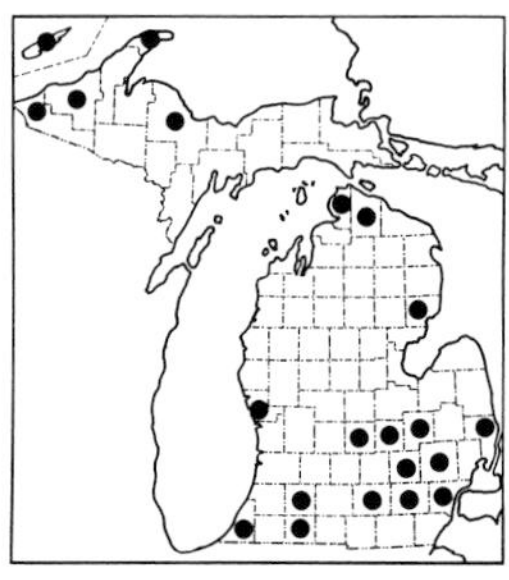
899. Oenothera villosa

900. Oenothera biennis

to mention intuition) is helpful in interpreting this character. The petals in the *biennis* group tend to be longer (ca. 14–20 mm in our material) than in the *parviflora* group (ca. 10–17 mm), but there is no practical distinction. Leaves in the *biennis* group are generally broader than in *O. oakesiana,* which is rather distinctively narrow-leaved (although *O. villosa* may approach it).

Three old specimens (Ingham, Washtenaw, & Genesee cos., 1879–1911) in poor condition and/or with scanty data have been identified by Dietrich as *O. nutans* Atkinson & Bartlett [= *O. biennis* ssp. *austromontana* Munz]. They are glabrous or glabrate throughout the inflorescence and lack glandular hairs (except on the calyx). This taxon has slightly larger flowers (ca. 2 cm or more) than *O. biennis*. It is not separately mapped or keyed here. The specimens probably represent scattered introductions from the Appalachian region and the southeast, if not from cultivation.

HALORAGACEAE — Water-milfoil Family

A small family, largely of the southern hemisphere and largely aquatic (including all of our species). The name is sometimes spelled Haloragidaceae, but the spelling used here is conserved. The submersed leaves in all our species (except one essentially leafless *Myriophyllum*) are pectinate, i. e., with a straight central axis and simple filiform lateral segments on both sides—like a 2-sided comb. Other aquatics with dissected leaves may have them forked dichotomously (*Ceratophyllum*) or more irregularly branched (*Ranunculus, Utricularia, Megalodonta*), but the only other species with a straight central axis is *Armoracia aquatica*. However, it has alternate leaves with the lateral segments usually further divided.

KEY TO THE GENERA

1. Leaves all clearly alternate, serrate to deeply pectinate; flowers perfect, 3-merous, in axils of emersed alternate leaves . . . . 1. **Proserpinaca**
1. Leaves all or mostly whorled (or nearly so), deeply pectinate, or (in one species) essentially absent (but alternate); flowers mostly unisexual, 4-merous, in axils of submersed deeply pectinate leaves or (most species) of emersed alternate or whorled bracts . . . . 2. **Myriophyllum**

1. **Proserpinaca** — Mermaid-weed

REFERENCE

Schmidt, Barbara L., & W. F. Millington. 1968. Regulation of Leaf Shape in Proserpinaca palustris. Bull. Torrey Bot. Club 95: 264–286.

320. *Proserpinaca palustris* ×1/2; emersed leaf ×2
321. *Myriophyllum spicatum* ×2/5; inflorescence & immature fruit ×3; whorl of leaves ×1 1/2

KEY TO THE SPECIES

1. Flowers in the axils of pectinate leaves (i. e., *all* leaves pectinate); species very rare . . . . 1. **P. pectinata**
1. Flowers only in the axils of merely serrate leaves (i. e., plant heterophyllous); species common . . . . 2. **P. palustris**

1. **P. pectinata** Lam.

Map 901. A species of the Coastal Plain, from Texas to Nova Scotia; apparently the only locality in the Great Lakes region is along a sandy ditch in Ottawa County, where a copious collection of fruiting material was made in 1941 (*Bazuin 4180,* MICH, ALM). The site was rediscovered in 1984 (*Reznicek et al. 7432,* MICH, MSC): an old peaty excavation, where *P. palustris* also occurs. Intermediate plants, in fruit, were found where the two species grow together; these apparent hybrids may presumably be called *P.* ×*intermedia* Mack.

The 3-angled fruit in the axils of stiff emersed alternate pectinate leaves will distinguish this species from any *Myriophyllum.*

2. **P. palustris** L. Fig. 320

Map 902. Wet shores and banks, shallow water of ponds and streams, interdunal pools and swales, sedge mats, ditches; often in rather alkaline habitats.

This is a remarkable species in its variable leaf forms, which have been the subject of physiological investigation since the beginning of this century. Most recently, both photoperiod and submergence have been demonstrated to play major roles in its heterophylly. Submersed shoots always produce dissected leaves, while aerial shoots produce dissected leaves under long days but simple leaves under short days (e. g., as summer progresses). High light intensity and high temperature, however, can at least partly counteract the effects of submergence. Since these plants tend to grow in areas of fluctuating water level (often shallow ponds, swales, and ditches), the series of leaf forms, including transitional states, can be quite varied on a single stem.

901. Proserpinaca pectinata

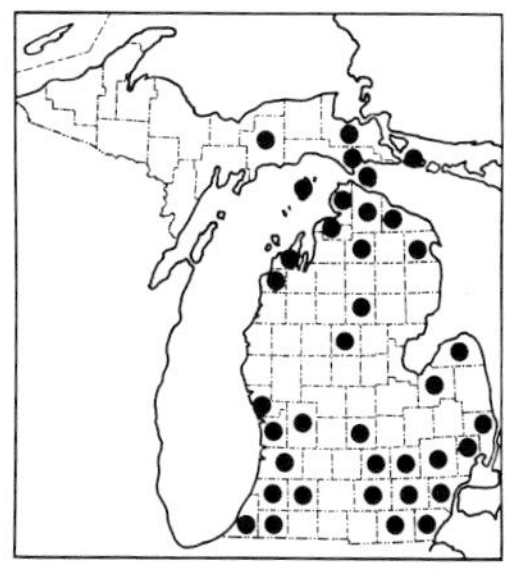

902. Proserpinaca palustris

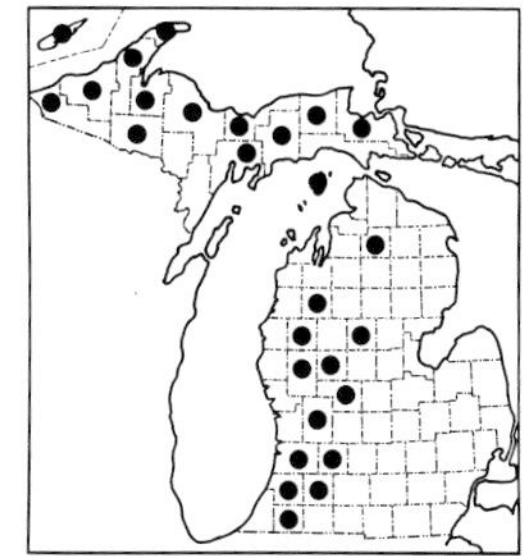

903. Myriophyllum tenellum

Even sterile, completely submersed, dissected-leaved plants can be readily distinguished from *Myriophyllum* by the consistently alternate arrangement of the leaves and their relatively broader rachis; and the internodes usually bear glandular-looking blackish flecks or spicules.

2. **Myriophyllum** Water-milfoil

Usually considered a difficult genus, because aquatic biologists often wish to identify sterile specimens and there is great phenotypic variation in leaves. The species are distinctive when fertile. Some suggestions for identification of sterile plants follow the listing of species. Like many aquatics, unusual forms develop, at least in some species, when plants are partly or entirely stranded by a lowering of water levels—or resubmersed upon a later rise in levels. But no local species displays quite as dramatic heterophylly as *Proserpinaca palustris,* except perhaps for the stranded terrestrial form of *M. heterophyllum.*

M. aquaticum (Vellozo) Verdc. [long called *M. brasiliense* Cambess.], parrot's-feather, is a dioecious South American species of which pistillate plants are adventive in the southern United States as well as on the Pacific coast. It is often grown as a greenhouse or aquarium plant for its attractive foliage, which usually rises several cm out of the water.

Anatomical studies (England & Tolbert 1964) have shown that the leaves are basically alternate in *M. heterophyllum,* though they appear (and are said to be) whorled as a result of extremely short internodes. This species often produces a few alternate leaves on its stems, and *M. farwellii* characteristically has some alternate leaves. In *M. alterniflorum* and *M. tenellum,* the flowers are alternate, and in the latter so also are the rudimentary leaves. So an alternate tendency seems widespread in the genus. Some species produce winter buds or turions—specialized shoots with very short internodes and compact leaves. Such structures aid in winter survival and dispersal. Extensive studies on Michigan material of *M. verticillatum* by Weber and Noodén have shown that formation, dormancy, and germination of its turions are controlled by changes in temperature and photoperiod which affect the relevant growth hormones. The species with emersed spikes are wind-pollinated, the pistillate (lower) flowers beginning to mature just before the staminate (upper) flowers (fig. 321).

REFERENCES

Aiken, S. G., P. R. Newroth, & I. Wile. 1979. The Biology of Canadian Weeds. 34. Myriophyllum spicatum L. Canad. Jour. Pl. Sci. 59: 201–215.

Aiken, S. G., & J. McNeill. 1980. The Discovery of Myriophyllum exalbescens Fernald (Haloragaceae) in Europe and the Typification of M. spicatum L. and M. verticillatum L. Jour. Linn. Soc., Bot. 80: 213–222.

Aiken, Susan G. 1981. A Conspectus of Myriophyllum (Haloragaceae) in North America. Brittonia 33: 57–69.

Aiken, Susan G. 1984. The Water-milfoils (Myriophyllum) of the Ottawa District and Ottawa River, Canada. Trail & Landscape 18: 35–52. [Same species as in Michigan, five of them well illustrated.]

Ceska, Oldriska. 1977. Studies on Aquatic Macrophytes Part XVII Phytochemical Differentiation of Myriophyllum Taxa Collected in British Columbia. B. C. Min. Environ. Water Invest. Br. File No. 0316533. 32 pp.

Coffey, Brian T., & Clarence D. McNabb. 1974. Eurasian Water-milfoil in Michigan. Mich. Bot. 13: 159–165.

England, Wayne H., & Robert J. Tolbert. 1964. A Seasonal Study of the Vegetative Shoot Apex of Myriophyllum heterophyllum. Am. Jour. Bot. 51: 349–353.

Faegri, Knut. 1982. The Myriophyllum spicatum Group in North Europe. Taxon 31: 467–471.

Weber, James A. 1972. The Importance of Turions in the Propagation of Myriophyllum exalbescens (Haloragidaceae) in Douglas Lake, Michigan. Mich. Bot. 11: 115–121.

Weber, James A., & Larry D. Noodén. 1974. Turion Formation in Myriophyllum verticillatum: Phenology and Its Interpretation. Mich. Bot. 13: 151–158.

Weber, James A., & Larry D. Noodén. 1976. Environmental and Hormonal Control of Turion Formation in Myriophyllum verticillatum. Pl. Cell Physiol. 17: 721–731.

Weber, James A., & Larry D. Noodén. 1976. Environmental and Hormonal Control of Turion Germination in Myriophyllum verticillatum. Am. Jour. Bot. 63: 936–944.

KEY TO THE SPECIES

1. Leaves all alternate, reduced to minute scales (less than 1 mm long) or bumps on the stem; floral bracts alternate, ca. 2–3 mm long or less, entire . . . . . . 1. **M. tenellum**

1. Leaves all or mostly whorled, developed and pectinate; floral bracts whorled, or if alternate then often toothed

 2. Flowers and fruit in the axils of normal submersed leaves; body of fruit ca. 1.5–2 mm long, sharply tuberculate in ridges on the back; leaves partly alternate (or irregularly whorled), extremely limp, the foliage becoming a shapeless mass upon removal from water . . . . 2. **M. farwellii**

 2. Flowers and fruit in the axils of small bracts in an emersed terminal spike; body of fruit various in length (often longer or shorter than in *M. farwellii*), but smoothly rounded on the back or irregularly roughened; leaves normally all whorled (a few often alternate in *M. heterophyllum*), usually retaining at least some shape on removal from water

  3. Well developed leaves 4–12 mm long; uppermost flowers in axils of alternate bracts . . . . 3. **M. alterniflorum**

  3. Well developed leaves ordinarily 10–40 mm long; uppermost flowers in axils of whorled bracts

   4. Middle and upper bracts much longer than the flowers and fruits, with distinct and merely toothed blades; leaves crowded, the internodes less than 8 (10) mm long; stamens 4 . . . . 4. **M. heterophyllum**

   4. Middle and upper bracts about as long as the flowers and fruits or else deeply pectinate; leaves usually less crowded, the internodes often 10–27 (40) mm long (and only rarely all less than 5 mm); stamens usually 8

    5. Bracts of middle and upper flowers deeply pectinate (more than halfway to the rachis); stems greenish or brownish; leaves, or many of them, essentially sessile (the lowest lateral segments arising next to the stem) . . . . .5. **M. verticillatum**

    5. Bracts of middle and upper flowers (not necessarily ones transitional to leaves) merely toothed or entire; stems (at least when dry) pale, whitish or light pink (except very near surface of the water); leaves all distinctly petioled

6. Lateral segments of leaves 5–11 (13) on a side; turions produced late in the season . . . . 6. **M. exalbescens**
6. Lateral segments of most well developed leaves (13) 14–17 (20) on a side; turions never produced . . . . 7. **M. spicatum**

1. **M. tenellum** Bigelow

Map 903. One of the most widespread of our interesting aquatics of softwater lakes and bays. Often forms dense turfs in deep water, the essentially leafless stems arising singly along buried rhizomes. In shallow water, at the sandy to mucky edges of lakes and ponds, the emersed tips produce typical *Myriophyllum* flowers.

Flowering stems can attain a height of 30 cm or more, but they are usually not over half that tall. This is an absolutely distinctive species, even when sterile, but is generally overlooked unless one has it in mind. I have myself made over 20 collections in 13 different counties, and noted the species at additional stations. Yet it was once considered a great rarity in the state (see Mich. Bot. 4: 21. 1965).

2. **M. farwellii** Morong

Map 904. Small ponds and lakes, marsh borders. Farwell described the type locality, 2 miles northwest of Cliff Mine, in a letter written in 1890 to S. Watson (GH) as "a small pond about 12 ft. in diameter, well shaded by evergreens & shrubs, and well supplied with water which averages about $1^1/_2$ ft. in depth."

Named for the prolific collector of Michigan plants, O. A. Farwell, who collected specimens in 1884 (*191*, BLH) and later years (TL: Keweenaw Co.). A very distinctive species when fertile, but sterile specimens may be impossible to separate from *M. heterophyllum* unless turions are present late in the summer (or old turion leaves are at the base). Fruiting material is known from all counties mapped except for Houghton, from which sterile material is referred here on the basis of very slender stem and extremely delicate foliage with minute dark spicules where lateral segments of the leaves

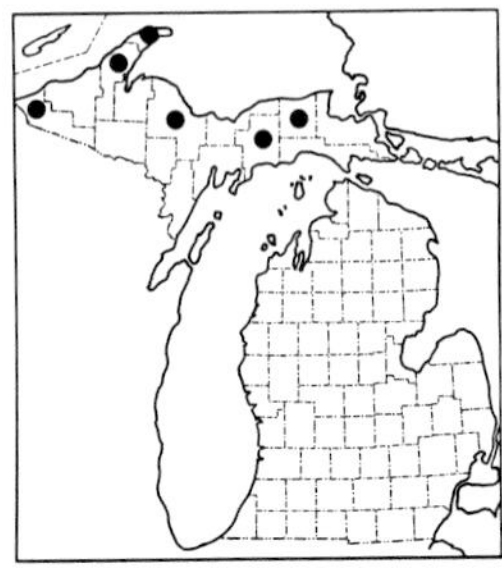
904. Myriophyllum farwellii

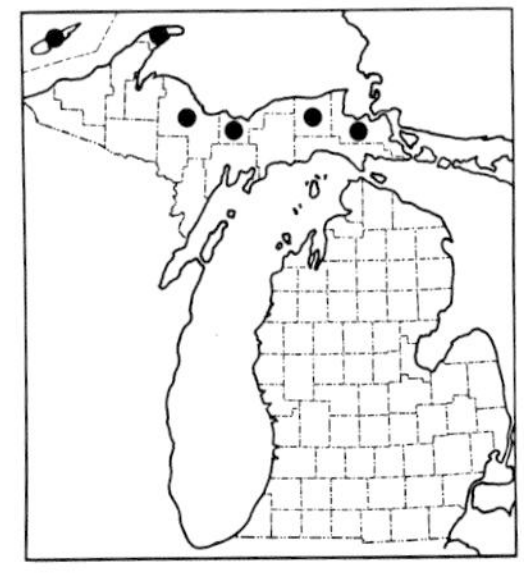
905. Myriophyllum alterniflorum

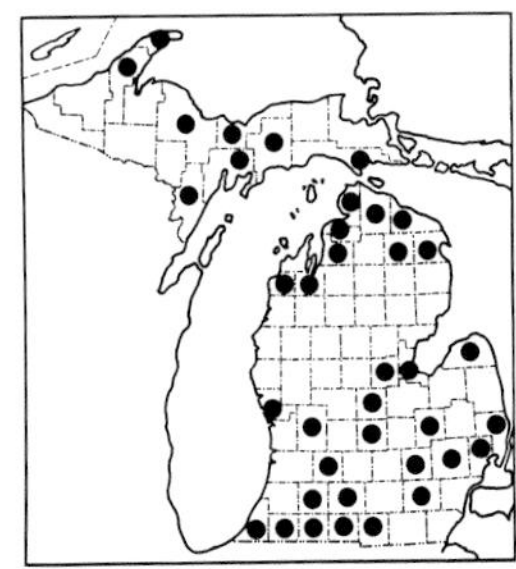
906. Myriophyllum heterophyllum

meet the rachis. These seem almost always to be present in this species (at least on some leaves), but they are usually lacking in *M. heterophyllum*. The specimen previously cited tentatively from Alger County (Mich. Bot. 4: 20. 1965) is certainly *M. heterophyllum*, of which fertile specimens with identical foliage were later collected from the same lake; that tentatively cited from Chippewa County (plus later collections from the same site, det. Aiken) is *M. verticillatum*.

3. **M. alterniflorum** DC.

Map 905. A northern species of softwater lakes and bays of Lake Superior and the St. Mary's River. A specimen from Algonac, St. Clair County (*W. S. Cooper* in 1903, MSC) is so far out of range as to suggest a mislabeling of one of Cooper's Isle Royale collections; it is not mapped.

A neat, slender plant, with rather dense very short leaves on an often sinuous, much-branched stem ca. 0.3–1 (1.5) mm thick.

4. **M. heterophyllum** Michaux

Map 906. Lakes and rivers, in water up to 15 feet deep.

There is a rather frequent tendency for some of the leaves to be alternate in this species, so that sterile fragments may resemble *M. farwellii* (see above). This is a variable species in other respects as well. Stranded plants on damp shores may produce deeply serrate (or shallowly pectinate) "terrestrial" leaves or bracts; the tips of these or other aerial shoots if they become inundated may again produce typical dissected leaves. Such unusual plants are probably the basis for reports from our area of the southern and eastern *M. pinnatum* (Walter) BSP. (including those in Mich. Bot. 4: 20. 1965).

The very short internodes and often more than 4 leaves at a node give this plant a very dense appearance; the long shoots look like green ropes under the water.

5. **M. verticillatum** L.

Map 907. Lakes, ponds (including bog pools), and rivers, usually in rather quiet water.

This species produces distinctive club-shaped turions late in the season (and hence not usually collected), when the temperature is about 15° C. These have very short internodes and leaves similar in color and texture to those of normal branches; they expand to full size when the turion germinates. Some of the differences from the next two species are described and illustrated by Aiken et al. (1979).

6. **M. exalbescens** Fern.

Map 908. Rivers, ponds, and lakes, in very shallow water or at depths as great as 18 feet; often in calcareous waters, with marly bottoms.

The turions produced by this species are more cylindrical than in the preceding, or even somewhat tapered toward the apex rather than clavate; they have very distinctive short, dark, stiff leaves, some of which often remain above the characteristic U-shaped base of the plant when the turion germinates (see Weber 1972). Rarely an otherwise normal mature stem will bear only such leaves (distinguished from *M. alterniflorum* by their very stiff nature and extremely dark green color); plants of this sort have been collected in cold running water and are doubtless similar to the dwarfed form "very like *M. alterniflorum*" reported from northern Canada by Aiken and McNeill (1980).

There has not been agreement on whether this species is truly distinct from the next, but recent evidence of differences in pollen (Faegri 1982) and chromatography of flavonoid compounds (which will separate other species of *Myriophyllum* as well; Ceska 1977) supports its recognition. If this is included in the next, it may be called *M. spicatum* ssp. *squamosum* Hartman f., or var. *squamosum* (Hartman f.) Hartman f.—names based on material from Europe (where this taxon was only recently again realized to occur) and antedating infraspecific names based on *M. exalbescens*.

7. **M. spicatum** L. Fig. 321 Eurasian Water-milfoil

Map 909. Ponds and lakes, including the Great Lakes and connecting waters. Reports from several counties additional to those mapped (see Coffey & McNabb 1974) are presumably reliable—as far north as Little Bay de Noc (Delta County) and as far west as Muskegon and Ottawa counties, but specimens have not been encountered in herbaria.

This, our only non-indigenous species of the genus, was first found in North America late in the 19th century, but did not become a serious aquatic weed until the mid-20th century. It was first recognized in Michigan in 1970 but had doubtless been in the state for at least a decade, confused with *M. exalbescens*. It is now an aggressive weed in many lakes and streams (dubbed "superweed" in the popular press), where it crowds out other species (Coffey & McNabb 1974). The plant spreads primarily by fragmentation, and

907. Myriophyllum verticillatum

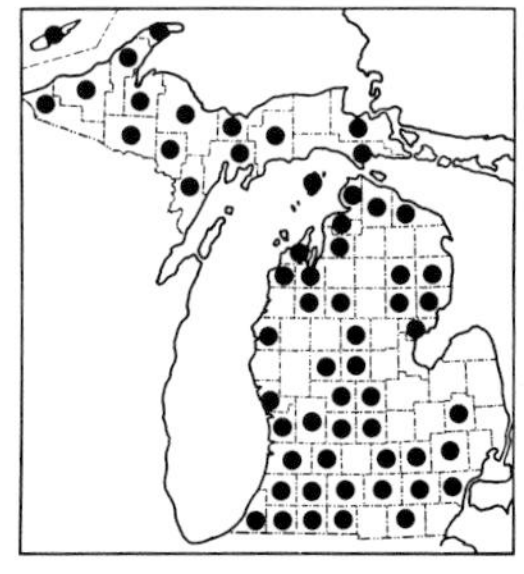

908. Myriophyllum exalbescens

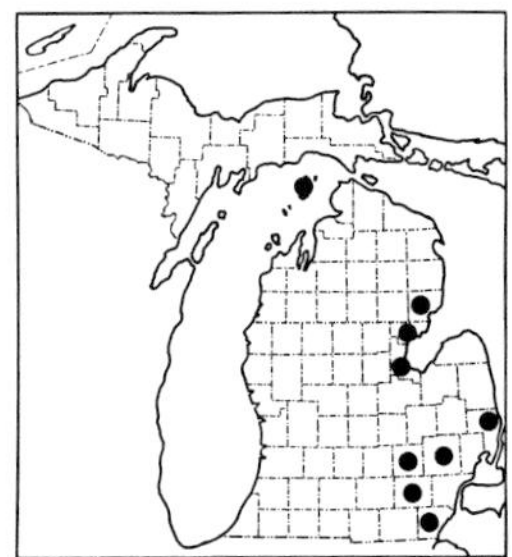

909. Myriophyllum spicatum

indeed it has been suggested that all North American populations may represent a single clone. The species may have been introduced at some sites through dumping of aquarium rubbish. In Monroe County marshes, it was reportedly planted, having been bought as "coontail" (*Ceratophyllum*), an important duck food.

Many characters have been proposed to distinguish this species from the preceding closely related one, but of the easily seen morphological ones, only the number of leaf segments (on well developed leaves) seems reliable. In the field, the two look quite different. *M. spicatum* tends to branch profusely near the surface of the water, forming a tangle of foliage that shades out plants below. The leaves are strongly feather-like, the segments all in one plane; the impression is well given in fig. 321.

NOTES ON STERILE MYRIOPHYLLUM

M. tenellum poses no problem. The small-leaved *M. alterniflorum* could be confused only with a form of *M. exalbescens* with reduced leaves (discussed above under that species). *M. farwellii* is extraordinarily limp and grows rarely in northern softwater lakes, nowhere else, and could be confused only with lax sterile *M. heterophyllum*. The following notes may help with the other four widespread species:

1. If there is a definite tendency to alternate leaves somewhere on the stem, the plant is *M. heterophyllum* (if not *M. farwellii*). But if *all* leaves are whorled, it may be any species (except *M. farwellii*).
2. If many internodes exceed 10 mm, it is *not M. heterophyllum*.
3. If the foliage is dense (very short internodes, sometimes with 6 leaves) and quite delicate, it is probably *M. heterophyllum*.
4. If the stems are whitish (with scattered yellow-orange dots), it is *not M. verticillatum*.
5. Only *M. spicatum* has as many as 14–17 segments on each side of some or all leaves.
6. *M. exalbescens* and *M. verticillatum* have fewer than 14 segments on each side of the leaf, but differ in stem color (brownish or greenish in the latter), and the former has the leaves all petioled. Plants with fewer than 8 or 9 segments are *M. exalbescens* rather than *M. verticillatum*.
7. If the base of the plant is present, in *M. verticillatum* and *M. exalbescens* roots may arise from a strong U-shaped bend in the stem, representing the axis of a turion, which only these species (and *M. farwellii*) produce.

Note: All four of these species may produce a strongly thickened stem (1.5–2 times the width lower down) shortly below the inflorescence, so this is not a reliable character, as sometimes stated, for *M. spicatum*.

HIPPURIDACEAE — Mare's-tail Family

This family, consisting of only one or two species, has traditionally been classified very near the Haloragaceae, or even included in it. However, recent systems of classification (e. g., by Cronquist, Takhtajan) place Hippuridaceae in the subclass Asteridae (where they are associated with Callitrichaceae by Cronquist) while the Haloragaceae are retained in the Rosidae. The flowers are very much reduced, consisting of little or no perianth (the calyx at best a mere rim on the ovary), 1 stamen, and 1 carpel containing 1 ovule; they are presumably wind-pollinated, and form in the axils of ± firm aerial leaves, quite unlike the limp submersed leaves which the plant produces under water (where light intensity is lower).

REFERENCE

McCully, Margaret E., & Hugh M. Dale. 1961. Heterophylly in Hippuris, a Problem in Identification. Canad. Jour. Bot. 39: 1099–1116.

1. **Hippuris**

1. **H. vulgaris** L. Plate 8-B — Mare's-tail

Map 910. A circumpolar species, formerly known south in the Great Lakes region as far as the head of Lake Michigan in Indiana. Lakes, streams, and rivers, especially in marshy borders (but in water as deep as 3 feet), on mucky to gravelly substrates and in cold spring-fed water; in temporary pools and marshes at the edge of the Great Lakes; ditches; the terrestrial form on muddy to sandy or gravelly banks and shores. The Kalamazoo County collection was made by the First Survey in 1838; the species has not been found in southern Michigan since.

Our only other true aquatic with definitely whorled simple leaves is *Elodea,* which has 3 (rarely 6) short leaves in a whorl. In *Hippuris,* submersed leaves are mostly 6–12 (often 9) in a whorl, about 12–25 times as long as wide, and all on unbranched stems. These leaves are very limp and dense (suggesting the generic and common names), but those on emersed tips are firm and average shorter. Such more rigid aerial leaves are also found on terrestrial plants on damp shores and banks, where they may look at first glance like some odd *Lycopodium*.

ARALIACEAE — Ginseng Family

Several species of this family are grown indoors or outdoors as ornamentals, including the true European ivy, *Hedera helix* L., which may appear to run wild. A few authors unite this family with Umbelliferae (Apiaceae), in which case the name of the latter family is to be used.

REFERENCE

Graham, Shirley A. 1966. The Genera of Araliaceae in the Southeastern United States. Jour. Arnold Arb. 47: 126–136.

KEY TO THE GENERA

1. Leaves simple, palmately lobed; plant a shrub, spiny throughout, of Isle Royale .... 1. **Oplopanax**
1. Leaves compound; plant herbaceous, or woody only at the base
  2. Leaves in 1 whorl, once-palmately compound; umbel solitary; carpels 2 or 3 .... 2. **Panax**
  2. Leaves alternate or basal, mostly 2–3 times compound, the ultimate leaflets pinnate; umbels normally 2 or more; carpels 5 .... 3. **Aralia**

1. Oplopanax

A species very similar to ours is *O. elatus* (Nakai) Nakai of eastern Asia. The generic name has been variously considered as masculine, feminine, or neuter, but for half a century the Code has recommended (or required) that compound generic names ending in *-panax* be treated as masculine, the classical gender of *Panax* itself. The purpose of such an explicit provision is to promote uniformity, and these names should uniformly be associated with adjectival epithets of masculine form, such as *horridus, trifolius*.

1. **O. horridus** (J. E. Smith) Miq. Plate 7-E, 7-F Devil's-club

Map 911. Primarily a northwestern species, ranging from southern Alaska and southwestern Yukon to Oregon and northwestern Montana. It is then strikingly disjunct to islands in northern Lake Superior, including Porphyry Island off the Sibley Peninsula of Ontario and (recently discovered) the Slate Islands. In Michigan, devil's-club thrives at the northeast end of Isle Royale (Blake Point), several of the nearby offshore islands, and Passage Island (see Mich. Bot. 20: 59. 1981). It is found especially between rock ridges with fir and paper birch in wooded ravines and openings.

The stems (which are often 2 m tall) and leaf veins are all decorated with

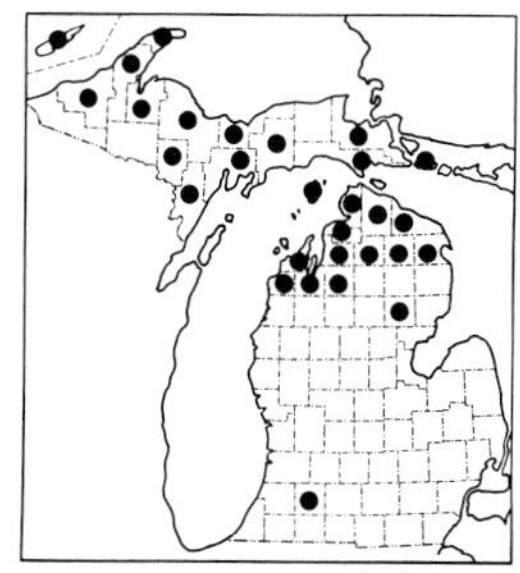

910. Hippuris vulgaris

911. Oplopanax horridus

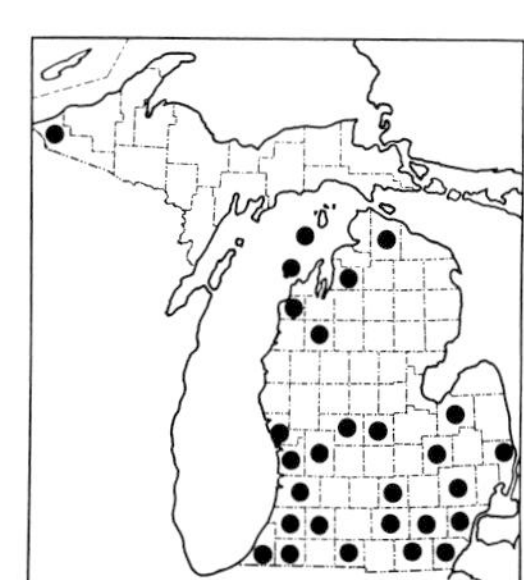

912. Panax quinquefolius

sharp spines which penetrate clothes and leave a lasting impression on anyone who walks through a thicket of this aptly named species. Even the moose of Isle Royale do not browse it. The flowers are rather fragrant, in an impressive panicle of umbels, and the bright red fruit held above the densely prickly stem and enormous leaves look dramatically out of place in the boreal forest. Flowers are at their peak about mid-July and fruit is ripe a month later.

2. **Panax** Ginseng

Panax as a Latin noun for a plant with all-healing properties was of masculine gender. When taken over as a generic name, it should retain its classical gender—even though some authors, including Linnaeus himself, have treated it as neuter. The flowers are perfect or unisexual. Functionally staminate flowers may have a single style. The roots of the Asian *P. ginseng* C. A. Meyer have for centuries been used medicinally by the Chinese. The American ginseng, *P. quinquefolius,* has similar often forked roots, suggesting a two-legged person, and hence is similarly reputed to have curative and tonic properties; it is therefore under considerable commercial pressure, especially for export to the Orient, and has become quite rare throughout its range in eastern North America. Considerable information on the life history and biology of the two native species is in the references cited below.

REFERENCES

Lewis, Walter H., & Vincent E. Zenger. 1982. Population Dynamics of the American Ginseng Panax quinquefolium (Araliaceae). Am. Jour. Bot. 69: 1483–1490.

Lewis, Walter H., & Vincent E. Zenger. 1983. Breeding Systems and Fecundity in the American Ginseng, Panax quinquefolium (Araliaceae). Am. Jour. Bot. 70: 466–468.

Philbrick, C. Thomas. 1983. Contributions to the Reproductive Biology of Panax trifolium L. (Araliaceae). Rhodora 85: 97–113.

Williams, Llewelyn, & James A. Duke. 1978. Growing Ginseng. U.S. Dep. Agr. Farmer's Bull. 2201, rev. 8 pp.

KEY TO THE SPECIES

1. Leaflets all distinctly stalked (petiolules (0.8) 1–2.5 cm long), at least the larger ones 3.5–6 (7) cm wide, ± obovate, abruptly acuminate; styles (in perfect flowers) all or mostly 2; root elongate, often forked; ripe fruit bright red .... 1. **P. quinquefolius**
1. Leaflets sessile or subsessile, 0.5–2.5 cm wide, narrowly elliptic to lanceolate, obtuse or rounded to acute (but not acuminate); styles (in perfect flowers) all or mostly 3; root globose; ripe fruit yellow-green .... 2. **P. trifolius**

1. **P. quinquefolius** L. Fig. 322 Ginseng; "Sang"

Map 912. Rich even swampy hardwoods (beech, sugar maple, hemlock), especially on slopes or ravines (including wooded dunes). Already noted by Dodge in 1917 as becoming scarce (in St. Clair County); intensely sought for its reputedly medicinal roots.

322. *Panax quinquefolius* $\times 1/4$
323. *Aralia hispida* $\times 1/2$
324. *Bupleurum rotundifolium* $\times 1/2$

2. **P. trifolius** L. Dwarf Ginseng

Map 913. Rich deciduous or hemlock woods, often low or swampy or even on floodplains.

A much smaller plant than the preceding, less than 30 (usually less than 20) cm tall. The round tuberous root has not suggested the medicinal merit attributed to true ginseng.

3. **Aralia**

Tall erect prickly shrubs or small trees in this genus are frequently cultivated, and there are records from Kalamazoo, Kent, Oakland, and Washtenaw counties of somewhat dubious status but which might indicate escapes from cultivation. Specimens are universally labeled as the Hercules-club, *A. spinosa* L., a native species from farther south in the United States. However, most if not all of them are probably *A. elata* (Miq.) Seem., a native of eastern Asia and more hardy in our climate. Herbarium specimens of a plant with such large leaves and inflorescences are very incomplete. In *A. spinosa* the umbels are in a very elongate panicle (as long as a meter), while in *A. elata* the inflorescence is very broad. The leaflets are petioluled in *A. spinosa* and more remotely toothed than the serrate ± sessile leaflets of *A. elata*.

KEY TO THE SPECIES

1. Inflorescence a large spreading panicle of numerous small umbels—often axillary as well as terminal . . . . 1. **A. racemosa**
1. Inflorescence of 3–13 umbels, terminal on erect peduncles (the umbels on spreading rays)
   2. Plant essentially stemless, the petioles and peduncles arising from the base, not prickly; leaves twice compound, the first division palmate (usually ternate), the second pinnate . . . . 2. **A. nudicaulis**
   2. Plant with erect leafy stem prickly toward the base; leaves twice-pinnately compound . . . . 3. **A. hispida**

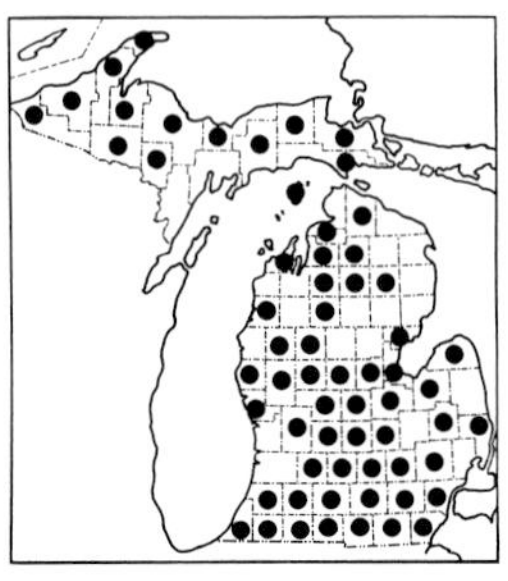

913. Panax trifolius

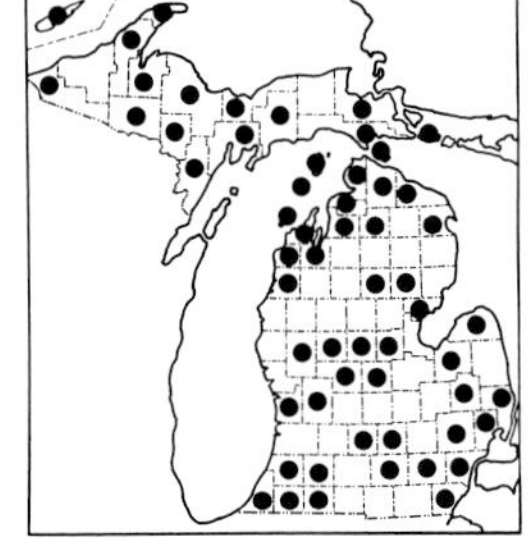

914. Aralia racemosa

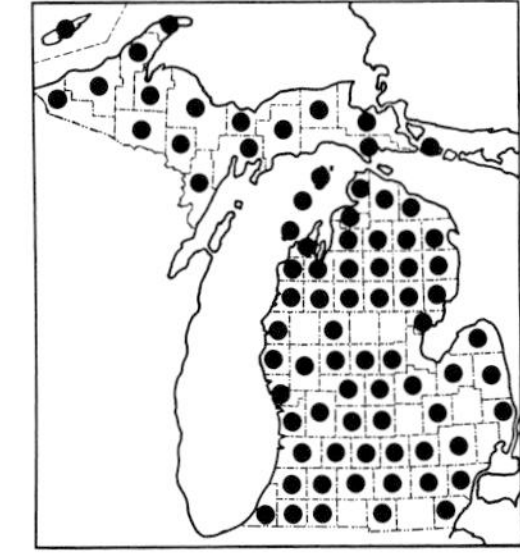

915. Aralia nudicaulis

1. **A. racemosa** L. Spikenard

Map 914. Rich usually moist beech-maple and hemlock-hardwoods, especially along edges and clearings and below bluffs; less often in oak woods; conifer (mostly cedar) swamps.

This is one of the largest herbaceous species in our flora, perhaps not quite as tall as some attain, but widely spreading and bushy, with enormous compound leaves and a massive inflorescence of countless umbels of purple-black fruit. There are often smaller axillary inflorescences as well. The individual umbels in this species are usually less than 2.5 cm broad, though occasionally almost 3 cm in fruit.

2. **A. nudicaulis** L. Wild Sarsaparilla

Map 915. In almost all sorts of dryish to moist deciduous, coniferous, and mixed woods (including oak-hickory, beech-maple, northern hemlock-hardwoods, pine); swamps (cedar, tamarack, elm); common with aspens and birch and on stabilized dunes.

The long horizontal aromatic roots are of reputed medicinal value and have contributed to patent medicines as well as herbal teas and aboriginal remedies. The larger flowering umbels in this species and the next are usually ca. 2–3 cm broad, becoming 3–4 (6) cm in fruit.

3. **A. hispida** Vent. Fig. 323 Bristly Sarsaparilla

Map 916. Dry sandy woodland (pine, aspen) on dunes or plains; sandy banks and bluffs, upper beaches, rock outcrops; barren open burned- or cut-over ground, gravel pits, clearings, newly made roadsides, recently bull-dozed areas; sometimes in damp, especially disturbed, ground or at borders of swamps. Often forms large colonies from an extensive system of horizontal roots in newly disturbed sandy soils, and persists for some years after disturbance.

UMBELLIFERAE (APIACEAE) Carrot or Parsley Family

This is one of the easiest of all families to recognize, but one of the most difficult as far as identification of species is concerned. The flowers are small and 5-merous except that the sepals are tiny or even absent and the carpels are 2. The inferior ovary ripens into a unique schizocarp in which 2 indehiscent, 1-seeded mericarps typically separate from a forking central axis that bears a mericarp suspended from the tip of each fork. Each mericarp usually has 5 longitudinal ribs, which may be relatively inconspicuous, or raised, or expanded into distinct wings. The flowers are in a head or simple umbel in a few genera (*Eryngium, Hydrocotyle, Sanicula*) but in the great majority of our genera the inflorescence is a compound umbel, the primary branches, termed *rays,* each bearing an *umbellet* of pediceled flowers—some of which in some species may be staminate rather than perfect. There may be bracts (an involucre) at the base of the primary umbel and/or there may be bractlets (an involucel) at the base of each umbellet. Only the Araliaceae (sometimes included in this family) have similar flowers and inflorescences, but they have a fleshy indehiscent fruit of usually 5 carpels.

Vegetatively, the Umbelliferae are ordinarily characterized by hollow internodes, petioles with expanded ± sheathing bases, and resin canals which are readily sensed in the aromatic foliage and often seen as prominent oil tubes on the fruit. Many pungent and fragrant species are used in flavoring foods and medicines, while others provide familiar vegetables, and most of these escape from cultivation at least occasionally. Other species, however, are deadly poisonous if taken internally. The leaves of Umbelliferae are usually compound to deeply dissected and are often difficult to describe, especially since they may vary from the base to the summit of the stem. Ordinarily, the best results with the key will be obtained by using leaves from the middle of the stem (unless others are specified).

Technical classification within the family depends heavily on characters of the fruit. Many keys rely so much on fruit that identification of flowering specimens is impossible. Fortunately, by restricting coverage to the species of a single state and utilizing vegetative characters as well as those visible on young fruit, a workable key is possible without requiring mature fruit. Since plants of most species have umbels of mixed ages, partly if not fully mature fruit is often available along with flowers. Measurements of the styles, unless stated to the contrary, include the expanded style base or *stylopodium,* and refer to the length that the style quickly attains by the time the petals are all shed. (Styles at anthesis may be minute, but on even the young fruit, elongate.) Measurements of fruit do not include any persistent style or stylopodium. The fruit is sometimes flattened "dorsally" (parallel to the adjoining faces of the mericarps) and sometimes "laterally" (at a right angle to the adjoining faces).

A few species have been repeatedly attributed to Michigan in the literature, but are not included here as no documenting specimens have been found in any herbarium examined. Specimens of *Angelica atropurpurea* are the basis for reports from Michigan of the European *Imperatoria ostruthium* L. from Muskegon County and of *Ligusticum canadense* (L.) Britton from Monroe County. *Perideridia americana* (Nutt.) Reichb. would be at the northeastern end of its range if in the state; it is said in recent manuals and monographs to occur in Michigan, though on what basis I have been unable to determine.

REFERENCES

Crawford, Daniel J. 1970. The Umbelliferae of Iowa. Univ. Iowa Stud. Nat. Hist. 21(4). 35 pp.

Kenoyer, L. A. 1924. Distribution of the Umbellales in Michigan. Pap. Mich. Acad. 3: 131–165. [Primarily of historical interest, with a number of old reports.]

Mathias, Mildred E., & Lincoln Constance. 1944–1945. Umbelliferae. N. Am. Fl. 28B: 43–295.

KEY TO THE GENERA

1. Margins of cauline leaves and bracts of the inflorescence with stiff spines; basal leaves simple, unlobed; cauline leaves simple (at most 3–7-lobed), on stiff erect stem; flowers in a dense bracted head . . . . 1. **Eryngium**

1. Margins of leaves and bracts without spines (or soft spines in *Sanicula*); basal leaves deeply lobed to compound or dissected, or if simple, the cauline leaves either compound or on a prostrate stem; flowers in umbels, at least some clearly pediceled

2. Leaves all simple, at most with merely crenate margins

3. Leaf blades distinctly longer than broad, entire, the upper ones perfoliate . . . . 2. **Bupleurum**

3. Leaf blades orbicular to reniform, crenulate, petioled . . . . 3. **Hydrocotyle**

2. Leaves (at least the cauline) mostly compound, dissected, or deeply lobed (basal or uppermost leaves sometimes simple); plants mostly of erect habit with compound umbels

4. Ovary and fruit pubescent, papillose, or bristly, or upper part of stem and peduncles with spreading or tomentose pubescence, or both conditions present

5. Bristles of ovary and fruit all strongly hooked (with incurved tips); leaves once-palmately compound or deeply lobed; umbellets consisting of 1 or more sessile or subsessile perfect flowers and pediceled staminate flowers, or entirely of staminate flowers . . . . 4. **Sanicula**

5. Bristles or hairs (if any) of ovary and fruit only partly or weakly and slightly if at all curved at the tips; leaves various (mostly pinnate or deeply divided) but not once-palmately lobed; umbels in most species consisting of all perfect flowers

6. Cauline leaves (at least the upper ones) pinnately or bipinnately divided into long linear-lanceolate lobes, grading into simple, at most shallowly lobed, basal leaves . . . . 5. **Pimpinella**

6. Cauline and basal leaves both finely dissected or with definite leaflets mostly ca. 1 cm or more wide

7. Leaves pinnate and with leaflets (or segments if dissected) less than 1 cm broad; plant introduced, mostly in waste places

8. Bracts (at least the larger ones) subtending the rays over (10) 15 mm long (except sometimes when very young), pinnate with narrow lobes . . . . 6. **Daucus**

8. Bracts all less than 15 mm long, simple

9. Leaves once-pinnately compound, the leaflets deeply lobed, with flat segments; ovaries and fruit with some bristles . . . . 7. **Torilis**

9. Leaves 2–3 times compound, with filiform segments; ovaries and fruit merely papillose-warty . . . . 8. **Trachyspermum**

7. Leaves definitely ternate, with 3 principal (and petioluled) divisions at the summit of the petiole, and/or with ± clearcut leaflets at least 1 cm broad

10. Primary rays of umbel 1–7; fruit, even when young, ± club-shaped, more than twice as long as broad, glabrous or with appressed hairs or bristles more than 0.5 mm long

11. Ovary and fruit glabrous; plant a low, weak-stemmed annual . . . . 21. **Chaerophyllum**

11. Ovary and fruit pubescent, the fruit with stiff appressed hairs especially toward the base; plant an erect perennial .... 9. **Osmorhiza**

10. Primary rays of umbel at least 15; fruit ± broadly elliptic-oblong and flat or winged, with spreading hairs less than 0.5 mm long

12. Leaflets mostly ca. 1–2 cm broad (rarely narrower), with regular obtuse teeth (occasionally lobed near the base); fruit less than 8 mm long, with conspicuous thin wings, the central part with ribs the entire length more prominent than the oil tubes .... 10. **Angelica (venenosa)**

12. Leaflets much larger, with acute lobes and teeth; fruit ca. 8–12 (14) mm long, not winged, with ribs less prominent than the dark oil tubes (which extend less than the full length) .... 11. **Heracleum**

4. Ovary and fruit smooth (except for ribs or wings) and glabrous; upper part of stem glabrous (minutely pubescent in *Pastinaca*)

13. Principal cauline leaves (unless submersed) clearly once-pinnately compound with 5 or more definite flat leaflets (entire or toothed) (figs. 335–337)

14. Flowers yellow; stem and leaves ± finely pubescent (very rarely glabrous); plant a common weed, often in dry places .... 12. **Pastinaca**

14. Flowers white; stem and leaves completely glabrous; plant native, mostly in damp or wet places

15. Leaflets entire or with a few irregularly spaced teeth (usually on apical half only); fruit ca. 4–5.5 mm long, strongly flattened dorsally .... 13. **Oxypolis**

15. Leaflets all toothed the full length of their margins; fruit ca. 1.5–3 mm long, rounded or flattened laterally

16. Most (or all) cauline leaves with leaflets less than 3 (4) times as long as broad, with teeth of very irregular size; fruit ca. 1.5–2 mm long, obscurely ribbed .... 14. **Berula**

16. Most (or all) cauline leaves with leaflets (3.5) 4–10 (14) times as long as broad, with numerous fine teeth of ± uniform size; fruit ca. 2–3 mm long, with prominent corky ribs .... 15. **Sium**

13. Principal cauline leaves trifoliolate, or palmate, or more than once-compound to deeply lobed, or finely divided into numerous flat to filiform toothed or entire segments

17. Plant bearing vegetative bulblets, subtended by broad-based acuminate bracts, on upper parts .... 33. **Cicuta (bulbifera)**

17. Plant without bulblets

18. Principal cauline leaves twice-pinnate (sometimes appearing ternate) with ± irregularly toothed ultimate segments ca. 3–10 mm broad (fig. 338); stem terete (or finely ridged), not grooved; umbellets subtended by narrow bractlets (scarcely if at all broader, at the base, than the rays); flowers yellow; fruit with corky marginal wing thicker than the body but with only obscure ribs on back; plant a rare (extirpated?) taprooted prairie species .. 16. **Polytaenia**

18. Principal cauline leaves more than twice-compound (sometimes clearly ternate) with ultimate segments various but if more than 3 mm broad and irregularly toothed, the stem grooved, the umbellets with broader bractlets, and/or the fruit with wing none or thin (though ribs may be prominent on back); plant of various habit

19. Cauline leaves (at least of middle and upper part of stem) finely divided, the ultimate lobes or segments ± filiform to linear-lanceolate and less than 5 mm wide

20. Ultimate segments of at least the upper leaves slenderly linear-filiform, less than 1 mm wide and several times as long, and all entire; plant escaped from cultivation, at least the ripe fruit with spicy or aromatic quality

21. Primary rays of umbel 3–7; styles ca. 1.5–2 mm long; fruit nearly globose, hardly splitting into mericarps; petals very unequal in size, the larger ca. 3–4 mm long, white (to pink) .... 17. **Coriandrum**

21. Primary rays of umbel (6) 8–50; styles less than 1.2 mm long; fruit longer than broad, readily splitting; petals essentially uniform, not over 1.5 (2) mm long, white or yellow

22. Petals white; styles (excluding base) becoming at least 0.5 mm long; primary rays of umbel (6) 8–14 (16); rays and pedicels often subtended by 1 or more bracts .... 18. **Carum**

22. Petals yellow (-green); styles (excluding base) less than 0.5 mm long; primary rays of umbel ca. (10) 17–50; rays and pedicels without subtending bracts

23. Petiolar sheaths ca. 1–2.5 cm long; mericarps bordered by a narrow wing at maturity; styles (including base) ca. 0.5 mm long on maturing fruit; plant annual .... 19. **Anethum**

23. Petiolar sheaths (at least the larger ones) ca. 3–9 cm long; mericarps without wing; styles (including relatively long base) ca. 1 mm long on maturing fruits; plant perennial .... 20. **Foeniculum**

20. Ultimate leaflets or segments of leaves (upper as well as lower) linear-oblong, mostly ca. 1–5 mm wide, less than 5 times as long as wide, sometimes toothed; plant native or escaped, with rank or little "umbelliferous" odor but not spicy-aromatic

24. Umbellets subtended by conspicuous bractlets much broader than the thickness of the rays

25. Bractlets blunt or rounded at the tip, symmetrical beneath the umbellet; primary rays of umbel 1–4; leaves ternate above the sheathing petiole (divided into 3 ± equal axes); plant a low native blooming in the spring

26. Fruit elongate (ca. 3.5–10 times as long as broad); anthers yellow; styles minute; plant usually lightly hairy (petioles, bractlets, stems, rays), annual .... 21. **Chaerophyllum**

26. Fruit no longer than broad; anthers deep maroon; styles elongate, exceeding the ovary and almost or quite as long as the fruit; plant glabrous, perennial from a subglobose tuber .... 22. **Erigenia**

25. Bractlets acute (cuspidate to acuminate or subulate) at the tip, all on one side of the umbellet; primary rays of umbel ca. (8) 10–20; leaves pinnate (central axis more prominent than the lateral ones); plant a tall erect introduction (usually near habitation), blooming in the summer

27. Stem strongly corrugated or angled, minutely pubescent toward base, unspotted; petiolar sheath densely pubescent at least at summit; fruit elongate, ca. 3 times as long as wide .... 23. **Anthriscus**

27. Stem nearly terete (slightly ridged), glabrous throughout, spotted with red-purple; petiolar sheath glabrous, fruit ovoid, less than twice as long as wide.. 24. **Conium**

24. Umbellets with bractlets scarcely (toward base) if at all broader than the thickness of the primary rays

28. Nodes pubescent with short stiff hairs; petals pale yellow; leaflets ± ovate-lanceolate and coarsely toothed or lobed .... 32. **Thaspium (barbinode)**

28. Nodes glabrous; petals white (yellow-green in *Petroselinum*); leaflets deeply divided into ± linear to lanceolate or oblong lobes

29. Styles becoming 1.3–2 mm long; fruit strongly flattened dorsally, with 3 prominent ribs in the middle of each winged mericarp; plant a tall native of swampy and springy places .... 25. **Conioselinum**

29. Styles less than 1.3 mm long; fruit ± rounded or slightly flattened laterally, with 5 prominent ribs on each wingless mericarp; plant an escape from cultivation, in waste ground or near gardens

30. Bractlets subtending umbellets not extending beyond flowers; ribs on fruit much narrower than the intervals between them; leaves usually ± crisped; pedicels and rays of umbel smooth .... 26. **Petroselinum**

30. Bractlets (or some of them) extending beyond the flowers; ribs on fruit broader than the intervals (nearly concealed) between them; leaves flat; pedicels and rays of umbel ± scabrous .... 27. **Aethusa**

19. Cauline leaves with definite leaflets mostly over 1 cm broad, entire or (usually) toothed but not dissected into narrow lobes

31. Cauline leaves with 3 leaflets (though these may be ± deeply lobed)

32. Petals yellow; fruit less than 7 mm long *and* less than twice as long as wide .... [go to couplet 34]

32. Petals white; fruit over twice as long as broad or over 8 mm long

33. Leaflets usually sessile or nearly so, glabrous; fruit less than 7 mm long, more than twice as long as broad; stem slender (less than 8 mm thick) .... 28. **Cryptotaenia**

33. Leaflets all on distinct petiolules (terminal one longer), pubescent; fruit 8–12 mm long, less than twice as long as broad; stem stout, (8) 10–30 mm thick (or even thicker). .... 11. **Heracleum**

31. Cauline leaves (except sometimes the uppermost) with more than 3 leaflets

34. Leaflets entire or with ca. 1–5 coarse and irregular lobes or teeth (mostly beyond the middle); petals yellow

35. Margins of all leaflets strictly entire; umbel and umbellets without bracts or bractlets, the rays and pedicels smooth; plant a widespread native in (usually) sandy soils .... 29. **Taenidia**

35. Margins of most or all leaflets with a few teeth or small lobes; umbel and/or umbellets subtended by bracts, the rays and pedicels ± scabrous or papillose; plant a rare native or escape from cultivation

36. Nodes glabrous; fruit ribbed but not winged, with evident large style base; umbels and umbellets both subtended by well developed bracts and bractlets broader than the rays and pedicels respectively; plant an escape from cultivation, near habitation .... 30. **Levisticum**

36. Nodes pubescent with very short stiff hairs; fruit strongly winged, without enlarged style base; umbels and umbellets with inconspicuous slender bracts or bractlets narrower than the rays and pedicels; plant a very rare native .... 32. **Thaspium (barbinode)**

34. Leaflets with numerous ± uniform and regularly spaced small teeth; petals yellow or white

37. Plant very stout, with terete purple main stem ca. 8–40 mm (or more) in diameter; expanded petiolar sheaths ca. 1–6 cm broad at the middle and often (6) 8–15 cm long; mature fruit ca. 6–8 mm long......... ....10. **Angelica (atropurpurea)**

37. Plant of more modest size, the stem (mostly not purple) less than 8 (15) mm in diameter (or larger in water); expanded petiolar sheaths narrower and shorter; mature fruit less than 6 (7) mm long

38. Petals yellow

39. Leaflets ± minutely pubescent or hispidulous, at least beneath; fruit without raised ribs or wings on the broad side, ca. 5–6 (7) mm long; style (including a large base) becoming ca. 1 mm or less long .... (depauperate) 12. **Pastinaca**

39. Leaflets glabrous (or ± scabrous-papillose only on main veins); fruit with elevated ribs or wings, ca. 3.5–4.5 mm long; style without a large base, slender throughout and becoming ca. 1–1.7 mm long

40. Central flower (if perfect) and fruit of each umbellet sessile or nearly so (if staminate, then pediceled); fruit with prominent ribs but not winged (fig. 343) .... 31. **Zizia**

40. Central flower of each umbellet staminate and pediceled; fruit pediceled, with several distinct thin wings on each side (fig. 345)—evident even when half-mature .... 32. **Thaspium**

38. Petals white, or plants with only immature to mature fruit

41. Leaflets with veins directed to sinuses between the teeth (with branch to tooth), or plant with vegetative bulblets on upper portion (or with both conditions); petals white .... 33. **Cicuta**

41. Leaflets with veins directed to or into the teeth; plant without bulblets; petals white or yellow

42. Style base not at all developed; petals yellow .... [go to couplet 40]

42. Style base becoming ca. 0.5 mm long; petals white ... 34. **Aegopodium**

1. **Eryngium** Eryngo

KEY TO THE SPECIES

1. Leaves all parallel-veined, simple and elongate (yucca-like); bracts subtending heads less than half their length; petals white; plant a rare native .... 1. **E. yuccifolium**

1. Leaves all netted-veined, of various shape (basal petioled with broad subcordate blades, upper and midcauline nearly sessile and lobed); bracts more than half as long as the head; petals blue; plant a local escape from cultivation ....2. **E. planum**

1. **E. yuccifolium** Michaux Fig. 325 Rattlesnake-master

Map 917. Prairies (often wet), boggy ground, open borders of marshes and swamps.

The iris-like (though not equitant) or yucca-like leaves, with remote spines along their margins, are unlike those of any other native plant in this area.

2. **E. planum** L.

Map 918. A garden perennial of Eurasian origin, occasionally escaped and established, e. g., in lawns and fields.

2. **Bupleurum**

1. **B. rotundifolium** L. Fig. 324 Thoroughwax

Map 919. An Old World annual of weedy habit, collected on ballast at Detroit in 1906 (*Farwell 2004,* WUD).

A very distinctive little plant with few-flowered umbels of yellowish flowers, subtended by relatively broad bractlets.

3. **Hydrocotyle** Water-pennywort

KEY TO THE SPECIES

1. Petioles marginal, attached at sinus of a reniform blade; umbels few-flowered, sessile or subsessile . . . . 1. **H. americana**
1. Petioles attached near center of leaf blade (leaves peltate); umbels usually 15–35-flowered, on long peduncles . . . . 2. **H. umbellata**

1. **H. americana** L. Fig. 326

Map 920. Mossy coniferous swamps and woods, especially along stream banks; marshy ground, shores, and river margins; damp rich woods, especially in springy places or depressions.

2. **H. umbellata** L.

Map 921. Sandy or marshy borders of lakes and creeks, often in shallow water; tamarack bogs.

Rarely a second umbel will be developed on a peduncle arising from the primary umbel.

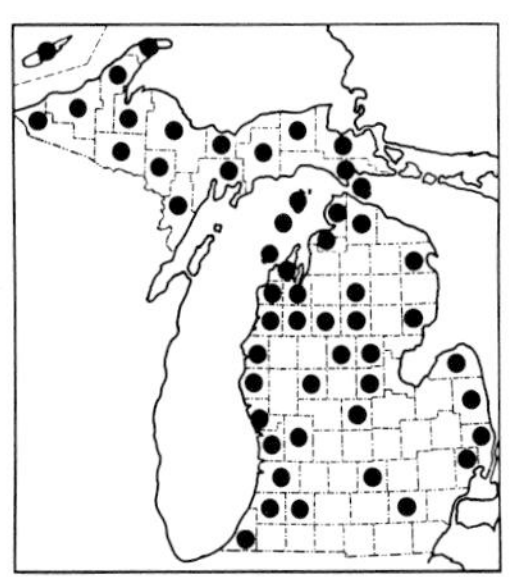

916. Aralia hispida

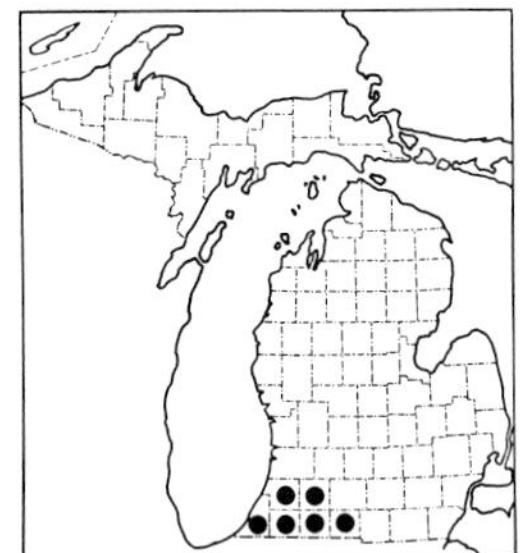

917. Eryngium yuccifolium

918. Eryngium planum

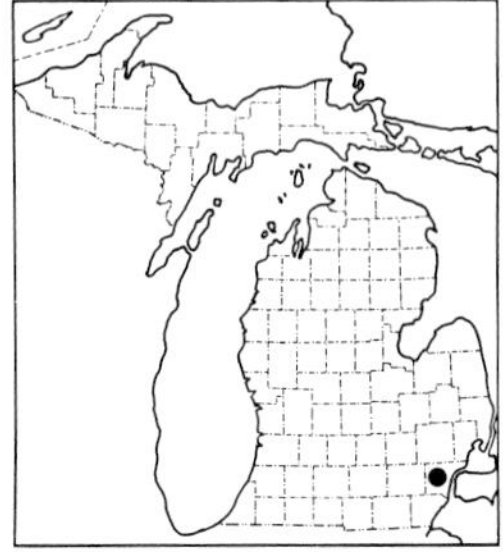

919. Bupleurum rotundifolium

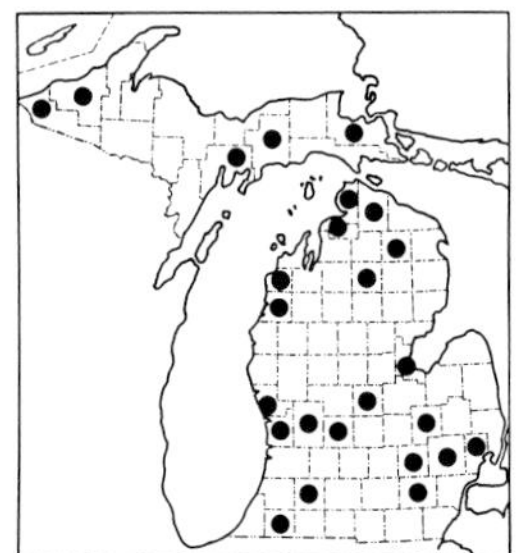

920. Hydrocotyle americana

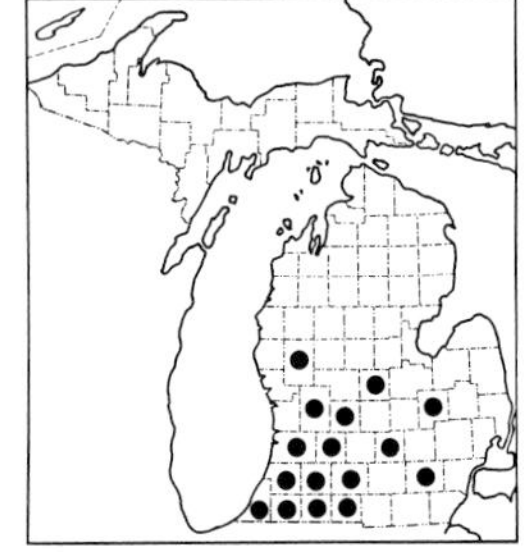

921. Hydrocotyle umbellata

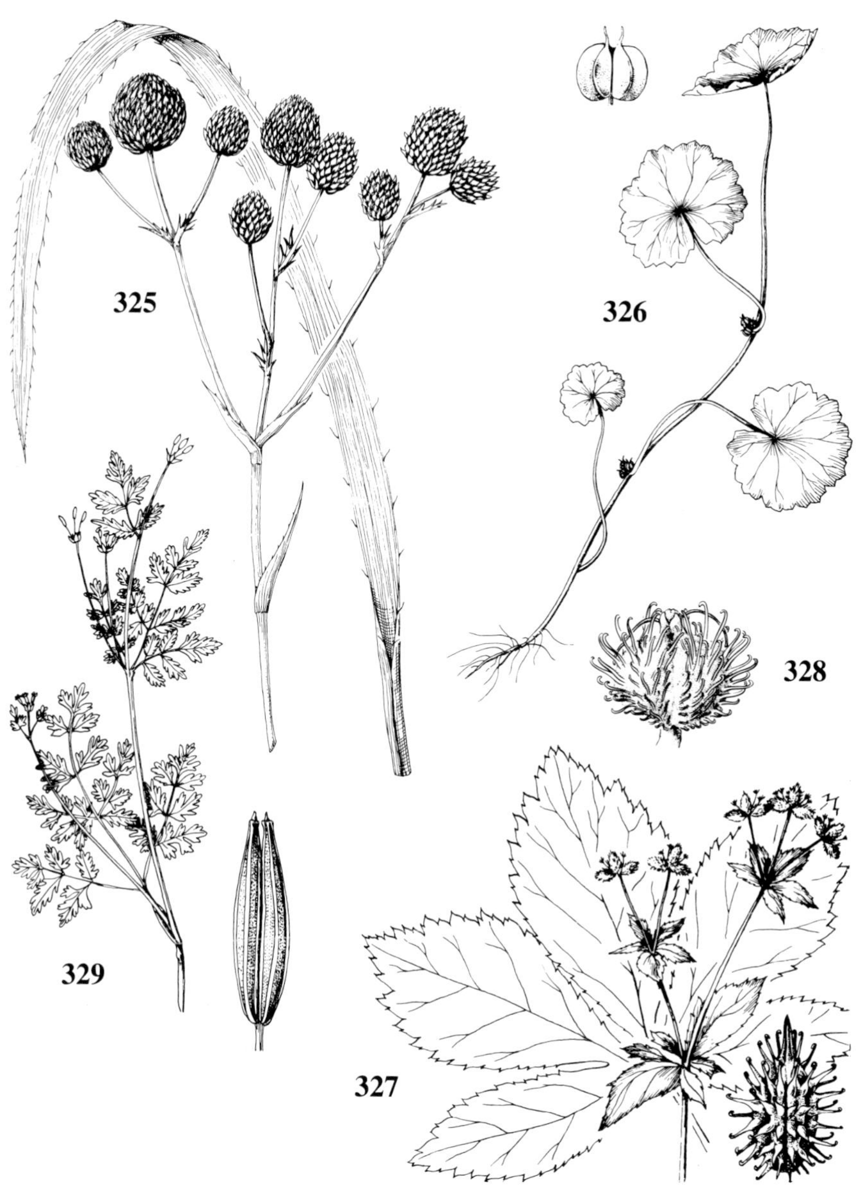

325. *Eryngium yuccifolium* $\times^1/_2$
326. *Hydrocotyle americana* $\times^1/_2$; fruit $\times 6$
327. *Sanicula trifoliata* $\times^1/_2$; fruit $\times 3$
328. *S. gregaria,* fruit $\times 4$
329. *Chaerophyllum procumbens* $\times^1/_2$; fruit $\times 4$

4. Sanicula — Black Snakeroot

REFERENCE

Shan, Ren Hwa, & Lincoln Constance. 1951. The Genus Sanicula (Umbelliferae) in the Old World and the New. Univ. California Publ. Bot. 25: 1–78.

KEY TO THE SPECIES

1. Styles shorter than the calyx of perfect flowers, scarcely visible (fig. 327); staminate flowers only 2–5, in the same umbels as perfect flowers
 2. Calyx lobes of perfect flowers conspicuous, equaling or exceeding the bristles; pedicels of staminate flowers ca. 2–4 times as long as their sepals . . . . . 1. **S. trifoliata**
 2. Calyx lobes of perfect flowers inconspicuous, shorter than the bristles; pedicels of staminate flowers twice as long as their sepals, or shorter . . . . 2. **S. canadensis**
1. Styles elongate (several times as long as calyx), arching conspicuously over (or among) the bristles (fig. 328); staminate flowers more numerous per umbel and often in separate umbels as well as in the same ones as perfect flowers
 3. Calyx lobes of staminate flowers soft, less than 1 mm long (usually ca. 0.5 mm) and less than twice as long as broad; mature fruit (including bristles) ca. 3–4 mm long (not including a short stipe) . . . . 3. **S. gregaria**
 3. Calyx lobes of staminate flowers rigid (even subulate-tipped), 1–1.6 mm long and 2–3 times as long as broad; mature fruit (including bristles) ca. 4.5–6.5 mm long . . . . 4. **S. marilandica**

1. **S. trifoliata** Bickn. Fig. 327
Map 922. Deciduous woods, especially beech-maple and oak-hickory forests.

2. **S. canadensis** L.
Map 923. Deciduous woods and thickets.

3. **S. gregaria** Bickn. Fig. 328
Map 924. Deciduous woods, including oak-hickory and rich beech-maple stands; river bottomland woods and thickets.

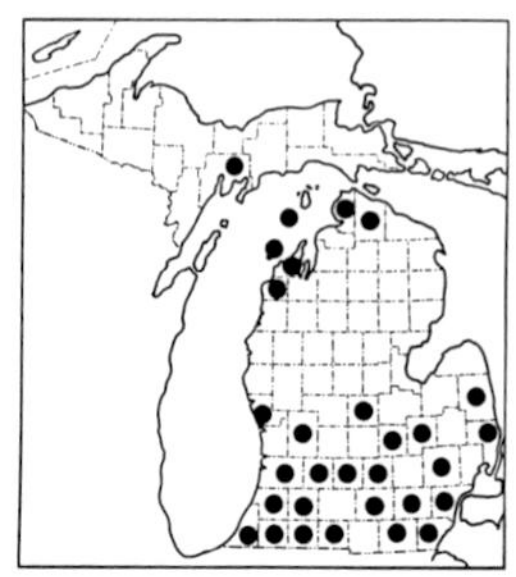
922. Sanicula trifoliata

923. Sanicula canadensis

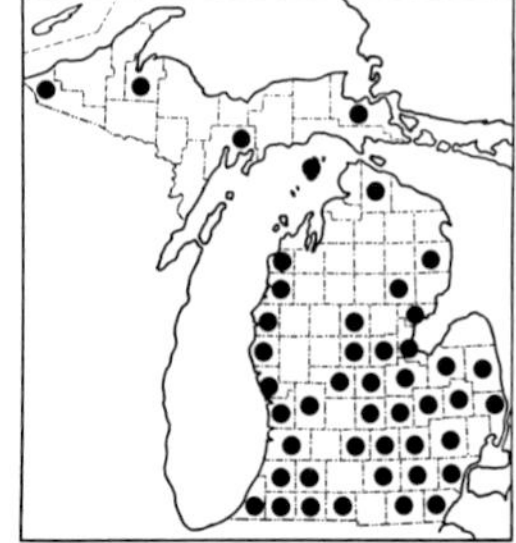
924. Sanicula gregaria

4. **S. marilandica** L.

Map 925. Deciduous, mixed, and coniferous (especially cedar) woods and swamps, often at borders and in openings and moist hollows, thriving with second-growth aspen-birch-fir cover; river-bank thickets.

The bristles on the fruit have more prominently bulbous bases than in the other species.

5. Pimpinella

1. **P. anisum** L. Anise

Map 926. A plant of Asian origin, whose seeds yield the licorice-like flavor of anise-seed and oil, widely used in many beverages, baked goods, candies, drugs, and other products. It rarely escapes from cultivation to waste grounds, and was collected by Farwell at Detroit (Rep. Mich. Acad. 20: 184. 1919).

6. Daucus

REFERENCES

Dale, Hugh M. 1974. The Biology of Canadian Weeds 5. Daucus carota. Canad. Jour. Pl. Sci. 54: 673–685.

Lacey, Elizabeth P. 1981. Seed Dispersal in Wild Carrot (Daucus carota). Mich. Bot. 20: 15–20.

Small, Ernest. 1978. A Numerical Taxonomic Analysis of the Daucus carota Complex. Canad. Jour. Bot. 56: 248–276.

1. **D. carota** L. Fig. 330 Wild Carrot; Queen-Anne's-lace

Map 927. An Old World native, now a well established weed of waste places, roadsides, and fields, invading woods, shores, and other habitats. Evidently established in Michigan in the 1880's.

The garden carrot, with enlarged, orange, brittle roots, is derived from this complex and variable species, and may be called *D. carota* var. *sativus*

925. Sanicula marilandica

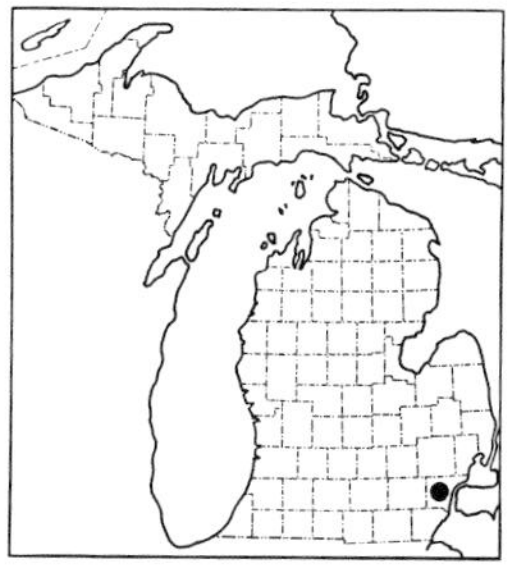

926. Pimpinella anisum

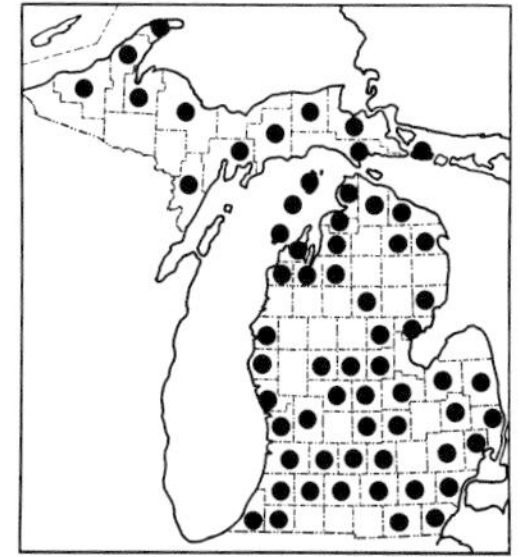

927. Daucus carota

Hoffman or ssp. *sativus* (Hoffman) Arcang. Wild carrots have whitish, often branched, relatively slender and tough roots; they may smell like carrots, and are not poisonous, but they are unpalatable. (And one should never confuse the poisonous tuberous roots of *Cicuta* with a carrot!)

There is usually a dark purple flower in the center of each inflorescence, and very rarely all the corollas are pink or purple.

7. **Torilis** — Hedge-parsley

KEY TO THE SPECIES

1. Peduncles not over 1 cm long; mericarps not uniform, some bristly and others merely warty . . . . 1. **T. nodosa**
1. Peduncles all or mostly 2–9 cm long; mericarps uniformly and densely bristly . . . . 2. **T. japonica**

1. **T. nodosa** (L.) Gaertner

Map 928. A native of the Mediterranean area, found as a weed in Ann Arbor in 1982 (*Reznicek 6979,* MICH) and to be expected elsewhere. (Farwell's notes indicate that he collected this species on waste ground at Detroit in 1906, but no specimen seems to be extant.)

2. **T. japonica** (Houtt.) DC. Fig. 331

Map 929. Roadsides, trails, clearings, and other disturbed areas of upland woods and low swamps. A European species, rather recently established as a weed in our area. First collected in Michigan in 1952 by Parmelee in oak woods in Clinton and Eaton counties, but not reported from the state until 1977 (Mich. Bot. 16: 136), by which time it was abundant in central Emmet County and the Lansing area, and known from a number of other counties.

Scandix pecten-veneris L., Venus'-comb, a native of Europe, was supposedly collected by Farwell in waste ground at Detroit in 1906, but no specimens have been found. The leaves are very finely dissected and the fruit is many times as long as broad, whereas in *Daucus* and *Torilis,* with somewhat similar leaves, the fruit is less than twice as long as broad.

8. **Trachyspermum**

1. **T. ammi** (L.) Sprague — Ajowan

Map 930. Native from northern Africa to the East Indies, cultivated in India, Asia Minor, and elsewhere for its aromatic fruits, used in cooking and medicines. Collected in waste ground at Detroit in 1894 (*Farwell 1479,* MICH; see Asa Gray Bull. 4: 46. 1896—where reported as *Carum copticum*). Not again recorded from the state—nor likely to be.

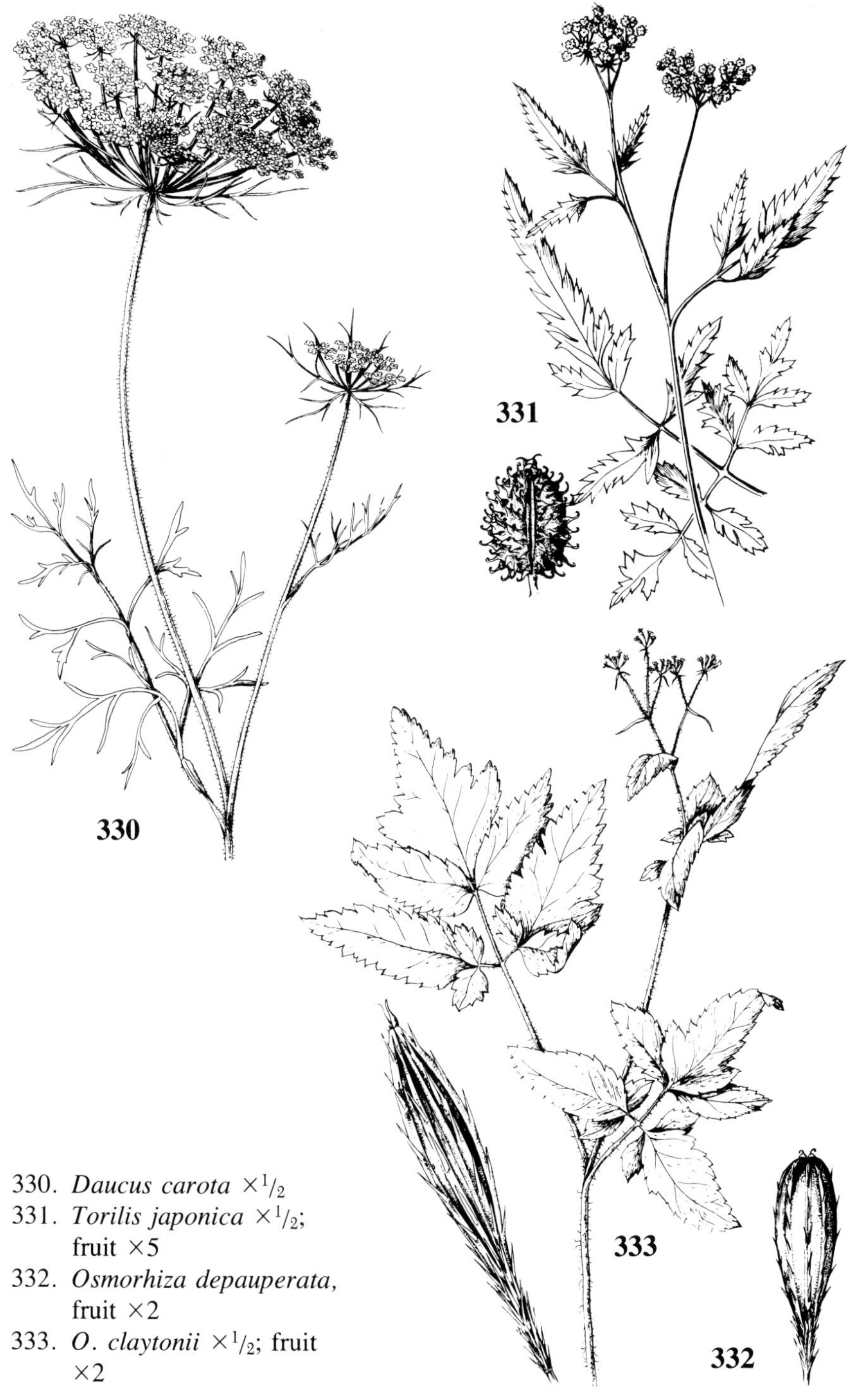

330. *Daucus carota* ×$^1/_2$
331. *Torilis japonica* ×$^1/_2$; fruit ×5
332. *Osmorhiza depauperata,* fruit ×2
333. *O. claytonii* ×$^1/_2$; fruit ×2

9. **Osmorhiza** Sweet-cicely

The base of the fruit in our species is prolonged into a pointed appendage covered with antrorse bristles which enable it to stick tenaciously to socks and other clothing—not to mention animals which may serve as agents of dispersal.

REFERENCES

Constance, Lincoln, & Ren Hwa Shan. 1948. The Genus Osmorhiza a Study in Geographic Affinities. Univ. California Publ. Bot. 23: 111–155.

Lowry, Porter P., II, & Almut G. Jones. 1979. Biosystematic Investigations and Taxonomy of Osmorhiza Rafinesque Section Osmorhiza (Apiaceae) in North America. Am. Midl. Nat. 101: 21–27.

Ostertag, Carol P., & Richard J. Jensen. 1980. Species and Population Variability of Osmorhiza longistylis and Osmorhiza claytonii. Ohio Jour. Sci. 80: 91–95.

KEY TO THE SPECIES

1. Inflorescence lacking involucels (bractlets at base of the umbellets); flowers all perfect
   2. Mature fruit ± convex toward apex, which is blunt, not beak-like (fig. 332); young fruit without any subapical constriction, but straight to slightly convex on the sides beneath the truncate apex . . . . 1. **O. depauperata**
   2. Mature fruits clearly concave-tapered to a short beak-like apex; young fruits with at least a slight constriction beneath the truncate apex . . . . 2. **O. chilensis**
1. Inflorescence with involucels; flowers both perfect and staminate
   3. Styles becoming 2–4 mm long; flowers (including withering staminate ones) (6) 9–18 per umbellet; plant (especially root) with strong licorice odor when bruised . . . . 3. **O. longistylis**
   3. Styles (even in fruit) not over 1.5 mm long; flowers 4–8 (10) per umbellet; plant with at most a weak licorice odor . . . . 4. **O. claytonii**

1. **O. depauperata** Phil. Fig. 332

Map 931. Mixed woods (insofar as our very few specimens bear habitat data). Long known as *O. obtusa* (Coulter & Rose) Fern., this species also

928. Torilis nodosa

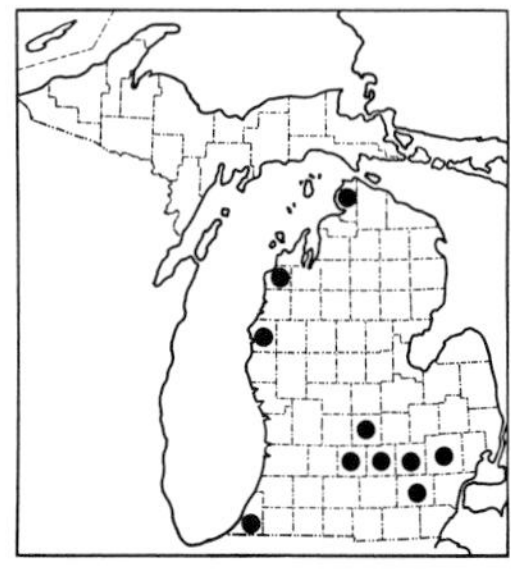

929. Torilis japonica

930. Trachyspermum ammi

occurs in southern South America; it ranges north from Lake Superior to James Bay and west to the Pacific coast—not as disjunct as the next species. The specimen on which published reports from Marquette County (Negaunee) are based has been reidentified as *O. chilensis*.

2. **O. chilensis** Hooker & Arn.

Map 932. Deciduous (beech-maple and northern hardwood), mixed, and coniferous forests and thickets, frequently on thin soil over limestone or other rock, or along trails. Often growing with the next two species. Originally described from Chile, this is, like *O. depauperata,* disjunct to North America; and it is also a classic disjunct between western North America, the Black Hills, the northern Great Lakes region, and the east (Newfoundland to New Hampshire). (See Mich. Bot. 20: 74–75. 1981.)

The bruised roots often have at least a weak licorice odor. The Alpena County material (Thunder Bay Island) is quite immature but looks much like *O. depauperata*.

3. **O. longistylis** (Torrey) DC.

Map 933. Rich, often moist (even swampy) deciduous woods; ± persistent after clearing and hence sometimes in fine stands along woodland roads. Frequently growing with the next, but usually restricted to rich moist sites.

Ordinarily one can readily distinguish this species in the field from *O. claytonii* by its taller, stiffer habit with redder, more glabrous stems; the flowers tend to be larger (petals often ca. 1.5–2.2 mm long, at least on perfect flowers) and, as they are more numerous in the umbellets (due mostly to a larger number of staminate ones), the plant is more showy. Bracts of the involucel tend to be broader than in *O. claytonii* (often over 1 mm). Rarely the stem is densely villous [var. *villicaulis* Fern.].

4. **O. claytonii** (Michaux) C. B. Clarke   Fig. 333

Map 934. Deciduous woods, including beech-maple, oak and oak-hick-

931. Osmorhiza depauperata

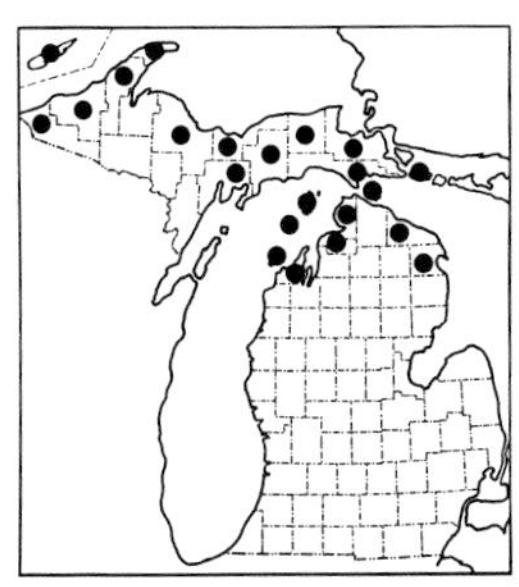

932. Osmorhiza chilensis

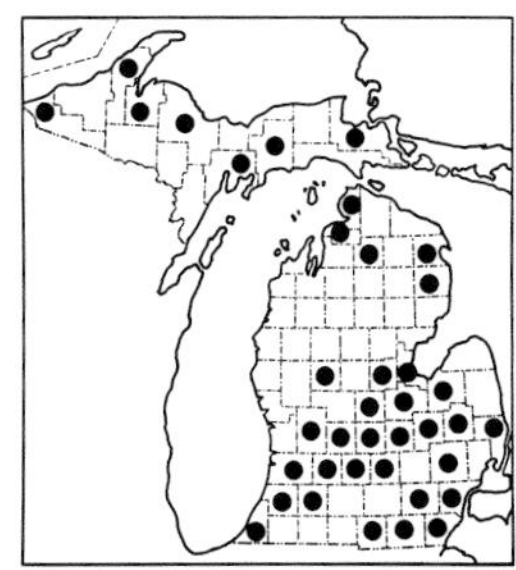

933. Osmorhiza longistylis

ory, northern hemlock-hardwoods; very rarely in strictly coniferous woods; sometimes on rocky or alluvial sites; ± persistent after clearing. Much more common than the preceding, and often in drier sites.

Rarely a plant of *O. longistylis* with very young flowers will have styles less than 1.5 mm long, but such can be distinguished from *O. claytonii* by the congested umbellets with broader bractlets in the involucels, as well as generally longer petals and licorice odor.

10. **Angelica** — Angelica

KEY TO THE SPECIES

1. Ovary, fruit, and upper part of stem finely pubescent; mature inflorescence ± hemispherical or flat-topped; leaflets 1–2.5 (3) cm broad (rarely narrower), acute to obtuse or rounded at the tip . . . . 1. **A. venenosa**
1. Ovary, fruit, and stem glabrous; mature inflorescence ± spherical; leaflets (at least the larger ones) usually 3–7 cm broad, acute . . . . 2. **A. atropurpurea**

1. **A. venenosa** (Greenway) Fern.

Map 935. Upland woods, especially oak; borders of woods and thickets; sandy open ground, prairie-like areas.

Although the specific epithet means "very poisonous" and some Umbelliferae are indeed toxic, there seems to be no record of poisoning by this species.

2. **A. atropurpurea** L. Plate 8-C

Map 936. Marshes and wet shores, sedge meadows and edges of tamarack swamps, stream and river banks, wet hollows or openings in mixed woods. Especially characteristic of cold, springy, seepy habitats, either shaded or sunny.

This is probably our tallest native herbaceous plant, frequently attaining a height of 2–3 m, with hollow purple-red stems as thick as 3–4 cm.

934. Osmorhiza claytonii

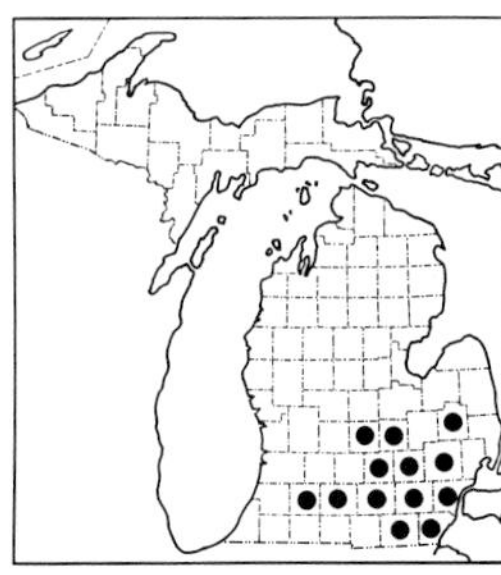

935. Angelica venenosa

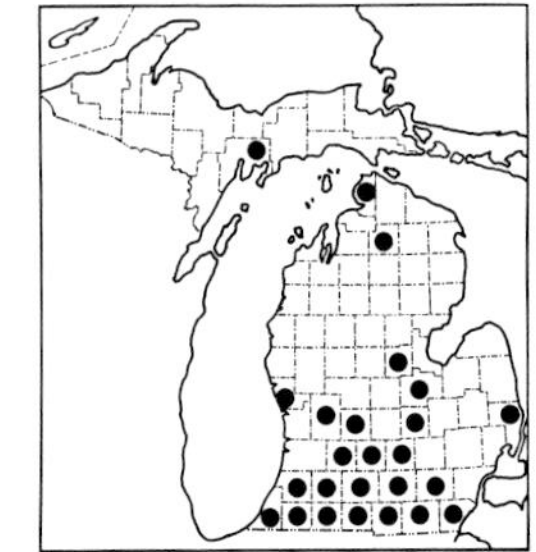

936. Angelica atropurpurea

11. **Heracleum**

REFERENCES

Brummitt, R. K. 1971. Relationship of Heracleum lanatum Michx. of North America to H. sphondylium of Europe. Rhodora 73: 578–584.

Morton, J. K. 1978. Distribution of Giant Cow Parsnip (Heracleum mantegazzianum) in Canada. Canad. Field-Nat. 92: 182–185.

1. **H. maximum** Bartram Plate 8-D; fig. 334 Cow-parsnip

Map 937. Floodplains and river banks, wet meadows and thickets; ± open hardwoods and clearings (spruce-fir at Isle Royale), borders of woods.

Individuals are often as tall as 2 m, with broad (at most ± 20 cm) flat-topped, white-flowered umbels—a striking sight. Widely known as *H. lanatum* Michaux, but current opinion treats the older Bartram name as validly published. A different taxonomic position is to treat our plant as *H. sphondylium* ssp. *montanum* (Gaudin) Briq., considering it to belong to the same species as the common European cow-parsnip. The distinctions from *H. sphondylium* are indeed not easily recognized. The latter is generally said to differ in having pinnate (rather than ternate) cauline leaves (although Brummitt notes that this character is not always reliable); broader, stiffer, straighter hairs on petioles, leaf rachises, and stems; and the fruit less strongly notched at the apex. Such a plant, with clearly pinnate leaves, was collected as a weed in Houghton in 1982 (*Gereau 1059,* MSC, MICH, BLH, AUB). It appears indeed to be the European form of cow-parsnip, whatever the best taxonomic disposition of these plants may be.

A more distinctive species in this genus, *H. mantegazzianum* Sommier & Levier, may be expected some day in Michigan as it has become a weed in the Bruce Peninsula of Ontario (known there since about 1950, as observed by George W. Thomson) and elsewhere in that province including Manitoulin Island, as well as in New York state and other places. This species is endemic to the Caucasus Mountains and is occasionally cultivated as a curiosity, becoming escaped locally in North America, as in Europe, to become a noxious weed. *H. mantegazzianum* is usually described in such terms

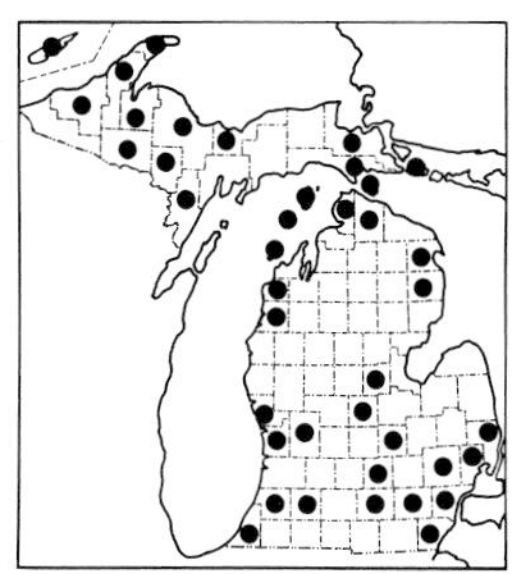

937. Heracleum maximum

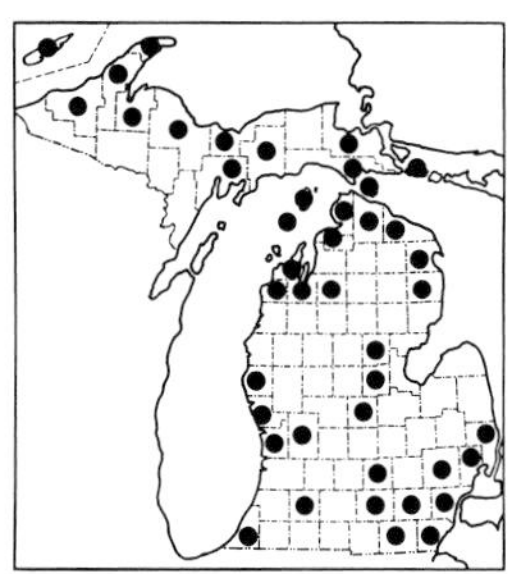

938. Pastinaca sativa

939. Oxypolis rigidior

as "enormous," as it ranges to 4 or 5 meters in height, with umbels as broad as 5 dm. The stiff, pustulate-based stem bristles are irritating, but the sap is more so, causing blistering and burning, even permanent scarring or brown staining of the skin of sensitive persons who have been exposed to the sap in sunlight. The oil tubes (at least the larger ones) on the fruit are ca. 0.8 mm or more in width, extending ca. 75% of the way from apex to base of the fruit, while in *H. maximum* the oil tubes on the fruit (fig. 334) are ca. 0.5 (0.8) mm or less in width, extending ca. 50–60 (70)% of the way from apex to base of the fruit.

12. **Pastinaca**

1. **P. sativa** L. Fig. 335 Wild Parsnip

Map 938. A Eurasian native, very well established (though usually ignored by collectors) throughout the state. Damp to dry roadsides, fields, clearings, shores; invading woodlands and thickets. Collected in the 1880's in counties as far apart as Keweenaw, Emmet, and Ingham; although no earlier collections have been seen, the species was listed by the First Survey so presumably it was in the state as early as 1838.

The common weed has presumably (in part, at least) reverted to the wild from cultivation, where strains are grown for their fleshy whitish edible roots. "Wild" plants have the same parsnip odor; the root is not poisonous to eat, despite a dubious reputation, but the foliage may irritate especially sensitive skins. The stems in this species are deeply furrowed and strongly angled. Very depauperate specimens may have leaves with only 3 leaflets.

13. **Oxypolis**

1. **O. rigidior** (L.) Raf. Cowbane

Map 939. Damp woods, especially with tamarack (and poison sumac); marshes, fens, and wet (rarely dry) prairies; swampy streamside thickets and shores.

The foliage and tuberous roots are reputed to be poisonous to cattle. Plants with entire leaflets are easily recognized. Those with toothed leaflets are sometimes confused with *Sium suave,* which is more finely and densely toothed and has a corrugated stem. The stem in *Oxypolis* is terete (but striate).

14. **Berula**

1. **B. erecta** (Hudson) Cov. Fig. 336

Map 940. Cold streams, marshes, and tamarack swamps; usually in calcareous areas.

A rare, rather delicate plant easily overlooked. The basal leaves are less incised-toothed than the midcauline ones. The styles are very short, barely 0.5 mm long including the stylopodium. Those who consider the American plant to be distinct from the Eurasian species have called ours *B. pusilla* (Nutt.) Fern.

15. **Sium**

1. **S. suave** Walter Fig. 337 Water-parsnip

Map 941. Marshes, swales, and ditches; swamps, potholes in forests; shores and borders of rivers, ponds, and lakes; often in shallow water (even to 2 feet deep).

The lower leaves if they are (or have been) submersed may be slightly to very extensively dissected—quite unlike the usual leaves; and the submersed stem may be very much inflated. The stem is corrugated, though not so deeply furrowed as in *Pastinaca*. The styles are at least 0.5 mm long beyond the stylopodium, and strongly recurved. An occasional plant of *Cicuta bulbifera* might key here, but the slender leaflets are very deeply toothed, the styles very short, and the stem terete (though striate)—apart from the usual presence of bulblets in the inflorescence.

It is not clear that this plant is poisonous, but so many white-flowered Umbelliferae are toxic that prudence suggests avoiding any risk of confusion or of providing a documented case of poisoning.

16. **Polytaenia**

1. **P. nuttallii** DC. Fig. 338 Prairie-parsley

Map 942. Collected August 2, 1837, on the Sturgis Prairie by Dr. Abram Sager of the First Geological Survey of Michigan, under Douglass Houghton. The species is (or was) here at the northeastern edge of its range. The only specimen (NY) was labeled by John Torrey, and the species has not been recognized in Michigan since.

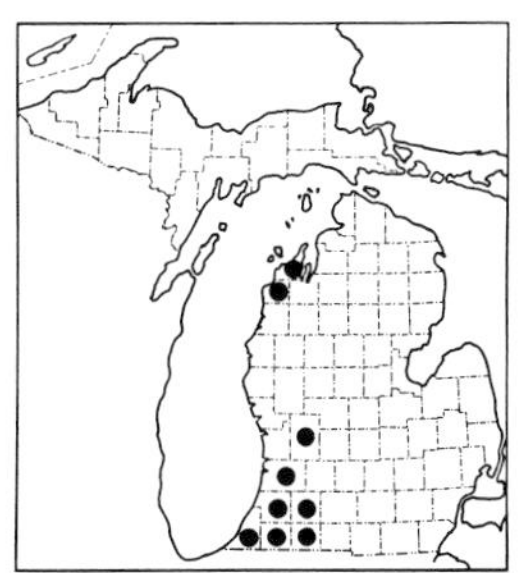
940. Berula erecta

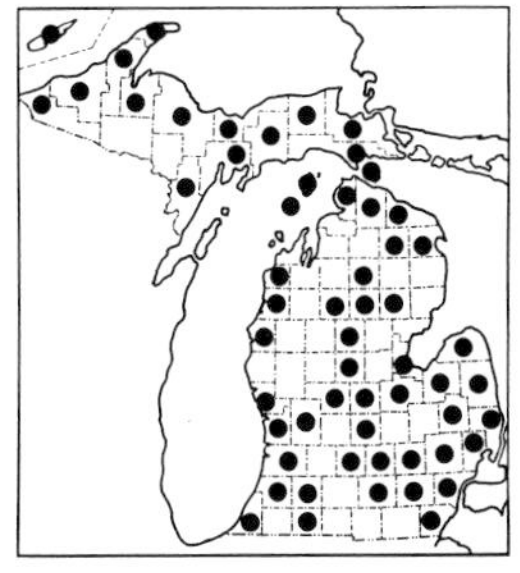
941. Sium suave

942. Polytaenia nuttallii

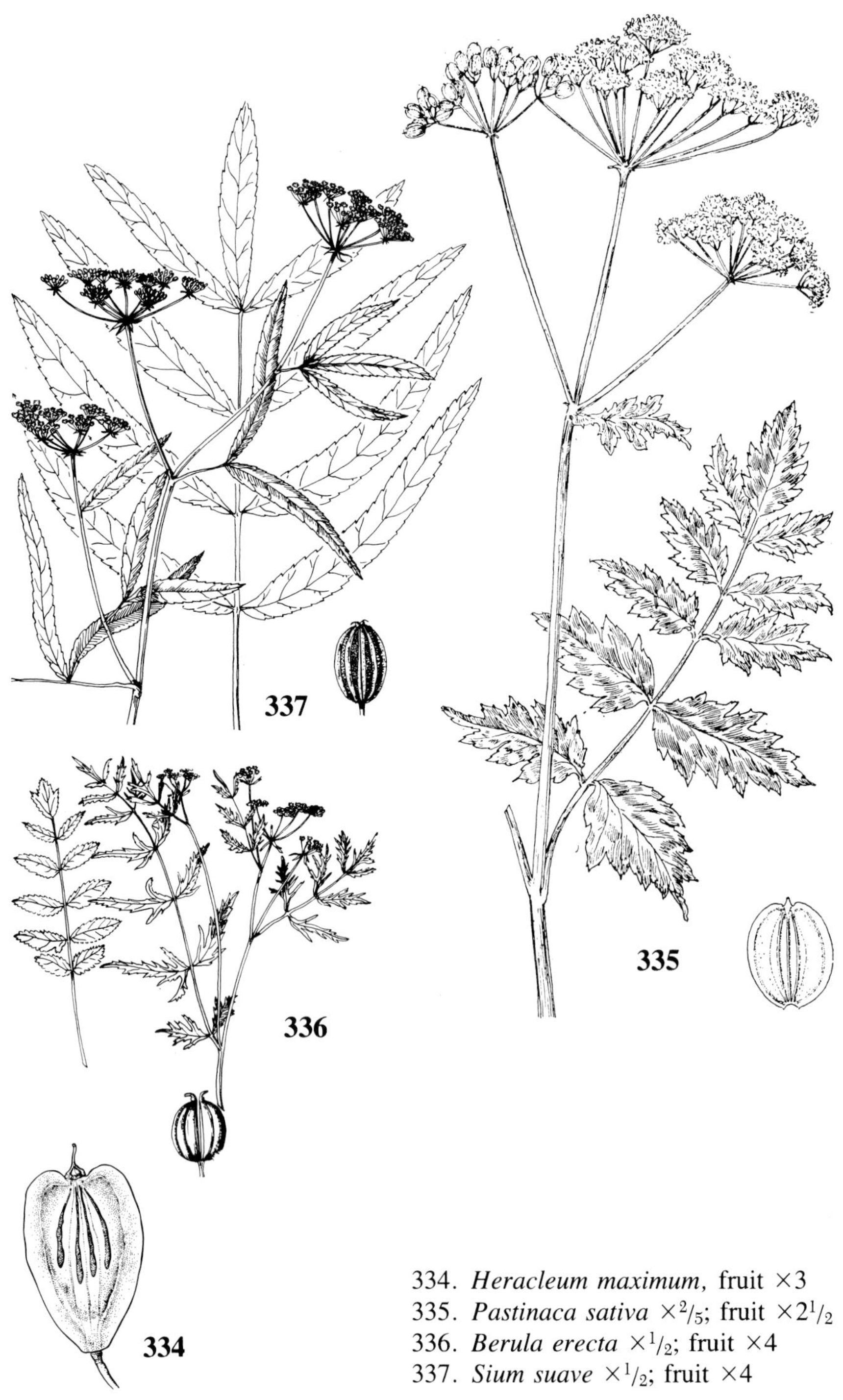

334. *Heracleum maximum,* fruit ×3
335. *Pastinaca sativa* $\times ^2/_5$; fruit $\times 2^1/_2$
336. *Berula erecta* $\times ^1/_2$; fruit ×4
337. *Sium suave* $\times ^1/_2$; fruit ×4

The foliage is rather variable, making difficulties in a key stressing vegetative characters, but this is really quite a distinct species, as described in the key. It may be noted, in addition, that *Polytaenia* blooms in June, so that the characteristic thick-margined fruit is ripe by August. There is no stylopodium, and the slender styles attain a length of 1.5–2 mm.

17. **Coriandrum**

1. **C. sativum** L. Coriander

Map 943. A native of the Mediterranean region, cultivated for the fragrant, spicy fruits (less pleasant when immature) and foliage, which are used as seasoning and flavoring. Rarely escaped, but collected by Farwell from waste ground in Detroit (*1394* in 1893, BLH; Rep. Mich. Acad. 20: 184. 1919).

18. **Carum**

1. **C. carvi** L. Caraway

Map 944. A Eurasian species, widely cultivated for its "seeds," used for flavoring in rye bread and many other products. Locally established along roadsides and in fields, clearings, farmyards, shores, and waste places.

Sometimes confused with *Daucus carota,* but blooms much earlier in the summer and lacks the conspicuous bracts at the base of the umbel. The rays of the umbel are quite unequal in length and are ± strongly ascending in fruit.

19. **Anethum**

1. **A. graveolens** L. Fig. 341 Dill

Map 945. Roadsides, dumps, fields, vacant lots, and other waste places. Originally native in southwestern Asia, extensively grown for its flavorful fruit and foliage.

943. Coriandrum sativum

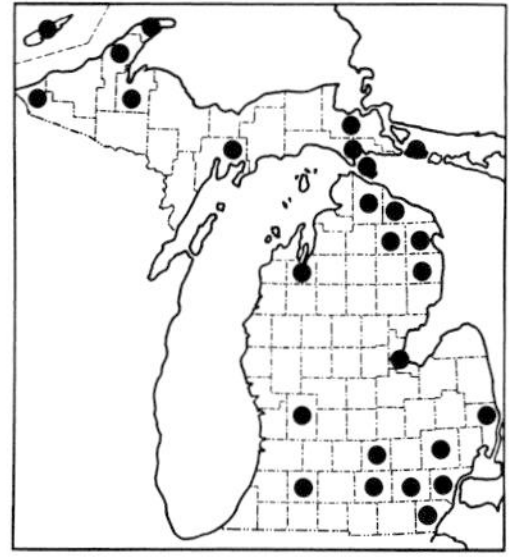

944. Carum carvi

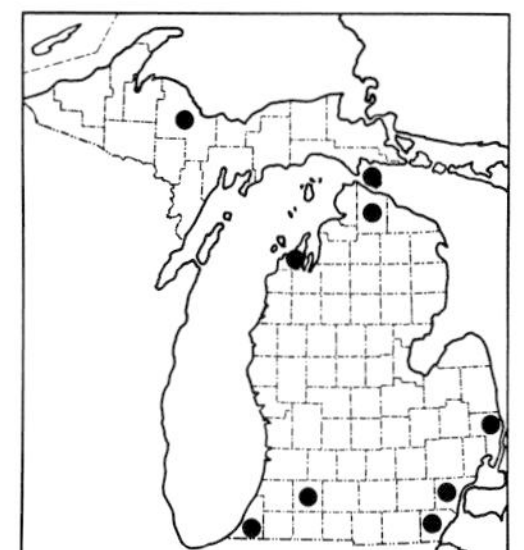

945. Anethum graveolens

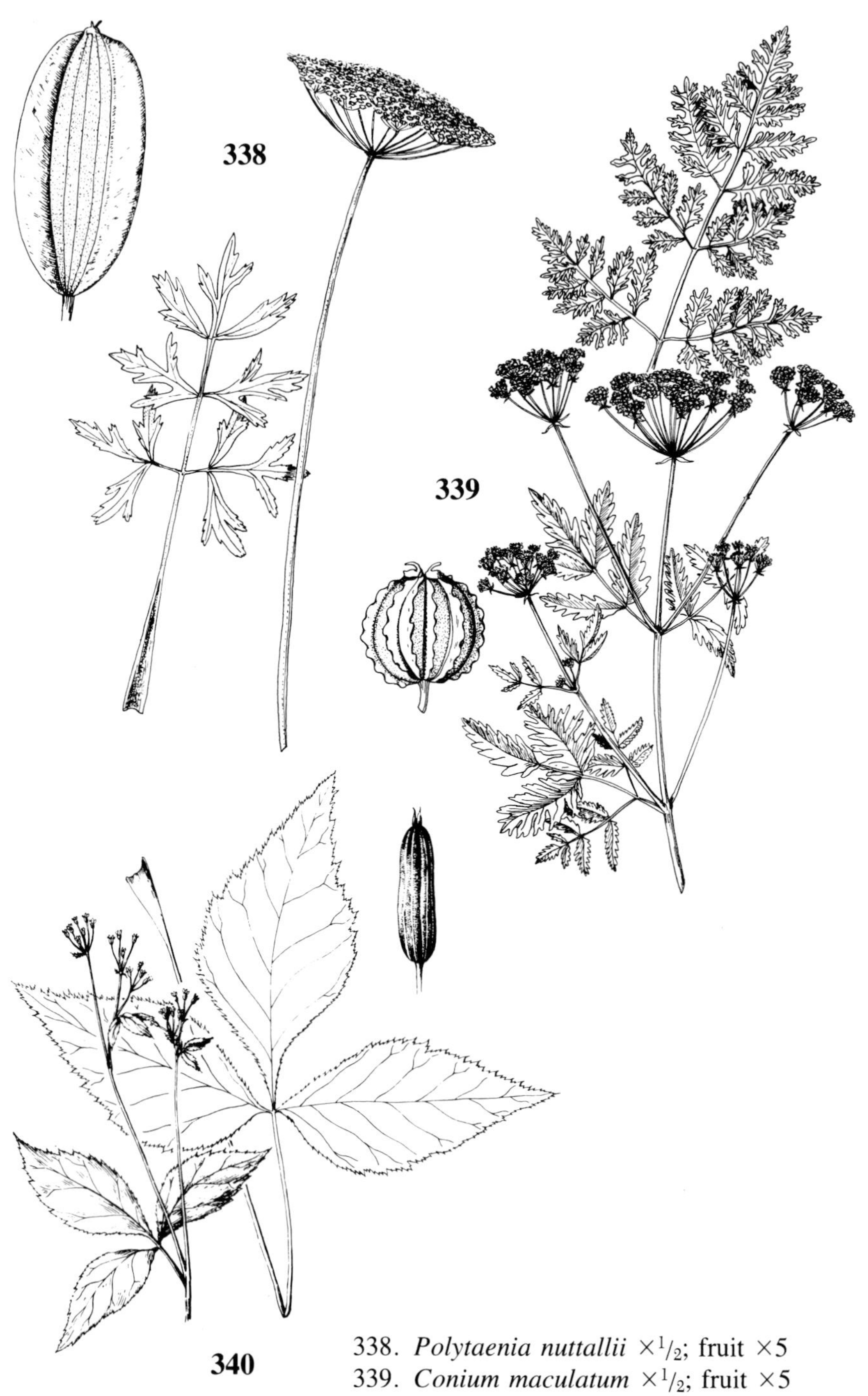

338. *Polytaenia nuttallii* $\times ^1/_2$; fruit $\times 5$
339. *Conium maculatum* $\times ^1/_2$; fruit $\times 5$
340. *Cryptotaenia canadensis* $\times ^1/_2$; fruit $\times 3$

20. **Foeniculum**

1. **F. vulgare** Miller — Fennel

Map 946. A native of the Mediterranean region and, like the preceding (to which it is closely related), cultivated for similar use of its fruit and foliage as a flavoring; escaped along railroads and other waste places.

Fennel tends to have fewer primary rays in the umbel (rarely as many as 30, while the large umbels of dill often have at least 30). The large stylopodium on the fruit is longer than the rest of the style.

21. **Chaerophyllum**

1. **C. procumbens** (L.) Crantz Fig. 329 — Wild-chervil

Map 947. Rich woods, especially on floodplains. Blooms in late April or early May, and withers to the ground by the end of June.

22. **Erigenia**

REFERENCE

Buddell, George F., II, & John W. Thieret. 1982. Harbinger-of-spring. Explorer 24(1): 24–26.

1. **E. bulbosa** (Michaux) Nutt. Fig. 346 — Harbinger-of-spring

Map 948. Rich, often moist deciduous woods, including floodplains and river banks. Blooms usually from early to late April, depending on the season and latitude.

The dark anthers and conspicuous contrasting white petals have led to a common name of "pepper-and-salt."

23. **Anthriscus** — Chervil

1. **A. sylvestris** (L.) Hoffm.

Map 949. Roadsides, fields, disturbed ground; invading woods. A Eurasian species, locally established in North America.

24. **Conium**

1. **C. maculatum** L. Fig. 339 — Poison-hemlock

Map 950. Roadsides, fields, waste places, clearings, banks and bluffs, shores. Established at a number of places across the state by the 1890's. A European species, once grown for medicinal purposes but too risky to rely on!

Although toxicity varies among plants and with the season, as well as the part of the plant, this is a very poisonous species, promptly producing symptoms when ingested—and often death from respiratory failure. Both livestock and humans are affected. Socrates is the best known victim of hemlock.

The plant has an unpleasant odor and so, fortunately, is not attractive to those seeking "wild foods." It can grow to 2 m or even more in height, the purple-spotted, slightly glaucous stem, lacy foliage, and white umbels making it very conspicuous.

25. **Conioselinum**

1. **C. chinense** (L.) BSP. Fig. 342 Hemlock-parsley

Map 951. Swampy places with deciduous trees, cedar, tamarack; springy river banks, creek borders.

The pedicels and usually rays of the umbel are ± scabrous or minutely hispidulous (as in *Aethusa*).

26. **Petroselinum**

1. **P. crispum** (Miller) A. W. Hill Parsley

Map 952. A species of Eurasian origin, commonly cultivated (especially

946. Foeniculum vulgare

947. Chaerophyllum procumbens

948. Erigenia bulbosa

949. Anthriscus sylvestris

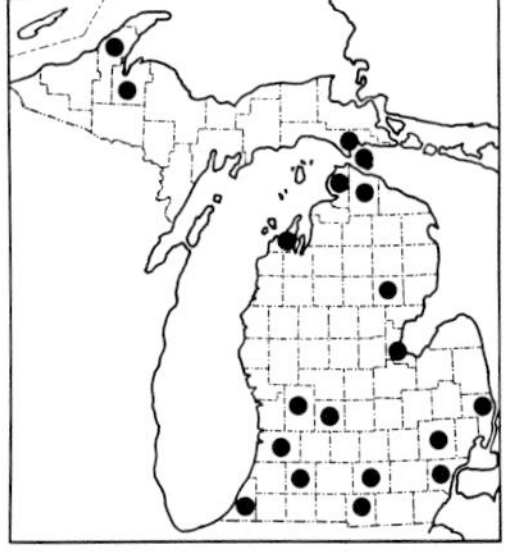

950. Conium maculatum

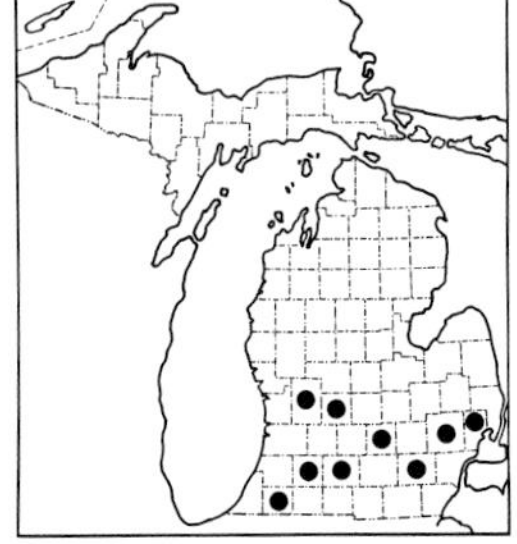

951. Conioselinum chinense

in a variety with curled, crisped leaves) as a familiar garnish and flavoring plant. Rarely escapes from cultivation, but found around uncultivated gardens and on waste ground (see Pap. Mich. Acad. 26: 17. 1941).

27. **Aethusa**

1. **A. cynapium** L. Fool's-parsley

Map 953. A European weed, local in North America. The only Michigan records are from Brighton in 1862 and from waste ground in Galien, Berrien County, in 1965.

Reputed to be poisonous if eaten.

28. **Cryptotaenia**

1. **C. canadensis** (L.) DC. Fig. 340 Honewort

Map 954. Beech-maple and other rich woods; swamp forests.

29. **Taenidia**

1. **T. integerrima** (L.) Drude Yellow-pimpernel

Map 955. Jack pine plains and other sandy (or rocky) woodlands on plains or low dunes, with oak, pines, and/or aspen; deciduous woods, especially along borders and in clearings; river- and stream-bank thickets; invading old fields.

An easily recognized, smooth (even glaucous) species, our only umbellifer with yellow flowers and strictly entire leaflets.

30. **Levisticum**

1. **L. officinale** Koch Lovage

Map 956. A native of Europe, cultivated as an ornamental and for the

952. Petroselinum crispum

953. Aethusa cynapium

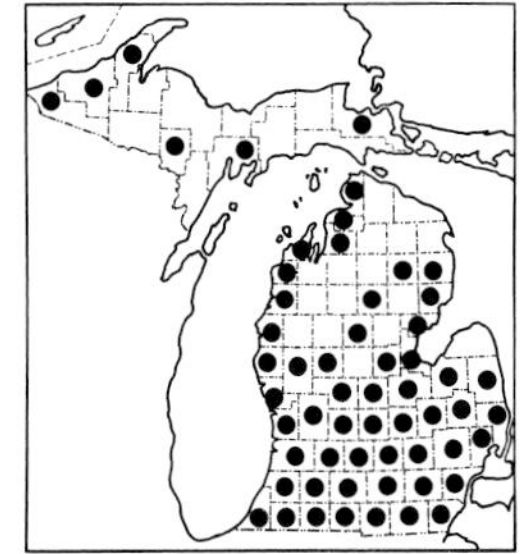

954. Cryptotaenia canadensis

fruits and leaves, which can be used much as celery. Rarely established near old gardens. The only Michigan collection seen is from Emmet County in 1948 (*McVaugh 9708*, MICH, UMBS), but the species was observed 23 years earlier in Cheboygan County (see Pap. Mich. Acad. 42: 26. 1957).

Celery, *Apium graveolens* L., is another southern European native widely cultivated as a vegetable, seasoning, and salad plant. It is rather similar in foliage to *Levisticum*, and might key here although the leaves tend to have more teeth. Commercial growing of celery in the United States began in Michigan over a century ago, and the state now ranks third in the nation in size of that crop. Kalamazoo was once dubbed "Celery City" and was a center for alleged medicinal preparations based on the plant (see Larry B. Massie, "Celebrated Celery City Quacks," Mich. History 66(5): 16–23. 1982). There seem to be no reports of celery escaping from cultivation in Michigan, however.

31. **Zizia**

Frequently confused with *Thaspium*, but material with young or mature fruit is easily distinguished not only by the presence of ribs (rather than definite thin wings) on the mericarps but also by the presence of a sessile central fruit in at least some umbellets. In *Thaspium*, the central flowers in all umbellets are pediceled and staminate, leaving only the marginal and pediceled perfect flowers to develop fruit. (This feature is described by Morley in *Spring Flora of Minnesota*, p. 203. 1966.) See further comments under the next genus.

KEY TO THE SPECIES

1. Basal leaves simple, with cordate blades; cauline leaves with 3 leaflets . . . 1. **Z. aptera**
1. Basal leaves compound, like the cauline, with [3] 5–11 leaflets . . . . 2. **Z. aurea**

1. **Z. aptera** (A. Gray) Fern.

Map 957. Very rare and local, on dry shaded bluffs.

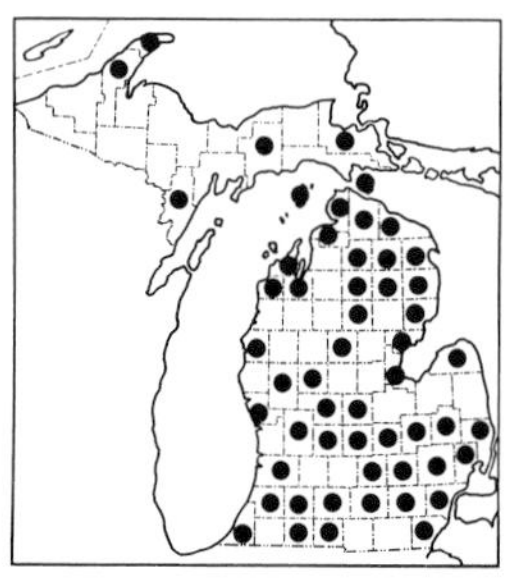
955. Taenidia integerrima

956. Levisticum officinale

957. Zizia aptera

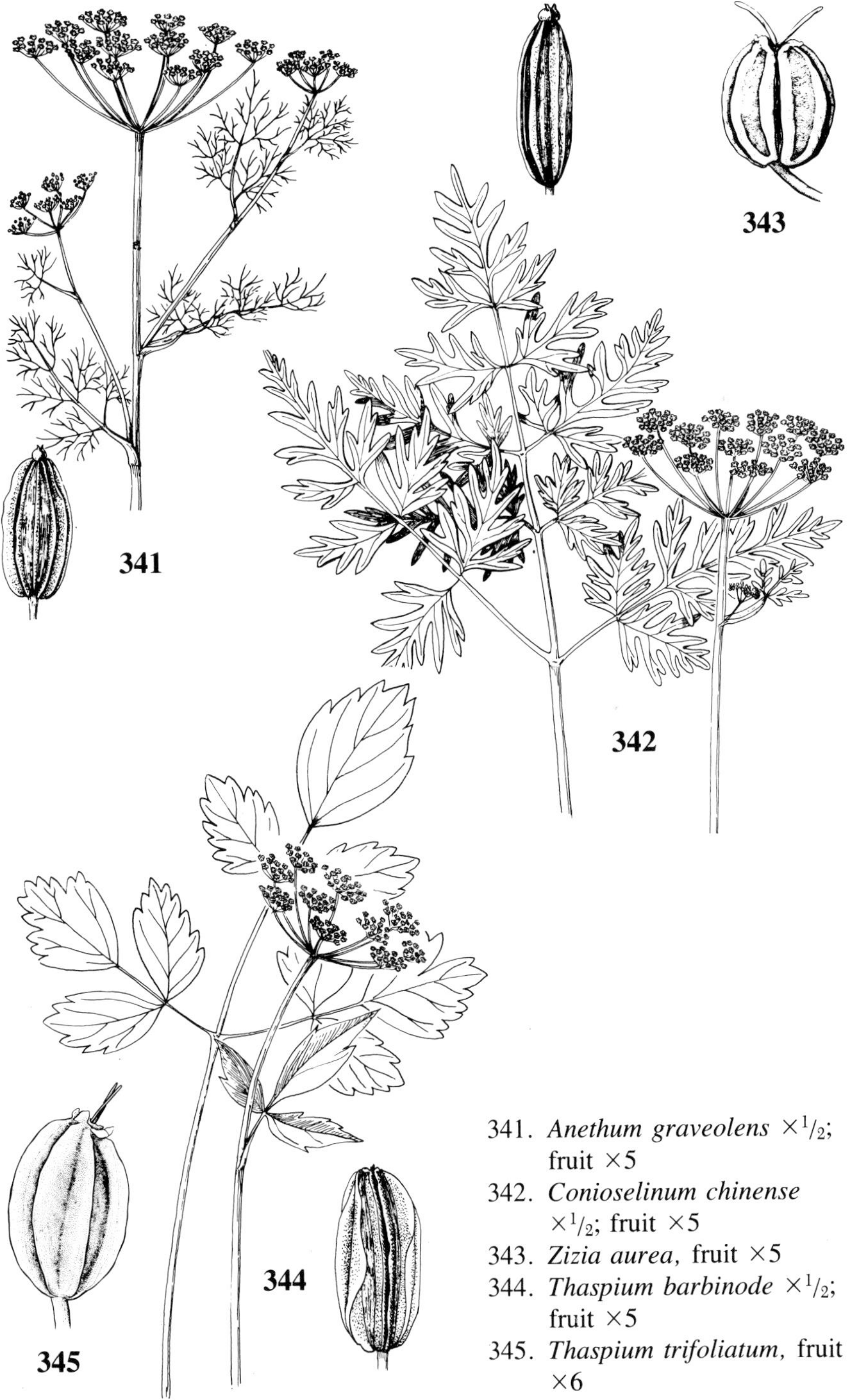

341. *Anethum graveolens* $\times{}^1/_2$; fruit $\times 5$
342. *Conioselinum chinense* $\times{}^1/_2$; fruit $\times 5$
343. *Zizia aurea*, fruit $\times 5$
344. *Thaspium barbinode* $\times{}^1/_2$; fruit $\times 5$
345. *Thaspium trifoliatum*, fruit $\times 6$

2. **Z. aurea** (L.) Koch Fig. 343 Golden Alexanders

Map 958. Woods, especially low swampy ones; more often in boggy ground (especially fens), with tamarack and poison sumac, in openings and thickets on river banks; meadows and fencerows.

32. Thaspium Meadow-parsnip

REFERENCE

Ball, P. W. 1979. Thaspium trifoliatum (Meadow-parsnip) in Canada. Canad. Field-Nat. 93: 306–307.

KEY TO THE SPECIES

1. Basal leaves at least twice-compound; margins of leaflets minutely ciliolate, with ca. 3–5 (7) coarse and irregular teeth or lobes on a side .... 1. **T. barbinode**
1. Basal leaves simple or trifoliolate; margins of leaflets smooth and glabrous, the principal ones with ca. 10–40 fine and regular teeth on a side .... 2. **T. trifoliatum**

1. **T. barbinode** (Michaux) Nutt. Fig. 344

Map 959. In dry to moist soil of oak and beech-maple forest and woodland, river bluffs, borders of wetlands.

2. **T. trifoliatum** (L.) A. Gray Fig. 345

Map 960. Oak-hickory and oak woods and thickets, prairie remnants.

Frequently confused with the much more common *Zizia aurea*—which usually grows in wetter places. When a sessile perfect flower is evident in the center of an umbellet, the plant is clearly *Zizia*; however, this is not always easily seen in very young infloresences, or if the central flowers are staminate, as they are in some umbellets of *Zizia* and all umbellets of *Thaspium*. The prominent broad wings on all sides of the fruit in *Thaspium* become evident (more than mere ribs) in even young fruit (at a stage when it is nearly smooth in *Zizia*). A comparative character, difficult to describe, is

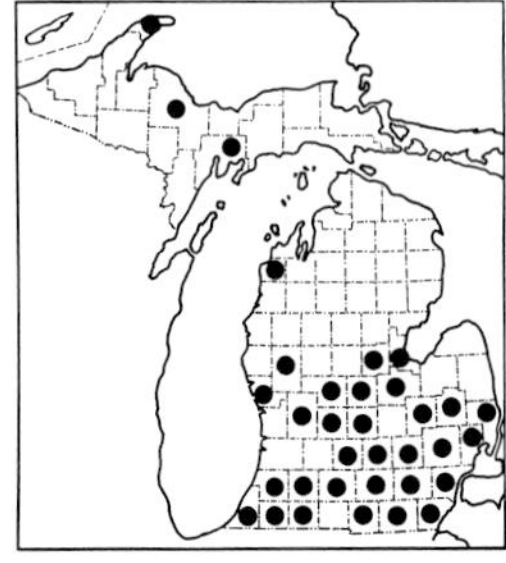

958. Zizia aurea

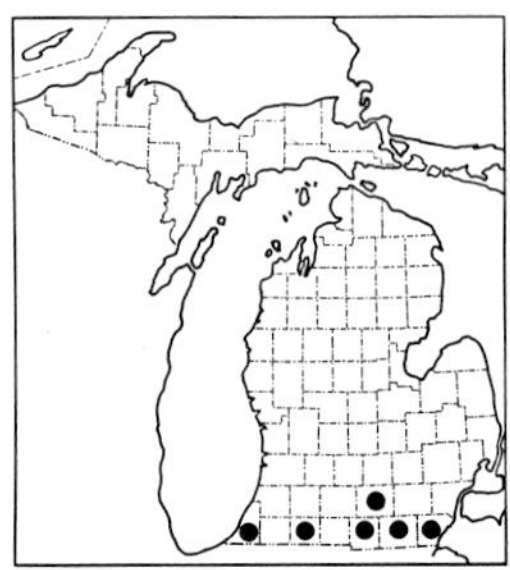

959. Thaspium barbinode

960. Thaspium trifoliatum

that the teeth on the leaflets of *T. trifoliatum* are obtuse with a prominently thickened pale border and callous tip, while in *Zizia* the teeth tend to be acute with less well developed border and callus.

This species closely resembles the rare *Z. aptera*, also of dry ground, but the petiolar sheaths are narrower at any given level on the stem: sheaths on midcauline leaves, for example, are less than 5 mm broad in *Thaspium trifoliatum* and more than 5 mm broad in *Zizia aptera*. A form of *T. trifoliatum* with flowers purple rather than yellow occurs south of our area.

33. **Cicuta** Water-hemlock

REFERENCES

Mulligan, Gerald A. 1980. The Genus Cicuta in North America. Canad. Jour. Bot. 58: 1755–1767.

Mulligan, Gerald A., & Derek B. Munro. 1981. The Biology of Canadian Weeds. 48. Cicuta maculata L., C. douglasii (DC.) Coult. & Rose and C. virosa L. Canad. Jour. Pl. Sci. 61: 93–105.

KEY TO THE SPECIES

1. Plant bearing numerous vegetative bulblets in the axils of broad-based, abruptly acuminate bracts, especially on upper portion of the plant; umbels very few or none; central part of leaflets (excluding teeth or lobes) less than 3 (5) mm wide; styles becoming ca. 0.5–0.7 mm long; fruit ca. 1.5–2 mm long . . . . 1. **C. bulbifera**
1. Plant without bulblets or acuminate bracts; umbels well developed; central part of leaflets (excluding teeth) mostly 5–25 mm wide; styles becoming ca. 1–2 mm long; fruit ca. 3–5 mm long . . . . 2. **C. maculata**

1. **C. bulbifera** L.

Map 961. Marshes and shores, borders of lakes and streams, bogs and tamarack swamps, swales and ditches, moist thickets and river margins.

This species is supposedly less poisonous than the next, but nevertheless toxic if ingested.

2. **C. maculata** L. Fig. 347

Map 962. Marshes and shores, swamps, moist to wet woods and thickets, stream banks, meadows, ditches.

Usually considered to be the most violently poisonous plant in temperate North America, for both livestock and humans. Poison is concentrated in the clustered tuberous roots, which one will not mistake for parsnips more than once, although all parts of the plant are poisonous. Symptoms appear promptly and include abdominal pain, violent convulsions, fever, paralysis, and respiratory failure, followed by death—sometimes within 15 minutes, or as long as 8 hours later.

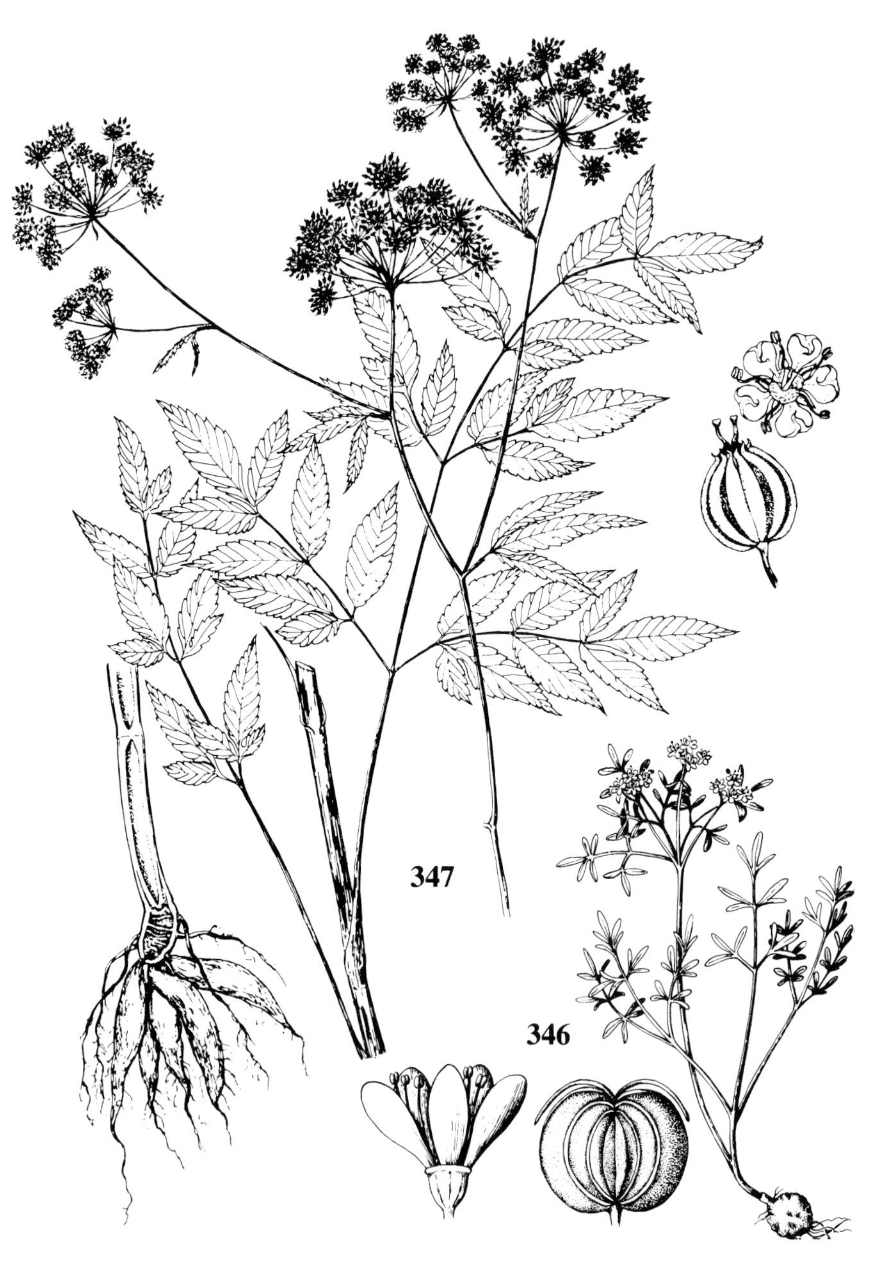

346. *Erigenia bulbosa* $\times^{1}/_{2}$; flower & fruit $\times 4$
347. *Cicuta maculata* $\times^{2}/_{5}$; flower $\times 6$; fruit $\times 4$

34. **Aegopodium**

1. **A. podagraria** L. Goutweed

Map 963. A native of Europe, widely grown as a foliage plant or ground-cover and readily escaping, forming colonies from rhizomes and also from seed near homesites, along roadsides, on banks, and at borders of woods.

Usually cultivated as cv. Variegatum, with the doubly serrate leaflets margined with white; the leaves revert to plain green on plants established from seed. The delicately ribbed fruit has a large stylopodium with slender style ca. 1–2 mm long.

CORNACEAE Dogwood Family

Flowers in this family are typically 4-merous, with inferior ovary and very small to obsolete calyx lobes. Only *Cornus* is found in Michigan.

1. **Cornus** Dogwood

While there is good agreement on the taxa in our flora, authors and check-lists vary considerably regarding the ranks at which several of these taxa should be recognized and the names, at any rank, which apply. Even the genus is divided into three by the more ardent "splitters."

The leaves of *Cornus* make the genus easily recognizable. They are opposite in most species, entire, with parallel lateral veins arching strongly as they approach the margins. Combinations of size, shape, texture, venation, and pubescence are rather distinctive for most species, but they do vary depending on conditions, and blades tend to be larger and less papillose beneath in the shade. Almost any character in the keys is open to exception, but identification is easier than the apparently overlapping statements might suggest. The first four species are quite easily recognized. Some hints for identification of the others are given after the last species, and may help

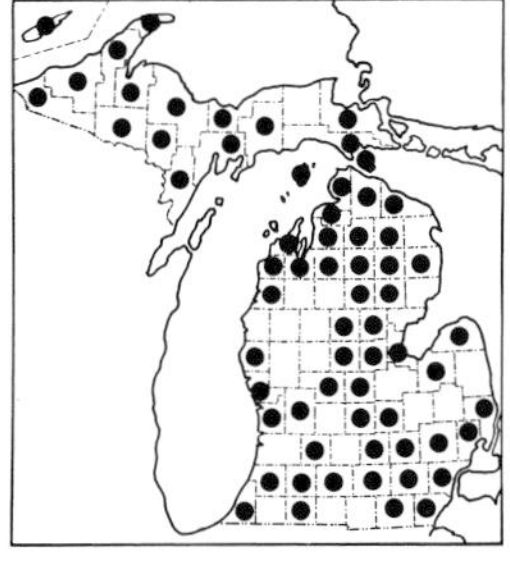

961. Cicuta bulbifera

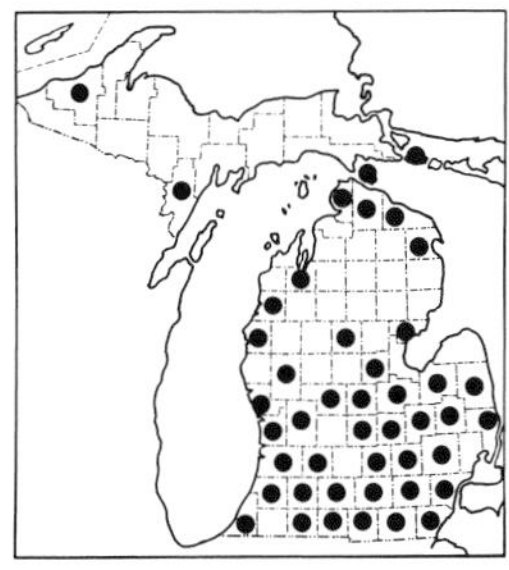

962. Cicuta maculata

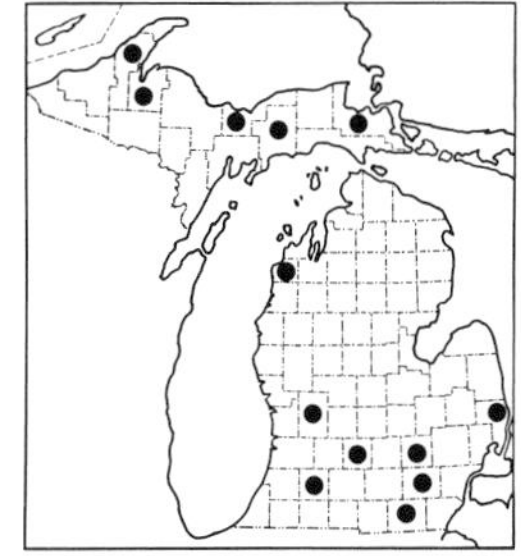

963. Aegopodium podagraria

when one prefers to apply characters in a different sequence than they are employed in a key or to contrast different pairs of species.

Several species are used horticulturally, including *Cornus mas* L., Cornelian-cherry, a native of eastern Asia with dense umbels of bright yellow flowers opening before the leaves (March–April) and juicy red edible fruits generally stolen by birds before they can be used by humans. Two cultivated species besides *C. florida* have similar large white bracts beneath the flowers: *C. nuttallii* Audubon, of the West Coast of North America, and *C. kousa* Hance, of eastern Asia.

REFERENCES

Ferguson, I. K. 1966. The Cornaceae in the Southeastern United States. Jour. Arnold Arb. 47: 106–116.

Rickett, H. W. 1944. Cornus stolonifera and Cornus occidentalis. Brittonia 5: 149–159.

Wilson, James S. 1965. Variation of Three Taxonomic Complexes of the Genus Cornus in Eastern United States. Trans. Kansas Acad. 67: 747–817. [Includes a key to all species and subspecies.]

KEY TO THE SPECIES

1. Leaves alternate or so crowded as to appear whorled, but not all distinctly opposite
  2. Plant herbaceous; leaves all or mostly in one whorl; inflorescence subtended by large white bracts; fruit bright red . . . . 1. **C. canadensis**
  2. Plant a shrub or small tree; leaves all alternate (or crowded at ends of branches); inflorescence without subtending bracts; fruit dark blue . . . . 2. **C. alternifolia**
1. Leaves clearly all opposite (or in pairs at ends of shoots)
  3. Flowers in a dense head-like cluster subtended by 4 large showy white (or pinkish) bracts; fruit dark red, in tight heads . . . . 3. **C. florida**
  3. Flowers in a compound cyme without subtending bracts; fruit white to blue, in an open infructescence
    4. Leaf blades rough to the touch above (from tiny projecting stiff hairs)
      5. Lateral veins of leaves 3–5 on a side; pith usually brown in old twigs; plant rare, only in southeastern Lower Peninsula . . . . 4. **C. drummondii**
      5. Lateral veins of leaves 6–8 on a side; pith white; plant frequent throughout most of the state . . . . 5. **C. rugosa**

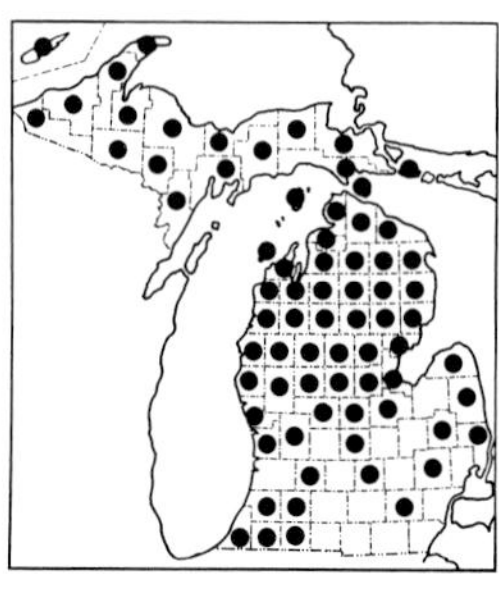

964. Cornus canadensis

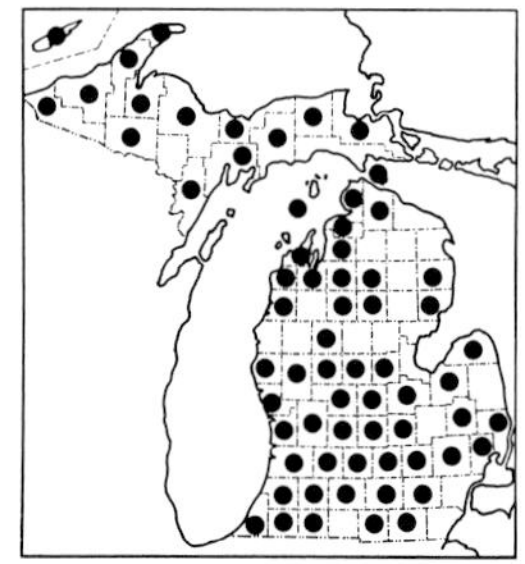

965. Cornus alternifolia

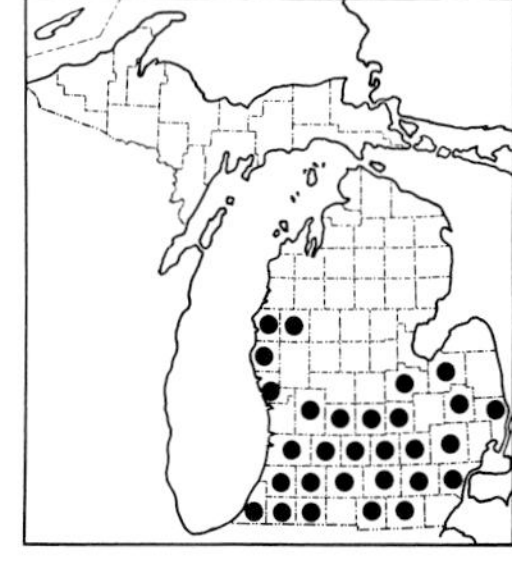

966. Cornus florida

4. Leaf blades smooth to the touch above (glabrous or with soft closely appressed hairs)
  6. Pith of 2-year-old twigs white (paler than surrounding wood); lateral veins of leaves 4–9 on a side
    7. Branchlets green, flecked with purplish streaks; leaf blades often less than 1.5 times as long as wide, with 6–8 (9) lateral veins per side .... 5. **C. rugosa**
    7. Branchlets unspotted except for paler lenticels, the older ones usually red (rarely green); leaf blades ca. 1.5 times as long as wide or longer, with 4–5 (6) lateral veins per side .... 6. **C. stolonifera**
  6. Pith of 2-year-old twigs light brown (darker than surrounding wood) or often white in *C. foemina*; lateral veins of leaves 3–5 on a side
    8. Calyx lobes (0.6) 0.8–1.3 (2) mm long; styles abruptly swollen for ca. 0.5–1 mm below the stigma; inflorescence flat or slightly convex; ripe fruit dark blue (with pale patches) on yellow-brown to maroon pedicels; bark of older branchlets reddish .... 7. **C. amomum**
    8. Calyx lobes less than 0.8 mm long; styles at most slightly expanded (no broader than the stigma); inflorescence strongly convex or pyramidal; ripe fruit pale blue to white, on bright red pedicels; bark of older branchlets gray .... 8. **C. foemina**

1. **C. canadensis** L. Plate 8-E, 8-F Bunchberry; Dwarf Cornel

Map 964. Coniferous and mixed woods and swamps (especially cedar); jack pine plains (except in the driest sites); not often in strictly deciduous woods (except for young aspen-birch stands); very local (bogs and tamarack swamps) at the southern edge of its range in the state.

Often there is a pair of small, opposite leaves on the stem below the conspicuous single whorl, which normally consists of 6 leaves on fertile stems and usually 4 leaves on sterile ones. Bunchberry blooms usually in late spring, but may be held back as long as a month at cold exposed sites, such as Whitefish Point, where there can be striking displays of plants in full bloom as late as the first of July. And every once in a while there is a truly late-bloomer looking fresh at the end of the summer. Ordinarily, by summer's end one sees the distinctive dense clusters of red fruit, which suggest the common name. One of the charming sights of the north woods is to watch a chipmunk sitting up to feast upon these little red morsels that nature has set forth for him at exactly the right level.

2. **C. alternifolia** L. f. Alternate-leaved or Pagoda Dogwood

Map 965. Deciduous and mixed forests (rarely in spruce-fir stands), either as an understory shrub or along borders; floodplains and cedar swamps; banks and thickets above lakes and streams.

When full-grown, a large shrub or small tree. The branches tend to be horizontal, curved upwards toward their tips, giving a pagoda-like aspect. The inflorescences are rather large as well as numerous; it is an especially handsome species in flower. The alternate arrangement of the leaves at the ends of new twigs is not always easy to see, but the branching pattern is clearly alternate, as are leaves on older shoots. The pith is white.

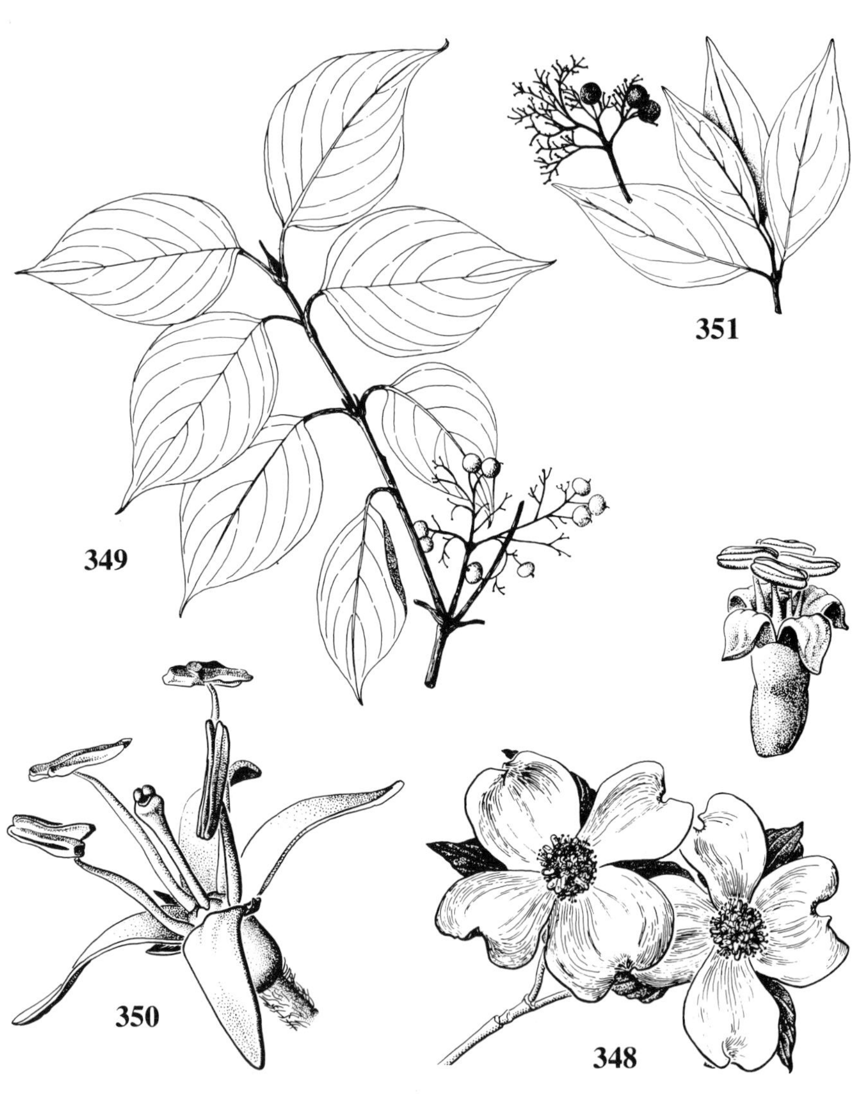

348. *Cornus florida* $\times{}^1/_2$; flower $\times 4$
349. *C. stolonifera* $\times{}^1/_2$
350. *C. amomum,* flower $\times 6$
351. *C. foemina* $\times{}^1/_2$

3. **C. florida** L. Fig. 348 Flowering Dogwood

Map 966. Dry (usually oak) to rich deciduous woods, especially on hillsides and river banks; rarely recorded with tamaracks.

One of our most attractive native shrubs, the large bracts (actually enlarged bud scales) usually ca. 3–5.5 cm long, the whole inflorescence simulating a single flower.

4. **C. drummondii** C. A. Meyer

Map 967. Banks and thickets, especially along rivers and borders of woods.

Vigorous plants from fencerows in Monroe County with leaves intermediate between this species and *C. foemina* have been considered by Wilson to be hybrids; they have high pollen viability and fruit set. A collection (*Wilson 261*, MICH) from along a ditch in the same county was considered to be a hybrid with *C. amomum* ssp. *obliqua*; it has longer and narrower leaf blades than in either parent and produces almost no viable pollen.

5. **C. rugosa** Lam. Round-leaved Dogwood

Map 968. Deciduous and mixed woods, especially at borders near lakes, rivers, and clearings; especially well developed with fir, aspen, paper birch, and/or cedar in thickets along the shores of lakes Michigan and Huron and on dunes; also on rock outcrops and talus slopes; fencerows, wooded bluffs.

6. **C. stolonifera** Michaux Fig. 349 Red-osier

Map 969. In almost all sorts of damp situations: marshes, swamps (even coniferous, with tamarack, spruce, and/or cedar), wet shores, sides of rivers and streams; on rock outcrops and talus slopes; coniferous and mixed thickets on shores and common on sand dunes, displaying there its "stoloniferous" habit by suckering from buried stems.

Until the name *C. sericea* L., which originally included both this species and the next, has been satisfactorily typified (or officially rejected as ambiguous), the least confusing course is not to use it, although some recent works have done so. Another position is that our plant is not a different species from the Old World *C. alba* L., in which case that name antedates both *C. sericea* and *C. stolonifera* for it.

The underside of the leaves varies from ± sparsely pubescent with a few appressed often 2-pronged hairs to having dense curly hairs. Plants with the latter may be recognized as f. *baileyi* (Coulter & Evans) Rickett, and occur often on sand dunes and shores, though widespread elsewhere in our area. Although the branches are typically bright red, occasionally they are green (but may turn red in drying); a form with yellow twigs is cultivated—as are some other red-stemmed species.

C. foemina may have white pith but has gray bark, never curly pubescence, and often fewer as well as less bold leaf veins compared to *C. stolonifera*; see comments under that species below.

7. **C. amomum** Miller Fig. 350 Pale Dogwood

Map 970. Wet (very rarely upland) sites: marshes, swamps (including cedar-tamarack), bogs (fens); margins of ponds, lakes, and streams and on banks of creeks and rivers; often forming dense thickets at the edges of swampy woods and bodies of water.

Most of our specimens have long been called *C. obliqua* Raf., which Wilson recognizes as *C. amomum* ssp. *obliqua* (Raf.) J. S. Wilson. True *C. amomum* ssp. *amomum* has been collected in the state only along fencerows in Monroe County; it has the underside of the leaf blades not papillose and the veins with mostly rust-colored hairs. In ssp. *obliqua*, the lower leaf surface is usually very minutely and densely papillose [use 20×–30× lens] and the hairs along the veins are often mostly if not entirely whitish. There are also differences in leaf shape, the blades of ssp. *obliqua* being lance-ovate (relatively narrower and more cuneate at the base); they are also pale beneath, whence the common name. New young shoots of *C. amomum* (especially the petioles and branchlets) have mostly rust-colored hairs, while in the next species the new shoots (and the inflorescence) are glabrous or with very sparse mostly or entirely whitish hairs. Hybrids between this species and the next have been found where their ranges overlap in several counties, and are intermediate in length of calyx lobes, pubescence, fruit color, inflorescence shape, and other characters—as well as having mostly inviable pollen.

8. **C. foemina** Miller Fig. 351 Gray Dogwood

Map 971. Along rivers, streams, and lakes, often forming dense thickets on banks and shores; marshes and swamps; sandy oak and pine woods; fencerows and borders of woods; less often in hardwood forests or prairie remnants. The Marquette County record may represent planted shrubs (at Ives Lake).

Typical *C. foemina* ranges well south of Michigan. Our plants are all *C. foemina* ssp. *racemosa* (Lam.) J. S. Wilson, long known as *C. racemosa* Lam. The young branchlets are ± 2-angled or ridged and are nearly or quite

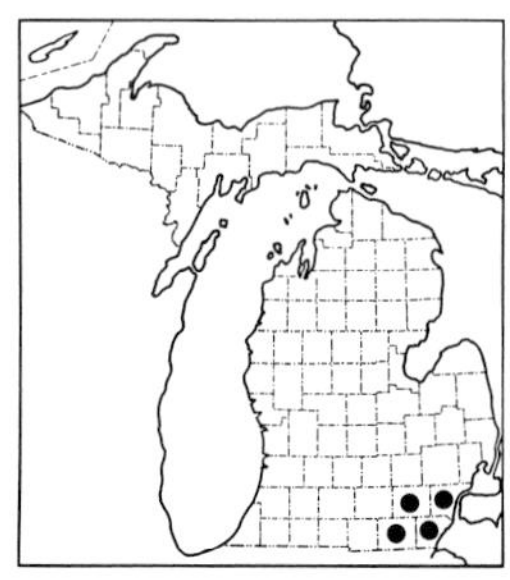

967. Cornus drummondii

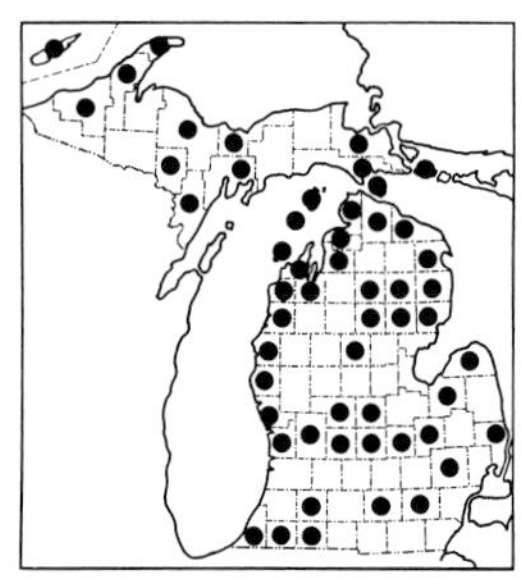

968. Cornus rugosa

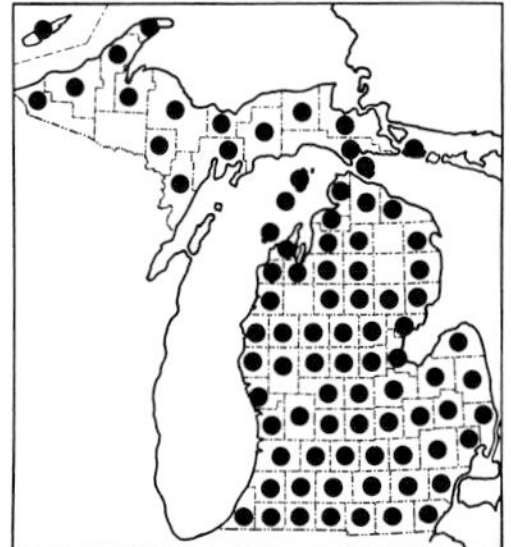

969. Cornus stolonifera

glabrous. Fruiting plants, with a more paniculate inflorescence than our other dogwoods and bright red stalks, are easily recognized.

The pith is often white even in 2-year-old twigs, but these have distinctly gray bark, which will separate *C. foemina* from *C. stolonifera*; the leaves also have only 3–4 (5) lateral veins per side (not so strong or parallel as in *C. stolonifera*) and are nearly glabrous beneath, especially on the veins. The leaves of *C. stolonifera* often have curly pubescence beneath and have the same number of veins as *C. amomum*—usually 5. Although the latter typically has reddish twigs (not so bright as in *C. stolonifera*), it is readily distinguished from all our other species by the strongly expanded style immediately below the stigma (usually slightly exceeding the stigma; fig. 350) and long calyx lobes. (Slight expansion of the style is often seen in *C. foemina*.) The consistently brownish pith in *C. amomum* is helpful to distinguish it from all species other than *C. drummondii* and *C. foemina*. Fresh young branchlets of *C. rugosa* are distinctive in the purple flecks on green bark, and these are visible as darker spots even on dried specimens when the bark darkens; it also usually has very broad, ± rotund leaves, often somewhat rough on the upper surface, with more lateral veins, on the average, than other species.

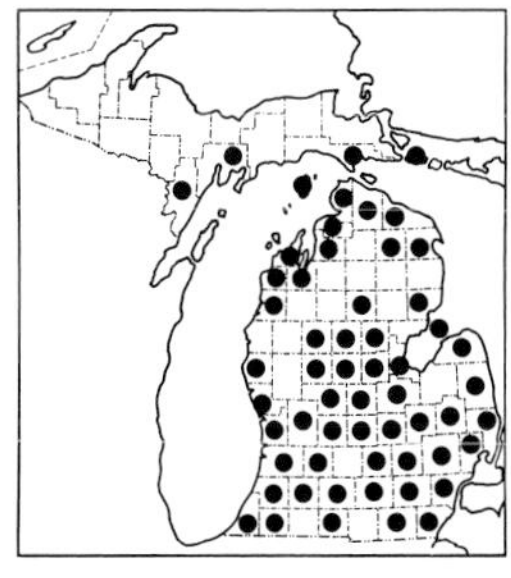

970. Cornus amomum

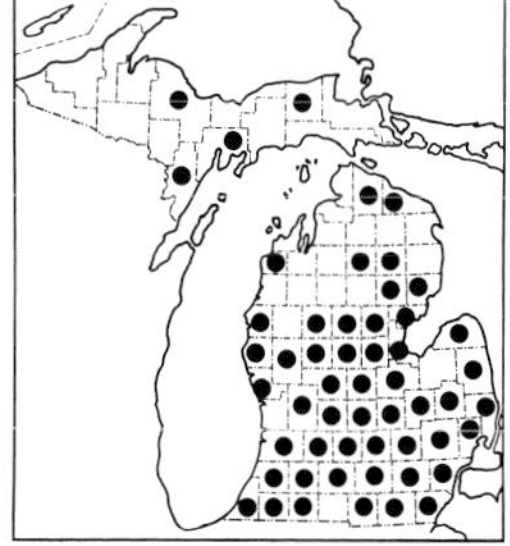

971. Cornus foemina

Glossary

Some frequently used terms for habitats are included here; for further discussion, see Part I of this Flora, pp. 17–23.

When this glossary lists a word, usually a noun, derivative words are usually not listed separately: e. g., *whorled* means "in a whorl"; *petioled* means "with a petiole"; *mucronate* means "with a mucro"; *papillose* (or *papillate*) means "with papillae"; *stipitate* means "having a stipe"; *involucral* means "pertaining to the involucre"; and so forth.

Likewise, negatives are usually not defined: *apetalous* means "without petals"; *eciliate* and *eglandular* mean "without cilia" and "without glands"; etc.

Some specialized terms used in certain families and genera are explained in the remarks for the group concerned and are not necessarily repeated here. Other terms not listed here are probably not specialized botanical ones; if they are unfamiliar, consult a good dictionary.

For many terms, representative figures are cited where the application of the term may be seen. When it may not be evident from the definition or the figure legend itself to what part of a figure the term applies, a brief suggestion is added.

Abaxial. Away from the axis; e. g., the "lower" or dorsal surface of a leaf. Cf. adaxial.

Achene. A dry indehiscent fruit, strictly speaking one derived from a single superior carpel, but generally used for similar fruits ("nutlets") derived from more than one carpel.

Acuminate. Prolonged into a very acute point (often slightly concave below the point). [Fig. 7, leaf tips]

Acute. With the sides or margins meeting at less than a 90° angle. [Fig. 16, leaf tips]

Adaxial. Toward the axis; e. g., the "upper" or ventral surface of a leaf. Cf. abaxial.

Adherent. Sticking (but not fused) to parts of a different kind. Cf. adnate.

Adnate. United (fused) to parts of a different kind; e. g., stamens to petals, stipule to blade. Cf. connate.

Adventive. Spreading from a native or naturalized source but not [yet] well established.

Albino. Lacking normal color; i. e., white—usually in reference to flowers, at least the perianth.

Alternate. Arranged singly at the nodes, as leaves on a stem or branches in an inflorescence; neither opposite nor whorled. [Figs. 2, 7, leaf arrangement]

Ament. A spike or spike-like inflorescence consisting of reduced (usually apetalous and unisexual) flowers and deciduous as a unit; also called a "catkin." [Figs. 3, 18; plate 1-A]

Amphiploid. An autopolyploid: a taxon of hybrid origin with double (or more) sets of chromosomes.

Anastomose. To merge (as veins in a leaf blade) so as to form a network. [Figs. 113, 223]

Annual. Living for one year; i. e., germinating, flowering, and setting seed in a single growing season (lacking perennial roots, rhizomes, or other such parts). A winter annual germinates in the fall and completes its cycle the following spring.

Anther. The pollen-bearing part of a stamen. [Figs. 314, 350; plate 8-A]
Anthesis. The time at which a flower is fully expanded and functional.
Antrorse. Directed toward the apex or upward. Cf. retrorse. [Figs. 118, bristles on ovary, 205, bristles on fruit]
Apiculus. A very small sharp beak-like tip. [Figs. 23, 24, at apex of acorn]
Apomixis. As used here, reproduction by seed without fertilization; i. e., a form of asexual reproduction.
Appressed. Oriented in a parallel or nearly parallel manner to the surface or axis to which attached. [Fig. 135, pedicels & fruit]
Aril. An appendage arising from or near the scar (hilum) on a seed marking its point of attachment. [Fig. 125]
Ascending. Directed strongly upward (or forward, in relation to the site of attachment) but not fully erect (or at right angles). [Figs. 68, branches, 137, fruiting pedicels]
Attenuate. Drawn out to a slender tapering apex or base. [Fig. 333, base of fruit]
Auricle. A lobe or appendage, often small and ear-like, typically projecting at the base or summit of an organ (as on a leaf blade). [Fig. 138]
Awn. A terminal appendage or elongation, often bristle-like. [Figs. 243, 244, on fruits]
Axil. The angle where a leaf or branch joins a stem or main axis, or where a lateral vein joins the midrib of a leaf.
Basal. At the base; i. e., unless the context indicates otherwise, at the base of the plant, or at ground level.
Beak. A comparatively slender prolongation (sometimes of firmer texture) on a broader organ. [Figs. 134, 135, at end of fruit]
Berry. A fleshy indehiscent several-seeded fruit derived from a single ovary.
Biennial. Living for two years. Such plants often produce a rosette of leaves the first year and a flowering stem the second year.
Bi-. Two- or twice- (as a prefix).
Bifid. Cleft in two. [Fig. 152, petals]
Bilaterally symmetrical. Capable of division into similar (mirror-image) halves on only one plane; zygomorphic. Cf. regular. [Figs. 121, 122, flowers; plate 7-B]
Blade. The expanded portion of a leaf or other flat structure.
Bog. A peatland typically with vegetation in ± concentric zones of increasing maturity from open water of a pond or small lake to surrounding swamp forest or upland.
Bract. A reduced leaf-like, sometimes scale-like, structure, often subtending a flower, inflorescence, branch, etc.
Bracteole. A secondary bract; bractlet.
Calcareous. Limy—rich in calcium carbonate, as from limestone (or dolomite) or marl.
Calyx. The outer series of perianth parts (or the only one); the sepals, collectively. [Fig. 152]
Campanulate. Bell-shaped. [Fig. 311, flower]
Capillary. Hair-like; very slender.
Capitate. Like a pin-head (as certain stigmas on the style). [Fig. 311, stigma]
Capsule. A fruit which dehisces along two or more sutures (derived from 2 or more carpels), usually several- or many-seeded. [Fig. 301]
Carpel. The basic female structural unit of the flower, homologous to a sepal, petal, or stamen; in a compound pistil, the carpels are united (connate), but the number can often be determined from the number of styles, stigmas, or locules.
Caudate. Bearing a prolonged tail-like appendage; or prolonged into such a shape. [Plate 1-D, sepal tips]
Cauline. On or pertaining to the stem—often in contrast to basal.
Ciliate. With hairs along the margin or edge. [Figs. 11, on leaf, 30, on fruit]
Ciliolate. Minutely ciliate.
Circumscissile. Dehiscing by a circular line. [Fig. 62, fruit]

Clavate. Club-shaped; i. e., with a ± prolonged and narrow base. [Figs. 318, 332, fruit]

Claw. A ± abruptly narrowed basal portion of some blades (e. g., of petals or tepals). [Fig. 126, on petals]

Cleistogamous. Fertilized and setting seed without opening.

Cline. A "character gradient"; i. e., ± regular or continuous change in a character across a geographic area.

Complex-heterozygote. In *Oenothera* and related genera, a race containing two different chromosomal complexes, each acting as a single linkage group (as a result of the ring-forming behavior peculiar to the chromosomes of such plants).

Compound. Composed of more than one part, or branched; e. g., a leaf with two or more blades (leaflets), a pistil of more than one carpel, a branched inflorescence. Cf. simple.

Connate. United (fused) to other parts of the same kind; e. g., petals to petals, leaf margin to leaf margin. Cf. adnate.

Cordate. Broadly 2-lobed; heart-shaped. [Figs. 2, 304, leaf blades]

Coriaceous. Leathery in texture; firm.

Corm. A short thick underground stem lacking the thick fleshy leaves that characterize a bulb.

Corolla. The inner series of perianth parts (when there are two series); the petals, collectively. [Fig. 152]

Corymb. An inflorescence of racemose or paniculate type but with the lower pedicels or branches longer than the upper so that the inflorescence is relatively short, broad, and flat-topped. [Fig. 193; plate 5-D]

Crenate. With very rounded teeth; scalloped. [Fig. 164, leaf margin]

Crenulate. Finely crenate. [Fig. 299, leaf margin]

Crisped. More or less puckered. [Fig. 41, leaf margin]

Cuneate. Wedge-shaped; i. e., with straight but not parallel sides. [Fig. 104, bases of the principal leaf divisions]

Cuspidate. With a sharp, firm point. [Fig. 54, bract tips]

Cyme. A type of inflorescence in which the terminal (rather than the lower) flower matures first. [Fig. 74]

Deciduous. Falling off naturally at the end of its period of function.

Decumbent. Prostrate basally but ascending toward the tip. [Fig. 311]

Decurrent. Extending downward and along, as a leaf base on a stem or a leaf blade on a petiole.

Dehiscent. Splitting open naturally at maturity at one or more definite points. Cf. indehiscent.

Deltoid. Broadly triangular. [Fig. 309, calyx lobes]

Dentate. With ± outward-pointing (often coarse and/or obtuse) marginal teeth. [Figs. 52, 285]

Denticulate. With minute, usually ± remote, marginal teeth. [Fig. 3]

Depauperate. Stunted or otherwise poorly developed.

Dichotomous. Forking into two ± equal branches. [Fig. 87]

Dimorphic. Of two forms.

Dioecious. Having the sexes on separate plants; i. e., all flowers on a single plant either staminate or pistillate. Cf. monoecious. [See fig. 271]

Diploid. (See *n*)

Dissected. So finely divided (as in some leaf blades) that the blade tissue is nearly restricted to bordering the main veins (definite leaflets not evident). [Figs. 87, 102, 321]

Distal. At or toward the apex; i. e., toward the opposite end from that at which a structure is attached.

Divaricate. Stongly divergent; spreading or forking at about a 90° angle or more. [Fig. 241, branches of inflorescence]

Divergent. Spreading away from the surface or axis to which attached. [Fig. 134, pedicels & fruit]

Dorsal. Pertaining to the surface (e. g., of a leaf or seed) away from the axis to which a structure is attached; abaxial. Cf. ventral. [Figs. 263, right seed, 258 & 259, left seed]

Double. (Of a flower) with extra cycles of perianth parts (morphologically derived usually from stamens and carpels converted to petals). (Of serrations) with primary teeth again toothed (doubly serrate). Cf. single. [Fig. 21, leaf margin]

Drupe. A fleshy indehiscent fruit with the seed (or seeds) enclosed in a hard tissue (endocarp) forming a central pit (or "stone"), as in a cherry.

Elliptic(al). Longer than wide, broadest at the middle, and tapering ± equally toward both ends. [Figs. 91, leaf blades, 222, leaflets]

Emarginate. With a shallow notch at the apex. [Figs. 78, 81, petals]

Emersed. Normally extending above the water. Cf. submersed.

Entire. Without teeth; with a continuous margin. [Fig. 208, leaf blade]

Erose. Irregular (of a margin), as if chewed or nibbled. [Fig. 271, petals]

Excurrent. Running beyond, as a vein prolonged beyond the margin of a leaf or other structure. [Figs. 22, 223]

Exserted. Protruding beyond the surrounding structure(s). [Fig. 223, stamens]

Farinose. Covered with a fine granular or powdery (mealy) coating.

Fen. An "alkaline bog"—a peatland so nourished by calcareous groundwater as to have more sedges and less sphagnum than the typical true acid bog. (Some parts of a single peatland may be termed bog and other parts, fen, depending on the distinctions made by the describer!)

Fertile. Normally reproductive; e. g., a fertile stamen produces pollen, a fertile flower bears seed (or at least reproductive parts), a fertile shoot bears flowers.

Filament. The stalk of a stamen, usually thread-like but sometimes flattened or expanded. [Figs. 314, 350]

Filiform. Thread-like: very slender and approximately as broad as thick.

Flexuous. More or less loose and sinuous, bent or curved (usually several times in alternate directions); zigzag. [Fig. 57, stem]

Floral tube. The usually saucer- or cup-shaped structure formed by the adnate portions of perianth and stamens, on which the free portions of these organs are inserted. (In some works, = "calyx tube" or "hypanthium.")

Follicle. A fruit which dehisces along a single suture (derived from a single carpel).

Forest. Vegetation dominated by trees closely enough spaced to provide a ± continuous or closed canopy.

Fruit. A ripened ovary and any closely associated structures.

Fusiform. Thickest at the middle and tapering to both ends; spindle-shaped.

Glabrate. Nearly without hairs.

Glabrous. Without hairs.

Gland. A secretory structure; any protuberance (often of different texture, e. g., shiny or sticky in appearance) resembling such a structure.

Glaucous. Covered with a pale (gray to blue-green) waxy coating or "bloom."

Globose. Spherical.

Hastate. Shaped like an arrowhead, but with basal lobes diverging. Cf. sagittate. [Fig. 46]

Head. A compact inflorescence of sessile flowers crowded on a receptacle. [Figs. 210, 325] Loosely used for compact clusters of fruits from a single flower. [Figs. 109, 205]

Heterophylly. The phenomenon of producing two kinds of leaves, quite different morphologically, on the same individual, either simultaneously or in the course of development. [Figs. 155, 320]

Hirsute. With rather coarse or stiff hairs.

Hispid. With stiff hairs or bristles.

Hispidulous. Minutely hispid.

Hyaline. Thin and translucent.

Hypanthium. (See floral tube)

Illegitimate. Contrary to one or more Articles of the International Code of Botanical Nomenclature, under which names in violation of certain rules are not available for use under any circumstances.

Imbricate. With the edges overlapping, like shingles on a roof. Cf. valvate. [Fig. 23, bud scales and cup scales]

Impressed. Slightly sunken, as the veins at the surface of some leaves.

Indehiscent. Not splitting open naturally. Cf. dehiscent.

Indurated. Firm and hardened.

Inferior. (Of an ovary) below the perianth. Cf. superior. [Figs. 170, 314]

Inflorescence. An entire flower cluster, including pedicels and bracts; often used to cover clusters of fruit as well.

Infructescence. A fruiting inflorescence.

Inserted. Attached to or on; appearing to arise from (often applied to the free portion of an adnate structure).

Introgression. The gradual infiltration of genes from one taxon into another, as the result of hybridization and back-crossing with the parent(s).

Involucel. Bracts at the base of a unit in a compound inflorescence, in contrast to the involucre at the base of the entire inflorescence. [Fig. 337, at base of umbellets]

Involucre. The bract or bracts (or even leaves) at the base of an inflorescence. [Figs. 107, 337, at base of umbel]

Involute. With the margins rolled in (i. e., adaxially). Cf. revolute.

Internode. The portion of a stem or axis between nodes.

Keel. A ridge ± centrally located on the long axis of a structure, such as a sepal or an achene. [Fig. 105] The pair of connate lowermost petals in a papilionaceous flower.

Lacerate. Ragged, irregularly cleft, appearing as if torn.

Laciniate. Deeply and ± narrowly lobed or slashed.

Lanceolate. Narrow and elongate, broadest below the middle. [Fig. 4]

Leaflet. One of the blades of a compound leaf.

Lenticel. A corky, porous spot on bark (especially noticeable and large on *Betula* and *Prunus*).

Linear. Narrow and elongate with ± parallel sides. [Figs. 172, petals, 233, leaflets; plate 4-B, leaves]

Lobe. A projection or extension, set off by an indentation (or sinus). [Figs. 26, 32, 193, all on leaves, 152, on petals]

Locule. A compartment or cavity, as in an anther or ovary (often termed a "cell").

Lyrate. Pinnatifid but with a relatively large terminal lobe. [Fig. 302]

Marsh. A wet (at least seasonally) and treeless area. Cf. swamp.

Meadow. A loosely defined term for a treeless area (including one with many introduced species), often less level and hence less wet than a marsh though usually with many grasses (or sedges).

Medifixed. Attached at the middle; two-pronged, as in some kinds of hairs. [Fig. 153]

Mericarp. One of the (usually indehiscent, 1-seeded) parts into which a schizocarp splits. [Figs. 243, 244]

-merous. -parted; i. e., with parts in the number cited or a multiple thereof.

Monoecious. Having the sexes in separate flowers but on the same individual. Cf. dioecious.

Mucro. A short, sharp, slender point. [Fig. 62, at tepal tips]

n. The haploid or gametic number of chromosomes; ordinary cells of a seed plant have this number of *pairs* of chromosomes. Many plants have more than the basic two sets or complements of chromosomes (diploid) and the number of these is indicated with the suffix *-ploid*: triploid (3n) = 3 sets; tetraploid (4n) = 4 sets; octoploid (8n) = 8 sets; etc.

Nerve. A vein or ridge, usually a relatively weak or less strong one.

Node. The point on a stem (extended to include the axis of an inflorescence) at which a leaf or branch arises.

Nut. A dry, hard, indehiscent (usually 1-seeded) fruit, often larger than normally termed an achene (or nutlet).

Nutlet. An achene; also used for the stony carpels embedded in the pome of *Crataegus.*

Ob-. A prefix signifying inversion, usually with adjectives indicating shape; e. g., obovate or obcordate (with the small end basal).

Oblong. Longer than wide and ± parallel-sided (but not as elongate as "linear").

Obtuse. With the sides or margins meeting at more than a 90° angle. [Fig. 186, leaf tip]

Ocrea. The tubular sheathing stipule peculiar to the Polygonaceae. [Figs. 48, 49]

Opposite. Two at a node (and ± 180° apart), as in some leaves. [Figs. 84, 349] Centered upon (rather than alternating with), as stamens opposite the sepals.

Orbicular. Circular in outline, or nearly so.

Ovary. The lower portion of a pistil, usually ± expanded, in which the seed or seeds are produced; ripens into a fruit. [Figs. 96, 118]

Ovate. Shaped in general outline like a longitudinal section of an egg; i. e., broadest below the middle (but broader than lanceolate). [Fig. 5]

Ovoid. Egg-shaped.

Palmate. Radiating from a common point, as veins or leaflets in a leaf. [Figs. 173, principal veins, 279, leaflets]

Panicle. A "branched raceme"; i. e., an inflorescence in which the pedicels arise from a branched axis rather than a simple central axis and the lowermost flowers mature first—although often inflorescences technically cymose are said to be paniculate if they resemble panicles. [Fig. 267]

Papilionaceous. Literally, "butterfly-like"; (of a flower) bilaterally symmetrical with 2 usually spreading lateral petals or wings, a lower keel (of 2 connate petals), and 1 upper (usually the largest) petal or standard (see Leguminosae). [Figs. 219, 231]

Papilla. A minute blunt or rounded projection on a surface.

Pedicel. The stalk of an individual flower.

Peduncle. The stalk of an entire inflorescence (or of a solitary flower when there is only one).

Peltate. With the stalk attached to the surface of a blade-like structure (rather than at the margin). [Figs. 91, 115, leaves]

Pendent, pendulous. Hanging or drooping. [Fig. 147, fruit; plate 5-A, racemes]

Perennial. Living 3 or more years.

Perfect. Containing both stamen(s) and pistil(s).

Perfoliate. With the stem (or other stalk) appearing to pass through the leaf (or other blade); i. e., the blade sessile and its tissue at the base surrounding the stalk. [Fig. 324]

Perianth. All of the calyx and corolla together insofar as these are present, in contrast to the reproductive organs of the flower.

Pericarp. The wall of a fruit; i. e., excluding the seeds.

Petal. One of the divisions of the corolla.

Petiole. The stalk portion of a leaf.

Petiolule. The stalk of a leaflet.

Pilose. With soft, usually long and ± straight, hairs.

Pinnate. Arranged in two rows, one on each side of a common axis, as veins in a leaf [Fig. 15] or leaflets in a compound leaf [Figs. 226, 269]. In an odd-pinnate leaf, there is a terminal leaflet [Fig. 224]; in an even-pinnate one, there is no terminal leaflet [Fig. 220]. Twice-pinnate: with the primary divisions again pinnate [Fig. 222].

Pinnatifid. Deeply lobed or cleft in a pinnate pattern. [Fig. 131, leaves] Bipinnatifid: with the primary divisions again pinnatifid.

Pinnatisect. Very deeply cleft in a pinnate pattern (often to the midrib). [Plate 4-E, petals]

Pistil. One of the female or seed-producing structures of a flower, whether composed of a single carpel or two or more carpels; usually consisting of an ovary and one or more styles and stigmas. [Fig. 118]

Pit. The seed and its stony covering of pericarp tissue in a drupe; a "stone." A small depression on a surface.

Plicate. Folded (along veins) like a fan or pleats of an accordion.

-ploid. (See *n*)

Pollen. The grains (microspores, containing male gametes) produced in the anther.

Polygamous. Bearing perfect and unisexual flowers on the same individual. Polygamo-monoecious: polygamous with unisexual flowers of both sexes. Polygamo-dioecious: polygamous with unisexual flowers of only one sex.

Pome. A fleshy fruit derived from an inferior ovary, the fleshy tissue developed chiefly from the floral tube (adnate to the ovary which forms a papery or cartilaginous core), as in apples and pears and the rest of their subfamily of the Rosaceae.

Prairie. A naturally treeless grassland (drier than a marsh, into which a "wet prairie" grades).

Puberulent. Minutely or finely pubescent.

Pubescent. With hairs (of whatever size or texture).

Pulvinus. A swelling or pad of tissue at the base of a petiole or petiolule, as in many Leguminosae. [Fig. 216, on petiolules]

Punctate. Dotted with tiny pits or glands.

Puncticulate. Very minutely punctate.

Pustulate. With blister-like swellings.

Pyriform. Pear-shaped.

Raceme. A type of inflorescence in which each flower is on an unbranched pedicel attached to a ± elongate unbranched central axis; the flowering sequence is from the base to the apex. [Figs. 64, 278]

Rachis. The central axis of an inflorescence or a compound leaf.

Ray. A branch of an inflorescence such as a compound umbel.

Receptacle. The surface on which the parts of a flower are inserted, or on which the flowers in a head or other dense inflorescence are inserted.

Reflexed. Bent backward or downward. [Fig. 209, flowers]

Regular. With radial symmetry (capable of division into similar halves on more than one plane); actinomorphic. Cf. bilaterally symmetrical. [Fig. 64, flower]

Remote. Relatively far apart.

Reniform. Shaped in general outline like a longitudinal section of a kidney; i. e., broader than long, ± shallowly cordate at base and otherwise ± rounded (obtuse at apex). [Fig. 300, leaf blades]

Reticulate. Having the appearance of a net. [Fig. 144, seed surface]

Retrorse. Directed toward the base or downward. Cf. antrorse. [Fig. 44, prickles on stem]

Revolute. With the margins rolled back or under (i. e., abaxially). Cf. involute.

Rhizome. An underground stem, usually ± elongate and growing horizontally (distinguishable from a root by the presence of nodes). [Figs. 92, 93, 140, 240]

Rosette. A ± dense and circular cluster of leaves. [Figs. 147, 150 (basal), 264 (terminal)]

Rugose. Wrinkled or puckered in appearance. [Plate 5-A, leaves]

Sagittate. Arrowhead-shaped, with basal lobes pointing downward (not divergent, often ± parallel). Cf. hastate. [Figs. 44, 229]

Samara. A dry indehiscent nut-like fruit with a well developed wing. [Figs. 30, 249, 275]

Scabrous. Rough (to the touch).

Scale. A small bract, such as the one subtending an individual flower in an ament. One of the structures covering a bud.

Scape. A peduncle arising from the *base* of a plant (directly from the root, rhizome, etc.)—a "leafless stem." [Figs. 98, 121, 240]

Scarious. Thin and dry, papery in texture.

Schizocarp. A fruit that splits at maturity into 2 or more (usually indehiscent and 1-seeded) parts which are dispersed as separate units (mericarps). [Figs. 288, 329, 343]

Sepal. One of the divisions of the calyx.

Serrate. With sharp, ± forward-pointing marginal teeth. In a doubly serrate margin, there are teeth on the primary teeth. [Figs. 16, 269; plate 5-D]

Serrulate. Minutely serrate. [Fig. 185]

Sessile. Attached without a stalk. [Figs. 292, 295, leaves]

Setaceous. Bristle-like.

Setose. Bristly.

Silique. A 2-carpellate fruit which dehisces from the base upward, leaving a septum between the locules, characteristic of the Cruciferae. Usually elongate; a short silique is called a silicle. [Figs. 131, 147, siliques, 154, 160, silicles]

Simple. Composed of a single or unbranched part; e. g., a leaf with one blade, a pistil of one carpel, an unbranched inflorescence, an unbranched hair. Cf. compound.

Single. (Of flowers) with one cycle of showy perianth parts. Cf. double. (Not the same as solitary, which means *only one*.) (Of serrations), without additional teeth on the primary teeth.

Sinuate. Broadly scalloped, with ± open sinuses and low teeth; coarsely dentate or wavy. [Fig. 57; plate 3-D, leaf margins]

Sinuous. With one or more wavy bends. [Figs. 264, 321, stems] Sinuate.

Sinus. The space or cleft between 2 lobes.

Spatulate. Broad and flat distally, contracted or tapered toward the base. [Fig. 96, petal]

Spicule. A minute sharp slender point, as on the margins of some leaves.

Spike. An elongate unbranched inflorescence in which the flowers are sessile; loosely, a dense elongate spike-like inflorescence with crowded flowers. [Fig. 47]

Spinulose. With a minute spine.

Stamen. One of the male or pollen-producing structures of a flower, usually consisting of a filament and an anther. [Figs. 247, 276]

Staminodium. A sterile stamen (so determined by its location), without an anther but sometimes cleft, glandular, or otherwise modified.

Standard. The usually upright larger petal opposite the keel in a papilionaceous flower.

Stellate. (Of a hair) ± radially branched.

Sterile. Lacking flowers or fruit (= vegetative); not fertile.

Stigma. The part of a pistil which is receptive to pollen, usually distinguished by a sticky, papillose, or hairy surface. [Figs. 96, 256]

Stipe. A stalk (generally used when no more precise term such as petiole is applicable); e. g., the short stalk on which some ovaries are elevated above the receptacle. [Fig. 133]

Stipel. The stipule-like structure at the base of a leaflet. [Fig. 218]

Stipule. An appendage on the stem at the base of a leaf (sometimes partly or wholly adnate to the petiole). [Figs. 5, 210, 218]

Stolon. An elongate above-ground (or at-ground) stem, growing ± horizontally and rooting at the nodes and/or apex. [Fig. 200]

Striate. With slender lines or stripes or low ridges.

Strigose. With short, straight, strongly appressed hairs.

Style. The portion of a pistil between the ovary and the stigma—often narrow and elongate. [Figs. 118, 314]

Sub-. A prefix meaning almost, not quite, just below; e. g., subterminal, just below the end; subglobose, almost spherical.

Submersed. Normally occurring under water. Cf. emersed.

Subtend. Occur immediately below, as a bract below a flower or a pedicel.

Subulate. Awl-shaped, very slender, firm, and sharp-pointed. [Fig. 69, leaves]

Superior. (Of an ovary) with the perianth and stamens inserted beneath it. Cf. inferior. [Fig. 96]

Suture. The line or joint along which 2 parts are fused (and along which they may separate).

Swamp. A wet (at least seasonally) and wooded area. Cf. marsh

Taxon. Any taxonomically recognized unit, regardless of rank; e. g., genus, species, variety.

Tendril. A slender coiling or twining organ, as on some vines. [Fig. 281]

Tepal. One of the divisions of a perianth when the sepals and petals are similar in color, texture, and (usually) size (though usually distinguishable by position, the sepals being the outer series and the petals the inner one). [Figs. 49, 62]

Terete. Round in cross section.

Ternate. Basically divided into 3 ± equal portions (as a compound or dissected leaf [Fig. 342]). Twice- or thrice-ternate structures have the first divisions again divided in a similar fashion (whereas sometimes each of the three primary divisions is pinnate [Fig. 323]).

Tetraploid. (See *n*)

Thicket. A loosely defined term for usually small areas (or narrow ones, as along a stream) with ± dense shrubs or small trees.

Tomentose. More or less densely covered with curly, matted hairs; woolly.

Trifoliolate. With 3 leaflets.

Truncate. Ending abruptly (at base or apex), as if cut off squarely (neither tapered nor lobed). [Fig. 194, base of leaf blades]

Tube. The fused portion of a cycle of perianth parts (beyond which the calyx lobes or corolla lobes extend). [Figs. 30, 69]

Tuber. A thickened portion of rhizome or root, usually a starch-storing organ. [Figs. 107, 347]

Type. A specimen which fixes the application of a name; a name of a species, for example, applies at least to its type—and to all other specimens deemed to belong to the same species. If a taxon is divided into two or more, the name remains with that element which includes its type, and a new name is required for the other element(s).

Umbel. An inflorescence in which the pedicels arise from the same point or nearly so; in a compound umbel, each primary ray bears an umbellet. [Figs. 337, 347; plate 8-C (all compound umbels)]

Undulate. Wavy, sinuate.

Unisexual. (Of a flower) containing only stamen(s) or pistil(s); imperfect. Cf. perfect. [Figs. 271, 276]

Valvate. Meeting at the edges without overlapping. Cf. imbricate. [Fig. 271, carpels in fruit]

Vein. A bundle of vascular tissue; the external ridge marking the location of an underlying vein.

Ventral. Pertaining to the surface (e. g., of a leaf, seed, etc.) toward the axis to which a structure is attached; adaxial. Cf. dorsal. [Figs. 263, left seed, 258 & 259, right seed]

Villous. With soft, not necessarily straight, hairs—practically synonymous with pilose.

Viscid. Sticky, glutinous.

Waste places, waste ground. Seriously disturbed areas that are not cultivated, such as roadsides, vacant lots, dumps, and construction sites.

Whorl. A ring of 3 or more similar structures around a stem or other axis (i. e., at the same node). [Fig. 65, leaves]

Wing. A flat, ± thin extension or appendage on the edge or surface of an organ (as on a seed or stem). [Figs. 52, 148, 344] One of the lateral petals in a papilionaceous flower.

Woodland. Vegetation with trees less dense than in a forest (often a savana).

Woods. Forest. A "mixed woods" is one of both coniferous and deciduous trees.

Index

Names of species followed by a map number (M-000) in parentheses are accepted ones for species mapped as part of our flora. An asterisk indicates that the species is illustrated. References to the color plates (pages 17–24) and the figures (numbered consecutively throughout the volume) may be found at the main entry for each species—which is the first page cited. Additional pages cited refer to discussion or significant observations elsewhere in the volume; page numbers where species are compared within the same genus are usually not indexed, nor are pages where taxa appear in the keys. Species and genera, with their descriptive key characters, can generally be found easily in the keys as they are numbered in sequence.

Names of species not followed by a map number represent synonyms of accepted names, most hybrids, minor species which may be keyed but not mapped in *Crataegus,* and species mentioned incidentally (including those not considered established and those erroneously reported from the state).

Varieties and other infraspecific taxa are not included in the index. Accepted names for families and for groups in *Crataegus* are in capitals. When only one species is mentioned in a genus, the name of the genus is not separately indexed. Common names consisting of two elements are ordinarily indexed only under the second one (e. g., Rock *Cress,* Sour *Dock,* Red *Maple*), especially if the elements are not hyphenated—unless the second element is such a broad or unseparated word as "berry," "plant," or "weed" (e. g., *straw*berry, *pie* plant, *mermaid*-weed). Hyphens (or spaces) and multiplication signs (designating hybrids) are ignored in alphabetizing.

Abbreviations, 27
Abelmoschus esculentus (M-804), 574
Abutilon theophrasti (M-792), 569*
Acacia, Rose-, 477
Acalypha, 519
  gracilens, 520
  rhomboidea (M-722), 520*
  virginica (M-721), 519
Acer, 545
  ginnala, 545
  negundo (M-763), 547
  nigrum, 549
  pensylvanicum (M-770), 551*
  platanoides (M-764), 547
  pseudoplatanus (M-768), 550, 336
  rubrum (M-767), 549*
  saccharinum (M-766), 549*
  saccharum (M-765), 547
  spicatum (M-769), 550*
ACERACEAE, 545
Acnida altissima, 145
Aconitum napellus, 199
Actaea, 207
  alba, 208
  pachypoda (M-260), 208*
  rubra (M-259), 207
  spicata, 207
Adlumia fungosa (M-321), 248*
Aegopodium podagraria (M-963), 675
Aesculus, 551
  ×carnea, 552
  glabra (M-772), 554*
  hippocastanum (M-771), 552
  octandra, 554
Aethusa cynapium (M-953), 669
Agrimonia, 440
  bicknellii, 443
  gryposepala (M-583), 443*

[Agrimonia]
parviflora (M-581), 442
pubescens (M-584), 443
rostellata (M-582), 443
striata (M-585), 444
Agrimony, 440
Agrostemma githago (M-209), 172*
Aiken, Susan G., xiv
Ailanthus altissima (M-709), 512*
Aizoaceae, 151
Ajowan, 656
Akebia quinata (M-301), 233*
Alcea rosea (M-803), 573
Alder, 63
Black, 64
Black-, 540
Green, 64
Mountain, 64
Smooth, 63
Speckled, 64
Tag, 64
Alexanders, Golden, 672
Alfalfa, 457
Alfileria, 502
Alliaria, 308
officinalis, 308
petiolata (M-416), 308*
Almond, Flowering, 365
Alnus, 63
crispa (M-43), 64*
glutinosa (M-41), 64
incana, 64
rugosa (M-42), 64*
serrulata, 63
viridis, 64
Althaea officinalis (M-802), 573
Alum-root, 321
Alyssum, 276
alyssoides (M-363), 276*
Hoary, 296
murale (M-364), 277
Pale, 276
saxatile (M-365), 277
Sweet, 298
Amaranth, 143
Family, 142
Green, 148
Purple, 148
AMARANTHACEAE, 142
Amaranthus, 143
albus (M-159), 145
arenicola (M-160), 145
blitoides (M-158), 145*
[Amaranthus]
caudatus (M-167), 148
cruentus (M-168), 148
gracilis (M-162), 146
graecizans, 145
hybridus (M-169), 148
hypochondriacus (M-166), 148
leucocarpus, 148
powellii (M-165), 146
retroflexus (M-164), 146*
spinosus, 143
tamariscinus, 143
tricolor (M-163), 146
tuberculatus (M-161), 145
viridis, 146
Amelanchier, 379
alnifolia, 379, 384
arborea (M-515), 382
bartramiana (M-514), 382
canadensis, 383
gaspensis, 385
humilis, 385
huronensis, 385
interior (M-517), 384
intermedia, 384
laevis (M-516), 383*
sanguinea (M-519), 385*
spicata (M-518), 384
stolonifera, 384
Ammannia, 610
coccinea, 610
robusta (M-864), 610*
Amorpha, 474
canescens (M-640), 474
fruticosa (M-639), 474*
Amphicarpaea bracteata (M-608), 460*
ANACARDIACEAE, 531
Anemone, 227
blanda, 227
Canada, 228
canadensis (M-291), 228
cylindrica (M-293), 229*
False Rue-, 224
hepatica, 209
multifida (M-290), 227*
nemorosa, 228
parviflora, 227
patens, 227
quinquefolia (M-292), 228
Red, 227
riparia, 229
Rue-, 224
virginiana (M-294), 229*

[Anemone]
Wood, 228
Anemonella thalictroides (M-286), 224*
Anethum graveolens (M-945), 665*
Angelica, 660
atropurpurea (M-936), 660*, 646
venenosa (M-935), 660
Angell, Virginia, 389
Anise, 655
ANNONACEAE, 235
Antenoron virginianum, 118
Anthriscus sylvestris (M-949), 667
Anthyllis vulneraria (M-674), 493
APIACEAE, 645
Apios americana (M-677), 493*
Apium graveolens, 670
Apomicts, 8
Appel, Heidi, xi
Apple, 418
American Crab, 419
Bechtel Crab, 419
Crab, 418
Flowering Crab, 418
Prairie Crab, 419
Sweet Crab, 419
Wild Crab, 419
Apricot, 370
AQUIFOLIACEAE, 539
Aquilegia, 202
canadensis (M-250), 202*
chrysantha, 202
vulgaris (M-251), 202
Arabidopsis thaliana (M-394), 295*
Arabis, 289
alpina, 292
canadensis (M-383), 290
caucasica (M-388), 292
dentata, 292
divaricarpa (M-392), 294
drummondii (M-385), 291
glabra (M-386), 292
hirsuta (M-387), 292*
holboellii (M-384), 291*
laevigata (M-391), 294
lyrata (M-382), 290*
missouriensis (M-390), 294
perstellata (M-389), 292
procurrens (M-381), 290
Aralia, 644
elata, 644
hispida (M-916), 645*
nudicaulis (M-915), 645
racemosa (M-914), 645
[Aralia]
spinosa, 644
ARALIACEAE, 640
Arceuthobium pusillum (M-87), 101*
Arenaria, 170
lateriflora (M-208), 171*
leptoclados, 171
macrophylla (M-207), 171
serpyllifolia (M-206), 171*
stricta (M-205), 170
Argemone, 240
alba, 242
albiflora (M-310), 242
intermedia, 242
mexicana (M-311), 242
Argentina anserina, 427
Argus, George W., xiv, 31
Aristolochia, 102
clematitis (M-89), 103
durior, 102
macrophylla (M-88), 102
serpentaria (M-90), 103*
ARISTOLOCHIACEAE, 102
Armoracia, 280
aquatica (M-369), 282*
lapathifolia, 282
rusticana (M-370), 282
Aronia, 377
arbutifolia, 379
atropurpurea, 379
melanocarpa, 379
prunifolia (M-513), 377*
Aruncus, 433
dioicus (M-568), 433
sylvester, 433
Asarum, 103
canadense (M-91), 103*
ypsilantense, 103
Ash,
Mountain-, 363
Prickly-, 510
Wafer-, 509
Asimina triloba (M-304), 235*
Aspen,
Bigtooth, 53
European, 51
Largetooth, 53
Quaking, 53
Astragalus, 494
canadensis (M-680), 495
cicer (M-679), 495
cooperi, 495
neglectus (M-681), 495*

Atriplex, 129
argentea, 129
hortensis (M-135), 129
littoralis, 130
patula (M-136), 129*
prostrata, 130
rosea (M-134), 129
triangularis, 130
Aurinia saxatilis, 277
Avens, 433
Awlwort, 307

Baby's-breath, 175
Ball, C. R., 31
Ballard, Harvey E., Jr., xiv, 589
Balloon-vine, 554
Balm-of-Gilead, 52
Balsam, 555
Garden, 555
BALSAMINACEAE, 554
Bamboo, Mexican, 118
Baneberry, 207
Red, 207
White, 208
Baptisia, 458
australis, 458
bracteata, 460
lactea (M-607), 460
leucantha, 460
leucophaea (M-606), 460
tinctoria (M-605), 460
Barbarea, 259
orthoceras (M-334), 260
stricta, 260
verna (M-333), 260
vulgaris (M-332), 259*
Barberry, 230
Common, 231
Family, 230
Japanese, 231
Barnes, Burton V., xiv
Basswood, 567
Bay, 236
Bayberry, 55
Family, 54
Beach-heath, 584
Beal, Ernest O., xiv
Bean, 460
Castor-, 517
Common, 461
Green, 461
Kidney, 461
[Bean]
Snap, 461
Soy, 472
String, 461
Wax, 461
Wild, 461, 493
Beech, 84
Blue-, 71
European, 84
Family, 72
Gray, 84
Red, 84
White, 84
Beet, 133
Beggars-tick, 462
Benzoin, 237
BERBERIDACEAE, 230
Berberis, 230
aquifolium, 231
thunbergii (M-295), 231
vulgaris (M-296), 231
Berteroa incana (M-396), 296*
Berula, 662
erecta (M-940), 662*
pusilla, 663
Beta vulgaris (M-141), 133
Betula, 65
alba, 68
alleghaniensis (M-44), 66
cordifolia, 68
lenta, 66
lutea, 66
murrayana, 68
nigra, 66
papyrifera (M-47), 68*
pendula (M-46), 68
populifolia, 68
pubescens, 68
pumila (M-45), 67*
×purpusii, 67
×sandbergii, 67
BETULACEAE, 61
Bilderdykia, 112
Bindweed, Black-, 119
Birch, 65
Black, 66
Bog, 67
Canoe, 68
Cherry, 66
Dwarf, 67
European White, 68
Family, 61

[Birch]
Gray, 68
Paper, 68
River, 66
Sweet, 66
White, 68
Yellow, 66
Birthwort, 102
Family, 102
Bishop's-cap, 323
Bistort, Alpine, 120
Bistorta vivipara, 120
Bitternut, 59
Bittersweet, 541
American, 543
Climbing, 543
Family, 541
Oriental, 543
Black-alder, 540
Blackberry, 340
Common, 353
Cutleaf, 348
Black-gum, 612
Black Medick, 457
Blackthorn, 374
Bladdernut, 545
Family, 545
Bleeding-heart, 245
Wild, 247
Blite,
Sea-, 133
Strawberry, 138
Blitum capitatum, 140
Bloodroot, 240
Blue-beech, 71
Bocconia cordata, 239
Boehmeria, 96
cylindrica (M-82), 96*
nivea, 96
Bois blanc, 566
Boston-ivy, 560
Boufford, David E., xiv
Bouncing Bet, 177
Bowman's Root, 421
Box-elder, 547
Boysenberry, 343
Bramble, 340
Brasenia schreberi (M-241), 190*
Brassica, 262
alba (M-340), 264
campestris, 263
hirta, 264
[Brassica]
juncea (M-343), 266
kaber (M-341), 264*
napus (M-339), 264
nigra (M-342), 265*
oleracea, 264
rapa (M-338), 263
BRASSICACEAE, 251
Braya humilis (M-395), 296*
Bridal-wreath, 375
Brier,
Sensitive, 477
Sweet, 358
Broccoli, 264
Broom, Scotch, 458
Brussels Sprouts, 264
Buckeye, 551
Family, 551
Ohio, 554
Red, 552
Sweet, 554
Yellow, 554
Buckthorn, 558
Alder-leaved, 559
Common, 559
Family, 556
Glossy, 558
Buckwheat, 111
False, 119
Fringed False, 119
Buffalo-berry, 608
Bugseed, 130
Bunchberry, 677
Bunias orientalis (M-344), 266
Bupleurum rotundifolium (M-919), 651*
Bur-clover, 456
Burnet,
American, 440
Garden, 440
Salad, 440
Burning-bush, 544
Bush-clover, 468
Korean, 469
Buttercup, 212
Bulbous, 221
Common, 221
Creeping, 221
Early, 221
Family, 199
Lapland, 219
Prairie, 220
Small-flowered, 219

[Buttercup]
Swamp, 223
Tall, 221
Butternut, 56

Cabbage, 264
Cabomba caroliniana (M-240), 190*
Cabombaceae, 189, 190
CACTACEAE, 603
Cactus Family, 603
Cakile, 308
edentula (M-417), 308*
maritima, 309
Callirhoë involucrata (M-801), 573
CALLITRICHACEAE, 528
Callitriche, 528
hermaphroditica (M-741), 528
heterophylla (M-742), 529
palustris, 530
verna (M-743), 529*
Caltha, 223
natans, 223
palustris (M-284), 224*
Caltrop, 508
Family, 508
Calylophus serrulatus (M-888), 624
Camelina, 273
microcarpa (M-356), 274*
sativa (M-355), 273
Campion,
Bladder, 182
Starry, 182
White, 183
Candytuft,
Globe, 309
Rocket, 309
CANNABACEAE, 90
Cannabis sativa (M-74), 92*
Caper Family, 249
CAPPARACEAE, 249
Capparis spinosa, 249
Capsella bursa-pastoris (M-380), 287*
Caragana arborescens (M-634), 472*
Caraway, 665
Cardamine, 282
bulbosa (M-372), 284
douglassii (M-371), 284
flexuosa (M-375), 284
hirsuta (M-376), 285
impatiens (M-373), 284
parviflora (M-377), 285
pensylvanica (M-378), 285*, 287
pratensis (M-374), 284
Cardaria, 300
draba (M-405), 302*
pubescens (M-404), 302
Cardiospermum halicacabum (M-773), 554
Carnation, 174
Carpetweed, 151
Family, 151
Carpinus caroliniana (M-50), 71*
Carrot, 655
Family, 645
Wild, 655
Carum, 665
carvi (M-944), 665
copticum, 656
Carya, 56
cordiformis (M-37), 59
glabra (M-39), 61
illinoinensis, 58
laciniosa (M-40), 61
ovalis, 61
ovata (M-38), 59*
tomentosa, 61
CARYOPHYLLACEAE, 155
Cashew Family, 531
Cassia, 479
chamaecrista (M-649), 480
fasciculata, 480
hebecarpa (M-651), 481*
marilandica, 481
nictitans (M-648), 480
obtusifolia, 481
tora (M-650), 480
Castanea dentata (M-64), 85*
Castor-bean, 517
Catchfly,
Night-flowering, 183
Nottingham, 183
Sleepy, 181
Sweet-William, 181
Cat-claw, 477
Catgut, 494
Caulophyllum, 231
giganteum, 232
thalictroides (M-298), 231*
Ceanothus, 557
americanus (M-779), 557
herbaceus (M-780), 558*
ovatus, 558
Redstem, 557
sanguineus (M-778), 557
Celandine, 244
Lesser-, 218
CELASTRACEAE, 541

Celastrus, 541
orbiculata (M-755), 543
scandens (M-756), 543*
Celery, 670
Celtis, 87
occidentalis (M-69), 88
tenuifolia (M-70), 88*
Cerastium, 165
arvense (M-197), 166
fontanum (M-201), 167*
glomeratum (M-199), 166
nutans (M-198), 166
semidecandrum (M-200), 167
tomentosum (M-196), 166
viscosum, 167
vulgatum, 167
Ceratocystis ulmi, 87
CERATOPHYLLACEAE, 186
Ceratophyllum, 186
demersum (M-238), 187*
echinatum (M-239), 187*
muricatum, 187
Cercis canadensis (M-586), 449*
Chaenomeles, 418
Chaerophyllum procumbens (M-947), 667*
Chamaecrista,
fasciculata, 480
nictitans, 480
Chamaerhodos, 433
erecta, 433
nuttallii (M-567), 433*
Chamaesyce, 520
Chamerion angustifolium, 620
Chard, Swiss, 133
Charlock, 264
Cheeses, 572
Chelidonium majus (M-317), 244*
CHENOPODIACEAE, 125
Chenopodium, 135
album (M-156), 141*
ambrosioides (M-143), 136
aristatum (M-145), 137*
berlandieri, 142
bonus-henricus (M-149), 138
botrys (M-144), 136
bushianum, 142
capitatum (M-151), 138*
desiccatum, 137
gigantospermum, 140
glaucum (M-147), 137
hybridum (M-152), 140*
lanceolatum, 141
leptophyllum, 137
[Chenopodium]
murale (M-155), 141*
paganum, 142
pratericola, 137
rubrum (M-150), 138
standleyanum (M-153), 140
subglabrum (M-146), 137
urbicum (M-154), 140
vulvaria (M-148), 138
Cherry, 365
Choke, 369
Cornelian-, 676
European Bird, 370
Fire, 371
Nanking, 370
Perfumed, 371
Pie, 372
Pin, 371
Sand, 370
Sour, 372
Sweet, 372
Wild Black, 369
Chervil, 667
Wild-, 667
Chestnut, 85
Horse-, 552
Chickweed, 162, 165
Field, 166
Forked, 157
Giant, 165
Jagged, 167
Mouse-ear, 167
Nodding, 166
China Fleece Plant, 120
Chokeberry, 377
Christmas-rose, 225
Chrysosplenium americanum (M-435), 320*
Cicely, Sweet-, 658
Cicuta, 673
bulbifera (M-961), 673, 663
maculata (M-962), 673*
Cimicifuga racemosa (M-258), 207*
Cinquefoil, 424
Common, 429
Marsh, 427
Old-field, 429
Prairie, 428
Rough, 429
Rough-fruited, 432
Shrubby, 427
Silvery, 432
Tall, 428
Three-toothed, 429

Circaea, 615
alpina (M-871), 617
×intermedia, 617
lutetiana (M-872), 617*
quadrisulcata, 617
CISTACEAE, 584
Citrus, 508
Clammy-weed, 250
Claytonia, 154
caroliniana (M-179), 155*
virginica (M-178), 154*
Clearweed, 96
Clematis, 201
occidentalis (M-248), 201*
verticillaris, 201
virginiana (M-249), 202*
Cleome, 250
dodecandra, 250
hassleriana (M-327), 251
serrulata (M-328), 251
spinosa, 251
Cloudberry, 420
Clover, 451
Alsike, 453
Bur-, 456
Bush-, 468
Crimson, 453
Hop, 453
Little Hop, 455
Low Hop, 455
Prairie-, 491
Rabbitfoot, 453
Red, 452
Sweet-, 449
White, 453
CLUSIACEAE, 575
Cockle,
Corn-, 172
White, 183
Coffee-tree, Kentucky, 472
Cohosh,
Black, 207
Blue, 231
Cole, Emma J., 388, 402
Columbine, 202
Garden, 202
Wild, 202
Colutea arborescens, 476
Comandra, 98
livida, 98
richardsiana, 98
umbellata (M-86), 98
Comarum palustre, 427
Common names, 10
Complexes, 8
Comptonia peregrina (M-33), 55*
Conioselinum chinense (M-951), 668*
Conium maculatum (M-950), 667*
Conringia orientalis (M-357), 274*
Consolida, 202
ajacis, 203
ambigua (M-252), 202
orientalis, 203
regalis, 203
Constance, Lincoln, xiv
Coontail, 186
Coptis, 208
groenlandica, 209
trifolia (M-261), 208*
Coriander, 665
Coriandrum sativum (M-943), 665
Corispermum, 130
americanum, 133
hyssopifolium (M-139), 132*
leptopterum, 133
nitidum, 132
orientale (M-138), 132
Corn-cockle, 172
CORNACEAE, 675
Cornel, Dwarf, 677
Cornelian-cherry, 676
Cornus, 675
alba, 679
alternifolia (M-965), 677
amomum (M-970), 680*
canadensis (M-964), 677*
drummondii (M-967), 679
florida (M-966), 679*
foemina (M-971), 680*
kousa, 676
mas, 676
nuttallii, 676
obliqua, 680
racemosa, 680
rugosa (M-968), 679
sericea, 679
stolonifera (M-969), 679*
Coronilla varia (M-676), 493*
Corydalis, 248, 138
aurea (M-325), 249*
flavula (M-324), 249
Golden, 249
Pale, 249
Pink, 249
sempervirens (M-323), 249*
Corylaceae, 61

Corylus, 69
americana (M-48), 69
cornuta (M-49), 71*
Cotinus coggygria, 531
Cotton, 567
Cottonweed, 143
Cottonwood, 52
Swamp, 52
Cowbane, 662
Cow Herb, 177
Cow-parsnip, 661
Cowslip, 224
Crab (see Apple)
Crane's-bill, 502
CRASSULACEAE, 315
Crataegus, 386
albicans, 389, 412
allecta, 416
ambitiosa, 416
apiomorpha, 413
arnoldiana, 408
asperata, 413
ater, 407
attenuata, 399
basilica, 414
bealii, 398
beata, 417
bellula, 408
borealis, 388
BRAINERDIANAE, 402
brainerdii (M-530), 402
brockwayae, 396
brumalis, 417
brunetiana, 405
caesa, 411
calpodendron (M-525), 398*
celsa, 401
chrysocarpa (M-532), 405
coccinea (M-539), 410, 412*
COCCINEAE, 410
coleae, 402
compacta, 406
comparata, 416
compta, 416
CORDATAE, 396
corusca, 411
CRATAEGUS, 396
crus-galli (M-527), 399*
CRUS-GALLI, 399
desueta, 401
dilatata (M-538), 410
DILATATAE, 410
disperma (M-528), 400
[Crataegus]
dissona, 409
diversifolia, 398
dodgei (M-531), 405*
douglasii (M-523), 396
DOUGLASII, 396
ellwangeriana, 412
fallax, 388
farwellii, 400
faxonii, 405
filipes, 416
flabellata (M-541), 414*
flavida, 405
foetida, 398
fontanesiana, 400
fretalis, 413
fulleriana, 412
gattingeri, 410
gemmosa, 398
glareosa, 410
gravis, 416
hillii, 411, 408
holmesiana, 411
honesta, 403
horridula, 409
immanis (M-535), 406
incerta, 402
intricata (M-526), 398
INTRICATAE, 398
iracunda, 417
irrasa (M-533), 405*
jesupii, 408
laevigata (M-522), 396
laurentiana, 405
laxiflora, 397
leiophylla, 409
lenta, 411
levis, 416
lucorum (M-540), 413
macracantha, 397
MACRACANTHAE, 397
macrosperma, 414*
margaretta (M-534), 406*
merita, 415
meticulosa, 398
michiganensis, 389, 397
miranda, 414
MOLLES, 407
mollipes, 408
mollis (M-536), 407*
monogyna (M-521), 396
multifida, 414
nitidula, 401

[Crataegus]
nutans, 408
opulens, 417
otiosa, 414
oxyacantha, 396
Oxyacanthae, 396
parvula, 407
pascens, 412
pedicellata, 412*
perampla, 409
perlaeta, 418
phaenopyrum (M-520), 396
pinguis, 403
populnea, 418
porteri, 410
pringlei, 411
prona, 414, 417
pruinosa (M-537), 409
PRUINOSAE, 408
punctata (M-529), 401
PUNCTATAE, 400
pura, 412
pusilla, 398
remota, 409, 417
roanensis, 414
ROTUNDIFOLIAE, 403
rubella, 399
rugosa, 409
scabrida, 403
schuettei, 413
SILVICOLAE, 415
sitiens, 409, 389
stolonifera, 418
streeterae, 414
structilis, 398
submollis, 408
suborbiculata, 401, 397
Suborbiculatae, 401
succulenta (M-524), 397
superata, 409, 389
taetrica, 414
tenax, 400
TENUIFOLIAE, 412
tortilis, 413
uber, 413
urbana, 403
wheeleri, 398
wisconsinensis, 401
Creeper, Virginia, 560
Cress, 252
Bitter, 282
Garden, 304
Hoary, 302
[Cress]
Hungarian, 290
Lake, 282
Mouse-ear, 295
Penny, 303
Pink Spring, 284
Rock, 289
Sand, 290
Spring, 284
Wall Rock, 292
Water, 286
Winter, 259
Yellow, 266
Cronartium,
comandrae, 97
comptoniae, 54
ribicola, 327
Cronquist, Arthur, xiv
Crotalaria sagittalis (M-587), 449
Croton, 519
glandulosus (M-719), 519*
monanthogynus (M-720), 519
Crowberry,
Black, 530
Family, 530
Crowfoot, 212
Bristly, 220
Cursed, 219
Hooked, 219
White Water, 214
Yellow Water, 216
CRUCIFERAE, 251
Cryptotaenia canadensis (M-954), 669*
Cuckoo-flower, 284
Cucumber-tree, 235
Cultivated taxa, 7
Currant, 327
Alpine, 328
Black, 332
Buffalo, 332
Golden, 332
Mountain, 328
Northern Black, 332
Red, 333
Skunk, 333
Swamp Black, 331
Swamp Red, 334
Wild Black, 332
Custard-apple Family, 235
Cycloloma atriplicifolium (M-142), 133*
Cydonia oblonga, 418
Cypress, Summer-, 130
Cytisus scoparius (M-604), 458

Dalea, 489
alopecuroides, 491
foliosa, 491
leporina (M-673), 491
purpurea (M-672), 491*
villosa, 489
Dalibarda repens (M-546), 420*
Daphne mezereum, 605
Darlingtonia californica, 310
Daucus carota (M-927), 655*, 665
Decodon verticillatus (M-865), 610*
Delphinium, 203
ajacis, 203
consolida, 203
elatum (M-253), 203
tricorne, 203
Dentaria, 277
diphylla (M-368), 280*
laciniata (M-366), 279
maxima (M-367), 280
Descurainia, 257
pinnata (M-329), 257*
sophia (M-330), 257
Desmodium, 462
canadense (M-621), 466*
canescens (M-618), 466
ciliare (M-615), 464
cuspidatum (M-622), 467*
dillenii, 467
glutinosum (M-613), 464
illinoense (M-619), 466
marilandicum (M-616), 464
nudiflorum (M-612), 463*
nuttallii, 467
obtusum (M-620), 466
paniculatum (M-623), 467
pauciflorum, 464
perplexum, 467
rigidum, 466
rotundifolium (M-614), 464
sessilifolium (M-617), 464
viridiflorum, 467
Deutzia, 319
Devil's-club, 641
Dewberry, 340
Northern, 349
Swamp, 348
Dewdrop, 420
Dianthus, 174
armeria (M-212), 174
barbatus (M-213), 174*
carthusianorum (M-214), 174
caryophyllus, 174
[Dianthus]
chinensis, 174
deltoides (M-215), 174*
plumarius (M-216), 175
sylvestris (M-217), 175
Dicentra, 245
canadensis (M-320), 247*
cucullaria (M-319), 247*
eximia (M-318), 247
spectabilis, 245
Dictamnus albus (M-706), 509
Dill, 665
Dionaea muscipula, 312
Diplotaxis, 272
muralis (M-352), 272*
tenuifolia (M-353), 272*
Dirca palustris (M-859), 605*
Dock, 105
Bitter, 108
Curly, 110
Great Water, 110
Sour, 110
Water, 110
Dodge, C. K., 388, 389
Dogwood, 675
Alternate-leaved, 677
Family, 675
Flowering, 679
Gray, 680
Pagoda, 677
Pale, 680
Red-osier, 679
Round-leaved, 679
Doll's-eyes 208
Draba, 298
arabisans (M-403), 300*
cana (M-402), 300
incana (M-401), 299
lanceolata, 300
nemorosa (M-399), 299
reptans (M-400), 299
verna, 298
Drosera, 313
×anglica (M-425), 315
intermedia (M-424), 314
linearis (M-423), 314*
longifolia, 315
rotundifolia (M-422), 314
DROSERACEAE, 312
Dryas, 435
Drymocallis arguta, 429
Duchesnea indica (M-550), 422*
Duncan, Thomas O., xiv

Dutchman's-breeches, 247

Easterly, N. William, xi
ELAEAGNACEAE, 606
Elaeagnus, 606
angustifolia (M-860), 606
commutata, 608
umbellata (M-861), 608
ELATINACEAE, 582
Elatine minima (M-822), 582*
Elm, 85
American, 87
Chinese, 86
Cork, 87
Family, 85
Red, 86
Rock, 87
Siberian, 86
Slippery, 86
White, 87
EMPETRACEAE, 530
Empetrum nigrum (M-744), 530*
Enchanter's-nightshade, 615
Epilobium, 618
adenocaulon, 622
angustifolium (M-877), 620
ciliatum (M-883), 622
coloratum (M-884), 622*
glandulosum, 622
hirsutum (M-880), 621*
latifolium, 620
leptocarpum, 622
leptophyllum (M-882), 622
obscurum, 618
palustre (M-881), 621
parviflorum (M-879), 620*
strictum (M-878), 620
×wisconsinense, 622
Erigenia bulbosa (M-948), 667*
Erodium cicutarium (M-695), 502*
Erophila verna (M-397), 298*, 296
Eruca, 262
sativa, 262
vesicaria (M-337), 262
Erucastrum gallicum (M-331), 259*
Eryngium, 651
planum (M-918), 651
yuccifolium (M-917), 651*
Eryngo, 651
Erysimum, 274
arkansanum, 275
asperum, 275
capitatum (M-358), 275
[Erysimum]
cheiranthoides (M-359), 275*
hieraciifolium (M-362), 276
inconspicuum (M-361), 276
repandum (M-360), 276
Eschscholzia californica (M-308), 239
Euonymus, 543
alata (M-758), 544
americana, 544
atropurpurea (M-760), 544*
europaea (M-761), 544
fortunei (M-757), 543
obovata (M-759), 544*
Winged, 544
Euphorbia, 520
commutata (M-737), 527*
corollata (M-732), 526*
cyparissias (M-739), 527*
dentata (M-733), 526*
esula (M-740), 527
geyeri (M-729), 525
glyptosperma (M-728), 524*
helioscopia (M-734), 526*
hirta (M-723), 523
humistrata, 520
hypericifolia, 524
maculata (M-724), 523*
marginata (M-731), 525
nutans (M-726), 524
obtusata (M-736), 527
peplus (M-738), 527*
pilulifera, 523
platyphylla (M-735), 526
polygonifolia (M-730), 525*
preslii, 524
pulcherrima, 516
serpyllifolia (M-727), 524*
supina, 524
vermiculata (M-725), 524
EUPHORBIACEAE, 516
Evening-primrose, 624
Family, 614

FABACEAE, 444
FAGACEAE, 72
Fagopyrum, 111
esculentum (M-106), 111*
sagittatum, 111
tartaricum, 111
Fagus, 84
grandifolia (M-63), 84*
sylvatica, 84

Fallopia, 112
cilinodis, 119
convolvulus, 119
False Mermaid, 531
Family, 531
Fanwort, 190
Farwell, O. A., 388
Fennel, 667
Ficus,
carica, 89
sycomorus, 336
Fig, 89
Filbert, 69
Filipendula, 439
rubra (M-578), 439
ulmaria, 439
Fireweed, 620
First Survey, 4
Five-finger, 424
Flax, 496
Common, 497
Fairy, 497
False, 273
Family, 496
Perennial, 497
Floerkea proserpinacoides (M-745), 531*
Flower-of-an-hour, 574
Foamflower, 325
Foeniculum vulgare (M-946), 667
Fool's-parsley, 669
Four-o'clock, 149
Family, 149
Wild, 150
Fragaria, 422
×ananassa, 423
chiloënsis, 423
vesca (M-551), 423
virginiana (M-552), 424*
Frangula alnus, 559
Freudenstein, John V., xi
Froelichia gracilis (M-157), 143
Frostweed, 584
Fumaria officinalis (M-322), 248
FUMARIACEAE, 245, 237
Fumitory, 248
Climbing, 248
Family, 245

Gale, Sweet, 55
Galega officinalis, 494
Garlitz, Russell, xi
Gas-plant, 509
Gaura, 622
biennis, (M-886), 623*
coccinea (M-885), 623
filiformis, 624
longiflora (M-887), 623
Gay-wings, 513
Gendlin, Judy, xi
Genista tinctoria, 449
Geocaulon lividum (M-85), 98*
GERANIACEAE, 502
Geranium, 502
bicknellii (M-703), 507*
carolinianum (M-702), 506
columbinum (M-704), 507
dissectum, 507
Family, 502
maculatum (M-696), 505*
molle (M-699), 505
pratense, 505
pusillum (M-700), 506
pyrenaicum (M-701), 506
robertianum (M-698), 505*
rotundifolium, 502
sanguineum (M-697), 505
Wild, 502
Gereau, Roy E., xi
Geum, 433
aleppicum (M-577), 439
canadense (M-572), 437
laciniatum (M-573), 437
macrophyllum (M-575), 438
rivale (M-570), 435*
triflorum (M-569), 435
urbanum (M-576), 439
vernum (M-571), 437
virginianum (M-574), 438
Gillenia, 421
Gillis, William T., xi, 553
Ginger, Wild-, 103
Ginseng, 642
American, 642
Dwarf, 644
Family, 640
Glasswort, 127
Glaucium flavum (M-316), 244
Gleditsia triacanthos (M-636), 472*
Globe-flower, 199
Glossary, 683
Glycine max (M-633), 472
Goatsbeard, 433
Goats-rue, 494
Goatweed, 579
Goff, F. Glenn, xi

Golden Alexanders, 672
Goldenseal, 212
Goldentuft, 277
Goldthread, 208
Good-King-Henry, 138
Gooseberry, 327
  Family, 327
  Garden, 330
  Northern, 331
  Prickly, 329
  Swamp, 330
  Wild, 329
Goosefoot, 135
  Family, 125
  Stinking, 138
Gossypium, 567
Goutweed, 675
Graham, Shirley T., xiv
Grape, 563
  Family, 560
  Fox, 563
  Frost, 564
  Oregon-, 231
  River-bank, 564
  Summer, 564
  Wine, 563
Grass-of-Parnassus, 321
Greenweed, Dyer's, 449
GROSSULARIACEAE, 327
Groundnut, 493
Gum,
  Black-, 612
  Sour-, 612
Gumbo, 574
GUTTIFERAE, 575
Gymnocladus dioicus (M-635), 472*
Gymnosporangium, 379
Gypsophila, 175
  elegans (M-220), 176
  muralis (M-221), 176
  paniculata (M-219), 176
  scorzonerifolia (M-218), 176

Habitats, 4, 681
Hackberry, 87
  Dwarf, 88
HALORAGACEAE, 631
HAMAMELIDACEAE, 334
Hamamelis, 334
  japonica, 334
  mollis, 334
  vernalis, 334
[Hamamelis]
  virginiana (M-459), 334*
Hanes, Clarence R., & Florence N., xv
Harbinger-of-Spring, 667
Hardhack, 375
Harlequin,
  Rock, 249
  Yellow, 249
Hawthorn, 386
  Black, 396
  Dotted, 401
  English, 396
  Washington, 396
Hazel, 69
  Witch-, 334
Hazelnut, 69
  Beaked, 71
Hazlett, Brian T., xi
Heart's-ease, 124
Heather, False, 584
Hedera helix, 640
Hedge-parsley, 656
Hedysarum alpinum (M-682), 496
Helianthemum, 584
  bicknellii (M-825), 585*
  canadense (M-826), 585
Hellebore, 225
  Green, 225
Helleborus, 225
  niger (M-289), 225
  viridis (M-288), 225
Hemlock,
  Poison-, 667
  Water-, 673
Hemlock-parsley, 668
Hemp, 92
  Family, 90
Hen-and-chickens, 315
Henson, Don, xi
Hepatica, 209
  acutiloba (M-262), 211*
  americana (M-263), 211
  nobilis, 209
Heracleum, 661
  lanatum, 661
  mantegazzianum, 661
  maximum (M-937), 661*
  sphondylium, 661
Herb Robert, 505
Herbaria, consulted, xii
Hercules-club, 644
Herniaria glabra (M-182), 157
Hesperis matronalis (M-393), 295

Heuchera, 321
americana (M-440), 322
richardsonii (M-439), 322
sanguinea, 321
Hevea brasiliensis, 516
Hibiscus, 574
esculentus, 574
laevis (M-807), 575
militaris, 575
moscheutos (M-806), 574*
palustris, 574
syriaca, 574
trionum (M-805), 574*
Hickory, 56
Bitternut, 59
Kingnut, 61
Mockernut, 61
Pignut, 61
Red, 61
Shagbark, 59
Shellbark, 59
HIPPOCASTANACEAE, 551
HIPPURIDACEAE, 640
Hippuris vulgaris (M-910), 640*
Hoch, Peter C., xiv
Hog-peanut, 460
Holly, 540
Family, 539
Michigan, 540
Mountain, 540
Hollyhock, 573
Holosteum umbellatum (M-202), 167
Honesty, 307
Honewort, 669
Hop-hornbeam, 71
Hops, 92
Common, 93
Japanese, 93
Hop-tree, 509
Hornbeam, 71
Hop-, 71
Hornwort Family, 186
Horse-chestnut, 552
Horseradish, 282
Hudsonia tomentosa (M-824), 584*
Humulus, 92
americanus, 93
japonicus (M-75), 93
lupulus (M-76), 93
Hybanthus concolor (M-832), 589*
Hybrids, 8
Hydrangea, 319
Hydrastis canadensis (M-264), 212*
Hydrocotyle, 652
americana (M-920), 652*
umbellata (M-921), 652
Hymenophysa, 302
Hypericaceae, 575
Hypericum, 577
ascyron (M-812), 579
boreale (M-819), 581
canadense (M-820), 582
ellipticum (M-816), 581
gentianoides (M-813), 579
kalmianum (M-811), 579*
majus (M-821), 582*
mutilum (M-818), 581
perforatum (M-815), 579*
prolificum (M-810), 578
punctatum (M-814), 579*
pyramidatum, 579
spathulatum, 578
sphaerocarpum (M-817), 581
virginicum, 576

Iberis, 309
amara (M-418), 309
sempervirens, 309
umbellata (M-419), 309
Identification, 25
Ilex verticillata (M-754), 540*
Illustrations,
acknowledged, xv
explained, 27
Impatiens, 554
balsamina (M-774), 555
biflora, 556
capensis (M-777), 556*
glandulifera (M-775), 556
pallida (M-776), 556
Imperatoria ostruthium, 646
Indian-potato, 493
Indigo, False, 458
Inkberry, 151
Ipecac, American, 421
Ironwood, 71
Islands, listed, 5
Isopyrum biternatum (M-287), 224*
Ivy,
Boston-, 560
European, 640
Poison-, 533

Jeffersonia diphylla (M-299), 232*
Jerusalem-oak, 136

Johnny-jump-up, 595
Jointweed, 111
Joseph's Coat, 146
Judas Tree, 449
JUGLANDACEAE, 55
Juglans, 55
cinerea (M-35), 56*
nigra (M-36), 56*
regia, 55
Jumpseed, 118
Juneberry, 379
Mountain, 382
Northern, 382

Kale, 264
Kentucky Coffee-tree, 472
Keys, use of, 25
Kingnut, 61
Kiss-me-over-the-garden-gate, 123
Klamath Weed, 579
Knawel, 160
Knotweed, 111, 116
Giant, 118
Japanese, 118
Kochia scoparia (M-137), 130
Kohlrabi, 264
Kruschke, Emil P., xiv, 387

Lady's-thumb, 124
Lambs-quarters, 141
Laportea canadensis (M-79), 95
Lardizabala Family, 233
LARDIZABALACEAE, 233
Larkspur, 202, 203
Lathyrus, 483
hirsutus (M-660), 486
japonicus (M-653), 483
latifolius (M-662), 486
maritimus, 484
ochroleucus (M-654), 484*
odoratus (M-659), 485
palustris (M-656), 484*
pratensis (M-657), 485*
sylvestris (M-661), 486
tuberosus (M-658), 485
venosus (M-655), 484
LAURACEAE, 236
Laurel, 236
Family, 236
Laurus nobilis, 236
Lead-plant, 474
Leatherwood, 605
Lechea, 585
intermedia (M-830), 587*
leggettii, 586
minor (M-828), 586
mucronata, 586
pulchella (M-829), 586
stricta (M-831), 587
villosa (M-827), 586
LEGUMINOSAE, 444
Lepidium, 303
apetalum, 305
campestre (M-408), 304
densiflorum (M-411), 304*
montanum (M-413), 305
perfoliatum (M-407), 303*
ruderale (M-410), 304*
sativum (M-409), 304
virginicum (M-412), 305*
Les, Donald H., xiv, 186
Lespedeza, 468
capitata (M-628), 470
cuneata (M-626), 469
hirta (M-627), 470
intermedia (M-631), 471
×nuttallii, 471
procumbens (M-629), 471
stipulacea (M-625), 469
thunbergii, 468
violacea (M-630), 471*
virginica (M-632), 471
Leucophysalis grandiflora, 138, 248
Levisticum officinale (M-956), 669
Ligusticum canadense, 646
Lilac, Wild-, 557
Lily,
Cow-, 195
Lotus-, 192
Pond-, 195
Water-, 193
LIMNANTHACEAE, 531
LINACEAE, 496
Linden, 567
Family, 566
Lindera benzoin (M-306), 237*
Linnaeus, Carl, 566
Linum, 496
catharticum (M-683), 497
lewisii, 497
medium (M-687), 498
perenne (M-685), 497
striatum (M-689), 498
sulcatum (M-686), 498*
usitatissimum (M-684), 497

[Linum]
virginianum (M-688), 498
Liquidambar styraciflua, 334
Liriodendron tulipifera (M-303), 235*
Live-forever, 317
Lizard's-tail, 29
Family, 29
Lobularia maritima (M-398), 298*
Locust, 476
Black, 476
Bristly, 477
Clammy, 477
Honey, 472
Loganberry, 343
Loosestrife, 611
False, 617
Family, 609
Purple, 611
Swamp, 610
Whorled, 610
Loranthaceae, 100
Lotus, 192
American, 192
corniculata (M-675), 493*
Sacred, 192
Lovage, 669
Love-in-a-mist, 224
Love-lies-bleeding, 148
Ludwigia, 617
alternifolia (M-874), 618
palustris (M-873), 617*, 610
polycarpa (M-876), 618
sphaerocarpa (M-875), 618
Lunaria, 307
annua (M-414), 307*
rediviva, 307
Lupine, 477
Wild, 479
Lupinus, 477
perennis (M-646), 479*
polycarpus (M-645), 479
polyphyllus (M-647), 479
Lychnis, 185
alba, 183
chalcedonica (M-237), 186
coronaria (M-236), 185*
drummondii, 178
Scarlet, 186
LYTHRACEAE, 609
Lythrum, 611
alatum (M-868), 612*
dacotanum, 612
hyssopifolia (M-867), 612
[Lythrum]
salicaria (M-866), 611
virgatum, 612

Macleaya cordata (M-307), 239
Maclura pomifera (M-71), 89
Magnolia,
acuminata, 235
Family, 235
MAGNOLIACEAE, 235
Mahonia aquifolium (M-297), 231
Mallow, 570
Common, 572
Curled, 571
Family, 567
High, 571
Marsh, 573
Musk, 571
Poppy, 573
Rose, 574
Swamp, 574
Vervain, 571
Maltese-cross, 186
Malus, 418
coronaria (M-544), 419
domestica, 419
ioënsis (M-543), 419
×platycarpa, 420
pumila (M-542), 418
sylvestris, 419
Malva, 570
alcea (M-795), 571
borealis, 572
crispa (M-798), 571
moschata (M-796), 571
neglecta (M-800), 572*
pusilla, 572
rotundifolia (M-799), 572*
sylvestris (M-797), 571
verticillata, 572
MALVACEAE, 567
Mandrake, 232
Manihot esculenta, 516
Map, Michigan, 6
Maple, 545
Amur, 545
Ash-leaved, 547
Black, 549
Family, 545
Goosefoot, 551
Hard, 547
Manitoba, 547
Mountain, 550

[Maple]
Norway, 547
Red, 549
Silver, 549
Striped, 551
Sugar, 547
Sycamore, 550
Maps, explained, 4
Mare's-tail, 640
Family, 640
Marigold, Marsh-, 224
Marijuana, 92
Marsh-marigold, 224
Matthiola, 295
bicornis, 295
incana, 295
longipetala, 295
May-apple, 232
Meadow-beauty, 614
Meadow-parsnip, 672
Meadow-rue, 204
Early, 204
Purple, 206
Meadowsweet, 376
Medicago, 456
denticulata, 456
falcata, 457
hispida, 456
lupulina (M-602), 457*
minima, 456
nigra, 456
polymorpha (M-601), 456
sativa (M-603), 457
×varia, 457
Medick, Black, 457
Meibomia, 462
Melandrium album, 183
MELASTOMATACEAE, 614
Melastome Family, 614
Melilotus, 449
alba (M-588), 450
altissima (M-590), 450
officinalis (M-589), 450*
MENISPERMACEAE, 233
Menispermum canadense (M-302), 233*
Mercury, Three-seeded, 519
Merkle, John, xi
Mermaid, False, 531
Mermaid-weed, 631
Mexican-tea, 136
Mezereum Family, 605
Mignonette, 310
Family, 310
Milfoil, Water-, 634
Milk-vetch, 494
Canada, 495
Chick-pea, 495
Cooper's, 495
Milkwort, 512
Family, 512
Minuartia michauxii, 171
Mirabilis, 149
albida (M-172), 150
hirsuta (M-173), 150
jalapa, 149
linearis (M-170), 150
nyctaginea (M-171), 150*
Mistletoe, 100
Dwarf, 101
Family, 100
Mitella, 323
diphylla (M-441), 323*
nuda (M-442), 325*, 420
Miterwort, 323
False, 325
Naked, 325
Moehringia,
lateriflora, 171
macrophylla, 171
MOLLUGINACEAE, 151
Mollugo verticillata (M-175), 151*
Money-plant, 307
Monkshood, 199
Moonseed, 233
Family, 233
Moosewood, 551
MORACEAE, 88
Morus, 89
alba (M-72), 89*
rubra (M-73), 90
Mountain-ash, 363
European, 364
Mulberry, 89
Family, 88
Red, 90
Russian, 89
White, 89
Mustard, 252
Ball, 273
Black, 265
Brown, 266
Chinese, 266
Dog, 259
Family, 252
Field, 263
Garlic, 308

[Mustard]
Hare's-ear, 274
Hedge, 270
Indian, 266
Tansy, 257
Tower, 292
Treacle, 276
Tumble, 270
White, 264
Wild, 264
Wormseed, 275
Mustard, Timothy S., xi
Myosoton aquaticum (M-195), 165
Myrica gale (M-34), 55*
MYRICACEAE, 54
Myriophyllum, 634
alterniflorum (M-905), 637
aquaticum, 634
brasiliense, 634
exalbescens (M-908), 637
farwellii (M-904), 636
heterophyllum (M-906), 637
pinnatum, 637
spicatum (M-909), 638*
tenellum (M-903), 636
verticillatum (M-907), 637

Nasturtium, 286, 266
lacustre, 282
microphyllum, 286
officinale (M-379), 286*
Nelumbo, 192
lutea (M-242), 192*
nucifera, 192
pentapetala, 193
Nelumbonaceae, 189
Nemopanthus mucronatus (M-753), 540*
Neslia paniculata (M-354), 273*
Nettle, 95
False, 96
Family, 93
Stinging, 95
Wood, 95
Nigella damascena (M-285), 224
Nightshade, Enchanter's-, 615
Ninebark, 374
Nomenclature, 9
Nuphar, 195
advena (M-247), 197*
lutea, 195
microphylla, 197
pumila (M-245), 196
[Nuphar]
×rubrodisca, 197
variegata (M-246), 197
NYCTAGINACEAE, 149
Nymphaea, 193, 196
lotus, 192
odorata (M-244), 194*
tetragona (M-243), 194
tuberosa, 195
NYMPHAEACEAE, 189
Nyssa sylvatica (M-869), 612*
NYSSACEAE, 612

Oak, 72
Black, 77
Blackjack, 73
Bur, 81
Chestnut, 82
Chinquapin, 82
Dwarf Chestnut, 82
Dwarf Chinquapin, 82
Hill's, 79
Jerusalem-, 136
Northern Pin, 79
Pin, 78
Poison-, 536
Red, 77
Rock Chestnut, 82
Scarlet, 79
Shingle, 76
Swamp White, 81
White, 80
Yellow Chestnut, 82
Oenothera, 624
biennis (M-900), 630*
clelandii (M-894), 627
erythrosepala, 629
fruticosa (M-891), 626
glazioviana (M-896), 629
laciniata (M-893), 627
lamarckiana, 629
nutans, 631
nuttallii (M-892), 627
oakesiana (M-898), 629
parviflora (M-897), 629
perennis (M-889), 626*
pilosella (M-890), 626
rhombipetala (M-895), 627
serrulata, 624
strigosa, 630
tetragona, 626
villosa (M-899), 629
Okra, 574

Oleaster, 606
  Family, 606
Olive, Russian-, 606
ONAGRACEAE, 614
Oplopanax, 641
  elatus, 641
  horridus (M-911), 641*
Opuntia, 603
  compressa, 605
  fragilis (M-857), 604*
  humifusa (M-858), 604
  macrorhiza, 604
  rafinesquei, 605
Orache, 129
  Garden, 129
Orange-grass, 579
Oregon-grape, 231
Orpine, 315
  Family, 315
Osage-orange, 89
Osier,
  Common, 44
  Purple, 43
  Red-, 679
  Silky, 44
Osmorhiza, 658
  chilensis (M-932), 659
  claytonii (M-934), 659*
  depauperata (M-931), 658*
  longistylis (M-933), 659
  obtusa, 658
Ostrya virginiana (M-51), 71*
Overlease, William R., xi
OXALIDACEAE, 498
Oxalis, 498
  acetosella (M-691), 501*
  corniculata (M-694), 502
  dillenii, 501
  europaea, 501
  fontana (M-692), 501*
  montana, 501
  stricta (M-693), 501
  violacea (M-690), 499
Oxybaphus, 149
Oxypolis rigidior (M-939), 662

Palmer, Ernest J., xiv, 386
Panax, 642
  ginseng, 642
  quinquefolius (M-912), 642*
  trifolius (M-913), 644
Pansy, 595
  Field, 595
Papaver, 242
  argemone, 242
  dubium, 243
  orientale (M-312), 242
  pseudoörientale, 243
  rhoeas (M-314), 243*
  somniferum (M-313), 243
  ×strigosum, 244
PAPAVERACEAE, 237
Parietaria, 94
  diffusa (M-77), 94
  pensylvanica (M-78), 95
Parnassia, 321
  glauca (M-436), 321*
  palustris (M-437), 321
  parviflora (M-438), 321
Parnassiaceae, 319
Paronychia, 157
  canadensis (M-180), 157*
  fastigiata (M-181), 157
Parrot's-feather, 634
Parsley, 668
  Family, 645
  Fool's-, 669
  Hedge-, 656
  Hemlock-, 668
  Prairie-, 663
Parsnip, 662
  Cow-, 661
  Meadow-, 672
  Water-, 663
  Wild, 662
Parthenocissus, 560
  inserta (M-786), 561*
  quinquefolia (M-785), 561*
  tricuspidata, 560
  vitacea, 561
Pastinaca sativa (M-938), 662*
Patience, 109
Pawpaw, 235
Pea,
  Beach, 483
  Common, 481
  Everlasting, 486
  Family, 444
  Field, 481
  Garden, 481
  Green, 481
  Marsh, 484
  Partridge-, 480
  Perennial, 486
  Rabbit-, 494
  Snow, 481

[Pea]
Sugar, 481
Peach, 370
Pear, 420
Pearlwort, 168
Pea-tree, 472
Pecan, 58
Pelargonium, 502
Pellitory, 94
Pennywort, Water-, 652
Pentaphylloides,
floribunda, 427
fruticosa, 427
PENTHORACEAE, 318, 315
Penthorum sedoides (M-434), 319*
Pepper-and-salt, 667
Pepper-grass, 303
Pepperidge, 612
Perideridia americana, 646
Persicaria, 112
amphibia, 121
hydropiper, 124
hydropiperoides, 124
lapathifolia, 123
orientalis, 123
pensylvanica, 121
punctata, 124
vulgaris, 124
Petalostemon,
foliosum, 491
purpureum, 491
Petrorhagia, 172
prolifera (M-210), 172
saxifraga (M-211), 172*
Petroselinum crispum (M-952), 668
Phaseolus, 460
polystachios (M-609), 460
vulgaris (M-610), 460
Philadelphus, 319
Phipps, James B., xiv
Phoradendron flavescens, 100
Phoradendron serotinum, 100
Physocarpus opulifolius (M-507), 374*
Phytolacca americana (M-174), 151*
PHYTOLACCACEAE, 150
Pie Plant, 105
Pignut, 61
Sweet, 61
Pigweed, 141, 143
Winged, 133
Pilea, 96
fontana (M-84), 97
microphylla, 96
[Pilea]
pumila (M-83), 97
Pimpernel, Yellow-, 669
Pimpinella anisum (M-926), 655
Pink, 174
Cluster-head, 174
Deptford, 174
Family, 155
Fire, 182
Garden, 175
Grass, 175
Maiden, 174
Mullein, 185
Pinkweed, 121
Pinweed, 585
Pipe, Dutchman's, 102
Pisum sativum (M-652), 481
Pitcher, Zina, 299
Pitcher-plant, 312
Family, 310
Plane-tree,
Family, 336
London, 336
Oriental, 336
PLATANACEAE, 336
Platanus, 336
×acerifolia, 336
×hybrida, 336
occidentalis (M-460), 336*
orientalis, 336
Plum, 365
Alleghany, 372
Canada, 373
Cherry, 373
Common, 374
Wild, 372
Podophyllum peltatum (M-300), 232*
Poinsettia, 516, 520, 526
Poison-hemlock, 667
Poison-ivy, 533
Poison-oak, 536
Poison Sumac, 533
Poke, 151
Pokeweed, 151
Family, 150
Polanisia, 250
dodecandra (M-326), 250*
graveolens, 250
Polygala, 512
cruciata (M-711), 514*
Fringed, 513
incarnata (M-715), 516
paucifolia (M-710), 513*

[Polygala]
polygama (M-714), 514
sanguinea (M-717), 516
senega (M-716), 516
verticillata (M-712), 514
vulgaris (M-713), 514
POLYGALACEAE, 512
POLYGONACEAE, 104
Polygonella articulata (M-105), 111*
Polygonum, 111
achoreum (M-114), 118*
amphibium (M-122), 120*
arenastrum, 116
arifolium (M-107), 115*
aubertii, 120
aviculare (M-112), 116
buxiforme, 116
careyi (M-125), 123
cespitosum (M-131), 125
cilinode (M-118), 119*
coccineum, 120
convolvulus (M-119), 119
cristatum, 120
cuspidatum (M-117), 118
douglasii (M-110), 115
dumetorum, 120
erectum (M-113), 118
hydropiper (M-128), 124
hydropiperoides (M-129), 124
lapathifolium (M-124), 123*
natans, 120
opelousanum, 124
orientale (M-126), 123
patulum, 115
pensylvanicum (M-123), 121
persicaria (M-130), 124
punctatum (M-127), 123*
ramosissimum (M-111), 115
robustius, 124
sachalinense (M-116), 118
sagittatum (M-108), 115*
scandens (M-120), 119
setaceum, 124
tenue (M-109), 115
virginianum (M-115), 118*
viviparum (M-121), 120
Polytaenia nuttallii (M-942), 663*
Poplar, 50
Balsam, 52
Carolina, 53
Lombardy, 52
Silver, 51
White, 51
[Poplar]
Yellow-, 235
Poppy, 242
California, 239
Corn, 243
Family, 237
Field, 243
Flanders, 243
Horned, 244
Opium, 243
Oriental, 242
Plume, 239
Prickly, 240
Wood, 244
Populus, 50
alba (M-26), 51
balsamifera (M-28), 52
×canadensis, 53
canescens, 51
deltoides (M-29), 52
×gileadensis, 52
grandidentata (M-32), 53
heterophylla (M-27), 52
×jackii, 52
nigra (M-30), 52
×smithii, 53
tremuloides (M-31), 53*
Porteranthus, 421
stipulatus (M-548), 421
trifoliatus (M-547), 421*
Portulaca, 153
grandiflora (M-176), 154
oleracea (M-177), 154*
PORTULACACEAE, 153
Potentilla, 424
anserina (M-555), 427*
argentea (M-564), 432*
arguta (M-559), 428
bipinnatifida, 427
blaschkeana, 430
canadensis, 429
canescens, 432
fruticosa (M-553), 427
gracilis (M-563), 430
hippiana (M-557), 428
inclinata (M-566), 432
intermedia, 432
monspeliensis, 429
norvegica (M-561), 429
palustris (M-554), 427
paradoxa (M-558), 428
pensylvanica (M-556), 427
pumila, 430

[Potentilla]
recta (M-565), 432*
simplex (M-562), 429
tridentata (M-560), 429*
Prairie-clover, Purple, 491
Prairie-parsley, 663
Prairie-tea, 519
Prickly-pear, 603
Primrose, Evening-, 624
Prince's Feather, 123, 148
Proserpinaca, 631
×intermedia, 633
palustris (M-902), 633*
pectinata (M-901), 633
Prunus, 365
alleghaniensis (M-502), 372
americana (M-501), 372
armeniaca, 370
avium (M-499), 372
besseyi, 370
cerasifera (M-503), 373
cerasus (M-500), 372
domestica (M-506), 374
glandulosa, 365
mahaleb (M-497), 371
nigra (M-504), 373
padus, 370
pensylvanica (M-498), 371
persica (M-495), 370
pumila (M-496), 370*
serotina (M-492), 369*
spinosa (M-505), 374
susquehanae, 370
tomentosa (M-494), 370
virginiana (M-493), 369
Psoralea psoralioides (M-624), 468
Ptelea trifoliata (M-707), 509*
Purslane, 153
Family, 153
Water-, 617
Pyrus, 420, 363, 377, 418
americana, 365
aucuparia, 364
communis (M-545), 420
coronaria, 418
decora, 365
floribunda, 379
malus, 419
microcarpa, 365
prunifolia, 379

Quassia Family, 510
Queen-Anne's-lace, 654
Queen-of-the-meadow, 439
Queen-of-the-prairie, 439
Quercus, 72
alba (M-57), 80
×bebbiana, 80
bicolor (M-59), 81*
coccinea (M-56), 79*
×deamii, 80, 81
ellipsoidalis, 79
×faxonii, 80
×hawkinsiae, 78
×hillii, 81
imbricaria (M-52), 76*
×jackiana, 80
×leana, 76
lyrata, 81
macrocarpa (M-58), 81
marilandica, 73
montana, 82
muehlenbergii (M-61), 82
×palaeolithicola, 80
palustris (M-55), 78*
prinoides (M-60), 82
prinus (M-62), 82
rubra (M-54), 77*
×runcinata, 76
×schuettei, 81
shumardii, 78
stellata, 81
velutina (M-53), 77*
Quince, 418

Rabbit-pea, 494
Rabeler, Richard K., xi, xiv, 155
Radicula, 266, 282
Radish, 260
Horse, 282
Wild, 260
RANUNCULACEAE, 199
Ranunculus, 212
abortivus (M-275), 219
acris (M-279), 221*
ambigens (M-270), 218
aquatilis, 214
bulbosus (M-281), 221*
carolinianus, 223
circinatus, 215
cymbalaria (M-268), 218
fascicularis (M-282), 223
ficaria (M-269), 218
flabellaris (M-266), 216*
flammula, 218
gmelinii (M-267), 216

[Ranunculus]
hispidus (M-283), 223*
lapponicus (M-272), 219
longirostris (M-265), 214*
macounii (M-277), 220
pensylvanicus (M-278), 220
recurvatus (M-273), 219
repens (M-280), 221
reptans (M-271), 218
rhomboideus (M-276), 220
sceleratus (M-274), 219*
septentrionalis, 223
subrigidus, 215
trichophyllus, 215
Rape, 264
Raphanus, 260
raphanistrum (M-335), 260
sativus (M-336), 262
Raspberry, 340
Black, 347
Dwarf, 347
Flowering, 345
Red, 348
Wild Red, 347
Rattlebox, 449
Rattlesnake-master, 651
Raven, Peter H., xiv, 614
Read, Robert H., xi
Redbud, 449
Red-osier, 679
References, 10
Reseda, 310
alba, 310
lutea (M-420), 310
RESEDACEAE, 310
Reynoutria, 112
cilinodis, 119
convolvulus, 119
japonica, 118
sachalinensis, 118
scandens, 120
Reznicek, A. A., xi
RHAMNACEAE, 556
Rhamnus, 558
alnifolia (M-782), 559*
cathartica (M-783), 559
davurica, 560
frangula (M-781), 558
utilis (M-784), 560
Rheum, 105
rhabarbarum, 105
rhaponticum (M-92), 105
Rhexia virginica (M-870), 614*
Rhizobium, 444
Rhubarb, 105
Rhus, 536
aromatica (M-748), 536
×borealis, 537
copallina (M-749), 537
glabra (M-751), 537*
×pulvinata (M-752), 539, 537
radicans, 533
typhina (M-750), 537*
vernix, 533
Ribes, 327
alpinum, 328
americanum (M-453), 332*
aureum, 332
cynosbati (M-448), 329
glandulosum (M-456), 333*
grossularia, 330
hirtellum (M-449), 330
hudsonianum (M-454), 332
lacustre (M-451), 331
missouriense, 330
nigrum (M-455), 332
odoratum (M-452), 332
oxyacanthoides (M-450), 331
reclinatum, 330
rubrum (M-457), 333
sativum, 333
setosum, 331
triste (M-458), 334*
uva-crispa, 330
vulgare, 333
Richweed, 96
Ricinus communis (M-718), 517
Robinia, 476
hispida (M-642), 477
neomexicana, 476
pseudoacacia (M-641), 476*, 510
viscosa (M-643), 477
Rocket,
Dame's, 295
Garden, 262
London, 272
Sea-, 308
Turkish, 266
Wall, 272
Yellow, 259
Rocket-salad, 262
Rockrose, 584
Family, 584
Rocky Mountain Bee Plant, 251

Rogers, C. Marvin, xiv
Rollins, Reed C., xiv
Rorippa, 266
curvipes (M-346), 268
islandica, 269
nasturtium-aquaticum, 286
palustris (M-347), 269*
sylvestris (M-345), 268
Rosa, 354
acicularis (M-486), 362*
×alba, 355
arkansana (M-485), 361
×bifera, 355
blanda (M-487), 362
canina, 355
carolina (M-484), 361
centifolia (M-475), 358
cinnamomea (M-481), 360
damascena, 355
eglanteria (M-476), 358
gallica (M-478), 360
humilis, 361
×michiganensis, 363
micrantha, 358
multiflora (M-474), 357*
×palustriformis, 363
palustris (M-482), 360*
pimpinellifolia, 361
rubiginosa, 358
rugosa (M-480), 360
×schuetteana, 363
setigera (M-479), 360*
spinosissima (M-483), 361
villosa (M-477), 358
virginiana, 361
ROSACEAE, 336
Rose, 354
Cabbage, 358
Cinnamon, 360
Damask, 355
Dog, 355
Family, 336
French, 360
Japanese, 357
Multiflora, 357
Pasture, 361
Prairie, 360
Scotch, 361
Swamp, 360
Wild, 362
Rose-acacia, 477
Rose-moss, 154
Rotala ramosior (M-863), 609*
Rowan, 364
Rubus, 340
abactus, 354
acaulis (M-463), 346*
allegheniensis (M-471), 353*
angustifolius, 351
arcticus, 346
arundelanus, 349
associus, 354
attractus, 353
avipes, 354
baileyanus, 350
bellobatus, 354
besseyi, 353
canadensis (M-472), 353
cauliflorus, 354
centralis, 350
chamaemorus, 420
complex, 350
compos, 351
conabilis, 351
darlingtonii, 353
dissensus, 351
distinctus, 349
enslenii, 349
exutus, 350
flagellaris (M-469), 349*
florenceae, 350
×fraseri, 345
frondosus, 354
geophilus, 350
glandicaulis, 351
hanesii, 354
hispidus (M-468), 348*
idaeus, 348
ithacanus, 350
jejunus, 351
junceus, 351
kalamazoensis, 349
laciniatus (M-467), 348
laetabilis, 353
licens, 354
limulus, 354
localis, 354
mediocris, 351
meracus, 350
michiganensis, 350
michiganus, 347
missouricus, 351
multiformis, 350
×neglectus, 347

[Rubus]
notatus, 351
occidentalis (M-465), 347
odoratus (M-461), 345
orarius, 353
×paracaulis, 346
parviflorus (M-462), 346*
pauper, 350
pensilvanicus (M-473), 354
peracer, 350
perdebilis, 351
pergratus, 354
permixtus, 349
perspicuus, 351
phoenicolasius, 348
plicatifolius, 350
plus, 349
potis, 351
pubescens (M-464), 347
rappii, 353
recurvans, 354
regionalis, 351
repens, 420
roribaccus, 350
rosa, 353
schoolcraftianus, 350
setosus (M-470), 350
signatus, 349
spectatus, 351
stipulatus, 351
strigosus (M-466), 347
superioris, 351
tantulus, 350
tenuicaulis, 350
ursinus, 343
uvidus, 354
vagus, 350
variispinus, 351
vermontanus, 351
wheeleri, 351
wisconsinensis, 351
Rue,
Family, 508
Goats-, 494
Rue-anemone, 224
False, 224
Rumex, 105
acetosa (M-94), 106
acetosella (M-93), 106*
altissimus (M-99), 108*
britannica, 110
conglomeratus, 105
[Rumex]
crispus (M-104), 110*
domesticus, 108
longifolius (M-98), 108
maritimus (M-96), 108*
mexicanus, 109
obtusifolius (M-97), 108
orbiculatus (M-103), 110
patientia (M-100), 109
×pratensis, 111
stenophyllus, 111
thyrsiflorus (M-95), 108
triangulivalvis (M-101), 109*
verticillatus (M-102), 110
Russian-olive, 606
Russian-thistle, 127
Rust,
Comandra Blister, 97
Sweetfern, 54
Wheat, 231
White Pine Blister, 327
Rutabaga, 264
RUTACEAE, 508
Rye, 665

Sagina, 168
nodosa (M-204), 168*
procumbens (M-203), 168
Saint (see St.)
SALICACEAE, 29
Salicornia europaea (M-132), 127*
Salix, 30
adenophylla, 41
alba (M-16), 46, 31
amygdaloides (M-21), 48
babylonica, 47
bebbiana (M-11), 44
candida (M-10), 44
caprea, 46
cinerea (M-15), 46
×clarkei, 44
cordata (M-5), 41*
discolor (M-13), 45
eriocephala (M-4), 41
exigua (M-2), 40*
fragilis (M-17), 47
×glatfelteri, 40
glaucophylloides, 49
humilis (M-14), 46
interior, 40
lucida (M-20), 48*
myricoides (M-23), 49

[Salix]
nigra (M-3), 40*, 31
occidentalis, 46
pedicellaris (M-6), 43*
pellita (M-8), 43
pentandra (M-18), 47
petiolaris (M-25), 49
planifolia (M-12), 45
purpurea (M-7), 43
pyrifolia (M-22), 48*
rigida, 41
×rubens, 46
sericea (M-24), 49
serissima (M-19), 48
subsericea, 49
syrticola, 41
tristis, 46
viminalis (M-9), 44
Salsola kali (M-133), 127*
Saltbush, 129
Salt-cedar, 583
Sand-spurrey, 159
Sandalwood Family, 97
Sandwort, 170
Rock, 170
Sang, 642
Sanguinaria canadensis (M-309), 240*
Sanguisorba, 440
canadensis (M-580), 440
minor (M-579), 440*
Sanicula, 654
canadensis (M-923), 654
gregaria (M-924), 654*
marilandica (M-925), 655
trifoliata (M-922), 654*
SANTALACEAE, 97
SAPINDACEAE, 554
Saponaria, 177
ocymoides (M-223), 177
officinalis (M-224), 177*
vaccaria, 177
Sarracenia purpurea (M-421), 312*
SARRACENIACEAE, 310
Sarrazin, Michel, 312
Sarsaparilla,
Bristly, 645
Wild, 645
Saskatoon, 379
Sassafras albidum (M-305), 236*
Sauer, Jonathan D., xiv, 143
SAURURACEAE, 29
Saururus cernuus (M-1), 29*
Saxifraga, 325
aizoön, 326
paniculata (M-444), 326
pensylvanica (M-447), 326
tricuspidata (M-445), 326*
virginiensis (M-446), 326*
SAXIFRAGACEAE, 319
Saxifrage, 325
Early, 326
Family, 319
Golden, 320
Lime-encrusted, 326
Prickly, 326
Swamp, 326
Scandix pecten-veneris 656
Schrankia nuttallii (M-644), 477
Scleranthus, 160
annuus (M-187), 160*
perennis (M-188), 160
Sea-blite, 133
Sea-rocket, 308
Sedum, 315
acre (M-432), 318*
album (M-431), 318
hispanicum (M-430), 317
purpureum, 317
sarmentosum (M-427), 317
sexangulare (M-433), 318
spurium (M-428), 317
telephium (M-429), 317
ternatum (M-426), 317
Sempervivum tectorum, 315
Senega Root, 516
Senna, 479
hebecarpa, 481
tora, 481
Wild, 481
Sensitive-plant, 444
Wild, 480
Sericea, 469
Serviceberry, 379
Shadblow, 379
Shadbush, 379
Shagbark, 59
Shamrock, 451, 498
Shepherd's-purse, 287
Shepherdia canadensis (M-862), 608*
Sibbaldiopsis tridentata, 429
Sickle-pod, 290, 480
Sida, 570
hermaphrodita (M-793), 570
spinosa (M-794), 570

Silene, 178
  alba, 183
  antirrhina (M-225), 181*
  armeria (M-226), 181*
  conica (M-230), 182
  csereii (M-227), 182
  cucubalus, 182
  dichotoma (M-232), 182*
  dioica, 185
  drummondii, 178
  gallica, 178
  latifolia, 182, 185
  noctiflora (M-234), 183
  nutans (M-233), 183
  pendula, 178
  pratensis (M-235), 183
  stellata (M-229), 182
  virginica (M-231), 182
  vulgaris (M-228), 182
Silverweed, 427
SIMAROUBACEAE, 510
Sinapis, 263, 265
  alba, 264
  arvensis, 265
Sisymbrium, 270
  altissimum (M-349), 270*
  irio (M-350), 272
  loeselii (M-351), 272
  officinale (M-348), 270
Sium suave (M-941), 663*, 662
Sloe, 374
Smartweed, 111
  Bigseed, 121
  Family, 104
  Nodding, 123
  Water, 120
Snakeroot,
  Black, 207, 654
  Sampson's, 468
  Seneca, 516
  Virginia-, 103
Snow-in-summer, 166
Snow-on-the-mountain, 525
Soapberry, 608
  Family, 554
Soapwort, 177
Sorbaria sorbifolia (M-488), 363
Sorbus, 363
  americana (M-491), 365*
  aucuparia (M-489), 364
  decora (M-490), 364*
  dumosa, 364
  sambucifolia, 364
[Sorbus]
  scopulina, 365
  sitchensis, 365
  subvestita, 365
Sorrel,
  Garden, 106
  Green, 106
  Red, 106
  Sheep, 106
  Wood-, 498
Sour-gum, 612
Soybean, 472
Spatterdock, 195
Spearscale, 129
Spearwort, Creeping, 218
Spergula arvensis (M-183), 157*
Spergularia, 159
  marina (M-186), 159*
  maritima, 159
  media (M-184), 159
  rubra (M-185), 159
  salina, 160
Spicebush, 237
Spider Plant, 250
Spikenard, 645
Spinach, 138
Spindle Tree, 544
Spiraea, 375
  alba (M-511), 376
  douglasii, 375
  False, 363
  Japanese, 375
  japonica (M-509), 375
  latifolia, 376
  salicifolia (M-512), 376
  tomentosa (M-510), 375*
  ×vanhouttei (M-508), 375
Spring-beauty, 154
Spurge, 520
  Cypress, 527
  Family, 516
  Flowering, 526
  Leafy, 527
  Seaside, 525
Spurrey, 157
  Sand-, 159
Squirrel-corn, 247
St. John's-wort, 577
  Common, 579
  Family, 575
  Giant, 579
  Kalm's, 579
  Marsh, 575

[St. John's-wort]
Shrubby, 578
Spotted, 579
Staphylea trifolia (M-762), 545*
STAPHYLEACEAE, 545
Steeplebush, 375
Stellaria, 162
aquatica, 165
calycantha (M-191), 163, 171
crassifolia (M-190), 163
glauca, 164
graminea (M-194), 164*
longifolia (M-193), 164
longipes (M-192), 163
media (M-189), 163*
palustris, 164
Stitchwort, 162
Stonecrop, 315
Ditch, 319
Mossy, 318
Stork's-bill, 502
Strawberry, 422
Barren-, 422
Indian-, 422
Wild, 424
Woodland, 423
Strawberry-bush, Running, 544
Strophostyles helvula (M-611), 460*
Stylophorum diphyllum (M-315), 244*
Styrax, 237
Suaeda, 133
calceoliformis, (M-140), 133
depressa, 133
Subularia aquatica (M-415), 307*
Sugarplum, 379
Sumac, 536
Dwarf, 537
Fragrant, 536
Poison, 533
Shining, 537
Smooth, 537
Staghorn, 537
Summer-cypress, 130
Sundew, 313
Family, 312
Sundrops, 624, 626
Sweetbrier, 358
Sweet-cicely, 658
Sweet-clover, 449
White, 450
Yellow, 450
Sweetfern, 55
Sweet William, 174
Swiss Chard, 133
Sycamore, 336

Taenidia integerrima (M-955), 669
TAMARICACEAE, 583
Tamarisk, 583
Family, 583
Tamarix parviflora (M-823), 583
Taxonomy, 8
Tea,
Mexican-, 136
New Jersey, 558
Prairie-, 519
Tear-thumb, 115
Tephrosia virginiana (M-678), 494*
Thalictrum, 204
confine, 206
dasycarpum (M-257), 206*
dioicum (M-254), 204
polygamum, 207
pubescens, 207
revolutum (M-256), 206
venulosum (M-255), 206
Thaspium, 672, 670
barbinode (M-959), 672*
trifoliatum (M-960), 672*
Thimbleberry, 346
Thimbleweed, 229
Thistle, Russian-, 127
Thlaspi arvense (M-406), 303*
Thompson, Paul W., xi
Thornapple, 386
Thoroughwax, 651
THYMELAEACEAE, 605
Tiarella cordifolia (M-443), 325*
Tick-trefoil, 462
Tilia americana (M-791), 567*
TILIACEAE, 566
Tiniaria, 112
Toadflax,
Bastard-, 98
Star-, 98
Tooth-cup, 609
Toothwort, 277
Cut-leaved, 279
Two-leaved, 280
Torilis, 656
japonica (M-929), 656*
nodosa (M-928), 656
Touch-me-not,
Family, 554
Pale, 556
Spotted, 556

Tovara, 118
Toxicodendron, 532
radicans (M-747), 533*
rydbergii, 536
vernicifluum, 533
vernix (M-746), 533*
Trachyspermum ammi (M-930), 656
Tree-of-Heaven, 512
Trefoil,
Birdfoot, 493
Tick-, 462
Triadenum, 575
fraseri (M-809), 576
virginicum (M-808), 576*
Tribulus terrestris (M-705), 508*
Trifolium, 451
agrarium, 455
arvense (M-594), 453
aureum (M-598), 453*
campestre (M-599), 455
depauperatum (M-591), 452
dubium (M-600), 455
fucatum (M-592), 452
hybridum (M-597), 453
incarnatum (M-595), 453*
pratense (M-593), 452
procumbens, 455
repens (M-596), 453*
Trollius laxus, 199
Tulip-poplar, 235
Tulip-tree, 235
Tunica, 172
Tupelo, 612
Family, 612
Turnip, 263
Twinleaf, 232

ULMACEAE, 85
Ulmus, 85
americana (M-67), 87*
fulva, 86
glabra, 85
parvifolia, 86
pumila (M-65), 86
rubra (M-66), 86
thomasii (M-68), 87
UMBELLIFERAE, 645
Umbrellawort, 149
Urtica, 95
dioica (M-81), 95
gracilis, 96
urens (M-80), 95
[Urtica]
viridis, 96
URTICACEAE, 93

Vaccaria, 177
hispanica (M-222), 177*
pyramidata, 177
segetalis, 177
Velvet-leaf, 569
Venus'-comb, 656
Venus' Fly-trap, 312
Vetch, 486
American, 488
Bird, 489
Common, 488
Crown-, 493
Hairy, 488
Hedge, 487
Milk-, 494
Pale, 488
Sparrow, 488
Spring, 488
Wood, 488
Vetchling,
Pale, 484
Tuberous, 485
Yellow, 485
Vicia, 486
americana (M-666), 488
angustifolia, 488
caroliniana (M-669), 488
cracca (M-671), 489*
dasycarpa, 488
grandiflora (M-664), 487
hirsuta (M-667), 488
sativa (M-665), 488
sepium (M-663), 487
tetrasperma (M-668), 488
villosa (M-670), 488*
Vine,
Alleghany, 248
Balloon-, 554
Chocolate, 233
Pipe, 102
Puncture, 508
Viola, 589
adunca (M-840), 597
affinis (M-853), 601
arvensis (M-833), 595*
blanda (M-845), 599
×brauniae, 596
canadensis (M-835), 595
conspersa (M-839), 596

[Viola]
cucullata (M-852), 601*
epipsila (M-854), 602
eriocarpa, 595
fimbriatula, 601
incognita, 599
labradorica, 597
lanceolata (M-841), 597
macloskeyi (M-844), 598*
×malteana, 596
nephrophylla (M-855), 602
novae-angliae, 602
obliqua, 601
odorata (M-843), 598
pallens, 598
palmata (M-850), 600
palustris, 602
papilionacea, 603
pedata (M-848), 600*
pedatifida (M-847), 599
pensylvanica, 595
×primulifolia (M-842), 598
pubescens (M-836), 595
rafinesquii, 595
renifolia (M-846), 599*
rostrata (M-838), 596*
sagittata (M-851), 601
selkirkii (M-849), 600
septentrionalis, 602
sororia (M-856), 602
striata (M-837), 595
tricolor (M-834), 595
triloba, 601
×wittrockiana, 595
VIOLACEAE, 589
Violet, 589
Arrow-leaved, 601
Birdfoot, 600
Canada, 595
Common Blue, 602
Cream, 595
Dog, 596
English, 598
False, 420
Family, 589
Great-spurred, 600
Green, 589
Kidney-leaved, 599
Lance-leaved, 597
Long-spurred, 596
Marsh, 601
Prairie, 599
Sand, 597
[Violet]
Smooth White, 598
Sweet White, 599
Sweet, 598
Wood, 600
Yellow, 595
Virginia Creeper, 560
Virgin's Bower, 201
VISCACEAE, 100
Viscum album, 100
VITACEAE, 560
Vitis, 563
aestivalis (M-788), 564*
cordifolia 566
labrusca (M-787), 563
riparia (M-789), 564*
rupestris, 564
vinifera, 563
vulpina (M-790), 564

Wafer-ash, 509
Wagner, Warren H., Jr., xiv, 73
Wagner, Warren L., xiv
Wahoo, 544
Waldsteinia fragarioides (M-549), 422*
Wallflower, Western, 275
Walnut,
Black, 56
English, 55
Family, 55
Watercress, 286
Water-lily, 193
Family, 189
Water-milfoil, 634
Eurasian, 638
Family, 631
Water-parsnip, 663
Water-pennywort, 652
Water-pepper, 124
Mild, 124
Water-purslane, 617
Water-shield, 190
Water-starwort, 528
Family, 528
Waterwort, 582
Family, 582
Wax-myrtle, 55
Wells, James R., xi
Wheeler, Charles F., 40
Wheeler, O. B., 516
White-top, 302
Whitlow-grass, 298
Whitlow-wort, 157

Wild-chervil, 667
Wild-lilac, 557
Willow, 30
  Autumn, 48
  Balsam, 48
  Basket, 43, 44
  Bay-leaved, 47
  Beaked, 44
  Bebb's, 44
  Black, 40
  Blueleaf, 49
  Bog, 43
  Brittle, 47
  Crack, 47
  Family, 29
  Florist's, 46
  Furry, 41
  Goat, 46
  Golden, 46
  Gray, 46
  Heart-leaved, 41
  Hoary, 44
  Laurel, 47
  Meadow, 49
  Peach-leaved, 48
  Prairie, 46
  Pussy, 45
  Sage, 44
  Sandbar, 40
  Sand-dune, 41
  Satiny, 43
  Shining, 48
  Silky, 49
  Slender, 49
  Upland, 46
  Weeping, 47
  White, 46
Willow-herb, 618
  Great Hairy, 621
  Great, 620
Willow-weed, 123
Windflower, 227
Wineberry, 348
Winterberry, 540
Wintercreeper, 543
Wintergreen, Flowering-, 513
Wisteria, 473
  Chinese, 474
  frutescens (M-637), 473
  macrostachya, 473
  sinensis (M-638), 474
Witch-hazel, 334
  Family, 334
Witch's brooms, 97, 101
Woodbine, 201, 560
Woodland, Dennis W., xiv
Wood-sorrel, 498
  Family, 498
Wormseed, 136
Woundwort, 493
Wyeomyia smithii, 312

Xanthoxylum, 510

Yellow-pimpernel, 669
Yellow-poplar, 235
Yellowtuft, 277

Zanthoxylum, 510
  americanum (M-708), 510*
  fraxineum, 510
Zizia, 670, 672
  aptera (M-957), 670
  aurea (M-958), 672*
ZYGOPHYLLACEAE, 508

Note to the Second Printing: Corrigenda

Omitting self-evident typographical errors and slips in formatting, the following corrections should be made to avoid misleading the reader. (These are not new information, but represent problems at the time of publication, such as gremlins that lost data as the manuscript was typed or maps made, consultation of an incomplete set of the Gray Index to new scientific names, and/or faulty proofreading.)

p. xi, line 7 from bottom: insert Robert W. Smith, Lenawee County

p. 14, insert in citations: Swink, Floyd, & Gerould Wilhelm. 1979. Plants of the Chicago Region, ed. 3. 1979. Morton Arboretum, Lisle, Ill. lxxiii + 922 pp.

p. 35, line 7 up: for "couplet 6" read "couplet 16"

p. 40, line 7 under *S. exigua*: authors of combination should be (Rowlee) C. Reed

p. 50, add to References: Wagner, Warren H., Thomas F. Daniel, & Joseph M. Beitel. 1980. Studies on Populus heterophylla in Southern Michigan. Michigan Bot. 19: 269–275.

p. 120, parenthetical clause at end of first paragraph should read: "or *Fallopia scandens* (L.) Holub"

p. 124, name at the end of the last paragraph should read: "*P. maculata* (Raf.) Á. & D. Löve."

p. 165: *Myosoton* should be numbered 7 (not 6) and all genera for the rest of the family be given one number higher as well.

p. 231, Loconte reference: Phytologia 49 is correct volume

p. 252, couplet 4 in key: 31 is correct number of *Lepidium* (also couplet 33 on p. 254)

p. 253, couplet 18 in key: 9 is correct number for *Sisymbrium*

p. 254, couplet 26 in key: 28 is correct number for *Draba*

p. 259, Map 331: add dot in Emmet Co.

p. 304, line 10: insert "as" between "round" and "in"

p. 335, legend for fig. 173, add: "fruiting head on twig $\times 1/2$"

p. 346, line 18 under *R. parviflorus*: epithet should be *lacer*

p. 347, Map 465: the following block of counties was mysteriously lost in transferring data and all need dots: Hillsdale, Ingham, Isabella, Kalamazoo, Kent, Lapeer, Leelanau, Leelanau (Manitou Is.), Lenawee, Livingston, Mackinac (islands), Macomb, Monroe, Montcalm, Newaygo, Oakland

p. 351, first name in list should be *R. angustifoliatus*

p. 433, Map 567: the dot belongs on mainland of Keweenaw Co., not Isle Royale

p. 439, *Geum urbanum* author should be L.

p. 457, Map 601: dot in southern Michigan belongs in Wayne Co., not Washtenaw

p. 499, Lourteig reference: Phytologia 42 is correct volume

p. 502, *E. cicutarium* authors should be (L.) L'Hér.

p. 549, line 13: authors for var. *viride* should be (Schmidt) E. Murray

p. 574, line 2 under *Hibiscus*: epithet should be *syriacus*

p. 578, Map 811: delete dot in St. Joseph Co. and add Schoolcraft Co.

p. 581, Map 816: add dot in Houghton Co.

p. 598, *V. odorata* author should be L.

p. 633, second paragraph under *P. palustris*: at end of line 5 and middle of line 6, switch "short" and "long"

p. 696: insert Betula glandulosa, 67

p. 705: second page number for Gillis should be 533; insert Gillman, Henry, 208

p. 721: page number for Strophostyles helvula should be 461; insert Syringa, 319

Note to the Second Printing: Addenda

While updating of the distribution maps is not feasible at this time, the following list of additional species can be presented. These belong to families covered in this volume and have been documented (at least in the herbaria consulted) since 1985 as occurring outside of cultivation in Michigan. Species marked with an asterisk are not indigenous (or only questionably so) in this state. Recorded counties (or island groups) are in brackets.

*Myrica pensylvanica** [Monroe, Washtenaw]
*Juglans regia** [Leelanau (Manitou Is.)]
Quercus shumardii [Macomb, Monroe]
*Ulmus glabra** [Chippewa]
Rumex occidentalis [Marquette]
*Salsola collina** [Ottawa]
Froelichia floridana [Van Buren]
*Claytonia sibirica** [Houghton]
*Cerastium pumilum** [Calhoun, Ingham, Livingston, Oakland]
*Stellaria pallida** [Berrien, Lenawee]
Myosurus minimus [Hillsdale]
Thalictrum pubescens? [Kalkaska, Otsego, Wayne]
*Chorispora tenella** [Hillsdale]
*Coincya monensis** [Berrien]
Draba glabella [Keweenaw (I. Royale)]
*Malus baccata** [Lenawee, Mackinac]
*Rhodotypos scandens** [Washtenaw]
*Rosa canina** [Lenawee]
Rosa virginiana [Lenawee]
*Galega officinalis** [Saginaw]
*Lespedeza bicolor** [Macomb]
*Lespedeza thunbergii** [Clinton, Washtenaw]
*Pueraria lobata** [Allegan]
Callitriche terrestris [Washtenaw]
*Cotinus coggygria** [Benzie, Leelanau, St. Joseph]
*Acer ginnala** [Calhoun, Hillsdale, Mackinac, Washtenaw]
*Ampelopsis aconitifolia** [Wayne]
*Ampelopsis brevipedunculata** [Wayne]
*Daphne mezereum** [Ontonagon]
Shepherdia argentea (*?) [Alpena]
Rhexia mariana [Allegan, Ottawa]
*Heracleum mantegazzianum** [Ingham]
*Myrrhis odorata** [Leelanau; Mackinac (islands)]
*Pimpinella saxifraga** [Delta, Dickinson, Mackinac, Menominee]

MICHIGAN
Scale of Miles
0 10 50
UMMZ-1957-WLB
ONT.
ONTARIO
WISCONSIN
INDIANA
OHIO
LAKE SUPERIOR
LAKE MICHIGAN
LAKE HURON
LAKE ERIE
ISLE ROYALE
KEWEENAW
HOUGHTON
ONTONAGON
BARAGA
GOGEBIC
MARQUETTE
ALGER
LUCE
CHIPPEWA
IRON
SCHOOLCRAFT
MACKINAC
DICKINSON
DELTA
MENOMINEE
MACKINAC IS
BOIS BLANC IS
DRUMMOND IS
BEAVER IS
FOX IS
MANITOU IS
EMMET
CHEBOYGAN
PRESQUE ISLE
CHARLEVOIX
ANTRIM
OTSEGO
MONT-MORENCY
ALPENA
LEELANAU
BENZIE
GRAND TRAVERSE
KALKASKA
CRAWFORD
OSCODA
ALCONA
MANISTEE
WEXFORD
MISSAUKEE
ROS-COMMON
OGEMAW
IOSCO
CHARITY IS
ARENAC
MASON
LAKE
OSCEOLA
CLARE
GLADWIN
HURON
OCEANA
NEWAYGO
MECOSTA
ISABELLA
MIDLAND
BAY
TUSCOLA
SANILAC
MUSKEGON
MONTCALM
GRATIOT
SAGINAW
OTTAWA
KENT
IONIA
CLINTON
SHIA-WASSEE
GENESEE
LAPEER
ST. CLAIR
ALLEGAN
BARRY
EATON
INGHAM
LIVINGSTON
OAKLAND
MACOMB
VAN BUREN
KALAMAZOO
CALHOUN
JACKSON
WASHTENAW
WAYNE
BERRIEN
CASS
ST. JOSEPH
BRANCH
HILLS-DALE
LENAWEE
MONROE
89
87
85
83
48
46
44
42